Photochemistry

Volume 10

Specialist Periodical Report

Photochemistry

Volume 10

A Review of the Literature published between
July 1977 and June 1978

Senior Reporter
D. Bryce-Smith, *Department of Chemistry,
University of Reading*

Reporters
N. S. Allen, *Manchester Polytechnic*
M. D. Archer, *University of Cambridge*
G. Beddard, *The Royal Institution, London*
H. A. J. Carless, *Birkbeck College, University of London*
A. Cox, *University of Warwick*
R. B. Cundall, *University of Salford*
A. Gilbert, *University of Reading*
W. M. Horspool, *University of Dundee*
J. M. Kelly, *Trinity College, University of Dublin*
J. F. McKellar (late), *University of Salford*
D. Phillips, *The University of Southampton*
S. T. Reid, *The University of Kent*
M. A. West, *The Royal Institution, London*
M. Wyn-Jones, *University of Salford*

The Chemical Society
Burlington House, London, W1V 0BN

British Library Cataloguing in Publication Data

Photochemistry. –
(Chemical Society. Specialist periodical reports).
Vol. 10.
1. Photochemistry
I. Bryce-Smith, Derek   II. Series

541'.35   QD708.2   73-17909
ISBN 0-85186-590-9
ISSN 0556-3860

Organic formulae composed by Wright's Symbolset method

PRINTED IN GREAT BRITAIN BY JOHN WRIGHT AND SONS. LTD, AT THE STONEBRIDGE PRESS BRISTOL BS4 5NU

Contents

Part I Physical Aspects of Photochemistry

Part II   *Photochemistry of Inorganic and Organometallic Compounds*

Part III   Organic Aspects of Photochemistry

Part V   *Photochemical Aspects of Solar Energy Conversion*
By M. D. Archer

Part VI   Chemical Aspects of Photobiology
By G. Beddard

Introduction and Review of the Year

It is a pleasure to welcome Dr. Alan Cox to the team of Reporters for 'Photochemistry', Volume 10. He is contributing a chapter on transition-metal complexes in Part II. In recent Volumes it has become necessary to cover certain sections on the subject on a biennial basis. Thus this year's developments in the spectroscopic and theoretical fields will form part of a biennial review in Volume 11, whereas the present Volume includes two-year reviews on 'Instrumentation and Techniques' (M. West) and 'Photochemical Aspects of Photobiology' (G. Beddard).

The citation of some 1600 references in Dr. West's Report testifies to the intense activity in the field of photochemical instrumentation and techniques. Great interest continues in new laser systems, especially those using rare gas fluorides, bromides, and chlorides (Sze and Scott, among many others). Other fields of strong activity include photoacoustic spectroscopy, the use of vidicon detectors, the 'thermal lens' technique for measuring weak absorption and fluorescence, and of course the laser separation of isotopes. New procedures for generating tunable ultraviolet sources continue to be described (Bilt et al., Ferguson et al.).

Bigio's interesting observation that irradiation with an α-source (^{241}Am) trebled the energy output from a XeF laser may have wider applications. The sulphur dimer S_2 appears to provide a promising new molecular system for pulsed gas lasers (Leone and Kosnik); but the number of atoms and molecules known to show laser action is in fact increasing very rapidly. Nevertheless, the well-tried nitrogen laser continues to be further improved (see Bergmann, von Bergman among many others). In an ingenious and potentially important development, Green et al. have employed the opto-galvanic effect to lock the frequency of a CW dye laser to the very narrow bandwidth of 0.003 nm over the range 550—640 nm. Holographic defraction gratings are being increasingly used. It is possible to extend the sensitivity in laser flash photolysis by using an interference pattern for excitation so as to produce an amplitude grating of absorption coefficient distributions which is detected by Bragg diffraction of a second low-intensity laser beam (Rondelez et al.).

The notorious background fluorescence interference in Raman spectroscopy, thought to arise from widespread contamination, has now been quantified as an intrinsic property by Hirshfeld. Practical suggestions to avoid this interference have been proposed.

Adamson et al. have described a new type or calorimetry termed photocalorimetery by which enthalpy changes occurring during photoprocesses may be measured. Despite the intrinsic low precision for processes of low quantum yield, we are likely to hear much more of this in the future.

Strong interest is developing in multiphoton infrared reactions involving the use of lasers: SF_6 and SiF_4 act as 'infrared sensitizers' for such processes. Although many of the applications are being directed towards isotope enrichment (*e.g.* 100% enrichment of $^{12}CH_2 = {}^{12}CH_2$ has been reported by Hall *et al.*), the list of reactions which can now be accomplished by the so-called 'ground state photochemistry' now extends to isomerization processes such as hexafluorocyclobutene $\rightarrow$ hexafluorobutadiene (Yogev and Benmair).

In the field of practical applications, photochemical techniques are being increasingly applied to studies of atmospheric chemistry, particularly those relating to the chemistry of Freons in the upper atmosphere. Concerning the important question of ozone depletion, the evidence on balance is tending to indicate that the problem may be a little less serious than some workers have previously believed; but there is certainly a case for the use of Freons having abstractable hydrogens since most of these (but not CHF_2Cl) have short tropospheric lifetimes as a result of preferred hydrogen abstraction by hydroxyradicals. The residence times of some hydrogen-free Freons in the upper atmosphere have been claimed to range up to 550 years (Chou *et al.*).

Attention is drawn to interesting work on dual molecular emission following pulsed excitation of mercury vapour (Stock *et al.*, among others) and the lithium isotope separation by two-step photoionization of Li_2 (Rothe *et al.*).

Benzene continues to be a molecule of great interest to both physical and organic photochemists, although cross-fertilization between the 'physical' and 'organic' findings continues to be rather rare. (The need to excite photochemists by interaction with aspects of photochemistry lying outside their special fields has encouraged the Reporters to resist the occasional suggestions that 'Photochemistry' should be split into several parts.) On the physical side Parmenter and Tang have contributed a notable study in which they have followed mode-to-mode vibrational relaxation of S_1 benzene under 'single-collision' conditions, and have shown that this follows highly specific patterns which predominantly favour four particular channels. An interesting anomaly has cropped up from Durnick and Kalantar's findings on polarization of the b_{2g} phosphorescence bands in benzene; these seem inconsistent with the normal assignment of T_1 benzene as $^3B_{1u}$. Wassam and Lim have provided an interesting study of factors influencing radiationless decay of S_1 states, particularly vibronic interactions between S_1 and S_2 states, and they have applied this with some success to nitrogen heterocycles. Schroder *et al.* have reported an elegant study of relaxation in triplet naphthalene which is likely to require reconsideration of previous work in this area.

Kammula *et al.* have observed the unusually efficient formation of a 'Dewar' isomer from [6]paracyclophane. Chambers *et al.* have prepared thermally stable azaprismanes from the corresponding pyridines *via* 'Dewar' isomers, which may themselves apparently undergo 'walk'-type photoisomerization prior to cyclization to the azaprismanes. Ring transposition reactions of pyrroles (*e.g.* 2-cyanopyrrole $\rightarrow$ 3-isomer) evidently proceeds *via* 2,5-cyclization followed by a 'walk' reaction (Barltrop, Day, and Ward). It is interesting in this connection that ring transposition processes have also been shown to occur in cyclopentadiene (Andrews and Baldwin).

Fluorescence from pyridine has at last been observed, but it is very weak (Yamazaki and Baba). The interesting photoisomerization of pyridines (1) to anilines (2) is now confirmed to involve 3,6-bonding followed by a 1,3-hydrogen shift and formation of the biradical (3) (Takagi and Ogata).

(1)   (2)   (3)

The photoaddition of ethylenes to aromatic rings constitutes an active area of research by several groups. The orientation in *meta*-cycloadditions to substituted benzenes has been found to depend not only on the benzene substituent(s) but also more surprisingly on the nature of the alkene in some, but not all, cases (Bryce-Smith, Dadson, Gilbert, Orger, and Tyrrell). Gilbert and Taylor have described some most interesting types of intramolecular cycloadditions of an ethylenic bond to the benzene ring. Pentafluoropyridine, unlike pyridine itself, has been found to undergo photocycloaddition of ethylene, and in this respect more resembles benzene (Barlow, Brown, and Haszeldine).

There is growing interest, both physical and chemical, in amine–aromatic interactions, notably the role of exciplexes, encounter complexes, and the like. It is notable in this connection that Lapouyade and his co-workers have discovered that certain amines, notably triethylamine, can strongly promote certain intramolecular cycloadditions of ethylenic bonds to aromatic rings. Although many photoreactions of amines appear to involve electron transfer, Amouyal and Bensasson have provided evidence that photoreduction by tertiary amines such as triethylamine may sometimes occur *via* initial hydrogen-extraction. Exciplex fluorescence from amine–aromatic pairs has only previously been observed in solution, but Hirayama, Abbott, and Phillips have now observed this in the gas phase.

Full details of the photoreaction of aliphatic amines with benzene have been described. The products of 1,2-, 1,3-, and 1,4-addition appear to arise directly or indirectly from S_1 benzene in every case (Bellas and co-workers). Kaupp has described a photoaddition to anthracene which, most unusually, appears to be hindered rather than assisted by the formation of an exciplex.

Rare examples of cycloadditions of C=C to (*a*) C=N, (*b*) N=N, and (*c*) −C≡N have been reported: (*a*) Oe, Tashiro, and Tsuge; (*b*) Berning and Hünig; (*c*) Cantrell. A most unusual photoaddition of butadiene to *N*-methylphthalimide giving the benzazepine dione (4) has been described by

(4)

Mazzochi *et al.* Some observations by Caldwell and Creed on the effects of di-t-butyl nitroxide and oxygen in promoting the cyclization of diradical intermediates may prove to be of significance and utility in a wider context, particularly in relation to mechanistic and synthetic studies. In this connection it is interesting that Nagel and co-workers have observed that the widely used oxidative photocyclizations of stilbenes to phenanthrenes can be very advantageously and specifically performed in the presence of strong electron acceptors such as tetracyanoethylene and p-chloranil.

Javaheripour and Neckers have provided evidence that a heavy-atom effect can operate in photoprocesses which occur in KBr discs.

Lenz has described an example in which the photo-Diels–Alder reaction occurs in a $4s + 2a$ stereochemical sense. An interesting [6 + 2]cycloaddition of an aldehyde carbonyl group to cycloheptatriene has been reported by Yang and Chiang. Prinzbach and co-workers have described what appears to be a a photochemical $\pi2_s + \pi2_a + \pi2_a + \sigma2_s$ process. Possibly the most remarkable intermolecular photoprocess to be reported during the year is the photodimerization of LiC≡CLi to tetralithiotetrahedrane in liquid ammonia (Rauscher *et al.*).

Fewer papers have appeared than previously on the photochemistry of simple carbonyl compounds in solution, in continuation of a trend noted in previous years: the principal emphasis is on physical aspects. (These remarks do not apply to the photochemistry of enones, *etc.* which continues to attract a strong band of devotees.) Norrish Type II elimination in a crown ether bearing keto-groups has been shown to be enhanced strongly by sodium or potassium ions, but to only a small extent by lithium ions (Hautala and Hastings). The photochemistry of some carbonyl compounds can be influenced by the presence of catalytic amounts of sodium ethoxide to promote enolization: Padwa and Owens have described an interesting example this year, using a chromanone. Although the effects of Brønsted acids on the photochemistry of carbonyl compounds has been known for some years, effects of Lewis acids have only more recently been attracting attention. Childs and Hor have reported an interesting example of the latter with eucarvone. Complexing with boron trifluoride or protons appears to reverse the order of $n\pi^*$ and $\pi\pi^*$ state energies, the latter becoming the lower (see also Snyder and Testa).

Much interest continues in the photochemistry of formaldehyde and glyoxal. Lucchese and Schaefer have made the provocative suggestion that some aspects of the photochemistry of formaldehyde could be explained in terms of an initial isomerization to hydroxycarbene HOCH. Kuttner *et al.* have reported that the previously described effect of a magnetic field on the fluorescence quenching of glyoxal is only observed in the presence of collision partners. The previously reported existence of an 'unquenchable' excited singlet acetaldehyde species has not been confirmed in a study by Gandini and Hackett.

The manifold complexities of ketone photoreduction may provide an exception to the earlier remarks concerning a lower level of interest in the photochemistry of unconjugated carbonyl compounds. Attention is drawn to some puzzling concentration effects observed in the photoreduction of 3,3,5-trimethyl-cyclohexanone (Despax *et al.*), the specific functionalization at C-5 and C-6 achieved in the *solid state* photoreductive addition of acetone to deoxycholic acid in a 1,2 complex of the reactants (Lahav *et al.*), and the application by Breslow and co-workers to studies of the conformations of long-chain alkyl groups. The work of Ampo and Kubokawa provides an important reminder that hydrogen-abstraction in the photoreduction of a ketone can be followed by retro-transfer of a (different) hydrogen to give an enol of the original ketone, with diminution of the apparent quantum yield: *cf.* Blank *et al.* The photoreduction of carbohydrate acetates and pivalates in aqueous hexamethylphosphoric triamide provides an efficient synthetic route to deoxy-sugars (Pete and Portella; Collins and Munasinghe). Rousseau *et al.* have described a useful partly photochemical procedure for the conversion of unconjugated carbonyl compounds into $\alpha\beta$-unsaturated aldehydes and esters: for example, methyl cyclohexene-1-carboxylate is obtained from cyclohexanone.

Van der Westhuizen *et al.* have described a photoprocess which is somewhat analogous to the benzil–benzilic acid rearrangement. Epling and Jones have demonstrated convincingly that photodecarboxylation of phenylacetic acid is not a concerted process, but involves formation of the benzyl radical. Givins *et al.* have used an [18]O label to demonstrate that oxygen scrambling can occur on photolysis of esters. The photolability of carboxylic esters is exemplified by the photoreduction of carbohydrate acetates mentioned above, and by Binkley's mild photochemical procedure for oxidation of carbohydrates *via* pyruvate photolysis.

The photochemistry of carbonium ions is still a comparatively blank area on the map. Some idea of the riches awaiting discovery is provided by Cornell and Filipescu's report that irradiation of the octalone (5) in concentrated sulphuric acid yields (6) at 300 nm and (7) at 254 nm. Steroid chemists please note! Pavlik *et al.* have described the photochemical formation of a furyl cation derivative by irradiation of the corresponding pyrylium cation: an 'oxygen-walk' mechanism is probably involved.

(5)   (6)   (7)

Natural products synthesized by photochemical procedures during the year include the mould metabolite terrein (Barton and Hulshof), dimethylcrocetin (Quinkert, Barton, and their co-workers), β-apolignan (Salomon and Sinha), and compounds related to the antibiotic adriamycin (Jung and Lowe). The terrein synthesis involved isomerization of a γ-pyrone to a zwitterionic cyclopentadienone epoxide intermediate, a type of process also investigated by Barltrop, Day, and Samuel for a simpler system.

The use of surfactants appears able to promote the photodimerization of acenaphthylene in micelles: this technique may well have wider applications.

Cyclobutadienes still attract interest, and the formation of tetrakis(trifluoromethyl)cyclobutadiene (8) by irradiation of the ozonide (9) in a matrix at 77 K is noteworthy (Masamune *et al.*).

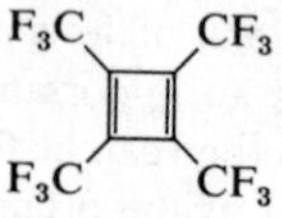

(structures 8 and 9 shown at top)

F_3C, CF_3, F_3C, CF_3 (8)

F_3C, O, CF_3, F_3C—O—O—CF_3, F_3C, CF_3 (9)

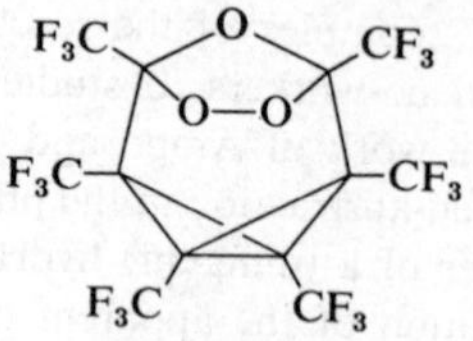

Interest continues on the photochemical generation of nitrile ylides from azirines, and interesting intermolecular and intramolecular examples of 1,3-dipolar cyclization of these ylides on to ethylenic and acetylenic bonds have been described by Padwa and his co-workers.

The photoisomer of pyridazine 1,2-dioxide has been reformulated as (10) by Arai *et al.* The formation process seems to be unprecedented. The photo-Beckmann rearrangement of oximes seems established in some cases as a fully concerted singlet process, the chirality of the migrating groups being retained (Suginome and Yagihashi), but non-Beckmann cleavage products have been reported by reactions in methanol: the solvent participates as a proton donor (Okazaki *et al.*). Suschitsky and his co-workers have observed an interesting 'nitrogen walk' reaction on irradiation of *p*-tolyl azide in ethane thiol to give the *m*-toluidine derivative (11). Work by Chapman and Le Roux indicates that the singlet photochemistry of phenyl azide may need to be reinterpreted in terms of the transitory formation (*via* phenylnitrene ?) of 1-azacyclohepta-1,2,4,6-tetraene (12). Schneider and Csacsko have shown that the photodecomposition of certain pyrazolines can differ from that of other azo-compounds in that it involves the intermediate formation of an isomeric diazo-compound. (Pyrazoles themselves are already known to react in this way.)

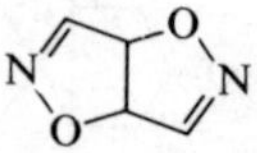

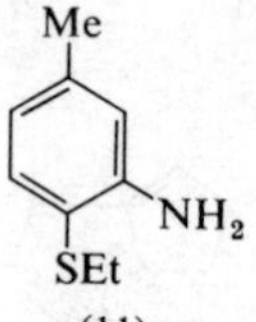

(structures 10, 11, 12 shown)

(10)

Me, NH_2, SEt (11)

N (12)

The long-standing controversy over the question of perepoxide intermediates in the addition of singlet oxygen to alkenes has continued unabated during this review year, and now shows promise of rivalling the classic debate over non-classical carbonium ions. For the latest lengthy episode, see Part III, Chapter 5. It is particularly interesting that Jefford and Boschung have observed that the products from dye-photosensitized photo-oxidation of hindered alkenes can depend critically on the choice of sensitizer. Thus methylene blue gives mainly the dioxetan, whereas erythrosin or Rose Bengal largely give the epoxide, the latter possibly because of their ability to form relatively stable radical-cations.

Interesting work on silacyclopropenes has been reported this year by Sakurai and Seyferth, *inter alia*. Kobayashi *et al.* have observed the formation of a diphosphabenzvalene from the aromatic species (13). This represents the first example of a stable benzvalene-type species having two hetero-atoms in the

$$\begin{array}{c} F_3C \quad\quad P \quad\quad CF_3 \\ \\ F_3C \quad\quad P \quad\quad CF_3 \end{array}$$

(13)

ring. Inous and Takeda have reported the photo-insertion of carbon dioxide into a C—Al bond, sensitized by 1-methylimidazole.

The photolysis of metallocenes Cp_2MR_2 has previously been considered to give the corresponding R· radicals plus Cp_2M, but studies by Rausch *et al.* *inter alia* have indicated that the reaction is more complex than this and may not always give free radicals to any great extent. On the other hand, Samuel *et al.* have reported the detection of such radicals (methyl) as spin-trap adducts. The photodecomposition of chromium hexacarbonyl appears to provide a rare example of an adiabatic photoprocess in that the dichromium nonacarbonyl product is obtained in an excited state (Turner *et al.*). Quantum mechanical calculations support this proposal (Burdett *et al.*, Hay). The triatomic metal atom clusters Cr_2Mo and $CrMo_2$ have been prepared by a photochemical procedure which may have applications in the production of heterogeneous catalysts (Klotzbücher and Ozin). The clusters Ag_2 and Ag_3 have also been prepared (Ozin and Hubek).

Nesmeyanov *et al.* have reported an unusual periodic oscillation of optical density on irradiation of octamethylferrocene or its ferricinium ion in ethanol. Barton *et al.* have shown experimentally that the matrix-isolated species $Fe(CO)_4$ has a triplet ground state, as earlier predicted by Burdett.

The new carbaborane species $Me_4C_4B_4H_4$ has been prepared by photo-reaction of but-2-yme with ferraborane $(B_4H_8)Fe(CO)_3$ (Fehlner).

The irradiation of certain vanadium alkoxide chelate complexes (VOL_2OR; L = 8-quinolyloxo, R = alkyl) appears to provide a useful new route to RO· radicals (Aliwi and Bamford).

Despite the failure to confirm the previous report of the photodissociation of water by $(Ru\ bipy_3)^{2+}$ complexes, these species continue to attract wide attention. Lehn and Sauvage have described a particularly interesting system containing mixed Ru and Rh complexes together with colloidal platinum and triethanolamine which evolves hydrogen on irradiation with *visible* light. Mann and co-workers have likewise reported the photochemical production of hydrogen using visible light (546 nm) by irradiation of the rhodium complex (14) in 12M-HCl.

$$[Rh_2(NCCH_2CH_2CH_2CN)_4]^{2+}(BF_4)_2^{2-}$$

(14)

Donohue has described an interesting photochemical procedure for separating europium from lanthanide mixtures in aqueous isopropanol. Photo-oxidation reactions of aqueous Eu^{2+} ions, which produce hydrogen, are attracting increased attention because of their possible use for solar energy conversion (Davis *et al.*, Ryason).

In the field of SO_2 photochemistry, the most notable development during the year has been the recognition of a pressure saturation effect in the phosphorescence quenching of $SO_2(^3B_1)$ by quenchers such as CO_2, CO, N_2, and C_2H_2. Thus the idea of three triplet states of sulphur dioxide now appears to be redundant (Su *et al.*, Rudolph and Strickler: *cf.* Partymiller and Heicklen).

One often hears it said that photochemistry lacks any really important industrial applications, other than photography. Although there may be an element of truth in this, it is also true to say that the great upsurge of activity in the field of polymer photochemistry has arisen in large measure from the interest shown by the industrial sector. For example, photochromic systems are beginning to look attractive for information storage. Allen and McKellar's study of photostabilizing effects in nylon 66 of metal salts and other species in relation to luminescence quenching provides a good example of an academic investigation having strong relevance to problems in applied chemistry. Bamford *et al.* have reported an interesting technique whereby dibenzazepine groups are built in at each end of a polystyrene or poly(methyl methacrylate) chain: these permit further chain extension *via* photodimerization of the dibenzazepine units on irradiation in the presence of a sensitizer such as benzil.

Crivello *et al.* have reported that $Ar_2I^+ SbF_6^-$ species are very efficient initiators for photocationic polymerization. Rather surprisingly, aqueous *N*-laurylpyridinium azide has been reported to be an effective initiator for the photopolymerization of styrene (Takeishi *et al.*).

Debate continues on the relative importance of hydroperoxides and $\alpha\beta$-unsaturated carbonyl moieties in the photo-oxidation of polypropylene *etc.* (Scott and co-workers; Allen and McKellar). Any idea that the former are generally of sole or predominant importance is no longer tenable.

Finally, attention is drawn to the theoretical study by Giles, Walsh, and Sinclair for the light fading of dyes on polymer substrates. This suggests that for large and small dye particles, the fading rate is proportional to $1/a^2$ and $1/a$, respectively, where a = radius; but for the smallest particles, the fading rate tends to be independent of a.

D. BRYCE-SMITH.

Part I

PHYSICAL ASPECTS OF PHOTOCHEMISTRY

1
Developments in Instrumentation and Techniques

BY M. A. WEST

1 Introduction

Instrumentation and techniques in photochemistry have probably advanced more in the last two years than over any other comparable period. Developments published during the period July 1976 to June 1978 are covered in this Report and papers cited are designed to be representative of the mass of literature covering photochemistry and related subjects, though not comprehensive. It has often been said that advances in photochemical techniques would have been impossible without parallel advances in electro-optics and electronics. The developments in lasers, without doubt the physical photochemists' most useful and applicable excitation source, have enabled unprecedented advances to be made in many areas. For example, in high resolution spectroscopy, i.r. diode lasers now offer spectral resolution of 10^{-4} cm^{-1}; fast pulsed techniques now employ tunable picosecond dye lasers for both transient absorption and emission measurements; and selective photochemistry enables high power tunable lasers to be employed effectively for isotope separation and preparative photochemistry.

The reader is referred to two particular issues of *Physics Today*[1,2] which include introductory review articles in i.r.–laser-induced-unimolecular reactions (by N. Bloembergen and E. Yablonovitch), rare-gas halide lasers (J. J. Ewing), and sub-picosecond spectroscopy (E. P. Ippen and C. V. Shank);[1] also physics and photochemistry (V. S. Letokhov), high-resolution spectroscopy of atoms, molecules (T. W. Hansch), and coherent Raman spectroscopy (M. D. Levenson).[2] A compilation of papers reviewing the state-of-the-art of lasers in chemistry included sections on laser Raman and other scattering, pollution and combustion, atomic and molecular spectroscopy, isotope separation and selective excitation, fast pulsed techniques, and developments in lasers and laser techniques.[3]

Although little information appears in the open literature on laser separation of uranium isotopes, there is no doubt that this photochemical process will have a significant impact on fuel processing costs if it can be made economically viable. Equally important in the energy context, of course, is the separation of deuterium for production of heavy water, and there are many laboratory examples (see *Laser Focus*, July 1978, p. 22) of laser-induced enrichment of this particular isotope.

[1] *Physics Today*, May 1978.
[2] *Physics Today*, May 1977.
[3] 'Lasers in Chemistry', ed. M. A. West, Elsevier, Amsterdam, 1977.

Apart from advances in laser applications, the reader will notice that great strides have been made in photoacoustic spectroscopy and the applications of vidicon detectors: *cf.* Vols. 6 and 8. The thermal lens effect offers a new way of determining weak absorption and the quantum yield of fluorescence. Several methods have also appeared for high sensitivity atomic detection at levels of 1 atom cm^{-3}.

2 Plasma Sources

Despite the advances in reducing the wavelength of lasers in the vacuum-u.v., electric discharge and microwave-powered lamps still offer the most useful continuous sources for photochemical applications. For example, a miniature cold-cathode discharge source using flowing neon with emission lines at 73.6 and 74.4 nm maintained its flux to *ca.* $\pm 2\%$ with continuous use over 6 months.[4] A commercially available d.c.-excited cold-cathode design of housing provided Lyman-α radiation of high spectral purity by using uranium hydride in a sealed tube as the source of molecular hydrogen.[5] A flowing hollow-cathode lamp produced intense ion resonance radiation (from Ca, Ba, Zn, Mg, Sr, Yb, or Eu) at *ca.* 1 mW into $4\pi/25$ steradians.[6] Sulphur hexafluoride at 8 K has been photolysed with a microwave-excited argon discharge in a windowless vacuum-u.v. photolysis lamp with principal lines at 105 and 107 nm.[7] The emission spectra of a similarly excited bromine lamp (used for the gas-phase photolysis of pent-1-ene)[8] and a krypton supersonic jet excited by an electron beam[9] have also been published.

Experiments have shown that the coaxial orifice discharge tube is an effective windowless resonance lamp for argon and neon photolysis of matrix-isolated molecules such as CCl_4.[10] A current-regulated power supply for cold cathode discharge lamps has maintained a current deviation of $< \pm 0.08\%$ with hydrogen.[11]

A vacuum-u.v. and soft *X*-ray radiation source has been proposed, which was based on spontaneous anti-Stokes scattering from an atomic population stored in a metastable level. The brightness of this source was predicted to be higher than any other laboratory-scale source and could be of picosecond durations.[12] Strong quasi-uniform continua covering the wavelength region 4—200 nm were produced by focusing the output of a Q-switched ruby laser (1 J) on to the rare earth metals suggesting use as a reliable background source for absorption spectroscopy in the vacuum-u.v. and soft *X*-ray region.[13]

[4] S. F. Somerstein, *Appl. Optics*, 1976, **15**, 2969.
[5] A. L. Buck, *Appl. Optics*, 1977, **16**, 2634.
[6] F. H. K. Rambow and L. D. Schearer, *Rev. Sci. Instr.*, 1977, **48**, 92.
[7] R. R. Smardzewski, *Appl. Spectroscopy*, 1977, **31**, 332.
[8] A. Wieckowski and G. J. Collin, *Canad. J. Chem.*, 1978, **56**, 1435.
[9] E. T. Verkhovtseva, E. A. Katrunova, A. E. Ovechkin, and Y. M. Fogel, *Chem. Phys. Letters*, 1977, **50**, 463.
[10] F. T. Prochaska and L. Andrews, *J. Chem. Phys.*, 1977, **67**, 1091.
[11] J. C. Traeger and G. W. Haertel, *J. Phys. (E)*, 1977, **10**, 678.
[12] S. E. Harris, *Appl. Phys. Letters*, 1977, **31**, 498.
[13] P. K. Carroll, E. T. Kennedy, and G. O'Sullivan, *Optics Letters*, 1978, **2**, 72.

The successful application of a commercially available radiometric voltage-to-frequency converter has resulted in the reduction of light source fluctuations with tungsten–halogen and xenon arc sources.[14] Modulated d.c. currents through compact arc lamps (such as xenon and krypton) have been shown to induce arc instabilities at discrete frequencies which are sufficiently severe to extinguish the lamp.[15] An arc lamp pulsing unit incorporating thyristor switching has been described which provides pulses variable from 10 to 990 µs.[16] The emission spectra of two identically-labelled types of fluorescent blacklight lamp (low pressure mercury lamps with u.v. phosphor coatings) have been reported to be significantly different with emission peaks at 350 and 365 nm.[17] The medium pressure mercury lamp has been shown to be an excellent source of far-i.r. radiation, with an effective blackbody temperature of 3000 K[18] and a stabilized d.c. supply for a 125 W mercury lamp has been shown to eliminate frequency modulation in a far-i.r. spectrometer.[19] Intense visible emission from Ca and Mg hydrides (620—640 and 470—610 nm, respectively) in an r.f.-excited heat pipe oven was produced with electrical efficiencies as high as 5% suggesting possible use as an efficient narrow band light source.[20]

3 Laser Sources

It is again appropriate to comment on safety codes regarding eye protection against laser radiation. There are two principal guides which are recommended and which provide information on maximum permissible levels (MPEs) under a variety of laser exposure conditions. The American National Standard published in 1976[21] contains a collection of tables and figures for a variety of experimental conditions. The earlier published (1972) British Standards guide also gives corneal MPEs for common lasers under a narrower range of operating conditions.[22] Although there is a mass of accumulated data on threshold values for laser damage, it must be stressed that only a few measurements of threshold values for the human eye have been determined and since the publication of both guides, reports have appeared of much lower threshold values for certain of the newer types of lasers, especially mode-locked and u.v. lasers. Picosecond pulses in particular have been shown to be potentially hazardous[23] and the revised British standard (to be published in 1978) requires that no greater than 10^{-9} J cm^{-2} be incident on the cornea for a time duration of some picoseconds. A comprehensive bibliography of laser radiation hazards in biological systems

14 A. de Sa and D. G. McCartan, *J. Phys. (E)*, 1976, **9**, 725.
15 C. F. Gallo and W. L. Lama, *Appl. Optics*, 1977, **16**, 819.
16 J. J. Meyer and J. Aubard, *Rev. Sci. Instr.*, 1978, **49**, 695.
17 P. D. Forbes, R. E. Davies, L. C. D'Aloisio, and C. Cole, *Photochem. and Photobiol.*, 1976, **24**, 613.
18 R. J. Emery and H. A. Gebbie, *Infrared Physics*, 1977, **17**, 231.
20 W. D. Slafer and D. J. Bernard, *Appl. Phys. Letters*, 1978, **32**, 654.
21 'American National Standard for the Safe Use of Lasers', American National Standards Institute, Report Z 136.1, 1976.
22 'Guide on Protection of Personnel against Hazards from Laser Radiation', British Standards Institution, BS 4803, 1972.
23 D. C. Winburn, *Electro-Optical Systems Design*, November 1977.

has been published [24] and a useful collection of articles related to laser measurements for a regulatory standard includes information on types of detectors, calibration, and measurements of beam properties. [25]

CW Lasers.—Sources of coherent u.v. radiation, emitting more than 1 W CW power are attractive for many applications such as spectroscopy, dye-laser pumping, photochemistry, and isotope separation. At present, tunable u.v. is best obtained from pulsed sources (although recent work with intracavity SHG in dye lasers is very encouraging) but CW operation offers better amplitude stability and beam quality and is usually not accompanied by electromagnetic interference. At present noble gas ion lasers still represent the highest average power sources in the u.v. region even though excimer lasers of the noble gas monohalide type hold great promise for the future. CW laser action of up to 16 W from the 351.1 and 363.8 nm lines of Ar^{III}, 6.7 W from the 351 and 356 nm lines of Kr^{III} and 1.8 W from the 375 and 378 nm lines of Xe^{III} has been observed in purely wall-confined noble gas discharges with a report that the maximum output is limited only by the rapid optical degradation of the u.v. cavity mirrors. [26] Higher powers from a highly ionized, low pressure discharge [27, 28] of up to 61 W from the combined 351 and 364 Ar lines have been obtained with an axial homogeneous magnetic field of 20 G. [28]

Twenty new laser transitions have been measured from a helium discharge in a gold-plated hollow cathode spanning the region 253—763 nm. [29] In general, lasers of this type require lower threshold currents (<4 A) than the noble gas ion lasers. A multiline output of 125 mW has been demonstrated in the 250—290 nm region. [29] CW laser transitions in Ne–Ag and He–Ag mixtures extended to 224.3 nm, with the strongest output at 318.1 nm of more than 350 mW. [30] A preliminary report has been made of CW laser action in Ne^{II} producing 6 W at 332.4 nm. [31] New u.v. ion laser transitions in F_2, Cl_2, I_2, Br_2, and S_2 look promising for future CW operation with three new transitions found in Br_2 below 280 nm. [32] The strongest line at 236.2 nm produced a peak power of 60 W. Generation of tunable CW u.v. radiation was reported earlier (Vol. 4, p. 87; Vol. 5, p. 81) using intracavity SHG of a CW rhodamine 6G (Rh6G) dye laser but over a limited tuning range (285—315 nm). A convenient method has now been developed for generating broadly tunable u.v. from 285 to 400 nm by SHG and sum frequency mixing. [33] Two pump lasers (5 W Ar and 5 W Kr),

[24] P. W. Crockett, 'Biological Effects of Laser Radiation; a bibliography with abstracts', NTIS/PS–77/0028, 1977.

[25] 'National Conference on Measurements of Laser Emissions for Regulatory Purposes', Bureau of Radiological Health, Report DHEW/PUB/FDA–76/8037, 1976.

[26] T. K. Tio, H. H. Luo, and S.-C. Lin, *Appl. Phys. Letters*, 1976, **29**, 795.

[27] H. R. Luthi, W. Seelig, and J. Steinger, *Appl. Phys. Letters*, 1977, **31**, 670; H. R. Luthi, *J. Appl. Phys.*, 1977, **48**, 664.

[28] H. R. Luthi and W. Seelig, *Z. Angew. Maths. Phys.*, 1977, **28**, 1168; *J. Appl. Phys.*, 1977, **49**, 4922; H. R. Luthi, W. Seelig, J. Steinger, and W. Lobsiger, *I.E.E.E. J. Quantum Electron.*, 1977, QE-13, 404.

[29] R. D. Reid, D. R. McNeil, and G. J. Collins, *Appl. Phys. Letters*, 1976, **29**, 666.

[30] J. R. McNeil, W. L. Johnson, G. J. Collins, and K. B. Persson, *Appl. Phys. Letters*, 1976, **29**, 172.

[31] H. R. Luthi and J. Steinger, *J. Opt. Soc. Amer.*, 1978, **68**, 703.

[32] J. Marling, *I.E.E.E. J. Quantum Electron.*, 1978, QE-14, 4.

[33] S. Bilt, E. G. Weaver, T. A. Rabson, and F. K. Tittle, *Appl. Optics*, 1978, **17**, 721.

a single dye laser (either a Rh6G or Oxazine 1) and three non-linear crystals (ADA, ADP, and RDP) were used. The output of the dye laser was either used directly in extracavity SHG or combined with the Kr laser by means of a dichroic mirror. The maximum power at 313 nm was reported to be 0.75 mW with a linewidth of 0.016 nm and powers in excess of 5 μW over the range 290 to 390 nm. Over 30 mW of CW radiation in the 285—315 nm range was produced by doubling the CW Rh6G laser with either ADA[34] or ADP[35] in the cavity. The conversion efficiency was shown to be limited by the onset of thermally-induced phase mismatching. Peak u.v. output was increased to 85 mW if the radiation from the laser was chopped.[35] U.v. laser lines having characteristics close to those presented in the visible by CW tunable single mode dye lasers were produced by superposition of both a CW Rh6G and pulsed (doubled Nd : YAG) laser. The recorded linewidth of less than 20 MHz was shown to be suitable for a high resolution study of rubidium Rydberg states.[36] Laser action at five lines between 358 (maximum power 7 mW) and 747 nm (1 mW) has been observed by exciting neon buffer gas in an aluminium hollow cathode.[37] A similar arrangement with a silver cathode produced 18 lines from 408.6 to 585.2 nm.[38]

A long-life He–Cd laser tube (greater than 1000 h) has been described in which sputter protection of the cathode was achieved by a cadmium film coverage.[39] The additions of small amounts of iodine to this laser discharge was shown to increase the power of the 441.6 nm line by 30—40%.[40] The He–Cd laser was shown to exhibit self mode-locking at 325 nm and retinal damage levels with this laser (0.36 J cm^{-2}) seem anomalously low when contrasted with similar wavelengths for Ar and Kr lasers where retinal damage is not reproducibly incurred at corneal thresholds of 67 J cm^{-2}.[41] It is not known if this sensitivity is related to the mode-locked nature of the laser.

Laser oscillation on transitions of Cd^{2+} in He–CdI$_2$,[42,43] He–CdCl$_2$, and He–CdBr$_2$ and similar transitions in Zn^{2+} compounds[44] have been reported over the range 442—888 and 491—759 nm, respectively.[42] The hollow cathode He–Cd laser operating on the red (636, 635.5 nm), green (537.8, 533.7 nm), and blue (441.6 nm) offers the possibility of obtaining efficient blue, green, or white light oscillation from a single laser tube.[45] A hollow cathode discharge with internal anodes has been shown to operate at significantly higher discharge voltages than conventional hollow cathode discharges allowing use of lower

[34] A. I. Ferguson and M. H. Dunn, *Optics Comm.*, 1977, **23**, 177.
[35] A. I. Ferguson, M. H. Dunn, and A. Maitland, *Optics Comm.*, 1976, **19**, 10.
[36] J. Pinard and S. Liberman, *Optics Comm.*, 1977, **20**, 344.
[37] D. C. Gerstenberger, R. D. Reid, and G. J. Collins, *Appl. Phys. Letters*, 1977, **30**, 466.
[38] W. L. Johnson, J. R. McNeil, G. J. Collins, and K. B. Persson, *Appl. Phys. Letters*, 1976, **29**, 101.
[39] K. G. Hemqvist, *I.E.E.E. J. Quantum Electron.*, 1978, **QE-14**, 129.
[40] T. Arai and T. Yabumoto, *I.E.E.E. J. Quantum Electron.*, 1977, **QE-13**, 405; T. Arai, T. Yabumoto, and T. Goto, *ibid.*, 1978, **QE-14**, 374.
[41] J. A. Zuclich and J. Taboada, *Appl. Optics*, 1978, **17**, 1482.
[42] J. A. Piper and M. Brandt, *J. Appl. Phys.*, 1977, **49**, 4486.
[43] J. A. Piper, *Optics Comm.*, 1976, **19**, 189.
[44] P. A. Gill and J. A. Piper, *Optics Comm.*, 1977, **22**, 280.
[45] L. Csillag, C. Z. Nam, M. Janossy, and K. Rozsa, *Optics Comm.*, 1977, **21**, 39.

discharge currents and thereby increasing the output of noble gas mixtures. CW laser oscillation was reported at 531.4 and 486.3 nm from Xe^{II}.[46]

Up to 52 W of dye laser power was obtained from a Rh6G jet stream dye laser pumped by all the lines of a high power Ar laser (175 W). No thermo-optical problems were reported with water as the solvent with a viscosity-raising additive (polyvinyl alcohol).[47] (Other dye laser studies are reported later.) Other CW lasers in the visible include molecular iodine pumped by the 514 nm Ar line (wavelength range 583—1338 nm at a power conversion efficiency of 8%),[48] a practical sealed-off He–I$^+$ laser (emission at 541, 568, 613, and 658 nm with a total power of 20—30 mW)[49] and quasi-CW operation of three laser lines (479.7, 521.0, and 650.1 nm with respective output powers of 25, 5, and 40 W) in a pure mercury low pressure discharge.[50] The first visible laser having nuclear energy as the only source of excitation was reported to be He–Hg (lasing at 615 nm) pumped by $a^1n(^{10}B, a)$ ^{7}Li nuclear reactor.[51]

In the near-i.r., Ar^{III} was shown to lase at 692 and 747 nm with a slot cathode discharge[52] and Cu^{II} produced laser oscillations at 740 and 790 nm in a hollow anode–cathode discharge.[53] Up to 300 mW of true CW laser power was reported from ruby at 77 K pumped by the 514 nm Ar laser line.[54]

Alkali halide host crystals containing F-centres, which have wide fluorescence bandwidths, have been optically pumped to produce tunable emission in the 0.8—3 μm band.[55] The advantage of these lasers is the long fluorescence life-times, the coincidence of the pump bands with Nd : YAG and GaAs laser wavelengths, and the strong refractory nature of the host crystals.[56]

CW laser action at room temperature between 1.06 and 1.12 μm has been obtained from double-heterojunction structures of $In_xGa_{1-x}/In_yGa_{1-y}P$[57] and other developments in semiconductor lasers has been reported in a collection of 34 papers.[58] Laser emission has even been found with a single crystal Nd : YAG fibre end pumped by a single high radiance LED.[59]

Chemical lasers operate on a population inversion produced directly or indirectly in the course of an exothermic chemical reaction and since their

[46] K. Rozsa, M. Janossy, J. Bergou, and L. Csillag, *Optics Comm.*, 1977, **23**, 15.
[47] P. Anliker, H. R. Luthi, W. Seelig, J. Steinger, H. P. Weber, S. Leutwyler, E. Schumacher, and L. Woste, *I.E.E.E. J. Quantum Electron.*, 1977, **QE-13**, 547.
[48] B. Wellegehausen, K. H. Stephen, D. Friede, and H. Welling, *Optics Comm.*, 1977, **23**, 157.
[49] T. Goto, H. Kao, N. Yoshino, J. K. Mizeraczyk, and S. Hattori, *J. Phys. (E)*, 1977, **10**, 292.
[50] P. Burkhard, H. R. Luthi, and W. Seelig, *Optics Comm.*, 1976, **18**, 485.
[51] M. A. Akerman, G. H. Miley, and D. A. McArther, *Appl. Phys. Letters*, 1977, **30**, 409.
[52] W. K. Schuebel, *Appl. Phys. Letters*, 1977, **30**, 516.
[53] K. Rozsa, M. Janossy, L. Schillag, and J. Bergou, *Optics Comm.*, 1977, **23**, 162.
[54] T. N. C. Venkatesan and S. L. McCall, *Rev. Sci. Instr.*, 1977, **48**, 539; M. Birnbaum, A. W. Tucker, and C. L. Fincher, *I.E.E.E. J. Quantum Electron.*, 1977, **QE-13**, 808.
[55] L. F. Mollenauer, *J. Opt. Soc. Amer.*, 1978, **68**, 636; *Optics Letters*, 1977, **1**, 164; G. Liftin, R. Beigang and H. Welling, *J. Phys. (E)*, 1977, **31**, 381; *I.E.E.E. J. Quantum Electron.*, 1977, **QE-13**, 31D.
[56] R. Beigang, G. Litkin, and H. Welling, *Optics Comm.*, 1977, **22**, 269.
[57] C. J. Neuss, G. H. Olsen, M. Ettenberg, J. J. Gannon, and T. J. Zamerowski, *Appl. Phys. Letters*, 1976, **29**, 807.
[58] *I.E.E.E. J. Quantum Electron.*, 1977, **QE-13**, No. 8, pp. 555—725.
[59] J. A. Stone, C. A. Burrus, A. G. Dentai, and B. I. Miller, *Appl. Phys. Letters*, 1976, **29**, 37.

discovery (in 1964), numerous studies have been made[60] largely on the high power HF and DF,[61] CO and I* lasers. Although currently reported devices are i.r. emitters, the lowest reported wavelength is close to 800 nm and it is likely that very high power lasers operating in the visible region will be produced, probably from metal-oxide systems.[62] A purely chemical HCl laser in which Cl atoms were produced by a branched chain reaction between NO and ClO_2 produced a multiline power of 4 W and a chemical efficiency of 6.6%.[63] The first example of CW laser action on a transition between two distinct electronic states which was pumped only by a chemical reaction and requiring no external sources of power was achieved on the $^2P_{\frac{3}{2}} \rightarrow {}^2P_{\frac{1}{2}}$ transition of the iodine atom by energy transfer from the $^1\Delta_g$ metastable state of O_2.[64] The excited oxygen was generated chemically by flowing chlorine gas through a basic solution of 90% H_2O_2 and the effluent mixed with molecular iodine at the entrance to a longitudinal flow laser cavity.

A detailed description of a liquid N_2-cooled CW CO laser was given in a significant publication describing monitoring techniques for CO produced by the u.v. photolysis of formaldehyde.[65] A CO chemical laser pumped by the O + CS reaction was shown to produce nearly 1 W of power oscillating on the $1 \rightarrow 0$ vibrational band.[66]

There are numerous examples of pulsed CO_2 lasers and their applications elsewhere in this Report (see Sections 6 and 8) but specific papers on CW CO_2 lasers have included reports of 36 new lines in the 9—11 μm region,[67] an optically-pumped gas-dynamic laser operating on the 14 and 16 μm transitions of $^{12}CO_2$,[68] glow discharge stabilization of transverse flow lasers, [69] a description of the high repetition rate photoionized TEA laser techniques to CW CO_2 lasers which may also be suitable for discharge-pumped noble gas halide lasers,[70] and a single CO_2 laser amplifier producing 40 W of CW power suitable for pumping far-i.r. lasers.[71]

Carbon dioxide laser pumping has been used to produce CW far-i.r. emission

[60] 'Handbook of Chemical Lasers', ed. R. W. F. Gross and J. F. Bott, Wiley, New York, 1977.
[61] K. O. Tan, D. E. Rothe, J. A. Nilson, and D. J. James, *I.E.E.E. J. Quantum Electron.*, 1977, **QE-13**, 27D; L. Bertrand, J. M. Gagne, R. G. Bosision, and M. Moisan, *ibid.*, 1977, **QE-14**, 8; J. Munch, M. A. Kolpin, and J. Levine, *ibid.*, 1978, **QE-14**, 17; L. Bertrand, J. M. Gagne, B. Mongeau, B. Lapointe, Y. Conturie, and M. Moisan, *J. Appl. Phys.*, 1977, **48**, 224.
[62] C. R. Jones and H. P. Broida, *Laser Focus*, March 1974.
[63] S. J. Arnold, K. D. Foster, D. R. Snelling, and R. D. Suart, *I.E.E.E. J. Quantum Electron.*, 1978, **QE-14**, 293.
[64] W. E. McDermott, N. R. Pchelkin, D. J. Bernard, and R. R. Bousek, *Appl. Phys. Letters*, 1978, **32**, 469.
[65] P. L. Houston and C. B. Moore, *J. Chem. Phys.*, 1976, **65**, 757.
[66] W. Q. Jeffers, H. Y. Ageno, C. E. Wiswall, and J. D. Kelley, *Appl. Phys. Letters*, 1976, **29**, 242.
[67] J. Reid and K. Siemsen, *Appl. Phys. Letters*, 1976, **29**, 250.
[68] T. J. Manuccia, J. A. Stregack, N. W. Harris, and B. L. Wexler, *Appl. Phys. Letters*, 1976, **29**, 360.
[69] Z. Bauman, F. Dothan, and S. Yatsiv, *J. Phys. (E)*, 1978, **11**, 189.
[70] H. J. J. Sequin, A. K. Nam, and J. Tulip, *Appl. Phys. Letters*, 1978, **32**, 418.
[71] B. W. Jolliffe, P. T. Woods, and D. J. Ravenhill, *Optics Laser Technol.*, 1978, **10**, 86.

from HCHO and DCDO (195—949 m)[72] and NH_3.[73] Other lasers operating in the far-i.r. include HCN (at 337 μm),[74] DCN (190 and 195 μm),[75] and an efficient waveguide laser (771—1222 μm).[76] Far-i.r. laser lines in optically-pumped $MeCHF_2$ were first discovered by use of the opto-acoustic effect.[77]

Pulsed Gas Lasers.—*Excimer Lasers.* The gas-phase excimer lasers reported earlier (Vol. 8, p. 6) are now well developed commercially and offer a new approach to high power tunable u.v. laser output. (Strictly speaking, rare gas halide molecules in their excited state are exciplexes and excimers refer only to excited homonuclear species. In this Report, the term 'excimer' will be used to cover both species.) Reviews on these lasers are given elsewhere.[78]

Of the lasers demonstrated to date, ArF (193 nm), KrF (248 nm), and XeF 351 and 353 nm) yield the highest efficiencies and highest single pulse energies. Simple electric discharge devices have produced energies of 1 J per pulse on the rare gas fluorides and energies in excess of 0.1 J on XeCl, KrCl, and XeBr.

The ArF and KrF systems have homogeneous emission bands while the XeF spectrum is a line structure that is only discretely tunable at reduced power levels. The rare gas excimer lasers in the vacuum-u.v. offer broader bands than the rare gas monohalides and the Xe_2^* laser at 172 nm[79] has been tuned. The practical advantages of excimer lasers (short wavelength, high power) are unfortunately accompanied by high cost of operation and the requirement for inert flow lines and a gas containment system. A closed-cycle recycling system has been employed in which a flowing gas is continuously purified in a high temperature titanium gettering furnace where halogens are removed by reaction to form volatile but condensable metal halides.[80] Degradation of the laser gas mixture in the discharge is also a limitation to laser operation at high pulse repetition frequencies. A 1 litre volume laser head used in a Blumlein-driven XeF excimer laser at pulse rates as high as 200 Hz was connected to a reclamation and recycling process for Xe and NF_3.[81]

Excimer cross-sections are in the range 10^{-17}—5×10^{-16} cm^{-2}, whereas similar atomic and molecular transitions at comparable oscillator strengths lie in the range 10^{-15}—10^{-12} cm^{-2}. Excimer lasers must therefore be vigorously pumped, usually by an electron beam. It has been shown that the small cross-sections for photoionization of metastable rare gas atoms (He*–Xe*) in some gas mixtures and optical windows in the 200—300 nm range are of major

[72] D. Dangoisse, A. Deldalle, J. P. Splingard, and J. Bellet, *I.E.E.E. J. Quantum Electron.*, 1977, **QE-13**, 730.
[73] A. Tanaka, A. Tanimoto, N. Murata, M. Yamanaka, and H. Yoshinaga, *Optics Comm.*, 1977, **22**, 17.
[74] D. D. Bicanic and A. Dymanus, *Infrared Physics*, 1976, **16**, 601; J. Vanderkooy and C. S. Kang, *ibid.*, 1976, **16**, 627.
[75] J. L. Bruneau, P. Belland, and D. Veron, *Optics Comm.*, 1978, **24**, 259.
[76] D. T. Hodges, F. B. Foote, and R. D. Reel, *Appl. Phys. Letters*, 1976, **29**, 663.
[77] G. Busse and R. Thurmaier, *Appl. Phys. Letters*, 1977, **31**, 194.
[78] C. A. Brau, in 'Laser 77 Opto-Electronics Conference Proceedings', ed. W. Waidelich, I.P.C., London, 1977, p. 49; C. W. Werner, E. V. George, P. W. Hoff, and C. K. Rhodes, *I.E.E.E. J. Quantum Electron.*, 1977, **QE-13**, 769.
[79] C. E. Turner, *Appl. Phys. Letters*, 1977, **31**, 659; F. Collier and P. Cottin, *Optics. Comm.*, 1978, **25**, 89.
[80] P. M. Johnson, N. Keller, and R. E. Turner, *Appl. Phys. Letters*, 1978, **32**, 291.
[81] C. P. Christensen, *Appl. Phys. Letters*, 1977, **30**, 483.

importance in laser development.[82] In the vacuum-u.v., electron-beam pumping of molecular Ar produced laser action at 126 nm [83] and of molecular F_2 at 157.5 nm.[84] Electric discharge excitation pulses of 500 ns duration and peak current densities of up to 14 kA cm^{-2} produced lasing on five Ar and Kr ion transitions below 200 nm. In this case, the two strongest emissions showed 0.1—1 kW peak powers from Kr^{IV} at 195.027 and 175.641 nm, which are clearly of interest in photochemistry.[85] A novel pulse generator used for travelling wave excitation in hydrogen provided laser peak powers in the Lyman bands of 45 kW with a pulse of 1 ns duration.[86] Considerable interest has been shown in developing a viable discharge alternative to the electron-beam-pumped excimer laser with an emphasis on pumping KrF and XeF lasers. A simple photoionization-stabilization discharge procedure with a Blumlein-type transverse discharge was used to produce 20—25 ns pulses with XeF and KrF lasers at pressures up to 2×10^5 N m^{-2} and specific energy loadings of up to 0.5 kJ 1^{-1} atm^{-1}.[87] A mechanism has been proposed to explain the improvement in discharge uniformity observed in electric-discharge-pumped lasers with u.v. preionization.[88] Improvements to the Tachisto double discharge excimer laser have been made by alterations to the unstable cavity and gas mixture.[89]

There have been numerous studies of rare gas monohalides especially ArF, KrF, and XeF. Measurement of the radiative lifetimes of the halide transitions is important for optimization studies of these lasers. Excitation of atomic krypton in the presence of F_2 by a wavelength-selective 2.5 ns argon excimer photolysis pulse was used to initiate the reaction sequence resulting in the formation of excited KrF with a radiative lifetime (at the high pressure limit) of 9 ns.[90] As with all excimer lasers, KrF* is potentially scalable to high average powers and signal gain and absorption measurements have been determined by probing the excited medium with a frequency-doubled dye laser.[91] The discharge-pumped KrF* laser has generated 10 W of average power at a repetition rate of 1 kHz. At a more modest pulse rate (2 Hz), a similar device has generated output pulse energies of 100 mJ at 1 % efficiency.[92] Energy outputs of 0.25 J from this laser have been reported for a cable-fed u.v. preionized discharge device [93] and 50 μJ from a fast-pin discharge using transmission line excitation techniques (as well as 4.5 mJ from a nitrogen laser).[94] A constant voltage excitation technique used in a u.v. preionized multiatmosphere KrF* laser has demonstrated scalability to larger volumes and higher operating pressures with,

[82] K. J. McCann and M. R. Flannery, *Appl. Phys. Letters*, 1977, **31**, 599.
[83] D. J. Bradley, M. H. R. Hutchinson, and C. C. Ling, *J. Opt. Soc. Amer.*, 1978, **68**, 702.
[84] J. A. Rice, A. K. Hays, and J. R. Woodworth, *Appl. Phys. Letters*, 1977, **31**, 31.
[85] J. B. Marling and D. B. Lang, *Appl. Phys. Letters*, 1977, **31**, 181.
[86] W. Ross, S. Florek, and J. Gatzke, *Optics Comm.*, 1977, **23**, 29.
[87] V. Hasson, C. M. Lee, R. Exberger, K. W. Billman, and P. D. Rowley, *Appl. Phys. Letters*, 1977, **31**, 167.
[88] J. Hsia, *Appl. Phys. Letters*, 1977, **30**, 101.
[89] D. L. Barker and T. R. Loree, *Appl. Optics*, 1977, **16**, 1792.
[90] G. P. Quigley and W. H. Hughes, *Appl. Phys. Letters*, 1978, **32**, 627, 649.
[91] A. M. Hawryluk, J. A. Mangano, and J. H. Jacob, *Appl. Phys. Letters*, 1977, **31**, 164.
[92] T. S. Fahlen, *J. Appl. Phys.*, 1978, **49**, 455.
[93] R. C. Sze and P. B. Scott, *Rev. Sci. Instr.*, 1978, **49**, 772.
[94] R. C. Sze and P. B. Scott, *J. Appl. Phys.*, 1976, **47**, 5492.

apart from the problem of gas recycling, no inherent repetition rate limitations.[95] Operational characteristics investigated have included the influence of F_2 dissociation and vibrational excitation,[96] the effect of long laser resonators,[97] fluorescence efficiencies as a function of gas composition,[98] spectral widths,[99] and properties of gas mixtures of He : Kr containing F_2, N_2F_4, and SF_6.[100] Studies with electron-beam-pumped mixtures included a comparision of a theoretical model of lasing with experimental results[101] and measurements of the instability onset.[102] Diffraction-limited performance has been achieved in discharge-excited KrF and XeF lasers by using an unstable resonator configuration[103] with peak powers of 5 MW produced by a Blumlein circuit with a stored energy of only 5 J. Simple prism spectral tuning of a double-discharge excimer laser resulted in non-continuous tuning ranges better than 2 nm in both KrF and ArF with interruptions due to gain loss in regions of self-absorption.[104] Electron-beam excitation of Ar–Kr–Cl$_2$ mixtures resulted in laser emission from KrCl* at 222 nm.[105]

Fast Blumlein discharge circuits used for XeF–KrF lasers[106—108] are reported to produce peak powers of 1 MW for XeF[107] and overall efficiencies of 1% (XeF) and 0.3% (KrF).[108] Double-discharge circuits have been used with KrF and ArF lasers.[109] Preionization is essential in many gaseous laser systems for the initiation of stable self-sustained discharges with the minimum of streamers and arcs and although it is usual to prepulse the discharge or to use u.v. photons from an arc source, an α-source of ^{241}Am was shown to be effective with the XeF laser resulting in a tripling of energy output.[110] Kinetic studies of the XeF laser have been made following irradiation of XeF$_2$–gas mixtures with ArCl (175 nm) radiation[111] and in a flowing afterglow apparatus.[112] The XeF efficiency has been increased dramatically (up to 1.8% efficiency) by adding neon as a diluent in place of argon and explained following measurements of

[95] W. J. Sarjeant, A. J. Alcock, and K. E. Leopold, *I.E.E.E. J. Quantum Electron.*, 1978, **QE-14**, 177.
[96] W. L. Nighan, *Appl. Phys. Letters*, 1978, **32**, 297, 424.
[97] R. S. Taylor, W. J. Sarjeant, A. J. Alcock, and K. E. Leopold, *Optics Comm.*, 1978, **25**, 231.
[98] J. A. Mangano, J. H. Jacob, M. Rokni, and A. Hawryluk, *Appl. Phys. Letters*, 1977, **31**, 26; J. H. Jacob, M. Rokni, J. A. Mangano, and R. Brochu, *Appl. Phys. Letters*, 1978, **32**, 109.
[99] J. Goldhar and J. R. Murray, *Optics Letters*, 1977, **1**, 199.
[100] A. J. Andrews, A. J. Kearsley, C. E. Webb, and S. C. Haydon, *Optics Comm.*, 1977, **20**, 265.
[101] W. B. Lacina and D. B. Cohn, *Appl. Phys. Letters*, 1978, **32**, 106.
[102] R. T. Brown and W. L. Nighan, *Appl. Phys. Letters*, 1978, **32**, 730.
[103] T. J. McKee, B. P. Stoicheff, and S. C. Walkee, *Appl. Phys. Letters*, 1977, **30**, 278.
[104] T. R. Loree, K. B. Butterfield, and D. L. Barker, *Appl. Phys. Letters*, 1978, **32**, 171.
[105] J. R. Murray and H. T. Powell, *Appl. Phys. Letters*, 1976, **29**, 252; J. G. Eden and S. K. Searles, *ibid.*, 1976, **29**, 350.
[106] M. Matera and R. Salimbeni, *Optics Comm.*, 1978, **25**, 251.
[107] C. P. Wang, *Appl. Phys. Letters*, 1976, **29**, 103.
[108] R. Burnham, F. X. Powell, and N. Djeu, *Appl. Phys. Letters*, 1976, **29**, 30.
[109] T. R. Loree, R. C. Sze, and R. Begley, *I.E.E.E. J. Quantum Electron.*, 1977, **QE-13**, 62D; S. Watanabe, T. Sato, and H. Kashiwagi, *Optics Comm.*, 1977, **22**, 143.
[110] I. J. Bigio, *I.E.E.E. J. Quantum Electron.*, 1978, **QE-14**, 75.
[111] J. G. Eden and R. W. Waynant, *J. Chem. Phys.*, 1978, **68**, 2850.
[112] J. E. Valazco, J. H. Kolts, D. W. Setser, and J. A. Coxon, *Chem. Phys. Letters*, 1977, **46**, 99.

u.v. absorption spectra and quenching efficiencies.[113] A simple flashlamp-pumped cell in which Xe was pumped at 147 nm and caused to react with NF_3 to form XeF* provided a means of examining the emission from rare gas halides, oxides, and sulphides.[114] XeF fluorescence emission spectra have also been studied following photolysis of XeF_2 with resonance radiation from Kr, Hg, and CO discharge lamps.[115] A scalable u.v.-preionized transverse discharge XeF laser operated at pressures up to 10^5 N m^{-2} has resulted in a u.v. output of >250 mJ at <20 ns from an active volume of 0.18 l.[116] Long pulse operation (1 μs) of this laser with Ar–Xe–NF_3 produced a similar power following electron-beam pumping.[117] Laser action at 500 Hz was reported in a fast flow system with a 3 mJ pulse output.[118] The short gain duration of rare gas halide lasers (<50 ns) prevents effective amplification by either active or passive mode-locking techniques.[119] However, the emission spectrum of XeF is sufficiently broad to permit amplification of pulses a few ps in duration. By using XeF as an amplifier, restrictions placed on the minimum attainable pulse duration by the gain duration are effectively circumvented and 200 ps third harmonic (353 nm) Nd : YAG laser pulses have been amplified by >6000 times in a high pressure XeF discharge[120] attaining peak powers of 50 kW.[120,121] The spectral output of the XeF laser can be controlled by injection locking to provide high spectral power densities on transitions within a single vibronic band.[122] Other studies on XeF have included two and three body quenching with Ar and Xe,[123] electron-beam pumping results with Xe–Ar–NF_3 mixtures,[124] measurements of spontaneous and stimulated emission spectra,[125] and measurements of radiative lifetimes following electron-beam pumping of XeF_2 plasmas.[126] Long pulse laser action has been achieved at several wavelengths near 350 nm in an electron-beam excited supersonic flow with mixtures of Ne–Xe–NF_3 aerodynamically-cooled to 120 K producing 1 mJ in 400 ns FWHM pulses at pressures of 5×10^4 N m^{-2}.[127]

[113] W. R. Wadt, D. C. Cartwright, and J. S. Cohem, *Appl. Phys. Letters*, 1977, **31**, 672; L. F. Champagne and N. W. Harris, *Appl. Phys. Letters*, 1977, **31**, 513; M. Rokni, J. H. Jacob, and J. A. Mangano, *ibid.*, 1978, **32**, 622.

[114] A. Mandl and J. J. Ewing, *Rev. Sci. Instr.*, 1977, **48**, 1434.

[115] H. C. Brashears, D. W. Setser, and D. Desmarteau, *Chem. Phys. Letters*, 1977, **48**, 84.

[116] W. J. Sarjeant, A. J. Alcock, and K. E. Leopold, *Appl. Phys. Letters*, 1977, **30**, 635.

[117] L. F. Champagne, J. G. Eden, N. W. Harris, N. Djeu, and S. K. Searles, *Appl. Phys. Letters*, 1977, **30**, 160.

[118] C. P. Wang, *Appl. Phys. Letters*, 1978, **32**, 260.

[119] C. P. Christense, L. W. Braversman, W. H. Steier, and C. Wittig, *Appl. Phys. Letters*, 1976, **29**, 424.

[120] I. V. Tomov, R. Fedosejevs, M. C. Richardson, W. J. Serjeant, A. J. Alcock, and K. E. Leopold, *Appl. Phys. Letters*, 1977, **30**, 146.

[121] I. V. Tomov, R. Fedosejevs, M. C. Richardson, W. J. Serjeant, A. J. Alcock, and K. E. Leopold, *Appl. Phys. Letters*, 1977, **31**, 747.

[122] J. Golderher, J. Dickie, L. P. Bradley, and L. D. Pleasance, *Appl. Phys. Letters*, 1977, **31**, 677.

[123] J. G. Eden and S. K. Searles, *Appl. Phys. Letters*, 1976, **29**, 356; M. Rokni, J. H. Jacob, J. A. Mangano, and R. Brochu, *ibid.*, 1977, **30**, 458.

[124] J. A. Mangano, J. H. Jacob, and J. B. Dodge, *Appl. Phys. Letters*, 1976, **29**, 426.

[125] J. Tellinghuisen, P. C. Tellinghuisen, G. C. Tisone, J. M. Hoffman, and A. K. Hays, *J. Chem. Phys.*, 1978, **68**, 5177; J. Tellinghuisen, P. C. Tellinghuisen, J. A. Coxon, J. E. Velazco, and D. W. Setser, *J. Chem. Phys.*, 1978, **68**, 5187.

[126] J. G. Eden and S. K. Searles, *Appl. Phys. Letters*, 1977, **30**, 287.

[127] B. Forestier and B. Fontaine, *Appl. Phys. Letters*, 1978, **32**, 569.

Lasing has also been demonstrated in XeCl at 308 nm,[128, 129] in KrCl (222 nm), and XeBr (282 nm),[130] with hydrogen halides and halocarbons as donor molecules in electrical discharges[128] and in XeOH (at 234 nm)[131] following electron-beam pumping. A large volume (6 l) electron-beam-controlled discharge of *ca.* 1 μs duration in Xe : Ar mixtures with HCl has been used to achieve XeCl laser action at 308 nm with an efficiency approaching the best reported for XeF and KrF lasers.[132] U.v. radiation at 248 nm from a KrF* laser absorbed in atomic Xe gas *via* a two-photon process was transferred by collisions with N_2O to XeO* which in turn was identified (with photon counting) by the characteristic green emission band.[133]

Observations have been made of laser action in ArCl excimers (at 175 nm) pumped by a Blumlein discharge circuit[134] and the formation and quenching kinetics of ArF.[135]

Nitrogen Lasers. Further designs and performance figures have been reported for the ever-useful nitrogen laser. A novel coaxial design involved the introduction of an adjustable self-inductance into the pulse generator circuit so as to synchronize the time evolution of the laser discharge voltage with the gas breakdown delay period in such a way that the laser output and stability (better than 3%) were maximized.[136] The output power of only 50 kW at 400 Hz was smaller than many TEA systems but the coaxial design also minimized electromagnetic interference associated with each laser pulse. A travelling wave pulsing system for transverse discharge excitation of nitrogen had an effective inversion time of <1 ns and produced a beam power (at 337.1 nm) of >2 MW.[137] The discharge capacitor in a parallel plate transmission line consisted of aluminium foils separated by Mylar and held in contact by atmospheric pressure.[138] In this case a long output pulse (10 ns) was produced by avoiding a travelling-wave discharge. A compact TEA laser operating at 10^5 N m^{-2} used a laser channel only 25 cm long and produced 1 ns pulses with output powers of 0.5 MW.[139] Other designs included a 125 Hz, 0.6 ns duration TEA laser,[140] a 1 MW peak-power, high stability laser with graphite electrodes,[141] a sealed-off miniature atmospheric pressure laser producing sub-nanosecond pulses with peak powers of 0.2 MW at repetition rates of 1.7 kHz,[142] and a further sealed-off

[128] R. C. Sze and P. B. Scott, *J. Opt. Soc. Amer.*, 1978, **68**, 704.
[129] V. N. Ischchenko, V. N. Lisitsyn, and A. M. Razhev, *Optics Comm.*, 1977, **21**, 30; R. Burnham, *ibid.*, 1978, **24**, 161; R. Shuker, *Appl. Phys. Letters*, 1976, **29**, 785.
[130] R. C. Sze and P. B. Scott, *Appl. Phys. Letters*, 1978, **32**, 479.
[131] M. H. R. Hutchinson, *Chem. Phys. Letters*, 1978, **54**, 359; *J. Opt. Soc. Amer.*, 1978, **68**, 715.
[132] J. I. Lavatter, J. H. Morris, and S.-C. Lin, *Appl. Phys. Letters*, 1978, **32**, 630.
[133] D. Kligler, D. Pritchard, W. K. Bischel, and C. K. Rhodes, *J. Appl Phys.*, 1978, **49**, 2219.
[134] R. W. Waynant, *Appl. Phys. Letters*, 1977, **30**, 234.
[135] M. Rokni, J. H. Jacob, J. A. Mangano, and R. Brochu, *Appl. Phys. Letters*, 1977, **31**, 79.
[136] W. W. Wladimiroff and H. E. B. Anderson, *J. Phys. (E)*, 1977, **10**, 361.
[137] H. M. von Bergmann and V. Hasson, *J. Phys. (E)*, 1976, **9**, 982.
[138] A. J. Schmidt, *J. Phys. (E)*, 1977, **10**, 453.
[139] E. E. Bergmann, *Rev. Sci. Instr.*, 1977, **48**, 545.
[140] R. K. Bauer and A. Kowalczyk, *Optics Comm.*, 1977, **23**, 169.
[141] A. Fisher, J. Peacock, and E. K. C. Lee, *Rev. Sci. Instr.*, 1978, **49**, 395.
[142] H. M. von Bergmann and A. J. Penderis, *J. Phys. (E)*, 1977, **10**, 602.

laser giving powers of 150 kW with pulse lengths from 0.3 to 1 ns.[143] In the latter case, no deterioration in output was reported after 10^8 shots at 30 Hz (1000 h operation). A small nitrogen laser with discrete $BaTiO_3$ ceramic discs as storage elements in a Blumlein circuit with a 15 cm long discharge circuit yielded 170 kW pulses of 5.5 ns FWHM duration at 17 kV charging voltage.[144] Copper–Mylar sandwiches were used as discharge capacitors in a simple design of nitrogen laser.[145] By charging the flat plate pulse-forming line only a few tens of nanoseconds before each laser pulse, it was possible to produce 700 kW of 6 ns FWHM duration at 1 Hz.[146] A novel circuit was used to provide fast risetime, large volume electrical discharge excitation[147] and a master-oscillator and power amplifier techniques were used to synchronize two TEA lasers to obtain spatially coherent 1 ns pulses at 337 nm with an energy per pusle of 0.3 mJ.[148] Transverse gas flow was shown to improve the performance of an earlier reported laser (P. Schenck, *Appl. Optics*, 1973, **12**, 183)[149] and an output increase of 1.6 times found with N_2–SF_6 gas mixtures.[150] With a long duration transverse discharge of 300 ns and preionization by a short pulse electron-beam gun (10 ns), laser emission was obtained from a N_2–CF_4 mixture with a pulse duration of 70 ns.[151] A 60% increase in laser output was reported on increasing the cathode temperature in a nitrogen laser by 75 K.[152]

The nitrogen laser is normally superfluorescent with little or no spatial coherence so the brightness is low. Spatially coherent radiation was obtained by placing the discharge inside an unstable resonator of high magnification with a pulse energy of 0.25 mJ in a diffraction-limited spot.[153]

Details have recently been published of a useful electronic delay trigger to synchronize a low frequency nitrogen laser (100 Hz) with electro-optic instrumentation operating at much higher modulation frequencies.[154]

Copper Lasers. There have been further studies on this potentially high power laser which produces visible emission at 510.6 and 578.2 nm (see Vol. 8, p. 8). Short, fast rising current pulses for pumping copper atoms in the CuCl laser are needed for efficient dissociation of CuCl and recent work has shown that the performance is improved by decreasing the inductance of the pulse circuit.[155] A modification of the Blumlein circuit in which two networks are charged in parallel and discharged in series to provide 5 kW at 15 Hz has been shown to be suitable for pumping at frequencies of up to 20 kHz.[156]

[143] H. M. von Bergmann, *J. Phys. (E)*, 1977, **10**, 1210.
[144] C. L. Sam, *Appl. Phys. Letters*, 1976, **29**, 505.
[145] T. Baer, *Appl. Optics*, 1976, **15**, 2953.
[146] R. Polloini, *Optics and Quantum Electron.*, 1976, **8**, 565.
[147] J. S. Levine, K. Bezdjian, and R. Rappaport, *I.E.E.E. J. Quantum Electron.*, 1976, **QE-12**, 437.
[148] E. E. Bergmann, *Appl. Phys. Letters*, 1977, **31**, 661.
[149] M. Feldman, P. Lebow, F. Raab, and H. Metcalf, *Appl. Optics*, 1978, **17**, 774.
[150] Y. Ishida and T. Yajima, *Oyo Butsuri*, 1977, **46** (10), 996; CA 89 : 68213.
[151] F. Collier, G. Thiell, and P. Cottin, *Appl. Phys. Letters*, 1978, **32**, 739.
[152] I. Baltog and C. B. Collins, *Optics Comm.*, 1976, **18**, 282.
[153] G. C. Thomas, G. Chakrapani, and C. M. L. Kerr, *Appl. Phys. Letters*, 1977, **30**, 633.
[154] W. L. Switzer, W. L. McCullen, and A. F. Schreiner, *Rev. Sci. Instr.*, 1977, **48**, 356.
[155] A. A. Vetter and N. M. Nerheim, *I.E.E.E. J. Quantum Electron.*, 1978, **QE-14**, 73.
[156] J. L. Pack, C. S. Liu, D. W. Feldman, and L. A. Weaver, *Rev. Sci. Instr.*, 1977, **48**, 1047.

A comparison of CuCl, CuBr, and CuI lasants showed that CuBr produced somewhat higher power than the other salts[157] and other copper salts investigated included a mixture of copper nitrate and copper acetylacetonate,[158] copper acetate,[159] and copper oxide.[160] In the latter case, average laser powers of up to 2 W were reported at 12 kHz repetition rate. Addition of small amounts of HCl to the buffer gas of a double-pulsed CuCl laser increased the maximum power by 15%.[161] Other studies of copper chloride lasers[162,163] included the description of a long-lived sealed-off laser.[163] Time-resolved fluorescence studies enabled copper atom concentrations in both ground and metastable states of pulsed copper chloride to be determined.[164] Copper lasers have been shown to be attractive pump sources for dye oscillator–amplifier systems operating in the range 560—700 nm[165] and have been mode-locked at pulse rates of up to 150 kHz.[166]

Mercury Lasers. Mercury halide molecules are proving to be attractive candidates as visible lasers offering efficiencies comparable with those of the rare gas halides, and there have been a number of reports of laser emission (*ca.* 558 nm) following electron-beam pumping of Hg–Cl$_2$–Ar and Hg–CCl$_3$Br–Ar mixtures[167] and HgBr$_2$ (at 502 nm).[168] Dissociation of HgBr$_2$ to attain laser action on the $B^2\Sigma^+ \rightarrow X^2\Sigma^+$ transition of HgBr has also been reported following excitation with a pulsed ArF laser[169] and by electron collision in a transverse electric discharge.[170]

Other Pulsed Gas Lasers. Lasing has been reported on a promising new molecular system, the sulphur dimer, S$_2$, on its $B^3\Sigma_u^- – X^3\Sigma_g^-$ transition following optical pumping with a frequency-doubled dye laser or nitrogen laser.[171] Initial experiments show laser action at 365—570 nm and indicate that the S$_2$ system is capable of non-degradable operation, is scalable, and has potential for efficient operation. Superfluorescent laser action has been observed in atomic

[157] S. Gabey, I. Smilanski, L. A. Levin, and G. Erez, *I.E.E.E. J. Quantum Electron.*, 1977, **QE-13**, 364.
[158] A. J. Andrews, C. E. Webb, R. C. Tobin, and R. G. Denning, *Optics Comm.*, 1977, **22**, 272.
[159] G. Chakrapani, T. A. Praseda Rao, A. A. N. Murty, and D. R. Rao, *Appl. Phys. Letters*, 1977, **31**, 832.
[160] R. S. Anderson, B. G. Bricks, and T. W. Karras, *Appl. Phys. Letters*, 1976, **29**, 187.
[161] A. A. Vetter and N. M. Nerheim, *Appl. Phys. Letters*, 1977, **30**, 405.
[162] J. A. Piper, *I.E.E.E.J. Quantum Electron.*, 1978, **OE-14**, 405; R. S. Anderson, B. G. Bricks, and T. W. Karras, *ibid.*, 1977, **QE-13**, 115; C. S. Liu, D. W. Feldman, J. L. Pack, and L. A. Weaver, *ibid.*, 1977, **QE-13**, 744; I. Smilanski, A. Kerman, L. A. Levin, and G. Erez, *ibid.*, 1977, **QE-13**, 24.
[163] D. W. Feldman, C. S. Liu, J. L. Pack, and J. A. Weaver, *I.E.E.E. J. Quantum Electron.*, 1977, **QE-13**, 64D.
[164] J. Tenenbaum, I. Smilanski, S. Gabay, G. Erez, and L. A. Levin, *J. Appl. Phys.*, 1978, **49**, 2662.
[165] R. S. Hargrove and T. Kan, *I.E.E.E. J. Quantum Electron.*, 1977, **QE-13**, 28D.
[166] T. S. Fahlen, *I.E.E.E. J. Quantum Electron.*, 1977, **QE-13**, 546.
[167] J. G. Eden, *Appl. Phys. Letters*, 1977, **31**, 448; K. Y. Tang, R. O. Hunter, J. Oldenettel, C. Howton, D. Huetis, D. Eckstrom, B. Perry, and M. McCusker, *ibid.*, 1978, **32**, 226; W. T. Whiting, *ibid.*, 1978, **32**, 239; J. H. Parks, *I.E.E.E. J. Quantum Electron.*, 1977, **QE-13**, 103D.
[168] J. H. Parks, *Appl. Phys. Letters*, 1977, **31**, 297.
[169] E. J. Schimitschek, J. E. Celto, and J. A. Trias, *Appl. Phys. Letters*, 1977, **31**, 608.
[170] E. J. Schimitschek and J. E. Celto, *Optics Letters*, 1978, **2**, 64.
[171] S. R. Leone and K. G. Kosnik, *Appl. Phys. Letters*, 1977, **30**, 346.

mercury at 365 and 546 nm following pumping with 266 nm Nd radiation.[172] A preionized tranverse discharge in nitrogen–helium mixtures produced emission at 427.8 nm (from N_2^+) at peak powers of 0.5 MW[173] and 180 kW.[174] Other pulsed lasers in the visible region included atomic indium at 451 nm (pumped by an ArF laser),[175] Br_2 which was tunable from 550 to 750 nm and capable of extension to 3.5 μm (pumped by a 532 nm Nd laser),[176] F_2 which was tunable from 635 to 780 nm (with a double discharge system)[177] and Bi which provided an emission tunable from 650 to 710 nm and attributed to Bi_2 molecules.[178] Several watts at 704 nm have been obtained from a He–F_2 laser suggesting use as a pump source for i.r. laser dyes.[179] Fluorescence of TlHg molecules induced by Tl resonant radiation at 459 and 656 nm showed that these bands have great promise for excimer laser action.[180]

Lasing has also been demonstrated in atomic iodine (at 1.315 μm) following flashlamp-pumped-dye-laser-pumping (FLPD laser pumping) of iodine vapour[181] and CF_3I.[182] Prompt laser emission of 100 ns duration from the iodine transition has also been produced by a transverse electrical discharge in i-C_3F_7I–buffer gas mixtures.[183] A convenient saturable absorber for mode-locking this laser was shown to be ground state atomic iodine which is easily saturated at moderate energy densities.[184] A closed-cycle iodine laser used cooling to remove molecular iodine and thereby maintain laser output.[185]

There are various designs for CO lasers emitting at 5.3 μm[186—189] including a simple sealed TEA variety yielding >600 mJ at 77 K,[187] a waveguide laser operating near room temperature,[188] and a u.v. preionized TE arrangement operated sealed-off for 10^5 pulses.[189]

Photoionization is well known to improve the operating characteristics of CO_2 lasers[190] and the addition of trace quantities of low ionization threshold vapours also shown to improve performance significantly.[191] Photon absorption and photoionization spectra in a CO_2 laser mixture sealed with selected additives revealed that tripropylamine exhibited a longer wavelength photoionization effect, which is essentially independent of the primary host laser gases.[191] (Amine additives were also shown to produce potentially lethal

172 H. Komine and R. L. Byer, *J. Appl. Phys.*, 1977, **48**, 2505.
173 D. E. Rothe and K. O. Tan, *Appl. Phys. Letters*, 1977, **30**, 152.
174 J. B. Laudenslager, T. J. Pacala, and C. Wittig, *Appl. Phys. Letters*, 1976, **29**, 580.
175 R. Burnham, *Appl. Phys. Letters*, 1977, **30**, 132.
176 F. J. Wodarczyk and H. R. Schlossberg, *J. Chem. Phys.*, 1977, **67**, 4476.
177 T. R. Loree and R. C. Sze, *Optics Comm.*, 1977, **21**, 255.
178 W. P. West and H. P. Broida, *Chem. Phys. Letters*, 1978, **56**, 283.
179 L. O. Hocker and T. B. Phi, *Appl. Phys. Letters*, 1976, **29**, 493.
180 D. Drummond and L. A. Schlie, *J. Chem. Phys.*, 1976, **65**, 3454.
182 S. J. Davis, *J. Opt. Soc. Amer.*, 1978, **68**, 690.
181 S. J. Davis, *Appl. Phys. Letters*, 1978, **32**, 656.
183 M. C. Wong and R. E. Beverly, *Optics Comm.*, 1977, **20**, 19, 23.
184 E. Fill and K. Hohla, *Optics Comm.*, 1976, **18**, 431.
185 W. Fuss and K. Hohla, *Optics Comm.*, 1976, **18**, 427.
186 D. G. Lim, P. J. Mendoza, and D. B. Cohn, *Rev. Sci. Instr.*, 1977, **48**, 1430.
187 P. J. M. Peters, *Optics Comm.*, 1978, **25**, 241.
188 E. K. Gorton, R. M. Jenkins, and D. R. Hall, *J. Phys. (E)*, 1977, **10**, 1234.
189 A. L. S. Smith and B. Norris, *Optics Comm.*, 1977, **23**, 183.
190 N. S. Kopeika, *I.E.E.E. J. Quantum Electron.*, 1977, **QE-13**, 968.
191 H. J. J. Sequin, D. McKen, and J. Tulip, *Appl. Optics*, 1977, **16**, 77; *Appl. Phys. Letters*, 1976, **29**, 110.

cyanides.)[192] U.v. preionized CO_2 TEA[193] and waveguide[194] arrangements provided output pulses of 1.3 MW and 0.5 MW with pulse durations of 30—35 and 20 ns, respectively. A high power (10 mJ) photopreionization-stabilized waveguide laser was found to operate at gas pressures of up to 1.3×10^6 N m^{-2}.[195] Double-discharge techniques[196,197] have been used to provide compact laser sources operating at frequencies up to 100 Hz.[197] Other designs included longitudinal excitation,[198] a compact cylindrical arrangement,[199] a BeO waveguide laser maintaining constant power for at least 2500 h,[200] and a 15 cm resonator operating on only one axial mode with a Blumlein-type discharge.[201] A sealed TEA laser producing $>10^6$ pulses without failure allowed economic use of isotopic mixtures of CO_2 as laser media.[202] Up to three different rotational–vibrational transitions were observed from a TEA laser with a helical distribution of electrodes.[203] A line-tunable laser operating in the 2280—2360 cm^{-1} region was pumped by energy transfer from $Br(4^2P_{\frac{1}{2}})$ which in turn was produced by the flash photolysis of Br_2.[204] A cold cathode electron gun introduced into a high pressure gas laser system enabled multiatmospheric laser operation on a number of molecular gases including CO_2, N_2O, CO, and C_2H_2.[205]

Although CW chemical lasers have been briefly mentioned earlier (p. 8), there are some useful pulsed chemical lasers including HF (lasing between 2.5 and 3.1 μm),[206] HCl,[207] HBr,[208] and DF.[209] The reader is referred elsewhere for a detailed introduction to the subject.[60]

In recent years optically-pumped i.r. lasers have greatly extended the range and number of lasing i.r. wavelengths.[210] In the middle i.r. it is desirable to have intense sources in the 14—17 μm for photochemical experiments with molecules

[192] H. N. Ruth and J. M. Butcher, *J. Phys. (E)*, 1977, **10**, 844.
[193] D. S. Stark, P. H. Cross, and M. R. Harris, *J. Phys. (E)*, 1978, **11**, 311.
[194] D. J. Brink and V. Hasson, *J. Appl. Phys.*, 1978, **49**, 2250.
[195] D. J. Brink, V. Hasson, and T. I. Salamon, *J. Phys. (E)*, 1977, **10**, 370.
[196] M. Lyszk, F. Herlemont, and J. Lemaire, *J. Phys. (E)*, 1977, **10**, 1110; P. W. Pace and M. Lacombe, *I.E.E.E. J. Quantum Electron.*, 1978, **QE-14**, 263.
[197] P. Pace and M. Lacombe, *J. Phys. (E)*, 1977, **10**, 208.
[198] M. M. T. Loy and P. A. Roland, *Rev. Sci. Instr.*, 1977, **48**, 554.
[199] G. R. Osche and H. E. Sonntag, *I.E.E.E. J. Quantum Electron.*, 1976, **QE-12**, 752.
[200] L. M. Laughman, *Rev. Sci. Instr.*, 1976, **47**, 1411.
[201] G. J. Ernst, *Rev. Sci. Instr.*, 1977, **48**, 1281.
[202] C. Willis, R. A. Back, and J. G. Purdon, *Appl. Phys. Letters*, 1977, **31**, 84.
[203] K.-H. Krahn and K. H. Finken, *Appl. Optics*, 1977, **16**, 1137.
[204] A. B. Petersen and C. Wittig, *J. Appl. Phys.*, 1977, **48**, 3665.
[205] N. W. Harris, F. O'Neill, and W. T. Whitney, *Rev. Sci. Instr.*, 1977, **48**, 1042.
[206] B. K. Deka, P. E. Dyer, and D. J. James, *Optics Comm.*, 1976, **18**, 462; J. E. Brandelik and R. F. Paulson, *I.E.E.E. J. Quantum Electron.*, 1977, **QE-13**, 933; A. F. Gibson, T. A. Hall, and C. B. Hatch, *I.E.E.E. J. Quantum Electron.*, 1977, **QE-13**, 801; R. L. Kerber and J. S. Whittier, *Appl. Optics*, 1976, **15**, 2358; G. J. Simonis, *Appl. Phys. Letters*, 1976, **29**, 42; D. B. Nichols, R. B. Hall, and J. D. McClure, *J. Appl. Phys.*, 1976, **47**, 4026.
[207] C. C. Badcock, W. C. Hwang, J. F. Kalsch, and R. F. Kameda, *Appl. Phys. Letters*, 1978, **32**, 363; Z. B. Alfassi, R. Giniger, E. Huler, and H. Reisler, *Chem. Phys.*, 1978, **31**, 263; S. J. Arnold, K. D. Foster, D. R. Snelling, and R. D. Suart, *Appl. Phys. Letters*, 1977, **30**, 637.
[208] H. Oodate and T. Fujioka, *I.E.E.E. J. Quantum Electron.*, 1978, **QE-14**, 227.
[209] R. Turner and T. O. Poehler, *J. Appl. Phys.*, 1976, **47**, 3038.
[210] T. Y. Chang, 'Optical pumping in gases', in 'Topics in Applied Physics', Vol. 16, Springer-Verlag, Berlin, 1976.

such as SF_6, UF_6, and OsO_4 (see Section 6). Large numbers of laser transitions in the 11—17 μm range have been obtained by optically pumping CF_4, NOCl, CF_3I, and NH_3[211] with simple line tunable CO_2 TEA lasers.[212] Many of these lasers have respectable conversion efficiencies and can be scaled to high energies in spectral regions where selective excitation of molecules is a prerequisite to performing selective photodissociation of laser-induced chemistry experiments. Laser oscillation has been demonstrated in $^{14}NH_3$ over the range 6—35 μm following two-photon excitation with a pair of CO_2 lasers[213] and in $^{15}NH_3$ over 14—18 μm following pumping with a pulsed HF laser.[214] Electron-beam pumping has enabled a continuously tunable 9—12.5 μm range to be obtained with multiatmospheric lasers (of $^{12}CS_2$, $^{13}CS_2$, N_2O, $^{13}CO_2$, and $^{12}CO_2$).[215] Stimulated Raman scattering in liquid nitrogen pumped by 3.4 μm coherent light occurred between 15 and 18 μm.[216] Fifty new laser lines (in the 10 μm region) were observed in the continuous emission from an N_2O laser by placing a hot N_2O absorber cell inside the laser cavity to suppress the regular laser lines.[217]

A great deal of interest in the development of 16 μm lasers is due to the proposed isotopic separation scheme for uranium.[218] One laser of particular interest in this region[219] was derived from potassium atoms which were pumped simultaneously by a gas discharge and by a tunable dye laser (at 534.3 nm) to a high-lying Rydberg state from which lasing at 15.95 μm occurred.[220] Most of the 16 μm lasers do not have the combined requirement of tunability and high energy, but an optically-pumped CO_2 laser excited by a 100 mJ HBr laser produced 1 mJ of multiline energy per pulse with an extensive stepwise tuning accomplished by an intracavity grating and the use of various isotopic forms of CO_2.[221]

The spin-flip laser is proving to be a useful high resolution tunable i.r. source for chemical spectroscopy and photochemistry and a number of applications have recently been reviewed.[222] Typical specifications of this laser are: tuning range

211 R. G. Harrison and F. A. Al-Watban, *Optics Comm.*, 1977, **20**, 225; B. Walker, G. W. Chantry, and D. G. Moss, *Optics Comm.*, 1977, **23**, 8; T. Mochizuki, M. Yamanaka, M. Morikawa, and C. Yamanaka, *J. Opt. Soc. Amer.*, 1978, **68**, 672.
212 J. J. Tiee and C. Wittig, *J. Appl. Phys.*, 1978, **49**, 61.
213 R. R. Jacobs, D. Prosnitz, W. K. Bischel, and C. K. Rhodes, *Appl. Phys. Letters*, 1976, **29**, 710.
214 C. R. Jones, M. I. Buchwald, M. Gundersen, and A. H. Bushnell, *Optics Comm.*, 1978, **24**, 27.
215 F. O'Neill and W. T. Whitney, *Appl. Phys. Letters*, 1977, **31**, 270.
216 R. Frey, F. Pradere, J. Lukasik, and J. Ducuing, *Optics Comm.*, 1977, **22**, 355.
217 K. Siemsen and J. Reid, *Optics Comm.*, 1977, **20**, 284.
218 B. L. Wexler, T. J. Manuccia, and R. W. Waynant, *Appl. Phys. Letters*, 1977, **31**, 730; E. V. George and W. F. Krupke, *Proc. Soc. Photo-Opt. Instrum. Eng.*, 1976, **86**, 140; E. A. O'Hair and M. S. Piltch, *ibid.*, 1976, **86**, 147; J. Cahen, M. Clerc, and P. Rigny, *Optics Comm.*, 1977, **21**, 387.
219 A. M. F. Lau, W. K. Bischel, C. K. Rhodes, and R. M. Hill, *Appl. Phys. Letters*, 1976, **29**, 245.
220 D. R. Grischkowsky, J. R. Lankard, and P. P. Sorokin, *I.E.E.E. J. Quantum Electron.*, 1977, **QE-13**, 392.
221 R. M. Osgood, *Appl. Phys. Letters*, 1978, **32**, 564.
222 R. B. Dennis, in 'Laser 77 Opto-Electronics Conference Proceedings', ed. W. Waidelich, I.P.C., London, 1977, p. 120; A. McNeish, J. H. Carpenter, D. H. Whiffen, and J. J. Turner, in ref. 3, p. 389.

1896—1820 cm^{-1} (5.27—5.49 m), linewidth <0.007 cm^{-1}, and power >50 mW (see also ref. 910).

Far-i.r. lasers are useful for laser magnetic resonance experiments (see Section 5) and almost invariably pumped by the pulsed or CW CO_2 laser.[223] A tabulation of optically-pumped far-i.r. laser lines (from 37 to 1.8 mm) included lasing medium, output power, pump laser, pump transition, wavelength, and power.[224] Common lasants included methanol,[225] D_2O,[226] HCN,[227] CH_3F,[228] and NH_3.[229]

Dye Lasers.—A special issue of *Optica Acta* included reviews on the principles of dye laser operation, pulsed narrow band dye laser operation, research in dye vapour lasers, simmer-mode high efficiency flashlamps, and thermal effects in FLPD lasers.[230] Other reviews on this important and useful class of lasers are given elsewhere.[231]

Although laser emission from dye laser vapours pumped by lasers and electric discharges has been known for some time, the first observation has been reported of super-radiant emission from POPOP in the vapour phase pumped by an electron beam with a conversion efficiency from electronic energy to coherent light of 5%.[232] Under similar conditions, broadband fluorescence at 400 nm was observed from POPOP and α-NPO in Xe and Ar.[233] The optimum conditions for achieving intense dye vapour fluorescence of a xanthene dye under short duration electron pulse excitation have also been established.[234] Other studies with dye vapours included discharge excitation,[235] and pumping POPOP, pyrene, and perylene with a nitrogen laser.[236]

Radiation from a CO_2 laser-produced plasma has been used to pump short wavelength laser dyes such as *p*-terphenyl, PPD, and *p*-quaterphenyl.[237] An

[223] P. Mathieu and J. R. Izatt, *I.E.E.E. J. Quantum Electron.*, 1977, **QE-13**, 465; G. Duxbury and H. Herman, *J. Phys. (E)*, 1978, **11**, 419; T. Y. Chang, *I.E.E.E. Trans. Microwave Theory and Technol.*, 1974, **22**, 983.

[224] J. J. Gallagher, M. D. Blue, B. Beau, and S. Perkowitz, *Infrared Physics*, 1977, **14**, 43.

[225] E. J. Danielewicz and P. D. Coleman, *I.E.E.E. J. Quantum Electron.*, 1977, **QE-13**, 485.

[226] Z. Drozdowicz, P. Woskoboinikow, K. Isobe, D. R. Cohn, R. J. Temkin, K. J. Button, and J. Waldman, *I.E.E.E. J. Quantum Electron.*, 1977, **QE-13**, 413; M. R. Green, P. D. Morgan, and M. R. Siegrist, *J. Phys. (E)*, 1978, **11**, 389; K. Lipton, J. P. Nicholson, and R. Illingworth, *Optics Comm.*, 1978, **21**, 42; K. Lipton and J. P. Nicholson, *I.E.E.E. J. Quantum Electron.*, 1977, **QE-13**, 811.

[227] C. Sturzenegger, B. Adam, and F. Kneubahl, *I.E.E.E. J. Quantum Electron.*, 1977, **QE-13**, 473; H. Lengfellner, G. Pauli, W. Heisel, and K. F. Renk, *Appl. Phys. Letters*, 1976, **29**, 566.

[228] F. Brown, P. D. Hislop, and J. O. Tarpinian, *I.E.E.E. J. Quantum Electron.*, 1977, **QE-13**, 445; T. K. Plant and T. A. DeTemple, *J. Appl. Phys.*, 1976, **47**, 3042.

[229] E. J. Danielewicz and C. O. Weiss, *I.E.E.E. J. Quantum Electron.*, 1978, **QE-14**, 222.

[230] *Optica Acta*, 1976, No. 10, p. 23.

[231] R. B. Green, *J. Chem. Educ.*, 1977, **54**, A365, A407.

[232] G. Marowsky, R. Cordray, F. K. Tittel, and W. L. Wilson, *Appl. Phys. Letters*, 1978, **32**, 561.

[233] S. A. Edelstein, H. H. Nakano, T. F. Gallagher, and D. C. Lorents, *Optics Comm.*, 1977, **21**, 27.

[234] G. Marowsky, R. Cordray, F. K. Tittel, B. Wilson, and J. W. Keto, *J. Chem. Phys.*, 1977, **67**, 4845; *J. Opt. Soc. Amer.*, 1978, **68**, 701.

[235] P. W. Smith, P. F. Liao, and P. J. Maloney, *I.E.E.E. J. Quantum Electron.*, 1976, **QE-12**, 539.

[236] T. Sakurai, A. Ogishima, and M. Sugawara, *Optics Comm.*, 1978, **25**, 75.

[237] O. R. Wood, L. H. Szeto, and W. T. Silvfast, *J. Appl. Phys.*, 1977, **48**, 1956.

optical conversion efficiency of 28% was obtained at 340 nm by pumping *p*-terphenyl in cyclohexane solution with a KrF laser.[238] Laser emission was tunable from 323 to 364 nm. Almost identical studies reported a tuning range from 335 to 346 nm from *p*-terphenyl in *p*-dioxane solution.[239]

The more instrumental aspects of dye lasers have received considerable attention in the last few years, particularly in terms of spectral narrowing. Fabry–Perot etalons, gratings, and prisms have been considered as tuning elements in a theoretical study of spectral narrowing in pulsed dye lasers[240] and a detailed model evaluated of the spectral and temporal properties of an evolving dye laser pulse.[241] A new single cavity design for nitrogen-laser-pumped dye (NLPD) lasers used a diffraction grating near grazing incidence with an additional mirror in order to obtain a large angular dispersion and linewidths of the order of $0.08\ cm^{-1}$.[242] A two-fold diffraction-limited beam was obtained for a high power Rh6G transverse flow dye laser using suitably correcting optics and a sound wave parallel to the flow to control beam divergence.[243] Rutile (TiO_2) has been shown to be a durable material for use as a tuning element producing linewidths smaller than those obtained with a 1800 line mm^{-1} grating.[244] Strontium titanate prisms (with dispersions as high as $74 \times 10^{-5}\ nm^{-1}$) used in a CW dye ring laser resulted in spectral bandwidths of 2 pm.[245] A peak power of 100 kW in the visible at a linewidth of 60 MHz was generated using a CW dye laser oscillator followed by three single pass dye amplifier stages pumped by a single nitrogen laser.[246] This high output power may be used to produce coherent u.v. pulses by sum or SHG in non-linear crystals.

Rapid wavelength tuning devices can be exploited with the broad lasing bands of dye lasers for numerous time-resolved spectral studies. For example, tuning of a NLPD laser was achieved by electronic coupling of a piezo-driven grating with a piezo-driven etalon resulting in a tuning range of 180 GHz with a laser linewidth of 250 MHz.[247] Electrochemical reactions in the dye solution are reported to shift the dye laser maximum from 408 to 494 nm in 4-methylumbelliferone.[248] A somewhat simpler method involved changing the concentration of an additional dye in the lasing solution.[249] Wavelengths shifts in dye lasers have been analysed from gain curves derived from transmission characteristics of tuning elements.[250]

[238] B. Gordard and O. de Witte, *Optics Comm.*, 1976, **19**, 325.
[239] D. G. Sutton and G. A. Capelle, *Appl. Phys. Letters*, 1976, **29**, 563.
[240] P. Flamant, *Appl. Optics*, 1978, **17**, 955.
[241] P. Juramy, P. Flamant, and Y. H. Meyer, *I.E.E.E. J. Quantum Electron.*, 1977, **QE-13**, 855.
[242] I. Shoshan, N. N. Danon, and U. P. Oppenheim, *J. Appl. Phys.*, 1977, **49**, 4495.
[243] D. B. Northam, M. E. Mack, and I. Itzkan, *Appl. Optics*, 1978, **17**, 931.
[244] J. Krasinski and W. Majewski, *Rev. Sci. Instr.*, 1976, **47**, 1293.
[245] G. Marowsky, *Rev. Sci. Instr.*, 1976, **47**, 843.
[246] M. M. Salour, *Optics Comm.*, 1977, **22**, 202.
[247] A. Yamagishi and A. Szabo, *Appl. Optics*, 1977, **16**, 691.
[248] V. A. Zhivnov, I. Y. Rumyantsev, V. I. Tomin, and A. N. Rubinov, *Optics Comm.*, 1977, **23**, 33.
[249] R. Konjevic and N. Konjeric, *Optics Comm.*, 1977, **23**, 187.
[250] I. J. Hodgkinson and J. I. Vukusic, *J. Phys. (E)*, 1978, **11**, 121; L. Singer, Z. Singer, and S. Kimel, *Appl. Optics*, 1976, **15**, 2678.

The opto-galvanic effect has been ingeniously used to frequency-lock a CW dye laser to a bandwidth of 0.003 nm over the range 550—640 nm with a tilting 0.5 nm etalon to provide fine tuning.[251] The laser was locked to excited state transition frequencies of an emission lamp. This constant frequency provision suggests potentially wide use on photochemistry and spectroscopy. Frequency sweeping ('chirping') in dye lasers has been investigated with a cavity-pumped dye laser.[252] Decreasing the pump power resulted in a shift towards longer wavelengths as the overall excited state population was increased and increasing the pumping pulse width (and hence average power) was shown to increase the gain at longer wavelengths resulting in a wide wavelength range approaching the CW laser-pumped case.[252]

A combination of a FLPD laser and acousto-optic output coupler was used to produce any pulse width from <12 ns to about 1 μs with energies from 0.1 to 0.5 mJ.[253]

Several designs of flashlamp-pumped dye lasers have been evolved. An efficient flashlamp-pumped u.v. dye laser has been described in which a four-stage Marx-bank generator provided a fast risetime.[254] An output energy of 25 mJ (420 kW) for an 81 J input was obtained with *p*-terphenyl solution at 342.6 nm (as well as 0.22 MW for BBD at 377 nm and 1.22 MW for DAMC at 456 nm).[254] A novel arrangement with four flashlamps to pump a single dye cell had a reported efficiency as high as 0.35% with Rh6G for an input power of 34 kW (at 50 Hz).[255] Simmer-mode operation of flashlamps in a double elliptical reflector resulted in reliable operation at repetition rates up to 100 Hz with an average power of 4 W, again with Rh6G.[256] By evaporating aluminium on the outside of a borosilicate elliptical cavity, a highly reflecting surface was obtained which was unaffected by cooling water and light below 300 nm (which will decompose dyes) was absorbed by the glass.[257] The performance of a FLPD laser was improved by cooling the lamps by forced gaseous convection allowing higher operating temperatures, filtering the u.v. and i.r. from the lamps (thus increasing the dye life), and minimizing thermal distortions of the dye tube.[258]

A short cavity is required for laser pumping with short duration excitation pulses (3—5 ns) and a new design with four prisms provided easily-achievable single axial mode pulses.[259]

Simultaneous laser operation at two or more tunable wavelengths in the visible and i.r. region has received attention lately because of the increasing interest in fields of application including differential absorption spectroscopy, excited state, and two-photon spectroscopy. Three dye laser cavity arrangements

[251] R. B. Green, R. A. Keller, G. G. Luther, P. K. Schenck, and J. C. Travis, *I.E.E.E. J. Quantum Electron.*, 1977, **QE-13**, 63.
[252] J. H. Richardson, L.-J. Steinmetz, and B. W. Wallin, *Appl. Optics*, 1977, **16**, 1133.
[253] F. E. Lytle and J. M. Harris, *Appl. Spectroscopy*, 1976, **30**, 633.
[254] M. Maeda, O. Vehino, E. Doi, K. Watanabe, and Y. Mijazoe, *I.E.E.E. J. Quantum Electron.*, 1977, **QE-13**, 65.
[255] J. Jethwa, F. P. Schaefer, and J. Jasny, *I.E.E.E. J. Quantum Electron.*, 1977, **QE-13**, 29D; *ibid.*, 1978, **QE-14**, 119.
[256] A. Hirth and H. Fagot, *Optics Comm.*, 1977, **21**, 318.
[257] S. Chu and E. Commins, *Appl. Optics*, 1977, **16**, 2619.
[258] D. J. Derc and S. W. Wallace, U.S.P. 3 967 212.
[259] G. K. Klauminzer, *I.E.E.E. J. Quantum Electron.*, 1977, **QE-13**, 92D.

have been used to generate two or more wavelengths simultaneously with bandwidths of 1 cm^{-1}.[260] Other arrangements described have included use of a single diffraction grating with two plane mirrors,[261] a NLPD laser producing two coaxial beams,[262] a Hansch-type tunable laser,[263] two adjacent cells,[264] and a double compartment cell[265] pumped by a pulsed laser. An arrangement for generating two wavelengths with orthogonal polarizations from a NLPD laser has also been described.[266] Two independently tunable dye lasers, synchronously pumped in common by the split output of a frequency-doubled mode-locked Nd-glass laser provided pulses which were perfectly synchronized in time. By overlapping the pulses in strontium vapour, tunable picosecond pulses in the vacuum-u.v. (190—200 nm) were produced.[267] Synchronized trains of picosecond pulses (5 ps pulsewidth) at two independent wavelengths were also generated by synchronous pumping of two Rh6G lasers.[268] Autocorrelation measurements indicated jitter times between pulse trains of <10 ps.

An important paper on the bleaching of laser dyes by a nitrogen laser showed that dye performance under non-flowing conditions was limited by bleaching even at low repetition rates and especially in the case of longitudinal pumping.[269] Bleaching in Rh6G probably involved excitation to a higher excited state through multiple photon excitation. The same dye pumped by u.v. or copper laser excitation also showed a strong bleaching dependence on solvent, the highest stability being achieved in alcoholic solvents.[270] Although viscosity-raising additives for jet-stream lasers are claimed to offer substantial advantages in terms of high stability and high power applications,[271] Triton X-100 (a commonly used deaggregating agent and triplet quencher) considerably increased the bleaching rate of Rh6G in reported experiments.[270] In contrast with existing ideas, the addition of cyclo-octatetriene was claimed to increase the triplet lifetimes (of Rh6G)[272] and high concentrations of this supposedly triplet quencher were reported to induce a negative effect because of competition with the dye for available u.v. excitation.[273]

Among the numerous studies of dye lasers are included reports of copper-vapour-pumped dyes,[274] excitation of dyes in liquid crystals,[275] xenon ion laser pumping (5 kW at 364.5 nm) of blue dyes with broadband efficiencies of 2%,[276] further calculations on electrochemical pumping,[277] and a comparison of

260 C. Kittrell and R. A. Bernheim, *Optics Comm.*, 1976, **19**, 5.
261 P. Burlamacchi, R. Coisson, R. Pratesi, and D. Pucci, *Appl. Optics*, 1977, **16**, 1553.
262 T. Mita and N. Nagasawa, *Optics Comm.*, 1978, **24**, 345.
263 B. R. Marx, G. Hollaway, and L. Allen, *Optics Comm.*, 1976, **18**, 437.
264 J. A. Buck and Z. A. Yasa, *I.E.E.E. J. Quantum Electron.*, 1977, **QE-13**, 935.
265 R. Dorsinville and M. M. Denariez-Roberge, *Optics Comm.*, 1978, **24**, 31.
266 E. Winter, G. Veith, and A. J. Schmidt, *Optics Comm.*, 1978, **25**, 87.
267 T. R. Royt and C. H. Lee, *Appl. Phys. Letters*, 1977, **30**, 332.
268 R. K. Jain and J. P. Heritage, *Appl. Phys. Letters*, 1978, **32**, 41.
269 E. Saher and D. Treves, *Optics Comm.*, 1977, **21**, 20.
270 I. Rosenthal, *Optics Comm.*, 1978, **24**, 164.
271 S. Leutwyler, E. Schumacher, and L. Woste, *Optics Comm.*, 1976, **19**, 197.
272 A. N. Runinov and M. M. Asimov, *J. Luminescence*, 1977, **15**, 429.
273 H. H. L. Wang and C. H. Taji, *I.E.E.E. J. Quantum Electron.*, 1978, **QE-14**, 85.
274 A. A. Pease and W. M. Pearson, *Appl. Optics*, 1977, **16**, 57.
275 S. Kuroda and K. Kubota, *Appl. Phys. Letters*, 1976, **29**, 737.
276 D. W. Fahey and L. D. Schearer, *I.E.E.E. J. Quantum Electron.*, 1978, **QE-14**, 220.
277 C. A. Heller and J. L. Jernigan, *Appl. Optics*, 1977, **16**, 61.

flashlamp and 530 nm Nd : YAG laser pumping of dyes in order to obtain high average power frequency doubling.[278]

Information concerning dye lasers emitting in the i.r. includes the use of an i.r. molecular nitrogen laser to pump a simple grating tuned dye laser operating in the range 915—1040 nm,[279] a reported 44% conversion efficiency at 700 nm for carbazine-122 pumped by 530 nm radiation,[280] and similar findings concerning tunable i.r. radiation produced over the range 1081—1216 nm.[281] A flashlamp-pumped oscillator amplifier arrangement was used to produce laser emission at 810 nm for a solution of DOTC in DMSO.[282]

Laser Dyes.—The low efficiency of FLPD lasers is mainly caused by the small spectral width of the long wavelength absorption band of the dye as compared with the broad spectral emission of the excitation lamp. Intramolecular energy transfer between dyes made up of two or more conventional fluorophoric systems has been suggested as a general means of improving pumping efficiency.[283] A considerable improvement in output power was obtained by using a bifluorophoric dye formed by linking *p*-terphenyl and dimethyl-POPOP by one CH_2 group. The fluorescence of this compound was almost identical with that of dimethyl-POPOP indicating practically complete energy transfer, and picosecond studies indicated a transfer time of *ca.* 1 ps.[284] Energy transfer between dyes is discussed elsewhere.[285]

One other significant advance in the area of new laser dyes has been the discovery of a class of bis-styryl dyes (one of which is now known commercially as Stilbene 3) which show excellent lasing properties in water and methanolic solutions and are far superior to coumarins as CW laser dyes to 403 nm.[286]

The photochemical stability of rare gas halide laser-pumped dyes (blue-green fluorophores) was shown to be better than with flashlamp-pumping and explained in terms of a shorter pulse length and faster risetime with the excimer laser pulse resulting in smaller populations in the reactive triplet states (and elimination of the far u.v. decomposing wavelengths of the flash).[287] The laser output power for fluorescein was shown to be very dependent on pH, and gains over neutral solutions could be as high as 26-fold.[288]

New investigations of dyes lasing in the u.v. included nine phenyl-oxadiazole derivatives shown to lase between 346.5 and 358.5 nm on pumping with a KrF

[278] D. T. Hon, H. W. Bruesselbach, and E. J. Woodbury, *Proc. Soc. Photo-Opt. Instrum. Eng.*, 1977, **122**, 95.
[279] L. O. Hocker, *I.E.E.E. J. Quantum Electron.*, 1977, **QE-13**, 548.
[280] K. Kato, *I.E.E.E. J. Quantum Electron.*, 1976, **QE-12**, 442; *Optics Comm.*, 1976, **18**, 447.
[281] K. Kato, *I.E.E.E. J. Quantum Electron.*, 1978, **QE-14**, 7.
[282] C. Loth and P. Flamant, *Optics Comm.*, 1977, **21**, 13.
[283] F. P. Schaefer, Z. Bor, W. Lüttke, and B. Liphardt, *Chem. Phys. Letters*, 1978, **56**, 455.
[284] B. Kopainsky, W. Kaiser, and F. P. Schaefer, *Chem. Phys. Letters*, 1978, **56**, 458.
[285] Y. Kusumoto, H. Sato, K. Maeno, and S. Yahiro, *Chem. Phys. Letters*, 1978, **53**, 388; C. T. Ryan and T. K. Gustafson, *Chem. Phys. Letters*, 1976, **44**, 241; P. Burlamacchi and D. Cutter, *Optics Comm.*, 1977, **22**, 283.
[286] J. Kuhl, H. Telle, R. Schieder, and U. Brinkmann, *Optics Comm.*, 1978, **24**, 251; H. Telle, U. Brinkman, and R. Raue, in 'Laser 77 Opto-Electronics Conference Proceedings', ed. W. Waidelich, I.P.C., London, 1977, p. 82; *ibid.*, *Optics Comm.*, 1978, **24**, 33, 248.
[287] E. A. Stappaerts, *Appl. Optics*, 1977, **16**, 3079.
[288] Y. Matsunagu, H. Ikeda, K. Matsumoto, and T. Fujioka, *J. Appl. Phys.*, 1977, **48**, 842.

laser,[289] benzoxazole and its phenyl derivatives which lase at 330 nm,[290] and in the blue included a study of common brightening agents.[291] The properties necessary to obtain efficient laser action from dyes have been reviewed showing that there is a shortage of booster dyes capable of increasing the output of dyes lasing in the blue and green portions of the spectrum.[292]

An interesting study on the power scalability of eight dyes excited with a Q-switched doubled Nd laser showed that conversion efficiencies for some dyes (*e.g.* IR-140, IR-144) became limited at pump-power densities as low as 14 MW cm^{-2} due to excited state absorption of pump radiation.[293] Rhodamine 6G was the least affected by increases in pump power densities.

New dyes shown to lase in the near i.r. included Ga, Mg, and Zn phthalo-cyanines (at around 700 nm)[294] and the phenoxazones.[295] DDI dye in DMSO pumped by a giant pulse ruby laser was shown to have a wide generation band, narrow bandwidth (0.1 cm^{-1}), and high conversion efficiency (30%) with a tuning range from 742 to 1050 nm.[296]

Picosecond Pulse Generation from Dye Lasers.—In comparison with solid state lasers, dye lasers, and especially those pumped by a CW laser, are relatively easy to mode-lock giving a convenient means of producing tunable pulses (in both the visible and u.v.) with picosecond pulse durations.[297] The com-monest example has again been found to be Rh6G pumped by a mode-locked argon laser (100 ps pulses) and generating wavelength tunable (562—620 nm) pulses of 0.6—1.0 ps duration.[298] (The minimum pulsewidth for Rh6G has been shown to be limited by the spectral bandwidth of the dye to about 0.5 ps[299] although pulses as short as 0.3 ps have been measured.[300]) The pulse-forming processes in a forced mode-locked system have been analysed by optical up-conversion in an ADP crystal[301] and pulsewidths (reputably as short as 200 fs) measured with a highly stable interferometer.[302] Tunable mode-locked pulses at two wavelengths have been obtained by using Rh6G and cresyl violet dye jets[303] or by inserting plane mirrors at convenient positions in the dispersion plane of a tuning prism.[304] Other studies with mode-locked systems included development of a cavity dumper to provide picosecond pulses at 50 MHz[305]

[289] C. Rulliere, J. P. Morand, and O. de Witte, *Optics Comm.*, 1977, **20**, 339.
[290] C. Rulliere and J. Joussot-Dubien, *Optics Comm.*, 1978, **24**, 38.
[291] W. Majewski and J. Krasinski, *Optics Comm.*, 1976, **18**, 255.
[292] T. G. Pavlopoulos, *Optics Comm.*, 1978, **24**, 170.
[293] C. A. Moore and C. D. Decker, *J. Appl. Phys.*, 1978, **49**, 47.
[294] R. Kugel, A. Svirmickas, J. J. Katz, and J. C. Hindman, *Optics Comm.*, 1977, **23**, 189.
[295] D. Basting, D. Ouw, and F. P. Schaefer, *Optics Comm.*, 1976, **18**, 260.
[296] J. S. Bakos and Zs. Sorlei, *Optics Comm.*, 1977, **23**, 258.
[297] C. V. Shank and E. P. Ippen, *Laser Focus*, July 1977, p. 44.
[298] J. C. Diels, E. Van Stryland, and G. Benedict, *J. Opt. Soc. Amer.*, 1978, **68**, 666; C. P. Ausschnitt, R. K. Jain, and J. P. Heritage, *ibid.*, 1978, **68**, 666; R. K. Jain and C. P. Ausschnitt, *Optic Letters*, 1978, **2**, 117; A. Scavennec, *Optics Comm.*, 1977, **20**, 335; T. Kurobori, Y. Cho, and Y. Matsuo, *ibid.*, 1978, **24**, 41.
[299] C. P. Ausschnitt and R. K. Jain, *Appl. Phys. Letters*, 1978, **32**, 727.
[300] I. S. Ruddock and D. J. Bradley, *Appl. Phys. Letters*, 1976, **29**, 296.
[301] N. J. Frigo, T. Daly, and H. Mehr, *I.E.E.E. J. Quantum Electron.*, 1977, **QE-13**, 101.
[302] J. C. Diels, E. Van Stryland, and G. Benedict, *Optics Comm.*, 1978, **25**, 93.
[303] H. Mahr, *I.E.E.E. J. Quantum Electron.*, 1976, **QE-12**, 554.
[304] B. Conillard and A. Ducasse, *Appl. Phys. Letters*, 1976, **29**, 665.
[305] K. G. Spears and J. Larsen, *Rev. Sci. Instr.*, 1977, **48**, 1333.

and an arrangement of a doubled-pulsed dye laser to produce simultaneous 0.5 ps pulses from Rh6G and 1.0 ps pulses from cresyl violet.[306] A mode-locked Kr laser was used to pump a series of dyes with an output continuously tunable from 685 to 900 nm with a reported conversion efficiency of 30%.[307] A similar arrangement used to pump carbocyanine dyes (range 685—965 nm) resulted in maximum output powers of 1.5 W with a 5 W pump power.[308] This efficiency is considerably higher than that reported for other dyes in the visible (including the rhodamines) and was aided by certain experimental precautions including use of a triplet quencher, a reduced oxygen partial pressure, the addition of surfactant, and increased jet velocity. With frequency doubling (to 343—482 nm), this picosecond arrangement would be most attractive for photochemistry. A pulse train from a Kr laser passed through a dye laser amplifier pumped by a nitrogen laser resulted in one pulse from the input train being amplified[309] by about 100 times but non-amplified pulses being attenuated by a factor of 10^{-1} or less.[310] A mode-locked Rh6G laser (10 ps pulsewidth) provided even shorter pulses (<1 ps) on pumping rhodamine B.[311]

An arrangement for actively mode-locking a FLPD laser by bleaching a saturable absorber with picosecond pulses from a master oscillator was used to provide independently tunable and highly synchronous pulses in the 550—643 nm range.[312] Direct measurements of pulse durations (with a streak camera) have shown that single pulses of 2—6 ps duration are obtained in the middle and latter stages of an actively mode-locked flashlamp-pumped dye laser but the initial pulses have a complicated structure and should not be used for kinetic measurements.[313] Flashlamp pumping was also used to provide 5 ps pulses from Fluorol 7GA[314] and 10 ps pulses from coumarin 314.[315] A short cavity dye laser pumped by picosecond pulses at 354 nm (from a Nd laser) produced pulsewidths of <14 ps with tuning over the range 396—470 nm with various dyes.[316]

Solid-state Lasers.—Streak camera measurements have recently revealed some interesting results on the performance characteristics of Nd-doped matrices used for picosecond pulse generation. Transform-limited pulses of *ca.* 5 ps duration were found throughout the whole pulse train of a mode-locked Nd : phosphate glass laser operating in the TEM_{00} mode but spectral and temporaral broadening became apparent as the higher-power mode-locked trains evolved.[317] Frequency-doubled pulses which were not transform limited

[306] Z. A. Yasa, A. Dienes, and J. R. Whinnery, *Appl. Phys. Letters*, 1977, **30**, 24.
[307] J. Kuhl, R. Lambrich, and D. von der Linde, *Appl. Phys. Letters*, 1977, **31**, 657.
[308] K. M. Romanek, O. Hildebrand, and E. Gobel, *Optics Comm.*, 1977, **21**, 16.
[309] B. Bolger, L. Baede, and H. M. Gibbs, *Optics Comm.*, 1976, **19**, 346.
[310] T. Urisu and K. Kajiyama, *Optics Comm.*, 1977, **20**, 34.
[311] R. K. Jain and J. P. Heritage, *J. Opt. Soc. Amer.*, 1977, **67**, 1425.
[312] E. Lill, S. Schneider, and F. Dorr, *Optics Comm.*, 1977, **22**, 107.
[313] E. Lill, S. Schneider, F. Dorr, S. F. Bryant, and J. R. Taylor, *Optics Comm.*, 1977, **23**, 318.
[314] E. Lill, S. Schneider, and F. Dorr, *Optics Comm.*, 1977, **20**, 223.
[315] J. C. Mialocq and P. Goujon, *Optics Comm.*, 1978, **24**, 255.
[316] A. J. Cox, G. W. Scott, and L. D. Talley, *Appl. Phys. Letters*, 1977, **31**, 389; G. W. Scott and A. J. Cox, *Proc. Soc. Photo-Opt. Instrum. Eng.*, 1977, **113**, 25.
[317] J. R. Taylor, W. Sibbett, and A. J. Cormier, *Appl. Phys. Letters*, 1977, **31**, 184.

($\Delta \nu = 46$ cm^{-1}, $\Delta t = 6$ ps) were observed with streak camera detection from the same glass.[318] The phosphate glass appeared to offer considerably higher conversion efficiencies for both second and further harmonic generation. Minimum bandwidth pulses of <15 ps (some two—three times shorter than Nd : YAG) were measured with Nd : BEL (neodymium : lanthanum beryllate).[319]

Neodymium : YAG unstable oscillators have recently become of growing importance owing mainly to the high energies which may be obtained in a uniphase output.[320,321] A 20 Hz system has been described in which Q-switched output energies of 180 mJ in 10 ns pulses have been obtained in a diffraction-limited beam with a divergence of <0.25 mR.[321] Nd : YAG rods have been shown to be effective amplifiers (100-times) when used with mode-locked Nd : glass oscillators and to provide pulses at 1061 and 1064 nm each less than 3 cm^{-1} bandwidth of 6 ps duration.[322] A reliable synchronizable single 1060 nm pulse has been produced in an actively mode-locked Nd : glass ring laser with a pulsewidth variable from 1.3 ns to the spectral transform limit of 15 ps.[323] Other Nd lasers described include the LED end-pumped varieties,[59,324] a simple room temperature NdP_5O_{14} laser excited by a small xenon flashlamp,[325] and a CW Nd : YAG laser pumped by a 5 kW krypton arc which could be tuned to 19 transitions from 1052 to 1444 nm.[326]

Fluorescence from the $5d \rightarrow 4f$ transitions in tervalent rare-earth ions (such as Nd^{3+}, Er^{3+}, and Tm^{3+}) in solid hosts (such as LaF_3, YF_3, and LuF_3) is characterized by broad bandwidths and large Stokes shifts–features which are particularly attractive for the development of tunable lasers.[327] Pumping these crystals with an electron beam, hydrogen laser or excimer laser could lead to the development of lasers tunable from 156 to 322 nm.[327,328]

There have been several useful developments in electro-optic switching for single pulse extraction. A coaxial silicon photoconductive switch capable of connecting 10 kV pulses with a risetime of <50 ps and jitter of a few picoseconds has found considerable use in driving a travelling-wave Kerr cell and fast Pockels cell.[329] The need for a photodiode-driven avalanche circuit has been eliminated by use of a krytron, triggered by focusing the output of the laser on to the grid in order to switch out a single pulse with a Pockels cell.[330] A light-triggered diode array used for a similar switching requirement has been used to

[318] G. R. Fleming, I. R. Harrowfield, A. E. W. Knight, J. M. Morris, R. J. Robbins, and G. W. Robinson, *Optics Comm.*, 1977, **20**, 36.
[319] L. S. Goldberg and J. N. Bradford, *Appl. Phys. Letters*, 1976, **29**, 585.
[320] R. L. Herbst, H. Komine, and R. L. Byer, *Optics Comm.*, 1977, **21**, 5; A. Owyoung and E. D. Jones, *Rev. Sci. Instr.*, 1978, **49**, 266.
[321] D. Andreou, *Rev. Sci. Instr.*, 1978, **49**, 586.
[322] D. Huppert and P. M. Rentzepis, *Appl. Phys. Letters*, 1978, **32**, 241.
[323] I. V. Tomov, R. Fedosejevs, and M. C. Richardson, *Appl. Phys. Letters*, 1977, **30**, 164.
[324] F. W. Ostermayer, *I.E.E.E. J. Quantum Electron.*, 1977, QE-13, 1; K. Washio, K. Iwamoto, K. Inoue, I. Hino, S. Matsumoto, and F. Saito, *Appl. Phys. Letters*, 1976, **29**, 720.
[325] S. R. Chinn and W. K. Zwicker, *Appl. Phys. Letters*, 1977, **31**, 178.
[326] J. Marling, *I.E.E.E. J. Quantum Electron.*, 1978, QE-14, 56.
[327] K. H. Yang and J. A. DeLuca, *Appl. Phys. Letters*, 1977, **31**, 594.
[328] K. H. Yang and J. A. DeLuca, *Appl. Phys. Letters*, 1976, **29**, 499.
[329] A. Antonetti, M. M. Malley, G. Mourou, and A. Orszay, *Optics Comm.*, 1977, **23**, 435.
[330] R. L. Hyde, D. Jacoby, and S. A. Ramsden, *J. Phys. (E)*, 1977, **11**, 1106.

extract two pulses with a delay from 10 to 70 ns.[331] Optimum risetime characteristics of a laser-triggered spark gap have been determined.[332] A Pockels cell
mounted between parallel polarizers was shown to be an effective optical shutter
to separate laser pulses from a laser-produced plasma with an opening time of
0.37 μs.[333]

Published work on passive Q-switches has included a useful review of phthalo-
cyanines and carbocyanines as both Q-switches and laser dyes,[334] a report of a
saturable absorber in plastic solution to mode-lock a Nd : glass laser,[335] and
means for producing high energy 40 ns pulses from Nd : glass with an exploding
PbS film.[336]

Frequency Conversion.—Second harmonic generation (SHG) by means of non-
linear crystals is the most common means of providing tunable u.v. radiation
from the numerous high power visible lasers. CW radiation tunable from 257 to
400 nm has been generated by frequency-doubling and sum frequency mixing
of a dye laser and selected lines in ion lasers.[33—35,337] A detailed analysis of
SHG with intracavity non-linear crystals such as ADA has shown the importance
of cavity length on efficiency and the need for crystals with very low absorptions.[338] Tunable u.v. radiation has been obtained from laser-pumped coumarin
dyes extending from 246 to 265 nm (with a 90° phase-matched and temperature-
tuned ADP crystal) at an efficiency of 8.4%[339] and from 217 to 250 nm (with
potassium phentaborate) at an efficiency of 5%.[340] Conversion efficiencies of
9.4 and 8.4%, respectively, have been reported for ADP and KDP crystals
employed with a flashlamp-pumped Rh6G laser[341] and up to 40% with CD*A
for Nd : YAG.[342] With short pathlength crystals at higher intensities ($>10^9$ W
cm^{-2}), up to 85% conversion was found for picosecond pulse conversion from
532 to 266 nm with ADP and KD*P.[343] Absolute measurements of the two-
photon absorption coefficients of these two crystals at 266.1 nm emphasized that
a crystal length exists for each input pump intensity giving optimum conversion
and that excessive crystal lengths can reduce SHG efficiencies.[344]

Spectral phase-matching bandwidths in several of the more useful non-linear
crystals ($LiIO_3$, $LiNbO_3$, CDA, and KDP) for SHG at 1060 nm have been
accurately measured with a tunable line-narrowed Nd : laser[345] and the optimum
conditions for third harmonic generation in this laser determined.[346] The
performance of an angle-tuned cell in which an ADP crystal was rotated inside

[331] I. K. Pearce and D. D. McLeod, *J. Phys. (E),* 1977, **10**, 961.
[332] J. C. Scott and A. W. Palmer, *J. Phys. (E),* 1978, **11**, 495.
[333] J. Watson and A. Maitland, *J. Phys. (E),* 1977, **10**, 1227.
[334] A. D. Britt and W. B. Moniz, *Appl. Spectroscopy,* 1977, **31**, 104.
[335] J. R. Taylor, W. Sibbett, and A. J. Cormier, *Appl. Phys. Letters,* 1977, **31**, 732.
[336] M. J. Landry, *Appl. Optics,* 1978, **17**, 635.
[337] S. Blit, F. K. Tittel, and E. G. Weaver, *J. Opt. Soc. Amer.,* 1977, **67**, 1384.
[338] A. I. Ferguson and M. H. Dunn, *I.E.E.E. J. Quantum Electron.,* 1977, **QE-13**, 751.
[339] R. K. Jain and T. K. Gustafson, *I.E.E.E. J. Quantum Electron.,* 1976, **QE-12**, 555.
[340] H. Zacharias, A. Anders, J. B. Halpern, and K. H. Welge, *Optics Comm.,* 1976, **19**, 116.
[341] A. Hirth, K. Vollrath, and J. Y. Allain, *Optics Comm.,* 1977, **20**, 347.
[342] K. Kato and R. S. Adhav, *I.E.E.E. J. Quantum Electron.,* 1976, **QE-12**, 443.
[343] J. Reintjes and R. C. Eckardt, *Appl. Phys. Letters,* 1977, **30**, 91.
[344] J. Reintjes and R. C. Eckardt, *I.E.E.E. J. Quantum Electron.,* 1977, **QE-13**, 791.
[345] Y. S. Liu, *Appl. Phys. Letters,* 1977, **31**, 187.
[346] H. B. Puell and C. R. Vidal, *I.E.E.E. J. Quantum Electron.,* 1978, **QE-14**, 364.

an index matching fluid has been assessed for the Rh6G laser[347] and a feedback device used to control the precise orientation of an angle-tuned crystal to ensure maximum SHG efficiency.[348] Phase matching in a fixed SHG crystal by changing the angle of the incident beam with a dispersive element such as gratings or prisms provided a means of generating tunable coherent radiation from 230 to 350 nm with a fast scanning rate.[349] A calculator-based system provided automatic phase-matching for u.v. scanning covering the range 217—360 nm.[350]

Among other papers on non-linear crystals are included the relative merits of RDA and ADP for SHG with a ruby laser,[351] use of a KD*P–ADP tandem arrangement with 266 nm generation from a Nd : YAG laser,[352] harmonic generation in potassium pentaborate[353] and ammonium malate,[354] and a $CdGeAs_2$ doubler for a pulsed CO_2 laser.[355] A variable temperature Dewar flask, with a temperature stability of ± 0.02 K over the range 100—393 K has been described for use with ADP, ADA, and RDP crystals.[356]

Frequency up-conversion using non-linear optical interactions has proved to be an important source of coherent radiation in the vacuum-u.v. and examples of this technique have included seventh harmonic generation in helium to produce 38 nm radiation from laser pulses at 266 nm,[357] four-wave mixing from dye lasers in producing coherent radiation in the 130—152 nm range[358] and third- and fifth-harmonic generation from a mode-locked Nd laser in sodium–xenon vapour.[359] Efficient vacuum-u.v. radiation at 172 nm by four-wave mixing with low power dye laser beams in strontium vapour has led to the prediction that it may be possible to generate low power CW vacuum-u.v. radiation in this way.[360] With the evolution of simple, reliable excimer lasers, high power coherent sources in the 120—130 nm region would appear realistic.[361] The conversion of visible wavelength radiation from two NLPD lasers into vacuum-u.v. radiation (135—160 nm) by a third-order non-linear process in atomic magnesium has been shown to be remarkably efficient with more than 10^{10} photons per pulse in the 155 nm region with a bandwidth of 0.3 cm^{-1}.[362]

[347] S. Blit and F. K. Tittel, *I.E.E.E. J. Quantum Electron.*, 1978, **QE-14**, 329.
[348] R. Bernheim, C. Kitrell, and D. K. Veirs, *Chem. Phys. Letters*, 1977, **51**, 325.
[349] S. Saikan, *Optics Comm.*, 1976, **18**, 439.
[350] G. K. Klauminzer, in 'Laser 77 Opto-Electronics Conference Proceedings', ed. W. Waidelich, I.P.C., London, 1977, p. 93.
[351] S. Sullivan and E. L. Thomas, *Optics Comm.*, 1978, **25**, 125.
[352] Y. S. Liu, W. B. Jones, and J. P. Chernoch, *Appl. Phys. Letters*, 1976, **21**, 32.
[353] T. S. Chen and W. P. White, *I.E.E.E. J. Quantum Electron.*, 1976, **QE-12**, 436; K. Kato, *Appl. Phys. Letters*, 1976, **29**, 562.
[354] K. Betzler, H. Hesse, and P. Loose, *Optics Comm.*, 1977, **22**, 66.
[355] N. Menyuk, G. W. Iseler, and A. Mooradian, *Appl. Phys. Letters*, 1976, **29**, 422.
[356] K. G. Spears, L. Hoffiand, and R. Loyd, *Appl. Optics*, 1977, **16**, 1172.
[357] J. Reintjes, C. Y. She, R. C. Eckardt, N. E. Karangelen, R. A. Andrews, and R. C. Elton, *Appl. Phys. Letters*, 1977, **30**, 480.
[358] K. K. Innes, B. P. Stoicheff, and S. C. Wallace, *Appl. Phys. Letters*, 1976, **29**, 715.
[359] V. L. Doitcheva, V. M. Mitev, L. I. Pavlov, and K. M. Stamenov, *Optics Quantum Electron.*, 1978, **10**, 131; D. I. Metchov, V. M. Mitev, L. I. Pavlov, and K. V. Stamenov, *Optics Comm.*, 1977, **21**, 391; K. N. Drabovich, D. I. Metchov, V. M. Mitev, L. I. Pavlov, and K. V. Stamenov, *ibid.*, 1977, **20**, 350.
[360] G. C. Bjorklund, J. E. Bjorkholm, R. R. Freeman, and P. F. Liao, *Appl. Phys. Letters*, 1977, **31**, 330.
[361] S. C. Wallace, T. J. McKee, and B. P. Stoicheff, *J. Opt. Soc. Amer.*, 1978, **68**, 677.
[362] A. C. Provorov, B. P. Stoicheff, and S. C. Wallace, *J. Chem. Phys.*, 1977, **67**, 5393.

This intensity is $>10^5$ that from synchrotron sources (see Section 9), and has been used to obtain the first lifetime measurements of individual rotational levels of the (0,0) band of the lowest allowed transition in CO ($A^1\pi - X^1\Sigma$). Tunable radiation from mixing experiments has also been reported from 208 to 217 nm (from two NLPD lasers in potassium pentaborate),[363] from 240 to 248 nm (from a ruby laser and ruby-laser-pumped DPTC and HITC laser dye in ADP KDP),[364] from 207 to 217 nm (by mixing fundamental and harmonics of a dye laser in potassium pentaborate),[365] from 208 to 234 nm (by mixing a dye laser harmonic with 1064 nm radiation),[366] and to 196 nm (by mixing Nd harmonics and a near-i.r. dye laser in potassium pentaborate).[367] Emission at 50 discrete wavelengths was observed from 231 to 803 nm when BaI$_2$ was excited with a dye laser tuned to the 791 nm intercombination line and a ruby laser, most of the lines being attributed to four-wave mixing.[368]

High efficiency tunable-frequency shifting in pulsed i.r. lasers has many applications in photochemistry, isotope separation, and high resolution spectroscopy, and four-wave|mixing|of CO$_2$ laser beams in germanium, for example, has been used to provide thousands of high power lines of discrete frequencies in the 8—13 μm region at conversion efficiencies of about 37%.[369] Four-wave sum and frequency difference frequency-generation in the outputs of two CO$_2$ TEA lasers has been demonstrated in liquid CO—CO$_2$ mixtures at 77 K[370] and a useful 16 μm radiation source obtained by mixing ruby and CO$_2$ laser beams in para-hydrogen.[371] Two dye lasers pumped by a single nitrogen laser have been used to generate i.r. radiation tunable from 5.5 to 18.3 μm in a single crystal of AgGaS$_2$.[372] An efficient linear electro-optical effect in CdTe has been used to shift the output of a CO$_2$ laser to cover the entire 1—30 μm range.[373] Tunable radiation at lower wavelengths (0.68—1.1 μm) has been generated by mixing a nitrogen laser and dye laser pulses in ADP.[374]

Although the excimer lasers provide powerful new pulses in the u.v. at discrete frequencies, it is desirable for many photochemical applications to have available a greater selection of wavelength. This has now been achieved in the 190—360 nm range by multiple Raman scattering of ArF and KrF laser wavelengths in high pressure hydrogen gas.[375] For the optimized case of KrF, more than 50% energy conversion was observed for the first Stokes lines at 276.9nm. Similarly, near-unit photon conversion of the XeF laser output at 351—585 nm

[363] F. B. Dunning and R. E. Stickel, *Appl. Optics*, 1976, **15**, 3131.
[364] R. E. Stickel and F. B. Dunning, *Appl. Phys. Letters*, 1978, **17**, 1313.
[365] K. Kato, *I.E.E.E. J. Quantum Electron.*, 1977, **QE-13**, 544.
[366] G. A. Massey and J. C. Johnson, *I.E.E.E. J. Quantum Electron.*, 1976, **QE-12**, 721.
[367] K. Kato, *Appl. Phys. Letters*, 1977, **30**, 583.
[368] C. H. Skinner and H. P. Palenius, *Optics Comm.*, 1976, **18**, 335.
[369] N. Lee, R. L. Aggarwal, and B. Lax, *I.E.E.E. J. Quantum Electron.*, 1977, **QE-13**, 59D; *J. Appl. Phys.*, 1977, **48**, 2470.
[370] H. Kildal and S. R. J. Brueck, *Appl. Phys. Letters*, 1978, **32**, 173.
[371] P. P. Sorokin, M. M. T. Loy, and J. R. Lankard, *I.E.E.E. J. Quantum Electron.*, 1977, **QE-13**, 871.
[372] R. J. Seymour and F. Zernike, *Appl. Phys. Letters*, 1976, **29**, 705.
[373] G. M. Carter, *Appl. Phys. Letters*, 1978, **32**, 810; *J. Opt. Soc. Amer.*, 1978, **68**, 635.
[374] C. L. Sam and M. M. Choy, *Appl. Phys. Letters*, 1977, **30**, 199.
[375] T. R. Loree, R. C. Sze, J. Ackerhalt, and D. L. Barker, *J. Opt. Soc. Amer.*, 1978, **68**, 677; T. R. Loree, R. C. Sze, and D. L. Barker, *Appl. Phys. Letters*, 1977, **31**, 37.

was obtained in Ba vapour.[376] Very broad stimulated emission of light covering almost the entire visible spectrum has been provided by a fused silica fibre Raman laser in which a 30—50 m length of 50 μm diameter fibre was pumped by a 60 kW 532 nm laser pulse (see also Vol. 8, pp. 46—47).[377] At longer wavelengths Raman scattering in Cs vapour of a visible dye laser provided tunable i.r. radiation.[378] In high pressure hydrogen, a Nd : YAG laser-pumped $LiNbO_3$ parametric oscillator produced i.r. radiation continuously tunable from 3.5 to 13 μm.[379]

Pulses covering the range 0.72—7.7 μm with powers between 9 and 1000 MW have been generated by using an amplified ruby-laser-pumped dye laser and its Stokes component in hydrogen gas[380] and superfluorescent far-i.r. lasers near 50 μm in rubidium vapour have been tuned over 800 MHz by using the dynamic Stark effect induced by a Q-switched CO_2 laser.[381]

Optical parametric oscillators (OPO) offer great promise as means for providing coherent light tunable from 300 nm to 30 μm.[382] A recent development in crystal technology enabled lithium niobate crystals to be grown along a direction only a few degrees from the direction needed for phase matching to a Nd-pumped optical parametric oscillator. As a result, high power parametric oscillations could be sustained over the 1.4—4.4 μm range[383] with an efficiency of 40%.[384] Ultrashort pulses tunable from 240 nm to 3.6 μm have been reported with a $LiNbO_3$ OPO on pumping with a repetitive mode-locked Nd : YAG laser.[385] Gallium selenide has recently been introduced as a crystal transparent from 0.65 to 18 μm with a large birefringence. By combining the signal and idler outputs from a $LiNbO_3$ OPO in this crystal, coherent light tunable from 4 to 12 μm has been obtained.[386] A tunable source at 16 μm with 500 W peak power has used a 20 Hz Nd : YAG driven $LiNbO_3$ OPO with non-linear mixing in CsSe.[387] Nearly bandwidth-limited i.r. pulses (from 1.4 to 4 μm) of 3—5 ps duration have been produced by single-path parametric amplification in two properly aligned $LiNbO_3$ crystals.[388] Other examples of OPOs have included 0.7—2.2 μm production in $LiIO_3$ with a nitrogen laser and dye laser[389] and high efficiency (50%) narrow bandwidth difference frequency generation in $LiNbO_3$ with two dye lasers.[390] Continuously tunable i.r. radiation over the

376   N. Djeu and R. Burnham, *Appl. Phys. Letters*, 1977, **30**, 473.
377   J. Botineau, F. Gires, A. Saissy, C. Vanneste, and A. Azema, *Appl. Optics*, 1978, **17**, 1209.
378   R. T. V. Kung and I. Itzkan, *I.E.E.E. J. Quantum Electron.*, 1978, **QE-14**, 73.
379   S. J. Brosnan, R. N. Fleming, R. L. Herbst, and R. L. Byer, *Appl. Phys. Letters*, 1977, **30**, 330.
380   M. Bierry, R. Frey, and F. Pradeve, *Rev. Sci. Instr.*, 1977, **48**, 733.
381   T. Y. Chang, T. C. Damen, V. T. Nguyen, J. D. McGee, and T. J. Bridges, *Appl. Phys. Letters*, 1978, **32**, 633.
382   D. S. Chelma, *Phys. Bull.*, 1978, **29**, 214.
383   R. B. Wiseman and S. A. Rice, *Optics Comm.*, 1976, **19**, 28.
384   R. Baumgartner and R. L. Byer, *J. Opt. Soc. Amer.*, 1978, **68**, 688; *Proc. Soc. Photo-Opt. Instrum. Eng.*, 1977, **122**, 89; R. L. Byer, *ibid.*, 1976, **95**, 92.
385   Y. Tanaka, T. Kushida, and S. Shionoya, *Optics Comm.*, 1978, **25**, 273.
386   A. Bianchi, A. Ferrario, and M. Musci, *Optics Comm.*, 1978, **25**, 256.
387   D. Andreou, *Optics Comm.*, 1977, **23**, 37.
388   A. Seilmeier, K. Spanner, A. Laubereau, and W. Kaiser, *Optics Comm.*, 1978, **24**, 237.
389   R. Koening, A. Rosenfeld, N. Tam, and S. Mory, *Optics Comm.*, 1978, **24**, 190.
390   R. J. Seymour and M. M. Choy, *Optics Comm.*, 1977, **20**, 101.

range 4.5—13 μm was obtained with a $LiNbO_3$ parametric oscillator followed by a hydrogen cell for Raman shifting.[391]

An improved i.r. detection capability has been attained by means of an i.r. quantum counter in which i.r. photons are converted into easily detectable visible photons. By mixing signals with an i.r. pump source in a medium which has real electronic transitions resonant at both signal and pump wavelengths (such as sodium vapour), high sensitivity detection was confirmed and a predicted NEP for the device put at 10^{-17} W $Hz^{\frac{1}{2}}$ in the 1.5—5 μm spectral region.[392] Proustite crystals have been used successfully for the detection of 75 ps pulsed CO_2 laser pulses with 1060 nm Nd radiation to generate 960 nm sum frequency radiation (detected by a streak camera)[393] and for 10 μm emission from astronomical objects (by mixing with a 0.25 W, 752.5 nm Kr laser beam).[394] In the latter case, the conversion efficiency for image conversion was *ca.* 2×10^{-7}. Similarly $AsGaS_2$ crystals were used in the conversion of mid-i.r. radiation into green light[395] and 3.39 μm radiation was up-converted into 446.7 nm light using a 514.5 Ar laser pump and a $LiNbO_3$ crystal.[396] Resonant two-photon pumping in the Cs $6s^2S–7s^2S$ transition has been used for 2.9 μm into 455 nm image up-conversion.[397]

Laser Measurements.—Calorimetric methods are commonly used for absolute energy measurements of both CW and pulsed lasers and the precision and accuracy of such measurements have been shown to depend critically on the alignment, beam width, and beam intensity distribution.[398] An absolute reference calorimeter for energy measurements of high power lasers (range 0.4—15 J) was based on volume absorption in a solid and provided a random error of $\pm 0.2\%$ and a systematic error of $\pm 2.3\%$ over the visible-to-near i.r. range.[399] (A simple calorimeter to measure spark energies in the range 200 μJ to 50 mJ has also been described.[400]) Circulated liquid absorption calorimeters have been described for use with 100 J, 50 ns KrF laser pulses[401] and <5 W CO_2 laser pulses.[402] Various disc calorimeters have also been described for CO_2 laser measurements[403] as well as a thermoelectric detector.[404] A technique for calibrating calorimeters has also been published.[405] A broadband power meter suitable for use from 0.4 to at least 500 μm with a sensitivity from 1 mW

[391] S. J. Brosnan and R. L. Byer, *I.E.E.E. J. Quantum Electron.*, 1977, **QE-13**, 58D.
[392] J. A. Gelbwachs, C. F. Klein, and J. E. Wessel, *I.E.E.E. J. Quantum Electron.*, 1978, **QE-14**, 77.
[393] A. C. Walker and A. J. Alcock, *Rev. Sci. Instr.*, 1976, **47**, 915.
[394] R. W. Boyd and C. H. Townes, *Appl. Phys. Letters*, 1977, **31**, 440.
[395] W. Jantz, P. Koidl, and R. Schatzle, *I.E.E.E. J. Quantum Electron.*, 1977, **QE-13**, 55D; W. Jantz and P. Koidl, *Appl. Phys. Letters*, 1977, **31**, 99.
[396] J. Falk and Y. C. See, *Appl. Phys. Letters*, 1978, **32**, 100.
[397] E. A. Stappaerts, S. E. Harris, and J. F. Young, *Appl. Phys. Letters*, 1976, **29**, 669.
[398] G. F. Tietz, *Appl. Optics*, 1977, **16**, 1136.
[399] D. L. Franzen and L. B. Schmidt, *Appl. Optics*, 1976, **15**, 3115.
[400] L. R. Merritt, *J. Phys. (E)*, 1978, **11**, 193.
[401] G. A. Fisk and M. A. Gusinow, *Rev. Sci. Instr.*, 1977, **48**, 118.
[402] D. C. Emmony and J. C. S. Bunn, *J. Phys. (E)*, 1976, **9**, 621.
[403] Z. Jurela and T. Jojic, *Rev. Sci. Instr.*, 1978, **49**, 392; D. F. Grosjean, R. A. Olson, B. Sanka, and D. C. Rabe, *Rev. Sci. Instr.*, 1978, **49**, 778; R. W. Zimmerer, *Optical Spectra*, 1978, **12** (2), 34.
[404] J. M. Moynault, G. Bachet, and R. Coulon, *J. Phys. (E)*, 1977, **10**, 703.
[405] P. D. Thacher, *Appl. Optics*, 1976, **15**, 1815.

to 1 W and an accuracy of better than 10% employed a copper cone lined with colloidal graphite paint as the absorber.[406] An internal-cavity calorimeter for CW DF lasers used a variable rectangular mode geometry, while containing the radiation inside the resonator.[407] Optical pulse energies (at 1060 nm) as small as 1 nJ could be measured to an electronic accuracy of 1% by means of a simple photodiode, amplifier, and peak follower-and-hold circuit.[408]

Changes in the temporal profile of i.r. laser pulses have been recorded with a low cost pyroelectric detector, which used a technique of chopping to eliminate heating of the detector element.[409]

Laser beam profiles (for laser powers <0.4 mW cm^{-2} CW or 6.0 μJ cm^{-2} for 10 ns pulses) in the spectral range 400 to 1100 nm have been measured using an array of 50 FET switches to access the elements of either a linear array of silicon photodiodes or an array of pyroelectric detectors formed on a single sheet of plastic.[410]

Laser-induced opto-galvanic spectra have been used to provide a simple yet highly accurate means of wavelength calibration for tunable lasers.[411] Bandwidths may be determined down to 0.01 nm and laser wavelengths for single frequency excitation experiments to 10^{-4} nm. The frequency-drift in picosecond pulses resulting from non-linearity of index of refraction of the amplifying medium has been measured.[412] A two-beam interferometer allowed ratios of an unknown wavelength to that of a standard to be measured to an accuracy of 6 parts in 10^8.[413]

4 Monochromators and Light Filters

Comparative studies of conventional ruled and holographic (interferometrically-produced) diffraction gratings have recently been published[414—416] in order to assist the selection of a particular grating for scattered light[414], efficiency,[415] and spectral resolution.[416] A loss in blaze efficiency of 20% for both planes of polarization found with a holographic grating (compared with that for triangular groove gratings) must be balanced against the other spectroscopic properties of holographic gratings including the absence of ghosts and low scattered light.[417] In a review on diffraction gratings, it was pointed out that in the visible region, ruled gratings are probably the best for low dispersion work and holographic

[406] K. M. Evenson, D. A. Jennings, F. R. Petersen, J. A. Mucha, J. J. Jimenez, R. M. Charlton, and C. J. Howard, *I.E.E.E. J. Quantum Electron.*, 1977, **QE-13**, 442.
[407] R. A. Chodzko, S. B. Mason, R. R. Giedt, and D. A. Durran, *Appl. Optics*, 1976, **15**, 2367.
[408] A. L. Smirl, R. L. Shoemaker, J. B. Hambenne, and J. C. Matter, *Rev. Sci. Instr.*, 1978, **49**, 672.
[409] D. C. Elbers, W. H. Thomason, and J. D. Macomber, *Appl. Optics*, 1978, **17**, 308.
[410] J. G. Edwards, H. R. Gallantree, and R. M. Quilliam, *J. Phys. (E)*, 1977, **10**, 699.
[411] D. S. King, P. K. Schenck, K. C. Smyth, and J. C. Travis, *Appl. Optics*, 1977, **16**, 2617.
[412] W. Zinth, A. Laibereau, and W. Kaiser, *Optics Comm.*, 1977, **22**, 161; J. P. Gex, C. Sauteret, P. Vallet, H. Tourbez, and M. Schelev, *Optics Comm.*, 1977, **23**, 430
[413] F. V. Kowalski, R. T. Hawkins, and A. W. Schawlow, *J. Opt. Soc. Amer.*, 1976, **66**, 965.
[414] J. F. Verrill, *Optica Acta*, 1978, **25**, 531.
[415] E. G. Loewen, M. Niviere, and D. Maystre, *Appl. Optics*, 1977, **16**, 2711.
[416] M. C. Hutley, *J. Phys. (E)*, 1976, **9**, 513.
[417] B. J. Brown and I. J. Wilson, *Optics Comm.*, 1977, **20**, 418.

gratings best for high spectral purity and high dispersion work.[416] For u.v. applications, either blazed holographic or unblazed gratings with high groove densities were recommended as suitable. Stray light problems in optical systems have also been dealt with in a series of papers on the light-scattering characteristics of optical surfaces.[418]

Holographic gratings have been used in a number of spectrometers including a double monochromator for measuring the ratio of diffuse to direct flux in the u.v. near to the atmospheric limit,[419] in a similar polychromator having a spectral range of ± 20 nm width centred at 180, 220, and 260 nm,[420] and in a new monochromator for far-u.v. spectroscopy (10—50 eV).[421] In the latter instrument, a holographically-fabricated transmission grating consisted of a fine gold grating supported on a random gold mesh. Other vacuum-u.v. monochromators included a simple compact grazing incidence arrangement for isolating emission lines of He, Ne, and Ar,[422] a high-flux instrument for synchrotron radiation covering the range 30—600 nm,[423] three new monochromators employing concave gratings,[424] and a normal incidence Wadsworth-type instrument for the range 27—103 nm.[425] A high resolution 3 m normal incidence spectrometer for the 300—3000 nm range has also been described.[426] Excellent image quality with high numerical apertures were features claimed for a new class of diffraction grating spectrographs.[427]

A hydrogen discharge lamp used in combination with a vacuum-u.v. monochromator has been the conventional source for monochromatic radiation below 200 nm. There are problems associated with lamps of this type including solarization of windows by the intense radiation and the need for mirrors with non-vulnerable coatings to focus the radiation onto the entrance slit of the monochromator. These inconveniences have been circumvented in a simple manner by integrating a 1 kW hydrogen lamp with a vacuum monochromator so that the entrance slit of the monochromator served as the lamp anode.[428] With this arrangement, the output at the exit slit of the monochromator was reported to be $>1.5 \times 10^{12}$ quanta s^{-1} at 121.6 nm (600 W input). A somewhat similar arrangement used a tunable u.v. flashlamp as the source slit for a vacuum-u.v. monochromator.[429]

A pre-monochromator for stray light rejection in Raman spectroscopy used two concave gratings mounted in a non-dispersive way and a sharp-edged movable mirror to reflect all the spectrum except Rayleigh lines to the second

[418] *Proc. Soc. Photo-Opt. Instrum. Eng.*, 1977, **107**.
[419] L. M. Garrision, L. E. Murray, D. D. Doda, and A. E. S. Green, *Appl. Optics*, 1978, **17**, 827.
[420] E. Pitz, C. Leinert, A. Schulz, and H. Link, *Appl. Optics*, 1978, **17**, 730.
[421] S. A. Flodstrom and R. Z. Bachrach, *Rev. Sci. Instr.*, 1976, **47**, 1464.
[422] N. V. Shevchik, *Rev. Sci. Instr.*, 1976, **47**, 1028.
[423] M. Lavollee and R. Lopez-Delgado, *Rev. Sci. Instr.*, 1977, **48**, 816.
[424] M. Singh, *Appl. Optics*, 1978, **17**, 1815.
[425] M. Howells, C. Norris, and G. R. Williams, *J. Phys. (E)*, 1977, **10**, 259.
[426] V. Saile, P. Gurtler, E. E. Koch, A. Kozevnikov, M. Skibowski, and W. Steinmann, *Appl. Optics*, 1976, **15**, 2559.
[427] L. Mertz, *Appl. Optics*, 1977, **16**, 3122.
[428] B. J. Mulder, *J. Phys. (E)*, 1977, **10**, 490.
[429] G. E. Miller, J. Halpern, and W. M. Jackson, *J. Opt. Soc. Amer.*, 1978, **68**, 641.

grating.[430] Spectral measurements close to a ruby laser line were made possible with a high rejection (10^9-10^{12}), high transmission triple grating monochromator with crossed dispersion.[431] Removal of undesired emssions from laser beams encountered in Raman and fluorescence spectroscopy was also accomplished with a high dispersion Amici-type dispersing prism used as a tunable optical filter.[432]

The polarized emission spectrum of 9-cyanoanthracene crystals over the range 2—300 K was determined with a conventional grating spectrometer in which a Fresnel rhomb before the spectrometer was used to rotate the polarization so that light falling on the grating was always polarized parallel to the grating.[433] In this way, the same system response was obtained for light polarized perpendicular and parallel to the crystal *c*-axis, and Wood's anomalies were eliminated. A very useful compact interference filter polychromator and narrow band rejection filter was described which used polarizing beam-splitting cubes.[434] With two cubes, four stages of light rejection were possible with an attainable extinction ratio of 5×10^{-6}. A spectrometer based on 16 interference filters was designed to scan the range 340—640 nm in 125 ms.[435]

Tunable acousto-optic filters have been described previously (Vol. 6, p. 77) and several reviews on their properties have been published.[436] Non-colinear acousto-optic tunable filters using TeO_2 crystals have been used as rapid scanning devices giving a separation of 3.2 nm at 400 nm in a period of 0.5 ms.[437] By refining the scanning method, it is claimed that scan speeds of 1 nm μs^{-1} would be possible with bandwidths <2 nm.

A useful review on factors affecting the stability of evaporated thin film filters commented that it is virtually impossible to prevent water penetration and that u.v. degradation of the epoxy adhesive (and laser damage) can be significant.[438] Commercial interference filters have been found to have widely varying wavelength transmission profiles over their areas.[439] Mixed semiconductor evaporated films can be used to provide sharp cut-off characteristics in the visible region[440] and thin films of silver have been deposited on plastic or glass surfaces by photolysing high molecular weight silver compounds.[441]

An invaluable paper on the spectral characteristics of 800 coloured glasses (which include all the common bandpass and cut-off filters) made by 13 manufacturers provides the internal spectral transmittance of typical glasses plotted as a function of wavelength in a series of 44 diagrams.[442] A series of bandpass

430 J. Fellman and P. Lindblom, *Appl. Optics*, 1977, **16**, 1085.
431 M. Greenwald and W. I. B. Smith, *Appl. Optics*, 1978, **17**, 587.
432 D. W. Collins, R. E. Cookingham, and A. Lewis, *Appl. Optics*, 1977, **16**, 252.
433 S.-T. Lee, S. Suzer, and D. A. Shirley, *Chem. Phys. Letters*, 1976, **41**, 25.
434 M. A. Mahdavi, *Appl. Optics*, 1977, **16**, 1765.
435 J. K. Bailiff, D. A. Morris, and M. J. Aitken, *J. Phys. (E)*, 1977, **10**, 1156.
436 I. C. Chang, *Proc. Soc. Photo-Opt. Instrum. Eng.*, 1976, **90**, 12; J. D. Feichtner, M. Gottlieb, and J. J. Conroy, *ibid.*, 1976, **82**, 106.
437 T. Yano and Watanabe, *Appl. Optics*, 1976, **15**, 2250.
438 T. M. Christmas and D. Richmond, *Optics and Laser Technology*, 1977, **9**, 109.
439 K. H. Elliot, *Appl. Optics*, 1977, **16**, 2799.
440 W. Kottler, D. Bonnet, and G. Ganson, *Appl. Optics*, 1978, **17**, 164.
441 Y. Konishi, H. Saijo, H. Hada, and M. Takura, *Nature*, 1977, **268**, 709.
442 J. A. Dobrowolski, G. E. Marsh, D. G. Charbonneau, J. Eng, and P. D. Josephy, *Appl. Optics*, 1977, **16**, 1491.

filter glasses for the visible region has been reported[443] in which up to four colouring oxides are added to four basic glass types to provide glasses with bandwidths from 58 to 168 nm and peak transmittances from 20 to 86%.[444] Cut-on glass filters from the 300—650 nm range containing Ce–Ti have also been reported.[445] Formulae for the response of an optical detector to the transmitted component of a tilted absorbing glass plate have been derived in terms of the angle of incidence, plate thickness, absorption coefficient, and index of refraction of the glass.[446] A liquid transmission filter, with 70% transmission at 250 nm, using a solution of rubeanic acid in ethanol has been proposed for wavelength isolation in the 235—310 nm range.[447]

The characteristics of interference filters for use in the far-i.r. (50 µm) have been reviewed with an emphasis on metallic mesh reflecting interfaces[448] and various narrow band far-i.r. filters described.[449] Black polyethylene and various salts mixed with black polyethylene have also been proposed as low cost transmission filters for use in this spectral region.[450]

A light trap has been proposed based on the principle that a polarized beam with its polarization appropriately oriented will be transmitted without reflection across a dielectric interface set at Brewster's angle with respect to the beam.[451] With a 3 mm glass u.v. transmitting filter set at 56° to a He–Ne laser the ratio of the incident beam intensity to the reflected beam intensity was 1000 : 1 and further attenuation of the reflected beam could be achieved with a simple matt black surface or second absorbing dielectric. Absorption cells containing iodine vapour have been shown to be narrow band spectral filters with very large solid angles and were especially effective in filtering out one of the two sodium lines which are separated by 0.6 nm.[452]

A solution of $CuCl_2$ in aqueous HCl has been reported as an effective bandpass filter for flashlamps in the region 460—580 nm with a maximum transmittance of 80% at 510 nm.[453] A variable pathlength water cell has been shown to be a simple attenuator for the iodine laser (at 1315 nm) with a typical range of 10^3.[454] Damage-resistant (15 MW cm^{-2}) rotators for polarization of u.v. and visible light consisted simply of aqueous solutions of sucrose.[455] The angle of rotation was governed by the concentration of sucrose and the optical pathlength.

[443] M. A. Res and C. J. Kok, *J. Phys.* (*D*), 1974, **7**, L196; M. A. Res, C. J. Kok, J. Bednarik, and K. Kroeger, *J. Phys.* (*D*), 1978, **11**, 735; M. A. Res, F. Hengstberger, and J. Bednarik, *J. Phys.* (*D*), 1977, **10**, 1275; M. A. Res, C. J. Kok, and J. Bednarik, *J. Phys.* (*D*), 1976, **9**, L119; M. A. Res, J. Bednarik, C. J. Kok, and K. Kroeger, *Optik*, 1977, **49**, 277.
[444] M. A. Res, C. J. Kok, J. Bednarik, and K. Kroeger, *Appl. Optics*, 1977, **16**, 1908.
[445] K. Kroeger, M. A. Res, and J. Bednarik, *Optik*, 1977, **49**, 347.
[446] R. L. Walraven, *Rev. Sci. Instr.*, 1978, **49**, 537.
[447] A. Passner, S. L. McCall, and M. Leventhal, *Rev. Sci. Instr.*, 1976, **47**, 1221.
[448] G. D. Holah and S. D. Smith, *J. Phys.* (*E*), 1977, **10**, 101.
[449] G. D. Holah and N. Morrison, *J. Opt. Soc. Amer.*, 1977, **67**, 971;
[450] P. S. Nayar and C. K. So, *Appl. Optics*, 1977, **16**, 289.
[451] P. B. Elterman, *Appl. Optics*, 1977, **16**, 2352.
[452] K. H. Liao and R. Gupta, *Rev. Sci. Instr.*, 1978, **49**, 867.
[453] K.-A. Engdahl and G. Ramme, *Chem. Phys. Letters*, 1976, **41**, 100.
[454] S. Ariga and B. Ahlborn, *Rev. Sci. Instr.*, 1978, **49**, 403.
[455] B. A. Baretz, G. E. Hall, and G. A. Kenney-Wallace, *Appl. Phys. Letters*, 1977, **31**, 387; *J. Opt. Soc. Amer.*, 1977, **67**, 1436.

5 Spectroscopic techniques

U.V.–Visible Absorption Spectrometry.—The apparently ever-present problem of standardization in absorption (and luminescence) spectrometry has been succintly reviewed in a collection of 15 papers which deal with the problems of accurate measurements in u.v. absorption spectrometry, fluorescence quantum yield determinations, diffuse reflectance spectrometry, and suggested standard reference materials.[456] The distinction has been made between the terms photometric accuracy, photometric linearity, and adherence to Beer's law and it has been shown that spectrometric instruments showing acceptable 'analytical linearity' are, nevertheless, photometrically inaccurate.[457] There have been a number of proposed reference materials for checks on the wavelength accuracy,[458–460] absorption precision,[459–462] and linearity,[458,461,463] which have been either solutions[459,460,463] or solid glasses (such as nichrome on glass).[461,462] The latter-mentioned transfer standard was found to be durable enough not to require a protective cover glass (unlike inconel).[461] A novel approach for wavelength calibration in the u.v.–visible involved the use of interference fringes induced in a demountable i.r. cell containing a 25 μm PTFE spacer.[464] The spacing of the fringes varied from about 1 nm at 220 nm to 14 nm at 850 nm.

The literature on u.v. and visible spectrometry for the period from December 1975 to November 1977 has been reviewed with an emphasis placed on analytical aspects.[465] Although there have been great strides made in certain branches of spectrometry such as photoacoustic and diode laser, there have been few major developments in the more conventional u.v.–visible field with the possible exception of data processing and computer control.[466] For example, microprocessor-control of a dye laser spectrometer for absorption analysis enabled any wavelength to be selected from 360 to 650 nm on command from a keyboard.[467] Wavelength selection was obtained by adjusting the angle of a diffraction grating and moving one of 16 dye cells into the laser cavity. An intracavity telescope provided spectral bandwidths as narrow as 0.06 nm. A rotating disc sample compartment enabled simultaneous recording of u.v.–visible absorption spectra of up to 14 samples with a precision of 0.1—0.0001 absorbance units to be determined by a computer-controlled spectrometer.[468]

[456] K. D. Mielenz, R. A. Valapoldi, and R. Mavrodineanu, in 'Standardisation in Spectrophotometry and Luminescence Measurements', N.B.S. special publication 466, 1977.
[457] H. J. Sloane and W. S. Gallaway, *Appl. Spectroscopy*, 1977, **31**, 25.
[458] C. Burgess, *UV Spectrometry Group Bulletin*, 1977 (5), 77.
[459] A. Knowles, *UV Spectrometry Group Bulletin*, 1977 (5), 94.
[460] M. A. West and D. R. Kemp, *Amer. Lab.*, 1977 (3), 37.
[461] F. J. J. Clarke, M. J. Downs, and W. McGivern, *UV Spectrometry Group Bulletin*, 1977 (5), 104.
[462] B. P. Popplewell, *UV Spectrometry Group Bulletin*, 1977 (5), 90.
[463] G. Milazzo, S. Caroli, M. Palumbo-Doretti, and N. Violante, *Analyt. Chem.*, 1977, **49**, 711.
[464] G. J. Buist, *J. Chem. Educ.*, 1976, **53**, 727.
[465] J. A. Howell and L. G. Hargis, *Analyt. Chem.*, 1978, **50**, 243R.
[466] D. E. Metzler, C. M. Harris, R. L. Reeves, W. H. Lawton, and M. S. Maggio, *Analyt. Chem.*, 1977, **49**, 864A.
[467] J. A. Perry, M. F. Bryant, and H. V. Malmstadt, *Analyt. Chem.*, 1977, **49**, 1702.
[468] J. P. Avery and H. V. Malmstadt, *Analyt. Chem.*, 1976, **48**, 1309.

The absorbances of highly absorbent samples (up to 7.5) were determined with an optically-delayed nitrogen laser pulse[469] and, in the case of liquids, in a wedge-shaped cell.[470] A differential measurement technique enabled changes in laser beam intensities resulting from absorption to be measured by means of an internally modulatable laser (absorption resolution of 10^4).[471] A simple interface allowed computer control of a Cary 118 C absorption spectrometer.[472]

With the few exceptions of spectrometers employing oscillating mirrors[473,474] or prisms[475] for wavelength selection, rapid scanning spectrometers described recently have incorporated diode arrays[476] or vidicon detectors[477—479] for either time-resolved[477,478] or derivative spectroscopy.[479] Second derivative spectroscopy is useful for qualitative enhancement of spectral data and for quantitative analysis. For u.v. absorption and other spectroscopic methods where additive low frequency noise may be dominant, the wavelength modulation techniques appear to give better signal-to-noise ratios than do electronically or numerically-obtained derivative spectra. An improved response for a wavelength-modulated spectrometer was obtained by combining first and second derivative spectra.[480] A combination of a second derivative u.v. spectrometer (using a vibrating quartz plate) and a gas chromatograph has been used for specific detection and analysis.[481] Wavelength modulation has also been used for reflectivity measurements between 150 and 900 nm.[482]

Experimental arrangements to measure reflectivity have included those for thin films,[483] aluminium coatings on mirrors[484] and two spectrometers for measurements in the vacuum-u.v.[485] The absorption spectra of unimolecular layers of a surface-active dye at a H_2O–CCl_4 interface have been determined using a multiple reflection technique with polarized and unpolarized light.[486] Diffuse reflectance measurements in the 1—3 μm range were determined on a Cary 17 I spectrometer fitted with a sulphur-coated integrating sphere.[487]

The experimental procedure associated with thermo-optical spectroscopy has been described and applied to determinations of absorption spectra of thin

[469] J. Taboada and W. J. Fador, *Appl. Optics*, 1977, **16**, 1132.
[470] I. L. Tyler, G. Taylor, and M. R. Querry, *Appl. Optics*, 1978, **17**, 960.
[471] D. J. Spencer in 'Laser 77 Opto-Electronics Conference Proceedings', ed. W. Waidelich, I.P.C., London, 1977, p. 164.
[472] J. C. Sutherland and T. T. Boles, *Rev. Sci. Instr.*, 1978, **49**, 853.
[473] R. Szentirmay and T. Kuwana, *Analyt. Chem.*, 1977, **49**, 1348.
[474] K. Ogan, A. Essig, K. W. Beach, and S. R. Caplan, *Rev. Sci. Instr.*, 1977, **48**, 142.
[475] A. Matsui and K. Tomioka, *J. Phys. (E)*, 1976, **9**, 529.
[476] M. J. Milano and K.-Y. Kim, *Analyt. Chem.*, 1977, **49**, 555; J. A. Haas, L. J. Perko, and D. E. Osten, *Industrial Research*, 1977 (5), 67.
[477] J. C. Moulder and A. F. Clark, *Proc. Soc. Photo-Opt. Instrum. Eng.*, 1976, **82**, 66.
[478] J. M. Pasachoff, D. F. Muzyka, and J. A. Schierer, *Appl. Optics*, 1976, **15**, 2884.
[479] T. E. Cook, R. E. Santini, and H. L. Pardue, *Analyt. Chem.*,1977, **49**, 871.
[480] A. R. Hawthorne and J. H. Thorngate, *Appl. Optics*, 1978, **17**, 724.
[481] P. M. Houpt and G. H. W. Baalhuis, *Appl. Spectroscopy*, 1977, **31**, 473.
[482] D. Theis and W. Busse, *J. Phys. (E)*, 1977, **10**, 57.
[483] P. Gadenne and G. Vuye, *J. Phys. (E)*, 1977, **10**, 733.
[484] F. M. Melsheimer and D. M. Rank, *Rev. Sci. Instr.*, 1977, **48**, 482.
[485] L. C. Emerson, J. T. Cox, G. L. Ostrom, L. R. Painter, and G. H. Cunningham, *Rev. Sci. Instr.*, 1976, **47**, 1065; E. Nicklaus, *J. Phys. (E)*, 1977, **11**, 1104.
[486] T. Ohnishi and H. Tsubomura, *Chem. Phys. Letters*, 1976, **41**, 77.
[487] J. B. Gillespie and J. D. Lindberg, *Appl. Spectroscopy*, 1977, **31**, 463.

films on surfaces of polycrystalline materials at low temperatures.[488] The essential component of the equipment was a special thermal detector with a response time faster than 10 μs which was also capable of measuring very small temperature changes. Laser excitation matrix-isolation spectroscopy has been reviewed elsewhere.[489] Other spectroscopic techniques for the solid phase included photon correlation spectroscopy applied to molecular relaxation in amorphous PMMA,[490] photoconductivity in dye polymer films,[491] and optical detection of magnetic resonance (ODMR) in the photoexcited triplet state of benzo(a)pyrene bound to DNA at 2 K.[492] Photochromic disproportion-ation of Cu^+ into Cu^{2+} and Cu^0 has been studied in the sytem $CuCl(s)-H_2O(l)$ by obtaining Auger spectroscopy measurements of Cu.[493]

High resolution optical spectra have been obtained by repetitive photoelectric scanning with a linear optical encoder and data accumulation in an MCA.[494] A photomultiplier was moved across the image plane of a 9.1 m spectrometer and its position determined by means of a Moiré fringe technique (accurate to 1 μm). Hyperfine splitting and broadening have been measured for the series of seven doublet laser lines of fluorine with the aid of a 10.9 m Czerny–Turner spectrograph (having a resolution of 1 part in 7×10^5) using a vidicon detec-tor.[495] Infrared up-conversion (see p. 29) – a technique which allowed detec-tion of i.r. radiation with sensitive detectors – has been effectively employed to convert 3.2—5.0 μm radiation into 800—880 nm radiation and thereby attain a spectral resolution of *ca.* 0.0003 μm at 4 μm wavelength.[496] The technique has been applied to measurements of the absorption spectra of polystyrene and CH_4-N_2O mixtures.

It has been found that the voltage between electrodes of a gas discharge plasma will change when illuminated at wavelengths corresponding to transitions in gas atoms. This is called the opto galvanic effect and promises to be a very sensitive analytical technique,[497] having already been employed to examine absorption transients in neon, [498]Ba^+, and Eu^+,[499] and numerous other metal ions.[500—502] See ref. 251 for an ingenious application of the effect in frequency-locking. Using a mechanically-scanned CW dye laser, absorption spectra of atomic transitions at high signal-to-noise ratios (SNR) were recorded for Cs, Li, Eu, U, Ne, Ar, He, and Hg plasmas.[501,502] The largest effect observed

488 H. Parker, K. W. Hipps, and A. H. Francis, *Chem. Phys.*, 1977, **23**, 117.
489 L. Andrews, *Appl. Spectroscopy Rev.*, 1976, **11**, 125.
490 T. A. King and M. F. Treadaway, *Chem. Phys. Letters*, 1977, **50**, 494.
491 D. B. Freeston, C. H. Nicholls, and M. T. Pailthorpe, *J. Phys. (E)*, 1976, **9**, 884.
492 P. A. Chiha, R. H. Clarke, and E. Kramer, *Chem. Phys. Letters*, 1977, **50**, 61.
493 B. Carlsson, C. Leygref, and G. Huttguist, *J. Photochem.*, 1977, **7**, 51.
494 G. F. Fulop, Chun-Sing O, and H. H. Stroke, *Rev. Sci. Instr.*, 1977, **48**, 1550.
495 L. O. Hocker, *J. Opt. Soc. Amer.*, 1978, **68**, 262.
496 T. R. Gurski, H. W. Epps, and S. P. Maran, *Appl. Optics*, 1978, **17**, 1238.
497 R. B. Green, R. A. Keller, G. G. Luther, P. K. Schenck, and J. C. Travis, *Appl. Phys. Letters*, 1976, **29**, 727.
498 K. C. Smyth and P. K. Schenck, *Chem. Phys. Letters*, 1978, **55**, 466.
499 P. K. Schenck and K. C. Smyth, *J. Opt. Soc. Amer.*, 1978, **68**, 626.
500 P. K. Schenck, D. S. King, K. C. Smyth, J. C. Travis, and G. C. Turk, in ref. 3, p. 431.
501 W. B. Bridges, *J. Opt. Soc. Amer.*, 1977, **67**, 1417.
502 W. B. Bridges, *J. Opt. Soc. Amer.*, 1978, **68**, 352.

was a 30% decrease in the operating voltage of a Cs discharge on illumination at 610 nm corresponding to the $6p^2P_{\frac{1}{2}} \rightarrow 8d^2D_{\frac{3}{2}}$ transition.[501]

Among the numerous papers on atomic absorption spectroscopy are included descriptions of a dual wavelength instrument using a 1 kHz pulsed hollow cathode lamp,[503] a background correction spectrometer,[504] and the use of resonance lamps to study the kinetics of F atoms[505] and I atoms[506] (see also pp. 96 and 111).

Two useful collections of papers have dealt with the chemical applications of linear dichroism and polarized spectroscopy[507] and recent advances (to 1976) in polarization techniques.[508] Polarized photochromism in solid solutions of pyrene (triplet absorption), salicylidene aniline (*cis–trans* isomerization), and spiropyrans (heterolytic cleavage) has been measured in a steady-state irradiation system using a chopper for modulation.[509] CD spectra of short-lived photo-induced transient species, in which circularly polarized light was introduced into the monitoring beam by using a quarter-wave plate and linear polarizer, have been published[510] but the accuracy of the method has been questioned.[511] An optical CD resolution of 2×10^{-6} for a 300 μs time constant and absorbance resolution of $\pm 4 \times 10^{-5}$ was reported for a temperature-jump apparatus which allowed simultaneous measurements to be made of both CD and optical transmission.[512] Other publications on CD spectroscopy included descriptions of vacuum-u.v. instruments,[513,514] in one case using reflecting optics (>135 nm),[514] a polarization attachment for a Cary spectrometer,[515] a pressure-jump technique using both optical rotation and CD detection,[516] and observation of laser-induced birefringence in neon gas using a circularly polarized probe beam and an analyser behind the cell.[517] A hydrogen discharge lamp has been used as a source in an apparatus for producing vacuum-u.v. circularly polarized radiation.[518] Polarization measurements of aromatic molecules in glassy solutions at 203 K employed Glan polarizing prisms for u.v. transmission[519] and a simple Fourier photopolarimeter employed two polarizers rotated at different speeds.[520] Simultaneous measurements of birefringence and dichroism induced in liquids or colloidal dispersions by a magnetic field

[503] T. Araki, T. Uchida, and S. Minami, *Appl. Spectroscopy*, 1977, **31**, 150.
[504] R. J. Sydor, J. T. Sinnamon, and G. M. Hieftje, *Analyt. Chem.*, 1976, **48**, 2030.
[505] M. A. A. Clyne and W. S. Nip, *Internat. J. Chem. Kinetics*, 1978, **10**, 367.
[506] I. Burak and M. Eyal, *Chem. Phys. Letters*, 1977, **52**, 534.
[507] *Spectroscopy Letters*, 1977, **10**, No. 6.
[508] *Proc. Soc. Photo-Opt. Instrum. Eng.*, 1976, vol. 88.
[509] C. Pasternak, M. A. Slifkin, and M. Shinitzky, *J. Chem. Phys.*, 1978, **68**, 2669.
[510] M. A. Slifkin and T. I. Y. Allos, *J. Phys. (E)*, 1977, **10**, 508.
[511] P. M. Bayley and M. Anson, *J. Phys. (E)*, 1977, **10**, 1299; M. A. Slifkin and T. I. Y. Allos, *ibid.*, 1977, **10**, 1300.
[512] M. Anson, S. R. Martin, and P. M. Bayley, *Rev. Sci., Instr.* 1977, **48**, 953.
[513] A. Gedanken and M. Levy, *Rev. Sci. Instr.*, 1977, **48**, 1661.
[514] K. P. Gross and O. Schnepp, *Rev. Sci. Instr.*, 1977, **48**, 362.
[515] T. A. Dessent and R. A. Palmer, *Rev. Sci. Instr.*, 1976, **47**, 1193.
[516] B. Gruenewald and W. Knoche, *Rev. Sci. Instr.*, 1978, **49**, 797.
[517] J.-C. Keller and C. Delsart, *Optics Comm.*, 1977, **20**, 147.
[518] U. Heinzmann, *J. Phys. (E)*, 1977, **10**, 1001.
[519] E. Leroy and H. Lami, *Chem. Phys. Letters*, 1976, **41**, 373.
[520] R. M. A. Azzan, *Optics Comm.*, 1978, **25**, 137.

have been reported[521] and a device described for controlling the photomultiplier output of a combined CD–MCD spectrometer.[522]

A variety of publications on data analysis in spectrometry have appeared including methods for determining the number of detectable species in a rapid scanning kinetic experiment,[523] for estimating the spectral reflectance curves from multi-spectral image data,[524] and for applying non-negative least squares and simplex optimization algorithms to the analysis of multi-component spectrophotometric data.[525] Model systems containing soluble dyes in scattering media have been considered in a study dealing with quantitative diffuse reflectance spectroscopy in highly absorbing media.[526] A theoretical assessment of precision in dual wavelength spectrometric measurements predicted that for high precision, a high intensity source was needed in conjunction with suitable modulation.[527]

Among other spectrometric instruments described have been reports of a fibre optics refractometer having a sensitivity of 10^{-5} refractive index units,[528] a beam transmittance meter for turbidity measurements in sea water,[529] a fully automated stopped-flow spectrometer in which a microcomputer controlled several operations including reagent proportions,[530] and an instrument to measure gas phase electrochromism spectra between 165 and 300 nm.[531]

The size, shape, and position of a light beam in a spectrometer could be determined by using a piece of light-sensitive material such as a diazonium compound in paper to fit into the cuvette holder, but exposure times at 425 nm were 1—24 h.[532]

Spectrometric studies of materials under non-ambient conditions have been conducted in cells designed to operate at high pressure only (up to 2.5×10^9 N m^{-2}),[533] at both low temperatures (2 K) and high pressures (2×10^8 N m^{-2}),[534] at variable temperatures (173—370 K) and high pressures (0—6 $\times 10^6$ N m^{-2}),[535] at high temperatures (100—700 K) and variable pressures (10^{-4}–10^7 N m^{-2}),[536] for fluorescence measurements at high temperatures (293—503 K) and high pressures (2×10^8 N m^{-2}),[537] and for reflectance measurements at high pressures.[538] A modification of a continuous flow

[521] J. V. Champion, D. Downer, G. H. Meeten, and L. F. Gate, *J. Phys.* (*E*), 1977, **10**, 1137.
[522] J. H. Obbink and A. M. F. Hezemans, *J. Phys.* (*E*), 1977, **10**, 769.
[523] R. N. Cochrane and F. H. Horne, *Analyt. Chem.*, 1977, **49**, 846.
[524] S. K. Park and F. O. Huck, *Appl. Optics*, 1977, **16**, 3017.
[525] D. J. Leggett, *Analyt. Chem.*, 1977, **49**, 276.
[526] H. G. Hecht, *Analyt. Chem.*, 1976, **48**, 1775.
[527] K. L. Ratzlaff and D. F. S. Natusch, *Analyt. Chem.*, 1977, **49**, 2170.
[528] D. J. David, D. Shaw, H. Tucker, and F. C. Unterleitner, *Rev. Sci. Instr.*, 1976, **47**, 989.
[529] L. Basano, P. Ottonello, and L. Papa, *Rev. Sci. Instr.*, 1977, **48**, 528.
[530] G. E. Mieling, R. W. Taylor, L. G. Hargis, J. English, and H. L. Pardue, *Analyt. Chem.*, 1976, **48**, 1686.
[531] G. C. Causley, J. D. Scott, and B. R. Russell, *Rev. Sci. Instr.*, 1977, **48**, 264.
[532] S. D. Rains, *Analyt. Chem.*, 1978, **50**, 680.
[533] E. Whalley, A. Lavergne, and P. T. T. Wong, *J. Phys.* (*E*), 1976, **9**, 845.
[534] J. C. Stryland and S. M. Till, *Rev. Sci. Instr.*, 1977, **48**, 1350.
[535] E. Forte and H. von den Bergh, *Chem. Phys.*, 1978, **30**, 325.
[536] H. Forster, V. Meyn, and M. Schuldt, *Rev. Sci. Instr.*, 1978, **49**, 74.
[537] A. Chalzel, *J. Phys.* (*E*), 1977, **10**, 633.
[538] J. Kalinowski and R. Jankowiak, *Chem. Phys. Letters*, 1978, **53**, 56; B. Welber, *Rev. Sci. Instr.*, 1977, **48**, 395.

cryostat enabled variable temperature (3—300 K), high resolution absorption spectra to be measured on mg quantities of powdered metal compounds in a mull.[539] Drawings have been published for a thermostated holder for four 10 mm pathlength cuvettes.[540] Multipass cells for laser excitation have also been described.[541] Other miscellaneous publications have included a description of a quartz boat placed in an e.s.r. cavity for u.v. irradiation studies of thin films on water,[542] the preparation of ultrathin mica windows for vacuum-u.v. studies to 160 nm,[543] a report of unwanted birefringence in low temperature silica windows,[544] and the optical properties of room temperature vulcanizing (RTV) silcone rubber.[545]

I.R. Spectroscopy.—There seems to have been an emphasis on the reflectance side of conventional i.r. spectroscopy, presumably because of the influence of diode lasers and other lasers in i.r. absorption spectrometry. I.r. spectrometers which have been described have included a double beam instrument for specular reflectance measurements at near-normal incidence in the 1.2—25 μm range,[546] an arrangement for matrix isolation studies by i.r. reflectance of alkyl radicals following the vacuum-u.v. photolysis of diacyl peroxides,[547] a specular reflectance accessory,[548] and a two-channel spectrometer operating in both reflectance (400—2500 nm) and thermal emissive (6—15 μm) modes.[549] Reviews in this area have included one on i.r. internal reflection spectroscopy[550] and the application of near-i.r. reflectance spectroscopy to the analysis of agricultural products (*e.g.* moisture content).[551]

A new technique for i.r. spectroscopy of molecular beams (using CO molecules) was based on detection of power absorbed, as a modulated increase of power carried by molecules to a microcalorimeter (a cryogenic bolometer) after the molecular beam had been crossed with the modulated output of a semiconductor diode laser.[552] Kinetics of the N_2O_5-catalysed decomposition of O_3 has been studied by a multiple-pass i.r.-absorption arrangement using a modulation technique.[553] A simple multipass accessory has been described to enhance absorption intensities and eliminate optical interference effects in examinations of polymer films[554] and a White-type multiple reflection cell (pathlength variable from 0.3 to 2.1 m) used for i.r. measurements in liquid argon solution.[555]

[539] C. D. Flint and A. P. Matthews, *J. Phys. (E)*, 1976, **9**, 727.
[540] R. J. Threlfall, *Lab. Practice*, 1976, **25**, 691.
[541] A. Owyoung, *Optics Comm.*, 1977, **22**, 323; K. P. Koch and W. Lahmann, *Appl. Phys. Letters*, 1978, **32**, 289.
[542] H. D. Gesser, T. A. Wildman, and Y. B. Tewari, *Environ. Sci. Technol.*, 1977, **11**, 605.
[543] B. J. Mulder, *J. Phys. (E)*, 1976, **9**, 724.
[544] G. Baldacchini U. M. Grassano, and A. Tanga, *Rev. Sci. Instr.*, 1978, **49**, 677.
[545] F. M. Smoka and H. A. Hill, *Appl. Optics*, 1977, **16**, 292.
[546] G. Cappucio and S. D'Angelo, *J. Phys. (E)*, 1978, **11**, 298.
[547] J. Pacansky, D. E. Horne, G. P. Gardini, and J. Bargon, *J. Phys. Chem.*, 1977, **81**, 2149.
[548] M. J. Gelten, A. van Oosterom, and C. van Es, *Infrared Physics*, 1976, **16**, 661.
[549] T. L. Barnett and R. D. Juday, *Appl. Optics*, 1977, **16**, 967.
[550] H. G. Tompkins, *Appl. Spectroscopy*, 1976, **30**, 377.
[551] C. A. Watson, *Analyt. Chem.*, 1977, **49**, 835A.
[552] T. E. Gough, R. E. Miller, and G. Scholes, *Appl. Phys. Letters*, 1977, **30**, 338.
[553] R. A. Graham and H. S. Johnson, *J. Phys. Chem.*, 1978, **82**, 254.
[554] J. M. Chalmers and B. J. Stay, *J. Phys. (E)*, 1977, **10**, 741.
[555] A. C. Jeannotte and J. Overend, *Spectrochim. Acta*, 1977, **33a**, 849.

Digitization systems for i.r. spectrometers[556, 557] included methods to allow computer subtraction and to provide fingerprint spectra of components.[556]

The increased importance of chemical lasers has led to a demand for high purity window materials for use at the DF (3.8 μm) and HF (2.7 μm) laser wavelengths.[558] Alkaline earth fluorides and alkali halides are probably the best materials for minimum absorption but a cautionary note added that common organic solvents used for cleaning have strong absorption bands in this region.[558] Absorbed contaminants on windows have been studied by internal reflection spectroscopy.[559] The transmission characteristics of various common materials (plastic, wood, rubber *etc.*) in the far-i.r. region (50—1000 m) have been examined with HCN and various waveguide lasers using a liquid helium-cooled GaAs detector.[560] The general characteristics of i.r. transmissive materials[561] and techniques of solid sampling with the alkali halide pelleting technique[562] are reviewed elsewhere.

Other publications in i.r. spectroscopy have discussed the problems associated with high pressure measurements using a diamond anvil cell.[563] and have described a portable helium cryostat for far-i.r. applications,[564] a double cylindrical cell for transmission measurements of solid sorption systems,[565] and high pressure cells for high[566] and low[567] temperature studies.

Specific applications of i.r. spectroscopy to photochemistry have concentrated on qualitative analysis of products and intermediates. For example, new i.r. absorption bands produced following the u.v. and laser photolysis of dilute Ar matrices containing O_3 and H_2S at 8 K have been ascribed to the previously unobserved HSOH intermediate.[568] Similar low temperature studies have enabled the HCCN radical to be identified following photolysis of C_2H_3N,[569] the $CFCl_3^+$ ion from $CFCl_3$, CF_2Cl_2, and CF_3Cl[570] and searches have been made for the HF_2 and HClF species following photolysis of F_2 in Ar–HF or Ar–HCl matrices.[571] Laser intracavity studies have been made on vinyl chloride, ethylene, and propylene[572] (see also ref. 754 *et seq.*) and insecticides.[573] In the far-i.r., a CO_2-laser-pumped-MeOH laser has been used with a Golay detector to determine absorption coefficients of water at 96, 163, 232, and 570 m.[574]

556 B. W. Tattershall, *Analyt. Chem.*, 1977, **49**, 772.
557 Y. Miyashita, K. Oikawa, H. Abe, and S.-I. Sasaki, *Analyt. Chem.*, 1977, **49**, 1875.
558 J. A. Harrington, D. A. Gregory, and W. F. Otto, *Appl. Optics*, 1976, **15**, 1953.
559 E. D. Palik, J. W. Gibson, R. T. Holm, M. Hass, M. Braunstein, and B. Gargia, *Appl. Optics*, 1978, **17**, 1776.
560 T. S. Hardwick, D. T. Hodges, D. H. Barker, and F. B. Foote, *Appl. Optics*, 1976, **15**, 1919.
561 F. M. Lussier, *Laser Focus*, 1976 (12), 47.
562 O. Y. Ataman and H. B. Mark, *Appl. Spectroscopy Rev.*, 1977, **13**, 1.
563 D. M. Adams and S. K. Sharma, *J. Phys. (E)*, 1977, **10**, 838.
564 J. V. Radostitz, I. G. Nott, P. Kittel, and R. J. Donnelly, *Rev. Sci. Instr.*, 1978, **49**, 86.
565 K. Moller and D. Kunath, *J. Phys. (E)*, 1977, **10**, 962.
566 J. M. Alberigs, A. J. Kaper, and W. M. A. Smith, *Appl. Spectroscopy*, 1977, **31**, 175.
567 R. K. Horne and G. Birnbaum, *Infrared Physics*, 1977, **17**, 174.
568 R. R. Smardzewski and M. C. Lin, *J. Chem. Phys.*, 1977, **66**, 3197.
569 A. Dendramis and G. E. Leroi, *J. Chem. Phys.*, 1977, **66**, 4334.
570 F. T. Prochaska and L. Andrews, *J. Chem. Phys.*, 1978, **68**, 5568, 5577.
571 B. S. Ault, *J. Chem. Phys.*, 1978, **68**, 4012.
572 J. W. Robinson and D. Nettles, *Spectroscopy Letters*, 1978, **11**, 73.
573 N. Konjevic, M. Orlov, and M. Trtica, *Spectroscopy Letters*, 1977, **10**, 311.
574 B. L. Bean and S. Perkowitz, *Appl. Optics*, 1976, **15**, 2617.

Accurate absorption spectra of liquids in the far-i.r. range have been determined with a two-compartment chamber fitted with a movable septum.[575]

A review on the analysis of ambient air[576] included instrumental techniques for fluorescence (of SO_2), chemiluminescence (of O_3 reactions), and gas filter correlation spectroscopy.[577] Differential absorption techniques enabled depth-resolved absorption measurements to be made to 3 km over the spectral range 1.4—4 μm[578] and have been used with a DF laser to measure HCl, CH_4, and N_2O[579] and a continuously tunable optical parametric oscillator has been used to determine SO_2.[580] The CO_2 laser has again been employed to measure ethylene,[581] a He–Ne laser used to determine humidity variations over a 300 m pathlength[582] and a CW dye laser employed to measure NO_2, I_2, and Br_2 concentrations.[583] Differential absorption techniques have also been used with a double wavelength u.v. laser to determine SO_2.[584] Atmospheric measurements of OH by absorption using a CW tunable u.v. laser appeared to be a promising alternative to fluorescence techniques and should have a sensitivity of 10^5 molecules cm^{-3}.[585]

Diode Laser Spectroscopy.—Tunable diode lasers fabricated from lead-salt semiconductors (*e.g.* PbCdS, PbSSe, PbSnSe) offer narrow bandwidth (10^{-4} cm^{-1}), high brightness sources, which may be tuned over selected portions (typically 100 cm^{-1}) of the i.r. from 3 to 25 μm. Both power and spectral resolution are orders of magnitude better than for conventional i.r. spectrometers and the use of these compact sources for high resolution spectroscopy is obvious (for reviews see ref. 586).

Quantitative determinations of atmospheric pollutants has naturally received much attention.[587] A White-multipass cell was employed to measure absorption coefficients as low as 3×10^{-8} cm^{-1} with sensitivity limits of 1 p.p.b. (for SO_2 and NO_2) and <0.1 p.p.b. (for NH_3).[588] Individual gaseous pollutants,

[575] H. Kilp, D. C. Barnes, F. W. J. Clutterbuck, M. N. Afsar, and G. W. Chantry, *Infrared Physics*, 1978, **18**, 11.

[576] D. J. Kroon, *J. Phys.* (*E*), 1978, **11**, 497.

[577] J. M. Russell, J. H. Park, and S. R. Drayson, *Appl. Optics*, 1977, **16**, 607.

[578] R. Baumgartner, D. Wolfe, and R. L. Byer, *I.E.E.E. J. Quantum Electron.*, 1977, **QE-13**, 71D.

[579] E. R. Murray, J. E. Van der Laan, and J. G. Hawley, *Appl. Optics*, 1976, **15**, 3140.

[580] R. A. Baumgartner and R. L. Byer, *Optics Letters*, 1978, **2**, 163.

[581] S. O. Kanstad, A. Bjerkestrand, and T. Lund, *J. Phys.* (*E*), 1977, **10**, 998; E. R. Murray and J. E. van der Laan, *Appl. Optics*, 1978, **17**, 814.

[582] M. Bertolotti, M. Carnevale, and D. Sette, *Appl. Optics*, 1978, **17**, 285.

[583] Y. Fujii and T. Masamura, *Optical Engineering*, 1978, **17**, 147.

[584] H. Inomata and A. I. Carswell, *Optics Comm.*, 1977, **22**, 278.

[585] D. K. Killinger and C. C. Wang, *Chem. Phys. Letters*, 1977, **52**, 374; M. Hanabusa, C. C. Wang, S. Japar, D. K. Killinger, and W. Fischer, *J. Chem. Phys.*, 1977, **66**, 2118.

[586] 'Laser Monitoring of the Atmosphere', ed., E. D. Hinckley, Springer-Verlag, Heidelberg, 1976; E. D. Hinckley, *Environ. Sci. Technol.* 1977, **11**, 564; K. W. Nill, *Laser Focus*, 1977 (2), 32; J. F. Butler, *Electro-Optic Systems Design*, 1977 (7), 33.

[587] E. D. Hinckley, R. T. Ku, K. W. Nill, and J. F. Butler, *Appl. Optics*, 1976, **15**, 1653.

[588] J. Reid, B. K. Garside, J. Shewchun, M. El-Sherbiny, and E. A. Ballik, *Appl. Optics*, 1978, **17**, 1806; J. Reid, M. El-Sherbiny, B. K. Garside, and E. A. Ballik, *J. Opt. Soc. Amer.*, 1978, **68**, 637; J. Reid, J. Shewchun, B. K. Garside, and E. A. Ballik, *Appl. Optics*, 1978, **17**, 300; J. Reid, J. Shewchun, B. K. Garside, and E. A. Ballik, *Optical Engineering*, 1978, **17**, 56.

such as SO_2,[589] H_2SO_4,[590] C_2H_4,[591] and NH_3[592] have been examined in detail by diode laser spectroscopy and spectra in some cases compared with those obtained with normal i.r. spectrometers (*e.g.* ref. 590).

High resolution studies which have been reported have included matrix-isolated SO_2,[593] CH_3F,[594] HNO_3 at a spectral resolution of $<10^{-3}$ cm^{-1},[595] H_2SO_4 (resolution of 10^{-4} cm^{-1}),[596] and SF_6 (resolution of 5×10^{-5} cm^{-1}).[597] By passing a diode laser output through one cell containing $^{14}C^{16}O_2$ and another containing a reference gas ($^{12}C^{16}O_2$ or $^{13}C^{16}O_2$), it was possible to determine all the absorption characteristics of $^{12-14}CO_2$ at around the ν_3 absorption centred at *ca.* 4.5 μm.[598]

Fundamental absorption strengths of the lines in the ClO radical were measured using sensitive derivative techniques.[599] Line frequencies were measured with respect to the $^{14}C^{16}O_2$ laser frequency at 853.181 cm^{-1} by observing the beat frequency when a diode laser was tuned through the 853.1 cm^{-1} region.[600] Because of the small concentration of ClO generated in the apparatus and the relatively short pathlength, derivative spectroscopy was to improve the SNR.

Diode lasers have also been used for studies on isotope separation mechanisms in UF_6 and other materials,[601] spectroscopy in molecular beams,[602] and the attainment of a fully resolved spectrum of ozone in the upper atmosphere using heterodyne techniques at ground level.[603]

Laser sideband spectroscopy, in which an electro-optically stabilized CO_2 laser was mixed with a tunable millimetre-wave signal, has been applied to SiH_4 and claimed to give a frequency better than 10^{-3} cm^{-1} and a resolution of 10^{-4} cm^{-1}.[604]

Laser Magnetic Resonance.—Laser magnetic resonance (LMR) has been used with considerable success at far-i.r. frequencies to study the pure rotational spectra of chemically unstable species.[605] In LMR, the Zeeman effect is utilized to shift molecular or atomic energy levels through resonance with fixed frequency

[589] E. Max and S. T. Eng, *Optics and Quantum Electron.*, 1977, **9**, 411; *I.E.E.E. J. Quantum Electron.*, 1977, **QE-13**, 74D.
[590] G. P. Montgomery and R. F. Majkowski, *Appl. Optics*, 1978, **17**, 173.
[591] R. T. Ku and D. L. Spears, *I.E.E.E. J. Quantum Electron.*, 1977, **QE-13**, 73D.
[592] J. P. Sattler and K. J. Ritter, *J. Mol. Spectroscopy*, 1978, **69**, 486.
[593] M. Dubs and H. H. Gunthard, *Chem. Phys. Letters*, 1977, **47**, 421.
[594] J. P. Sattler and G. J. Simonis, *I.E.E.E. J. Quantum Electron.*, 1977, **QE-13**, 461.
[595] P. Brockman, C. H. Bair, and F. Allario, *Appl. Optics*, 1978, **17**, 91.
[596] R. S. Eng, G. Petagna, and K. W. Nill, *Appl. Optics*, 1978, **17**, 1723.
[597] R. S. McDowell, H. W. Galbraith, C. D. Cantrell, N. G. Nereson, and E. D. Hinckley, *J. Mol. Spectroscopy*, 1977, **68**, 288.
[598] R. S. Eng, K. W. Nill, and M. Wahlen, *Appl. Optics*, 1977, **16**, 3072; M. Wahlen, R. S. Eng, and K. W. Nill, *Appl. Optics*, 1977, **16**, 2350.
[599] R. T. Menzies, J. S. Margolis, E. D. Hinckley, and R. A. Toth, *Appl. Optics*, 1978, **17**, 523; J. S. Margolis, R. T. Menzies, and E. D. Hinckley, *Appl. Optics*, 1978, **17**, 1680.
[600] N. Nereson, *J. Mol. Spectroscopy*, 1978, **69**, 489.
[601] D. N. Travis, J. C. McGurk, D. McKeown, and R. G. Denning, *Chem. Phys. Letters*, 1977, **34**, 287.
[602] T. E. Gough, *Appl. Phys. Letters*, 1977, **30**, 338.
[603] M. A. Frerking and D. J. Muehlner, *Appl. Optics*, 1977, **16**, 526.
[604] G. Magerl, E. Bonek, and W. A. Kreiner, *Chem. Phys. Letters*, 1977, **52**, 473.
[605] P. B. Davies, D. K. Russell, B. A. Thrush, and H. E. Radford, *Proc. Roy. Soc.*, 1977, **A353**, 299.

i.r. or far-i.r. lasers. Rotation and vibration–rotation spectra of free radicals[606] and spin–orbit transitions of free atoms can be measured with high precision and detected with high sensitivity.

Spectrometers using a discharge laser (D_2O, H_2O, HCN, or DCN)[607] are useful for rotational spectroscopy of the lighter free radicals (such as NH_2),[608] but more complete spectral coverage can be achieved by using a CO_2 laser to pump the many molecular lasers in the far-i.r. region (*e.g.* MeF, MeOH, HCOOH).[606—609] Among the many species which have been examined this way are included HO_2,[610] NH,[609] O,[611] MeO,[612] and SeH.[613] Rate constants for the reaction of OH radicals with ethylenic compounds have been determined using LMR to detect OH[614] and the reaction of OH_2 and NO_2 studied similarly in a discharge flow system with LMR detection of HO_2.[610] LMR has also been used to detect and measure the $^3P_1 \leftarrow {}^3P_0$ transitions of mercury by flowing excited Hg atoms through an absorption cell located inside the cavity of a CO laser and Zeeman shifting some of the components of the transition into resonance with the laser.[615] Selective quantitative detection of NO_2 in p.p.m. concentrations in N_2 has been determined using magnetic field modulation of the molecular absorption of a 1616 cm^{-1} CO laser line.[616] A magnetic field of $<$500 G was sufficient to shift the i.r. transition of interest into coincidence with the laser. In an investigation of the laser-enhanced reaction between NO and O_3, $^{14}N^{16}O$ molecules pumped by a CO laser were tuned into resonance by a magnetic field.[617]

Fourier Transform Spectroscopy.—An excellent review has described the recent advances in instrumentation, measurement techniques, and methods of data reduction for Fourier spectroscopy in the i.r.[618] and a comparison made between Fourier transform (FT) spectrometers and grating i.r. spectrometers on the basis of their optical and digital advantages.[619] Further reviews in this area have dealt with the impact of FT i.r. spectrometers on industrial spectroscopy[620] and the evaluation of commercial FT spectrometers.[621] New spectrometers described in the literature have included a high resolution (0.005 cm^{-1}) instrument for the 0.6—100 μm range employed to measure the absorption spectra of

[606] J. W. C. Johns, A. R. W. McKellar, and M. Riggin, *J. Chem. Phys.*, 1978, **68**, 3957; 1977, **67**, 2427; J. M. Brown, J. Buttershaw, A. Carrington, and C. R. Parent, *Mol. Phys.*, 1977, **33**, 589.
[607] J. P. Burrows, G. W. Harris, and B. A. Thrush, *Nature*, 1977, **267**, 233; also ref. 406.
[608] P. B. Davies, D. K. Russell, D. R. Smith, and B. A. Thrush, in ref. 3, p. 97.
[609] F. D. Wayne and H. E. Radford, *Mol. Phys.*, 1976, **32**, 1407.
[610] C. J. Howard, *J. Chem. Phys.*, 1977, **67**, 5258.
[611] P. B. Davies, B. J. Handy, E. K. Murray Lloyd, and D. R. Smith, *J. Chem. Phys.*, 1978, **68**, 1135.
[612] H. E. Radford and D. K. Russell, *J. Chem. Phys.*, 1977, **66**, 2222.
[613] P. B. Davies, B. J. Handy, E. K. Murray Lloyd, and D. K. Russell, *J. Chem. Phys.*, 1978, **68**, 3377.
[614] C. J. Howard, *J. Chem. Phys.*, 1976, **65**, 4771.
[615] J. W. C. Johns, A. R. W. McKellar, and M. Riggin, *J. Chem. Phys.*, 1977, **67**, 3962.
[616] S. M. Freund, D. M. Sweger, and J. C. Travis, *Analyt. Chem.*, 1976, **48**, 1944.
[617] J. C. Stephenson and S. M. Freund, *J. Chem. Phys.*, 1976, **65**, 4303.
[618] G. W. Chantry, *J. Mol. Structure*, 1978, **45**, 307.
[619] J. W. Griffiths, H. J. Sloane, and R. W. Hannel, *Appl. Spectroscopy*, 1977, **31**, 485.
[620] J. G. Grasselli and L. E. Wolfram, *Appl. Optics*, 1978, **17**, 1387.
[621] S. T. Dunn, *Appl. Optics*, 1978, **17**, 1367.

gases over a 40 m pathlength,[622] a lamellar-grating instrument for high resolution and general measurements in the submillimetre region,[623] and a versatile Michelson interferometer system for measurements in the lower wavelengths (u.v. to mid i.r.).[624] Diffuse reflectance measurements from 5000 to 500 cm^{-1} (2—20 μm) as well as normal transmittance measurements were reported on a FT i.r. spectrometer[625] and a system was described for improving the sensitivity of absorption spectrometry with a double-beam FT spectrometer.[626] (FT spectroscopy of transient absorbing species is dealt with elsewhere, p. 95.)

Studies at high resolution, both in emission (*e.g.* the hyperfine patterns of ^{237}NP) and in absorption (*e.g.* the iodine absorption spectrum from 18 520 to 18 529 cm^{-1}), have shown that the performance achieved by FT spectroscopy in the u.v. and visible was comparable with that reached in the i.r.[627] Time-resolved i.r. emission of electron-beam irradiated CO_2–N_2 mixtures has been detected by FT spectroscopy with a prediction that this technique could be extended to give spectral resolutions of 0.1 cm^{-1} and time resolutions of <1 μs.[628] Other examples of FT spectroscopy have included an analysis of the rotational absorption spectrum of $ClONO_2$ in the region 5—15 cm^{-1} at 0.07 cm^{-1} resolution,[629] i.r. chemiluminescence from F + HBr,[630] an analysis of chlorophyll-a and -b in cyclohexane solution in the 10—200 cm^{-1} region[631] and studies on peroxynitric acid.[632] The opto-acoustic effect was used in a rapid scan FT spectrometer to obtain absorption lines of CH_3OH vapour.[633]

Photoacoustic Spectroscopy.—The technique of photoacoustic spectroscopy (PAS) has recently received much attention and there have been a considerable number of publications recently on both instrumental developments and applications (see also Vol. 8, p. 22), with a considerable emphasis on the use of laser sources.

A number of papers have dealt with the theory of PAS[634] and the various criteria involved in the selection of light sources, sample chambers, and data acquisition systems.[635] Unlike in conventional absorption spectrometry, cell design in PAS is crucial for high sensitivity studies and there have been descrip-

[622] A. A. Balashov, V. S. Bukreev, N. G. Kultepin, I. N. Nesteruk, E. B. Perminov, V. A. Vagin, and G. N. Zhizhin, *Appl. Optics*, 1978, **17**, 1717.
[623] K. Sakin, H. Masumoto, K. Ichimura, and H. Kojima, *Appl. Optics*, 1978, **17**, 1709.
[624] G. Horlick and W. K. Yuen, *Appl. Spectroscopy*, 1978, **32**, 38.
[625] R. R. Willey, *Appl. Spectroscopy*, 1976, **30**, 593.
[626] D. Kuehl and P. R. Griffiths, *Analyt. Chem.*, 1978, **50**, 418.
[627] P. Luc and S. Gerstenkorn, *Appl. Optics*, 1978, **17**, 1327.
[628] H. Sakui and R. E. Murphy, *Appl. Optics*, 1978, **17**, 1342.
[629] J. W. Fleming, *Chem. Phys. Letters*, 1977, **50**, 107.
[630] J. P. Sung and D. W. Setser, *Chem. Phys. Letters*, 1977, **48**, 413.
[631] S. Pekowitz and B. L. Bean, *J. Chem. Phys.*, 1977, **66**, 2231.
[632] R. A. Graham, A. M. Winer, and J. N. Pitts, *Chem. Phys. Letters*, 1977, **51**, 215; S. Z. Levine, W. M. Uselman, W. H. Chan, J. C. Calvert, and J. H. Shaw, *Chem. Phys. Letters*, 1977, **48**, 528.
[633] G. Burse and B. Bullemer, *Infrared Physics*, 1978, **18**, 255.
[634] F. A. McDonald and G. C. Wetsel, *J. Appl. Phys.*, 1978, **49**, 2313; A. Rosencwaig, *ibid.*, 1978, **49**, 2905; M. A. Afromowitz, P.-S. Yeh, and S. Yee, *ibid.*, 1977, **48**, 209.
[635] G. F. Kirkbright and M. J. Adams, *European Spectroscopy News*, 1977 (14), 22; A. Rosenswaig, *Rev. Sci. Instr.*, 1977, **48**, 1133; J. F. McClelland and R. N. Kniseley, *Appl. Optics*, 1976, **15**, 2658.

tions of resonantly-enhanced cells,[636] discussions of size considerations,[637] and choice criteria of buffer gases.[638] Scattered light has been shown to contribute significantly to the PAS signal in regions of low absorbance and/or high reflectivity.[639] Although conventional sample cells commonly use a microphone and amplifier,[640] piezoelectric detectors[641] and strain transducers[642] have also been used. Piezoelectric detectors, in particular, have been shown to be capable of operation over a wide temperature range (4—700 K) with a frequency response as high as 100 MHz.[641] A method for calibrating PAS cells directly in terms of absolute energy units has used an electrically-conducting metal block deposited on an insulating substrate.[643] The block served as a spectrally flat optical absorber and acted as a uniform source for resistance heating. With this arrangement, the time-response of PAS cells to pulsed excitation was examined theoretically and experimentally.[644]

Absolute values of optical absorption coefficients for strongly absorbing liquids and solids have been derived from measurements of the dependence of the photoacoustic signal on chopping frequency.[645] Other papers have discussed PAS measurements of small quantities of solids (5 mg) at a spectral resolution of 15 nm[646] and the experimental details of a low temperature (77 K) apparatus.[647] Applications of conventional PAS have included UF_4 powder,[648] studies of transport properties in semiconductors,[649] microstructures of solid surfaces,[650] and radiationless transitions of Eu^{2+}[651] and Cr^{3+} ions[652] in crystals.

The detection of weak molecular absorption spectra has traditionally been achieved by the use of multiplass cells with total path lengths of several hundred metres or more. For unstable and/or scarce materials (*e.g.* with isotopic labelling), multipass spectroscopy is not practicable. PAS, on the other hand, can be made in a very sensitive method by using appropriate laser sources. For example, the detection of forbidden transitions in thioformaldehyde and a previously unreported transition in S_2O were discovered with a 7 cm pathlength cell and a CW tunable dye laser (1 cm^{-1} bandwidth) in a photoacoustic arrange-

[636] R. S. Quimby, P. M. Selzer, and W. M. Yen, *Appl. Optics*, 1977, **16**, 2630.
[637] L. C. Aamodt, J. C. Murphy, and J. G. Parker, *J. Appl. Phys.*, 1977, **48**, 927; W. G. Ferrell and Y. Haven, *ibid.*, 1977, **48**, 3984; E. M. Monahan and A. W. Nolle, *ibid.*, 1977, **48**, 3519.
[638] L. J. Thomas, M. J. Kelly, and N. M. Amer, *Appl. Phys. Letters*, 1978, **32**, 736; K. Y. Wong, *J. Appl. Phys.*, 1978, **49**, 3033.
[639] J. F. McClelland and R. N. Kniseley, *Appl. Optics*, 1976, **15**, 2967.
[640] R. C. Gray, V. A. Fishman, and A. J. Bard, *Analyt. Chem.*, 1977, **49**, 697.
[641] M. M. Farrow, R. K. Burnham, M. Auzanneau, S. L. Olsen, N. Purdie, and E. M. Eyring, *Appl. Optics*, 1978, **17**, 1093.
[642] A. Hordvik and H. Schlossberg, *Appl. Optics*, 1977, **16**, 101.
[643] J. C. Murphy and L. C. Aamodt, *Appl. Phys. Letters*, 1977, **31**, 728.
[644] L. C. Aamodt and J. C. Murphy, *J. Mol. Phys.*, 1978, **49**, 3036.
[645] G. C. Wetsel and F. A. McDonald, *Appl. Phys. Letters*, 1977, **30**, 252.
[646] H. E. Eaton and J. D. Stuart, *Analyt. Chem.*, 1978, **50**, 587.
[647] J. C. Murphy and L. C. Aamodt, *J. Appl. Phys.*, 1977, **48**, 3502.
[648] H. R. Eaton, D. R. Anton, and J. D. Stuart, *Spectroscopy Letters*, 1977, **10**, 847.
[649] C. C. Ghizoni, M. A. A. Sigueira, H. Vargas, and L. C. M. Miranda, *Appl. Phys. Letters* 1978, **32**, 554.
[650] Y. H. Wong, R. L. Thomas, and G. F. Hawkins, *Appl. Phys. Letters*, 1978, **32**, 539.
[651] L. D. Merkle and R. C. Powell, *Chem. Phys. Letters*, 1977, **46**, 303.
[652] R. G. Peterson and R. C. Powell, *Chem. Phys. Letters*, 1978, **53**, 366.

ment.[653] The system was calibrated with thiophosgene and could measure absorbance values of 10^{-6} with a SNR of 20. An argon laser operating at both 488 and 514 nm alternatively allowed the background signal in PAS to be reduced considerably.[654] When applied to the assay of small amounts of non-fluorescent absorbers (such as β-carotene) detection limits of 9×10^{10} molecules cm^{-3} (12 p.p.t.) were obtained corresponding to absorption coefficients of 2.2×10^{-5} cm^{-1}.[654] Other visible laser studies have included the absorption spectrum of NO_2 over the 480—625 nm range[655] and measurements of dichroism spectra of polymers, crystals, and biological samples with a linearly polarized CW dye laser.[656] Preliminary results have been reported which show the application of PAS to measurements of excited state absorption $(S_n \leftarrow S_1)$ at 474 nm for an opaque sample of Rose Bengal with nanosecond time resolution.[657]

Photoacoustic measurements of heat energy yields for vapour phase benzene excited at 248 nm have also been used to determine the parameters associated with the photo-oxidation of benzene[658] and oxygen quenching of both singlet and triplet states of benzene and naphthalene.[659] Molecular dephasing and radiationless decay measurements have been made by a new technique known as laser-acoustic diffraction spectroscopy.[660]

Photoacoustic (or better known as optoacoustic) spectroscopy in the i.r. is well established for use with gaseous,[661,662] liquid, and solid samples, especially with laser sources (such as the CO_2 laser).[663] Carbon dioxide laser excitation has been employed with a resonant cell to measure linear absorptions of SF_6[664,665] at concentrations as low as 1 in 10^{11},[665] i.r. absorptivities of mirrors and window materials[666] and thin films,[667] ethylene of 0.3 p.p.b.,[662] and in laser opto-acoustic detection in gas chromatography.[668] A subtle source of contamination in spectrophones was shown to be traces of methanol used to clean cell windows.[669] Detailed measurements of absorptivity of mixtures of water vapour in air have been made using resonant optoacoustic detection with 49 wavelengths from 9.2 to 10.7 m from a $^{12}C^{16}O_2$ laser.[670] *In situ* measurements of aerosol absorption with a resonant CW laser enabled the spectral dependence of quartz dust from 9.2 to 10.8 μm to be found.[671] A new opto-

653 R. N. Dixon, D. A. Hamer, and C. R. Webster, *Chem. Phys.*, 1977, **22**, 199.
654 W. Lahmann, H. J. Ludewig, and H. Welling, *Analyt. Chem.*, 1977, **49**, 549.
655 P. C. Claspy, C. Ha, and Y.-H.-Pao, *Appl. Optics*, 1977, **16**, 2972.
656 D. Fournier, A. C. Boccara, and J. Badoz, *Appl. Phys. Letters*, 1978, **32**, 640.
657 M. G. Rockley and J. P. Devlin, *Appl. Phys. Letters*, 1977, **31**, 24.
658 T. F. Hunter, D. Rumbles, and M. G. Stock, *Chem. Phys. Letters*, 1977, **45**, 145.
659 L. M. Hall, T. F. Hunter, and M. G. Stock, *Chem. Phys. Letters*, 1976, **44**, 145.
660 T. E. Orlowski, K. E. Jones, and A. H. Zewail, *Chem. Phys. Letters*, 1978, **54**, 197.
661 R. D. Kamm, *J. Appl. Phys.*, 1976, **47**, 3550.
662 S. Shtrikman and M. Slatkine, *Appl. Phys. Letters*, 1977, **31**, 830.
663 P.-E. Nordal and S. O. Kanstad, *Optics Comm.*, 1977, **22**, 185.
664 D. M. Cox, *Optics Comm.*, 1978, **24**, 336.
665 E. Nodov, *Appl. Optics*, 1978, **17**, 1110.
666 R. Gibbs and K. L. Lewis, *J. Phys. (E)*, 1978, **11**, 304.
667 T. J. Moravec and E. G. Bernal, *Appl. Optics*, 1978, **17**, 1939.
668 L. B. Kreuzer, *Analyt. Chem.*, 1978, **50**, 597A.
669 J.-C. Peterson, R. J. Nordstrom, and R. K. Long, *Appl. Optics*, 1976, **15**, 2974.
670 M. S. Shumate, R. T. Menzies, J. S. Morgolis, and L.-G. Rosengren, *Appl. Optics*, 1976 **15**, 2480.
671 C. W. Bruce and R. G. Pinnick, *Appl. Optics*, 1977, **16**, 1762.

acoustic cell suitable for use with diode lasers enabled absorption coefficients as low as 10^{-10} cm^{-1} to be determined together with studies on the effects of temperature and electric and magnetic fields.[672] Sensitive and unambiguous detection of CO has been reported with a CW CO laser and both resonant and non-resonant optoacoustic cells.[673]

A technique related to optoacoustic spectroscopy has used a super-conducting bolometer to track heat-pulse temperature profiles with temperature changes as low as 10^{-5} K.[674] The detection method bears some similarity to fluorescence except that phonons rather than photons are detected. In the example quoted, the relaxation profiles of Cr^{III} ions excited in Al$_2$O$_3$ and other host matrices were determined following pumping with a 3 ns dye laser pulse. The thermal response time of the bolometer was 10—20 ns but r.f. interference and the finite size of the detector limited the actual time resolution to a few hundred ns.[674]

The thermal lens effect in liquids has now been shown to form the basis of a very sensitive technique for absorption spectrometry capable of measuring absorption of the order of 10^{-6} cm^{-1}. A liquid sample placed in the beam of a laser will be heated by the absorbed power and so create in the liquid a transverse temperature profile matching the intensity profile of the laser. Since most liquids have positive coefficients of thermal expansion, the thermal lens thus formed is divergent. In the single-beam method, the thermal lens is detected by monitoring the divergence of the transmitted laser beam. In the 'double-beam' method a second, much weaker, 'probe' laser can be directed along the same path through the liquid. The double-beam approach has been used to detect the relatively narrow (200 cm^{-1}) absorption peak of 600 nm attributed to the fourth overtone of the non-hydrogen-bonded OH stretching vibration[675] and CH stretchings in benzene.[676] At longer wavelengths, the thermal lens effect has been used to provide data complementary to that obtained from i.r. fluorescence studies on the magnitude of vibrational energy transfer processes.[677]

Two-photon Spectroscopy.—An excellent introductory review to two-photon spectroscopy emphasized the instrument arrangements available including both single- and double-beam instruments.[678] NLPD lasers are commonly employed for excitation with observation of fluorescence from various excited states.[679—681] In a representative study of naphthalene in the gas phase, a NLPD laser with a peak power of 100 kW at 10 Hz provided spectral bandwidths of 0.02 nm at 600 nm and various dyes were used to cover the range 575—

[672] C. K. N. Patel and R. J. Kerl, *Appl. Phys. Letters*, 1977, **30**, 578.
[673] R. Gerlach and N. M. Amer, *Appl. Phys. Letters*, 1978, **32**, 228.
[674] M. B. Robin and N. A. Kuebler, *J. Chem. Phys.*, 1977, **66**, 169.
[675] R. L. Swofford, M. E. Long, M. S. Burberry, and A. C. Albrecht, *J. Chem. Phys.*, 1977, **66**, 664.
[676] R. L. Swofford, M. E. Long, and A. C. Albrecht, *J. Chem. Phys.*, 1976, **65**, 179; R. L. Swofford, M. S. Burberry, J. A. Morrell, and A. C. Albrecht, *J. Chem. Phys.*, 1977, **66**, 5245; M. E. Long, R. L. Swofford, and A. C. Albrecht, *Science*, 1976, **191**, 183.
[677] R. T. Bailey, F. R. Cruickshank, D. Pugh, and W. Johnstone, in ref. 3, p. 257.
[678] D. Frohlich and M. Sondergeld, *J. Phys. (E)*, 1977, **10**, 761.
[679] U. Boesl, H. J. Neusser, and E. W. Schlag, *Chem. Phys.*, 1976, **15**, 167.
[680] A. Bree, M. Edelson, and C. Taliani, *J. Phys. (E)*, 1978, **11**, 541.
[681] N. Mikami and M. Ito, *Chem. Phys.*, 1977, **23**, 141.

600 nm.[679] Two-photon spectrometers with FLPD lasers[682—684] had somewhat similar operating characteristics and used photon counting[682] for sensitive fluorescence determinations and stabilized flashlamp pumping[683] to minimize long term fluctuations in laser output. Two independently tunable FLPD lasers allowed the two-photon absorption (TPA) spectrum of naphthalene in benzene to be determined at a spectral resolution of 20 cm^{-1}.[684] The technique has been further extended in terms of additional degrees of freedom by using two tunable dye lasers with different wavelengths and polarizations.[685] In this case, simultaneous absorption of two photons from one laser beam was precluded by choosing appropriate incident wavelengths. The arrangement was used to record the TPA spectrum of benzene in the range 470—545 nm.

A two-photon absorption study of the fine structure of excited *P*-states in CdS involved use of a tunable dye laser beam and pulsed CO_2 laser beam which excited the sample simultaneously.[686] Consecutive u.v. laser excitation was used to populate high-lying singlet states of rhodamine dyes by two-photon absorption followed by fluorescence observation from several states.[687] There are now numerous other reports of TPA spectra with measurements in the gas and vapour phase for NO,[688] benzene and [$^2H_6$]benzene,[689] *p*-difluorobenzene;[690] in the liquid phase for umbelliferone,[691] all-*trans* diphenylhexatriene,[692] and benzene at high resolution (bandwidth <0.03 cm^{-1});[693] and in the solid phase for fluorene single crystals at 6 K,[694] ground state xenon,[695] and biphenyl and naphthalene crystals at 77 and 1.8 K.[696]

Two-photon excitation of cyclohexane and decalin by a pulsed nitrogen laser was reported on the basis of an observed emission spectrum, decay time, and quadratic dependence of emission intensity on laser beam intensity.[697] High resolution laser spectroscopy of CO and N_2 was undertaken by two- or three-photon excitation with a dye laser oscillator–amplifier system with excitation detected by direct or indirect fluorescence in the vacuum-u.v.[698] Laser-excited emission from the Ca_2 van der Waals dimer isolated in solid krypton was detected following one-or consecutive two-photon excitation at 12 K.[699]

682 R. R. Birge, J. A. Bennett, B. M. Pierce, and T. M. Thomas, *J. Amer. Chem. Soc.*, 1978, **100**, 1533.
683 H. Gattermann and M. Stockburger, *Chem. Phys.*, 1977, **24**, 327.
684 V. SethuRaman, G. J. Small, and E. S. Yeung, *Rev. Sci. Instr.*, 1977, **48**, 1436.
685 W. Hampf, H. J. Neusser, and E. W. Schlag, *Chem. Phys. Letters*, 1977, **47**, 405.
686 V. T. Nguyen, T. C. Damen, and E. Gornik, *Appl. Phys. Letters*, 1977, **30**, 33.
687 H.-B. Lin and M. R. Topp, *Chem. Phys. Letters*, 1977, **47**, 442.
688 H. Zacharias, J. B. Halpern, and K. H. Welge, *Chem. Phys. Letters*, 1976, **43**, 41.
689 L. Wunsch, F. Metz, H. J. Neusser, and E. W. Schlag, *J. Chem. Phys.*, 1977, **66**, 386.
690 M. J. Robey and E. W. Schlag, *Chem. Phys.*, 1978, **30**, 9.
691 L. Parma and N. Omenetto, *Chem. Phys. Letters*, 1978, **54**, 541.
692 G. R. Holtom and W. M. McClain, *Chem. Phys. Letters*, 1976, **44**, 436.
693 J. R. Lombardi, R. Wallenstein, T. W. Hansch, and D. M. Friedrich, *J. Chem. Phys.* 1976, **65**, 2357.
694 A. Bree, M. Edelson, and C. Taliani, *Chem. Phys.*, 1978, **30**, 343.
695 B. A. Bushaw and T. J. Whitaker, *Proc. Soc. Photo-Opt. Instrum. Eng.*, 1976, **82**, 92.
696 R. M. Hochstrasser and H. N. Sung, *J. Chem. Phys.*, 1977, **66**, 3265, 3276.
697 S. Dellonte, E. Gardini, F. Barigelletti, and G. Orlandi, *Chem. Phys. Letters*, 1977, **49**, 596.
698 R. Wallensteing and H. Zacharias, *J. Opt. Soc. Amer.*, 1978, **68**, 677.
699 J. C. Miller and L. Andrews, *J. Chem. Phys.*, 1978, **68**, 1701.

During two-photon experiments with *trans*-butadiene, a focused laser beam caused a transient thermal (density) gradient enabling the refractive index gradient to be probed with a low power He–Ne laser.[700] Further work with this compound in which multiphoton ionization techniques were used for detection (see below) confirmed the usefulness of thermal lensing for studying weak u.v.–visible spectra of molecular gases.[701] Thermal lensing effects have also been applied to determine the TPA spectrum of benzene[702] and in a search for the low-lying ^1A state of 1,3,5-hexatriene.[703]

Among the errors in conventional TPA experiments which have been discussed were the degree of overlapping of the two dye laser beams (which are often of poor beam quality) and other uncertainties in the properties of the laser beam.[704] Such errors are less important in parametric mixing measurements (using three-wave mixing or CARS) and these have been recommended for more precise measurements and demonstrated in the case of NO and SO_2[705] and *trans*-stilbene.[706] In the former case, two tunable dye lasers (ω_1 and ω_2) were combined in a gas cell and a dispersing element used to select the ω_3 beam (where $\omega_3 = 2\omega_2 - \omega_1$).[705]

Doppler-free two-photon spectroscopy involved irradiation of a gas sample with oppositely directed optical beams at a frequency which corresponds to one half of the molecular transition centre frequency. Simultaneous absorption of one photon from each of the counterpropagating beams yielded an absorption spectrum free from Doppler broadening.[707] An investigation into the feasibility of this technique for *in situ* detection of stratospheric gases using fluorescence detection indicated that detection limits of the order of 10^5—10^7 molecules cm^{-3} were possible for molecular species with strong two-photon u.v. transitions, which undergo efficient radiative relaxation from the excited state.[707] One of the first examples of Doppler-free measurements of two-photon absorptions was naphthalene with a CW tunable dye laser.[708] A sensitive variation of the above method to select only those molecules moving perpendicularly to the beam direction used the technique of intermodulated fluorescence in which two laser beams were chopped at different frequencies and the saturation detected as a modulation of the fluorescence at the same frequency.[709] This technique has been applied to examine regions of the (0, 0) band of the $A^2\pi–X^2\Sigma^+$ system of CaCl with linewidths of 100 MHz.[709]

Two-step sequential absorption processes are a natural extension to conventional TPA measurements where simultaneous excitation occurs and are somewhat analogous to flash photolysis. For example, two independently tunable high resolution dye lasers, excited simultaneously by a pulsed nitrogen laser,

[700] G. C. Nieman and S. D. Colson, *J. Chem. Phys.*, 1978, **68**, 2994.
[701] V. Vaida, R. E. Turner, J. L. Casey, and S. D. Colson, *Chem. Phys. Letters*, 1978, **54**, 25.
[702] A. J. Twarowski and D. S. Kliger, *Chem. Phys.*, 1977, **20**, 253, 259.
[703] A. J. Twarowski and D. S. Kliger, *Chem. Phys. Letters*, 1977, **50**, 36.
[704] R. T. Lynch and H. Lotem, *J. Chem. Phys.*, 1977, **66**, 1905.
[705] R. M. Hochstrasser, G. R. Meredith, and H. P. Trommsdorf, *Chem. Phys. Letters*, 1978, **53**, 423.
[706] R. J. M. Anderson, G. R. Holtom, and W. M. McClain, *J. Chem. Phys.*, 1977, **68**, 3832.
[707] J. Gelbwachs, *Appl. Optics*, 1976, **15**, 2654.
[708] K.-M. Chen, I. C. Khoo, L. E. Steenhoek, and E. S. Yeung, *Optics Comm.*, 1977, **23**, 90.
[709] J. M. Brown, H. Martin, and F. D. Wayne, *Chem. Phys. Letters*, 1978, **55**, 67.

have been used to excite selective rotational band structure in iodine and to probe for absorptions to higher excited states.[710] Sequential two-photon absorption by two lasers (a FLPD laser and NLPD laser with 3 and 0.5 cm^{-1} bandwidths) of C_2H_2 and a rare gas on a Pt mirror at 4.2 K produced emission attributed to $C_2d^3\pi_g$–$a^3\pi_u$.[711] Intermolecular energy transfer from the short-lived S_2 singlet state of Rh6G to the ground state of BBOT was observed following sequential time-delayed two-photon excitation of the energy donor (530 nm pulse excitation followed by a 55 ps delayed 1060 nm pulse) and interrogation of the enhancement of the $S_1 \rightarrow S_0$ fluorescence of the acceptor.[712] Ultrafast two-step spectroscopy on a picosecond time scale using two intense 5 ps tunable pulses, one in the visible and the other in the near-i.r., was used to measure the energy states of S_0 and S_1 of a large polyatomic molecule.[713] The vibrational mode was first excited well above the thermal population level by the i.r. pulse and the second visible pulse used to excite molecules to an upper vibrational level of S_1 from which they fluoresced. The fluorescence signal was a measure of the population of vibrational states in S_0.[713] Repopulation times for various levels in sodium molecules were determined by using two dye laser pulses delayed by 0.1—0.5 μs[714] and sequential two-photon ionization with a CW Ar laser of the same molecule was shown to be useful for molecular beam diagnostics.[715]

Multiphoton ionization (MPI) is an extension of two-photon absorption spectroscopy except that ions rather than photons are detected. The signal-to-noise ratios in MPI are generally higher than that of a comparable laser-induced fluorescence experiment because the relative efficiency of counting collected ions *versus* the relative inefficiency of collecting only a small fraction of the emitted fluorescence. There is also no requirement for an atomic or molecular species to be luminescent. In laser-induced fluorescence, the ultimate sensitivity is set by scattered light levels, whereas in MPI, which is unaffected by scattered light, only the ion background (which can be suppressed) and fluctuations in the laser intensity determine this limit. The technique of MPI has been applied to studies of the Na_2 B–X spectrum (by crossing a molecular beam with a tunable dye laser),[716] a new state of NH_3 around 451 nm,[717] the symmetry of two-photon states in benzene with polarized excitation,[718] photoionization and Rydberg spectra of ten lanthanides,[719] and the spectrum of NO cooled by supersonic expansion.[720] Two-step photoionization of HCHO by simultaneous irradiation by hydrogen and nitrogen lasers has enabled the lifetime of the $^1A_2$ state to be

[710] M. D. Danyluk and G. W. King, *Chem. Phys.*, 1977, **22**, 59; 1977, **25**, 343; *Chem. Phys. Letters*, 1976, **43**, 1; 1976, **44**, 440.
[711] V. E. Bondybey, *J. Chem. Phys.*, 1976, **65**, 2296.
[712] I. Kaplan and J. Jortner, *Chem. Phys. Letters*, 1977, **52**, 202.
[713] W. Kaiser, A. Seilmeier, and A. Laubereau, *J. Opt. Soc. Amer.*, 1978, **68**, 680.
[714] R. Feinberg, R. E. Teets, J. Rubbmark, and A. L. Schawlow, *J. Chem. Phys.*, 1977, **66**, 4330.
[715] B. P. Mathur, E. W. Rothe, and G. P. Reck, *J. Chem. Phys.*, 1978, **68**, 2518.
[716] D. L. Feldman, R. K. Lengel, and R. N. Zare, *Chem. Phys. Letters*, 1977, **52**, 413.
[717] G. C. Nieman and S. D. Colson, *J. Chem. Phys.*, 1978, **68**, 5656.
[718] J. O. Berg, D. H. Parker, and M. A. El-Sayed, *J. Chem. Phys.*, 1978, **68**, 5661.
[719] E. F. Worden, R. W. Solarz, J. A. Paisner, and J. G. Conway, *J. Opt. Soc. Amer.*, 1978, **68**, 52.
[720] D. Zakheim and P. Johnson, *J. Chem. Phys.*, 1978, **68**, 3644.

determined.[721] Resonance ionization spectroscopy applied to CsI was used to measure diffusion of Cs atoms in Ar and the reactions of Cs with O_2 in Ar gas.[722] In this case, every CsI molecule in a small volume was photodissociated with a FLPD laser and a second pulsed dye laser was used to detect single Cs atoms through resonance ionization.

Double Resonance Spectroscopy.—The realm of laser excitation spectroscopy can be greatly extended in energy coverage with the use of two lasers to perform optical–optical double resonance so as to attain spectroscopic states which are not readily accessible from the ground state. For example, the complete rotational analysis of the $E^2–B^2$ system of CaCl has been accomplished using two tunable dye lasers. The first (an Ar^+-pumped dye laser) was tuned to a known $B^2–X^2$ transition and the second (a FLPD laser) was counterpropagated through the cell colinearly with the CW beam and the resulting photoluminescence detected photometrically.[723] There are numerous other examples of double resonance experiments with excitation sources varying from u.v. to microwave wavelengths. Energy transfer processes in biacetyl vapour were followed after stepwise molecular absorption of a pulsed CO_2 laser and a CW 441.6 nm He–Cd laser.[724] A chopped CW CO_2 laser was employed to induce an excited vibronic state transition in NO_2 following prior excitation with a CW dye laser to the $^2B_2$ and $^2B_1$ states.[725] Population inversion between ground and excited state iodine atoms (produced by injecting I_2 into a stream of chemically produced $O_2^*(^1\Delta_g)$) were determined by passing the stream of flowing gases through the optical cavity of a pulsed discharge I* laser and monitoring the I* concentration with a resonant lamp at 206.2 nm.[726] As the laser was fired, high radiation densities inside the laser cavity rapidly drove the $I^*(5^2P_{\frac{1}{2}})$ and $I(5^2P_{\frac{3}{2}})$ populations in the gas stream towards saturation.

I.r.–i.r. double resonance experiments have been undertaken to measure rotational transfer in CO with two CO lasers (pump on $\nu = 1 \to 0$, probe on $\nu = 2 \to 1$),[727] to examine rotational relaxation effects in HF (with pulsed and CW HF lasers)[728] and to determine vibrational relaxation times of CH_3F isolated in a krypton matrix excited by two CO_2 lasers.[729] V–V energy transfer studies[730] and direct measurements of collisionless intramolecular vibrational energy redistribution in SF_6[731] have been made following excitation with CW and pulsed CO_2 lasers. Energy transfer between isotopic species, $^{12}CH_3F$ and

[721] S. V. Andreyev, V. S. Antonov, I. N. Knyazev, and V. S. Letokhov, *Chem. Phys. Letters*, 1977, **45**, 166.

[722] L. W. Grossman, G. S. Hurst, M. G. Payne, and S. L. Allman, *Chem. Phys. Letters*, 1977, **50**, 70; L. W. Grossman, G. S. Hurst, S. D. Kramer, M. G. Payne, and J. P. Young, *ibid.*, 1977, **50**, 207.

[723] P. J. Domaille, T. C. Steimle, and D. O. Harris, *J. Chem. Phys.*, 1978, **68**, 4977.

[724] B. J. Orr, *Chem. Phys. Letters*, 1976, **43**, 446.

[725] I. P. Herman, A. Javan, and R. W. Field, *J. Chem. Phys.*, 1978, **68**, 2398.

[726] A. T. Pritt, R. D. Coombe, D. Pilipovich, R. I. Wagner, D. Bernard, and C. Dymek, *Appl. Phys. Letters*, 1977, **31**, 745.

[727] P. Brechignac, *Optics Comm.*, 1978, **25**, 53.

[728] J. J. Hinchen and R. H. Hobbs, *J. Chem. Phys.*, 1976, **65**, 2732.

[729] L. Abouaf-Marguin, B. Gauthier-Roy, and F. Legay, *Chem. Phys.*, 1977, **23**, 443.

[730] D. S. Frankel, *J. Chem. Phys.*, 1976, **65**, 1696.

[731] T. F. Deutsch and S. R. J. Brueck, *Chem. Phys. Letters*, 1978, **54**, 258.

$^{13}CH_3F$, was followed using a double-resonance experiment in which the $^{12}CH_3F$ v_3 mode was pumped by a Q-switched 9.6 μm CO_2 laser operating on the $P(20)$ line, while the $^{13}CH_3F$ v_3 absorption was monitored by a CW CO_2 laser operating on the $P(32)$ 9.6 μm transition.[732]

Other i.r. double resonance experiments included the observed production of visible luminescence of the tris-(2,2′-bipyridine)ruthenium(II) cation at 2 K[733] and measurements of vibrational energy relaxation of CH_3F dissolved in liquid O_2 and Ar.[734] Energy transfer was observed in H_2O–NH_3 mixtures excited in an i.r. (N_2O) laser – microwave double irradiation[735] and 11 transitions of the v_2 band of $^{15}NH_3$ were measured using an isotopic CO_2 laser and 20 mW microwave source.[736] A new method has been reported to detect molecular ion emission and measure lifetimes with a photon–photoion coincidence arrangement.[737]

The technique for photon echo measurements (see Vol. 8, p. 48) has provided a means of studying phase relaxation characteristics of excited molecules and usually employs two optical pulses (derived in general from dye lasers) with an optical delay between the pulses. Most of the recently reported studies have been with molecular mixed crystals at low-temperatures[738,739] although studies on the excited state manifolds in sodium vapour have also been made.[739] Synchronized picosecond excitation and probe pulses from mode-locked dye lasers have also enabled optical dephasing to be examined in organic mixed crystals of pentacene in *p*-terphenyl.[740] A third probe pulse has been used to convert the optical polarization induced by the laser pulses into a change in excited state population giving rise to spontaneous emission.[741]

Laser Spectroscopy.—Without doubt, the tunability and exceptionally narrow bandwidth of laser sources have revolutionized atomic and molecular spectroscopy, particularly in the areas of selective excitation and high resolution spectroscopy.[742] An overview, by A. L. Schawlow, on a few ways by which laser sources have extended knowledge of the interaction of light with matter has included saturated absorption and polarization spectroscopy and polarization labelling for spectrum identification.[743] Publications in laser spectroscopy have included a collection of 71 papers on fundamental physical applications,

732 J. M. Preses and G. W. Flynn, *J. Chem. Phys.*, 1977, **66**, 3112.
733 D. C. Baker, K. W. Hipps, and G. A. Crosby, *Chem. Phys. Letters*, 1978, **53**, 333.
734 S. R. J. Brueck, T. F. Deutsch, and R. M. Osgood, *Chem. Phys. Letters*, 1977, **51**, 339.
735 F. Herlemont, M. Lyszyk, and J. Lemaire, *Chem. Phys. Letters*, 1978, **54**, 603.
736 H. Jones, *J. Mol. Spectroscopy*, 1978, **70**, 279.
737 J. H. D. Eland, M. Devoret, and S. Leach, *Chem. Phys. Letters*, 1976, **43**, 97.
738 J. B. W. Morsink, T. J. Aartsma, and D. A. Wiersma, *Chem. Phys. Letters*, 1977, **49**, 34; T. J. Aartsma and D. A. Wiersma, *ibid.*, 1976, **42**, 520; 1978, **54**, 415; W. H. Hesselink and D. A. Wiersma, *ibid.*, 1977, **50**, 51; T. J. Aartsma, J. Morsink, and D. A. Wiersma, *ibid.*, 1977, **47**, 425.
739 T. J. Aartsma and D. A. Wiersma, *Chem. Phys. Letters*, 1976, **42**, 520.
740 W. Hesselink and D. A. Wiersma, *Chem. Phys. Letters*, 1978, **56**, 227.
741 A. H. Zewail, T. E. Orlowski, K. E. Jones, and D. E. Godar, *Chem. Phys. Letters*, 1977, **48**, 256.
742 T. Shimizu, *Appl. Spectroscopy Rev.*, 1976, **11**, 163; L. J. Kinsey, *Ann. Rev. Phys. Chem.*, 1977, **28**, 349; S. Kimel and S. Speiser, *Chem. Rev.*, 1977, **77**, 437.
743 A. L. Schawlow, *J. Opt. Soc. Amer.*, 1977, **67**, 140.

multiple photon dissociation, and high resolution studies,[744] and a further collection of papers on unconventional spectroscopy.[745] The application of tunable lasers and i.r. FT spectroscopy to the identification and quantification of a number of species in photochemical smog has also been reviewed.[746]

The isolation of individual rovibronic transitions in glyoxal vapour can be regarded as a representative example of high resolution spectroscopy.[747] The linewidth of the 454.5 nm line of an Ar^+ laser can be measured to be less than 10^{-4} cm^{-1} and an intracavity etalon allows tuning over about 0.2 cm^{-1}. By using fluorescence detection, the probing laser was used to examine 12 rovibronic levels and allow their identification. Narrow line lasers have also been used to examine several rotational lines in the $A^2\pi_{\frac{1}{2}}-X^2\pi_{\frac{3}{2}}$ system of ^{35}ClO,[748] resonance absorption in NO (with a line tunable CO laser),[749] absorption in OCS near 1890 cm,[750] fine structure in HeI,[751] the saturation spectrum of krypton isotopes,[752] and the Rydberg states of Rb atoms (using a space-charge-limited thermionic diode as a detector).[753]

Intracavity CW dye laser excitation has the distinct advantage of making available 20—150 times higher circulating power inside the laser cavity compared with outside and tunability over the entire visible spectrum and beyond. A novel excitation technique for the preparation of $O_2*(b^1\Sigma_g^+)$ in order to study its reactions with various olefins used a dye laser tuned to the O_2 transition at 760.6775 nm and monitored the photoacoustic signal from an intracavity detector.[754] More conventional intracavity spectroscopy measurements enabled extremely weak transitions (near 630 nm) in O_2 and HCl to be measured and estimated that band systems with oscillator strengths of 10^{-12} could be detected readily.[755] Further intracavity absorption techniques have been applied to time-resolved kinetics of the HCO radical,[756] chemical reactions between Ba, Ca, and Sr and N_2O and CO (which have potential as visible chemical lasers)[757] and have been used as a simple relative population probe.[758] Intracavity laser detection of NO_2 was shown to produce spectra with discrete absorption features and to exhibit a linear dependence on the concentration over more than an order of magnitude.[759] The experimental arrangement used an optical

[744] 'Laser Spectroscopy', ed. J. L. Hall and J. L. Carlsten, vol. III, Springer-Verlag, Heidelberg, 1977.
[745] *Proc. Soc. Photo-Opt. Instrum. Eng.*, 1976, vol. 82.,
[746] J. N. Pitts, B. J. Finlayson-Pitts, and A. M. Winer, *Environ. Sci. Technol.*, 1977, **11**, 569.
[747] B. F. Rordorf and C. S. Parmenter, *J. Mol. Spectroscopy*, 1978, **69**, 365.
[748] P. H. Wine, A. R. Ravishankara, D. L. Philen, D. D. Davis, and R. T. Watson, *Chem. Phys. Letters*, 1977, **50**, 101.
[749] B. K. Garside, E. A. Ballik, M. Elsherbiny, and J. Shewchun, *Appl. Optics*, 1977, **16**, 398.
[750] P. G. Buckley, J. H. Carpenter, A McNeish, J. D. Muse, J. J. Turner, and D. H. Whiffen, *J.C.S. Faraday II*, 1978, **74**, 129.
[751] Ph. Cahuzac and R. Damaschini, *Optics Comm.*, 1977, **20**, 111.
[752] H. Gerhardt and A. Timmermann, *Optics Comm.*, 1977, **21**, 343.
[753] A. I. Ferguson and M. H. Dunn, *Optics Comm.*, 1977, **23**, 227.
[754] K. V. Reddy and M. J. Berry, *J. Opt. Soc. Amer.*, 1978, **68**, 694.
[755] R. G. Bray, W. Henke, S. K. Liu, K. V. Reddy, and M. J. Berry, *Chem. Phys. Letters*, 1977, **47**, 213.
[756] J. H. Clark, C. B. Moore, and J. P. Reilly, *Internat. J. Chem. Kinetics*, 1978, **10**, 427.
[757] D. J. Eckstrom, J. R. Barker, J. G. Hawley, and J. P. Reilly, *Appl. Optics*, 1977, **16**, 2102.
[758] J. P. Reilly and G. C. Pimentel, *Appl. Optics*, 1976, **15**, 2372.
[759] G. H. Atkinson, T. N. Heimlich, and M. W. Schuyler, *J. Chem. Phys.*, 1977, **66**, 5005.

multichannel analyser to subtract the output of a FLPD laser with and without absorber and is clearly applicable to transient species. Intracavity CW dye laser excitation to photoactivate methyl isocyanide around 730 nm revealed that the photoacoustic spectrum had identical features to the first-order reaction rate constant for isomerization as a function of wavelength.[760]

High sensitivity laser wavelength-modulation spectroscopy has been put forward as a technique to provide both derivative and direct absorption spectra directly.[761] Sample and reference beams are measured continuously in phase quadrature. The derivative spectrum reported corresponded to the extremely weak fifth overtone of the CH stretching vibration in the electronic state of benzene and was observed in a 1 cm pathlength cell at about 608 nm.

6 Chemical Techniques

Low and medium pressure mercury lamps are probably still the most common irradiation sources in photochemistry and many of the conventional preparations and kinetic studies are carried out with simplex immersion well or 'merry-go-round' reactors. For example, the irradiation of adenosine 5'-phosphate was undertaken in an immersion-well type of reactor with 15 W low pressure mercury and 75 W medium pressure ZnCdHg lamps.[762] High intensities at 313 nm (up to 1.3×10^{17} photons s^{-1} cm^{-2} were obtained from two 400—750 W high pressure mercury lamps and used to irradiate bromodeoxyuridine-substituted DNA.[763] By surrounding a cylindrical cell with a toroidal lamp, it was possible to irradiate and simultaneously monitor the u.v. absorption spectrum of a small volume (14 cm^3) of solution.[764] An immersion-well reactor with a single lamp and a number of concentric reaction flasks enabled a number of reactions to be carried out simultaneously. Each reaction is designed to use different regions of the lamp spectrum and act as a filter for the adjacent reaction.[765]

A simple modification to the 'merry-go-round' apparatus with a single central lamp used variable slits to attenuate the photolysis light and enable reactions which differ in quantum yield by a factor of eight or less to be irradiated for equal periods of time with acceptable conversion.[766] Degassable and re-usable tubes using O-ring sealing have been suggested for use with this reactor.[767]

An efficient gas–liquid reactor has been designed to ensure rapid reaction between ozone and acidified KI solution.[768] Ozone was produced by the irradiation of oxygen with 184 nm light and a key feature of the reactor was an inverted centrifugal stirrer in the form of an inverted funnel with two diametrically opposite outputs. The ozone was passed to this funnel from the bottom of the vessel. In studies of the photochemical reactions of mercuric sulphide in aqueous

[760] K. V. Reddy and M. J. Berry, *Chem. Phys. Letters*, 1977, **52**, 111.
[761] E. I. Moses and C. L. Tang, *Optics Letters*, 1977, **1**, 115.
[762] J. T. H. Goosen and J. G. Kloosterboer, *Photochem. and Photobiol.*, 1978, **27**, 703.
[763] F. E. Ahmed, D. C. Robb, and R. W. Hart, *Rev. Sci. Instr.*, 1977, **48**, 1442.
[764] H. A. Friedman and L. M. Toth, *J. Phys. (E)*, 1977, **10**, 796.
[765] J. J. Bloomfield, U.S.P. 4 087 342.
[766] J. S. Keute and T. H. Koch, *Mol. Photochem.*, 1977, **8**, 371.
[767] G. G. Wubbels, *J. Chem. Educ.*, 1977, **54**, 49.
[768] A. K. Davies, K. A. Khan, J. F. McKellar, and G. O. Phillips, *Mol. Photochem.*, 1976, **7**, 343.

solutions, a suspension of the mercuric sulphide was formed by stirring, and powdered quartz added to prevent adhesion of the suspended solids to the walls during the reaction.[769] A flat plate reactor has been designed for solid state irradiations at low temperatures[770] and a product yield study following the flash photolysis of azomethane used a quartz vessel inside a perspex-body flash lamp.[771]

A high speed rotor technique has been used to provide a pulsed accelerated beam of xenon atoms (travelling at <2 km s^{-1}) which subsequently reacted with Br_2 in a crossed-beam chemiluminescence arrangement.[772]

Large scale irradiations of alkane–NO_x systems in air were carried out in a 5800 l evacuable, Teflon-coated environmental chamber using a 25 kW solar simulator[773] with an air purification system used to produce clean reaction conditions.[774] A 720 l continuous flow reactor used fluorescent sunlamps and incandescent lamps to simulate irradiation in a marine community.[775] Photochemical reactions between Cl_2, NO_2, and $HCHO$ were undertaken in a 690 l glass irradiation cell using 96 u.v. fluorescent lamps and simultaneously analysed by FT i.r. techniques using folded light paths of up to 504 m.[776] A similar large photolysis cell (6.3 m long, 445 l volume) also provided with multipass optics for *in situ* i.r. analysis was used to study the reaction $HO + CO \rightarrow H + CO_2$ with irradiation (from blacklight fluorescence lamps) designed to mimic the solar intensity and wavelength distribution in the photochemically important region, 300—430 nm.[777] Ozone production was investigated in a cyclohexene–NO_2–air mixture flowing inside a 9 m long $\times$ 15 cm i.d. borosilicate tube and irradiated by u.v. lamps.[778] A concentric photolysis flash-sample compartment was used for aerosol studies of sulphur and iodoform.[779]

Various theoretical and experimental studies have been undertaken to describe the rate of photochemical reactions in solid materials[780,781] in terms of diffuse reflectance,[782] in the absence of diffusion,[780] and as a dependence on reactant concentration and radiation concentration and intensity.[783] The kinetic characteristics of reverse thermal reactions in powdered photochromic samples

[769] H. Akagi, Y. Fujita, and E. Takabatake, *Photochem. and Photobiol.*, 1977, **26**, 363.
[770] J. R. Scheffer and A. A. Dzakpasu, *J. Amer. Chem. Soc.*, 1978, **100**, 2163.
[771] M. J. Pilling, J. A. Robertson, and G. J. Rogers, *Internat. J. Chem. Kinetics*, 1976, **8**, 883.
[772] M. R. Levy, C. T. Rettner, and J. P. Simons, *Chem. Phys. Letters*, 1978, **54**, 120; P. B. Moon, C. T. Rettner, and J. P. Simons, *J.C.S. Faraday II*, 1978, **74**, 630.
[773] W. P. L. Carter, K. R. Darnall, A. C. Lloyd, A. M. Winer, and J. N. Pitts, *Chem. Phys. Letters*, 1976, **42**, 22.
[774] G. J. Doyle, P. J. Bekowies, A. M. Winer, and J. N. Pitts, *Environ. Sci. Technol.*, 1977, **11**, 45.
[775] R. C. Worrest, H. van Dyke, and B. E. Thomson, *Photochem. and Photobiol.*, 1978, **27**, 471.
[776] P. L. Hanst and B. W. Gay, *Environ. Sci. Technol.*, 1977, **11**, 1105; P. L. Hanst, J. W. Spence, and M. Miller, *ibid.*, 1977, **11**, 403.
[777] W. H. Chan, W. M. Uselman, J. G. Calvert, and J. H. Shaw, *Chem. Phys. Letters*, 1977, **45**, 240.
[778] C.-H.-Shen, G. S. Springer, and D. H. Stedman, *Environ. Sci. Technol.*, 1977, **11**, 151.
[779] P. G. Whitkop and P. L. Goodfriend, *Photochem. and Photobiol.*, 1976, **24**, 303.
[780] E. L. Simmons, *J. Chem. Phys.*, 1977, **66**, 1413.
[781] B. Claudel, M. Feve, J. P. Puaux, and H. Sautereau, *J. Photochem.*, 1978, **8**, 117.
[782] E. L. Simmons, *J. Phys. Chem.*, 1977, **81**, 1592.
[783] E. L. Simmons, *J. Chem. Phys.*, 1976, **65**, 5357.

have been tested by studying the reactions of 2-(2,4-dinitrobenzyl)pyridine.[784] A review of surface photochemistry (defined as chemistry which photo-excited molecules undergo as a result of interaction with a surface) and photodesorption (defined as the optically selective excitation of molecules absorbed on a surface) has also been published.[785] Spectra of partially oriented rigid solutions of azides and nitrenes have been obtained by a stretched film technique at 77 K.[786] The nitrenes were generated by irradiation of frozen solutions of azides with a mercury lamp.

A novel technique of on-line photochemical derivatization has been described which enhanced both the sensitivity and specificity of detection in h.p.l.c.[787] The material eluting from the column was irradiated with a high flux of u.v. light, which may induce a reaction to form fluorescence or highly u.v. absorbing products, which may be detected. An example chosen was the detection of cannabinol which was converted into a highly fluorescent compound on irradiation. A somewhat similar method involved exposure of several phenothiazine drugs to intense u.v. irradiation resulting in simultaneous photo-oxidation and fluorescence exitation.[788]

Irradiation vessels for photochemistry which have been described include a miniature capillary flow tube originally designed for CARS,[789] a small volume cell for non-aqueous spectroelectrochemistry made from a machinable glass-ceramic which offers a low coefficient of expansion, inertness, and a low gas permeability,[790] a reaction cell for end-on irradiations permitting rapid and reproducible sampling by gas chromatography,[791] and a divided cell for photo-electrolysis.[792] A cell design for photoconductivity or photovoltaism of solid discs or flakes has been developed employing a variety of electrical contacts.[793] KBr discs have been suggested for visible and i.r. spectroscopy as suitable matrices for inorganic co-ordination compounds.[794] Gas pump systems described include an improved design of Toepler pump with a simple internal stopcock[795] and a closed system circulating pump offering circulation rates from 1 ml h^{-1} to 20 ml min^{-1}.[796] A miniature helical stirrer has been used with 1 cm flow through cuvettes.[797] Modifications to a glass–PTFE stopcock has allowed their use at 130 °C and gas pressures up to 1.3 $\times$ 10^8 N m^{-2}.[798] A new method for sealing optically flat windows onto glass cells involved the

[784] E. L. Simmons and J. Pierrus, *J. Chem. Phys.*, 1977, **67**, 4557.
[785] T. E. Gangwer, *Optics and Laser Technol.*, 1978, **10**, 81.
[786] R. Alvarado, J.-P. Grivet, C. Igier, J. Barcelo, and J. Rigaudy, *J.C.S. Faraday II*, 1977, **73**, 844.
[787] P. J. Twitchett, P. L. Williams, and A. C. Moffat, *J. Chromatog.*, 1978, **149**, 683.
[788] V. R. White, C. S. Frings, J. E. Villafranca, and J. M. Fitzgerald, *Analyt. Chem.*, 1976, **48**, 1314.
[789] L. B. Rogers, J. D. Stuart, L. P. Goss, T. B. Malloy, and L. A. Carreira, *Analyt. Chem.*, 1977, **49**, 959.
[790] F. Hawkridge, J. E. Pemberton, and H. N. Blount, *Analyt. Chem.*, 1977, **49**, 1646.
[791] A. E. Ledofrd and W. Braun, *Rev. Sci. Instr.*, 1977, **48**, 537.
[792] P. D. Fleischauer and J. K. Allen, *J. Phys. Chem.*, 1978, **82**, 432.
[793] D. R. Rosseinsky, R. E. Malpass, and T. E. Booty, *J. Phys. (E)*, 1977, **10**, 1236.
[794] J. T. Wrobleski and G. T. Long, *Appl. Spectroscopy*, 1977, **31**, 177.
[795] A. G. Sherwood, *J. Phys. (E)*, 1977, **10**, 204.
[796] E. Schumacher, *Rev. Sci. Instr.*, 1977, **48**, 767.
[797] D. S. Bright, *Analyt. Chem.*, 1977, **49**, 191.
[798] P. J. Robinson and G. G. Skelhorne, *J. Phys. (E)*, 1976, **9**, 1064.

use of glass transfer tapes[799] and the use of Stycast 1266 epoxy resin described for low temperature (1 K) fabrication.[800] A window assembly for a ultra-high vacuum system using Irtran 4 was sealed in place with a suspension of Pyroceram in nitrocellulose followed by rapid heating to 450 °C.[801] Lead fluoride was used to seal LiF windows helium-tight to copper or gold cylinders.[802] The seals could be heated to 400—500 °C to cure the effect of u.v. solarization on LiF.

A useful summary of results of viscosity measurements covers various organic solvents over the temperature range 80—333 K, but specifically excludes EPA and solutions containing decalin, which are known to undergo a phase separation at intermediate temperatures.[803] Various temperature controllers have been described, for the range −200 to +500 °C with a precision of 10^{-5} °C,[804] and for the range 280—350 K (in a water-bath) with temperature control to better than ± 0.002 K.[805] Gallium arsenide phosphide chips retrieved from ordinary light-emitting diodes can be employed as inexpensive, reproducible, and durable temperature sensors over the range 4—300 K.[806] Their low thermal mass results in a response time (at <25 K) of 1 ms. A heating-rate controller for thermoluminescence studies provided a heating rate adjustable from 3 to 60 °C min^{-1}[807] and an optical vacuum furnace has been described which is capable of temperatures of 1650 K using a low current source of only 200 W.[808]

Photochemical experiments in the vacuum-u.v. included 149 nm irradiation of hexafluoroacetone by a xenon resonance lamp in a cell containing a pinhole leading to a mass spectrometer,[809] the photolysis of acetic acid in Ar, N_2, CO, and |CO-doped Ar matrices at 10 K by exciting into specific absorption bands with various resonance lamps and cut-off filters,[810] and the photolysis of SF_6 and SF_5X (where X = Cl, Br, or|SF_5)|in dilute Ar matrices at 8 K with 106.7 and 104.8 nm radiation.[811] In the latter case, a new i.r. absorption was attributed to the $SF_5 \cdot$ radical. Absolute quantum yields for the dissociation of glyoxal into CO following excitation to seven vibronic levels by a NLPD laser was determined by monitoring CO by resonance emission following excitation in the vacuum-u.v. near 147 nm.[812]

CW laser excitation (Ar$^+$ or dye) has been used to stimulate the reaction of NO_2 with CO to produce CO_2[813] and lines from an Ar$^+$ laser (at intensities up to 380 W cm^{-2}) have been used for irradiation in a photoelectrolysis experi-

[799] R. A. L. Mason, *J. Phys.* (*E*), 1976, **9**, 816.
[800] G. Armstrong, A. S. Greenberg, and J. R. Sites, *Rev. Sci. Instr.*, 1978, **49**, 345.
[801] P. E. Wierenga, G. J. Mollenhorst, and A. T. B. Wikkerink, *Rev. Sci. Instr.*, 1978, **49**, 408.
[802] B. J. Mulder, *J. Phys.* (*E*), 1977, **10**, 591.
[803] G. Fischer and E. Fischer, *Mol. Photochem.*, 1977, **8**, 279.
[804] M. Lewandowski and S. Randzio, *J. Phys.* (*E*), 1977, **10**, 903.
[805] J. D. B. Featherstone and N. A. Dickinson, *J. Phys.* (*E*), 1978, **11**, 334.
[806] B. F. Griffing and S. A. Shivashanker, *Rev. Sci. Instr.*, 1977, **48**, 1225.
[807] K. Inabe, *J. Phys.* (*E*), 1976, **9**, 931.
[808] R. T. Harley, D. I. Manning, and J. F. Ryan, *J. Phys.* (*E*), 1978, **11**, 517.
[809] G. G. A. Perkins, E. R. Austin, and F. W. Lampe, *J. Chem. Phys.*, 1978, **68**, 4357.
[810] J. L. Wilkinson and W. A. Guillory, *J. Photochem.*, 1977, **7**, 251.
[811] R. R. Smardzewski and W. B. Fox, *J. Chem. Phys.*, 1977, **67**, 2309.
[812] G. H. Atkinson, M. E. McIlwain, and C. G. Venkatesh, *J. Chem. Phys.*, 1978, **68**, 726.
[813] I. P. Herman, R. P. Mariella, and A. Javan, *J. Chem. Phys.*, 1976, **65**, 3792.

ment of water with TiO_2 and $SrTiO_3$ photoanodes.[814] A He–Cd laser emitting at 325 nm has been used for excitation of colchicine at low absorbances.[815]

In the gas phase, photobromination of MeF at low pressures has been monitored by mass spectrometry[816] and absolute quantum yields for photolysis of H_2CO and D_2CO have been determined in a careful study using a frequency-doubled dye laser as the excitation source.[817] (Films of H_2CO–D_2CO mixtures at 77 K have also been photolysed at 123.6 and 147 nm.[818])

In the solid phase, conversion of Fe^{II}–diamine complexes to the Fe^{III} form has been achieved on irradiation at room temperature in a molten salt of $AlCl_3$ and ethylpyridinium bromide.[819] At low temperatures, a single-beam arrangement has been used for the photolysis and determination of quantum yields for the photolysis of $Mo(CO)_6$ in CH_4, Ar, and CO matrices,[820] photoisomerization of azoethane induced by 313—390 nm radiation has been studied in a hydrocarbon glass at 77 K,[821] and the photolysis of ClNO in solid oxygen at 10 K shown to produce chlorine nitrate in high yield.[822] Partially oriented samples of $Fe(CO)_5$ in solid CO at 20 K have been prepared by irradiation with polarized u.v. radiation[823] and the i.r. spectrum of CF_2Br free radical determined following photolysis of matrix-isolated CF_2Br_2.[824] The u.v. absorption spectrum of Ar–Xe–O_3 mixtures on a Cary 17 spectrometer subjected to mercury arc photolysis showed a marked decrease in the O_3 absorption at 250 nm and the appearance of a weak new band at 227 nm (attributed to a charge-transfer transition of XeO).[825]

Quantum yields of fluorescence and of triplet molecule formation have been determined for the xylenes and the ethyltoluenes using the biacetyl method.[826] The photoaddition of an aldehyde to an α-diketone gave a product mixture which was affected considerably by a variation in the light intensity.[827] An expression derived for the treatment of quenching data has shown that under appropriate conditions, plots of Φ_0/Φ_q for product formation, measured at constant ratios of concentration, can give information from which the rates of excitation transfer from a sensitizer to a substrate and to a quencher can be derived.[828]

Flow Experiments.—Flow methods for chemical reaction coupled with photometric techniques such as chemiluminescence, atomic absorption *etc.* are

[814] A. B. Bocarsly, J. M. Bolts, P. G. Cummins, and M. S. Wrighton, *Appl. Phys. Letters*, 1977, **31**, 568.
[815] R. Croteau and R. M. Leblanc, *J. Luminescence*, 1977, **15**, 353.
[816] L. N. Krasnopyorov and V. N. Panfilov, *Chem. Phys.*, 1977, **25**, 375.
[817] J. H. Clark, C. B. Moore, and N. S. Nogar, *J. Chem. Phys.*, 1978, **68**, 1264.
[818] S. Glicker, *J. Chem. Phys.*, 1976, **65**, 4426.
[819] H. L. Chum, D. Koran, and R. A. Osteryoung, *J. Amer. Chem. Soc.*, 1978, **100**, 310.
[820] M. Poliakoff, *J.C.S. Faraday II*, 1977, **73**, 569.
[821] V. I. Pergushov, O. N. Bormot'ko, and V. S. Gurman, *Chem. Phys. Letters*, 1977, **51**, 269.
[822] D. E. Tevault and R. R. Smardewski, *J. Phys. Chem.*, 1978, **82**, 375.
[823] J. K. Burdett, J. M. Grzybowski, M. Poliakoff, and J. J. Turner, *J. Amer. Chem. Soc.*, 1976, **98**, 5729.
[824] M. E. Jacox, *Chem. Phys. Letters*, 1978, **53**, 192.
[825] B. S. Ault and L. Andrews, *Chem. Phys. Letters*, 1976, **43**, 350.
[826] S. A. Lee, *J. Chem. Phys.*, 1978, **68**, 602.
[827] M. B. Rubin and S. Inbar, *J. Amer. Chem. Soc.*, 1978, **100**, 2266.
[828] C. S. Ilenda, R. J. Daughenbaugh, and S. J. Cristol, *Mol. Photochem.*, 1976, **7**, 287.

extremely useful for studies on small molecules in the gas phase. The following examples are representative of recent studies in this general area.

A mixture of NO, H_2, and Hg vapour illuminated in a flow system by intensity-modulated 253.7 nm radiation produced luminescence from $HNO('A'')$ which was measured by conventional phase methods.[829] The modulation amplitudes and phase angles of luminescence were found to be functions of wavelength, [NO], and the modulation frequency and indicated the involvement of two distinct $HNO('A'')$ precursors, H atoms, and HgH radicals. NO fluorescence resulting from reaction of NO with ozone excited with a pulsed CO_2 laser was detected in a rapidly flowing mixture.[930] A discharge flow system was used to detect NO chemiluminescence resulting from the reaction of $N(^4S)$ with $O(^3P)$.[831] I.r. emission from NO was observed from $N(^4S) + O_2$ with a PbS detector.[832] A similar flow system used with resonance fluorescence detection enabled reactions between OH radicals and various halocarbons to be followed.[833] Chemiluminescence observed in the reaction between hydrogen and molecular oxygen showed emission peaks in the i.r. between 1.2 and 1.5 μm (which were ascribed to transitions involving HO_2), while that at 762 nm was due to an O_2 transition.[834]

The cross-sections of ClO produced in a flow reactor have been determined by conventional u.v. spectrometry[835] and a discharge flow reactor used to examine the kinetics of formation of $ClONO_2$ from reaction between ClO and NO_2.[836] Rate constants have been determined for the reactions of Cl atoms with O_2 and NO using resonance fluorescence detection at 134.7 nm[837] and H_2O_2, H_2O, O, CH_4, and HNO_3 with mass spectrometric detection.[838] $E \rightarrow V$ energy transfer from $Br(4^2P_{\frac{1}{2}})$ to H_2O has been studied after pulsed dye laser photolysis of a flowing gas mixture of Br_2, H_2O, and Ar[839] and the Br (or Cl) atom resonance radiation emitted following collision of Br_2 (or Cl_2) with Ar* metastable atoms has also been investigated.[840] In the latter case, both resonance fluorescence and absorption measurements were made on atomic species at concentrations of 10^{13} cm^{-3}. Emission arising from the reaction of O with tetrafluoroethylene in a fast flow reactor has been tentatively assigned to the $CF_2(^3B_1) \rightarrow CF_2(^1A_1)$ phosphorescence transition.[841]

Decay rates of $Ar(^3P_2)$, $Ar(^3P_2)$, and $Xe(^3P_2)$[842] and $Ne(^3P_2)$[843] have been measured in a flowing afterglow apparatus and the gas phase reactions of

[829] K. Oka, D. L. Singleton, and R. J. Cvetanovic, *J. Chem. Phys.*, 1977, **66**, 713.
[830] J. Moy, E. Bar-Ziv, and R. J. Gordon, *J. Chem. Phys.*, 1977, **66**, 5439.
[831] I. M. Campbell and R. S. Mason, *J. Photochem.*, 1976, **5**, 383.
[832] M. E. Whitson, L. A. Darnton, and R. J. McNeal, *Chem. Phys. Letters*, 1976, **41**, 552.
[833] J. S. Chang and F. Kaufman, *J. Chem. Phys.*, 1977, **66**, 4989.
[834] J. R. Hislop and R. P. Wayne, *J.C.S. Faraday II*, 1977, **73**, 506.
[835] P. Rignaud, B. Leroy, G. LeBras, G. Poulet, J. L. Jourdain, and J. Combourieu, *Chem. Phys. Letters*, 1977, **46**, 161.
[836] M. S. Zahniser, J. S. Chang, and F. Kaufman, *J. Chem. Phys.*, 1977, **67**, 997.
[837] M. S. Zahniser and F. Kaufman, *J. Chem. Phys.*, 1977, **68**, 3673.
[838] M. T. Leu and W. B. Demore, *Chem. Phys. Letters*, 1976, **41**, 121.
[839] A. Haviri and C. Wittig, *J. Chem. Phys.*, 1978, **68**, 2109.
[840] M. A. A. Clyne and D. J. Smith, *J.C.S. Faraday II*, 1978, **74**, 263.
[841] S. Koda, *Chem. Phys. Letters*, 1978, **55**, 353.
[842] J. H. Kolts and D. W. Setser, *J. Chem. Phys.*, 1978, **68**, 4848.
[843] J. M. Brom, J. H. Kolts, and D. W. Setser, *Chem. Phys. Letters*, 1978, **55**, 44.

$Ge(^3P_{0,1})$ and N_2O studied by atomic absorption spectroscopy in a flow tube system at 350 K.[844]

Laser-induced Chemistry.—Since its discovery in the early 1970's (N. R. Isenor and M. C. Richardson, *Appl. Phys. Letters*, 1971, **18**, 225), laser-induced dissociation of polyatomic molecules has been the subject of many investigations (for reviews see refs. 1 and 845). Interest in this dissociation became widespread when it was shown to be isotopically selective.[846] The high average power, tunability, and narrow spectral linewidth of available lasers makes them ideal for exciting a specific absorption line of a single isotope or molecule in a mixture and the species may be removed selectively by chemical reaction or some other process (for a review on laser isotope separation see ref. 847). Other appropriate reviews in this area include sources of atomic and molecular data for applications of laser chemistry,[848] lasers for uranium isotope separation,[849] and theoretical aspects of dissociation.[850]

With few exceptions, most of the investigations in laser-induced chemistry are concerned with two broad types of reaction: three-centre systems (A + B*C → AB + C) in which the diatom is vibrationally excited by a relatively small number of quanta and with unimolecular conversions of polyatomics, which are superexcited to states well above the critical level for dissociation isomerization (the so-called multiphoton processes). In one example, a bimolecular reaction has been reported between a highly excited polyatomic (at a level of excitation below that required for fragmentation) and attacking radicals, viz. SF_6 and SiH_4. A single unfocused pulse from a CO_2 laser was used to initiate an explosion in the mixture ($\rightarrow S_2^* + SiH_4 + HF + H_2$).[851] Krypton resonance radiation at 123.58 nm was shown to excite the ^{13}CO transition selectively in a natural isotopic composition of CO resulting in ^{13}C enrichment of 10-fold and 60-fold in the photolysis products CO_2 and C_3O_2, respectively.[852] The overall efficiency of laser-induced separation in deuterium using a one-step separation process has been considered in an arrangement using a 1 km path

844 P. M. Swearengen, S. J. Davis, and T. M. Niemcyzk, *Chem. Phys. Letters*, 1978, **55**, 274; P. M. Swearengen, S. J. Davis, S. G. Hadley, and T. M. Niemcyzk, *ibid.*, 1977, **49**, 571.
845 V. S. Letokhov, *Physics Today*, 1977 (5), 23; S. R. Leone, *Proc. Soc. Photo-Opt. Instrum. Eng.*, 1976, **92**, 130; E. K. C. Lee, *Accounts Chem. Res.*, 1977, **10**, 319; R. V. Ambartzumian and V. S. Letokhov, in 'Chemical and Biological Applications of Lasers', ed. C. B. Moore, vol. 3, Academic Press, New York, 1977.
846 R. V. Ambartzumian, V. S. Letokhov, E. A. Ryabov, and N. V. Chekalin, *J.E.T.P. Letters*, 1974, **20**, 597; J. L. Lyman, R. J. Jensen, J. Rink, C. P. Robinson, and S. D. Rockwood, *Appl. Phys. Letters*, 1974, **27**, 273.
847 V. S. Letokhov and C. B. Moore, in 'Chemical and Biological Applications of Lasers', ed. C. B. Moore, vol. 3, Academic Press, New York, 1977.
848 D. R. Lide, *Laser Focus*, 1977 (2), 53.
849 E. V. George and W. F. Krupke, *Proc. Soc. Photo-Opt. Instrum. Eng.*, 1976, **86**, 140.
850 D. P. Hodgkinson and J. S. Briggs, *Chem. Phys. Letters*, 1976, **43**, 451; C. D. Cantrell and K. Fox, *J. Opt. Soc. Amer.*, 1978, **68**, 693.
851 S. H. Bauer and J. A. Haberman, *I.E.E.E. J. Quantum Electron.*, 1978, **QE-14**, 233; S. H. Bauer, E. Bar-Ziv, and J. A. Haberman, *ibid.*, 1978, **QE-14**, 238.
852 A. C. Vikis, *Chem. Phys. Letters*, 1978, **53**, 565.

length.[853] Two-photon ionization has been employed for isotope separation in some rare earth elements[854] and Li_2 beams.[855]

Large molecules such as SF_6, OsO_4,[856] and BCl_3 have been dissociated with intensely focused, high power i.r. lasers (see ref. 1). The initial few photons are absorbed on specific vibrational transitions, the first of which can be isotopically specific. When the molecule is excited to the third or fourth fundamental vibration, the density of states becomes so high that there is nearly a broad band continuum of vibrational levels and in an intense field the molecule can quickly absorb 50 or 100 more photons. The highly excited molecule has now more than enough energy to dissociate eventually along one of its bonds.

The most studied unimolecular reaction induced by i.r. has been the dissociation of SF_6. There have been numerous studies [857] of isotopic distribution [858] and product analysis [859] as well as the effects of laser intensity,[860] wavelength,[861] pressure,[862] and the number of photons absorbed.[863] The dense and complex absorption features of this (and other large molecules) have only been recorded recently following the application of tunable diode lasers (see earlier p. 44 and ref. 864). I.r.-photosensitized decomposition of SF_6 mixtures with hydrocarbons [865] and NO and O_3 [866] has also been observed, in the last example, by an increase in the visible NO_2 emission when the laser was tuned to an SF_6 absorption line.[866] Ions produced by photolysis of flowing SF_6–hydrocarbon mixtures have been attributed to a reactive mechanism involving F atoms and

[853] J. C. Vanderleeden, *Appl. Optics*, 1978, **17**, 785.
[854] N. V. Karlov, B. B. Krynetskii, V. A. Mishin, and A. M. Prokhov, *Appl. Optics*, 1978, **17**, 857; N. V. Karlov, B. B. Krynetskii, V. A. Mishin, A. M. Prokhorov, A. D. Savelev, and V. V. Smirnov, *Optics Comm.*, 1977, **21**, 384.
[855] E. W. Rothe, B. P. Mathur, and G. P. Reck, *Chem. Phys. Letters*, 1978, **53**, 74; B. P. Mathur, E. W. Rothe, G. P. Reck, and A. J. Lightman, *ibid.*, 1978, **56**, 336.
[856] R. V. Ambartzumian, V. S. Letokhov, G. N. Makarov, and A. A. Puretzky, *Optics Comm.*, 1978, **25**, 69; R. V. Ambartzumian, Y. A. Gorokhov, G. N. Makarov, A. A. Puretzky, and N. P. Furzikov, *Chem. Phys. Letters*, 1977, **45**, 231.
[857] A. S. Akhmanov, V. Y. Baranov, V. D. Pismenny, V. N. Bagratasvili, Y. R. Kolomiisky, V. S. Letokhov, and E. A. Ryabov, *Optics Comm.*, 1977, **23**, 357; Y. J. Kaufman and U. P. Oppenheim, *Appl. Optics*, 1977, **16**, 1931; J. L. Lyman, S. D. Rockwood, and S. M. Freund, *J. Chem. Phys.*, 1977, **67**, 4545.
[858] J. L. Wyman, J. W. Hudson, and S. M. Freund, *Optics Comm.*, 1977, **21**, 112; M. C. Gower and K. W. Billman, *ibid.*, 1977, **20**, 123.
[859] F. Brummer, T. P. Cotter, K. L. Kompa, and D. Proch, *J. Chem. Phys.*, 1977, **67**, 1547; E. R. Grant, M. J. Coggiola, Y. T. Lee, P. A. Schulz, A. S. Sudbo, and Y. R. Shen, *Chem. Phys. Letters*, 1977, **52**, 595; S. T. Lin and A. M. Ronn, *ibid.*, 1978, **56**, 415.
[860] J. D. Campbell, G. Hancock, and K. H. Welge, *Chem. Phys. Letters*, 1976, **43**, 581; D. R. Keefer, J. E. Allen, and W. B. Person, *ibid.*, 1976, **43**, 394; H. Stafast, W. E. Schmid, and K. L. Kompa, *Optics Comm.*, 1977, **21**, 121; F. Brunner and D. Proch, *J. Chem. Phys.*, 1978, **68**, 4936; P. Kolodner, C. Winterfield, and E. Yablonovitch, *Optics Comm.*, 1977, **20**, 119; J. Dupre, J. Dupre-Maquaire, P. Pinson, and C. Meyer, *Infrared Physics*, 1978, **18**, 185.
[861] W. Fub, D. Proch, and W. Schmid, *J. Opt. Soc. Amer.*, 1978, **68**, 710.
[862] P. Bado and H. van der Bergh, *J. Chem. Phys.*, 1978, **68**, 4188; S. T. Lin, S. M. Lee, and A. M. Ronn, *Chem. Phys. Letters*, 1977, **49**, 526.
[863] D. O. Ham and M. Rothschild, *Optics Letters*, 1977, **1**, 28; M. Rothschild, W.-S. Tsay, and D. O. Ham, *Optics Comm.*, 1978, **24**, 327.
[864] R. S. McDonald, *Proc. Soc. Photo-Opt. Instrum. Eng.*, 1977, **113**, 160.
[865] M. Rothschild, W. S. Tsay, and D. O. Ham, *J. Opt. Soc. Amer.*, 1978, **68**, 683; J. Tardieu de Maleissye, F. Lempereur, C. Marsal, and R. I. Ben-Aim, *Chem. Phys. Letters*, 1976, **42**, 46; K. Nagai and M. Katayama, *ibid.*, 1977, **51**, 329; B. J. Orr and M. V. Keentok, *ibid.*, 1976, **41**, 68.
[866] E. Bar. Ziv and R. J. Gordon, *Chem. Phys. Letters*, 1977, **52**, 355.

hydrocarbon radicals.[867] Studies on related compounds to SF_6 included SF_5Cl and S_2F_{10}.[868] Multiphoton absorption has also been reported in the ν_3 band of UF_6 at 16 μm[869] and isotopically selective photodissociation of SeF_6 reported using the output of an optically pumped NH_3 laser at 337 μm.[870]

In addition to CO_2 lasers, pulsed HF and DF lasers can produce sufficiently strong i.r. fields to induce unimolecular decomposition of formic acid with the added advantages for separation of hydrogen isotopes in higher photon energies and larger isotope shifts at this lower excitation frequency.[871] In a recent example, equimolar mixtures of HCOOH and HCOOD irradiated by a pulsed HF laser provided a physically separated product (H_2) which had a 25-fold enrichment in H *versus* D as compared with the formic acid mixture.[871] Isotopically enriched ozone (in ^{17}O and ^{18}O) was found following irradiation of an oxygen cell with an ArF laser in which the $^{16}O_2$ Schumann–Runge lines had been removed from the broad laser emission by simple absorption through air.[872] New u.v. photodissociation of HCHO with strong ion laser transitions in Xe^{III} at 345 nm and Ne^{II} at 332 and 228 nm was shown to be a highly selective method for enrichment of rare stable isotopes (^{17}O, ^{18}O, ^{13}C, and D) contained in this molecule.[873] For example, ^{18}O was enriched up to nine-fold using natural neon lasing at 332.374 nm, while substituting ^{22}Ne permitted up to nine-fold enrichment of ^{17}O at 332.371 nm.

Analysis of isotopically selective induced reactions of CF_2Cl_2 and Br_2CF_2 with olefins[874] first suggested the idea of CF_2 as an isotopically specific reactive intermediate for organic synthesis.[875] Chemical reactions of excited species from CH_2F_2 also showed evidence of the involvement of CF_2.[876] Significant enrichment of the ^{35}Cl and ^{37}Cl isotopes in chlorine produced by the photolysis of CF_2Cl_2 occurred at pressures up to several hundred Torr.[877] Other studies have included product analysis from the photolysis of CF_2Cl_2,[878, 879] $CFCl_3$,[879] CCl_3F_3,[880] CF_3I,[881] alkyl halides in helium,[882] and an octafluorocyclobutane[883] together with evidence for photoelimination from vinyl compounds,[884]

[867] F. F. Crim, G. H. Kwei, and J. L. Kinsey, *Chem. Phys. Letters*, 1977, **49**, 526.
[868] K. M. Leary, J. L. Lyman, L. B. Asprey, and S. M. Freund, *J. Chem. Phys.*, 1978, **68**, 1671; J. L. Lyman and K. M. Leary, *J. Opt. Soc. Amer.*, 1978, **68**, 683.
[869] A. Kaldor, P. Rabinowitz, D. M. Cox, J. A. Horsley, and R. Brickman, *J. Opt. Soc. Amer.*, 1978, **68**, 684.
[870] J. J. Tiee and C. Wittig, *Appl. Phys. Letters*, 1978, **32**, 236.
[871] D. K. Evans, R. D. McAlpino, and F. K. McClusky, *Chem. Phys.*, 1978. **32**, 81.
[872] R. K. Sanders, T. R. Loree, S. D. Rockwood, and S. M. Freund, *Appl. Phys. Letters*, 1977, **30**, 150.
[873] J. Marling, *J. Chem. Phys.*, 1977, **66**, 4200.
[874] J. J. Ritter, *I.E.E.E. J. Quantum Electron.*, 1977, **QE-13**, 97D.
[875] J. J. Ritter, *J. Amer. Chem. Soc.*, 1978, **100**, 2441.
[876] S.-T. Lin and A. M. Ronn, *Chem. Phys. Letters*, 1978, **53**, 255.
[877] R. E. Huie, J. T. Herron, W. Braun, and W. Tsang, *Chem. Phys. Letters*, 1978, **56**, 193.
[878] G. Folcher and W. Braun, *J. Photochem.*, 1978, **8**, 341; V. Slezak, J. Caballero, A. Burgos, and E. Quel, *Chem. Phys. Letters*, 1978, **54**, 105.
[879] J. W. Hudgens, *J. Chem. Phys.*, 1978, **68**, 777.
[880] D. F. Dever and E. Grunwald, *J. Amer. Chem. Soc.*, 1976, **98**, 5055.
[881] S. Bittenson and P. L. Houston, *J. Chem. Phys.*, 1977, **67**, 4819.
[882] W. Braun and W. Tsang, *Chem. Phys. Letters*, 1976, **44**, 354.
[883] J. M. Preses, R. E. Weston, and G. W. Flynn, *Chem. Phys. Letters*, 1977, **46**, 69.
[884] C. R. Quick and C. Wittig, *Chem. Phys.*, 1978, **32**, 75; F. M. Lussier and J. I. Steinfeld, *Chem. Phys. Letters*, 1977, **50**, 175; C. Willis, A. Gandini, and R. A. Back, *J. Opt. Soc. Amer.*, 1978, **68**, 692.

and selective isomerization in dichloroethylene.[885] The isotopic selectivity of CH_3F photobromination with a CO_2 laser was found to be 2.5 if the laser excited $^{13}CH_3F$ and 1.4 on excitation of $^{12}CH_3F$.[886] Laser-induced reactions between a number of methyl halides and chlorine occurring at frequencies far removed from the methyl halides resonances indicated that Stark broadening plays an important role in the absorption process.[887] The ratio [HCl] : [DCl] for the decomposition of CH_2DCH_2Cl was found to be independent for changes in beam energy, beam geometry, rotational lines, irradiation time, and pressure.[888]

By selecting an appropriate laser frequency, enrichment of either of two bromine isotopes could be obtained by vibrationally enhancing a bimolecular step in the radical chain reaction of CH_3Br with atomic chlorine.[889] Rates of the $Br + Br_2$ halogen atom exchange to the $Br + HI$ reaction have been determined following excitation of $^{81}Br_2$ with a spectrally narrowed (1.2 GHz) Q-switched Nd : YAG laser at 558 nm[890] and elementary processes in an ortho–para molecular iodine mixture were studied during selective excitation with a 514.5 nm Ar^+ laser.[891] By exciting HBr to $\nu > 2$ with a dye laser, it was possible to increase the rate of reaction with I ($\rightarrow$ HI + Br) by a factor of 10^9.[892]

In the collisionless multiphoton dissociation of MeOH, the appearance of CH and OH radicals was monitored in time by resonance fluorescence and it was noted that the OH appeared at 50 ± 20 ns after the laser pulse (and independently of the initial MeOH pressure), whereas the CH appeared after a further 70 ± 20 ns.[893] Molecular decomposition products (CO and H_2) have also been identified from pulsed irradiation of MeOH.[894] Photofragment concentrations of H_2, HD, and D_2 from HCHO, HCDO, and DCDO have been measured under various irradiation conditions with 20 MW pulses from a CO_2 laser[895, 896] and highly selective photodissociation found to occur. A deuterium enrichment factor of 40 was discovered after 300 laser pulses.[895]

Final product analysis of chemical reactions of ethylene, chloroethylenes, and chloroethanes showed that the principal reaction mechanism involved elimination of HCl, H_2, or Cl_2.[897] The mechanisms responsible for the production of $C_2(a^3\pi_u)$ following the i.r. photolysis of 97.2% $^{12}CH_2^{13}CH_2$ were elucidated

[885] R. V. Ambartzumian, N. V. Chekalin, V. S. Doljikov, V. S. Letokhov, and V. N. Lokhman, *J. Photochem.*, 1976, **6**, 55; R. V. Ambartzumian, N. V. Chekalin, V. S. Doljikov, V. S. Letokhov, and V. N. Lokhman, *Optics Comm.*, 1976, **18**, 400.
[886] Y. N. Molin, V. N. Panfilov, and V. P. Strunin, *Chem. Phys. Letters*, 1978, **56**, 557.
[887] B. L. Earl and A. M. Ronn, *Chem. Phys. Letters*, 1976, **41**, 29.
[888] A. J. Colussi, S. W. Benson, R. J. Hwang, and J. J. Tiee, *Chem. Phys. Letters*, 1977, **52**, 349.
[889] T. J. Manuccia, M. D. Clark, and E. R. Lory, *J. Chem. Phys.*, 1978, **68**, 2271.
[890] F. Zaraga, S. R. Leone, and C. B. Moore, *Chem. Phys. Letters*, 1976, **42**, 275.
[891] V. I. Balykin, V. S. Letokhov, V. I. Mishin, and V. A. Semchishen, *Chem. Phys.*, 1976, **17**, 111.
[892] C. C. Badcock, W. C. Hwang, and J. F. Kalsch, *Chem. Phys. Letters*, 1977, **50**, 381.
[893] S. E. Bialkowski and W. A. Guillory, *J. Chem. Phys.*, 1978, **68**, 3339.
[894] S. E. Bialkowski and W. A. Guillory, *J. Chem. Phys.*, 1977, **67**, 2061.
[895] G. Koren, U. P. Oppenheim, D. Tal, M. Okon, and R. Weil, *Appl. Phys. Letters*, 1976, **29**, 40.
[896] G. Koren, M. Okon, and U. P. Oppenheim, *Optics Comm.*, 1977, **22**, 351.
[897] K. Nagai and M. Katayama, *Bull. Chem. Soc., Japan*, 1978, **51**, 1269.

using tunable dye laser induced fluorescence of the various C_2 isotopes to characterize various reactions.[898] The CO_2 laser was also shown to induce polymerization[899] and pyrolysis[900] of ethylene. An important paper on the photolysis of cyclopropane and propylene showed that significant decomposition occurred when the reaction cell windows were not scrupulously clean.[901] Under these conditions, photolysis products appeared even when the radiation was tuned far from resonance.

An impressive example reported of photochemistry occurring in the electronic ground state was the isomerization of hexafluorocyclobutene induced by a tunable CO_2 laser to the thermodynamically less stable hexafluorobutadiene.[902] Quantitative conversion resulted from unfocused laser light but focused conditions caused the reaction to be contaminated with decomposition products. With a 400 mJ pulse, 90% conversion was achieved after 2200 pulses.[902]

Removal of $COCl_2$ in BCl_3 (which is commonly used in the electronics industry) was accomplished most effectively using CO_2 laser radiation.[903] Decomposition of phosgene occurred with no detectable BCl_3 dissociation and it was believed that the boron compound acts as an i.r. sensitizer. A similar laser-sensitized reaction was shown to occur in mixtures of BCl_3 and C_2Cl_4 in which the main product, C_6Cl_6, resulted from a trimerization of C_2Cl_4 with elimination of Cl_2 (88% yield) and the BCl_3 remained unchanged.[904] Selective CO_2 laser photochemistry of $^{11}BCl_3$–H_2 mixtures with Ti and Pd catalysts revealed that it was possible to produce a new class of compounds which could not be produced by a thermal reaction or by a catalysis-free photochemical process.[905] Selective photolysis with an ArF laser enabled PH_3, AsH_3, and B_2H_6 to be removed from SiH_4 with quantum yields (for removal) of 0.35, 0.42, and 0.15, respectively.[906]

While much of the work in laser-isotope separation has been directed towards gas phase studies, the problems in separating lanthanides and actinides in solution are similar to those found in isotope separation since the chemical similarity within each group leads to difficult separation by conventional techniques. The first selection photoreduction following absorption of radiation from a nitrogen laser, CW argon laser, and mercury lamp was applied to oxalato complexes of Fe and Co in solution.[907] Later studies enabled mixtures of lanthanides to be separated photochemically with the output of an ArF laser (193 nm) or resonance lines from a low pressure mercury lamp.[908]

Unlike multiphoton dissociation reactions in the gas phase, reactions in low

[898] J. H. Hall, M. L. Lesiecki, and W. A. Guillory, *J. Chem. Phys.*, 1978, **68**, 2247.
[899] S. L. Chin, *Canad. J. Chem.*, 1976, **54**, 2341.
[900] J. Tardieu de Maleissye, F. Lempereur, and C. Marsal, *Chem. Phys. Letters*, 1976, **42**, 472.
[901] Z. Karny and R. N. Zare, *Chem. Phys.*, 1977, **23**, 321.
[902] A. Yogev and R. M. J. Benmair, *Chem. Phys. Letters*, 1977, **46**, 290.
[903] J. A. Merritt and L. C. Robertson, *J. Chem. Phys.*, 1977, **67**, 3545.
[904] H. R. Bachman, R. Rinck, H. Noth, and K. L. Kompa, *Chem. Phys. Letters*, 1977, **45**, 169.
[905] C. T. Lin and T. D. Z. Atvars, *J. Chem. Phys.*, 1978, **68**, 4233.
[906] J. H. Clark and R. G. Anderson, *Appl. Phys. Letters*, 1978, **32**, 46.
[907] T. Donohue, *Chem. Phys. Letters*, 1977, **48**, 119.
[908] T. Donohue in ref. 3, p. 216; *J. Chem. Phys.*, 1977, **67**, 5402; F. J. Gustafson and J. C. Wright, *Analyt. Chem.*, 1977, **49**, 1680.

temperature matrices often have low energy barriers for reaction which can be overcome by individual molecules absorbing only one i.r. photon. Furthermore, i.r. lasers can introduce a very high degree of selectivity into matrix chemistry allowing not only the chemical reaction but also the stereochemistry of a reaction to be followed. For example, i.r. laser irradiation of $^{13}C^{18}O$ enriched $Fe(CO)_4$ isolated in Ar, Xe, and CH_4 matrices at 20 K revealed not only isotopic and stereochemical selectivity but also selectivity between molecules isolated in different substitutional sites in the matrix.[909] With a tunable CW spin-flip Raman laser providing a resolution of *ca.* 0.01 cm^{-1}, complete selectivity was observed between molecules with absorption bands separated by only 0.72 cm^{-1}.[910] Other low temperature studies included isotopically selective photochemistry of *sym*-tetrazine in crystals and rare gas matrices at 4 K,[911] isotope separation of boron isotopes in BCl_3 by selective migration on a liquid nitrogen-cooled plate,[912] and strong H/D isotopic effects observed in the photochemical decomposition of acetaldehyde, acetone, and ethyl bromide at 77 and 4 K.[913] A mercury lamp enabled isotopic effects of C and N to be observed at both low temperature (4 and 77 K) and room temperature following photolysis of benzoyl peroxide and diazoaminobenzene.[914]

Photodissociation Spectroscopy.—The commercial development of rare gas excimer lasers has opened up a new realm in photochemistry in which the high laser intensities are capable of driving photochemical systems. States that normally predissociate now may have the opportunity to absorb one or more photons and can be driven to a higher state. Multiphoton dissociation following irradiation with a 193 nm ArF excimer laser has been observed in C_2H_2, C_2N_2, C_2H_4, CH_3OH, and C_2H_5OH.[915] In all these cases fluorescence from small free radicals such as CN, CH, and OH has been observed (see also p. 84). More conventional product analysis (by mass spectrometer) was used to follow laser photodissociation of molecular beams of IBr,[916] AsI_3,[917] and photodissociation reactions of Ar_2^+, Ne_2^+, and $(CO_2)_2^+$.[918] A photofragment spectrometer designed to probe the internal energy distributions of fragments directly by laser-induced fluorescence[919] has been applied to the decomposition

909   B. Davies, A. McNeish, M. Poliakoff, and J. J. Turner, *J. Amer. Chem. Soc.*, 1977, **99**, 7573; M. Poliakoff, B. Davies, A. McNeish, M. Tranquille, and J. J. Turner, *Ber. Bunsengesellschaft Phys. Chem.*, 1978, **82**, 121; A. McNeish, M. Poliakoff, K. P. Smith, and J. J. Turner, *J.C.S. Chem. Comm.*, 1976, 859; M. Poliakoff, B. Davies, A. McNeish, K. P. Smith, and J. J. Turner, *J. Opt. Soc. Amer.*, 1978, **68**, 693.
910   M. Poliakoff, N. Breedon, B. Davies, A. McNeish, and J. J. Turner, *Chem. Phys. Letters*, 1978, **56**, 474.
911   R. M. Hochstrasser and D. S. King, *J. Amer. Chem. Soc.*, 1976, **98**, 5443; D. S. King, B. Dellinger, R. M. Hochstrasser, and A. B. Smith, *I.E.E.E. J. Quantum Electron.*, 1977, **QE-13**, 98D.
912   K. Wada, *I.E.E.E. J. Quantum Electron.*, 1977, **QE-13**, 9D.
913   R. Z. Sagdeev, A. B. Doktorov, V. V. Pervukhin, A. A. Obynochny, S. V. Camyshan, and V. M. Moralyov, *Chem. Phys.*, 1978, **29**, 311.
914   R. Z. Sagdeev, A. A. Obynochny, V. V. Pervukin, Y. N. Molin, and Y. M. Moralyov, *Chem. Phys. Letters*, 1977, **47**, 292.
915   W. M. Jackson, J. B. Halpern, and C.-S. Lin, *Chem. Phys. Letters*, 1978, **55**, 254.
916   M. S. De Vries, N. J. A. Van Veen, and A. E. De Vries, *Chem. Phys. Letters*, 1978, **56**, 15.
917   M. Kawaski and R. Bersohm, *J. Chem. Phys.*, 1978, **68**, 2105.
918   M. L. Verstal and G. H. Mauclaire, *Chem. Phys. Letters*, 1976, **43**, 499; *J. Chem. Phys.*, 1977, **67**, 3758, 3767.
919   A. P. Baronavski and J. R. McDonald, *Chem. Phys. Letters*, 1977, **45**, 172.

of ICN and HN_3[920] following laser photolysis at 266 nm. Other studies with ICN photodissociation concentrated on the observation of $CN^+(X^2\Sigma^+)$ by laser-induced fluorescence.[921]

Crossed-beam geometries have been used with a laser-pumped OPO and beam of C^- ions for photodetachment experiments,[922] with pulsed polarized light and C_2H_5ONO to obtain the photofragment spectrum,[923] and with a ruby laser and tunable dye laser to obtain the translational energy spectrum of O^+ ions from the photodissociation of $^{16}O_2^+$ and $^{18}O_2^+$.[924]

The kinetics of radicals produced by photodissociation from pyrocatechol and pyrogallol have been studied using modulated u.v. excitation and either i.r. or e.s.r. spectroscopy for analysis.[925] From measurements of amplitude and phase of transient concentrations as a function of modulation frequency, it was possible to determine the kinetics of photoinduced reactions, lifetimes, and approximate i.r. extinction coefficients of transient radicals.

The combination of ion cyclotron resonance (ICR) spectrometry[926] and tunable laser excitation has been used with great effect to produce photodissociation spectra of hexatriene cations[927] and CH_3Cl^+,[928] and to study the process $C_7H_8^+ + h\nu \rightarrow C_7H_7^+ + H$.[929] An ICR spectrometer fitted with a monochromator and synchronous photon counting equipment has also been used to study luminescence in near-thermal charge exchange of He^+ with H_2O and D_2O.[930]

Undergraduate Experiments.—It is pleasing to see a variety of undergraduate experiments in photochemistry being suggested. Publications have included several straightforward preparative reactions such as the photobromination of cinnamic acid at 436 nm[931] and the photohydration of pyridine in aqueous solution with a low pressure mercury lamp.[932] Stern–Volmer plots were introduced in quenching experiments of anthracene by CCl_4[933] and uranyl ions by alcohol.[934] The fluorescence of biacetyl in CCl_4 solution formed the basis of a study of absorption and self-absorption quenching.[935] Other luminescence

[920] J. R. McDonald, R. G. Miller, and A. P. Baronavski, *Chem. Phys. Letters*, 1977, **51**, 57.
[921] M. J. Sabety-Dzvonik and R. J. Cody, *J. Chem. Phys.*, 1977, **66**, 125; G. A. West and M. J. Berry, *Chem. Phys. Letters*, 1978, **56**, 423.
[922] D. Feldman, *Chem. Phys. Letters*, 1977, **47**, 338.
[923] A. F. Tuck, *J.C.S. Faraday II*, 1977, **73**, 689.
[924] A. Tabche-Fouhaille, J. Durup, J. T. Moseley, J. B. Ozenne, C. Pernot, and M. Tadjeddine, *Chem. Phys.*, 1976, **17**, 81.
[925] M. Forster, K. Loth, M. Andrist, U. P. Fringeli, and H. H. Gunthard, *Chem. Phys.*, 1976, **17**, 56.
[926] 'Ion Cyclotron Resonance Spectrometry', by T. A. Lerman and M. M. Bursey, Wiley, New York, 1976.
[927] R. C. Dunbar and H. Ho-I. Teng, *J. Amer. Chem. Soc.*, 1978, **100**, 2279. R. C. Dunbar, *ibid.*, 1976, **98**, 4671.
[928] R. G. Orth and R. C. Dunbar, *J. Chem. Phys.*, 1978, **68**, 3254.
[929] J. R. Eyler, *J. Amer. Chem. Soc.*, 1976, **98**, 6831.
[930] T. R. Govers, M. Gerald, and R. Marx, *Chem. Phys.*, 1976, **15**, 185.
[931] M. R. F. Bazley and G. R. Wooley, *J. Chem. Educ.*, 1977, **54**, 771.
[932] J. C. Andre, M. Niclause, J. Joussot-Dubien, and X. Deglise, *J. Chem. Educ.*, 1977, **54**, 387.
[933] M. W. Legenza and C. J. Marzzacco, *J. Chem. Educ.*, 1977, **54**, 183.
[934] H. D. Burrows and S. J. Formosinho, *J. Chem. Educ.*, 1978, **55**, 125.
[935] G. Henderson, *J. Chem. Educ.*, 1977, **54**, 57.

demonstrations have included triboluminescence from uranyl salts [936] and room temperature phosphorescence.[937]

It is perhaps a sign of the sophisticated techniques now being used in photochemistry that undergraduate experiments on N-atom recombination [937] and isotope separation in SF_6 (including the construction of a pulsed CO_2 laser) [938] are now being suggested.

7 Light Detection and Measurement

Photomultipliers.—High sensitivity light measurements in the u.v. and visible regions are still best undertaken by photomultipliers used with photon counting techniques and a commendable review on this combination discusses current detectors and their use.[939] The RCA C31034 and C31034A photomultiplier tubes with gallium arsenide photocathodes have been assessed for use in photon correlation spectroscopy and details given of their pulse height distributions, overall quantum efficiency, and dark-count properties.[940] An improved InGaAsP material developed for near-i.r. photocathode applications was reported to have a quantum efficiency of 9% per incident photon at 1.06 μm at room temperature,[941] but the same material used in Varian photomultipliers was reported to show a deterioration with time unless tubes were refrigerated continuously.[942] In this case, the quantum efficiency was flat at 10% from 170 to 450 nm but fell to 0.02% at 1.1 μm.[942] The u.v. response of a semi-transparent visible sensitive photocathode showed a response at the Lyman-α line (121.6 nm) essentially equal to the peak response in the visible.[943] The quantum efficiency of caesium telluride semi-transparent photocathodes from 120 to 800 nm was shown to be enhanced by about 16% by application of high electric fields.[944]

One of the major problems in using photon counting systems which accurately determine the relative number of photons per pulse is the existence of inevitable gain shifts that can occur during the course of a long experiment (*e.g.* hours), which will lead to an uncertainty and/or smearing in the photon number distribution. A stable pulsed light source such as an LED is required [945,946] and one has been designed to monitor the counting system.[945] Used with a temperature-compensating current pump, the temperature stability of 200 ns pulses was found to be 0.05% °C^{-1}. Measurements of transient noise in a photomultiplier tube resulting from illumination by a short pulse of low intensity light have been reported earlier (Vol. 8, p. 27) and recent work [947] has found that transient noise in either the EMI 9558 or 9569 tubes are more than two-orders of magnitude smaller than that reported for a Philips 56TUVP tube. Abrupt pulse height

[936] J. Martin Gill and F. J. Martin Gill, *J. Chem. Educ.*, 1978, **55**, 340.
[937] E. M. Schulman, *J. Chem. Educ.*, 1976, **53**, 522.
[938] C. R. Quick and C. Wittig, *J. Chem. Educ.*, 1977, **54**, 705.
[939] J. W. Longworth, *Photochem. and Photobiol.*, 1977, **26**, 665.
[940] E. Gulari and B. Chu, *Rev. Sci. Instr.*, 1977, **48**, 1560.
[941] J. S. Escher, G. A. Antypas, and J. Edgecumbe, *Appl. Phys. Letters*, 1976, **29**, 153.
[942] W. A. Feibelman, *Appl. Optics*, 1977, **16**, 800.
[943] S. Sobieski, *Appl. Optics*, 1977, **16**, 2298.
[944] C. I. Coleman, *Appl. Optics*, 1978, **17**, 1789.
[945] E. C. Hager and P. C. Eklund, *Rev. Sci. Instr.*, 1976, **47**, 1144.
[946] D. R. Andrews and L. J. Wallis, *J. Phys. (E)*, 1977, **10**, 95.
[947] B. R. Clemesha, *J. Phys. (E)*, 1977, **10**, 814.

changes and different steady state pulse heights observed with the 8850 and 8575 tubes have been shown to be identical in nature with hysteresis effects.[948] An electronic gate has been described which will pass only the Poisson distributed pulses from noise bursts on a real time basis.[949] Further papers on pulse counting deal with the relationship between optimum and maximum count rates and overall precision, deadtime uncertainty and count rate for high precision work,[950] a circuit of a low cost preamplifier/discriminator for photon counting with an EMI 9813 photomultiplier,[951] and stepping motor programmer designed to control a monochromator during photon counting.[952]

It was noted that the addition of ferrite beads concentric with the active dynode leads and direct connection to dynode 5 in the RCA IP28 photomultiplier tube resulted in a very fast response and a significant improvement in the signal-to-noise ratios in laser flash photolysis.[953]

A photomultiplier gating circuit was required in order to reduce the effect of substantial scattered light from a coaxial flashlamp used in studies of $P(4^4S_{\frac{3}{2}})$ atom monitoring.[954] A pretrigger pulse, fired 75 μs before the flashlamp, effectively switches off the output by defocusing a pair of dynodes in the photomultiplier tube (an EMI 9816QB) operating normally at a gain of 10^9. A gating circuit for an EMI 9598 tube in which dynodes 1, 3, and 5 were driven negative showed that stable gain returned within 700 ns after the removal of the gating pulse.[955] On a much slower time scale, a light-controlled stepping motor was used to insert a set of neutral density filters in front of a photomultiplier to protect it from damaging laser levels.[956] A photomultiplier was left in operation all the time during a phosphorescence experiment but the signal fed to a lock-in amplifier was recorded only during a fully variable viewing time.[957] A useful circuit for automatically cutting off the high voltage to a photomultiplier tube was designed to prevent accidental overload.[958]

A photomultiplier power supply (1 kV at 5 mA) with a long term stability of $\pm 0.01\%$ over an 8 h run has been described[959] as well as a log ratio device for a low cost absorption spectrometer.[960]

Other Photodetectors.—The detection of a weak laser beam in the presence of ordinary incoherent light as background (*e.g.* sunlight) is a problem of considerable topical importance and an ingenious technique, based on laser coherence, used a frequency-modulated tunable acousto-optical filter to detect CW radiation (about 5×10^{-7} mW) in the presence of an incident background (of 0.1 W).[961]

948 M. Yamashita, *Rev. Sci. Instr.*, 1978, **49**, 499.
949 M. Arnow, *Rev. Sci. Instr.*, 1977, **48**, 1354.
950 J. M. Hayes and D. A. Schoeller, *Analyt. Chem.*, 1977, **49**, 306.
951 H. Guttinger, M. Gautschi, and E. Serrallach, *J. Phys. (E)*, 1976, **9**, 936.
952 D. A. Jackson and P. Schoen, *J. Phys. (E)*, 1978, **11**, 403.
953 R. D. Small and J. C. Scaiano, *J. Phys. Chem.*, 1977, **81**, 2126.
954 D. Husain and P. E. Norris, *J.C.S. Faraday II*, 1977, **73**, 415.
955 M. R. Groves, *Photochem. and Photobiol.*, 1978, **27**, 491.
956 J. S. Rayside, W. B. McLendon, and W. H. Fletcher, *Chem. Instrumentation*, 1976, **7**, 271.
957 F. Gruneis, S. Schneider, and F. Dorr, *J. Phys. (E)*, 1976, **9**, 1013.
958 W. Gough, *J. Phys. (E)*, 1977, **10**, 867.
959 T. Gonda and K. Kimoto, *Rev. Sci. Instr.*, 1977, **48**, 195.
960 A. L. Crumbliss, M. E. McCabe, J. A. Dilts, and H. B. Herman, *J. Chem. Educ.*, 1976, **53**, 529.
961 I. C. Chang, *I.E.E.E. J. Quantum Electron.*, 1978, **QE-14**, 108.

A comparative evaluation of the sensitivity characteristics of u.v.-sensitized silicon vidicon and silicon-intensified target vidicon tubes in the u.v.–visible region reported that both devices had lower sensitivities than a photomultiplier tube.[962] Several photomultipliers and image devices (silicon vidicons, SIT vidicons *etc.*) have been compared theoretically with particular respect to SNR and their use in atomic and molecular fluorescence spectrometry.[963] The conclusion was reached that in the visible (>350 nm) the intensified image devices possess considerable analytical potential combining a multichannel time advantage with a SNR approaching that of a photomultiplier. Other papers on vidicon tubes have reported on their gating characteristics,[964] to produce optical gate widths of 5 ns, their use as multiwavelength detectors for liquid chromatography,[965] and a comparison with high speed i.r. film.[966] Observation of bioluminescence phenomena, which is often limited by the small number of photons emitted, has now been shown to be possible by using an image intensifier coupled to a TV vidicon.[967] The entire visible spectrum can be dispersed by a fast spectrograph, recorded on film, and analysed by means of a digitized densitometer. An integrating camera based on a cooled vidicon detector has been described and treatment given of the competitive lag which limits the performance for small signals.[968] Luminescence spectra of solids at low temperature have been recorded on an optical multichannel analyser equipped with a synchronization interface to allow background subtraction with a repetitive mode-locked laser.[969]

A 1024 element spectroscopic detector known as the self-scanned digicon consisted of a magnetically-focused image intensifier tube with a Reticon silicon photodiode array serving as the photoelectron image detector, amplifier, and intermediate storage device.[970] With this combination, the instrument was shown to have shot noise limited performance and a relative quantum efficiency near to the maximum of 80% for 700—900 nm radiation.[970] A self-scanned photodiode array has also been fibre-optically-coupled to an image intensifier tube.[971] The noise characteristics and linear range of a Reticon 256 photodiode array have been assessed for use in liquid chromatography over the range 227—300 nm.[972] A new imaging optical detector called a Photicon has been described, which is sensitive to single photons over a two-dimensional area.[973]

Detectors for the vacuum–u.v. have been reviewed[974] and descriptions given

[962] N. G. Howell and G. H. Morrison, *Analyt. Chem.*, 1977, **49**, 106.
[963] R. P. Cooney, G. D. Boutilier, and J. D. Winefordner, *Analyt. Chem.*, 1977, **49**, 1048.
[964] J. L. Weber, *Proc. Soc. Photo-Opt. Instrum. Eng.*, 1976, **82**, 60; R. W. Simpson and Y. Talmi, *Rev. Sci. Instr.*, 1977, **48**, 1295.
[965] A. McDowell and H. L. Pardue, *Analyt. Chem.*, 1976, **48**, 1815; R. P. Cooney, T. Vo-Dinh, G. Walden, and J. D. Winefordner, *Analyt. Chem.*, 1977, **49**, 939.
[966] A. R. Lang and A. P. W. Makepeace, *J. Phys. (E)*, 1977, **10**, 1292.
[967] G. T. Reynolds, *Photchem. and Photobiol.*, 1978, **27**, 405.
[968] D. M. Hunten and C. J. Stump, *Appl. Optics*, 1976, **15**, 3105.
[969] E. Ostertag, *Rev. Sci. Instr.*, 1977, **48**, 18.
[970] S. S. Vogt, R. G. Tull, and P. Kelton, *Appl. Optics*, 1977, **16**, 574.
[971] B. R. Sandel and A. L. Broadfoot, *Appl. Optics*, 1976, **15**, 3111.
[972] M. J. Milano, S. Lam, M. Savonis, D. B. Paulter, J. W. Pav, and E. Grushka, *J. Chromatog.*, 1978, **149**, 599.
[973] E. Kellog, S. Murray, U. Briel, and D. Bardas, *Rev. Sci. Instr.*, 1977, **48**, 550.
[974] W. R. Hunter, *Electro-Optical Systems Design*, 1976 (7), 44.

of ceramic channel electron multipliers[975] and microchannel plates[976] for use in this region. The temporal characteristics of microchannel plate detectors, measured with a 530 nm pulse, revealed a response time of 0.2—0.3 ns at a gain of up to 10^6.[977]

New processes and materials for u.v. detection with solid state devices have also been reviewed.[978] By using d.c. biases higher than those recommended by the manufacturers, higher u.v. sensitivities to radiation with gas-filled phototubes have been reported by exploiting the photoionization of excited atoms.[979] A u.v.–visible phosphor (magnesium tungstate having an excitation spectrum similar to the erythema action spectrum) has been used with a 1P39 photodiode as a 'sun-burning' u.v. meter.[980] A simple monitor for phosphor-coated lamps consisted of a cadmium sulphide photoconductor with a u.v. filter (UG 1 glass).[981] A sensitive u.v. detector with an NEP of 10^{-13} W Hz$^{-\frac{1}{2}}$ has been described, which utilized photoionization of excited atoms.[982]

A direct method for measuring absorbance by a linear relationship to an indicating voltage has been developed by combining the law of absorptivity of light with the photovoltaic relationship of a barrier layer cell.[983] A meter adjusted for a f.s.d. of 2 absorbance units could be read consistently with an accuracy of 2% over the entire range 2—0.01 absorbance units. A pair of silicon junction photovoltaic cells have been used in a temperature-independent logarithmic light meter covering the range 10^{-1}—10^4 lux with an accuracy of 1%.[984] A battery-operated radiant energy meter also using a silicon cell was shown to have good linearity over five decades.[985] A photon irradiance meter for measuring the energy available for photosynthesis (*i.e.* total quanta between 350 and 700 nm) used a silicon photodiode with a rhodamine B quantum counter.[986] A low-noise feedback amplifier for high impedance photoconductive detectors[987] and an optical isolation amplifier to measure currents from 10^{-11} to 10^{-6} A at ± 1000 V[988] have also been developed. A digitally-controlled shutter allowed exposure times from ms to h to be set with a jitter time of ± 0.1 ms.[989]

The relative line intensities in a spectrograph have been estimated visually using colour reversal film as the recording medium. Moderate overexposure of such a film becomes apparent as a colour change and the colour of a line, relative to its position in the visible spectrum, can be used as a measure of its

975 T. Masuoka, *Rev. Sci. Instr.*, 1977, **48**, 1284.
976 P. J. K. Langendam and M. J. Van Der Wiel, *J. Phys. (E)*, 1977, **10**, 870.
977 J. P. Boutol, J. C. Delmotte, J. A. Miehe, and B. Sipp, *Rev. Sci. Instr.*, 1977, **48**, 1405.
978 D. Chopra, *Proc. Soc. Photo-Opt. Instrum. Eng.*, 1977, **116**, 10.
979 N. S. Kopeika, R. Gellman, and A. P. Kushelevsky, *Appl. Optics*, 1977, **16**, 2470.
980 D. S. Berger, *Photochem. and Photobiol.*, 1976, **24**, 587.
981 B. L. Diffey and A. Miller, *Phys. Med. Biol.*, 1978, **23**, 514.
982 G. Eytan, N. S. Kopeika, R. Gellman, and A. P. Kushelevsky, *Optics and Quantum Electron.*, 1977, **9**, 354.
983 P. Kundu and A. Kar, *J. Phys. (E)*, 1976, **9**, 974.
984 P. Kundu and A. Kar, *Appl. Optics*, 1978, **17**, 138.
985 N. B. McLaughlin and J. R. Allan, *J. Phys. (E)*, 1976, **9**, 651.
986 D. Spitzer and M. R. Wernand, *Appl. Optics*, 1977, **16**, 12.
987 E. L. Dereniak, R. R. Joyce, and R. W. Capps, *Rev. Sci. Instr.*, 1977, **48**, 392.
988 R. N. J. Chou and A. Castellano, *Rev. Sci. Intr.*, 1977, **48**, 569.
989 G. Beenen, J. R. Rentner, and J. P. Walters, *Appl. Spectroscopy*, 1977, **31**, 444.

intensity.[990] A photographic technique has been described to record u.v.-activated spots on a t.l.c. plate.[991]

Other techniques and/or light-measuring devices have included a means for sensitizing poly(vinyl alcohol) with various organic dyes to visible light,[992] a reflective image intensifier based on the twisted electro-optic effect which can be read out with intense visible light,[993] a KF–LiF crystal showing photoinduced dichroism and birefringence,[994] and a field-averaging spectrograph for remote sensing.[995]

Two useful reviews have covered the state of the art in i.r. detectors[996] and the specific properties of HgCdTe detectors.[997]

A useful comparison of ten commercial PIN silicon photodiodes for noise characteristics was made in order to select an appropriate one for oxygen night glow measurements at 762 nm.[998] An accompanying circuit was designed for current signals as low as 10^{-14} A. Although there is a report that the responsivity of certain types of silicon detectors will increase on exposure to u.v. (by 25% using 5 mV cm^{-2} of 313 nm radiation),[999] fatigue was reported for vacuum-u.v. exposures.[1000] High speed photodiodes[1001]−[1003] were reported to offer risetimes as short as 50—60 ps when used in a matched holder[1001] and operated at 6 kV.[1003] Sampling techniques used with a Nd : YAG laser have enabled the time response and risetimes of two fast photodiodes to be determined,[1004] and a trigger with picosecond time dispersion was claimed for a tunnel diode.[1005] A silicon photovoltaic cell has proved its usefulness as a pyrometer in an ultra-high vacuum system, which could be baked out at 375 °C.[1006]

Avalanche photodiodes covering a range up to 1.6 μm have also been shown to have fast risetimes,[1007, 1008] in one case less than 50 ps.[1007] Germanium diodes have been used for near-i.r. detectors (0.5—1.5 μm) for nanosecond flash photolysis[1008] and gallium-doped Ge detector with a 1 ns risetime has been used to record time-resolved far-i.r. pulses between 50 and 114 μm.[1009] The photocapacitive effect in metal-insulator-semiconductor slabs has been

990   H. M. Cartwright, *J. Phys. (E)*, 1976, **9**, 1022.
991   P. W. Rulon and M. J. Cardone, *Analyt. Chem.*, 1977, **49**, 1640.
992   A. Bloom and W. J. Burke, *Appl. Optics*, 1977, **16**, 2614.
993   W. E. L. Haas, G. A. Dir, J. E. Adams, and I. P. Gates, *Appl. Phys. Letters*, 1976, **29**, 631.
994   W. C. Collins and M. O. Greev, *Appl. Optics*, 1978, **17**, 1550.
995   H. Genda and H. Okayama, *Appl. Optics*, 1978, **17**, 601.
996   H. Levinstein, *Proc. Soc. Photo-Opt. Instrum. Eng.*, 1977, **124**, 52.
997   M. B. Reine and R. M. Broudy, *Proc. Soc. Photo-Opt. Instrum. Eng.*, 1977, **124**, 80.
998   R. C. Schaeffer, *Appl. Optics*, 1976, **15**, 2902.
999   A. R. Schaefer, *Appl. Optics*, 1977, **16**, 1539.
1000 R. L. Ohlhaber, *Appl. Optics*, 1977, **16**, 14.
1001 S. I. Green, *Rev. Sci. Instr.*, 1976, **47**, 1083.
1002 H. Kanbe, Y. Mizushima, T. Kimura, and K. Kajiyama, *J. Appl. Phys.*, 1976, **47**, 3749.
1003 G. Beck, *J. Phys. (E)*, 1976, **9**, 849.
1004 A. Antonetti, A. Migus, M. M. Malley, and G. Mourou, *Optics Comm.*, 1977, **21**, 211.
1005 B. Cunin, J. A. Miehe, B. Sipp, and J. Thebault, *Rev. Sci. Instr.*, 1976, **47**, 1435.
1006 B. K. Lunde and F. A. Schmidt, *Appl. Optics*, 1976, **15**, 2625.
1007 H. D. Law, L. R. Tomasetta, and K. Nakano, *J. Opt. Soc. Amer.*, 1978, **68**, 632.
1008 B. B. Craig, M. A. J. Rodgers, and B. Wood, *J.C.S. Faraday II*, 1977, **73**, 349.
1009 D. E. Evans, R. A. Guinee, D. A. Huckridge, and G. Taylor, *Optics Comm.*, 1977, **22**, 337.

reported and a measured detectivity of about $3 \times 10^{12}\,W^{-1}\,cm\,Hz^{\ddagger}$ measured in a prototype silicon device.[1010] Other i.r. diode detectors described have included a low capacitance PbTe photodiode with a background detectivity of typically $1.7 \times 10^{11}\,W^{-1}\,cm\,Hz^{\ddagger}$ for a peak wavelength near 5 µm,[1011] an InGaAs–InP detector having a high quantum efficiency in the 1.0—1.7 µm range[1012] and which can be composition-tuned from 3.7 to 6.9 µm at 77 K.[1013] Exposure of InSb detectors to visible light at 77 K was reported to induce a slow component to the response time.[1014] A cooled (77 K) intrinsic germanium detector was recommended for high sensitivity measurements of HF emission in the 1.1—1.6 µm range.[1015] A large area pyroelectric detector was reported to have a response flat to $\pm 1\%$ over the entire visible to 12 µm range.[1016] Type IZ ammonia-hypersensitized photographic plates have been used up to 1.315 µm in order to characterize a single transverse mode iodine laser.[1017] A tellurium detector for CO_2 laser radiation was reported to be superior to available photon drag detectors.[1018]

Radiometry and Actinometry.—A brief account of the standard and calibration services at the NPL for measurements of spectral emission of sources and spectral sensitivities of detectors in the u.v., visible, and i.r.[1019] has been accompanied by a corresponding description of work at the NBS on standard tungsten lamps.[1020] Deuterium lamps are commonly recommended as reference sources in the u.v. and an interlaboratory comparison has been made for spectral emission in the 165—300 nm region.[1021] Black-body limited lines of C and N from an argon-wall stabilized arc have been used to determine the spectral radiance of specially developed deuterium lamps at selected wavelengths between 165 and 193 nm, and a further comparison made with a calibration based on synchrotron radiation.[1022] Thirty of the more intense Cu^{II} vacuum-u.v. lines from a liquid nitrogen-cooled hollow-cathode lamp have been recommended for use as wavelength standards.[1023] A simple electronic actinometer using a IP28 photomultiplier tube and voltage-to-frequency converter, which has been proposed as a device to monitor light outputs in an irradiation apparatus,[1024] appears to offer little advantage over the quantum counter device of Schaffner and co-workers (Vol. 8, p. 31). Fatigue and drift in a photomultiplier occur at

[1010] A. Shev, R. K. Crouch, S. S.-M. Lu, W. E. Miller, and J. A. Moriarty, *Appl. Phys. Letters*, 1978, **32**, 713.

[1011] H. Holloway and K. F. Yeung, *Appl. Phys. Letters*, 1977, **30**, 210.

[1012] K. J. Bachmann and J. L. Shay, *Appl. Phys. Letters*, 1978, **32**, 446.

[1013] R. B. Schooler, J. D. Jensen, and G. M. Black, *Appl. Phys. Letters*, 1977, **31**, 620.

[1014] W. A. Seddon, E. B. Selkirk, and J. W. Fletcher, *Radiation Phys. Chem.*, 1978, **11**, 53.

[1015] P. R. Poole and I. W. M. Smith, *J.C.S. Faraday II*, 1977, **73**, 1435.

[1016] G. W. Day, C. A. Hamilton, and K. W. Pyatt, *Appl. Optics*, 1976, **15**, 1865.

[1017] J. C. Farcy and D. Beaupere, *Rev. Sci. Instr.*, 1977, **48**, 975.

[1018] G. Ribakovs and A. A. Gundjian, *I.E.E.E. J. Quantum Electron.*, 1978, **QE-14**, 42.

[1019] E. J. Gillham, *Appl. Optics*, 1977, **16**, 301.

[1020] R. D. Saunders, W. R. Ott, and J. M. Bridges, *Appl. Optics*, 1977, **16**, 593.

[1021] J. M. Bridges, W. R. Ott, E. Pitz, A. Schulz, D. Einfeld, and D. Stuck, *Appl. Optics*, 1977, **16**, 1788.

[1022] P. J. Key and R. C. Preston, *Appl. Optics*, 1976, **16**, 2477.

[1023] G. H. C. Freeman and W. H. King, *J. Phys. (E)*, 1977, **10**, 894.

[1024] H. E. Zimmerman, T. P. Cutler, V. R. Fitzgerald, and T. J. Weigt, *Mol. Photochem.*, 1977, **8**, 379.

anode currents above about 10 μA thereby restricting the dynamic range without the use of neutral density filters.

Two useful papers on the ferrioxalate actinometer discussed a 10—15-fold improvement by using 0.15 M-potassium ferrioxalate solution and the need for deoxygenation,[1025] and warned that 1,10-phenanthroline solutions can be photo-oxidized under fluorescent room lights.[1026] By adding buffer solution before the phenanthroline solution and keeping the latter in the dark, many of the erratic and irreproducible actinometric results can be eliminated.[1026] The photoisomerization of 2,2′,4,4′-tetra-isopropylazobenzene in n-heptane has formed the basis of an oxygen-insensitive actinometer for use in the range 350—390 nm.[1027] The quantum output of a low pressure Hg lamp has been measured in the wavelength range 180—200 nm by using two actinometers – one based on the gas phase conversion of oxygen into ozone and the other on the liquid phase decomposition of ethanol.[1028] Although there is a reasonable agreement (within an order of magnitude) between the two methods, ethanol is preferred because the total output from 180 to 200 nm can be obtained (oxygen does not absorb the 194 nm line), liquid can surround much of the lamp and simulate actual irradiation conditions, and hydrogen is easier to measure than ozone.[1028]

Two dosimeters for the mid-u.v. region have implied practical use. A thin polymer film containing 8-methoxy psoralen as the chromaphore showed a decrease in absorption at 254 and 305 nm on exposure to u.v. radiation (<380 nm).[1029] A 'sunburn dosimeter' used the irreversible photochemical conversion of 3′-[(p-dimethylamino)phenyl]spiro(fluorene-9,4′oxazolidine)-2′,5′-dione in a polycarbonate film to a red product on exposure to light below 350 nm.[1030]

8 Fluorescence and Phosphorescence Spectrometry

Instruments and Methods.—Although the biennial review on fluorescence and phosphorescence which appeared in *Analytical Chemistry*[1031] emphasized the analytical aspects, brief citations are given to many of the developments in instrumentation for quantitative studies. Recent advances (since 1970) in instrumentation for the study of electronic emission spectra have been outlined briefly in a review which covered steady-state and time-resolved measurements.[1032]

The earlier Sections of this Chapter describe some of the specific developments in plasma sources and lasers, many of which are used as excitation sources for fluorescence and phosphorescence measurements. Pulsed source excitation-gated detection (with or without time resolution) and CW excitation–CW

[1025] D. E. Nicodem, M. L. P. F. Cabral, and J. C. N. Ferreira, *Mol. Photochem.*, 1977, **8**, 213.
[1026] W. D. Bowman and J. N. Demas, *J. Phys. Chem.*, 1976, **80**, 2434.
[1027] R. Frank and G. Gauglitz, *J. Photochem.*, 1977, **7**, 355.
[1028] A. K. Davies, K. A. Khan, J. F. McKellar, and G. O. Phillips, *Mol. Photochem.*, 1976, **7**, 389.
[1029] B. L. Diffey and A. Davis, *Phys. Med. Biol.*, 1978, **23**, 318.
[1030] A. Zweig and W. A. Henderson, *Photochem. and Photobiol.*, 1976, **24**, 543.
[1031] C. M. O'Donnell and T. N. Solie, *Analyt. Chem.*, 1978, **50**, 189R.
[1032] K. R. Naqvi, A. P. Holzwarth, and U. P. Wild, *Appl. Spectroscopy Rev.*, 1976, **12**, 131.

detection have been compared theoretically with respect to signal-to-noise ratios in atomic and molecular fluorescence spectrometry.[1033] Calculations indicate that pulsed source excitation with delayed gated detection results in a considerable improvement in SNR as long as the major source of noise is unrelated to the source decay. A pulsed xenon flashlamp (2.8 J at 13 Hz) was used for excitation in a fast response fluorescence monitor for NO_2.[1034] Pulsed laser sources are commonly used for both spectral and lifetime measurements of fluorescence and the relatively low cost and versatile nitrogen laser pumped dye laser shown to be more flexible and generally applicable as an excitation source in comparison with the cavity-dumped or mode-locked Ar^+ laser.[1035] With a boxcar detector, this laser source was shown to be quite capable of detecting rhodamine B fluorescence at concentration limits of 1 p.p.t. More conventional plasma sources used for fluorescence excitation have included a Lyman-α lamp using heated uranium hydride as a hydrogen source for OH production and measurement[1036] and a new design of an atomic hydrogen resonance fluorescence lamp incorporating a simple blackened collimator.[1037] When used to follow the reaction H + NO, the lamp reduced the scattered light by a factor of 100 and the corresponding fluorescence by a factor of 10 and could be used to detect H atoms at concentrations greater than 5×10^9 cm^{-3} with an SNR of 1.[1037] (The lamp was less satisfactory than a wide bore lamp for following H + Cl_2 reactions.[1038])

Until recently no method existed for quantitative detection of atomic and molecular species at concentrations less than 10^8 cm^{-3}. The detection of low concentrations of sodium atoms reported earlier (Vol. 8, p. 38) has now been extended by the combination of saturated optical absorption and non-resonant emission spectroscopy (abbreviated to SONRES) so that concentrations below 10 atoms cm^{-3} can be determined.[1039] An alternative new analytical technique which does not require the use of a tunable laser has been claimed to provide a simple quantitative detection of particles in a gas stream.[1040] The method involves injecting excess of an unstable metastable species, $N_2(A^3\Sigma_u^+)$ in the example, into the gas stream. Energy transfer from the metastable to the sample specimen (Bi) resulted in light emission at the wavelength characteristic of the sample; bismuth vapour concentrations were 1.5×10^4 cm^{-3}.[1040] Resonance ionization spectroscopy (RIS) in which all of a given quantum species are converted into ion pairs has been used to develop a detector for a single Cs atom.[1041] By using tunable dye laser excitation, the method has been shown to be highly selective (1 atom in 10^{19} atoms or molecules of another kind).

[1033] G. D. Boutilier, J. D. Bradshaw, S. J. Weeks, and J. D. Winefordner, *Appl. Spectroscopy*, 1977, **31**, 307.
[1034] C. L. Fincher, A. W. Tucker, M. Birnbaum, R. J. Paur, and W. A. McClenny, *Appl. Optics*, 1977, **16**, 1359.
[1035] J. H. Richardson and S. M. George, *Analyt. Chem.*, 1978, **50**, 616.
[1036] D. Kley and E. J. Stone, *Rev. Sci. Instr.*, 1978, **49**, 691.
[1037] M. A. A. Clyne and P. B. Monkhouse, *J.C.S. Faraday II*, 1977, **73**, 298.
[1038] P. B. Bemand and M. A. A. Clyne, *J.C.S. Faraday II*, 1977, **73**, 394.
[1039] J. A. Gelbwachs, C. F. Klein, and J. E. Wessel, *Proc. Soc. Photo-Opt. Instrum. Eng.*, 1977, **113**, 68; *I.E.E.E. J. Quantum Electron.*, 1977, QE-13, 11D; *ibid.*, 1978, QE-14, 121; *Appl. Phys. Letters*, 1977, **30**, 489.
[1040] G. A. Capelle and D. G. Sutton, *Appl. Phys. Letters*, 1977, **30**, 407.
[1041] G. S. Hurst, M. H. Nayfeh, and J. P. Young, *Appl. Phys. Letters*, 1977, **30**, 229.

A commendable design of fluorescence spectrometer has been reported, which uses easily interchangeable optical components (*e.g.* prism and grating monochromators, sample chambers) of high optical speed ($f/3$) with conventional electronic processing for measurements of excitation and emission spectra and their anisotropies as well as their low temperature spectra.[1042] A computer-controlled photon-counting fluorescence spectrometer employing a conventional optical design has been described.[1043] A hybrid system combining both photon counting and pulse sampling was designed for use with a pulse-picked Ar^+ laser source.[1044] Counting rates up to 5 MHz could be accepted on a fast digital lock-in system employing a 36 Hz modulation frequency to present alternate positive and negative pulses for subtraction.[1045] A bipolar averaging circuit enabled SNR of fluorescence spectra to be enhanced.[1046] A digital counting device, which also generates serial code to punch a paper tape and produce formatted count listings on a teletype, is a useful component for single-channel photon counting.[1047]

The effect of excited beam absorption on measured values of fluorescence has been studied with a computer-controlled fluorescence spectrometer which is capable of simultaneous measurements of both absorption and fluorescence.[1048] By applying corrections for solute absorbance, it is possible to produce an absorbance-corrected fluorescence curve which is linear up to absorbances as high as two. An ingenious arrangement of a differential spectrometer was essentially a single-beam instrument in which fluorescence from two samples was shared in time. A cylindrical fused spectrosil quartz cell consisting of two separate half chambers with illumination from the bottom flat surface was rotated and emission was viewed at 90° to the direction of excitation.[1049] Other differential fluorescence spectrometers employed two synchronously rotating mirrors for alternate illumination of sample and reference cuvettes.[1050] A differential phase spectrometer was employed in a comparative study of aromatic hydrocarbons and molecules with similar skeletons but with several groups which can form hydrogen bonds with the solvent in order to demonstrate strongly anisotropic rotations.[1051]

The technique of synchronous fluorescence spectrometry has been further refined[1052] in order to improve selectivity and compared with conventional fluorescence spectrometry in a study of a series of standard oils.[1053] A method

[1042] J. Jasny, *J. Luminescence*, 1978, **17**, 149.
[1043] D. M. Jameson, R. D. Spencer, and G. Weber, *Rev. Sci. Instr.*, 1976, **47**, 1034.
[1044] J. M. Harris and F. E. Lytle, *Rev. Sci. Instr.*, 1977, **48**, 1469.
[1045] E. Siefast, *Appl. Optics*, 1977, **16**, 1163.
[1046] J. A. Wehrly, J. F. Williams, D. M. Jameson, and D. A. Kolb, *Analyt. Chem.*, 1976, **48**, 1424.
[1047] G. J. Perreault, R. E. Cookingham, J. P. Spoonhower, and A. Lewis, *Appl. Spectroscopy*, 1976, **30**, 614.
[1048] J. F. Holland, R. E. Teets, P. M. Kelly, and A. Timnick, *Analyt. Chem.*, 1977, **49**, 707.
[1049] K. A. Bostian, *Analyt. Biochem.*, 1977, **82**, 353.
[1050] T. J. Porro and D. A. Terhaar, *Analyt. Chem.*, 1976, **48**, 1103A; B. Bablouzian and G. D. Fasman, *Analyt. Biochem.*, 1977, **77**, 79.
[1051] G. Weber, *J. Chem. Phys.*, 1977, **66**, 4081; W. M. Mantulin and G. Weber, *J. Chem. Phys.*, 1977, **66**, 4092.
[1052] J. B. F. Lloyd and I. W. Evett, *Analyt. Chem.*, 1977, **49**, 1710; T. Vo-Dinh, *ibid.*, 1978, **50**, 396.
[1053] S. F. Wakeham, *Environ. Sci. Technol.*, 1977, **11**, 272.

for analysis of fluorescence samples containing multiple components has been described for use with a mechanically-scanned computer-controlled spectrometer.[1054] Other qualitative analyses of mixtures have been applied to polycyclic hydrocarbons by spectral decomposition of molecular fluorescence[1055] and to other fluorescent mixtures using methods of least squares or linear programming.[1056]

Arrangements described for fluorescence polarization measurements have included a T-format optical lay-out used with photon counting[1057] and instruments to measure spectra of single crystals of $Ba[Pt(CN)_4],4H_2O$ at high pressures,[1058] polarization ratios in stretched polyethylene films,[1059] and polarization properties of perylene in dodecyl- and hexadecyl-trimethylammonium bromide micelles.[1060] A fluorescence polarization instrument enabled heterogeneous populations of biological cells to be characterized according to the fluidity of their plasma membranes.[1061] Corrections have been shown to be necessary for polarization measurements when light is detected into a finite solid angle.[1062] Molecular realignment in an electric field, which is accompanied by changes in the fluorescence properties of solutions of macromolecules, has been detected in an apparatus which measured both the amplitudes and rates of change in the polarized components of the fluorescence on laser excitation.[1063]

Detectors for fluorescence spectrometers are described in Section 7, and comments have already been made on the performance characteristics of the vidicon detector. A silicon-intensified–OMA combination detector for measurements of molecular fluorescence in flames was shown to cover the range 200—1100 nm with a linear response over three decades and to have a high quantum efficiency ($>50\%$ in the range 350—800 nm).[1064] On the other hand, little or no use of a SIT vidicon tube was predicted following a study of atomic fluorescence from 15 elements in flames excited with a Xenon source.[1065] A combination of an OMA and Eimac xenon lamp applied to an Amico fluorescence spectrometer was shown to be suitable for fluorescence studies but the observed limits of detection were attributed mostly to the Eimac lamp, which increased the normal photon flux by a factor of two—three times.[1066] Increased sensitivity with a Farrand spectrometer[1067] employed a low light level TV camera as a multi-

1054 I. M. Warner, G. D. Christian, E. R. Davidson, and J. B. Callis, *Analyt. Chem.*, 1977, **49**, 564.
1055 H. S. Gold, C. E. Rechsteiner, and R. P. Buck, *Analyt. Chim. Acta*, 1978, **103**, 167.
1056 I. M. Warner, E. R. Davidson, and G. D. Christian, *Analyt. Chem.*, 1977, **49**, 2155.
1057 B. Valeur and G. Weber, *Photochem. and Photobiol.*, 1977, **26**, 441; D. M. Jameson, G. Weber, R. D. Spencer, and G. Mitchell, *Rev. Sci. Instr.*, 1978, **49**, 510.
1058 M. Stock and H. Yersin, *Chem. Phys. Letters*, 1976, **40**, 423.
1059 L. Margulies and A. Yogev, *Chem. Phys.*, 1978, **27**, 89.
1060 R. C. Dorrance, T. F. Hunter, and J. Philp, *J.C.S. Faraday II*, 1977, **73**, 89.
1061 A. Bruck, E. Sahar, E. Agmon, and M. Shinitzky, *Appl. Optics*, 1978, **17**, 564.
1062 P. E. Zinsli, *J. Phys. (E)*, 1978, **11**, 17.
1063 P. J. Ridler and B. R. Jennings, *J. Phys. (E)*, 1977, **10**, 558.
1064 H. Haraguchi, W. K. Fowler, D. J. Johnson, and J. D. Winefordner, *Spectrochim. Acta*, 1976, **32a**, 1539.
1065 T. L. Chester, H. Haraguchi, D. O. Knapp, J. D. Messman, and J. D. Winefordner, *Analyt. Spectroscopy*, 1976, **30**, 410.
1066 T. Vo-Dinh, D. J. Johnson, and J. D. Winefordner, *Spectrochim. Acta*, 1977, **33a**, 341.
1067 H. Steinhart and J. Sandmann, *Analyt. Chem.*, 1977, **49**, 950.

channel photometer.[1068] Three-dimensional plots of fluorescence spectra, *i.e.* of excitation and emission spectra and intensity, were obtained from a specially-designed plotter.[1069]

Several new approaches to the measurement of fluorescence quantum yields have now been described. Conventional photoacoustic spectrometry applied to quinine sulphate solution obtained a quantum yield of 0.53 ± 0.2, which agrees well with the literature.[1070] The corrected excitation spectrum (relative to carbon black) and quantum yield of benzene vapour have been measured on a photoacoustic spectrometer and a warning given that since Φ_f falls off rapidly at wavelengths shorter than 255 nm, any determinations with a high pressure mercury lamp and broad excitation filter could be subject to error.[1071] Quantum yield determinations with a chopped Ar^+ laser in which a piezoelectric microphone was built into a cell were shown to be limited to >0.2 where conventional calorimetry was admitted to be a better method.[1072] A photoacoustic determination of Φ_f for cresyl violet (where $\Phi_f = 0.00$—0.10) relied on the use of the $S_n \leftarrow S_1$ transition probability as an internal standard for comparision with S_1 radiationless transition probability, both probabilities being measured.[1073] It was shown that a pulsed 347 nm laser irradiating a cell containing a barium titanate piezoelectric transducer could be used to measure quantum yields of fluorescence.[1074] A calorimetric method using filtered light from a 75 W Hg lamp for excitation has been used to determine Φ_f for DPA, 9,10-diphenyl-anthracene (0.955 in cyclohexane), POPOP (0.975 in cyclohexane, 0.91 in ethanol), and to confirm that both Φ_f values and lifetimes of DPA varied significantly from one solvent to another.[1075] By comparing an unknown with a non-luminescent reference, a thermal blooming method was shown to offer considerable advantages over conventional photometric techniques for determinations of fluorescence quantum yields (see p. 50).[1076] The method was simple and inexpensive to implement, and the only restriction was the need for an appropriate laser of >1 mW of the correct wavelength (which is little problem in the visible but more difficult in the u.v.). The study revealed how only slow progress is being made to refine luminescence quantum yield standards. *NN*-Dimethylaniline vapour ($\Phi_f = 0.18 \pm 0.03$, F. Kraus *et al.*, *Z. Physik Chem.*, 1972, **NF 82**, 139) was used as a standard reference material in a pressure-dependent study of the fluorescence of pyrimidine vapour.[1077] It was shown that the formula ordinarily taken to compute the quantum yield of luminescence relative to that of a 'standard' substance should be changed in order to take account of polarization effects in the excitation and emission

[1068] J. B. Callis, I. M. Warner, G. D. Christian, and E. R. Davidson, *Industrial Research*, 1976 (11), 44.

[1069] J. H. Rho and J. L. Stuart, *Analyt. Chem.*, 1978, **50**, 620.

[1070] M. J. Adams, J. G. Highfield, and G. F. Kirkbright, *Analyt. Chem.*, 1977, **49**, 1850.

[1071] M. G. Rockley, *Chem. Phys. Letters*, 1977, **50**, 427.

[1072] W. Lahmann and H. J. Ludewig, *Chem. Phys. Letters*, 1977, **45**, 177.

[1073] M. G. Rockley and K. M. Waugh, *Chem. Phys. Letters*, 1978, **54**, 597.

[1074] T. K. Razumova and I. O. Starobogatov, *Optics and Spectroscopy*, 1977, **42**, 274; I. O. Starobogatov, *ibid.*, 1977, **42**, 172.

[1075] M. Mardelli and J. Olmsted, *J. Photochem.*, 1977, **7**, 277.

[1076] J. H. Brannon and D. Magde, *J. Phys. Chem.*, 1978, **82**, 705.

[1077] H. Reineccius and H. von Weyssenhoff, *Chem. Phys. Letters*, 1977, **52**, 34.

spectrometers as well as the non-zero emission anisotropy of the emitting systems.[1078] An automatic system to measure quantum efficiency as a function of photon energy has also been described in which a standard photocell measured part of the excitation beam.[1079]

The sensitivity of chemical analysis using fluorescent-tagged reagents is usually limited by concentration quenching at high reagent concentrations and by photochemical bleaching if high illumination intensities are used. Studies have now shown by both theoretical analysis and experimental verification that concentration quenching lengthens the bleaching lifetimes leading to very sensitive fluorescence determinations.[1080] In laser-induced fluorescence spectroscopy, the fluorescence signal has been found to be independent of quenching rates or laser power when operating in the saturation mode.[1081]

A differential thermoluminescence technique has been described in which undesirable black-body radiation is cancelled out to give a signal from the luminophor alone.[1082] In this way, the temperature range of thermoluminescence can be extended considerably, thereby allowing high temperature investigations to be made in the red region of the spectrum. A double integrating sphere described in a thermoluminescence apparatus has been effectively used to establish linearity of photomultipliers at light levels corresponding to 10^{-11} that of the lamp intensity.[1083] A diaphragm slide connecting the spheres was fitted with a series of accurately sized pinholes.

Other fluorescent instruments developed have included a dual-beam microspectrofluorimeter used for photoreceptor pigments,[1084] an image intensifier spectroscope for spectral studies on triboluminescent materials,[1085] and adaptations for high pressure liquid chromatography.[1086] Short wavelength excitation (<250 nm) was shown to have particular merits in h.p.l.c.–fluorescence determinations of indole peptides and other compounds.[1087] An optical scanning technique with a spatial resolution of 10 m, developed for measurements of surface luminance, used Ar^+ laser excitation, a high resolution spectrometer, and photon counting for detection.[1088]

A useful apparatus has been described for low temperature studies (4—30 K) of emission spectra of excited states (from SO_2, benzene) produced by u.v. photolysis and trapped in rigid matrices.[1089] In this case, photolysis was achieved using rare gas lamps through a LiF window and emission measurements could be made over the range 180—600 nm. Low temperature matrix isolation measurements of polycyclic hydrocarbons, with detection limits of

[1078] J. A. Poole and A. Findeisen, *J. Chem. Phys.*, 1977, **67**, 5338.
[1079] A. Schmidt-Ott and F. Meier, *Rev. Sci. Instr.*, 1977, **48**, 524.
[1080] T. Hirschfeld, *Appl. Optics*, 1976, **15**, 3135.
[1081] J. W. Daily, *Appl. Optics*, 1978, **17**, 568.
[1082] E. P. Manche, *Rev. Sci. Instr.*, 1978, **49**, 715.
[1083] A. A. Mills, D. W. Sears, and R. Hearsey, *J. Phys. (E)*, 1977, **10**, 51.
[1084] P. A. Benedetti and F. Lenci, *Photochem. and Photobiol.*, 1977, **26**, 315.
[1085] A. J. Walton and P. Botos, *J. Phys. (E)*, 1978, **11**, 513.
[1086] W. Slavin, A. T. Rhys-Williams, and R. F. Adams, *U.V. Spectrometry Group Bull.*, 1977 (5), 21; J. R. Jadmec, W. A. Saner, and Y. Talmi, *Analyt. Chem.*, 1977, **49**, 1316.
[1087] G. J. Krol, C. A. Mannan, R. E. Pickering, D. V. Amoto, B. T. Kho, and A. Sonnenschein, *Analyt. Chem.*, 1977, **49**, 1836.
[1088] M. J. Luciano and D. L. Kingston, *Rev. Sci. Instr.*, 1978, **49**, 718.
[1089] J. Fornier, J. Deson, and C. Vermeil, *J. Phys. (E)*, 1976, **9**, 879.

10^{-11} gm, has been shown to be a useful qualitative and quantitative technique with working curves linear over five decades.[1090] A cryostat containing a thin film of a simple hydrocarbon was provided with both a titanium film window for electron irradiation and a quartz window for optical measurements of fluorescence.[1091] An X-type cell in a glass jacket was used for observation of fluorescence spectra over the temperature range 77—373 K,[1092] and a variable temperature (77—450 K), high pressure 10^{10} N m^{-2} cell enabled efficiency studies to be made of tungstate and molybdate phosphors.[1093] Other high pressure cells have been employed for polarization studies,[1094] sensitized luminescence using KCl–Ag–Tl,[1095] and fluorenone emission in plastic media.[1096] An efficient, easy-to-construct long-path cell for laser-induced fluorescence measurements consisted of an integrating volume with seven flat, highly polished specularly reflecting walls meeting at irregular angles so that both pumping and fluorescence photons could make many quasi-random traversals of the cell.[1096]

A review article on fluorescence analysis on solid surfaces dealt with analysis of paper chromatograms, t.l.c. plates *etc.*,[1097] and theoretical fluorescence calibration curves have been compared with experimental findings for solid-surface fluorescence analysis.[1098] An apparatus for irradiation and fluorescence monitoring of solid samples has been used to examine uranyl formate mono-hydrate[1099] and comments made on the effect of a bandpass intensity distribution spectrum on measurements of solid samples using fluorescence spectrometry.[1100] Two scattering materials commonly used in integrating spheres, namely Eastman White reflectance standard[1101] and barium sulphate,[1102] have been reinvestigated and revised figures published for their absolute reflectance values.

Fluorescence Applications.—It would be quite impossible to cover in a short space even a small fraction of the publications dealing with fluorescence, which have appeared in the past two years, and the following Section (which is similar in many ways to that for transient emission spectroscopy) has simply taken representative examples across the whole of chemistry which have interest from the point-of-view of instrumentation or technique. The reader is referred elsewhere for the more quantitative fluorescence assays.[1031,1103]

A NLPD laser with an etalon to give a tuning range of 5 nm with a bandwidth of less than 10^{-3} nm has been used to obtain fluorescence excitation spectra of

[1090] R. C. Stroupe, P. Takousbalides, R. B. Dickinson, E. L. Wehry, and G. Mamantov, *Analyt. Chem.*, 1977, **49**, 701.
[1091] I. Sugawara and Y. Tabata, *Chem. Phys. Letters*, 1976, **41**, 357.
[1092] M. D. Swords, M. Caplin, and D. Phillips, *Rev. Sci. Instr.*, 1978, **49**, 669.
[1093] C. E. Tyner and H. G. Drinkamer, *J. Chem. Phys.*, 1977, **67**, 4103.
[1094] G. S. Chryssomallis, H. G. Drinkamer, and G. Weber, *J. Appl. Phys.*, 1978, **49**, 3085.
[1095] K. W. Bieg and H. G. Drinkamer, *J. Chem. Phys.*, 1977, **66**, 1437.
[1096] D. J. Mitchell, G. B. Schuster, and H. G. Drinkamer, *J. Chem. Phys.*, 1977, **67**, 4832.
[1097] G. G. Guilbault, *Photochem. and Photobiol.*, 1977, **25**, 403.
[1098] R. J. Hurtibise, *Analyt. Chem.*, 1977, **49**, 2160.
[1099] B. Claudel, M. Feve, J. P. Puaux, and H. Sautereau, *J. Photochem.*, 1977, **7**, 113.
[1100] M. G. King, *J. Phys. (E)*, 1977, **10**, 284.
[1101] F. Grum and T. E. Wightman, *Appl. Optics*, 1977, **16**, 2775.
[1102] E. M. Patterson, C. E. Shelden, and B. H. Stockton, *Appl. Optics*, 1977, **16**, 729.
[1103] P. Froehlich, *Appl. Spectroscopy Rev.*, 1976, **12**, 83.

various interhalogen compounds including IF and ICl[1104] and BrF.[1105] Strong fluorescence emission at 340 nm arising from flashlamp-pumped iodine vapour has indicated the possibility of constructing a molecular iodine laser.[1106] Laser-induced fluorescence has also been observed from the 0–0 band of the $B \leftarrow X$ visible spectrum of I_2[1107] and from the $B^3\pi(0_u{}^+)$ state of Cl_2.[1108] Iodine atom reactions with bromine[1109] and HF[1110] can be conveniently followed by observing I* fluorescence at 1.315 µm. High I* concentrations can be produced by the laser photolysis of CF_3I using a 265 nm Nd[1110,1111] or doubled-dye laser.[1109] Strong u.v. fluorescence originating from radiative decay of I* has also been observed following dye laser photolysis of iodine vapour and excited I* assumed to be formed by two-photon absorption in I_2 followed by saturated photodissociation.[1112] Deactivation rates of $Br^*(^2P_{\frac{1}{2}})$ atoms by halogens and interhalogen compounds[1113] and relative quantum yields for production from Br_2 and IBr[1114] have been measured by observing the i.r. fluorescence (at 2.715 µm) with a suitable detector, such as InSb.

The application of laser-induced fluorescence techniques to studies of multi-photon dissociation has been shown to provide a very sensitive diagnostic technique to monitor ground state or low-lying excited state fragments.[1115] A clear demonstration of the sensitivity of the technique was the identification of C_2 ($d^3\pi_g \leftarrow a^3\pi_u$) following the pulsed CO_2 laser photolysis of C_2H_5CN, C_2H_5OH, and $C_2H_5N_2$. A dye laser was tuned to the (0,0) band of C_2 at 516.5 nm and fluorescence was monitored on the ($d \leftarrow a$) system at 563.5 nm with a 0.7 nm FWHM interference filter. Ground state $NH_2(^2B_1)$ fragments produced by the pulsed CO_2 laser photolysis of NH_3 have also been observed by laser-induced fluorescence.[1116]

Infrared fluorescence studies are now extensively used to measure relaxation rates of vibrational energy transfer. Although these emissions are often weak, with state-of-the-art detectors, excellent signal-to-noise ratios can be achieved from electronic and vibrational states of all kinds including isotopic species or single excited states of atoms and molecules.[1117] V–V transfer rates and V–T relaxation rates have been measured for a wide variety of diatomics and small polyatomic molecules.[1118] Rotational relaxation of HF and DF[1119] following excitation by an HF laser, E–V energy transfer from $O_2(^1\Delta_g)$ to HF,[1120] colli-

1104 M. A. A. Clyne and I. S. McDermid, *J.C.S. Faraday II*, 1976, **72**, 2252.
1105 M. A. A. Clyne, A. H. Curran, and J. A. Coxon, *J. Mol. Spectroscopy*, 1976, **63**, 43.
1106 A. B. Callear and M. P. Metcalfe, *Chem. Phys. Letters*, 1976, **43**, 197.
1107 R. E. Smalley, L. Wharton, and D. H. Levy, *Chem. Phys. Letters*, 1977, **51**, 392.
1108 R. E. Huie, N. J. T. Long, and B. A. Thrush, *Chem. Phys. Letters*, 1976, **44**, 608.
1109 P. L. Houston, *Chem. Phys. Letters*, 1977, **47**, 137.
1110 R. D. Coombe and A. T. Pritt, *J. Chem. Phys.*, 1977, **66**, 5214.
1111 A. T. Pritt and R. D. Coombe, *J. Chem. Phys.*, 1976, **65**, 2097.
1112 C. Tai and F. W. Dalby, *Canad. J. Phys.*, 1978, **56**, 183.
1113 H. Hofmann and S. R. Leone, *Chem. Phys. Letters*, 1978, **54**, 314.
1114 A. B. Petersen and I. W. M. Smith, *Chem. Phys.*, 1978, **30**, 407.
1115 J. D. Campbell, M. H. Yu, and C. Wittig, *Appl. Phys. Letters*, 1978, **32**, 413.
1116 J. D. Campbell, G. Hancock, J. B. Halpern, and K. H. Welge, *Chem. Phys. Letters*, 1976, **44**, 404.
1117 S. R. Leone, *Proc. Soc. Photo-Opt. Instrum. Eng.*, 1976, **82**, 152.
1118 N. C. Long, J. C. Polanyi, and J. Wanner, *Chem. Phys.*, 1977, **24**, 219.
1119 J. K. Hancock and A. W. Saunders, *J. Chem. Phys.*, 1976, **65**, 1275.
1120 S. Madronich, J. R. Wiesenfeld, and G. J. Wolga, *Chem. Phys. Letters*, 1977, **46**, 267.

sional energy rates from SO_2^* to $O(^3P)$,[1121] $H^{35}Cl$–$H^{37}Cl$ vibrational energy transfer,[1117] and vibrational chemiluminescence from the ion–molecule reaction O^- + CO[1122] are only a few examples of this extensive field of study. The extensive range of i.r. lasers now available allows selective excitation to be undertaken and has allowed measurements of vibrational relaxation rates in N_2–CO–OCS mixtures by monitoring CO fluorescence following excitation of N_2 with an HBr laser (4.04 μm),[1123] and similar measurements in $H_2^{18}O_2$ following excitation of various V–R lines of the ν_1 and ν_2 stretching levels by a tunable OPO.[1124] A CW CO laser was used to measure the extent of CO product vibrational relaxation following pulsed dye laser excitation of I_2, ICl, and NO_2–CO mixtures by i.r. fluorescence measurements with a Ge : Au detector at 77 K.[1125] Strong laser-induced fluorescence has been observed in COF_2 following excitation with over 30 different CO_2 laser lines.[1126]

Hydroxy-radical concentrations have been determined by resonance fluorescence excitation at 282.5 nm in numerous kinetic experiments following flash photolysis (see p. 97) and there are several other examples of using fluorescence characterization including the determination of trace impurities of H_2O and D_2O in rare gas mixtures at 4.2 K,[1127] sonoluminescence spectra of argon-saturated water,[1128] simulated stratospheric studies,[1129] and rotational state populations in the reaction H + $NO_2 \rightarrow$ OH + NO.[1130] Quantum yields for $O(^1D)$ from the photolysis of ozone[1131] has been determined from measurements of NO_2 chemiluminescence produced by the reaction $O(^1D)$ + N_2O. An oxygen-induced luminescence burst on admission of O_2 to a phosphorescing sample (*e.g.* naphthalene in a polystyrene fluff)[1132] has been attributed to observation of exciplex emission: T_1 + $O_2 \rightarrow S_0$ + $O_2(^3\Sigma)$ + $h\nu$.[1133] Other studies of fluorescence from molecular oxygen have included the pressure dependence of 1.27 μm emission from $O_2(^1\Delta)$,[1134] reactions with O_2, CO_2, NH_3, and H_2 in a 220 m^3 reactor,[1135] and Ar^+ laser-pumped dye laser excitation of fluorescence from the $^1\Sigma^1\Delta$ dimer state of O_2.[1136] Small quantities of $O_2(b^1\Sigma_g^+)$ ($\nu = 0$) produced by direct excitation of gas phase O_2 in the for-

[1121] G. A. West, R. E. Weston, and G. W. Flynn, *J. Chem. Phys.*, 1977, **67**, 4873.

[1122] V. M. Bierbaum, G. B. Ellison, J. H. Futrell, and S. R. Leone, *J. Chem. Phys.*, 1977, **67**, 2375.

[1123] S. R. J. Brueck and R. M. Osgood, *J. Chem. Phys.*, 1978, **68**, 4941.

[1124] J. Finzi, F. E. Hovis, V. N. Panfilov, P. Hess, and C. B. Moore, *J. Chem. Phys.*, 1977, **67**, 4053.

[1125] D. S. Y. Hsu and M. C. Lin, *Chem. Phys. Letters*, 1978, **56**, 79.

[1126] K. H. Casleton and G. W. Flynn, *J. Chem. Phys.*, 1977, **67**, 3133.

[1127] J. Goodman and L. E. Brus, *J. Chem. Phys.*, 1977, **67**, 4858.

[1128] C. Sehgal, R. P. Steer, R. G. Sutherland, and R. E. Verrall, *J. Phys. Chem.*, 1977, **81**, 2618.

[1129] C. H. Wu, C. C. Wang, S. M. Japar, L. I. Davis, M. Hanabusa, D. Killinger, H. Niki, and B. Weinstock, *Internat. J. Chem. Kinetics*, 1976, **8**, 765.

[1130] J. A. Silver, W. L. Dimpfl, J. H. Brophy, and J. L. Kinsey, *J. Chem. Phys.*, 1975, **65**, 1811.

[1131] D. L. Philen, R. T. Watson, and D. D. Davis, *J. Chem. Phys.*, 1977, **67**, 3316; J. Arnold, F. J. Comes, and G. K. Moortgat, *Chem. Phys.*, 1977, **24**, 211.

[1132] R. D. Kenner and A. U. Khan, *J. Chem. Phys.*, 1977, **67**, 1605.

[1133] R. D. Kenner and A. U. Khan, *Chem. Phys. Letters*, 1977, **45**, 340.

[1134] D. J. Bernard and N. R. Pehelkin, *Rev. Sci. Instr.*, 1978, **49**, 794.

[1135] A. Leiss, U. Schurath, K. H. Becker, and E. H. Fink, *J. Photochem.*, 1978, **8**, 211.

[1136] S. E. Novick and H. P. Broida, *J. Chem. Phys.*, 1978, **67**, 5975.

bidden atmospheric band at 762.8225 nm (with a tunable dye laser) have been detected by fluorescence in spite of the small line strength of 1.56×10^{-4} cm^{-2} atm^{-1}.[1137]

Vacuum-u.v. excitation with microwave-powered rare gas lamps has been used to obtain fluorescence excitation spectra of H_2O and D_2O.[1138] CO fluorescence has been identified following the photodissociation of CO_2 with an atomic nitrogen lamp at 92.8 nm.[1139] In this study, signal levels measured with a solar-blind photomultiplier were low and gating photon counting (3—13 μs after the flash) was used to minimize the effect of noise and dark count. Pulsed u.v. excitation was used to obtain the fluorescent spectra of XeF in Ar and Ne matrices at 4.2 K,[1140] and the intrinsic fluorescence band at 172 nm of solid xenon was induced by pulsed electron irradiation.[1141] Emission following vacuum-u.v. photolysis of OCS and CS_2 in rare gas matrices has been attributed to $S(^1S)$ atoms.[1142] Laser-induced fluorescence has also been found for matrix-isolated CNN (produced by the photolysis of N_3CN)[1143] and CH_3–N_2 mixtures[1144] and CCl_2 produced by the Lyman-α decomposition of CH_2Cl_2.[1145] Other products identified by fluorescence following vacuum-u.v. photodissociation have included $PH(b^1\Sigma^+)$ from PH_3,[1146] CN from HCN and DCN,[1147] and CH_3CN[1148] and CH_3S from CH_3SCH_3.[1149] A detailed analysis of the primary photodissociative products and their reactions following 266-nm laser photolysis of HN_3 enabled identification to be made of NH_2 (by chemiluminescence) and NH (by fluorescence).[1150] In this case, NH was monitored at right angles to a tunable dye laser exciting beam in a 30 l sphere and detected with an RCA 31034 photomultiplier tube using a gated boxcar integrator. Photodissociation of ketene (CH_2CO) by 337 nm laser radiation produced CH_2, which was detected by laser-induced fluorescence.[1151] There are extensive publications on the laser-induced fluorescence of NO_2, including detection at sub-p.p.m. levels in flames,[1152] emission spectra on excitation with a conventional[1153] and single isotope[1154] He–Cd laser, resolved hyperfine structure from

[1137] L. R. Martin, R. B. Cohen, and J. F. Schatz, *Chem. Phys. Letters*, 1976, **41**, 394.
[1138] M. T. McPherson and J. P. Simons, *Chem. Phys. Letters*, 1977, **51**, 261.
[1139] E. Phillips, L. C. Lee, and D. L. Judge, *J. Chem. Phys.*, 1976, **65**, 3118; 1977, **68**, 3688.
[1140] J. Goodman and L. E. Brus, *J. Chem. Phys.*, 1976, **76**, 3809.
[1141] U. Hahn, N. Schwentner, and G. Zimmerer, *Optics Comm.*, 1977, **21**, 237.
[1142] C. Lalo, *J. Mol. Structure*, 1978, **45**, 403; J. Fornier, C. Lalo, J. Deson, and C. Vermeil, *J. Chem. Phys.*, 1977, **66**, 2656; J. M. Brom and E. J. Lepak, *Chem. Phys. Letters*, 1976, **41**, 185.
[1143] J. L. Wilkerson and W. A. Guillory, *J. Mol. Spectroscopy*, 1976, **62**, 188.
[1144] V. E. Bondybey and J. H. English, *J. Chem. Phys.*, 1977, **67**, 664.
[1145] V. E. Bondybey, *J. Mol. Spectroscopy*, 1977, **64**, 180.
[1146] G. Di Stefano, M. Lenzi, A. Margani, and C. N. Xuan, *J. Chem. Phys.*, 1978, **68**, 959.
[1147] I. Stein and A. Gedanken, *J. Chem. Phys.*, 1978, **68**, 2982.
[1148] M. L. Lesiecki and W. A. Guillory, *J. Chem. Phys.*, 1977, **66**, 4239; *Chem. Phys. Letters*, **49**, 93.
[1149] K. Ohhbayashi, H. Akimoto, and I. Tanaka, *Chem. Phys. Letters*, 1977, **52**, 47.
[1150] A. R. Baronavski, R. G. Miller, and J. R. McDonald, *Chem. Phys.*, 1978, **30**, 119; J. R. McDonald, R. G. Miller, and A. R. Baronavski, *ibid.*, 1978, **30**, 133.
[1151] J. Danon, S. V. Filseth, D. Feldman, H. Zacharias, C. H. Dugan, and K. H. Welge, *Chem. Phys.*, 1978, **29**, 345.
[1152] R. H. Barnes and J. F. Kircher, *Appl. Optics*, 1978, **17**, 1099.
[1153] M. Birnbaum, C. L. Fincher, and A. W. Tucker, *J. Photochem.*, 1977, **6**, 237.
[1154] G. I. Senum and S. E. Schwartz, *J. Mol. Spectroscopy*, 1977, **64**, 75.

a single mode Ar^+ laser,[1155] and depolarization effects of resonance fluorescence by application of a magnetic field (the Hanle effect).[1156] Room-temperature gas phase reactions of NO_2 with CO on excitation with various CW visible lasers was shown to produce CO_2 (identified by laser-induced fluorescence) at a rate far exceeding that of an associated thermal reaction.[1157] Vibrationally-excited NO_2 was observed by fluorescence following photodissociation of CH_2NO_2 and $C_3H_7NO_2$.[1158] Laser-induced fluorescence has also been reported for CO_2^+,[1159] CO^+,[1160] and two-photon excited CO and N_2.[1161] HNO*(A) luminescence in gaseous mixtures of H_2, NO, and Hg has been induced by intensity-modulated 253.7 nm radiation.[1162] The $v' = 3$, v'' Lyman band progression from argon-sensitized fluorescence of H_2 has been monitored in fluorescence with a channel electron multiplier by photon counting.[1163]

An arrangement to measure high resolution absolute quantum yields over the rotational envelope of several vibronic levels of naphthalene used doubled FLPD laser for excitation and a well-baffled fluorescence cell ahead of a multiple path absorption cell.[1164] Experimental techniques to measure absolute quantum yields of CO from a single vibronic level of glyoxal excited by a NLPD laser employed vacuum-u.v. resonance emission for CO determinations.[1165] Photon-counting techniques were employed in the determination of anomalous $S_3 \rightarrow S_0$ fluorescence from azuleno(5,6,7-*cd*)phenalene,[1166] $S_2 \rightarrow S_0$ fluorescence of [18]annulenes,[1167] emission spectra of radical cations of 1,3,5-hexatriene,[1168] $T_2 \rightarrow T_1$ fluorescence in substituted anthracenes,[1169] and quantum yields of fluorescence from S_2 vibronic levels of thiophosgene.[1170] Emission spectra from two vibronic levels of HCHO have been obtained following excitation with a dye laser.[1171] The weak fluorescence signal ($\Phi_f < 0.005$) was detected by an OMA and averaged for 20—30 min. Low quantum yield fluorescence spectra of common polycyclic hydrocarbons in solution at 300 K over the region 240—370 nm has been determined following direct laser excitation.[1172] Laser two-photon excited fluorescence detection has been successfully applied to the characterization of oxadiazoles PPD, PBD, and BBD using an Ar^+ laser excitation source in h.p.l.c. with detection limits generally comparable to u.v.

[1155] R. Schmiedl, I. R. Bonilla, F. Paech, and W. Demtroder, *J. Mol. Spectroscopy*, 1977, **68**, 236.

[1156] H. Figger, D. L. Monts, and R. N. Zare, *J. Mol. Spectroscopy*, 1977, **68**, 388.

[1157] I. P. Herman, R. P. Mariella, and A. Javan, *J. Chem. Phys.*, 1978, **68**, 1070.

[1158] K. G. Spears and S. P. Brugge, *Chem. Phys. Letters*, 1978, **54**, 373.

[1159] V. E. Bondybey and T. A. Miller, *J. Chem. Phys.*, 1977, **67**, 1790.

[1160] T. A. Miller and V. E. Bondybey, *Chem. Phys. Letters*, 1977, **50**, 275.

[1161] S. V. Filseth, R. Wallenstein, and H. Zacharias, *Optics Comm.*, 1977, **23**, 231.

[1162] K. Oda, D. L. Singleton, and R. J. Cvetanovic, *J. Chem. Phys.*, 1977, **67**, 4681.

[1163] D. J. McKenrey and R. N. Dubinsky, *Chem. Phys.*, 1977, **26**, 141.

[1164] W. E. Howard and E. W. Schlag, *Chem. Phys.*, 1976, **17**, 123.

[1165] G. H. Atkinson, M. E. McIlwain, C. G. Venkatesh, and D. M. Chapman, *J. Photochem.*, 1978, **8**, 307.

[1166] A. R. Holzwarth, K. Razi Naqvi, and U. P. Wild, *Chem. Phys. Letters*, 1977, **46**, 473.

[1167] U. P. Wild, H. J. Griesser, Vo Dinh Tuan, and J. F. M. Oth, *Chem. Phys. Letters*, 1976, **41**, 450.

[1168] M. Allan and J. P. Maier, *Chem. Phys. Letters*, 1976, **43**, 94.

[1169] G. D. Gilliespie and E. C. Lim, *J. Chem. Phys.*, 1976, **65**, 2022.

[1170] T. Oka, A. R. Knight, and R. P. Steer, *J. Chem. Phys.*, 1977, **66**, 699.

[1171] K. Y. Tang and E. K. C. Lee, *Chem. Phys. Letters*, 1976, **43**, 232.

[1172] H. B. Lin and M. R. Topp, *Chem. Phys. Letters*, 1977, **48**, 251.

detection methods.[1173] Two-photon excited single vibronic level fluorescence of benzene vapour was also recorded after excitation with a 4 kW FLPD laser pulse at 504.2 nm[1174] (see also the Section on Two-photon Excitation). Detection limits for polycyclic hydrocarbons with a NLPD laser and boxcar detection were found to be in the sub-p.p.t. range[1175] and for fluorescence excited with an esculin dye laser, 2 p.p.t. – a factor of 35 times less than that reported with a Perkin–Elmer MPF-4 fluorescence spectrometer.[1176]

Quasilinear 'Shpol'skii' fluorescence spectra of polycyclic aromatic hydrocarbons have been observed by matrix isolation employing an n-alkane as the matrix if a vapour-deposited sample was annealed at an elevated temperature (140 K) prior to observation at a lower temperature.[1177] The Shpol'skii fluorescence spectrum of acenaphthene in n-alkane matrices has been studied as a function of temperature.[1178] Other low temperature studies included observation of u.v.–visible fluorescence of DNA at 77 K following excitation with a BRV vacuum-u.v. source[1179] and the use of site selection laser excitation to study the fluorescence of the cyclohexadienyl radical at 4.2 K.[1180]

Resonance fluorescence from isoquinoline vapour has been measured following excitation into each of eight bands in the 310 nm absorption spectrum with a 0.2 cm^{-1} linewidth doubled FLPD laser.[1181] Single vibronic level excitation has also been applied to cyclobutanone[1182] and p-difluorobenzene.[1183] The latter example employed a xenon source–monochromator arrangement with photon counting and the quantum yield values for fluorescence were found to be significantly different for continuous excitation to those obtained by flash excitation.

There are numerous examples of high resolution fluorescence excitation spectra of alkali metals, and recent publications have included studies on Na,[1184] K (and Mg),[1185] Cs,[1186] and Rb.[1187] In the latter case, a nozzle beam of rubidium under collision-free conditions was excited by a CW laser beam at 476 nm and the fluorescence spectrum used to give evidence for $Rb_2^*(c^1\pi_u) \rightarrow Rb^*(^2P_{\frac{3}{2}}) + Rb(^2S_{\frac{1}{2}})$. The vibrational–rotational structure of the c^1 states of the NaK molecule, excited by a single-mode CW dye laser, was detected by observing the atomic potassium fluorescent D-lines, which result from dissocia-

[1173] M. J. Sepaniak and E. S. Yeung, *Analyt. Chem.*, 1977, **49**, 1554.
[1174] A. E. W. Knight and C. S. Parmenter, *Chem. Phys. Letters*, 1976, **43**, 399.
[1175] J. H. Richardson and M. E. Ando, *Analyt. Chem.*, 1977, **49**, 955.
[1176] T. Imasaka, H. Kadone, T. Ogawa, and N. Ishibashi, *Analyt. Chem.*, 1977, **49**, 667.
[1177] P. Takousbalides, E. L. Wehry, and G. Mamantov, *J. Phys. Chem.*, 1977, **81**, 1769.
[1178] J. J. Dekkers, G. P. Hoornweg, C. MacLean, and N. H. Velthorst, *J. Mol. Spectroscopy*, 1977, **68**, 56.
[1179] S. Basu, R. B. Cundall, M. W. Jones, and G. O. Phillips, *Chem. Phys. Letters*, 1978, **53**, 439.
[1180] S. J. Sheng, *J. Phys. Chem.*, 1978, **82**, 442.
[1181] G. Fischer and A. E. W. Knight, *Chem. Phys.*, 1976, **17**, 327.
[1182] K. Y. Tang and E. K. C. Lee, *J. Phys. Chem.*, 1976, **80**, 1833.
[1183] L. J. Volk and E. K. C. Lee, *J. Chem. Phys.*, 1977, **67**, 236.
[1184] J. P. Woerdman, *Chem. Phys. Letters*, 1976, **43**, 279.
[1185] D. J. Bernard and W. D. Slafer, *Chem. Phys. Letters*, 1978, **56**, 438.
[1196] T. Yabuzaki, A. C. Tam, M. Hou, W. Happer, and S. M. Curry, *Optics Comm.*, 1978, **24**, 305; R. T. M. Su, J. W. Beavan, and R. F. Curl, *Chem. Phys. Letters*, 1976, **43**, 162.
[1187] D. L. Feldman and R. N. Zare, *Chem. Phys.*, 1976, **15**, 415.

tion of the molecule in collisions with a buffer gas.[1188] Fluorescence excited in a [139]La atomic beam by a single-frequency dye laser in the region of 593.06 nm has been examined with Doppler-free geometry.[1189] The concentrations of $Hg(^3P_0)$ in mixtures of quenching gases with nitrogen have been monitored in a repetitive pulsed laser system by measuring the intensity of the 546.1 nm fluorescence which accompanied laser excitation at 404.7 nm[1190] (see also ref. 1550). Collisions between $Hg(6^3P_1)$ and $Tl(6^2P_{\frac{1}{2}})$ atoms induced by 253.7 nm radiation resulted in the population of various Tl states whose decay gave rise to sensitized fluorescence at 352.0 and 535.2 nm.[1191] Mercury-sensitized luminescence of NH_3 was used in a specific analytical technique,[1192] and the optical and collisional properties of the ZnHg excimer have been examined by laser-induced fluorescence.[1193]

Laser excitation has been used in fluorescence studies on YO at 1800 K,[1194] MgO,[1195] CrO_2Cl_2,[1196] and Pb_2 in rare gas solids.[1197] A tunable dye laser (0.04 nm linewidth) has been used to obtain over 50 new transitions in BO_2 between 489 and 547 nm[1198] and a CW dye laser used to obtain single vibronic level fluorescence from $^{11}BO_2$.[1199] Luminescence spectra of toluene–chromium hexacarbonyl in gas matrices at 10 K have been measured in a cryostat with a Farrand fluorescence spectrometer turned on its back to enable the light paths for excitation and analysis to be directed onto the central cold windows.[1200] Visible emission spectra from $(Y, Yb, Er)F_3$ and $(Y, Yb, Ho)F_3$ have been reported following excitation at 970 nm.[1201] Detection limits of 2.5×10^{-14} gm for Tl and 1.2×10^{-12} gm of Cs, achieved in aqueous solutions for dye laser excitation, were shown to be substantially lower than those obtained with atomic absorption.[1202] Boron and phosphorus atom concentrations, with detection limits of 10^{11} cm^{-3} and 5×10^{11} cm^{-3}, respectively, in silicon crystals have been measured by a novel photoluminescence technique at 4 K using excitation from an Ar$^+$ laser at 514 nm.[1203]

In crystalline matrices, a new method for performing trace analysis by selective laser excitation of probe ions[1204] has been shown to be specific for Eu^{3+} whose crystal field levels were perturbed by the presence of PO_4 in a $BaSO_4$ lattice. The fluorescence intensity resulting from dye laser excitation was a direct

[1188] M. Allegrini, L. Moi, and E. Arimondo, *Chem. Phys. Letters*, 1977, **45**, 245.
[1189] W. J. Childs and L. S. Goodman, *J. Opt. Soc. Amer.*, 1977, **67**, 1230.
[1190] L. F. Phillips, *J.C.S. Faraday II*, 1977, **73**, 97.
[1191] M. K. Wade, M. Czajkowski, and L. Krause, *Canad. J. Phys.*, 1978, **56**, 891.
[1192] W. Ho and A. B. Harker, *Analyt. Chem.*, 1976, **48**, 1780.
[1193] J. G. Eden, *Optics Comm.*, 1978, **25**, 201.
[1194] C. Linton, *J. Mol. Spectroscopy*, 1978, **69**, 351.
[1195] T. Ikeda, N. B. Wong, and D. O. Harris, *J. Mol. Spectroscopy*, 1977, **68**, 452.
[1196] R. N. Dixon and C. R. Webster, *J. Mol. Spectroscopy*, 1976, **62**, 271.
[1197] V. E. Bondybey and J. H. English, *J. Chem. Phys.*, 1977, **67**, 3405.
[1198] A. Fried and C. W. Mathews, *Chem. Phys. Letters*, 1977, **52**, 363.
[1199] R. N. Dixon, D. Field, and M. Noble, *Chem. Phys. Letters*, 1977, **50**, 1.
[1200] A. J. Rest and J. R. Sodeau, *J.C.S. Faraday II*, 1977, **73**, 1691.
[1201] D. G. Etlinger and T. M. Niemczyk, *J. Chem. Phys.*, 1978, **68**, 872.
[1202] P. J. Hargis and J. P. Hohimer, *I.E.E.E. J. Quantum Electron.*, 1977, **QE-13**, 10D.
[1203] M. Tajima, *Appl. Phys. Letters*, 1978, **32**, 719.
[1204] M. P. Miller, D. R. Tallant, F. J. Gustafson, and J. C. Wright, *Analyt. Chem.*, 1977, **49**, 1474.

measure of the PO_4 concentration.[1205] Delayed fluorescence in pure and doped $NaNO_2$ single crystals was used to give evidence of triplet excitons.[1206]

Phosphorescence.—There has certainly been much interest recently in establishing the conditions necessary for observation of phosphorescence at room temperature – a subject which has now been reviewed.[1207] Phosphorescence in fluid media (including common and useful solvents such as acetonitrile) was reported to be readily observable requiring only nitrogen purging, attention to concentration, and conventional purification schemes (*e.g.* halonaphthalenes in micellar solutions).[1208] Barium(II) was shown to promote $T_1 \leftarrow S_0$ transitions in solutions of quinoline. Lifetimes of simple arenes (pyrene, naphthalene *etc.*) and naphthalene[1209] in aqueous micellar solutions have been shown to be very sensitive to the purity of nitrogen used for deoxygenating, the quality of the water, and surfactant and general cleanliness of glassware and other apparatus used to make up solutions.[1210] Phosphorescence from solid substrates at room temperature has been reported previously (Vol. 8, p. 41) and recent studies have demonstrated the enhancement produced by heavy-atom compounds such as silver nitrate and sodium iodide with polycyclic hydrocarbons[1211] and alkali halides on adsorbed dyes.[1212] The effect of moisture and oxygen on organic phosphors adsorbed onto paper has been studied[1213] and sodium acetate reported to have analytical advantages of selectivity and insensitivity to moisture.[1214] The construction and use of a thin-layer phosphorimeter have been described, which permitted flexible chromatography media to be scanned at 77 K.[1215] Phosphorescence intensities could be enhanced by spraying the medium with solvent immediately before examination.

An analogue switch phosphoroscope was shown to be comparable in performance to a rotating can variety in the rejection of scattered or stray light and fluorescence but could be used at much higher chopping frequencies (250 Hz).[1216] An Osram 150-watt xenon lamp operated in a free-running mode at 16 kV produced surprisingly short (less than 1 μs) pulses suitable for phosphorescence studies with benzoylpyridines.[1217] Pesticidal carbamates could be examined with a more conventional 2 ms closing shutter and signal averager.[1218]

Following measurements by Saltiel and co-workers (J. Saltiel *et al.*, *J. Amer.*

1205 J. C. Wright, *Analyt. Chem.*, 1977, **49**, 1690.
1206 L. Schmidt, H. Port, and D. Schmid, *Chem. Phys. Letters*, 1977, **51**, 413.
1207 T. Vo Dinh and J. D. Winefordner, *Appl. Spectroscopy Rev.*, 1977, **13**, 261.
1208 N. J. Turro, K.-C. Liu, M.-F. Chow, and P. Lee, *Photochem. and Photobiol.*, 1978, **27**, 523.
1209 A. M. P. C. de Amorim, H. D. Burrows, S. J. Formosinho, and A. M. da Silva, *Spectrochim. Acta*, 1977, **33a**, 245.
1210 K. Kalyanasundaram, F. Grieser, and J. K. Thomas, *Chem. Phys. Letters*, 1977, **51**, 501.
1211 T. Vo Dinh, E. L. Yen, and J. D. Winefordner, *Talanta*, 1977, **24**, 146.
1212 W. White and P. G. Seybold, *J. Phys. Chem.*, 1977, **81**, 2035.
1213 E. M. Schulman and R. T. Parker, *J. Phys. Chem.*, 1977, **81**, 1932.
1214 R. M. A. von Wandruszka and R. J. Hurtubise, *Analyt. Chem.*, 1976, **48**, 1784; 1977, **49**, 2164.
1215 J. N. Miller, *U.V. Spectrometry Group Bull.*, 1977 (5), 8; J. N. Miller, D. L. Phillips, D. T. Burns, and J. W. Bridges, *Analyt. Chem.*, 1978, **50**, 613.
1216 E. L. Yen, G. B. Boutilier, and J. D. Winefordner, *Canad. J. Spectroscopy*, 1977, **22**, 120.
1217 G. Favaro and F. Masetti, *J. Phys. Chem.*, 1978, **82**, 1213.
1218 J. B. Addison, G. P. Semeluk, and I. Unger, *J. Luminescence*, 1977, **15**, 323.

Chem. Soc., 1970, **92**, 410), solutions of benzophenone in CCl_4 are recommended as reference materials for phosphorescence quantum yields since a measure of the triplet lifetime is a simple and convenient means of determining the quality of the sample and properly degassed solutions are reported to be stable for years.[1219] There is a need to correct for a refractive index difference between CCl_4 and water which results in an inner filter effect (R. B. Cundall and D. A. Robinson, *J.C.S. Faraday II*, 1972, **68**, 1133).

Two-photon excited phosphorescence of triphenylene in PMMA at 77 K has been measured using a tunable FLPD laser over an effective range 278—312 nm.[1220] The main advantage of phosphorescence detection was shown to lie in the long decay time which allows easy shuttering of scattered laser radiation and detection of minute signals. Two-photon fluorescence measurements always involve spectral rejection which is never absolute and leaves a considerable background. Reports of other laser-induced phosphorescence included gas phase thiophosgene with the 528.7 nm Ar^+ line[1221] and a high resolution study of anthracene single crystals with a tunable CW dye laser.[1222]

Studies of phosphorescence as a function of temperature[1223-1225] included arrangements to monitor spectra of polycyclic hydrocarbons, different solvents from 120 to 273 K as a magnetic field was varied from 0—6000 G,[1223] of benzene and [$^2H_6$]benzene from 2 to 30 K,[1224] and of dinaphthylalkanes at several temperatures.[1225] Other publications on phosphorescence dealt with the detection of slow rotational motions of proteins by steady-state phosphorescence anisotropy[1226] and studies of intramolecular relaxation processes within the lowest triplet state of 1,4-dibromonaphthalene by phosphorescence microwave photoexcitation spectroscopy.[1227]

Chemiluminescence.—Chemiluminescence observed in condensed phases can usually be detected with a conventional fluorescence spectrometer providing the lifetime of the emitting species is long or some steady-state mixing arrangement (such as a flow cell)[1228] is used. For example, an HCl detector has been described with a lower detection limit of $<5 \times 10^{-3}$ p.p.m. (by volume), which relies on HCl in air reacting with bromide–bromate solution to create bromine which, in turn, is quantitated by the chemiluminescent reaction with luminol.[1229] The visible light generated was found to be proportional to the HCl concentration of the sample air stream but the process was found to be subject to interference from strong acids and the halogen gases. Simultaneous and selective analyses of chlorine (as hypochlorous acid) and chlorine dioxide have been accomplished by measuring the chemiluminescence of reactions with luminol

[1219] M. A. Winnik and A. Lemire, *Chem. Phys. Letters*, 1977, **46**, 283.
[1220] L. Singer, Z. Baram, A. Ron, and S. Kimel, *Chem. Phys. Letters*, 1977, **47**, 372.
[1221] J. R. Lomardi, J. B. Koffend, R. A. Gottscho, and R. W. Field, *J. Mol. Spectroscopy*, 1977, **65**, 446.
[1222] H. Post and D. Rund, *Chem. Phys. Letters*, 1978, **54**, 474.
[1223] J. Spichtig, H. Bulska, and H. Labhart, *Chem. Phys.*, 1976, **15**, 279.
[1224] N. G. Kilmer and A. H. Kalantar, *Chem. Phys.*, 1978, **27**, 355.
[1225] S. Okajima, P. C. Subudhi, and E. C. Lim, *J. Chem. Phys.*, 1977, **67**, 4611.
[1226] G. B. Strambini and W. C. Galley, *Nature*, 1976, **260**, 554.
[1227] P. N. Prasad, A. I. Attia, and A. H. Francis, *Chem. Phys. Letters*, 1977, **46**, 125.
[1228] S. Stieg and T. A. Nieman, *Analyt. Chem.*, 1978, **50**, 401.
[1229] G. L. Gregory and R. H. Moyer, *Rev. Sci. Instr.*, 1977, **48**, 1464.

and H_2O_2 at two sites on the flow line.[1230] Chemiluminescence yields of luminol and related compounds have been determined by comparision with fluorescein under electron-beam excitation,[1231] and the versatile luminol has been used for assay of vitamin B_{12}.[1232]

A method developed to measure the absolute light output from a gas or liquid in a cylindrical tube can be applied to any source of chemiluminescence or fluorescence which will radiate uniformly within a cylindrical volume.[1233] In the arrangement described, a luminous surface disc was used as a piston source in a cylinder. An alternative procedure to measure absolute photon yields for chemiluminescence reactions under single collision conditions used the reaction between Sm and NO_2 as a standard.[1234]

Visible chemiluminescence has been observed on warming a u.v.-photolysed inert gas matrix containing O_3–H_2S, H_2S–O_2 or O_3–H_2S,[1235] or O_2–OCS,[1236] mixtures and analysed, in one case,[1235] by optical multichannel techniques.

Flow tubes are commonly employed for gas phase reactions in which chemilumescence techniques are employed for quantitative or qualitative analysis (see p. 61). Reactions of ozone with olefins, for example, were studied with a stopped-flow gas phase reactor.[1237] Nitrogen dioxide chemiluminescence produced from the reaction O + NO has become a useful indicator for reactions of either O or NO in arrangements as different as a crossed-beam[1238] or a low temperature argon matrix.[1239] Quantum yields of $O(^1D)$ from the near-u.v. photolysis of O_3 have been determined as a function of wavelength by a chemiluminescence technique.[1240]

Crossed-beam techniques are frequently used to characterize reaction between atomic beams of metals (such as Mg or Sr[1241] or Hg[1242] with halogens—see also ref. 1545). The chemiluminescent reaction of holmium atoms (produced by the ruby laser photolysis of a 1 μm thick holmium film) with N_2O has been studied by observing the emission of N_2^* with an OMA.[1234]

Raman and CARS.—The occasional humorous assertion of Raman spectroscopists concerning the near-universal presence of fluorescence interference in their samples has now been quantified by Hirschfeld who has revealed the grim fact that fluorescence is an intrinsic property of all samples at cross-sections similar to those for Raman spectroscopy.[1244] The only consolation is that the fluorescence may occur in a different spectral region. Practical solutions to

1230 U. Isacsson and G. Wettermark, *Analyt. Letters*, 1978, **A11**, 13.
1231 E. Wurzberg and Y. Haas, *Chem. Phys. Letters*, 1978, **55**, 250.
1232 T. L. Sheehan and D. M. Hercules, *Analyt. Chem.*, 1977, **49**, 446.
1233 P. H. Lee, G. A. Woolsey, and W. D. Slafer, *Appl. Optics*, 1976, **15**, 2825.
1234 C. R. Dickson, S. M. George, and R. N. Zare, *J. Chem. Phys.*, 1977, **67**, 1024.
1235 R. R. Smardzewski, *J. Chem. Phys.*, 1978, **68**, 2878.
1236 S. R. Long and G. C. Pimentel, *J. Chem. Phys.*, 1977, **66**, 2219.
1237 D. A. Hansen, R. Atkinson, and J. N. Pitts, *J. Photochem.*, 1977, **7**, 379.
1238 T. Kasai, T. Masui, H. Nakane, I. Hanazakiz, and K. Kuwata, *Chem. Phys. Letters*, 1978, **56**, 84.
1239 J. Fournier, J. Deson, and C. Vermeil, *J. Chem. Phys.*, 1978, **67**, 5688.
1240 G. K. Moortgat, E. Kudszus, and P. Warneck, *J.C.S. Faraday II*, 1977, **9**, 1216.
1241 A. Kowakski and J. Heldt, *Chem. Phys. Letters*, 1978, **54**, 240.
1242 S. Hayashi, T. M. Mayer, and R. B. Bernstein, *Chem. Phys. Letters*, 1978, **53**, 419.
1243 S. P. Tang, B. G. Wicke, and J. F. Friichtenicht, *J. Chem. Phys.*, 1978, **68**, 5471.
1244 T. Hirschfeld, *Appl. Spectroscopy*, 1977, **31**, 328.

eliminate fluorescence have included modulation of the optical frequency of a CW laser[1245] and the use of two different laser lines to excite alternatively a sample.[1246]

A double-beam Raman difference spectrometer for measurement of small changes (due to temperature effects, concentration difference *etc.*) uses a modulated Ar^+ laser and a synchronous photon-counting system to determine counts passing through sample cells.[1247] Pulsed Raman spectrometers employing mode-locked Ar^+ lasers and gated detection with time-to-amplitude techniques have been assembled in attempts to suppress unwanted fluorescence interference,[1248,1249] in one case to obtain high quality Raman spectra of benzene doped with the fluorophors acridine orange and rubrene.[1249] A general discussion covered the techniques of time-resolved Raman spectroscopy over time scales from 10 ns to 100 ms,[1250] and the use of laser Raman and fluorescence techniques in combustions diagnostics, with particular reference to many small molecules, has also been reviewed.[1251] The more instrumental aspects of resonance Raman scattering including details of laser sources and techniques for gas liquid and solid phase studies are covered in a comprehensive review.[1252] Other instrumental papers have dealt with obtaining polarization spectra by modulation techniques[1253] and a method for determining the polarization of scattered light.[1254] Applications have included measurement of the vibrational relaxation time of the $^nH_2$ molecule at low temperatures[1255] and detection of oil spills at sea using laser harmonics from a Nd laser and a gated OMA as a high speed detection system.[1256]

Coherent anti-Stokes Raman spectroscopy (CARS) generally offers solutions to the particular problems of Raman spectroscopy, namely the fluorescent or thermal background radiation and low sensitivity, and several reviews on the instrumental and theoretical aspects have been published,[1257] including an experimental technique for high temperature low pressure inorganic systems.[1258] The fluorescence rejection capability of CARS has been shown in a study of β-

[1245] F. L. Galeener, *Chem. Phys. Letters*, 1977, **48**, 7.
[1246] J. F. Morhange and C. Hirlimann, *Appl. Optics*, 1976, **15**, 2969.
[1247] M. Moskovits and K. Michaelian, *Appl. Optics*, 1977, **16**, 2044.
[1248] S. Burgess and I. W. Shepherd, *J. Phys. (E)*, 1977, **10**, 617.
[1249] J. M. Harris, R. W. Christian, F. E. Lytle, and R. S. Tobias, *Analyt. Chem.*, 1977, **48**, 1937.
[1250] A. Campion, M. A. El-Sayed, and J. Terner, *Proc. Soc. Photo-Opt. Instrum. Eng.*, 1977, **113**, 128.
[1251] A. C. Eckbreth, P. A. Bonczyk, and J. F. Verdieck, *Appl. Spectroscopy Rev.*, 1977, **13**, 15.
[1252] A. Compaan, *Appl. Spectroscopy Rev.*, 1977, **13**, 295.
[1253] L. I. Horvath and A. J. McCaffery, *J.C.S. Faraday II*, 1977, **73**, 562.
[1254] H. Engstrom, *Rev. Sci. Instr.*, 1976, **47**, 928.
[1255] C. Dellande and G. M. Gale, *Chem. Phys. Letters*, 1977, **50**, 339.
[1256] T. Sato, Y. Suzuki, H. Kashiwagi, M. Nanjo, and Y. Kakui, *J. Opt. Soc. Amer.*, 1977, **67**, 1363.
[1257] W. M. Tolles, J. W. Nibler, J. R. McDonald, and A. B. Harvey, *Appl. Spectroscopy*, 1977, **31**, 253; G. K. Klauminzer and J. Weber, in 'Laser 77 Opto-Electronics Conference Proceedings', ed. W. Waidelich, I.P.C., London, 1977, p. 183; M. D. Levenson, *Physics Today*, 1977 (5), 45.
[1258] I. R. Beattie, J. D. Black, T. R. Gilson, D. Greenhalgh, D. Hanna, and L. Laycock, in ref. 3, p. 7; J. D. Black, T. R. Gilson, D. A. Greenhalgh, and L. C. Laycock, *Laser Focus*, 1978 (3), 84.

naphthol,[1259] CARS in the u.v. with a frequency-doubled pulsed dye laser demonstrated with benzene and methanol[1260] and high resolution spectra (0.03 cm^{-1}) obtained for O_2, N_2, C_2H_2, and NH_3.[1261] The detection of β-carotene in benzene solution at *ca.* $5 \times 10^{-7} \text{ mol l}^{-1}$ was demonstrated with a CARS arrangement in which a single nitrogen laser was used to pump two dye lasers.[1262] Low laser powers were required to avoid saturation effects and avoid losing the CARS signal in the background emission. Negative peaks in CARS spectra were shown to arise from a cross-term between the background susceptibility and the imaginary part of the Raman susceptibility.[1263] A picosecond CARS experiment in which two incident pumping beams with their frequency difference equal to a vibrational frequency of calcite enabled a measured dephasing time of 7 ± 1 ps to be compared with the Raman linewidth measured on the same sample.[1264]

9 Transient Absorption Spectrometry

Conventional Flash Photolysis.—A versatile pulsed plasma source operating in the vacuum-u.v. to visible region consisted of two flat electrodes mounted on a dielectric surface (polythene or PTFE) and contained in a vacuum.[1265] In the wavelength region between 10 and 100 nm the spectrum, although discrete with powers of $10^{17} \text{ W sr}^{-1} \text{ m}^{-3}$ at the centre of the optically thick lines, was shown to match or exceed the spectral power levels obtained from high energy electron synchrotron devices. The pulse duration could be varied from 1 to 10 μs and by seeding the dielectric with inorganic compounds selected resonance lines could be produced. Ablative flashlamps normally use a low pressure gas to initiate a discharge, which vaporizes material from the tube wall leading to a high pressure discharge of ionized wall material. Two useful papers on high energy short-pulse flashlamps[1266, 1267] discuss the operating parameters of this lamp type. In one case, a 2 μs discharge in tubes of 1.4, 4.4, and 8 mm diameter was unaffected by gas type and pressure above loadings of 150 J cm^{-3} although xenon was always superior for light emission.[1267] A demountable quartz flashlamp containing a quartz rod reduced the 1% decay time for a 1000 J discharge from 40 to 15 μs, but at the expense of useful lamplife ($>10^4$ to *ca.* 100 shots).[1268] A conventional spark-gap-triggered linear lamp offered a choice of fast or slow flashlamp operation, 400 J dissipated in 146 μs when operated at 2 kV and 263 J dissipated in 2.7 μs when operated at 14.5 kV.[1269] A transverse discharge lamp used two 20 cm aluminium electrodes, machined to a $\pi/2$ Rogowski profile and filled with Hg vapour, H_2 at 0.5 Torr, and Ne at 320 Torr

1259 L. A. Carriera, L. P. Goss, and T. B. Malloy, *J. Chem. Phys.*, 1978, **68**, 280.
1260 L. A. Carriera, L. P. Goss, and T. B. Malloy, *J. Chem. Phys.*, 1977, **66**, 2762.
1261 W. Nitsch and W. Kiefer, *J. Mol. Structure*, 1978, **45**, 343.
1262 L. A. Carriera, T. C. Maguire, and T. B. Malloy, *J. Chem. Phys.*, 1977, **66**, 2621.
1263 L. A. Carriera, L. P. Goss, and T. B. Malloy, *J. Chem. Phys.*, 1977, **66**, 4360.
1264 C. H. Lee and D. Ricard, *Appl. Phys. Letters*, 1978, **32**, 168.
1265 B. A. Norton and E. R. Wooding, *J. Phys. (E)*, 1977, **10**, 493.
1266 T. Efthymiopoulos and B. K. Garside, *Appl. Optics*, 1977, **16**, 70.
1267 Y. Levy, G. Neumann, and D. Treves, *Appl. Optics*, 1977, **16**, 2293.
1268 A. Dunne, M. F. Quinn, and J. K. Taaffe, *J. Phys. (E)*, 1976, **9**, 918.
1269 C. C. Davis and R. J. Pirkle, *J. Phys. (E)*, 1976, **9**, 580.

to produce a FWHM duration of 0.5 μs at 0.125 J for a Hg-photosensitized experiment.[1270] The output pulses reported were shorter and more intense than the multipin lamps described earlier (Vol. 4, p. 105).

U.v. spectral efficiencies of surface-spark discharges were made in an attempt to find an improved pumping source for the iodine photodissociation laser and an electrical-to-optical conversion efficiency of $9.4 \pm 1.1\%$ for u.v. emission in the 250—290 nm region was obtained from discharges across a Cr_2O_3-Al_2O_3 substrate in Ar gas.[1271] The spectral distributions from 400 to 1300 nm were determined for pulsed and d.c. linear arc discharges containing alkali metal vapours and a general conclusion reached that the effective radiation temperature increases with increasing current density and decreasing pressure.[1272] A prepulse discharge (amounting to 10% of the available energy) through a flashlamp, which resulted in a much higher current than simmer operation, produced a 20% increase in laser efficiency (for a Nd:YAG rod) with better reproducibility ($+1\%$ compared with $\pm 6\%$) and quieter operation.[1273] An optically isolated HV trigger system has been built to eliminate ground loop and e.m.i. problems.[1274]

Other papers associated with flashlamp construction and operation have discussed the use of electro-optical methods (Faraday and Kerr effects) to measure rapidly varying high voltages and currents,[1275] the choice of reflector material in the 250—300 nm region (optimum found to be MgF_2-coated aluminium or $BaSO_4$ powder),[1276] switching an array of lamps with the use of a homopolar generator to supply 100 kA current[1277] and a fast flashlamp for T-jump (35 J in 4 μs).[1278] A description of a modulated excitation d.c. monitoring lamp was given in a study on the spectra of intermediates in the dye-sensitized oxidation of phenols.[1279]

An optical channel analyser was used to record the absorption spectrum of transient species of L-N-acetyl-tryptophanamide by means of a 6 μs spectroflash following conventional flash photolysis.[1280] It was noted that low dispersion gratings (100 lines mm^{-1}) produced considerable total stray light ($7 \pm 2\%$) distributed homogeneously over all the OMA channels. A procedure was described for using an on-line computer to acquire and process digital data from a transient decay over the time scale 1 μs to 1 s and to display data on a logarithmic time scale with signal-to-noise ratios optimized. The example chosen was the reassociation of myoglobin with CO following laser flash photolysis at a low temperature.[1281]

[1270] A. B. Callear, D. R. Kendall, and L. Krause, *Chem. Phys.*, 1978, **29**, 415.
[1271] R. E. Beverly, R. H. Barnes, C. E. Moeller, and M. C. Wong, *Appl. Optics*, 1977, **16**, 1572.
[1272] W. F. Hug, C. M. Hains, and C. J. Marlett, *J. Opt. Soc. Amer.*, 1978, **68**, 62.
[1273] J. A. Mroczkowski and R. H. Milburn, *Rev. Sci. Instr.*, 1977, **48**, 1555.
[1274] I. Henins and M. S. Kelly, *Rev. Sci. Instr.*, 1977, **48**, 168.
[1275] W. Botticher, Ch. Homann, and H. Hubner, *J. Phys. (E)*, 1978, **11**, 248.
[1276] R. E. Palmer, T. D. Padrick, and R. B. Pettit, *J. Appl. Phys.*, 1977, **48**, 3125.
[1277] G. B. Gillman, *J. Phys. (E)*, 1977, **10**, 876.
[1278] F. Survey and H. Strehlow, *J. Phys. (E)*, 1977, **10**, 1272.
[1279] M. J. Thomas and C. S. Foote, *Photochem. and Photobiol.*, 1978, **27**, 683.
[1280] R. F. Evans, W. A. Volkert, R. R. Kuntz, and C. A. Ghiron, *Photochem. and Photobiol.*, 1976, **24**, 3; R. F. Evans, C. A. Ghiron, W. A. Volkert, R. R. Kuntz, R. Santus, and M. Bazin, *Chem. Phys. Letters*, 1976, **42**, 39.
[1281] M. Sharrock, *Rev. Sci. Instr.*, 1977, **48**, 1202.

In a technique analogous to flash photolysis, minority carrier lifetimes in semiconductors have been determined from observation of photoconductive decay following the generation of excess carrier density by flash illumination. Secondary effects such as trapping and uneven sample illumination produced deviations from single exponential behaviour.[1282] Instrumentation for conductimetric detection of charged species formed in the pulse photolysis of solutions could be used for transient species having lifetimes from 1 μs to 100 ms.[1283] Absolute CIDNP enhancement factors were determined by the pulse photolysis of cyclododecanone contained in a cell within a nuclear spin echo–n.m.r. spectrometer.[1284] Flash spectroscopy in the 220—650 nm region has been accomplished following a stopped-flow mixing experiment.[1285]

The first results have been reported of an i.r. study of transient species using a continuously scanning Michelson interferometer to provide both temporal and spectral multiplex techniques.[1286, 1287] The time-resolved spectrum following the flash photolysis of acetone[1286] clearly showed an absorption at 2145 cm^{-1} attributable to CO. Later work on the photolysis products of acetaldehyde revealed HCHO features in emission and CH_4 with its strong Q-branch in absorption.[1288]

Conventional flash spectroscopy has been employed to study the gas phase reactions of $CBr(\tilde{X}^2\pi)$ produced from CHBr with a variety of paraffins, olefins *etc.*,[1289] transient species in gas phase thiophene attributed to the C_4H_3 radical,[1290] the absorption spectrum of HgH in the mercury photosensitized decomposition of water vapour,[1291] and ethanol,[1292] and to obtain extinction coefficients of the 2-methylallyl radical in the flash photolysis of 2-methylbut-1-ene and isobutene.[1293] Flash dissociation of methyl ketene and acrolein near 200 nm has been investigated using CO laser probing[1294] (see also ref. 1329—1331). Line broadening in the absorption spectrum of HNO was studied at high resolution using a conventional flash technique with a flashlamp-pumped Rh6G laser for the background continuum.[1295] A mixture of Xe–UF$_6$ was reported to yield the B–X transition of XeF in absorption following flash photolysis.[1296] Further studies in the gas phase included the kinetics of NH_2 reactions

1282 W. F. Richardson, J. M. Meese, and R. D. Westbrook, *Rev. Sci. Instr.*, 1978, **49**, 329.
1283 S. G. Ballard, *Rev. Sci. Instr.*, 1976, **47**, 1157.
1284 A. V. Dushkin, Y. A. Grishin, and R. Z. Sagdeev, *Chem. Phys. Letters*, 1978, **55**, 174.
1285 P. J. Benson and P. Moore, *J. Phys. (E)*, 1976, **9**, 787.
1286 A. W. Mantz, *Appl. Spectroscopy*, 1976, **30**, 459.
1287 A. W. Mantz, *Proc. Soc. Photo-Opt. Instrum. Eng.*, 1976, **82**, 54.
1288 A. W. Mantz, *Appl. Optics*, 1978, **17**, 1347.
1289 R. S. McDaniel, R. Dickson, F. C. James, O. P. Strausz, and T. N. Bell, *Chem. Phys. Letters*, 1976, **43**, 130.
1290 S. L. N. G. Krishnamachari and T. V. Venkitachalam, *Chem. Phys. Letters*, 1978, **55**, 116.
1291 S. L. N. G. Krishnamachari and R. Venkatasubramanian, *Mol. Photochem.*, 1977, **8**, 419.
1292 S. L. N. G. Krishnamachari and R. Venkatasubramanian, *Mol. Photochem.*, 1976, **7**, 295.
1293 F. Bayrakceken, *Chem. Phys. Letters*, 1976, **43**, 183.
1294 M. E. Umstead, R. G. Shortridge, and M. C. Lin, *J. Phys. Chem.*, 1978, **82**, 1455.
1295 P. A. Freedman, *Chem. Phys. Letters*, 1976, **44**, 605.
1296 A. L. Smith and P. C. Kobrinsky, *J. Mol. Spectroscopy*, 1978, **69**, 1.

with propylene following the flash photolysis of NH_3 mixtures.[1297] The NH_2 radicals were detected in a multipass arrangement by a CW single-mode dye laser. Flash photolysis of dicyanoacetylene (C_4N_2) at 160 nm with a 0.6 µs argon flashlamp was shown to produce $CN(X^2\Sigma^+)$, which could be determined, after a time delay, by laser-induced fluorescence.[1298] Chemical laser techniques were employed to study the photodissociative excitation of the cyanide radical generated by the flash photolysis of methyl isocyanide.[1299] Optical gain was observed on the 470 nm continuum of CdHg following flash excitation of Cd–Hg mixtures by monitoring with an Ar^+ laser.[1300] Population measurements of the $^3P_0$ and $^3P_1$ states of Hg (produced by the pulse photolysis of N_2–Hg gas mixtures) indicated that in the design of a mercury laser, any increase in power by increasing the optical pumping would be offset by the concomitant heating of the nitrogen vibration.[1301] Modulation kinetic spectroscopy has been used to investigate reaction of $Hg(6^3P)$ with O_2 in the presence of energy acceptors.[1302]

Characteristics of the triplet states of various rhodamine dyes[1303] and carbonyl-containing compounds[1304] have been studied by conventional flash photolysis. Benzenes in the gas phase[1305] and 23 polycyclic hydrocarbons in high resolution[1306] have also been examined by modulation excitation spectrophotometry.

The combination of flash photolysis with either atomic absorption or, to a lesser extent, atomic fluorescence is undoubtably a powerful one for studying atomic reactions, for example (of Group III—VI elements) and determining quenching cross-sections and rate constants. The monitoring sources of atomic emission required for atomic absorption studies are usually microwave-powered, which are usually more intense than the conventional hollow cathode lamps used for routine atomic absorption spectrometers. Among atomic reactions studied using atomic absorption were those of Ge,[1307] Sb,[1308] Si,[1309] P,[1310] Se,[1311] Sn,[1312] Pb,[1313] and Bi.[1314] For Bi, time-resolved atomic fluorescence

[1297] R. Leclaux, J. C. Soulignac, and P. V. Khe, *Chem. Phys. Letters*, 1976, **43**, 520.
[1298] M. J. Sabety-Dzvonik, R. J. Cody, and W. M. Jackson, *Chem. Phys. Letters*, 1976, **44**, 131.
[1299] J. T. Knudston and M. J. Berry, *J. Chem. Phys.*, 1978, **68**, 4419.
[1300] M. W. McGeoch and G. R. Fornier, *J. Appl. Phys.*, 1978, **49**, 2659.
[1301] J. Degani, E. Rosenfeld, and S. Yatsiv, *J. Chem. Phys.*, 1978, **68**, 4041.
[1302] H. Hippler, H. R. Wendt, and H. E. Hunziger, *J. Chem. Phys.*, 1978, **68**, 5103.
[1303] V. E. Korobov, V. V. Shubin, and A. K. Chibisov, *Chem. Phys. Letters*, 1977, **45**, 498.
[1304] M. V. Alfimov, N. Y. Buben, V. L. Glagolev, E. S. Kugumdzhi, Y. V. Pomazan, and V. N. Shamshev, *Optics and Spectroscopy*, 1977, **42**, 267.
[1305] M. E. Sime and D. Phillips, *Chem. Phys. Letters*, 1978, **56**, 138.
[1306] M. A. Slifkin and A. O. Al-Chalabi, *Spectrochim. Acta*, 1977, **33a**, 1091.
[1307] M. A. Chowdhury and D. Husain, *J. Photochem.*, 1977, **7**, 41; *J.C.S. Faraday II*, 1977, **73**, 1805.
[1308] D. Husain and N. K. H. Slater, *J. Photochem.*, 1977, **7**, 59; D. Husain, L. Krause, and N. K. H. Slater, *J.C.S. Faraday II*, 1977, **73**, 1706.
[1309] D. Husain and P. E. Norris, *Chem. Phys. Letters*, 1978, **53**, 474; 1977, **51**, 206; *J.C.S. Faraday II*, 1978, **74**, 93, 106, 335.
[1310] D. Husain and P. Norris, *J.C.S. Faraday II*, 1977, **73**, 1107, 1815.
[1311] R. J. Donovan and D. J. Little, *Chem. Phys. Letters*, 1978, **53**, 394.
[1312] J. R. Wiesenfeld and M. J. Yuen, *Chem. Phys. Letters*, 1976, **42**, 293.
[1313] P. J. Cross and D. Husain, *J. Photochem.*, 1977, **7**, 157; 1978, **8**, 183.
[1314] D. W. Trainor, *J. Chem. Phys.*, 1977, **67**, 1206, 3094.

techniques were also used to study ground-state atoms using an integral end-on capillary source coupled to a borosilicate cell with a Wood's horn used to reduce scattered excitation light.[1315] Interest has also been shown in halogen atoms including the reactions of Br,[1316] Cl,[1317] and iodine.[1318,1319]

A detailed description of an apparatus has been given to monitor the time-resolved atomic fluorescence of $I(5^2P_{\frac{1}{2}})$ produced by the flash photolysis of CD_3I and C_3F_7I under isothermal conditions.[1318] Flash spectroscopy has been used to study the mechanism of formation of IO in I_2–O_2 mixtures.[1319]

Time-resolved atomic absorption of N following the photoionization of buffered NO (e + $NO^+ \rightarrow$ N + O) used an arrangement in which the image of a 160 J nitrogen flashlamp in a cell was crossed with radiation from a nitrogen resonance lamp at 120 nm.[1320] Multichannel recording using up to 100 shots was required to obtain a reasonable SNR. Flash photolysis with resonance fluorescence was used to obtain rate constants for the reaction between Cl, NO, and N_2,[1321] O + NO + M (by resonance fluorescence detection of O),[1322] and to study the kinetics of $NH(X^3\Sigma^-)$ (by resonance fluorescence of NH) following the vacuum-u.v. flash photolysis of NH_3–NO mixtures.[1323]

A technique involving flash photolytic production and resonance fluorescence detection of oxygen atoms coupled with CW production of vibrationally excited reactant molecules (with a CO_2 laser) has been used to assess any effect of reactant vibrational energy on the dynamics of the reactions O + $C_2H_4^+$ and O + OCS^+.[1324] Rate constants for the reaction of $O(^3P)$ with C_2H_2 and C_2H_3Cl,[1325] NO,[1326] CH_3CN and CF_3CN,[1327] and $ClONO_2$[1328] have been determined following flash photolysis by resonance fluorescence detection of O. A CO laser resonant technique was used to study product formation following the flash-induced reactions of $O(^3P)$ with C_3O_2,[1329] with C_2F_4,[1330] and with but-1- and but-2-yne.[1331]

Chemiluminescence from NO_2 (produced directly or by titration with NO)

[1315] D. Husain, L. Krause, and N. K. H. Slater, *J.C.S. Faraday II*, 1977, **73**, 1678.
[1316] W. L. Ebenstein, J. R. Wiesenfeld, and G. L. Wolk, *Chem. Phys. Letters*, 1978, **53**, 185, J. V. Michael, J. H. Lee, W. A. Payne, and L. J. Stief, *J. Chem. Phys.*, 1978, **68**, 4093; H. Hippler, S. H. Luu, H. Teitelbaum, and J. Troe, *Internat. J. Chem. Kinetics*, 1978, **10**, 155.
[1317] J. V. Michael, D. A. Whytock, J. H. Lee, W. A. Payne, and L. J. Stief, *J. Chem. Phys.*, 1977, **67**, 3533; I. S. Fletcher and D. Husain, *Chem. Phys. Letters*, 1977, **49**, 516; R. G. Manning and M. J. Kurylo, *J. Phys. Chem.*, 1977, **81**, 291; D. A. Whytock, J. H. Lee, J. V. Michael, W. A. Payne, and L. J. Stief, *J. Chem. Phys.*, 1977, **66**, 2690.
[1318] R. J. Donovan, H. M. Gillispie, and R. H. Strain, *J.C.S. Faraday II*, 1977, **73**, 1553.
[1319] A. B. Callear and M. P. Metcalfe, *Chem. Phys.*, 1977, **20**, 233.
[1320] D. Kley, G. M. Lawrence, and E. J. Stone, *J. Chem. Phys.*, 1977, **66**, 4157.
[1321] J. H. Lee, J. V. Michael, W. A. Payne, and L. J. Stief, *J. Chem. Phys.*, 1978, **68**, 5410.
[1322] R. Atkinson, R. A. Perry, and J. N. Pitts, *Chem. Phys. Letters*, 1977, **47**, 197.
[1323] I. Hansen, K. Hoinghaus, C. Zetzsch, and F. Stuhl, *Chem. Phys. Letters*, 1976, **42**, 370.
[1324] R. G. Manning, W. Braun, and M. J. Kurylo, *J. Chem. Phys.*, 1976, **65**, 2609.
[1325] A. A. Westenberg and N. deHaas, *J. Chem. Phys.*, 1977, **66**, 4900.
[1326] D. A. Whytock, J. V. Michael, and W. A. Payne, *Chem. Phys. Letters*, 1976, **42**, 466.
[1327] R. J. Bonanno, R. B. Timmons, L. J. Stief, and R. B. Klemm, *J. Chem. Phys.*, 1977, **66**, 92.
[1328] M. J. Kurylo, *Chem. Phys. Letters*, 1977, **49**, 467; M. J. Kurylo and R. G. Manning, *Chem. Phys. Letters*, 1977, **48**, 279.
[1329] D. S. Y. Hsu and M. C. Lin, *J. Chem. Phys.*, 1978, **68**, 4347.
[1330] D. S. Y. Hsu and M. C. Lin, *Chem. Phys.*, 1977, **21**, 235.
[1331] M. E. Umstead and M. C. Lin, *Chem. Phys.*, 1977, **25**, 353.

was used to follow O atom reactions with n-butane and NO[1332] and aliphatic amines.[1333] Reactions of $O(2^1D_2)$ (produced by the flash photolysis of O_3 or N_2O) with halomethanes have been studied in a flow tube flash unit coupled to a nozzle beam mass spectrometer.[1334, 1335] Products such as ClO of primary and secondary radical reactions were monitored with time with the mass spectrometer set to a preselected mass number followed by signal averaging.[1334] Somewhat similar flash photolysis–resonance fluorescence techniques have been used to study hydrogen atom reactions[1336] and the reactions of hydroxy-radicals with a multitude of small molecules.[1337, 1338] The substantial interest in OH reactions is, of course, stimulated by their importance in atmospheric chemistry, particularly at stratospheric altitudes. The reaction between $OH + NO_2 + M$ is of special interest,[1338] since it connects the cycles based on HO_x and NO_x species, which destroy O_3 and converts the two reactive constituents of the stratosphere into HNO_3, which is relatively unreactive. Flash photolysis of O_3–H_2–Ar mixtures was used to observe a pure rotational collisionally-pumped OH laser emitting in the far-i.r.[1339]

A gas phase chemi-ionization process (*i.e.* reaction with atomic samarium) was used to detect O atoms produced by the vacuum-u.v. photolysis of O_2 and N_2O.[1340]

Nanosecond Flash Photolysis.—An elegant tandem laser experiment to follow the radiationless formation of triplet states from naphthalene in the gas phase used two laser pulses produced from the same nitrogen laser.[1341, 1342] The first pulse derived by frequency-doubling of a dye laser (0.2 cm⁻¹ bandwidth) was used to excite single vibronic levels and a second pulse, from another dye laser (bandwidth 60 cm⁻¹), was delayed and used to probe the excitation volume at various wavelengths. The naphthalene vapour was contained in a 5 m pathlength cell and the probe beam split into two equal and parallel beams, one of which overlapped the excitation beam.[1341] This double-pulse arrangement showed some similarity to one described previously (Vol. 6, p. 101), which has been used more recently to determine excited singlet state absorption cross-sections for a number of laser dyes in the pumping and lasing regions.[1342] The wavelength dependence of ISC quantum yield for all-*trans* and all-*cis* retinal

[1332] R. Atkinson, R. A. Perry, and J. N. Pitts, *Chem. Phys. Letters*, 1977, **47**, 197.
[1333] R. Atkinson and J. N. Pitts, *J. Chem. Phys.*, 1978, **68**, 911.
[1334] R. J. Donovan, K. Kaufmann, and J. Wolfrum, *Nature*, 1976, **262**, 204.
[1335] H. M. Gillespie, J. Garraway, and R. J. Donovan, *J. Photochem.*, 1977, **7**, 29.
[1336] J. H. Lee, J. V. Michael, W. A. Payne, and L. J. Stief, *J. Chem. Phys.*, 1978, **68**, 1817; J. H. Lee, L. J. Stief, and R. B. Timmons, *ibid.*, 1977, **67**, 1705; D. A. Whytock, J. V. Michael, W. A. Payne, and L. J. Stief, *ibid.*, 1976, **65**, 4871; D. A. Whytock, W. A. Payne, and L. J. Stief, *ibid.*, 1976, **65**, 191; J. H. Lee, J. V. Michael, W. A. Payne, D. A. Whytock, and L. J. Stief, *ibid.*, 1976, **65**, 3280.
[1337] For recent papers see R. Overend and G. Paraskevopoulos, *J. Chem. Phys.*, 1977, **67**, 674; R. A. Perry, R. Atkinson, and J. N. Pitts, *ibid.*, 1978, **67**, 5577; *Chem. Phys. Letters*, 1978, **54**, 14; R. Atkinson and J. N. Pitts, *J. Chem. Phys.*, 1978, **68**, 3581.
[1338] C. Anastasi and I. W. M. Smith, *J.C.S. Faraday II*, 1976, **72**, 1459.
[1339] J. H. Smith and D. W. Robinson, *J. Chem. Phys.*, 1978, **68**, 5474.
[1340] E. J. Stone, G. M. Lawrence, and C. E. Fairchild, *J. Chem. Phys.*, 1976, **65**, 5083.
[1341] H. Schroeder, H. J. Neusser, and E. W. Schlag, *Chem. Phys. Letters*, 1977, **48**, 12; 1978, **54**, 4.
[1342] E. Sahar and D. Treves, *I.E.E.E. J. Quantum Electron.*, 1977, **QE-13**, 962.

was determined from absorption measurements with two dye laser pulses also derived from a single nitrogen laser pulse.[1343] The probe beam was directed into a double monochromator enabling low beam intensities to be used. A boxcar integrator enabled the noise between the pulses to be minimized by gating the pump and probe signals. Fluorescence following $T_n \leftarrow T_1$ absorption in substituted anthracenes was detected using a double excitation method with a delay beteeen the first exciting flash and a laser pulse at 694 nm.[1344]

Since all two-photon mechanisms will show a quadratic dependence on the intensity of a single laser pulse, double-pulse laser experiments are required to distinguish between a consecutive two-photon and double-quantum absorption. Proof of a two-photon mechanism in the photoionization of pyrene came from laser flash photolysis studies using two 347 nm pulses separated by 40 ns. An intermediate state was identified as the vibrationally relaxed, lowest excited singlet of pyrene.[1345]

Transient absorption spectra of excited doublet states of thiocyanate complexes of Cr^{III} were determined using a nitrogen laser, which simultaneously pumped adjacent cells containing the sample and a fluorophor (quinine sulphate).[1346] Fluorescence passing through the sample was detected with a monochromator–OMA combination and covered the range 420—570 nm. The lower limit on wavelength was a consequence of a cut-off filter to eliminate Raman scattered light from water. A more conventional nanosecond laser kinetic spectrometer used an off-axial arrangement for a xenon monitoring source with a nitrogen laser to determine polarized and spectral data for the single state absorption of 1,4-diphenylnaphthalene.[1347] A boxcar integrator with a 350 ps window was employed to scan the time profile averaging as it scans or to average the data at a fixed time delay. Stimulated emission near 300 nm was observed following the KrF laser excitation of $Fe(CO)_5$.[1348]

An important method for extending the detection sensitivity of laser flash photolysis has involved excitation of a sample by an optical fringe pattern obtained by interfering two light sources from a high-power pulsed laser. Absorption coefficient distributions resulting from this excitation produce an amplitude grating, which is easily observed through the Bragg diffraction of a second, low-intensity CW laser beam. Preliminary experiments on the *cis–trans* isomerization of an azo-dye showed that absorbance changes of less than 10^{-4} could be detected.[1349]

There are again numerous examples of widely differing chemical systems which have been examined in absorption following laser flash photolysis. The nitrogen laser, for example, has been used for excitation and in studies on free radical

[1343] R. M. Hochstrasser and D. L. Marva, *Photochem. and Photobiol.*, 1977, **26**, 595.
[1344] S. Kobayashi, K. Kituchi, and H. Kokubun, *Chem. Phys.*, 1978, **27**, 399; *Chem. Phys. Letters*, 1976, **42**, 494.
[1345] G. A. Kenney-Wallace and K. Serantidis, *Chem. Phys. Letters*, 1978, **53**, 495; G. E. Hall and G. A. Kenney-Wallace, *Chem. Phys.*, 1978, **28**, 205.
[1346] S. C. Pyke, M. Ogasawara, L. Kevan, and J. F. Endicott, *J. Phys. Chem.*, 1978, **82**, 302.
[1347] E. L. Russell, A. K. Twarowski, D. S. Kliger, and E. Switzes, *Chem. Phys.*, 1977, **22**, 167.
[1348] D. W. Trainor and S. A. Mani, *J. Chem. Phys.*, 1978, **68**, 5481.
[1349] F. Rondelez, H. Hervet, and W. Urbach, *Chem. Phys. Letters*, 1978, **53**, 138.

formation in crystalline cysteine,[1350] hydrated electron formation in aqueous β-naphthol,[1351] photoionization and isothermal recombination luminescence of NN-disubstituted dihydrophenazines at 77 K,[1352] *cis–trans* isomerization of thioindigo,[1353] exciplex formation in fumaronitrile,[1354] and transient absorption of pyrene–diethylaniline mixtures.[1355] 1,4-Biradicals generated during the laser photolysis of γ-methyl valerophenone–oxygen mixtures were monitored in absorption using a He–Ne laser.[1356] The lifetime of $O(^1\Delta_g)$ was determined from observations of the decay of 1,3-diphenylisobenzofuran in micelles following dye laser[1357] and nitrogen laser excitation.[1358] The voltage waveforms from the monitoring photomultiplier were transferred to a sequence of 2048 voltage-time co-ordinates in a Biomation digitizer on-line to a computer system, thereby enabling immediate transformation to absorbance and ease of data analysis.[1358] The frequency-quadrupled pulse from a Nd laser at 265 nm has been used for nanosecond studies of substituted benzenes in the gas phase[1359] and unimolecular isomerization of substituted cycloheptatrienes.[1360]

Dye laser excitation has been used in studies on the photoreduction of methylene blue by amines[1361] and the properties of $I^*(^2P_{\frac{1}{2}})$ following photolysis of C_2H_5I at 295 nm[1362] and I_2.[1363] Atom products of dye laser photolysis of Br_2 have been monitored in the vacuum-u.v. by resonance absorption.[1364]

Transient absorption spectra induced in silver halide photochromic glasses by 347 nm doubled ruby laser radiation have been monitored with a flash and monochromator–vidicon arrangement.[1365] Other examples of chemical systems examined following 347 nm photolysis have included photochromism in 2,2'-dinitrophenylmethane,[1366] biradicals in γ-methylvalerophenone[1367] (see also ref. 1356), the decay rate of 1,8-ANS in aqueous and ethanolic solutions,[1368] and photoisomerization processes in IR-140 laser dye.[1369] Trapped electrons in 9.5 M-LiCl and 3.1 M-$MgCl_2$ aqueous glasses at 77 K were bleached by single pulses of light at 694 nm from a Q-switched ruby laser to produce an absorption in the i.r. with an associated luminescence in the visible and near-u.v.

[1350] P. I. Richter, J. E. Johnson, L. I. Finch, and G. C. Moulton, *J. Chem. Phys.*, 1976, **65**, 5527.
[1351] U. Lachish, M. Ottolenghi, and G. Stein, *Chem. Phys. Letters*, 1977, **48**, 402.
[1352] U. Bruhlmann and J. R. Huber, *J. Phys. Chem.*, 1977, **81**, 386.
[1353] K. H. Grellmann and P. Hentzschel, *Chem. Phys. Letters*, 1978, **53**, 545.
[1354] H. Hayashi and S. Nagakura, *Chem. Phys. Letters*, 1978, **53**, 201.
[1355] I. P. Bell and M. A. J. Rodgers, *Chem. Phys. Letters*, 1976, **44**, 249.
[1356] R. D. Small and J. C. Scaiano, *Chem. Phys. Letters*, 1977, **48**, 354.
[1357] I. B. C. Matheson, J. Lee, and A. D. King, *Chem. Phys. Letters*, 1978, **55**, 49.
[1358] A. A. Gorman and M. A. J. Rodgers, *Chem. Phys. Letters*, 1978, **55**, 52.
[1359] R. Bonneau, M. E. Sime, and D. Phillips, *J. Photochem.*, 1978, **8**, 239.
[1360] H. Hippler, K. Luther, J. Troe, and R. Walsh, *J. Chem. Phys.*, 1978, **68**, 323.
[1361] R. H. Kayser and R. H. Young, *Photochem. and Photobiol.*, 1976, **24**, 395, 403.
[1362] F. J. Comes and S. Pionteck, *Chem. Phys. Letters*, 1976, **42**, 558.
[1363] D. H. Burde and R. A. McFarlane, *Chem. Phys.*, 1976, **16**, 295.
[1364] T. G. Lindemann and J. R. Wiesenfeld, *Chem. Phys. Letters*, 1977, **50**, 364.
[1365] C. L. Marquardt, J. F. Giuliani, and R. T. Williams, *J. Appl. Phys.*, 1976, **47**, 4915.
[1366] D. Klemm, E. Klemm, A. Graness, and J. Kleinschmidt, *Chem. Phys. Letters*, 1978, **55**, 503.
[1367] R. D. Small and J. C. Scaiano, *Chem. Phys. Letters*, 1977, **50**, 431.
[1368] G. R. Fleming, G. Porter, R. J. Robbins, and J. A. Synowiec, *Chem. Phys. Letters*, 1977, **52**, 228.
[1369] J.-P. Fouassier, D.-J. Loughnot, and J. Faure, *Optics Comm.*, 1977, **23**, 393.

(due to an unidentified species).[1370] A laser flash photolysis was also employed to study the spectral properties of benzanthrone dyes on poly(ethylene terephthalate) fabric,[1371] and various tricarbocyanine dyes.[1372]

Transient absorption measurements following CO_2 laser photolysis of N_2F_4 with both i.r. and 254 nm monitoring beams enabled initial dissociation rates into NF_2 to be followed.[1373] A strong Q-switched N_2O laser capable of a 60 MHz frequency sweep was used to induce population inversion by optical adiabatic rapid passage on the ν_2 transition of $^{14}NH_3$.[1374] The fast inversion and subsequent return to equilibrium were detected by a weak counterpropagating N_2O laser pulse.

Sub-nanosecond Photophysical Techniques.—An extremely useful compilation of reviews on picosecond techniques and applications contains chapters by D. J. Bradley (picosecond pulse production), E. P. Ippen and C. V. Shank (techniques of pulsewidth measurements and measurements of picosecond events), D. H. Auston (non-linear optics), D. von der Linde (picosecond interactions in liquids and solids), and K. B. Eisenthal (picosecond relaxation measurements in biology).[1375] Each chapter has been written by a known authority and the whole volume is commended to the picosecond researcher. A further review has covered sub-nanosecond dynamic processes in polyatomic molecules in liquids.[1376] An equally useful collection of 66 papers presented at a conference in May 1978 from an excellent review of the current progress in experimental techniques used to measure picosecond phenomena.[1377]

Before dealing with the more common direct methods for picosecond spectroscopy, it is worth noting that i.r. line broadening has been suggested as a means for estimation of sub-picosecond lifetimes of collisional complexes[1378] and that a fast signal averager combining the time resolution of a sampling oscilloscope, signal averaging capabilities of an MCA, and processing power of a minicomputer is claimed to be capable of sub-nanosecond time resolution.[1379] The technique of picosecond hole-burning has been utilized to measure sub-nanosecond lifetimes of excited states in crystals at 2 K[1380] and transient grating methods discussed in relation to lifetime measurements.[1381]

Probe Technique. A brief description of this simple arrangement for transient absorption measurements has been given previously (Vol. 8, p. 49) and more recent work has perhaps concentrated on using different wavelengths for

[1370] H. A. Gillis and D. C. Walker, *J. Chem. Phys.*, 1976, **65**, 4590.
[1371] P. Bentley and J. F. McKellar, *J. Photochem.*, 1976, **5**, 377.
[1372] J.-P. Fouassier, D. J. Loughnot, and J. Faure, *Optics Comm.*, 1976, **18**, 263.
[1373] P. Lavigne, J. L. Lachambre, and G. Otis, *Optics Comm.*, 1977, **22**, 75.
[1374] S. M. Hamadani, N. A. Kurnit, and A. Javan, *Chem. Phys. Letters*, 1977, **49**, 277.
[1375] 'Ultrashort Light Pulses. Picosecond Techniques and Applications', ed. S. L. Shapiro, Springer-Verlag, Heidelberg, 1977.
[1376] A. Laubereau and W. Kaiser, in 'Chemical and Biological Applications of Lasers', ed. C. B. Moore, vol. 2, Academic Press, New York, 1976.
[1377] 'Picosecond Phenomena', ed. C. V. Shank, E. P. Ippen, and S. L. Shapiro, Springer-Verlag, Heidelberg, 1978.
[1378] S. Weiss, *J. Chem. Phys.*, 1978, **67**, 5735.
[1379] I. C. Plumb, G. H. Cooper, and D. G. Heap, *J. Phys. (E)*, 1977, **10**, 744.
[1380] H. de Vries and D. A. Wiersma, *Chem. Phys. Letters*, 1977, **51**, 565.
[1381] H. J. Eichler, *Optica Acta*, 1977, **24**, 631; A. E. Siegman, *Appl. Phys. Letters*, 1977, **30**, 21.

excitation and monitoring. For example, tunable i.r. pulses (7000—2500 cm^{-1}) produced by parametric three-photon processes in $LiNbO_3$ with a Nd : glass laser has been used to excite vibrational modes of a dye vapour such as coumarin 6 followed by delayed 530 nm pulses for interrogation.[1382] The green pulses promote the excited molecules to the fluorescence first singlet state and thereby enable vibrational relaxation times (of 4 ± 1 ps for an overtone at 5950 cm^{-1}) to be measured. Excess population in CHI-stretching modes in molecules such as $CHCl_3$ and CH_3CH_2OH generated by pumping with picosecond 1057 nm pulses were determined by anti-Stokes scattering of subsequent ultrashort 530 nm probe pulses.[1383] Transient absorptions induced in laser dyes by 347 nm pulses were measured by a 347 nm-induced fluorescence probe beam.[1384] Further studies of transient absorptions in DODCI used a single pulse extracted from a mode-locked Rh6G dye laser for both excitation and monitoring[1385] and a kinetic model to simulate fast processes in this molecule compared with picosecond bleaching experiments.[1386] Kinetics of energy transfer and photo-oxidation in photosynthetic bacteria were determined with a delayed probe technique by using a 530 nm actinic and 1242 nm (from stimulated Raman scattering) probe pulse.[1387] Among other investigations reported using probe techniques were singlet state populations in stilbene,[1388] photoselection in gallium phthalocyanine,[1389] excited state build-up in anthrone,[1390] and studies of azulene fluorescence[1391] and erythrosin absorption.[1392] A picosecond arrangement for monitoring $I_\perp$ and $I_\parallel$ transmissions at 694 nm following photo-excitation at 347 nm of the model compound anthryl-$(CH_2)_3$-NN-dimethyl-aniline was used to study geometrical requirements and the effects of solvent polarity on excited-state charge-transfer complexes.[1393] Absorption anisotropy of rhodamine dye molecules has been measured using a delayed polarized probe pulse following partial optical bleaching.[1394] The excess population in individual CH_3-stretching modes excited by a single tunable i.r. pulse was monitored by subsequent probe pulses using spontaneous anti-Stokes Raman scattering.[1395] Absorption at the 530 nm excitation wavelength has been shown to recover in a rapid two-state process following bleaching of tris-(1,10-phenanthroline)-iron(II).[1396]

Light Gate Technique. Although the CS_2 shutter is capable of offering high time

[1382] J. P. Maier, A. Seilmeier, A. Laubereau, and W. Kaiser, *Chem. Phys. Letters*, 1977, **46**, 527.
[1383] A. Laubereau, S. F. Fischer, K. Spanner, and W. Kaiser, *Chem. Phys.*, 1978, **31**, 335.
[1384] J.-P. Fouassier, D. J. Loughnot, F. Wielder, and J. Faure, *J. Photochem.*, 1977, **7**, 17.
[1385] J. Jaraudias, P. Goujon, and J. C. Mialocq, *Chem. Phys. Letters*, 1977, **45**, 107.
[1386] G. L. Olson, K. S. Greve, and G. E. Busch, *J. Chem. Phys.*, 1978, **68**, 1474.
[1387] E. Moskowitz and M. M. Malley, *Photochem. and Photobiol.*, 1978, **27**, 55.
[1388] H. Neumann, W. Triebel, R. Uhlmann, and B. Wilhemi, *Chem. Phys. Letters*, 1977, **45**, 425.
[1389] D. Magde, *J. Chem. Phys.*, 1978, **68**, 3717.
[1390] G. W. Scott and L. D. Talley, *Chem. Phys. Letters*, 1977, **52**, 431.
[1391] J. P. Heritage and A. Penzkofer, *Chem. Phys. Letters*, 1977, **45**, 76.
[1392] A. Al-Obaidi and G. A. Oldershaw, *J. Photochem.*, 1976, **6**, 153.
[1393] K. Gnaedig and K. B. Eisenthal, *Chem. Phys. Letters*, 1977, **46**, 339.
[1394] A. Penzkofer and W. Falkenstein, *Chem. Phys. Letters*, 1976, **44**, 547.
[1395] K. Spanner, A. Laubereau, and W. Kaiser, *Chem. Phys. Letters*, 1976, **44**, 88.
[1396] A. J. Street, D. M. Goodall, and R. C. Greenhow, *Chem. Phys. Letters*, 1978, **56**, 326.

resolution,[1397] it is necessary to use a large photon input to the sample (10^{14} photons cm^{-2}), and measurements can only be made over a low dynamic range. One of the few alternatives to CS_2 proposed has been cryptocyanine in glycerin, which was reported to be more transmissive at giant-pulse excitation levels up to 10 MW cm^{-2}.[1398] A more extreme variation as a shutter consisted of a thin film of aluminium which has an opening time of a few ps on irradiation by a 100 ps duration, 1 J Nd pulse.[1399] Applications of the optical gate technique to fluorescence lifetime determinations have included studies on pseudoisocyanine,[1400] azobenzene, [1401] various polymethine dyes,[1402] xanthione,[1403] and *sym*-tetrazine.[1404] A reflection echelon and optical gate technique was used to obtain fluorescence risetimes of *p*-dimethylaminobenzonitrile in various solvents,[1405] and time delays in stimulated Raman emission from ethanol.[1406] Relaxation kinetics of supercooled liquid phenyl salicylate have been investigated using the opticao Kerr effect with intense 1060 nm pulses to induce birefringence of the sample between crossed polarizers and time-delayed 530 nm pulses to probe the time evolution of the transitory birefringence.[1407] Other relaxation studies with this arrangement have been described elsewhere.[1408]

A method giving higher signal-to-background ratios than a standard Kerr light gate has used sum and difference frequency-conversion in a type-II phase-matched ADP crystal.[1409] Harmonic beams from a Nd laser were used to excite the fluorescence (in dyes such as rhodamine, erythrosin, and xanthione) and both the fluorescence signal and the laser fundamental were passed through a KDP crystal for detection. The main limitation of the technique was found to be scattered light arising from the laser harmonics but the time resolution is <10 ps. Fluorescence below 400 nm (the CS_2 wavelength cut-off) may be detected with this technique.

There are now numerous descriptions of picosecond spectrometers employing SPM continuum generation for transient absorption measurements,[1410] to which the reader can be referred. Alternative materials to CCl_4 and phosphoric

1397 W. S. Struve, *Optics Comm.*, 1977, 1977, **21**, 215.
1398 D. W. Vahey, *J. Appl. Phys.*, 1976, **47**, 3057.
1399 B. H. Ripin, U. Feldman, and G. A. Doschek, *Rev. Sci. Instr.*, 1977, **48**, 935.
1400 F. Fink, E. Klose, K. Teuchner, and S. Dahne, *Chem. Phys. Letters*, 1977, **45**, 548.
1401 W. S. Struve, *Chem. Phys. Letters*, 1977, **46**, 15.
1402 H. Tashiro and T. Yajima, *Chem. Phys. Letters*, 1976, **42**, 553.
1403 R. W. Anderson, R. M. Hochstrasser, and H. J. Pownall, *Chem. Phys. Letters*, 1976, **43**, 224.
1404 R. M. Hochstrasser, D. S. King, and A. C. Nelson, *Chem. Phys. Letters*, 1976, **42**, 8.
1405 W. S. Struve and P. M. Rentzepis, *J. Mol. Structure*, 1978, **47**, 273.
1406 E. Pochon and M. Bourene, *J. Chem. Phys.*, 1976, **65**, 2056.
1407 P. P. Ho and R. R. Alfano, *J. Chem. Phys.*, 1977, **67**, 1004.
1408 P. P. Ho and R. R. Alfano, *Chem. Phys. Letters*, 1977, **50**, 74; *J. Chem. Phys.*, 1978, **68**, 4551; *J. Opt. Soc. Amer.*, 1978, **68**, 678.
1409 L. A. Halliday and M. R. Topp, *Chem. Phys. Letters*, 1977, **46**, 8; 1977, **48**, 40.
1410 T. Kobayashi and S. Nagakura, *Chem. Phys. Letters*, 1976, **43**, 429; R. M. Hochstrasser and A. C. Nelson, *Optics Comm.*, 1976, **18**, 361. H. Shizuka, K. Matsui, Y. Hirata, and I. Tanaka, *J. Phys. Chem.*, 1976, **80**, 2070; A. Matsuzaki, T. Kobayashi, and S. Nagakura, *ibid.*, 1978, **82**, 1201; V. Sundstrom, P. M. Rentzepis, and E. C. Lim, *J. Chem. Phys.*, 1977, **66**, 4287; G. E. Busch, D. Huppert, and P. M. Rentzepis, *Proc. Soc. Photo-Opt. Instrum. Eng.*, 1976, **82**, 80; P. Avouris and P. M. Rentzepis, *ibid.*, 1977, **113**, 1; D. Huppert, P. M. Rentzepis, and D. S. Kliger, *Photochem. and Photobiol.*, 1977, **25**, 193.

acid for spectral broadening have included a mixture of 60 % H_2O–40 % D_2O [1411] and neat D_2O. [1412] In the latter case, the spectrum showed practically no structure and extended from at least 380 to 800 nm. [1412] It would be inappropriate to list many of the chemical systems which have been studied in absorption by picosecond techniques, and the following examples are not necessarily representative of the extensive studies in this field. It has been possible to observe the build-up of triplet absorption of acridine and phenazine [1413] and carbonyl derivatives of anthracene, [1414] to study exciplex kinetics in anthracene-*NN*-diethylaniline mixtures, [1415] excited state lifetimes and spectra of dimethyl-aminobenzaldehyde, [1416] DTTC and HITC, [1417] NO_2, [1418] and intermediate photoproducts of rhodopsin. [1419]

Streak Cameras. Streak cameras offer 0.5—5 ps time resolution and a dynamic range of 100, but require a relatively high excitation intensity at the sample (10^{12} photons cm^{-2}). A compilation of papers on high speed optical techniques includes reports on developments in streak cameras and their use in studying transient phenomena in photosynthetic material. [1420] An image converter streak camera with a streak record time ranging from 50 ns to 10 µs has been described [1421] together with a comment on the development of a sub-picosecond camera system. [1422] The dynamic range of two streak cameras has been shown to depend on the time resolution employed. [1423] For events up to *ca.* 2 ps, a useful dynamic range of 30 was reported, increasing to 180 for 30 ps events. Picosecond fluorescence decays of dyes have been measured using a streak camera operating repetitively at up to 140 MHz. [1424] Successive streaks were precisely superimposed on the tube phosphor at *ca.* 10^8 pulses s^{-1} to produce an essentially steady-state image. The time resolution was reported to be <10 ps and non-linear effects were avoided by keeping individual pulse powers very low. A streak camera–OMA detection system used for time-dependent fluorescence

[1411] M. J. Colles and G. E. Walrafen, *Appl. Spectroscopy*, 1976, **30**, 463.
[1412] D. K. Sharma, R. W. Yip, D. F. Williams, S. E. Sugamori, and L. L. T. Bradley, *Chem. Phys. Letters*, 1976, **41**, 460.
[1413] Y. Hirata and I. Tanaka, *Chem. Phys.*, 1977, **25**, 381; L. J. Noe, E. O. Degenkolb, and P. M. Rentzepis, *J. Chem. Phys.*, 1978, **68**, 4435; Y. Hirata and I. Tanaka, *Chem. Phys. Letters*, 1976, **41**, 336; 1976, **43**, 568.
[1414] S. Hirayama and T. Kobayashi, *Chem. Phys. Letters*, 1977, **52**, 55.
[1415] T. Nishimura, N. Nakashima, and N. Mataga, *Chem. Phys. Letters*, 1977, **46**, 334; H. Fujiware, N. Nakashima, and N. Mataga, *ibid.*, 1977, **47**, 185.
[1416] S. Dahne, E. Heumann, I. Kapp, W. Triebel, and B. Wilhelmi, *J. Opt. Soc. Amer.*, 1978, **68**, 683; E. Heumann, I. Kapp, W. Triebel, and B. Wilhelmi, *J. Mol. Structure*, 1978, **45**, 395.
[1417] T. Kobayashi and S. Nagakura, *Chem. Phys.*, 1977, **23**, 153.
[1418] W. L. Faust, L. S. Goldberg, T. R. Royt, J. N. Bradford, E. J. Friebele, J. M. Schnur, P. G. Stone, and R. G. Weiss, *J. Opt. Soc. Amer.*, 1978, **68**, 639.
[1419] V. Sundstrom, P. M. Rentzepis, K. Peters, and M. L. Applebury, *Nature*, 1977, **267**, 645; Y. Shichida, T. Kobayashi, H. Ohtani, T. Yoshizawa, and S. Nagakura, *Photochem. and Photobiol.*, 1978, **27**, 335.
[1420] *Proc. Soc. Photo-Opt. Instrum. Eng.*, 1976, vol. 94.
[1421] D. G. Lewis, M. R. Barrault, and R. Smith, *J. Phys. (E)*, 1978, **11**, 409.
[1422] A. J. Lieber, H. D. Sutphin, C. B. Webb, and A. H. Williams, *Electro-Optical System Design*, 1976 (9), 26; A. J. Lieber, H. D. Sutphin, R. C. Hyer, J. S. McGurn, and K. R. Winn, *Proc. Soc. Photo-Opt. Instrum. Eng.*, 1977, **122**, 29.
[1423] D. J. Bradley, S. F. Bryant, J. R. Taylor, and W. Sibbett, *Rev. Sci. Instr.*, 1978, **49**, 215.
[1424] M. C. Adams, D. J. Bradley, M. M. Salour, and W. Sibbett, *J. Opt. Soc. Amer.*, 1978, **68**, 665.

depolarization measurements on Rose Bengal and eosin revealed that excess total light produced stray electron signals in the streak camera causing a distortion in the fluorescence decay curves.[1425] With this arrangement, the linearity was also affected by a lag problem in the OMA (for single scans), which could only be held to within $\pm 5\%$ of the pulsed light intensity in the range 100—2000 counts by adjustment of the cathode voltage.[1426] Linearity was clearly essential in order to observe non-exponential decays of tryptophan, ANS, and TNS in mixed solvents.[1427] A signal averager coupled to a streak camera–OMA combination allowed 50—100 single short measurements of fluorescence decays of solid tetracene to be summed in a study of the decay dependence on excitation intensity.[1428] Fluorescence lifetimes of *trans*-stilbene excited at 265 nm were determined with a streak camera with digitization provided by means of a TV camera and microcomputer.[1429]

This technique has also been used to study lifetimes of laser dyes,[1430] energy transfer between Rh6G and malachite green,[1431] the risetime of acetone fluorescence following the photolysis of tetramethyldioxetane at 264 nm,[1432] and rotational diffusion time constants of tetrachlorotetraiodofluorescein in various low viscosity, hydrogen-bonding solvents.[1433]

CW Mode-locked Lasers. The continuous mode-locked dye laser offers certain unique properties for picosecond diagnostics, namely extremely short pulse duration (<0.2 ps), wide wavelength tunability (typically 530—900 nm for the fundamental and 265—450 nm for the second harmonic), and repetition rates from 200 MHz to a few Hz. Measurements of transient absorption in samples with these lasers have used a variation of the delay line technique in which the absorption (or gain) of a weak probe pulse has been measured as a function of the delay between the excitation and probe pulse. The probe wavelength may be the same as that of the excitation or a harmonic (as in, for example, ref. 1434) or could originate from a second dye laser synchronously pumped by the same ion laser as the excitation laser, but operating at a different wavelength: see ref. 303. In one simple example of a probe technique, the second harmonic at a dye laser was used to excite very short-lived upper singlet states of coronene and the fundamental at 615 nm was used to probe the excited state absorptions

1425 G. W. Robinson, J. M. Morris, R. J. Robbins, and G. R. Fleming, *Proc. Soc. Photo-Opt. Instrum. Eng.*, 1977, **113**, 13; G. R. Fleming, J. M. Morris, and G. W. Robinson, *Chem. Phys.*, 1976, **17**, 91.
1426 G. R. Fleming, A. E. W. Knight, J. M. Morris, R. J. Robbins, and G. W. Robinson, *Chem. Phys. Letters*, 1977, **49**, 1.
1427 G. W. Robinson, J. M. Morris, R. J. Robbins, and G. R. Fleming, *J. Mol. Structure*, 1978, **45**, 221.
1428 A. J. Campillo, R. C. Hyer, S. L. Shapiro, and C. E. Swenberg, *Chem. Phys. Letters*, 1977, **48**, 495; A. J. Campillo, S. L. Shapiro, and C. E. Swenberg, *ibid.*, 1977, **52**, 11.
1429 M. Sumitani, N. Nakashima, K. Yoshihara, and S. Nagakura, *Chem. Phys. Letters*, 1977, **51**, 183.
1430 G. R. Fleming, A. E. W. Knight, J. M. Morris, R. J. Robbins, and G. W. Robinson, *Chem. Phys. Letters*, 1977, **51**, 399; *Chem. Phys.*, 1977, **23**, 61.
1431 G. Porter and C. J. Tredwell, *Chem. Phys. Letters*, 1978, **56**, 278.
1432 K. K. Smith, J.-Y. Koo, G. B. Schuster, and K. F. Kaufman, *Chem. Phys. Letters*, 1977, **48**, 267.
1433 G. Porter, P. J. Sadkowski, and C. J. Tredwell, *Chem. Phys. Letters*, 1977, **49**, 416.
1434 H. E. Lessing, A. von Jena, and M. Reichert, *Chem. Phys. Letters*, 1976, **42**, 218.

produced (S_3).[1435] A similar arrangement was used to measure optically induced conformational changes in 1,1'-binaphthyl in different liquids [1436] and to study the dynamics of intramolecular motion in the singlet states of *cis*- and *trans*-stilbene.[1437] Further applications of the synchronously-pumped dye laser for sub-nanosecond fluorescence lifetimes are given in Section 10.

10 Transient Emission Spectroscopy

Although time-correlated photon counting offers the advantages of analysis of non-exponential decays and time-resolved spectra of short-lived fluorophors, the resolution of this technique with plasma discharge excitation sources is usually limited to about 0.5 ns, depending on the fluorophor and experimental arrangement used. Although laser sources have now extended the time resolution of photon counting to a fraction of a nanosecond, two significant papers have revealed how plasma sources can be used in certain experimental arrangements to measure lifetimes to 90 ps.[1438, 1439] Two key experimental changes were made, first the operating characteristics of the lamp were adjusted to minimize the spread and light pulse profile and second, a new constant fraction timing discriminator was used, which had a time walk of no more than ± 35 ps over a 50 mV to 5 V input.[1438] An important evaluation of a typical lifetime system employing a deuterium gated lamp relied on self-quenching of concentrated solutions of N-methylpiperidine in n-hexane to generate solutions fluorescing with known sub-nanosecond lifetimes.[1440] A lower limit for reliable measurements was placed at *ca.* 0.2 ns for flashlamp having a FWHM of 2.6 ns and 1/e value of 0.8 ns. Reports of more conventional arrangements included a description of a high pressure deuterium or deuterium–neon flashlamp operated at 10 kHz with a monochromator (13—16 nm bandpass) set to select 270 nm for excitation of tyrosyl peptides.[1441] Emission at 315 nm (recorded with a second monochromator having a 4—6 nm bandwidth) was detected with a XP2020 12-stage photomultiplier (replacing the now obsolete 56DUVP tube). Data collection of 10^6 counts (at a count rate of 2% of the flash frequency) was achieved in less than two hours.[1441] A free-running (10—20 kHz) nanosecond lamp was used to dissociate alkali iodides at high temperature for measurement of lifetimes of $Na(3p^2 P)$ and fluorescence detected with a Bendix Channeltron photon counter tube. The detector had a typical gain of 5×10^7, a single electron response distribution of only 40% FWHM, and a dark count of 3 cps at room temperature (0.8 mm² photocathode area); decay curves were not reported to show any satellite peaks.[1442] Time-correlated photon counting was also undertaken with a conventional two-monochromator arrangement using a

1435 C. V. Shank, E. P. Ippen, and O. Teschke, *Chem. Phys. Letters*, 1977, **45**, 291.
1436 C. V. Shank, E. P. Ippen, O. Teschke, and K. B. Eisenthal, *J. Chem. Phys.*, 1977, **67**, 5547.
1437 O. Teschke, E. P. Ippen, and G. R. Holtom, *Chem. Phys. Letters*, 1977, **52**, 233.
1438 B. Leskovar, C. C. Lo, P. R. Hartig, and K. Sauer, *Rev. Sci. Instr.*, 1976, **47**, 1113.
1439 P. R. Hartig, K. Sauer, C. C. Lo, and B. Leskovar, *Rev. Sci., Instr.*, 1976, **45**, 1122.
1440 D. K. Wong and A. M. Halpern, *Photochem. and Photobiol.*, 1976, **24**, 609.
1441 P. Gauduchon and Ph. Wahl, *Biophys. Chem.*, 1978, **8**, 87.
1442 J. R. Barker and R. E. Weston, *J. Chem. Phys.*, 1976, **65**, 1427.

50 kHz lamp.[1443] Details have also been given of a coaxial spark lamp filled with a 200 psi mixture of 15% H_2–85% Ar,[1444] a multivibrator circuit and free-running avalanche transistor arrangement to operate an HY-6 thyratron[1445] and of a free-running (1—2 kHz) high pressure (2 × 10^5 N m^{-2}) nitrogen-filled lamp producing pulses of 4.7 ns FWHM.[1446] An arrangement for vacuum-u.v. studies employed a ceramic lamp housing filled with H_2, N_2, or CO and an EMR solar blind photomultiplier tube fitted with a MgF_2 window.[1447] Hydrogen was found to be a suitable filling gas for the 120—400 nm range. The RCA C31034 photomultiplier tube, which is suitable for measurements to 900 nm, produces output pulses of only a few mV making a manosecond spectrometer prone to e.m.i. pickup in the early time stages. Precautions outlined to minimize this effect have included use of a cooled photomultiplier housing and e.m.i. screening in order to measure phosphorescence decays of Cr^{III} complexes with lifetimes from ns to tens of μs.[1448] A related technique to photon counting, namely multichannel scaling, has been employed to determine long (500 μs) decays of NO_2 vapour following excitation with an 80 Hz doubled FLPD laser.[1449] A 512-channel photon counter with a resolution of 10 ns per channel has been specifically designed to measure laser-induced fluorescence in a weakly ionized plasma.[1450] A low repetition rate laser (few pps) is of little use for time-correlated photon counting where only one photon per pulse is collected. A multichannel scaler was used to overcome this difficulty by counting all detected photons in a number of channels which opened sequentially following a trigger pulse.[1450]

Before dealing with the extensive use of laser as excitation sources in fluorescence, it is appropriate to deal with the only broadband and intense pulsed plasma source, namely the radiation from a synchrotron storage ring (see Vol. 10, p. 54). This short duration (0.3—1 ns pulse) excitation source, which extends over the entire u.v. and vacuum-u.v. region, has been used to great effect for lifetime studies on gaseous[1451, 1452] and solid[1453] xenon over the spectral range 104—190 nm, on krypton[1451] and on tetracene crystals[1454] as well as time-resolved excitation studies of styrenes.[1455] The lifetimes of haem proteins excited into their tryptophan absorption bands[1456] have also been determined. Time-resolved studies have been made on the vacuum-u.v. and u.v.–visible

[1443] D. J. S. Birch and R. E. Imhof, *J. Phys. (E)*, 1977, **10**, 1044.
[1444] T. H. McGee and R. E. Weston, *J. Chem. Phys.*, 1978, **68**, 1736.
[1445] D. J. S. Birch, *Mol. Photochem.*, 1977, **8**, 273.
[1446] S. Yagi, T. Hikida, and Y. Mori, *Chem. Phys. Letters*, 1978, **56**, 113.
[1447] R. L. Lyke and W. R. Ware, *Rev. Sci. Instr.*, 1977, **48**, 320.
[1448] D. C. Stuart and A. D. Kirk, *Rev. Sci. Instr.*, 1977, **48**, 186.
[1449] V. M. Donnelly and F. Kaufman, *J. Chem. Phys.*, 1977, **66**, 4100.
[1450] M. Lawton, R. C. Bolden, and M. J. Shaw, *J. Phys. (E)*, 1976, **9**, 686.
[1451] E. Mathias, R. A. Rosenberg, E. D. Poliakoff, M. G. White, S.-T. Lee, and D. A. Shirley, *Chem. Phys. Letters*, 1977, **52**, 239.
[1452] M. Ghelfenstein, H. Szwarc, and R. Lopez-Delgado, *Chem. Phys. Letters*, 1977, **52**, 236; R. Brodman, G. Zimmerer, and U. Han, *ibid.*, 1976, **41**, 160; M. Ghelfenstein, R. Lopez-Delgado, and H. Schwarc, *ibid.*, 1977, **49**, 312.
[1453] K. Monahan, V. Rehn, E. Matthia, and E. Poliakoff, *J. Chem. Phys.*, 1977, **67**, 1784.
[1454] R. Lopez-Delgado, J. A. Miehe, and B. Sipp, *Optics Comm.*, 1976, **19**, 79.
[1455] K. P. Ghiggino, K. Hara, K. Salisbury, and D. Phillips, *J.C.S. Faraday II*, 1978, **74**, 607.
[1456] B. Alpert and R. Lopez-Delgado, *Nature*, 1976, **263**, 445.

radiation emitted by $Ar-N_2-SF_6$ mixtures[1457] and on the emission spectra of ArF^* and Ar_2F^* from $Ar-F_2$ mixtures[1458] following pulsed proton excitation.[1458]

The mode-locked ion and ion-pumped dye lasers have allowed extension of the photon-counting technique to measure subnanosecond fluorescence decay times with high precision. The limiting factor for time resolution of the technique when a short, essentially jitter-free excitation pulse is used is the spread of transit time for the photoelectrons in the STOP photomultiplier since jitter in the electronics (discriminators and time-to-amplitude converter) is usually 10 ps. Photomultipliers (such as the XP2020Q) wired correctly for minimum jitter have enabled instrumental response functions to be only a few hundred ps and should allow reliable deconvolution for fluorescence decays of a few tens of ps. Various arrangements employing synchronously-pumped dye laser excitation pulses have been described.[1459,1460] In one case, the high repetition rate and intensity of the excitation source allowed two-fold discrimination (low level and differentiation) in the stop channel, giving excellent time resolution (50 ps).[1460] Examples of systems studied with mode-locked (and occasionally frequency-doubled) lasers have included CF_3NO,[1461] Rose Bengal,[1462] and pyrimidine vapour.[1463] Fluorescence from S_2 of azulene has been reported following excitation with a mode-locked at 615 nm.[1464] Further studies with this material used a mode-locked dye laser to generate two-photon fluorescence patterns and establish a lifetime of <1 ps for excited vibronic levels of S_1.[1465]

Cavity-dumped laser sources offer high repetition rate pulses with durations of 5—7 ns and a $1/e$ decay time of 2—3 ns.[1466] A manageable 5 MHz counting rate with this laser has been used with a gated constant-fraction discriminator[1467] in one arrangement[1466] and fluorescent lifetime studies undertaken on benzene vapour,[1466] 1-phenylbut-2-ene,[1468] styrenes,[1469] and poly(1-vinylnaphthalene).[1470] Similar systems have also been employed to examine PBD in ethanol[1471] and heated TiO_2.[1472] A mode-locked Rh6G laser with a Pockels cell to isolate a single pulse at 610 nm has been used for fluorescence excitation

[1457] C. H. Chen, M. G. Payne, G. S. Hurst, and J. P. Judish, *J. Phys. Chem.*, 1976, **65**, 4028.
[1458] C. H. Chen and M. G. Payne, *Appl. Phys. Letters*, 1978, **32**, 358.
[1459] K. G. Spears, L. E. Cramer, and L. D. Hoffland, *Rev. Sci. Instr.*, 1978, **49**, 255; K. G. Spears, *Laser Focus*, 1978 (2), 96.
[1460] U. P. Wild, A. R. Holzwarth, and H. R. Good, *Rev. Sci. Instr.*, 1977, **48**, 1621.
[1461] K. G. Spears and L. Hoffland, *J. Chem. Phys.*, 1977, **66**, 1755; K. G. Spears, *Chem. Phys. Letters*, 1978, **54**, 139.
[1462] K. G. Spears and L. E. Cramer, *Chem. Phys.*, 1978, **30**, 1.
[1463] K. G. Spears and M. El-Manguch, *Chem. Phys.*, 1977, **24**, 65.
[1464] E. P. Ippen, C. V. Shank, and R. L. Woerner, *Chem. Phys. Letters*, 1977, **46**, 20.
[1465] P. Wirth, S. Schneider, and F. Dorr, *Chem. Phys. Letters*, 1976, **42**, 483.
[1466] M. D. Swords and D. Phillips, *Chem. Phys. Letters*, 1976, **43**, 228.
[1467] A. W. Sloman and M. D. Swords, *J. Phys. (E)*, 1978, **11**, 521.
[1468] R. P. Steer and K. Salisbury, *J. Photochem.*, 1977, **7**, 417.
[1469] K. P. Ghiggino, D. Phillips, K. Salisbury, and M. D. Swords, *J. Photochem.*, 1977, **7**, 141; R. P. Steer, M. D. Swords, P. M. Crosby, D. Phillips, and K. Salisbury, *Chem. Phys. Letters*, 1976, **43**, 461.
[1470] K. P. Ghiggino, R. D. Wright, and D. Phillips, *Chem. Phys. Letters*, 1978, **53**, 552.
[1471] J. M. Harris, L. M. Gray, M. J. Pelletier, and F. E. Lytle, *Mol. Photochem.*, 1978, **8**, 161.
[1472] J. Feinberg and S. P. Davis, *J. Mol. Spectroscopy*, 1978, **69**, 445.

of spinach chloroplasts as a function of intensity (10^{12}—10^{16} photons per pulse).[1473]

A fluorescence correlation technique offering considerable potential for lifetime measurements has been described[1474] in which the exciting laser pulse was mixed with laser-excited fluorescence in a non-linear crystal. As before, the time resolution of the technique is limited by the laser pulse duration to a few ps. In the example described, fluorescence (at 780 nm from a bacteriorhodopsin) was mixed with a 590 nm probe pulse in a lithium iodate crystal and the sum frequency (at 366 nm) was measured as a function of the delay between the exciting laser pulse and the probe pulse. The sensitivity of the technique was readily demonstrated by the reported quantum yield of between 1.2 and 2.5 × 10^{-4} and the decay time of 15 ± 3 ps. The signal was enhanced by scanning over 10^9 laser pulses.[1474]

An interesting alternative arrangement using an Ar^+ laser for lifetime measurements to 200 ps has been proposed, which requires neither photon counting equipment nor the laser to be mode-locked.[1475] In this case, mode noise in a free-running laser is used to produce variations in the excited-state population of a fluorophor and measurement of the r.f. power spectrum of the fluctuations in the resulting fluorescence can be used to determine the fluorescence lifetime. The time resolution of the described technique was limited to >0.2 ns (by the detector noise background) and to <7.0 ns by the large spacing of the laser mode-noise peaks,[1475] but these are not thought to be inherent restrictions.

Alternative near-i.r. laser sources to ion and dye lasers are GaAlAs double heterostructure devices which have produced up to 0.1 W of monochromatic radiation with pulse widths <1 ns over the range 660—900 nm.[1476] Lifetime measurements of semiconductors such as GaAs and InP were determined by time-correlated photon counting using an RCA C31034 C photomultiplier tube following excitation with a diode heterostructure laser emitting at 790 nm and with a peak power >40 nW and a pulse duration of 0.5 ns.

There have been several novel arrangements of phase fluorometers described including one intriguing system employing a CW laser and an AM-radio.[1477] In this case the CW laser is modulated with a standing-wave acousto-optic modulator and phase shifts were determined optically rather than electronically. With a spectrum analyser and 170 MHz light modulation frequency, a time resolution of 4 ps was reported. More conventional laser arrangements used a mode-locked cavity-dumped Ar^+ laser emitting at 514 nm to determine lifetimes of DNS in various solvents and various temperatures[1478] and a similarly modulated laser (5—250 MHz) to give a time resolution of 1 ps from a phase

[1473] N. E. Geacintov, J. Breton, C. E. Swenberg, and G. Paillotin, *Photochem. and Photobiol.*, 1977, **26**, 629.
[1474] M. D. Hirsch, M. A. Marcus, A. Lewis, H. Mahr, and N. Frigo, *Biophys. J.*, 1976, 1399; H. Mahr and M. D. Hirsch, *Optics Comm.*, 1975, **13**, 96.
[1475] G. M. Hieftje, G. R. Haugen, and J. M. Ramsey, *Appl. Phys. Letters*, 1977, **30**, 463.
[1476] R. J. Nelson, *Rev. Sci. Instr.*, 1978, **49**, 770.
[1477] E. R. Menzel and Z. D. Popovic, *Rev. Sci. Instr.*, 1978, **49**, 39; *Chem. Phys. Letters*, 1977, **45**, 537.
[1478] H. E. Lessing and M. Reichert, *Chem. Phys. Letters*, 1977, **46**, 111.

accuracy of $\pm 0.1°$.[1479] A phase fluorometer with a 150 W xenon arc source modulated at 15—60 MHz was shown to be capable of 100 ps time resolution and to be able to discriminate between single and multiple decay schemes by changing the modulation frequency.[1480]

Deconvolution.—It is disturbing to read that the various mathematical techniques of deconvolution have been used in some papers to determine fluorescence lifetimes which are so short that their values are meaningless. This is particularly true when cases of sub-nanosecond lifetimes are reported following excitation with a 2—3 ns lamp. Even under the most favourable experimental conditions (taking into account the wavelength response of the photomultiplier, lamp jitter, photomultiplier jitter *etc.*), a useful guide should be that the minimum lifetime which can be determined for a single exponential is about one tenth that of the full width of the instrumental response function (or one fifth for double exponentials). Actual real time measurements (with a streak camera or by up-conversion [1474]) should be compared with values obtained by deconvolution, wherever possible, to establish the credibility of the particular mathematical technique employed. In one case for example, a review of numerical techniques for deconvolution recommended interactive deconvolution [1481] but reported lifetimes with this technique for a double exponential decay of aqueous tryptophan (of typically 3.14 and 0.51 ns) showed published irreproducibilities of up to 100%.[1482] Other publications in deconvolution have included methods of modulating functions,[1483] a theoretical study of statistical accuracy of rotational correlation times,[1484] a comment on the 'moment index displacement' procedure used in the method of moments,[1485] and distortion and statistical accuracy of nanosecond spectrometers used with and without pile-up corrections.[1486] An extensive review of the various effects which may distort decay curves emphasised those due to escape out of the viewing region due to thermal motion or, in the case of ions, to electrostatic repulsion.[1487] Frequency fluctuations of pulsed dye lasers, used for the determination of atomic lifetimes, were shown to produce a small shift in estimated lifetimes due to the inability to correct completely for pile-up errors.[1488] A somewhat-related paper discussed methods for analysing decays which were distorted by flow effects, *e.g.* in the gas phase, by laser-induced fluorescence.[1489] It was noted that careful experimental techniques were required to measure unbiased fluorescence lifetimes with a boxcar integrator since systematic drifts in laser excitation source intensity (from a NLPD laser) or concentration of fluorescing molecules (ICl or IBr) led to non-exponential decays over a period of 20 min.[1490]

[1479] H. P. Haar, U. K. A. Klein, F. W. Hafner, and M. Hauser, *Chem. Phys. Letters*, 1977, **49**, 563; H. P. Haar and M. Hauser, *Rev. Sci. Instr.*, 1978, **49**, 632.
[1480] K. Schurer, P. G. F. Ploegaert, and P. C. M. Wennekes, *J. Phys.* (*E*), 1976, **9**, 821.
[1481] A. E. McKinnon, A. G. Szabo, and D. R. Miller, *J. Phys. Chem.*, 1977, **81**, 1564.
[1482] D. M. Rayner and A. G. Szabo, *Canad. J. Chem.*, 1978, **56**, 743.
[1483] B. Valeur, *Chem. Phys.*, 1978, **30**, 85.
[1484] Ph. Wahl, *Chem. Phys.*, 1977, **22**, 245.
[1485] J. Eisenfeld and D. J. Mishelevich, *J. Chem. Phys.*, 1976, **65**, 3384.
[1486] B. E. A. Saleh and B. K. Selinger, *Appl. Optics*, 1977, **16**, 1408.
[1487] L. J. Curtis and P. Erman, *J. Opt. Soc. Amer.*, 1977, **67**, 1218.
[1488] J. B. Atkinson, *J. Phys.* (*E*), 1977, **10**, 482.
[1489] C. C. Davis and M. E. Lewittes, *Internat. J. Chem. Kinetics*, 1977, **9**, 235.
[1490] M. A. A. Clyne and I. S. McDermid, *J.C.S. Faraday II*, 1977, **73**, 1094.

Applications.—There are simply so many publications now on the radiative properties of organic and inorganic materials that it is possible in the following section to illustrate only a few of the studies which have been made on various chemical systems, many of which have also been studied by steady-state methods. The emphasis has been directed towards those techniques which have been newly applied to a particular chemical study.

Atoms and Small Molecules. Halogen atoms and interhalogen compounds have been subjected to numerous fluorescence studies, in part because of their importance in atmospheric chemistry. A flashlamp-pumped dye laser was used to produce $Br(4^2P_{\frac{1}{2}})$ atoms in gas mixtures containing HCN. By monitoring the time-resolved fluorescence from the (001) state of HCN, it was possible to determine the rate constants for E–V energy transfer.[1491] Deactivation rates of $I^*(^2P_{\frac{1}{2}})$ by HCl, HBr, and NO were measured directly in emission at 1.315 μm following dye laser photolysis of I_2 mixtures.[1492] The lifetime of the $D^1\Sigma_u{}^+$ state of I_2 was measured using the high frequency deflection technique and found to be 15.5 ± 0.5 ns.[1493] Conventional dye laser excitation and analogue methods have been employed to measure fluorescence lifetimes of BrCl,[1494] IBr,[1495] and BrF.[1496] In the latter case, the decay curve obtained from a transient recorder was reported to be superior in terms of accuracy and sensitivity to a boxcar arrangement. Rates of HF removal by H atoms have been measured by a laser-induced fluorescence technique in which a small fraction of HF was pumped by a pulsed HF TEA laser to $\nu = 1$, 2, or 3 and fluorescence was detected with either a RCA 7102 or a C31034 photomultiplier tube.[1497] Time-resolved measurements of HF spontaneous emission following CO_2 laser excitation of SF_6H_2 mixtures were made with InSb[1498] and AuGe[1499] detectors.

Basic measurements of the characteristics of rare-gas–halide compounds are reported in the Section on Excimer Lasers (see p. 10). The photolysis of KrF_2 has been used to measure the rates of collisional quenching of KrF(*B*) excimers in two- and three-body collisions with Ar, F_2, and Kr_2.[1500] In the experimental arrangement, fluorescence of the ArF($B \to X$) band at 193 nm (produced by e-beam pumping of Ar–F_2 mixtures) was used to photolyse a fraction of the KrF_2 molecules in a diluent flow system.[1500,1501] The subsequent radiative decay of KrF* at 249 nm was monitored axially by a narrow band filter and a fast risetime (0.5 ns) S-5 biplanar photodiode.[1500] The lifetime of XeF was also obtained through dissociative photoexcitation of XeF_2 by a pulsed ArF laser at 193 nm.[1502]

1491 A. Hariri, A. B. Petersen, and C. Wittig, *J. Chem. Phys.*, 1976, **65**, 1872.
1492 A. J. Grimley and P. L. Houston, *J. Chem. Phys.*, 1978, **68**, 3366.
1493 A. B. Callear, P. Erman, and J. Kurepa, *Chem. Phys. Letters*, 1976, **44**, 599.
1494 J. J. Wright, W. S. Spates, and S. J. Davis, *J. Chem. Phys.*, 1977, **66**, 1566.
1495 J. J. Wright and M. D. Havey, *J. Chem. Phys.*, 1978, **68**, 864.
1496 M. A. A. Clyne and I. S. McDermid, *J.C.S. Faraday II*, 1978, **74**, 644.
1497 J. F. Bott and R. F. Heidner, *J. Chem. Phys.*, 1977, **66**, 2878.
1498 C. R. Quick and C. Wittig, *Chem. Phys. Letters*, 1977, **48**, 420.
1499 J. M. Preses, R. E. Weston, and G. W. Flynn, *Chem. Phys. Letters*, 1977, **48**, 425.
1500 J. G. Eden, R. W. Waynant, S. K. Searles, and R. Burnham, *Appl. Phys. Letters*, 1978, **32**, 733.
1501 R. Burnham and S. K. Searles, *J. Chem. Phys.*, 1978, **67**, 5967.
1502 R. Burnham and N. W. Harris, *J. Chem. Phys.*, 1977, **16**, 2742.

Transient species such as small radicals are often produced by photolysis immediately prior to examination by fluorescence techniques. For example, laser-induced fluorescence measurements of CN followed photolysis of C_2N_2 with a repetitive argon flashlamp (50 Hz, 0.8—1 J per pulse, 0.6 μs FWHM, *ca.* 10^{13} photons per pulse) through a sapphire window.[1503] Vibrational energy transfer from ND and NH($A^3\pi$) to CO and N_2 in solid Ar at 4 K was studied using frequency-doubled pulsed dye laser excitation followed by subsequent time and wavelength resonance fluorescence using a transient recorder.[1504]

A study similar to that in ref. 1504 was used to determine the time, wavelength, and polarization-resolved fluorescence of XeO in solid rare gas hosts.[1505] Dye laser excitation was used in studies on the decay characteristics of CFCl and CCl_2 in the gas phase[1506] and BO and BO_2.[1507]

I.r. fluorescence from small molecules has been discussed earlier and among lifetime measurements in this region are those of CO_2, C_2H_4, and C_2H_5Cl,[1508] tetramethyldioxetane,[1509] CD_3H,[1510] CO,[1511] and NOCl.[1512] The pulsed i.r. photolysis of cyclopropane using fluorescence detection with a boxcar detector for both spectra and time decays produced an emission attributed, in the main, to C_2^+.[1513] At the other end of the excitation scale, a pulsed ArF laser (at 193 nm) was used to induce multiphoton dissociation of CH_3Br in order to study time-resolved fluorescence of CH fragments.[1514] With excimer lasers, photochemistry involving two- and three-photon absorption processes is possible allowing excitation in the hard vacuum-u.v. at 97 and 64 nm. Relative $O(^1S)$ quantum yields from the photolysis of CO_2 between 106 and 117.5 nm using a 1 s duration hydrogen flashlamp and monochromator for excitation were determined by conventional emission techniques.[1515] Atomic resonance fluorescence was used to monitor the disappearance rate of $O(^3P)$ from vibrationally excited O_3 produced by a pulsed CO_2 laser.[1516]

Pulsed dye lasers tunable from 295 to 300 nm[1517] and 297 to 327 nm[1518] have been employed for fluorescence decay studies of SO_2 [1517—1519] and SO_2-sensitized biacetyl phosphorescence.[1517] Other sulphur compounds studied

[1503] R. J. Cody, M. J. Sabety-Dzvonik, and W. M. Jackson, *J. Chem. Phys.*, 1977, **66**, 2145.
[1504] J. Goodman and L. E. Brus, *J. Chem. Phys.*, 1977, **66**, 1156.
[1505] J. Goodman, J. C. Tully, V. E. Bondybey, and L. E. Brus, *J. Chem. Phys.*, 1977, **66**, 4803.
[1506] R. E. Huie, N. J. T. Long, and B. A. Thrush, *Chem. Phys. Letters*, 1977, **51**, 197.
[1507] R. E. Huie, N. J. T. Long, and B. A. Thrush, *Chem. Phys. Letters*, 1978, **55**, 404.
[1508] R. T. Bailey and F. R. Cruickshank, *J. Phys. Chem.*, 1976, **80**, 1596.
[1509] Y. Haas and G. Yahav, *Chem. Phys. Letters*, 1977, **48**, 63.
[1510] W. S. Drozdoski, A. Fakkr, and R. D. Bates, *Chem. Phys. Letters*, 1977 ,**47**, 309.
[1511] N. Legay-Sommaire and F. Legay, *Chem. Phys. Letters*, 1977, **52**, 213.
[1512] A. Hartford, *Chem. Phys. Letters*, 1977, **50**, 85.
[1513] M. L. Lesiecki and W. A. Guillory, *J. Chem. Phys.*, 1977, **66**, 4317.
[1514] A. P. Baronavski and J. R. McDonald, *Chem. Phys. Letters*, 1978, **56**, 369.
[1515] T. G. Slanger, R. L. Sharpless, and G. Black, *J. Chem. Phys.*, 1977, **67**, 5317.
[1516] G. A. West, R. E. Weston, and G. W. Flynn, *Chem. Phys. Letters*, 1978, **56**, 429.
[1517] F. Ahmed, K. F. Langley, and J. P. Simons, *J.C.S. Faraday II*, 1977, **73**, 1659.
[1518] F. Su, J. W. Bottenheim, H. W. Sidebottom, J. G. Calvert, and E. K. Damon, *Internat. J. Chem. Kinetics*, 1978, **10**, 125.
[1519] F. Su, J. W. Bottenheim, D. L. Thorsell, J. G. Calvert, and E. K. Damon, *Chem. Phys. Letters*, 1977, **49**, 305.

include $S_2(B^3\Sigma_u^-)$,[1520] CS_2 (following nitrogen laser excitation),[1521] and thio-phosgene.[1522] In the latter mentioned study, up to six hours' accumulation time with photon counting were required for decay measurements into single vibronic levels at low pressure of thiophosgene and great care was taken to ensure that the observed decays were genuinely non-exponential (consistency of lamp profile, reduction of scattered light, background subtraction). The produc-tion and decay of $N_2(A^3\Sigma_u^+)$ ($v = 0$, 1, or 2) states were monitored photo-electrically from sensitized fluorescence of NO-γ bands.[1523] These were produced by collisional transfer from excited Xe, which in turn was produced by radiative energy transfer *via* resonance radiation (147 nm) from a closely coupled Xe flashlamp. The small size of the flashlamp-reaction cell (two cubes 25 mm in length) enables repetitive flashing and boxcar integration to be used.[1523] Radiative lifetimes of N_2 following selective laser excitation were determined by time-correlated single photon counting.[1524]

Dramatic time-resolved fluorescence spectra from large organic molecules were well illustrated in a study on *N*-n-butyl-*NN*-di-(2-naphthylmethyl)amine, which exhibits both intramolecular excimer and exciplex formation.[1525] Single vibronic lifetimes of two-photon excited states in naphthalene at low pressures were measured by averaging very noisy decay curves over many laser shots with a transient digitizer.[1526] A similar arrangement was also used to examine fluorescence decays of glyoxal in a weak magnetic field.[1527] The ubiquitous nitrogen laser has been used for excitation in studies of time-resolved fluor-escence spectra of β-naphthol in water,[1528] the room temperature fluorescence of the diphenyl ketyl radical,[1529] and S_1 and S_2 lifetimes of fluoranthene in a rigid glass at 77 K.[1530] In order to study the radiative properties of the isolated benzophenone molecule, a molecular beam was excited by a nitrogen laser (or tunable dye laser) and subsequent emission intensities measured with time.[1531] It was noted that the decay lengthened and became biexponential as the back-ground pressure is increased. The kinetics of sensitized fluorescence in naph-thalene crystals doped with 2-methyl naphthalene and anthracene were studied as a function of temperature from 4 to 300 K following excitation with a doubled NLPD laser, and a signal averager[1532] and the influence of a magnetic field on the lifetimes of *p*-terphenyl crystals were determined by photon counting.[1533] Neodymium lasers have been used to record a fluorescence risetime of 21 ns for low pressure pyrene on excitation at 354 nm,[1534] and mode-locked systems used

[1520] T. H. McGee and R. E. Weston, *Chem. Phys. Letters*, 1977, **47**, 352.
[1521] S. J. Silvers and M. R. McKeever, *Chem. Phys.*, 1976, **18**, 333.
[1522] D. Phillips and R. P. Steer, *J. Chem. Phys.*, 1977, **67**, 4780.
[1523] A. Mandl and J. J. Ewing, *J. Chem. Phys.*, 1977, **67**, 3490.
[1524] K. H. Becker, H. Engels, and T. Tatarczyk, *Chem. Phys. Letters*, 1977, **51**, 111.
[1525] G. S. Beddard, R. S. Davidson, and T. D. Whelan, *Chem. Phys. Letters*, 1977, **56**, 54.
[1526] U. Boesl, H. J. Neusser, and E. W. Schlag, *Chem. Phys. Letters*, 1976, **42**, 16.
[1527] H. G. Kuttner, H. L. Selzle, and E. W. Schlag, *Chem. Phys. Letters*, 1977, **48**, 207.
[1528] T. Kishi, J. Tanaka, and T. Kouyama, *Chem. Phys. Letters*, 1976, **41**, 497.
[1529] K. Razi Naqvi and U. P. Wild, *Chem. Phys. Letters*, 1976, **41**, 570.
[1530] D. L. Philen and R. M. Hedges, *Chem. Phys. Letters*, 1976, **43**, 358.
[1531] R. Naaman, D. M. Lubman, and R. N. Zare, *Chem. Phys.*, 1978, **32**, 17.
[1532] A. Braun, H. Pfisterer, and D. Schmid, *J. Luminescence*, 1978, **17**, 15.
[1533] G. Klein and M. J. Carvalho, *Chem. Phys. Letters*, 1977, **51**, 409.
[1534] D. J. Ehrlich and J. Wilson, *J. Chem. Phys.*, 1977, **67**, 5391.

to measure the fluorescence lifetimes of crystalline tetracene[1535] and Rh6G.[1536] In the latter case, the laser pulse had sufficient flux density that fluorescence was induced by a two-photon absorption process at 530 nm. Time-resolved luminescence of benzophenone vapour[1537] and biacetyl at pressures as low as 6.6 $\times 10^{-3}$ N m^{-2}[1538] have been determined by photon counting following excitation with a 100 Hz NLPD laser.

Radiative lifetimes of the alkali metals have been determined following Na$_2$,[1541] and Cs[1542] in various vibrational levels with a variety of dye lasers. Specific atomic levels of metallic vapours of Mg, Ca, and Sr[1543] and Al, Ga, In, and Tl[1544] were excited by short pulse tunable lasers prior to fluorescence decay measurements using an analogue technique. The time-of-flight spectrum for different internal states of metastable barium atoms was determined from time-resolved chemiluminescence measurements following reaction of a pulsed atomic beam with an oxidant gas in a scattering chamber.[1545] A time-of-flight velocity selection method utilizing pulsed tunable laser fluorescence detection has been described. The flight time was determined by the delay between the time when a pulse of particles (NaCl beam and a Ca atomic beam) was let through a chopper slit and when a pulsed dye laser was fired.[1546]

There have been a number of attempts to understand the luminescence produced between mercury and ammonia.[1547—1549] Direct excitation of luminescence has been studied by a repetitive analogue technique in which a flashlamp containing 50 Torr Ar and a droplet of Hg was operated at 8—10 kV with a repetition rate of 1—10 kHz producing an output practically the same as a low pressure mercury lamp (*i.e.* at 184.9 and 253.7 nm).[1547] The FWHM duration of the lamp at 253.7 nm was 40 ns. In a phase study, the phase angles between resonance fluorescence at 253.7 nm and luminescence from the excited complex of Hg–NH$_3$ was found to vary with the wavelength of the emission band of the complex indicating the presence of two emitting species.[1549] The decay of Hg($^3P_0$) following pulsed excitation with a modulated microwave discharge lamp through a narrow band interference filter was followed by measuring the intensity of fluorescence (with a boxcar integrator), which accompanied pulsed dye laser excitation at 404.7 nm.[1550] Both phase shift[1551] and dye laser excita-

[1535] F. Heisel, J. A. Miehe, B. Sipp, and M. Schott, *Chem. Phys. Letters*, 1976, **43**, 534.
[1536] K. A. Selanger, J. Falnes, and T. Sikkeland, *J. Phys. Chem.*, 1977, **81**, 1960.
[1537] D. Zevenhuijzen and R. van der Werf, *Chem. Phys.*, 1977, **23**, 279.
[1538] R. van der Werf and J. Kommandeur, *Chem. Phys.*, 1976, **16**, 125.
[1539] S. Lemont, R. Giniger, and G. W. Flynn, *J. Chem. Phys.*, 1977, **66**, 4509.
[1540] P. H. Wine and L. A. Melton, *Chem. Phys. Letters*, 1977, **45**, 509.
[1541] T. W. Ducas, M. G. Littman, M. L. Zimmerman, and D. Kleppner, *J. Chem. Phys.*, 1976, **65**, 842.
[1542] G. Alessandretti, F. Chiarini, G. Gorini, and F. Petrucci, *Optics Comm.*, 1977, **20**, 289.
[1543] M. D. Havey, L. C. Balling, and J. J. Wright, *J. Opt. Soc. Amer.*, 1977, **67**, 488.
[1544] M. D. Havey, L. C. Balling, and J. J. Wright, *J. Opt. Soc. Amer.*, 1977, **67**, 491.
[1545] R. C. Estler and R. N. Zare, *Chem. Phys.*, 1978, **28**, 253.
[1546] L. Pasternack and P. J. Dagdigian, *Rev. Sci. Instr.*, 1977, **48**, 226.
[1547] T. Hikida, M. Santoku, and Y. Mori, *Chem. Phys. Letters*, 1978, **55**, 280.
[1548] A. B. Callear and C. G. Freeman, *Chem. Phys.*, 1977, **23**, 343.
[1549] H. Umemoto, S. Tsunashima, and S. Sato, *Chem. Phys. Letters*, 1978, **53**, 521.
[1550] L. F. Phillips, *J.C.S. Faraday II*, 1976, **72**, 2082.
[1551] J. Skonieczny, *J.C.S. Faraday II*, 1977, **73**, 1145.

tion with a transient digitizer[1552] were used to determine the decay rate of the Hg$_2$ fluorescence band. Information on the operating conditions of the low pressure HgBr laser was obtained from fluorescence decay studies following excitation with a NLPD laser.[1553] UF$_6$ fluorescence lifetimes were followed by analogue methods with dye laser excitation at 393 nm.[1554]

Reports of other inorganic studies included use of a pulsed hydrogen laser (emitting at 160 nm) for excitation of Tb^{3+}-activated phosphates,[1555] a PLPD laser to excite PbO vapour,[1556] a nitrogen laser to excite europium(III) hexa-fluoroacetylacetonate,[1557] and nanosecond spectrometry to measure energy transfer between Ce^{3+} and Tb^{3+} in rare-earth pentaphosphate crystals.[1558]

11 Signal Processing

Many of the photochemical techniques employed, for example, in fluorescence, absorption spectrometry, picosecond techniques, and especially transient emission spectroscopy have sophisticated signal- and data-processing facilities as mandatory. Only a few of the remaining signal-processing publications can be mentioned further here.

A review on the applications of digital techniques in laboratory automation and instrumentation is commended as a useful introduction to the subject.[1559] A tapeloop boxcar averager (accompanied by filtering) was shown to be capable of recovering a sinewave buried in 40 nB of noise[1560] and a signal averaging system, which used a TV camera as a buffer memory prior to insertion of 50 Hz signals into a normal signal averager (such as a CAT) was claimed to provide a faster improvement in SNR.[1561] Transient recorders described included one with a 1024 digital memory capable of subtracting a background signal,[1562] a multichannel system dissecting a 400 ns signal into 21 sampling channels,[1563] and a fast arrangement with a linearity and accuracy of better than 10% at recording rates up to 10 MHz.[1564] The low SNR found in i.r.-induced fluorescence experiments was found to be complicated by the characteristic high r.f. noise of a TEA CO$_2$ laser and was enhanced by multichannel signal averaging techniques, using a transient recorder and computer.[1565]

[1552] M. Stock, E. W. Smith, R. E. Drullinger, M. M. Hessel, and J. Pourcin, *J. Chem. Phys.*, 1978, **68**, 1785; M. Stock, E. W. Smith, R. E. Drullinger, and M. M. Hessel, *ibid.*, 1978, **68**, 4167; M. Stock, R. E. Drullinger, and M. M. Hessel, *Chem. Phys. Letters*, 1977, **45**, 592.
[1553] N. Djeu and C. Mazza, *Chem. Phys. Letters*, 1977, **46**, 172.
[1554] F. B. Wampler, R. C. Oldenborg, and W. W. Rice, *Chem. Phys. Letters*, 1978, **54**, 554, 557, 560; W. W. Rice, F. B. Wampler, and R. C. Oldenberg, *J. Opt. Soc. Amer.*, 1978, **68**, 687.
[1555] T. Fukuzawa and S. Tanimizu, *J. Luminescence*, 1978, **16**, 447.
[1556] W. H. Beattie, M. A. Revelli, and J. M. Brom, *J. Mol. Spectroscopy*, 1978, **70**, 163.
[1557] A. J. Twarowski and D. S. Kliger, *Chem. Phys. Letters*, 1976, **41**, 329.
[1558] B. Blanzat, J. P. Denis, and R. Reisfeld, *Chem. Phys. Letters*, 1977, **51**, 403.
[1559] E. J. Millett, *J. Phys. (E)*, 1976, **9**, 794.
[1560] G. Palmer, M. McKellar, and O. G. Fritz, *J. Phys. (E)*, 1976, **9**, 783.
[1561] W. Venhuizen and G. A. Sawatzky, *J. Phys. (E)*, 1976, **9**, 551.
[1562] K. R. Betty and G. Horlick, *Analyt. Chem.*, 1977, **49**, 343.
[1563] P. Watkinson, M. R. Barrault, and D. G. Lewis, *J. Phys. (E)*, 1976, **9**, 776.
[1564] T. A. Last and C. G. Enke, *Analyt. Chem.*, 1977, **49**, 19.
[1565] S. Bialkowski and W. A. Guillory, *Rev. Sci. Instr.*, 1977, **48**, 1445.

A simple and fast (0.2 μs risetime) photometric unit employing a subtractive method was reported to offer a common mode rejection ratio of 75 dB.[1566] Up to eight channels of analogue information with a frequency component not exceeding 1 Hz could be recorded on a low cost system employing an unmodified stereo cassette tape recorder.[1567] An integrating analogue-to-digital converter displaying high signal fidelity and noise immunity has a maximum sample rate of 1000 |s^{-1}.[1568]

[1566] M. Krizan and M. Grubie, *Rev. Sci. Instr.*, 1978, **49**, 63.
[1567] A. L. Greenhill and D. J. Winning, *J. Phys. (E)*, 1977, **10**, 717.
[1568] J. W. Frazer, G. M. Hieftje, L. R. Layman, and J. T. Sinnamon, *Analyt. Chem.*, 1977, **49**, 1869.

2
Photophysical Processes in Condensed Phases

BY R. B. CUNDALL AND M. W. JONES

1 Introduction

It is apparent again during this year under survey that the application of photo-chemistry to chemical analysis is ever increasing, as is the use of fluorescence in the biological field. The use of fluorescent probes as well as different types of polarization experiments has now assumed an extremely important role in characterizing biological molecules. The usefulness of the synchrotron in physical photochemistry has now been made clearer through the publication of several papers, and it is likely that this is only a beginning to its future exploitation.

Again the same format has been adhered to as in Volume 9, and comments on most papers have been kept to a minimum for reasons of space restrictions.

2 Excited Singlet-state Processes

Baird[1] reviews the present state of art of *ab initio* quantum-mechanical calculations for polyatomic molecules with particular reference to spectral predictions, substituent effects on energy gaps, and potential energy surfaces. The main conclusion drawn from this review is that the energetics for low-lying valence states of small polyatomic molecules can be obtained *via ab initio* calculations provided that a large basis set and extensive configuration interaction are employed. Such calculations can be useful in attempts to understand the nature of the changes in electronic structure which occur as the result of a reaction or an electronic transition. A treatment of polarization effects in the measurements of luminescence yields has been presented.[2] This paper takes account of the effect of polarization in the excitation and emission monochromators as well as the non-zero emission anisotropy of the emitting species. The analysis gives a 'correction factor', the magnitude of this factor being dependent on the nature of the sample. For highly viscous solutions this factor may be of the order of the experimental error, but for solids it might be even greater. Several papers have appeared on the possibility of characterizing radiationless transitions in competition with fluorescence and phosphorescence.[3-8] Birks[3] presents a

[1] N. C. Baird, *Pure Appl. Chem.*, 1977, **49**, 223.
[2] J. A. Poole and A. Findeisen, *J. Chem. Phys.*, 1977, **67**, 5338.
[3] J. B. Birks, *Nouveau J. de Chim.*, 1977, **1**, 453.
[4] I. H. Kuhn, D. F. Heller, and W. M. Gelbart, *Chem. Phys.*, 1977, **22**, 435.
[5] W. A. Wassam and E. C. Lim, *J. Chem. Phys.*, 1978, **68**, 433.
[6] E. S. Medvedev, V. I. Osherov, and V. M. Pschenichnikov, *Chem. Phys.*, 1977, **23**, 397.
[7] T.-S. Lee, *J. Amer. Chem. Soc.*, 1977, **99**, 3909.
[8] K. F. Freed, *Accounts Chem. Res.*, 1978, **11**, 74.

paper on the reversibility of photophysical processes and establishes general kinetic relations for these processes; consideration is given to evaluation of thermodynamic parameters. Relative rate calculations for intramolecular radiationless transitions which take account of anharmonicity (K), geometry shifts (ΔQ) and frequency changes ($\Delta \omega$) have been made[4] and the effect of vibronic interaction between the first and second excited states on the radiationless decay rate of the S_1 level has been considered.[5] Such analysis has been applied to the radiationless transitions of N-heterocyclics with close-lying $n\pi^*$ and $\pi\pi^*$ excited states. The decay rate was found to vary with the strength of the $n\pi^*$–$\pi\pi^*$ vibronic interaction, the $n\pi^*$ and $\pi\pi^*$ separation, the energy gap between the initial and final electronic states of the radiationless process, and isotropic substitution in accord with experimental observations. Medvedev *et al.*[6] give a mathematical treatment of radiationless transitions[7] in isolated large polyatomic molecules and show that the transition rate is dependent on the widths of vibronic levels involved in the transition. The pressure dependence, collision-induced intersystem crossing, and intramolecular vibrational relaxation are also discussed by Freed.[8] Calculations of the vibronic coupling integrals confirm the b_{1g} assignment of the 918 cm^{-1} vibration in the $n\pi^*$ transition of pyrazine.[9]

Vibronic analysis of the visible absorption and fluorescence spectra of the fluorescein dianion has been carried out[10] and a complete rate constant equation for intersystem crossing including rotational states established and applied to naphthalene.[11]

Several papers have appeared on the use of lasers in the field of interest.[12—17] Kimel and Speiser have provided a useful and balanced review of the use of lasers in chemistry and start the review with a brief description of the fundamentals of laser operation and a summary of existing laser devices and their properties. Further sections demonstrate the applications of lasers in areas of spectroscopy, photophysics, and photochemistry. Of more specialized nature are the papers by Kinsey[12] and Burland.[13] Kinsey deals specifically with laser-induced fluorescence (LIF) as a tool for the detection of minute amounts of atoms or molecules in specific quantum states where the role of the dye laser has become an obvious advantage. Burland has given an interesting account on the use of lasers in molecular crystal systems, with specific examples of the recent studies on the photodissociation of *s*-tetrazine (ST) and its dimethyl derivative (DMST) in molecular crystal hosts. Tetrazines are of current interest through the work of Hochstrasser and co-workers who have shown that the photodissociation of ST in a molecular crystal host and in an argon matrix can lead to an efficient method of separating $^{12}C/^{13}C$, $^{14}N/^{15}N$, and $^{1}H/^{2}H$ isotopes, and also through the work of de Vries and Wiersma on photochemical hole burning

[9] G. Orlandi and G. Marconi, *Chem. Phys. Letters*, 1978, **53**, 61.
[10] T. Kurucsev, *J. Chem. Educ.*, 1978, **55**, 128.
[11] W. E. Howard and E. W. Schlag, *Chem. Phys.*, 1978, **29**, 1.
[12] J. L. Kinsey, *Ann. Rev. Phys. Chem.*, 1977, **28**, 349.
[13] D. M. Burland, Advances in Laser Chemistry, *Proc. SPIE*, 1977, **113**, 151.
[14] S. Kimel and S. Speiser, *Chem. Rev.*, 1977, **77**, 437.
[15] P. Zoller and F. Ehlotzky, *Z. Phys. (A)*, 1978, **285**, 245.
[16] J. M. Harris, L. M. Gray, M. J. Pelletier, and F. E. Lytle, *Mol. Photochem.*, 1977, **8**, 161.
[17] J. M. Richardson and S. M. George, *Analyt. Chem.*, 1978, **50**, 616.

applied as a high-resolution spectroscopic tool for measuring processes associated with the homogeneous broadening of absorption lines, processes such as phonon scattering and energy transfer. An account of the use of the dye laser as a tunable u.v. source for measuring fluorescence decays has been presented[16] and a comparison of three laser excitation sources (nitrogen pumped dye laser, cavity-dumped argon ion laser, and externally pulse-picked mode-locked argon ion laser) obtained using rhodamine B as the fluorophor.[17] For many analytical problems the nitrogen pumped dye laser appears to be the most flexible and generally applicable laser excitation source.

Interest in benzene is sustained with two comprehensive reviews on benzene organic photochemistry by Bryce-Smith and Gilbert.[18, 19] These reviews cover photoisomerization, photoaddition, and substitution processes. An orbital symmetry treatment of the photoisomerization and photoaddition processes is presented. Mulder has discussed additional orbital symmetry aspects of the benzene–benzvalene interconversion.[20]

The effect of one-dimensional translational and free rotation of benzene on its π-electronic states has been analysed, and it has been concluded that translation causes mixing of various π-states whereas rotation brings about a small amount of $\sigma-\pi$-mixing and therefore explains the absence of phosphorescence in a rigid glass.[21] Singlet and triplet exciton percolation theory has been applied to benzene isotopic mixed crystals[22] and optical absorption properties of cluster states in substitutionally disordered molecular crystals, taking physical parameters based on the ν_{11} of benzene vibrational excitation, have been obtained.[23] The exciton band density of states of the first excited singlet state of hexamethylbenzene (HMB) crystal has been obtained experimentally.[24] The centre of the band is at 35 156 cm^{-1} and the $K = 0$ is at 35 134 cm^{-1}, the full width of the band being about 40 cm^{-1}.

The radiative and non-radiative decay constant for singlet excited states of halogen- and methyl-substituted benzenes have been obtained, the effect of halogen substitution on the non-radiative decay being much greater than that of methyl substitution.[25] Chlorobenzene photoreactivity has been studied,[26] together with energy migration and energy transfer studies in benzene matrices.[27]

Both monomer and dimer emission have been reported for benzoic acid at room temperature.[28, 29] Martin and Clarke[28] detail the monomer emission in aqueous and ethanolic solutions, the emission being very much dependent on pH, buffer, solute concentration, and solvents. In aqueous acidic media, where the $^1(\pi\pi^*)$ lies below the $^3(n\pi^*)$ and the $^1(n\pi^*)$, the fluorescence is assigned

18 D. Bryce-Smith and A. Gilbert, *Tetrahedron*, 1976, **32**, 1309.
19 D. Bryce-Smith and A. Gilbert, *Tetrahedron*, 1977, **33**, 2459.
20 J. J. C. Mulder, *J. Amer. Chem. Soc.*, 1977, **99**, 5177.
21 S. Basu and P. Sen, *Indian J. Chem.*, 1977, **15A**, 161.
22 S. D. Calson, S. M. George, T. Keyes, and V. Vaida, *J. Chem. Phys.*, 1977, **67**, 4941.
23 J. Hoshen and R. Kopelman, *Phys. Stat. Solidi (B)*, 1977, **81**, 479.
24 S. D. Woodruff and R. Kopelman, *Chem. Phys.*, 1977, **22**, 1.
25 D. M. Shold, *Chem. Phys. Letters*, 1977, **49**, 243.
26 N. J. Bunce and L. Ravanal, *J. Amer. Chem. Soc.*, 1977, **99**, 4150.
27 A. J. Rest and J. R. Sodeau, *Ber. Bunsengesellschaft phys. Chem.*, 1978, **82**, 20.
28 R. Martin and G. A. Clarke, *J. Phys. Chem.*, 1978, **82**, 81.
29 A. U. Acuna, A. Ceballos, and M. J. Molera, *An. Quim.*, 1977, **73**, 1261.

to the undissociated monomer of benzoic acid, as shown in Figure 1. On the other hand, Acuna *et al.*[29] have observed room-temperature fluorescence from benzoic acid in non-polar solvents and estimate a quantum yield of 0.018. The emission, however, was attributed to dimer benzoic acid emission. The differences are to be expected in view of the great variation in solvent conditions.

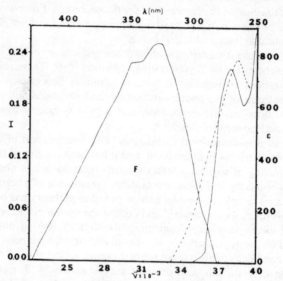

Figure 1 *Absorption and fluorescence spectra of a* 5.0×10^{-4} mol l^{-1} *aqueous benzoic acid solution, exciting at 250 nm. Excitation spectrum* (λ_{em} 330 nm) *shown by dashed curve* (---)
(Reproduced by permission from *J. Phys. Chem.*, 1978, **82**, 85)

Photoreduction of substituted nitrobenzenes in propan-2-ol has been examined by CIDNP,[30] and quantum yield measurements reported[31] for the solution-phase photolysis of some geminal chloro-nitroso-compounds.

Computational analysis of the vibronic intensities in the electronic transitions of naphthalene, in particular the singlet transition ($B_{2u}-A_g$), has been carried out.[32] The $n-\pi^*$ solvent blue shifts of ketones and azo-compounds have been investigated and it has been found that the enthalpy of transfer of the ground and excited state into the blue-shifted solvent could be either exothermic or endothermic.[33]

Absorption and fluorescence spectra of the ion pairs of 1,2,4,5-tetracyano-benzene and several methyl-substituted benzenes have been measured.[34]

[30] N. Levy and M. D. Cohen, *Mol. Photochem.*, 1977, **8**, 155.
[31] D. Forrest, B. G. Gowenlock, and J. Pfab, *J. C.S. Perkin II*, 1978, 12.
[32] M. J. Robey, I. G. Ross, R. V. Southwood-Jones, and S. J. Strickler, *Chem. Phys.*, 1977, **23**, 207.
[33] P. Haberfield, M. S. Lux, and D. Rosen, *J. Amer. Chem. Soc.*, 1977, **99**, 6828.
[34] S. Matsumoto, S. Nagakura, and T. Kobayashi, *Bull. Chem. Soc. Japan*, 1976, **49**, 2967.

A review of emission spectroscopy by Barnes[35] has recently appeared. The review gives a very useful account of various aspects of instrumentation, including standards and excitation sources, which are of interest to photochemists.

Fluorescence has now become an important tool in the assay of substances adsorbed on solid surfaces, such as paper chromatograms and thin-layer chromatography (t.l.c.) plates. Guilbault[36] gives a concise account of the use of fluorescence analysis on solid surfaces. Reports have also appeared which describe the experimental and theoretical calibration curves in solid-surface fluorescence analysis[37] and a thin-layer chromatographic method for the estimation of procainamide and its major metabolite, *N*-acetylprocamamide.[38] Uchiyama *et al.*[39] describe the technique of spraying non-volatile and viscous organic solvents such as liquid paraffin or glycerol on thin-layer plates after development. This results in a considerable increase in fluorescence intensity and stability of the fluorescent spots. The angular distribution of light emitted by molecules adsorbed on metal surfaces has been analysed for both i.r. emission from vibrational transitions and visible emission from electronic transitions.[40]

Several papers have appeared on the use of fluorescence in high-pressure liquid chromatography.[41-43] Thacker[42] describes the development of a miniature flow fluorometer capable of detecting 1—10 ng quantities. Applications of the technique to indole peptide, naphthyl, and phenolic compounds are also described.[43] A fluorescence detection system is also described for colorimetric titrations of microequivalent amounts of acids.[44]

Numerous papers have continued to appear on the photochemistry of anthracene and its derivatives. Fluorescence spectra and lifetimes of ultra-pure anthracene crystals subjected to photodimerization and photo-oxidation have been obtained.[45] Photo-oxidation does not change the spectral characteristics of anthracene even at 4 K, whereas photodimerization was found to modify both spectra and fluorescence lifetimes. A broad unstructured 'a' polarized emission was seen at 475 nm in photodimerized samples together with the characteristic peaks at 406, 421, and 444 nm, which are essentially unpolarized. These emissions are attributed respectively to traps which are interpreted as anthracene 'incipient' dimer and single, displaced anthracene molecules in close proximity to the stable photodimer. For the singlet exciton, the two trap depths are assigned to be at $\sim$4000 cm^{-1} and 300 cm^{-1} respectively with a distribution for the latter extending to $\sim \pm 50$ cm^{-1}. Excitation of a thin surface layer of pure anthracene crystal ($\lambda_{ex} = 366$ nm) gives a decay which is not in agreement

[35] R. M. Barnes, *Analyt. Chem.*, 1978, **50**, 100R.
[36] G. G. Guilbault, *Analyt. Chem.*, 1977, (Essays in memory of Anders Ringbom. Edited by E. Wanninen), p. 435.
[37] R. J. Hurtubise, *Analyt. Chem.*, 1977, **49**, 2160.
[38] R. N. Gupta, F. Eng, and D. Lewis, *Analyt. Chem.*, 1978, **50**, 197.
[39] S. Uchiyama, T. Kondo, and M. Uchiyama, *Bunseki Kagaku*, 1977, **26**, 762.
[40] R. G. Greenler, *Surface Sci.*, 1977, **69**, 647.
[41] S. Marsh and C. Grandjean, *J. Chromatog.*, 1978, **147**, 411.
[42] L. H. Thacker, *J. Chromatog.*, 1977, **136**, 213.
[43] G. J. Krol, C. A. Mannan, R. E. Pickering, D. V. Amato, B. T. Kho, and A. Sonnenschein, *Analyt. Chem.*, 1977, **49**, 1836.
[44] G. E. Cadwegan and D. J. Curran, *Mikrochim. Acta*, 1977, **11**, 461.
[45] J. O. Williams, D. Donati, and J. M. Thomas, *J.C.S. Faraday II*, 1977, **73**, 1169.

with delayed fluorescence.[46] This emission has been interpreted as a manifold exponential decay due to traps on the crystal surface. Triplet exciton trapping characteristics have been investigated using both steady-state and temporal

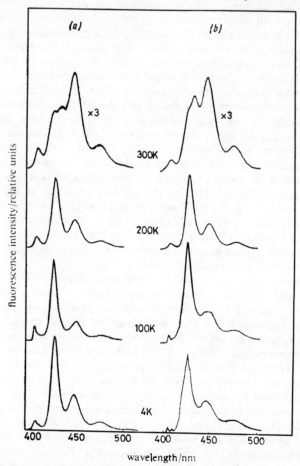

Figure 2 *Emission spectra of acridine + anthracene single crystals containing* (a) 50 p.p.m., (b) 200 p.p.m. *acridine as a function of temperature*
(Reproduced by permission from *J.C.S. Faraday II*, 1977, **73**, 1371)

dependencies of delayed fluorescence (and phosphorescence) as a function of temperature.[47] The results indicate that even at high temperatures the triplet exciton is localized in deep structural traps of low concentration $\sim10^{-11}$ mol dm³ per mol dm³. Fluorescence emission from anthracene doped with both acridine and carbazole has been investigated and compared with the emission from pure anthracene crystals.[48] Carbazole-doped crystals give an anthracene

[46] A. Schmillen and H. Wolff, *Z. Naturforsch.*, 1977, **32a**, 798.
[47] J. O. Williams and Z. Zboinski, *J.C.S. Faraday II*, 1978, **74**, 618.
[48] J. O. Williams and B. P. Clarke, *J.C.S. Faraday II*, 1977, **73**, 1371.

excitonic emission and a 'defect' spectrum with a progression based on a broad origin at $\sim$275 cm^{-1}. The intensity of the defect emission was found to increase with increasing carbazole concentration and decrease with increasing temperature. Acridine-doped crystals show the absence of intrinsic anthracene emission at all concentrations of acridine and the emission is structured with a broad origin and varies in both half-width and position with acridine concentration, corresponding to defect anthracene emission (Figure 2).

Two-photon excited singlet excitons are found to be quenched by triplet excitons generated by the ruby laser.[49] This causes a linear increase of the first-order decay rate of the fluorescence, yielding a singlet–triplet annihilation rate of 3 × 10^{-8} cm^3 s^{-1}. The use of reflection studies has been extended to the study of polaritons in anthracene and naphthalene single crystals.[50] The use of two-photon excitation in the 'transparent' regions of these crystals gives a better insight into their bulk characteristics.

Triplet excitation spectra of delayed fluorescence in anthracene crystals doped with tetracene have been investigated.[51] Below 150 K the quantum yield ratio of delayed host and guest fluorescence is dependent on excitation wavelength within the triplet absorption band. This is consistent with direct excitation of traps below 150 K. In addition to the homofusion of free host triplet excitons, a further component in the delayed host fluorescence is due to the annihilation of a free exciton with another localized at a direct excited trap.

Trace analysis of anthracene and phenanthrene in carbazole has been investigated by fluorescence lifetimes and time-resolved fluorescence spectra.[52] Concentrations of anthracene in carbazole less than 10^{-8} mol/mol could be measured by time-resolved spectra. Investigations of fluorescence light polarizations of anthracene crystals with particle impact excitation[53] have been reported.

Fluorescence spectra of anthracene and pyrene in water and in aqueous surfactant solution [sodium dodecyl sulphate (SDS)] have been obtained and compared with spectra of anthracene in ethanol (Figure 3). The spectra of anthracene show large broadenings and red shifts, whereas with pyrene intensity reductions of the vibronic bands are seen, indicating a weak polarity of the environment in the SDS micelles.[54]

The relationship between fluorescence and molecular geometry and the role of solvents in formation of intramolecular exciplexes from anthronyl-substituted anthracenes has been presented.[55] A centrosymmetric excimer emission at 490 nm has been observed from 9-cyanoanthracene solutions, glasses, polycrystalline films, and single crystals at 298 and 77 K.[56] Rate constants, k_q, for fluorescence quenching of some 9- and 9,10-polar-substituted anthracenes have

[49] M. E. Michel-Beyerle, R. Haberkorn, J. Kinder, and H. Seidlitz, *Phys. Stat. Solidi* (B), 1978, **85**, 45.
[50] R. M. Hochstrasser and G. R. Meredith, *J. Chem. Phys.*, 1977, **67**, 1273.
[51] N. Graber and I. Zschokke-Granacher, *Phys. Stat. Solidi* (B), 1978, **85**, 505.
[52] S. Iwashima, T. Sawada, and J. Aoki, *Bunseki Kagaku*, 1977, **26**, 401.
[53] A. E. Wohanka, *Z. Phys.* (A), 1978, **285**, 267.
[54] A. Nakajima, *J. Luminescence*, 1977, **15**, 277.
[55] H.-D. Becker and K. Sandros, *Chem. Phys. Letters*, 1978, **53**, 228.
[56] E.-Z. M. Ebeid, S. E. Morsi, M. A. El-Bayoumi, and J. O. Williams, *J.C.S. Faraday I*, 1978, **74**, 1457.

124   *Photochemistry*

been measured in dioxan.[57] Benzenes monosubstituted with OMe, CN, or Br
or 1,4-disubstituted with OMe, Cl; OMe, Br; Me, Br; CN, Cl; or CN, Br have
been used as quenchers. Only in the case of 9,10-dicyanoanthracene–methoxy-
benzene was excimer emission observed.

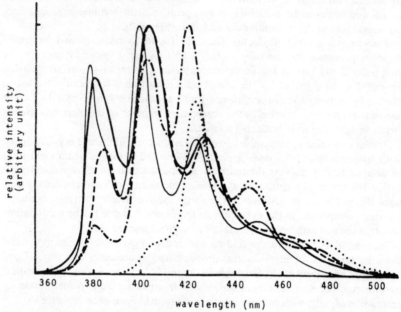

Figure 3 *Fluorescence spectra of anthracene in water (——), in aqueous SDS solution
(– – –), in a solution by the dilution method
(–.–.–), in ethanol (——), and for microcrystals (. . . .)*
(Reproduced by permission from *J. Luminescence*, 1977, **15**, 277)

Aperiodic and periodic variations in fluorescence intensity have been observed
for chlorocarbon solutions of anthracene and 9,10-dimethylanthracene.[58] No
detailed interpretation of the results has been put forward to date. Electro-
optical experiments on various exciplexes of 9,10-dicyanoanthracene and of
NN-dimethylaniline allow accurate determination of dipole moments μ_e in the
fluorescent state as well as information on the direction of the transition moment
P with respect to the dipole moment μ_e. The exciplex fluorescence was found to
be polarized parallel to the molecular planes of donor and acceptor.[59] Time-
dependent emission anisotropy has been applied to intermolecular energy
transfer on 9-methylanthracene in foils of acetylcellulose and a critical transfer
distance of 24.2 Å obtained.[60]

[57] D. Schulte-Frohlinde and H. Hermann, *Ber. Bunsengesellschaft phys. Chem.*, 1977, **81**, 562.
[58] R. J. Bose, J. Ross, and M. S. Wrighton, *J. Amer. Chem. Soc.*, 1977, **99**, 6119.
[59] E. J. J. Groenen and P. N. T. Th. van Velzen, *Mol. Phys.*, 1978, **35**, 19.
[60] J. Kaminski, A. Kawski, and A. Schmillen, *Z. Naturforsch.*, 1977, **32a**, 1335.

Two-photon excitation has been used to obtained low quantum yield fluorescence spectra for a number of polycyclic aromatic hydrocarbons,[61] the emission corresponding to fluorescence from highly excited singlet states directly to the ground state. Evidence is presented to show that the emission occurs from vibrationally unrelaxed states. Table 1 shows the data obtained for the hydrocarbon studied.

Table 1 *Hydrocarbon $S_{n>1} \rightarrow S_0$ spectral data*

Molecule	$\lambda_{irr}{}^a$	λ_{eff}	λ_F	λ_{abs}	τ_{rel}
Anthracene	366	186	259	252	—
1,2-Benzanthracene	366	188	303	287 (298)	—
1,2,3,4-Dibenzanthracene	366	185	300	286	—
1,2,5,6-Dibenzanthracene	366	190	313	297	—
Rubrene	515	259	313	298	—
Tetracene	474	238	285	273	1
			(~305)	293	~6
Chrysene	366	182	282	267	1
			330	318	~8
Pyrene	376	187	263	240	1
			285	272	1.1
			343	334	6
Coronene (ethanol)	410	208	310	302	1
			347	338 (344)	9
Perylene	437	218	258	252	1
			(~305)	292	~15

a λ_{irr} = irradiation wavelength (nm). λ_F, λ_{abs} = fluorescence and absorption maxima. τ_{rel} = estimated relative decay time, internally referenced to the shortest-wavelength transition for each molecule. λ_{eff} = effective irradiation wavelength calculated from the sum of incident photon energy and the 0–0 band. For rubrene the first vibronic maxima for fluorescence and absorption do not coincide and λ_{eff} for rubrene is calculated from the absorption peak.

Delayed fluorescence spectra of 1,2-benzanthracene, fluoranthene, pyrene, and chrysene in methylcyclohexane from upper excited singlet states, S_n ($n > 1$), have also been reported.[62] With all four compounds a weak emission from $S_n \rightarrow S_0$ was obtained corresponding to the energy level associated with triplet–triplet annihilation. Laser-induced molecular fluorescence has been shown to be a sensitive and selective means for detecting representative polycyclic aromatic hydrocarbons (PAH) in aqueous solutions.[63] The limits of detection of benzene, naphthalene, anthracene, fluoranthene, and pyrene were found to be 19, 1.3×10^{-3}, $<4.4 \times 10^{-3}$, and 0.5×10^{-3} $\mu g l^{-1}$ respectively. Improvements in these detection limits should result with advances in second harmonic generation and developments in u.v. lasers.

The lineshape function for the $S_0 \rightarrow T_1$ absorption in 1,4-dibromonaphthalene

[61] H.-B. Lin and M. R. Topp, *Chem. Phys. Letters*, 1977, **48**, 251.
[62] B. Nickel, *Helv. Chem. Acta*, 1978, **61**, 198.
[63] J. H. Richardson and M. E. Ando, *Analyt. Chem.*, 1977, **49**, 955.

has been analysed in terms of exchange theory[64] and the exciton percolation theory has been tested for the migration of the lowest singlet exciton in binary isotopic mixed crystals of naphthalene with an added exciton sensor of 2-methylnaphthalene.[65]

Temperature and concentration dependencies of the delayed luminescence in naphthalene (guest)–biphenyl (host) mixed crystals have been carried out.[66] Temperature-independent delayed fluorescence (TIDF) was obtained for this system and the significance of TIDF and the nature of the decay process accounted for.

U.v. fluorescence spectra have been obtained for 3,4-benzpyrene, 1,12-benzperylene, and 3,4,9,10-dibenzpyrene by means of a single-laser two-photon excitation.[67] Fluorescence line-narrowing in the spectrum of dibenzpyrene at 300 K in hexane solution has been observed and interpreted in terms of optical site-selection. Fluorescence spectra of polycyclic aromatic hydrocarbons (PAH) in n-alkane matrices have been reported.[68] Special care in annealing procedures is required, and specific examples are given for benz[a]anthracene. Several papers have appeared on the interaction and migration of excitons in tetracene single crystals.[69—71] Campillo *et al.*[69] have studied the fluorescence efficiency of crystalline tetracene as a function of intensity. Singlet–singlet annihilation rates of $\gamma_{ss} = 10^7$ cm^3 s^1 have been obtained and a single- diffusion rate D_s, of 4×10^2 cm^2 s^1 has been postulated. The energy-transfer kinetics for tetracene doped with pentacene have also been investigated[70] and a diffusion model is inferred for singlet migration from host to guest at low pentacene concentrations.

Rather few studies have to date been made on the luminescence of aromatic hydrocarbons in aqueous solution. Nakajima[72] reports on the fluorescence spectra of naphthalene, phenanthrene, triphenylene, 1,2-benzanthracene, 1,12-benzoperylene, coronene, benz[a]pyrene ,and benz[e]pyrene in water, aqueous sodium dodecyl sulphate (SDS), and ethanol, and for microcrystals. Examples of the fluorescence spectra of 1,12-benzoperylene are given in Figure 4. The emission in water is relatively weak and broad and composed primarily of microcrystalline and molecular spectra. In aqueous SDS solution, the fluorescence occurs more strongly from the solubilized molecules and the spectrum is blue-shifted relative to the microcrystalline spectrum, thereby indicating some interaction probably caused by aggregation in the micelles.

Time-resolved fluorescence of tetracene crystals has been studied and a singlet exciton lifetime of 300 ± 30 ps at 293 K obtained.[73] The rate constant for singlet–singlet exciton annihilation was found to be 3×10^{-9} cm^3 s^{-1}. Lumin-

[64] C. B. Harris, *Chem. Phys. Letters*, 1977, **52**, 5.
[65] R. Kopelman, E. M. Monberg, and F. W. Ochs, *Chem. Phys.*, 1977, **21**, 373.
[66] Y. Kusumoto, Y. Gondo, H. Sato, and Y. Kanda, *Bull. Chem. Soc. Japan*, 1977, **50**, 1698.
[67] M. R. Topp and H. B. Lim, *Chem. Phys. Letters*, 1977, **50**, 412.
[68] P. Tokousbalides, E. L. Wehry, and G. Mamantov, *J. Phys. Chem.*, 1977, **81**, 1769.
[69] A. J. Campillo, R. C. Hyer, S. L. Shapiro, and C. E. Swenberg, *Chem. Phys. Letters*, 1977, **48**, 495.
[70] A. J. Campillo, S. L. Shapiro, and C. E. Swenberg, *Chem. Phys. Letters*, 1977, **52**, 11.
[71] P. Schlotter, J. Kalinowski, and H. Bassler, *Phys. Stat. Solidi (B)*, 1977, **81**, 521.
[72] A. Nakajima, *Photochem. and Photobiol.*, 1977, **25**, 593.
[73] G.-R. Fleming, D. P. Miller, G. C. Morris, J. M. Morris, and G. W. Robinson, *Austral. Chem.*, 1977, **30**, 2353.

escence and decay measurements have also been reported for pentacene in *p*-terphenyl[74] and for 4,5-iminophenanthrene.[75]

Continued interest is shown in azulene even after many years of research into the nature of the $S_2 \rightarrow S_0$ and $S_1 \rightarrow S_0$ emissions. The pressure dependence of the fluorescence of azulene and several of its derivatives has been investigated

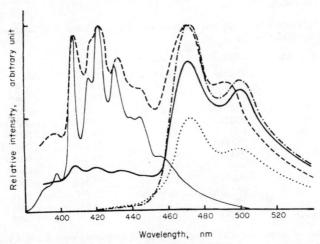

Figure 4 *Fluorescence spectra of 1,12-benzoperylene in water (——), in the solution prepared by dilution method (–.–.–.), in aqueous SDS solution (– – –), in ethanol (——), and for microcrystals (. . . .). The microcrystalline spectrum is shown for clarity, with the intensity reduced to half relative to the solution spectra*
(Reproduced by permission from *Photochem. and Photobiol.*, 1977, **25**, 593)

over a 130 kbar range.[76] The efficiency of fluorescence from both S_2 and S_1 states was pressure dependent, as was the relative energy of these states. A summary of the results for the azulenes (1)—(7) is shown in Table 2.

(1) R = X¹ = X² = H
(2) R = X² = H, X¹ = Cl
(3) R = X² = H, X¹ = I
(4) R = CO²Et, X¹ = Cl, X² = H
(5) R = CO²Et, X¹ = Cl, X² = Br
(6) R = CO²Et, X¹ = NH², X² = H
(7) R = CO²Et, X¹ = NH², X² = Br

74 H. De Vries, P. De Bree, and D. A. Wiersma, *Chem. Phys. Letters*, 1977, **52**, 399.
75 J. O. Williams, B. P. Clarke, J. M. Thomas, and G. J. Sloan, *Chem. Phys. Letters*, 1977, **48**, 560.
76 D. J. Mitchell, H. G. Drickamer, and G. B. Schuster, *J. Amer. Chem. Soc.*, 1977, **99**, 7489.

Quantum yields and lifetimes of 2-halogeno- and 1,3-dihalogeno-azulenes have also been reported.[77] A distinct heavy-atom effect was found in the radiationless deactivation, and it is possible that the intersystem crossing rates cannot be neglected in comparison with $S_2 \to S_1$ internal conversion. Interesting results have also been obtained for cryptocyanine[78] and N-(dialkylaminomethyl)phthalimides[79] where the fluorescence emission has been attributed to $S_2 \to S_0$ transition.

Table 2 *Quantum yields Φ_f and emission maxima of azulene derivatives at 1 atm in plastic media*

Compound	Medium	Transition	ν_{max}/cm^{-1}	$\Phi_f{}^a$
(1)	PMMA	$S_2 \to S_0$	26 670	0.027
(1)	PS	$S_2 \to S_0$	26 520	0.028
(2)	PMMA	$S_2 \to S_0$	26 330	0.0097
(3)	PMMA	$S_2 \to S_0$	25 510	0.0061
(4)	PMMA	$S_2 \to S_0$	24 810	0.000079
(4)	PS	$S_2 \to S_0$	24 790	0.000048
(5)	PMMA	$S_2 \to S_0$	24 640	0.000052
(4)	PMMA	$S_1 \to S_0$	15 230	0.00045
(4)	PS	$S_1 \to S_0$	15 210	0.00055
(5)	PMMA	$S_1 \to S_0$	15 160	0.00044
(6)	PMMA	$S_1 \to S_0$	16 440	0.0036
(6)	PS	$S_1 \to S_0$	15 946	0.0065
(7)	PMMA	$S_1 \to S_0$		0.0008

[a] All yields are determined relative to that of 9,10-diphenylanthracene in PMMA, which is taken to be 0.99.

The role of the low-lying singlet state in the *cis–trans* photochemical isomerization of stilbene is still unclear, but recently Stachelek *et al.* have detected and assigned the 'phantom' photochemical singlet of *trans*-stilbene by two-photon excitation.[80] Birks[81] discusses horizontal radiationless transitions (HRT) due to intramolecular rotation in stilbene and polyene derivatives. HRT's do not normally change the electronic state order of the S_0, T_1, S_1 states but change its steric configuration, zero-point energy, and the energy separation of adjacent electronic states. In his paper, Birks proposes a semi-empirical model of the dependence of the potentials of the two lowest excited singlet states on the angle of internal rotation.

Absorption, fluorescence, and phosphorescence spectra of *trans*-stilbene and derivatives thereof are shown to be functions of excitation wavelength, solvent viscosity, and temperature.[82] The results are explained by the existence of comparatively stable conformers which are formed by rotation of aryl fragments around single chemical bonds. Additional information has been obtained on *trans*-stilbene through fluorescence quantum yield and lifetime measurements.[83]

[77] G. Eber, S. Schneider, and F. Dorr, *Chem. Phys. Letters*, 1977, **52**, 59.
[78] V. Rehak, A. Novak, and M. Titz, *Chem. Phys. Letters*, 1977, **52**, 39.
[79] J. D. Coyle, G. L. Newport, and A. Harriman, *J.C.S. Perkin II*, 1978, 133.
[80] T. M. Stachelek, T. A. Pazoha, W. M. McClain, and R. P. Drucker, *J. Chem. Phys.*, 1977, **66**, 4540.
[81] J. B. Birks, *Chem. Phys. Letters*, 1978, **54**, 430.
[82] Y. B. Sheck, N. P. Kovalenko, and M. V. Alfimov, *J. Luminescence*, 1977, **15**, 157.
[83] J. L. Charlton and J. Saltiel, *J. Phys. Chem.*, 1977, **81**, 1940.

The derived parameters indicate that the majority of the fluorescence decays with a very short lifetime at room temperature and above (73%, 0.07 ns at 25 °C) whereas the measured value for fluorescence lifetime is much longer. Further information relating to the singlet state of *trans*-stilbene has been obtained through measurements of lifetime as a function of temperature[84] and through picosecond absorption measurements[85] in the excited singlet state of stilbene. The lifetime measurements reveal a radiative rate k_f, essentially constant at 6.0×10^8 s^1 in the temperature range 77—295 K. This is in contrast with results obtained previously using a single photon counting technique, and these measurements reveal only one exponential decay throughout the temperature range rather than the two exponential decays suggested by single photon counting. Quantum yields for stilbene *trans*-to-*cis* isomerization (Φ_t) and dimerization (Φ_{dim}) are shown in Table 3.[86] Stilbene dimerization competes

Table 3 *Quantum yields for* trans-*stilbene isomerization and dimerization*

[trans-*Stilbene*]/mol l^{-1}	Φ_t	Φ_{dim}
0.01	0.45 ± 0.02	<0.01
0.05	0.45	0.03 ± 0.01
0.10	0.44	0.08 ± 0.02
0.25	0.38	0.19
0.40	0.31	0.26
0.55	0.27	0.33

efficiently with isomerization at concentrations above 0.1 mol l^{-1}. The increase in dimerization with increasing stilbene concentration is proportional to the decrease in isomerization according to:

$$\Phi_{dim} = 1.6(\Phi_t^0 - \Phi_t)$$
$$\quad\quad\quad {}_{t \to c} \quad {}_{t \to c}$$

Interesting results have appeared on the quenching of *trans*-stilbene fluorescence by several secondary and tertiary amines.[87] The temperature dependence of fluorescence quenching by secondary and tertiary amines in polar and nonpolar solvents is indicative of reversible exciplex formation. The behaviour of exciplexes formed from *trans*-stilbene with secondary and tertiary amines and conjugated dienes is compared.

Fluorescence quantum yields and lifetimes of excited ion pairs formed by 1-naphthol and 2-naphthol with aliphatic amines have been studied as a function of temperature in a number of aprotic solvents.[88] Phenylethylamine, benzylamine, and amine-terminated polystyrene have been investigated,[89] and the quantum yields and lifetimes of fluorescence and phosphorescence given for 2-, 3-, and 4-aminopyridines, 2- and 4-(dimethylamino)pyridines in an ethanol–methanol mixture and in a mixture of triethylanine and 2-methyltetrahydrofuran at 300 and at 77 K.[90]

[84] M. Sumitani, N. Nakashima, K. Yoshihara, and S. Nagakura, *Chem. Phys. Letters*, 1977, **51**, 183.
[85] D. Teschke, E. P. Ippen, and G. R. Holtom, *Chem. Phys. Letters*, 1977, **52**, 233.
[86] F. D. Lewis and D. E. Johnson, *J. Photochem.*, 1977, **7**, 421.
[87] F. D. Lewis and T.-I. Ho, *J. Amer. Chem. Soc.*, 1977, **99**, 7991.
[88] I. Y. Martyunov, B. M. Uzhinov, and M. G. Kuzmin, *High Energy Chem.*, 1977, **11**, 36.
[89] L. K. Tan, R. A. Patsiga, *Analyt. Letters*, 1977, **10**, 437.
[90] K. Kimura, H. Takaoka, and R. Nagai, *Bull. Chem. Soc. Japan*, 1977, **50**, 1343.

No excimer emission at low temperatures is observed for perylene crystals, this being replaced by a weakly structured emission at higher energy which is attributed to a new crystal state (*Y*-state) populated independently of the high-temperature excimer (*E*-state).[91]

Interest has also developed recently in the nature of the first excited state of styrene, principally owing to uncertainty about the $^1A_1 \leftrightarrow {}^1A_1$ transition ($\sim C_{2v}$). Fluorescence analyses of [$^1H_8$]-, [$^2H_3$]-, [$^2H_3$]-, and [$^2H_8$]-styrene have been carried out in polycrystalline methylcyclohexane at 77 K, with the aim of establishing the position and nature of the 0–0 band for $S_0 \leftrightarrow S_1$ transition in styrene.[92] Investigation of the $S_1 \rightarrow S_0$ 0–0 transition in free-base porphin in n-octane matrix has also been presented.[93]

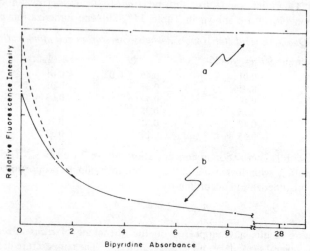

Figure 5 *Fluorescence intensity of* p-*terphenyl as a function of chromophore absorbance. Intensity from* (a) *two-photon and* (b) *one-photon excitation. The broken line indicates theoretical predicted intensity*
(Reproduced by permission from *Analyt. Chem.*, 1977, **49**, 2054)

Fluorescence data have been obtained for 2,2-bipyridyl,[94] and the two-photon absorption spectrum of hexa-1,3,5-triene[95] has been investigated using thermal blooming techniques.

An interesting paper has appeared on the technique of two-photon excited fluorimetry to quantitate molecular species in the presence of a high and/or varying matrix absorbance.[96] This technique has been applied successfully to the problem of interfering chromophores, and an example is given in Figure 5 for a model system with p-terphenyl as the fluorophor and 2,2'-bipyridyl as the chromophore and cyclohexane as the solvent. From the Figure it can be seen

[91] E. V. Freydorf, J. Kinder, and M. E. Michel-Beyerle, *Chem. Phys.*, 1978, **27**, 199.
[92] D. A. Condirston and J. D. Laposa, *J. Luminescence*, 1978, **16**, 47.
[93] S. Voelker, R. M. Macfarlane, and J. H. van der Waals, *Chem. Phys. Letters*, 1978, **53**, 8.
[94] A. Harriman, *J. Photochem.*, 1978, **8**, 205.
[95] A. J. Twarowski and D. S. Kliger, *Chem. Phys. Letters*, 1977, **50**, 36.
[96] M. J. Wirth and F. E. Lytle, *Analyt. Chem.*, 1977, **49**, 2054.

that the intensity due to two-photon excited *p*-terphenyl is not affected by the variation of 2,2'-bipyridyl concentration.

Fluorescence has also been applied to liquid crystal display,[97] reaction rate counters,[98] and identification of fuel oils.[99]

Many papers have appeared on various aspects of dye photochemistry, and therefore only a few comments will be made on each. Interest in dye photochemistry has been partly stimulated through the use of tunable dye lasers in photochemistry. The decay processes of rhodamine 6G in methanol,[100] and of its aggregates,[101] have been presented. The fluorescence lifetime in methanol decreases linearly from 3.7 ns at 2×10^{-4} mol l^{-1} to about 2.5 ns at 9×10^{-3} mol l^{-1}, thereby indicating quenching by excimer formation. Above 9×10^{-3} mol l^{-1}, there is a rapid decrease to about 0.9 ns at 2×10^{-2} mol l^{-1} which is attributed to quenching of excited monomers by aggregates. The molecular fluorescence lifetime of rhodamine 6G is given as 3.7 ± 0.4 ns. Fluorescence decay data have also been presented for rhodamine 6G, rhodamine B, acridine orange, eosin blue, PPO, and the 7-diethylamino-4-methyl derivative of marin.[102] In particular, the fluorescence decay as a function of aggregation was investigated. These results are summarized in Table 4.

Energy transfer investigations of rhodamine 6G in aqueous Triton X-100 have also been reported.[103] Rate constants for triplet-state quenching of rhodamine 6G by electron donors and acceptors have been obtained by a flash photolysis method.[104] Intermolecular electronic energy transfer (EET) from the short-lived S_2 state of rhodamine 6G to the ground state of 2,5-bis-(5-t-buty-1,2-benzoxazolyl) thiophen has been investigated using sequential time-delayed two-photon excitation.[105] Further studies of energy transfer have been made on mixtures of rhodamine 6G and cresyl violet.[106] Highly efficient *N*-de-ethylation of rhodamine B adsorbed on CdS has been studied by illumination, and the experimental results are consistent with electron transfer from the excited singlet state of rhodamine B to the conduction band of CdS.[107]

Picosecond lifetime measurements using laser excitation of acridine,[108] fluorescein and three of its derivatives (eosin, erythrosin, and rose bengal),[109] and indigo, merocyanine, and triphenylmethane,[110] have been reported. It has also been shown that intramolecular hydrogen-bonding can give rise to very fast non-radiative deactivation of the S_1 singlet state of indigo dyes.[111] Solvent and temperature are found to have marked effects on S_1 lifetime.

97 G. Baur and W. Grenbel, *Applied Phys. Letters*, 1977, **31**, 4.
98 R. L. Wilson and J. D. Ingle, *Analyt. Chem.*, 1977, **49**, 1060.
99 S. H. Fortier and D. Eastwood, *Analyt. Chem.*, 1978, **50**, 334.
100 K. A. Selanger, J. Falnes, and T. Sikkeland, *J. Phys. Chem.*, 1977, **81**, 1960.
101 A. K. Chibisov and T. D. Slavnova, *J. Photochem.*, 1978, **8**, 285.
102 J. Knof, F.-J. Theiss, and J. Weber, *Z. Naturforsch.*, 1978, **33a**, 98.
103 Z. Konefal, E. Lisicki, and T. Marszalek, *Acta Phys. Polon.*, 1977, **A52**, 149.
104 V. L. Pugachev, T. D. Slavnova, and A. K. Chibisov, *High Energy Chem.*, 1977, **11**, 296.
105 I. Kaplan and J. Jortner, *Chem. Phys. Letters*, 1977, **52**, 202.
106 Y. Kusumoto, H. Sato, and K. Malno, *Sci. Rep. Kagoshima Univ.*, 1977, **26**, 151.
107 T. Watanabe, T. Takizawa, and K. Honda, *J. Phys. Chem.*, 1977, **81**, 1845.
108 V. Sundstrom, P. M. Rentzepis, and E. C. Lim, *J. Chem. Phys.*, 1977, **66**, 4287.
109 G. R. Fleming, A. W. E. Knight, J. M. Morris, R. J. S. Morrison, and G. W. Robinson, *J. Amer. Chem. Soc.*, 1977, **99**, 4306.
110 P. Wirth, S. Schneider, and F. Forr, *Ber. Bunsengesellschaft phys. Chem.*, 1977, **81**, 1127.
111 W. Windhager, S. Schneider, and F. Dorr, *Z. Naturforsch.*, 1977, **32a**, 876.

Table 4 Dependence of fluorescence decay times on temperature (solvent mixture of 10% methanol + 90% ethanol) and Arrhenius parameters E_A and K_{NR}^0

Temperature/K	300	275	255	235	215	195	175	155	135	115	E_A/cal mol^{-1}	K_{NR}^0/s^{-1}
PPO 1×10^{-3} m	2.1	2.1	2.1	2.2	2.2	2.2	2.2	2.2	2.2	2.2	—	—
1×10^{-4} m	2.2	2.2	2.2	2.2	2.2	2.2	2.2	2.2	2.2	2.2	—	—
Eosin blue 1×10^{-3} m	3.6	3.6	3.7	3.8	3.8	3.9	4.0	4.0	4.1	4.1		
1×10^{-4} m	3.8	3.8	3.9	3.9	4.0	4.0	4.1	4.1	4.1	4.1	1261	1.5×10^8
7-Diethylamino-4-methylmarin 1×10^{-3} m	3.6	3.9	4.2	4.4	4.5	4.7	4.8	4.9	4.9	4.9		
1×10^{-4} m	3.6	3.8	4.0	4.2	4.5	4.7	4.8	4.9	4.9	4.9	2440	4.3×10^8
Rhodamine 6G 1×10^{-3} m	4.3	4.3	4.4	4.5	4.6	4.7	4.7	4.7	4.7	4.7		
1×10^{-5} m	4.7	4.7	4.7	4.7	4.7	4.7	4.7	4.7	4.8	4.8	—	—
Rhodamine B 1×10^{-3} m	2.9	3.3	3.6	3.9	4.1	4.4	4.6	4.7	4.8	4.8		
1×10^{-5} m	3.2	3.6	4.0	4.3	4.4	4.6	4.7	4.7	4.8	4.8	2675	9×10^8
2×10^{-3} m	5.3	5.6	6.1	6.4	6.9	5—6 10—12	5—6 12—14	5—6 14—16	6—7 16—18	6—7 18—20		
Acridine orange 1×10^{-3} m	4.7	5.0	5.5	5.6	5.9	6.3	6.6	5—6 9—10	5—6 10—15	6—7 15—20	—	—
1×10^{-4} m	4.7	4.9	5.1	5.3	5.5	5.6	5.8	5.9	5.9	5.9	2256	19×10^8

Of interest are the absorption and fluorescence properties of a variety of cyanine dyes in dimethyl sulphoxide.[112] The wavelengths of absorption and fluorescence maxima and quantum yield have been ascertained and the results discussed in relation to molecular structure. Spectra of monomeric and dimeric species of xanthene dyes[113] in aqueous solution have been obtained and compared with theoretical spectra calculated from two exciton interaction theories. Excited-state kinetics have been applied to singlet pK determination of acridine, xanthone, and thioxanthone.[114]

Flash photolysis has been applied to *cis–trans* photochemical isomerization of azo-dyes[115] and azomethine dyes.[116] Preliminary results for azo-dyes show that optical density changes of less than 10^{-4} can be detected.

Some preliminary results for spectra of organic dye molecules on thin films have recently appeared.[117] Fluorescence labelling of organic acidic compounds with 4-bromomethyl-7-methoxycoumarin has been reported,[118] and fluorescence and decay times of 4-methylum belliferone in ethanol–water–acid solutions have been investigated.[119]

A survey of spectral sensitization of silver halide emulsions in terms of spectral absorption of dyed emulsions and relative quantum efficiency of sensitization has been presented with the aim of examining an electron-transfer theory of sensitization.[120] Further aspects of sensitization of dichromated-poly(vinyl alcohol) are presented.[121] Certain aspects of dye-sensitization on the photocurrent at zinc oxide electrodes[122] and the use of covalently bound dyes on semiconductor surfaces used in photocells[123] have been investigated.

An interesting account has been given on the industrial use of fluorescent pigments.[124] Various methyl- and/or methoxy-substituted 1,3-diphenyl-2-pyrazolines have been synthesized and evaluated as solutes in liquid scintillation counting systems.[125] Reports of absorption and luminescence spectra of anthraquinone adsorbed on silica gel and microporous glass,[126] energy degradation processes in carbocyanine dyes,[127] and circular dichroism of the acridine orange-sulphated amylopectin complex[128] have also appeared.

112 R. C. Benson and H. A. Kues, *J. Chem. and Eng. Data*, 1977, **22**, 379.
113 L. P. Gianneschi, A. Cant, and T. Kurucsev, *J.C.S. Faraday II*, 1977, **73**, 664.
114 F. Celardin, *Z. Phys. Chem.*, 1977, **106**, 25.
115 F. Rondelez, H. Hervet, and W. Urbach, *Chem. Phys. Letters*, 1978, **53**, 138.
116 V. A. Kuz'min, A. M. Vinogradov, M. A. Al'perovich, and I. I. Levkoev, *High Energy Chem.*, 1977, **11**, 27.
117 J. D. Swalen, M. Tacke, R. Santo, K. E. Rieckhoff, and J. Fischer, *Helv. Chim. Acta*, 1978, **61**, 32.
118 W. Dunges, A. Meyer, K.-E. Muller, M. Muller, R. Pietschmann, C. Plachetta, R. Sehr, and H. Tuss, *Z. analyt. Chem.*, 1977, **288**, 361.
119 R. K. Bauer, A. Kowalczyk, and A. Balter, *Z. Naturforsch.*, 1977, **32a**, 560.
120 B. H. Carroll, *Photographic Sci. Engn.*, 1977, **21**, 151.
121 A. Bloom and W. J. Burke, *Appl. Optics*, 1977, **16**, 2614.
122 M. Matsumura, Y. Nomura, and H. Tsubomura, *Bull. Chem. Soc. Japan*, 1977, **50**, 2533.
123 M. Fujihira, N. Ohishi, and T. Osa, *Nature*, 1977, **268**, 226.
124 C. D. Dane, *Chem. in Britain*, 1977, **13**, 335.
125 H. Gusten, P. Schuster, and W. Seitz, *J. Phys. Chem.*, 1978, **82**, 459.
126 V. Ya. Oginets, *High Energy Chem.*, 1977, **11**, 141.
127 A. M. Vinogradov, V. A. Kuz'min, M. A. Al'perovich, I. I. Zil'berman, and I. I. Levkoev, *High Energy Chem.*, 1977, **11**, 113.
128 K. Nishida, K. Shibata, and H. Watanabe, *Colloid and Polymer Sci.*, 1977 **255**, 1008.

The interaction of the polyanions poly(potassium vinyl sulphate) (PVS), poly(sodium acrylate) (PANa), and poly(potassium *p*-styrenesulphonate) (PSS) on the formation of mixed dimers of methylene blue and tryptaflavine, methylene blue and phenosafranine, and methylene blue and pyronine[129] have been investigated spectrophotometrically, together with the alkaline fading reaction of methylene blue.[130]

Lifetime and polarization measurements have now become a fairly routine method of obtaining decay kinetics of molecules. This has come about with a greater knowledge of the use of photon counting apparatus[131] and dye laser spectroscopy.[132] Several papers deal with the accuracy of such measurements[133—135] and provide a useful insight into the various methods available of extracting lifetime values from decay curves.

Time-resolved emission spectroscopy of styrenes excited in the ^1L_a band have been obtained.[136,137] Measurements have been obtained using both laser excitation at 257.25 nm and synchrotron radiation.[138,139] The latter provides a tunable source of light with pulses of <1 ns and excitation wavelengths from the soft *X*-rays to the far visible (see Figure 6). Dual fluorescence is observed and is explained on the basis of a short-lived component arising from the high vibrational levels of the S_1 (^1L_b) state produced from internal conversion from S_2 and a long-lived component from a state which arises through twisting of the ^1L_a state about the olefinic bond.

Picosecond time-dependent fluorescence depolarization[140] has been used to measure the rotational diffusion time of 3,3′-diethyloxadicarbocyanine iodide (DODCI) in solution at 293 K.[141] The stable form of DODCI, the *cis,cis*-1,5-conformation, has a rotational correlation time of 160 ± 30 ps in ethanol and 320 ± 40 ps in propan-2-ol, the corresponding fluorescence lifetimes being 1.2 ± 0.1 and 1.6 ± 0.1 ns. The all-*trans*-photoisomer obtained by irradiating DODCI in solution has a rotational correlation time >700 ps in propan-2-ol and a lifetime of 0.42 ± 0.05 ns. These results followed closely those predicted on the Stokes–Einstein hydrodynamic model and thereby indicated a lack of any solvent attachment to DODCI in solution.

Decay of the excited state of 1,8-anilinonaphthalene sulphonate (ANS) in aqueous and ethanolic solutions has been studied by laser flash photolysis.[142]

Steady-state and time-resolved fluorescence spectroscopy of excited *p*-(9-

[129] M. Shirai, T. Nagatsuka, and M. Tanaka, *Makromol. Chem.*, 1978, **179**, 173.
[130] M. Shirai, M. Yamashita, and M. Tanaka, *Makromol. Chem.*, 1978, **179**, 747.
[131] J. M. Kurepa, B. M. Panic, and B. J. Levi, *Spectrochim. Acta*, 1977, **32B**, 313.
[132] E. P. Ippen and C. V. Shank, *Phys. Today*, May 1978, p. 41.
[133] A. L. Hinde, B. K. Selinger, and P. R. Nott, *Austral. J. Chem.*, 1977, **30**, 2383.
[134] I. M. Warner, E. R. Davidson, and G. D. Christian, *Analyt. Chem.*, 1977, **49**, 2155.
[135] L. J. Curtis, *Topics Current Phys.*, 1976, **1**, 63.
[136] K. P. Ghiggino, D. Phillips, K. Salisbury, and M. D. Swords, *J. Photochem.*, 1977, **7**, 141.
[137] K. P. Ghiggino, K. Hara, K. Salisbury, and D. Phillips, *J.C.S. Perkin II*, 1978, 88.
[138] R. E. Watson and M. L. Perlman, *Science*, 1978, **199**, 1295.
[139] O. Aita, K. Ichikawa, H. Nakamura, Y. Iwasaki, and K. Tsutsumi, *Japan J. Appl. Phys.*, 1978, **17**, 595.
[140] E. Frehland, *Biophys. Struct. Mechanism*, 1976, **2**, 243.
[141] G. R. Fleming, A. E. W. Knight, J. M. Morris, R. J. Robbins, and G. W. Robinson, *Chem. Phys. Letters*, 1977, **49**, 1.
[142] G. R. Fleming, G. Porter, R. J. Robbins, and J. A. Synowiec, *Chem. Phys. Letters*, 1977, **52**, 228.

anthryl)-*NN*-dimethylaniline derivatives in polar solvents has revealed the presence of two emitting states in all the compounds studied.[143] The influence of solvent polarity,[144—149] temperature,[150,151] pressure,[152—155] magnetic fields,[156—161] and electric field[162] on fluorescence spectra and lifetimes have also been investigated in some considerable detail.

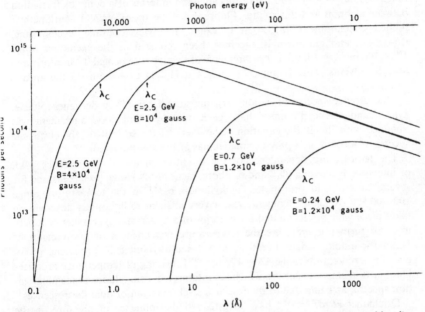

Figure 6 *Synchrotron radiation spectra for several typical electron energies and bending magnet fields. Intensities are given for 1% bandwidth, 100 mA, and 10 mrad horizontal slice of the radiation*

(Reproduced by permission from *Science*, 1978, **199**, 1295)

143 A. Siemiarczuk, Z. R. Grabowski, A. Krowczynski, M. Asher, and M. Ottolenghi, *Chem. Phys. Letters*, 1977, **51**, 315.
144 C. Huggenberger and H. Labhart, *Helv. Chim. Acta*, 1978, **61**, 250.
145 N. Vlahovici and A. Vlahovici, *J. Luminescence*, 1977, **15**, 421.
146 K.-C. Wu and S. Lipsky, *J. Chem. Phys.*, 1977, **66**, 5614.
147 L. E. Cramer and K. G. Spears, *J. Amer. Chem. Soc.*, 1978, **100**, 221.
148 A. Kawski and T. Wasniewski, *Z. Naturforsch.*, 1977, **32a** 1368.
149 E. Schmidt, H. Loeliger, and R. Zürcher, *Helv. Chim. Acta*, 1978, **61**, 488.
150 S. Suzuki, T. Fujii, A. Imal, and H. Akahori, *J. Phys. Chem.*, 1977, **81**, 1592.
151 W. Heinzelmann, *Helv. Chim. Acta*, 1978, **61**, 234.
152 L. A. Brey, G. B. Schuster, and H. G. Drickamer, *J. Chem. Phys.*, 1977, **67**, 2648.
153 L. A. Brey, G. B. Schuster, and H. G. Drickamer, *J. Chem. Phys.*, 1977, **67**, 5763.
154 D. J. Mitchell, G. B. Schuster, and H. G. Drickamer, *J. Chem. Phys.*, 1977, **67**, 4832.
155 J. Kalinowski and R. Jankowiak, *Chem. Phys. Letters*, 1978, **53**, 56.
156 R. C. Johnson, *Chem. Phys. Letters*, 1977, **49**, 85.
157 H. Bouchriha, V. Ern, J. L. Fave, C. Guthnann, and M. Schott, *J. Phys.*, 1978, **39**, 257.
158 H. Bouchriha, V. Ern, J. L. Fave, C. Guthmann, and M. Schott, *Chem. Phys. Letters*, 1978, **53**, 288.
159 G. Klein and M. J. Carvalho, *Chem. Phys. Letters*, 1977, **57**, 409.
160 F. W. Birss, D. A. Ramsay, and S. M. Till, *Chem. Phys. Letters*, 1978, **53**, 14.
161 B. M. Rumyantsev, V. I. Lesin, and E. L. Frankevich, *High Energy Chem.*, 1977, **11**, 104.
162 G. N. R. Padmaker and D. R. Dubey, *J. Luminescence*, 1978, **16**, 69.

The fluorescence of 9-fluorenone[144] in non-hydrogen-bonding solvents produced an emission at $19.3 \leqslant \bar{\nu}_{max} \leqslant 21.4 \times 10^3$ cm^{-1} whereas in hydrogen-bonding solvents the fluorescence suffers a red shift to $16.7 \leqslant \bar{\nu}_{max} \leqslant 17.5 \times 10^3$ cm^{-1}. A strong temperature dependence was also seen. 3-Amino-N-methylphthalimide[145] has been investigated in polar and non-polar solvents in the temperature range $+25$ to -50 °C and the results analysed in terms of a complex formation between alcohols and this solute. Fluorescence spectra of $NNN'N'$-tetramethyl-p-phenylenediamine (TMPD) in n-hexane, methylcyclohexane, cyclohexane, iso-octane, and tetramethylsilane have been reported in the excitation range 170—300 nm.[146] Lifetime measurements of the dye rose bengal[147] in protic and aprotic solvents ranged from 77 ps in CF_3CH_2OH to ~ 2.5 ns in the aprotic solvents.

Fluorescence spectra of various organic scintillators[148] in benzene, toluene, xylene, and xylene–n-butanol have been measured and good agreement was obtained with the Bothe equation. The effect of the solvent on the electronic spectra of a solute in a polymer matrix has also been discussed.[149]

The temperature dependence of the quantum yields of fluorescence (Φ_f), of intersystem crossing (Φ_t), and of photoreactions (Φ_R) have been measured for several neutral and protonated 2-alkylindazoles[151] in the temperature range 100–300 K. At higher temperatures, a very efficient radiationless deactivation takes place which is accounted for in terms of a hypersurface crossing or touching. Of further interest are the temperature dependences of fluorescence of 1-naphtholindole, 1-naphthonitrile, and 1-naphthylamine in propylene glycol over the temperature range 320—170 K.[150] Lowering of temperature produces blue shifts in emission spectra which are interpreted, using fluorescence polarization spectra, as being due to fluorescence level inversion of dual fluorescences.

Drickamer *et al.*[152—154] have investigated the influence of pressure on the fluorescence efficiency of crystal violet and auramine O[152] and fluorenone[153] and on the rate of quenching of naphthalene fluorescence by biacetyl in degassed methylcyclohexane at room temperature.[154] For fluorenone, the application of high pressure induces changes of quantum yield (Φ_f) which range from 0.001 at low pressure to a maximum at 0.1 at high pressure in hydrocarbon plastics. These results are interpreted as arising from a decrease in the energy of the lowest $\pi\pi^*$ excited singlet state relative to the other relevant states as the external pressure is increased. The effect of pressure on the 0–0 component of the first singlet transition of tetracene single crystals has also been measured as a function of pressure in the range 1 atm to 6.5 kbar.[155] The pressure-induced spectral red shifts are found to be larger in the low-pressure phase than those in the high-pressure crystal structure.

Interesting results have been obtained by Johnson[156] on the effect of magnetic fields on the delayed fluorescence intensity from pyrene and methylenephenanthrene crystals. It is shown that the rate at which triplet excitons interact and annihilate to produce singlet excitons depends on the strength and direction of an applied magnetic field. The magnetic field influences the triplet exciton spin states through interplay between their Zeeman interaction and zero-field splitting energies. Investigations of the effect of magnetic field on triplet–triplet exciton annihilation in a quasi-one-dimensional crystal 1,4-dibromonaphthalene

(DBN)[158] and on singlet exciton fission in crystalline tetracene[157] have also been presented. Other compounds investigated under applied magnetic fields are *p*-terphenyl,[159] formaldehyde,[160] and excited complexes of rubrene with oxygen.[161]

Electroluminescence excited by an a.c. electric field has been reported at temperatures of 273, 283, and 293 K from oxalic acid dehydrate dispersed in the epoxy-resin 'araldite'.[162] The voltage and frequency dependence of the emitted light flux have been reported.

The importance of photochemistry in biological research is shown by the ever increasing number of publications in this field. Comments on most are therefore kept to a minimum. The dependence of DNA luminescence on excitation wavelengths in the u.v. and vacuum u.v. region has been reported and the emission was found to cover the region 310—490 nm.[163] An investigation of polarization and rotational depolarization of fluorescence of DNA bases in aqueous solution at room temperature[164] and a pulse fluorimetry study in polarized light of DNA–ethidium bromide complexes have been carried out.[165] Of particular interest are the fluorescence properties (quantum yield, lifetime, and polarization) of acridine orange (AO) and proflavine (PF) bound to DNA[166] as a function of nucleotide to dye (P/D) ratio. Lifetime and quantum yield of acridine orange increased with increasing (P/D) ratio, there being no parallel relation between the lifetime and quantum yield for proflavine. Fluorescence polarization measurements suggested the presence of Förster energy transfer between molecules bound to the fluorescence site (AT pair) and bound to the quenching site (GC pair) and critical transfer distances of 26.4 and 37.0 Å were obtained for bound proflavine and acridine orange respectively. Figure 7 shows the results for lifetimes and quantum yields of AO and PF bound to DNA as function of P/D.

Nanosecond fluorescence anisotropy measurements of DNA–acridine complexes give values of between 21 and 31 ns for the rotational relaxation times of these complexes.[167] Energy-transfer studies in fluorescent derivatives of uracil and thymine[168] and a new emission band of tyrosine induced by interaction with phosphate ion[169] have been recently reported.

The photochemical reactions of nucleic acids and their constituents[170] between furocoumarins and various base compositions of DNA[171] and the fluorimetric determination of 4′,5′-cycloadducts in the DNA–psoralen photoreaction[172] have been investigated recently. A yearly review on protein–nucleic acid photo-interaction has been presented and deals with the utilization of

[163] S. Basu, R. B. Cundall, M. W. Jones, and G. O. Phillips, *Chem. Phys. Letters*, 1978, **53**, 439.
[164] J. P. Morgan and M. Daniels, *Photochem. and Photobiol.*, 1978, **27**, 73.
[165] D. Genest and Ph. Wahl, *Biophys. Chem.*, 1978, **7**, 1317.
[166] Y. Kubota and R. F. Steiner, *Biophys. Chem.*, 1977, **6**, 279.
[167] Y. Kubota and R. F. Steiner, *Bull. Chem. Soc. Japan*, 1977, **50**, 1502.
[168] Y. J. Lee, W. A. Summers, and J. G. Burr, *J. Amer. Chem. Soc.*, 1977, **99**, 7679.
[169] O. Shimizu and K. Imakubo, *Photochem. and Photobiol.*, 1977, **26**, 541.
[170] D. Elad, *Pure Appl. Chem.*, 1977, **49**, 503.
[171] F. Dall'Acqua, D. Vedaldi, and M. Recher, *Photochem. and Photobiol.*, 1978, **27**, 33.
[172] F. Dall'Acqua, S. Caffieri, and G. Rodighiero, *Photochem. and Photobiol.*, 1978, **27**, 77.

photochemistry for the determination of packing arrangements and recognition domains in protein–nucleic acid complexes.[173]

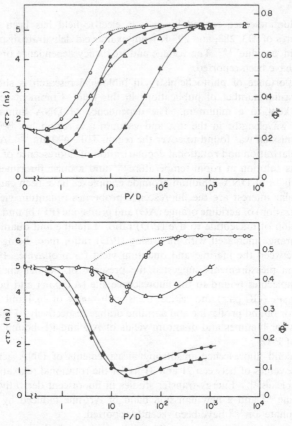

Figure 7 *Mean fluorescence lifetimes ($\langle\tau\rangle$) and quantum yields (Φ_f) of (a) acridine orange (AO) and (b) proflavine bound to DNA as a function of P/D. τ: $\bigcirc$ nature DNA, $\triangle$ denatured DNA; Φ_f: $\bullet$ nature DNA, $\blacktriangle$ denatured DNA. Excitation wavelengths (a) 450 mn, (b) 400 nm*
(Reproduced by permission from *Biophys. Chem.*, 1977, **6**, 279)

The use of photochemistry in evaluating the conformational changes involved in the denaturation of tyrosine residues in ribonuclease-A (RNase) has also proved successful.[174] Further examples where photochemistry has proved successful are in evaluating the luminescence properties of pesticides,[175,176] measuring the rates of exchange of carcinogens from particulate matter to cell

[173] J. Sperling and A. Havron, *Photochem. and Photobiol.*, 1977, **26**, 661.
[174] N. Borboy and J. Flitelson, *Photochem. and Photobiol.*, 1977, **26**, 561.
[175] J. B. Addison, G. P. Semeluk, and I. Unger, *J. Luminescence*, 1977, **15**, 323.
[176] Y. Kumar, G. P. Semeluk, and I. Unger, *J. Luminescence*, 1977, **15**, 341.

membranes,[177] and characterizing the surface distribution of chemical carcinogens on particulate matter.[178] Again laser fluorimetry in conjunction with high-pressure liquid chromatography have enabled the carcinogens, aflatoxins B_1, B_2, G_1, and G_2 to be detected down to 7.5×10^{-13} g.[179] Continued interest has been shown in photophysical aspects of the storage of solar energy.[180—185] Chain and Arnon[180] have studied the quantum efficiency of photosynthetic energy conversion in isolated spinach chloroplasts by measurements of the quantum requirements of ATP formation by cyclic and non-cyclic photophosphorylation catalysed by ferridoxin. The possible use of organic materials, namely linked anthracenes, for photochemical energy conversion has also been considered.[182] Bolton[181, 183] has also discussed the efficiency of photochemical storage of solar energy and concludes that the maximum solar energy storage efficiency for any photochemical system is 15—16%. This compares favourably with photosynthesis which has an efficiency of *ca.* 9.5%.

An interesting review on the primary photochemistry of photosynthesis has been presented,[186] and several papers have appeared on various luminescent properties of chlorophyll.[187—190] Picosecond light pulses have been used to study the kinetics of energy transfer and photo-oxidation in *Rhodopseudomanas sphaeroides* R-26 reaction centres.[191] and to study the fluorescence yield (Φ_f) of spinach chloroplasts as functions of exciting intensity.[192] The fluorescence spectroscopy of P100–chlorophyll–protein complex has also been studied;[193] the picosecond results obtained by Peters *et al.*[194] suggest that the primary step in vision is not a *cis–trans* isomerization but rather proton translocation. Spectroscopic properties of solid films of retinyl polyenes[195, 196] and retinal and retinyl acetate[197] have been reported, and also a flash photolysis study of hindered isomers of retinal.[198] Absorption and emission spectra of all-*trans*-octatetra-1,3,5,7-ene[199] and radiative lifetimes in hexane of 220 and 190 ns at

[177] H. S. Black and J. T. Chan, *Photochem. and Photobiol.*, 1977, **26**, 183.
[178] J. R. Lakowicz, M. McNamara, and L. Steenson, *Science*, 1978, **199**, 305.
[179] G. J. Diebold and R. N. Zane, *Science*, 1977, **196**, 1439.
[180] R. K. Chan and D. I. Arnon, *Proc. Nat. Acad. Sci. U.S.A.*, 1977, **74**, 3377.
[181] J. R. Bolton, *Solar Energy*, 1978, **20**, 181.
[182] G. Jones, jun., T. E. Reinhardt, and W. R. Bergmark, *Solar Energy*, 1978, **20**, 241.
[183] J. R. Bolton, *J. Solid State Chem.*, 1977, **22**, 3.
[184] J.-M. Lehn and J.-P. Sauvage, *Nouveau J. Chim.*, 1977, **1**, 449.
[185] Yu, P. Chukova, *High Energy Chem.*, 1977, **11**, 100.
[186] P. A. Loach, *Photochem. and Photobiol.*, 1977, **26**, 87.
[187] R. P. H. Kooyman, T. J. Schaafsma, and J. F. Kleibeuber, *Photochem. and Photobiol.*, 1977, **26**, 235.
[188] J. C. Hindman, R. Kugel, A. Svirmickas, and J. J. Katz, *Chem. Phys. Letters*, 1978, **53**, 197.
[189] Govindjee, *Acta Phys. et Chem. Szeged.*, 1977, **23**, 49.
[190] K. Vacek, D. Wong, and Govindjee, *Photochem. and Photobiol.*, 1977, **26**, 269.
[191] E. Maskowitz and M. M. Malley, *Photochem. and Photobiol.*, 1978, **27**, 55.
[192] N. E. Geacintov, J. Breton, C. E. Swenberg, and G. Paillotin, *Photochem. and Photobiol.*, 1977, **26**, 629.
[193] J. S. Brown, *Photochem. and Photobiol.*, 1977, **26**, 519.
[194] K. Peters, M. L. Applebury, and P. M. Rentzepis, *Proc. Nat. Acad. Sci. U.S.A.*, 1977, **74**, 3119.
[195] S. Hotchandani, P. Paquin, and R. M. Leblanc, *Photochem. and Photobiol.*, 1977, **26**, 167.
[196] S. Hotchandani and R. M. Leblanc, *Photochem. and Photobiol.*, 1978, **24**, 59.
[197] S. Hotchandani, P. Paquin, and R. M. Leblanc, *Canad. J. Chem.*, 1978, **56**, 1985.
[198] A. Harriman and R. S. H. Liu, *Photochem. and Photobiol.*, 1977, **26**, 29.
[199] R. M. Gavin, C. Weisman, J. K. McVey, and S. A. Rice, *J. Chem. Phys.*, **68**, 1978, 522.

room temperature and 77 K respectively have been obtained. Appriou[200,201] gives a detailed analysis of absorption and emission spectra of different heterocyclic spiropyrans and considers the concept of spiroconjugation in relation to energy transfer from the heterocyclic to the chromene group.

Interest is sustained in indole, one of the most widely studied fluorescent heterocyclics. The role of the water structure in the quenching of indole fluorescence is still unclear, and Kleim and Tatischeff[202] have obtained further data on the temperature-dependent fluorescence quenching of indole, *N*-methylindole, 5-methoxyindole, and benzimidazole. Their results suggest a reaction between the excited indole chromophore and a monomeric non-hydrogen-bonded water molecule. Decay kinetics for the indole ring in water–alcohol glasses at 77 K have been obtained over a range from 10 ms to several seconds.[203] Fluorescence quenching of indoles by *NN*-dimethylformamide[204] and energy-transfer studies of 2-phenylindole with different concentrations of xylene[205] have also been carried out. Additional papers of biological interest are concerned with the phototoxicity of coumarin derivatives[206] and the absorption and emission spectra of 4-hydroxycoumarin.[207]

Applications of fluorescence techniques to protein systems[208—211] is also expanding, and the use of time-resolved fluorescence spectroscopy to aqueous tryptophan is of particular interest.[211] The fluorescence decay of aqueous tryptophan consists of a two-exponential decay function whose components have lifetimes of 3.14 and 0.57 ns. These components are assigned as the solvent-equilibrated ^1L_a and ^1L_b states respectively. The time-resolved emission spectra are resolved into two spectra with λ_{max} at 350 and 335 nm corresponding to these two states.

Considerable interest is shown in fluorescent labelling[212—216] of biological systems. Dreyfuss *et al.*[212] describe the application of fluorescence spectroscopy to the study of receptor sites in protein mixtures or supramolecular assemblies, using photoaffinity labelling. This approach combines the merits of photoaffinity labelling to attain site-directed reactivity with the probing power of fluorescent ligands. Chen[217] describes a novel fluorometric technique for

200 P. Appriou and R. Guglielmetti, *J. Photochem.*, 1978, **8**, 145.
201 P. Appriou, R. Guglielmetti, A. Botrel, and A. leBeuze, *J. Photochem.*, 1978, **8**, 73.
202 R. Klein and I. Tatischeff, *Chem. Phys. Letters*, 1977, **51**, 333.
203 M. Bazin, M. Aubailly, and R. Santus, *J. Chem. Phys.*, 1977, **67**, 5070.
204 P. M. Froehlich, D. Gantt, and V. Paramasigamani, *Photochem. and Photobiol.*, 1977, **26**, 639.
205 A. Kawski and T. Wasniewski, *Bull. Acad. polon. Sci., Ser. Sci. chim.*, 1977, **15**, 1197.
206 S. C. Shim, K. Y. Choi, and P.-S. Song, *Photochem. and Photobiol.*, 1978, **27**, 25.
207 O. S. Walfbeis and G. Uray, *Monatsh.*, 1978, **109**, 123.
208 H. Steinhart and J. Sandmann, *Analyt. Chem.*, 1977, **49**, 950.
209 V. Giancotti, M. Fonda, and C. Crane-Robinson, *Biophys. Chem.*, 1977, **6**, 379.
210 P. Sherline and K. Schiavone, *Science*, 1977, **198**, 1038.
211 D. M. Rayner and A. G. Szabo, *Canad. J. Chem.*, 1978, **56**, 743.
212 G. Dreyfuss, K. Schwartz, E. R. Blout, J. R. Barrio, F.-T. Liu, and N. J. Leonard, *Proc. Nat. Acad. Sci. U.S.A.*, 1978, **75**, 1199.
213 M. W. Geiger and N. J. Turro, *Photochem. and Photobiol.*, 1977, **26**, 221.
214 M. E. Switzer, *Sci. Progr. (Oxford)*, 1978, **65**, 19.
215 R. Takashi, Y. Tonomura, and M. F. Morales, *Proc. Nat. Acad. Sci. U.S.A.*, 1977, **74**, 2334.
216 D. Atlas and A. Levitzki, *Proc. Nat. Acad. Sci. U.S.A.*, 1977, **74**, 5290.
217 R. F. Chen, *Analyt. Letters*, 1977, **10**, 787.

enzyme assay and an account is given of the use of a dye laser in obtaining fluorescence characteristics of single, intact human red blood cells.[218] The high sensitivity of the system allows the detection of potential-dependent uptake and release of dye added to the suspension of living erythrocytes. Dye concentrations as low as 10^{-8} mol l^{-1} in water have been detected in a layer of 10 μm thickness and a cross-sectional area of 1 μm². Other investigations of interest are those on metalloporphyrins,[219] the effect of proteins on the photodecomposition of pyrazolines,[220] the photochemistry of organic chromophores incorporated into fatty-acid monolayers,[221] and absorption and emission studies of solubilization in micelles.[222]

Singlet Quenching by Energy Transfer and Exciplex Formation.—*Electron Donor–Acceptor Complexes and Related Charge-transfer Phenomena.* Turro[223] has given a very comprehensive and interesting account of energy-transfer processes in organic molecules with particular emphasis on polymer systems. He deals with simple emission and reabsorption mechanisms and goes on to more complex coulomb interaction mechanisms of energy transfer. Bojarski[224] extends the dipole–dipole mechanism of non-radiative energy transfer to systems containing dimers. Equations representing the decay kinetics of fluorescence of a donor luminophore in the presence of an acceptor in the same flexible molecule have been obtained, and molecular dimensions of the end-to-end distances evaluated from the decay kinetics.[225] Mehlhorn *et al.*[226—228] have attempted to find molecular structures suitable as energy acceptors or donors in Scheme 1 and therefore able to quench or sensitize the photo-Fries rearrangement, highly important as an initial step in the photodecomposition of polymers containing *N*-arylamido-groups. Of further general interest are the studies of Guillet[229] on energy transfer and molecular mobility in polymer systems. The Förster theory, taking diffusion into account, has been applied to the system pyrene (donor)–perylene (acceptor) in liquid solutions of various viscosities and has been shown to be valid under various conditions.[230]

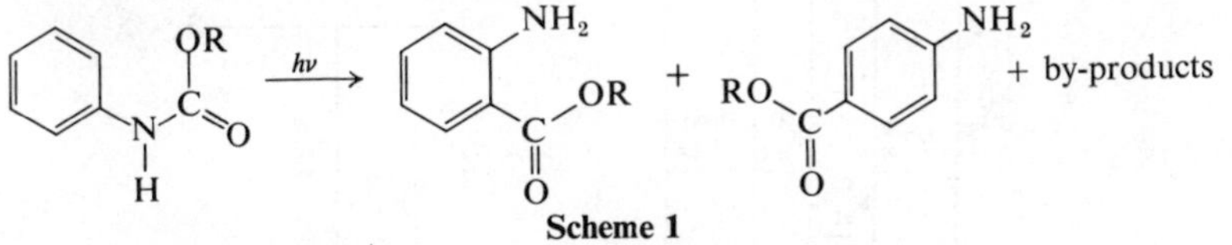

Scheme 1

[218] G. Marowsky, G. Cornelius, and L. Rensing, *Optics Comm.*, 1977, **22**, 361.
[219] B. B. Smith and C. A. Reberz, *Photochem. and Photobiol.*, 1977, **26**, 527.
[220] N. A. Evans, L. A. Holt, and B. Milligan, *Austral. J. Chem.*, 1977, **30**, 2277.
[221] D. G. Whitten, F. R. Hopf, F. H. Quina, G. Sprintschnik, and H. W. Sprintschnik, *Pure Appl. Chem.*, 1977, **49**, 379.
[222] R. C. Darrance and T. F. Hunter, *J.C.S. Faraday I*, 1977, **73**, 1891.
[223] N. J. Turro, *Pure Appl. Chem.*, 1977, **49**, 405.
[224] C. Bojarski, *Acta Physet. Chem. Szeged.*, 1977, **23**, 1.
[225] A. Englert and M. Leclerc, *Proc. Nat. Acad. Sci. U.S.A.*, 1978, **75**, 1050.
[226] A. Mehlhorn, B. Schwenzer, and K. Schwetlick, *Tetrahedron*, 1977, **33**, 1483.
[227] A. Mehlhorn, B. Schwenzer, and K. Schwetlick, *Tetrahedron*, 1977, **31**, 1489.
[228] A. Mehlhorn, B. Schwenzer, H.-J. Bruckner, and K. Schwetlick, *Tetrahedron*, 1978, **34**, 481.
[229] J. E. Guillet, *Pure Appl. Chem.*, 1977, **49**, 249.
[230] M. Hauser, R. Frey, U. K. A. Klein, and V. Gosele, *Acta Phys. et Chem. Szeged.*, 1977, **23**, 1.

A theory for the excitation energy transfer among homotransfer molecules in solution based on a shell model of the luminescent centre has been developed.[231] Analysis of decay curves of electronically excited molecules in the presence of quencher, taking into account 'static' quenching, has shown good agreement between the proposed model and experimental results.[232]

Fluorescence quenching of naphthalene, 2-methoxynaphthalene, 2,6-dimethoxynaphthalene, and 1,4-dimethoxynaphthalene by chloroacetonitrile in acetonitrile has been described.[233] The rate constants span almost three orders of magnitude and it is shown that triplet state formation occurs with three of the naphthalenes. Efficient energy transfer has been demonstrated from micellar sodium dodecyl sulphate (SDS)-solubilized naphthalene to terbium chloride.[234] The photophysical mechanism involves energy deposition in naphthalene, excitation into singlets, intersystem crossing into the triplet domain, and diffusion of naphthalene triplets to the micellar surface where energy transfer occurs to terbium chloride.

The quenching of the S_1 state of pyrene solubilized within the hydrophobic region of sodium dodecyl sulphate micelles has been studied as a function of cationic quencher.[235] At concentrations below *ca.* 10^{-3} mol dm^{-3} the fluorescence decay follows a complex rate law and the results are discussed in terms of a model which requires the water–hydrocarbon interface to penetrate the micelles in the region of the probe.

The fluorescence lifetime of coronene and the rate of dipole–dipole energy transfer from coronene to rhodamine 6G in PMMA matrices were found to be temperature dependent.[236] These results are analysed in terms of a model involving thermally activated energy transfer from excited states of coronene, the proposed scheme being shown in Figure 8.

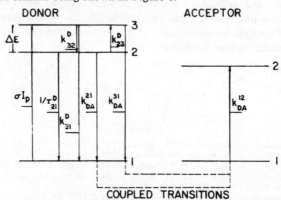

Figure 8 *Kinetic scheme for describing thermal activation of an energy-transfer process. The system is pumped by a short laser pump of intensity I_p at a rate σI_p*
(Reproduced by permission from *Chem. Phys. Letters*, 1977, **52**, 16)

[231] J. Kaminski and A. Kawski, *Z. Naturforsch.*, 1977, **32a**, 1339.
[232] J. C. Andre, I. M. Niclause, and W. R. Ware, *Chem. Phys.*, 1978, **28**, 371.
[233] F. H. Quina, Z. Hamlet, and F. A. Carroll, *J. Amer. Chem. Soc.*, 1977, **99**, 2240.
[234] J. R. Escabi-Perez, F. Nome, and J. H. Fendler, *J. Amer. Chem. Soc.*, 1977, **99**, 7749.
[235] M. A. J. Rodgers, M. F. Da Silva, and E. Wheeler, *Chem. Phys. Letters*, 1978, **53**, 165.
[236] R. Katraro, A. Ron, and S. Speiser, *Chem. Phys. Letters*, 1977, **52**, 16.

Fluorescence lifetime and quantum yield measurements of malachite green and erythrosin have been measured in a water–acetone mixture, and the lifetimes were found to decrease linearly as the mole concentration of water increases in the mixture.[237] The mechanism of fluorescence quenching by water is considered as a possible explanation for the experimental results.

Several papers have appeared on energy transfer in excited polymer systems.[238—240] Singlet and triplet lifetimes of anthracene incorporated as pendant groups in a terpolymer as well as rate constants of the bimolecular quenching by oxygen have been presented.[240] The results are summarized in Table 5.

Table 5 *Fluorescence lifetimes in aerated (τ_f) and deaerated (τ_f^0) solutions, triplet lifetimes (τ_T) in aerated solutions, quenching rate constants for singlet (K_s) and triplet states (K_q). Measurements made in benzenic solution; the oxygen concentrations in the aerated benzene solutions was taken as 1.8×10^{-3} mol l^{-1}*

Compound	τ_f^0/ns	τ_f/ns	$K_s \times 10^{-10}/$ mol^{-1}ls^{-1}	τ_T/ns	$K_q \times 10^{-9}/$ mol^{-1}ls^{-1}
Anthracene	4.5	3.7	2.7	154	3.6
EPDM-anthracene (8)	6.7	3.7	6.7	207	2.7
EPDM-anthracene (9)	5.3	3.7	4.6	156	3.6
EPDM-anthracene (10)	4.2	3.5	2.7	168	3.3
EPDM-anthracene (11)	5.3	3.7	4.6	156	3.6

Fluorescence quenching has been applied to investigations of equilibrium and kinetics of stacking associations between 2-aminopurine and thymidine.[241] Zander *et al.* report on the quenching of polycyclic aromatic hydrocarbons by nitromethane or nitrobenzene.[242] Photolysis of cyclohexane in the presence of solute results in energy transfer from excited cyclohexane to solute, the rate constants for energy transfer varying in the range $(5—14) \times 10^{10}$ lmol^{-1} s^{-1} at 15 °C.[243] The results are summarized in Table 6. It has also been shown that

237 W. Yu, F. Pellegrino, M. Grant, and R. R. Alfano, *J. Chem. Phys.*, 1977, **67**, 1766.
238 C. David, N. P. Lavareille, and G. Geuskens, *European Polymer J.*, 1977, **13**, 15.
239 I. Ohmine, R. Silbey, and J. M. Deutch, *Macromolecules*, 1977, **10**, 862.
240 J. Fouassier, D. Loughot, and J. Faure, *Makromol. Chem..*, 1978, **179**, 437.
241 A. Bierzynski, H. Kozlowska, and K. L. Wierzchowski, *Biophys. Chem.*, 1977, **6**, 223.
242 M. Zander, U. Breymann, H. Dreeskamp, and E. Koch, *Z. Naturforsch*, 1977, **32a**, 1561.
243 T. Wada and Y. Hatano, *J. Phys. Chem.*, 1977, **81**, 1057.

Table 6 *Energy transfer efficiency* α_t *and rate constant* k_t *for various solutes in the photolysis of cyclohexane*

Solute	$\alpha_t/l\ mol^{-1}$	$k_t \times 10^{-10}\ l\ mol^{-1}\ s^{-1}$
CCl_4	25	8.2
SF_6	16	5.4
CO_2	15	5.1
N_2O	31	10.2
CH_3I	43	14.3
CH_3Br	16	5.4
C_2H_5Br	22	7.2
C_6H_6	4	1.3

photolysis of the bond R_{ar}—NO_2 contributes to quenching of the fluorescence of aromatic nitro-compounds.[244]

The effects of geometry on intramolecular fluorescence quenching and exciplex formation have been studied in the (carbazole)—$(CH_2)_n$—(terephthalic acid methyl ester) systems where $n = 1$—10.[245]

Fluorimetry and time-resolved spectroscopy have been used to study proton-induced quenching of 6-(2-hydroxy-5-methylphenyl)-s-triazines,[246] 9,9′-bianthryl,[247] and naphthylamines[248] in a sulphuric acid–water mixture. In the case of 9,9′-bianthryl, efficient fluorescence quenching was observed in polar solvents ($^1k_q = 1.5 \times 10^9\ l\ mol^{-1}\ s^{-1}$ in MeOH at 298 K). The quenching involves intersystem crossing ($k_{isc}' \approx 5.4 \times 10^8\ l\ mol^{-1}\ s^{-1}$) and another radiationless process, probably internal conversion, ($k_n' \approx 1 \times 10^9\ l\ mol^{-1}\ s^{-1}$). Of further interest is the paper by MacCall and Berns on the study of light-gathering and energy transfer modes for various cryptomonad biliproteins (phycocyanin or phycoerythrins).[249] Although each cryptomonad biliprotein is composed of at least two distinct types of absorbing chromophore it is shown that, after radiationless energy transfer, emission is seen from one chromophore only. Reports of singlet quenching of photoexcited aromatics by carbonyl compounds,[250] singlet- and triplet-state photodecomposition of tetramethyl-cyclobutane-1,3-dione,[251] and interaction of organic molecules with oxygen and oxygen-containing acceptors[252] have also recently appeared.

The systems acridizinium[253] and N-methylacridinium ion[254] as energy donors and rhodamin B (tetrafluoroborate) as energy acceptors show multi-step energy transfer by resonance in solutions of low viscosity.

A solution of 1,2,4,5-tetracyanobenzene (TCNB) in 2-methyltetrahydrofuran (MTHF) in which a solute–solvent electron donor–acceptor (EDA) complex is formed has been studied at 77 K.[255] The radical-anion formation of TCNB

[244] E. Lippert and J. Kelm, *Helv. Chim. Acta*, 1978, **61**, 279.
[245] Y. Hatano, M. Yamamoto, and Y. Nishijima, *J. Phys. Chem.*, 1978, **82**, 367.
[246] H. Shizuka, K. Matsui, Y. Hirata, and I. Tanaka, *J. Phys. Chem.*, 1977, **81**, 2243.
[247] H. Shizuka, Y. Ishii, and T. Morita, *Chem. Phys. Letters*, 1977, **51**, 40.
[248] K. Tsutsumi and H. Shizuka, *Chem. Phys. Letters*, 1977, **52**, 485.
[249] R. MacCall and D. S. Berns, *Photochem. and Photobiol.*, 1978, **27**, 343.
[250] D. Busch, L. Dahm, B. Siwicke, and R. W. Ricci, *Tetrahedron Letters*, 1977, 4489.
[251] R. Spafford and M. Vala, *J. Photochem.*, 1978, **8**, 61.
[252] E. L. Frankevich, B. M. Rumyantsev, and V. I. Lesin, *High Energy Chem.*, 1977, **11**, 31.
[253] V. J. Bendig, S. Helm, and D. Kreysig, *J. prakt. Chem.*, 1977, **319**, 809.
[254] J. Bendig and D. Kreysig, *Z. Naturforsch.*, 1978, **33a**, 78.
[255] K. Kimura and Y. Achiba, *Chem. Phys. Letters*, 1977, **46**, 585.

suggests a sensitized biphotonic ionic dissociation of the EDA complex. Further studies have been made on 1,3-diphenyl-2-pyrazoline and on five of its isomeric pairs with donor–acceptor substituents in the 4,4'-positions of the phenyl rings,[256] and on fluorescence quenching of excited-state donor–acceptor pairs in surfactant micelles.[257]

Polarized fluorescence measurements have been made on the ethidium bromide (EB)–poly(rA-rU) complex.[258] A single-exponential decay lifetime of 27 ns was obtained. The absolute value of the EB transition moment defined by an angle ϕ is 38°.

Charge-transfer emission studies have been carried out on propellicene: the results indicate a lack of charge-transfer emission and probably imply that the non-radiative rate is too high for a charge-transfer state.[259] Further charge-transfer studies have been made on 1-benzyl-1,4-dihydronicotinamide (BNAH),[260] anthracene-dimethylpyromellitimide,[261] and on photo-induced ternary electron-transfer reactions.[262]

Considerable interest is still shown in the formation and decay of exciplexes.[263,264] Normally one regards the exciplex being formed directly from the excited state A and the ground state Q, precluding the formation of any encounter complex:

$$A^1 + Q \underset{k_{-1}}{\overset{k_1}{\rightleftharpoons}} (AQ)^1 \xrightarrow{\ k_3\ } A + Q$$

An alternative mechanism exists, however, whereby the exciplex is formed *via* a variety of collisional complexes termed encounter complexes in which the components are separated by distances of 5—7 Å:

$$A^1 + Q \underset{k_{-q}}{\overset{k_{diff}}{\rightleftharpoons}} (A^1/Q) \xrightarrow{\ k_3\ } A^- + Q^+$$

White and Buckles[264] discuss the role of encounter complexes in singlet quenching and study the fluorescence quenching of 2,6-dimethoxynaphthalene (DMN) by a series of benzyl chlorides in acetonitrile.

The role of exciplex intermediates in the abstraction of α-hydrogen from substituted amines by aldehydes and ketones has been discusesd.[265] Fluorescence and laser photolysis studies have been carried out on intramolecular exciplex systems consisting of pyrene and *NN*-dimethylaniline in micellar solutions.[266] Pulsed laser excitation has also been used to study the decay of

[256] H. Gusten, G. Heinrich, and H. Fruhbeis, *Ber. Bunsengesellschaft phys. Chem.*, 1977, **81**, 810.
[257] B. K. Selinger, *Austral. J. Chem.*, 1977, **30**, 2087.
[258] Ph. Wahl, D. Genest, and J. L. Tichadou, *Biophys. Chem.*, 1977, **6**, 311.
[259] E. M. Kosower, H. Dodiuk, B. Thulin, and O. Wennerstrom, *Acta Chem. Scand.*, 1977, **B31**, 526.
[260] F. M. Marteus, J. W. Verhoeven, R. A. Gase, U. K. Pandit, and Th. J. DeBaer, *Tetrahedron*, 1978, **34**, 443.
[261] E. L. Frankevich, M. M. Triebel, I. A. Sokolik, and B. V. Kotov, *Phys. Stat. Solidi (A)*, 1977, **40**, 655.
[262] S. Tazuke and N. Kitamura, *J.C.S. Chem. Comm.*, 1977, 515.
[263] R. C. White and R. E. Buckles, *J. Photochem.*, 1977, **7**, 359.
[264] R. C. White and R. E. Buckles, *J. Photochem.*, 1978, **8**, 67.
[265] K. Morokuma, G. H. Neems, and S. Yamake, *Tetrahedron Letters*, 1977, 767.
[266] H. Masuhara, K. Kaji, and N. Mataga, *Bull. Chem. Soc. Japan*, 1977, **50**, 2084.

aromatic hydrocarbon-substituted dimethylaniline exciplexes. It was shown that the lifetime of the exciplexes varies with the reduction potential of the aromatics as well as the ionization energy of the dimethylaniline.[267] Further studies have been made on exciplex formation with *NN*-diethylaniline[268] and on 1,4-dicyanonaphthalene with alkylbenzenes and alkylnaphthalenes.[269] Interest has also been shown in exciplex formation and decay in styrene–amine systems.[270, 271]

Fluorescence Quenching by Inorganic Species. The fluorescence of the complex of tetracycline with calcium ions (Ca^{2+}) and sodium diethylbarbiturate (sodium barbitol), extracted into ethyl acetate, has been studied in the concentration range 10^{-5}—1.3×10^{-4} mol l^{-1}. The results show the fluorescence emission to be independent of time and exciting wavelength and tetracylcine concentration.[272]

5-Sulpho-8-quinolinol reacts with magnesium, zinc, cadmium, aluminium, gallium, and indium to form water-soluble chelates which emit green fluorescence.[273] Nishikawa *et al.* report on the fluorescence properties of these metallochelates: fluorescence lifetime values of 1 : 1 chelates ranged from 11.3 to 4 ns.

Addition of high concentrations of Br^- during the flash photolysis of aqueous solutions containing either tryptophan or *N*-acetyl-L-tryptophanamide produces an increased yield of the cation, neutral radicals, and triplet. Examination of the mechanism by which Br^- induces these effects indicates that the initial event involves an interaction between the fluorescent state and Br^- with formation of a long-lived radical precursor that may be the triplet. It is further shown that the induced neutral- and cation-radical formation originate from this long-lived state.[274]

Miller *et al.*[275] describe the quenching of pyrenesulphonate fluorescence in inverse micelles by bromide ions. The quenching constant is shown to be virtually independent of water concentrations, a consequence of its inability to dilute the ionic layer.

Heavy-atom Quenching. The use of heavy-atom additives to enhance intersystem crossing has assumed some importance in recent years with emphasis being placed on its role in the elucidation of photophysical and photochemical kinetics. Carroll[276] in an interesting paper derives kinetic expressions which allow determination of the reactivity of excited singlet and triplet states. He further discusses applications and limitations of the technique in photochemistry.

[267] U. P. Agarwal, H. Jagannath, and D. R. Rao, *J.C.S. Faraday II*, 1977, **73**, 1020.
[268] P. J. Harman, P. R. Nott, and B. K. Selinger, *Austral. J. Chem.*, 1977, **30**, 1875.
[269] M. Itoh, S. Furuya, and T. Okamoto, *Bull. Chem. Soc. Japan*, 1977, **50**, 2509.
[270] K. P. Ghiggino, K. Hara, G. R. Mant, D. Phillips, K. Salisbury, R. P. Steer, and M. D. Swords, *J.C.S. Perkin II*, 1977, 88.
[271] R. L. Brentnall, P. M. Crosby, and K. Salisbury, *J.C.S. Perkin II*, 1977, 2002.
[272] J. Grzywacz, J. Widuchowska, and A. Regosz, *Acta Phys. et Chem. Szeged.*, 1977, **23**, 1.
[273] Y. Nishikawa, K. Hiraki, K. Morishige, and T. Katagi, *Bunseki Kagaku*, 1977, **26**, 365.
[274] W. A. Volkert, R. R. Kuntz, C. A. Chiron, R. F. Evans, R. Santus, and M. Bazin, *Photochem. and Photobiol.*, 1977, **26**, 3.
[275] D. J. Miller, U. K. A. Vilen, and M. Hauser, *J.C.S. Faraday I*, 1977, **73**, 1654.
[276] F. A. Carroll, *Mol. Photochem.*, 1977, **8**, 133.

The effect of added alkali-metal halide salts on the room-temperature fluorescence and phosphorescence of naphthalene-2-sulphonate adsorbed on filter paper has been reported.[277] The luminescence is seen to increase strongly in the order NaF < NaCl < NaBr < NaI, and the presence of the heavy atom increases the radiative triplet decay constant k_p more than the competing nonradiative constant k_{qp}. The effect of NaI on the luminescence is shown in Figure 9.

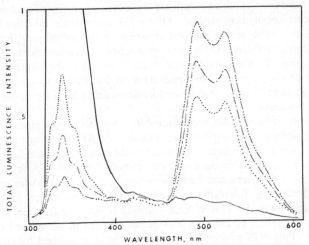

Figure 9 *Room-temperature luminescence of naphthalene-2-sulphonate adsorbed on filter paper with (——) no iodide (peak intensity ~5 × that of the 0.30-NaI sample); (· · · ·) 0.3M-NaI; (– · –) 0.42M-NaI; and (– · · · –) 1.0M-NaI* (Reproduced by permission from *J. Phys. Chem.*, 1977, **81**, 2035)

The photophysical properties of a series of charge-transfer complexes of hexamethylbenzene with various electron acceptors containing heavy halogen atoms have been determined.[278] Again the $T_1 \rightarrow S_0$ transition is strongly enhanced, and it is concluded that increasing spin–orbit mixing of the radiative $T_1 \rightarrow S_0$ charge-transfer transition with singlet–singlet is responsible for the observed effects.

Ruzo *et al.*[279] report photoreactions of various halogenated naphthalenes and biphenyls in methanol at 290—310 nm in the presence of lithium salts. The triplet decay rate appears to be increased relative to the reaction rate.

Excimer Formation and Decay. Several papers have appeared on excimer formation and decay. The geometrical requirements for the fluorescent intramolecular excimer formation for several α,ω-dinaphthylalkanes[280] and polymorphic dianthronylidene ethane[281] have been reported. Such analysis helps to reveal

277   W. White and P. G. Seybold, *J. Phys. Chem.*, 1977, **81**, 2035.
278   I. Deperasinska, J. Dresner, B. Kozankiewicz, K. Luczak, and J. Prochorow, *J. Luminescence*, 1978, **16**, 89.
279   L. O. Ruzo, G. Sundstrom, O. Hutzinger, and S. Safe, *Rec. Trav. chim.*, 1977, **96**, 249.
280   R. S. Davidson and T. D. Whelan, *J.C.S. Chem. Comm.*, 1977, 361.
281   H. Becker, K. Sandros, B. Karlsson and A. Pilotti, *Chem. Phys. Letters*, 1978, **53**, 232.

whether excimer formation takes placebe fore the sandwich configuration is attained or whether it is due to radiationless decay of the excimer.

Energy migration and intramolecular excimer formation have been studied in a series of co-polymers comprising 1-vinylnapthalene,[282] 2-vinylnaphthalene, and styrene with methyl methacrylate.[283] The results show that it is necessary to take into account the role of energy migration in population of the excimer sites. An excitation hopping model is proposed and luminescence depolarization techniques have been applied to estimate the extent of energy migration in the vinyl aromatic copolymers studied. Of further interest are investigations of solvent effects on the absorption and emission spectra of poly(vinylnaphthalenes).[284] In poor solvents such as cyclohexane, new absorption and fluorescence spectra are seen which are attributable to the dimer formed between naphthyl side-groups of the polymer. The dependence of band intensity on solvent as well as temperature indicates that dimer formation depends on the conformation of the polymer chain.

Excimer formation has also been studied in three polystyrene model systems, 2,4-diphenylpentane, 2,4,6-triphenylheptane, and 2,4,6,8-tetraphenylnonane.[285] The results show that isotactic conformations are more favourable for excimer formation than syndiotactic ones. Co-polymers of 2-phenyl-5-(p-vinylphenyl)-oxazole (POS) with styrene and the homopolymer poly-POS have been prepared and characterized by fluorescence decay times and examination of excimer formation and decay.[286] Fluorescence lifetimes of 0.6 ($\pm$0.4) ns and 9 ($\pm$0.2) ns have been obtained at 298 K in toluene solution for monomer and excimer respectively. Poly-POS exhibits excimer formation characterized by an activation energy of 5.9 ($\pm$0.5) kJ mol^{-1} and a binding energy of 17.0 ($\pm$1.0) kJ mol^{-1}.

A model for the long-range mechanism for the temperature dependence of the ratio of delayed monomer and delayed excimer fluorescence following triplet–triplet annihilation has been prescribed by Butler and Pilling.[287]

Excimer formation of fluoradene in several solvents has been examined.[288] In hexane a broad, structureless band appears at 440 nm and is attributed to excimer emission. Powell and Martel[289] have recently reported preliminary neutron diffraction studies of dimer formation in polycrystalline 1-methylthymine. This system is attractive in that it simulates the structural features of a nucleoside.

3 Triplet-state Processes

The rapid formation of triplet states having comparatively long lifetimes from initially formed excited singlet states means that triplet involvement is potentially

[282] R. F. Reid and I. Soutar, *Polymer Letters*, 1977, **15**, 153.
[283] R. F. Reid and I. Soutar, *J. Polymer Sci.*, 1978, **16**, 231.
[284] M. Irie, T. Kamijo, M. Aikawa, T. Takemura, K. Hayashi, and H. Baba, *J. Phys. Chem.*, 1977, **81**, 1571.
[285] L. Bokobza, B. Jasse, and L. Monnerie, *European Polymer J.*, 1977, **13**, 921.
[286] I. McInally, I. Soutar, and W. Steedman, *J. Polymer Sci.*, 1977, **15**, 2511.
[287] P. R. Butler and M. J. Pilling, *J.C.S. Faraday II*, 1977, **73**, 886.
[288] T. E. Hogen-Esch, J. M. Plodinec, M. Vala, and S. G. Shulman, *Chem. Phys. Letters*, 1977, **50**, 120.
[289] B. M. Powell and P. Martel, *Photochem. and Photobiol.*, 1977, **26**, 305.

possible in all aspects of photochemistry. During the period of this Report, progress in many areas has been apparent, and the participation of triplet exciplexes is becoming recognized as particularly widespread. As with other sections of this chapter the survey is based on types of compounds; theory and techniques are discussed within this framework.

Very detailed investigations continue to be made with benzene. Polarization and Zeeman measurements of the ^1A_g–$^3B_{1u}$ transition in crystalline benzene have been used to assign factor group states;[290] the results do not agree with recent calculations and the latter require further development. Kalantar and his co-workers have continued their group's work on the low-temperature phosphorescence of benzene. The rotational relaxation times for C_6H_6 and C_6D_6 in 3-methylpentane–2-methylbutane glasses at 77 K have been measured and found to be independent of the isotopic differences, with values greater than 200 s.[291] Polarization of the weak b_{2g} bands in phosphorescence has 15—20% out-of-plane character, paralleling the in-plane character in the e_{2g} region, contrary to the $^3B_{1u}$ assignment for the lowest triplet state.[292] This is in disagreement with other experiments, and requires radiative decay *via* the A_{2g} spin sublevel. This possibility has been disproved by measurements on the phosphorescence lifetime as a function of temperature[293] and the explanation of the anomaly is unknown. Precise lifetime and relative intensity measurements in several glassy and crystalline media have been made between 2 and 30 K. When spin–lattice relaxation is fast, the increase in τ_p below 10 K is due to changes in the Boltzmann population of spin sublevels. In cyclohexane slower spin–lattice relaxation also increases τ_p.[294] Coupling between the $^3B_{1u}$ and $^3E_{1u}$ states of benzene under the combined influence of substituents and crystal field requires a model to explain the e.s.r. data derived by Vergragt and van der Waals.[295] The phosphorescent triplet of *p*-xylene in isotopically mixed crystals at 1.2 K has been examined by phosphorescence spectroscopy, e.s.r., and microwave-induced delayed phosphorescence. A consistent model in which crystal field and methyl substituents influence the pseudo-Jahn–Teller structure of triplet benzene is described.[296] The phosphorescence lifetime of benzene complexed with $CHCl_3$ and $CDCl_3$ in 3-methylpentane in the range 10—90 K shows non-exponential decays attributed to formation of 1 : 1 and 1 : 2 $CHCl_3$: C_6H_6 complexes.[297] The temperature dependence of the two lifetimes is interpreted as involving thermal production of chloroform- or solvent-substituted hexatrienes from the benzene triplet. A precursor of the substituted hexatriene is described as an exterplex, a complex formed between chloroform, triplet benzene, and solvent. The authors suggest that features of benzene photochemistry under other conditions can also be understood in terms of exciplex formation.

290 H. P. Trommsdorff and G. Castro, *J. Chem. Phys.*, 1977, **67**, 501.
291 T. J. Durnick and A. H. Kalantar, *J. Chem. Phys.*, 1977, **66**, 1914.
292 T. J. Durnick and A. H. Kalantar, *Chem. Phys.*, 1977, **20**, 347.
293 N. G. Kilmer and A. H. Kalantar, *Chem. Phys. Letters*, 1977, **48**, 274.
294 N. G. Kilmer and A. H. Kalantar, *Chem. Phys.*, 1978, **27**, 355.
295 P. J. Vergragt and J. H. van der Waals, *Mol. Phys.*, 1977, **33**, 1507.
296 P. J. Vergragt, J. A. Kooter, and J. H. van der Waals, *Mol. Phys.*, 1977, **33**, 1523.
297 W. Moehle and M. Vala, *J. Phys. Chem.*, 1977, **81**, 1082.

Spin–lattice relaxation of triplet naphthalene in durene is shown to be due to phonon absorption to nearly vibronic states,[298] and spin–lattice relaxation times of metastable triplet states in naphthalene crystals have been measured at high magnetic fields.[299] A detailed interpretation of magnetic field-induced level crossing in triplet–triplet annihilation in naphthalene and anthracene-doped phenanthrene has recently appeared.[300] Further characterizations of the zero-field splittings of triplet [$^2H_{10}$]phenanthrene and [$^2H_8$]naphthalene in biphenyl at 40 K have been reported.[301] Evidence has appeared for a strong exchange-resonance interaction between triplet states of a number of aromatic hydrocarbons and aromatic solvents which results in the broadening of triplet–triplet absorption spectra.[302] Triplet–triplet energy-transfer studies between like molecules using isotopic mixtures have been further improved and applied to naphthalene and phenanthrene molecules. Critical transfer distances of 15 and 13 Å are reported.[303] The method has the advantage of not being complicated by singlet transfer effects.

Lim and his group[304] have extended their studies on triplet excimers by investigating excimer phosphorescence, delayed fluorescence, and triplet–triplet absorption of α,α-dinaphthylalkanes in iso-octane. The intramolecular triplet excimer conformations are significantly different from those of the singlet excimers. In the case of 1,3-diphenylpropane, the binding energy of the triplet excimer is less than for the dinaphthylalkanes. Pilling and Russell[305] find that triplet phenanthrene in propan-1,2-diol is quenched by the ground state by a process which is diffusion controlled below 200 K. The influence of iodide ion on the triplet lifetimes of naphthalene, phenanthrene, and chrysene was found to fit a Dexter-type relation.[306] Deactivation rate constants and their temperature dependence for the quenching of triplet phenanthrene by metal ions have also been determined in methanol–water solution.[307] At temperatures below 140 K, the quenching showed diffusion-controlled behaviour whereas between 200 and 140 K the activation energies were small and attributed to exchange-mediated energy transfer. Above 200 K the activation energy increases owing to ligand displacement for Pr^{3+}, Nd^{3+}, and Ho^{3+}, whereas electron transfer is suggested for Cu^{2+} and Eu^{3+}. Chloride ions increase the rate constants, possibly owing to formation of complexes of the type $[CuCl_n]^{(2-n)-}$.

Trapping of triplet excitons at dislocations produced by bending of anthracene crystals shows a trap depth of 0.3 eV.[308] The room-temperature e.p.r. spectrum of triplet excitons in crystalline pyrene showed no evidence of an excimeric contribution to the triplet exciton.[309] The nature of triplet coronene molecules

[298] P. J. F. Verbeek, C. A. Van't Hof, and J. Schmidt, *Chem. Phys. Letters*, 1977, **51**, 292.
[299] K. F. Penk, H. Sixl, and H. Wolfrum, *Chem. Phys. Letters*, 1977, **52**, 98.
[300] S. H. Tedder, K. W. Otto, and S. E. Webber, *Chem. Phys. Letters*, 1977, **23**, 357.
[301] A. S. Cullick and R. E. Gerkin, *Chem. Phys.*, 1977, **23**, 217.
[302] M. W. Alfimov, V. I. Gerko, L. S. Popov, and V. F. Razumov, *Chem. Phys. Letters*, 1977, **50**, 398.
[303] A. Inoue, H. Chuman, and N. Ebara, *Bull. Chem. Soc. Japan*, 1978, **51**, 345.
[304] S. Okajima, P. C. Subudhi, and E. C. Lim, *J. Chem. Phys.*, 1977, **67**, 4611.
[305] M. J. Pilling and M. A. Russell, *Chem. Phys. Letters*, 1977, **49**, 343.
[306] J. Najbar, J. B. Birks, and T. D. S. Hamilton, *Chem. Phys.*, 1977, **23**, 281.
[307] E. J. Marshall and M. J. Pilling, *J.C.S. Faraday II*, 1977, **74**, 579.
[308] A. Sasaki and S. Hayakawa, *Japan J. Appl. Phys.*, 1978, **17**, 283.
[309] W. Bizzaro, L. Yarmus, J. Rosenthal, and N. F. Berk, *Chem. Phys. Letters*, 1978, **53**, 49.

in different Shpol'skii sites in n-heptane single crystals at 1.6 K has been investigated by polarized microwave double-resonance techniques.[310] An interesting paper clarifies the nature of the rubrene triplet.[311] A transient absorption assigned to the T_1 state of rubrene can be produced by energy transfer from flash-excited anthracene or benzophenone. Delayed fluorescence from rubrene was observed and its intensity was proportional to the square of the triplet absorbance. By the use of various electron donors as quenching agents the rubrene triplet energy was located between 1.04 and 1.29 eV.

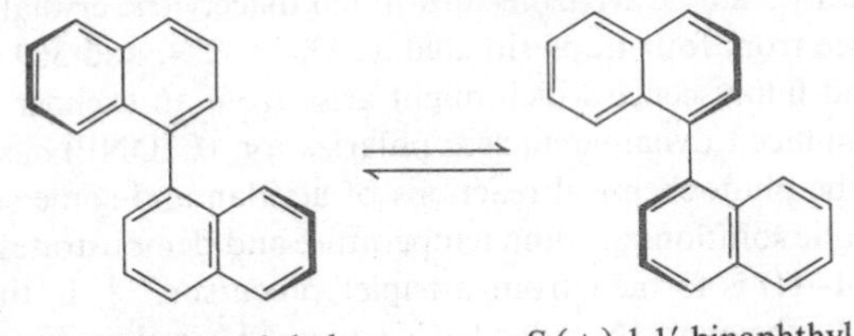

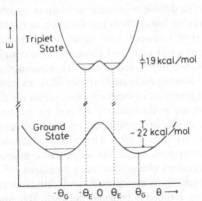

R-(−)-1,1'-binaphthyl S-(+)-1,1'-binaphthyl

Scheme 2

A particularly significant paper on the photoracemization (Scheme 2) of 1,1'-binaphthyl shows that the triplet state is involved.[312] This was observed directly by laser flash photolysis and through the effect of additives. Figure 10 shows the interpretation placed upon the activation energies for the thermal (22 kcal mol^{-1}) and photo- (1.9 kcal mol^{-1}) racemizations.

Figure 10 *Schematic diagram of the ground and triplet states of* 1,1'-*binaphthyl* (Reproduced by permission from *J. Phys. Chem.*, 1977, **81**, 969)

An *ab initio* study of the phosphorescence and intersystem crossing of the T_1 state of pyrazine has been carried out.[313] The computed radiative lifetime (10.2 ns) of the short-lived component of $^3B_{3u}$ ($n\pi^*$) state agrees well with experimental values. Connors and Walsh[314] find that the phosphorescence

310 A. M. Merle, W. M. Pitts, and M. A. El-Sayed, *Chem. Phys. Letters*, 1978, **54**, 211.
311 D. K. K. Liu and L. R. Faulkner, *J. Amer. Chem. Soc.*, 1977, **99**, 4594.
312 M. Irie, K. Yoshida, and K. Hayashi, *J. Phys. Chem.*, 1977, **81**, 969.
313 G. L. Bendazzoli, G. Orlandi, and P. Palmieri, *J. Chem. Phys.*, 1977, **67**, 1948.
314 R. E. Connors and P. S. Walsh, *Chem. Phys. Letters*, 1977, **53**, 436.

spectrum and lifetime of xanthone in polycrystalline n-hexane changes drastically between 4.2 and 77 K. The emitting state is predominantly $^3(\pi\pi^*)$ at 4.2 K and $^3(n\pi^*)$ at 77 K.

An e.p.r. study of the azulene triplet by energy transfer from phenazine by using azulene as dopant in single crystals of the latter has recently been carried out.[315] A detailed vibrational analysis of the phosphorescence $(^3B_2-S_0)$ of quinoxaline in Shpol'skii-type hydrocarbon matrices as well as the $S_1(^1B_1) \rightarrow S_0$ transition[316] reveals the possibility of out-of-plane distortion of the quinoxaline molecule in the T_1 state. Carbazole-substituted diacetylene crystals show strong phosphorescence from four traps situated at 23, 71, 224, and 369 cm^{-1}, as well as a broad-band fluorescence which might arise from an excimer emission.[317]

Chemically induced dynamic nuclear polarization (CIDNP) has been applied to investigate the photochemical reactions of acridan and some substituted acridans in alcoholic solutions at room temperature and demonstrates that the transient (N-radical--·H) is formed from a triplet precursor.[318] In the presence of CCl_4 or $CHCl_3$ the pair [N-radical--·CCl_3(or ·$CHCl_2$)] is formed from the singlet state. Cautionary remarks on difficulties involved in using CIDNP for assigning the electronic multiplicity of the intermediate state in producing radical pairs are illustrated by results on pyrene–NN-diethylaniline in [$^2H^3$]-acetonitrile.[319] Redox potentials and excited-state energy values must also be considered.

Interest in heavy-atom effects on triplet-state properties continues and the studies become more sophisticated. Latas and Nishimura[320] have used phosphorescence microwave double-resonance techniques to observe the effects of inter- and intra-molecular heavy-atom perturbation in naphthalenes and quinolines. Relative radiative rate and total decay rate constants from the triplet state are interpreted in terms of out-of-plane vibronic coupling to the radiative process. All the monohalogen-substituted naphthalenes and phenanthrenes have been studied[321] and the decrease in triplet lifetimes with substitution varies with the square of the atomic spin–orbit coupling factor for the attached halogen. The position dependence is qualitatively related to the unpaired spin density distribution, and the effects of substituents on the 0–0 band shifts in the phosphorescence spectra are discussed. The phosphorescence of 1,2,4,5-tetrachlorobenzene in durene host under the influence of magnetic fields has been studied and cross-relaxation effects have been measured.[322] The rate constants for bimolecular quenching, hopping rate, and diffusion coefficient of triplet excitons in crystalline p-dibromobenzene between 160 and 77 K have been measured.[323] These were determined by observing the phosphorescence decay curve at high- and low-density excitations using an intense nitrogen laser source.

[315] S. S. Kim, *J. Chem. Phys.*, 1978, **68**, 333.
[316] K. Brenner and Z. Ruziewicz, *J. Luminescence*, 1977, **15**, 235.
[317] V. Enkelmann, G. Schleier, G. Wegner, H. Eichele, and M. Schwoerer, *Chem. Phys. Letters*, 1977, **52**, 1314.
[318] W. Schwarz, P. Hesse, and F. Dorr, *Ber. Bunsengesellschaft phys. Chem.*, 1977, **81**, 1231.
[319] G. L. Closs and M. S. Czeropski, *J. Amer. Chem. Soc.*, 1977, **99**, 6127.
[320] K. J. Latas and A. M. Nishimura, *J. Phys. Chem.*, 1978, **82**, 491.
[321] J. C. Miller, J. S. Meek, and S. J. Strickler, *J. Amer. Chem. Soc.*, 1977, **99**, 8175.
[322] W. M. Pitts and M. A. El-Sayed, *Chem. Phys.*, 1977, **25**, 315.
[323] T. Kobayashi and N. Hirota. *Bull. Chem. Soc. Japan*, 1977, **50**, 1743.

The electronic configuration of the lowest triplet state for tetrachlorophthalic anhydride has been shown to be B_2 ($\pi\pi^*$) by means of Zeeman spectroscopy.[324] Shpol'skii-type quasi-line phosphorescence spectra of *p*-chlorobenzaldehyde (PCB) and its deuteriated derivatives have been studied in a methylcyclohexane polycrystalline matrix at 4.2 K.[325] The spectrum is composed of two subspectra separated by 44 cm^{-1} in [^{2}H$_5$]PCB, a low-energy short-lived ($\tau \approx 1$ ms) component and a higher-energy longer-lived component ($\tau \approx 12$ ms). The inadequacy of a two-state or two-site model is brought out in the analysis.

Nanosecond laser photolysis studies of 4-nitro-*NN*-dimethylnaphthylamine (4-NDMNA) in non-polar and polar solvents at room temperature shows a transient species with an absorption maximum in the 500—510 nm region assigned to the lowest triplet state of 4-NDMNA.[326] The hydrogen abstraction reactivity and lack of shift with increasing solvent polarity indicate this state to be $n\pi^*$ under all conditions. The triplet state of 3-nitropyrene has also been characterized in some detail by Scheerer and Henglein.[327] The triplet–triplet absorption spectra of seven highly dipolar aromatic nitro-compounds in EPA glass at 77 K have been observed and compared with theoretical predictions.[328] Two nitrostyrenes have also been studied by nanosecond flash photolysis in fluid media at room temperature and their triplet lifetimes found to increase substantially with increase in solvent polarity and charge-transfer character of the compound. Phosphorescence and e.s.r. resonance spectra of methyl mesitoate and mesitonitrile have been studied to assess steric effects in the former.[329] The triplets generated by laser pulse excitation of *N*-methyl-2-*N*-phenylamino-naphthalene-6-sulphonate and its *C*-protonated form do not interconvert, nor do these from its *N*-hydroxyethyl derivatives.[330] Metal-containing compounds investigated by the Zeeman effect are palladium porphin[331] and platinum phthalocyanine,[332] and optical detection of magnetic resonance has been achieved with the zinc aetioporphyrin triplet[333] in n-octane.

Triplet-state quenching may involve physical quenching or chemical interaction. Several examples of studies on such mechanisms have appeared during the year. Flash photolysis has been used to study the reactions of the thionine triplet and its protonated form with a variety of electron donors.[334] The results are shown in Table 7. The different yields with various donors are due to partial physical quenching of triplets in the electron-transfer reaction. A triplet exciplex as an intermediate is favoured in the electron-transfer reaction. Electron transfer from aromatic compounds to the methylene blue triplet has been studied in

[324] M. Sano, T. Narisawa, M. Matsuka, and Y. J. I'Haya. *Bull. Chem. Soc. Japan*, 1977, **50**, 2266.
[325] O. S. Shalil and L. Goodman, *J. Amer. Chem. Soc.*, 1977, **99**, 5924.
[326] C. Capellos and F. Lang, *Internat. J. Chem. Kinetics.*, 1977, **9**, 943.
[327] R. Scheerer and A. Henglein, *Ber. Bunsengesellschaft Phys. Chem.*, 1977, **81**, 1234.
[328] D. J. Cowley, *Helv. Chim. Acta*, 1978, **61**, 184.
[329] D. R. Arnold, J. R. Bolton, G. E. Palmer, and K. V. Prablu, *Canad. J. Chem.*, 1977, **55**, 2728.
[330] H. Dodiuk, E. M. Kosomer, M. Ottolenghi, and N. Orbach, *Chem. Phys. Letters*, 1977, **49**, 174.
[331] J. A. Kooter, G. W. Canters, and J. H. Van der Waals, *Mol. Phys.*, 1977, **33**, 1545.
[332] K. Kaneto, J. Yamamoto, K. Yoshino, and Y. Inuishi, *J. Chem. Phys.*, 1977, **67**, 1807.
[333] W. R. Leenstra, M. Gouterman, and A. L. Kwiram, *J. Chem. Phys.*, 1978, **68**, 327.
[334] U. Steiner, G. Winter, and H. E. A. Kramer, *J. Phys. Chem.*, 1977, **81**, 1104.

Table 7 *Total quenching constants ($k_Q + k_R$) and radical yields [$\Phi_R = kq/(k_R + k_R)$] of different electron donors in their reactions with basic and acid thionine triplet $^3TH^+$ and $^3TH_2^{2+}$*

Donors	Reaction with $^3TH^+$		Reaction with $^3TH_2^{2+}$	
	$(k_Q + k_R)$[a]	Φ_R	$(k_Q + k_R)$[a]	Φ_R
9,10-Dimethylanthracene	3.5×10^9	0.80 ± 0.08	4.0×10^9	0.80 ± 0.08
Azulene	2.0×10^9	0.75 ± 0.08	4.0×10^9	0.85 ± 0.09
1,3-Diphenylisobenzofuran	4.5×10^9	0.80 ± 0.08	4.0×10^9	0.85 ± 0.09
4-Methoxydimethylaniline	5.0×10^9	0.80 ± 0.1		
Hydroquinone	2.5×10^9[b]	0.85 ± 0.09[b]	2.5×10^9	0.80 ± 0.08
Diazabicyclo[2,2,2]octane	8.0×10^8	0.85 ± 0.09		
4-Bromoaniline	3.0×10^9	0.40 ± 0.05	4.0×10^9	0.45 ± 0.05
Bromobenzene	$<4 \times 10^4$		$<4 \times 10^4$	
Thiourea	7.5×10^6	0.20 ± 0.03	8.0×10^8	0.16 ± 0.02
N-Allylthiourea	8.0×10^6	0.31 ± 0.03	7.0×10^8	0.30 ± 0.03
Selenourea	4.5×10^9	0.04 ± 0.01	3.5×10^9	0.10 ± 0.02

[a] Units: l mol^{-1} s^{-1}.   [b] Presumably hydrogen transfer.

detail.[335] It is shown that if the Gibbs free energy is positive electron transfer does not occur, in contrast with the behaviour of the excited singlet state. Details of the deactivation of triplet 1-anthrol and 2-naphthol by pyridine and quinoline show that in the hydrogen-bonded species hydrogen atom transfer is less effective than in the encounter of the free species.[336] Ruthenocene quenches triplet states of organic molecules having energies greater than 24 000 cm^{-1} at a diffusion-controlled rate. For energies below this the efficiency fall-off is less than expected and is consistent with distortion of the excited state; with benzil there is evidence of charge-transfer interaction.[337] Transfer from triplet triphenylene to ferrocene and ruthenocene shows that the former is predominantly a dipole–dipole mechanism whereas the latter is largely an exchange interaction.[338] Triplet–triplet energy transfer from acetophenone to naphthalene has been studied in rigid EPA glass at 77 K by measuring the decay of donor phosphorescence and rise of acceptor T–T absorption over a wide range of time.[339] A very detailed comparison with the theory of Inokuti and Hirayama has been made.

Geuskens and David[340] have reviewed electronic energy transfer in polymer systems and have also analysed the consequence of triplet–triplet migration in styrene–vinylbenzophenone co-polymers.[341] Polarization measurements have been used to measure the extent of sensitized phosphorescence in poly-(*m*-phenylene-isophthalamide).[342] Phosphorescence depolarization has also been used to measure macromolecular segmental relaxation times on samples of poly(methyl acrylate) incorporating phosphorescent probes. Transient depolarization confirms the absence of motion below the glass transition temperature.[343]

The interception of a triplet biradical by oxygen or di-t-butyl nitroxide (N) enhances photocycloaddition by the process

$$N\uparrow + {}^3SS \xrightarrow{} N\downarrow + {}^1SS \xrightarrow{fast} \text{products}$$

This provides a useful kinetic probe as well as having potential synthetic utility.[344] A theoretical analysis of the stereoselection rules for spin inversion in triplet photochemical reactions has been carried out and three classes of triplet photoreaction complexes are distinguished.[345]

Chan and his co-workers have studied the photophysics of the small α-dicarbonyls [$^1H_2$]- and [$^2H_2$]-glyoxal[346] and biacetyl[347] in the crystalline

335 K. Kikuchi, S. I. Tamura, C. Iwanaga, and H. Kokubun, *Z. phys. Chem. (Frankfurt)*, 1977, **106**, 17.
336 S.-A. Yamamoto, K. Kikuchi, and H. Kokubun. *J. Photochem.*, 1977, **7**, 177.
337 A. P. Chapple, J.-P. Vikesland, and F. Wilkinson, *Chem. Phys. Letters*, 1977, **50**, 81.
338 J. P. Vikesland and F. Wilkinson, *J.C.S. Faraday II*, 1977, **73**, 1.
339 K. Yamamoto, T. Takemura, and H. Baba, *J. Luminescence*, 1977, **15**, 445.
340 G. Geuskens and C. David, *Pure Appl. Chem.*, 1977, **49**, 479.
341 C. David, D. Baeyens-Volant, P. Macedo de Abreu, and G. Geuskens, *European Polymer J.*, 1977, **13**, 841.
342 T. Werner and H. E. A. Kramer, *European Polymer J.*, 1977, **13**, 501.
343 H. Rutherford and I. Soutar, *J. Polymer Sci.*, 1977, **15**, 2213.
344 R. A. Caldwell and D. Creed, *J. Amer. Chem. Soc.*, 1977, **99**, 8360.
345 S. Shaik and N. D. Epiotis, *J. Amer. Chem. Soc.*, 1978, **100**, 18.
346 I. Y. Chan and K. R. Walton, *Mol. Phys.*, 1977, **34**, 65.
347 I. Y. Chan and S. Hsi, *Mol. Phys.*, 1977, **34**, 85.

state. Many aspects of these papers are of considerable interest in view of the remarkable behaviour of biacetyl. The existence of several low-lying $n\pi^*$ states facilitates the radiative process from the triplet state. Perturbality of the states gives rise to a variety of intersystem crossing patterns for populating the triplet and non-radiative relaxation. Zero-field splittings and phosphorescence decay rate constants from the triplet states of cycloalkanones in several hosts have been observed.[348] The photolysis of acetone in various solvents shows that the rate of photolysis is decreased by addition of methanol or increasing concentration of acetone.[349] This is taken as compelling evidence for the role of intermolecular hydrogen abstraction in quenching of triplet acetone (Scheme 3). This mechanism is supported by spectroscopic evidence for enol formation.

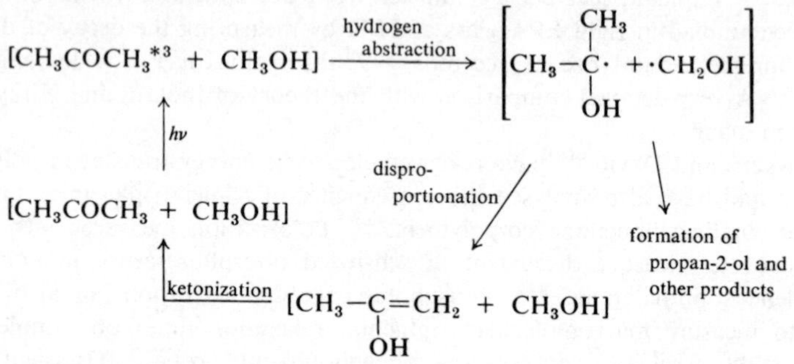

$$[CH_3COCH_3^{*3} + CH_3OH] \xrightarrow[\text{abstraction}]{\text{hydrogen}} \left[\begin{array}{c} CH_3 \\ | \\ CH_3-C\cdot \ +\cdot CH_2OH \\ | \\ OH \end{array} \right]$$

hv ↑

$[CH_3COCH_3 + CH_3OH]$

dispro-portionation

formation of propan-2-ol and other products

ketonization

$$[CH_3-C{=}CH_2 + CH_3OH]$$
$$| \atop OH$$

Scheme 3

Benzophenone still stimulates interest. Strong radiofrequency transitions between pairs of anti-crossing magnetic sublevels of $^3(n\pi^*)$ BP and three isotopically labelled derivatives have been observed using optical detection of magnetic resonance methods.[350] Magnetically induced changes in the phosphorescence intensities of triplet-state benzophenone, [*carbonyl*-^{13}C]benzophenone, and three 4,4'-dihalogenobenzophenones in single crystals of 4,4'-dibromodiphenyl ether at 1.6 K have been examined in detail by Mucha and Pratt.[351] Triplet–triplet annihilation of benzophenone in benzene is essentially diffusion controlled whereas hopping encounter is shown by comparison of experiment and computer simulation to be secondary even at 1 mol l^{-1}.[352] Measurements of photocurrents after laser flashing solutions of benzophenone and its derivatives with unsaturated compounds and compounds alone in solution showed transient currents, indicating the existence of exciplexes formed by triplet excited benzophenone compounds and ground-state solvent molecules.[353]

Phosphorescence spectra of benzophenone at 77 K have been measured in aqueous sodium acetate solutions at pH 6. The spectral changes are attributed to a straightforward acid–base equilibrium with $pK(T_1) = 1.5$ and not to the

[348] R. K. Power and A. M. Nishimura, *J. Photochem.*, 1978, **8**, 263.
[349] M. Anpo and Y. Kubokawa, *Bull. Chem. Soc. Japan*, 1977, **50**, 1913.
[350] G. Kothandaraman, P. F. Brode, and D. W. Pratt, *Chem. Phys. Letters*, 1977, **51**, 137.
[351] J. A. Mucha and D. W. Pratt, *J. Chem. Phys.*, 1977, **66**, 5356.
[352] T.-S. Fang, R. Fukuda, R. E. Brown, and L. A. Singer, *J. Phys. Chem.*, **82**, 1978, 246.
[353] R. Kuhlmann and W. Schnakel, *J. Photochem.*, 1977, **7**, 287.

benzophenone–hydroxonium complex suggested by aqueous photochemistry.[354] The quenching of triplet benzophenone in ethanol at low temperature is governed by a hydrogen atom abstraction reaction which results in a transient radical pair in the solvent cage.[355] The various activation energies are also determined. A linear relation between the log of the quenching rate constant for triplet acetone and the IP of the electron-rich quenching olefin is consistent with the formation of triplet exciplexes between the triplet ketone and olefin.[356] A CDNO treatment has been made of carbonyl–olefin charge-transfer complexes, *p*-benzoquinone and formaldehyde being selected as illustrative examples.[357] Information on the quenching of aromatic carbonyl triplets by norbornadienes and quadricyclenes has been obtained by pulse radiolysis.[358] The relative extents of energy transfer and decay to the ground state depend on the triplet energy and electron accepting properties of the triplet state involved. Self-quenching of triplets in a series of *para,para'*-disubstituted benzophenones has been examined in detail and correlated with the Hammett functions and the reversibility of exciplex formation.[359] Amongst other systems which have been investigated are the triplet state traps of 1-indanone and acetophenone by zero-field optically detected magnetic resonance and phosphorescence at low temperature.[360] The reversibility of exciplex formation in the benzil–anisole system has been experimentally demonstrated and the kinetic parameters have been evaluated.[361] The involvement of triplet exciplexes in the mechanism of enone photocycloadditions to olefins provides an illustration of the significance of these intermediates.[362] Preliminary work has been done on the excited-state absorption kinetics of anthrone at 533 nm.[363] A rapid build-up (<20 ps) of excited state absorption at 533 nm follows excitation at 355 nm in benzene at room temperature. There is probably rapid intersystem crossing followed by slow internal conversion and/or slow vibrational relaxation within the triplet manifold. The phosphorescence spectra, polarization, and lifetimes of a series of linear polynuclear *p*-quinones have been measured and interpreted.[364] Evidence for a solvent-dependent barrier to torsional motion in cyclic ketones which can lead to a 'freezing out' of non-radiative decay at low temperatures has been obtained from phosphorescence spectra of two rigid cyclic $\alpha\beta$-enones.[365] Intrinsic barriers are of much less importance in determining the pattern of photophysical behaviour.

The present rapid growth and progress in photobiology have led to the further recognition of the importance of the triplet state as an intermediate. Only a few topics of direct photophysical interest will be mentioned here.

[354] D. M. Rayner, P. K. Tolg, and A. G. Szabo, *J. Phys. Chem.*, 1978, **82**, 86.
[355] H. Murai, M. Jinguji, and K. Obi, *J. Phys. Chem.*, 1978, **82**, 38.
[356] R. O. Loutfy, R. W. Yip, and S. K. Dogra, *Tetrahedron Letters*, 1977, 2843.
[357] R. M. Wilson, R. Outcalt, and H. H. Jaffé, *J. Amer. Chem. Soc.*, 1978, **100**, 301.
[358] A. J. G. Barwise, A. A. Gorman, R. L. Leyland, P. G. Smith, and M. A. J. Rodgers, *J. Amer. Chem. Soc.*, 1978, **100**, 1814.
[359] M. W. Wolf, R. E. Brown, and L. A. Singer, *J. Amer. Chem. Soc.*, 1977, **99**, 526.
[360] S. Niijuma and N. Hirota, *J. Phys. Chem.*, 1978, **82**, 453.
[361] R. E. Brown, T.-S. Fang, C. L. Kwan, and L. A. Singer, *Chem. Phys. Letters*, 1977, **51**, 526.
[362] R. L. Loutfy and P. de Mayo, *J. Amer. Chem. Soc.*, 1977, **99**, 3559.
[363] G. W. Scott and L. D. Talley, *Chem. Phys. Letters*, 1977, **53**, 431.
[364] M. Nepras and A. Novak, *Coll. Czech. Chem. Comm.*, 1977, **42**, 2343.
[365] S. W. Beavan and D. Phillips, *Mol. Photochem.*, 1977, **8**, 311.

Theoretical calculations have been made on indole, and observed $T-T$ transitions are considered to involve excitation to very high lying triplet states.[366] Phosphorescence of ribonuclease T in neutral aqueous solution at 293 K is assigned to emission from a single tryptophan residue in a hydrophobic environment.[367] The formation of triplet acetone by the horseradish peroxidase–O_2 oxidation of isobutyraldehyde is shown by the sensitized emission from 9,10-dibromoanthracene-2-sulphonate.[368]

The advances in molecular biology inspire interest in the electronic states of pyrimidine. The phosphorescence spectrum of pyrimidine in cyclohexane has been examined in order to ascertain the vibrational structure of the ground state.[369] Many of the photophysical properties of riboflavin and several other N-10-substituted isoalloxazines have been measured at 77 and 298 K[370] and the data correlated with photochemical processes known to involve the triplet state in aqueous solution. The photosensitizing properties of the furocoumarins have stimulated investigation of their photophysical properties. The triplet–triplet absorption spectra of psoralen, xanthotoxin, angelicin, and bergapten have been determined and used to determine the $S_1 \rightarrow T_1$ crossing yields.[371] The yields vary considerably with solvent and the triplet reactivities measured. The high reactivities favour the suggestion that furocoumarin triplets are involved in the damage induced in DNA. The efficiency of psoralen and angelicin triplets in reacting with amino-acids suggests that such triplets *in vivo* may occur with protein components of the cell as well as nucleic acids (see Table 8).

Zeeman fine structure of the triplet states of the pheophytins has been reported.[372] The role of the triplet states of the chlorophylls in photosyntheses is speculative. The phosphorescence properties have proved difficult to establish because of the poor sensitivity of detectors in the infrared. A study of chlorophyll samples with a system of good response has been published by Mau and Puga (Figure 11).[373] Triplet-state e.s.r. of chlorophylls *a* and *b* in poly(methyl methacrylate) and methyltetrahydrofuran has been reported in some detail.[374,375]

The triplet and singlet excited states of protoporphyrin IX dimethyl ester (PPIXDME) have been identified by pulse radiolysis in benzene solution.[376] The absorption spectrum, extinction coefficients, energy level, and rate of O_2 quenching of the triplet and energy transfer properties have been measured. The triplet level ($\sim$150 kJ mol^{-1}) for PPIXDME means that if triplet quenching is the important step in carotenemia therapy of erythropoietic porphyria then any polyene with a triplet level below this value is likely to be effective. If

[366] E. M. Evleth, O. Chalvet, and P. Bamière, *J. Phys. Chem.*, 1977, **81**, 1913.
[367] K. Imakubo and Y. Kai, *J. Phys. Soc. Japan*, 1977, **42**, 1431.
[368] N. Duran, O. M. M. Faria Oliveira, M. Haun, and G. Cilento, *J.C.S. Chem. Comm.*, 1977, 442.
[369] A. Kawski, Z. Kofro, I. Gryczynski, and P. Baluk, *Bull. Acad. polon. Sci.*, *Sér. Sci. Chim.*, 1977, **25**, 1183.
[370] W. M. Moore, J. C. McDaniels, and J. A. Hen, *Photochem. and Photobiol.*, 1977, **25**, 505.
[371] R. V. Bensasson, E. J. Land, and C. Salet, *Photochem. and Photobiol.*, 1978, **27**, 73.
[372] J. F. Kleibeuker, R. J. Platenkamp, and J. J. Schaafsma, *Chem. Phys.*, 1978, **27**, 51.
[373] A. W. H. Mau and M. Puza, *Photochem. and Photobiol.*, 1977, **25**, 601.
[374] W. Hagele, D. Schmid, and H. C. Wolf, *Z. Naturforsch.*, 1978, **33a**, 83.
[375] W. Hagele, S. Schrud, and H. C. Wolf, *Z. Naturforsch.*, 1978, **33a**, 94.
[376] S. J. Chantrell, C. A. McAuliffe, R. W. Munn, A. C. Pratt, and E. J. Land, *J.C.S. Faraday I*, 1977, **73**, 858.

Table 8   *Rate constants for psoralen and angelicin triplet quenching by nucleic acid bases and amino-acids*

Quencher	Rate constant/l mol^{-1} s^{-1}	
	Psoralen	Angelicin
Thymine	7.5×10^8	1.1×10^9
Thymidine	1.9×10^8	3.5×10^8
Thymidine-5'-monophosphate	1.3×10^8	2.4×10^8
Uracil	1.1×10^8	2.1×10^8
Uridine	1.4×10^8	2.0×10^8
Uridine-5'-monophosphate	1.0×10^8	2.0×10^8
Cytosine	1.7×10^7	2.6×10^7
Cytidine	3.4×10^7	5.5×10^7
Cytidine-5'-monophosphate	1.8×10^7	4.8×10^7
Adenine	1.0×10^8	1.6×10^8
Adenosine	4.1×10^7	4.7×10^7
Guanine[a]	$<10^9$	$<10^9$
Guanosine	7.3×10^8	7.2×10^8
Tryptophan	3.1×10^9	2.1×10^9
Tyrosine	2.1×10^8	1.6×10^8
Histidine	3.3×10^7	3.7×10^7
Phenylalanine	$\sim 3 \times 10^6$	$\sim 5 \times 10^6$
Arginine	$\sim 2 \times 10^6$	$<10^5$
Methionine	$<10^5$	$<10^5$
Cysteine	$<10^5$	$<10^5$
Glycine	$<10^5$	$<10^5$

[a] Only a limit could be obtained, owing to the low solubility of guanine in water (~ 20 mol l^{-1}).

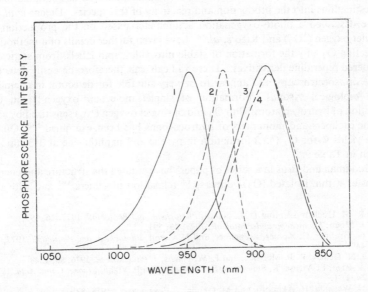

Figure 11   *Phosphorescence spectra of 1, chlorophyll* a; *2, pheophytin* a; *3, chlorophyll* b; *and 4, pheophytin* b *in EPA glass at 77 K*
(Reproduced by permission from *Photochem. and Photobiol.*, 1977, **25**, 601)

$O_2(^1\Delta_g)$ quenching is the important step then only polyenes with $E_T < 94$ kJ mol^{-1} will be effective.

The relative triplet-formation quantum yields of all-*trans*- and 11-*cis*-retinal at 77 K are found to be independent of wavelength, showing that previous interpretations of photophysical decay processes in all-*trans*-retinal are unacceptable.[377]

Photoionization of aromatic molecules in low-temperature glasses has been the subject of numerous investigations. From the comparative longevity of the triplets, most workers consider that ionization should be biphotonic involving the triplet state. Moan[378] has found that the radical yield does not correlate with the concentration of triplet states and this is accounted for in the way that the recombination probability of ejected electrons increases with radiation time for tryptophan and tyrosine glasses. The formation of hydrated electrons from excited phenol has been measured and apparently requires three alternative pathways to account for the observation.[379] The processes are (i) a monophotonic process involving the fluorescent state, (ii) a monophotonic process from a higher state competing with internal conversion to S_1, and (iii) a consecutive biphotonic pathway in which the lowest triplet state absorbs the second photon (it is concluded that this occurs only at intensities too high to be of biological significance).

4 Physical Aspects of Some Photochemical Studies

Photo-oxidation.—The possible role of singlet oxygen ($^1\Delta_g$) as an intermediate in chemical reactions, biological processes, and air pollution has initiated many investigations into the production and reactivity of this species. Demas *et al.*[380] have described a photo-oxygenation actinometer based on the production of singlet oxygen ($^1O_2$) and Rigo *et al.*[381] have given further details of a method of detecting $O_2{}^-$ by the formation of stable nitroxides, spin labels, from sterically hindered piperidine derivatives. These radicals can therefore be easily detected by e.s.p. spectroscopy and appear to be very suitable for detection in chemical and biological systems. Formation of singlet molecular oxygen[382] in the reaction of hydrogen atoms with O_2 and of singlet oxygen ($^1\Delta_g$) sensitized by aromatic excimers and monomers of hydrocarbons has been examined.[383] Quantum yields Φ for O_2 ($^1\Delta_g$) formation in pyrene and naphthalene at 313 nm are given in Table 9.

Goodman and Brus in a series of papers have studied the structure and energy transfer within isolated $(O_2)_2$ dimers,[384] relaxation of dimers,[385] and with the

[377] R. M. Hochstrasser and D. L. Narva, *Photochem. and Photobiol.*, 1977, **26**, 595.
[378] J. Moan, *Photochem. and Photobiol.*, 1977, **25**, 591.
[379] G. Grabner, G. Kohler, J. Zechner, and N. Getoff, *Photochem. and Photobiol.*, 1977, **26**, 449.
[380] J. N. Demas, R. P. McBride, and E. W. Harris, *J. Phys. Chem.*, 1976, **80**, 2248.
[381] A. Rigo, E. Argese, R. Stevanato, E. F. Orsega, and P. Viglino, *Inorg. Chim. Acta*, 1977, **24**, L71.
[382] N. Washida, H. Akimoto, and M. Okuda, *J. Phys. Chem.*, 1978, **82**, 18.
[383] D. M. Shold, *J. Photochem.*, 1978, **8**, 39.
[384] J. Goodman and L. E. Brus, *J. Chem. Phys.*, 1977, **67**, 4398.
[385] J. Goodman and L. E. Brus, *J. Chem. Phys.*, 1977, **67**, 4408.

Table 9 Quantum yields Φ for O_2 ($^1\Delta_g$) formation of 313 nm

Naphthalene concentration (mol l^{-1})	0.01	0.10	0.20	0.30	0.40	0.50
(Methanol)[a] Φ	0.41 ± 0.01	0.45 ± 0.02	0.50 ± 0.01	0.50 ± 0.03	0.55 ± 0.02	0.70 ± 0.05
(Hexanes) Φ	0.77 ± 0.07					
Pyrene concentration (mol l^{-1})	0.001	0.004	0.008	0.013	0.018	0.020
(Methanol) Φ	0.74 ± 0.01	0.75 ± 0.01	0.67 ± 0.02	0.65 ± 0.04	0.65 ± 0.04	
(Hexanes) Φ	0.83 ± 0.05					0.90 ± 0.08
Phenanthrene concentration (mol l^{-1})	0.01	0.13				
(Methanol) Φ	0.50 ± 0.02	0.51 ± 0.03				
1-Bromonaphthalene concentration (mol l^{-1})	0.01					
(Methanol) Φ	0.86 ± 0.03					

[a] The solvent is given in parentheses.

spectroscopy and dynamics of the low-lying $A^3\Sigma_u^+$, $C^3\Delta_u$, and $C^1\Sigma_u^-$ states of O_2 in solid hosts.[386]

Interest has been shown in the rapid photochemical generation of singlet oxygen in natural water,[387] and data are given to show that $O_2\cdot^-$ (or $HO_2\cdot$) is formed during the near-u.v. photo-oxidation of tryptophan and has at least two fates, one of which is H_2O_2 formation.[388] The role of singlet oxygen in the oxidation of arylazonaphthols,[389] hexamethylbenzene,[390] and N-allylthiourea[391] has also been investigated. A comparative study has been made on triplet and singlet molecular oxygen in the oxidation process of fully substituted alkoxyoxazoles[392] and also with dihydroflavins.[393]

The thermal regeneration of O_2 ($^1\Delta_g$) and the parent hydrocarbon from 9,10-diphenylanthracene (DPA) and rubrene peroxides has been investigated in several solvents.[394] The photo-oxidation of biacetyl produces biacetyl peroxide and carbon dioxide, the formation of which has been directly shown to involve chemical deactivation of T_1 biacetyl by O_2.[395]

Rate constants for the quenching of O_2 ($^1\Delta_g$), produced during the photolysis of ozone in the Earth's upper atmosphere, by atmospheric gases show that for ground-state molecular oxygen the rate constant of $1.7 \pm 0.1 \times 10^{-18}$ cm^3 mol^{-1} s^{-1} is several orders of magnitude greater than that for molecular nitrogen.[396] This is in good agreement with the rate constant value of $(1.57 \pm 0.05) \times 10^{-18}$ obtained by Borrell et al.[397]

Measurements of the efficiency of rubrene-sensitized photo-oxidation of 1,3-diphenylisobenzofuran show the formation of singlet oxygen via oxygen quenching to be inefficient.[398] This inefficiency is rationalized in terms of the spin-allowed decay of an initially formed $^3(T_1 + {}^1\Delta)$ complex state to a lower-energy $^3(T_1 + {}^3\Sigma)$ state prior to complex dissociation.

Reaction rates of singlet oxygen with olefins[399,400] and with aliphatic amines[401] have also been given. In agreement with earlier studies, a correlation of quenching rate with amine ionization potential was found for amines having unbranched alkyl groups.

Papers have appeared on dye-sensitized photo-oxidation reactions of chlor-

[386] J. Goodman and L. E. Brus, J. Chem. Phys., 1977, 67, 1482.
[387] R. G. Zepp, N. L. Wolfe, G. L. Baughman, and R. C. Hollis, Nature, 1977, 267, 421.
[388] J. P. McCormick and T. Thomason, J. Amer. Chem. Soc., 1978, 100, 312.
[389] J. Griffiths and C. Hawkins, J.C.S. Perkin II, 1977, 747.
[390] C. J. M. van den Heuvel, H. Steinberg, and Th. J. de Boer, Rec. Trav. chim., 1977, 96, 157.
[391] I. Kralfic and H. E. A. Kramer, Photochem. and Photobiol., 1978, 27, 9.
[392] M. L. Graziano, A. T. Corotenuto, M. R. Iesce, and R. Scarpati, J. Heterocyclic Chem., 1977, 14, 1215.
[393] C. Kemal, T. W. Chan, and T. C. Bruice, J. Amer. Chem. Soc., 1977, 99, 7272.
[394] B. Stevens and R. D. Small, J. Phys. Chem., 1977, 81, 1605.
[395] M. Bouchy and J. C. Andre, Mol. Photochem., 1977, 8, 345.
[396] A. Leiss, U. Schurath, K. H. Becker, and E. H. Fink, J. Photochem., 1978, 8, 211.
[397] P. Borrell, P. M. Borrell, and M. D. Pedley, Chem. Phys. Letters, 1977, 51, 300.
[398] P. B. Merkel and W. G. Herkstroeter, Chem. Phys. Letters, 1978, 53, 350.
[399] B. M. Monroe, J. Phys. Chem., 1978, 82, 15.
[400] L. B. Harding and W. A. Goddard, Tetrahedron Letters, 1978, 747.
[401] B. M. Monroe, J. Phys. Chem., 1977, 81, 1861.

promazine,[402] *NN*-dimethylbenzylamine,[403] spirocyclic vinylcyclopropanes,[404] tyramine,[405] biadamantylidene,[406] and in aqueous micellar solutions.[407]

Photolysis.—Only those papers which have a photophysical interest are referred to in this section. Laser flash photolysis is now widely applied to the spectroscopic investigation of molecules. Faure[408] uses laser flash photolysis to investigate the primary photodegradation processes on polymeric materials such as poly(vinyl phenyl ketone) and has observed triplets and biradicals by studying their decay and absorption spectra.

A laser photolysis study of the reactions of excited chloranil with acrylonitrile, methyl methacrylate, and styrene has revealed an interaction between the excited triplet state of chloranil and vinyl monomers and/or solvents. The mechanism of triplet quenching and transient formation has been accounted for by taking the relaxed triplet-state complex with charge-transfer character as a precursor leading to the production of semiquinone radicals, chloranil anions, and other intermediates.[409]

Laser flash excitation of rhodopsin (involved in the primary process of vision) produces initial transient absorption peaks at ~570 and ~420 nm and a subsequent transient species with a maximum absorption at 480 nm.[410] These results have been interpreted in terms of initial formation of bathorhodopsin (570 nm) followed by conversion into lumirhodopsin (470 nm). The peak at 420 nm in the first transient has been attributed to either hypsorhodopsin or isorhodopsin. Schichida *et al.*[411] in a similar experiment have established hypsorhodopsin as a physiological intermediate in the photobleaching process of rhodopsin and a precursor of bathorhodopsin.

Singlet excited acetone has been found as one of the observed products in the photolysis of tetramethyldioxetan at 264 nm. The rise time of acetone fluorescence was found to be less than 10 ps.[412]

Further examples of laser flash photolysis are found in investigations of intramolecular charge transfer and heteroexcimer formation processes in the excited state of $Me_2NC_6H_4(CH_2)_7$-(1-pyrenyl),[413] in electron transfer from anilines to excited pyrene near a micelle water interface,[414] and in the primary processes of the photodegradation of benzoins.[415]

Triplet excited states of carotenoids, believed to be involved in the protection

402 I. Rosenthal, T. Bercouici, and A. Frimer, *Jl Heterocyclic Chem.*, 1977, **14**, 355.
403 K. Inoue, I. Saito, and T. Matsuura, *Chem. Letters*, 1977, 607.
404 T. Hatsui and H. Takeshita, *Chem. Letters*, 1977, 603.
405 G. R. Seely, *Photochem. and Photobiol.*, 1977, **26**, 115.
406 C. W. Jefford and A. F. Boschung, *Helv. Chim. Acta*, 1977, **60**, 2673.
407 Y. Usui, M. Tsukada, and H. Nakamura, *Bull. Chem. Soc. Japan*, 1978, **51**, 379.
408 J. Faure, *Pure Appl. Chem.*, 1977, **49**, 487.
409 M. Kobashi, H. Gyoda, and T. Morita, *Bull. Chem. Soc. Japan*, 1977, **50**, 1731.
410 R. Bensasson, E. J. Land, and T. G. Truscott, *Photochem. and Photobiol.*, 1977, **26**, 601.
411 Y. Schichida, T. Yoshizawa, T. Kobayashi, H. Ohtani, and S. Nagakura, *F.E.B.S. Letters*, 1977, **80**, 214.
412 K. K. Smith, J. Koo, G. B. Schuster, and K. J. Kaufmann, *Chem. Phys. Letters*, 1977, **48**, 267.
413 M. Migita, M. Kawai, N. Mataga, Y. Sakata, and S. Misumi, *Chem. Phys. Letters*, 1978, **53**, 67.
414 Y. Waka, K. Hamanoto, and N. Mataga, *Chem. Phys. Letters*, 1978, **53**, 242.
415 J. Fouassier, D. Lougnot, and J. Faure, *Compt. rend.*, 1977, **284**, C, 643.

of photosynthetic membranes against the damaging effects of singlet oxygen, have been investigated.[416] The results indicate that singlet–triplet intersystem crossing quantum yields are very low (0.001) and suggest that cartenoid triplets are formed in a triplet–triplet energy transfer or via $^1O_2$ quenching.

Flash photolysis of aqueous tryptophan produces the Trp^+ radical cation (λ_{max} 580 nm) with quantum yield 0.12 ± 0.01 and e_{aq}^- with quantum yield 0.10 ± 0.01 at pH 4—8.[417] The neutral Trp radical (λ_{max} 510 nm) is formed by protonation with the rate constant $1.5 \times 10^6 \, s^{-1}$, while the disappearance of e_{aq}^- follows a complex decay inconsistent with homogeneous kinetics. The results are discussed in terms of a recombination model in which the electron diffuses through the medium as a quasi-free entity. Energy-transfer processes are shown to be involved in the photoionization of two dipeptides, tryptophanyltyrosine (TrpTyr) and tryptophanyltryptophan (TrpTrp). The study also confirmed that indole ring singlet states are involved in the photoejection process.[418]

Of further interest are the photolysis studies on polyethylene films,[419] penta-1,3-dienes (where excimer formation has been suggested[420]), and on Hg-photosensitized production of trapped radicals in 3-methylpentane glass.[421] The reversible photochemical reaction between thionine and iron(II) still arouses considerable interest because of its use as a possible photogalvanic system.[422] The photoredox reactions of thionine and iron(II) using flash photolysis have been extended to include reactions with cobalt(II) and manganese(II) and it is shown that the semithionine radical anion is not formed in the case of cobalt(II) or manganese(II), but that the reaction leads to enhanced deactivation of triplet thionine. Excitation of the metal ion directly leads to full electron transfer with concomitant formation of radical ions.

Photoisomerization.—Photoisomerization and emission characteristics of linked anthracenes with increasing molecular constriction have been presented (Scheme 4).[423] Fluorescence yields and lifetimes are significantly a function of the bridging group for the linked anthracenes whereas the quantum yields for photoisomerization (Φ_{12}) appear to be more a function of the 9'-substituent group Y (see Table 10). Evidence is presented to show weakly stabilized excimer emission from (12a), whereas similar experiments on (12c—e) in which sandwich dimers are produced reveal only normal anthracene fluorescences.

Kutal et al.[424] describe a potential solar energy storage system, based upon the valence isomerization of norbornadiene to quadricyclane, which possesses several attractive features including high specific energy storage capacity, kinetic stability of the energy-rich photoproduct in the absence of suitable catalysts, and relatively inexpensive reactants. Use of transition-metal compounds provides a

[416] R. Bensasson, E. A. Dawe, D. A. Long, and E. J. Land, *J.C.S. Faraday I*, 1977, **73**, 1319.
[417] J. F. Baugher and L. I. Grossweiner, *J. Phys. Chem.*, 1977, **81**, 1349.
[418] M. P. Pileni, M. Bazin, and R. Santus, *Chem. Phys. Letters*, 1977, **51**, 61.
[419] G. R. A. Johnson and A. Wilson, *Radiat. Phys. Chem.*, 1977, **10**, 89.
[420] P. Vanderlinden and S. Boue, *Bull. Soc. chim. belges*, 1977, **86**, 785.
[421] B. J. Brown and J. E. Willard, *J. Phys. Chem.*, 1977, **81**, 977.
[422] M. I. C. Ferreira and A. Harriman, *J.C.S. Faraday I*, 1977, **73**, 1085.
[423] W. R. Bergmark and G. Jones, *Nouveau J. Chim.*, 1977, **1**, 271.
[424] C. Kutal, D. P. Schwendiman, and P. Grutsch, *Solar Energy*, 1977, **19**, 651.

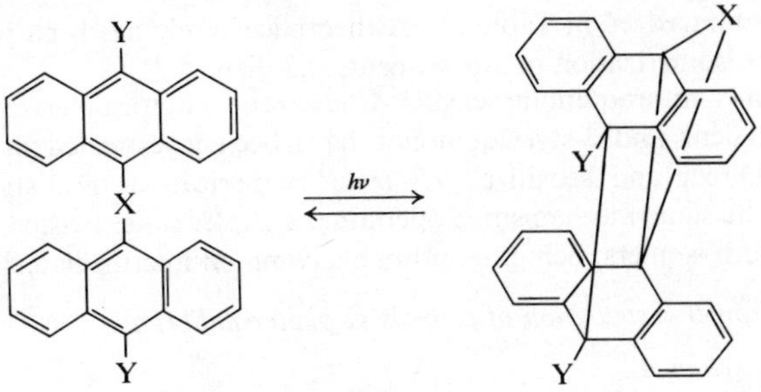

(12)   (13)

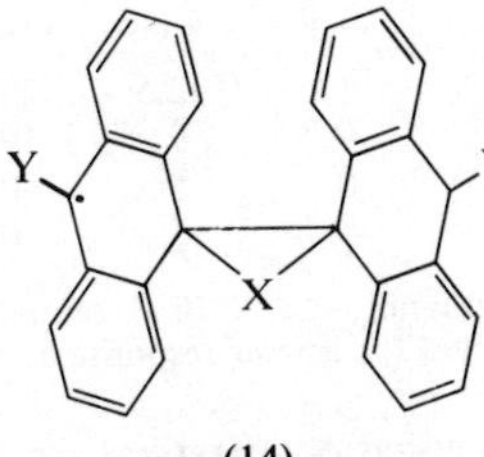

(14)

Scheme 4

Table 10 *Photochemical and photophysical data for linked anthracenes* (8)

	X	Y	$\Phi_{12}{}^a$	$\Phi_{21}{}^a$	$\Phi_f t$	τ_f/ns^b
(a)	CH_2CH_2	H, H	0.26	0.55	0.16	1.7
(b)	CH_2CH_2	CH_3, CH_3	0.04		0.14	2.0
(c)	CH_2	H, H	0.15	0.76	0.06	1.1
(d)	CHOH	H, H	0.29	0.81	0.02	0.3
(e)	CHOH	H, OCH_3	0.05	—	0.02	0.3
(f)	CH_2CH_2	CH_2CH_2	0.36	0.60	<0.001	$(\sim0.01)^c$

^{a} Isomerization quantum yield, benzene, 25 °C: (12) → (13), 366 nm, (13) → (12), 285 nm. ^{b} Fluorescence yield and lifetime in benzene, 25 °C (single photon counting measurements except where noted). ^{c} From dimethylaniline quenching of photoisomerization [k_q assumed $\sim 10^{10}$ l mol^{-1} s^{-1} as shown for (12a, c, d)].

potentially useful route to sensitize the desired energy storage step. Further aspects of light-energy storage systems utilizing valence isomerization between dienes and cage compounds have been reported.[425]

Various models for the primary photoevent in vision have been critically discussed. Rosenfeld *et al.*[426] conclude that the classic picture of a single *cis–trans* isomerization step is the only one which satisfactorily accounts for the available experimental data. Experimental results show that the yields of the 11-*cis* → all-*trans* isomerization of the free chromophore are small, exhibiting a marked dependence on the excitation wavelength. Quantum yields and products of the *trans–cis* photoisomerization of all-*trans*-retinal in hexane and ethanol have also been determined as a function of excitation wavelength.[427] These

[425] T. Mukai and Y. Yamashita, *Tetrahedron Letters*, 1978, 357.
[426] T. Rosenfeld, B. Honig, and M. Ottolenghi, *Pure Appl. Chem.*, 1977, **49**, 341.
[427] W. H. Waddell and D. L. Hopkins, *J. Amer. Chem. Soc.*, 1977, **99**, 6457.

results are summarized in Table 11. A theoretical study has been presented for the *cis–trans* isomerization of *s-trans*-penta-1,3-diene.[428]

Direct and anthraquinone-sensitized *cis–trans* photoisomerizations of 3-styrylnaphthalene and 3-styrylquinoline have been investigated in benzene at 25 °C.[429] Direct and sensitized *cis–trans* isomerization of 2-styrylthiophen shows that the same mechanism is operating, a triplet state, twisted by 90° with respect to both isomers, being the probable common intermediate.[430]

Table 11 *Isomer distribution of retinals at photoequilibrium*

Solvent[a]	Excitation wavelength/ nm[b]	all-trans	7-cis	9-cis	11-cis	13-cis	9,13-di-cis
				Percentage retinal isomer[c]			
Ethanol	430	45	1	12.5	20	21.5	1
	390	52.5	0.5	7	19.5	20.5	1
Hexane	430	62	0	5	0	31	2
	390	70	0	5	0	24	1

[a] 5.0×10^{-5} mol l^{-1} retinal in ethanol; 4.6×10^{-5} retinal in hexane. [b] 10 nm band pass, 150 W Xe lamp, 2 h. [c] Corrected for isomer response at 365 nm detection.

Biacetyl-photosensitized isomerization of styrylpyridines produced evidence for two competitive pathways.[431] Thirteen selected sensitizers have been used in the triplet sensitized *cis–trans* isomerization of 4-nitro-4'-methoxystilbene (NMS) in polar and non-polar solvents.[432,433]

Reports on the photochemical valence bond isomerization of some benzo-C_9H_{20} hydrocarbons[434] and on the mechanism of photoisomerization of thioindigo dyes in non-polar solvents[435] have recently appeared.

Photochromism.—A number of *N*-salicyldieneanilines (15) have been shown to be photochromic both in the crystal and in rigid and fluid solutions. Previous

(15)

studies have indicated that photochromic change occurs through hydrogen transfer followed by a geometrical rearrangement of the molecule around the C-1—C-7 bond. Nakagaki *et al.*[436] have further investigated the primary process of photochromism of *N*-salicylideneanilines and related compounds by using

[428] I. Baraldi, M. C. Bruni, F. Momicchiolo, J. Langlet, and J. P. Malrieu, *Chem. Phys. Letters*, 1977, **5**, 493.

[429] G. Gennari, G. Cauzzo, and G. Caliazzo, *J. Phys. Chem.*, 1977, **81**, 1551.

[430] L. L. Costanzo, S. Pistara, G. Condorelli, and G. Scarlata, *J. Photochem.*, 1977, **7**, 297.

[431] G. Bartocci, G. Favaro, and P. Bortolus, *Z. phys. Chem.* (*Frankfurt*), 1977, **105**, 13.

[432] H. Gorner and D. Schulte-Frohlinde, *J. Photochem.*, 1978, **8**, 91.

[433] H. Gorner and D. Schulte-Frohlinde, *Ber. Bunsengesellschaft phys. Chem.*, 1977, **81**, 713.

[434] M. Kato, T. Chikamoto, and T. Miwa, *Bull. Chem. Soc. Japan*, 1977, **50**, 1082.

[435] T. Karsteus, K. Kobs, R. Memming, and F. Schroppel, *Chem. Phys. Letters*, 1977, **48**, 540.

[436] R. Nakagaki, T. Kobayashi, J. Nakamura, and S. Nagakura, *Bull. Chem. Soc. Japan*, 1977, **50**, 1909.

time-resolved spectroscopy techniques to clarify the molecular structure of the photochromic coloured species by means of Fourier-transform i.r. spectroscopy. The results indicate that the photochromic behaviour occurs through the scheme

$$\text{enol imine} \xrightarrow{h\nu} \text{enol imine*} \longrightarrow \textit{cis}\text{-keto-amine*} \longrightarrow$$
$$\text{intermediate} \longrightarrow \textit{trans}\text{-keto-amine}$$

2,2,4,6-Tetraphenyl-1,2-dihydro-1,3,5-triazine has been shown to undergo photochromism only in the solid state and at temperatures above *ca.* $-70\ °C$ under excitation with 320—380 nm light. Such behaviour has not hitherto been reported.[437]

The photochromism of bis-spiropyrans has been studied,[438] and a novel steady-state method for measuring the degree of polarization of photochromic processes has been developed and applied to the photochromic behaviour of pyrene, salicylidene-aniline, diphenylbutadiyne, and spiropyrans.[439] Figure 12 shows an example of the technique applied to 1,3′,3′-trimethylspiro-(2*H*-1-benzopyran-2,2′-indoline) (BIPS) (16). The negative value of polarization indicates

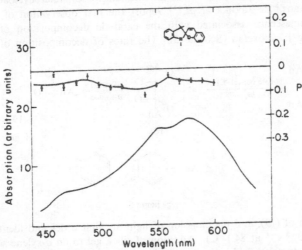

Figure 12 *Absorption and polarization spectra of the phototransient of BIPS induced by excitation at 365 nm*
(Reproduced by permission from *J. Chem. Phys.*, 1978, **68**, 2669)

(16)

that the angle between the $S_0 \rightarrow S_1$ transitions of the parent and the transient compounds is equal to or slightly greater than, 60°.

[437] T. Hayashi, *Bull. Chem. Soc. Japan*, 1077, **50**, 2489.
[438] Ya. N. Malkin, V. A. Kuz'min, and F. A. Mikhailenko, *Bull. Acad. Sci. U.S.S.R., Div. Chem. Sci.*, 1977, **26**, 70.
[439] C. Pasternak, M. A. Slifkin, and M. Shinitsky, *J. Chem. Phys.*, 1978, **68**, 2669.

Chemiluminescence.—Mendenholl[440] in an interesting article details the analytical applications of chemiluminescence, *e.g.* for trace analysis of materials, polymer degradation, and service-life predictions.

The role of dioxetans as possible intermediates in visible chemiluminescence is of great interest, the results so far suggesting that triplet states are virtually exclusive products of the dioxetans so far characterized. McCapra *et al.*[441—443] report studies made of three dioxetans which show remarkable singlet excited state yields of 25% and an abnormally short lifetime. An alternative mechanism to the 'concerted' decomposition of dioxetan whereby simultaneous cleavage of both the C—O and C—C bonds occurs is given by McCapra. It is suggested that for dioxetans either substituted, or by interaction with compounds with strongly electron-donating groups decompose *via* ions and that excitation occurs by electron transfer.

The effect of deuterium substitution on the thermal decomposition of *trans*-3,4-diphenyl-1,2-dioxetan has been investigated and the results provide further convincing evidence for the biradical mechanism for reterocycloaddition of 1,2-dioxetans.[444] Zaklika *et al.*[445] report the first observation of enhanced chemiluminescence associated with the catalytic decomposition of a stable crystalline 1,2-dioxetan (Scheme 5). The rates of decomposition of (18), the

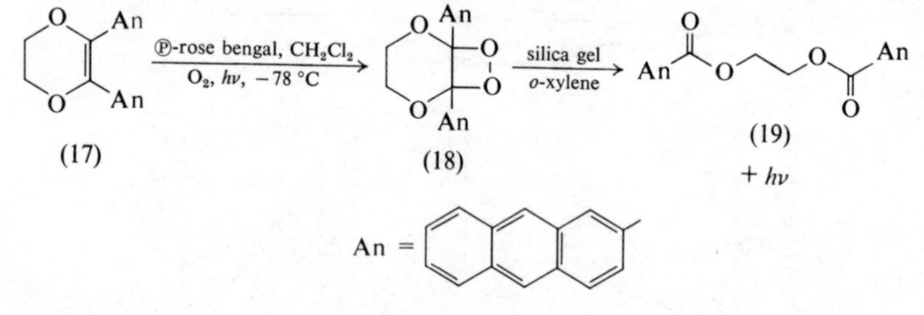

Scheme 5

formation of (19), and the emission of light are first order and identical ($k = 6.56 \times 10^{-3}\,\text{s}^{-1}$ at 84.1 °C). Addition of silica gel to an *o*-xylene solution of (18) leads to a 10^4-fold increase in light intensity and a corresponding 150-fold increase in rate.

Chemiluminescence has been observed from base-catalysed decomposition of hydroperoxy-ketones in the presence of fluorescer. The efficiency of fluorescers appears to follow the order dibromoanthracene (DBA) $\gg$ diphenylanthracene (DPA), eosin > fluorescein. Most peroxides exhibit chemiluminescence in the presence of DBA although the quantum yields are low, *ca* 10^{-5}. The results suggest that the chemiluminescent dioxetan mechanism is operating to produce

[440] G. D. Mendenhall, *Angew. Chem. Internat. Edn.*, 1977, **16**, 225.
[441] F. McCapra and A. Burford, *J.C.S. Chem. Comm.*, 1977, 874.
[442] F. McCapra, I. Beheshti, A. Burford, R. A. Hann, and K. A. Zaklika, *J.C.S. Chem. Comm.*, 1977, 944.
[443] F. McCapra, *J.C.S. Chem. Comm.*, 1977, 946.
[444] J. Koo and G. B. Schuster, *J. Amer. Chem. Soc.*, 1977, **99**, 5403.
[445] K. A. Zaklika, P. A. Burns, and A. P. Schaap, *J. Amer. Chem. Soc.*, 1978, **100**, 318.

triplet ketone.[446] Of further interest is the paper by Sawaki and Ogata[447] on the base-catalysed decomposition of α-hydroperoxy-nitriles, the results being explained by a cyclic mechanism containing a rate-determining fragmentation of dioxetanimine.

The chemiluminescence behaviour of tetramethyl-1,2-dioxetan (TMD)–ergostatetraenone in acidified acetonitrile has been investigated and a chemiluminescence yield (Φ_{CL}) of 12×10^{-8} obtained.[448] Chemiluminescence has been observed in a reaction between the tranquillizer imipramine and human polymorphonuclear leukocytes.[449] Luminol-dependent chemiluminescence has been employed in analysis of cellular and humoral defects of phagocytosis,[450] and of bacterial luciferase.[451]

Kamiya reports on the chemiluminescence from air oxidation of 9,10-di-isobutyrylanthracene[452] and of ketones and carboxylic acids[453] in alkaline aprotic solvents and of 3-methylbutan-2-one and isopropyl phenyl ketone in a dimethyl sulphoxide solution containing potassium t-butoxide.[454]

Several papers have appeared on electrogenerated chemiluminescence (ECL),[455–458] and a recent publication has appeared on the acoustic field effect on electrochemiluminescence.[459] In electroluminescence, ground-state molecules are excited by an electron-transfer reaction from the anion radical to the radical cation:

$$A^- + D^+ \longrightarrow A^* + D \ \text{ or } \ A + D^*$$

ECL, electrode potentials, and fluorescence of 14 substituted naphthalenes have been investigated[455] and a comparison has been made between electrochemical and ECL of substituted 9-cyanophenanthrenes.[456]

The chemiluminescent reaction between the thianthrene cation-radical and the anion-radical of 2,5-diphenyl-1,3,4-oxadiazole[457] gives an emission efficiency, Φ_{ECL}, of the order 10^{-4}, the singlet state being produced *via* a pure triplet route. Of further interest is a recent report on the electrochemical oxidation of oxalate ion in the presence of luminescers in acetonitrile solutions, whereby, emission characteristic of the fluorescer only occurs during the simultaneous oxidation of additive and oxalate. Zhivnov *et al.*[459] have investigated ECL under applications of ultrasonic fields. Application of acoustic vibrations produces variations in the peak power, quantum efficiency, duration of the emission pulse, time of

446 Y. Sawaki and Y. Ogata, *J. Amer. Chem. Soc.*, 1977, **99**, 5412.
447 Y. Sawaki and Y. Ogata, *J. Amer. Chem. Soc.*, 1977, **99**, 6313.
448 W. Adam, G. Cilento, L. O. Rodriguez, O. Rodrigues, A. S. Sarma, and K. Zinner, *Photochem. and Photobiol.*, 1977, **26**, 299.
449 M. A. Trush, K. V. Dyke, M. E. Wilson, and M. J. Reasor, *Res. Comm. Chem. Pathol. Pharmacol.*, 1977, **18**, 645.
450 K. V. Dyke, M. Trush, M. Wilson, P. Stealey, and P. Miles, *Microchem. J.*, 1977, **22**, 463.
451 C. Kemal and T. C. Bruice, *J. Amer. Chem. Soc.*, 1977, **99**, 7064.
452 I. Kamiya and T. Sugimoto, *Chem. Letters*, 1978, 335.
453 I. Kamiya and T. Sugimoto, *Bull. Chem. Soc. Japan*, 1977, **50**, 2442.
454 I. Kamiya, *Bull. Chem. Soc. Japan*, 1977, **50**, 2447.
455 S. Park, M. T. Paffett, and G. H. Daub, *J. Amer. Chem. Soc.*, 1977, **99**, 5393.
456 S. Park and R. A. Caldwell, *J. Electrochem. Soc.*, 1977, **124**, 1861.
457 P. R. Michael and L. R. Faulkner, *J. Amer. Chem. Soc.*, 1977, **99**, 7754.
458 M. Chang, T. Saji, and A. J. Bard, *J. Amer. Chem. Soc.*, 1977, **99**, 5399.
459 V. A. Zhionov, I. Yu, Rumyantsev, and V. I. Tomin, *Spectroscopy Letters*, 1977, **10**, 763.

its build-up, and the redox potential of the molecule. The investigators studied the behaviour of solutions of 1,5-diphenyl-3-styrylpyrazoline, rubrene, 9,10-diphenylanthracene (DPA), 9,10-dimethylanthracene, and perylene. Figure 13 illustrates the effect of applied voltage on the relative ECL peak intensities I/I_0 for DPA and rubrene, where I_0 is the ECL intensity without acoustic field. As can be seen from Figure 13, switching of an acoustic field decreases the intensity of the emission at the threshold potentials (potentials corresponding to the ECL appearance) while the growth of the voltage amplitude and the switching of an acoustic field lead to an increase of the ECL intensity.

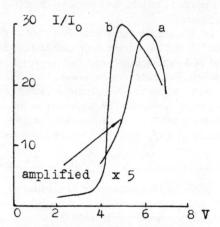

Figure 13 *Dependence of ECL peak intensities I/I_0 on applied voltage for* (a) *DPA and* (b) *rubrene*
(Reproduced by permission from *Spectroscopy Letters*, 1977, **10**, 763)

Triboluminescence.—Hardy *et al.*[460] report on the triboluminescence of crystals of a variety of compounds containing aromatic groups and compare this with the corresponding photoluminescence spectra of the crystals. The triboluminescence of the aromatic molecules investigated has five distinct origins, namely fluorescence and phosphorescence of the aromatic crystals, phosphorescence from metal centres in complexes, luminescence from charge transfer complexes, and fluorescence from the second positive group of molecular nitrogen. In general, triboluminescence spectra are rather similar to the crystal photoluminescence spectra: the authors discuss the physical aspects of the observed similarities and differences.

Bioluminescence.—Galt[461] reports on the dual mechanism of bioluminescence from a planktonic tunicate: emission arises both spontaneously and on mechanical stimulation. Field observations suggest that such emission may contribute substantially to the surface coastal displays in marine waters.

[460] G. E. Hardy, J. C. Baldwin, J. I. Zink, W. C. Kaska, P. Lui, and L. Dubois, *J. Amer. Chem. Soc.*, 1977, **99**, 3552.
[461] C. P. Galt, *Science*, 1978, **200**, 70.

3
Gas-phase Photoprocesses

BY D. PHILLIPS

1 Alkanes, Alkenes, and Alkynes

The photodissociation of the CH_4^+ cation using exciting radiation of 3 cm^{-1} bandpass gives the products shown in reactions (1) and (2) with respective relatively low cross-sections of 2.5×10^{-20} and 4×10^{-21} cm^2.[1] The process corresponds to direct dissociation or rapid predissociation. The lifetimes of CH_4^+, CD_4^+ ions have been given as 10^{-7}—10^{-5} s, and the predissociation decay process has been identified as tunnelling through a rotational barrier.[2] Other photodissociation studies on positive ions have been carried out on $C_6H_6^+$,[3,4] $C_7H_7^+$,[5] ions of cyclo-octatetraene, styrene, and related molecules,[6]

$$CH_4^+ + h\nu \longrightarrow CH_3^+ + H \qquad (1)$$

$$\longrightarrow CH_2^+ + H_2 \qquad (2)$$

and CH_3Cl^+.[7] In the last study it was shown that the average kinetic energy of the CH_3^+ fragment is 0.58 eV, a reult which suggests that some excess energy is partitioned into the internal modes of the fragment. The transition moment was found to be parallel to the C—Cl bond, suggesting a $^2E \leftarrow {}^2E$ optical transition.

Photoabsorption cross-sections in CH_4, CF_4, CF_3Cl, SF_6, and C_2F_6 in the 17.5—77.5 nm region have been reported; their spectra exhibit many Rydberg transitions.[8] The two-photon excitation of saturated hydrocarbons by a pulsed N_2 laser gave fluorescence decay times for cyclohexane, decalin, and n-pentadecane of 0.91, 2.3, and 5.2 ns, respectively.[9] These are remarkably long, but undoubtedly correct.

Photolysis at 7.1 eV (nitrogen resonance lines) of cis-[10] and trans-but-2-ene[11] has been investigated. For the former compound, the fragmentations shown in

[1] D. C. McGilvery, J. D. Morrison, and D. L. Smith, J. Chem. Phys., 1978, 68, 3949.
[2] J. P. Flamme, H. Wankenne, R. Locht, J. Momigny, P. J. C. M. Nowak, and J. Los, Chem. Phys., 1978, 27, 45.
[3] M. Allan, J. P. Maier, O. Marthaler, and E. Kloster-Jensen, Chem. Phys., 1978, 29, 331.
[4] R. C. Dunbar and H. H.-I. Teng, J. Amer. Chem. Soc., 1978, 100, 2279.
[5] D. A. McCrery and B. S. Freleer, J. Amer. Chem. Soc., 1978, 190, 2902.
[6] E. W. Fu and R. C. Dunbar, J. Amer. Chem. Soc., 1978, 100, 2283.
[7] R. G. Orth and R. C. Dunbar, J. Chem. Phys., 1978, 68, 3254.
[8] L. C. Lee, E. Phillips, and D. L. Judge, J. Chem. Phys., 1977, 67, 1237.
[9] S. Dellonte, E. Gardini, F. Barigelletti, and G. Orlandi, Chem. Phys. Letters, 1977, 49, 596.
[10] A. Wieckowski and G. J. Collin, J. Phys. Chem., 1977, 81, 2592.
[11] A. Wieckowski and G. J. Collin, Canad. J. Chem., 1977, 55, 3636.

processes (3) and (4) were detected, with the quantum yields Φ shown. Subsequent reactions of the C_3H_5 radical were as shown in (5) and (6). The photolysis

$$C_4H_8{}^{**} \longrightarrow C_4H_6 + 2H\cdot \qquad \Phi = 0.40 \qquad (3)$$

$$\longrightarrow CH_3CH=CH\cdot\dagger + \cdot CH_3 \quad \Phi = 0.38 \qquad (4)$$

$$CH_3CH=CH\cdot\dagger \longrightarrow \cdot CH_2CH=CH_2 \text{ allene} + H\cdot \qquad (5)$$

$$CH_3CH=CH\cdot \longrightarrow CH_3\cdot + C_2H_2 \qquad (6)$$

of but-1-ene and isobutene at the same wavelengths has been investigated.[12] The mercury-photosensitized decomposition of 3-methylbut-1-ene can be represented by the reactions shown in Scheme 1.[13]

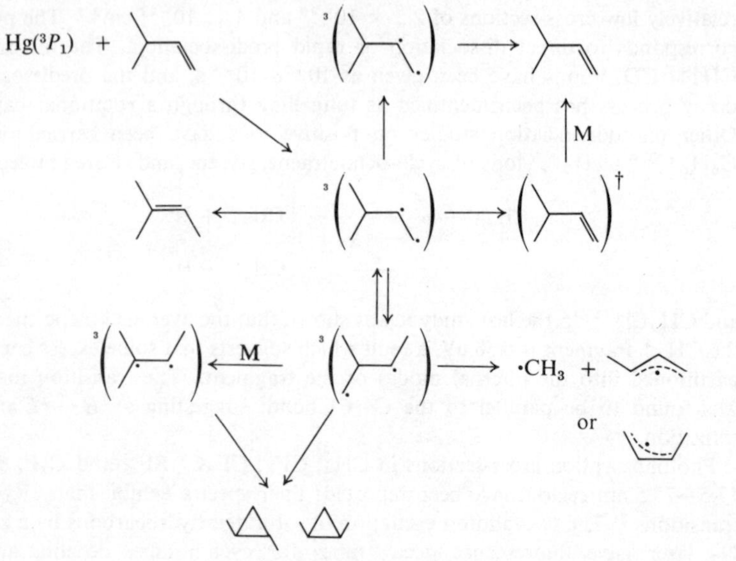

Scheme 1

The gas-phase photolysis of ethylcyclopropane at 123.6 nm and 147.0 nm,[14] and of buta-1,2-diene at the latter wavelength,[15] have been reported. In the latter study a complete analysis of all primary processes was carried out, and these are shown in reactions (7)—(15) with corresponding quantum yields.

In the two-photon absorption of hexa-1,3,5-triene using the thermal blooming effect in the 4.1—6.5 eV region no evidence was found for the $^1A_g{}^-$ state supposedly present,[16] in contrast with reports on *trans*-stilbene in solution, where

[12] G. J. Collin and A. Wieckowski, *J. Photochem.*, 1978, **8**, 103.
[13] D. C. Montague, *J.C.S. Faraday I*, 1978, **74**, 262, 277.
[14] E. Lopez and R. D. Doepker, *J. Phys. Chem.*, 1978, **82**, 753.
[15] Z. Diaz and R. D. Doepker, *J. Phys. Chem.*, 1977, **81**, 1442.
[16] A. J. Twarowski and D. S. Kliger, *Chem. Phys. Letters*, 1977, **50**, 36.

such a state was observed.[17] The unimolecular isomerization of cycloheptatrienes through the ground state populated by u.v. laser excitation to an excited
singlet state followed by internal conversion has permitted testing of unimolecular reaction rate theory.[18] The mercury-photosensitized decomposition of
1,1,2,2-tetrafluorocyclobutane[19] and HF elimination in the photodissociation of
3,3,3-trifluoropropyne at 200 nm[20] have been reported.

$$1,2\text{-}C_4H_6 + h\nu \longrightarrow H_2C{=}CH{-}C{\equiv}CH + 2H \qquad \Phi = 0.3 \qquad (7)$$

$$\longrightarrow H_2C{=}C{=}C{=}CH_2 + 2H \qquad \Phi = 0.08 \qquad (8)$$

$$\longrightarrow CH_2 + C_3H_4 \qquad \Phi = 0.02 \qquad (9)$$

$$\longrightarrow C_2H_2 + C_2H_4 \qquad \Phi = 0.02 \qquad (10)$$

$$\longrightarrow 2C_2H_2 + H_2 \qquad \Phi = 0.05 \qquad (11)$$

$$\longrightarrow H + C_2H_4 + C_2H \qquad \Phi = 0.02 \qquad (12)$$

$$\longrightarrow C_2H_2 + C_2H_3 + H \qquad \Phi = 0.03 \qquad (13)$$

$$\longrightarrow CH_3 + C_3H_3 \qquad \Phi = 0.36 \qquad (14)$$

$$\longrightarrow H + C_4H_5 \qquad \Phi = 0.09 \qquad (15)$$

Extensive studies on styrene and related compounds[21—24] have revealed
further good evidence of dual emission in these compounds upon excitation into
the second excited singlet (^1L_a) band in the isolated molecule limit. A weak,
blue-shifted, long-lived emission seen upon laser excitation at 257.25 nm was
attributed to that arising from a state in which some twisting motion had
occurred about the olefinic double-bond.

Table 1 *Dual fluorescence in styrenes* [22]

Compound	Short component Decay time/ns	Long component Decay time/ns
Styrene	1.5	50.0
trans-1-Phenyl propene	2.0	45.0
1-Phenylcyclobutene	3.0	70.0
2-Phenylnorborn-2-ene	4.5	Absent

Restriction of this motion prolonged the fluorescence decay times of both
long and short components, as shown in Table 1, and complete restriction of
the motion, as in the compound 2-phenylnorborn-2-ene (1), removed the long-
lived 'twisted' component completely. The short-lived component was attributed
to $S_1 \rightarrow S_0$ fluorescence originating in high vibrational levels of S_1 produced
upon internal conversion from the S_2 state.

[17] T. M. Stachelek, T. A. Pazoha, W. M. McClain, and R. P. Drucker, *J. Chem. Phys.*, 1977, **66**, 4540.
[18] H. Hippler, K. Luther, J. Troe, and R. Walsh, *J. Chem. Phys.*, 1978, **68**, 323.
[19] A. R. Ravishankara and R. J. Hanrahan, *J. Photochem.*, 1977, **7**, 201.
[20] L. J. Colcord and M. C. Lin, *J. Photochem.*, 1978, **8**, 337.
[21] R. P. Steer and K. Salisbury, *J. Photochem.*, 1977, **7**, 417.
[22] K. P. Ghiggino, K. Hara, G. R. Mant, D. Phillips, K. Salisbury, R. P. Steer, and M. D. Swords, *J.C.S. Perkin II*, 1978, 88.
[23] K. P. Ghiggino, K. Hara, K. Salisbury, and D. Phillips, *J.C.S. Faraday II*, 1978, **74**, 607.
[24] K. P. Ghiggino, K. Hara, K. Salisbury, and D. Phillips, *J. Photochem.*, 1978, **8**, 267.

(1)

2 Aromatic Molecules

Because of their highly fluorescent properties, aromatic hydrocarbons have long occupied a place of special interest in photophysics. Many of the current theoretical models for use in describing radiative and non-radiative decay characteristics have been devised around, and tested upon, rate data for aromatic hydrocarbons. It is thus of some importance that experimental data should be accurate, and absolute quantum yield data are notoriously difficult to obtain. A new method based upon the photoacoustic technique has been devised which is capable of great accuracy[25] and has been used to remeasure the benzene fluorescence quantum yield for 24 Torr, 253.7 nm excitation as 0.19 ± 0.2. Accurate decay data are also a requirement, and it has been demonstrated recently that synchronously-pumped mode-locked dye lasers with cavity-dumping provide ideal tunable fast repetition rate light sources for time-correlated single-photon counting decay measurements.[26] The analysis of such data can also pose problems, particularly when multiexponential decays are to be analysed and alternative methods have been proposed to accomplish this, including analysis of 'residuals',[27] the method of modulating functions,[28] and analysis of covariance ellipsoids.[29] A phase fluorimeter has been designed for picosecond processes,[30] and details of sub-picosecond spectroscopy have been given.[31] Many details of novel techniques and results are to be found in the proceedings of a conference held in the past year.[32]

In a noteworthy study[33] highly resolved fluorescence spectra from low pressures of benzene with nine added gases have been used to follow mode-to-mode vibrational relaxation in the S_1 state of benzene under 'single-collision' conditions. CW pumping of the S_1 fundamental 6^1 ($v_6' = 522$ cm^{-1}) allowed the study of collisional vibrational energy flow into each of four channels shown in Figure 1.

The mode-to-mode transfer has highly specific patterns, with roughly 70% of the transfer going into the four channels in spite of many other nearby levels. The largest cross-sections are always to a level, D, 237 cm^{-1} above the initial

[25] M. G. Rockley, *Chem. Phys. Letters*, 1977, **50**, 427.
[26] K. G. Spears, L. E. Cramer, and L. D. Hoffland, *Rev. Sci. I*, 1977, **48**, 71.
[27] D. M. Rayner, A. E. McKinnon, A. G. Szabo, and P. A. Hackett, *Canad. J. Chem.*, 1976, **54**, 3246.
[28] B. Valeur, *Chem. Phys.*, 1978, **30**, 85.
[29] A. L. Hinde, B. K. Selinger, and P. R. Nott, *Austral. J. Chem.*, 1977, **30**, 2383.
[30] H. P. Haar and M. Hauser, *Rev. Sci. I*, 1978, **49**, 632.
[31] E. P. Ippen and C. V. Shank, *Phys. Today*, 1978, **31**, 41.
[32] 'Lasers in Chemistry', ed. M. A. West, Elsevier, Amsterdam, 1977.
[33] C. S. Parmenter and K. Y. Tang, *Chem. Phys.*, 1978, **27**, 127.

level rather than to a level nearly resonant ($\Delta E = 7$ cm^{-1}) with the initial level. A common pattern of flow occurs for the four gases transferring energy by V–T,R processes alone, and another common pattern is established for the five gases which can also use V–V transfers. With the exception of one channel, V–V resonances with vibrationally complex partners increase cross-sections by less than a factor of two over that provided by the V–T,R path. V–V transfers

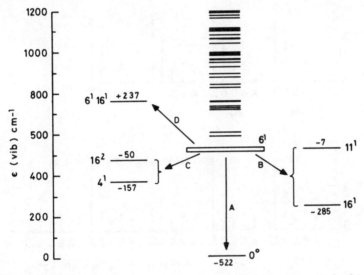

Figure 1 *Benzene vibrational levels up to* 1200 cm^{-1} *above the zero-point level in* S_1. *Level* 6^1 *is initially pumped, and flows via collisions into the four channels* A, B, C, *and* D (Reproduced from *Chem. Phys.*, 1978, **27**, 127)

have a similarly small effect on the overall vibrational relaxation rate out of the initial level 6^1. Both the flow patterns and the V–V *versus* V–T,R competitions are accounted for with an extremely simple and general set of propensity rules based on the degeneracies of the final levels, the number of vibrational quantum changes, and the amount of energy exchanged between vibrational and translational/rotational degrees of freedom. The cross-sections for collision partners SF$_6$ and perfluorohexane are small relative to those for other partners with similar vibrational complexity and mass. The results are summarized in Figure 2.

It is evident that in benzene energy flow from the initially populated level in isolation does not occur,[34] and this problem of unimolecular vibrational redistribution has exercised other authors.[35, 36] The Channel III process in benzene continues to attract attention. A close analysis of apparent anomalies in the one-photon spectrum of benzene $^1B_{2u} \leftarrow {}^1A_{1g}$, according to a recent publication, can only be explained if there exists an electronic state of benzene with a sparse set of levels close to that of the $^1B_{2u}$ level, and which is itself

[34] P. J. Nagy and W. L. Hase, *Chem. Phys. Letters*, 1978, **54**, 73.
[35] G. V. Maier, *Izvest. Akad. Nauk S.S.S.R. Ser. fiz.*, 1978, **42**, 366.
[36] P. Russegger, *Chem. Phys.*, 1977, **22**, 41.

coupled to a quasi-continuum, which could be provided by the ground state.[37] However, the sudden onset of non-radiative decay in isolated benzene as a function of excess vibrational energy, compared with a rather smoother increase observed in other substituted benzenes, has now been shown to be a consequence of the highly selective excitation possible in the case of benzene, which is not

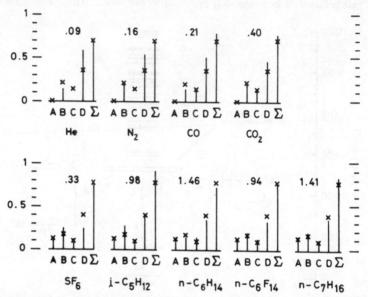

Figure 2 *Comparison of experimentally determined fractions of energy flow in channels A, B, C, and D in S_1 benzene (see Figure 1), solid vertical lines, with values calculated from simple propensity rules, crosses*
(Reproduced from *Chem. Phys.*, 1978, **27**, 127)

realizable in other more complex molecules.[38] It has been demonstrated convincingly that the onset of a Channel III process is *general* to simple aromatic hydrocarbons, and not specific to benzene. The authors suggest that the Channel III is most likely an internal conversion brought about by multiple excitations in C—C stretching frequencies, and that each mode has its own threshold for the internal conversion to occur.[39] Several papers have calculated single vibronic level decay rates for large molecules.[40,41a,41b] In one, an analytical expression was shown to give good agreement with benzene decay data.[40] A common approximation in such calculations of a single saddle point has been shown to be capable of leading to error.[41a] Extensive calculations on the one- and two-photon absorption spectra of benzene have been carried out.[42,43]

[37] G. Fischer, S. Jakobsen, and R. Naaman, *Chem. Phys. Letters*, 1977, **49**, 427.
[38] M. Jacon, C. Lardeaux, R. Lopez-Delgado, and A. Tramer, *Chem. Phys.*, 1977, **24**, 145.
[39] M. Jacon, *Chem. Phys. Letters*, 1977, **47**, 466.
[40] M. Pagitsas and K. F. Freed, *Chem. Phys.*, 1977, **23**, 387.
[41] (a) E. S. Medvedev, V. I. Osherov ,and V. M. Pschenichnikov, *Chem. Phys.*, 1977, **23**, 397; (b) I. H. Kuhn, D. F. Heller, and W. M. Gelbart, *Chem. Phys.*, 1977, **22**, 435.
[42] F. Metz, M. J. Robey, E. W. Schlag, and F. Dorr, *Chem. Phys. Letters*, 1977, **51**, 8.
[43] M. J. Robey and E. W. Schlag, *J. Chem. Phys.*, 1977, **67**, 2775.

The radiative and non-radiative decay rates of a number of methyl- and fluoro-substituted benzenes have been determined in solution and results compared with the vapour phase.[44] Confirmation that the radiative rate is mostly influenced by the second-order inductive effect exerted by the fluorine atom was obtained.

A high resolution fluorescence excitation study[45] on the fluorescence quantum yields of *p*-difluorobenzene vapour has given results at variance with earlier data,[46] and with a lower resolution study.[47] Thus the zero-point level yield of the $^4B_{3u}$ state was found to be 0.7 compared with earlier estimates of unity.[46] Relevant data are summarized in Table 2. The discrepancies between the studies may be due to the differences in bandpass of excitation used. Volk and Lee's study clearly resolved the sequence bands as is shown in Figure 3, and as Table 2 shows these have quantum yields which can differ markedly from those of the adjacent progression in mode 1. Fluorescence decay times do not show such large deviations. These findings are thus similar to those obtained for aniline and toluene discussed above.[38]

Table 2 *Fluorescence quantum yields and decay times for excitation of p-difluorobenzene*[45—47]

Band excited	λ/nm	Φ_f	τ_f/ns	k_r /10^7 s^{-1}	k_{nr} /10^7 s^{-1}	Ref.
0–0	271.3	1.00	11.3	8.85	—	45
0–0	271.3	0.70	—	6.2	2.7	46
0–0	271.3	1.00 ($\pm$0.2)	11.4	9.0	1.8	47
$16a_1^1$	271.6	0.85	11.1	7.66	1.35	45
$16a_1^1$	271.6	0.69	—	6.21	2.8	46
$16a_2^2$	271.9	0.80	11.0	7.30	1.8	45
$16a_2^2$	271.9	0.65	—	5.9	3.2	46
1_0^1	265.4	0.94	10.0	9.4	0.6	45
1_0^1	265.4	0.58	—	5.8	4.2	46
1_0^1	265.4	0.91	9.9	9.2	0.9	47
$1_0^1 16a_1^1$	265.7	0.56	—	—	—	46
$1_0^1 16a_2^2$	266.0	0.50	—	—	—	46
1_0^2	259.8	0.69	8.6	8.0	3.6	45
1_0^2	259.8	0.42	—	4.9	6.7	46
1_0^2	259.8	0.50	7.6	6.6	6.6	47
1_0^3	254.4	0.16	6.5	2.5	12.9	45
1_0^3	254.4	0.25	—	3.8	12.3	46
1_0^3	254.4	0.19	4.6	4.1	17.6	47

Collisional quenching of fluorescence and vibrational relaxation in the singlet manifold of *p*-difluorobenzene has also been studied.[45, 47] In both cases it was shown that the rate constant for *apparent* vibrational relaxation of the initially populated zero-point level by ground state *p*-difluorobenzene was very high, greater even than gas-kinetic, being 5.8 and 7×10^{11} l mol^{-1} s^{-1} (refs. 47 and

[44] D. M. Shold, *Chem. Phys. Letters*, 1977, **49**, 243.
[45] L. J. Volk and E. K. C. Lee, *J. Chem. Phys.*, 1977, **67**, 236, 242.
[46] C. Guttman and S. A. Rice, *J. Chem. Phys.*, 1974, **61**, 661.
[47] M. D. Swords, R. P. Steer, and D. Phillips, *Chem. Phys.*, 1978, **34**, 95; D. Phillips, M. G. Rockley and M. D. Swords, *Chem. Phys.*, 1979, **38**, 301.

45, respectively). In the latter study, a stochastic model was used to fit vibrational relaxation for a number of initially pumped levels. Not surprisingly, *p*-difluorobenzene was found to be a better vibrational relaxer than fluoroform.[47] The two-photon excitation spectra of the $^1B_{2u} \leftarrow {}^1A_g$ system in *p*-difluorobenzene has been recorded,[48] paving the way for studies on decay characteristics.

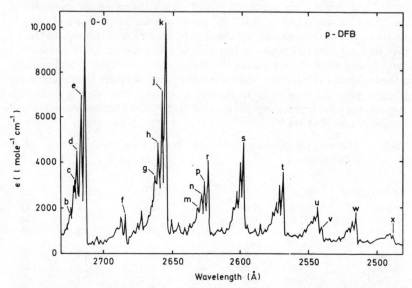

Figure 3 *High resolution absorption spectrum of p-difluorobenzene vapour, showing progression in v_1, bands k, $1_0{}^1$; s, $1_0{}^2$; and u, $1_0{}^3$; together with complicating sequence bands, for example b, $(16a)_4{}^4$; c, $(16a)_3{}^3$; d, $(16a)_2{}^2$; and e, $(16a)_1{}^1$ and others not identified here*
(Reproduced from *J. Chem. Phys.*, 1977, **67**, 236)

The triplet decay characteristics of a number of fluorinated benzenes in the vapour phase have been recorded, using a neodymium laser flash system,[49] and in all cases decay times shown to be at least two orders of magnitude shorter than that for benzene, being in the 1 μs → 100 ns range. No systematic variation in pressure was possible in these experiments, but a technique using a modulated mercury-resonance lamp for excitation has yielded the results shown in Table 3, where the rate constants refer to processes (16) and (17). The dramatic effect of

$$^3A \longrightarrow A \qquad (16)$$

$$^3A + A \longrightarrow \text{quenching} \qquad (17)$$

fluorine substitution on both k_{16} and k_{17} is apparent, and tentative explanations were offered. The collisional quenching is of interest, since the data on benzene showed a non-linear Stern–Volmer plot, indicating a complex quenching

[48] M. J. Robey and E. W. Schlag, *Chem. Phys.*, 1978, **30**, 9.
[49] R. Bonneau, M. E. Sime, and D. Phillips, *J. Photochem.*, 1978, **8**, 239.

Table 3 Rate constants for non-radiative (k_{16}) and bimolecular (k_{17}) decay of triplet states of substituted benzenes in the vapour phase

Compound	$k_{nr}{}^a$/s^{-1}	$k_q{}^b$/l mol^{-1} s^{-1}	$(P)^c$/s	Ref.	$(P)^d$/s
Benzene	5.6×10^{3e}	6.2×10^7	18×10^{-6} (20)	f	26×10^{-6} (20)[g]
Benzene	1.7×10^3	6.6×10^7	—	h	—
Benzene	3.3×10^3	—	—	i	—
[²H₆]Benzene	1.8×10^3	1.9×10^7	—	f	—
[²H₆]Benzene	1.9×10^3	2.1×10^7	—	h	—
[²H₆]Benzene	2.0×10^3	—	—	i	—
Toluene	3.4×10^3	2.2×10^7	—	f	—
Toluene	2.3×10^3	—	—	i	—
1,3-bis(Trifluoromethyl)benzene	1.8×10^5	4.0×10^8	4.0×10^{-6} (15)	f	2×10^{-6} (15)[49]
1,4-bis(Trifluoromethyl)benzene	4.0×10^5	9.6×10^8	800×10^{-9} (15)	f	560×10^{-9} (15)[49]
1,3-Difluorobenzene	2.2×10^6	2.5×10^9	240×10^{-9} (15)	f	340×10^{-9} (15)[49]
1,4-Difluorobenzene	4.15×10^6	5.5×10^9	160×10^{-9} (10)	f	200×10^{-9} (10)[49]

[a] Obtained by extrapolation of pressure-dependent lifetimes to zero pressure; [b] from slopes of pressure-dependent lifetimes; [c] value of decay time obtained by extrapolation to pressure (in Torr) shown in parentheses; [d] corresponding value of lifetime at pressure (in Torr) shown; [e] at pressures below 0.5 Torr, the benzene results showed a strong deviation from the results reported here, which are thus high pressure results comparable to other workers'; [f] ref. 50; [g] C. S. Parmenter and B. L. Ring, *J. Chem. Phys.*, 1967, **46**, 1998; [h] T. F. Hunter and M. G. Stock, *J.C.S. Faraday II*, 1974, **70**, 1028; [i] H. E. Hunziker, *J. Res. Development*, 1971, **15**, 10.

mechanism at low pressures. The enhancement of k_{17} in fluoro compounds is compatible with quenching being due to triplet excimer formation, but other mechanisms, including crossing to an electronic state close in energy to that of $^3B_{1u}$, cannot be discounted, particularly since such a state (but not the same one) has been invoked to explain singlet state absorption spectra.

The radiative relaxation of the $\tilde{B}(\Pi^{-1})$ state of radical cations of a variety of fluorinated benzenes has been reported,[51] and two-photon ionic dissociation in bromobenzene cations discussed.[52] In the reactions shown in Scheme 2 the cross-sections for absorption of photons 1 and 2 were found to be 4.4×10^{-18} cm^2, and 7.9×10^{-18} cm^2, respectively. The value of k_q for deactivation by the bromobenzene neutral was very high, being 1.1×10^{-9} cm^3 molecule^{-1} s^{-1}. The photophysics of aryl-substituted olefins has been discussed in the preceding section.[21—24]

$$C_6'H_5Br^+ \underset{k_q}{\overset{h\nu,\ \sigma_1}{\rightleftharpoons}} (C_6H_5{}^+Br)^* \xrightarrow{h\nu,\ \sigma_2} C_6H_5{}^+ + Br\cdot$$

Scheme 2

Two-photon excitation spectra of naphthalene vapour[53] and crystals[54] at 4.2 K have been recorded. The first study shows definitely that vibronic levels of $B_{3u} \times b_{3u}$ in the lowest singlet state are the predominantly excited species (Figure 4). The non-resolvable background in the fluorescence spectrum was

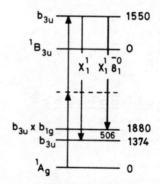

Figure 4 *Principal fluorescence bands observed following two-photon excitation of a $B_{3u} \times b_{3u}$ vibronic level of [$^2H_8$]naphthalene*
(Reproduced from *Chem. Phys.*, 1977, **24**, 327)

attributed to subsequent excitation of the two-photon state by the further stepwise excitation summarized in Figure 5.

A further study on the effects of rotational excitation on the radiative and non-radiative decay of isolated naphthalene vapour has appeared.[55]

[50] M. E. Sime and D. Phillips, *Chem. Phys. Letters*, 1978, **56**, 138.
[51] M. Allan, J. P. Maier, and O. Marthaler, *Chem. Phys.*, 1977, **26**, 131.
[52] R. C. Dunbar and E. W. Fu, *J. Phys. Chem.*, 1977, **81**, 1531.
[53] H. Gattermann and M. Stockburger, *Chem. Phys.*, 1977, **24**, 327.
[54] N. Mikami and M. Ito, *Chem. Phys.*, 1977, **23**, 141.
[55] W. E. Howard and E. W. Schlag, *Chem. Phys.*, 1978, **29**, 1.

An extremely elegant tandem laser system has been used to investigate the time evolution of the triplet state spectrum of naphthalene vapour following singlet state excitation.[56] It was shown (Figure 6) that the triplet state initially formed has a very much broader distribution than the vibrationally relaxed triplet state, and this undoubtedly has led to erroneous conclusions in previous studies where only a narrow wavelength region was scanned in absorption. It was shown that the vibrational relaxation occurs with almost hard sphere collision efficiency for naphthalene as relaxer. The loss rate of unrelaxed triplet state, probably *via* intersystem crossing to a high vibrational level of the ground state, has an estimated rate of $1.7 \times 10^5 \text{ s}^{-1}$.

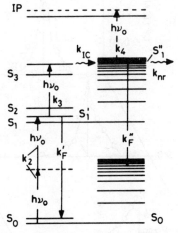

Figure 5 *Relaxation processes following two-photon excitation of naphthalene vapour* (Reproduced from *Chem. Phys.*, 1977, **24**, 327)

The fluorescence from pyrene vapour at 170 °C excited to the second excited singlet state is not completely quenched by O_2 at pressures up to 1000 Torr.[57] The quantum yield of the non-quenchable fraction was 5×10^{-6}. From these effects it was possible to estimate that the rate constant for $S_2 \rightsquigarrow S_1$ internal conversion was $3 \times 10^{13} \text{ s}^{-1}$, independent of excitation energy, whereas the reverse internal conversion $S_1 \rightsquigarrow S_2$ had a rate constant of 10^9 s^{-1} for 33 600 cm^{-1} excitation, and $3 \times 10^{10} \text{ s}^{-1}$ for 42 700 cm^{-1} excitation. Delayed fluorescence has been studied in anthracene vapour.[58]

Following 300-ps excitation from the third harmonic of a neodymium laser (28 460 cm^{-1}), pyrene vapour exhibited a 21-ns rise-time in fluorescence, whereas this was totally absent in fluoranthene vapour.[59] The two compounds differ in that in the excitation of pyrene, with a zero–zero band at 26 837 cm^{-1}, the initial state has an excess energy of some 1500—2000 cm^{-1} accommodated in a_g and/or b_{1g} modes, in a sparse level density, whereas for fluoranthene excitation is to a

[56] H. Schroder, H. J. Neusser, and E. W. Schlag, *Chem. Phys. Letters*, 1978, **54**, 4.
[57] K. Chihara and H. Baba, *Chem. Phys.*, 1977, **25**, 299.
[58] V. T. Pavlova, *Izvest. Akad. Nauk S.S.S.R. Ser. fiz.*, 1978, **42**, 353; A. V. Dorokhin and A. A. Kotov, *ibid.*, p. 349.
[59] D. J. Ehrlich and J. Wilson, *J. Chem. Phys.*, 1977, **67**, 5391.

dense set of levels. The authors propose that the rise-time observed corresponds to the intramolecular redistribution of energy into these vibronic levels in pyrene most strongly coupled radiatively to the ground state, and if this is true, the

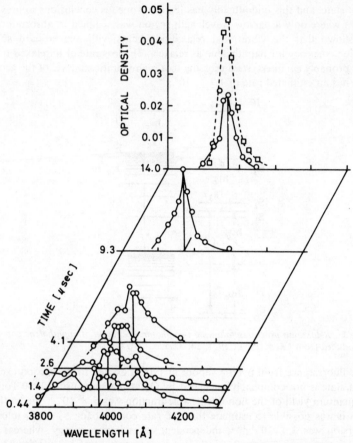

Figure 6 *The time evolution of the $T^* \leftarrow T_1$ absorption spectrum in $[^2H_8]$naphthalene, subsequent to ISC at low pressure ($\bigcirc$) (p = 68 mTorr N). The molecules were optically excited to the $8(b_{1g})$ vibronic band of S_1. For comparision, a 'high pressure' T–T spectrum ($\square$) (p = 68 mTorr N + 2.5 Torr Ar) is given at 14 μs. The time scale is related to the onset of the exciting laser pulse*
(Reproduced from *Chemical Physics Letters*, 1978, **54**, 4)

process would have a rate constant $\sim 10^8$ s^{-1}. Should this interesting observation be substantiated, it will mark the first unambiguous evidence for the redistribution process. The decay time of the fluoranthene vapour was found to be 24 $\pm$ 3 ns. A further study has appeared comparing the ps decay of azulene in solution and gas phase.[60]

[60] D. Huppert, J. Jortner, and P. M. Rentzepis, *Israel J. Chem.*, 1977, **16**, 277.

3 Carbonyl and Other Oxygen-containing Compounds

The photolysis and photophysics of formaldehyde are of especial interest since the molecule is the prototype carbonyl compound and is capable of excitation to single vibronic (and even single rotational) levels, may be of importance in photochemical smog formation, and could provide a means to photochemical separation of hydrogen isotopes. Not surprisingly, many papers have appeared in the past year concerned with this species.

A detailed theoretical paper has considered the absorption of formaldehyde,[61] and an *ab initio* computation on the non-radiative decay rate of formaldehyde $^1A_2$ has been performed.[62] A review by a pioneer in studies on laser-induced photochemistry in selected rovibronic states has appeared.[63] Excited formaldehyde decomposes *via* two routes (18) and (19). Two recent evaluations of the

$$H_2CO + h\nu \longrightarrow H_2 + CO \qquad (18)$$
$$\longrightarrow H + HCO \qquad (19)$$

quantum yields of these reactions are in some disagreement. Thus the quantum yields of radical production measured by but-2-ene scavenging give Φ_{19} as 0.68 at 275.4 nm excitation, 0.6 at 288 nm, and 0.68 at 303.5 nm.[64] The corresponding high energy excitation yield from NO scavenging studies, however, was 0.4.[65] The threshold energy for radical production in H_2CO was found to be 330—339 nm, while that for D_2CO was 326—320 nm. The Φ_{19} decreases steeply to zero above the threshold wavelength, where Φ_{18} becomes unity for both isotopes. A further paper on this subject has appeared very recently.[66] Many authors have invoked an intermediate between the excited electronic singlet state of formaldehyde produced upon excitation, and the appearance of products in reaction (18). The identity of the intermediate is not entirely compatible with either the triplet state or a high vibrational level of the ground state produced by internal conversion. It has been suggested that hydroxy carbene $:C{<}^{OH}_{H}$ an isomer of formaldehyde, could be the intermediate species.[67]

Very extensive studies on the fluorescence spectra of single *rotational* levels and rotational relaxation of the first excited singlet state of glyoxal have been reported,[68] and single *vibronic* level decay data for the ^1A_u and ^3A_u states of this compound have also been measured.[69] A reinvestigation of the effect of magnetic field on the quenching of glyoxal fluorescence[70—72] has shown that the

[61] P. W. Langhoff, S. R. Langhoff, and C. T. Corcoran, *J. Chem. Phys.*, 1977, **67**, 1722.
[62] J. M. F. van Dijk, M. J. H. Kemper, J. H. M. Kerp, H. M. Buck, and G. J. Visser, *Chem. Phys. Letters*, 1978, **54**, 353.
[63] E. K. C. Lee, *Accounts Chem. Res.*, 1977, **10**, 319.
[64] R. S. Lewis and E. K. C. Lee, *J. Phys. Chem.*, 1978, **82**, 249.
[65] J. H. Clark, C. B. Moore, and N. S. Nogar, *J. Chem. Phys.*, 1978, **68**, 1264.
[66] G. K. Moortgat, F. Slemr, W. Seiler, and P. Warneck, *Chem. Phys. Letters*, 1978, **54**, 444.
[67] R. R. Lucchese and H. F. Schaefer, *J. Amer. Chem. Soc.*, 1978, **100**, 298.
[68] C. S. Parmenter and B. F. Rordorf, *Chem. Phys.*, 1978, **27**, 1; B. F. Rordorf, A. E. W. Knight and C. S. Parmenter, *ibid.*, p. 11.
[69] C. Cossart-Magos, A. Frad, and A. Tramer, *Spectrochim. Acta*, 1978, **34**, 195.
[70] H. G. Kuttner, H. L. Selzle, and E. W. Schlag, *Chem. Phys. Letters*, 1977, **48**, 207.
[71] H. G. Kuttner, H. L. Selzle, and E. W. Schlag, *Israel J. Chem.*, 1977, **16**, 264.
[72] H. G. Kuttner, H. L. Selzle, and E. W. Schlag, *Chem. Phys.*, 1978, **28**, 1.

collision-free decay time is unaffected by the magnetic field, and thus the magnetic effect is observed only in the presence of collision partners. At a pressure of 200 mTorr, a 20% reduction in both yield and decay time are observed at 1 kG, a field at which saturation of the effect also occurs. It was argued that the results show the collision partner acts solely as an energy transfer agent, and does not induce degeneracy of singlet and triplet states or influence spin–orbit coupling, since helium and argon were as useful as collision partners as glyoxal itself, or methanol with high dipole moment and methyl iodide, containing a heavy atom.

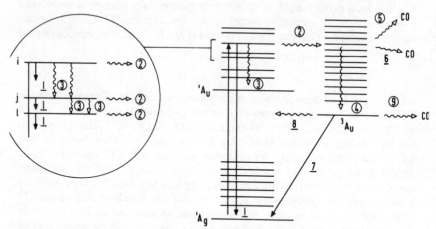

Figure 7 *Dissociation mechanism for the production of* CO *from excitation of single vibronic levels of* ^1A_u *glyoxal, with key to processes as follows. Circled processes are collisionally induced:* (1) $S_1^*(i) \rightarrow h\nu + S_0$, *fluorescence;* (2) $S_0 + S_1^*(i) \rightarrow T_1^*(i) + S_0$, *intersystem crossing;* (3) $S_0 + S_1^*(i) \rightarrow S_1^*(j) + S_0$, *vibrational relaxation;* (4) $S_0 + T_1^*(i) \rightarrow T_1(0) + S_0$, *vibrational relaxation;* (5) $S_0 + T_1^*(i) \rightarrow$ CO, *dissociation;* (6) $T_1^*(i) \rightarrow$ CO, *dissociation;* (7) $T_1(0) \rightarrow h\nu + S_0$, *phosphorescence;* (8) $T_1(0) \rightarrow S_0^*$, *intersystem crossing;* (9) $S_0 + T_1(0) \rightarrow$ CO, *dissociation, where* $S_0 \equiv {}^1A_g$ (ground) *state,* $S_1^*(i) \equiv {}^1A_u$ SVL *populated by absorption,* $T_1^*(i) \equiv$ *vibrationally excited* ^3A_u, $S_1^*(j) \equiv {}^1A_u$ SVL *not populated by absorption,* $T_1(0) \equiv$ *zero point* ^3A_u *state*
(Reproduced from *J. Chem. Phys.*, 1978, **68**, 726)

Some absolute quantum yields for the production of CO from glyoxal photolysis have been carefully measured.[73,74] The data are too extensive to be considered here in detail, but the scheme proposed is summarized in Figure 7. Levels excited were the 0^0, 7^1, 5^1, 8^1, $8^1 7^2$, 2^1, $8^1 4^1$ and significant vibronic and pressure dependencies were observed for all levels examined, some of which are indicated in Figure 8. The photolysis of methyl glyoxal vapour at 436 nm has also been reported.[75]

[73] G. H. Atkinson, M. E. McIlwain, C. G. Venkatesh, and D. M. Chapman, *J. Photochem.*, 1978, **8**, 307.
[74] G. H. Atkinson, M. E. McIlwain, and C. G. Ventatesh, *J. Chem. Phys.*, 1978, **68**, 726.
[75] E. Kyle and S. W. Orchard, *J. Photochem.*, 1977, **7**, 305.

A reinvestigation[76] of the fluorescence of acetaldehyde vapour at total pressures of up to 2200 Torr for 250—330 nm excitation has failed to reveal any sign of an 'unquenchable' species reported earlier by other workers. Singlet acetaldehyde thus appears to behave normally in that the initially prepared singlet state undergoes radiative and non-radiative relaxation in competition with collisional relaxation. A sharp discontinuity in the triplet state surface of hexafluoroacetone HFA has been revealed such that above this threshold all

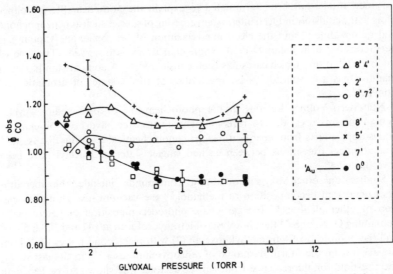

Figure 8 *Absolute quantum yields for* CO *production following excitation of the single vibronic levels of* ^1A_u *glyoxal shown in the inset*
(Reproduced from *J. Chem. Phys.*, 1978, **68**, 726)

triplet HFA molecules decompose, whereas below it phosphorescence is a possibility.[77] A scheme capable of fully explaining the results is outlined in processes (20)—(28) for the system at low pressures. At higher pressures collisional relaxation in the singlet state manifold also occurs. A similar scheme was

$$A_{th} + h\nu \longrightarrow {}^1A_n \tag{20}$$

$$^1A_n \longrightarrow A + h\nu_f \tag{21}$$

$$^1A_n \longrightarrow A \tag{22}$$

$$^1A_n \longrightarrow {}^3A_q \tag{23}$$

$$^3A_q \longrightarrow products \tag{24}$$

$$^3A_q + A_{th} \longrightarrow {}^3A_{th} \tag{25}$$

$$^3A_{th} \longrightarrow CF_3CO + CF_3 \tag{26}$$

$$^3A_{th} \longrightarrow A + h\nu_p \tag{27}$$

$$^3A_{th} \longrightarrow A \tag{28}$$

[76] A. Gandini and P. A. Hackett, *Chem. Phys. Letters*, 1977, **52**, 107.
[77] P. A. Hackett and K. O. Kutschke, *J. Phys. Chem.*, 1977, **81**, 1245.

shown to explain the results for a related molecule, 1,1,1-trifluoroacetone, TFA.[78] The quantum yield for production of biacetyl, easily observable in phosphorescence from TFA photolysis at 302.5 nm, was found to be 0.15 ± 0.03.[78] Relative rate constants for quenching of triplet TFA by but-2-ene in the absence and presence of mercury vapour showed the triplet state of TFA to be quenched by mercury since the ratio k_q (Hg free)/k_q(Hg) was found to be 1.5.[78]

A new model has been formulated to explain the non-linear Stern–Volmer plots in the collision-induced intersystem crossing of species such as benzophenone and quinoxaline,[79] and the electronic relaxation of benzophenone vapour has been considered in detail;[80] dual exponential decay is observed. The role of exciting radiation in such cases has been considered.[81] A comparision has been made of gas and solution phase reactivities of triplet states of aromatic carbonyls.[82]

Non-Stern–Volmer kinetics in the photochemistry of pentanedione–cyclopentadiene mixtures, due to reversible energy transfer have been observed,[83] and isotope effects from energy transfer in photochemical reactions discussed.[84] Radiationless transitions between excited singlet states of biacetyl have been investigated.[85]

Studies on other oxygen-containing compounds include the mercury-photosensitized decomposition of methanol,[86] the vacuum u.v. photodissociation of other alcohols,[87] the gas phase photodecomposition of heterocyclics containing O, N, or S,[88] the photolysis of tetrahydrofuran at 147 and 123.6 nm,[89] and the photolysis of tetramethyldioxetan (2) with picosecond pulses of 264 nm radiation from the fourth harmonic of a neodymium laser.[90] In the last study, the rise-time for fluorescence from singlet acetone was shown to be less than

10 ps, and thus no long-lived intermediate exists between excited oxetan and electronically excited acetone product. The photochemical reaction thus possibly proceeds along a reaction pathway different from that for the thermal

[78] S. W. Beavan, H. Inoue, D. Phillips, and P. A. Hackett, *J. Photochem.*, 1978, **8**, 247.
[79] R. Naaman, V. Yakhot, and G. Fischer, *J. Chem. Phys.*, 1977, **67**, 5472.
[80] D. Zevenhuijzen and R. van der Werf, *Chem. Phys.*, 1977, **26**, 279.
[81] C. A. Langhoff, *Chem. Phys.*, 1977, **20**, 357.
[82] M. Berger, R. N. Camp, I. Demetrescu, L. Giering, and C. Steel, *Israel J. Chem.*, 1977, **16**, 311.
[83] A. W. Jackson and A. J. Yarwood, *Mol. Photochem.*, 1977, **8**, 255.
[84] R. Z. Sagdeev, A. B. Doktorov, V. V. Pervukhin, A. A. Obynochny, S. V. Camyshan, Yu. N. Molin, and V. M. Moralyov, *Chem. Phys.*, 1978, **29**, 311.
[85] C. Richard, M. Bouchy, J. C. Andre, and M. Niclause, *Internat. J. Chem. Kinetics*, 1978, **10**, 213.
[86] S. L. N. G. Krishnamachari and R. Venkatasubramanian, *Mol. Photochem.*, 1977, **8**, 419.
[87] I. P. Vinogradov and F. I. Vilesov, *High Energy Chemistry*, 1977, **11**, 17.
[88] S. Braslavsky and J. Heicklen, *Chem. Res.*, 1977, **77**, 473.
[89] Z. Diaz and R. D. Doepker, *J. Phys. Chem.*, 1978, **82**, 10.
[90] K. K. Smith, J. Y. Koo, G. B. Schuster, and K. J. Kaufman, *Chem. Phys. Letters*, 1977, **48**, 267.

reaction, since the radical intermediate (3) has been implicated in the latter process, although there is some doubt about the feasibility of the energetics of this.

4 Ketene, Methylene, and Radical Reactions

A laser-induced fluorescence study (a review of such techniques has appeared [91]) on $CH_2(\tilde{a}^1A_1)$ produced from ketene photolysis at 337 nm has been used to estimate the $\tilde{a}$–$\tilde{X}$ energy separation as 6.3 ± 0.8 kcal mol^{-1},[92] which can be compared with calculated values.[93, 94a] It should be noted that a thorough study[94b] on the production of $CH_2(^1A_1)$ from the photolysis of CH_2N_2 gives results for the $CH_2(\tilde{a}$–$\tilde{X})$ splitting, which are incompatible with the estimate from photodetachment measurements of 19.5 kcal mol^{-1}; the kineticists' value of $\sim$8 kcal mol^{-1} being preferred. There still exists, therefore, a serious discrepancy between values obtained kinetically and the spectroscopic measurement. At 313, 334, and 336 nm, ketene photolysis has been explained by the mechanism shown in reactions (29)—(39) which involves association of an excited ketene

$$S_0 + h\nu \longrightarrow S_1{}^* \tag{29}$$

$$S_1{}^* + S_0 \longrightarrow (S_1S_0)^* \tag{30}$$

$$(S_1S_0)^* \longrightarrow C_2H_4 + 2CO \tag{31}$$

$$(S_1S_0)^* + S_0 \longrightarrow (S_1S_0) + S_0 \tag{32}$$

$$S_1{}^* \longrightarrow T_1{}^* \tag{33}$$

$$T_1{}^* \longrightarrow {}^3CH_2 + CO \tag{34}$$

$$S_1{}^* + S_0 \longrightarrow 2S_0 \tag{35}$$

$${}^3CH_2 + CH_2CO \longrightarrow CH_3 + CHCO \tag{36}$$

$$2CH_3 \longrightarrow C_2H_6 \tag{37}$$

$$2CHCO \longrightarrow C_2H_2 + 2CO \tag{38}$$

$$CH_3 + CHCO \longrightarrow C_2H_4 + CO \tag{39}$$

with a ground state species.[95] At limiting low pressures, the quantum yield for CO production approached 2. This scheme may be compared with that shown by reactions (40)—(62) derived from a flash photolysis study in the 160 nm region, from which many rate constants (shown in parentheses in units of cm^3 molecule^{-1} s^{-1}) were derived.[96] A notable feature in this study is that products derived from $CH_2(^1A_1)$ persist at pressures when quenching of this

[91] J. K. Kinsey, *Ann. Rev. Phys. Chem.*, 1977, **28**, 349.
[92] J. Danon, S. V. Filseth, D. Feldmann, H. Zacharias, C. H. Dugan, and K. H. Welge, *Chem. Phys.*, 1978, **29**, 345.
[93] S.-K. Shih, S. D. Peyerimhoff, R. J. Buenker, and M. Peric, *Chem. Phys. Letters*, 1978, **55**, 206.
[94] (a) L. B. Harding and W. A. Goddard, *Chem. Phys. Letters*, 1978, **55**, 217; (b) T. H. Richardson and J. W. Simons, *J. Amer. Chem. Soc.*, 1978, **100**, 1062.
[95] W. L. Hase and P. M. Kelley, *J. Chem. Phys.*, 1977, **66**, 5093.
[96] M. J. Pilling and J. A. Robertson, *J.C.S. Faraday I*, 1977, **73**, 968.

species to $^3B_1$ through reaction (41) should have been exclusive. The results can be explained if the $CH_2(^1B_1)$ which is formed is comparatively stable to collisional quenching through reaction (52), but reacts with H_2, reaction (51) and CH_2CO, reaction (53). Reactions of CHF with O_2 and NO leading to HF laser action have been reported.[97]

$$CH_2CO + h\nu \longrightarrow {}^1CH_2 + CO \tag{40}$$

$$^1CH_2 + M \longrightarrow {}^3CH_2 + M \quad (6.7 \times 10^{-13})* \tag{41}$$

$$^1CH_2 + H_2 \longrightarrow CH_3 + H \quad (7 \times 10^{-12}) \tag{42}$$

$$^3CH_2 + {}^3CH_2 \longrightarrow C_2H_2 + H_2 \quad (5.3 \times 10^{-11}) \tag{43}$$

$$CH_3 + CH_3 \longrightarrow C_2H_6 \quad (4.2 \times 10^{-11}) \tag{44}$$

$$CH_3 + H \longrightarrow CH_4 \tag{45}$$

$$H + CH_2CO \longrightarrow CH_3 + CO \quad (1.3 \times 10^{-13}) \tag{46}$$

$$^3CH_2 + CH_3 \longrightarrow C_2H_4 + H \quad (5.0 \times 10^{-11}) \tag{47}$$

$$^1CH_2 + CH_2CO \longrightarrow C_2H_4 + CO \quad (2.0 \times 10^{-11}, 3.5 \times 10^{-12}) \tag{48}$$

$$C_2H_4 + h\nu \longrightarrow C_2H_2 + H_2 \tag{49}$$

$$CH_3 + h\nu \longrightarrow {}^1CH_2 + H \tag{50}$$

$$CH_2(^1B_1) + H_2 \longrightarrow CH_3 + H \quad (2.0 \times 10^{-11})\dagger \tag{51}$$

$$CH_2(^1B_1) \longrightarrow {}^1CH_2 + h\nu \quad (10^6 \, s^{-1}) \tag{52}$$

$$CH_2(^1B_1) + CH_2CO \longrightarrow C_2H_4 + CO \quad (3.0 \times 10^{-11})\dagger \tag{53}$$

$$H + C_2H_4 \longrightarrow C_2H_5 \quad (1.6 \times 10^{-12}) \tag{54}$$

$$H + C_2H_2 \longrightarrow C_2H_3 \quad (1.6 \times 10^{13}) \tag{55}$$

$$H + H + M \longrightarrow H_2 + M \quad (4 \times 10^{-13}) \text{ at 800 Torr} \tag{56}$$

$$H \longrightarrow \text{wall} \tag{57}$$

$$^3CH_2 + H_2 \longrightarrow CH_3 + H \quad (<5 \times 10^{-15}) \tag{58}$$

$$^3CH_2 + C_2H_2 \longrightarrow C_3H_4 \quad (7.5 \times 10^{-12}, 4.0 \times 10^{-12}) \tag{59}$$

$$^3CH_2 + NO \longrightarrow CH_2NO \quad (1.6 \times 10^{-11}, 1.0 \times 10^{-11})$$
$$\text{200 Torr Ar} \tag{60}$$

$$^3CH_2 + O_2 \longrightarrow \text{products} \quad (1.5 \times 10^{-12}, 1.2 \times 10^{-12}) \tag{61}$$

$$^3CH_2 + CO \longrightarrow \text{products} \quad (<10^{-15}) \tag{62}$$

* For M = Ar.   † Dependent on k_{53}.

Reactions of CH_3 of importance in combustion,[98] and with hydrogen atoms [99] have been reported, the absolute rate constant for methoxy radical reacting with

[97] M. C. Lin, *J. Chem. Phys.*, 1978, **68**, 2004.
[98] A. C. Baldwin and D. M. Golden, *Chem. Phys. Letters*, 1978, **55**, 350.
[99] J.-T. Cheng and C.-T. Yeh, *J. Phys. Chem.*, 1977, **81**, 1982.

acetaldehyde has been measured,[100] and a Fourier transform i.r. spectroscopic study on the reaction of peroxy radicals with NO_2 described.[101]

5 Sulphur-containing Compounds

The electron photodetachment cross-section for S^- ions in the presence of magnetic fields has been measured,[102] and the rate constant for reaction of $S(^3P)$ with NO determined.[103] The u.v. ion-laser spectroscopy of S_2, F_2, Cl_2, and Br_2 has been investigated,[104] and the collisional quenching of fluorescence from $S_2(\tilde{B}^3\Sigma_u^-)$ excited at 292 nm by He, Ar, Xe, N_2, S_2, CF_4, or C_2F_6 has been measured. Two components of fluorescence were observed, with quenching cross-sections (in Å), shown in Table 4.[105] The two components were attributed to either effects of interacting electronic states such as $\tilde{B}''(^3\Pi_u)$, or to vibronic effects within the $^3\Sigma_u^-$ manifold. The analysis is complicated by the fact that broad-band excitation conditions were used.

Table 4 *Quenching cross-sections of fluorescence, from* $S_2(\tilde{B}^3\Sigma_u^-)$

Quencher	$(\sigma_S)l^a$	$(\sigma_L)q^b$
He	11.2	0
Ar	13.7	0.041
Xe	24.6	0.20
N_2	36.3	0.021
S_2	81.3	—
CF_4	21.4	0.052
C_2F_6	72.4	0.088

a Quenching of short-lived component;   *b* quenching of long-lived component.

The $S_3 \rightarrow S_0$ fluorescence from CS_2 vapour has been observed following excitation in the 200—220 nm region with a maximum quantum yield of $\sim 10^{-3}$.[106] The quantum yield of fluorescence from the isolated molecule decreases rapidly with increase in excitation energy; the non-radiative rate constant for decay being $\sim 10^{11}$ s^{-1} for excitation of the 1^1 level, rising to 8.2×10^{11} s^{-1} for the 1^9 level. The non-radiative decay process competing with fluorescence is certainly not $S_3 \rightsquigarrow S_2$, S_1 internal conversion, since no fluorescence from these lower levels was observed, and is most likely dissociation to ground-state S and CS fragments. Hanle effects in CS_2 vapour following pulsed excitation with a N_2 laser have been reported.[107]

Strongly non-exponential fluorescence decay curves have been observed following the excitation of thiophosgene Cl_2CS vapour to the second excited singlet state under isolated molecule conditions.[108] It was suggested that the

[100] N. Kelly and J. Heicklen, *J. Photochem.*, 1978, **8**, 83.
[101] H. Niki, P. D. Maker, C. M. Savage, and L. P. Breitenbach, *Chem. Phys. Letters*, 1978, **55**, 289.
[102] W. A. M. Blumberg, R. M. Jopson, and D. J. Larson, *Phys. Rev. Letters*, 1978, **40**, 1320.
[103] A. Vanroodselaar, K. Obi, and O. P. Strausz, *Internat. J. Chem. Kinetics*, 1978, **10**, 76.
[104] J. Marling, *I.E.E.E.J. Quantum Electronics*, 1978, **14**, 4.
[105] T. H. McGee and R. E. Weston, jun., *J. Chem. Phys.*, 1978, **68**, 1736.
[106] K. Hara and D. Phillips, *J.C.S. Faraday II*, 1978, **74**, 1441.
[107] S. J. Silvers and M. R. McKeever, *Chem. Phys.*, 1979, **27**, 27.
[108] D. Phillips and R. P. Steer, *J. Chem. Phys.*, 1977, **67**, 4780.

data are best explained by a reversible intersystem crossing to the T_2 or T_1 levels of Cl_2CS, which have vibronic level densities of 2×10^2 and 3×10^3, respectively, isoenergetic with the zero-point level of S_2. Recurrence times calculated from these densities are of the same order of magnitude as the estimated radiative lifetime of S_2 of 17 ns. The decay times measured in these experiments support the revised absorption band assignments proposed earlier by the same group. The longest decays are observed for transitions which populate either the zero-point level, e 35, 112 cm^{-1} above S_0, or e $+147$ cm^{-1}. Other transitions populate levels which decay faster, and therefore must produce higher energy final states. The excitation spectrum of thiophosgene has been further analysed.[109]

Fluorescence from the radical species CH_3S has been observed following photodissociation of CH_3SSCH_3 by photons of wavelengths 147 and 200 nm.[110] The threshold for the production of the radical species in its excited state was 202 ± 3 nm. The decay time was estimated as ~ 2 μs, and quenching parameters for N_2, H_2, D_2, CH_4, and CH_3SSCH_3 collision partners were reported. Isotope effects in the photolysis of methanethiol,[111] and the mercury-photosensitized decomposition of thietane[112] have been discussed.

6 Nitrogen-containing Compounds

The excitation and u.v. spectra of N_2, CO, HCN, C_2N_2, CO_2, and N_2O have been reported.[113] HCN and related cyanogen halides have proved to be fruitful molecules for the study of energy partitioning in product fragments. Several theoretical papers,[114—119] including a review article,[117] have appeared on the subject; Table 5 shows some of the results for populations in the $CN(\tilde{B}^2\Sigma)$ state produced upon photodissociation of HCN, DCN at 123.6,[120] 116.5,[120] and 121.6[121] nm. Although the trends in the two studies are similar, exact agreement is not realized. Reactions of $CN(v)$ with $O(^3P)$ have been studied and rotational energy distributions in $CN(\tilde{B}^2\Sigma^+)$ from HCN, ClCN, BrCN, and ICN photolysis under collision-free conditions and at high resolution over a range of wavelengths in the vacuum u.v. region have been obtained.[123] Results were rationalized in terms of Franck–Condon considerations, the constraints introduced by geometry changes, and requirements of energy and angular momentum conservation

[109] P. F. Bernath, P. G. Cummins, J. R. Lombardi, and R. W. Field, *J. Mol. Spectroscopy*, 1978, **69**, 166.
[110] K. Ohbayashi, H. Akimoto, and I. Tanaka, *Chem. Phys. Letters*, 1977, **52**, 47.
[111] D. Kamra and J. M. White, *J. Photochem.*, 1977, **7**, 171.
[112] D. R. Dice and R. P. Steer, *Canad. J. Chem.*, 1978, **56**, 114.
[113] L. Asbrink, C. Fridh, and E. Lindholm, *Chem. Phys.*, 1978, **27**, 159; C. Fridh, L. Asbrink, and E. Lindholm, *ibid.*, p. 169.
[114] Y. B. Band and K. F. Freed, *J. Chem. Phys.*, 1977, **67**, 1462.
[115] M. D. Morse, K. F. Freed, and Y. B. Band, *Chem. Phys. Letters*, 1977, **49**, 399.
[116] M. S. Child and R. Lefebvre, *Chem. Phys. Letters*, 1978, **55**, 213.
[117] O. Atabek and R. Lefebvre, *Chem. Phys. Letters*, 1977, **52**, 29.
[118] E. J. Heller, *J. Chem. Phys.*, 1978, **68**, 3891.
[119] W. M. Gelbart, *Ann. Rev. Phys. Chem.*, 1977, **28**, 323.
[120] I. Stein and A. Gedanken, *J. Chem. Phys.*, 1978, **68**, 2982.
[121] S. Tatematsu and K. Kuchitsu, *Bull. Chem. Soc. Japan*, 1977, **56**, 2896.
[122] K. J. Schmatjko and J. Wolfrum, *Ber. Bunsengesellschaft. Phys. Chem.*, 1978, **82**, 419.
[123] M. N. R. Ashfold and J. P. Simons, *J.C.S. Faraday II*, 1978, **74**, 280.

Table 5 *Product state distributions in* HCN, DCN *photolysis*

CN($\tilde{B}$)	Molecule photolysed		
vibrational level	HCN	DCN	*Ref.*
0	(1.00)	(1.00)	120, 121
1	0.54	0.56	121
1	0.90	0.93	120
2	0.20	0.28	121
2	0.48	0.68	120
3	0.10	0.17	121
4	0.06	0.09	121

following photodissociation.[123—125] Similar studies on H_2O and D_2O have been carried out.[126]

An *ab initio* study on the photolysis of NH_3 has been reported.[127] Electron impact spectroscopy has not revealed any low-lying triplet electronic states of ammonia.[128] A mechanism for the photolysis of NH_3, which predissociates from the $\tilde{A}^1A_2''$ state in which NH_2 fragments are formed in competition in the $\tilde{A}^2A_1$ and $\tilde{X}^2B_1$ states[129] has been challenged on the basis of an excitation study.[130] The zero pressure lifetime of NH_2 $\tilde{A}(^2A_1)$ has been estimated to be 15 μs,[131] with a rate-constant for quenching by NH_3 of $1.6 \pm 0.3 \times 10^{-10}$ cm^3 molecule^{-1} s^{-1}.[131] Details of the reaction of NH_2 with O_2 have been given.[132] The photolysis of HN_3 at 266 nm gives the product HN principally in the $\tilde{a}^1\Delta$, $v = 0$ state with 900 cm^{-1} rotational energy, $\leqslant 5000$ cm^{-1} translational energy.[133] The rate constant for the new chemiluminescent reaction (63) was measured as 9.25×10^{-11} cm^3 molecule^{-1} s^{-1}. Reactions of HN($^1\Delta$) with HCl and hydrocarbons were also reported.[134]

$$NH(^1\Delta) + HN_3 \longrightarrow NH_2(^2A_1) + N_3 \qquad (63)$$

Emission from complexes between electronically excited triplet mercury and simple amines, including ammonia, has been known for some time, but has recently been investigated at low pressures.[135] Under these conditions two emission peaks are seen, *e.g.* in 0.005 Torr ND_3, peaks at 305 and 340 nm are observed.[135] At high pressures, only the long wavelength emission is observable, and this clearly corresponds to a relaxed emitting state. A similar situation is

124 M. N. R. Ashfold, M. T. Macpherson, and J. P. Simons, *Chem. Phys. Letters*, 1978, **55**, 84.
125 M. N. R. Ashfold and J. P. Simons, *J.C.S. Faraday II*, 1977, **73**, 858.
126 M. T. McPherson and J. P. Simons, *Chem. Phys. Letters*, 1977, **51**, 261.
127 R. Runau, S. D. Peyerimhoff, and R. J. Buenker, *J. Mol. Spectroscopy*, 1977, **68**, 253.
128 K. E. Johnson and S. Lipsky, *J. Chem. Phys.*, 1977, **66**, 4719.
129 R. A. Back and S. Koda, *Canad. J. Chem.*, 1977, **55**, 1387.
130 G. D. Stefano, M. Lenzi, A. Margani, and C. Nguyen Xuan, *J. Chem. Phys.*, 1977, **67**, 3832.
131 S. Koda, *Bull. Chem. Soc. Japan*, 1977, **50**, 1683.
132 R. Lesclaux and M. Demissy, *Nouveau J. Chim.*, 1977, **1**, 443.
133 J. R. McDonald, R. G. Miller, and A. P. Baronavski, *Chem. Phys. Letters*, 1977, **51**, 57.
134 A. P. Baronavski, R. G. Miller, and J. R. McDonald, *Chem. Phys.*, 1978, **30**, 119; J. R. McDonald, R. G. Miller, and A. P. Baronavski, *ibid.*, p. 133.
135 T. Hikida, T. Ishihara, and Y. Mori, *Chem. Phys. Letters*, 1977, **52**, 43; A. B. Callear and C. G. Freeman, *Chem. Phys.*, 1977, **23**, 343.

observed with the mercury–t-butylamine system, for which experimental results were explained on the basis of the scheme[136,137] shown in reactions (64)—(70).

$$Hg(^3P_0) + A \longrightarrow HgA\dagger \quad (64)$$

$$HgA\dagger \longrightarrow HgH + R\cdot \quad (65)$$

$$HgA\dagger \longrightarrow h\nu + Hg + A \quad (66)$$

$$HgA\dagger + M \longrightarrow HgA^* + M \quad (67)$$

$$HgA^* \longrightarrow Hg + A + h\nu \quad (68)$$

$$HgA^* + A \rightleftharpoons HgA_2^* \quad (69)$$

$$HgA_2^* \longrightarrow h\nu + Hg + 2A \quad (70)$$

The short wavelength emission observed arises from a vibrationally excited complex HgA† through reaction (66); the lifetime of this species being of the order of 1 ns. The relaxed state, HgA* has a decay time of ~2.75 µs, and mainly decays by emission with $\lambda_{max} = 370$ nm. The removal of the Hg($^3P_0$) was found to occur solely *via* reaction (64), with a rate constant of $k_{64} = 3.9 \times 10^{-10}$ cm^3 molecule^{-1} s^{-1}.

Although these complexes have been well documented for several years, and exciplex emission between amine–aromatic partners is similarly well known in solution; no reports of exciplex emission in the *vapour phase* in which the excited partner is a complex polyatomic molecule had appeared until very recently. Now, however, exciplex fluorescence at 515 and 490 nm has been seen for the 9-cyanoanthracene (9-CNA)-triethylamine and 9-CNA-dimethylaniline (DMA) systems, respectively,[138,139] and the 9-CNA-tri-n-butylamine system shows similar emissions.[139] Exciplex fluorescence decay times were much shorter than in solution, *e.g.* the 9-CNA-DMA system had a decay time of 27.1 ns in the vapour phase at 231 °C compared with 60.2 ns in cyclohexane solution. Results were explained on the basis of Scheme 3. It was shown that in order to rationalize the higher exciplex emission efficiencies in the gas phase than in solution, yet the shorter decay times, the rate-constant for radiative decay of the exciplex k_{re} must be much greater in the gas phase than in solution, even in non-polar solvents.[138]

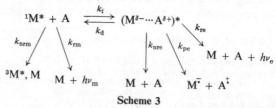

Scheme 3

The surprisingly strong fluorescence of tertiary amines in the vapour phase has been shown to be quenched through a dynamic mechanism by simple

[136] A. B. Callear, D. R. Kendall, and L. Krause, *Chem. Phys. Letters*, 1977, **49**, 29; *Chem. Phys.*, 1978, **29**, 415.
[137] T. Hikida, M. Santoku, and Y. Mori, *Chem. Phys. Letters*, 1978, **55**, 280.
[138] S. Hirayama and D. Phillips, *J.C.S. Faraday II*, 1978, **74**, 2035.
[139] S. Hirayama, G. D. Abbott, and D. Phillips, *Chem. Phys. Letters*, 1978, **56**, 497.

fluorocarbons, usually thought of as inert.[140] Thus the gas phase rate constant for quenching of triethylamine by perfluoromethylcyclohexane was 2.5×10^{11} l mol^{-1} s^{-1}, and in cyclohexane solution the corresponding value was 1.3×10^{10} l mol^{-1} s^{-1}. Charge transfer interactions from amine to halocarbon is a likely quenching mechanism, and although *ground-state* complex formation is a possibility in these systems, there was no spectroscopic evidence for strong interactions between ground states. Similar results were obtained for the quenching of aromatic amines by carbon tetrachloride in solution,[141] with quenching rate constants of 1.4×10^{10} and 9.7×10^{9} l mol^{-1} s^{-1} being measured for diphenylamine in methanol, and n-hexane, respectively. Charge transfer was proven in this case by the observation of the $Ph_2\overset{\bullet}{N}H^+$ ion in absorption.

The photolysis of CH_2N_2 was mentioned earlier.[94b] In the photochemistry of the azoalkanes: *trans*-azomethane, *trans*-azoethane, *trans*-azo-n-propane, and *trans*-azo-isopropane at high pressures of up to 140 atmospheres, the simple mechanism of reactions (71)—(73) was found to be inadequate to explain the observed dissociation.[142] *Two* reactive channels were required to rationalize the

$$A \xrightarrow{\;h\nu\;} A^* \tag{71}$$

$$A^* + M \longrightarrow A + M \tag{72}$$

$$A^* \longrightarrow \text{products} \tag{73}$$

data, the slower of the two being associated with the triplet state excited azoalkane, the faster with the singlet manifold. Rate constants for the slow decomposition were reported as 4×10^{10}, 6×10^{9}, 4×10^{7}, and 5×10^{8} s^{-1}, respectively. Vibrationally excited NO_2 production from the photolysis of CH_3NO_2 and $C_3H_7NO_2$ has been observed,[143] and a stochastic model of vibrational relaxation in the excited singlet state of CF_3NO (a blue gas) successfully fitted to fluorescence decay data.[144] The photolysis of 3-methyl-3-chlorodiazirine has been studied.[145]

Although aza-aromatics have been known for some years to fluoresce, no fluorescence from pyridine has been observed until recently.[146] Since the observed emission with $\bar{\nu}_{max}$ at 30 000 cm^{-1} had a maximum quantum yield of only 10^{-4} for excitation in the $n \to \pi^*$ zero–zero band, falling to 3×10^{-6} at the onset of $S_0 \to S_2$ ($\pi \to \pi^*$) absorption, it is not surprising that the fluorescence was overlooked until single-photon-counting methods were employed. A normal deuterium isotope effect was noted in that emission from perdeuterio pyridine was of higher yield by a factor of approximately 1.5. Other studies on aza-aromatics have included the study of non-radiative processes in pyridine,

140 P. C. Alford, C. G. Cureton, R. A. Lampert, and D. Phillips, *Chem. Phys. Letters*, 1978, **55**, 100.
141 T. Iwasaki, T. Sawada, M. Okuyama, and H. Kamada, *J. Phys. Chem.*, 1978, **82**, 371.
142 S. Chervinsky and I. Oref, *J. Phys. Chem.*, 1977, **81**, 1967.
143 K. G. Spears and S. P. Brugge, *Chem. Phys. Letters*, 1978, **54**, 373.
144 K. G. Spears, *Chem. Phys. Letters*, 1978, **54**, 139.
145 H. M. Frey and D. E. Penny, *J.C.S. Faraday I*, 1977, **73**, 2010.
146 I. Yamazaki and H. Baba, *J. Chem. Phys.*, 1977, **66**, 5826.

pyrazine, and pyramidine by energy transfer to biacetyl,[147] the dual exponential decay of pyrimidine in the $^1B_1$ state,[148] the ps excitation of acridine and phenazine and build-up of triplet–triplet absorption,[149] and the pressure-dependence of triplet formation in vapours of acridine and phenazine.[150] A scheme involving reversible intersystem crossing is required to fit the data, and rate constants from one of the studies[148] are indicated in reactions (74)—(79).

$$P + h\nu \longrightarrow {}^1P^* \tag{74}$$

$$^1P^* \longrightarrow P + h\nu_f \quad k = 4.4 \times 10^6 \, s^{-1} \tag{75}$$

$$^1P^* \longrightarrow \text{products (irreversible)} \tag{76}$$

$$^1P^* \longrightarrow {}^3P^* \quad k = 3.7 \times 10^8 \, s^{-1} \tag{77}$$

$$^3P^* \longrightarrow {}^1P^* \quad k = 1.1 \times 10^8 \, s^{-1} \tag{78}$$

$$^3P^* + M \longrightarrow {}^3P + M \quad k = 9.5 \times 10^7 \, Torr^{-1} \, s^{-1} \tag{79}$$

Elegant studies on the spectroscopy of s-tetrazine and its van der Waals complexes with He, H_2, and Ar have been reported,[151, 152] and the role of van der Waals molecules in vibrational relaxation has been discussed.[153]

7 Halogenated Compounds

Ground-state reactions of halogen atoms cannot be discussed here, but recent studies include the following subjects: oscillator strengths for the $F[(3s)^2 P_J \leftarrow (2p)^2 P_J]$ transition,[154] the F + tetrafluoroethylene,[155] F + HBr,[156] and $F + O_2 + M$ reactions;[157] the $Cl[(4s)^2 D_J \leftarrow (3p)(^2 P_J)]$ transition oscillator strength,[154] reactions of Cl with H_2,[158] Br_2,[159] CH_4,[160] C_2H_4,[161,162] NH_3,[163] H_2O_2,[164] $ClONO_2$,[165] and other substrates,[166] C_2H_2[167] and the Cl atom-

147 K. Aizawa, H. Igarashi, and K. Kaya, *Chem. Phys.*, 1977, **23**, 273.
148 H. Reineccius and H. Von Weyssenhoff, *Chem. Phys. Letters*, 1977, **52**, 34.
149 Y. Hirata and I. Tanaka, *Chem. Phys.*, 1977, **25**, 381.
150 J. L. Baptista, S. J. Formosinho, and M. de Fatima Leitao, *Chem. Phys.*, 1978, **28**, 425.
151 R. E. Smalley, L. Wharton, D. H. Levy, and D. W. Chandler, *J. Chem. Phys.*, 1978, **68**, 2487.
152 R. E. Smalley, L. Wharton, D. H. Levy, and D. W. Chandler, *J. Mol. Spectroscopy*, 1977, **66**, 375.
153 G. Ewing, *Chem. Phys.*, 1978, **29**, 253.
154 M. A. A. Clyne and W. S. Nip, *J.C.S. Faraday II*, 1977, **73**, 1308.
155 N. I. Buktovskaia, M. N. Larichev, I. O. Leipunskii, I. I. Morozov, and V. L. Talroze, *Doklady Akad. Nauk S.S.S.R.*, 1978, **240**, 366.
156 J. P. Sung and D. W. Setser, *Chem. Phys. Letters*, 1977, **48**, 413.
157 H.-L. Chen, D. W. Trainor, R. E. Center, and W. I. Fyfe, *J. Chem. Phys.*, 1977, **66**, 5513.
158 J. H. Lee, J. V. Michael, W. A. Payne, L. J. Stief, and D. A. Whytock, *J.C.S. Faraday I*, 1977, **73**, 1530.
159 J. J. Valentini, Y. T. Lee, and D. J. Anerbach, *J. Chem. Phys.*, 1977, **67**, 4866.
160 M. S. Zahniser, B. M. Berquist, and F. Kaufman, *Internat. J. Chem. Kinetics*, 1978, **10**, 15.
161 D. J. Stevens and L. D. Spicer, *J. Phys. Chem.*, 1977, **81**, 1217.
162 F. S. C. Lee and F. S. Rowland, *J. Phys. Chem.*, 1977, **81**, 1222, 1229, and 1235.
163 A. A. Westenberg and N. de Haas, *J. Chem. Phys.*, 1977, **67**, 2388.
164 J. V. Michael, D. A. Whytock, J. H. Lee, W. A. Payne, and L. J. Stief, *J. Chem. Phys.*, 1977, **67**, 3533.
165 M. J. Kurylo and R. G. Manning, *Chem. Phys. Letters*, 1977, **48**, 279.
166 J. V. Michael and J. H. Lee, *Chem. Phys. Letters*, 1977, **51**, 303.
167 G. Poulet, G. Le Bras and J. Combourieni, *J. Phys. Chem.*, 1977, **81**, 2303.

sensitized oxidation of $HCCl_3$, HCF_2Cl, and HCF_3;[168] Br atom recombinations,[169,170] reactions of Br with O_3.[171] $OClO$,[172] bromine atom resonance radiation and atom concentration measurements;[173a] iodine atom recombinations[170] and reactions of I atoms with olefins.[173b]

Table 6 *Reactions of excited halogen atoms*

Excited atom	Substrate	Rate constant cm^3 molecule^{-1} s^{-1}	Ref.
$Cl(^2P_{\frac{1}{2}})$	He	3.8×10^{-15}	172
$Cl(^2P_{\frac{1}{2}})$	Ne	4.0×10^{-14}	172
$Cl(^2P_{\frac{1}{2}})$	Ar	1.1×10^{-12}	172
$Cl(^2P_{\frac{1}{2}})$	Kr	1.4×10^{-12}	172
$Cl(^2P_{\frac{1}{2}})$	Xe	1.8×10^{-11}	172
$Cl(^2P_{\frac{1}{2}})$	CCl_4	2.0×10^{-10}	173
$Cl(^2P_{\frac{1}{2}})$	Cl_2	4.5×10^{-11}	173
$Cl(^2P_{\frac{1}{2}})$	O_2	2.1×10^{-11}	173
$Br(^2P_{\frac{1}{2}})$	H_2O	6.2×10^{-11}	174
$Br(^2P_{\frac{1}{2}})$	Br_2	9.3×10^{-13}	174
$I(^2P_{\frac{1}{2}})$	HBr	1.5×10^{-13}	175
$I(^2P_{\frac{1}{2}})$	HBr	1.1×10^{-13}	176
$I(^2P_{\frac{1}{2}})$	DBr	3.1×10^{-14}	175
$I(^2P_{\frac{1}{2}})$	HCl	1.4×10^{-14}	176
$I(^2P_{\frac{1}{2}})$	NO	1.2×10^{-13}	176
$I(^2P_{\frac{1}{2}})$	CH_4	9.7×10^{-14}	175
$I(^2P_{\frac{1}{2}})$	C_3H_8	1.6×10^{-13}	175
$I(^2P_{\frac{1}{2}})$	C_2D_4	3.1×10^{-15}	178
$I(^2P_{\frac{1}{2}})$	C_3D_8	4.6×10^{-15}	178
$I(^2P_{\frac{1}{2}})$	CD_3I	4.2×10^{-15}	178

Table 6 gives some *selected* rate constant data for reactions of electronically excited halogen atoms with various substrates, including $Cl(^2P_{\frac{1}{2}})$ with noble gases,[173c] and other substrates,[173d] $Br(^2P_{\frac{1}{2}})$ with water and Br_2,[174] $I(^2P_{\frac{1}{2}})$ with HBr,[175-177] CH_4 and C_3H_8[175] and deuteriated alkenes and CD_3I;[178] $Br(^2P_{\frac{1}{2}})$ deactivation by CO_2, COS, CS_2,[179] $Br(^2P_{\frac{1}{2}})$ with interhalogens,[180] and with CH_3F.[181] A notable feature of the reaction of $Br(^2P_{\frac{1}{2}})$ with H_2O[174] is the remarkable high mode specificity of the vibrational excitation of the water resulting, since all of the available energy was found in the 100 or 001 states of

168 E. Sanhueza, *J. Photochem.*, 1977, **7**, 325.
169 H. Hippler, S. H. Luu, H. Teitelbaum, and J. Troe, *Internat. J. Chem. Kinetics*, 1978, **10**, 155.
170 N. S. Snider and D. G. Leaist, *J. Phys. Chem.*, 1977, **81**, 1033.
171 M.-T. Leu and W. B. DeMore, *Chem. Phys. Letters*, 1977, **48**, 317.
172 M. A. A. Clyne and R. T. Watson, *J.C.S. Faraday I*, 1977, **73**, 1169.
173 (a) M. A. A. Clyne and D. J. Smith, *J.C.S. Faraday II*, 1978, **74**, 263; (b) K.-C. To, M. E. Berg, and E. P. Rack, *J. Phys. Chem.*, 1977, **81**, 1239; (c) I. S. Fletcher and D. Husain, *J.C.S. Faraday II*, 1978, **74**, 203; (d) I. S. Fletcher and D. Husain, *Chem. Phys. Letters*, 1977, **49**, 516.
174 A. Hariri and C. Wittig, *J. Chem. Phys.*, 1978, **68**, 2109.
175 J. R. Wiesenfeld and G. L. Wolk, *J. Chem. Phys.*, 1977, **67**, 509.
176 A. J. Grimley and P. L. Houston, *J. Chem. Phys.*, 1978, **68**, 3366.
177 C. Fotakis and R. J. Donovan, *Chem. Phys. Letters*, 1978, **54**, 91.
178 R. J. Donovan, H. M. Gillespie, and R. H. Strain, *J.C.S. Faraday II*, 1977, **73**, 1553.
179 A. Hariri and C. Wittig, *J. Chem. Phys.*, 1977, **67**, 4454.
180 H. Hofmann and S. R. Leone, *Chem. Phys. Letters*, 1978, **54**, 314.
181 L. N. Krasnopyorov and V. N. Panfilov, *Chem. Phys.*, 1977, **25**, 375.

H_2O. An important paper has pointed out the difficulties in measuring the lifetimes of atoms such $I(^2P_{\frac{1}{2}})$ due to radiation imprisonment, and the authors conclude that values of τ_r published for this species are unreliable.[182] A study of reaction (80) has shown that 50% of the iodine atoms are formed in the excited electronic state.[183] Calculations have been carried out on the emission of

$$F + HI \longrightarrow HF + I(^2P_{\frac{3}{2}}) \text{ or } I^*(^2P_{\frac{1}{2}}) \qquad (80)$$

$I(^2P_{\frac{1}{2}})$ in the presence of argon and xenon.[184]

The gas-phase photodissociation spectrum of $Cl_2{}^-$ arises from the broad $^2\Sigma_g{}^+ \leftarrow {}^2\Sigma_u{}^+$ transition centred at 350 nm, but extending from 220 to 700 nm.[184] It has been shown that upon excitation by electron impact, $Cl_2{}^-$ is formed in a vibrationally excited state.[185] Resonance fluorescence from Cl_2 excited by single mode lasers has been recorded,[186, 187] and the excitation of Cl_2 in active nitrogen reported.[188]

Relative absorption cross-sections for the excitations $\tilde{B}^3\Pi O_u{}^+ \leftarrow \tilde{X}^1\Sigma_g{}^+$ and $^1\Pi_{1u} \leftarrow \tilde{X}^1\Sigma_g{}^+$ in Br_2 at 18 350, 21 010, and 22 125 cm^{-1} have been reported.[189] U.v. laser transitions in Br_2 at 290 nm and I_2 at 340 nm have been assigned as due to the 1432 $2g^3\Pi \rightarrow$ 2431 $2u^3\Pi$ transition, contradicting all previous assignments.[190] The $\tilde{B} \leftarrow X(0,0)$ transition in I_2, despite its very small Franck Condon factors, has been observed recently,[191] and the energy levels of this molecule near the $\tilde{B}$ state dissociation limit have been probed.[192] Non-linear photodissociation,[193] laser-induced fluorescence,[194] and high resolution time-resolved spectroscopy in effusive beams of I_2 [195] have been studied. In the last study it was found that the optical dephasing time in a 10^{-4} cm^{-1} molecular packet is related directly to the zero-pressure T_1 obtained from static Stern–Volmer measurements of I_2 in a bulb.

There have been many studies on the interhalogens during the past year. Using 0.04 cm^{-1} bandwidth excitation, fluorescence from the $v' = 8$ and $v' = 8$ levels of ^{79}BrF and ^{81}BrF in the $\tilde{B}^3\Pi(O^+)$ state has been observed.[196] Rotational levels lower than $J' = 23$ of $v' = 7$ were stable with an extrapolated zero pressure lifetime of 22.4 ± 3.3 μs. Predissociation occurred at $J' = 28$, with decay time shortening down to 370 ± 10 ns for $J' = 34$. All rotational levels of $v' = 8$ were predissociated, with lifetimes shortening systematically

[182] D. Husain, N. K. H. Slater, and J. R. Wiesenfeld, *Chem. Phys. Letters*, 1977, **51**, 201.
[183] I. Burak and M. Eyal, *Chem. Phys. Letters*, 1977, **52**, 534.
[184] (a) M. Krauss and P. S. Julienne, *J. Chem. Phys.*, 1977, **67**, 669; (b) S. A. Sullivan, B. S. Freiser, and J. L. Beauchamp, *Chem. Phys. Letters*, 1977, **48**, 294.
[185] O. I. Asubiojo, H. L. McPeters, W. N. Olmstead, and J. J. Brauman, *Chem. Phys. Letters*, 1977, **48**, 127.
[186] J. S. Choe, H. Chang, and D. M. Hwang, *J. Chinese Chem. Soc.*, 1977, **24**, 1.
[187] D.-M. Hwang and H. Chang, *J. Mol. Spectroscopy*, 1978, **69**, 11.
[188] G. Black and I. M. Campbell, *J. Photochem.*, 1978, **8**, 49.
[189] T. G. Lindeman and J. R. Wiesenfeld, *Chem. Phys. Letters*, 1977, **50**, 364.
[190] J. Tellinghuisen, *Chem. Phys. Letters*, 1977, **49**, 485.
[191] R. E. Smalley, L. Wharton, and D. H. Levy, *Chem. Phys. Letters*, 1977, **51**, 392.
[192] M. D. Danyluk and G. W. King, *Chem. Phys.*, 1977, **25**, 343.
[193] C. Tai and F. W. Dalby, *Canad. J. Phys.*, 1978, **56**, 183.
[194] R. Clark and A. J. McCaffery, *Mol. Phys.*, 1978, **35**, 609, 617.
[195] A. H. Zewail, T. E. Orlowski, R. R. Shah, and K. E. Jones, *Chem. Phys. Letters*, 1977, **49**, 520.
[196] M. A. A. Clyne and I. S. McDermid, *J.C.S. Faraday II*, 1978, **68**, 644.

from 1.74 μs for $J' = 1$—3 down to 110 ns for $J' = 28$. Non-exponential decay curves were observed for J' levels of the $v' = 7$ state just above or just below the onset of predissociation, due to collisional upward and downward rotational energy transfer. Results showed that multiquantum $\Delta J \leqslant 10$ transfers occurred with high probability. The quenching of the $J' = 17, 20$ levels of BrF $\tilde{B}^3\Pi(O^+)$ $v' = 6$ by BrF, Br_2, and He had rate constants of 2.6×10^{-10}, 1.5×10^{-10}, and $\leqslant 2 \times 10^{-12}$ cm³ molecule⁻¹ s⁻¹, respectively.[197]

In BrCl $\tilde{B}^3\Pi(O^+)$, all rotational levels of $v' = 7$ are predissociated,[198] with lifetimes varying systematically from 910 ns for $J' = 7$ down to 240 ns for $J' = 16$. In $I^{35}Cl$ $\tilde{B}^3\Pi(O^+)$ the decay time for $v' = 1,2$ at various J' levels was 520 ns, and the second-order rate-constant for deactivation by the ICl ground-state was measured as $2.1 \pm 0.4 \times 10^{-10}$ cm³ molecule⁻¹ s⁻¹.[197] The rate constant for reaction (80a) was given as 9×10^{-14} cm³ molecule⁻¹ s⁻¹.[199]

$$ICl(\tilde{A}^3\Pi_i) + H_2 \longrightarrow HCl + HI \qquad (80a)$$

The decay time of IF $\tilde{B}^3\Pi(O^+)$ $v' = 9$ lies between the limits of 10 and 3 μs,[197] whereas that for the corresponding state of IBr $v' = 2, 3$, or 4 is 540 ns.[200] The spectroscopy of the IBr $(\tilde{B}^3\Pi(O^+) \rightarrow \tilde{X}^1\Sigma^+)$ system has been reported.[201] The enrichment of chlorine isotopes by selective photoaddition of ICl to acetylene will be discussed later.[202]

Emission from rare-gas halides is of interest since this forms the basis of the rare gas 'excimer' lasers.[203, 204] The radiative lifetime of the excited state of Ar_2F has been given as 180 ns,[205] and the rate constant for the quenching of this state by F_2 as 2.1×10^{-10} cm³ molecule⁻¹ s⁻¹.[205] Two values of the zero-pressure decay time of KrF* have been given, viz. 9 ± 0.5 ns[206] and 6.8 ± 0.2 ns,[207] with the latter being considered the more reliable. Quenching of the $\tilde{B}$ state of KrF by Ar, F_2, KrF, and (termolecular) 2Ar has the rate constants shown in Table 7:[206—208] the closely similar results for quenching of XeF* are also given.[209] The flash photolysis of XeF and vibrational analysis of the $\tilde{B} \leftrightarrow \tilde{X}$ transition[210] and studies on reaction (81)[211] have been reported. A computer simulation of reactive collisions of Kr $5s(\frac{3}{2})_2$ and Kr $5s(\frac{3}{2})_1$ atoms

$$Xe(^3P_{0,2}) + Br_2 \longrightarrow XeBr* + Br \qquad (81)$$

leading to KrF* emission has been carried out.[212]

197 M. A. A. Clyne and I. S. McDermid, *J.C.S. Faraday II*, 1977, **73**, 1094.
198 M. A. A. Clyne and I. S. McDermid, *J.C.S. Faraday II*, 1978, **74**, 798, 807.
199 S. J. Harris, *J. Amer. Chem. Soc.*,1977, **99**, 5798.
200 J. J. Wright and M. D. Havey, *J. Chem. Phys.*, 1978, **68**, 864.
201 E. M. Weinstock and A. Preston, *J. Mol. Spectroscopy*, 1978, **70**, 188.
202 M. Stuke and F. P. Schafer, *Chem. Phys. Letters*, 1977, **48**, 271.
203 M. Krauss, *J. Chem. Phys.*, 1977, **67**, 1712.
204 J. J. Ewing, *Phys. Today*, 1978, **31**, 32.
205 C. H. Chen and M. G. Payne, *Appl. Phys. Letters*, 1978, **32**, 358.
206 R. Burnham and S. K. Searles, *J. Chem. Phys.*, 1977, **67**, 5967.
207 J. G. Eden, R. W. Waynant, S. K. Searles, and R. Burnham, *Appl. Phys. Letters*, 1978, **32**, 733.
208 G. P. Quigley and W. M. Hughes, *Appl. Phys. Letters*, 1978, **32**, 627.
209 H. C. Brashers, jun., D. W. Setser, and D. Des Marteau, *Chem. Phys. Letters*, 1977, **48**, 84.
210 A. L. Smith and P. C. Kobrinsky, *J. Mol. Spectroscopy*, 1978, **69**, 1.
211 M. R. Levy, C. T. Rettner, and J. P. Simons, *Chem. Phys. Letters*, 1978, **54**, 120.
212 K. Tamagake and D. W. Setser, *J. Chem. Phys.*, 1977, **67**, 4370.

Table 7 *Rate constants for quenching of* KrF* $\tilde{B}$·, XeF*[206—209]

Collision partner	Excited state	Rate constant cm^3 molecule^{-1} s^{-1}	Ref.
Ar	KrF*	1.8×10^{-12}	206
Ar + Ar (termolecular)	KrF*	1.1×10^{-31a}	207
Kr + Kr (termolecular)	KrF*	2.9×10^{-31a}	208
F$_2$	KrF*	4.8×10^{-10}	206
F$_2$	KrF*	5.7×10^{-10}	208
KrF	KrF*	3.7×10^{-10}	207
N$_2$	XeF*	4.6×10^{-12}	209
Xe	XeF*	2.9×10^{-11}	209
CO$_2$	XeF*	1.6×10^{-10}	209
NF$_3$	XeF*	1.5×10^{-11}	209
F$_2$	XeF*	3.3×10^{-10}	209
Ar	XeF*	1.6×10^{-12}	209

^{a} Units of cm^6 molecule^{-2} s^{-1}

Table 8 *Relative efficiencies for production of* Br*($^2P_{\frac{1}{2}}$) *from photolysis of alkyl bromides*[214]

Compound	Φ
CH$_3$Br	0.2
C$_2$H$_5$Br	<0.1
CF$_3$Br	2.0
C$_2$F$_5$Br	0.9
n-C$_3$F$_7$Br	<0.1

$$\Phi = \frac{\text{Yield Br*}(^2P_{\frac{1}{2}})}{\text{Br}(^2P_{\frac{3}{2}})}$$

Photofragmentation in CH$_3$Cl$^+$ [7] and bromobenzene cations[52] was reported earlier. In the photolysis of CH$_3$I, CH$_3$Br, and CH$_3$Cl at 184.9 nm, and CH$_3$I at 253.7 nm, the quantum yield of methylene production was found to be less than 10^{-3} in all cases.[213] In the photodissociation of the alkyl bromides shown in Table 8, values for the production of Br*($^2P_{\frac{1}{2}}$) relative to Br($^2P_{\frac{3}{2}}$) were as shown.[214] The results show that alkyl bromides are relatively inefficient sources of excited bromine, and only CF$_3$Br and C$_2$F$_5$Br could find use in Br photochemical lasers. In the photolysis of neopentyl bromide at 478—564 K in the vapour phase, the major products were HBr and 1,1-dimethylcyclopropane formed by cyclization of intermediate radicals.[215] The photolysis of the following fluorocarbons has also been studied: CHFCl$_2$,[216] 1,1-dichloroethane at 147 nm,[217] 1,1,1-trichloroethane at 147 nm,[218] CF$_2$ClCH$_3$ and CH$_2$FCH$_2$Cl[219]

[213] G. A. Takacs and J. E. Willard, *J. Phys. Chem.*, 1977, **81**, 1343.
[214] W. L. Ebenstein, J. R. Wiesenfeld, and G. L. Wolk, *Chem. Phys. Letters*, 1978, **53**, 185.
[215] S. G. Bayliss, R. L. Failes, J. S. Shapiro, and E. S. Swinbourne, *J.C.S. Faraday I*, 1978, **74**, 776.
[216] R. E. Rebbert, S. G. Lias, and P. Ausloos, *J. Photochem.*, 1978, **8**, 17.
[217] A. W. Kirk and E. Tschuikow-roux, *Internal. J. Chem. Kinetics*, 1977, **9**, 619.
[218] D. Salomon, A. W. Kirk, and E. Tschuikow-roux, *J. Photochem.*, 1977, **7**, 345.
[219] T. Ichimura, A. W. Kirk, and E. Tschuikow-roux, *Internat. J. Chem. Kinetics*, 1977, **9**, 697, 743.

at 147 nm; 1,1,1-trifluoro-2-chloroethane,[220] and 1,2-fluorochloroethane and 1,1,1-difluorochloroethane at 123.6 nm.[221] The photolysis of the last-named compound at 123.6 nm was virtually unchanged from that at 147 nm, but the 1,2-fluorochloroethane had a much increased quantum yield ($\Phi = 0.41$) for the HCl elimination reaction (82) at 123.6 nm compared with 147 nm photolysis.[221]

$$CH_2FCH_2Cl + h\nu_{123.6} \longrightarrow C_2H_3F + HCl \qquad (82)$$

The laser-induced fluorescence from the carbenes CFCl and CCl$_2$ in the gas phase has been observed,[222] and radiative lifetimes measured as 644 ns and 3.81 μs respectively. The rate constant for quenching of CFCl was 5.5×10^{-12} cm^3 molecule^{-1} s^{-1}, independent of excitation wavelength, whereas that for CCl$_2$ varied between 2.5×10^{-11} for 515 nm excitation and 3.8×10^{-11} for 437 nm excitation.

8 Atom Reactions

Halogen atom reactions were considered in the preceding section.

Vacuum u.v. laser action from highly ionized noble gases has been observed,[223] and studies reported on the inelastic collision of metastable He and Hg atoms,[224] and the quenching of He(2^3S) and He$_2$($2^3\Sigma$) by N$_2$ and O$_2$.[225] Rate constants for the quenching of Ne($^3P_2$)[226] have been given, and the collision of Ar*($4s$) atoms with nitrogen shown to give rise to laser transitions at 357.7 nm *and* 337.1 nm (associated with the $\tilde{c}^3\Pi_u$ state of N$_2$).[226,227] The radiative decay of the 2^3P_2 and 2^3P_0 levels of argon[228] and rate constants for reaction of these states with H(^2S),[229] H$_2$,[230] PCl$_3$,[231] and Kr[232] have been studied. The reaction with PCl$_3$ gives rise to emission in the PCl ($\tilde{A}^3\Pi \rightarrow \tilde{X}^3\Sigma^-$) and PCl($\tilde{b}^1\Sigma^+ \rightarrow \tilde{X}^3\Sigma^-$) systems,[231] whereas that with Kr results in krypton resonance emission from the $5s[\frac{3}{2}]_1$ level at 123.6 nm, although the Kr $5p\,[\frac{3}{2}]_{2,1}$ states are the initial reaction products.[232]

Vacuum u.v. fluorescence from excited krypton and xenon has also been seen.[233-236] In one of these studies,[236] the following decay times were measured for the excited species shown: Kr $5s[\frac{3}{2}]_1$, $\tau = 3.18$ ns; Kr $5s[\frac{1}{2}]_1$, $\tau = 3.11$ ns;

[220] J. L. Jourdain, G. Lebras, and J. Combourieu, *J. Chimie Physique*, 1978, **75**, 318.
[221] T. Ichimura, A. W. Kirk, and E. Tschuikow-roux, *J. Phys. Chem.*, 1977, **81**, 2040.
[222] R. E. Huie, N. J. T. Long, and B. A. Thrush, *Chem. Phys. Letters*, 1977, **51**, 197.
[223] J. B. Marling and D. B. Lang, *Appl. Phys. Letters*, 1977, **31**, 181.
[224] V. A. Kartazaev, Yu. A. Piotrovsky, and Yu. A. Tolmachev, *Optika i Spektroskopiya*, 1977, **43**, 196.
[225] G. Myers and A. J. Cunningham, *J. Chem. Phys.*, 1977, **67**, 3352.
[226] J. M. Brom jun., J. H. Kolts, and D. W. Setser, *Chem. Phys. Letters*, 1978, **55**, 44; V. Puech, F. Collier, and P. Cottin, *J. Chem. Phys.*, 1977, **67**, 2887.
[227] J. Krenos and J. Bel. Bruno, *Chem. Phys. Letters*, 1977, **49**, 447.
[228] W. A. Davis and R. Marrus, *Phys. Rev. (A)*, 1977, **15**, 1963.
[229] M. A. A. Clyne, P. B. Monkhouse, and D. W. Setser, *Chem. Phys.*, 1978, **18**, 447.
[230] D. J. McKenney and R. N. Dubinsky, *Chem. Phys.*, 1977, **26**, 141.
[231] J. A. Coxon and M. A. Wickramaaratchi, *J. Mol. Spectroscopy*, 1977, **68**, 372.
[232] L. G. Piper, *J. Chem. Phys.*, 1977, **67**, 1795.
[233] R. Brodmann and G. Zimmerer, *J. Phys. (B)*, 1977, **10**, 3395.
[234] E. Matthias, M. G. White, E. D. Poliakoff, R. A. Rosenberg, S.-T. Lee, and D. A. Shirley, *Chem. Phys. Letters*, 1978, **54**, 43.
[235] M. Chenevier, N. Sadeghi, and J. C. Pebay-Peyroula, *J. Phys. Letters*, 1978, **39**, L105.
[236] E. Matthias, R. A. Rosenberg, E. D. Poliakoff, M. G. White, S.-T. Lee, and D. A. Shirley, *Chem. Phys. Letters*, 1977, **52**, 239.

Xe $6s[\frac{3}{2}]_1$, $\tau = 3.46$ ns; Xe$[\frac{1}{2}]_1$, $\tau = 3.44$ ns; Xe $5p^5[\frac{3}{2}]5d[\frac{3}{2}]J = 1$, $\tau = 1.40$ ns (note *jl* coupling notation is used here and elsewhere in this Chapter). Energy transfer in reactions of Xe *n,f* Rydberg atoms with NH$_3$,[237] molecular beam chemiluminescence from the reaction of tetramethyldioxetan collisionally dissociated by a fast Xe beam,[238] and emission from XeOH exciplexes[239] have been reported.

A theoretical study of the effect of bandwidth of laser radiation upon two-photon absorption cross-sections has been checked against the experimental data for the Na atom,[240] and the ultra-high resolution two-photon spectrum of this species has been discussed.[241] Ionization of sodium vapour has also been investigated.[242, 243] Other studies include the CW laser excitation of Na atomic fluorescence,[244] the relaxation of Na $4D_{\frac{3}{2}}$ and $4D_{\frac{5}{2}}$ by noble gases,[245] and energy transfer between Na* and NO[246] and Na$_2$.[247] The rate constant for (83) was measured as 3.4×10^{-9} cm^3 molecule^{-1} s^{-1}, and thus corresponds to a long-range process.

$$\text{Na}(3p) + \text{Na}_2(\tilde{X}\,^1\Sigma) \longrightarrow \text{Na}(3s) + \text{Na}_2{}^*\,(\tilde{A}\,^1\Sigma)$$
$$\text{and Na}_2{}^*\,(\tilde{a}\,^3\Pi) \qquad (83)$$

There have been many studies on chemiluminescence and from atom plus molecule reactions, often in crossed molecular beams, and probed by laser-induced fluorescence, a technique which has been reviewed recently. Thus Ca and Sr atoms in reaction with HF and DF, $v' = 1$ give rise to CaF and SrF whose internal energy states were probed by laser-induced fluorescence.[248] Chemiluminescence arising from the reactions of Ca(^3P) with N$_2$O,[249] NaCl,[250] Sr(^3P) with Cl$_2$,[251] Ba(^3P) with N$_2$O,[252] and in Ba–O$_2$ flames[253] has been studied. Possible chemiluminescent species in the last system are shown in reactions (84)—(89). The state of BaO* formed in reaction (85) is either the $\tilde{A}^1\Sigma^+$ state or a reservoir state.[253] Other chemiluminescence studies have been reported,[254, 255] including that from Al + O$_3$,[256] Al + O$_2$,[257] and Pb[258] and

237 K. A. Smith, F. G. Kellert, R. D. Rundel, F. B. Danning, and R. F. Stevvings, *Phys. Rev. Letters*, 1978, **40**, 1362.
238 J. C. Brown and M. Menzinger, *Chem. Phys. Letters*, 1978, **54**, 235.
239 M. H. R. Hutchinson, *Chem. Phys. Letters*, 1978, **54**, 359.
240 B. R. Marx, J. Simons, and L. Allen, *J. Phys. (B)*, 1978, **11**, L273.
241 M. M. Salour, *Ann. Physics*, 1978, **111**, 364.
242 E. W. Rothe, B. P. Mathur, and G. P. Reck, *Chem. Phys. Letters*, 1977, **51**, 71.
243 R. M. Measures, *J. Appl. Phys.*, 1977, **48**, 2673.
244 B. Smith, J. D. Winefordner, and N. Omenetto, *J. Appl. Phys.*, 1977, **48**, 2676.
245 F. Biraben, K. Beroff, E. Giacobino, and G. Grynberg, *J. Phys. Letters*, 1978, **39**, L108.
246 J. A. Silver, N. C. Blais, and G. H. Kwei, *J. Chem. Phys.*, 1977, **67**, 839.
247 L. K. Lam, T. Fujimoto, A. C. Gallagher, and M. M. Hessel, *J. Chem. Phys.*, 1978, **68**, 3553.
248 Z. Karny and R. N. Zare, *J. Chem. Phys.*, 1978, **68**, 3360.
249 P. J. Dagdigan, *Chem. Phys. Letters*, 1978, **55**, 239.
250 P. J. Dagdigan, *Chem. Phys.*, 1977, **21**, 453.
251 A. Kowalski and J. Heldt, *Chem. Phys. Letters*, 1978, **54**, 240.
252 A. Siegel and A. Shultz, *Chem. Phys.*, 1978, **28**, 265.
253 S. A. Edelstein, B. E. Perry, D. J. Eckstrom, and T. F. Gallagher, *Chem. Phys. Letters*, 1977, **49**, 293.
254 R. C. Estler and R. N. Zare, *Chem. Phys.*, 1978, **28**, 253.
255 J. L. Gole, D. R. Preuss, and C. L. Chalek, *J. Chem. Phys.*, 1977, **67**, 850.
256 M. J. Sayers and J. L. Gole, *J. Chem. Phys.*, 1977, **67**, 5442.
257 L. Pasternack and P. J. Dagdigian, *J. Chem. Phys.*, 1977, **67**, 3854.
258 B. G. Wicke, S. P. Tang, and J. F. Friichtenicht, *Chem. Phys. Letters*, 1978, **53**, 304.

$$Ba + O_2 \longrightarrow BaO(\tilde{X}^1\Sigma^+) + O \tag{84}$$

$$O + Ba + Ar \longrightarrow BaO^* + Ar \tag{85}$$

$$BaO + O + Ar \longrightarrow BaO_2^* + Ar \tag{86}$$

$$BaO + Ba + Ar \longrightarrow Ba_2O^* + Ar \tag{87}$$

$$Ba + O + O \longrightarrow Ba^* + O_2 + Ar \tag{88}$$

$$BaO + O + O \longrightarrow BaO^* + O_2 + Ar \tag{89}$$

$Ge(^3P_0)$[259,260a] with N_2O, which procedes *via* reaction (90). For Pb, the PbO*

$$M(^3P) + N_2O \longrightarrow MO^* + N_2 \tag{90}$$

is in the $\tilde{B}$, $v' = 0$ state, whereas in the case of Ge, the $\tilde{a}^3\Sigma^+$ state of GeO was formed.[259] The rate constant for this process was 6.7×10^{-13} cm^3 molecule^{-1} s^{-1},[259] Metal oxide rather than metal hydroxide formation in the reactions of La, Se, and Y with the alcohols H_2O, D_2O, CH_3OH, C_2H_5OH, iso-C_3H_7OH, and n-C_3H_7OH has been observed through reactions analogous to (91).[260b] Ground-state reactions of $Bi(^4S_{\frac{3}{2}})$,[261] $Sb(^4S_{\frac{3}{2}})$,[261] $As(^4S_{\frac{3}{2}})$,[262] $P(^4S_{\frac{3}{2}})$,[263] and $Si(^3P_J)$[260a,264] have been studied using atomic resonance radiation in the vacuum u.v. Among excited state reactions studied are those of $Bi(6p^3\,^2D_{\frac{3}{2}})$[265]

$$M + H_2O \longrightarrow MO + H_2 \tag{91}$$

and $Bi(6p^3\,^2D_{\frac{5}{2}})$,[265] $Si(^1D_2)$,[266] and $Si(^1S_0)$,[267] $Ge(^1S_0)$,[268] $Pb(^3P_0)$,[269,270] and $Se(^1D_2)$.[271] *Selected* rate constants are shown in Table 9. The lifetimes of some levels of Sm have been determined.[272] An oscillating concentration of particulate matter, probably CsD, has been observed in the laser irradiation of Cs–D_2 mixtures.[273]

There is continuing interest in the mercury laser.[274] Thus molecular luminescence at 485 and 335 nm has been observed following pulsed excitation of mercury vapour.[275,276] Both bands were shown to decay at the same rate at

[259] P. M. Swearengen, S. J. Davis, S. G. Hadley, and T. M. Niemczyk, *Chem. Phys. Letters*, 1977, **49**, 571.

[260] P. M. Swearengen, S. J. Davis, and T. M. Niemzyk, *Chem. Phys. Letters*, 1978, **55**, 274.

[260a] K. Liu and J. M. Parson, *J. Chem. Phys.*, 1978, **68**, 1794.

[261] D. Husain, L. Krause, and N. K. H. Slater, *J.C.S. Faraday II*, 1977, **73**, 1678, 1706.

[262] D. Husain and P. E. Norris, *J.C.S. Faraday II*, 1977, **73**, 1815.

[263] D. Husain and P. E. Norris, *J.C.S. Faraday II*, 1977, **73**, 1107.

[264] D. Husain and P. E. Norris, *J.C.S. Faraday II*, 1978, **74**, 93, 106.

[265] D. W. Trainor, *J. Chem. Phys.*, 1977, **67**, 1206.

[266] D. Husain and P. Norris, *Chem. Phys. Letters*, 1978, **53**, 474.

[267] D. Husain and P. E. Norris, *Chem. Phys. Letters*, 1977, **51**, 206.

[268] M. A. Chowdhury and D. Husain, *J.C.S. Faraday II*, 1977, **73**, 1805.

[269] P. J. Cross and D. Husain, *J. Photochem.*, 1978, **8**, 183.

[270] P. J. Cross and D. Husain, *J. Photochem.*, 1977, **7**, 157.

[271] R. J. Donovan and D. J. Little, *Chem. Phys. Letters*, 1978, **53**, 394.

[272] J. Marek and P. Munster, *Astronomy and Astrophys.*, 1978, **62**, 245.

[273] A. C. Tam, W. Happer, and D. Siano, *Chem. Phys. Letters*, 1977, **49**, 320.

[274] H. Komine and R. L. Byer, *J. Appl. Phys.*, 1977, **48**, 2505.

[275] M. Stock, E. W. Smith, R. E. Drullinger, M. M. Hessel, and J. Pourcin, *J. Chem. Phys.*, 1978, **68**, 1785.

[276] J. Skonieczny, *J.C.S. Faraday II*, 1977, **73**, 1145; H. Komine and R. L. Byer, *J. Chem. Phys.*, 1977, **67**, 2536; M. Stock, E. W. Smith, R. E. Drullinger, and M. M. Hessel, *J. Chem. Phys.*, 1977, **67**, 2463.

Table 9 *Rate constants for excited atom reactions*[a]

Atom	Collision partner	Rate constant /cm³ molecule⁻¹ s⁻¹	Ref.
$Bi(^2D_{3/2})$	N_2	$<0.0 \times 10^{-16}$	265
		$<3.4 \times 10^{-16}$ (450 K)	265
		1.4×10^{-15} (550 K)	265
$Bi(^2D_{5/2})$	N_2	$<6.2 \times 10^{-16}$	265
		9.7×10^{-15} (450 K)	265
		2.8×10^{-14} (550 K)	265
$Bi(^2D_{3/2})$	H_2	7.3×10^{-15}	
		1.9×10^{-14} (450 K)	265
		2.9×10^{-14} (550 K)	
$Bi(^2D_{5/2})$	H_2	1.0×10^{-11}	
		1.6×10^{-11} (450 K)	265
		2.2×10^{-11} (550 K)	
$Si(^1D_0)$	H_2	8.1×10^{-11}	266
$Si(^1S_0)$	H_2	$\leqslant 10^{-14}$	267
$Si(^1D_0)$	O_2	2.3×10^{-11}	266
$Si(^1S_0)$	O_2	1.5×10^{-11}	267
$Si(^1D_0)$	He	$\leqslant 10^{-15}$	266
$Si(^1S_0)$	He	1.3×10^{-15}	267
$Se(^1D_2)$	He	3×10^{-14}	271
$Se(^1D_2)$	Ne	4×10^{-13}	271
$Se(^1D_2)$	Ar	7×10^{-12}	271
$Se(^1D_2)$	Kr	2×10^{-11}	271
$Se(^1D_2)$	Xe	6×10^{-11}	271

[a] At 300 K unless otherwise specified.

high densities, suggesting that whereas at low densities there are two reservoir states, the 6^3P_0 atomic level and the 0_g^+ state of Hg_2, at high densities the lifetime of the former is very small, and overall decay characteristics are those of the metastable Hg_2 state.[275] Studies on mercury vapour in the presence of CO and NO have given cross-sections for the $6^3P_1 \rightarrow 6^3P_0$ relaxation of 22.4 and 7.1 Å² for CO and NO respectively, and values of 1.8 Å² and 37 Å² for CO and NO for the $6^3P_0 \rightarrow 6^1S_0$ relaxation.[277] Inelastic collisions between metastable He and mercury atoms have been referred to earlier,[224] and the excitation of $Hg(^3P_1)$ by low energy alkali ions[278] has been observed. Chemi-

[277] H. Horiguchi and S. Tsuchiya, *Bull. Chem. Soc. Japan*, 1977, **50**, 1657, 1661.
[278] V. Aquilanti, P. Casavecchia, and G. Grossi, *J. Chem. Phys.*, 1978, **68**, 1499.

luminescence has been seen in a cross-beam experiment represented by (92),[279] for which the cross-section for collision energies of 0.33 eV was 3 Å^2. Asymptotic

$$Hg(^3P_0) + Br_2 \longrightarrow HgBr(\tilde{B}) + Br \qquad (92)$$

yields in triplet-sensitized decompositions have been discussed.[280] The quenching of $Cd(^3P_1)$ by deuteriated methanols gave the results summarized in Table 10.[281] Luminescence from excited complexes was also observed and quenching

Table 10[281]   *Deuterium isotope effects in quenching of* $Cd(^3P_1)$

Methanol quencher	*Relative* k_q	*Relative intensity of complex emission*
CH_3OH	1	1
CH_3OD	0.3	3.1
CD_3OH	0.9	1.1
CD_3OD	0.3	3.1

was ascribed to reactions (93) and (94) and their analogues. The kinetic isotope effect associated with (93) was estimated to have a value of D/H = 0.3.

$$Cd(^3P_1) + CH_3OH \longrightarrow Cd\cdots H\cdots OCH_3 \qquad (93)$$

$$\longrightarrow Cd\cdots O \begin{matrix} H \\ \diagdown \\ CH_3 \end{matrix} \qquad (94)$$

Emission from an excited complex of $Mg(^3P)$ and potassium arising from reaction (95) has been observed,[282] and chemiluminescence from the reaction of

$$K_2 + Mg(^3P) \longrightarrow KMg^* + K \qquad (95)$$

magnesium atoms with Cl_2 described.[251] Many papers on laser-induced fluorescence, state selected kinetic studies, and chemiluminescence have been published collectively.[283]

9 Miscellaneous

The cross-sections for quenching of $He_2(^3\Sigma_u)$ bimolecularly and termolecularly with Ar atoms have been measured as 1.5×10^{-10} cm^3 molecule^{-1} s^{-1} and 2.4×10^{-30} cm^6 molecule^{-2} s^{-1}, respectively.[284] Quenching of the same state by N_2 and O_2 has also been reported.[225]

Photodissociation cross-sections for Ar_2^+, Kr_2^+; Xe_2^+ at 3.0 and 3.5 eV have been given.[285] In photoexcited Xe_2, two short decay times of 2 ns and 60 ns have been attributed to the *radiative* decay of the $0u^+(^1\Sigma_u^+)$ and $1u(^3\Sigma_u^+)$ states, respectively,[286, 287] whereas longer components at 150 ns and 500 ns

[279] S. Hayashi, T. M. Mayer, and R. B. Bernstein, *Chem. Phys. Letters*, 1978, **53**, 419.
[280] I. Oref, *Internat. J. Chem. Kinetics*, 1977, **9**, 751.
[281] S. Tsunashima, K. Morita, and S. Sato, *Bull. Chem. Soc. Japan*, 1977, **50**, 2283.
[282] D. J. Benard, P. J. Love, and W. D. Slafer, *Chem. Phys. Letters*, 1977, **48**, 321.
[283] 'State-to-state Chemistry', ACS Symposium Series, Volume 56, ed. P. R. Brooks and E. F. Hayes, American Chemical Society, Washington, 1977.
[284] F. W. Lee, C. B. Collins, L. C. Pitchford, and R. Deloche, *J. Chem. Phys.*, 1978, **68**, 3025.
[285] J. A. Vanderhoff, *J. Chem. Phys.*, 1978, **68**, 3311.
[286] M. Ghelfenstein, R. Lopez-Delgado, and H. Szwarc, *Chem. Phys. Letters*, 1977, **49**, 312.
[287] M. Ghelfenstein, H. Szwarc, and R. Lopez-Delgado, *Chem. Phys. Letters*, 1977, **52**, 236.

must correspond to the *overall* depopulation of the $(^3\Sigma_g{}^+)1_g$ and $(^1\Sigma_g{}^+)0_g{}^+$ states.[287] Time-resolved fluorescence in Cs–Xe systems has also been studied,[288] and a high pressure (6 atmospheres) efficient vacuum u.v. ($\lambda = 172$ nm) xenon excimer laser proposed.[289] New rare-gas excimers operating on transitions in NaXe* and NaKr* in the 600—700 nm region have been discussed.[290]

Radiative lifetimes of the $\tilde{B}^1\Pi_u$ state of Li_2[291] and isotope separation by the two-step photoionization of this species,[292] and two-step photoabsorption in Na_2,[293] have been discussed. The $\tilde{D}^1\Pi \rightarrow \tilde{X}^1\Sigma^+$ and $\tilde{D}^1\Pi \rightarrow \tilde{a}^3\Sigma^+$ emissions of NaK in the 625—710 nm region have been explored,[294] and absorption in NaI has been discussed.[295] The cross-section for reaction (96) has been measured

$$BaCl(\tilde{C}^2\Pi_{\frac{1}{2}}, v = 0) + Ar \longrightarrow BaCl(\tilde{C}^2\Pi_{\frac{1}{2}}, v = 0, 1, 2) \qquad (96)$$

as 16 Å^2.[296] Complete saturation of photodissociation (*i.e. all* molecules dissociated) has been achieved in CsI.[297] The time-resolved fluorescence of GeF ($\tilde{A}^2\Sigma$) has been studied, with lifetimes of the $v' = 0$, 1, or 2 levels being measured as 990 ± 100 ns,[298] Calculated lifetimes of the $\tilde{B}$ states of HgCl, HgBr, and HgI of 20, 16, and 12 ns were in fair agreement with the only experimental measurement available of 23 ns for HgBr $\tilde{B}$.[299] Laser action at 557.6 nm and 501.8 nm has been achieved in the $\tilde{B}^2\Sigma_{\frac{1}{2}}{}^+ \rightarrow \tilde{X}^2\Sigma_{\frac{1}{2}}{}^+$ bands of HgCl[300] and HgBr,[301] respectively.

The vacuum u.v. photodissociation of phosphine is represented in equation (97).[302,303] The excited PH gave rise to emission at 697.3 nm. Laser spectro-

$$PH_3(\tilde{X}^1A_1) + h\nu \longrightarrow PH(\tilde{b}^1\Sigma^+) + 2H(^2S) \qquad (97)$$

scopy of Cu_2[304] and BO_2[305,306] has been reported. The chemiluminescence arising from the dissociation of tetramethyldioxetan induced by molecular beams of Xe[238] can be compared with the luminescence observed upon photoexcitation at 265 nm[90] and that from a high powered CO_2 laser.[307]

[288] J. Marek, *J. Phys. (B)*, 1977, **10**, L235.
[289] F. Collier and P. Cottin, *Optics Commun.*, 1978, **25**, 89.
[290] W. P. West, P. Shuker, and A. Gallagher, *J. Chem. Phys.*, 1978, **68**, 3864.
[291] T. Uzer, D. K. Watson, and A. Dalgarno, *Chem. Phys. Letters*, 1978, **55**, 6.
[292] E. W. Rothe, B. P. Mathur, and G. P. Reck, *Chem. Phys. Letters*, 1978, **53**, 74.
[293] J. P. Woerdman, *Chem. Phys. Letters*, 1977, **50**, 41.
[294] E. J. Breford and F. Engelke, *Chem. Phys. Letters*, 1978, **53**, 282.
[295] W. R. Anderson, B. M. Wilson, and T. L. Rose, *Chem. Phys. Letters*, 1977, **48**, 284.
[296] S. A. Edelstein, B. E. Perry, and T. F. Gallagher, *Chem. Phys. Letters*, 1977, **51**, 552.
[297] L. W. Grossman, G. S. Hurst, M. G. Payne, and S. L. Allman, *Chem. Phys. Letters*, 1977, **50**, 70.
[298] R. A. Anderson, L. Hanko, and S. J. Davis, *J. Chem. Phys.*, 1978, **68**, 3286.
[299] C. Duzy and H. A. Hyman, *Chem. Phys. Letters*, 1977, **52**, 345.
[300] J. H. Parks, *Appl. Phys. Letters*, 1977, **31**, 192.
[301] J. H. Parks, *Appl. Phys. Letters*, 1977, **31**, 297.
[302] G. di Stefano, M. Lenzi, A. Margani, and C. Nguyen Xuan, *J. Chem. Phys.*, 1978, **68**, 969.
[303] G. di Stefano, M. Lenzi, A. Margani, A. Mele, and C. Nguyen Xuan, *J. Photochem.*, 1977, **7**, 335.
[304] J. Lochet, *J. Phys. (B)*, 1978, **11**, L55.
[305] R. N. Dixon, D. Field, and M. Noble, *Chem. Phys. Letters*, 1977, **50**, 1.
[306] A. Fried and C. W. Mathews, *Chem. Phys. Letters*, 1977, **52**, 363.
[307] Y. Haas and G. Yahov, *Chem. Phys. Letters*, 1977, **48**, 63.

Other papers of interest have discussed the reactions of CF_3 radicals in the photolysis of mercury compounds containing CF_3 groups[308] and reactions and decay of fluorescence in UF_6.[309,310] The quenching rate-constant of UF_6^* was found to follow the Arrhenius expression with a critical energy close to 80 kcal mol^{-1}, the known dissociation energy for (98).[311] The quenching pro-

$$UF_6 \longrightarrow UF_5 + F \qquad (98)$$

cess is thus most likely to be collision-induced dissociation. The use of van der Waals molecules to study band origins in spectroscopy has been applied to the chromyl chloride system.[312]

10 Laser-enhanced, Multiphoton I.R. Reactions and Isotope Enrichment

The large number of papers included in this section this year bears testimony to the activity in the field. There have been reviews on lasers and chemistry,[313] i.r. laser-driven unimolecular reactions,[314] theories proposed for the latter process[315] and trajectory studies of the phenomenon,[316] and evidence presented for heterogeneous decomposition[317] during i.r. laser reactions.

SF_6 continues to be studied widely.[318—329] Several papers show that the initial dissociation produces fluorine atoms.[319—322] In the reactions with hydrocarbons CH_4 was found to lead to ionization 'more efficiently than C_2H_6 or C_3H_8,[321] the ionization being chemi-ionization of energetic radicals produced through reactions of F atoms with the hydrocarbon. SF_6 acts as a 'photosensitizer' of i.r. laser reactions, since the rate of formation of the laser-enhanced reaction (99) is faster in the presence of SF_6 than in its absence,[328] and the

$$NO + O_3 \longrightarrow NO_2^* + O_2 \qquad (99)$$

308 N. D. Kagramanov, V. B. Kazanski, A. K. Maltsev, O. M. Nefedov, B. N. Shelimov, and A. J. Shteinshneider, *Doklady Akad. Nauk S.S.S.R.*, 1977, **237**, 140.
309 F. B. Wampler, R. C. Oldenborg, and W. W. Rice, *Chem. Phys. Letters*, 1978, **54**, 554, 557, 560.
310 F. B. Wampler, R. C. Oldenborg, and W. W. Rice, *Chem. Phys. Letters*, 1978, **54**, 560.
311 A. Andreoni, R. Cubeddu, S. de Silvestri, and F. Zaraga, *Chem. Phys. Letters*, 1977, **48**, 431.
312 R. A. Blazy and D. H. Levy, *Chem. Phys. Letters*, 1977, **51**, 395.
313 S. Kimel and S. Speiser, *Chem. Revs.*, 1977, **77**, 437.
314 N. Bloembergen and E. Yablonovitch, *Phys. Today*, 1978, **31**, 23.
315 J.-M. Yuan and T. F. George, *J. Chem. Phys.*, 1978, **68**, 3040.
316 D. W. Noid, M. L. Koszykowski, R. A. Marcus, and J. D. McDonald, *Chem. Phys. Letters*, 1977, **51**, 540.
317 Z. Karny and R. N. Zare, *Chem. Phys.*, 1977, **23**, 321.
318 D. Tal, U. P. Oppenheim, G. Koren, and M. Okon, *Chem. Phys. Letters*, 1977, **48**, 67.
319 J. M. Preses, R. E. Weston, jun., and G. W. Flynn, *Chem. Phys. Letters*, 1977, **48**, 425.
320 C. R. Quick, jun. and C. Wittig, *Chem. Phys. Letters*, 1977, **48**, 420.
321 F. F. Crim, G. H. Kwei, and J. L. Kinsey, *Chem. Phys. Letters*, 1977, **49**, 526.
322 G. J. Diebold, F. Engelke, D. M. Lubman, J. C. Whitehead, and R. N. Zare, *J. Chem. Phys.*, 1977, **67**, 5407.
323 T. F. Deutsch and S. R. J. Brueck, *Chem. Phys. Letters*, 1978, **54**, 258.
324 Iu. I. Arkhangelskii, V. D. Klimov, V. A. Kuzmenko, V. A. Legasov, and S. L. Nedoseev, *Doklady Akad. Nauk S.S.S.R.*, 1977, **235**, 1075.
325 F. Brunner, T. P. Cotter, K. L. Kompa, and D. Proch, *J. Chem. Phys.*, 1977, **67**, 1547.
326 M. J. Shultz and E. Yablonovitch, *J. Chem. Phys.*, 1978, **68**, 3007.
327 S. T. Lin, S. M. Lee, and A. M. Ronn, *Chem. Phys. Letters*, 1978, **53**, 260.
328 E. Bar-Ziv and R. J. Gordon, *Chem. Phys. Letters*, 1977, **52**, 355.
329 K. Nagai and M. Katayama, *Chem. Phys. Letters*, 1977, **51**, 329.

isomerization of 1,2-dichloroethylene by a CO_2 laser is enhanced sixfold by SF_6.[329] A similar result was found for SiF_4 isomerization of allene to methyl-acetylene.[330] CO_2 multiphoton laser chemistry of SF_5Cl has also been reported.[331] Studies on reaction (99) have shown that excitation of the stretching and bending modes in O_3 contributes equally to the enhancement of the reaction.[332—334a] The lowering of the activation energy for reaction (99) was 1.3 kcal mol^{-1},[332] and that for the corresponding reaction leading to ground-state NO_2 is 1.3 kcal mol^{-1}[332] (1.6 kcal mole^{-1}).[333] The rate constant for the laser-stimulated reaction (100) was measured as 3.2×10^{-15}.[334b] For reaction

$$NO_2{}^* + CO \longrightarrow NO + CO_2 \qquad (100)$$

(101), deactivation of the excitation in the ν_1 and ν_3 stretches was found to be very rapid, some 30—40 times that found for collision of $SO_2\dagger$ with rare gas atoms.[335] Energy transfer between modes was very inefficient with respect to this deactivation. Such V–V transfer processes have been further discussed,[336—338] and the use of the local mode model to describe excitation has been reviewed.[339]

$$SO_2\dagger + O(^3P) \longrightarrow SO_2 + O(^3P) \qquad (101)$$

Other multiphoton i.r. laser (usually CO_2 laser)-induced reactions which have been studied include those of propane,[340] ethylene,[341] common monomers,[342] acrolein,[343] *trans*-butadiene,[344] methanol[345, 346] (reactions 102—104), and many

$$CH_3OH + nh\nu \longrightarrow CH_2O^* + H_2 \quad 90\% \qquad (102)$$

$$CH_2O^* \longrightarrow CO + H_2 \qquad (103)$$

$$CH_3OH + nh\nu \longrightarrow CH_3{\cdot} + {\cdot}OH \qquad (104)$$

[330] C. Cheng and P. Keetin, *J. Amer. Chem. Soc.*, 1977, **99**, 5808.
[331] K. M. Leary, J. L. Lyman, L. B. Asprey, and S. M. Freund, *J. Chem. Phys.*, 1978, **68**, 1671.
[332] K.-K. Hui and T. A. Cool, *J. Chem. Phys.*, 1978, **68**, 1022.
[333] E. Bar-Ziv, J. Moy, and R. J. Gordon, *J. Chem. Phys.*, 1978, **68**, 1013.
[334] (a) J. May, E. Bar-Ziv, and R. J. Gordon, *J. Chem. Phys.*, 1977, **66**, 5439; (b) I. P. Herman, R. P. Mariella, jun., and A. Javan, *J. Chem. Phys.*, 1978, **68**, 1070.
[335] G. A. West, R. E. Weston, jun., and G. W. Flynn, *J. Chem. Phys.*, 1977, **67**, 4873.
[336] R. E. McNair, S. F. Fulghum, G. W. Flynn, M. S. Feld, and B. J. Feldman, *Chem. Phys. Letters*, 1977, **48**, 241.
[337] D. N. Shigorin and G. G. Konoplev, *Doklady Akad. Nauk S.S.S.R.*, 1977, **235**, 1118.
[338] (a) S. Mukamel and J. Ross, *J. Chem. Phys.*, 1977, **66**, 5235; (b) A. N. Oraevskii, A. V. Pankratov, A. N. Skachkov, and V. M. Shabarshin, *High Energy Chemistry*, 1977, **11**, 121.
[339] B. R. Henry, *Accounts Chem. Res.*, 1977, **10**, 207.
[340] J. A. McMordie and G. W. Sentance, *J. Phys.* (*D*), 1977, **19**, L127.
[341] J. H. Hall, jun., M. L. Lesiecki, and W. A. Guillory, *J. Chem. Phys.*, 1978, **68**, 2247.
[342] A. Gandini, *Canad. J. Chem.*, 1977, **55**, 4045.
[343] S. I. Blinov, G. A. Zalesskaya, and A. A. Kotov, *Izvest. Akad. Nauk S.S.S.R. Ser. Fiz.*, 1978, **42**, 383.
[344] V. Vaida, R. E. Turner, J. L. Casey, and S. D. Colson, *Chem. Phys. Letters*, 1978, **54**, 25.
[345] S. E. Bialkowski and W. A. Guillory, *J. Chem. Phys.*, 1978, **68**, 3339.
[346] S. E. Bialkowski and W. A. Guillory, *J. Chem. Phys.*, 1977, **67**, 2061.

halocarbons.[329,347-360] Among points of interest in these studies are the fact that excitation of different vibrational modes, for example the 921 cm^{-1} CCl_2 stretch and 1088 cm^{-1} CF_2 stretch in CF_2Cl_2 gives exactly the *same* products,[352] *e.g.* reaction (105), and this appears to be general.[353] Two different decomposi-

$$CF_2Cl_2 + nh\nu \longrightarrow CF_2ClCF_2Cl + CF_3Cl \qquad (105)$$

tion modes have been proposed for CH_2F_2, one at high pressures, reactions (106)—(108), the other at low pressures, reactions (109)—(111).[349]

$$CH_2F_2 + nh\nu \longrightarrow CHF_2 + H\cdot \qquad (106)$$

$$CHF_2\cdot + Cl_2 \longrightarrow CHF_2Cl + Cl\cdot \qquad (107)$$

$$H\cdot + Cl\cdot \longrightarrow HCl \qquad (108)$$

$$CH_2F_2 + nh\nu \longrightarrow CH_2F_2{}^* \qquad (109)$$

$$CH_2F_2{}^* + Cl_2 \longrightarrow CH_2F_2 + 2Cl\cdot \qquad (110)$$

$$CH_2F_2 + Cl\cdot \longrightarrow CHF_2Cl\cdot + H \qquad (111)$$

Nitrogen compounds have also been subjected to multiphoton dissociation. Thus NH_3,[361] N_2F_4,[362] methyl cyanide,[363] methyl isocyanide,[364,365] and methyl nitrite[366] have been studied.

The laser-induced separation of isotopes has been reviewed,[367] and among successful techniques recorded this year are the use of selective electronic excitation of $I^{37}Cl$, and scavenging by H_2,[199] reaction (112), bromobenzene,[368] reaction (113), the latter process having an enrichment factor of 75%, and

$$ICl^*(\tilde{A}^3\Pi_i) + H_2 \longrightarrow HCl + HI \qquad (112)$$

$$I^{37}Cl^*(\tilde{A}^3\Pi_i) + C_6H_5Br \longrightarrow C_3H_5{}^{37}Cl + IBr \qquad (113)$$

347 H. Pummer, J. Eggleston, W. K. Bischel, and C. K. Rhodes, *Appl. Phys. Letters*, 1978, **32**, 427.
348 V. Slezak, J. Caballero, A. Burgos, and E. Quel, *Chem. Phys. Letters*, 1978, **54**, 44.
349 S. T. Lin and A. M. Ronn, *Chem. Phys. Letters*, 1977, **49**, 255.
350 D. S. King and J. C. Stephenson, *Chem. Phys. Letters*, 1977, **51**, 48.
351 M. Kneba and J. Wolfrum, *Ber. Bunsengesellschaft Phys. Chem.*, 1977, **81**, 1275.
352 E. Grunwald, K. J. Olszynn, D. F. Dever, and B. Knishkowy, *J. Amer. Chem. Soc.*, 1977, **99**, 6515; G. A. Hill, E. Grunwald, and P. Keehn, *ibid.*, p. 6521.
353 R. N. Zitter and D. F. Koster, *J. Amer. Chem. Soc.*, 1977, **99**, 5491.
354 I. Oref and B. S. Rabinovitch, *J. Phys. Chem.*, 1977, **81**, 2587.
355 J. J. Ritter, *J. Amer. Chem. Soc.*, 1978, **100**, 2441.
356 K. Nagai and M. Katayama, *Bull. Chem. Soc. Japan*, 1978, **51**, 1269.
357 R. L. Swofford, M. S. Burberry, J. A. Morrell, and A. C. Albrecht, *J. Chem. Phys.*, 1977, **66**, 5245.
358 T. H. Richardson and D. W. Setser, *J. Phys. Chem.*, 1977, **81**, 2301.
359 V. V. Vizhin, Yu. N. Molin, A. K. Petrov, V. N. Lisitsyn, and A. R. Sorokin, *Reaction Kinet. Catal. Letters*, 1977, **7**, 303.
360 A. J. Colussi, S. W. Benson, R. J. Hwang, and J. J. Tiee, *Chem. Phys. Letters*, 1977, **52**, 349.
361 V. A. Dudkin, V. V. Panin, and V. B. Rukhin, *Khim. Vysok. Energii*, 1976, **10**, 439.
362 P. Lavigne, J. L. Lachambre, and G. Otis, *Optics Comm.*, 1977, **22**, 75.
363 M. L. Lesiecki and W. A. Guillory, *Chem. Phys. Letters*, 1977, **49**, 92.
364 C. Kleinermanns and H. G. Wagner, *Ber. Bunsengesellschaft Phys. Chem.*, 1977, **81**, 1283.
365 K. V. Reddy and M. J. Berry, *Chem. Phys. Letters*, 1977, **52**, 111.
366 A. Hartford, jun., *Chem. Phys. Letters*, 1978, **53**, 503.
367 V. S. Letokhov, *Ann. Rev. Phys. Chem.*, 1977, **28**, 133.
368 D. M. Brenner, S. Datta, and R. N. Zare, *J. Amer. Chem. Soc.*, 1977, **99**, 4554.

$$I^{37}Cl(\tilde{A}^3\Pi_i) + HC{\equiv}CH \longrightarrow \underset{H}{\overset{H}{\diagdown}}C{=}C\underset{^{37}Cl}{\overset{H}{\diagup}} \qquad (114)$$

acetylene, reaction (114) for which 94% enrichment is claimed.[202] 100% enrichment of $^{12}CH_2{}^{12}CH_2$ in the i.r. photolysis of C_2H_4 was experienced.[341] In mixtures of BCl_3 and phosgene, the BCl_3 acts as an 'i.r. photosensitizer' like SF_6.[369] Two-step photoionization leading to fractionation in Li_2[292] and other multiphoton ionizations in Na_2, K_2[370,371] have been reported, and other aspects of isotope separation discussed.[372] In the CO_2 laser-induced decomposition of tetramethyl dioxetan[308] it was suggested that the intermediate first formed was neither singlet nor triplet acetone, but possibly an excimer of acetone. Luminescence in metal flames under CO_2 laser irradiation,[373] multiphoton u.v. photochemistry,[374] laser-induced two-photon fluorescence detection in high pressure liquid chromatography,[375] and photo-selection with intense laser pulses[376] have been reported.

11 Atmospheric Chemistry

There have been several discussions of the chemical species observed in interstellar space,[377] and the chemistry of their evolution.[378—380] The distribution of ammonia and its photochemical production on Jupiter have been investigated,[381] and ionic photochemistry in the thermosphere as revealed by Explorer satellite measurements has been reported.[382]

Amongst papers relating to the stratosphere of general interest are included those on the production and release of Fluorocarbons 11 and 12 in 1975,[383] the environmental effects of halocarbons,[384,385] the photochemistry of the stratosphere[386] including the ozone layer,[386,387] the effects of NO_x emission

[369] J. A. Merritt, and L. C. Robertson, *J. Chem. Phys.*, 1977, **67**, 3545.
[370] A. Herrmann, S. Leutwyler, E. Schumacher, and L. Woste, *Chem. Phys. Letters*, 1977, **52**, 418.
[371] D. L. Feldman, R. K. Lengel, and R. N. Zare, *Chem. Phys. Letters*, 1977, **52**, 413.
[372] A. V. Oraevskii, N. B. Rodionov, A. A. Stepanov, and V. A. Sticheglov, *High Energy Chemistry*, 1977, **11**, 223.
[373] J. C. Moulder and A. F. Clark, *Chem. Phys. Letters*, 1977, **49**, 471.
[374] W. M. Jackson, J. B. Halpern, and C. S. Lin, *Chem. Phys. Letters*, 1978, **55**, 254.
[375] M. J. Sepaniak and E. S. Yeung, *Analyt. Chem.*, 1977, **49**, 1554.
[376] D. Magde, *J. Chem. Phys.*, 1978, **68**, 3717.
[377] C. H. Townes, *The Observatory*, 1977, **97**, 52.
[378] W. D. Watson, *Accounts Chem. Res.*, 1977, **10**, 221.
[379] W. T. Huntress, jun., *Chem. Soc. Revs.*, 1977, **6**, 295.
[380] I. Iglesias, *Atrophys. J.*, 1977, **218**, 697.
[381] S. K. Atreya, T. M. Donohue, and W. R. Kuhn, *Icarus*, 1977, **31**, 348.
[382] M. Oppenheimer, E. R. Constantinides, K. Kirbydocken, G. A. Victor, A. Dalgarno, and J. H. Hoffman, *J. Geophys. Res.*, 1977, **82**, 5485.
[383] R. L. McCarthy, F. A. Bower, and J. P. Jesson, *Atmos. Environment*, 1977, **11**, 491.
[384] P. J. Crutzen, I. S. A. Isaksen and J. R. McAfee, *J. Geophys. Res.*, 1978, **83**, 345.
[385] R. S. Scorer, *Atmos. Environment*, 1977, **11**, 655.
[386] D. F. Strobel, *Geophys. Res. Letters*, 1977, **4**, 424.
[387] R. G. Prinn, F. N. Alyea, and D. M. Cunnold, *Ann. Rev. Earth and Planetary Sci.*, 1978, **6**, 43.

from aircraft,[388] and a proposal for NO emission abatement in jet engines.[389]

Subjects of relevance to tropospheric photochemistry include papers on the modelling of photochemical reactions,[390] the photochemistry of tropospheric ozone[391] and NO_x,[392] the effects of non-methane hydrocarbons in the atmosphere,[393] atmospheric CO_2 and the effects of forest clearance,[394] SO_2 emission in the UK,[395] weekend and weekday differences in photochemical air pollution,[396] and kinetic photochemistry downwind over water from urban areas.[397] A review of the uses of kinetic spectroscopy for the study of atmospheric reactions has appeared.[398]

Among papers dealing with remote sensing of atmospheric constituents are those on O_2 monitoring with an i.r. heterodyne radiometer,[399,400] O_3 tropical stratospheric measurements from backscatter u.v. measurements,[401] measurements on vertical and horizontal fluxes of O_3 at the tropopause,[402] and of O_3 concentrations in London,[403] $O(^1D)$ measured by O_3 photolysis in the troposphere,[404] vertical and lateral distributions of H_2 in the troposphere,[405] tropospheric N_2O mixing ratios,[406] and simultaneous determination of the ground-level abundances of N_2O, CO_2, CO, and H_2O,[407] the diode laser spectra of CF_2Cl_2 at 10.8 nm in air,[408] and the laser-induced i.r. fluorescence of organic pollutants.[409] A method of remote detection of oil pollution on water using electronic fluorescence detection from aircraft has been developed.[410]

Hydrogen Atoms, Molecules.—Reactions of hydrogen atoms with the following substrates have been studied; H_2 ($v'' = 1$),[411] HCl and DCl,[412] Cl_2,[166] ICl and

[388] H. Hidalgo and P. J. Crutzen, *J. Geophys. Res.*, 1977, **82**, 5833.
[389] R. A. Craig and H. O. Pritchard, *Canad. J. Chem.*, 1977, **55**, 1599.
[390] R. J. Gelinas and P. D. Skewes-Cox, *J. Phys. Chem.*, 1977, **81**, 2468; L. A. Farrow and T. E. Graedel, *ibid.*, p. 2480; A. C. Baldwin, J. R. Barker, D. M. Golden, and D. G. Henry, *ibid.*, p. 2483.
[391] R. W. Stewart, S. Hameed, and J. P. Pinto, *J. Geophys. Res.*, 1977, **82**, 3134.
[392] W. L. Chameides, *Geophys. Res. Letters*, 1978, **5**, 17.
[393] W. L. Chameides and R. J. Cicerone, *J. Geophys. Res.*, 1978, **83**, 128.
[394] P. D. Moore, *Nature*, 1977, **268**, 296.
[395] L. E. Reed, *Nature*, 1978, **273**, 334.
[396] B. Elkus and K. R. Wilson, *Atmosph. Environment*, 1977, **11**, 509.
[397] T. E. Gradel and L. A. Farrow, *J. Geophys. Res.*, 1977, **82**, 4943.
[398] R. J. Donovan, 'Molecular Spectroscopy', ed. A. R. West, Heyden and Son Ltd., Bellmawr, New Jersey, 1977, p. 519.
[399] R. T. Menzies and R. K. Seals, jun., *Science*, 1977, **197**, 1275.
[400] M. M. Abbas, T. Kostiuk, M. J. Mumma, D. Buhl, V. G. Kunde, and L. W. Brown, *Geophys. Res. Letters*, 1978, **5**, 317.
[401] J. E. Frederick, B. W. Guenther, P. B. Hays, and D. F. Heath, *J. Geophys. Res.*, 1978, **83**, 935.
[402] G. D. Nastrom, *J. Appl. Meterology*, 1977, **16**, 740.
[403] J. E. Thornes, *Nature*, 1978, **273**, 407; M. L. Williams and E. J. Sullivan, *ibid.*, p. 408.
[404] F. Bahe and U. Schurath, *Pure Appl. Geophys.*, 1978, **116**, 537.
[405] U. Schmidt, *J. Geophys. Res.*, 1978, **83**, 941.
[406] P. D. Goldan, Y. A. Bush, F. C. Fehsenfeld, D. L. Albritton, P. J. Gutzen, A. L. Schmeltekopf, and E. E. Ferguson, *J. Geophys. Res.*, 1978, **83**, 935.
[407] Y. S. Chang, J. H. Shaw, J. G. Calvert, and W. M. Uselman, *J. Phys. Chem.*, 1977, **81**, 589.
[408] D. E. Jennings, *Geophys. Res. Letters*, 1978, **5**, 241.
[409] J. W. Robinson, D. Nettles, and P. L. H. Jowett, *Analyt. Chim. Acta*, 1977, **92**, 13.
[410] D. M. Rayner and A. G. Szabo, *Appl. Optics*, 1978, **17**, 1624.
[411] M. J. Ovchinnikova, *Doklady Akad. Nauk S.S.S.R.*, 1977, **235**, 1103.
[412] F. S. Klein and I. Veltman, *J.C.S. Faraday II*, 1978, **74**, 17.

ClBr,[413] O_2,[414,415] NO, [416] NO_2,[417] N_2O,[418] ethylene[419] and fluoroethylenes,[420] C_3H_4,[166] carbon suboxide,[421] CF_3Br,[163] $CDCl_3$.[422] acetaldehyde,[166] silanes,[423,424] and germanes.[424,425] The H + O_2 reaction sequence is that shown in reactions (115) and (116).[415] In reaction (116) the fractions of O_2

$$H + O_2 + M \longrightarrow HO_2 + M \qquad (115)$$

$$HO_2 + H \longrightarrow H_2 + O_2(\tilde{X}, \tilde{a}^1\Delta_g, \tilde{b}^1\Sigma_g^+) \qquad (116)$$

formed in the $\tilde{X}$, $\tilde{a}$, and $\tilde{b}$ states were 0.60, 0.015, and 0.0003, respectively. The room-temperature rate constant for reaction (117) was found to be 1.55×10^{10} $l\ mol^{-1}\ s^{-1}$ and $1.91 \times 10^{10}\ l\ mol^{-1}\ s^{-1}$, respectively for H_2 and NO as the third body, and the temperature dependence found to be (M = H_2) $k_{117} = 4.6 \times 10^9 \exp(722/RT)\ l^2\ mol^{-2}\ s^{-1}$ (R in calories).[416] The temperature dependence of the ethylene reaction was found to be $k = 3.67 \times 10^{-11} \exp(-1040/T)$.[419] The primary reaction with carbon suboxide was represented

$$H + NO + M \longrightarrow HNO + M \qquad (117)$$

by reaction (118) with a rate constant of $k_{118} = 1.7 \times 10^{10} \exp(-1480/T)$.[421]

$$C_3O_2 + H \longrightarrow HC_2O + CO \qquad (118)$$

Rate constants for the reaction of H atoms with Cl_2, CH_3CHO, and C_3H_4 at room temperature were measured as 1.6×10^{-11}, 9.8×10^{-14}, and 6.3×10^{-13} cm^3 molecule$^{-1}\ s^{-1}$, respectively.[166]

The photodissociation spectra of D_2^+, HD^+ and N_2O^+,[426] and H_3^+ [427] ions have been measured. The photodissociation of H_2 near the threshold produces 57% of hydrogen atoms in the $^2S_{\frac{1}{2}}$ state,[428] with reaction (119) following.

$$H_2 + H(2s) \longrightarrow H_3^+ + e^- \qquad (119)$$

Predissociation in the $\tilde{c}^3\Pi_u$ state of H_2 [429] and quantum beats in the electric field quenching of metastable H_2 [430] have been observed.

[413] J. C. Polanyi and W. J. Skrlac, *Chem. Phys.*, 1977, **23**, 167.
[414] A. Gauss, jun., *J. Chem. Phys.*, 1978, **68**, 1689.
[415] N. Washida, H. Akimoto, and M. Okuda, *J. Phys. Chem.*, 1978, **82**, 18.
[416] K. Oka, D. L. Singleton, and R. J. Cvetanovic, *J. Chem. Phys.*, 1977, **67**, 4681.
[417] R. P. Mariella, jun., and A. C. Luntz, *J. Chem. Phys.*, 1977, **67**, 5388.
[418] A. M. Dean, D. C. Steiner, and E. E. Wang, *Combustion and Flame*, 1978, **32**, 102.
[419] J. H. Lee, J. V. Michael, W. A. Payne, and L. J. Stief, *J. Chem. Phys.*, 1978, **68**, 1817.
[420] W. E. Jones, G. Matinopoulos, J. S. Wasson, and M. H. T. Liu, *J.C.S. Faraday II*, 1978, **74**, 831.
[421] C. Faubel and H. Gg. Wagner, *Ber. Bunsengesellschaft Phys. Chem.*, 1977, **81**, 684.
[422] G. A. Oldershaw and E. A. Robinson, *Chem. Phys. Letters*, 1978, **54**, 527.
[423] D. Mihelcic, V. Schubert, R. N. Schindler, and P. Potzinger, *J. Phys. Chem.*, 1977, **81**, 1543.
[424] E. R. Austin and F. W. Lampe, *J. Phys. Chem.*, 1977, **81**, 1134.
[425] E. R. Austin and F. W. Lampe, *J. Phys. Chem.*, 1977, **81**, 1546.
[426] T. F. Thomas, F. Dale, and J. F. Paulson, *J. Chem. Phys.*, 1977, **67**, 793.
[427] K. C. Kulander and C. Bottcher, *Chem. Phys.*, 1978, **29**, 141.
[428] J. E. Mentall and P. M. Guyon, *J. Chem. Phys.*, 1977, **67**, 3845.
[429] M. Vogler and B. Meierjohann, *Phys. Rev. Letters*, 1977, **38**, 57.
[430] A. van Wijngaarden, E. Goh, G. W. F. Drake, and P. S. Farago, *J. Phys. (B)*, 1976, **9**, 2017.

Oxygen Atoms, Molecules, O_3, and HO_x.—The photodissociation and photodetachment of molecular ions have been widely studied,[431—437] including atmospheric negative ions in the 700—800 nm region,[431] O_2^+,[433] O_4^+ and O^+H_2O,[434] and O_3^-.[436] The reactions of O^- with CO_2[432] and D_2[437] have been described.

Reactions of $O(^3P)$ with the following substrates have been investigated: H_2,[438] OH,[439] H_2/CO,[440] HCO,[441] species of general type HO_x,[442] NO,[443] N_2O_5,[444] HCl,[445—447] DCl, DBr,[448] olefins,[116,449—452] alkynes,[453—456] benzenes,[457] amines,[458] alkyl nitrates,[459] nitroalkanes,[460] and $ClONO_2$.[461] *Selected* rate-constant data are shown in Table 12.

The yield of $O(^1D)$ atoms from the near-u.v. photolysis of O_3 has been shown to be temperature dependent, increasing with temperature.[462] The quantum yield of this process has been further investigated.[463] Theoretical studies on the reactions of $O(^1D)$ with rare gases[464] and N_2[465] have been carried out, and the experimental values for reaction of $O(^1D)$ with N_2O, H_2, CH_4, HCl, and NH_3 found to be 1.1×10^{-10}, 9.9×10^{-11}, 1.4×10^{-10}, 1.4×10^{-10}, and

[431] G. P. Smith, L. C. Lee, P. C. Cosby, J. R. Peterson, and J. T. Moseley, *J. Chem. Phys.*, 1978, **68**, 3818.
[432] M. L. Vestal and G. H. Mauclaire, *J. Chem. Phys.*, 1977, **67**, 3758, 3767.
[433] A. Carrington, P. G. Roberts, and P. J. Sarre, *Mol. Phys.*, 1977, **34**, 291.
[434] G. P. Smith, P. C. Cosby, and J. T. Moseley, *J. Chem. Phys.*, 1977, **67**, 3818.
[435] L. C. Lee, *J. Phys. (B)*, 1977, **10**, 3033.
[436] R. L. C. Wu, T. O. Tiernan, and C. Lifshitz, *Chem. Phys. Letters*, 1977, **51**, 211.
[437] S. G. Johnson, L. N. Kremer, C. J. Metral, and R. J. Cross, jun., *J. Chem. Phys.*, 1978, **68**, 1444.
[438] G. C. Light, *J. Chem. Phys.*, 1978, **68**, 2831.
[439] I. M. Campbell and B. J. Handy, *Chem. Phys. Letters*, 1977, **47**, 475.
[440] I. M. Campbell and B. J. Handy, *J.C.S. Faraday I*, 1978, **74**, 316.
[441] K. H. Becker, H. Lippmann, and U. Schurath, *Ber. Bunsengesellschaft Phys. Chem.*, 1977, **81**, 567.
[442] R. Shaw, *Internat. J. Chem. Kinetics*, 1977, **9**, 929.
[443] G. A. Woolsey, P. H. Lee, and W. D. Slafer, *J. Chem. Phys.*, 1977, **67**, 1220.
[444] E. W. Kaiser and S. M. Japar, *Chem. Phys. Letters*, 1978, **54**, 265.
[445] A. R. Ravishankara, G. Smith, R. T. Watson, and D. D. Davis, *J. Phys. Chem.*, 1977, **81**, 2220.
[446] W. Hack, G. Mex, and H. Gg. Wagner, *Ber. Bunsengesellschaft Phys. Chem.*, 1977, **81**, 677.
[447] R. G. Macdonald and C. B. Moore, *J. Chem. Phys.*, 1978, **66**, 513.
[448] R. D. H. Brown and I. W. M. Smith, *Internat. J. Chem. Kinetics*, 1978, **10**, 1.
[449] B. B. Krieger and R. H. Kummler, *J. Phys. Chem.*, 1977, **81**, 2493.
[450] R. Atkinson and J. N. Pitts, jun., *J. Chem. Phys.*, 1977, **67**, 2488, 2492.
[451] R. Atkinson and J. N. Pitts, jun., *J. Chem. Phys.*, 1978, **68**, 2992.
[452] R. Atkinson and J. N. Pitts, jun., *J. Chem. Phys.*, 1977, **67**, 38.
[453] C. Vinckier, M. P. Gardner, and K. D. Bayes, *J. Phys. Chem.*, 1977, **81**, 2137.
[454] A. A. Westenberg and N. de Haas, *J. Chem. Phys.*, 1977, **66**, 4900.
[455] M. E. Umstead and M. C. Lin, *Chem. Phys.*, 1977, **25**, 353.
[456] D. S. Y. Hsu, L. J. Colcord, and M. C. Lin, *J. Phys. Chem.*, 1978, **82**, 121.
[457] T. M. Sloane, *J. Chem. Phys.*, 1977, **67**, 2267.
[458] R. Atkinson and J. N. Pitts, jun., *J. Chem. Phys.*, 1978, **68**, 911.
[459] L. F. Salter and B. A. Thrush, *J.C.S. Faraday I*, 1977, **73**, 1098.
[460] L. F. Salter and B. A. Thrush, *J.C.S. Faraday I*, 1977, **73**, 2025.
[461] M. J. Kurylo, *Chem. Phys. Letters*, 1977, **49**, 467.
[462] G. K. Moortgat, E. Kudszus, and P. Warneck, *J.C.S. Faraday II*, 1977, **73**, 1216.
[463] I. Arnold, F. J. Comes, and G. K. Moortgat, *Chem. Phys.*, 1977, **24**, 211.
[464] S. R. Kinnersly, J. N. Murrell, and W. R. Rodwell, *J.C.S. Faraday II*, 1978, **74**, 600.
[465] G. Delgado-Barrio and J. A. Beswick, *Chem. Phys. Letters*, 1977, **48**, 358.

Table 12   *Rate constants for reactions of* $O(^3P)^a$

Substrate	Rate constant /cm³ molecule⁻¹ s⁻¹	Temperature range/K	Ref.
OH	4.4×10^{-11}	—	439
HCl	$5.2 \times 10^{-11} \exp(-7510/RT)^b$	—	445
HCl	$8.6 \times 10^{-12} \exp(-3230/T)$	293—718	446
C_2H_4	$9.23 \times 10^{-12} \exp(-1475/RT)^b$	298—439	452
C_3H_6	$1.05 \times 10^{-11} \exp(-515/RT)^b$	298—439	452
1-Butene	$1.39 \times 10^{-11} \exp(-665/RT)^b$	298—439	452
Isobutene	$1.76 \times 10^{-11} \exp(+85/RT)^b$	298—439	452
cis-But-2-ene	$1.21 \times 10^{-11} \exp(+235/RT)^b$	298—439	452
trans-But-2-ene	$2.26 \times 10^{-11} \exp(+20/RT)^b$	298—439	452
2-Methyl but-2-ene	4.76×10^{-11}	299.4	451
C_3H_6	3.95×10^{-12}	298	166
CH_3CHO	4.9×10^{-13}	298	166
C_2H_2	1.2×10^{-13}	297	454
C_2H_3Cl	6.8×10^{-13}	297	454
CH_3NH_2	$9 \times 10^{-12} \exp(-1650/RT)^b$	298—440	458
$C_2H_5NH_2$	$1.13 \times 10^{-11} \exp(-1275/RT)^b$	298—440	458
$(CH_3)_2NH$	$1.52 \times 10^{-11} \exp(-550/RT)^b$	298—440	458
$(CH_3)_3NH$	$1.08 \times 10^{-11} \exp(+415/RT)^b$	298—440	458
$ClONO_2$	$3.03 \times 10^{-12} \exp(-808/T)$	213—295	461

^{a} Room temperature unless otherwise specified; ^{b} R in cal mol⁻¹.

2.5×10^{-10} cm³ molecule⁻¹ s⁻¹, respectively.[466] The branching ratio for collisional deactivation of $O(^1S)$ by reaction (120) to (122) is dependent upon collision partner.[467] With N_2O and CO_2 there is no reaction, whereas N_2O

$$O(^1S) + M \longrightarrow O(^1D) + M \qquad (120)$$

$$\longrightarrow O(^3P) + M \qquad (121)$$

$$\longrightarrow \text{reaction} \qquad (122)$$

quenching of $O(^1D)$ proceeds exclusively by reaction. With H_2O the reactive pathway [for $O(^1S)$] is dominant, whereas with NO, reaction (120) is preferred. For $O(^1S)$ collisions with O_2 there is 31% production of $O(^1D)$, and 69% of $O(^3P)$.

Reactions of $O(^5S)$ with rare gases have been studied.[468] The quantum yield of $O(^1D)$ production from O_2 photolysis in the 116—177 nm region exhibits a cut-off at 175 nm, and below 139 nm has a value of unity.[469] The photoabsorption cross-sections of O_2 in the 5.5—35 nm region[470] and the vacuum u.v. absorption spectra from the $\tilde{b}^1\Sigma_g^+$ and $\tilde{a}^1\Delta_g$ states of O_2[471] have been recorded.

[466] J. A. Davidson, H. I. Schiff, G. E. Streit, J. R. McAfee, A. L. Schmeltekopf, and C. J. Howard, *J. Chem. Phys.*, 1977, **67**, 5021.
[467] T. G. Slanger and G. Black, *J. Chem. Phys.*, 1978, **68**, 989, 998.
[468] J. Windrinch, H. D. Wolf, and J. Fricke, *J. Phys. (B)*, 1978, **11**, 1235.
[469] L. C. Lee, T. G. Slanger, G. Black, and R. L. Sharpless, *J. Chem. Phys.*, 1977, **67**, 5602.
[470] G. Mehiman, D. L. Ederer, and E. B. Saloman, *J. Chem. Phys.*, 1978, **68**, 1862.
[471] D. H. Katayama, S. Ogawa, M. Ogawa, and Y. Tanaka, *J. Chem. Phys.*, 1977, **67**, 2132.

There have been several new measurements of rate constants for reactions of $O_2(^1\Delta_g)$.[472-474] Selected rate constants are given in Table 13.

Table 13 *Rate constants for reactions of $O_2(^1\Delta_g)$*

Substrate	Rate constant $l\,mol^{-1}\,s^{-1}$	Ref.
O_2	9.4×10^2	472
N_2O	4.5×10^7	473
D_2O	5.0×10^8	473
NH_3	1.2×10^9	473
C_3H_8	2.0×10^8	473
C_2H_2	2.3×10^8	473
HCl	4.4×10^7	473
DCl	3.5×10^6	473

O_3 has been photolysed in the 600 nm region,[475] and in the near u.v.[462,463,476] There was a sharp decrease in $O(^1D)$ production at 308 nn, but at wavelengths below 300, the quantum yield for $O(^1D)$ was unity.[476] Numerical model studies on the $NO + O_3$ stratospheric reaction have been carried out,[477] and chemiluminescence from the same reaction (123) has been studied in molecular beams.[478] Chemiluminescence from the reactions of O_3 with organics has also

$$NO + O_3 \longrightarrow NO_2{}^* + O_2 \qquad (123)$$

been observed.[479] Reaction (124) has been shown to have an upper limiting rate-constant value of 5×10^{-19} cm^3 molecule^{-1} s^{-1} at 226 K, 1×10^{-19} cm^3 molecule^{-1} s^{-1} at 300 K, and can thus be neglected in the stratopshere.[480]

$$O_3 + HNO_2 \longrightarrow HNO_3 + O_2 \qquad (124)$$

Numerous papers have again been concerned with the hydroxy-radical, particularly its reactions with atmospheric constituents. The cross-sections for the appearance of $OH(\tilde{A}^2\Sigma^+ \to \tilde{X}^2\Pi)$ emission following the photodissociation of water have been measured,[481] and a device produced capable of monitoring OH concentrations as low as 10^5 molecules cm^{-3} based upon absorption measurements.[482] Rotational energy transfer in OH $\tilde{A}^2\Sigma^+$,[483] predissociation of that state,[484] the OH airglow,[485,486] and the tropospheric OH concentrations,[487] have all been discussed. Reactions of OH with the following species

472 P. Borrell, P. M. Borrell, and M. D. Pedley, *Chem. Phys. Letters*, 1977, **51**, 300.
473 R. G. O. Thomas and B. A. Thrush, *Proc. Roy. Soc. (A)*, 1977, **356**, 287, 295, 307.
474 A. Leiss, U. Schurath, K. H. Becker, and E. H. Fink, *J. Photochem.*, 1978, **8**, 211.
475 Yu. M. Gershenzon and S. K. Chekin, *High Energy Chemistry*, 1977, **11**, 293.
476 D. L. Philen, R. T. Watson, and D. D. Davis, *J. Chem. Phys.*, 1977, **67**, 3316.
477 A. F. Tuck, *Proc. Roy. Soc. (A)*, 1977, **355**, 267.
478 A. E. Redpath, M. Menzinger, and T. Carrington, *Chem. Phys.*, 1978, **27**, 409.
479 D. A. Hansen, R. Atkinson, and J. N. Pitts, jun., *J. Photochem.*, 1977, **7**, 379.
480 E. W. Kaiser and S. M. Japar, *Chem. Phys. Letters*, 1977, **52**, 121.
481 L. C. Lee, L. Oren, E. Phillips, and D. L. Judge, *J. Phys. (B)*, 1978, **11**, 47.
482 D. K. Killinger and C. C. Wang, *Chem. Phys. Letters*, 1977, **52**, 374.
483 R. K. Lengel and D. R. Crosley, *J. Chem. Phys.*, 1977, **67**, 2085.
484 K. R. German, *J. Chem. Phys.*, 1977, **67**, 5411; H. B. Palmer, *ibid.*, p. 5413.
485 E. J. Llewellyn and B. H. Long, *Canad. J. Phys.*, 1978, **56**, 581.
486 V. I. Krassovsky and M. V. Shagaev, *Planetary Space Sci.*, 1977, **25**, 509.
487 P. J. Crutzen and J. Fishman, *Geophys. Res. Letters*, 1977, **4**, 321.

have been studied: $O(^3P)$,[439] OH,[488] CO,[439,489—492] C_3O_2,[493] HCl,[445,446] $HCHO$,[494] CH_3CHO,[494] COS, CS_2, CH_3SCH_3,[495] CH_4 and CH_3F,[496] halocarbons,[497] olefins,[498—500,451] ethers,[499] esters,[501,502] benzenoid hydrocarbons,[503—506] and amines.[507] Selected rate-constant data are shown in Table 14.

The results of studies on the reaction of OH with CO indicate that a complex mechanism [(125)—(128)] is required to explain the data.[489,490] As a result of these reactions the tropospheric concentration of OH is lower than expected,

$$OH + CO \; \underset{\longleftarrow}{\longrightarrow} \; HOCO \qquad\qquad (125)$$

$$HOCO \; \longrightarrow \; H + CO_2 \qquad\qquad (126)$$

$$HOCO + M \; \longrightarrow \; HOCO + M \qquad\qquad (127)$$

$$HOCO + O_2 \; \longrightarrow \; HO_2 + CO_2 \qquad\qquad (128)$$

a result of some importance since the reactions of Freons in the troposphere are principally with OH. Freons with abstractable hydrogens have short tropospheric lifetimes, however,[497] and of those listed in Table 14, only CHF_2Cl (Freon-22) is long-lived, with a lifetime of some 25 years. The study on the temperature dependence of $O(^3P)$ and OH reactions with HCl results in a prediction that stratospheric Cl concentrations are some 20% lower than currently accepted levels in the 13—30 km region.[445] The disproportionation reaction of OH, reaction (129), has been shown not to exhibit Arrhenius type behaviour, the data being better fitted to an expression of the type $k = AT^M$.[488]

The absorption spectra of H_2O[508] and H_2O_2,[509,510] cross-sections for the production of $OH(\tilde{A}^2\Sigma^+)$ in the dissociation of H_2O,[481] polarized photofragment spectroscopy in H_2O and D_2O,[126] and collisional energy transfer between

[488] J. Ernst, H. G. Wagner, and R. Zellner, *Ber. Bunsengesellschaft Phys. Chem.*, 1977, **81**, 1270.
[489] R. A. Perry, R. Atkinson, and J. N. Pitts, jun., *J. Chem. Phys.*, 1977, **67**, 5577.
[490] R. Overend and G. Paraskevopoulos, *Chem. Phys. Letters*, 1977, **49**, 109; I. W. M. Smith, *ibid.*, p. 112.
[491] W. C. Gardiner, jun., D. B. Olson, and J. N. White, *Chem. Phys. Letters*, 1978, **53**, 134.
[492] R. Butler, I. J. Solomon, and A. Snelson, *Chem. Phys. Letters*, 1978, **54**, 19.
[493] C. Faubel, H. Gg. Wagner, and W. Hack, *Ber. Bunsengesellschaft Phys. Chem.*, 1977, **81**, 689.
[494] R. Atkinson and J. N. Pitts, jun., *J. Chem. Phys.*, 1978, **68**, 3581.
[495] R. Atkinson, R. A. Perry, and J. N. Pitts, jun., *Chem. Phys. Letters*, 1978, **54**, 14.
[496] J. Ernst, H. Gg. Wagner, and R. Zellner, *Ber. Bunsengesellschaft Phys. Chem.*, 1978, **82**, 409.
[497] J. S. Chang and F. Kaufman, *J. Chem. Phys.*, 1977, **66**, 4989.
[498] R. Overend and G. Paraskevopoulos, *J. Chem. Phys.*, 1977, **67**, 674.
[499] R. A. Perry, R. Atkinson, and J. N. Pitts, jun., *J. Chem. Phys.*, 1977, **667**, 458, 611.
[500] R. Atkinson, R. A. Perry, and J. N. Pitts, jun., *J. Chem. Phys.*, 1977, **67**, 3170.
[501] I. M. Campbell and P. E. Parkinson, *Chem. Phys. Letters*, 1978, **53**, 385.
[502] A. M. Winer, A. C. Lloyd, K. R. Darnall, R. Atkinson, and J. N. Pitts, jun., *Chem. Phys. Letters*, 1977, **51**, 221.
[503] R. A. Perry, R. Atkinson, and J. N. Pitts, jun., *J. Phys. Chem.*, 1977, **81**, 1607.
[504] M. Hoshino, H. Akimoto, and M. Okuda, *Bull. Chem. Soc. Japan*, 1978, **51**, 718.
[505] R. A. Kenley, J. E. Davenport, and D. G. Hendry, *J. Phys. Chem.*, 1978, **82**, 1095.
[506] T. M. Sloane, *Chem. Phys. Letters*, 1978, **54**, 269.
[507] R. Atkinson, R. A. Perry, and J. N. Pitts, jun., *J. Chem. Phys.*, 1978, **68**, 1850.
[508] P. Gurtler, V. Saile, and E. E. Koch, *Chem. Phys. Letters*, 1977, **51**, 386.
[509] L. T. Molina, S. D. Schinke, and M. J. Molina, *Geophys. Res. Letters*, 1977, **4**, 580.
[510] C. L. Lin, N. K. Rohatgi, and W. B. Demore, *Geophys. Res. Letters*, 1978, **5**, 113.

H_2O and NH_3 [511] have been discussed. In the reactions of HO_2 with H, reaction (116), $O_2(X^3\Sigma_g^-)$ was the major product.[415] The kinetics of reactions of HO_2 with NO,[512,513] NO_2,[513,514] and HO_2 [515] have been studied. For the reaction with NO_2 and third body M, rate constants varied from 10^{-31} cm^6 molecule^{-2} s^{-1} for M = He to 6.6×10^{-31} cm^6 molecule^{-2} s^{-1} for M = NO_2. For the reaction of HCO with O_2 and NO, rate constants have been measured as 5.6×10^{-12} and 8.5×10^{-12} cm^3 molecule^{-1} s^{-1}, respectively at room temperature.[516]

Table 14 *Rate-constant data for reactions of OH*

Substrate	Rate constant /cm^3 molecule^{-1} s^{-1}	Temperature range/K	Ref.
C_3O_2	1.17×10^{-11} exp$(-620/T)$	—	493
HCl	3.3×10^{-12} exp$(-937/RT)^a$	—	445
HCl	6.7×10^{-13} exp$(-3230/T)$	293—718	446
H_2CO	1.25×10^{-11} exp$(-175/RT)^a$	299—426	494
CH_3CHO	6.87×10^{-12} exp$(+510/RT)^a$	299—426	494
CH_4	4.17×10^{-12}	1300	496
$CHFCl_2$	1.16×10^{-12} exp$(-1073/T)$	—	497
CHF_2Cl	1.2×10^{-12} exp$(-1657/T)$	—	497
CH_3CCl_3	1.95×10^{-12} exp$(-1331/T)$	—	497
C_2HCl_3	5.3×10^{-13} exp$(+445/T)$	—	497
C_2Cl_4	9.4×10^{-12} exp$(-1199/T)$	—	497
C_2H_4	$1 \times 10^{-11\,b}$	296	498
$CH_2{=}CHF$	1.48×10^{-12} exp$(+775/RT)^a$	299—426	499
$CH_2{=}CHCl$	1.14×10^{-12} exp$(+1045/RT)^a$	299—426	499
$CH_2{=}CHBr$	1.79×10^{-12} exp$(+805/RT)^a$	299—426	499
Dimethyl ether	1.29×10^{-11} exp$(-770/RT)^a$	299—427	499
2-Methyl-2-butene	8.73×10^{-11}	299.4	451
Allene	5.59×10^{-12} exp$(+305/RT)^a$	299—424	500
Methyl acetate	1.83×10^{-13}	292	501
Peroxyacetyl nitrate (PAN)	$\leqslant 1.67 \times 10^{-13}$	299	502
Methoxybenzene	1.96×10^{-11}	299	503
o-Cresol	3.41×10^{-11}	299	503
$(CH_3)_2NH$	2.89×10^{-11} exp$(+490/RT)^a$	298—426	507
$(CH_3)_3N$	2.62×10^{-11} exp$(+500/RT)^a$	298—426	507
$C_2H_5NH_2$	1.47×10^{-11} exp$(+375/RT)^a$	198—426	507

^{a} R in calories; ^{b} high pressure limiting value.

ClO$_x$ Reactions, Freon Question.—The important questions here are the residenec times of halocarbons in the troposphere, dictated largely by OH reactions; the photolysis of halocarbons in the stratosphere, and modelling of this in laboratory studies, and the fates of chlorine atoms in the stratosphere. There have been several papers concerned with the overall impact of halocarbon release into the atmosphere.[383—385] It has been shown that most fluorocarbons containing

511 F. Herlemont, M. Lyszyk, and J. Lemaire, *Chem. Phys. Letters*, 1978, **54**, 603.
512 C. J. Howard and K. M. Evenson, *Geophys. Res. Letters*, 1977, **4**, 437.
513 R. Simonaitis and J. Heicklen, *Internat. J. Chem. Kinetics*, 1978, **10**, 67.
514 C. J. Howard, *J. Chem. Phys.*, 1977, **67**, 5258
515 E. J. Hamilton and R. R. Lii, *Internat. J. Chem. Kinetics*, 1977, **9**, 875.
516 K. Shibuya, T. Ebata, K. Obi, and I. Tanaka, *J. Phys. Chem.*, 1977, **81**, 2292.

hydrogen atoms have extremely short tropospheric lifetimes due to their efficient reaction (through hydrogen abstraction) with hydroxy-radicals.[497] As such these pose no threat to the stratospheric ozone filter, but there are some species, such as Freon-22 (CHF_2Cl), which *are* long-lived in the troposphere ($\sim$25 years), and must thus be monitored carefully, since the stratospheric photodissociation rates of many Freons [F-13 ($CClF_3$), F-113 (CCl_2FCClF_2), F-114 (CF_2ClCF_2Cl), and F-115 ($CClF_2CF_3$)] have been shown to give rise to upper atmosphere residence times of 40—550 years.[517] Freons F-13 and F-115 are especially long-lived. The gas phase laboratory photolysis of other halogenated compounds, although usually at higher energies than of direct concern in the stratosphere, has nevertheless provided results of interest.[216, 219, 221] In the stratosphere, such photolyses produce chlorine atoms, and the rate constants for many reactions of atmospheric interest involving ClO_x species have been tabulated.[518] The fate of the species ClO is of major interest, and since one such fate is photolysis,[519] it is of importance to have accurate absorption cross-section measurements for this molecule. These have now been provided. [520] The third-body reaction of ClO with NO_2, reaction (129) has been shown to have a rate constant of 8.3×10^{-32} cm^6 molecule^{-2} s^{-1} at 300 K for M = He, and 1.52×10^{-31} cm^6 molecule^{-2} s^{-1} at 297 K for M = N_2.[521] From studies on the rates of reaction of $ClONO_2$ with $O(^3P)$,[461] OH,[521] NO,[523] ClNO,[522, 523] and

$$ClO + NO_2 + M \longrightarrow ClONO_2 + M \qquad (129)$$

$Cl(^2P)$ [165] it can be concluded that chemical reactions of $ClONO_2$ are insignificant loss mechanisms for the species compared with u.v.-photolysis, accounting for less than 15% of $ClONO_2$ removal in the stratosphere at 20—30 km.[461] $BrONO_2$ is removed from the atmosphere at a rate twenty times that of $ClONO_2$.[524] Other $Cl(^2P)$ atom reactions have been referred to earlier.[160, 162, 164, 167, 168]

N Atom, N_2.—NO_x *Reactions.* A time-of-flight determination of radiative lifetimes of high Rydberg states of nitrogen[525] and crossed molecular beam study of reactions of atomic nitrogen with halogens[526] have been reported. The radiative lifetime of the $\tilde{C}^3\Pi_u$ state of N_2 has been given as 40.8 ns,[527] near to those of the $v' = 0$ and $v = 4$ vibrational levels of the same state which were measured as 36.6 and 36.5 ns, respectively.[528] For the $x^1\Sigma_g^-$ and $y^1\Pi_g$ states

[517] C. C. Chou, R. J. Milstein, W. S. Smith, H. Vera Ruiz, M. J. Molina, and F. S. Rowland, *J. Phys. Chem.*, 1978, **82**, 1.
[518] R. T. Watson, *J. Phys. Chem. Reference Data*, 1977, **6**, 871.
[519] W. S. Smith, C. C. Chou, and F. S. Rowland, *Geophys. Res. Letters*, 1977, **4**, 517.
[520] P. H. Wine, A. R. Ravishankara, D. L. Philen, D. D. Davis, and R. T. Watson, *Chem. Phys. Letters*, 1977, **50**, 101.
[521] M. S. Zahniser, J. S. Chang, and F. Kaufman, *J. Chem. Phys.*, 1977, **67**, 997.
[522] H.-D. Knauth, *Ber. Bunsengesellschaft Phys. Chem.*, 1978, **82**, 428.
[523] H.-D. Knauth, *Ber. Bunsengesellschaft Phys. Chem.*, 1978, **82**, 212; R. Algasmi, H.-D. Knauth, and D. Rohlack, *ibid.*, p. 217.
[524] J. E. Spencer and F. S. Rowland, *J. Phys. Chem.*, 1978, **82**, 7.
[525] C. A. Kocher and C. E. Fairchild, *J. Chem. Phys.*, 1978, **68**, 1884.
[526] R. L. Love, J. M. Herremann, R. W. Bickes, jun., and R. B. Bernstein, *J. Amer. Chem. Soc.*, 1977, **99**, 8316.
[527] T. W. Carr and S. Dardes, *J. Phys. Chem.*, 1977, **81**, 2225.
[528] K. H. Becker, H. Engels, and T. Tatarczyk, *Chem. Phys. Letters*, 1977, **51**, 111.

of N_2, lifetimes are 23.1 and 19.9 ns.[529] The quenching of the $\tilde{A}^3\Pi_u{}^+$ state of N_2 by I_2 and NO has rate constants of 6.9×10^{-12} and 4.5×10^{-11} cm^3 molecule^{-1} s^{-1}, respectively,[530] and those for quenching of the $\tilde{C}^3\Pi_u$ state by He and N_2 were 7.8×10^{-14} and 7.1×10^{-12} cm^3 molecule^{-1} s^{-1}.[527]

The photodissociation of the positive ions $NO^+(NO)$ and $NO^+(H_2O)$[531] and N_2O^+ [426, 532] has been studied.

The two-photon absorption cross-section for the system NO $\tilde{C}^2\Pi$ ($v = 0$) ← $\tilde{X}^2\Pi$ ($v = 0$) has been measured,[533] and decay times of NO $\tilde{b}^4\Sigma^-$, $v = 1$ and $\tilde{b}^4\Sigma^-$, $v = 2$ reported as 6.43 and 5.77 μs, respectively.[534] The rate constant for the removal of NO in the $\tilde{b}$ state by the ground state was 1.01×10^{-10} cm^3 molecules^{-1} s^{-1},[534] and that for the reaction of NO, X with HCO was 8.5×10^{-12} cm^3 molecule^{-1} s^{-1} (at 298 K).[516] The kinetics of the β and γ afterglows in NO,[535] the reaction of NO with O_3 in a molecular beam[478] and with H atoms[416] have been studied.

There have been several studies on the fluorescence of NO_2,[536—540] including that from selectively excited hyperfine levels,[536] the anomalous magnetic fluorescence properties in the $^2B_2$ state,[538] laser NO_2 fluorescence measurements in flames,[539] and quenching studies.[540, 541] The quantitative intracavity determination of NO_2 has been discussed.[542] A study on the wavelength dependence of the quantum yield of photodissociation of NO_2 in the 375—420 nm region has shown that Φ declines significantly and anomalously between 380 and 390 nm,[543] leading to speculation that there may be a non-dissociative state of NO_2 in this region. The sensitized photo-oxidation of isobutane by photolysis of NO_2 at 366 nm has been studied from a mechanistic viewpoint,[544] and the role of NO in the combustion of hydrocarbons investigated.[545]

There have been continuing studies of NO_x reactions of interest to both tropospheric and stratospheric regions of the atmosphere. Numerical model studies on the $NO + O_3$ stratospheric reaction have been carried out,[477] and the sensitivity of the modelling to choice of rate constants has been discussed.[546] A paper has appeared concerned with the effect of attitude upon the rate of

529   C. R. Hummer and D. J. Burns, *J. Chem. Phys.*, 1977, **67**, 4062.
530   A. Mandl and J. J. Ewing, *J. Chem. Phys.*, 1977, **67**, 3490.
531   J. A. Vanderhoff, *J. Chem. Phys.*, 1977, **67**, 2332.
532   R. Frey, B. Gotcher, W. B. Peatman, H. Pollak, and E. W. Schlag, *Chem. Phys. Letters*, 1978, **54**, 411.
533   P. A. Freedman, *Canad. J. Phys.*, 1977, **55**, 1387.
534   G. R. Mohlmann and F. J. de Heer, *Chem. Phys. Letters*, 1977, **49**, 588.
535   I. M. Campbell and R. S. Mason, *J. Photochem.*, 1978, **8**, 321.
536   R. Schmiedl, I. R. Bonilla, F. Paech, and W. Demtroder, *J. Mol. Spectroscopy*, 1977, **68**, 236.
537   J. L. Hardwick, *J. Mol. Spectroscopy*, 1977, **66**, 248.
538   H. Figger, D. L. Monts, and R. N. Zare, *J. Mol. Spectroscopy*, 1977, **68**, 388.
539   R. H. Barnes and J. F. Kircher, *Appl. Optics*, 1978, **17**, 1099.
540   J. Heicklen, *J. Photochem.*, 1977, **7**, 219.
541   V. M. Donnelly and F. Kaufman, *J. Chem. Phys.*, 1977, **67**, 4768.
542   G. H. Atkinson, T. N. Heimlich, and M. W. Schuyler, *J. Chem. Phys.*, 1977, **66**, 5005.
543   A. B. Harker, W. Ho, and J. J. Ratto, *Chem. Phys. Letters*, 1977, **50**, 394.
544   G. Paraskevopoulos and R. J. Cvetanovic, *J. Phys. Chem.*, 1977, **81**, 2598.
545   W. M. Shaub and S. H. Bauer, *Combustion and Flame*, 1978, **32**, 35.
546   H. S. Johnston and H. Nelson, *J. Geophys. Res.*, 1977, **82**, 2593; W. H. Duewer, D. J. Wuebbles, H. W. Ellsaesser, and J. S. Chang, *ibid.*, p. 2599.

photodissociation of NO_2,[547] the photochemistry of atmospheric pollutants derived from NO_x has been discussed,[548] sources and sinks of atmospheric N_2O[549] and the reactions, including photoreactions[550] of this species, have been described.

Reactions (130) and (131) have been shown to have rate-constants of $2.5 \times 10^{-14} \exp(-1230/T)$ and 1.9×10^{-11} cm^3 molecule^{-1} s^{-1} (at 297 K),

$$NO_2 + NO_3 \longrightarrow NO + O_2 + NO_2 \qquad (130)$$

$$NO + NO_3 \longrightarrow 2NO_2 \qquad (131)$$

respectively.[551] The average values for the quantum yields of reactions (132) and (133) had a yield of only 0.07 in the 510—670 nm region.[551]

$$NO_3 + h\nu \longrightarrow NO + O_2 \qquad (132)$$

$$NO_3 + h\nu \longrightarrow NO_2 + O \qquad (133)$$

Electronic absorption in HNO_3,[552] HONO and N_2O_3[553] has been studied. The reaction of HNO_2 with O_3 has been shown to have such a low rate constant as to be of little importance in the atmosphere.[480] The ratio of rate constants k_{134}/k_{135} for reactions (134) and (135) is $\geqslant 10^3$ in the 254—283 K temperature range,[554] and the rate constant for reaction (136) is $1.4 \times 10^{14} \exp(-20700/RT)$ l mol^{-1} s^{-1} (R in cal).[554] The thermal decomposition lifetimes of pernitric

$$HO_2 + NO_2 \longrightarrow HO_2NO_2 \qquad (134)$$

$$HO_2 + NO_2 \longrightarrow HONO + O_2 \qquad (135)$$

$$HO_2NO_2 \xrightarrow{M} HO_2 + NO_2 \qquad (136)$$

acid are thus strongly temperature-dependent, ranging from 12 s at 298 K to 1 month at 220 K. The dissociation of nitrous acid has been shown to be a heterogeneous reaction (137).[555]

$$2HONO \longrightarrow NO + NO_2 + H_2O \qquad (137)$$

Other papers have been concerned with the tropospheric and stratospheric atmospheric composition perturbed by NO_x emission from aircraft,[388] NO_x emission abatement in jet engines,[389] enhancement of smog by NN'-diethyl-hydroxylamine in polluted air,[556] and the reaction of Cl atoms with NH_3.[163]

CO_x Reactions.—Anomalous fluorescence decay curves in CO excited in the vacuum u.v. due to reversible collision-induced singlet–triplet transitions have been observed,[557, 558] and the two-photon fluorescence excitation spectrum of CO has been recorded.[559] Reactions of CO with OH were discussed

[547] J. T. Peterson, *Atmos. Environment*, 1977, **11**, 689.
[548] C. W. Spicer, *Atmos. Environment*, 1977, **11**, 1089.
[549] S. C. Lin, R. J. Cicerone, T. M. Donahue, and W. L. Chameides, *Tellus*, 1977, **29**, 251.
[550] G. Selwyn, J. Podolske, and H. S. Johnston, *Geophys. Res. Letters*, 1977, **4**, 427.
[551] R. A. Graham and H. S. Johnston, *J. Phys. Chem.*, 1978, **82**, 254.
[552] G. D. T. Tejwani and E. S. Yeung, *J. Chem. Phys.*, 1978, **68**, 2012.
[553] W. R. Stockwell and J. G. Calvert, *J. Photochem.*, 1978, **8**, 193.
[554] R. A. Graham, A. M. Winer, and J. N. Pitts, jun., *Chem. Phys. Letters*, 1977, **51**, 215.
[555] E. W. Kaiser and C. H. Wu, *J. Phys. Chem.*, 1977, **81**, 1701.
[556] J. N. Pitts, jun., *Science*, 1977, **197**, 255.
[557] M. Lavollee and A. Tramer, *Chem. Phys. Letters*, 1977, **47**, 523.
[558] A. C. Provorov, B. P. Stoicheff, and S. Wallace, *J. Chem. Phys.*, 1977, **67**, 5393.
[559] R. A. Bernheim, C. Kittrell, and D. K. Veirs, *Chem. Phys. Letters*, 1977, **51**, 325.

earlier.[489—492] The $\tilde{A}^2\Pi_u \rightarrow \tilde{X}^2\Pi_g$ fluorescence of $CO_2{}^+$,[560] and predissociation in the $\tilde{C}^2\Sigma_g{}^+$ state of $CO_2{}^+$ [56] have been investigated.

At 147 nm and 130.2—130.6 nm, quantum yields of O atom production and CO yields are unity,[561] suggesting that previously reported CO deficiencies are in error. At 106—117.5 nm, CO_2 photolysis yields $O(^1S)$, the production of which has a quantum yield which dips sharply to $\leqslant 0.15$ at 108.9 nm.[562] This is the centre of a Rydberg transition in CO_2. At the same wavelength $O(^1D)$ production has a quantum yield of 0.65 ± 0.1.

The rate constants for reaction of HCO with NO and O_2 were given earlier as 8.5×10^{-12} and 5.6×10^{-12} cm^3 molecule^{-1} s^{-1}, respectively.[516]

SO_x Reactions.—The most important observation in the past year has been that there exists a pressure saturation effect in the phosphorescence quenching of $SO_2(^3B_1)$ when quenchers such as CO_2, CO, N_2, and C_2H_2 are used with excitation in the 266.2—363.1 nm region.[563—566] The important corollary of this study is that the three states postulated earlier to explain $^3SO_2$ photochemistry are no longer required,[566] and at zero-pressures, the quantum yield of phosphorescence from SO_2 becomes 0.95 ± 0.29,[565] resulting in the revised conclusion that no unimolecular non-radiative decay of $SO_2(^3B_1)$ exists, in satisfactory agreement with intuition. Rate constants for quenching of $^3SO_2$ by a variety of species are shown in Table 15.

Table 15 *Rate constants for quenching of* $SO_2(^3B_1)$

Substrate	Rate constant /l mol^{-1} s^{-1}	Ref.
SO_2	4.2×10^8	565
CO	2.6×10^6	566
$C_2H_2{}^a$	1.56×10^9	566
$C_2H_2{}^b$	8.0×10^9	566

[a] Reactive quenching; [b] physical quenching.

A further extensive study has produced no evidence of a low-lying $^3B_2$ state of SO_2,[567] and two studies of the singlet state of SO_2 have appeared.[567, 568] Results indicate a single fluorescent species,[567] with a common precursor state $(^1A_2)$ to both the emitting singlet $(^1B_1)$ and triplet $(^3B_1)$ states. The photolysis of SO_2 at 313 in the presence of acetylene and O_2,[569] and allene and O_2,[570] has been interpreted on the basis of three triplet states; but this proposal must now be reconsidered in the light of the above-mentioned pressure saturation effects.

560 V. E. Bondybey and T. A. Miller, *J. Chem. Phys.*, 1977, **67**, 1790.
561 J. H. D. Eland and J. Berkowitz, *J. Chem. Phys.*, 1977, **67**, 2782.
561a T. G. Slanger and G. Black, *J. Chem. Phys.*, 1978, **68**, 1844.
562 T. G. Slanger, R. L. Sharpless, and G. Black, *J. Chem. Phys.*, 1977, **67**, 5317.
563 F. Su, F. B. Wampler, J. W. Bottenheim, D. L. Thorsell, J. G. Calvert, and E. K. Damon, *Chem. Phys. Letters*, 1977, **51**, 150.
564 R. N. Rudolph and S. J. Strickler, *J. Amer. Chem. Soc.*, 1977, **99**, 3871.
565 F. Su, J. W. Bottenheim, D. L. Thorsell, J. G. Calvert, and E. K. Damon, *Chem. Phys. Letters*, 1977, **49**, 305.
566 F. Su and J. G. Calvert, *Chem. Phys. Letters*, 1977, **52**, 572.
567 F. Ahmed, K. F. Langley, and J. P. Simons, *J.C.S. Faraday II*, 1977, **73**, 1659.
568 F. Su, J. W. Bottenheim, H. W. Sidebottom, J. G. Calvert, and E. K. Damon, *Internat. J. Chem. Kinetics*, 1978, **10**, 125.
569 K. Partymiller and J. Heicklen, *J. Photochem.*, 1977, **7**, 221.
570 K. Partymiller and J. Heicklen, *J. Photochem.*, 1978, **8**, 167.

Part II

PHOTOCHEMISTRY OF INORGANIC AND ORGANOMETALLIC COMPOUNDS

1
Photochemistry of Transition-metal Complexes

BY A. COX

1 Introduction

A number of reviews on various aspects of inorganic photochemistry have appeared. These include such topics as a glossary of photochemical terms and definitions,[1] the luminescence properties of inorganic compounds,[2] luminescence analysis,[3] flash photolysis of co-ordination compounds including experimental techniques,[4] primary processes,[5,6] photochromic glasses,[7] co-ordination compounds as potential photocatalysts,[8] and a review discussing possible practical photochemical systems for the utilization of solar energy.[9] Finally, a book has appeared incorporating a chapter on photochemistry in low-temperature matrices.[10]

From theoretical considerations an expression has been derived[11] relating the observed rate of luminescence decay of labile complexes at equilibrium in solution to the types of complexes in excited states, the equilibrium concentration of ligand, and the stability constants of the complexes. A study has been made of the quenching of excited singlet states by several different metal ions. The results[12] show that in the absence of ground-state complexation, rapid dynamic quenching is only likely if M^{n+} is readily reduced or oxidized by a single-electron process, and/or if there is appreciable overlap between the fluorescence spectrum and spin-allowed absorptions of M^{n+}. Photochemical substitution reactions of six-co-ordinated transition-metal complexes have been analysed[13] according to a ligand field approach in which the model involves an additive contribution to the orbital energy by each ligand. The results for Cr^{III} and Co^{III} show highly satisfactory agreement between theory and experiment and provide an alternative

1 G. B. Porter, V. Balzani, and L. Moggi, *Adv. Photochem.*, 1974, 9, 147.
2 D. J. Robbins and A. J. Thomson, in 'Electronic Structure and Magnetism of Inorganic Compounds', ed. P. Day (Specialist Periodical Reports), The Chemical Society, London, 1977, vol. 5, p. 153.
3 D. P. Shcherbov and R. N. Plotnikova, *Zavod. Lab.*, 1976, 42, 1429.
4 Z. Stasicka and A. Marchaj, *Coord. Chem. Rev.*, 1977, 23, 131.
5 G. A. Shagisultanova, *Reakts. Sposobnost Koordinats. Soedin.*, 1976, 91.
6 C. R. Bock and E. A. Körner von Gustorf, *Adv. Photochem.*, 1977, 10, 221.
7 S. Sakka, *Kagaku Kogyo*, 1978, 29, 189.
8 H. Hennig, P. Thomas, R. Wagener, D. Rehorek, and K. Jurdeczka, *Z. Chem.*, 1977, 17, 241.
9 G. Stein, *Curr. State Knowl. Photochem. Form. Fuel Rep. Workshop*, 1974, 59.
10 J. K. Burdett and J. J. Turner, in 'Cryochemistry', ed. M. Moskovits and G. A. Ozin, Wiley, New York, 1976, p. 493.
11 V. P. Gruzdev, *Zhur. neorg. Khim.*, 1977, 22, 3208.
12 J. A. Kemlo and T. M. Shepherd, *Chem. Phys. Letters*, 1977, 47, 158.
13 L. G. Vanquickenborne and A. Ceulemans, *J. Amer. Chem. Soc.*, 1977, 99, 2208.

to Adamson's rules. A report has appeared[14] of a new type of calorimetry, *photocalorimetry*, in which the enthalpy of a light-induced reaction may be determined. The ΔH is calculated as $(F^0 - F)/n$, where F^0 is the light flux, F the observed rate of heat production, and n the moles of reaction per second. Photocalorimetry suffers in precision because it depends on $(F^0 - F)$ and is, therefore, insensitive if Φ is small.

2 Titanium

Irradiation of solutions of Ti^{III} salts in the presence of catalytic amounts of cuprous ions proceeds[15] with the stoicheiometry shown in equation (1).

$$Ti^{III} + H^+ \xrightarrow{\ h\nu\ } Ti^{IV} + \tfrac{1}{2}H_2 \qquad (1)$$

The photo-oxidation probably occurs by the mechanism detailed in steps (2)—(4).

$$Cu^{I} \xrightarrow{\ h\nu\ } Cu^{I*} \qquad (2)$$

$$Cu^{I*} + H^+ \longrightarrow Cu^{II} + \tfrac{1}{2}H_2 \qquad (3)$$

$$Cu^{II} + Ti^{III} \longrightarrow Cu^{I} + Ti^{IV} \qquad (4)$$

A method of determining titanium fluorimetrically involving the 1 : 2 complexes with the Schiff base benzal-2-amino-3-cyano-4,5-diphenylfuran and its dimethoxy derivative has been reported.[16]

3 Vanadium

Irradiation of solutions of $VOCl_3$ in aqueous ethanol leads to V^{III} and proceeds with evolution of hydrogen. The quantum yields of each stage in the reduction have been determined and the effects of C_{1-5} alcohols on the reaction investigated.[17] These results are supported by two other studies. In the first,[18] a single-electron photoreduction of V^V to V^{III} in two stages has been observed in deaerated solutions containing alcohols of equal and higher concentrations; this is accompanied by hydrogen evolution. The quantum yield of the photo-reduction increases with alcohol concentration and its oxidation products also appear to participate in the reaction. Secondly,[19] using 254-nm light, irradiation of $VOCl_3$ in ethanol solution shows the possibility of the stepwise reduction, $V^V \to V^{IV} \to V^{III} \to V^{II}$, the kinetics and quantum yields of which have been investigated.

A study has been made[20] of the photolysis of some V^{VI} chelates of the form VOL_2OR [reaction (5); L = 8-quinolyloxo, R = alkyl]. The products arise

[14] A. W. Adamson, A. Vogler, H. Kunkely, and R. Wachter, *J. Amer. Chem. Soc.*, 1978, **100**, 1298.

[15] K. L. Stevenson and D. D. Davis, *Inorg. Nuclear Chem. Letters*, 1976, **12**, 905.

[16] A. T. Tashkhodzhaev, L. E. Zel'tser, T. Sabirova, and L. A. Morozova, *Doklady Akad. Nauk Uzb. S.S.R.*, 1976, **4**, 32.

[17] A. I. Kryukov, S. Ya. Kuchmii, I. S. Shchegoleva, and A. V. Korzhak, *Teor. Eksp. Khim.*, 1977, **13**, 135.

[18] I. S. Shchegoleva and A. I. Kryukov, *Dopov. Akad. Nauk Ukr. R.S.R.*, Ser. B, 1977, **5**, 429.

[19] S. Ya, Kuchmii, A. V. Korzhak, and A. I. Kryukov, *Doklady Akad. Nauk S.S.S.R.*, 1977, **235**, 354.

[20] S. Aliwi and C. H. Bamford, *J.C.S. Faraday I*, 1977, **73**, 776.

$$L-\overset{\overset{O}{\|}}{\underset{\underset{OR}{|}}{V}}-L \quad \xrightarrow{h\nu} \quad L-\overset{\overset{O}{\|}}{V}-L \ + \ \dot{O}R \qquad (5)$$

by a first-order process and are V^{IV} chelates (VOL_2). It is concluded that the primary photochemical process is scission of the V—OR bond; thus photolysis of these readily accessible chelates should provide a convenient and widely applicable route to alkoxy-radicals. Since thermal decomposition of the chelates is very slow at temperatures below 100 °C, the technique can be used over a reasonably wide range of temperatures.

4 Chromium

The $^2E \rightarrow {}^4A_2(O)$ luminescence and $^4A_2 \rightarrow {}^2E, {}^2T_1$ absorption spectra of *cis*-[Cr(en)$_2F_2$]I, *cis*-[Cr(en)$_2X_2$]X,H$_2$O (X = Cl or Br), and NH$_4$[Cr(en)F$_4$] have been measured[21] between 80 and 300 K. A discrepancy is found between the experimentally determined separation of the two components of the ^2E state and that calculated from the complete weak-field matrices of the d^3-tetragonal crystal field-interelectron repulsion-spin–orbit coupling-Trees correction perturbations. A study has appeared[22] of the lifetimes for the doublet state emission of *trans*-[Cr(NH$_3$)$_2$(NCS)$_4$]$^-$ at room temperature for twenty solvents and also as a function of solvent composition in H$_2$O–MeCN mixtures. From this it appears that the emission lifetimes are determined primarily by the rate at which the doublet excited state undergoes chemical reactions. A related study has been carried[23] out on [Cr(NCS)$_6$]$^{3-}$, [Cr(NH$_3$)$_5$(NCS)]$^{2+}$, and Reineckate.

To account for the observation that the lifetime of the ^2E state of [Cr(bipy)$_3$]$^{3+}$ is prolonged in the presence of high concentrations of certain salts, a model is suggested[24] based upon extensive ion-pairing and the replacement of interligand water molecules by ClO_4^-. The effect is to decrease the vibrational freedom of the ligands and the efficiency of energy transfer between the metal-centred state and the vibrational modes of the ligands. The phosphorescence intensity, lifetime, and doublet absorption of [Cr(bipy)$_3$]$^{3+}$ are[25] independent of pH from 1.0 to 9.6, whereas the photoaquation quantum yield increases from $\leqslant 10^{-3}$ at pH 4 to ~ 0.1 at pH 9. Added I$^-$ causes complete phosphorescence quenching together with at least 95% quenching of the photoreaction, an observation which indicates that the transient absorption is due to the same excited state which is responsible for the phosphorescence emission. The phosphorescence quenching of *trans*-[Cr(en)$_2$(NCS)$_2$]$^+$ by a variety of substrates including HO$^-$ has been studied.[26] The results indicate that the interaction between the (^2E) state of *trans*-[Cr(en)$_2$(NCS)$_2$]$^+$ and HO$^-$ quenches the 'intrinsic' manifestations of the ^2E excited state but introduces a new path for

21 C. D. Flint, A. P. Matthews, and P. J. O'Grady, *J.C.S. Faraday II*, 1977, **73**, 655.
22 A. R. Guttierez and A. W. Adamson, *J. Phys. Chem.*, 1978, **82**, 902.
23 S. C. Pyke, M. Ogasawara, L. Kevan, and J. F. Endicott, *J. Phys. Chem.*, 1978, **82**, 302.
24 M. S. Henry, *J. Amer. Chem. Soc.*, 1977, **99**, 6138.
25 M. Maestri, F. Bolletta, L. Moggi, V. Balzani, M. S. Henry, and M. Z. Hoffmann, *J.C.S. Chem. Comm.*, 1977, 491.
26 D. Sandrini, M. T. Gandolfi, A. Juris, and V. Balzani., *J. Amer. Chem. Soc.*, 1977, **99**, 4523.

NCS$^-$ aquation from ^2E. Consequently, HO$^-$ should not be used as a doublet quencher in mechanistic studies involving Cr^{III}. Irradiation of *trans*-[Cr(en)$_2$-(NCS)$_2$]$^+$ leads[27] to release of NCS$^-$ and H$^+$ consumption. Sensitization experiments have also been carried out, and the results are discussed in terms of the spin–orbit coupling framework of Kane–Maguire.

The stereochemistry of some photoinduced substitution reactions of Cr^{III} involving ligand loss and isomerization, and followed by nucleophilic attack of an entering ligand, has been analysed[28] using correlation diagrams. This has led to the formulation of a set of electronic selection rules which can be restated in terms of either the FO concept or the conservation of orbital symmetry. The results of a study of the ability of leaving ligand to behave as an influential parameter in the photolysis of *trans*-[Cr(en)$_2$(NCS)Br]$^+$ have been compared[29] with those previously reported for *trans*-[Cr(en)$_2$(NCS)Cl]$^+$. In both ground and excited state, the behaviour of the leading ligands is the same, suggesting that Adamson's arguments based on the ligand field are not in themselves sufficient to explain the photochemistry of complexes of Cr^{III}. Studies of the CT photochemistry of [Cr(NH$_3$)$_5$N$_3$]$^{2+}$ under a variety of conditions have revealed[30] the operation of both redox and nitrene pathways, both of which show definite threshold modes for their quantum yields. The primary steps can be accounted for by the mechanistic scheme given in steps (6)—(8).

$$[(NH_3)_5Cr-N_3]^{2+} \xrightarrow{h\nu} \begin{array}{l} (NH_3)_5Cr^{III}-N^- + N_2 \\ \\ \{(NH_3)_5Cr^{2+}N_3\} \end{array} \qquad (6)$$

$$(NH_3)_5Cr^{III}-N^- \longrightarrow (NH_3)_5Cr^{III}-NH \qquad (7)$$

$$(NH_3)_5Cr^{III}-NH \begin{array}{l} \xrightarrow{HClO_4} [(NH_3)_5Cr-NH_2OH]^{2+} \\ \\ \xrightarrow{HCl} [(NH_3)_5Cr-NH_2Cl]^{3+} \end{array} \qquad (8)$$

Photoisomerization of [Cr^{III}(pfa)$_3$] on irradiation in the *d–d* region occurs[31] with an unexpectedly large difference in quantum yields for *cis–trans* and *trans–cis* isomerizations. This effect, also apparent in other complexes, has been interpreted in terms of the more efficient use by the *cis* isomer of photochemical energy for bond breaking as a result of greater steric strain. Other mechanistic and synthetic photochemistry of tris(β-diketonato)chromium(III) complexes has been investigated.[32] In the photolysis of *trans*-[Cr(en)$_2$NH$_3$F]$^{2+}$ (1) and [Cr(NH$_3$)$_5$F]$^{2+}$ (2) predominant loss of NH$_3$ from the position *trans* to the F ligand in (2) occurs[33] giving 85% *cis*-aquofluoro product, whereas (1) gives 99% *cis*-aquofluoro product. Equatorial loss of NH$_3$ from (2) can, therefore, be associated with stereochemical change and if F is accepted to be a strong

[27] D. Sandrini, M. T. Gandolfi, L. Moggi, and V. Balzani, *J. Amer. Chem. Soc.*, 1978, **100**, 1463.
[28] L. G. Vanquickenborne and A. Ceulemans, *J. Amer. Chem. Soc.*, 1978, **100**, 475.
[29] R. C. Bifano, *Acta Cient. Venez.*, 1977, **28**, 266.
[30] R. Spiram and J. F. Endicott, *Inorg. Chem.*, 1977, **16**, 2766.
[31] S. S. Minor and W. G. Everett, *Inorg. Chim. Acta*, 1976, **20**, L51.
[32] S. S. Minor, *Diss. Abs. Internat. B*, 1977, **38**, 677.
[33] C. F. C. Wong and A. D. Kirk, *Inorg. Chem.*, 1977, **16**, 3148.

σ and π donor, the results are in disagreement with the MO/σ-donor models of Cr^{III} photochemistry. The photochemical behaviour of *trans*-$[Cr(en)_2F_2]^+$ is different from that of the other *trans*-$[Cr(en)_2X_2]^+$ complexes. Regardless of the excitation wavelength, irradiation leads[34] to two photoreactions, *viz.* release of F^- and consumption of H^+ ions. This latter process is the dominant one and is the result of detachment from the metal of one *en* end from the metal. In a related investigation,[35] photolysis of aqueous $[Cr(NH_3)_5F]^{2+}$ at pH 2 gives *cis*-$[Cr(NH_3)_4(H_2O)F]^{2+}$ as the principal product. Changes in the pH of the medium from 2 to 1 do not appear to affect significantly the quantum yields of either NH_3 or F^- loss. However at pH 7, there is a noticeable increase in both quantum yields. In an attempt to provide data on the dependence of photosubstitution reactions on the entering ligand, the photoanation of $[Cr(Me_2SO)_6]^{3+}$ by N_3^- and CNS^- has been studied.[36] The results are interpreted in terms of an associative mechanism.

An investigation has been carried out[37] of formation conditions, spectrophotometric, and fluorescence characteristics of the ion association Cr^{VI} with safranine T cation (L^+). At pH 1, a fluorescent complex is formed, probably $Cr_2O_7^{2-}$: L^+ in the ratio 1 : 2, and since Beer's law is obeyed, this constitutes a useful luminescence determination of Cr^{VI}. It is reported[38] that citric, oxalic, and tartaric acids and 8-hydroxyquinoline can be determined by measurement of the luminescence of their Cr^{III} complexes.

5 Tungsten

Use has been made[39] of the charged micelle surface as a reaction field to investigate the complexation of W^{VI} with 3-hydroxyflavone. This fluorescent chelate complex has its maximum emission at 470 nm on excitation at 400 nm, but the emission is shifted to longer wavelength by about 10 nm in the presence of a micelle. The fluorescence intensity of the complex at the charged micelle surface increases with ligand concentration, but decreases after attaining the maximum owing to dynamic quenching. In the neighbourhood of the charged micelle, surface complexation is prevented because of restriction of the chelating group in the ligand by the micelle.

6 Manganese

Aqueous solutions of $Na[Mn(malonate)_2(H_2O)_2]$ and $Na_3[Mn(malonate)_3]$ have been photolysed[40] at 77 K. From an analysis of the i.r. and e.s.r. spectra of the product, a decarboxylation process appears to be occurring. Solutions in DMF of $K[Mn(C_2O_4)_2(H_2O)_2]$ in the concentration range 1×10^{-4}—2.5×10^{-3}M have been irradiated[41] in the LMCT band using 313-nm light.

34 M. F. Manfrin, D. Sandrini, A. Juris, and M. T. Gandolfi, *Inorg. Chem.*, 1978, **17**, 90.
35 R. E. Wright and A. W. Adamson, *Inorg. Chem.*, 1977, **16**, 3360.
36 C. H. Langford and J. P. K. Tong, *J.C.S. Chem. Comm.*, 1977, 138.
37 A. T. Pilipenko, T. L. Shevchenko, and A. I. Volkova, *Zhur. analit. Khim.*, 1977, **32**, 731.
38 E. A. Solov'ev, E. A. Bozhevol'nov, and S. L. Khacheryan, *Zhur. analit. Khim.*, 1977, **32**, 1459.
39 H. Kohara, *Kitakyushu Kogyo Koto Semmon Gakko Kenkyu Hokoku*, 1977, **10**, 9.
40 V. E. Stel'mashok, A. L. Poznyak, and S. I. Arzhankov, *Koord. Khim.*, 1977, **3**, 524.
41 T. Oncescu, S. G. Ionescu, and C. Andreescu, *Rev. Roumaine Chim.*, 1977, **22**, 697.

A marked dependence of quantum yields on the O_2 and complex concentrations is observed, and these data lend support to the photolysis proposed previously by Murgulescu *et al.* in 1972. Aqueous solutions of $CeMn(SO_4)_2,12H_2O$ containing HCl have been photolysed[42] using a Hg arc lamp and both Cl_2^- and Mn^{II} have been identified as electron transfer products by u.v. spectroscopy.

Irradiation[43] of the stable di-μ-oxo-dimanganese complex $[Mn_2^{IV,III}O_2(bipy)_4](ClO_4)_3$ in acidic solution using 313-nm light is reported to bring about photoreduction of the complex. On the other hand, the related complex $[Mn_2^{IV,IV}O_2(phen)_4](ClO_4)_4$ is reduced in the dark to the Mn^{IV}–Mn^{III} complex and further reduced upon irradiation. Both transformations are accompanied by one-electron oxidation of H_2O to generate $H\dot{O}$. A study is available[44] of phosphorescence in molten tetrabutylammonium tetrahalomanganates.

7 Iron

By use of Raman laser temperature-jump kinetics, the intersystem crossing rates for fourteen (low-spin) $\rightleftarrows$ (high-spin) spin-equilibrium metal complexes of six-co-ordinated Fe^{II}, Fe^{III}, and Co^{II} have been investigated.[45] The results are interpreted in terms of a model in which the spin multiplicity change is treated as an internal electron-transfer reaction.

An investigation[46] of the quenching of thionine by $FeCl_2$ and $FeSO_4$ has produced a linear Stern–Volmer plot; thus the Fe^{II}–thionine complex is probably not implicated in the quenching process. The lifetimes of the triplet excited states of thionine and methylene blue and the first-order rate constants for their reduction by Fe^{II} have been determined[47] in aqueous and 50 vol. % aqueous acetonitrile solutions acidified with $0.005M$-H_2SO_4 or CF_3SO_3H.

In solutions containing SO_4^{2-} the rate constant for the reaction of triplet dyes with Fe^{II} increases with increasing SO_4^{2-} concentration and with solvent change from water to aqueous acetonitrile. The results are explained in terms of the probability of association of the positively charged reactive ions with SO_4^{2-}.

Variable-temperature magnetic susceptibility measurements and Mössbauer spectroscopy have shown[48] that the bis(N-methylethylenediaminesalicylaldiminato)iron(III) complexes are new (low-spin, $S = \frac{1}{2}$) $\rightleftarrows$ (high-spin, $S = \frac{5}{2}$) spin-equilibrium compounds in the solid state, and from Mössbauer spectra an upper limit of $\sim 10^7\ s^{-1}$ has been established for the rate of the intersystem crossing (9) in the solid state.

$$(S = \tfrac{1}{2}) \xrightleftharpoons[k_{-1}]{k_1} (S = \tfrac{5}{2}) \qquad (9)$$

[42] A. L. Poznyak, V. E. Stel'mashok, and V. Rumas, *Khim. Vys. Energ.*, 1977, **11**, 374.

[43] Y. Otsuji, K. Sawada, I. Morishita, Y. Taniguchi, and K. Mizuno, *Chem. Letters*, 1977, 983.

[44] N. Presser, *Diss. Abs. Internat. B*, 1977, **38**, 714.

[45] E. V. Dose, M. A. Hoselton, N. Sutin, M. F. Tweedle, and L. J. Wilson, *J. Amer. Chem. Soc.*, 1978, **100**, 1141.

[46] M. D. Archer, M. I. C. Ferreira, G. Porter, and C. J. Tredwell, *Nouv. J. Chim.*, 1977, **1**, 9.

[47] P. D. Wildes, N. N. Lichtin, M. Z. Hoffman, L. Andrews, and H. Linschitz, *Photochem. Photobiol.*, 1977, **25**, 21.

[48] R. H. Petty, E. V. Dose, M. F. Tweedle, and L. J. Wilson, *Inorg. Chem.*, 1978, **17**, 1064.

In solution, laser Raman temperature-jump kinetics have been employed to measure directly the forward and reverse rate constants for ISC giving $2 \times 10^7 \, \text{s}^{-1} \lesssim k_1 \lesssim 2 \times 10^8 \, \text{s}^{-1}$.

Some theoretically predicted properties of certain hexacyano- and pentacyano-ferrate complexes with nitrogen-containing aromatic ligands have been obtained and the results compared[49] with experimental data. This suggests that a better understanding of the photochemistry can be gained by analysis of the low-energy CT absorption, and the spectroscopic properties of the pentacyanoferrate complexes with aromatic heterocycles are better described by adding the '$d-\pi^*a$' MLCT and the $\pi-\pi^*$ transitions.

Excitation of the CTTL band of pentacyanoferrate(II) complexes [(CN)$_5$-Fe^{II}L]$^{n-}$ (L = aromatic N-heterocyclic base) leads[50] to photosubstitution of L at room temperature in aqueous solutions. The excited state responsible for the photosubstitution is of LF character and efficient interconversion between these states occurs with the magnitude of the quantum yield dependent on whether the LF ($0.1 < \Phi_L < 1.0$) or CTTL ($\Phi < 0.05$) excited state is lower in energy.

Following excitation at 360 nm, the $^3A_2 \rightarrow {}^1A_1$ emission of [Fe(CN)$_5$CO]$^{3-}$ is centred at 18 900 cm^{-1}, and in agreement with the spin-forbidden character of the transition, the measured[51] lifetime is 81 ms. The lack of luminescence from the parent [Fe(CN)$_6$]$^{4-}$ complex is interpreted in terms of radiationless decay to the ground state with possible quenching due to photo-oxidation of the central metal ion.

A study[52] of the thermal reaction, quantum yield, and mechanism of the photochemical reaction of alkaline solutions of [Fe(CN)$_5$NO]$^{2-}$ shows that although irradiation at $\lambda = 576$ and $\lambda = 546$ leads to no photochemistry, irradiation at $\lambda = 436$ nm produces appreciable photodecomposition.

In aqueous solution, flash photolysis studies of the [Fe(CN)$_6$]$^{3-}$–N$_3$ and [Fe(CN)$_6$]$^{3-}$–SCN systems have shown[53] that the rate constants are dependent on ligand concentration; a photoexcited state mechanism is proposed. The quantum yields of the photolysis of [(NC)$_5$Fe(NO)]$^{2-}$ in aqueous solution have been determined[54] at 405, 366, 313, and 254 nm and a correlation established with the total interatomic overlap population calculated according to the Mulliken scheme of population analysis. These results are used to explain the changes in quantum yield of the primary reaction of nitroprusside ions with excitation energy of the ions.

Irradiation[55] of hexa-aquoiron(III) at 240 nm in the presence of alcohols leads to two processes. The first involves a CT transition from a water-centred orbital to an iron-centred orbital and leads to the production of HO· radicals which then abstract H· from the alcohols. The second reaction involves the CT

[49] G. Calzaferri and F. Franz, *Helv. Chim. Acta*, 1977, **60**, 730.
[50] J. E. Figard and J. D. Petersen, *Inorg. Chem.*, 1978, **17**, 1059.
[51] L. Vera and F. Zuloaga, *Inorg. Chem.*, 1978, **17**, 765.
[52] T. Jarzynowski, T. Senkowski, and Z. Stasicka, *Roczniki Chem.*, 1977, **51**, 2299.
[53] D. J. Kenney and H. M. Lin, *Bull. Inst. Chem. Acad. Sin.*, 1976, **23**, 53.
[54] R. Gogolin and M. Klobukowski, *Roczniki Chem.*, 1977, **51**, 207.
[55] J. H. Carey, E. G. Cosgrove, and B. G. Oliver, *Canad. J. Chem.*, 1977, **55**, 625.

excited states of the Fe^{III} and leads to outer-sphere oxidation of the alcohols. For aqueous ethylene glycol, these reactions have been studied in detail.

Preliminary studies of the photochemistry of iron(II)–di-imine complexes in a room temperature molten salt composed of $AlCl_3$ and ethylpyridinium bromide in a 2 : 1 molar ratio are reported.[56] The results indicate that upon irradiation most of the Fe^{II} complexes are converted into the corresponding Fe^{III} complexes. Two mechanisms have been suggested, fast static quenching, and electron transfer from the excited state. Sunlight irradiation of a methanolic Fe complex of 2,3,9,10-tetramethyl-1,4,8,11-tetra-azacyclotetradeca-1,3,8,10-tetraene (TIM), $[Fe^{III}(TIM)(MeOH)(OMe)]^{2+}$ leads[57] to oxidation of the methanol to formaldehyde and, in the absence of O_2, also to stoicheiometric reduction of the Fe^{III} complex to $[Fe^{II}(TIM)(MeOH)_2]^{2+}\cdot$. The initial step is photohomolysis of Fe—OMe giving MeO· and the Fe^{II} complex, which is re-oxidized to the Fe^{III} complex. This cyclic process continues until all available O_2 has been consumed. Irradiation of $[Fe(TIM)XY](PF_6)_2$, X and Y = MeCN, CO, imidazole, or $(EtO)_3P$, has been investigated[58, 59] in various organic solvents. For $[Fe(TIM)-MeCN(CO)](PF_6)_2$ two reaction modes have been defined, loss of MeCN in Me_2CO, and loss of CO in MeCN. The photoreactivity of the $Fe^{II}(TIM)$ complexes can be rationalized in terms of ligand field and CT excited states.

An investigation[60] of the effect of viscosity and co-ordinating ability of the medium and of type and concentration of the entering ligand on the rates and quantum yields of the CT photochemistry of tris(dibenzyldithiocarbamato)-iron(III) with halogenated hydrocarbons has been analysed quantitatively. A new green photosenstive compound $[phenH]_3[Fe(C_2O_4)_3]$ has been prepared[61] and characterized. Photoreduction of the complex gives $[Fe(phen)_3]^{2+}$ making it useful as an actinometric substance.

E.s.r. spin-trapping experiments have been carried out[62] on the photolysate of $[phenH][Fe(C_2O_4)_3]$, and two nitroxide radicals detected. It is concluded that free oxalate radicals are probably the primary products in the photolysis process, whereas the formate radicals, which are also detected, probably arise in a secondary reaction. Following bleaching of $[Fe^{II}(phen)_3]^{2+}$ by a single picosecond pulse at 530 nm, absorption at the excitation wavelength recovers in a sub-ns three-stage process.[63] Absorption coefficients and decay times suggest that one CT and two d–d excited states are consecutively occupied before ground-state repopulation.

Fe^{II} is produced[64] ($\Phi = 0.110 \pm 0.002$) on irradiation of $Fe(ClO_4)_3$ in the presence of t-amyl alcohol. Studies using other t-alcohols (but not t-butanol) show similar low yields, probably owing to the formation of γ-hydroxyalkyl radicals which are incapable of reducing Fe^{III} to Fe^{II}. The results support the

[56] H. L. Chum, D. Koran, and R. A. Osteryoung, *J. Amer. Chem. Soc.*, 1978 **100**, 310.
[57] D. W. Reichgott and N. J. Rose, *J. Amer. Chem. Soc.*, 1977, **99**, 1813.
[58] M. J. Incorvia and J. I. Zink, *Inorg. Chem.*, 1977, **16**, 3161.
[59] J. I. Zink and M. J. Incorvia, *J.C.S. Chem. Comm.*, 1977, 730.
[60] P.-H. Liu and J. I. Zink, *J. Amer. Chem. Soc.*, 1977, **99**, 2155.
[61] P. Thomas, M. Benedix, and H. Hennig, *Z. Chem.*, 1977, **17**, 114.
[62] D. Rehorek, M. Benedix, and P. Thomas, *Inorg. Chim. Acta*, 1977, **25**, L100.
[63] A. J. Street, D. M. Goodall, and R. C. Greenhow, *Lasers Chem. Proc. Conf.*, ed., M. A. West, Elsevier, Amsterdam, 1977, 311.
[64] J. H. Carey, B. G. Oliver, and C. H. Langford. *Canad. J. Chem.*, 1977, **55**, 1207.

suggestion that H on the C bearing O is necessary for direct reaction between the scavenger and the Fe^{III} CT excited state.

The mechanism and kinetics of the photopolymerization of acrylamide have been determined[65] in the presence of potassium trisoxalatoferrate as initiator. The rate of monomer consumption was found to be proportional to the monomer and initiator concentration, concentration of the formed complex, and light intensity. A selective kinetic method for the determination of trace amounts of Fe and Co has been described,[66] which is based upon the ability of these metals to catalyse the chemiluminescent oxidation of luminol by H_2O_2.

8 Ruthenium

A discussion of electron transfer reactions of CT excited states with particular reference to Ru^{II} has been presented[67] and the excited state processes of [Ru(bipy)$_3$]$^{2+}$ have been discussed.[68]

A flash photolysis investigation of [Ru(bipy)$_3$]$^{2+}$ has been undertaken[69] in the presence of NN-dimethylaniline and paraquat diperchlorate as quenchers. It is found that the transient spectral changes arise from back electron transfer reactions and it appears that disproportionation of *[Ru(bipy)$_3$]$^{2+}$ can be catalysed in the presence of two different one-electron redox couples. The same technique has also been used[70] to study reduction of the excited CT state of [Ru(bipy)$_3$]$^{2+}$ by other organic donors. Except for diphenylamine, for which a rate constant of 2.5×10^7 l mol^{-1} s^{-1} was obtained, the reduction is by electron transfer and the rate is diffusion controlled.

TMPD, dimethylaniline, p-anisidine, and other amines in MeCN will[71] all reductively quench *[Ru(bipy)$_3$]$^{2+}$. Support comes from the observation that the absorption spectrum of [Ru(bipy)$_3$]$^{+}$, generated by electrochemical reduction of [Ru(bipy)$_3$]$^{2+}$, agrees well with the spectrum obtained by flash photolysis.

A related study of excited state processes has also been reported.[72] Based on the redox quenching of the CT excited state of [Ru(bipy)$_3$]$^{2+}$, quenching and back electron transfer have been observed for a series of excited states including CT[Ru(phen)$_3$]$^{2+}$, π,π*[Pd(OEP)] (OEP = octaethylporphyrin), and f-f[Eu(phen)$_3$]$^{3+}$. It has been found that in suitable systems containing more than one redox reagent, the quenching and back electron transfer steps can be separated and a study made of reactions which are driven photocatalytically by excited states like [Ru(bipy)$_3$]$^{2+}$. Electron transfer reactions of complexes of the type [RuL$_3$]$^{2+}$ [L = compound (3), R = isopropyl, cyclohexyl, PhCH$_2$, α-naphthyl, or dihydrocholesteryl] have also been studied.[73] In a number of

[65] K. Sahul, L. V. Natarajan, and Q. Anwaruddin, *J. Polymer Sci. Polymer Letters Edition*, 1977, **15**, 605.

[66] V. I. Rigin and A. I. Blokhin, *Zhur. analit. Khim.*, 1977, **32**, 312.

[67] G. S. Laurence, *Mol. Rate Processes, Pap. Symp.*, R. Austral. Chem. Inst., Parkville, Australia, 1975, G2.

[68] J. C. Van Houten, *Diss. Abs. Internat. B*, 1977, **38**, 681.

[69] J. K. Nagle, R. C. Young, and T. J. Meyer, *Inorg. Chem.*, 1977, **16**, 3366.

[70] M. Maestri and M. Graetzel, *Ber. Bunsenges. Phys. Chem.*, 1977, **81**, 504.

[71] C. P. Anderson, D. J. Salmon, T. J. Meyer, and R. C. Young, *J. Amer. Chem. Soc.*, 1977, **99**, 1980.

[72] T. J. Meyer, *Israel J. Chem.*, 1977, **15**, 200.

[73] P. J. DeLaive, J. T. Lee, H. W. Sprintschnik, H. Abruna, T. J. Meyer, and D. G. Whitten, *J. Amer. Chem. Soc.*, 1977, **99**, 7094.

$$CO_2R \qquad CO_2R$$

(3)

cases, the hydrophobic barrier surrounding the Ru cation core reduces the rates of electron transfer reactions relative to that for $[Ru(bipy)_3]^{2+}$, so that in the quenching by neutral amines, the reactive radical-cation can dominate the energy-wasting back electron transfer step and allow formation of the high energy $[RuL_3]^+$ in good yield.

In contrast with previous claims,[74] it has now been reported[75] that highly purified samples of the dioctadecyl ester of (4,4'-dicarboxy-2,2'-bipyridine)bis-(2'2'-bipyridine)ruthenium(II) are inactive as catalysts for the photocleavage of water in monolayer assembles. This present conclusion is also supported by other workers[76] who report a detailed description of their reaction conditions.

A study[77] has been made of the photolysis and radiolysis of $[Ru(bipy)_3]^{2+}$ in sodium dodecyl sulphate micelles. The results indicate that the Ru complex resides in the negatively charged ionic atmosphere and that it interacts strongly with the hydrocarbon chain of the surfactant. Under the conditions used, essentially all of the complex is bound. Both $[Ru(2,2'-bipyridine)_2(2,2'-bipyridine-4,4'-dicarboxylic acid)]^{2+}$ and its deprotonated form emit in aqueous solution, so direct measurement of the excited state equilibrium is possible by means of a luminescence titration.[78] At pH 3.5, it is mainly the deprotonated form which is excited, but emission is also observed from the protonated form. This is the first observation involving a metal complex in which excited state proton transfer can occur without excited state deactivation.

A number of geometric isomers of β-diketonate complexes of Co^{III} having identical electronic spectra quench $[Ru(bipy)_3]^{2+}$ and $[Ru(bpic)_3]^{2+}$ at a diffusion-controlled limit: both *cis* and *trans* forms yield identical Stern–Volmer constants.[79] In contrast, Cr^{III} isomers quench at a rate which is an order of magnitude less than this. Related work[80] has revealed that *cis*-$[Cr(bipy)_2Cl_2]^+$ and *cis*-$[Cr(phen)_2Cl_2]^+$, alone among a group of Cr^{III} complexes studied, quench $[Ru(bipy)_3]^{2+}$ phosphorescence at nearly diffusion controlled rate, whereas the remainder quench significantly more slowly.

The rates of electron transfer between $[Ru(bipy)_3]^{3+}$ and $[Fe(phen)_3]^{2+}$ and between $[Ru(phen)_3]^{3+}$ and $[Ru(bipy)_3]^{2+}$ have been measured[81] by differential

[74] G. Sprintschnik, H. W. Sprintschnik, P. P. Kirsch, and D. G. Whitten, *J. Amer. Chem. Soc.*, 1976, **98**, 2337.
[75] G. Sprintschnik, H. W. Sprintschnik, P. P. Kirsch, and D. G. Whitten, *J. Amer. Chem. Soc.*, 1977, **99**, 4947.
[76] L. J. Yellowlees, R. G. Dickinson, C. S. Halliday, J. S. Bonham, and L. E. Lyons, *Austral. J. Chem.*, 1978, **31**, 431.
[77] D. Meisel, M. S. Matheson, and J. Rabani, *J. Amer. Chem. Soc.*, 1978, **100**, 117.
[78] P. J. Giordano, C. R. Bock, M. S. Wrighton, L. V. Interrante, and R. F. X. Williams, *J. Amer. Chem. Soc.*, 1977, **99**, 3187.
[79] S.-M. Y. Huang and H. D. Gafney, *J. Phys. Chem.*, 1977, **81**, 2602.
[80] K. Nakamaru and H. Nakagawa, *Sci. Rep. Hirosaki Univ.*, 1977, **24**, 22.
[81] R. C. Young, F. R. Keene, and T. J. Meyer, *J. Amer. Chem. Soc.*, 1977, **99**, 2468.

excitation flash photolysis. The net reactions are slightly below the diffusion-controlled limit and a slight barrier exists arising from reorganization of outer co-ordination sphere solvent molecules.

Flash photolysis and pulse irradiation of aqueous solutions of $[Ru(bipy)_3]^{2+}$ lead [82] to e_{aq}^- production probably by electron ejection from the first triplet of $[Ru(bipy)_3]^{2+}$ upon absorption of a second photon; the possibility of a biphotonic effect leading to H_2 formation is discussed. In related work, the reaction of e_{aq}^- with the excited states of $[Ru(bipy)_3]^{3+}$ has been studied. [83] About 38% of e_{aq}^- yields $(CT)[Ru(bipy)_3]^{2+}$, which decays to ground state both by first-order conversion and through quenching by $[Ru(bipy)_3]^{3+}$. However, more than 55% yields another excited state or co-ordinated radical complex which decays to $[Ru(bipy)_3]^{2+}$ almost solely through a reaction with $[Ru(bipy)_3]^{3+}$ without conversion into $(CT)[Ru(bipy)_3]^{2+\cdot}$.

Photoanation of $[Ru(bipy)_3](SCN)_2$ in DMF proceeds [84] by SCN^--dependent and SCN^--independent pathways. In view of the variation of the quantum yield with $[SCN^-]$, an ion-pair/ion-triplet mechanism seems probable.

Irradiation, using visible light, of aqueous solutions (pH $\sim$ 7.5) of a weak reducing agent (triethanolamine: TEA), $Ru(bipy)_3Cl_2$, a rhodium bipyridine complex, and a platinum catalyst results in generation of appreciable quantities of hydrogen. [85] The TEA/TEA–H^+ buffer probably serves as an electron donor and proton source, with the Ru complex being the photoactive species and the Rh complex the storage for electrons and protons by hydride formation.

The complexes $Ru(bipy)_2(CN)_2$ and $Ru(phen)_2(CN)_2$ are very useful sensitizers of transition-metal complexes because, in addition to their intense long-lived fluid solution emissions, they are free from ion-pairing problems and the large charge dependence of the quenching rates with the tris complexes. [86] Furthermore, varying the solvents enables a tuning of the energy of the excited states to be achieved. The quenching mechanism appears to be by energy and electron transfer.

A study has been made [87] of the quenching by oxygen of sixteen luminescent di-imine complexes of Ru^{II}, Os^{II}, and Ir^{II}. Singlet oxygen is produced by complexes of all three metals, probably from the CT excited states of Ru^{II} and Os^{II} and from the $\pi-\pi^*$ triplet states of the Ir^{III} complexes. Quantum yields for the solvation of $[Ru(NH_3)_5py-x]^{2+}$, where py-x is a substituted pyridine or related aromatic heterocycle, have been measured in several different solvents. [88] The results demonstrate that systematic parameters such as ligand substituents, solvent, and temperature can be used to tune the photochemical properties of these complexes.

An examination of the photochemistry at 265 nm of $[Ru(NN\text{-dialkyldithio-carbamato})_3]$ in chlorinated solvents has shown that only $[Ru(R_2dtc)_3Cl]$ and

[82] D. Meisel, M. S. Matheson, W. A. Mulac, and J. Rabani, *J. Phys. Chem.*, 1977, **81**, 1449.
[83] C. D. Jonah, M. S. Matheson, and D. Meisel, *J. Amer. Chem. Soc.*, 1978, **100**, 1449.
[84] P. E. Hoggard and G. B. Porter, *J. Amer. Chem. Soc.*, 1978, **100**, 1457.
[85] J. M. Lehn and J. P. Sauvage, *Nouv. J. Chim.*, 1977, **1**, 449.
[86] J. N. Demas, J. W. Addington, S. H. Peterson, and E. W. Harris, *J. Phys. Chem.*, 1977, **81**, 1039.
[87] J. N. Demas, E. W. Harris, and R. P. McBride, *J. Amer. Chem. Soc.*, 1977, **99**, 3547.
[88] G. Malouf and P. C. Ford, *J. Amer. Chem. Soc.*, 1977, **99**, 7213.

$a[Ru_2(R_2dtc)_5]^+Cl^-$ are formed.[89] Since the electronic absorption spectra of complexes with the MS_6 core are very rich, consisting primarily of intense CT bands, the photochemistry of these complexes is expected to be characteristic of reactions from CT excited states.

The dependence of photoelectron formation in aqueous solution on wavelength has been investigated[90] for $[Ru(CN)_6]^{4-}$ and $[W(CN)_8]^{4-}$. The general mechanism suggested for their photochemistry is given in steps (10) and (11).

$$[M(CN)_x]^{4-} \xrightarrow[\text{CT or CTTS}]{h\nu} [M(CN)_x]^{4-*} \longrightarrow [M(CN)_x]^{3-} + e_{aq} \quad (10)$$

$$[M(CN)_x]^{4-} \xrightarrow[\text{LF}]{h\nu} [M(CN)_x]^{4-*} \longrightarrow \text{photoaquation} \quad (11)$$
$$\text{products}$$

9 Osmium

A photochemically induced *cis–trans* isomerization of solid chloroiodo-osmates-(IV) has been observed[91] spectrophotometrically at $-50\ °C$. The quantum yields are in the range 10^{-1}—10^{-2} and the temperature dependence is low, the rearrangement occurring at nearly the same rate at $77\ K$; an intramolecular exchange mechanism is proposed.

In various polar solvents (H_2O, MeCN, or pyridine), the irradiation of hexahalo-osmates(IV) leads[92] to the products of monosubstitution, *e.g.* $[OsCl_5-(MeCN)]^-$, $[OsBr_5(MeCN)]^-$, and $[OsCl_5py]^-$; quantum yields are in the region 313—436 nm, are wavelength-dependent, and vary between 10^{-2} and 10^{-3}.

Photolysis of $[OsBr_6]^{2-}$ in CH_2Cl_2 and 1,2-dichloroethane leads to the mixed ligand complexes $[OsCl_nBr_{6-n}]^{2-}$ in which for $n = 2$, 3, or 4, the statistical ratio of the stereoisomers is observed following ionophoretic separation.

A study is reported[93] of the photoprocesses occurring between excited $Os(bipy)_3^{2+}$ in the presence of various $Co(NH_3)_5X^{n+1}$ [complexes of oxidation state (III)]. Luminescence quenching results in the formation of $[Os(bipy)_3]^{3+}$ and Co^{2+} in equimolar amounts, and taken together with the consistency of the reaction stoicheiometry as well as the high limiting yields of Co^{2+}, this suggests electron transfer as the quenching mechanism.

10 Cobalt

Photochemical decomposition of $[Co(EDTA)_4]^-$ proceeds[94] differently from that of analogous Fe^{III} complexes. The results suggest a mechanism in which primary photolysis yields Co^{II}, CO_2, and a radical which attacks the ligand in preference to reduction of the slowly reducible metal oxidation state, Co^{III}.

[89] K. W. Given, B. M. Mattson, M. F. McGuiggan, G. L. Miessler, and L. H. Pignolet, *J. Amer. Chem. Soc.*, 1977, **99**, 4855.
[90] O. Kalisky and M. Shirom. *J. Photochem.*, 1977, **7**, 215.
[91] W. Hasenpusch and W. Preetz, *Z. anorg. Chem.*, 1977, **432**, 101.
[92] W. Hasenpusch and W. Preetz, *Z. anorg. Chem.*, 1977, **432**, 107.
[93] E. Finkenberg, P. Fisher, S.-M. Y. Sze, and H. D. Gafney, *J. Phys. Chem.*, 1978, **82**, 526.
[94] C. H. Langford and G. W. Quance, *Canad. J. Chem.*, 1977, **55**, 3132.

Irradiation at $\lambda = 460$—585 of aqueous $[Co(en)_3]^{3+}$ leads[95] to ligand substitution with bis-(2-hydroxyethyl)dithiocarbamate ion (L) to give $[CoL_3]$, $\Phi = 0.343$—0.845, which is high compared with those of the photo-ligand-substitution reactions of Co^{III} complexes so far known.

In acidic solution, the photoaquation of *trans*-$[Co(en)_2(NH_3)Cl]^{2+}$ occurs[96] with essentially complete stereoretention, but the principal photoaquation paths for *cis*-$[Co(en)_2(NH_3)Cl]^{2+}$ are more complicated in that they yield mainly *cis*-$[Co(en)_2(NH_3)(H_2O)]^{3+}$ as the chloride photoaquation product, and mainly *trans*-$[Co(en)_2(H_2O)Cl]^{2+}$ as the NH_3 photoaquation product. These results can be accounted for in terms of Adamson's photolysis rules[97] with the added assumption that if one end of an ethylenediamine is labilized, it may undergo an edge displacement to detach an adjacent unidentate ligand.

On excitation ($\lambda = 254$) of $[Co(NH_3)_5(MeCN)]^{3+}$(CTTM) and $[Co(NH_3)_5-(PhCN)]^{3+}$ (IL, $\pi-\pi^*$), the dominant reaction in both complexes is photoreduction[98] to Co^{2+}. However, in the benzonitrile complex the $IL_{\pi\pi^*}$ state can undergo internal conversion to give a redox active CTTM state, but it may also have independent pathways for deactivation, perhaps including internal conversion into the lower energy ligand-field states. At 313 nm there is a reversal of the relative redox activity.

The effect of solvent viscosity on the extent of radical breakdown and, in the case of the NO_2 complex, linkage isomerization, has been studied[99] in the photolysis of $[Co(NH_3)_5L]^{2+}$ ($L = NO_2$ or N_3) in aqueous hydroxylic media at 365 nm. Addition of glycerol or glycol brings about $>30\%$ decrease in quantum yield but cyclohexanol is without effect, suggesting that a more specific role exists for the added solvent.

The products from photolysis[100] of *trans*-$[Co^{III}(en)_2(S_2O_3)_2]^-$ at 25 °C are $[Co(en)_2(S_2O_3)(H_2O)]^+$ (partly converted into $[Co(en)_2(S_2O_3)(OH)]$), $Co^{II}aq$, $(CH_2NH_2)_2$, and oxidation products of $S_2O_3^{2-}$. However, at 77 K in a $(CH_2OH)_2$ matrix, only minor spectral changes occur and the primary processes are electron transfer from $S_2O_3^{2-}$ to Co^{III}, and direct photolysis of co-ordinated $S_2O_3^{2-}$ in contact with water. E.p.r. studies[101] of aqueous $[Co(CN)_5X]^{3-}$ ($X = Cl^-, Br^-, N_3^-$, or CN^-) irradiated at 100 K and $\lambda = 254$ nm have shown that the pentacyanocobaltate ion is formed. Photolysis of $Co(acac)_3$ in benzene solution leads[102] to ligand-deficient products which undergo recombination processes. These reactions can be scavenged by addition of metal salts to form stable acetylacetonates which are transparent over the wavelength range of interest.

An investigation[103] of the thermal and photoredox reactions of $[Co(phen)_2-(C_2O_2S_2)]X$ ($X = Cl$, Br, I, NO_3, ClO_4, or AcO) has shown that the thermal

[95] M. Nakashima and S. Kida, *Chem. Letters*, 1977, 857.
[96] R. A. Pribush, R. E. Wright, and A. W. Adamson, *J. Amer. Chem. Soc.*, 1977, **99**, 2495.
[97] A. W. Adamson, *J. Phys. Chem.*, 1967, **71**, 798.
[98] A. W. Zanella, K. H. Ford, and P. C. Ford, *Inorg. Chem.*, 1978, **18**, 1051.
[99] P. Natarajan, *J.C.S. Dalton*, 1977, 1400.
[100] K. P. Balashev, I. P. Serova, and G. A. Shagisultanova, *Koord. Khim.*, 1977, **3**, 82.
[101] K. P. Balashev and G. A. Shagisultanova, *Khim. Vys. Energ.*, 1977, **11**, 371.
[102] T. Tominaga and Y. Nishi, *Radiochem. Radioanalyt. Letters*, 1977, **31**, 129.
[103] H. Hennig, W. Kumpf, K. Jurdeczka, and R. Benedix, *J. Prakt. Chem.*, 1977, **319**, 444.

reactions lead to the formation of $[Co(phen)_2X_2]$, $[Co(phen)_2(C_2O_2S_2)]$, and COS, except for X = NO_3 and AcO.

The photoinitiated polymerization of acrylamide using pentammineoxalato-cobalt(III) chloride is reported to depend[104] on the first power of monomer concentration in aqueous media at pH 4.05—5.23. A first-order relationship in monomer and in H_2O–glycol mixtures was also found for tetra-amminebis-oxalatocobalt(III) chloride, thereby eliminating the possibility of radical termination. The influence of viscosity is also studied.

A study has been made[105] of the effect of $Co(OAc)_2,4H_2O$ and other additives of the chemiluminescence in pseudocumene oxidation.

11 Rhodium

Photolysis of the lowest ligand field band of aqueous *cis*-$[Rh(NH_3)_4Cl_2]^+$ gives[106] *trans*-$[Rh(NH_3)_4(H_2O)Cl]^{2+}$ in which aquation of Cl takes place concomitantly with isomerization to the *trans* configuration. As inhibition of the stereo rearrangement appears to have little effect with Rh^{III} complexes (the quantum yield for this reaction is very close to that measured for chloride aquation from *cis*-$[Rh(cyclam)Cl_2]^+$), it appears that the rearrangement occurs subsequently to the chloride aquation, perhaps by the rearrangement of a five-co-ordinate intermediate formed by photodissociation of Cl^-. These results suggest strongly that the *trans*-configuration may be expected regardless of whether a *cis* or *trans* NH_3 were photoaquated.

Studies on the complex *trans*-$Rh(tfa)_3$ (tfa is the anion of 1,1,1-trifluoropentane-2,4-dione) have shown[107] the existence of two highly solvent-dependent photoinduced processes. Thus in cyclohexane, irradiation leads to *trans–cis* isomerization only, whereas in propan-2-ol, photodecomposition is the only observable process: in mixtures of these solvents, both processes occur. A wavelength dependence is apparent and oxygen is found to be a quencher for the system, both of which suggest the existence of more than one photoactive state in *trans*-$Rh(tfa)_3$. The enhanced photoreactivity in the presence of H atom donors suggests that at least one of these photoactive states imparts considerable radical character to the complex. These results stand in sharp contrast with most first row transition metal β-diketonate complexes, irradiation of which induces either stereochemical rearrangement (Cr) or redox decomposition (Mn, Fe, Co, Ni, Cu).

Ligand-field photolysis of $[Rh(en)_3]Cl_3,3H_2O$ in deoxygenated 1M-HCl leads[108] to *cis*-$[Rh(en)_2(enH)Cl]Cl_3,2H_2O$ together with a small amount of *trans*-$[Rh(en)_2Cl_2]^+$ resulting from a secondary reaction. The retention of stereochemistry in the photolysis of $[Rh(en)_3]^{3+}$ to form *cis*-$[Rh(en)_2(enH)Cl]^{3+}$ and the presence of *trans*-$[Rh(en)_2Cl_2]^+$ as a secondary photolysis product indicate that the photocatalysed exchange of Cl^- for enH$^+$ (second step) is

104   R. Bhaduri and S. Aditya, *Makromol. Chem.*, 1977, **178**, 1385.
105   S. Ivanov and M. Khinkova, *Doklady Bolg. Akad. Nauk*, 1977, **30**, 1145.
106   D. Strauss and P. C. Ford, *J.C.S. Comm.*, 1977, 194.
107   P. A. Grutsch and C. Kutal, *J. Amer. Chem. Soc.*, 1977, **99**, 7397.
108   J. D. Petersen and F. P. Jakse, *Inorg. Chem.*, 1977, **16**, 2845.

proceeding through an excited state rearrangement rather than a photodissociative pathway. Photolysis of [Rh(en)$_3$]Cl$_3$ appears to parallel the [Rh(NH$_3$)$_6$]$^{3+}$ system.

Following the discovery that in reducing conditions complexes of Rh with pyridine and halide (X) will combine with dioxygen to give first, transient red–orange species, then isolable blue dimers involving [XRh(py)$_4$O$_2$Rh(py)$_4$X]$^{3+}$, the observation was made that aqueous solutions of salts of the colourless *cis*-[Rh(en)$_2$(NO$_2$)$_2$]$^+$X (X = NO$_3$ or Cl) become red in daylight. Further investigations show[109] that this is due to the formation of monomeric paramagnetic superoxorhodium(III) species [Rh(en)$_2$(NO$_2$)(O$_2$)]$^+$X$^-$ which are relatively stable in water. In the presence of oxygen, the superoxo product from the resolved *cis*-dinitro compound is not optically active, so presumably, an intermediate is stereolabile.

Irradiation at 546 nm of [Rh$_2$(bridge)$_4$] (BF$_4$)$_2$ (bridge = 1,3-di-isocyanopropane) in 12M-HCl yields [Rh$_2$(bridge)$_4$Cl$_2$]$^{2+}$ and H$_2$.[110] It has been shown that the principal species undergoing the photoreaction is [Rh$_2$(bridge)$_4$H]$^{3+}$·Cl$^-$ and $\Phi = 0.0040 \pm 0.002$. The photoreaction is considered to be that given in equation (12). CT excitation $1a_{2u} \to 2a_{1g}$ leads to the photoactive state $^1A_{2u}$ or

$$[Rh_2(bridge)_4H]^{3+}\cdot Cl^- + H^+ + Cl^- \xrightarrow[\text{12M-HCl}]{\lambda\,=\,546nm} [Rh_2(bridge)_4Cl_2]^{2+} + H_2 \tag{12}$$

$^3A_{2u}$ which, it is proposed, is attacked by H$_2$O or H$_3$O$^+$ to give the products shown in steps (13) and (14). Possibly the axial ligand–metal interactions may

$$[(Cl^-)Rh^I \cdots Rh^I(H^+)]^{2+} \xrightarrow{\lambda\,=\,546} [(Cl) \cdots Rh^{I\frac{1}{2}} \cdots Rh^{I\frac{1}{2}} \cdots (H.)]^{2+*} \tag{13}$$

$$[(Cl^-) \cdots Rh^{I\frac{1}{2}} \cdots Rh^{I\frac{1}{2}} \cdots (H.)]^{2+*} \xrightarrow[\text{HCl}aq]{\text{fast}} [Cl—Rh^{II}—Rh^{II}—Cl]^{2+} + H_2 \tag{14}$$

operate in such a way as to reduce the fraction of excited species returning to the Rh^I–Rh^I ground state by back electron transfer. The photoreaction may be useful as a means of producing H$_2$ in solar energy conversion schemes as the thermal back reaction is relatively slow.

12 Iridium

Absorption and emission spectra and emission lifetimes of purified samples of *cis*-[Ir(phen)$_2$Cl$_2$]$^+$ and *cis*-[Ir(5,6-Me$_2$phen)$_2$Cl$_2$]$^+$ have been reported.[111] By photolysis of [Ir(bipy)$_2$Cl$_2$]$^+$ the complexes [Ir(bipy)$_2$Cl(H$_2$O)]$^{2+}$ and [Ir(bipy)$_2$(H$_2$O)$_2$]$^{3+}$ have been prepared[112] and converted by treatment with base into [Ir(bipy)$_2$Cl(OH)]$^+$ and [Ir(bipy)$_2$(OH)$_2$]$^+$. The contrasting effects of protonation of π-donating and π-accepting ligands on CTTL transitions are illustrated by comparision of the effects of protonation of these complexes on their visible

[109] R. D. Gillard, J. D. Pedrosa de Jesus, and L. R. H. Tipping, *J.C.S. Chem. Comm.*, 1977, 58.
[110] K. R. Mann, N. S. Lewis, V. M. Miskowski, D. K. Erwin, G. S. Hammond, and H. B. Gray, *J. Amer. Chem. Soc.*, 1977, **99**, 5525.
[111] R. Ballardini, G. Varani, L. Moggi, and V. Balzani, *J. Amer. Chem. Soc.*, 1977, **99**, 6881.
[112] R. J. Watts, J. S. Harrington, and J. Van Houten, *J. Amer. Chem. Soc.*, 1977, **99**, 2179.

absorption and emission spectra with the effects of protonation of the unidentate bipy complex. The luminescence lifetimes of the unidentate complexes are long in fluid solution ($\sim$10 μs), and their high luminescence quantum yields ($\sim$0.3) indicate that they may be useful as high-energy sensitizers and in flash-lamp-pumped dye lasers. The structures of these complexes provide a model for the structures of reactive intermediates formed in the photolysis of tris-2,2'-bipy complexes such as $[Ru(bipy)_3]^{2+}$. On the basis of slight shifts in the n.m.r. spectrum during photolysis, and the somewhat reversible nature of the reaction, it is suggested[113] that photoinduced association of solvent occurs [reaction (15)]

$$[Ir(MeCN)_4]^+ + S \; \xleftrightarrow{\; h\nu \;} \; [Ir(MeCN)_4(S)]^+ \quad (S = Me_2CO, MeCN, or MeOH)$$
$$(15)$$

to give a stable solvated complex. It is now shown[114] that in solution $[Ir(MeCN)_4]^+$ does not exist as a monomeric complex, but rather as the oligomeric $[Ir(MeCN)_4]_n^{n+}$. Contrary to previous suggestions, the photochemical reaction is not a photoinduced association but rather a photoinduced dissociation of the solution oligomers to generate monomeric $[Ir(MeCN)_4]^+$.

13 Nickel

At 530 nm Q-switched laser photolysis of $[NiEt_4dienCl]Cl$ in acetonitrile results[115] in a rapid ionic photoassociation reaction to give as principal absorbing species at the laser wavelength the planar ion-pair $NiEt_4dienCl^+ \| Cl^-$. Subsequent fast relaxation of the equilibrium between ions and ion-pair accounts for the observed rapid conductivity decrease. The probable reason for the photoassociation is formation of a tetrahedrally distorted excited state of the planar complex which, by lessening the repulsion between the filled dz^2 orbital in the planar $NiLCl^+$ and Cl^-, aids approach of this latter ion to form a five-co-ordinate complex.

14 Palladium and Platinum[72]

The laser flash photolysis technique has been used[116] to measure the quenching rate constants of five organic molecules by $PdCl_4^{2-}$, $PtCl_4^{2-}$, and $Ni(CN)_4^{2-}$. Both $PdCl_4^{2-}$ and $PtCl_4^{2-}$ have high triplet energies and show no quenching. This observation, taken together with the variations in quenching constants observed as a function of the triplet energy of the organic molecule being quenched, supports electronic energy transfer as the quenching mechanism. It has been shown that exothermic energy transfer to these complexes is almost diffusion controlled and much more efficient than triplet energy transfer to many octahedrally co-ordinated complexes in aqueous solution. This is attributed to more favourable orbital overlap in the case of the square planar complexes. In agreement with some theoretical estimates, variation in the quenching constant

[113] W. M. Bedford and G. Rouschias, *J.C.S. Chem. Comm.*, 1972, 1224.
[114] G. L. Geoffroy, M. G. Bradley, and M. E. Keeney, *Inorg. Chem.*, 1978, **17**, 777.
[115] L. Campbell and J. J. McGarvey, *J. Amer. Chem. Soc.*, 1977, **99**, 5809.
[116] K. C. Marshall and F. Wilkinson, *Z. Phys. Chem. (Frankfurt)*, 1976, **101**, 67.

as a function of the triplet energy of the donor supports the assignment of the weak absorption at 17 000 cm^{-1} at the lowest state of PtCl$_4$$^{2-}$ and indicates that the lowest triplet state of PdCl$_4$$^{2-}$ lies at an energy of less than 13 000 cm^{-1}. Irradiation of [Pt(NH$_3$)Cl$_3$]$^-$ in aqueous solution with yellow light leads[117] to chloride aquation with formation of *cis*- and *trans*-[Pt(NH$_3$)(H$_2$O)Cl$_2$] in the ratio 2:1 and a quantum yield and isomer ratio nearly independent of the wavelength of irradiating light and of temperature. On the other hand, irradiation of [Pt(NH$_3$)$_3$Cl]$^+$ in the near u.v. leads to aquation of both NH$_3$ and Cl. If the photoreaction had been an accelerated form of the thermal reaction, the ratio of *cis*- to *trans*-[Pt(NH$_3$)(H$_2$O)Cl$_2$] produced, as deduced from the rate constants for the thermal reactions, should have been *ca*. 10:1. From the fact that this is not the case, it follows that the photochemical reaction takes an entirely different path from the thermal.

Following irradiation at 25 °C using 254—460 nm light, aqueous solutions of *cis*- and *trans*-dichlorobis(propylenediamine)platinum(II) were analysed for free Cl$^-$, NH$_4$$^+$, and H$^+$, and compounds of bivalent Pt.[118] Quantum yields of substitution of chloride ligands by H$_2$O molecules were measured, and reduction of the central metal atom, and oxidation of propylenediamine to NH$_3$ and other products were observed. Irradiation of the *cis*- form yielded the *trans*-isomer, but no *cis*-isomer was produced by u.v. irradiation of the *trans*- form. In order to obtain further insight into the mechanism of ligand photosubstitution reactions of Pt^{II} complexes, a study has been made[119] of the photochemical behaviour of [Pt(Et$_4$dien)Br]$^+$. Irradiation of this complex ion in aqueous solution at 313 nm leads, in acid or neutral solutions, to [Pt(Et$_4$dien)H$_2$O]$^{2+}$ independently of the presence of Cl$^-$ (up to 1 × 10^{-2} M) or Br$^-$ (up to 1 × 10^{-2} M). In alkaline solutions (pH 12), [Pt(Et$_4$dien)OH]$^+$ is formed: the quantum yield is 4.0 × 10^{-3} in both acid and alkaline solutions independently of the presence of outer ligands.

15 Copper

Excitation in the CT band of the macrocyclic copper complex Cu([14]diene-N$_4$)$^{2+}$ can lead to two processes depending on the conditions.[120] In aqueous solution, ligand hydrolysis occurs. This is thought to involve formation of a ketone–amine intermediate, which has been characterized, and a precursor, possibly a carbinolamine species ($t_{0.5} \simeq 0.03$ s), observed by means of flash photolysis. In neat methanol, a copper(I) species is formed by a second-order reaction ($k = 2.2 \times 10^4$ l mol^{-1} s^{-1}). A mechanism has been proposed involving a common precursor with both ligand and metal centres. Among the series of copper(I) halide complexes with cyclic nitrogen bases (L), Cu(L)$_n$X (n = 1, 2, or 3) some monosolvates reversibly alter[121] their fluorescence colour with temperature (fluorescence thermochromism). For elucidation of this phenomenon, emission and excitation spectra of Cu$_4$(py)$_4$I$_4$, Cu$_4$(piperidine)$_4$I$_4$, and Cu$_4$-

[117] P. Natarajan and A. W. Adamson, *J. Indian Chem. Soc.*, 1977, **54**, 25.
[118] A. V. Loginov and G. A. Shagisultanova, *Koord. Khim.*, 1977, **3**, 1567.
[119] C. Bartocci, A. Ferri, V. Carassiti, and F. Scandola, *Inorg. Chim. Acta*, 1977, **24**, 251.
[120] G. J. Ferraudi and J. F. Endicott, *Inorg. Chem.*, 1977, **16**, 2762.
[121] H. D. Hardt and A. Pierre, *Inorg. Chim. Acta*, 1977, **25**, L59.

(morpholine)$_4$I$_4$ were recorded between 77 and 293 K. In the case of Cu$_4$(py)$_4$I$_4$ lowering of the temperature brings about a shift in the emission peak to longer wavelengths, and the results suggest that the emissions originate from Cu ions with different environments following the lower symmetries of the Cu$_4$(py)$_4$I$_4$ cluster. In the latter two complexes, the extent of the red shift of the emission peak with temperature depends more on the symmetry of the cluster than on the Cu—Cu distances. The higher the symmetry of the cluster, the less the red shift. These observations make it probable that at low temperatures, higher excited states are preferentially populated.

Irradiation of aqueous solutions containing both bis-(2,9-dimethyl-1,10-phenanthroline)copper(I) (4) and *cis*-bis-(iminodiacetato)cobaltate(III) (5) at 454 nm into the MLCT band of the former complex induces a redox reaction.[122] The quantum yield for loss of (4) was observed to increase with increasing concentration of (5). The observations do not require that an excited state of Cu(dmp)$_2$$^+$ be involved when the redox step occurs, but it is clear that the singlet MLCT is at least a precursor to reaction. There are two possible mechanisms, (i) direct electron transfer from an excited Cu(dmp)$_2$$^+$ ion or (ii) energy transfer from an excited Cu(dmp)$_2$$^+$ ion to *cis*-Co(ida)$_2$$^-$, which might undergo an intramolecular redox decomposition producing radicals that *could* oxidize Cu^I in a subsequent thermal step. The direct electron transfer mechanism appears the more likely since the 1 : 1 stoicheiometry observed would be improbable in the event that ligand-based radicals were formed. In related work[123] on the same compound the primary excited species is shown to be CTTL in nature.

A study has been made[124] of the photolysis of CuC$_2$O$_4$,2.5H$_2$O at low pressures and under normal conditions both in the presence and absence of Pd^{2+}, Bi^{3+}, Tl$^+$, and hypophosphite ion. The primary products are CO and CO$_2$ and a mechanism for their formation is discussed. The valence isomerization of norborna-2,5-diene to quadricyclane can be sensitized[125] by complexation of the olefin to Cu^I, several different compounds proving to be effective. Factors determining the ability of a transition-metal compound to function as a sensitizer *via* complexation mechanisms of this type are discussed. Spectral evidence[126] indicates that a 1 : 1 ClCu—NBD π-complex is present in solution and is in fact the photoactive species. Quantum yield data in several solvents and at different initial concentrations of norbornadiene support this premise. The photoreaction is thought to originate *via* population of an olefin-metal CT state of the complex.

16 Rare-earth Elements[72]

An investigation[127] of the fluorescent properties of complexes of some phenolcarboxylic acids of the triarylmethane series (6) has revealed that with La^{III} and

[122] D. R. McMillin, M. T. Buckner, and B. T. Ahn, *Inorg. Chem.*, 1977, **16**, 943.
[123] G. Ferraudi, *Inorg. Chem.*, 1978, **17**, 1370.
[124] G. V. Nysh, G. G. Savel'ev, and L. F. Trushina, *Deposited Doc.*, 1974, *VINITI*, 1963.
[125] C. Kutal, D. P. Schwendiman, and P. Grutsch, *Solar Energy*, 1977, **19**, 651.
[126] D. P. Schwendiman and C. Kutal, *J. Amer. Chem. Soc.*, 1977, **99**, 5677.
[127] L. M. Kudryavtseva, L. K. Sukhova, and R. K. Chernova, 'Org. Reagenty Anal. Khim., Tezisy, Doklady, Vses. Konf., 4th, Sarat. Gos. Univ. in. Chernyshevskogo, Saratov, U.S.S.R., 1976, vol. 2, p. 89.

(6)

other metals, freezing leads to an increase in the fluorescence yields. The use of the rare-earth elements in the production of light is discussed.[128]

Two interesting determinations of Ce^{III} have been reported.[129,130] These depend on redox potentiometic and amperometric titration with $Fe(CN)_6^{3-}$ in the presence of methylene blue, Zn–EDTA, excess Zn^{II}, and O_2, while the sample is being illuminated with a tungsten lamp to increase reaction rate. Based on theoretical information about the radiationless dipole–dipole energy transfer from rare-earth ions to the molecules of the environment, the possible use of solutions of $Nd(NO_3)_3,3Bu_3PO_4$ as an active medium for a Nd^{III} laser has been investigated.[131] The system efficiency was determined as 0.015%.

A study has been made[132] of the intermolecular energy transfer between tris(2,2,6,6-tetramethyl-3,5-heptanedionato)terbium(III) and tris(2,2,6,6-tetra-methyl-3,5-heptanedionato)europium(III) in different solvents. Quenching of TbL_3 by EuL_3 occurred in all solvents except DMSO, DMF, and pyridine and the results suggest that the dominant quenching mechanism is dynamic. No evidence is found for strong dimeric associations between the donor and acceptor chelate pairs.

Substantial changes in the lifetime of $Eu^{3+}(^5D_0)$ fluorescence are reported[133] for perchlorate concentrations much lower than had been previously assumed. The effect is thought to be due to a weakly bound complex, for which a formation constant is estimated. Intramolecular energy transfer from organic donor to tris(dibenzoylmethanato)europium leads[134] to the 612 nm emission from Eu^{3+}. By using donors having a known quantum yield of intersystem crossing to EuL_3, the quantum yield of intersystem crossing of the chelate has been determined in some cases. Separation of europium from lanthanide mixtures in aqueous solution has been achieved[135] by a photochemical technique. This consists in irradiating ($\lambda = 240$ nm) equimolar mixtures (0.01M) of binary or ternary combinations of lanthanides as perchlorates and 0.05M-K_2SO_4 in 10% aqueous propan-2-ol solution. The propan-2-ol solution acts as a scavenger for the hydroxy-radicals formed during photoreduction and Eu^{3+} is selectively reduced and precipitated as $EuSO_4$. The kinetics and activation energies for dissociation

128 V. T. Mishchenko, *Khim. Zhizn.*, 1977, 72.
129 F. Sierra, M. S. Garcia, and R. T. Perez, *An. Quim.*, 1977, **73**, 400.
130 F. Sierra, M. S. Garcia, and L. C. Martinez, *An. Quim.*, 1977, **73**, 405.
131 E. M. Goryaeva, A. V. Shablya, and A. Serov, *Zhur. priklad. Spektroskoppii*, 1978, **28**, 75.
132 H. G. Brittain and F. S. Richardson, *J.C.S. Faraday II*, 1977, **73**, 545.
133 J. F. Giuliani and T. Donohue, *Inorg. Chem.*, 1978, **17**, 1090.
134 G. Perichet and B. Pouyet, *J. Chim. Phys.-Chim. Biol.*, 1977, **74**, 373.
135 T. Donohue, *J. Chem. Phys.*, 1977, **67**, 5402.

of the excited (1 : 1) complexes of Eu^{3+} with phen and bipy have been measured[136] in D_2O at 295 K. The mechanism appears to involve replacement of the ligand by D_2O. The photoredox reactions of some metal ions, including Eu, in water proceeds[137] according to reaction (16). For Eu^{II} the reaction occurs

$$M^{n+} + H^+ \longrightarrow M^{(n+1)} + 0.5 H_2 \qquad (16)$$

in visible light with a minimum of photochemical complications and also appears to be potentially useful for energy storage. Photo-oxidation of Eu^{2+} has been observed[138] in aqueous solution with a sufficiently high quantum yield of hydrogen formation to justify further explorations of photoredox reactions for solar energy conversion.

The structures of compounds of rare earths with monocarboxylic acids and phenanthroline doped with Eu have been investigated[139] by studying their luminescence spectra. From the results, it appears that there are two types of compounds differing in the degree of hydration as well as the number of bidentate, bridging, and bidentate–bridging carboxy-groups.

Eu^{III} fluorescence has been used[140] as a spectroscopic tool to detect the presence of PO_4^{3+} ions coprecipitated with $BaSO_4$. The Eu^{III} ions close to PO_4^{3+} ions can be excited by means of a tunable dye laser highly selectively since their crystal field levels are perturbed by the PO_4^{3+} ions. This means that the resulting fluorescence intensity is a direct measure of the PO_4^{3+} concentration. Determination of Eu can be accomplished[141,142] by measurement of the luminescence of the 1 : 3 : 1 Eu-1-naphthoic acid-1,10-phenanthroline complex extracted by toluene from a solution containing urotropine. In some related work the luminescence spectra of Eu complexes with several dibasic acids have been determined.[143]

17 Terbium

Recently, the structural and conformational nature of nucleic acids has been probed by the emission of bound Tb^{III} and Eu^{III}. It has now been reported[144] that the photochemistry (λ 240—300 nm) of Tb^{III} complexes of polyuridylate (Mol. Wt. $> 10^5$) is not observed in uridine nucleotides, dinucleoside phosphates or the double-stranded helix of polyuridylate with polyadenylate.

Determination of terbium can be achieved[145] with both increased sensitivity

[136] V. P. Gruzdev and V. L. Ermolaev, *Zhur. neorg. Khim.*, 1977, **22**, 3022.
[137] D. D. Davis, G. K. King, K. L. Stevenson, E. R. Birnbaum, and J. H. Hageman, *J. Solid State Chem.*, 1977, **22**, 63.
[138] P. R. Ryason, *Solar Energy*, 1977, **19**, 445.
[139] V. I. Tsaryuk, V. F. Zolin, and L. G. Koreneva, *Koord. Khim.*, 1977, **3**, 183.
[140] J. C. Wright, *Analyt. Chem.*, 1977, **49**, 1690.
[141] V. T. Mishchenko, E. I. Tselik, and A. P. Koev, *Zhur. analit. Khim.*, 1977, **32**, 71.
[142] V. T. Mishchenko, E. T. Tselik, L. I. Ovchar, A. P. Koev, and N. S. Polvektova, in ref. 127, p. 62.
[143] V. I. Tsaryuk, L. G. Koreneva, and V. F. Zolin, *Koord. Khim.*, 1977, **3**, 465.
[144] C. Formoso, *Photochem. and Photobiol.*, 1977, **26**, 159.
[145] M. A. Tishchenko, N. S. Poluektov, G. I. Gerasimento, V. Ya. Temkina, I. I. Zheltvai, G. F. Yaroshenko, and L. M. Timakova, *Otkrytiya Izobret. Prom. Obraztsy Tovarnye Znaki*, 1977, **54**, 125.

and selectivity by using *o*-hydroxyphenyliminod!acetic acid. Luminescence has been observed from some hydroxyl-containing polycarboxylates of type (7).[146]

(7)

18 Uranium

From tunnel effect theory, it has been reasoned[147] that, kinetically, the most important quenching mechanism of excited UO_2^{2+} involves chemical quenching *via* H abstraction. In the case of alcohols and ethers, the experimental and theoretical results agree well and evidence is also presented to suggest that in the deactivation of $[UO_2^{2+}]^*$ by H_2O, hydrogen abstraction is the predominant process.

A number of photoreactions of compounds containing a carbon–oxygen or carbon–nitrogen multiple bond with alcohols have been studied[148] in the presence of UO_2Cl_2 or $TiCl_4$. In the UO_2Cl_2-catalysed reactions, bond formation occurs on a C=C—C atom of the substrates.

A comparision made[149] between Stern–Volmer data obtained for the quenching of UO_2^{2+} luminescence by MeOH and data for Ag^+ quenching[150] shows that both quenchers form transient exciplexes with the linear excited uranyl ion $[U_2O_4H^{4+}]^*$. The primary act is probably a simultaneous overlap between the π_uMO and one of the virtually atomic components $[f_{x,y,z}$ or $f_{z(x^2-y^2)}]$ of the σ_uMO of $[UO_2^{2+}]^*\pi_u^3(5f, \sigma_u)$ by an appropriate orbital of the quencher, *i.e.* d_{yz} of Ag^+ or the group $MO_{\pi Me}$ of MeOH. An examination[151] of negative Stern–Volmer deviations in the quenching of UO_2^{2+} luminescence by Ag^+ ions strongly suggests that $[UO_2^{2+}]^*$ is the quenched species. The most plausible quenching mechanism is $d(Ag^+) \rightarrow \pi_u[UO_2^{2+}]^*$ electron transfer and it cannot be excluded that Ag^+ acts as a two-way quencher, also perturbing $[UO_2^+]^*$ by a $S^0(Ag^+) \leftarrow \pi_u(UO_2^+)^*$ electron transfer.

A review has appeared[152] of the photochemical reduction of the uranyl ion with alcohols, and a discussion has been published[153] of the quenching mechanism of the UO_2^{2+} luminescence by metal ions. Electron transfer from metal

146 Y. G. Kryazhev, N. M. Snegovskikh, Y. P. Belousov, V. M. Kirilets, and A. A. Nechi-
      tailo, *Otkrytiya Izobret. Prom. Obraztsy Tovarnye Znaki*, 1977, **54**, 56.
147 H. D. Burrows and S. J. Formosinho, *J.C.S. Faraday II*, 1977, **73**, 201.
148 T. Sato, S. Yoshiie, T. Imamura, K. Hasegawa, M. Miyahara, S. Yamamura, and
      O. Ito, *Bull. Chem. Soc. Japan*, 1977, **50**, 2714.
149 M. D. Marcantonatos, *Inorg. Chim. Acta*, 1977, **24**, L37.
150 H. D. Burrows, S. J. Formosinho, M. da G. Miguel, and F. Pinto Coelho, *J.C.S. Faraday
      I*, 1976, **72**, 163.
151 M. D. Marcantonatos, *Inorg. Chim. Acta*, 1977, **25**, L87.
152 A. S. Brar, S. S. Sandhu, and A. S. Sarpal, *J. Chem. Sci.*, 1977, **2**, 8.
153 M. D. Marcantonatos, *Inorg. Chim. Acta*, 1977, **24**, L53.

ion to $[UO_2^{2+}]^*$ is probably the mechanism involved, and it is argued that an appropriate choice for the energies of the highest occupied MOs is the chemical ionization energies of the aquometallic cationic complexes. A mechanism is suggested which correlates at least some of the available experimental observations.

Irradiation of solutions of pyridine-2,6-dicarboxylic acid (8) containing $[UO_2^{2+}]$ results in decarboxylation to pyridine-2-carboxylic acid.[154] It is believed that the $[UO_2^{2+}]$ is chelated between the two carboxy-groups and decomposition results from energy transfer from the $\pi-\pi^*$ transition in (8) and the excitation of the complexed $[UO_2^{2+}]$.

The photochemical reaction between $UO_2(NO_3)_2$ and tri-n-butyl phosphate (TBP) solution has been investigated[155] in both the presence and absence of HNO_3 and H_2O. The products include U^{IV}, NO_2^-, and hydrolysis products of TBP. Since the overall photochemical results derive from at least three photochemical reactions and several thermal reactions, the details of the reduction of UO_2^{2+} to U^{IV} are not yet clear.

Photolysis of the uranyl oxalate system at pH 4.5 leads[156] to $[UO_2]^{2+}$, U^{IV}, and CO_2. A mechanism has been proposed based upon $[UO_2(C_2O_4)_2]^{2-}$ as the primary photoactivated species and is in agreement with the quantitative and qualitative results of this investigation showing similar tendencies to the mechanism suggested for the photolysis in solutions of pH ≤ 1.7. Flash photolysis of $0.5M$-HCl/U^{IV} leads[157] to Cl_2^- ($\lambda_{max}345$) which in the presence of U^{VI} no longer decays by second-order kinetics. The mechanism in steps (17) and (18) has been proposed.

$$Cl_2^- + U^{IV} \xrightleftharpoons{K_{eq}} UCl_2^{3+} \qquad (17)$$

$$UCl_2^{3+} + 2H_2O \xrightarrow{k} UO_2^+ + 2Cl^- + 4H^+ \qquad (18)$$

The spectrum, quantum efficiency, and decay time of the fluorescence of UF_6, excited in the A band, have been reported.[158] It was observed that the quantum efficiency decreases for an excitation <380 nm and the Stern–Volmer plot shows a two-slope behaviour. Irradiation into the ligand-to-metal CT bands of uranyl carboxylate complexes in aqueous solution leads[159] to oxidative decarboxylation of the ligands and in the presence of excess carboxylic acid gives the U^{IV} complexes: $U(HOCH_2CO_2)_4,2H_2O$; $U(O_2CCH_2OCH_2CO_2)_2,2H_2O$; $Na_2[U(O_2-CCH_2OCH_2CO_2)_3],2HO_2CCH_2OCH_2CO_2H,H_2O$; $U(MeOCH_2CO_2)_4,H_2O$; and $UO(O_2CCH_2NHCH_2CO_2),2H_2O$.

The kinetics of the solid-state photodecomposition of uranyl formate monohydrate has been interpreted[160] in terms of three different models. The first two consider the absorption of light by the photoproduct layer and the more satis-

[154] M. Munakata and S. Niina, *Kinki Daigaku Rikogakubu Kenkyu Hokoku*, 1977, 57.
[155] C. K. Rofer-DePoorter and G. L. DePoorter, *J. Inorg. Nuclear Chem.*, 1977, **39**, 631.
[156] A. G. Brits, R. Van Eldik, and J. A. Van den Berg, *J. Inorg. Nuclear Chem.*, 1977, **39**, 1195.
[157] P. K. Bhattacharyya and R. D. Saini, *Inorg. Nuclear Chem. Letters*, 1977, **13**, 479.
[158] O. De Witte, R. Dumanchin, M. Michon and J. Chatelet, in ref. 63, p. 228.
[159] G. Sbrignadello, G. Battiston, G. Tomat, and O. Traverso, *Inorg. Chim. Acta*, 1977, **24**, L43.
[160] B. Claudel, M. Feve, J. P. Puaux, and H. Sautereau, *J. Photochem.*, 1978, **8**, 117.

factory one allows the effective quantum yields of photodecomposition and fluorescence to be calculated. In the third model, CO_2 is considered as a kinetic inhibitor of the photochemical step. In order to explain the luminescence of the U^{VI} ion, MO LCAO calculations have been carried out using an ionic model to determine the characteristics of $[UO_2]^{2+}$ and $[UO_4]^{2-}$ in $SrMoO_4$ and $[UO_6]^{6-}$ in Ba_2MgWO_6.[161] The suggestion has been made that the first excited state of the uranyl ion might be useful in the conversion of solar energy for synthetical purposes.[162] This is in accord with the observation that many U^{IV} compounds such as $U(SO_4)_2,4H_2O$ and UF_4 are frequently prepared by the photochemical reaction of acidic uranyl with EtOH in sunshine.

19 Plutonium

Further work describing the Pu^{IV} disproportionation equilibrium under irradiation has appeared.[163] The results show that the disproportionation equilibrium is reversible, and the photodecomposition of time- and temperature-aged Pu^{IV} polymers in $HClO_4$ and HNO_3 solutions has been examined as a function of ageing conditions.

[161] D. E. Onopko, E. P. Pashnina, and N. V. Starostin, *Opt. Spektrosk.*, 1977, **43**, 901.
[162] C. K. Jörgensen, *Naturwiss.*, 1977, **64**, 37.
[163] H. A. Friedman, L. M. Toth, and J. T. Bell, *J. Inorg. Nuclear Chem.*, 1977, **39**, 123.

2

The Photochemistry of Transition-metal Organometallic Compounds, Carbonyls, and Low-oxidation-state Complexes

BY J. M. KELLY

1 Introduction

This chapter deals with the photochemistry of transition-metal complexes containing metal-to-carbon bonds, including carbonyls (but not cyanide complexes), and of other complexes in which the metal is in a low oxidation state. Reports pertaining to a particular compound are discussed in the section concerned with its constituent metal. A selection of recently reported photosubstitution reactions of metal carbonyls of preparative significance is presented in Table 1 at the end of the chapter (p. 267).

Recent reviews relevant to this area have considered the primary processes of organo-transition compounds,[1] the flash photolysis of co-ordination compounds including metal carbonyls,[2] and the matrix photochemistry of metal carbonyls.[3,4]

2 Titanium, Zirconium, and Hafnium

In recent years photolysis of Cp_2MR_2 (M = Ti, Zr, or Hf; Cp = η^5-cyclopentadienyl) in the presence of suitable substrates has been found to be a useful method for the production of other metallocene derivatives.[5] Although the photochemical decomposition of these complexes has been represented by equation (1), it is clear from more recent studies that the actual situation is more complex.[5-9] Thus irradiation of Cp_2TiMe_2 in benzene or hexane gives a solid

$$Cp_2MR_2 \xrightarrow{h\nu} Cp_2M + 2R\cdot \qquad (1)$$

('black titanocene') and methane.[6] Deuteriation experiments have revealed that the hydrogen abstracted in the formation of methane originated either in the cyclopentadienyl group or in the other methyl group but not in the solvent. In

[1] C. R. Bock and E. A. Koerner von Gustorf, *Adv. Photochem.*, 1977, **10**, 221.
[2] Z. Stasicka and A. Marchaj, *Coordination Chem. Rev.*, 1977, **23**, 131.
[3] J. J. Turner, J. K. Burdett, R. N. Perutz, and M. Poliakoff, *Pure Appl. Chem.*, 1977, **49**, 271.
[4] J. K. Burdett and J. J. Turner, in 'Cryochemistry', ed. M. Moskovits and G. A. Ozin, Wiley, New York, 1976, p. 493.
[5] M. D. Rausch, W. H. Boon, and H. G. Alt, *Ann. N.Y. Acad. Sci.*, 1977, **295**, 103.
[6] M. D. Rausch, W. H. Boon, and H. G. Alt, *J. Organometallic Chem.*, 1977, **141**, 299.
[7] E. Samuel, P. Maillard, and C. Giannotti, *J. Organometallic Chem.*, 1977, **142**, 289.
[8] M. Peng and C. H. Brubaker, *Inorg. Chim. Acta*, 1978, **26**, 231.
[9] G. Erker, *J. Organometallic Chem.*, 1977, **134**, 189.

the same study attempts to trap the methyl radical with anthracene indicated that the formation of free radicals is not the predominant photo-induced process, and the authors have suggested that the reactions proceed *via* caged radicals or other unspecified radical-like organometallic species. However, Samuel *et al.* have been able to obtain e.s.r. evidence for the methyl adducts of 5,5'-dimethyl-1-pyrroline-1-oxide and of nitrosodurene during irradiation of toluene solutions of Cp_2MMe_2 (M = Ti, Zr, or Hf) in the presence of the spin traps.[7] E.s.r. signals attributed to the tervalent Cp_2MMe were also recorded, and evidence has been adduced to demonstrate that this species undergoes further photolysis to yield the metallocene.

In contrast with the case of Cp_2TiMe_2, where only traces of ethane are formed, irradiation of Cp_2TiPh_2, $Cp_2Ti(C_6H_4Me)_2$,[5,8] or their zirconium analogues[9] yields predominantly the biphenyls. Whether these coupled products arise by direct reductive elimination from the photo-excited Cp_2MPh_2 or by radical recombination is not as yet clear. E.s.r. investigations of THF solutions of Cp_2TiPh_2 at $-60\,^\circ\text{C}$ and at room temperature have revealed a complex series of signals, amongst which those due to solvated titanocene and titanocene monohydride could be identified.[8] It is certain from these reports that further studies, possibly using CIDNP or flash photolysis, will be required before the detailed mechanism for these reactions can be elucidated.

The exchange reaction (2) previously recorded for the titanocene dichlorides has also been observed with the zirconium and hafnium analogues.[10,11] A

$$Cp_2MCl_2 + (C_5D_5)_2MCl_2 \xrightarrow{h\nu} 2(C_5D_5)CpMCl_2 \qquad (2)$$

mechanism such as that shown in Scheme 1 appears to be more likely for the titanium and zirconium derivatives than is the alternative involving initial dissociation of a cyclopentadienyl radical. Photo-induced cyclopentadienyl group

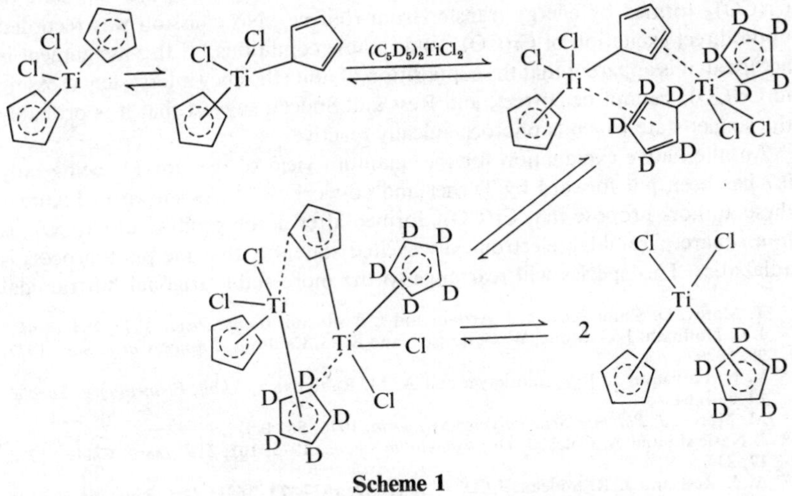

Scheme 1

10 M. H. Peng and C. H. Brubaker, *J. Organometallic Chem.*, 1977, **135**, 333.
11 J. G.-S. Lee and C. H. Brubaker, *Inorg. Chim. Acta*, 1977, **25**, 181.

exchange has also been observed for Cp_2V, Cp_2VCl, and Cp_2VCl_2.[11] The electronic spectra of substituted titanocene dichlorides have recently been discussed on the basis of CNDO calculations.[12]

Reduction of carbon monoxide by hydrogen to methane may be catalysed by $Cp_2Ti(CO)_2$ either upon irradiation or at high temperatures. The primary process under both sets of conditions appears to be CO-expulsion and creation of the free co-ordination site required for catalysis.[13]

Other authors have discussed the photodecomposition of diphenylzirconium,[14] and the initiation of the cationic polymerization of isobutylene by excitation of olefin–$TiCl_4$ complexes [equation (3)].[15]

$$(\text{olefin})\text{–}TiCl_4 \xrightarrow{\quad h\nu \quad} (\text{olefin})^+ \ TiCl_4{}^- \qquad (3)$$

3 Chromium, Molybdenum, and Tungsten

It is now clear from studies by Nasielski and Colas[16] that the quantum yield for photo-dissociation of the Group VI metal hexacarbonyls is substantially less than 1.0. Thus for reaction (4), both upon direct irradiation and upon benzo-phenone sensitization, a quantum yield of approximately 0.7 is obtained. On the basis of these results and those derived from investigations of the wavelength and solvent viscosity dependence, it has been proposed that the triplet state, formed from the singlet state with unit efficiency, is the photochemically active species, and further it has been assumed that approximately 30% of the triplet states undergo radiationless transition to the ground state. However, the role

$$M(CO)_6 + py \xrightarrow{\quad h\nu \quad} M(CO)_5py \qquad (4)$$
$$(M = \text{Cr or W})$$

of the triplet state in reaction (4) has been brought into question by the observa-tion of luminescence upon irradiation of mixtures of $Cr(CO)_6$ and toluene in argon matrices at 10 K.[17] This emission has been assigned to the $^3T_{1g}$ state of $Cr(CO)_6$ formed by energy transfer from toluene. No emission was recorded upon direct excitation of $Cr(CO)_6$ under these conditions. If this assignment is accepted, it is apparent that the supposition of unit efficiency intersystem crossing in $Cr(CO)_6$ cannot be correct, and Rest and Sodeau suggest that it is probably the singlet state which is photochemically reactive.

An alternative explanation for the quantum yield of reaction (4) being only 0.7 has been put forward by Turner and co-workers.[18] As shown in Figure 1 these authors propose that $Cr(CO)_5$ formed after decomposition of $Cr(CO)_6$ is in a square pyramidal, electronically excited state, *i.e.* that the photoprocess is adiabatic. This species will rearrange to the more stable trigonal bipyramidal

[12] T. Marey, D. Simon-leClerc, J. Arriau, and J. Besancon, *Compt. rend.*, 1977, **284**, C, 967.
[13] J. C. Huffman, J. G. Stone, W. C. Krusell, and K. G. Caulton, *J. Amer. Chem. Soc.*, 1977, **99**, 5829.
[14] V. N. Latyaeva, L. I. Vyshinskaya, and A. M. Rabinovich, *Khim. Elementoorg. Soedin.*, 1976, **4**, 64.
[15] M. Marek, *J. Polymer Sci.*, *Polymer Symposia*, 1976, **56**, 149.
[16] J. Nasielski and A. Colas, *J. Organometallic Chem.*, 1975, **101**, 215; *Inorg. Chem.*, 1978, **17**, 237.
[17] A. J. Rest and J. R. Sodeau, *J.C.S. Faraday II*, 1977, **73**, 1691: *Ber. Bunsengesellschaft phys. Chem.*, 1978, **82**, 20.
[18] J. K. Burdett, J. M. Grzybowski, R. N. Perutz, M. Poliakoff, J. J. Turner, and R. F. Turner, *Inorg. Chem.*, 1978, **17**, 147.

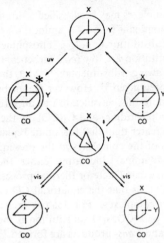

Figure 1 *Photochemical behaviour of* M(CO)₅ *in a mixed matrix containing X and* Y. *Reorientation* via *a trigonal bipyramidal intermediate.* (*Each rotation of the square pyramidal species should strictly be 120°, not 90°; 90° angles are used here for simplicity*). (*) *electronically excited state;* (≠) *electronically or vibrationally excited state* (Reproduced by permission from *Inorg. Chem.*, 1978, **17**, 152)

isomer, which then converts back into the ground state C_{4v} species. In the last step one-third of the pentacarbonyl species formed will have an orientation such that efficient recombination may occur, while in the other cases reaction with the matrix takes place and the overall quantum yield is hence predicted to be 0.67. The interesting postulate of the initial formation of $Cr(CO)_5$ in its electronically excited state is predicted by quantum mechanical calculations.[18,19] The experimental evidence for the above mechanism was obtained by studying the photochemical behaviour of $M(CO)_5L$ (L = Ar, Xe, N₂ *etc.*) upon excitation with polarized light. Thus, for example, irradiation of $Mo(CO)_5N_2$ in a mixed Ar–N₂ matrix [reaction (5)] caused dichroic photodepletion of the $Mo(CO)_5N_2$ (due to preferential absorption by correctly oriented molecules) but no dichroism was present in the $Mo(CO)_5Ar$ produced. These results require that some motion occurring during the photochemical act causes randomization of the photoproduct, and this is consistent with the inverse Berry twist represented in Figure 1.

$$Mo(CO)_5N_2 + Ar \xrightarrow{h\nu} M(CO)_5Ar + N_2 \qquad (5)$$

It has been demonstrated that the above isomerization of the five-co-ordinate species is not a feature unique to the photochemistry of the metal hexacarbonyls by showing that $W(CO)_3(^{13}CO)(CS)$ formed as an intermediate species upon irradiation of *trans*-$W(CO)_4(^{13}CO)(CS)$ also isomerizes in an argon matrix.[18,20] This observation has obvious implications for the solution photosubstitution reactions of complexes of the type $M(CO)_5L$. It appears, at least for the tungsten

[19] P. J. Hay, *J. Amer. Chem. Soc.*, 1978, **100**, 2411.
[20] M. Poliakoff, *Inorg. Chem.*, 1976, **15**, 2892.

derivatives, that these complexes may be classified as (*a*) those in which expulsion of L is efficient and predominant (*e.g.* L = amine); (*b*) those in which both CO and L undergo efficient substitution (*e.g.* L = phosphine); (*c*) those in which the quantum yields for substitution are low (*e.g.* carbenes, CS, and amines bearing strongly electron-withdrawing substituents).[21] MO theoretical treatments of these phenomena have been given.[21] However Darensbourg and Murphy have cautioned against extending this classification to all Group VI metals, because for $Cr(CO)_5(pip)$ (pip = piperidine) and $Cr(CO)_5py$ the quantum yield for CO substitution is equal or greater than that for amine replacement.[22] Interestingly in these cases photolysis of the complexes in the presence of ^{13}CO gives exclusively *cis*-$Cr(CO)_4(^{13}CO)$(amine), indicating either that the five-co-ordinate species rearranges (*via* an excited state or thermal process) to give the C_s isomer before reaction with ^{13}CO, or that the equatorial CO is preferentially labilized in these $Cr(CO)_5$(amine) complexes. For 366 nm excitation of $Mo(CO)_5PPh_3$ in the presence of PPh_3 or ^{13}CO, CO substitution is the predominant process, but in this case both *cis*- and *trans*- products are formed.[23] Again in this instance there is evidence for isomerization of the C_{4v} and C_s isomers (1) and (2).

(1)   (2)

$Mo(CO)_4(PPh_3)$ also appears to be an important intermediate in the *trans*- to *cis*- photoisomerization of $Mo(CO)_4(PPh_3)_2$, and if this reaction is carried out in the presence of ^{13}CO *cis*-$Mo(CO)_4(^{13}CO)(PPh_3)$ is the principal product.[23] Photolysis of bis(phosphine) complexes of type $M(CO)_4L_2$ (M = Cr, Mo, or W) in the presence of other donor ligands, such as amines, halide ions, or acetonitrile provides a useful route to mixed complexes of the type $M(CO)_4LL'$.[24]

Other reports on the photosubstitution of $M(CO)_6$ include the preparation of $M(CO)_5N_2$ species in low-temperature matrices and their characterization by Raman spectroscopy,[25] and the determination, by photo-calorimetry, that the substitution of CO by piperidine in $Cr(CO)_6$ is essentially thermoneutral.[26]

Further details on the use of $W(CO)_6$–CCl_4 as a catalyst for photo-activated olefin metathesis have been communicated.[27] The species $W(CO)_5Cl$ has been identified as one of the probable reaction intermediates. It has also been demonstrated that olefin metathesis takes place in the dark if the $W(CO)_6$–CCl_4 mixture is irradiated prior to addition of the olefin.[28] Metathesis of cyclopentene

[21] R. M. Dahlgren and J. I. Zink, *Inorg. Chem.*, 1977, **16**, 3154.
[22] D. J. Darensbourg and M. A. Murphy, *Inorg. Chem.*, 1978, **17**, 884.
[23] D. J. Darensbourg and M. A. Murphy, *J. Amer. Chem. Soc.*, 1978, **100**, 463.
[24] W. A. Schenk, *J. Organometallic Chem.*, 1977, **139**, C63.
[25] J. K. Burdett, A. J. Downs, G. P. Gaskill, M. A. Graham, J. J. Turner, and R. F. Turner, *Inorg. Chem.*, 1978, **17**, 523.
[26] A. W. Adamson, A. Vogler, H. Kunkely, and R. Wachter, *J. Amer. Chem. Soc.*, 1978, **100**, 1298.
[27] P. Krausz, F. Garnier, and J. E. Dubois, *J. Organometallic*, 1978, **146**, 125.
[28] C. Tanielian, R. Kieffer, and A. Harfouch, *Tetrahedron Letters*, 1977, **52**, 4589.

to give stereoregular polypentenamer has been induced by irradiation in the presence of WCl_6.[29]

Studies in argon matrices[30] and in hydrocarbon solutions[31] confirm that the principal primary photoprocess for $(arene)Cr(CO)_3$ complexes is carbon monoxide dissociation [reaction (6)] and that it is secondary reactions of $(arene)Cr(CO)_2$ which lead to formation of the free arene. For methyl meth-

$$(arene)Cr(CO)_3 \xrightarrow{\quad h\nu \quad} (arene)Cr(CO)_2 + CO \qquad (6)$$

acrylate (M) solution, however, it has been suggested that the formation of an exciplex $[M \cdots (arene)Cr(CO)_3]^*$ is the initial reaction step.[31] Polymerization of methyl methacrylate may be achieved by irradiation of mixtures of the $(arene)$-$Cr(CO)_3$, CCl_4, and the monomer, an initiating radical being produced for every molecule of the carbonyl complex which decomposes. Replacement of the arene in $(benzene)Cr(CO)_3$ or $(mesitylene)Cr(CO)_3$ by fulvenes to give (3) may be efficiently achieved photochemically.[32] Photo-excitation of (4) causes exclusively phosphine substitution to yield (5) as product.[33] This substantiates the belief

(3) a; R^1 = Me; R^2 = Ph
    b; R^1 = R^2 = p-ClC$_6$H$_4$

(4) a; X = O; L = PPh$_3$
    b; X = CH$_2$; L = P(OPh)$_3$, P(OEt)$_3$,
                        PPh$_3$, or PMe$_2$Ph

(5)

that cleavage of the Cr–L linkage is a general feature of the photochemistry of $(arene)Cr(CO)_2L$ complexes.

Irradiation of $CpM(CO)_3Me$ (M = Cr, Mo, or W) causes cleavage of the M—Me bond rather than rupture of the M—CO linkage.[34] In solvents such as pentane, benzene, or THF the corresponding dimeric species $[CpM(CO)_3]_2$ and methane are formed, whereas in chloroform or carbon tetrachloride, $CpM(CO)_3Cl$ (M = Mo or W) or $[CpCrCl_2]_2$ is produced. Labelling studies for irradiation in C_6D_6 show that the liberated methane is formed principally by hydrogen abstraction from the cyclopentadienyl ring (a process also observed with the

[29] M. K. Yakovleva, A. P. Sheinker, E. B. Kotin, and A. D. Abkin, *Vysokomol. Soedineniya* (*B*), 1977, **19**, 604.
[30] A. J. Rest, J. R. Sodeau, and D. J. Taylor, *J.C.S. Dalton*, 1978, 651.
[31] C. H. Bamford, K. G. Al-Lamee, and C. J. Konstantinov, *J.C.S. Faraday I*, 1977, **73**, 1406.
[32] F. Edelmann, D. Wormsbaecher, and U. Behrens, *Chem. Ber.*, 1978, **111**, 817.
[33] G. P. Donnini and A. Shaver, *Canad. J. Chem.*, 1978, **56**, 1477.
[34] M. D. Rausch, T. E. Gismondi, H. G. Alt, and J. A. Schwärzle, *Z. Naturforsch.*, 1977, **32b**, 998.

Group IV Cp$_2$MMe$_2$ derivatives[6]). Upon irradiation in the presence of acetylenes, CpM(CO)$_3$Me (M = Mo or W) species are converted into the alkenyl-ketone derivatives (6) and in smaller amounts into CpW(CO)(RC≡CR)Me.[35]

(6)

Analogous products are formed from the η^5-indenyl derivatives.[36] With ethylene, CpW(CO)$_2$(C$_2$H$_4$)Me and (η^5-C$_9$H$_7$)W(CO)$_2$(C$_2$H$_4$)Me are obtained.[37] Photolysis of CpMo(COCF$_3$)(CO)$_3$ in the presence of various disubstituted acetylenes gives the formally 16-electron complexes shown in equations (7) and (8).[38]

$$\text{CpMo(COCF}_3\text{)(CO)}_3 + \text{MeC} \equiv \text{CMe} \xrightarrow{h\nu} \text{CpMo(CF}_3\text{)(CO)(MeC} \equiv \text{CMe)}$$
(7)

$$\text{CpMo(COCF}_3\text{)(CO)}_3 + \text{RC} \equiv \text{CR} \xrightarrow{h\nu} \text{CpMo(CF}_3\text{)(RC} \equiv \text{CR)}_2$$
(8)
$$(\text{R} = \text{CF}_3 \text{ or CO}_2\text{Et})$$

Photosubstitution of CpW(CO)$_3$H by tributylphosphine may be initiated by u.v. irradiation ($\lambda < 400$ nm); the reaction apparently proceeding by cleavage of the W—H bond.[39] The 17-electron fragment then takes part in a chain reaction [equations (9—12)] similar to that previously proposed for Re(CO)$_5$. If small amounts of the dimer [CpW(CO)$_3$]$_2$ are added, the above photosubstitution may be initiated by visible light ($\lambda = 500$ nm), which causes rupture of the W—W bond of the dimer and production of the propagating metal-centred radical. In this case quantum yields of 1000 are obtained.

$$\text{CpW(CO)}_3\text{H} \xrightarrow{h\nu} \text{CpW(CO)}_3 + \text{H·}$$
(9)

$$\text{CpW(CO)}_3 \longrightarrow \text{CpW(CO)}_2 + \text{CO}$$
(10)

$$\text{CpW(CO)}_2 + \text{L} \longrightarrow \text{CpW(CO)}_2\text{L}$$
(11)

$$\text{CpW(CO)}_2\text{L} + \text{HCpW(CO)}_3 \longrightarrow \text{CpW(CO)}_2\text{LH} + \text{CpW(CO)}_3$$
(12)

Photolysis of [CpM(CO)$_3$]$_2$ (M = Mo or W) and Co$_2$(CO)$_8$ or [CpFe(CO)$_2$]$_2$ provides a convenient synthetic route to CpM(CO)$_3$Co(CO)$_4$ and CpM(CO)$_3$-FeCp(CO)$_2$.[40] On flash photolysis these species may be reconverted into the orginal dinuclear complexes [*e.g.* equation (13)]. In the presence of carbon tetrachloride the 17-electron intermediates are trapped, as illustrated by equation (14). From the results of these and similar experiments with heterodinuclear metal–metal bonded carbonyl complexes it has been possible to express the

[35] H. G. Alt, *Chem. Ber.*, 1977, **110**, 2862.
[36] H. G. Alt, *Z. Naturforsch.*, 1977, **32b**, 1139.
[37] H. G. Alt, J. A. Schwärzle, and C. G. Kreiter, *J. Organometallic Chem.*, 1978, **153**, C7.
[38] J. L. Davidson, M. Green, J. Z. Nyathi, F. G. A. Stone, and A. J. Welch, *J.C.S. Dalton*, 1977, 2246.
[39] N. W. Hoffman, and T. L. Brown, *Inorg. Chem.*, 1978, **17**, 613.
[40] H. B. Abrahamson and M. S. Wrighton, *Inorg. Chem.*, 1978, **17**, 1003.

relative reactivities of such metal carbonyl fragments in halogen-atom abstraction reactions as follows: $Re(CO)_5 > Mn(CO)_5 > CpW(CO)_3 > CpMo(CO)_3 > CpFe(CO)_2 > Co(CO)_4$.[40,41] Photocleavage of Mo—Mo bonds has also been reported for the complexes $Mo_2Cl_4L_4$ (L = PEt_3, PBu_3, or $PEtPh_2$) upon irradiation in chlorinated solvents.[42]

$$2Cp(CO)_3MFe(CO)_2Cp \xrightarrow{hv} [CpM(CO)_3]_2 + [CpFe(CO)_2]_2 \quad (13)$$

$$Cp(CO)_3MFe(CO)_2Cp \xrightarrow[CCl_4]{hv} CpM(CO)_3Cl + CpFe(CO)_2Cl \quad (14)$$

Photonitrosylation of $[CpCr(CO)_3]_2$, $[CpCr(CO)_2]_2$, and $CpCr(CO)_2NO$ has been achieved by irradiation of carbon tetrachloride or chloroform solutions in the presence of nitric oxide.[43] Other publications discuss the photoreactions of arsine-bridged complexes such as $CpMo(CO)_3AsMe_2Fe(CO)_4$,[44] the photodecomposition of $(\mu\text{-dppe})[CpCr(CO)(NO)]_2$ (dppe = $Ph_2PCH_2CH_2PPh_2$),[45] and the photodecarbonylation of complexes of type (7).[46-48] A full report has

(7)

appeared on the reduction of nitrogen to ammonia in complexes such as $Mo(N_2)_2(dppe)_2$.[49] A study of the MLCT spectra of $M(N_2)_6$ (M = Cr, V, or Ti) reveals that the CT states of these dinitrogen complexes lie lower than those of their carbonyl analogues. This gives support to the idea that the increased basicity of the dinitrogen in this excited state may make its fixation easier.[50]

In previous investigations, Green and co-workers[51] have noted that tungstenocene, formed upon irradiation of Cp_2WH_2, inserts readily into C—H bonds, and this behaviour has now been observed with tetramethylsilane. With propylene oxide, insertion into a C—O bond occurs, the complex so formed decomposing to yield propene.[52] Irradiation of Cp_2MoH_2 in benzene gives (8) in high yield, probably *via* the photolysis of (9), the dimerization product of molybdenocene.[52] Evidence for molybdenocene has also been obtained by trapping experiments with carbon monoxide, acetylene, and phosphines.[53] The gas formed upon photolysis of Cp_2MoH_2 in deuteriated toluene contained

[41] H. B. Abrahamson and M. S. Wrighton, *J. Amer. Chem. Soc.*, 1977, **99**, 5510.
[42] W. C. Trogler and H. B. Gray, *Nouveau J. Chim.*, 1977, **1**, 475.
[43] M. Herberhold, R. Klein, and H. G. Alt., *Israel J. Chem.*, 1977, **15**, 206.
[44] R. Mueller and H. Vahrenkamp, *Chem. Ber.*, 1977, **110**, 3910.
[45] D. Sellmann and E. Kleinschmidt, *Z. Naturforsch.*, 1977, **32b**, 1010.
[46] W. H. de Roode and K. Vrieze, *J. Organometallic Chem.*, 1978, **153**, 345.
[47] W. H. de Roode, M. L. Beekes, A. Oskam, and K. Vrieze, *J. Organometallic Chem.*, 1977, **142**, 337.
[48] B. Gaylani and M. Kilner, *J. Less-Common Metals*, 1977, **54**, 175.
[49] J. Chatt, A. J. Pearman, and R. L. Richards, *J.C.S. Dalton*, 1977, 1852.
[50] A. B. P. Lever and G. A. Ozin, *Inorg. Chem.*, 1977, **16**, 2012.
[51] M. L. H. Green, M. Berry, C. Couldwell, and K. Prout, *Nouveau J. Chim.*, 1977, **1**, 187.
[52] M. Berry, S. G. Davies and M. L. H. Green, *J.C.S. Chem. Comm.*, 1978, 99.
[53] G. L. Geoffroy and M. G. Bradley, *J. Organometallic Chem.*, 1977, **134**, C27.

(8)   (9)

predominantly H_2 and very little HD, indicating that free hydrogen atoms are not produced and suggesting that the elimination of hydrogen may be a concerted process.[53]

Diatomic metals Cr_2 and Mo_2, which have previously been observed upon flash photolysis of $Cr(CO)_6$ and $Mo(CO)_6$ in the gas phase,[54] can also be readily prepared by condensation of molybdenum or chromium atoms into argon matrices at 10 K.[55] Other small clusters can also be selectively prepared by mobilizing particular atoms using u.v. or visible light.[56] In this way clusters such as Cr_2Mo and $CrMo_2$ have been synthesized, and the relevance of this work for heterogeneous catalysis has been emphasized.[57]

4 Manganese and Rhenium

By consideration of the photofragment angular distribution and the translational energy of the $M(CO)_5$ fragments following photodissociation ($\lambda = 300$ nm) of $M_2(CO)_{10}$ (M = Mn or Re) in molecular beams, it has been possible to show (i) that the excited state of $M_2(CO)_{10}$ decomposes within a few picoseconds of excitation, (ii) that only the M—M bond and not the M—CO bonds are cleaved, and (iii) that the $M(CO)_5$ fragment contains about 125 kJ of vibrational energy.[58]

The photosubstitution of $Mn_2(CO)_{10}$ by tributylphosphine and by triethyl phosphite yields the diaxially substituted product $Mn_2(CO)_8L_2$ as the main product.[59] Quantum yields are of the order of 0.9, but the reactions may be suppressed by added carbon monoxide. The mechanism is that proposed in equations (15—17), an essential feature being the extreme lability of $Mn(CO)_5$. It has been suggested that this lability might be due to photo-excitation: but *cf.* $Cr(CO)_5$, ref. 18. Photolysis of the diphosphine complexes $Mn_2(CO)_8L_2$ in the presence of excess phosphine produces the paramagnetic species $Mn(CO)_3$-$(PR_3)_2$ (R = Bu or OEt) and these have been characterized by e.s.r.[59, 60] Spin trap adducts of $Mn(CO)_4PR_3$ have also been identified after thermal reaction of

[54] Y. M. Efremov, A. N. Samoilova, and L. V. Gurvich, *Chem. Phys. Letters*, 1976, **44**, 108.
[55] W. Klotzbucher, G. A. Ozin, J. G. Norman, and H. J. Kolari, *Inorg. Chem.*, 1977, **16**, 2871.
[56] W. E. Klotzbücher and G. A. Ozin, *J. Amer. Chem. Soc.*, 1978, **100**, 2262.
[57] G. A. Ozin and W. Klotzbücher, *J. Mol. Catal.*, 1977, **3**, 195.
[58] A. Freedman and R. Bersohn, *J. Amer. Chem. Soc.*, 1978, **100**, 4416.
[59] D. R. Kidd and T. L. Brown, *J. Amer. Chem. Soc.*, 1978, **100**, 4095.
[60] D. R. Kidd, C. P. Cheng, and T. L. Brown, *J. Amer. Chem. Soc.*, 1978, **100**, 4103.

$Mn_2(CO)_8(PR_3)_2$ with Me_3CNO.[61]

$$Mn_2(CO)_{10} \underset{\longleftarrow}{\overset{h\nu}{\longrightarrow}} \{2\,Mn(CO)_5\} \rightleftharpoons 2\,Mn(CO)_5 \quad (15)$$

$$Mn(CO)_5 \rightleftharpoons Mn(CO)_4 + CO \underset{\longleftarrow}{\overset{+L}{\longrightarrow}} Mn(CO)_4L \quad (16)$$

$$2Mn(CO)_4L \longrightarrow Mn_2(CO)_8L_2 \quad (17)$$

The salt $[fac\text{-}Mn(CO)_3(NH_3)_3]^+[Mn(CO)_5]^-$ is produced on passing ammonia through an irradiated pentane solution of $Mn_2(CO)_{10}$.[62] In the presence of substituted olefins [*e.g.* $(EtO_2C)_2C{=}C(CO_2Et)_2$][63] or *N*-bromoacetyl-dibenzazepine,[64] $Mn_2(CO)_{10}$ or $Re_2(CO)_{10}$ function as photo-initiators for the polymerization of methyl methacrylate. The standard heat of reaction ΔH^0 for process (18) has been determined as -217 kJ mol^{-1} by photocalorimetry.[26]

$$Re_2(CO)_{10} + I_2 \longrightarrow 2Re(CO)_4I + 2CO \quad (18)$$

Although CO-expulsion is the primary process for $[Mn(CO)_5]^-$ upon photo-excitation, for $[Mn(CO)_4(PPh_3)]^-$, it is phosphine elimination, with the substitution reaction (19) being observed.[65] Photo-excited $[Mn(CO)_5]^-$ also undergoes oxidative addition as illustrated in equations (20) and (21).

$$[Mn(CO)_4(PPh_3)]^- + P(OMe)_3 \xrightarrow{h\nu} [Mn(CO)_4P(OMe)_3]^- + PPh_3 \quad (19)$$

$$[Mn(CO)_5]^- + HSiPh_3 \xrightarrow{h\nu} cis\text{-}[HMn(CO)_4SiPh_3]^- + CO \quad (20)$$

$$[Mn(CO)_5]^- + Ph_4P^+ \xrightarrow{h\nu} cis\text{-}[PhMn(CO)_4PPh_3] + CO \quad (21)$$

Extended photolysis of the complex $[Mn(CO)_3(H_2O)_3]^+Cl^-$ in aqueous solution leads to the formation of glyoxal, probably *via* the protonation of the excited state [equation (22)].[66]

$$[Mn(CO)_3(H_2O)_3]^+ + H^+ \xrightarrow{h\nu} [Mn(CO)_2(CHO)(H_2O)_3]^{2+} \quad (22)$$

In argon matrices, irradiation of $Mn(NO)_3(CO)$ leads both to $Mn(NO)_3$ and also to a species represented as $Mn(CO)(NO)_2(NO^*)$, in which the metal centre is electron deficient and the unique nitrosyl ligand exhibits a particularly low value of $\nu(NO)$.[67] Under similar conditions $Mn(CO)_4(NO)$ yields $Mn(CO)_3$-(NO) and $Mn(CO)_4(NO^*)$.[68] It is suggested that both these species are active in the solution photochemistry of $Mn(CO)_4(NO)$, where the quantum yield for substitution has previously been found to be dependent on the nature of the nucleophile.[69]

Whereas both $XRe(CO)_3(4\text{-benzoylpyridine})_2$ and $XRe(CO)_3(3\text{-benzoyl-pyridine})_2$ ($X = Cl$, Br, or I) show emission from MLCT states at 298 K and 77 K, an additional luminescence band, apparently a ligand–ligand n–π^* tran-

[61] L. S. Benner and A. L. Balch, *J. Organometallic Chem.*, 1977, **134**, 121.
[62] M. Herberhold, F. Wehrmann, D. Neugebauer, and G. Huttner, *J. Organometallic Chem.*, 1978, **152**, 329.
[63] C. H. Bamford and S. U. Mullik, *J.C.S. Faraday I*, 1977, **73**, 1260.
[64] A. K. Alimoglu, C. H. Bamford, A. Ledwith, and S. U. Mullik, *Macromolecules*, 1977, **10**, 1081.
[65] R. A. Faltynek and M. S. Wrighton, *J. Amer. Chem. Soc.*, 1978, **100**, 2701.
[66] C. H. Bamford and M. Coldbeck, *J.C.S. Dalton*, 1978, 4.
[67] O. Crichton and A. J. Rest, *J.C.S. Dalton*, 1978, 202.
[68] O. Crichton and A. J. Rest, *J.C.S. Dalton*, 1978, 208.
[69] D. P. Keeton and F. Basolo, *Inorg. Chim. Acta*, 1972, **6**, 33.

sition, is observed for the latter complex at the lower temperature.[70] This observation of two types of emission presumably indicates that the MLCT and $n\pi^*$ excited states are not thermally equilibrated at 77 K, a consequence of their markedly different geometries.

$CpMn(CO)_2$ has been characterized in low temperature matrices,[30] and has been the subject of MO calculations which predict that the pyramidal structure is energetically favourable.[71] The metal–metal bonded compounds (10) are formed on irradiation of the corresponding $CpMn(CO)_2AsMe_2M(CO)_{n+1}$ (M = Mn or Re, n = 4; M = Co, n = 3).[72] Complex (11) is formed upon photoexcitation of $Mn(CO)_3(C_5H_4CH_2OCH_2CHCH_2)$.[73]

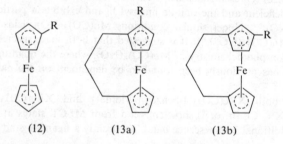

Me₂
As
／ ＼
$Cp(CO)_2Mn$—$M(CO)_n$

(10)   (11)

5 Iron, Ruthenium, and Osmium

Upon u.v. irradiation in the presence of nitrous oxide, substituted ferrocenes (12) [R = $(CH_2)_nCO_2^-$, n = 0 or 1,[74] n = 2—4;[75] R = $CH{=}CH(CH_2)_nCO_2^-$, n = 0 or 1;[76] R = $C{\equiv}CCO_2^-$ [76]] and (13)[77] undergo photo-oxidation to give the corresponding ferricenium zwitterions and nitrogen. From quenching studies with alcohols it is estimated that the excited state for complexes of type (12) [R = $(CH_2)_nCO_2^-$; n = 1—4] has a lifetime >1 μs; an interesting observation in view of the very short lifetimes of the unsubstituted ferrocene triplet state itself.

(12)   (13a)   (13b)

[70] P. J. Giordano, S. M. Fredericks, M. S. Wrighton, and D. L. Morse, *J. Amer. Chem. Soc.*, 1978, **100**, 2257.

[71] P. Hofmann, *Angew. Chem.*, 1977, **89**, 551; *Angew. Chem. Internat. Edn.*, 1977, **16**, 536.

[72] U. Richter and H. Vahrenkamp, *J. Chem. Research (S)*, 1977, 156.

[73] A. N. Nesmeyanov, M. I. Rybinskaya, and L. M. Korneva, *Doklady Akad. Nauk S.S.S.R.*, 1977, **237**, 1373.

[74] E. K. Heaney and S. R. Logan, *J.C.S. Perkin II*, 1978, 590.

[75] E. K. Heaney and S. R. Logan, *J.C.S. Perkin II*, 1977, 1353.

[76] E. K. Heaney, S. R. Logan, and W. E. Watts, *J. Organometallic Chem.*, 1978, **153**, 229.

[77] E. K. Heaney, S. R. Logan, and W. E. Watts, *J. Organometallic Chem.*, 1978, **150**, 309.

An unusual periodic variation in optical density has been observed upon irradiation of octamethylferrocene (OMF) or its ferrocenium ion in ethanol solution.[78] The oscillations are attributed to the formation and dissociation of aggregates of the type $[(OMF)_m(OMF^+)_n]$.

Photolysis of monosubstituted ferrocenes in CCl_4–EtOH solution produces the corresponding ester (14).[79] A study of substituent effects in this reaction reveals that they control the efficiency of the reaction by affecting the extent of formation of the charge transfer complex (15) rather than the rate of attack of $CCl_3\cdot$ on the ferrocenium ion (Scheme 2). Excitation of a charge transfer complex between ferrocene and $MeO_2CC\equiv CCO_2Me$ in aqueous acetic acid yields (16) as principal product.[80] Irradiation of the charge transfer complexes of ferrocene or ruthenocene with mercuric halides induces formation of the corresponding metallocenium ions and the production of mercurous ions.[81] Other

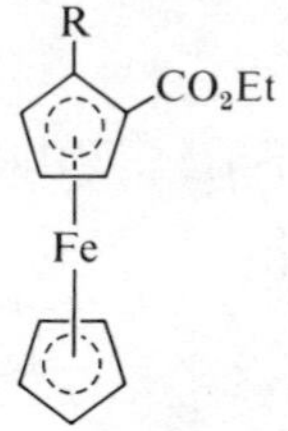

(14)

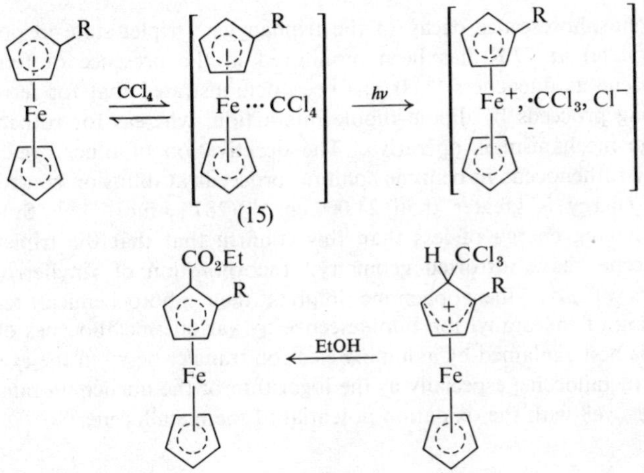

Scheme 2

[78] A. N. Nesmeyanov, N. S. Kochetkova, R. B. Materikova, I. R. Lyatifov, N. V. Fok, T. V. Burova, B. M. Yavorskii, and V. N. Trembovler, *J. Organometallic Chem.*, 1978, **145**, 69; *Doklady Akad. Nauk S.S.S.R.*, 1977, **236**, 624.

[79] T. Akiyama, T. Kitamura, T. Kato, H. Watanabe, T. Serizawa, and A. Sugimori, *Bull. Chem. Soc. Japan*, 1977, **50**, 1137.

[80] H. Watanabe, K. Ito, F. Nakanishi, T. Akiyama, and A. Sugimori, *Bull. Chem. Soc. Japan*, 1977, **50**, 1349.

[81] O. Traverso, C. Chiorboli, U. Mazzi, and G. L. Zucchini, *Gazz. Chim. Ital.*, 1977, **107**, 181.

(16)

authors have reported the polymerization of methyl methacrylate by ferrocene and carbon tetrachloride in micelles,[82] the photodecomposition of acetoacetyl-ferrocene,[83] the light-induced reaction of (17) to give (18),[84] and the use of ethanolic electrolytes containing ferrocene and ferrocenium ions in an n-type silicon electrochemical cell.[85]

(17)   (18)

The phosphorescence decay of the triphenylene triplet state in poly(methyl methacrylate) at 77 K has been monitored in the presence of ferrocene or ruthenocene as quenchers.[86] It has been demonstrated that for ferrocene the quenching proceeds by dipole–dipole interaction, whereas for ruthenocene an exchange mechanism is operative. The deactivation of other organic triplet states by ruthenocene in benzene solution proceeds at diffusion-controlled rates if their energy is greater than $24\,000$ cm^{-1} (287 kJ mol^{-1}).[87] Studies with triplets having energy of less than this confirm that that the triplet state of ruthenocene has a distorted geometry. Incorporation of vinylferrocene into polymers of *p*-vinylbenzophenone inhibits their photochemical reactions.[88] Quenching of the uranyl ion luminescence by various metallocenes of Fe, Ru, and Os is best explained by assuming electron transfer between the excited state and the metallocene, especially as the logarithm of the quenching rate constant correlates well with the oxidation potential of the metallocene.[89]

[82] M. Tsunooka and M. Tanaka, *J. Polymer Sci. Polymer Letters*, 1978, **16**, 119.
[83] T. Shiraiwa and H. Imai, *Technol. Rep. Kansai Univ.*, 1977, **18**, 83.
[84] S. Toma, M. Pariciova, E. Solcaniova, and K. Lesko, *Coll. Czech. Chem. Comm.*, 1978, **43**, 295.
[85] K. D. Legg, A. B. Ellis, J. M. Bolts, and M. S. Wrighton, *Proc. Natl. Acad. Sci., U.S.A.*, 1977, **74**, 4416.
[86] J. P. Vikesland and F. Wilkinson, *J.C.S. Faraday II*, 1977, **73**, 1082.
[87] A. P. Chapple, J. P. Vikesland, and F. Wilkinson, *Chem. Phys. Letters*, 1977, **50**, 81.
[88] G. Sanchez, W. Weill, and R. Knoesel, *Makromol. Chem.*, 1978, **179**, 131.
[89] O. Traverso, R. Rossi, L. Magon, A. Cinquantini, and T. J. Kemp, *J.C.S. Dalton*, 1978, 569.

In a series of papers[90-95a] and in recent reviews,[95b] Poliakoff and co-workers have further examined the properties of matrix-isolated $Fe(CO)_4$. It had previously been demonstrated that this species, which is produced by photolysis of $Fe(CO)_5$, had C_{2v} symmetry,[96] and MO calculations predicted that it would be a ground-state triplet.[97] This has been confirmed by a study of the temperature dependence of its magnetic circular dichroism spectrum.[90] $Fe(CO)_4$ isomerizes upon i.r. laser irradiation into its ν(C—O) bands.[91] Using labelled complexes $Fe(CO)_x(^{13}C^{18}O)_{4-x}$, it has been shown that isomerizations such as (23) proceed

$$\underset{(X\,=\,{}^{13}C^{18}O)}{} \qquad \underset{1880.5\ cm^{-1}}{\overset{1897.5}{\rightleftarrows}} \qquad \tag{23}$$

by a non-Berry pseudorotation. In nitrogen matrices, where isomerization is more rapid than the formation of $Fe(CO)_4N_2$,[92] examination of the polarized i.r. spectra reveals that the isomerization is orientationally selective and that the $Fe(CO)_4$ does not freely rotate during the experiment.[93] In the case of methane matrices, reaction proceeds more rapidly than the isomerization, possibly a consequence of the different structures of the methane and dinitrogen compounds (19) and (20).[92]

(19)　　　　　　　　　　(20)

The radicals formed upon irradiation of $Fe(CO)_5$ and CCl_4 [reaction (24)] have been trapped by Bu^tNO and $PhCH{=}N(O)Bu^t$, and identified by e.s.r.[98]

$$Fe(CO)_5 + CCl_4 \xrightarrow{\ h\nu\ } Fe(CO)_4Cl + CCl_3{\cdot} + CO \tag{24}$$

Presumably similar intermediates are involved in the photonitrosylation of $Fe(CO)_5$ in CCl_4 for which the final product is $Fe(NO)_3Cl$.[43]

Sarel has reviewed his group's work on rearrangements and insertions into cyclopropyl olefins upon irradiation in the presence of $Fe(CO)_5$,[99] and has

90　T. J. Barton, R. Grinter, A. J. Thomson, B. Davies, and M. Poliakoff, *J.C.S. Chem. Comm.*, 1977, 841.

91　B. Davies, A. McNeish, M. Poliakoff, and J. J. Turner, *J. Amer. Chem. Soc.*, 1977, **99**, 7573.

92　B. Davies, A. McNeish, M. Poliakoff, M. Tranquille, and J. J. Turner, *J.C.S. Chem. Comm.*, 1978, 36.

93　B. Davies, A. McNeish, M. Poliakoff, M. Tranquille, and J. J. Turner, *Chem. Phys. Letters*, 1977, **52**, 477.

94　J. K. Burdett, P. G. Buckley, B. Davies, J. H. Carpenter, A. McNeish, M. Poliakoff, J. J. Turner, D. H. Whiffen, R. L. Allwood *et al.*, *Springer Ser. Opt. Sci.*, 1977, **7**, 427.

95a　M. Poliakoff, B. Davies, A. McNeish, M. Tranquille, and J. J. Turner, *Ber Bunsengesellschaft Phys. Chem.*, 1978, **82**, 121.

95b　J. K. Burdett, *Chem. Soc. Rev.* 1978, **7**, 507; M. Poliakoff, *ibid.*, p. 527.

96　M. Poliakoff and J. J. Turner, *J.C.S. Dalton*, 1974, 2276.

97　J. K. Burdett, *J.C.S. Faraday II*, 1974, **70**, 1599.

98　R. G. Gasanov and R. K. Freidlina, *Doklady Akad. Nauk S.S.S.R.*, 1977, **235**, 1309.

99　S. Sarel, *Accounts Chem. Res.*, 1978, **11**, 204.

reported on the production of bicyclic octadienones from divinylcyclopropanes *via* this route.[100] The usefulness of the $Fe(CO)_5$ photolysis procedure is further exemplified by studies with tricyclic vinylcyclopropanes,[101] tricyclic heptenes,[102] tricyclic enones,[103] norbornene and acetylenes,[104] 7-oxanorbornene derivatives,[105] and mono-epoxidized dienes.[106]

(Cyclobutadiene)$Fe(CO)_2$ and its dinitrogen complex have been identified and characterized following photolysis of $(C_4H_4)Fe(CO)_3$ in low temperature matrices.[30] These studies substantiate the findings of earlier reports about the non-lability of cyclobutadiene under these conditions. It is probable that the corresponding dicarbonyl is involved in the reactions of (substituted cyclobutadiene)$Fe(CO)_3$ complexes with acetylenes, even though the products observed are arenes formed by the cycloaddition of the acetylene to the cyclobutadiene.[107, 108] With $(C_4Me_4)Fe(CO)_3$, organometallic products (21) and (22)

(21)   (22)

were isolated; (21) being the first known example of an (arene)$Fe(CO)_3$ complex.[107] Photolysis of $(C_4H_4)Fe(CO)_3$ with B_5H_9 gives a small amount of the metallocarbaborane $(BC_4H_5)Fe(CO)_3$.[109] Ferraborane $(B_4H_8)Fe(CO)_3$ may be regarded as an analogue of $(C_4H_4)Fe(CO)_3$ and its photoreaction with but-2-yne provides a useful route to the new carbaborane species Me_4C_4-B_4H_4.[110]

It is now well established that the 17-electron compound $CpFe(CO)_2\cdot$ abstracts halogen atoms readily from halocarbons.[40] Attempts to obtain evidence for hydrogen atom abstraction have, however, proved negative as the photolysis of $[CpFe(CO)_2]_2$ in solvents such as cyclohexane, toluene, or propan-2-ol leads to no reaction.[111] Only with benzaldehyde was a reaction observed [reaction (25)], but in this case too it is probable that the $CpFe(CO)_2$ does not abstract a hydrogen atom from the benzaldehyde, reaction (26), but rather forms a com-

100 S. Sarel and M. Langbeheim, *J.C.S. Chem. Comm.*, 1977, 827.
101 K. Kayakawa and H. Schmid, *Helv. Chim. Acta*, 1977, **60**, 1942.
102 R. Aumann, H. Woermann, and C. Krueger, *Chem. Ber.*, 1977, **110**, 1442.
103 J. Elzinga and H. Hogeveen, *J. Org. Chem.*, 1978, **43**, 745.
104 K. Hayakawa and H. Schmid, *Helv. Chim. Acta*, 1977, **60**, 2160.
105 A. A. Pinkerton, P. A. Carrupt, P. Vogel, T. Boschi, N. H. Thuy, and R. Roulet, *Inorg. Chim. Acta*, 1978, **28**, 123.
106 G. D. Annis and S. V. Ley, *J.C.S. Chem. Comm.*, 1977, 581.
107 A. Bond, M. Bottrill, M. Green, and A. J. Welch, *J.C.S. Dalton*, 1977, 2372.
108 P. L. Pruitt, E. R. Biehl, and P. C. Reeves, *J. Organometallic Chem.*, 1977, **134**, 37.
109 T. P. Fehlner, *J. Amer. Chem. Soc.*, 1978, **100**, 3250.
110 T. P. Fehlner, *J. Amer. Chem. Soc.*, 1977, **99**, 8355.
111 J. A. Labinger and S. Madhavan, *J. Organometallic Chem.*, 1977, **134**, 381.

plex, reaction (27). Photoreaction of $[(MeCp)Fe(CO)_2]_2$ with sulphur yields

$$[CpFe(CO)_2]_2 + 2PhCHO \xrightarrow{\ h\nu\ } 2CpFe(CO)_2Ph + 2CO + H_2 \qquad (25)$$

$$CpFe(CO)_2 + PhCHO \longrightarrow CpFe(CO)_2H + Ph\dot{C}O \qquad (26)$$

$$CpFe(CO)_2 + PhCHO \longrightarrow CpFe(CO)(PhCHO) + CO \qquad (27)$$

complex (23).[112] Species (24) is the product of photolysis of $[CpFe(CO)_2]_2$ and $(RCOCH=CHI)Fe(CO)_4$ (R = Me or OMe).[113] The insertion of $CF_3C\equiv CCF_3$ or of $MeO_2CC\equiv CCO_2Me$ into the metal-to-metal bond of $[FeX(CO)_3]_2$ (X = SMe, SPh, or PMe_2) is induced photochemically.[114]

(23)

(24)

In contrast with the reaction of the analogous alkyl complexes, photoreaction of $CpFe(CO)_2SiMe_3$ with cyclohexyl isocyanide causes CO-substitution and the production of the mono- and di-substituted complexes.[115] Photolysis of $CpFe(CO)_2\overline{Si(CH_2)_2}CH_2$ in hexane causes cleavage of the Fe—Si bond, although photosubstitution of the carbon monoxide by phosphines may be achieved by shorter term irradiation.[116] Photo-induced CO-elimination from $[CpFe(CO)_2]_2SiMeH$ yields complex (25).[117] $[CpFe(CO)(dppe)]^+Br^-$ undergoes efficient photosubstitution of the carbon monoxide, and dinitrogen and ammonia- and hydrazine-derivatives have been synthesized.[118]

Photolabilization of carbon monoxide in the ionic macrocyclic complexes $[Fe(TIM)(MeCN)CO]^{2+}$ [TIM = (26)] occurs in acetonitrile ($\Phi = 0.6$), but not in acetone, where expulsion of acetonitrile is observed.[119,120] Other reports consider CO-dissociation from Ru(porphyrin)(CO)(EtOH) complexes[121] and the phosphorescence of $[Fe(CN)_5CO]^{3-}$.[122]

[112] C. Giannotti, A. M. Ducourant, H. Chanaud, A. Chiaroni, and C. Riche, *J. Organometallic Chem.*, 1977, **140**, 289.
[113] A. N. Nesmeyanov, M. I. Rybinskaya, L. V. Rybin, E. A. Petrovskaya, and A. I. Lutsenko, *Izvest. Akad. Nauk S.S.S.R. Ser. khim.*, 1977, 904.
[114] R. Mathieu and R. Poilblanc, *J. Organometallic Chem.*, 1977, **142**, 351.
[115] T.-M. Chan, J. W. Connolly, C. D. Hoff, and F. Millich, *J. Organometallic Chem.*, 1978, **152**, 287.
[116] C. S. Cundy, M. F. Lappert, and C.-K. Yuen, *J.C.S. Dalton*, 1978, 427.
[117] W. Malisch and W. Ries, *Angew. Chem.*, 1978, **90**, 140; *Angew. Chem. Internat. Edn.*, 1978, **17**, 120.
[118] D. Sellmann and E. Kleinschmidt, *J. Organometallic Chem.*, 1977, **140**, 211.
[119] J. I. Zink and M. J. Incorvia, *J.C.S. Chem. Comm.*, 1977, 730.
[120] M. J. Incorvia and J. I. Zink, *Inorg. Chem.*, 1977, **16**, 3161.
[121] N. Farrell, D. H. Dolphin, and B. R. James, *J. Amer. Chem. Soc.*, 1978, **100**, 324.
[122] L. Vera and F. Zuloaga, *Inorg. Chem.*, 1978, **17**, 765.

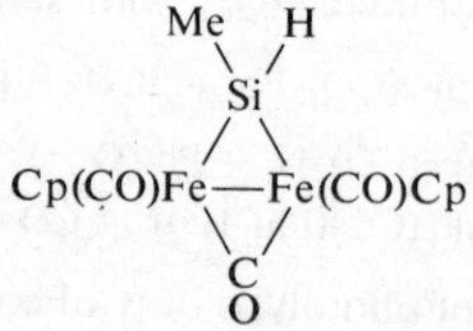

Me H
\
Si
/
Cp(CO)Fe——Fe(CO)Cp
\
C
‖
O

(25)

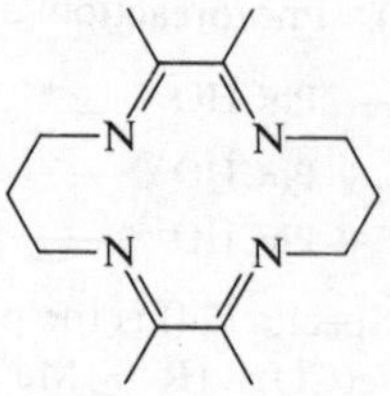

(26)

6 Cobalt, Rhodium, and Iridium

The photochemical cleavages of the Co—Me bonds in $[Co(A)(H_2O)Me]^{2+}$ [A = (27a) or (27b)] have particularly low threshold energies, and may be induced by irradiation of all wavelengths in the range 254—540 nm.[123] Interestingly although the excited state responsible for this reaction has not yet been identified it does not appear to be associated with any of the charge transfer

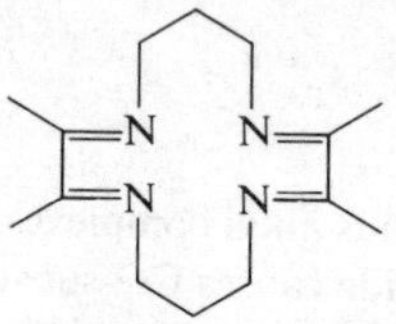

(27a)

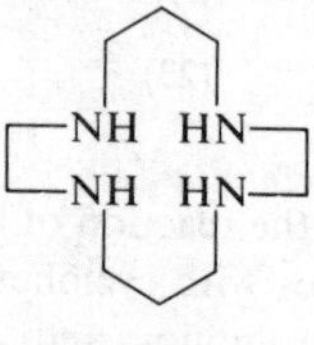

(27b)

transitions, all of which lie in the u.v. region of the spectrum. $\cdot CH_2CO_2^-$ radicals formed upon photolysis of $[Co(CN)_5(CH_2CO_2^-)]^{4-}$, react with nitroxide spin labels.[124] Recently this technique has been employed for the study of diffusion in phospholipid bilayers. Related publications consider light-induced bond cleavage reactions in some alkyl cobaloximes[125,126] and in other similar chelate systems.[127]

Transitions to the lowest excited states of the cluster compounds (28) correspond to population of metal-to-metal anti-bonding orbitals. Consistent with this proposal, it has been shown that irradiation of (28) causes declusterification, although CO-substitution by phosphines also occurs.[128] The conversion of the C_{3v} form of $Cp_3Rh(CO)_3$ into its C_s isomer can be achieved photochemically.[129] Other reports have described the isolation of (29) by irradiation of $CpIr(CO)_2$ in benzene,[130] the photochemical addition of bis(alkoxy)carbenes to $CpCo$-$(CO)_2$,[131] the formation of the novel rectangular cluster (30) from $Ir_4(CO)_{12}$

[123] C. Y. Mok and J. F. Endicott, *J. Amer. Chem. Soc.*, 1978, **100**, 123.
[124] J. R. Sheats and H. M. McConnell, *J. Amer. Chem. Soc.*, 1977, **99**, 7091.
[125] B. T. Golding, C. S. Sell, and P. J. Sellars, *J.C.S. Chem. Comm.*, 1977, 693.
[126] T. Funabiki, B. D. Gupta, and M. D. Johnson, *J.C.S. Chem. Comm.*, 1977, 653.
[127] G. Roewer, G. A. Shagisultanova, and I. V. Woyakin, *J. Prakt. Chem.*, 1977, **319**, 1031.
[128] G. L. Geoffroy and R. A. Epstein, *Inorg. Chem.*, 1977, **16**, 2795.
[129] R. J. Lawson and J. R. Sharpley, *Inorg. Chem.*, 1978, **17**, 772.
[130] M. D. Rausch, R. G. Gastinger, S. A. Gardner, R. K. Brown, and J. S. Wood, *J. Amer. Chem. Soc.*, 1977, **99**, 7871.
[131] M. L. Ziegler, K. Weidenhammer, and W. A. Herrmann, *Angew. Chem.*, 1977, **89**, 557; *Angew. Chem. Internat. Edn.*, 1977, **16**, 555.

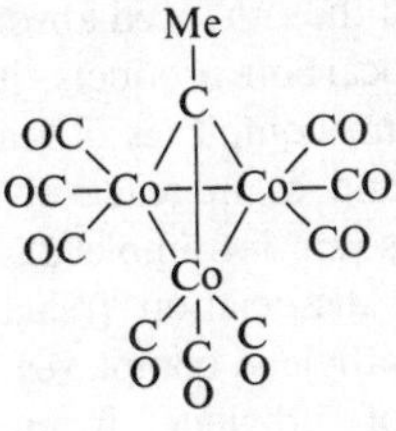

(28)

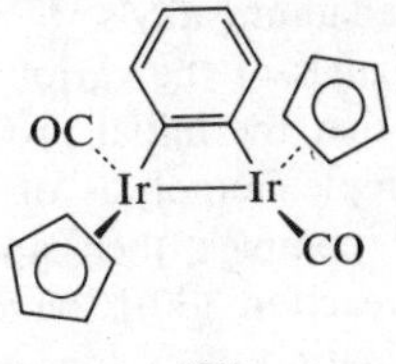

(29)

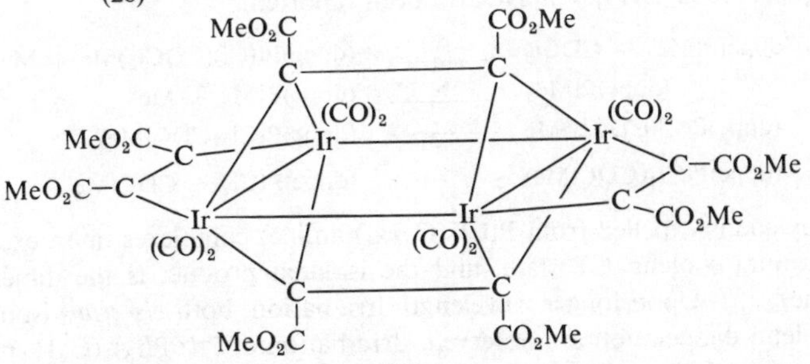

(30)

and $MeO_2CC\equiv CCO_2Me$,[132] and the photochemical oxygenation of $Cp(PPh_3)$- $\overline{CoCPh{=}CRCR}{=}CPh$ (R = Ph or Me).[133]

Upon photolysis of the polystyrene bonded complex (PS)-$CH_2C_5H_4Co(CO)_2$ no evidence was found for di- or poly-nuclear species such as are formed from $CpCo(CO)_2$ in homogeneous solution. Instead, the co-ordinatively unsaturated centre produced by CO-expulsion interacted with the aromatic rings of the polymer support.[134]

In a re-examination of the photochemistry of tetrakis(methylisocyanide)-iridium(I), Geoffroy *et al.* have demonstrated that the compound exists as an oligomer in solution, and that irradiation merely causes dissociation of the aggregate.[135] Similarly $[Rh(tol)_4]^+PF_6^-$ (tol = *p*-methylphenylisocyanide) exists partly as a dimer in solution, and excitation of this species leads both to a very short-lived ($\tau < 2$ ns) emission and a longer-lived ($\tau > 90$ ns) transient absorption.[136] Intense short-lived emission is also reported for the bridged complex $[Rh_2\{CN(CH_2)_3NC\}_4]^{2+}$ in acetonitrile, and a long-lived transient absorption ($\tau = 8$ µs) in this case has been assigned to a triplet state of the complex.

7 Nickel, Palladium, and Platinum

The complexities of the photochemistry of the metal–carbon bond cleavage reaction are illustrated by CIDNP investigations with various phosphine plati-

132 P. F. Heveldt, B. F. G. Johnson, J. Lewis, P. R. Raithby, and G. M. Sheldrick, *J.C.S. Chem. Comm.*, 1978, 340.
133 F. W. Grevels, Y. Wakatsuki, and H. Yamazaki, *J. Organometallic Chem.*, 1977, **141**, 331.
134 G. Gubitosa, M. Bolt, and H. H. Brintzinger, *J. Amer. Chem. Soc.*, 1977, **99**, 5174.
135 G. L. Geoffroy, M. G. Bradley, and M. E. Keeney, *Inorg. Chem.*, 1978, **17**, 777.
136 V. M. Miskowski, G. L. Nobinger, D. S. Kliger, G. S. Hammond, N. S. Lewis, K. R. Mann, and H. B. Gray, *J. Amer. Chem. Soc.*, 1978, **100**, 485.

num and palladium alkyls.[137] From an analysis of the enhanced absorption and emission signals of the various chlorinated hydrocarbon products, it has been postulated that the initial process for the platinum complexes is reaction (28) and not simple homolysis of the metal–alkyl bond as in reaction (29). The initial Pt^{III} complex then apparently undergoes an intramolecular carbene insertion [reaction (30)], which is followed by dissociation [reaction (31)]. Photo-elimination of cyclopropane from the trimethylene complexes (L—L)Pt-$(C_3H_6)X_2$ (L—L = bipy or phen),[138] and of ethylene from $(PPh_3)_2$-$NiCH_2CH_2CH_2CH_2$,[139] have also been reported.

$$(dppe)PtMe_2 + CDCl_3 \xrightarrow{h\nu} (dppe)Pt(Cl)(CDCl_2)Me + Me\cdot \quad (28)$$

$$(dppe)PtMe_2 \xrightarrow{h\nu} (dppe)PtMe + Me\cdot \quad (29)$$

$$(dppe)PtCl(CDCl_2)Me \longrightarrow (dppe)PtCl_2(CDClMe) \quad (30)$$

$$(dppe)PtCl_2(CDClMe) \longrightarrow (dppe)PtCl_2 + \dot{C}DClMe \quad (31)$$

Ethylene is expelled from $PtCl_2(C_2H_4)$(amine) complexes upon excitation to a platinum → olefin CT state, and the isolated product is the dimer Pt_2Cl_4-(amine)$_2$.[140] Upon longer wavelength irradiation, both *cis–trans* isomerization and olefin dissociation are observed. Irradiation of $Pt(PPh_3)_2(C_2H_4)$ in a band assigned to platinum → olefin charge transfer induces olefin dissociation and the resultant production of the highly reactive $Pt(PPh_3)_2$.[141] However, irradiation at shorter wavelengths ($\lambda = 254$ nm) excites a phosphine localized transition, and yields the orthometallated product (31).

$$Ph_2P\text{———}Pt(Et)PPh_3$$
(31)

8 Copper and Silver

Because of its high specific energy storage capacity, transition-metal sensitized isomerization of norbornadiene to quadricyclene is a potentially attractive system for photochemical solar energy conversion, if this can be achieved using visible light.[142] Irradiation ($\lambda < 350$ nm) of CuCl–norbornadiene complexes induces quadricyclene formation, although whether the reactive excited state is a charge-transfer one or a norbornadiene ligand localized (π–π^*) species is as yet not clear.[143] The pathway shown in reactions (32) and (33) *via* carbonium ion products is not favoured by the authors as there is no noticeable solvent effect. Phosphine complexes of Cu^I also sensitize the reaction, the quantum

[137] P. W. N. M. Van Leeuwen, C. F. Roobeek, and R. Huis, *J. Organometallic Chem.*, 1977, **142**, 233.

[138] G. Phillips, R. J. Puddephatt, and C. F. H. Tipper, *J. Organometallic Chem.*, 1977, **131**, 467.

[139] R. H. Grubbs, A. Miyashita, M. Liu, and P. Burk, *J. Amer. Chem. Soc.*, 1978, **100**, 2418.

[140] P. Courtot, R. Rumin, and A. Peron, *J. Organometallic Chem.*, 1978, **144**, 357.

[141] S. Sostero, O. Traverso, M. Lenarda, and M. Graziani, *J. Organometallic Chem.*, 1977, **134**, 259.

[142] C. Kutal, D. P. Schwendiman, and P. Grutsch, *Solar Energy*, 1977, **19**, 651.

[143] D. P. Schwendiman and C. Kutal, *J. Amer. Chem. Soc.*, 1977, **99**, 5677.

$$\text{(structure)} \quad \xrightarrow{h\nu} \quad \text{(structure)} \quad (32)$$

$$\text{(structure)} \quad \xrightarrow{} \quad \text{(structure)} \quad (33)$$

yield ($\lambda = 313$ nm) for $Cu(PPh_3)_2,BH_4$ being about 0.6 at high norbornadiene concentrations.[144] However, in this case there is no evidence for complex formation with the norbornadiene, and the sensitization apparently proceeds either *via* direct energy transfer or *via* exciplex formation.

Cu^I triflate complexes catalyse the cycloaddition of conjugated dienes to cyclohexene and cycloheptene.[145] Thus with butadiene and cyclohexene the major product is (32). The photo-induced step appears to be formation of *trans*-cyclohexene in the Cu^I–cyclohexene complex. In the absence of dienes and at low cyclohexene concentrations, irradiation of the cyclohexene–Cu^I complexes produces a mixture of isomers (33—35) probably *via* *trans*-cyclohexene.[146] Other recent publications dealing with Cu^I triflate sensitized

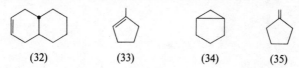

(32)   (33)   (34)   (35)

reactions are a full paper on the photofragmentation and rearrangement of methylenecyclopropanes,[147] the photochemical reactions of allyl alcohol and related compounds,[148] and the photocycloaddition of allyl alcohol to norbornene or dicyclopentadiene.[149]

Cationic polymerization of THF has been initiated by u.v. irradiation of Cu^I, Cu^{II}, and Ag^I triflate complexes.[150]

Selective excitation of atomic silver in low temperature matrices allows the preparation of small clusters such as Ag_2 and Ag_3.[151]

9 Mercury

The major decomposition route for $Hg[C(N_2)CO_2Et]_2$ is reaction (34).[152] Photolysis of $Hg(GeEt_3)_2$ in the presence of C_6F_5COR (R = CF_3, C_6F_5, or

[144] P. A. Grutsch and C. Kutal, *J. Amer. Chem. Soc.*, 1977, **99**, 6460.
[145] J. T. M. Evers and A. Mackor, *Tetrahedron Letters*, 1978, 2317.
[146] J. T. M. Evers and A. Mackor, *Tetrahedron Letters*, 1978, 2321.
[147] R. G. Salomon, A. Sinha, and M. F. Salomon, *J. Amer. Chem. Soc.*, 1978, **100**, 520.
[148] J. T. M. Evers and A. Mackor, *Tetrahedron Letters*, 1978, 821.
[149] R. G. Salomon and A. Sinha, *Tetrahedron Letters*, 1978, 1367.
[150] M. E. Woodhouse, F. D. Lewis, and T. J. Marks, *J. Amer. Chem. Soc.*, 1978, **100**, 996.
[151] G. A. Ozin and H. Hubek, *Inorg. Chem.*, 1978, **17**, 155.
[152] T. B. Patrick and T.-T. Wu, *J. Organometallic Chem.*, 1978, **43**, 1506.

Ph) initially produces the radical $C_6F_5(Et_3GeO)RC\cdot$.[153] The light-induced reaction of (cyclopentyl)HgBr and $Pt(PPh_3)_3$ yields $RPt(PPh_3)_2Br$.[154]

$$Hg[C(N_2)CO_2Et]_2 \xrightarrow{h\nu} N_2 + EtO_2C\ddot{C}HgC(N_2)CO_2Et \quad (34)$$

153 D. V. Gendin, O. A. Kruglaya, T. I. Vakul'skaya, L. P. Petukhov, and N. S. Vyazankin, *Zhur. obshchei Khim.*, 1977, **47**, 1427.
154 V. V. Bashlov, V. I. Sokolov, G. Z. Suleimanov, and O. A. Reutov, *Izvest. Akad. Nauk S.S.S.R., Ser. Khim.*, 1977, 2562.

Table 1 *Photochemical substitution reactions of metal carbonyl compounds*

Substrate	Reactant L	Products	Ref.
$[V(CO)_6]^-$	Ph_2PPPh_2	$\{(\mu\text{-}L)_2[V(CO)_4]_2\}^{2-}$ $\{(\mu\text{-}L)[V(CO)_5]_2\}^{2-}$	155
$CpV(CO)_4$	$PhP[(CH_2)_2PPh_2]_2$	$[V(CO)_4L]^-$ and $[V(CO)_3L]^-$	156
	MR_3 (M = As, Sb, or Bi)	$[V(CO)_5L]^-$	157
	Ph_2PPPh_2	$CpV(CO)_3L$	155
	$PhP[(CH_2)_2PPh_2]_2$	$CpV(CO)_2L$	156
	(bicyclic dione structure)	$(C_4H_4)CpV(CO)_2$	158
$M(CO)_6$ (M = Cr, Mo, or W)	SMe_2; SPh_2	$M(CO)_5L$	159
	$Ph(CH_2)_nCN$ ($n = 0\text{—}3$)	$M(CO)_5L$	160
	(Et, Et, Me, Me-substituted dithiaborole structure: $Et\text{—}C{=}C\text{—}Et$, B, B, S, Me)	$M(CO)_4L$ and $M(CO)_2L_2$	161
	$P(M^1Me_3)$ (M^1 = C, Si, Ge, or Sn)	$M(CO)_5L$	162

[155] D. Rehder, *J. Organometallic Chem.*, 1977, **137**, C25.
[156] I. Mueller and D. Rehder, *J. Organometallic Chem.*, 1977, **139**, 293.
[157] R. Talay and D. Rehder, *Chem. Ber.*, 1978, **111**, 1978.
[158] M. D. Rausch and A. V. Grossi, *J.C.S. Chem. Comm.*, 1978, 401.
[159] M. Herberhold and G. Suess, *J. Chem. Res. (S)*, 1977, 246.
[160] A. N. Nesmeyanov, V. S. Kaganovich, V. V. Krivykh, and M. I. Rybinskaya, *Izvest. Akad. Nauk S.S.S.R., Ser. khim.*, 1977, 156.
[161] K. Kinberger and W. Siebert, *Chem. Ber.*, 1978, **111**, 356.
[162] H. Schumann, L. Roesch, H. J. Kroth, J. Pickardt, H. Neumann, and B. Neudert, *Z. anorg. Chem.*, 1977, **430**, 51.

Table 1 (cont.)

Substrate	Reactant L	Products	Ref.
$(C_6H_5CO_2Me)Cr(CO)_3$	CN^-	$[(C_6H_5CO_2Me)Cr(CO)_2L]^-$	163
(phospholyl–Mn(CO)₃ complex: Me, Me ring with P, $Mn(CO)_3$)	PR_3	(phospholyl–Mn(CO)₂L complex: Me, Me ring with P, $Mn(CO)_2L$)	164
$Re(CO)_5$–$C(CF_3)_2$–C–CF_3 =CF_2	$P(OPh)_3$	$Re(CO)_4L$–$C(CF_3)_2$–C–CF_3 =CF_2	165
$CpRe(CO)_2$	NH_3	$CpRe(CO)_2L$	166
$Fe(CO)_5$	$(C_6H_{11}N=CH—)_2$	$Fe(CO)_3L$ and $Fe_2(CO)_6L$	167
	1-allylpyrazole	$Fe(CO)_4L$	168
	$HSiMe_2NEt_2$	$Fe(CO)_4(SiMe_2)\,(NHEt_2)$	169
$Fe_3(CO)_{12}$	$Me_2SiCH_2CH_2CH_2$	$(CO)_4FeCH_2CH_2CH_2SiMe_2$	170
$[CpFe(CO)_2]_2$	$MeN(PF_2)_2$	$Fe_2(CO)L_3(PF_2)$ (MeN=PF₂)	171
	$MeN(PF_2)_2$	$[CpFe(CO)]_2L$ and $Cp_2Fe_2L(PF_2)$ (MeN=PF₂)	172
$Fe(CO)_4PPh_3$	$PhCH=CHCOMe$	$Fe(CO)_2(PPh_3)L$	173
$CpFe(CO)_2$–$C(CF_3)_2$–C–CF_3 =CF_2	$CH_2=CHCH=CH_2$	$LCpFe$–$C(CF_3)(F_2)$=C–CF_3, CF_3	174
$(CO)_3Fe(\mu\text{-}AsMe_2)_2Fe(CO)_3$	$PPhMe_2$; $PPhMeCH_2Ph$	$(CO)_3Fe(\mu\text{-}AsMe_2)_2Fe(CO)_2L$	175

CpFe(CO)₂C(CN)=C(CN)CH₂CH=CH₂   CpFe(CO)C(CN)=C(CN)CH₂CH=CH₂   176

CpFe(CO)₂ ⟨ring⟩ Mn(CO)₃   CpFe(COL) ⟨ring⟩ Mn(CO)₃   177

Co₂(CO)₈   MeN(PF₂)₂   PPh₃   L₃Co₂(PF₂=NMeH)₂   178

163 P. Le Maux, B. Simmoneaux, G. Jaouen, L. Ouahab, and P. Batail, *J. Amer. Chem. Soc.*, 1978, **100**, 4312.
164 A. Breque and F. Mathey, *J. Organometallic Chem.*, 1978, **144**, C9.
165 A. N. Nesmeyanov, N. E. Kolobova, I. B. Zlotina, L. V. Ivanova, and K. N. Anisimov, *Izvest. Akad. Nauk S.S.S.R., Ser. khim.*, 1977, 707.
166 D. Sellmann and E. Kleinschmidt, *Z. Naturforsch.*, 1977, **32b**, 795.
167 H. W. Frühauf, A. Landers, R. Goddard, and C. Krüger, *Angew. Chem.*, 1978, **90**, 56; *Angew. Chem. Internat. Edn.*, 1978, **17**, 64.
168 K. Fukushima, T. Miyamoto, and Y. Sasaki, *Bull. Chem. Soc. Japan*, 1978, **51**, 499.
169 G. Schmid and E. Welz, *Angew. Chem.*, 1977, **89**, 823; *Angew. Chem. Internat. Edn.*, 1977, **16**, 785.
170 C. S. Cundy and M. F. Lappert, *J.C.S. Dalton*, 1978, 665.
171 M. G. Newton, R. B. King, M. Chang, and J. Gimeno, *J. Amer. Chem. Soc.*, 1978, **100**, 326.
172 M. G. Newton, R. B. King, M. Chang, and J. Gimeno, *J. Amer. Chem. Soc.*, 1978, **100**, 1632.
173 B. F. G. Johnson, J. Lewis, G. R. Stephenson, and E. J. S. Vichi, *J.C.S. Dalton*, 1978, 369.
174 N. E. Kolobova, I. B. Zlotina, and I. I. Ivanova, *Izvest. Akad. Nauk S.S.S.R., Ser. Khim.*, 1977, 2130.
175 T. C. Flood, F. J. Disanti, and K. D. Campbell, *Inorg. Chem.*, 1978, **17**, 1643.
176 J. P. Williams and A. Wojcicki, *Inorg. Chem.*, 1977, **16**, 3116.
177 A. N. Nesmeyanov, E. G. Perevalova, L. I. Leont'eva, and E. V. Shumilina, *Izvest. Akad. Nauk S.S.S.R., Ser. Khim.*, 1977, 1142.
178 M. G. Newton, R. B. King, M. Chang, N. S. Pantaleo, and J. Gimeno, *J.C.S. Dalton*, 1977, 531.

3
Photochemistry of Compounds of the Main Group Elements

BY J. M. KELLY

1 Introduction

Reports on the photochemistry of inorganic anions are discussed in the next section; those dealing with other compounds are considered under the particular element.

2 Anions

Irradiation ($\lambda > 340$ nm) has no effect on solutions of aqueous nitrite ion in the presence of thallous or sodium ions. However, $TlNO_2$ in DMF or DMSO solutions undergoes decomposition, presumably due to charge transfer in the ion pair.[1] The production of ammonia following u.v. irradiation of nitrite ion in the presence of hydrocarbons has been discussed.[2]

Phosphate radicals formed upon photolysis of peroxodiphosphate ions in aqueous solution [equation (1)] have been characterized by the e.s.r. spectra of their fumaric and maleic acid adducts.[3]

$$P_2O_8^{2-} \longrightarrow 2PO_4^{\cdot -} \qquad (1)$$

The quenching of the fluorescence of anionic pyrene derivatives by halide ions has been studied both in inverse micelles[4] and in sodium dodecyl sulphate micelles.[5]

3 Alkali and Alkaline Earth Metals

The quantum yield for the photoelimination of propene from the valerophenone derivative (1) of dibenzo-18-crown-6 is increased by five-fold and ten-fold, respectively, upon co-ordination with sodium or potassium ions.[6] It is not yet clear whether this effect is caused by an increase in the hydrogen abstraction rate or by a decrease in the rate of radiationless decay of the reactive excited state.

The non-exponential decay of the phosphorescence observed from ethanolic glasses of acetylacetonate complexes of alkali metals and of alkaline earth metals

[1] J. Cunningham, J. O'Shea, P. Reddan, and A. Walker, *J. Photochem.*, 1977, **7**, 185.
[2] G. L. Petriconi and H. M. Papee, *Water Air Soil Pollution*, 1977, **8**, 217, 225.
[3] P. Maruthamuthu and H. Taniguchi, *J. Phys. Chem.*, 1977, **81**, 1944.
[4] D. J. Miller, U. K. A. Klein, and M. Hauser, *J.C.S. Faraday I*, 1977, **73**, 1654.
[5] F. H. Quina and V. G. Toscano, *J. Phys. Chem.*, 1977, **81**, 1750.
[6] R. R. Hautala and R. H. Hastings, *J. Amer. Chem. Soc.*, 1978, **100**, 648.

(1)

at 77 K appears to be due to emission originating from both metal complexes and the free ion.[7] The observation of singlet-to-triplet absorption in naphthalene and quinoline in the presence of barium ions and the efficient quenching of naphthalene fluorescence by the same ion have been attributed to the heavy ion effect.[8]

Flash photolysis of the sodium salt of 1,1,4,4-tetraphenylbutane dianion leads to photoejection of an electron [equation (2)], but not to dissociation [equation (3)].[9, 10] The product formed decomposes within 100 μs as shown in equation (4). Photolysis of dilithioacetylene in liquid ammonia produces a compound which is believed to be tetralithiotetrahedrane.[11]

$$Na^+Ph_2\bar{C}CH_2CH_2\bar{C}Ph_2Na^+ \xrightarrow{h\nu} Na^+Ph_2\bar{C}CH_2CH_2\dot{C}Ph_2 + Na^+, e^-$$
(2)

$$Na^+Ph_2\bar{C}CH_2CH_2\bar{C}Ph_2Na^+ \xrightarrow{h\nu} 2Na^+[CPh_2{=}CH_2]^{\cdot-}$$
(3)

$$Na^+Ph_2\bar{C}CH_2CH_2\dot{C}Ph_2 \longrightarrow Na^+[CPh_2{=}CH_2]^{\cdot-} + CPh_2{=}CH_2$$
(4)

4 Boron

The conversion of alkyl- and cycloalkyl-substituted dienylboranes (2) into boracyclopentenes (3) proceeds by initial *trans*-to-*cis* isomerization and rapid thermal cyclization of the *cis*-dienylborane (4).[12] Isomerization *trans*-to-*cis* has also been observed for boronic esters (5).[13] Surprisingly the phenanthrene derivative

[7] J. A. Kemlo, J. D. Neilson, and T. M. Shepherd, *J. Inorg. Nuclear Chem.*, 1977, **39**, 1945.
[8] A. M. P. C. De Amorim, H. D. Burrows, S. J. Formosinho, and A. M. Da Silva, *Spectrochim. Acta*, 1977, **33A**, 245.
[9] H. C. Wang, E. D. Lillie, S. Slomkowski, G. Levin, and M. Szwarc, *J. Amer. Chem. Soc.*, 1977, **99**, 4612.
[10] M. Szwarc, H. C. Wang, and G. Levin, *Chem. Phys. Letters*, 1977, **51**, 296.
[11] G. Rauscher, T. Clark, D. Poppinger, and P. V. R. Scheyer, *Angew. Chem.*, 1978, **90**, 306; *Angew. Chem. Internat. Edn.*, 1978, **17**, 276.
[12] G. Zweifel, S. J. Backlund, and T. Leung, *J. Amer. Chem. Soc.*, 1977, **99**, 5192.
[13] A. Hassner and J. A. Soderquist, *J. Organometallic Chem.*, 1977, **131**, C1.

(6)

(7) R^1 = cyclopentyl, cyclohexyl,
         Bu^s, Pr^n, or $n-C_8H_{17}$
      R^2 = H or Me
      R^3 = H or Me

(6) is not formed upon photolysis of (2-pyridylamino)diphenylborane, although irradiation of mixtures of 2-aminopyridine and chlorodiphenylborane does indeed yield (6).[14]

Photolysis of trialkylboranes BR_3 in solutions of alcohols gives substituted dioxaboralanes (7), both in the presence and absence of air.[15]

The emission of the triplet state of benzophenone at 77 K is quenched by boron trichloride and replaced by an unstructured emission attributed to the phosphorescence of $Ph_2C=O^+ \cdots BCl_3$.[16] Similarly the photochemical properties of eucarvone are altered in the presence of BX_3 (X = F, Cl, or Br) due to formation of donor–acceptor complexes.[17]

The conversion of diborane into higher boranes using intense i.r. radiation from a CO_2 laser has been re-examined.[18,19] By use of a chopped beam method it has been demonstrated that $B_{20}H_{16}$ is produced *via* thermal reaction rather than by controlled vibration excitation.[18] Other publications consider mass spectroscopic photo-ionization evidence for the cleavage of a basal B—H bond during mercury-sensitized dissociation of B_5H_9,[20] the photo-isomerization of bis[π-(3)-1,2-dicarbollyl]nickel(IV),[21] and the i.r. laser-induced reaction of boron trichloride and silicon tetrafluoride.[22]

5 Aluminium

Upon irradiation in the presence of 1-methylimidazole, carbon dioxide inserts into the Al—C bond of (TPP)AlEt (TPP = tetraphenylporphin) giving (TPP)AlO$_2$CEt.[23]

6 Thallium

The radical polymerization of methyl methacrylate upon irradiation of thallous ions in sulphuric acid,[24] and the decomposition of $TlNO_2$ in non-aqueous solvents,[1] have been the subjects of recent publications.

[14] K. D. Mueller and K. Niedenzu, *Inorg. Chim. Acta*, 1977, **25**, L53.
[15] V. V. Chung, K. Inagaki, and M. Itoh, *J.C.S. Chem. Comm.*, 1977, 690.
[16] R. Snyder and A. C. Testa, *J. Phys. Chem.*, 1978, **82**, 842.
[17] R. F. Childs and Y.-C. Hor, *Canad. J. Chem.*, 1977, **55**, 3501.
[18] C. Riley, S. Shatas, and V. Arkle, *J. Amer. Chem. Soc.*, 1978, **100**, 658.
[19] S. Shatas, D. Gregory, R. Shatas, and C. Riley, *Inorg. Chem.*, 1978, **17**, 163.
[20] J.-S. Wang, A. J. DeStefano, and R. F. Porter, *Inorg. Chem.*, 1978, **17**, 1374.
[21] V. V. Volkov, S. Y. Dvurechenskaya, and V. A. Ershova, *Zhur. neorg. Khim.*, 1977, **22**, 3283.
[22] Y. A. Adamova, A. V. Pankratov, V. G. Sagitova, A. N. Skachkov, G. I. Stolyarova, and G. V. Shmerling, *Khim. vysok. Energii*, 1978, **12**, 89.
[23] S. Inoue and N. Takeda, *Bull. Chem. Soc. Japan*, 1977, **50**, 984.
[24] V. S. Balasubramanian, Q. Anwaruddin, and L. V. Natarajan, *J. Polymer Sci., Polymer Letters*, 1977, **15**, 599.

7 Silicon

Photocleavage of the silicion-to-silicon bond in organopolysilanes has proved a useful process for the preparation of silacyclopropanes, silacyclopropenes, silaethylenes, and silylenes. Ishikawa has recently summarized his group's previous efforts in this field.[25]

Silacyclopropenes of type (8) are produced upon irradiation of $(Me_3Si)_3SiPh$ in the presence of acetylenes $(RC{\equiv}CR)$.[26] Photolysis of $Me_3SiSiMe_2C{\equiv}CPh$ gives (9).[27] Although (9) is thermally stable, dimethylsilylene is expelled upon photo-excitation. In the presence of ketones (*e.g.* acetone) silacyclopropenes form addition products [*e.g.* compound (10)] probably *via* attack of the $n{-}\pi^*$ triplet of the ketone.[27, 28] With nitriles silacyclopropenes photoreact to give adducts (11), presumably after initial formation of (12).[29]

The intermediate species (13) formed upon irradiation of $SiMe_3SiMe_2Ph$ has been caused to react with arynes[30] and carbonyl compounds.[31] A new route to silaethylene derivatives, involving photo-induced 1,3-shifts of a silyl group to the carbon atom of the vinyl group in 1-alkenyldisilanes, has been reported.[32] Thus species (14) is converted into (15), which has been trapped by methanol. Silaethylenes are also presumed to be formed upon vacuum u.v. excitation ($\lambda = 147$ nm) of $Me_2\overline{SiCH_2CH_2CH_2}$ in the gas phase, because in the presence of methanol Me_3SiOMe is produced.[33] The reactions of silylenes generated by photolysis of polysilanes with carbonyl compounds,[34] siloxanes,[35] and olefins[36] have been described.

25 M. Ishikawa, *Pure Appl. Chem.*, 1978, **50**, 11.
26 M. Ishikawa, K. Nakagawa, and M. Kumada, *J. Organometallic Chem.*, 1977, **131**, C15.
27 H. Sakurai, Y. Kamiyama, and Y. Nakadaira, *J. Amer. Chem. Soc.*, 1977, **99**, 3879.
28 D. Seyferth, S. C. Vick, M. L. Shannon, T. F. O. Lim, and D. P. Duncan, *J. Organometallic Chem.*, 1977, **135**, C37.
29 H. Sakurai, Y. Kamiyama, and Y. Nakadaira, *J.C.S. Chem. Comm.*, 1978, 80.
30 M. Ishikawa, T. Fuchikami, and M. Kumada, *J. Organometallic Chem.*, 1977, **127**, 261.
31 M. Ishikawa, T. Fuchikami, and M. Kumada, *J. Organometallic Chem.*, 1977, **133**, 19.
32 M. Ishikawa, T. Fuchikami, and M. Kumada, *J. Organometallic Chem.*, 1978, **149**, 37.
33 S. Tokach, P. Boudjouk, and R. D. Koob, *J. Phys. Chem.*, 1978, **82**, 1203.
34 M. Ishikawa, K. Nakagawa, and M. Kumada, *J. Organometallic Chem.*, 1977, **135**, C45.
35 H. Okinoshima and W. P. Weber, *J. Organometallic Chem.*, 1978, **50**, C25.
36 M. Ishikawa, K.-I. Nakagawa, M. Ishiguro, F. Ohi, and M. Kumada, *J. Organometallic Chem.*, 1978, **152**, 155.

$$\text{(13)} \qquad \text{(14)} \qquad \text{(15)}$$

Evidence has been presented for photoreactions of disilanes in DMSO involving the nucleophilic attack of the oxygen atom of the DMSO on the excited state of the disilane. The products formed are substituted silanes, disiloxanes, dimethylsilicone oligomers, and dimethyl sulphide.[37]

Mercury-sensitized cleavage of the Si—Si linkage in Si_2F_6, Si_2Me_6, and related compounds provides a useful synthetic route to species of the type $Si_2Me_xF_{6-x}$.[38] Similarly in the presence of phosphorus trifluoride, mercury-sensitized decomposition of Si_2F_6 yields $P(SiF_3)_3$.[39]

In the photolysis ($\lambda = 185$ nm) of hexamethylsiloxane the predominant primary processes are cleavage of the Si—C bonds [equations (5) and (6)]; rupture of the Si—O linkage is not observed.[40] This behaviour contrasts with that of dibutyl ether (the carbon analogue), where breaking of the C—O bond is the principal primary photoreaction.

$$Me_3Si-O-SiMe_3 \xrightarrow{h\nu} Me_3Si-O-SiMe_2\cdot + Me\cdot \qquad (5)$$

$$Me_3Si-O-SiMe_3 \xrightarrow{h\nu} Me_3Si-O-SiMe=CH_2 + CH_4 \qquad (6)$$

Recent publications on i.r. laser-induced reactions of silicon compounds have described the decomposition of silane,[41] and the use of silicon tetrafluoride as a sensitizer for reactions such as the isomerization of allene and methylacetylene.[42,43]

8 Germanium, Tin, and Lead

Investigations of the formation of persistent tervalent germanium- and tin-centred amide radicals by irradiation of the corresponding bivalent amides [equation (7)] has now been extended to compounds having $R = GeEt_3$ or $GeMe_3$.[44] However, with bulkier amide ligands such as those where $R = SiEt_3$ or $GePh_3$, photolysis of the bivalent amide does not generate the tervalent amide, although the corresponding aminyl radical NR_2 could be detected. Related publications have described the photochemical generation of persistent triaryl-

[37] H. Okinoshima and W. P. Weber, *J. Organometallic Chem.*, 1978, **149**, 179.
[38] K. G. Sharp and P. A. Sutor, *J. Amer. Chem. Soc.*, 1977, **99**, 8046.
[39] K. G. Sharp, *J.C.S. Chem. Comm.*, 1977, 564.
[40] H.-P. Schuchmann, A. Ritter, and C. von Sonntag, *J. Organometallic Chem.*, **148**, 213.
[41] A. N. Oraevskii, A. V. Pankratov, A. N. Skachkov, and V. M. Shabarshin, *Khim. vysok. Energii*, 1978, **12**, 59.
[42] C. Cheng and P. Keehn, *J. Amer. Chem. Soc.*, 1977, **99**, 5808.
[43] K. J. Olszyna, E. Grunwald, P. M. Keehn, and S. P. Anderson, *Tetrahedron Letters*, 1977, 1609.
[44] M. J. S. Gynane, D. H. Harris, M. F. Lappert, P. P. Power, P. Riviere, and M. Riviere-Baudet, *J.C.S. Dalton*, 1977, 2004.

germanium compounds,[45] and the spin-trapping of other germyl radicals with nitrones.[46]

$$2M(NR_2)_2 \xrightarrow{h\nu} M(NR_2)_3 + {}^{\cdot}M(NR_2)' \qquad (7)$$

Tin(III) radicals are conveniently produced by photolysis of $(\eta^1\text{-}C_5H_5)$–tin(IV) complexes [equation (8)], and this procedure has been recommended as an alternative to the photocleavage of distannanes as a method for the generation

$$(\eta^1\text{-}C_5H_5)SnR_3 \longrightarrow C_5H_5{}^{\cdot} + SnR_3 \quad (R = Cp \text{ or } Bu) \qquad (8)$$

of such radicals.[47] Other reports consider the formation of SnR_2 species (R = Ph or Me) from the corresponding dimethyl–[48] or diphenyl–tin polymers.[49]

Photolysis of (16) gives perylene in high yield.[50] The photocyclization of $Me_3SnCPh{=}CPh_2$ in the presence of oxygen yields (17).[51]

(16)   (17)

Irradiation of the CT complexes of R_4Sn (R = Me, Et, or Ph) and tetracyanoethylene induces insertion of the olefin into the metal–carbon bond [equation (9)].[52] Similarly Sn_2R_6 or Pb_2Ph_6 form CT complexes with phenanthroquinone (PQ) and photolysis of these yields adducts of the kind R_3M,PQ, which may be detected by e.s.r.

$$R_4Sn + (NC)_2C{=}C(CN)_2 \xrightarrow{h\nu} R_3Sn{-}C(CN)_2{=}C(CN)_2R \qquad (9)$$

The quenching of the phosphorescence of MPh_4 (M = C, Si, Ge, Sn, or Pb) and of SnR_4 (R = Me, Bu^n, or $C_5H_{11}{}^n$) by oxygen in ethanol glasses has been investigated. The quenching process is attributed to complex formation between the organometallic compounds and oxygen.[53] Other publications deal with the

[45] M. J. S. Gynane, M. F. Lappert, P. Riviere, and M. Riviere-Baudet, *J. Organometallic Chem.*, 1977, **142**, C9.
[46] P. Riviere, S. Richelme, M. Riviere-Baudet, J. Satge, M. J. S. Gynane, and M. F. Lappert, *J. Chem. Res. (S)*, 1978, 218.
[47] A. G. Davies and M.-W. Tse, *J.C.S. Chem. Comm.*, 1978, 353.
[48] K. Kuno, K. Kobayashi, M. Kawanisi, S. Kozima, and T. Hitomi, *J. Organometalic Chem.*, 1977, **137**, 349.
[49] K. Kobayashi, K. Kuno, M. Kawanisi, and S. Kozima, *Bull. Chem. Soc. Japan*, 1977, **50**, 1353.
[50] J. Meinwald, S. Knapp, T. Tatsuoka, J. Finer, and J. Clardy, *Tetrahedron Letters*, 1977, 2247.
[51] C. J. Cardin, D. J. Cardin, J. M. Kelly, D. J. H. L. Kirwan, R. J. Norton, and A. Roy, *Proc. Roy. Irish Acad.*, 1978, **77**, 365.
[52] K. Mochida, J. K. Kochi, K. S. Chen, and J. K. S. Wan, *J. Amer. Chem. Soc.*, 1978, **100**, 2927.
[53] G. A. Razuvaev, A. N. Egorochkin, V. A. Kuznetsov, V. N. Glushakova, A. V. Shabanov, Y. A. Alexandrov, and Y. Y. Baryshnikov, *J. Organometallic Chem.*, 1978, **148**, 147.

quenching of the fluorescence of 9-cyanoanthracene by $Me_2N(CH_2)_3SnMe_3$,[54] the photoreduction of bridgehead halides with organotin hydrides,[55] the photodecomposition of dialkylgermyl- and dialkylstannyl-tetraphenylporphyrin complexes,[56] and the catalysed photoreduction of nitrobenzene using tin tetraphenylporphyrin complexes.[57]

9 Nitrogen, Phosphorus, Arsenic, and Antimony

The bis(trimethylsilyl)aminyl radical may be conveniently prepared by reaction (10), and its adducts with ethylene, triethylphosphite, and t-butylisocyanide have been characterized by e.s.r.[58]

$$(Me_3Si)_2NN(SiMe_3)_2 \xrightarrow{h\nu} 2\dot{N}(SiMe_3)_2 \qquad (10)$$

Other authors have reported the low-temperature synthesis of $[NF_4]^+$ by u.v. photolysis of $NF_3-F_2-AsF_5$ or $NF_3-F_2-BF_3$ mixtures,[59] the conversion of nitrosyl chloride to chlorine nitrate upon photoexcitation in oxygen matrices at 10 K,[60] the photoreaction of nitrogen dioxide on sodium chloride at 77 K,[61] and the photodissociation of dinitrogen tetrafluoride by CO_2 laser pulses.[62]

The phosphorescence of MPh_3 (M = N, P, As, or Sb) in ethanol glasses is quenched by oxygen, apparently through formation of a CT complex.[63]

E.s.r. studies have shown that photolysis of $Ph_3P=CHCO_2Et$ causes production of triphenylphosphine and the carbene. Secondary photoreactions then yield $\cdot PPh_2$ and $\cdot PPh_2=CHCO_2Et$.[64] The light-accelerated decomposition of 1,2-diphenyldiphosphine, the reaction of phosphorus pentafluoride and sulphur dioxide induced by intense i.r. radiation,[65] the photodissociation of arsenic tri-iodide in molecular beams,[66] the photosynthesis of arsenic pentachloride from its trichloride and chlorine,[67] the photosensitivity of silver-doped arsenic sulphide systems,[68] and the initiation of polymerization of methyl methacrylate by photolysis of pentaphenylantimony,[69] have been the subjects of recent papers.

10 Oxygen, Sulphur, Selenium, and Tellurium

Reports on the photo-oxidation of sulphur[70] and of ascorbic acid[71] by hydrogen peroxide have been published.

[54] D. E. Vander Donckt, M. R. Barthels, N. Antheunis, and M. Swinnen, *Mol. Photochem.*, 1977, **8**, 121.
[55] T.-Y. Luh and L. M. Stock, *J. Org. Chem.*, 1977, **42**, 2790.
[56] C. Cloutour, D. Lafargue, J. A. Richards, and J. C. Pommier, *J. Organometallic Chem.*, 1977, **137**, 157.
[57] Y. Harel and J. Manassen, *J. Amer. Chem. Soc.*, 1977, **99**, 5817.
[58] B. P. Roberts and J. N. Winter, *J.C.S. Chem. Comm.*, 1978, 545.
[59] K. O. Christe and I. B. Goldberg, *Inorg. Chem.*, 1978, **17**, 759.
[60] D. E. Tevault and R. R. Smardewski, *J. Phys. Chem.*, 1978, **82**, 375.
[61] H. D. Breuer and J. Krueger, *Ber. Bunsengesellschaft Phys. Chem.*, 1978, **82**, 97.
[62] P. Lavigne, J. L. Lachambre, and G. Otis, *Optics Comm.*, 1977, **22**, 75.
[63] M. Zanger and R. Poupko, *Spectroscopy Letters*, 1977, **10**, 737.
[64] M. Baudler, B. Carlsohn, D. Koch, and P. K. Medda, *Chem. Ber.*, 1978, **111**, 1210.
[65] V. D. Klimov, V. A. Kuz'menko, and V. A. Legasov, *Zhur. fiz. Khim.*, 1977, **51**, 949.
[66] M. Kawasaki and R. Bersohn, *J. Chem. Phys.*, 1978, **68**, 2105.
[67] K. Seppelt, *Z. anorg. Chem.*, 1977, **434**, 5.
[68] M. T. Kostyshin and V. I. Min'ko, *Ukrain. fiz. Zhur.*, 1977, **22**, 1560.
[69] H. Matsuda, T. Isaka, and N. Iwamoto, *Makromol. Chem.*, 1978, **179**, 539.
[70] I. N. Barshchevskii, *Zhur. fiz. Khim.*, 1977, **51**, 2269.
[71] E. V. Shtamm, A. Purmalis, and Y. I. Skurlatov, *Zhur. fiz. Khim.*, 1978, **52**, 117.

E.s.r. evidence has been presented for the cleavage of S—S bonds in arene thiosulphonates [*e.g.* reaction (11)], the radicals so produced reacting further to give arene sulphinyl radicals [equation (12)].[72] Presumably rupture of the

$$Ph—\overset{\overset{O}{\|}}{\underset{\underset{O}{\|}}{S}}—S—Ph \xrightarrow{h\nu} PhSO_2\cdot + \cdot SPh \qquad (11)$$

$$PhSO_2\cdot + PhS\cdot \longrightarrow 2PhSO\cdot - O\cdot \qquad (12)$$

S—S bond is also the primary process upon irradiation of PhS—SPh, which leads to (18), a versatile chelating ligand.[73] Dibenzylditelluride is photolysed to dibenzyltelluride, which upon irradiation in the presence of oxygen reacts further to give tellurium-free products.[74]

(18)

Photochemical syntheses of arylphenylselenides from arylhalides and phenyl-selenide anion in liquid ammonia have been described.[75] The primary process seems to involve an electron transfer step of the type shown in equation (13).

$$ArX + PhSe^- \longrightarrow ArX^{\cdot-} + PhSe\cdot \qquad (13)$$

Photoaddition of alkynes or alkenes to benzoselenophene,[76] the photochemical decomposition and isomerization of (19),[77] a study of the phosphorescence and intersystem crossing in compounds (20),[78] and the photopolymerization of S_2N_2[79] have been discussed recently.

At room temperature, substituted 1,2,4- and 1,2,5-selenadiazoles (21), (22) photodissociate to nitriles and selenium.[80, 81] At low temperatures or in nitrogen matrices, spectroscopic data indicated that the nitrile selenide RCNSe is formed

72 B. C. Gilbert, B. Gill, and M. D. Sexton, *J.C.S. Chem. Comm.*, 1978, 78.
73 S. D. Killops, S. A. R. Knox, G. H. Riding, and A. J. Welch, *J.C.S. Chem. Comm.*, 1978, 486.
74 H. K. Spencer and M. P. Cava, *J. Org. Chem.*, 1977, **42**, 2937.
75 A. B. Pierini and R. A. Rossi, *J. Organometallic Chem.*, 1978, **144**, C12.
76 Tran Quang Minh, L. Christaens, P. Grandclaudon, and A. Lablache-Combier, *Tetrahedron*, 1977, **33**, 2225.
77 G. Höhne, W. Lohner, K. Praefcke, U. Schulze, and H. Simon, *Tetrahedron Letters*, 1978, 613.
78 P. Jardon, *J. Chim. phys.*, 1977, **74**, 1177.
79 P. Love, H. I. Kao, G. H. Meyer, and M. M. Labes, *J.C.S. Chem. Comm.*, 1978, 301.
80 C. L. Pedersen, N. Harrit, M. Poliakoff, and I. Dunkin, *Acta Chem. Scand.*, 1977, **B31**, 848.
81 C. L. Pedersen and N. Hacker, *Tetrahedron Letters*, 1977, 3981.

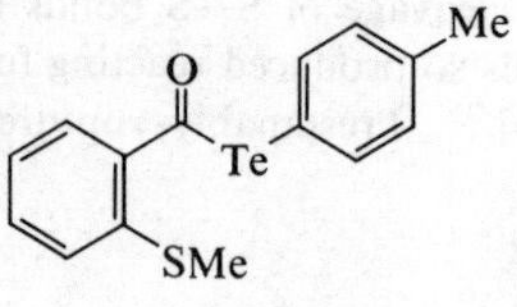

278 Photochemistry

(19)

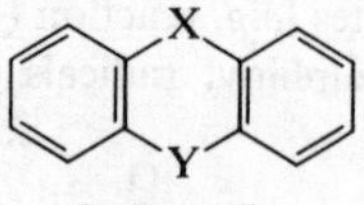

(20) X = O, S, or Se
Y = O, S, Se, Te, or CH₂

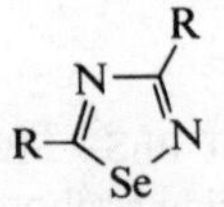

(21) 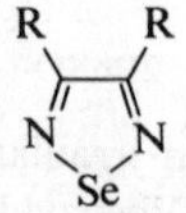   (22)   (23)    (24)

as the initial product. Irradiation of 1,2,3-thiadiazole (23) isolated in argon matrices at 8 K yields thiirene (24).[82]

The decomposition of SF_6,[83—88] SF_5Cl,[89] and SeF_6[90] induced by i.r. laser irradiation continues to attract attention.

11 Halogens

Photodecomposition of fluorine in low-temperature matrices containing other halogens gives rise to species of type XF_2, X_2F, and X_2F_2.[91] A study has been carried out on the properties of the excited state of xenon monofluoride formed upon vacuum u.v. photolysis of xenon difluoride.[92]

[82] A. Krantz and J. Laureni, *Ber. Bunsengesellschaft Phys. Chem.*, 1978, **82**, 13.
[83] R. V. Ambartsumyan, Y. A. Gorokhov, G. N. Makarov, A. A. Puretskii, and N. P. Furzikov, *Springer Ser. Opt. Sci.*, 1977, **7**, 439.
[84] Y. I. Arkhangel'skii, V. D. Klimov, V. A. Kuz'menko, V. A. Legasov, and S. L. Nedoseev, *Doklady Akad. Nauk S.S.S.R.*, 1977, **235**, 1075.
[85] N. Bloembergen and E. Yablonovitch, *Springer Ser. Opt. Sci.*, 1977, **7**, 86.
[86] G. J. Diebold, F. Engelke, D. M. Lubman, J. C. Whitehead, and R. N. Zare, *J. Chem. Phys.*, 1977, **67**, 5407.
[87] E. R. Grant, M. J. Coggiola, Y. T. Lee, P. A. Schulz, A. S. Sudbo, and Y. R. Shen, *Chem. Phys. Letters*, 1977, **52**, 595.
[88] M. Rothschild, W.-S. Tsay, and D. O. Ham, *Optics Comm.*, 1978, **24**, 327.
[89] K. M. Leary, J. L. Lyman, L. B. Asprey, and S. M. Freund, *J. Chem. Phys.*, 1978, **68**, 1671.
[90] J. J. Tiee and C. Wittig, *Appl. Phys. Letters*, 1978, **32**, 236.
[91] E. S. Prochaska, L. Andrews, N. R. Smyrl, and G. Mamantov, *Inorg. Chem.*, 1978, **17**, 970.
[92] H. C. Brashears, jun., D. W. Setser, and D. Desmarteau, *Chem. Phys. Letters*, 1977, **48**, 84.

Part III

ORGANIC ASPECTS OF PHOTOCHEMISTRY

1
Photolysis of Carbonyl Compounds

BY W. M. HORSPOOL

1 Introduction

The decline in interest in the solution-phase photochemistry of simple carbonyl compounds has continued during the past year.

More interest has been shown in the physical aspects of carbonyl photo-chemistry, as in the study of the quenching of triplet acetone by olefins[1] and the report by Tabushi *et al.*[2] of triplet energy transfer from the benzophenone-cyclodextrin molecule (1) to trapped host molecules such as 1-bromonaphthalene. The intersystem-crossing quantum yields for butyraldehyde have been measured in the gas and liquid phases.[3]

Kossanyi and Furth have published a two-part review on the photochemistry of carbonyl compounds.[4, 5]

2 Norrish Type I Reactions

The rate constants for the Norrish Type I cleavage of 4-methylbutan-2-one have been measured.[6] Encina and Lissi[7] have shown that the rate of α-cleavage in the photoreaction of several alkyl ketones is almost completely determined by the nature of the α-substituent. Norrish Type I photocleavage has been reported for the cyclopentanone (2) when it is irradiated in benzene solution. The fate of

(1)   (2)   (3)

[1] R. O. Loutfy, R. W. Yip, and S. K. Dogra, *Tetrahedron Letters*, 1977, 2843.
[2] I. Tabushi, K. Fujita, and L. C. Yuan, *Tetrahedron Letters*, 1977, 2503.
[3] M. Tolgyesi, A. Nacsa, and T. Berces, *Reaction Kinetics Catalysis Letters*, 1977, **7**, 45 (*Chem. Abs.*, 1977, **87**, 117 211).
[4] J. Kossanyi and B. Furth, *Actual. Chim.*, 1974, 3 (*Chem. Abs.*, 1978, **88**, 21 465).
[5] J. Kossanyi and B. Furth, *Actual. Chim.*, 1974, 7 (*Chem. Abs.*, 1978, **88**, 21 466).
[6] E. Abuin and E. A. Lissi, *J. Photochem.*, 1976, 6, 1 (*Chem. Abs.*, 1977, **86**, 139 046).
[7] M. V. Encina and E. A. Lissi, *J. Photochem.*, 1978, 8, 131 (*Chem. Abs.*, 1978, **88**, 104 369).

the biradical in this instance is disproportionation to yield the Z- and E-aldehydes (3) in high yield (77%).[8, 9] The cyclopentanone (4) also undergoes α-fission.[10] The biradical in this example reacts further by ring opening of the epoxide moiety, yielding (5). This biradical either ring-closes to the lactone (6) or else disproportionates competitively to the dialdehyde (7). Similar behaviour

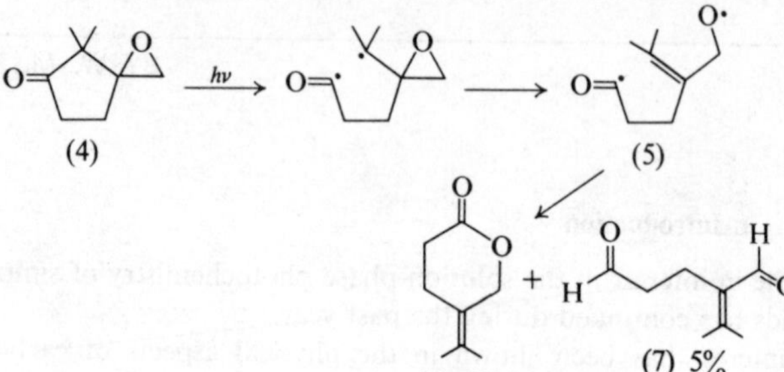

(4)   (5)

(6) 65%   +   (7) 5%

is encountered with the cyclohexanone derivative (8) which yields the analogous products (9) and (10).[10] Weiss [11] has studied quantitatively the photoreactions of the ketones (11). The results obtained show that α-cleavage in (11a) and (11b) occurs solely at the methoxycarbonyl-substituted carbon. The biradicals produced from each ketone by this mode of cleavage show very similar rates of reaction.

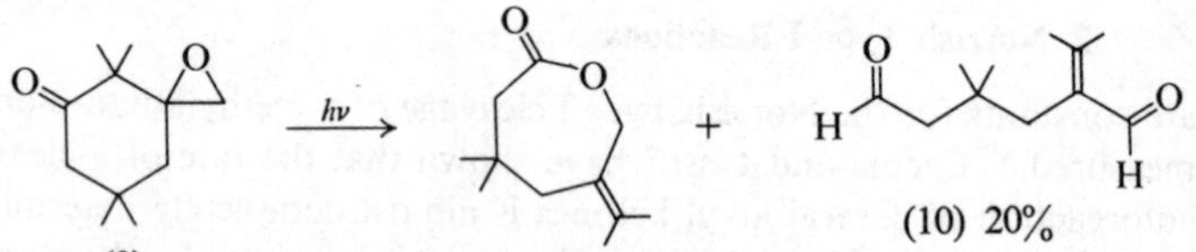

(8)   (9) 45%   (10) 20%

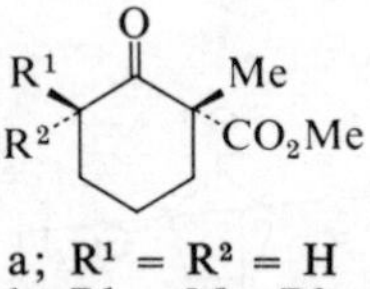

(11) a; $R^1 = R^2 = H$
     b; $R^1 = Me$, $R^2 = H$
     c; $R^1 = R^2 = Me$

Irradiation of norcamphor (12a) has been shown to yield only the ene-al (13a; $\Phi = 0.28$) as a result of Norrish Type I fission and disproportionation of the resultant biradical.[12] The introduction of substituents in the norcamphor skeleton as in (12b), (12c), and (12d) changes the outcome of the photochemical reaction. Thus the ketones (12b—d) undergo Norrish Type I cleavage to afford

[8] G. Bidan, J. Kossanyi, V. Meyer, and J.-P. Morizur, *Tetrahedron*, 1977, **33**, 2193.
[9] J. P. Morizur, G. Bidan, and J. Kossanyi, *Tetrahedron Letters*, 1975, 4167.
[10] R. K. Murray, jun. and C. A. Andruskiewicz, jun., *J. Org. Chem.*, 1977, **42**, 3994.
[11] D. S. Weiss, *Tetrahedron Letters*, 1978, 1039.
[12] A. G. Singer, S. Wolff, and W. C. Agosta, *J. Org. Chem.*, 1977, **42**, 1327.

the two aldehydes (13b—d) and (14b—d) in each case. The formation of the Δ^3-aldehyde (14) was in each case a minor process. The photochemical ring cleavage of the type shown by norcamphor is also exhibited by the cyclic ketone (15). Pyrex-filtered irradiation of this ketone results in the formation of the aldehyde (16) in good yield. This aldehyde was used as a precursor for the synthesis of a deoxyprostaglandin.[13] Another photochemical process was also observed in this system which resulted in isomerization of the side-chain double bond, yielding (17).[13]

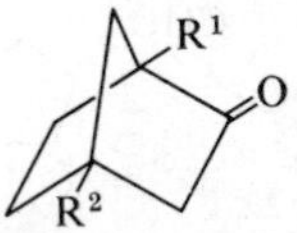

(12) a; $R^1 = R^2 = H$
b; $R^1 = H$, $R^2 = Me$
c; $R^1 = Pr^i$, $R^2 = H$
d; $R^1 = Pr^i$, $R^2 = Me$

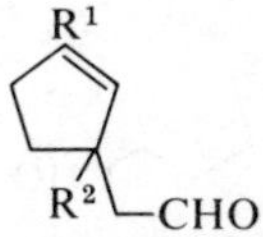

(13) a; $\Phi = 0.28$
b; $\Phi = 0.19$
c; $\Phi = 0.29$
d; $\Phi = 0.16$

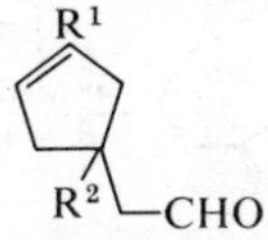

(14) a; $\Phi = 0$
b; $\Phi = 0.13$
c; $\Phi = 0.04$
d; $\Phi = 0.10$

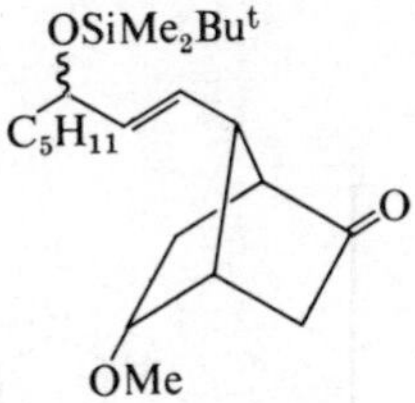

(15)

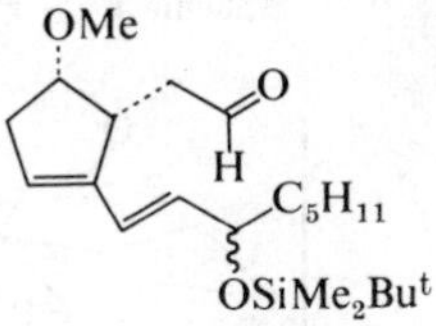

(16)

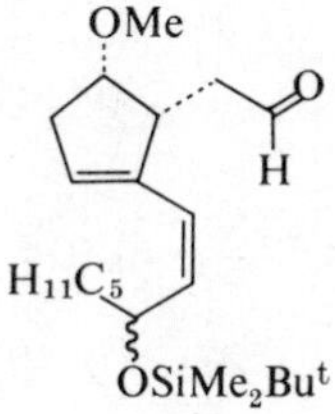

(17)

The bicyclic ketones (18) undergo photochemical α-cleavage in methanol solution to yield aldehydes and ketones in the ratios shown in Scheme 1.[14] In all the cases studied, the aldehyde was the predominant product. The reason for this is presented in terms of the conformation of the transition state which controls the hydrogen transfer step in the disproportionation of the biradical intermediate.

A full account of the photochemistry of bicyclo[3,2,1]octan-6-ones (19) has been published.[15, 16] This work has sought to analyse the problems of conformational change associated with the biradicals produced by the photochemical fission of (19). The conformational inversion can be followed by suitable deuterium labelling as shown in Scheme 2.

The gibbanes (20) undergo photochemical ring cleavage to afford the secogibbanes (21).[17] It is interesting that this work reports the Norrish Type I process without the decarboxylation of the gibbane. The androstene (22) also undergoes

13 N. M. Crossland, S. M. Roberts, and R. F. Newton, *J.C.S. Chem. Comm.*, 1977, 886.
14 S. C. Critch and A. G. Fallis, *Canad. J. Chem.*, 1977, **55**, 2845.
15 W. C. Agosta and S. Wolff, *J. Amer. Chem. Soc.*, 1977, **99**, 3355.
16 W. C. Agosta and S. Wolff, *J. Amer. Chem. Soc.*, 1976, **98**, 4316.
17 G. Adam and V. S. Tran, E.Ger.P., 118 875 (*Chem. Abs.*, 1977, **86**, 155 833).

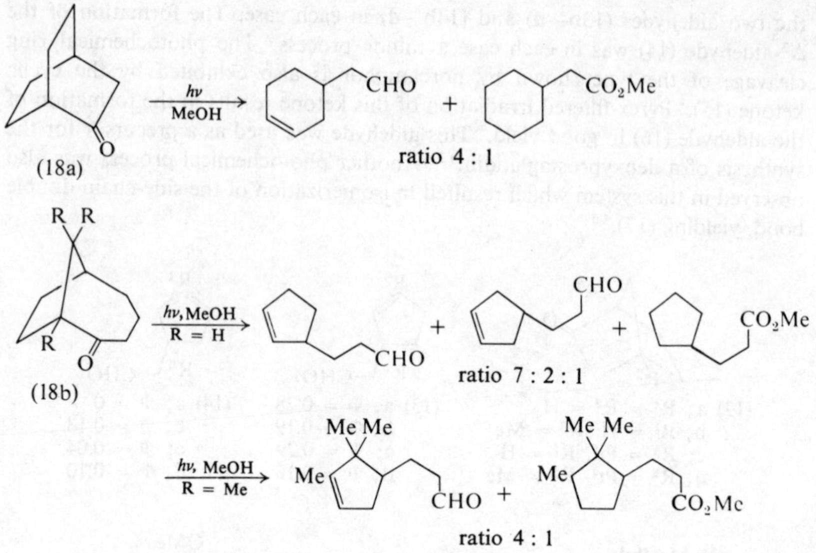

$\xrightarrow[\text{MeOH}]{h\nu}$

ratio 4 : 1

(18a)

R R

$\xrightarrow[R = H]{h\nu, \text{MeOH}}$

CHO

CHO

CO$_2$Me

(18b)   ratio 7 : 2 : 1

Me Me   Me Me

$\xrightarrow[R = Me]{h\nu, \text{MeOH}}$ Me   Me

CHO +   CO$_2$Me

ratio 4 : 1

Scheme 1

2%
CH$_2$CDO

H

88%

+

90%
CH$_2$CHO

(19) a; R^1 = H, R^2 = D

b; R^1 = D, R^2 = H

$\xrightarrow[C_6H_6-\text{MeOH}]{h\nu}$

D

10%

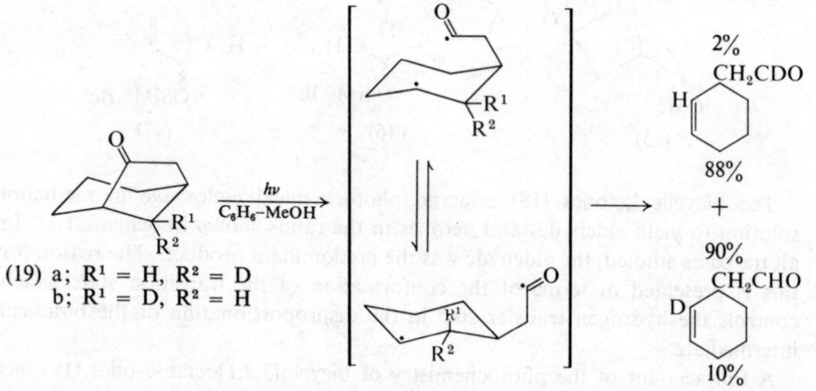

Scheme 2

Norrish Type I ring opening; this time opening of ring D occurs and yields the aldehyde (23).[18] A full account of the photochemical behaviour of friedelin (24) in diethyl ether has been made.[19, 20] One reaction exhibited by this compound under these conditions is quite novel in that the Norrish Type I product [the keten (25)] is attacked by the ketyl radical of acetone.

[18] J. Brussee and H. J. C. Jacobs, *Org. Photochem. Synth.*, 1976, **2**, 67 (*Chem. Abs.*, 1977, **87**, 23 607).

[19] H. Shirasaki, T. Tsuyuki, T. Takashi, and R. Stevenson, *Bull. Chem. Soc. Japan*, 1977, **50**, 921.

[20] H. Shirasaki, T. Tsuyuki, T. Takashi, and R. Stevenson, *Tetrahedron Letters*, 1975, 2271.

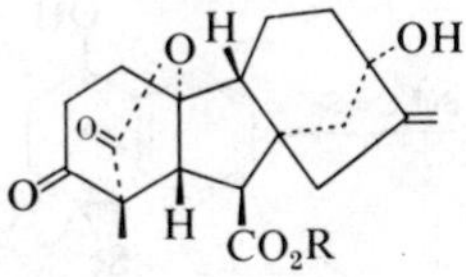

(20) 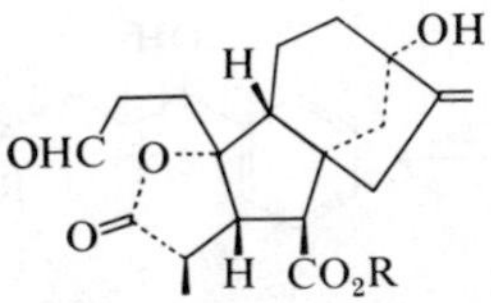   (21) 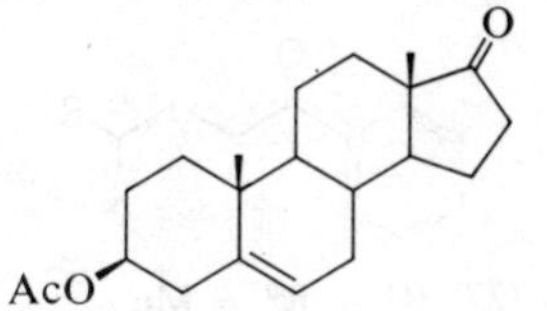   (22)

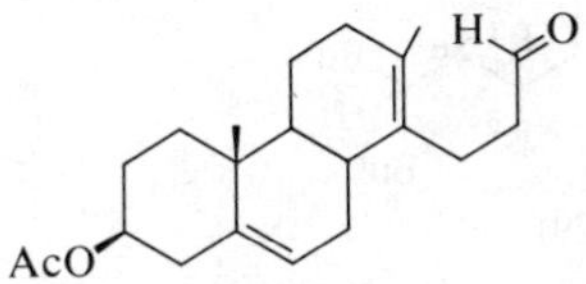

(23) 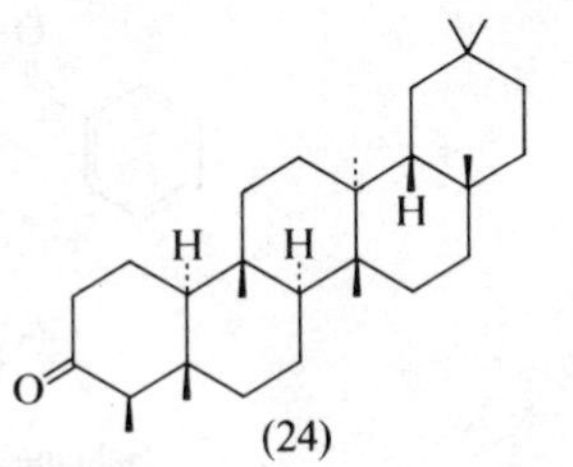   (24) 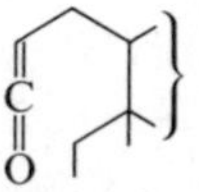   (25)

3 Norrish Type II Reactions

The lifetimes of the biradicals produced by the Norrish Type II reactions of the ketones (26) have been measured.[21] Lifetime studies have also been carried out on the biradical produced by the irradiation of 4′-methylvalerophenone (27) in

$$R^1CHR^2CH_2CH_2COPh$$

(26) R^1 = H, Me, or Et, R^2 = H
     R^1 = R^2 = Me, R^1 = Ph, R^2 = OH

a laser flash system. The biradicals in this instance were produced in the presence of paraquat ions.[22] Triphenylphosphine oxide has been used as an interactant with the biradicals produced by the photolysis of valerophenone (28a).[23] Scaiano has also studied [24] the kinetics of the trapping by di-t-butyl selenoketone of the biradicals produced upon photolysis of butyrophenone (28b), valerophenone (28a), and 4′-methylvalerophenone (27). The products from the reaction were identified as the ketones (29) which are formed as shown in Scheme (3).

Hautala and Mayer [25] have studied the influence of nitro-substituents upon the photochemical reactivity of the carbonyl group in the ketones (30). Their work has shown that deactivation is extremely important (thus excitation yields a carbonyl-excited state which undergoes rapid decay to a nitro-excited state) with concomitant loss of hydrogen abstracting ability.

A laser flash study of the photoenolization of 2-methylbenzophenone has been reported.[26] A flash photolytic study of the reaction of 2-methylacetophenone

[21] R. D. Small, jun. and J. C. Scaiano, *J. Phys. Chem.*, 1977, **81**, 2126.
[22] J. C. Scaiano, *J. Phys. Chem.*, 1977, **81**, 828.
[23] J. C. Scaiano, *J. Org. Chem.*, 1978, **43**, 568.
[24] J. C. Scaiano, *J. Amer. Chem. Soc.*, 1977, **99**, 1494.
[25] R. R. Hautala and T. Mayer, *Tetrahedron Letters*, 1977, 2499.
[26] K. Ujiie, K. Kikuchi, and H. Kokobun, *Chem. Letters*, 1977, 543 (*Chem. Abs.*, 1977, **87**, 52 447).

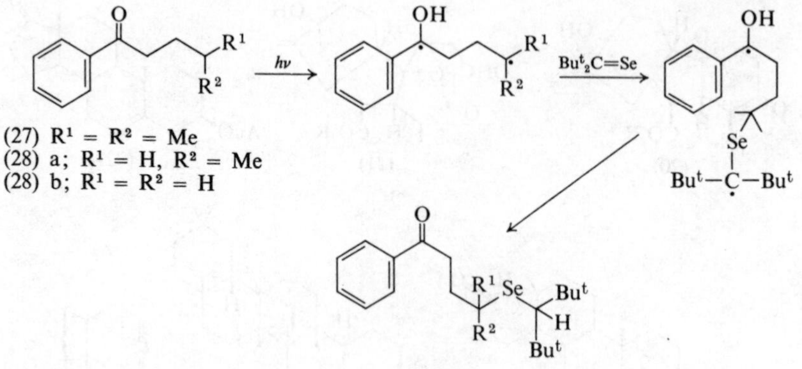

(27) $R^1 = R^2 = Me$
(28) a; $R^1 = H$, $R^2 = Me$
(28) b; $R^1 = R^2 = H$

Scheme 3

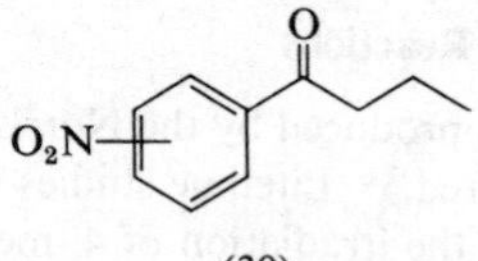

(30)

has shown that there are *syn-* (31) and *anti-* (32) conformers of the ketone triplet.[27] This work is in agreement with the interpretation put forward by Wagner and Chen.[28] In this earlier work two triplet states with decay rates of 3×10^7 and 5×10^9 s^{-1} had been identified. The shorter-lived triplet state is the *syn*-conformer, and it is from this state that photoenolization arises.

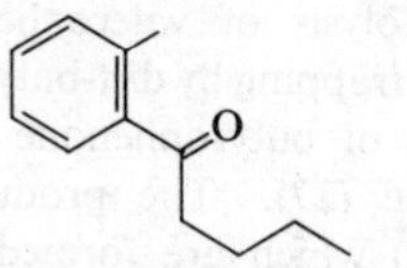

(31) 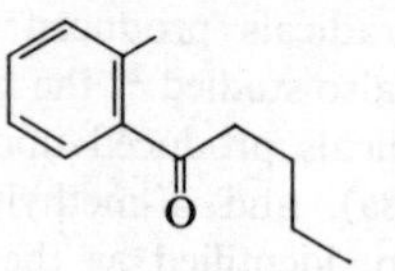   (32)

Bergmark[29] suggests that the irradiation of the chloro-ketones (33) in benzene results in the formation of a photoenol (34) by a Norrish Type II pathway. The products obtained from this reaction are the indanones (35) as a result of the cyclization of enol. When methanol is used as the solvent for the irradiation the formation of the indanone is suppressed and the solvolysis product (36) is obtained. Irradiation of the bromo-ketone (37) yields *o*-methylacetophenone by loss of the halogen.

The photolysis of the crown ether (38) has been studied in the presence and absence of complexed alkali metals.[30] The results show that the presence of

[27] R. D. Small, jun. and J. C. Scaiano, *J. Amer. Chem. Soc.*, 1977, **99**, 7713.
[28] P. J. Wagner and C.-P. Chen, *J. Amer. Chem. Soc.*, 1976, **98**, 239.
[29] W. R. Bergmark, *J.C.S. Chem. Comm.*, 1978, 61.
[30] R. R. Hautala and R. H. Hastings, *J. Amer. Chem. Soc.*, 1978, **100**, 648.

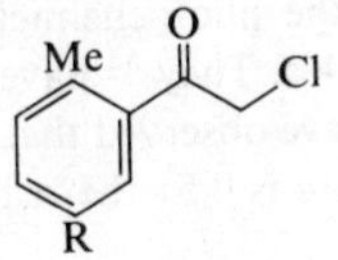

(33) R = Me or H 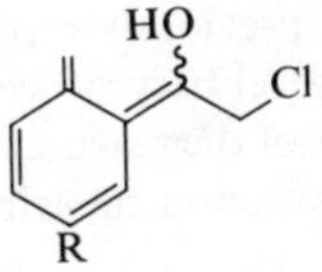   (34) 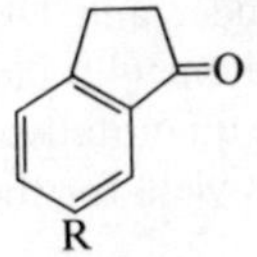   (35)

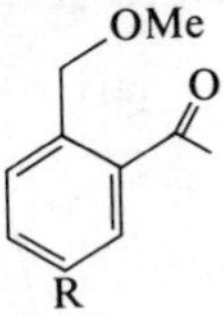

(36) 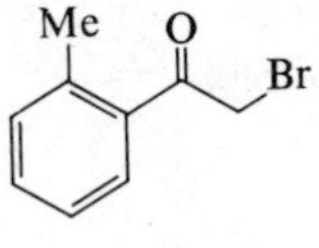   (37)

sodium ions enhances the elimination process by a factor of 5, affording (39) by a Norrish Type II mechanism. Potassium ions increase the process by a factor of 10, but lithium salts have little effect (Table). Quenching studies indicate that a portion of the reaction is not quenchable and therefore arises from the singlet state.

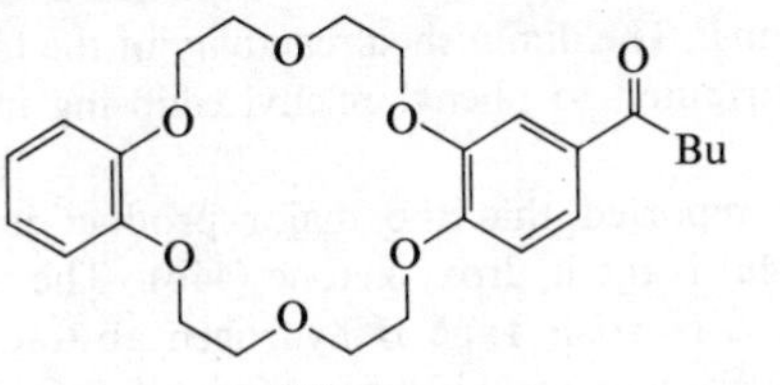

(38) 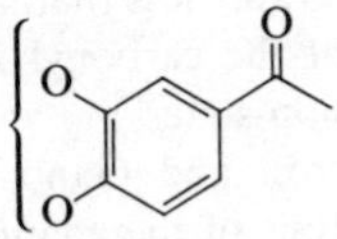   (39)

Table *Effects of salts on photoelimination reactions of ketone* (38) [30]

Ketone	Salt (1×10^{-3}M)	Φ_{rel}
(38)	—	1.0
(38)	LiOAc	1.4
(38)	LiCl	1.2
(38)	NaOAc	5.0
(38)	NaCl	4.3
(38)	KOAc	8.8
(38)	KCl	10.8

The synthetic utility of the Norrish Type II process has been demonstrated in the synthesis of methylenecyclopentane by the irradiation of the ketone (40).[31] Polymer supports have also been used in the synthetic applications of Norrish Type II eliminations. Thus a copolymer of styrene–divinylbenzene was acylated with the acid chlorides (41) (*n* = 7 or 8). The resulting products were then irradiated (254 nm) in diglyme to afford either hept-1-ene or oct-1-ene by a Norrish Type II process.[32]

[31] D. C. Neckers, R. M. Kellogg, W. L. Prins, and B. Scoustra, *Org. Photochem. Synth.*, 1976, **2**, 77 (*Chem. Abs.*, 1977, **87**, 5435).
[32] F. Camps, J. Castells, and J. Pi, *Anales de Quim.*, 1976, **72**, 483 (*Chem. Abs.*, 1977, **87**, 151 641).

Alexander and Uliana [33] previously reported the photochemical cyclization of the ketone (42) into the novel tricyclic system (43). They [34] have now studied the more quantitative aspects of the reaction and have observed that although the quantum yield for the cyclization in the triplet state is 0.51 the reactivity of the

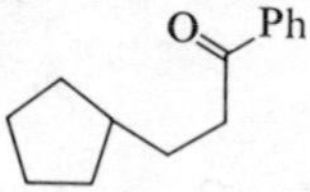

(40)

Me(CH$_2$)$_n$COCl

(41)

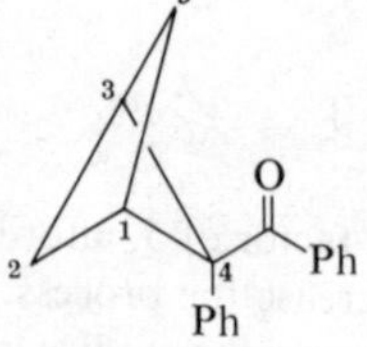

(42)

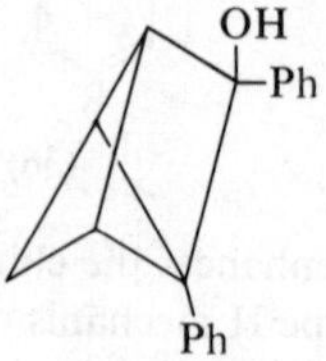

(43)

triplet state of the carbonyl group towards hydrogen abstraction from C-5 is considerably less than in model systems. The diminished reactivity of the triplet state of the carbonyl function is attributed to phenyl–phenyl eclipsing in the transition state.

Marron and Gano [35] previously reported that the major product in the photolysis of the adamantyl ester (44a) is the hydroxy-ketone (44b). The route to product in this reaction is again a Norrish Type II hydrogen abstraction, yielding the biradical (45) which cyclizes to the unstable oxetanol (46). This

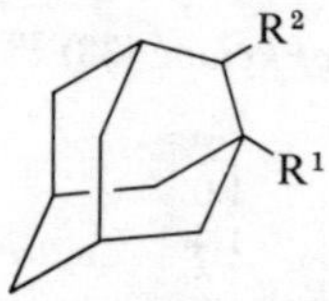

(44) a; R^1 = H, R^2 = CF$_3$CO$_2$
     b; R^2 = OH, R^1 = CF$_3$CO

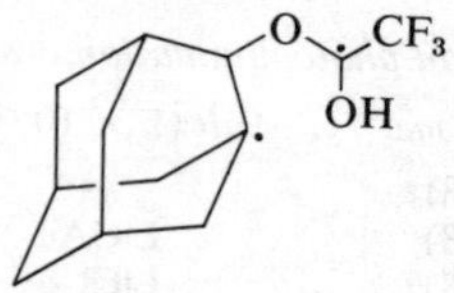

(45)

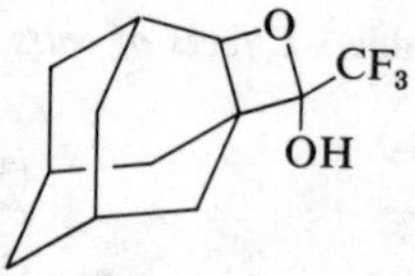

(46)

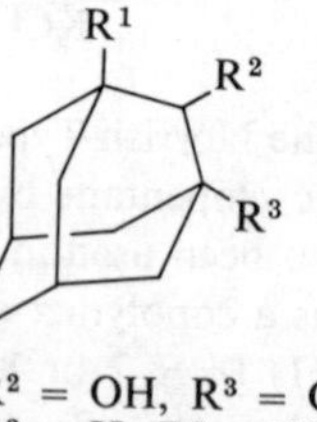

(47) a; R^1 = Me, R^2 = OH, R^3 = CF$_3$CO
     b; R^1 = Me, R^3 = H, R^2 = CF$_3$CO$_2$
     c; R^1 = MeCO$_2$ or CF$_3$CO$_2$, R^2 = R^3 = H
     d; R^1 = OH, R^2 = MeCO or CF$_3$CO, R^3 = H

[33] E. C. Alexander and J. Uliana, *J. Amer. Chem. Soc.*, 1976, **98**, 4324.
[34] E. C. Alexander and J. Uliana, *Tetrahedron Letters*, 1977, 1551.
[35] N. A. Marron and J. E. Gano, *J. Amer. Chem. Soc.*, 1976, **98**, 4653.

compound is not isolated and undergoes ring opening to the product (44b). The synthetic usefulness of the reaction has been probed and the authors [36] have shown that the adamantols (47a and d) can be prepared from the esters (47b and c).

The direct irradiation (313 or 366 nm) of the imino-ketone (48) affords two photochemical products (49a) and (49b).[37] The formation of the cyclobutanol products most likely arises *via* a Norrish Type II process affording the biradical (50) which subsequently cyclizes. The formation of the two isomers is a result of *syn–anti* isomerization of the oximino-group.

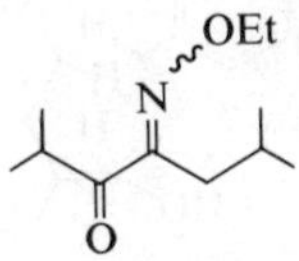

(48)

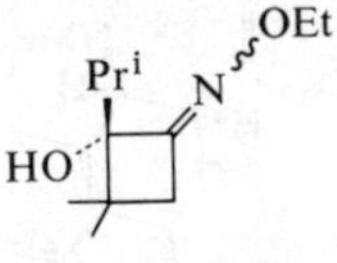

(49) a; α-OEt
     b; β-OEt

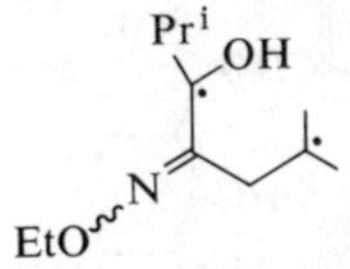

(50)

A biradical (51) is formed by Norrish Type II hydrogen abstraction in the photolysis of the 3*RS*-3,4-epoxy-1-phenylbutan-1-one (52). This biradical does not undergo fragmentation but ring closes to yield a 1.7 : 1 mixture of the epoxycyclobutanols (53a) and (53b).[38] Other epoxy-enones (54) have also been studied and the photolysis affords the corresponding epoxycyclobutanols (55)

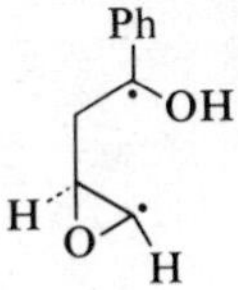

(51)

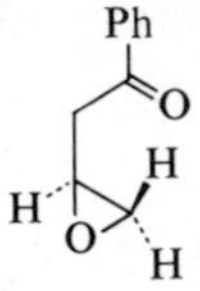

(52)

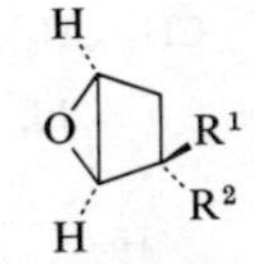

(53) a; R^1 = OH, R^2 = Ph
     b; R^1 = Ph, R^2 = OH

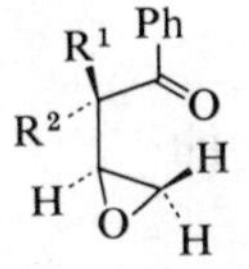

(54) a; R^1 = H, R^2 = Me
     b; R^1 = Me, R^2 = H

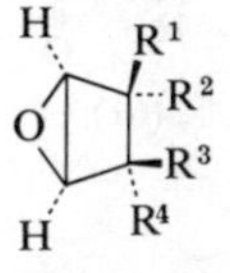

(55) a; R^1 = H, R^2 = Me, R^3 = Ph, R^4 = OH
     b; R^1 = Me, R^2 = H, R^3 = OH, R^4 = Ph

the stereochemistry of which is determined by steric interactions between the methyl on C-2 and the substituents on C-1.[39] Pyrex-filtered irradiation of solutions of the dialdehyde (56) in Bu^{t}OH results in the formation of the alkaloid (57).[40] A likely route to the formation of this product involves hydrogen abstraction to yield the photo-enol (58). This could be intramolecularly trapped by the other aldehyde function to yield the observed product.

[36] N. A. Marron and J. E. Gano, *Synth. Comm.*, 1977, **7**, 515 (*Chem. Abs.*, 1978, **88**, 136 185).
[37] P. Baas and H. Cerfontain, *Tetrahedron Letters*, 1978, 1501.
[38] J. M. Coxon and G. S. C. Hii, *Austral. J. Chem.*, 1977, **30**, 161.
[39] J. M. Coxon and G. S. C. Hii, *Austral. J. Chem.*, 1977, **30**, 835.
[40] S. Prabhakar, A. M. Lobo, and I. M. C. Oliveira, *J.C.S. Chem. Comm.*, 1977, 419.

Further details on the photochemical remote functionalization of benzo-phenone esters (59) have been published.[41]

The pterin (60) undergoes side-chain fission ($\Phi = 0.11$) when it is irradiated (365 nm) in phosphate buffer. The reaction, which affords acetaldehyde and the pterin aldehyde (61), is analogous to a Norrish Type II process.[42] However, in this instance the hydrogen transfer takes place to a ring nitrogen in much the same manner as that reported by Alexander and Jackson [43, 44] in their study of (62).

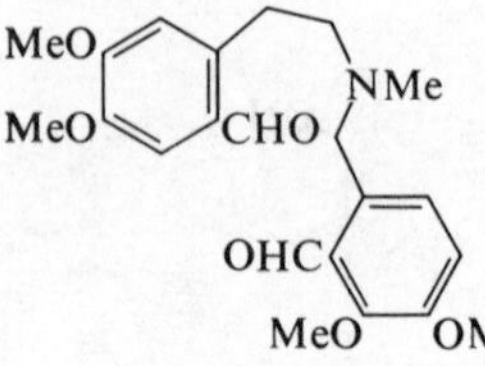

(56) 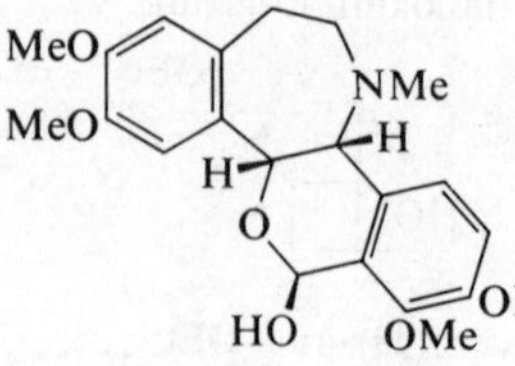   (57) 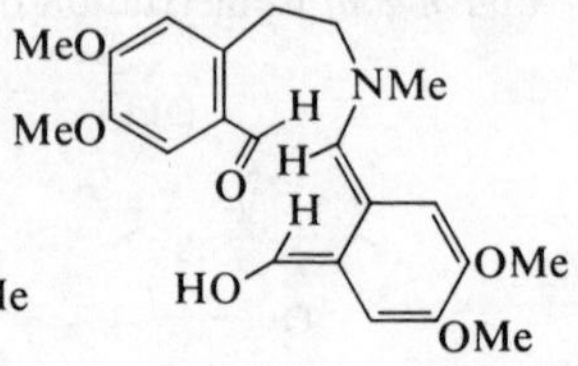   (58)

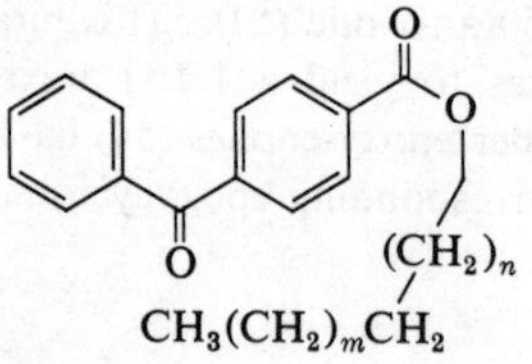

(59) $m + n = 10, 14, $ or 18 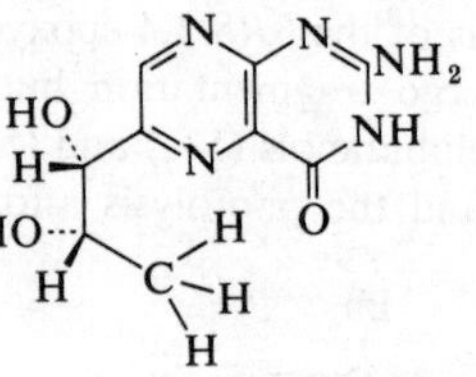   (60)

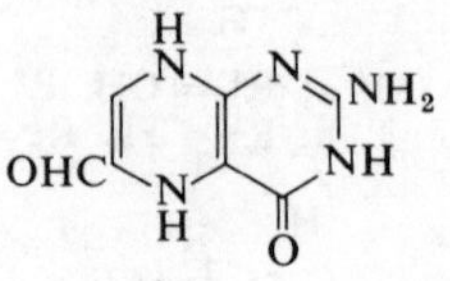

(61)

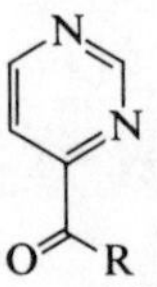

(62) R = Me, Et, Pr, or Bu

4 Rearrangement Reactions

Dewar and Sutherland [45] have re-examined the photochemistry of chalcone epoxide (63) and some derivatives and have shown that, contrary to an earlier report,[46] dibenzoylmethane is the major product, a result which is in agreement with the work of Kagan.[47] The present work [45] has identified that an $n\pi^*$ triplet state is responsible for the reaction. A hydrogen transfer mechanism is proposed

[41] R. Breslow, J. Rothbard, F. Herman, and M. L. Rodriguez, *J. Amer. Chem. Soc.*, 1978, **100**, 1213.
[42] P. Mengel, W. Pleiderer, and W.-R. Knappe, *Tetrahedron Letters*, 1977, 2817.
[43] E. C. Alexander and R. J. Jackson, *J. Amer. Chem. Soc.*, 1974, **96**, 5663.
[44] E. C. Alexander and R. J. Jackson, *J. Amer. Chem. Soc.*, 1976, **98**, 1609.
[45] D. J. Dewar and R. G. Sutherland, *J.C.S. Perkin II*, 1977, 1522.
[46] S. Bodforss, *Ber.*, 1918, **51**, 214.
[47] V. T. Ramakrishnan and J. Kagan, *J. Org. Chem.*, 1970, **35**, 2898.

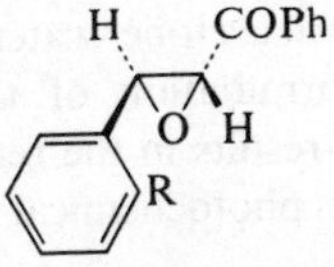

(63) R = H or OMe

to account for the formation of the dibenzoylmethane. In some experiments the formation of the *cis*-epoxide, possibly *via* a triplet $\pi\pi^*$ state, was observed. Photofragmentation products were also encountered.

Irradiation of (64) in benzene affords the isomeric product (65),[48] possibly as in Scheme 4. The photochemistry of (64a) is solvent-dependent and upon irradiation in methanol dihydrocoumarin (66) is produced. The methyl analogue (64b) does not exhibit similar solvent dependency. The reaction encountered in benzene arises from the singlet excited state. Padwa *et al.*[48] have also studied the photochemistry of the chroman derivatives (67) in which the formation of ring-opened products was found to be dominant (*e.g.* Scheme 5). Both of the reaction schemes involve fission of a C—O bond. It is tempting to suggest that biradical intermediates such as (68) and (69) are involved. However, there is always the possibility that the intermediates are polar as is implied by the solvent dependency of the photolysis of (64a).

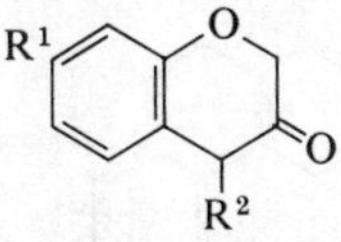

(64) a; $R^1 = H$, $R^2 = Ph$
b; $R^1 = R^2 = Me$

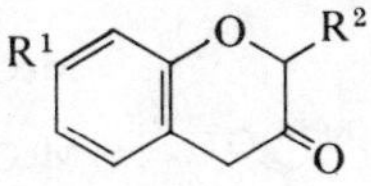

(65)

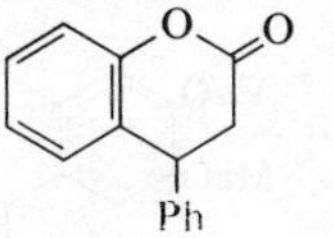

(66)

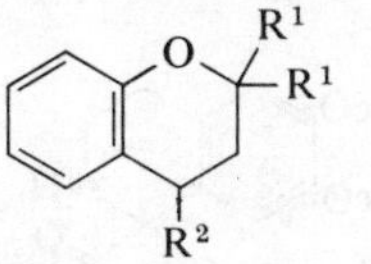

(67) a; $R^1 = Ph$, $R^2 = H$
b; $R^1 = R^2 = Ph$
c; $R^1 = Ph$, $R^2 = OH$

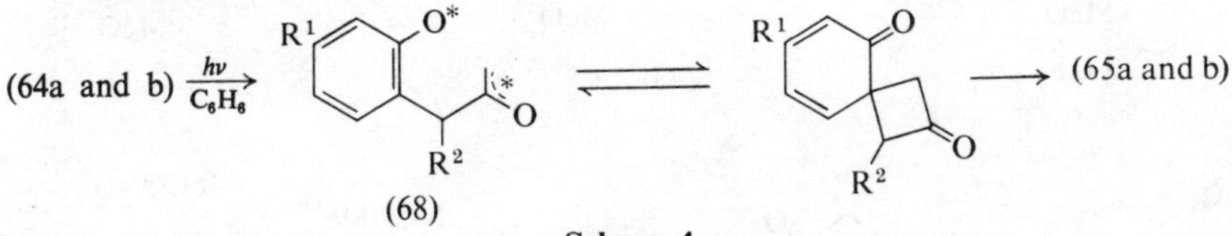

(64a and b) $\xrightarrow[C_6H_6]{h\nu}$

(68)

(65a and b)

Scheme 4

$Ph_2C=CH_2$
+

(67a) $\xrightarrow[MeOH]{h\nu}$

(69)

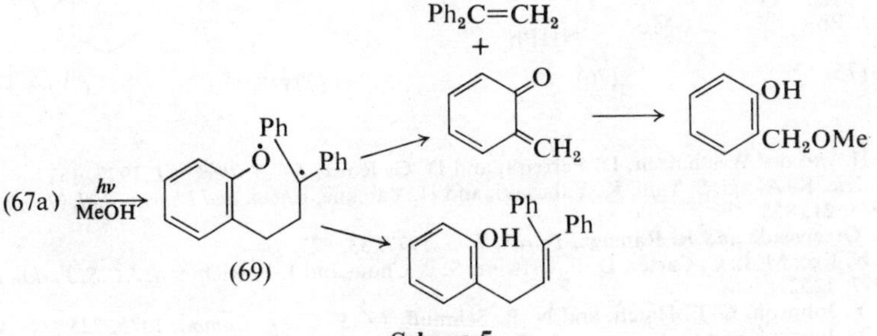

Scheme 5

[48] A. Padwa, A. Au, and W. Owens, *J. Org. Chem.*, 1978, **43**, 303.

Irradiation (350 nm) of (70a) in acetone–water results in its conversion into the keto-alcohol (70b). Further irradiation of this product (70b) at the same wavelength but in THF–water results in the rearrangement to (71). This transformation is described as the photochemical equivalent of the benzilic acid rearrangement.[49]

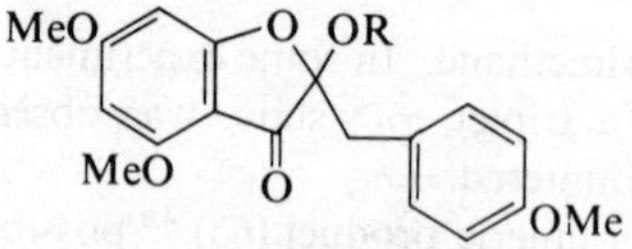

(70) a; R = Me
b; R = H

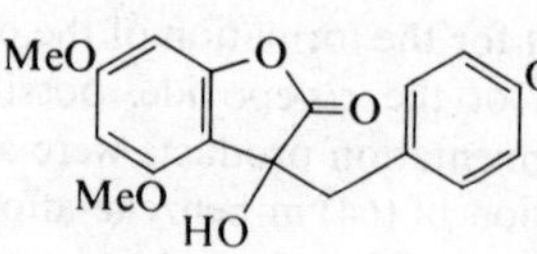

(71)

Irie *et al.*[50] have shown that the irradiation of (72) in THF yields (73), as the minor product, and (74). Greenslade and Ramage[51] have studied the influence of pH upon the reaction and have demonstrated that irradiation of (72) in ether affords (73; 30%), whereas under acidic conditions the major product is (74; 28%). Irradiation of the pyridazinone (75) in anhydrous methanol affords the ring-contracted product (76).[52] This ring contraction is reminiscent of the earlier work of Johnson and Hatch[53–56] who studied the ring-contraction reactions of pyrazolidinones [*e.g.* (77)] into azetidinones [*e.g.* (78)].

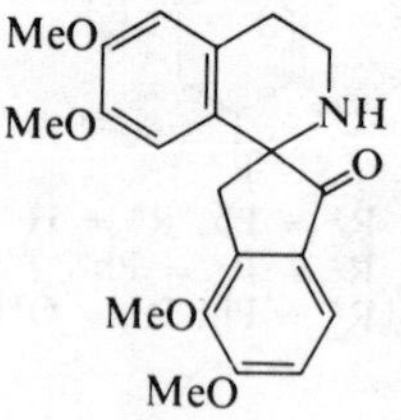

(72)

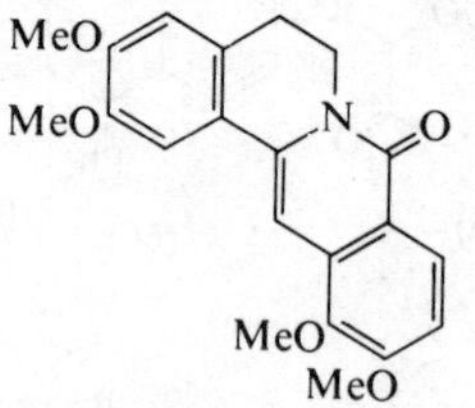

(73)

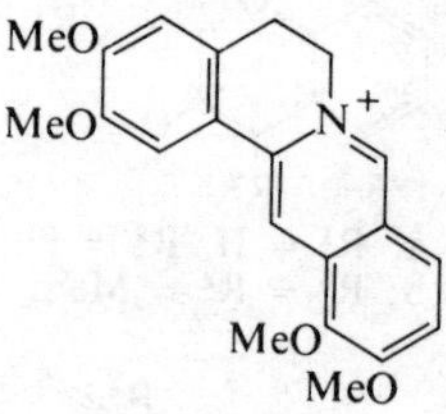

(74)

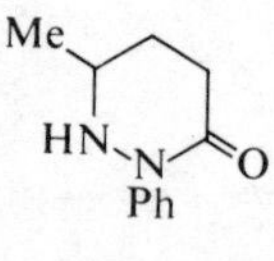

(75)

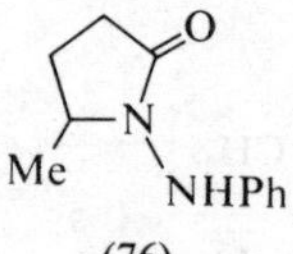

(76)

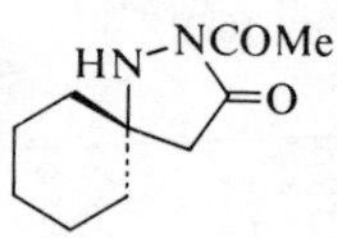

(77)

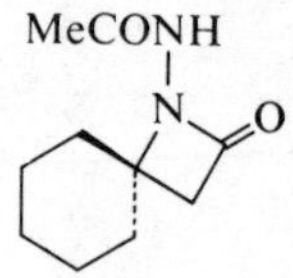

(78)

[49] J. H. van der Westhuizen, D. Ferreira, and D. G. Roux, *J.C.S. Perkin I*, 1977, 1517.
[50] H. Irie, K. Akagi, S. Tani, K. Yabusaki, and H. Yamane, *Chem. and Pharm. Bull. (Japan)*, 1973, **21**, 855.
[51] D. Greenslade and R. Ramage, *Tetrahedron*, 1977, **33**, 927.
[52] S. N. Ege, M. L. C. Carter, D. F. Ortwine, S. P. Chou, and J. F. Richman, *J.C.S. Perkin I*, 1977, 1252.
[53] P. Y. Johnson, C. E. Hatch, and N. R. Schmuff, *J.C.S. Chem. Comm.*, 1975, 725.
[54] P. Y. Johnson and C. E. Hatch, *J. Org. Chem.*, 1975, **40**, 3502.
[55] P. Y. Johnson and C. E. Hatch, *J. Org. Chem.*, 1975, **40**, 3510.
[56] P. Y. Johnson and C. E. Hatch, *J. Org. Chem.*, 1975, **40**, 909; *Tetrahedron Letters*, 1974, 2719.

5 Oxetan Formation

A polymer with oxetan units in the backbone has been prepared by the photochemical addition of benzophenone to polyisoprene.[57] Acetaldehyde adds photochemically to either *E*- or *Z*-penta-1,3-diene, yielding the oxetans (79—85), the relative ratios of which are shown below the appropriate structures.[58]

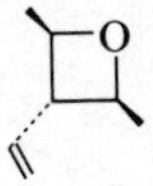

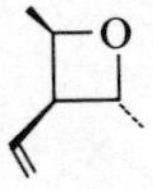

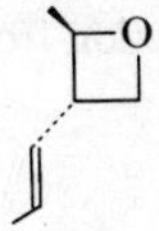

	(79)	(80)	(81)	(82)
E-Diene	0.51	0.16	<0.02	0.25
Z-Diene	<0.01	0.48	0.15	<0.01

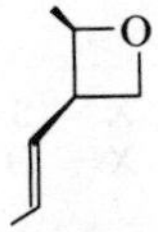

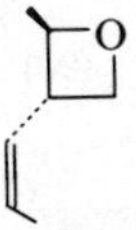

	(83)	(84)	(85)
	0.08	<0.005	<0.005
	<0.005	0.29	0.08

Several oxetans (86a—g) have been prepared by the photochemical addition of aldehydes and ketones to furan.[59] This process was developed for the synthesis of 3-substituted furans since it was discovered that the oxetans (86a—e) undergo easy ring opening in the presence of toluene-*p*-sulphonic acid. The oxetan (87) is obtained from the irradiation of a benzene solution of the dioxolone (88) and 1,3-diacetoxypropan-2-one.[60]

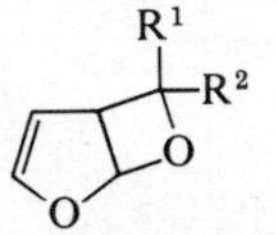

(86) a; $R^1 = H$, $R^2 = Me$
b; $R^1 = H$, $R^2 = CO_2Bu$
c; $R^1 = H$, $R^2 = CH_2OCOMe$
d; $R^1 = H$, $R^2 = (CH_2)Pr^i$
e; $R^1 = H$, $R^2 = Ph$
f; $R^1 = R^2 = CO_2Et$
g; $R^1 = H$, $R^2 = 2$-furyl

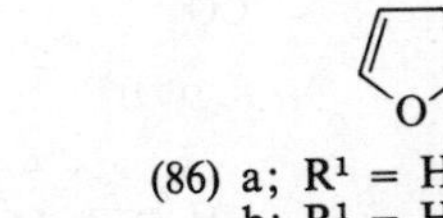

(87)

(88)

[57] H. Ng and J. E. Guillet, *Macromolecules*, 1977, **10**, 866.
[58] H. A. J. Carless and A. K. Maitra, *Tetrahedron Letters*, 1977, 1411.
[59] A. Zamojski and T. Kozluk, *J. Org. Chem.*, 1977, **42**, 1089.
[60] Y. Araki, J. Nagasawa, and Y. Ishido, *Carbohydrate Res.*, 1977, **58**, C4.

Cantrell [61] has studied the photochemical behaviour of the 3-acyl thiophens and furans towards 2,3-dimethylbut-2-ene. He has observed that the reactions are sluggish in comparison with the reactions of the 2-acyl thiophens. The acyl derivatives (89a, e, and f) yield oxetans (90) or products derived from subsequent reaction of the oxetans (89b, c, and d). However, the other derivatives do not show this mode of reactivity and only yield the product (91) following hydrogen abstraction from the olefin.

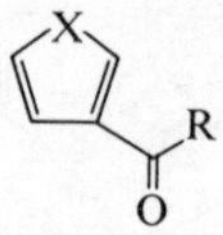

(89) a; R = Ph, X = S
     b; R = Ph, X = O
     c; R = Me, X = S
     d; R = Me, X = O
     e; R = H, X = S
     f; R = H, X = O

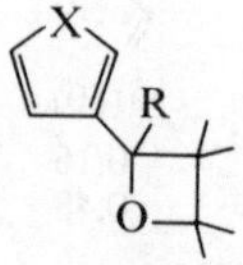

(90) a; 28%
     e; R = H, X = S
     f; R = H, X = O

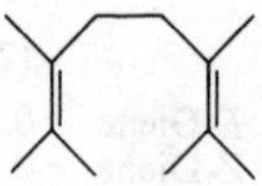

(91)

Adam and Sung [62, 63] have reported in a patent the formation of the two gibbanes (92) and (93) from photolysis of the gibberellin (94). It is tempting to interpret these reactions in terms of the two possible modes of addition of the carbonyl group to the ethylenic bond. Straightforward addition affords the oxetan (92) while the alternative mode affords a biradical intermediate (95) which does not ring-close but undergoes hydrogen transfer to yield (93).

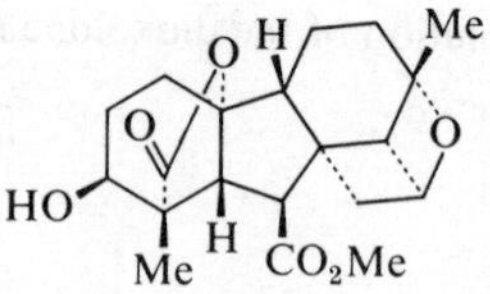

(92)

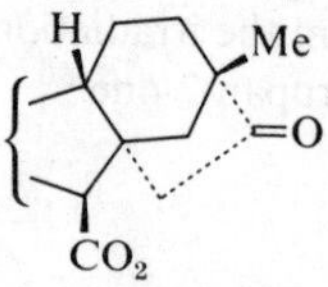

(93)

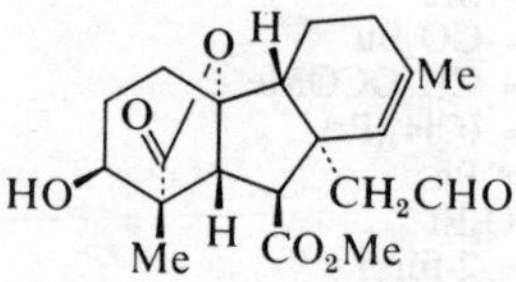

(94)

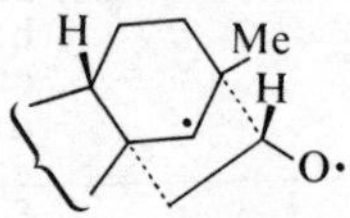

(95)

Yang and Chiang [64] have reported the photochemical addition of singlet excited pyrenyl-1-carboxaldehyde to cycloheptatriene to afford the [6 + 2] product (96) by a path which could be considered to be concerted. The more conventional [2 + 2] addition products (97a and b) and a product (98) from a

[61] T. S. Cantrell, *J. Org. Chem.*, 1977, **42**, 3774.
[62] G. Adam and T. V. Sung, E.Ger.P. 124 525 (*Chem. Abs.*, 1978, **88**, 89 393).
[63] G. Adam and T. V. Sung, E.Ger.P. 125 660 (*Chem. Abs.*, 1978, **88**, 62 488).
[64] N. C. Yang and W. Chiang, *J. Amer. Chem. Soc.*, 1977, **99**, 3163.

hydrogen abstraction path were also isolated. The photochemical addition of benzophenone to the diene (99a) affords the two adducts (100) and (101). It is interesting that only one adduct (102) was obtained when the alcohol (99b) was used.[65]

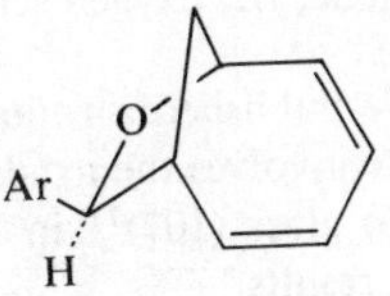

(96)

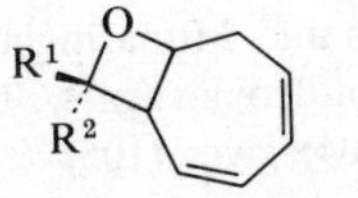

(97) a; R¹ = Ar, R² = H
      b; R¹ = H, R² = Ar

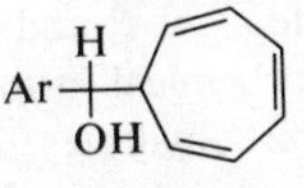

(98)

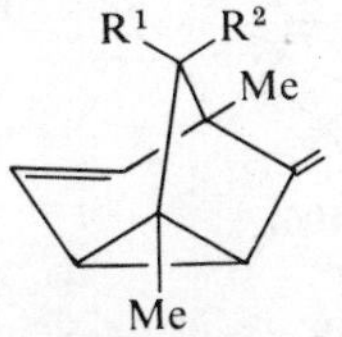

(99) a; R¹ = R² = O
      b; R¹ = OH, R² = H

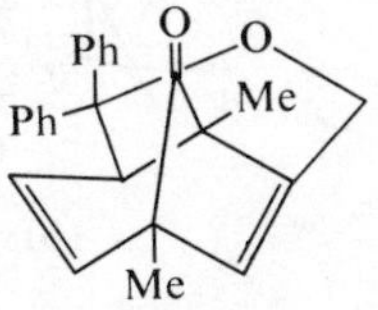

(100)

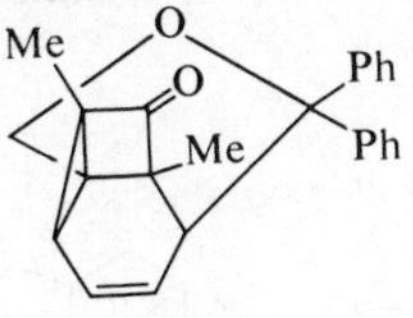

(101)

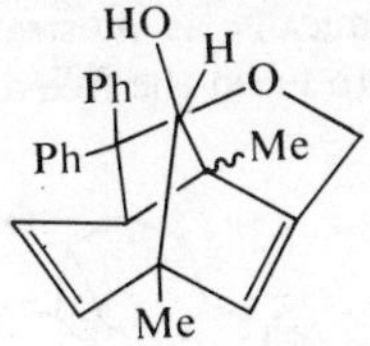

(102)

6 Fragmentation Reactions

Epling and Lopes [66] have reasoned from a series of experiments that the singlet-state photochemical decarboxylation of phenylacetic acid cannot be a concerted process. It is evident by performing the experiments in deuteriated solvents that a benzyl radical is produced from the photolysis of the acid. In experiments with sodium phenylacetate the path for decarboxylation was established as ionic. The photochemical decarboxylation of esters [*e.g.* (103)] is a reasonably well-investigated reaction and details of the mechanistic path have been obtained and

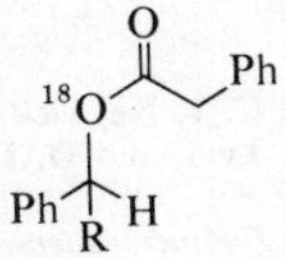

(103) R = H or Me

[65] M. Nitta, H. Sugiyama, and Y. Sekine, *Chem. Letters*, 1977, 55 (*Chem. Abs.*, 1977, **86**, 170 392).
[66] G. A. Epling and A. Lopes, *J. Amer. Chem. Soc.*, 1977, **99**, 2700.

reported in earlier years.[67] Givens and his co-workers [68, 69] have studied further the mechanistic aspects of the reaction by the use of labelling studies and have shown that the product composition is largely determined by the diffusion-controlled parameters of the solvent in which the reaction is carried out. The use of an ^{18}O label has identified another reaction mode, *viz.* oxygen scrambling within the ester.

Pete *et al.*[70] and Collins and Munasinghe [71] have published a route for the photochemical formation of deoxy-sugars. The route involves the irradiation of the carbohydrates: *e.g.* (104) gives (105)[70] or (106) gives (107)[71] in HMPA–H_2O, whereby elimination of the acetoxy side-chain results.

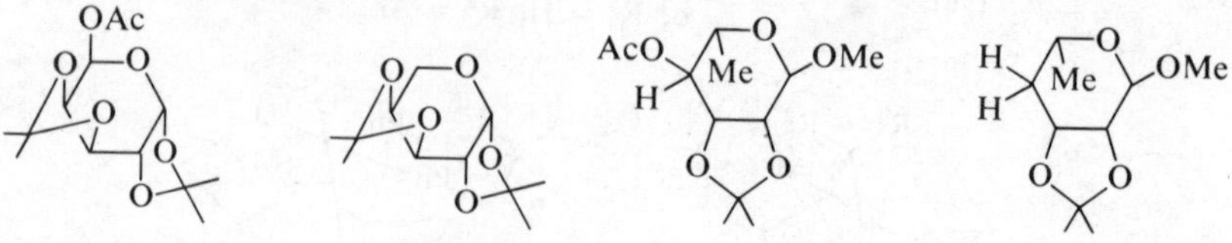

(104)   (105)   (106)   (107)

p-Methoxyphenacyl esters have been used to protect the carboxylic acid function of gibberellin derivatives.[72] The protecting group can be removed photochemically as illustrated in Scheme 6 where irradiation of the ester in ethanol produced the free acid (108).

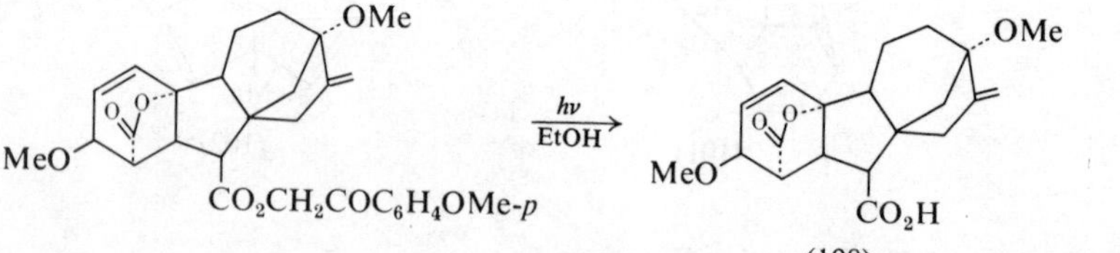

Scheme 6

Kametani *et al.*[73] report the photochemical epimerization and racemization of (109a) into (109b) by photochemical excitation.

Irradiation of the dihydrothiophenone (110) in methanol results in the fission of a S—C bond to yield the biradical intermediate (111).[74] The preference for this mode of fission is thought to be due to the stability of the thiyl radical. Hydrogen abstraction and dimerization of the resultant radical account for the formation of the major product (112).

[67] R. S. Givens, B. Matuszewski, and C. V. Neywick, *J. Amer. Chem. Soc.*, 1974, **96**, 5547.
[68] R. S. Givens, B. Matuszewski, N. Levi, and D. Leung, *J. Amer. Chem. Soc.*, 1977, **99**, 1896.
[69] R. S. Givens and B. Matuszewski, *J. Amer. Chem. Soc.*, 1975, **97**, 5617.
[70] J.-P. Pete, C. Portella, C. Monneret, J.-C. Florent, and Q. Khuong-Huu, *Synthesis*, 1977, 774.
[71] P. M. Collins and V. R. Z. Munasinghe, *J.C.S. Chem. Comm.*, 1977, 927.
[72] E. P. Serebryakov, L. M. Suslova, and V. F. Kucherov, *Tetrahedron*, 1978, **34**, 345.
[73] T. Kametani, H. Inoue, T. Sugahara, and K. Fukomoto, *J.C.S. Perkin I*, 1977, 374.
[74] P. Yates and Y. C. Toong, *J.C.S. Chem. Comm.*, 1978, 205.

(109) a; R¹ = OMe or H, R² = α-H
     b; R¹ = OMe or H, R² = β-H

(110)　　　　　(111)　　　　　(112)

Enone Cycloadditions and Rearrangements: Photoreactions of Cyclohexadienones and Quinones

BY W. M. HORSPOOL

1 Cycloaddition Reactions

Intramolecular.—Further interest in photochemical asymmetric synthesis has been reported by Green and his co-workers.[1] In this study they have examined the photochemical [2 + 2] dimerization of cinnamate esters attached to erythritol derivatives as in (1) and (2). [These workers had previously studied cinnamates attached to mannitol, *e.g.* (3).[2]] After irradiation and freeing the truxinates from the erythritol moiety it was shown that in the less rigid molecule (1)

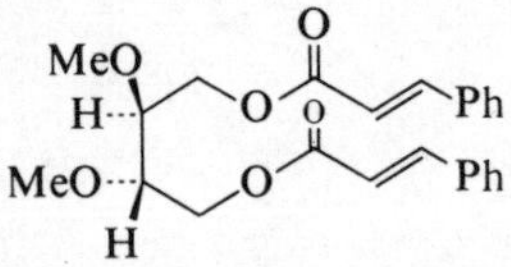

(1) 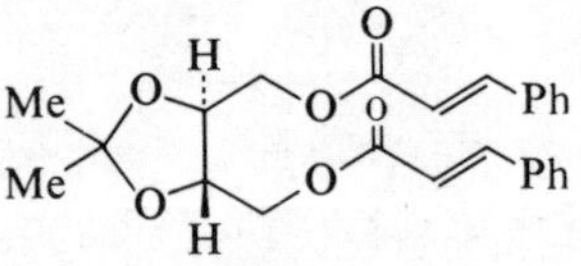   (2)

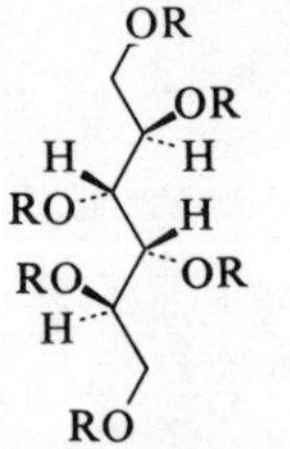

(3) R = cinnamate

asymmetric induction was 85% in the synthesis of (+)-dimethyl-γ-truxinate whereas the more rigid derivative (2) gave only 6% asymmetric induction and afforded (−)-dimethyl-γ-truxinate in excess. Ors and Srinivasan[3] have investigated the photochemical ring closure of the cinnamate ester (4) which yields the macrocycle (5). Direct irradiation is effective in this process although *cis–trans* isomerization also occurs efficiently. Interestingly, the efficiency of the cyclization decreases from compounds of type (6)[4] to that presently studied, an observation which is in line with a decreased probability of encounter of one end of the olefinic system with the other. Benzophenone-sensitized irradiation is more efficient since the longer lifetime of the triplet excited state enhances the number of encounters between the ends of the molecule.

The enones (7) undergo intramolecular [2 + 2] photoaddition to afford the tricyclic derivatives as shown in Scheme (1).[5] Cycloadditions of this type have

[1] B. S. Green, A. T. Hagler, Y. Rabinsohn, and M. Retjo, *Israel J. Chem.*, 1977, **15**, 124 (*Chem. Abs.*, 1978, **88**, 23 288).
[2] B. S. Green, Y. Rabinsohn, and M. Rejto, *J.C.S. Chem. Comm.*, 1975, 313; *Carbohydrate Res.*, 1975, **45**, 115.
[3] J. A. Ors and R. Srinivasan, *J. Amer. Chem. Soc.*, 1978, **100**, 315.
[4] J. Rennert, S. Soloway, I. Waltcher, and B. Leong, *J. Amer. Chem. Soc.*, 1972, **94**, 7242.
[5] M. Mellor, D. A. Otieno, and G. Pattenden, *J.C.S. Chem. Comm.*, 1978, 138.

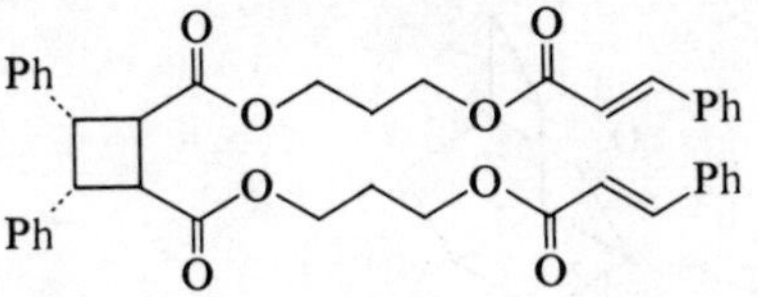

(4)

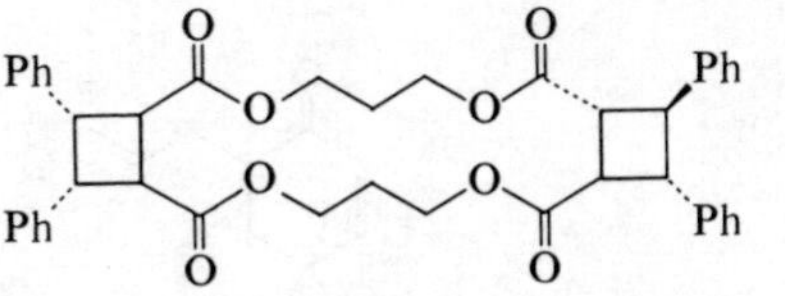

(5)

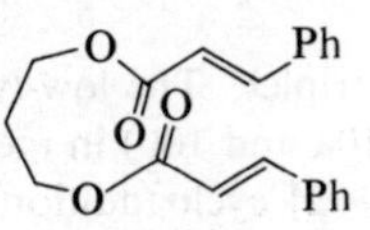

(6)

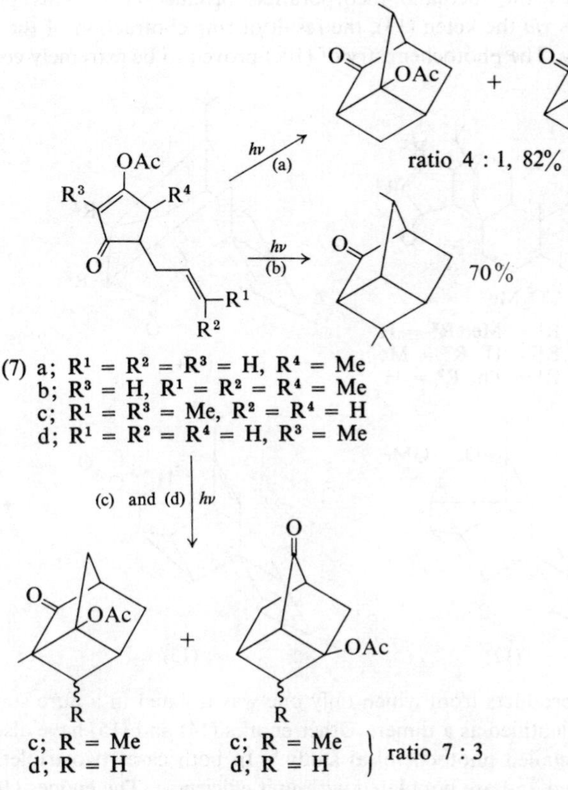
OAc
ratio 4 : 1, 82%
hν (a)
hν (b)
70%
(7) a; $R^1 = R^2 = R^3 = H, R^4 = Me$
b; $R^3 = H, R^1 = R^2 = R^4 = Me$
c; $R^1 = R^3 = Me, R^2 = R^4 = H$
d; $R^1 = R^2 = R^4 = H, R^3 = Me$
(c) and (d) hν
c; R = Me
d; R = H
c; R = Me
d; R = H
ratio 7 : 3

Scheme 1

previously been reported for the allyl-substituted cyclohexenone (8) which affords the tricyclic product (9).[6]

Herz and Iyer [7] have studied the photochemistry of some rigid cyclopentenones (10). The emission spectra of these compounds show that the excited state

[6] G. Jones and B. R. Ramachandran, *J. Photochem.*, 1976, **5**, 341.
[7] W. Herz and V. S. Iyer, *J. Org. Chem.*, 1977, **42**, 1573.

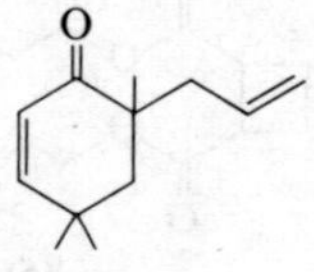

(8)

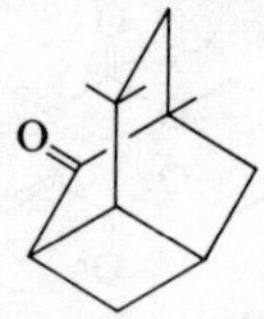

(9)

of both (10a) and (10b) is an $n\pi^*$ triplet. The low-lying triplet state of (10c) is $\pi\pi^*$ in character. Irradiation of (10a and 10b) in methanol using Pyrex-filtered light gave the intramolecular [2 + 2] cycloaddition products (11a) and (11b) respectively. In the case of (10b) this was the only reaction encountered, whereas (10a) also yielded the methanol-incorporation product (12). This product presumably arises *via* the keten (13), the result of ring contraction of the cyclo-pentenone moiety. The photochemistry of (10c) proved to be extremely complex

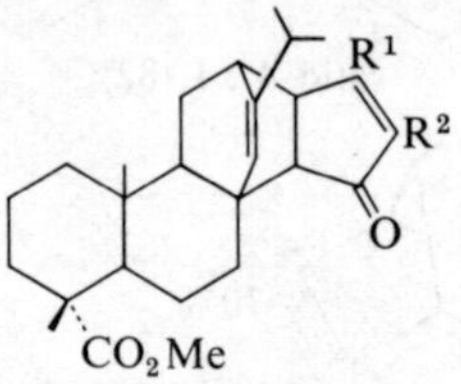

CO₂Me

(10) a; R^1 = Me, R^2 = H
b; R^1 = H, R^2 = Me
c; R^1 = Ph, R^2 = H

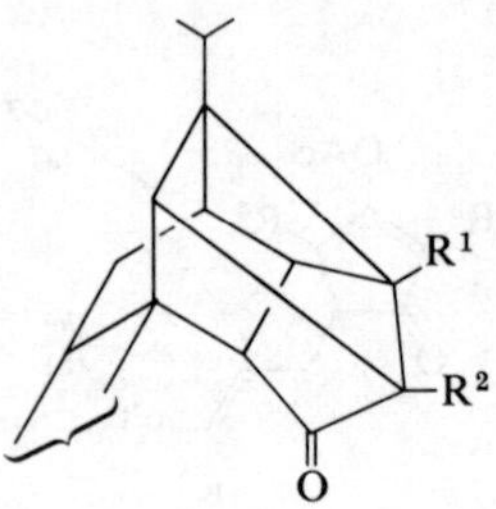

(11)

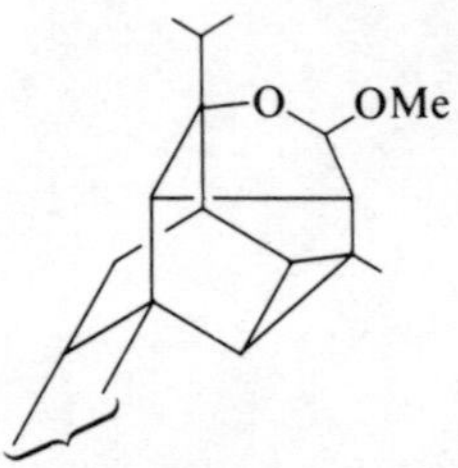

(12)

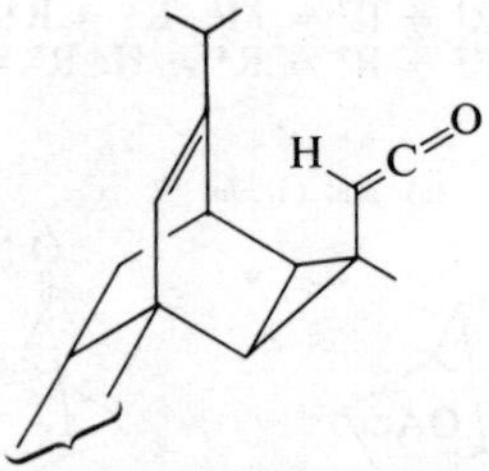

(13)

and gave many products from which only one was isolated in a pure state and was tentatively identified as a dimer. Other enones (14) and (15) have also been subjected to a detailed photochemical study.[8] In both cases two triplet states have been detected and are populated with unit efficiency. The enones (14) and (15) react through the ethylenic bond. Thus (14) is converted into the cage compound (16) while enone (15) undergoes hydrogen abstraction reactions to yield the products (17) and (18). The photochemical ring closure of the bis-enone (19) to yield the cage compound (20) has been reported.[9] The acrylamide (21) undergoes photochemical [2 + 2] addition when irradiated (Pyrex filter) in

[8] W. Herz, V. S. Iyer, M. G. Nair, and J. Saltiel, *J. Amer. Chem. Soc.*, 1977, **99**, 2704.
[9] P. E. Eaton, R. H. Mueller, G. R. Carlson, D. A. Cullison, G. F. Cooper, T.-C. Chou, and E.-P. Krebs, *J. Amer. Chem. Soc.*, 1977, **99**, 2751.

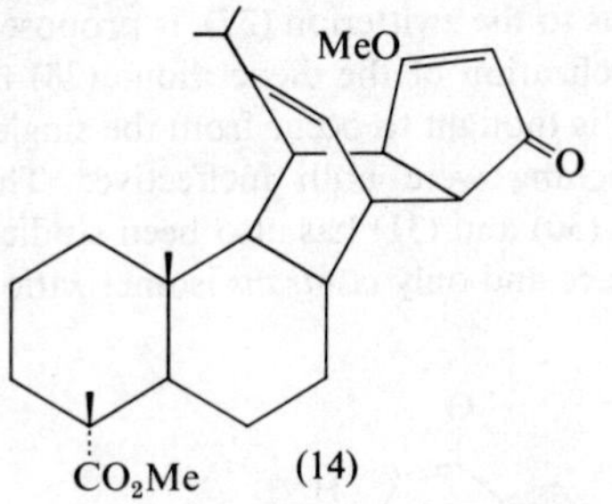

(14)

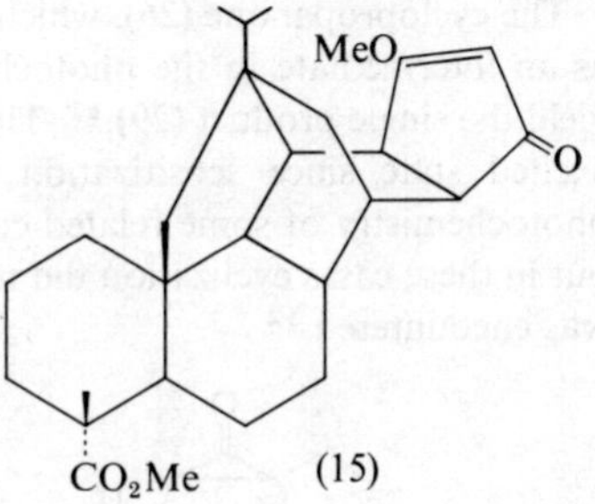

(15)

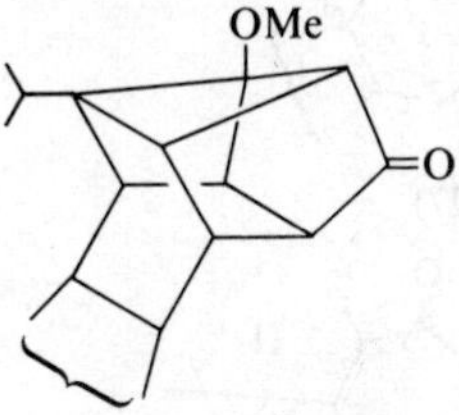

(16)

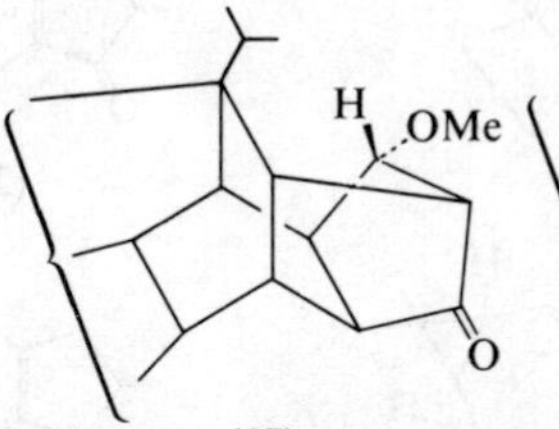

(17)

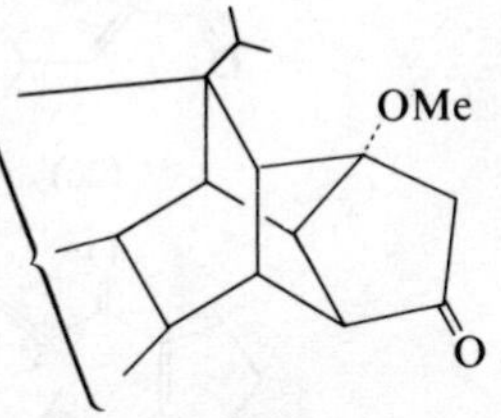

(18)

dioxan or pyridine to yield the lactam (22; 6%).[10] The enones (23) also undergo [2 + 2] photochemical addition to afford unstable intermediates (24). These readily ring-open under the condition of the reaction (methanol–Pyrex filter) to afford the tricyclic ketones (25).[11]

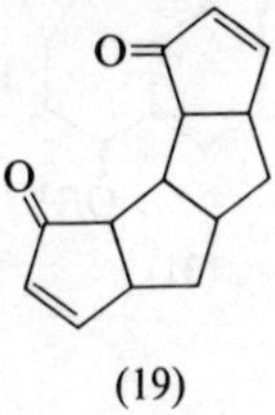

(19)

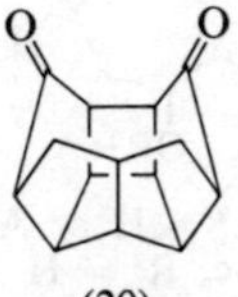

(20)

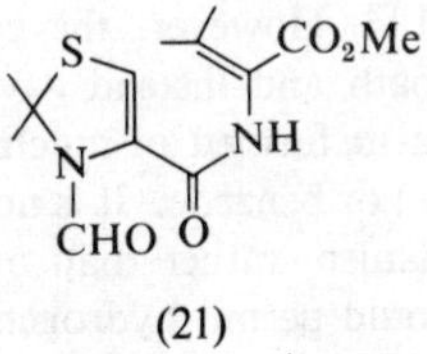

(21)

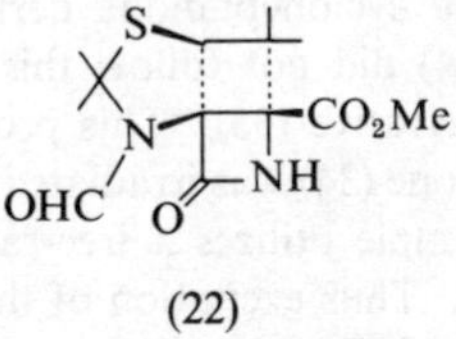

(22)

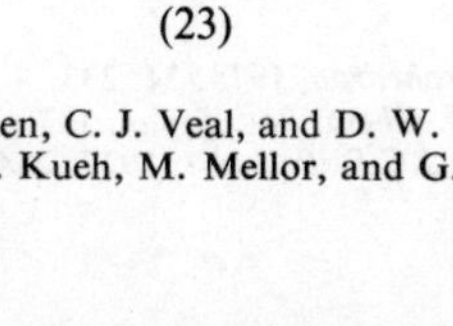

(23)

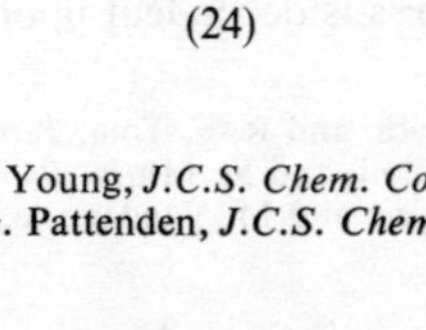

(24)

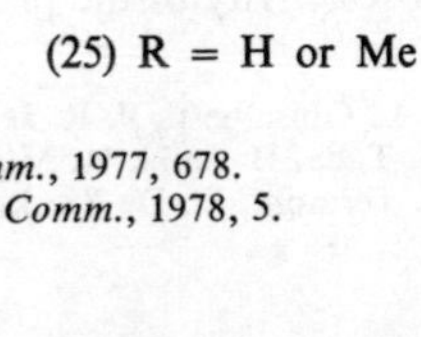

(25) R = H or Me

[10] P. K. Sen, C. J. Veal, and D. W. Young, *J.C.S. Chem. Comm.*, 1977, 678.
[11] J. S. H. Kueh, M. Mellor, and G. Pattenden, *J.C.S. Chem. Comm.*, 1978, 5.

The cyclopropanone (26), which ring-opens to the zwitterion (27), is proposed as an intermediate in the photochemical cyclization of the diene dione (28) to yield the single product (29).[12] The reaction is thought to occur from the singlet excited state since sensitization and quenching were both ineffective. The photochemistry of some related compounds (30) and (31) has also been studied but in these cases cyclization did not take place and only *cis–trans* isomerization was encountered.[12]

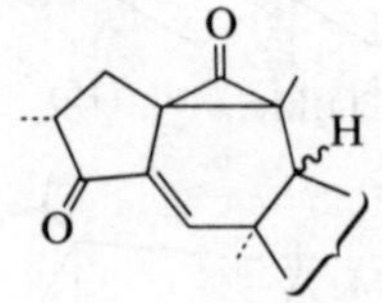

(26)

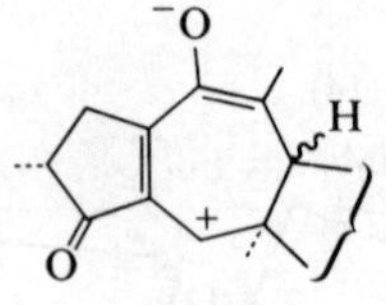

(27)

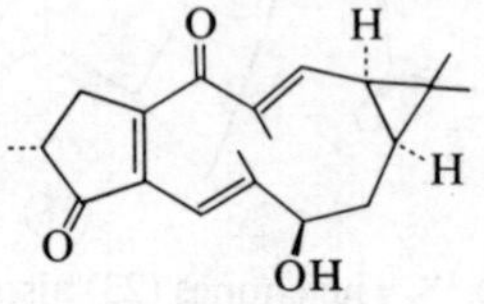

(28)

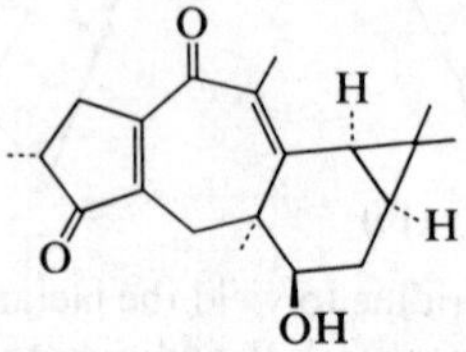

(29)

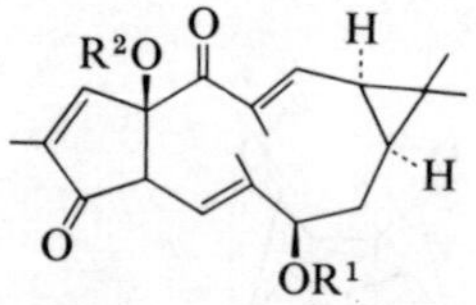

(30) $R^1 = R^2 = H$ or Ac
$R^1 = Ac, R^2 = H$

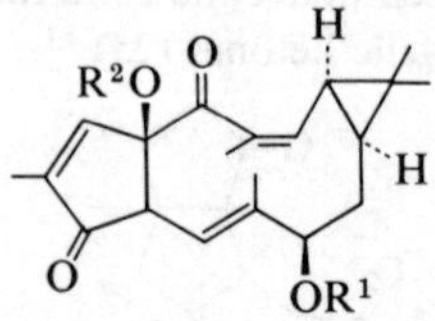

(31)

In the presence of an acid catalyst such as BF_3–Et_2O, irradiation of the enone (32) gave the cycloheptanone derivative (33).[13] However, the cyclopentenyl derivative (34) did not follow this reaction path and instead gave the cyclobutanone derivative (35). This product was also formed in much better yield when the ketone (34) was irradiated in methanol or benzene. It is not impossible that this example utilizes a free-radical mechanism rather than one involving ionic species. Thus excitation of the enone would permit hydrogen abstraction by the terminal C-atom of the enone, yielding the biradical (36) which subsequently cyclizes to (35).

Intermolecular.—The photochemical addition of cyclopent-2-enone to electron-rich olefins has been studied.[14] The reaction is regioselective and the regioselectivity increases with increasing nucleophilic character of the olefin. The stereoselectivity of the process is dependent upon the substituent pattern on the

[12] E. L. Ghisalberti, P. R. Jeffries, and R. F. Toia, *Tetrahedron*, 1978, **34**, 233.
[13] M. Tada, H. Saiki, H. Mizutani, and K. Miura, *Bull. Chem. Soc. Japan*, 1978, **51**, 343.
[14] D. Termont, D. De Keukeleire, and M. Vandewalle, *J.C.S. Perkin I*, 1977, 2349.

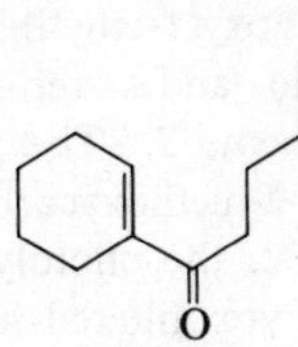

(32)

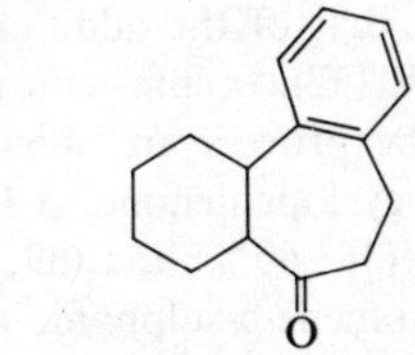

(33)

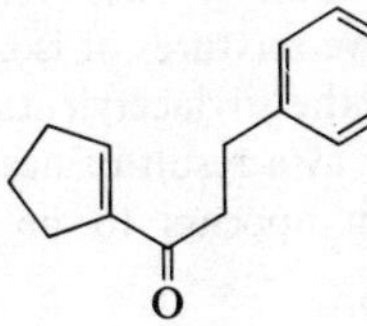

(34)

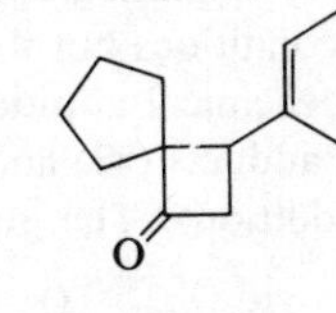

(35)

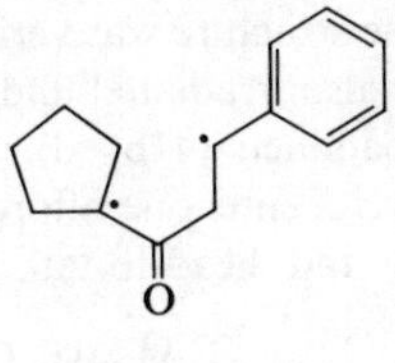

(36)

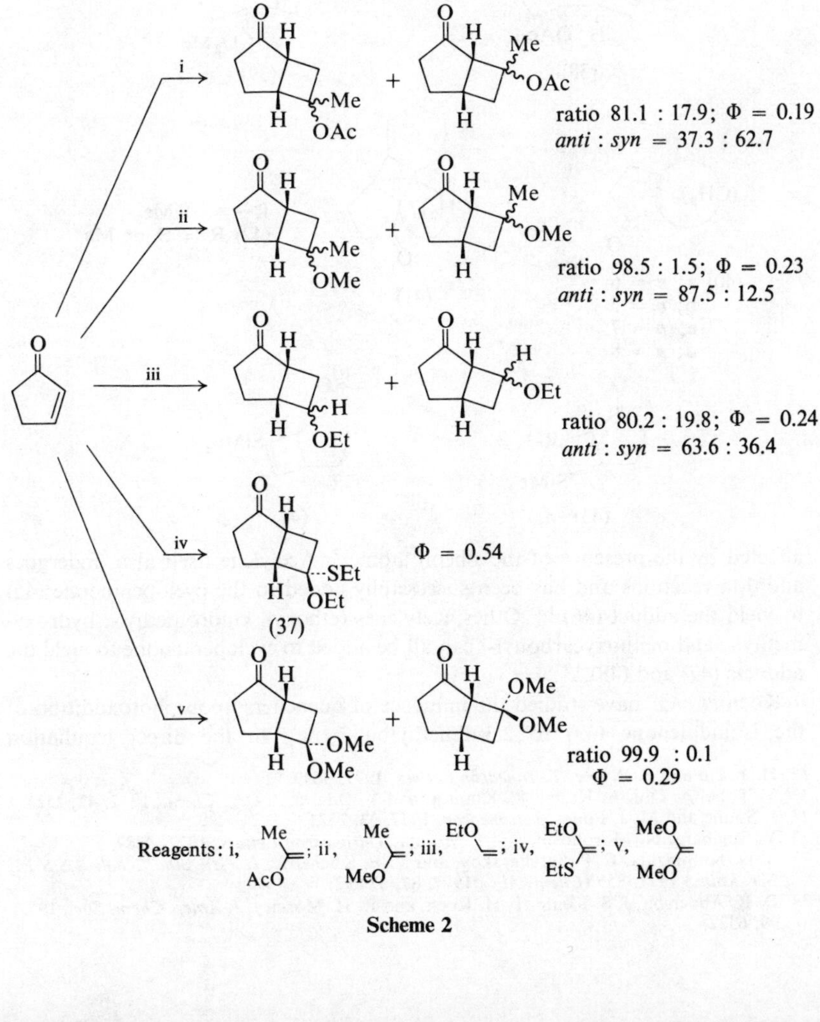

ratio 81.1 : 17.9; Φ = 0.19
anti : *syn* = 37.3 : 62.7

ratio 98.5 : 1.5; Φ = 0.23
anti : *syn* = 87.5 : 12.5

ratio 80.2 : 19.8; Φ = 0.24
anti : *syn* = 63.6 : 36.4

(37)   Φ = 0.54

ratio 99.9 : 0.1
Φ = 0.29

Reagents: i, ; ii, ; iii, ; iv, ; v,

Scheme 2

alkene. In one case, that of the addition to 1-ethoxy-1-ethylthioethylene, the addition, which yields (37), occurs with total regio- and stereo-selection. The other results from the process are shown in Scheme 2. The photochemical addition of 4-acetoxycyclopentenone to 1-acetoxy-2-methoxycarbonylcyclopentene gave the enones (38; 60%) and (39; 20%) after the photolysate had been treated briefly with toluene-*p*-sulphonic acid.[15] Pyrex-filtered irradiation of a solution of the enone (40a) in cyclohexene gave a single cycloadduct (41; 84%) whose structure was verified by *X*-ray crystallography.[16] Other enones (40b—d) were also irradiated under similar conditions but they gave mixtures of isomeric cycloadducts (41b—d). The photochemical addition of the silylacetylenes (42) to cyclopentenone affords the two adducts (43) and (44) as a result of head-to-head and head-to-tail [2 + 2] addition. The reaction appears to be little

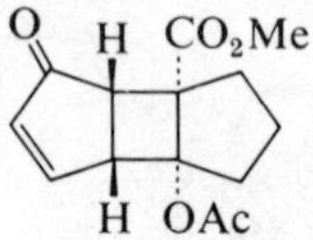

(38) 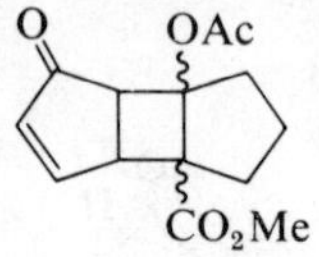   (39)

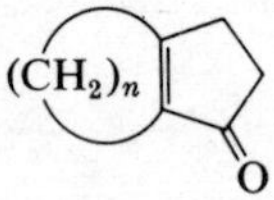

(40) a; *n* = 6
     b; *n* = 5
     c; *n* = 7
     d; *n* = 8

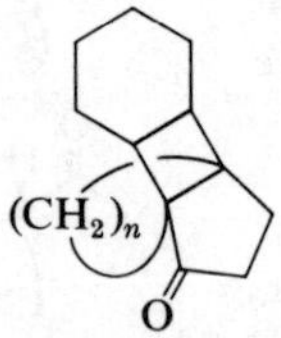

(41)

$R-\!\!\equiv\!\!-SiMe_3$
(42) R = H or Me

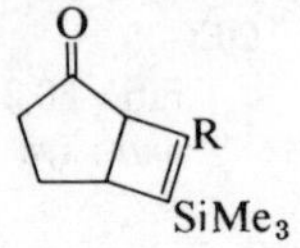

(43) 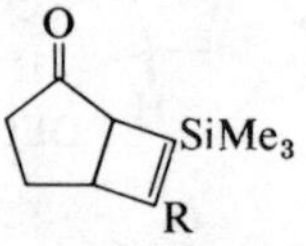   (44)

affected by the presence of the silicon atom.[17] Acetylene itself also undergoes addition reactions and has been successfully added to the cyclopentenone (45) to yield the adduct (46).[18] Other acetylenes (ethoxy-, chloromethyl-, hydroxymethyl-, and methoxycarbonyl-) can all be added to cyclopentenone to yield the adducts (47) and (48).[19]

Koch *et al.*[20] have studied the influence of quenchers upon photoaddition of the isoindolenone (49) to 2,3-dimethylbut-2-ene. In the direct irradiation

[15] H.-J. Liu and S. P. Lee, *Tetrahedron Letters*, 1977, 3699.
[16] Y. Tobe, A. Doi, A. Kunai, K. Kimura, and Y. Odaira, *J. Org. Chem.*, 1977, **42**, 2523.
[17] J. Soulie and M. J. Pouet, *Tetrahedron*, 1977, **33**, 2521.
[18] Y. Sugihara, N. Morokoshi, and I. Murata, *Tetrahedron Letters*, 1977, 3887.
[19] S. D. Kulomzina, E. P. Serebryakov, and V. F. Kucherov, *Izvest. Akad. Nauk S.S.S.R., Ser. khim.*, 1977, 855 (*Chem. Abs.*, 1977, **87**, 67 892).
[20] D. R. Anderson, J. S. Keute, T. H. Koch, and R. H. Moseley, *J. Amer. Chem. Soc.*, 1977, **99**, 6332.

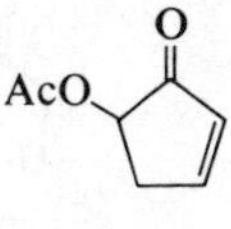

(45) 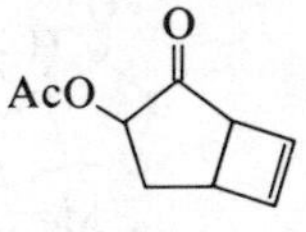   (46)

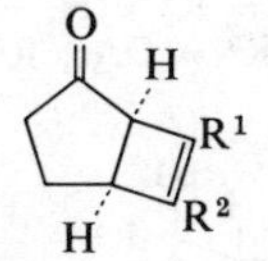

(47) R^1 = EtO, ClCH$_2$, HOCH$_2$, or MeO$_2$C, R^2 = H
(48) R^1 = H, R^2 = EtO, ClCH$_2$, HOCH$_2$, or MeO$_2$C

experiment the addition affords the products (50), (51), and (52) shown in Scheme 3. When di-t-butyl nitroxide is used as quencher, part of the quencher is destroyed, presumably by radical scavenging reactions. *cis*-Piperylene and biacetyl also quench the reaction.

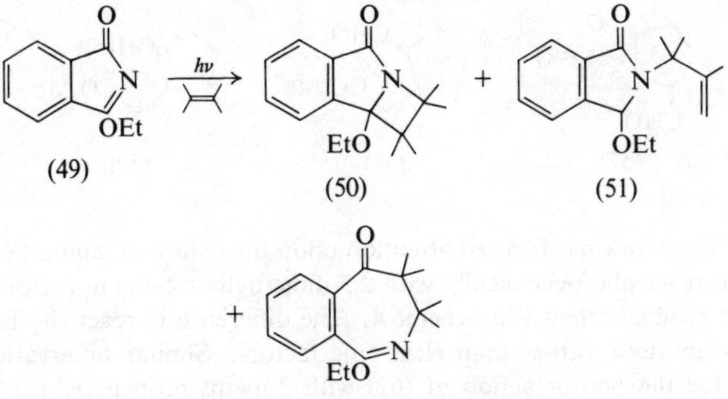

Scheme 3

The photochemical addition of methyl 3-acetoxyprop-2-enoate to the enone (53) affords four products (54) and (55) where the head-to-head products (54a) and (54b) predominate.[21—23] Addition of (53) to acetoxyethylene has also been studied and the [2 + 2] adducts (56) were isolated.[22, 23] Cyclopent-2-enone also undergoes photocycloaddition of the propenoate to afford a mixture of five isomeric 1 : 1 adducts. These were reduced with sodium borohydride and hydrolysed with sodium methoxide to afford the three products (57; 31%), (58; 60%), and (59; 4%).[21]

A further study of the influence of substituents upon the cyclobutane : oxetan ratio in the photochemical addition of olefins to cyclic enones has been pub-

[21] T. Ogino, K. Yamada, and K. Isogai, *Tetrahedron Letters*, 1977, 2445.
[22] T. Ogino, *Niigata Daigaku Kyiokugakubu Kiyo, Shizen Kagaku Hen*, 1976, **18**, 15 (*Chem. Abs.*, 1977, **87**, 84 136).
[23] T. Ogino, T. Kubota, and K. Manaka, *Chem. Letters*, 1976, 323 (*Chem. Abs.*, 1976, **85**, 20 982).

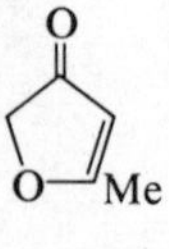

(53)

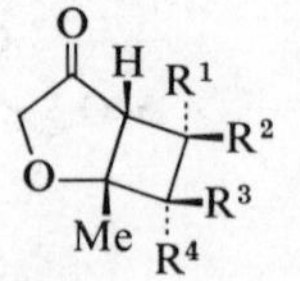

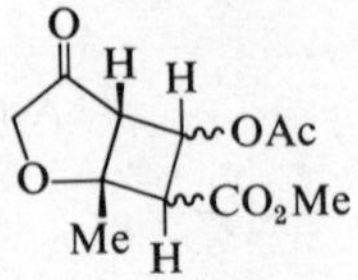

(54) a; R^4 = OAc, R^1 = R^3 = H, R^2 = CO_2Me
     b; R^1 = CO_2Me, R^3 = OAc, R^2 = R^4 = H

(55)

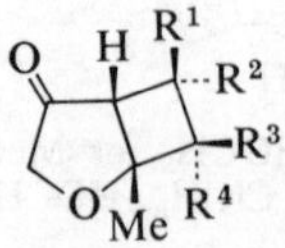

(56) R^1 = R^3 = R^4 = H, R^2 = OAc
     R^1 = R^2 = R^3 = H, R^4 = OAc
     R^1 = R^3 = R^4 = H, R^3 = OAc

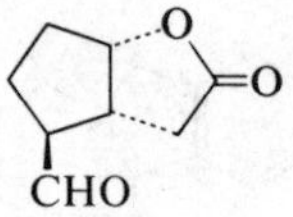

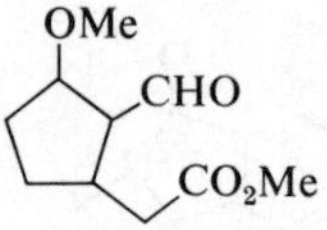

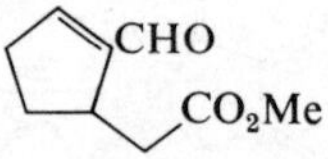

(57)   (58)   (59)

lished.[24] This work has focused attention upon the chloro-compounds (60) and (61) which react photochemically with 2,3-dimethylbut-2-ene in cyclohexane to afford the products shown in Scheme 4. The difference in reactivity is thought to be due to steric rather than electronic factors. Similar observations were recorded for the photoreaction of (62) with 2-methylpropene which gave the products shown in Scheme 5. Margaretha and Thi [25] had previously studied the cyclopentenone (63) and its photoreactions with 2,3-dimethylbut-2-ene. The photochemical [2 + 2] addition of 2-acetoxypropene to isophorone gave the adduct (64).[26] The irradiation of mixtures of *cis*-1,2-dimethoxy- or *cis*-1,2-diethoxy-ethylene or tetramethoxyethylene and either 3-acetoxy- or 3-chloro-cyclohex-2-en-1-one affords the [2 + 2] adducts (65).[27]

Kinetic data for the photoreaction of cyclohex-2-en-1-one with hex-1-yne have been obtained.[28] The photoaddition reaction yields two adducts (66a) and (66b) in ratios which are temperature dependent, *viz.* 8.58, 7.02, 5.55, and 4.42 at 250, 300, 350, and 400 K, respectively. Irradiation ($\lambda > 300$ nm) of methylene chloride solutions of the gibberellins (67) in the presence of acetylene yielded the

[24] I. Altmeyer and P. Margaretha, *Helv. Chim. Acta*, 1977, **60**, 874.
[25] G. V. Thi and P. Margaretha, *Helv. Chim. Acta*, 1976, **59**, 2236.
[26] Z. Valenta and H. J. Liu, *Org. Synthesis*, 1977, **57**, 113.
[27] R. K. Boeckman, jun., M. H. Delton, T. Nagasaka, and T. Watanabe, *J. Org. Chem.*, 1977, **42**, 2946.
[28] A. Kh. Margaryan, E. P. Serebryakov, and V. F. Kucherov, *Izvest. Akad. Nauk S.S.S.R., Ser. khim.*, 1977, 418 (*Chem. Abs.*, 1977, **87**, 38 476).

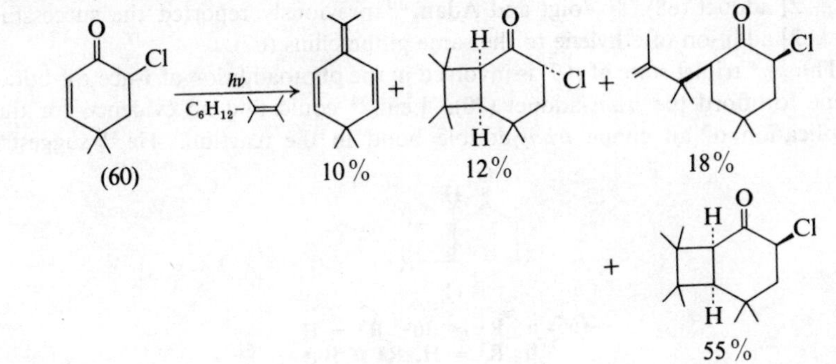

Scheme 4

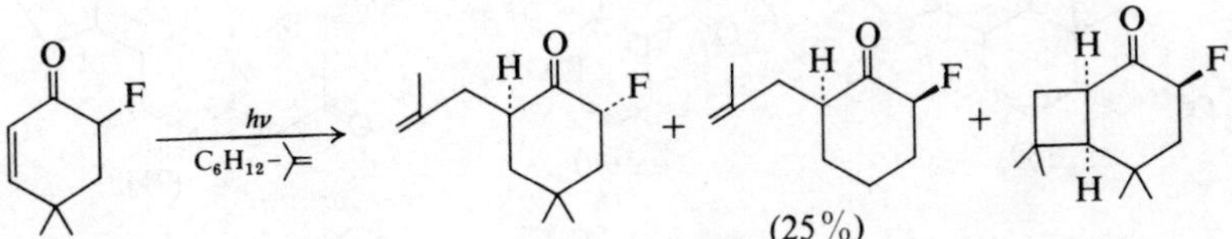

Scheme 5

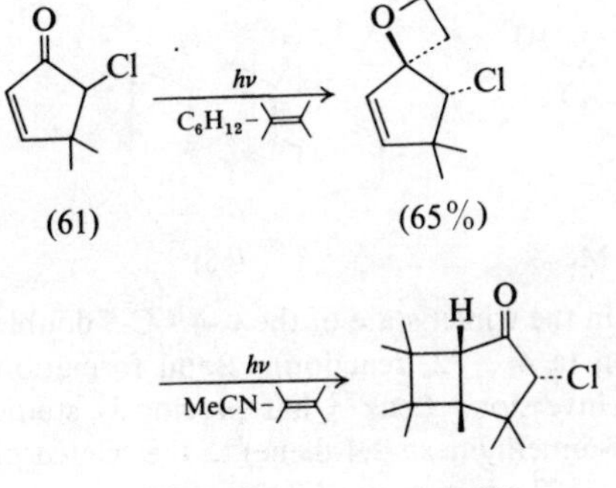

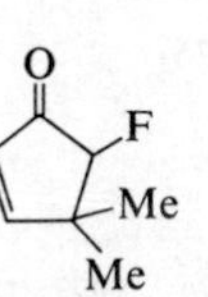

(63)

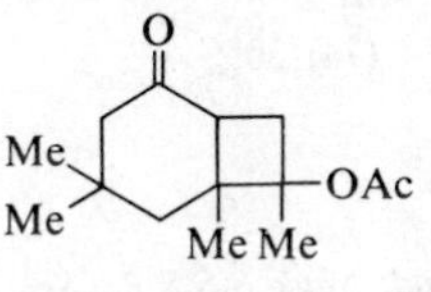

(64)

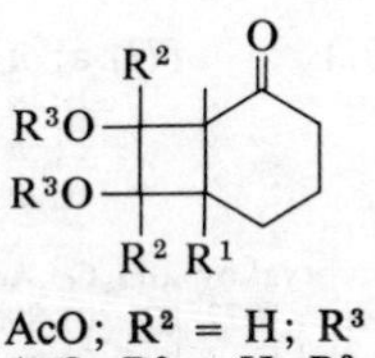

(65) a; R^1 = AcO; R^2 = H; R^3 = Me
b; R^1 = AcO, R^2 = H, R^3 = Et
c; R^1 = Cl, R^2 = H, R^3 = Me, or Et
d; R^1 = OAc, R^2 = R^3 = Me

[2 + 2] adduct (68).[29] Voigt and Adam [30] previously reported the successful
[2 + 2] addition of ethylene to the same gibberellins (67).

The $\pi\pi^*$ triplet state of (69) is involved in the photoaddition of 1-acetoxybuta-
diene to afford the *trans*-adduct (70). Lenz [31] could find no evidence for the
implication of an enone *trans*-double bond in the reaction. He [31] suggests,

(66) a; $R^1 = Bu^n$, $R^2 = H$
     b; $R^1 = H$, $R^2 = Bu^n$

(67) R = H or Me   (68)

however, that the slight twisting in the triplet state of the C-4—C-5 double bond
permits the Diels–Alder addition (a $4_s + 2_a$ reaction). Bond formation must
occur simultaneously with spin inversion. Lenz [32] has previously studied the
addition of other dienes (*e.g.* 2,5-dimethylhexa-2,4-diene) to the related dienone
(71) and in that example only [2 + 2] addition products were encountered.

(69)   (70)   (71)

(72)   (73) a; $n = 2$, 67%   (74) 30%   (75)
             b; $n = 1$, 51%
             c; $n = 3$, 55%
             d; $n = 4$, 58%

[29] E. P. Serebryakov and G. Adam, *Izvest. Akad. Nauk S.S.S.R., Ser. khim.*, 1977, 1831 (*Chem. Abs.*, 1977, **87**, 201 808).
[30] B. Voigt and G. Adam, *Tetrahedron*, 1976, **32**, 1581.
[31] G. R. Lenz, *Tetrahedron Letters*, 1977, 2483.
[32] G. R. Lenz, *Tetrahedron*, 1975, **31**, 1587.

The photochemical addition of the hydroxycoumarin (72) to cyclohexene in t-butyl alcohol yields the [2 + 2] adduct (73; $n = 2$) of uncertain stereochemistry.[33] Analogous adducts (73b—d) were obtained from cyclopentene, cycloheptene, and cyclo-octene. The adducts (74) and (75) were obtained with norbornene and 2,3-dimethylbutene.

Photodimerization of 3,4-dichlorocinnamic acid in the β-crystalline form affords dimers when a suspension in hexane is irradiated.[34] The dimer obtained is of the β-truxinic type but differs in physical properties from dimers obtained without dispersant. The difference in the dimer obtained from the hexane suspension is due to the incorporation of hexane into the dimer (0.4 mol hexane per mol of dimer). The authors [34] point out that the incorporation of the hexane occurs during the formation of the dimer and not subsequent to its formation. They demonstrated that other hydrocarbons could be incorporated in a like manner. The styryl dye (76) is reported to undergo solid-phase dimerization to afford (77).[35]

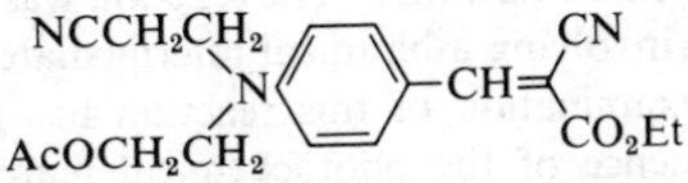

(76)

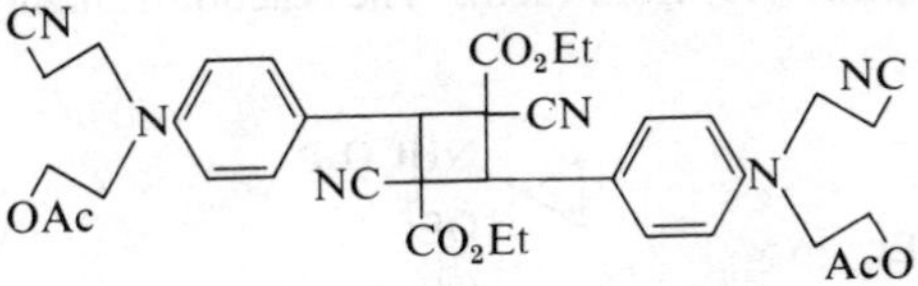

(77)

The results of empirical force-field calculations on the [2 + 2] cycloaddition process have been published.[36] Dilling [37] has published a review of the application of photocycloaddition to the synthesis of natural products.

2 Enone Rearrangements

The polarity of the solvent is important in the photochemical reactivity of methyl 2-naphthylacrylate in that intersystem crossing occurs more readily in polar solvents.[38] A study of the photochemical *cis–trans* isomerization of pent-2-en-3-one has indicated that a variety of mechanisms can be involved in the process.[39] Thus for example when carbonyl sensitizers (acetophenone, benzophenone, *etc.*) are used in reducing solvents, a radical chain mechanism can account for the isomerization. Classical triplet–triplet energy transfer can also take place when the reaction is carried out in a solvent such as benzene.

33 D. J. Haywood, R. G. Hunt, C. J. Potter, and S. T. Reid, *J.C.S. Perkin I*, 1977, 2458.
34 F. Nakanishi, S. Yamada, and H. Nakanishi, *J.C.S. Chem. Comm.*, 1977, 247.
35 A. T. Peters and M. S. Wild, *J. Soc. Dyers and Colourists*, 1976, **92**, 397 (*Chem. Abs.*, 1977, **86**, 141 601).
36 E. Osawa, K. Aigami, and Y. Inamoto, *J. Org. Chem.*, 1977, **42**, 2621.
37 W. L. Dilling, *Photochem. and Photobiol.*, 1977, **25**, 605.
38 H. Tanaka, K. Honda, and N. Suzuki, *J.C.S. Chem. Comm.*, 1977, 506.
39 A. Rioual, A. Deflandre, and J. Lemaire, *Canad. J. Chem.*, 1977, **55**, 3915.

Benzene itself can also induce the isomerization process at higher concentrations when an exciplex is involved. The photoisomerization of the enones (78) yielded mixtures of both the *cis*- and the *trans*-isomers regardless of the nature of the starting material.[40]

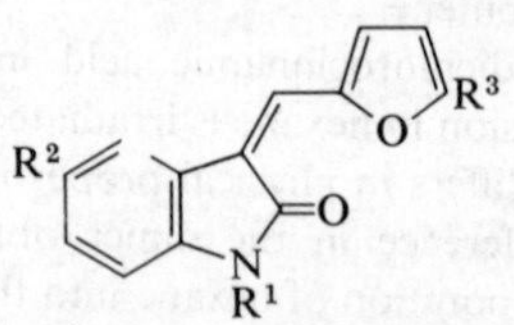

(78) R^1 = H or Ac; R^2 = H or NO_2;
R^3 = H, Me, Cl, Br, I, MeS, CO_2Me, Me_2N, NO_2,
o-$NO_2C_6H_4$ or m-$NO_2C_6H_4$

Koch *et al.*[41] previously reported the photochemical ring contraction of the imino-ether (79) to the carbamate (80). The reaction was established as a singlet-state process, perhaps involving a biradical intermediate (81) (a Norrish Type I process). A further examination of this reaction has sought to establish the sterochemical consequence of the photochemical rearrangement of the keto-imino-ethers (82).[42] The rearrangement affords the ring-contracted products (83) from (82a) and (84) from (82b). The reaction is highly stereospecific with

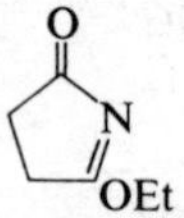

(79)

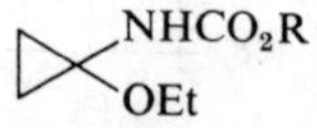

(80)

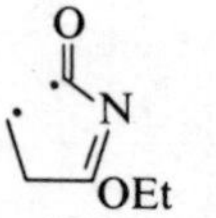

(81)

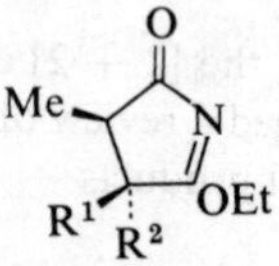

(82) a; R^1 = Me, R^2 = H   (83) a; R^1 = $NHCO_2Me$, R^2 = OEt
     b; R^1 = H, R^2 = Me   b; R^1 = OEt, R^2 = $NHCO_2Me$

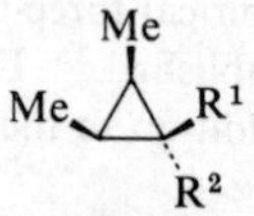

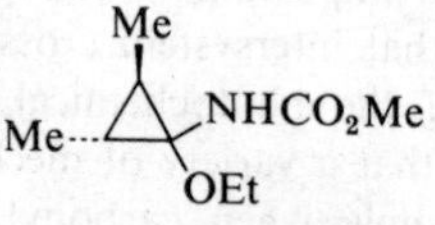

(84)

respect to C-4 of (82). The synthetic potential of this ring-contraction process has been further exploited.[43] Thus irradiation of the keto-imino-ethers (85) and (86) in THF affords the ring-contracted isocyanates (87) and (88).

[40] J. Stetinova, J. Kovac, J. Sura, and M. Dandarova, *Coll. Czech. Chem. Comm.*, 1977, **42**, 2201 (*Chem. Abs.*, 1977, **87**, 151 450).
[41] T. H. Koch, R. J. Sluski, and R. H. Moseley, *J. Amer. Chem. Soc.*, 1973, **95**, 3957.
[42] J. M. Burns, M. E. Ashley, G. C. Crockett, and T. H. Koch, *J. Amer. Chem. Soc.*, 1977, **99**, 6924.
[43] G. C. Crockett and T. H. Koch, *J. Org. Chem.*, 1977, **42**, 2721.

(85) a; $n = 1$ (86)   (87)   (88)
b; $n = 2$
c; $n = 3$

The lactone (89) undergoes photochemical rearrangement to afford the product (90) by a path which is formally akin to the rearrangement reactions of enones.[44] Two other products (91) and (92) are also formed in the reaction. The former product involves a straightforward [2 + 2] dimerization but the second is unusual since the reaction was carried out in anhydrous t-butyl alcohol.

(89)   (90)   (91)   (92)

The hydration of the enone could arise by the photoaddition of adventitious water. The enone (93) undergoes photochemical skeletal rearrangement to afford the normal lumi-product (94) by selective $n\pi^*$ ($\lambda > 300$ nm) excitation.[45] This rearrangement is also brought about by triplet sensitization. Other products are also produced under direct or sensitized irradiation. These have been identified as the isomeric enones (95) in which the allyl side-chain has also isomerized.[45] Excitation of enone (93) by selective irradiation of the $\pi\pi^*$ band with 254 nm light also yielded the products (94) and (95) previously mentioned, but an

(93)   (94)   (95) a; $R^1 = Me$, $R^2 = H$ (96)
b; $R^1 = H$, $R^2 = Me$

additional product (96) was also isolated. This compound arises by a hydrogen abstraction path involving abstraction by the α-C of the enone system (Scheme 6). Cyclizations of this type are not uncommon and have been reported previously, as for example in the conversion of (97) into (98).[46]

[44] V. Ferrer, J. Gomez, and J.-J. Bonet, *Helv. Chim. Acta*, 1977, **60**, 1357.
[45] F. Nobs, U. Burger, and K. Schaffner, *Helv. Chim. Acta*, 1977, **60**, 1607.
[46] J. Gloor, K. Schaffner, and O. Jeger, *Helv. Chim. Acta*, 1971, **54**, 1864.

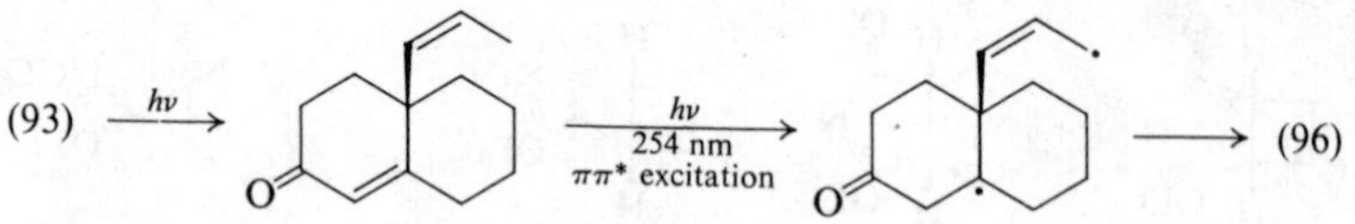

Scheme 6

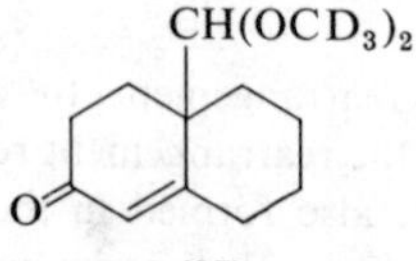

(97)

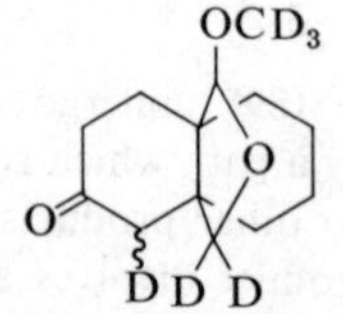

(98)

Direct irradiation of the lactone (99) in the presence of cyclohexane or toluene affords the two products (100) and (101) in equal quantities.[47] Although no definitive information concerning the pathway for the addition reaction was obtained it is not unreasonable to assume that addition occurs *via* a $\pi\pi^*$ excited state. The exact nature of the multiplicity of this state is uncertain since in the presence of quencher, although the disappearance of (99) is not quenched, the products (100) and (101) are not formed. Many products are produced in this reaction and it may be that the reaction of the S_1 state of (99) is faster with quencher than with the solvent. When (99) is irradiated under triplet sensitization conditions a different mechanism is involved. This is an example of chemical sensitization where the excited sensitizer (acetophenone or benzophenone) abstracts hydrogen from the solvent and then transfers a hydrogen from the ketyl radical (102) to the lactone. Independently, Toder *et al.*[48] have shown that the addition of cyclohexane to (99) gives the same two products (101) and (102)

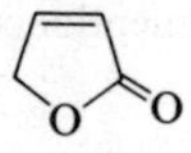

(99) 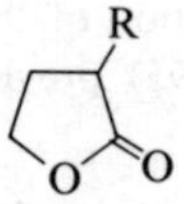   (100) 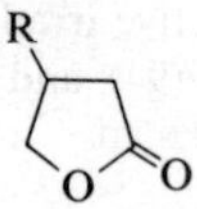   (101) 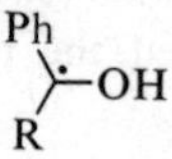   (102) R = Me or Ph

R = cyclohexyl or benzyl

in virtually equal amounts. They have, however, demonstrated by the use of deuteriated cyclohexane that the formation of (100) results from hydrogen (deuterium) abstraction by the β-carbon of the enone, followed by radical combination. Agosta and his co-workers have further examined the photochemical reactions of cyclopentenones where hydrogen abstraction occurs at the β-carbon of the excited-state cyclopentenone.[49] In particular they have examined the photochemistry of the cyclopentenones (103) and (104). The products from the irradiation are shown in Scheme 7. A detailed analysis of the stereochemical effects in play on the resultant biradicals [*e.g.* (105) and (106)] produced by the hydrogen abstraction step is reported.

[47] T. W. Flechtner, *J. Org. Chem.*, 1977, **42**, 901.
[48] B. H. Toder, S. J. Branca, and A. B. Smith, *J. Org. Chem.*, 1977, **42**, 904.
[49] S. Ayral-Kaloustian, S. Wolff, and W. C. Agosta, *J. Amer. Chem. Soc.*, 1977, **99**, 5984.

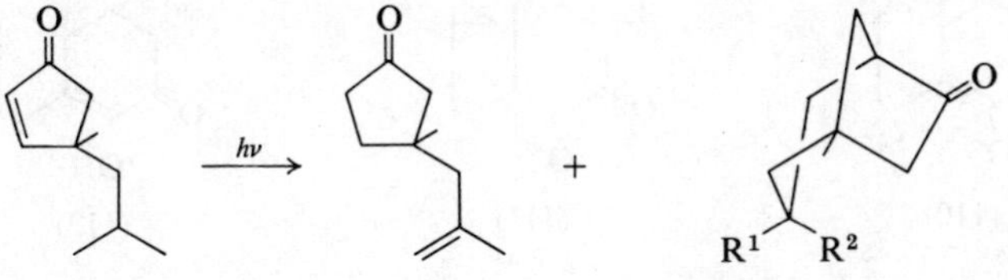

(103) (31%) +

a; R¹ = H, R² = Me (31%)
b; R¹ = Me, R² = H (3%)

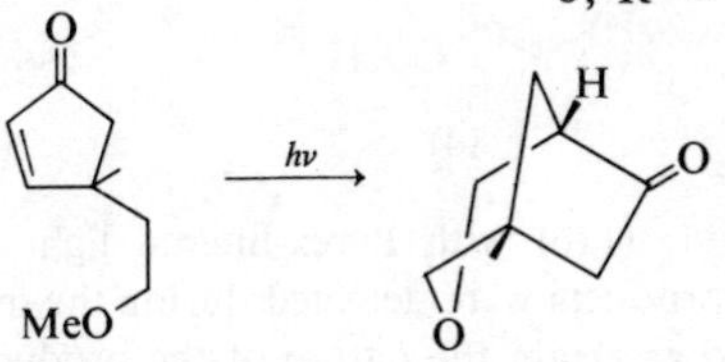

MeO
(104) (17%)

Scheme 7

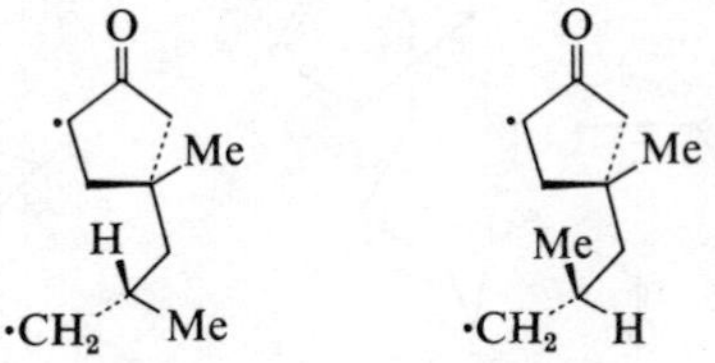

(105) (106)

The methoxycholestenone (107) undergoes an intramolecular hydrogen abstraction reaction to yield a biradical (108) which is capable of cyclizing either at the carbonyl carbon or at the β-carbon of the enone to yield the three principal products (109)—(111).[50] These products arise from the excited singlet state of the cholestenone (107). A fourth product (112) could be formed as a result of oxidation. The photoisomerization of *E*-4-methylpent-2-enoic acid yields the deconjugated isomer (113) by the normal route of intramolecular hydrogen abstraction by the carbonyl group.[51] Deconjugation is also reported for the acid (114), which yields (115).

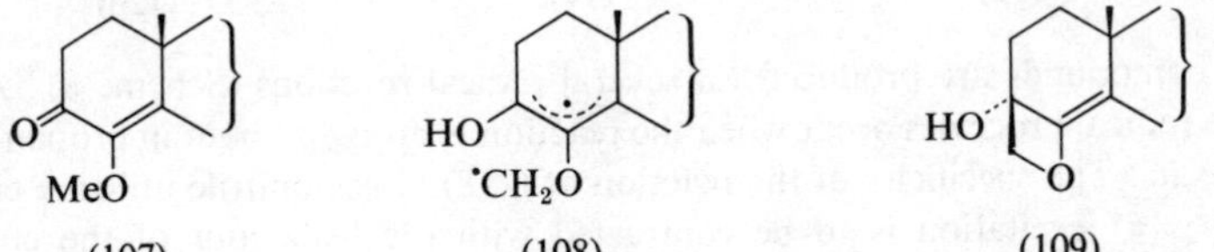

MeO ·CH₂O HO

(107) (108) (109)

[50] A. Feigenbaum and J.-P. Pete, *Bull. Soc. chim. France*, 1977, 351.
[51] J. M. Biot, D. De Keukeleire, and M. Verzele, *Bull. Soc. chim. belges*, 1977, **86**, 973 (*Chem. Abs.*, 1978, **88**, 120 573).

(110)   (111)   (112)

(113)   (114)   (115)

Irradiation of the ketone (116) with Pyrex-filtered light gave one isolable compound although two products were detected during the irradiation. Insufficient data are available to ascertain the nature of the product although a cage structure of the type (117) is thought likely.[52] When circularly polarized light was used to investigate the chiroptical properties of the ketone (116) no enantiomeric enhancement was detected in recovered starting material.

(116)   (117)

The epoxyacetylenic ketone (118) undergoes γ-C—O fission when excited in its $n\pi^*$ or $\pi\pi^*$ bands.[53] The biradical (119) produced undergoes rearrangement by methyl migration to yield the 1,5-diketone (120) when the reaction is carried out in acetonitrile or benzene. However, when cyclopentane is used as solvent,

(118)   (119)   (120)

many compounds are produced *via* several radical reactions (Scheme 8). Analogous radical processes occur when the reaction is brought about in propan-2-ol or dioxan. The specificity of the reaction of (118) in acetonitrile utilizing either $n\pi^*$ or $\pi\pi^*$ excitation is to be contrasted with the behaviour of the epoxyolefinic ketone (121). When this compound is excited by 254 nm irradiation, ($\pi\pi^*$) fission of the γ,δ-C—C bond is predominant yielding (122) whereas $n\pi^*$

[52] J. F. Nicoud, C. Eskenazi, and H. B. Kaga, *J. Org. Chem.*, 1977, **42**, 4270.
[53] H. Eichenberger, H. R. Wolf, and O. Jeger, *Helv. Chim. Acta*, 1977, **60**, 743.

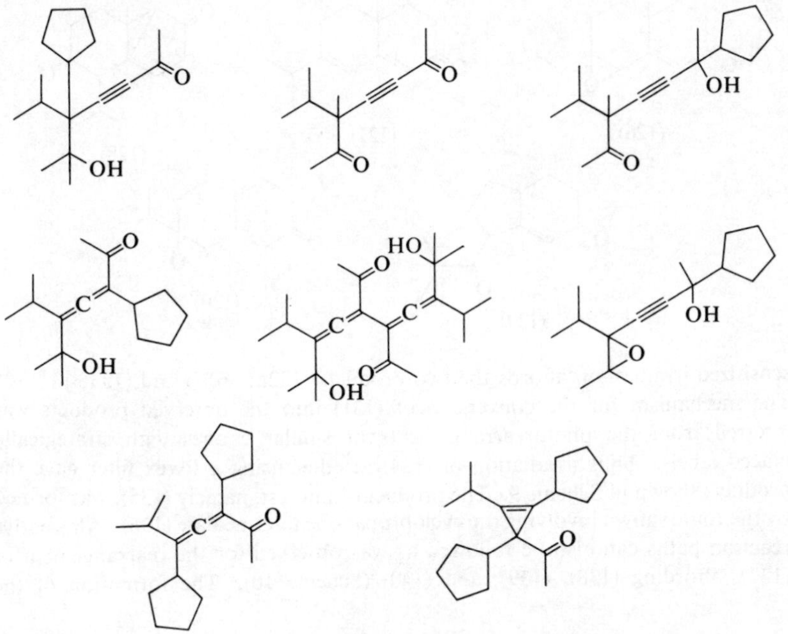

Scheme 8

excitation affords (123), (124), and (125).[54] Specific reactions are also reported for the epoxy-enone (126) depending upon the excitation mode. Thus $n\pi^*$ excitation of (126) results in deoxygenation with the formation of the dienone (127).[55] A different result is found when (126) is irradiated into the $\pi\pi^*$ band when isomerization to (128) and (129) occurs. The formation of (128) and (129) must involve ring opening of the epoxide to the biradical (130). Migration of the isopropyl group in the radical (130) affords an enone which on secondary photolysis can rearrange to (128). Migration of a ring C—C bond in (130) will yield (129). The triplet state of the enone (131a), populated by either direct or

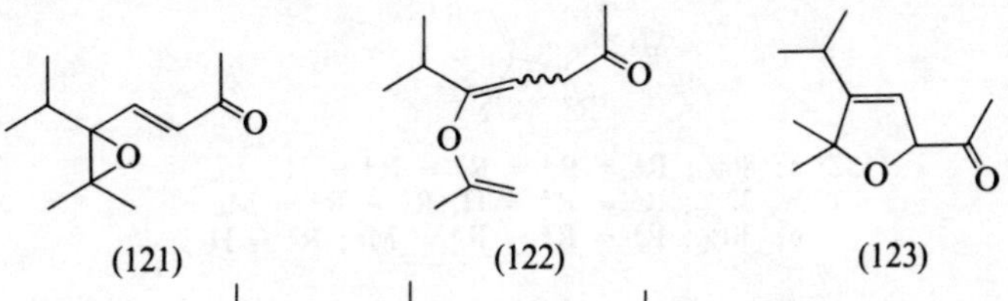

(121) (122) (123)

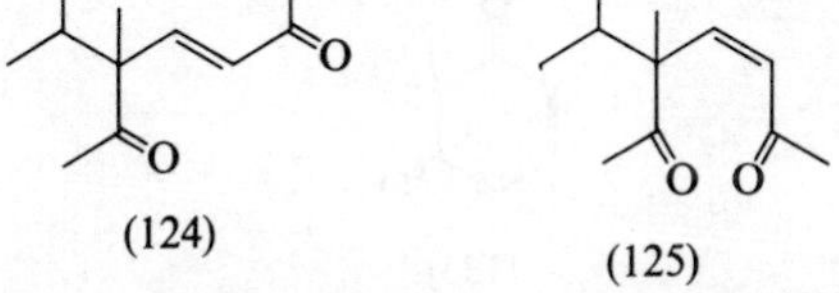

(124) (125)

[54] H. Eichenberger, H. R. Wolf, and O. Jeger, *Helv. Chim. Acta*, 1976, **59**, 1253.
[55] A. K. Dey and H. R. Wolf, *Helv. Chim. Acta*, 1978, **61**, 626.

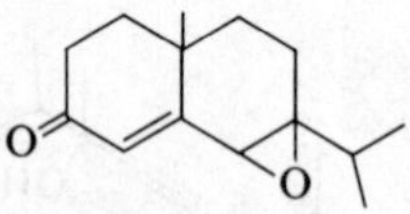

(126) 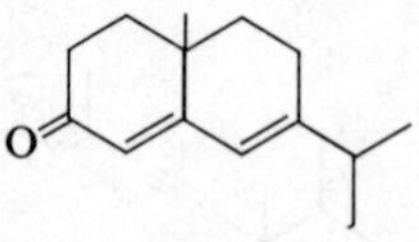   (127)

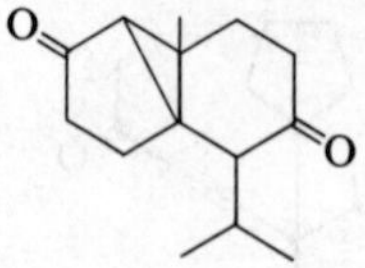

(128)

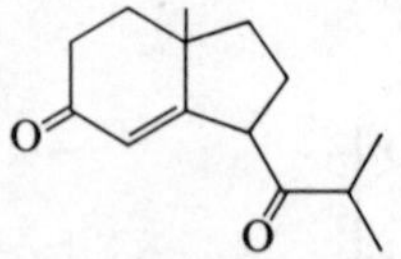

(129) 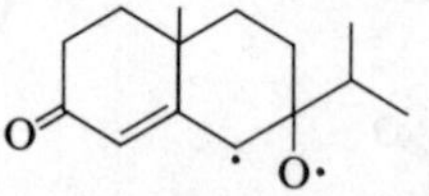   (130)

sensitized irradiation, affords the two products (132a; 86%) and (133; 14%).[56]
The mechanism for the conversion of (131) into the observed products was
inferred from the photorearrangements of similar enones with strategically
placed labels. Thus irradiation of (134) in ether using a Pyrex filter gave the
products shown in Scheme 9. The product of interest, namely (135), was formed
by the route shown involving the cyclopropanone intermediate (136). Alternative
reaction paths can also be followed as was observed for the rearrangement of
(137), affording (138), (139), and (140) (Scheme 10). The formation of the

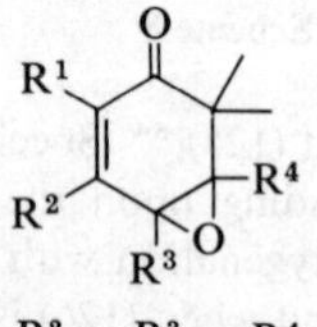

(131) a; $R^1 = R^2 = R^3 = R^4 = H$
b; $R^1 = R^4 = H; R^2 = R^3 = Me$
c; $R^1 = R^3 = R^4 = Me, R^2 = H$

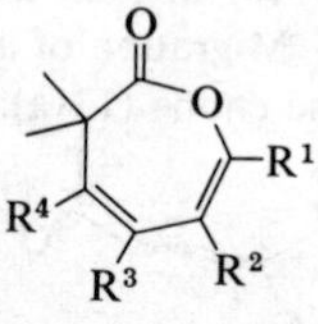

(132) a; 86%; $R^1 = R^2 = R^3 = R^4 = H$
b; 22%; $R^1 = R^4 = H, R^2 = R^3 = Me$
c; 30%; $R^1 = R^3 = R^4 = Me; R^2 = H$

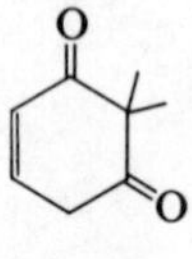

(133)

[56] H. Hart, C.-T. Peng, and E. Shih, *Tetrahedron Letters*, 1977, 1641.

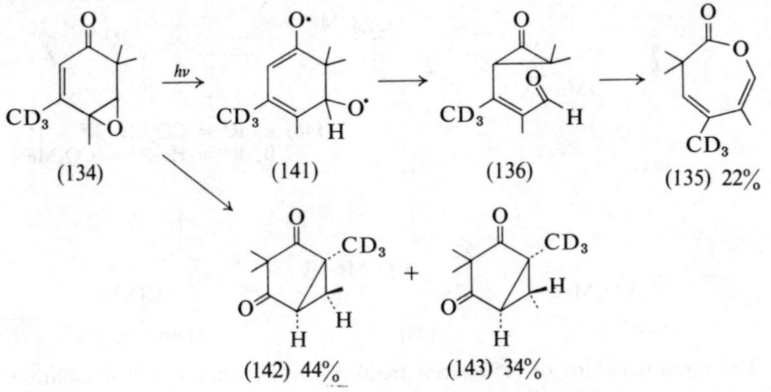

(134) (141) (136) (135) 22%

(142) 44% (143) 34%

Scheme 9

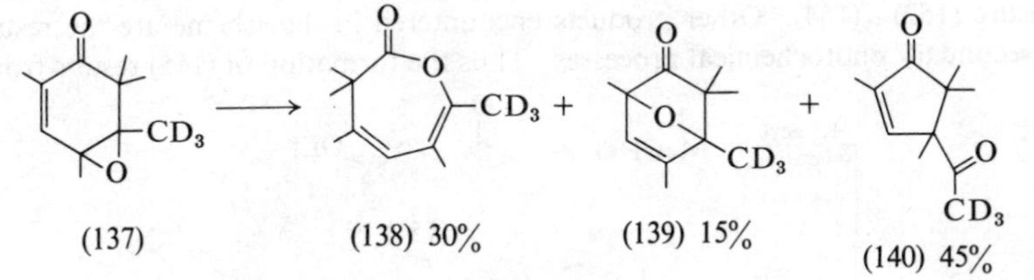

(137) (138) 30% (139) 15% (140) 45%

Scheme 10

cyclohexene-1,3-dione (133) is the result of group migration within the intermediate (141) which is also an intermediate in the formation of the lactone products. These diones [*e.g.* (133)] are themselves photolabile and can be converted by secondary irradiation into bicyclic diones. Products of this type [*e.g.* (142) and (143)] are encountered in the irradiation of (134). Other examples of the conversion of (131b and c) into the enol lactones (132b and c) were also reported.[57] Hart and Shih [58] have previously reported other reactions of enone epoxides closely related to these.

The di-π-methane reaction still excites considerable activity throughout the photochemical world. One interesting report [59] in the past year has dealt with the different modes of migration, dependent upon the multiplicity, which have been recognized in the irradiation of the ester (144) and (145) as a mixture of *cis*- and *trans*-isomers. Acetone sensitization (300 nm) leads to the formation of the products (146a) and (146b) where the biradical (147) has collapsed to afford the vinylcyclopropane in conjugation with the ester group.[60, 61] However, direct irradiation of the ester (144), (145) affords the chrysanthemate (148) as well as several other isomeric products, among which is (149). There is, however, no trace of (146a) or (146b) in the products of direct irradiation.

[57] H. Hart, C.-T. Peng, and E. Shih, *J. Org. Chem.*, 1977, **42**, 3635.
[58] H. Hart and E. Shih, *J. Org. Chem.*, 1976, **41**, 3377.
[59] P. Baeckström, *J.C.S. Chem. Comm.*, 1976, 476.
[60] M. J. Bullivant and G. Pattenden, *J.C.S. Perkin I*, 1976, 256.
[61] M. J. Bullivant and G. Pattenden, *Pyrethrum Post*, 1976, **13**, 110 (*Chem. Abs.*, 1977, **87**, 5118).

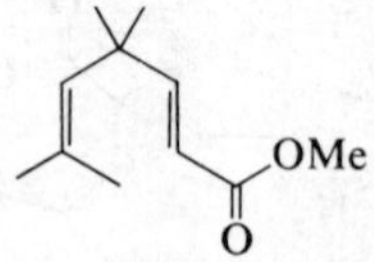

(144) 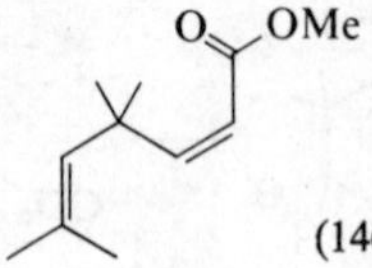   (145)

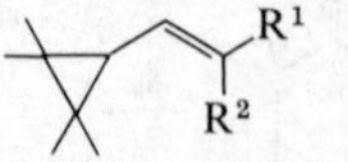

(146) a; R¹ = CO₂Me, R² = H
      b; R¹ = H, R² = CO₂Me

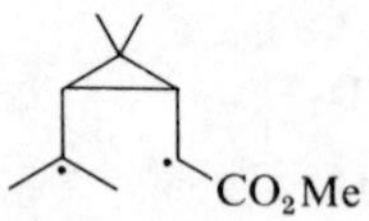

(147) 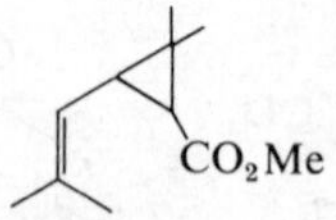   (148) 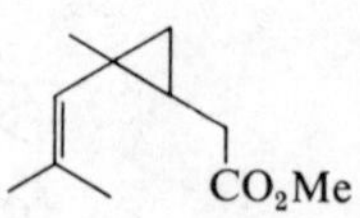   (149)

The photoreactions of (150) arise from S_1 or T_2 upon direct irradiation.[62] The products produced from such an irradiation are shown in Scheme 11 and arise from the 1,3-acyl migrations (151), Norrish Type II fission, or ring closure (152)—(154). Other products encountered in the scheme are the result of secondary photochemical processes. Thus the formation of (155) results from

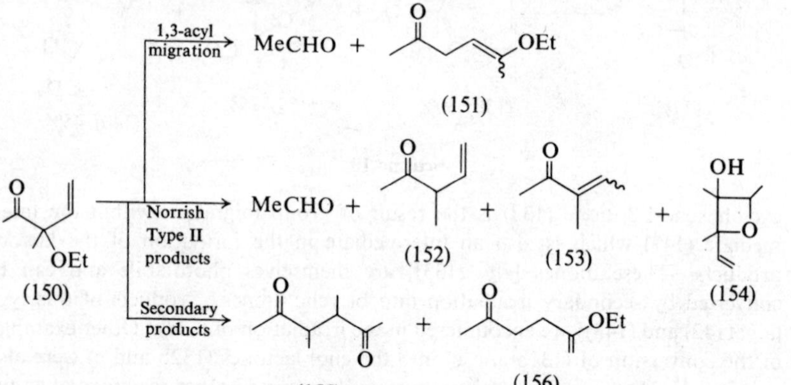

Scheme 11

the reaction of the primary photoproduct (152) with acetaldehyde (itself a primary product). The enone (156) also is a secondary product and arises from the further rearrangement of the product (151). The 1,3-acyl migration encountered in the irradiation of (150) is analogous to that reported for the alcohol (157) in its direct or xanthene-sensitized conversion into the diketone (158).[63, 64]

A photo-CIDNP study of the rearrangement of $\beta\gamma$-unsaturated ketones has shown that the α-cleavage process occurs from the excited $n\pi^*$ singlet state if the alkene part of the system is substituted with a phenyl group at the γ-position [*e.g.* (159)] whereas the cleavage reaction occurs from the triplet state if methyl groups are the substituents at that site [*e.g.* as in (160)]. The authors[65] suggest

[62] K. G. Hancock and P. L. Wylie, *J. Org. Chem.*, 1977, **42**, 1850.
[63] K. G. Hancock, P. L. Wylie, and J. T. Lan, *J. Amer. Chem. Soc.*, 1977, **99**, 1149.
[64] K. G. Hancock, P. L. Wylie, and J. T. Lan, *Tetrahedron Letters*, 1974, 4149.
[65] A. J. A. van der Weerdt, H. Cerfontain, J. P. M. van der Ploeg, and J. A. den Hollander, *J.C.S. Perkin II*, 1978, 155.

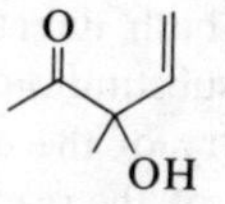

(157) 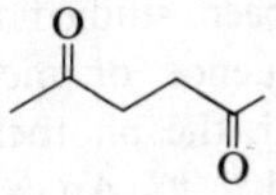 (158) 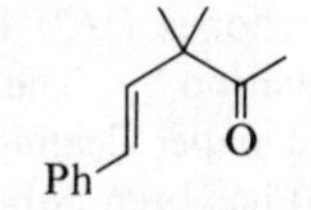 (159) 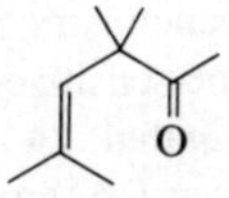 (160)

that the difference in excited-state reactivity arises as a result of intramolecular quenching. Another report by Cerfontain *et al.*[66] describes the influence of substitution upon the sensitized photoreactions of the enones (161).[66] All the enones undergo *Z–E* isomerization when they are triplet-sensitized with acetophenone. Some of the enones (161b, d, f, and g) and (162) also undergo the oxa-di-π-methane reaction and yield cyclopropane derivatives. Interestingly the authors [66] suggest that the influence of bulky groups could lead to a conformation of the enone from which an oxa-di-π-methane reaction can occur. The

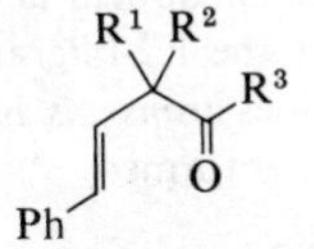

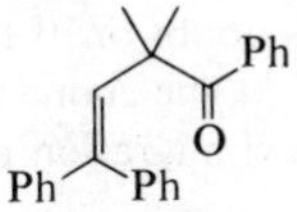

(161) a; $R^1 = R^2 = R^3 = Me$
b; $R^1 = R^2 = Pr^i$, $R^3 = Me$
c; $R^3 = Me$, $R^1-R^2 = (CH_2)_4$
d; $R^1 = H$, $R^2 = Pr^i$, $R^3 = Me$
e; $R^1 = H$, $R^2 = Me$, $R^3 = Bu^t$
f; $R^1 = R^2 = Me$, $R^3 = Bu^t$
g; $R^1 = R^2 = Me$, $R^3 = Ph$

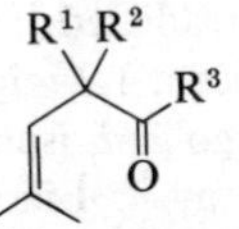

(162)

(163) $R^1 = R^3 = Me$, $R^2 = H$
$R^1 = R^2 = R^3 = Me$
$R^1 = R^2 = Me$, $R^3 = Pr^i$

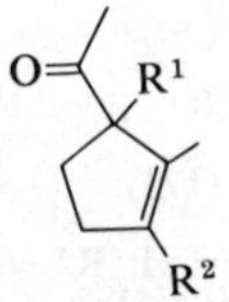

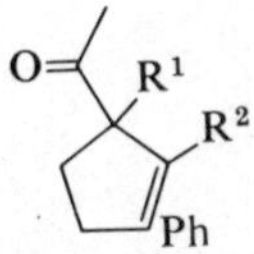

(164) a; $R^1 = R^2 = H$
b; $R^1 = Me$, $R^2 = H$
c; $R^1 = H$, $R^2 = Me$
d; $R^1 = R^2 = Me$

(165) a; $R^1 = R^2 = H$
b; $R^1 = Me$, $R^2 = H$

66 A. J. A. van der Weerdt and H. Cerfontain, *Rec. Trav. chim.*, 1977, **96**, 247.
67 A. J. A. van der Weerdt and H. Cerfontain, *J.C.S. Perkin II*, 1977, 1357.
68 H.-U. Gonzenbach, I.-M. Tegmo-Larsson, J.-P. Grosclaude, and K. Schaffner, *Helv. Chim. Acta*, 1977, **60**, 1091.
69 K. Schaffner, *Pure Appl. Chem.*, 1973, **33**, 329; *Tetrahedron*, 1976, **32**, 641; I.-M. Tegmo-Larsson, H.-U. Gonzenbach, and K. Schaffner, *Helv. Chim. Acta*, 1976, **59**, 1376.

photochemistry of the enones (163) has been studied using both direct and acetone-sensitized irradiation.[67] The influence of methyl substitution was investigated. A detailed paper dealing with the photochemistry of the cyclo-pentenes (164) and (165) has been published.[68, 69] An example of the reactions encountered in this sytem are outlined in Scheme 12. The chrysene-sensitized

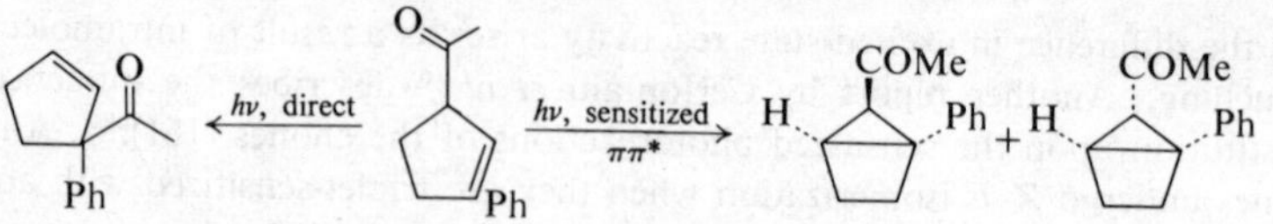

Scheme 12

irradiation of the keto-olefin (166) results in an oxa-di-π-methane reaction affording the two products (167a) and (167b) where loss of configuration has occurred at the methanic carbon. The lack of optical activity in the products clearly casts doubt on the concertedness of the 1,2-migration reaction in such systems.[70, 71] The enone (166) also undergoes _trans–cis_ isomerization to (168) and a 1,3-acyl migration product (169) is also formed.

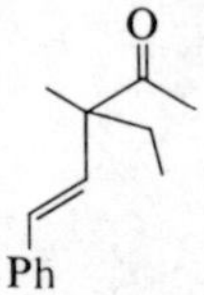

(+)-(166)

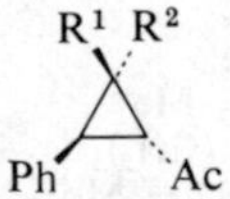

(167) a; R¹ = Et, R² = Me
b; R¹ = Me, R² = Et

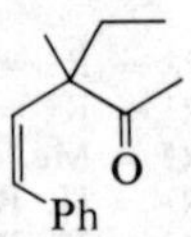

(168)

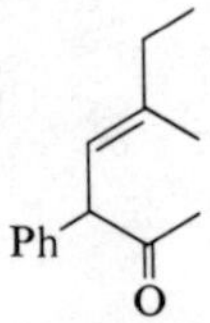

(169)

The vinylcyclopropanes (170a and b) are produced when the allylcyclo-pentenones (171a and b) are irradiated.[72, 73] The reactions occur _via_ a radical path where hydrogen undergoes a 1,2-migration. Interestingly, the substituted derivatives (171c and d) undergo _Z–E_ isomerization perhaps as a result of free-rotor dissipation of triplet energy.[74] Efficient cyclization also is reported when the central atom is _gem_-dimethyl substituted as in (171e).

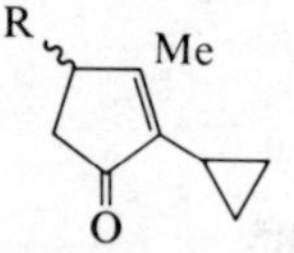

(170)

a; R = H, OMe, or OAc
b; R = H

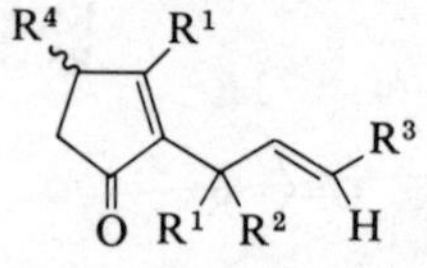

(171)

a; R¹ = Me, R² = R³ = H, R⁴ = H, OMe, or AcO
b; R¹ = R² = R³ = R⁴ = H
c; R¹ = Me, R² = H, R³ = Me, R⁴ = OMe or OAc
d; R¹ = Me, R² = H, R³ = Et, R⁴ = OMe or OAc
e; R¹ = R³ = R⁴ = H, R² = Me

[70] W. G. Dauben, G. Lodder, and J. D. Robbins, _Nouveau J. Chim._, 1977, **1**, 243 (_Chem. Abs._, 1977, **87**, 151 614).
[71] W. G. Dauben, G. Lodder, and J. D. Robbins, _J. Amer. Chem. Soc._, 1976, **98**, 3030.
[72] M. J. Bullivant and G. Pattenden, _Pyrethrum Post_, 1976, **13**, 100 (_Chem. Abs._, 1977, **86**, 188 775).
[73] M. J. Bullivant and G. Pattenden, _J.C.S. Perkin I_, 1976, 249.
[74] S. S. Hixson, P. S. Mariano, and H. E. Zimmerman, _Chem. Rev._, 1973, **73**, 531.

The enone (172), produced as the primary photochemical product from the bicyclic ketone (173) by irradiation with 254 nm light in acetonitrile solution, is converted by acetone-sensitized irradiation into the isomeric bicyclic ketone (174). It is likely that this second product arises by a di-π-methane process involving phenyl migration within the biradical intermediate (175).[75] The reactivity of the triplet state in this reaction is as a result of the absence of a free-rotor effect to dissipate triplet energy.[74] Hixson[76] had previously found that the triplet state of (176) was unreactive as a result of a free-rotor effect and only the singlet excited state gave rise to the di-π-methane product (177).

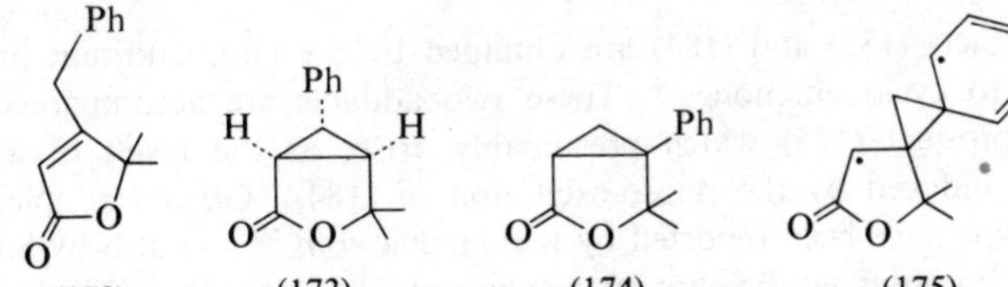

(172)   (173)   (174)   (175)

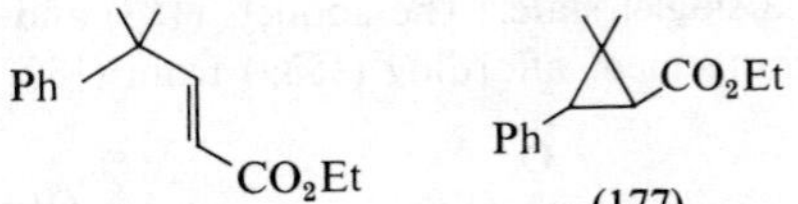

(176)   (177)

Previous studies by Padwa[77] and his group have shown that the photochemical reactions encountered in the furanone systems (178) arise from the triplet state when the reactions are carried out in benzene (Scheme 13). If the reactions were carried out in methanol the products (179) and (180) (Scheme 13) were isolated. The most striking feature of the change in solvent is the change in migratory aptitude encountered with the various aryl functions. The authors[78] interpret

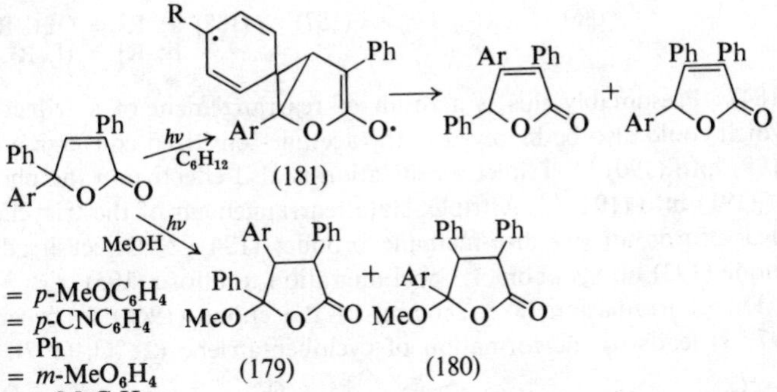

(178) a; Ar = p-MeOC$_6$H$_4$
      b; Ar = p-CNC$_6$H$_4$
      c; Ar = Ph
      d; Ar = m-MeOC$_6$H$_4$
      e; Ar = p-MeC$_6$H$_4$

(179)   (180)

Scheme 13

75 P. C. M. van Noort and H. Cerfontain, *Tetrahedron Letters*, 1977, 3899.
76 S. S. Hixson and J. C. Tausta, *Tetrahedron Letters*, 1974, 2007.
77 A. Padwa and D. Dehm, *J. Amer. Chem. Soc.*, 1975, **97**, 4779.
78 A. Padwa, T. Brookhart, D. Dehm, G. West, and G. Wubbels, *J. Amer. Chem. Soc.*, 1977, **99**, 2347.

this behaviour in terms of a biradical mechanism in benzene (a di-π-methane-like process) involving (181) and a zwitterionic mechanism in methanol involving (182). Such a zwitterion would be readily trapped by solvent to afford the observed products.

(182)

The adducts (183) and (184) are obtained by the photoaddition of ethoxy-acetylene to cyclopentenone.[19] These two adducts are accompanied by the isomeric product (185) which presumably arises as the result of a 1,3-aryl migration induced by the photo-excitation of (184). Other examples of this reaction type have been reported by Kulomzina *et al.*[19] and also by Soulie and Pouet [17] who observed the photoconversion of (186) into (187). These reactions most likely arise from a singlet state. The adducts (183) and (184) undergo an oxa-di-π-methane rearrangement affording (188a) from (183) and (188b) from

(183)   (184)   (185)

(186)   (187)   (188) a; R¹ = OEt, R² = H
                                       b; R¹ = H, R² = OEt

(184). Presumably this is a result of rearrangement of a triplet excited state, which could also be involved in the acetone-sensitized conversion of the adduct (189) into (190).[18] Triplet sensitization is also effective in the photoconversion of (191) into (192).[17] A triplet-state rearrangement of the tricyclic enone (193) also affords an oxa-di-π-methane product (194).[79] Direct irradiation of the enone (193) brings about 1,3-acyl migration to afford (195).

Direct irradiation (λ > 280 nm) of the enone (196a) in benzene–methanol (97:3) leads to the formation of cycloheptatriene (22%), (197), and (198) in

(189)   (190)   (191) R = H or Me   (192)

[79] K. Hirao, S. Unno, H. Miura, and O. Yonemitsu, *Chem. and Pharm. Bull. (Japan)*, 1977, **25**, 3345 (*Chem. Abs.*, 1978, **88**, 120 295).

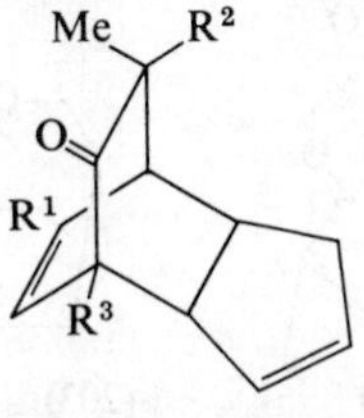

(193) $R^1 = R^3 = H$ or Me
$R^2 = H$ or OH

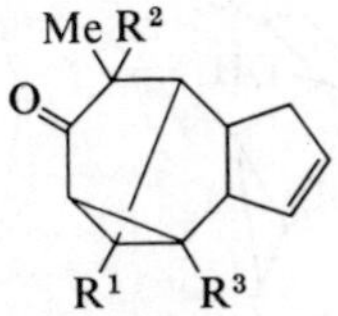

(194)

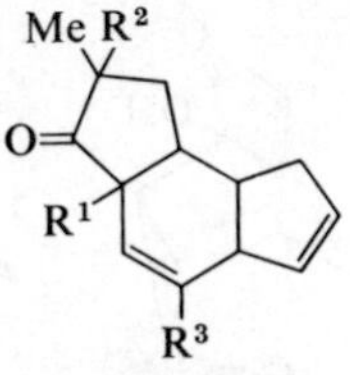

(195)

the yields shown.[80] The main product (198) arises from a 1,3-acyl migration, a reaction common in $\beta\gamma$-unsaturated ketones. Compound (197) arises from the decarbonylation of this principal product (198), whereas cycloheptatriene presumably arises *via* photoelimination of keten from (198); however, this is not the sole path for formation of the triene since it can be formed directly from the starting material (196a). Irradiation of (196b) under conditions identical with those used for (196a) leads to a high yield of a single product identified as (199) which again is the result of a 1,3-acyl migration. Although a biradical mechanism was considered for the formation of this product, one cannot rule out the possibility of a concerted process. However, the absence of fragmentation or other processes such as those encountered in the irradiation of (196a) suggests that other structural features may be important.[80]

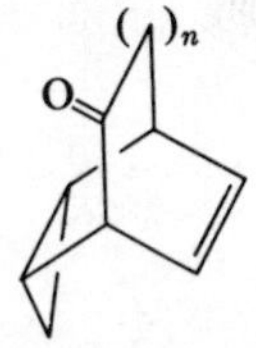

(196) a; $n = 1$
b; $n = 2$

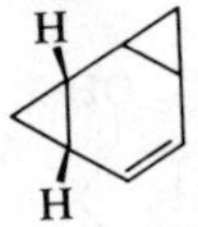

(197) 3—4%

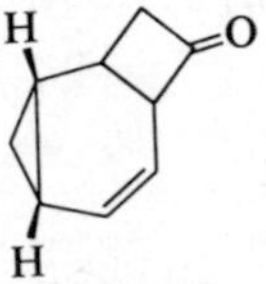

(178) 60%

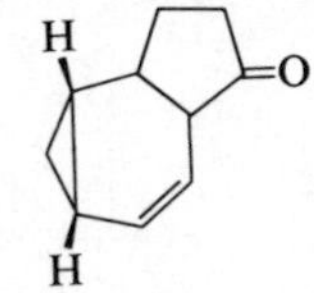

(199)

The dieneone (200) photorearranges in methanol by a path initially reminiscent of a di-π-methane process. However, divergence from the conventional path arises at the biradical intermediate (201) which ring-opens to a keten (202) trapped as the ester (203).[81] In acetic acid as reaction medium, the reaction follows a different path affording the diketone (204). This ketone is also photo-reactive in methanol and undergoes Norrish Type I fission, perhaps producing a biradical intermediate (205) which subsequently unzips to yield a keten trapped as (206).

The enone (207) undergoes photorearrangement and deoxygenation to the phenazine (208). The process by which this occurs could well be a variant of an oxa-di-π-methane reaction.[82] Full experimental details of the photochemical behaviour of (209) have been published.[83]

[80] I. M. Takakis and W. C. Agosta, *Tetrahedron Letters*, 1978, 531.
[81] T. Uyehara, Y. Kayama, M. Taniguchi, C. Kabuto, and Y. Kitahara, *Chem. Letters*, 1977, 569 (*Chem. Abs.*, 1977, **87**, 84 596).
[82] C. S. Panda, *Indian J. Chem.*, 1976, **14B**, 610 (*Chem. Abs.*, 1977, **86**, 139 932).
[83] J. R. Williams and H. Ziffer, *Org. Photochem. Synth.*, 1976, **2**, 65 (*Chem. Abs.*, 1977, **87**, 23 600).

(200)   (201)   (202)   (203)

(204)   (205)   (206)

(207)   (208)

(209)

A full account of the photochemistry of (210) and related compounds has been reported[84] supplementing a previous preliminary report.[85] The acetone-sensitized irradiation of (211) results in decarbonylation and the formation of the diene (212).[86] Irradiation of the oxazolinone (213) brings about decarbonylation and the formation of (214).[87] Changes in substitution affect the direction of the photochemical reaction and the oxazolinone (215) undergoes loss of CO_2 with the formation of a stabilized nitrile ylide. The photochemical double decarboxylation of the gibbane (216) affords the aromatic derivative (217) in an unspecified yield.[88] Voigt and Adam[89] report the useful photodecarboxylation of (218), also to yield (217; 50%).

[84]   J. R. Grunwell, D. L. Forest, and M. J. Sanders, *J. Org. Chem.*, 1977, **42**, 1142.
[85]   J. R. Grunwell, *Chem. Comm.*, 1969, 1437.
[86]   R. N. Warrener, R. A. Russell, and R. Y. S. Tan, *Tetrahedron Letters*, 1978, 1589.
[87]   M. R. Johnson and L. R. Sousa, *J. Org. Chem.*, 1977, **42**, 2439.
[88]   B. Voigt and G. Adam, E. Ger. P. 118 858, 1975 (*Chem. Abs.*, 1977, **86**, 171 662).
[89]   B. Voigt and G. Adam, E. Ger. P. 185 827, 1975 (*Chem. Abs.*, 1977, **86**, 190 277).

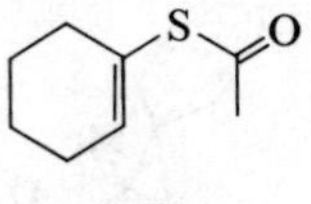

(210)

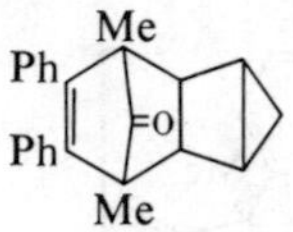

(211)

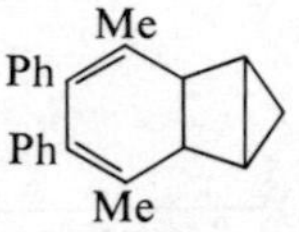

(212)

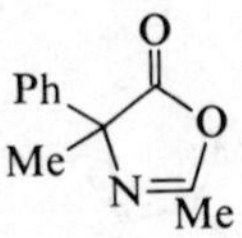

(213)

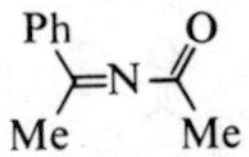

(214)

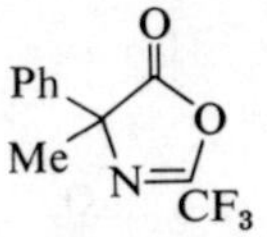

(215)

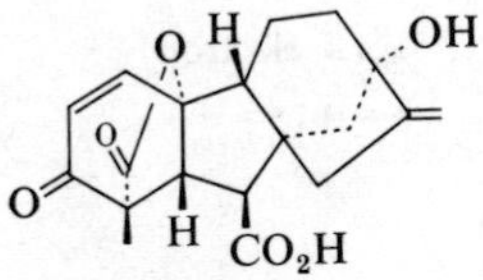

(216)

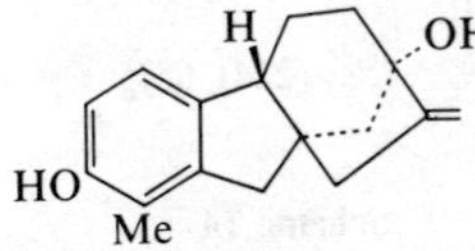

(217)

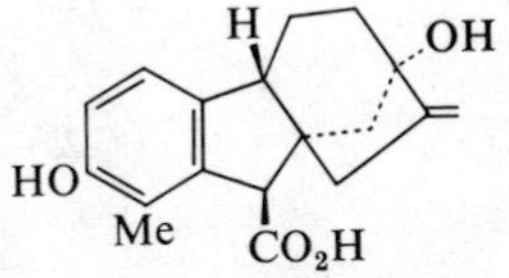

(218)

Hart *et al.*[57] have reported that the dienones (219) undergo valence-bond isomerization to yield (220) when irradiated (Corex filter) in ethereal solution. Sato *et al.*[90] have previously reported the valence-bond isomerization of (219a) into (220a), but under triplet-sensitized conditions. Direct irradiation at 254 nm of (221) in methylene chloride did, however, give (222).[90] Sensitized irradiation (β-acenaphthone) of (221) induces the molecule to follow a different path whereby a 1,4-phenyl migration takes place as shown in Scheme 14 to afford the two products (223) and (224).[90] This report has reappeared in another journal.[91]

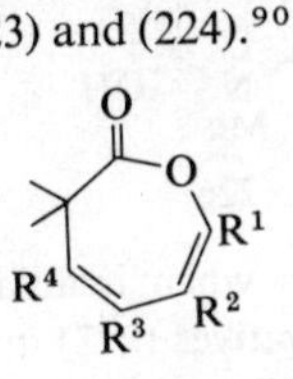

(219)

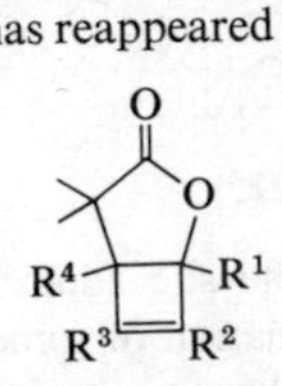

(220) a; $R^1 = R^2 = R^3 = R^4 = H$
b; $R^1 = R^4 = H$, $R^2 = R^3 = Me$
c; $R^1 = R^3 = R^4 = Me$, $R^2 = H$

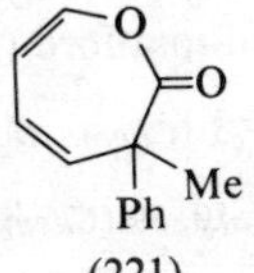

(221)

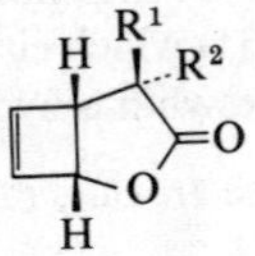

(222) a; $R^1 = Me$, $R^2 = Ph$ ⎫
b; $R^1 = Ph$, $R^2 = Me$ ⎬ 44% yield

[90] K. Sato, H. Hagiwara, H. Uda, M. Sato, and N. Horada, *J. Amer. Chem. Soc.*, 1976, **98**, 8281.

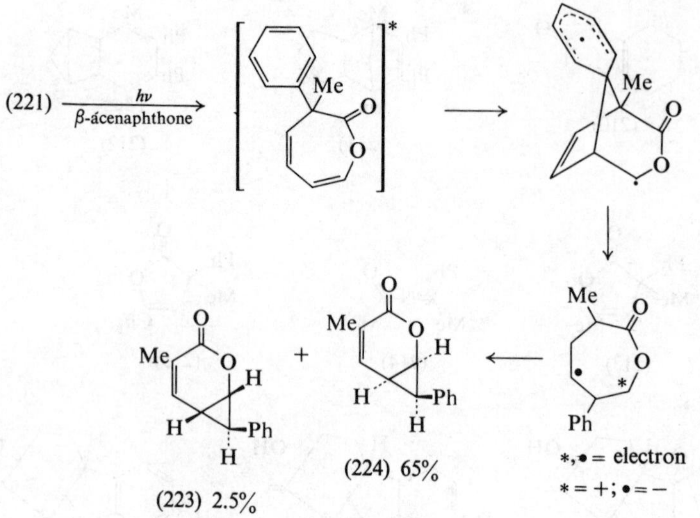

Scheme 14

3 Photoreactions of Thymines *etc.*

A review dealing with the photohydration of pyrimidines has been published.[92] The photohydration of 1,3-dimethyluracil (225) yielding (226) has been suggested as a useful actinometer for 254 nm irradiation. The quantum yield was measured, relative to potassium ferrioxalate, as 0.013.[93] *N*-Methylpyrid-4-one undergoes

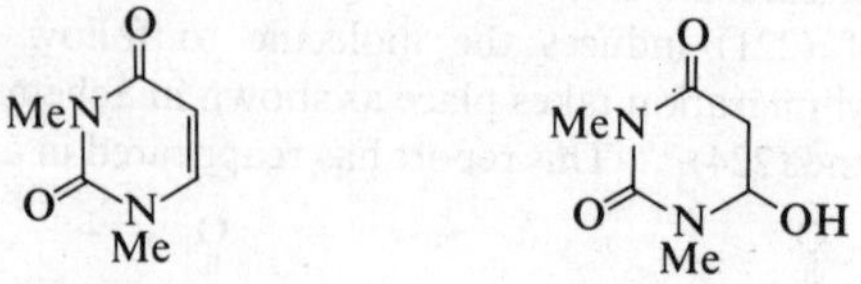

(225)   (226)

photohydration on the site adjacent to nitrogen when irradiated in aqueous solution.[94] The irradiation of some uridine derivatives (227) in methanol leads to the formation of three photoproducts.[95] The first (228) is produced by the photochemical addition of solvent in a route reminiscent of the cationic pathway involved in photohydration. The reduction products (229) and (230) could well arise by a triplet path whereby the photoexcited uridine abstracts hydrogen from solvent-yielding radical (231) which either abstracts a further hydrogen atom to yield (229) or else combines with a hydroxymethyl radical to afford (230).

[91] K. Sato, H. Hagiwara, and H. Uda, *Chem. Letters*, 1977, 175 (*Chem. Abs.*, 1977, **86**, 170 399).
[92] S. Y. Wang, *Jerusalem Symp. Quantum Chem. Biochem.*, 1977, **10**, 39 (*Chem. Abs.*, 1978, **88**, 120 121).
[93] N. Numao, T. Hamada, and O. Yonemitsu, *Tetrahedron Letters*, 1977, 1661.
[94] C. S. Shim and G. S. Hammond, *Hanguk Saenghwa Hakhoe Chi*, 1976, **9**, 181 (*Chem. Abs.*, 1977, **87**, 101 637).
[95] J.-L. Fourrey and P. Jouin, *Tetrahedron Letters*, 1977, 3397.

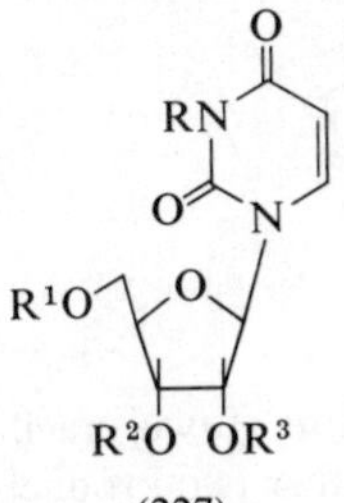

(227) 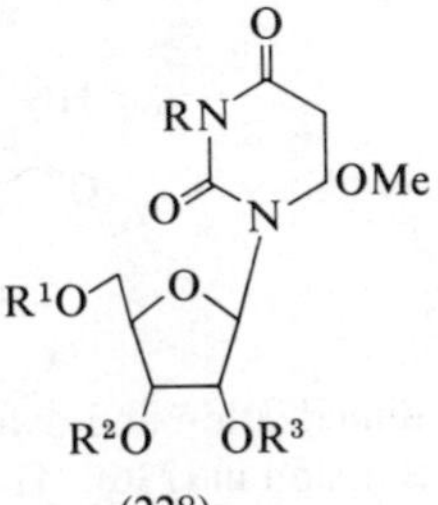   (228)

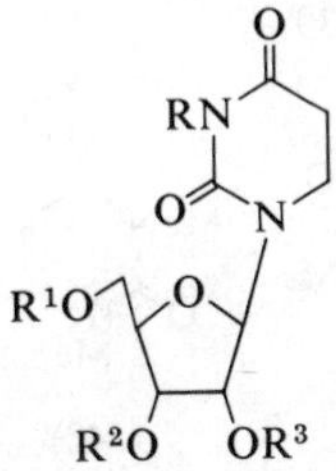

(229) 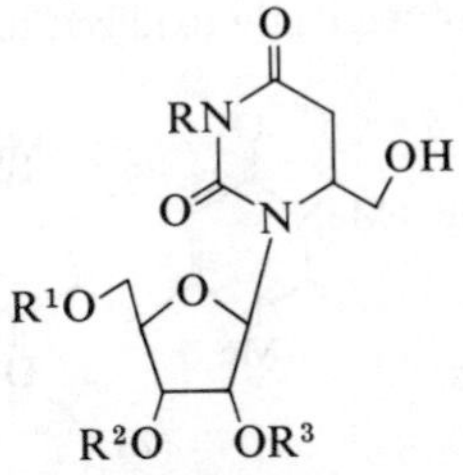   (230) 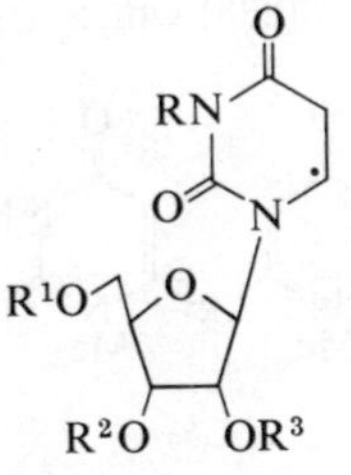   (231)

A study of the photochemical dimerization of cytosine and its derivatives has been published.[96] Tetramethyluracil (232) undergoes photochemical dimerization in aqueous solution to afford the [2 + 2] dimers (233) and (234).[97] The dimerization (Φ 1.5 $\times$ 10^{-2}) is not quenched by piperylene, and so is probably a singlet process. In non-aqueous media, dimerization also arises from the singlet state but the ratio of the dimers is independent of the solvent polarity.

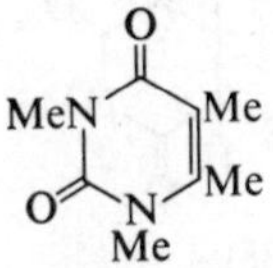

(232) 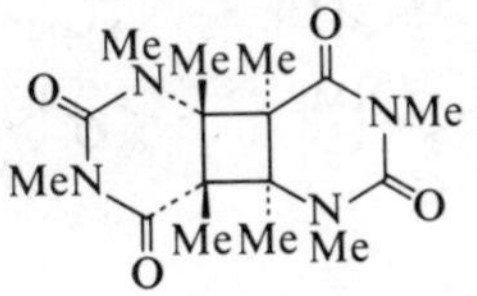   (233) 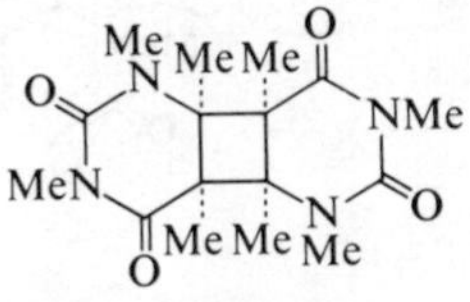   (234)

The quantum yield for the dimerization in methanol is 4.0 $\times$ 10^{-3}. Morrison *et al.*[98] have previously suggested that the dimerization of pyrimidines arises within a ground-state aggregate (stacking). The formation of dimers of uracil has been reported following the irradiation of uracil (235) in water under air at room temperature.[99] Dimers of uracil have been previously reported.[100]

Irradiation of the uracil derivative (236) yields three products (237; 18%), (238; 8.2%), and (239; 9%). The formation of the first product is most readily

[96] H. Taguchi, B.-S. Hahn, and S. Y. Wang, *J. Org. Chem.*, 1977, **42**, 4127.
[97] J. G. Otten, C. S. Yeh, S. Byrn, and H. Morrison, *J. Amer. Chem. Soc.*, 1977, **99**, 6353.
[98] H. Morrison, D. Clark, J. Otten, M. Pallmer, and C. S. Yeh, *Mol. Photochem.*, 1976, **7**, 301.
[99] H. Ishihara, *Nagoya Shiritsu Daigaku Kyoyobu Kiyo, Shizen Kagaku-Hen*, 1975, **21**, 1 (*Chem. Abs.*, 1977, **87**, 84 927).
[100] A. Smietanowska and D. Shugar, *Bull. Acad. polon. Sci., Ser. Sci. biol.*, 1961, **9**, 375; P. A. Swensen and R. B. Setlow, *Photochem. and Photobiol.*, 1963, **2**, 419.

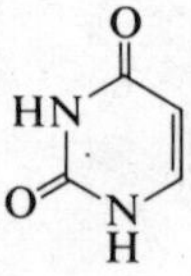

(235)

explained by the dimerization and oxidation of two uracil-6-thiol radicals (240) produced by C—S fission in (236). The other two products are thought to arise by the route shown in Scheme 15 whereby rearrangement yields the dihydro-compound (238) which is subsequently oxidized to (239).[101]

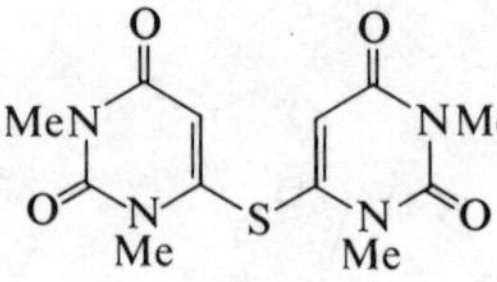

(236) 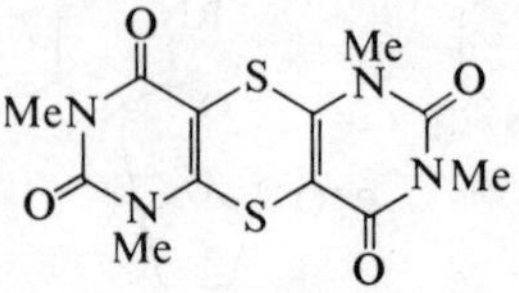   (237) 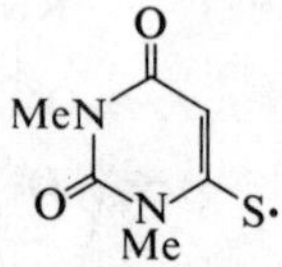   (240)

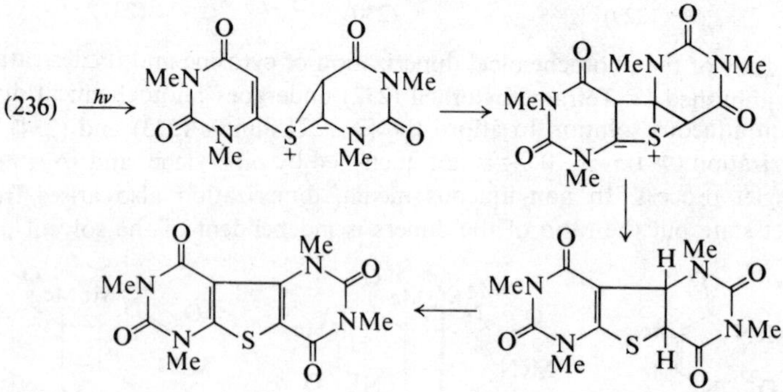

(239)   (238)

Scheme 15

Swenton and his co-workers [102] have compared the photochemical cyclo-addition reactions of thymine derivatives with those of cyclohexenones. They [102] have observed that although the substitution of methyl groups in cyclohexenones has a profound effect on the [2 + 2] cycloaddition, this effect is only minimal in the nitrogen-base analogues. Likewise [2 + 2] cycloaddition appears to be the predominant reaction in the thymine system whereas in cyclohexenones ene-reactions are also important. Unlike cyclohexenones where both *cis-* and *trans*-ring fusion is found in the cycloadducts, the cycloaddition to thymine systems tends to give *cis*-fusion. Typical examples of these are shown in Schemes 16 and 17. Triplet-sensitized ($E_T \geqslant 62$ kcal mol^{-1}, 260 kJ mol^{-1}) irradiation

[101] T. Itoh, H. Ogura, and K. A. Watanabe, *Tetrahedron Letters*, 1977, 2595.
[102] A. J. Wexler, J. A. Hyatt, P. W. Raynolds, C. Cottrell, and J. S. Swenton, *J. Amer. Soc.*, 1978, **100**, 512.

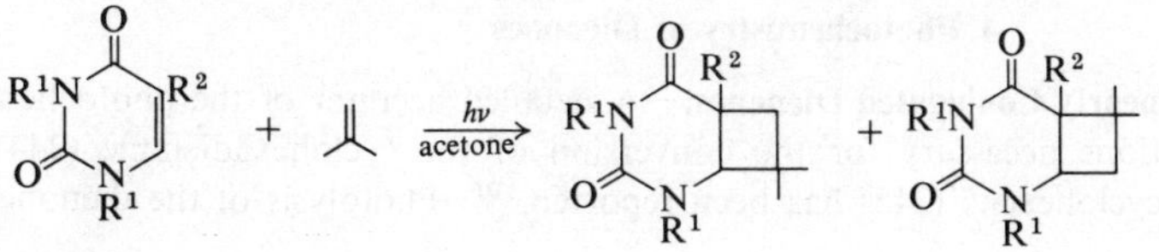

R^1 = Me, R^2 = H
R^1 = R^2 = H
R^1 = Me, R^2 = D
R^1 = R^2 = Me

Scheme 16

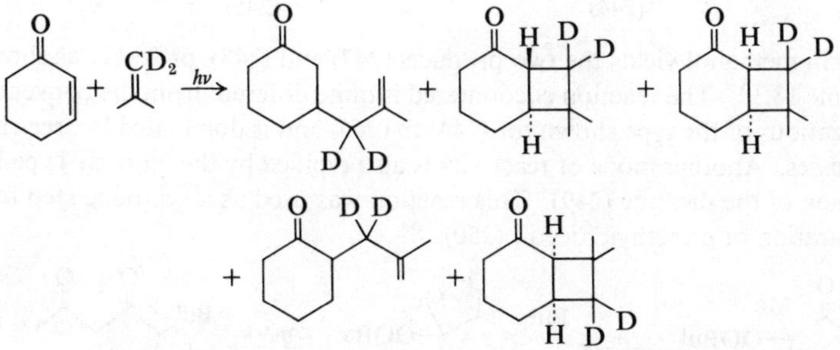

Scheme 17

of *N*-methyl-2-pyridone and methyl acrylate gave the two [2 + 2] adducts (241a) and (241b). The unsensitized irradiation afforded the adducts (242a) and (242b). A more unusual adduct (243) was also formed from the excited singlet state of the pyridone.[103—105]

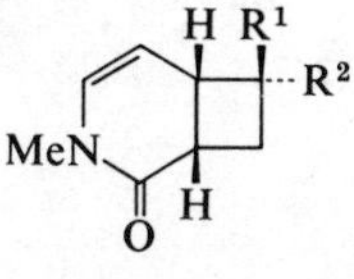

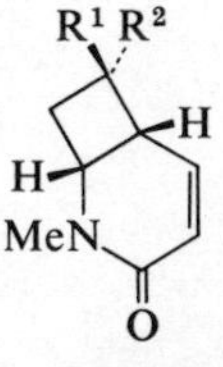

(241) a; R^1 = CO_2Me, R^2 = H; b; R^1 = H, R^2 = CO_2Me

(242) a; R^1 = CO_2Me, R^2 = H; b; R^1 = H, R^2 = CO_2Me

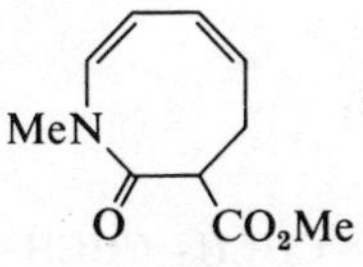

(243)

[103] K. Somekawa and S. Kumamoto, *Nippon Kagaku Kaishi*, 1977, 1489 (*Chem. Abs.*, 1978, **88**, 73 832).

[104] K. Somekawa, T. Shimou, H. Muta, and S. Kumamoto, *Nippon Kagaku Kaishi*, 1976, 1443 (*Chem. Abs.*, 1977, **86**, 88 708).

[105] K. Somekawa, T. Shimou, K. Tanaka, and S. Kumamoto, *Chem. Letters*, 1975, 45.

4 Photochemistry of Dienones

Linearly Conjugated Dienones.—A detailed account of the photochemical conditions necessary for the conversion of the cyclohexadienone (244) into the bicyclohexene (245) has been reported.[106] Photolysis of the dienone peroxide

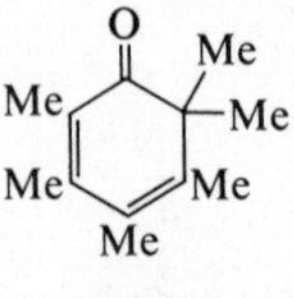

(244) 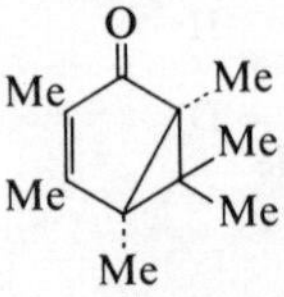   (245)

(246) in methanol yields the two products (247) and (248), probably as shown in Scheme 18.[107] The reaction encountered is quite different from the conventional cyclizations of the type shown for (244) to (245) and is dominated by free-radical processes. Another mode of reactivity is also typified by the Norrish Type I ring opening of the dienone (249). This reaction was used as a synthetic step for the preparation of dimethylcrocetin (250).[108]

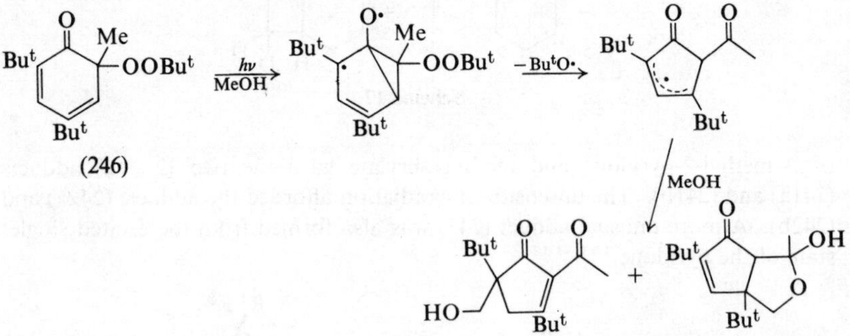

Scheme 18

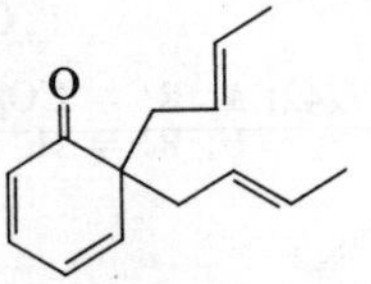

(249)

$$MeO_2CC(Me){=}CHCH{=}CHC(Me){=}CHCH{=}CHCH{=}C(Me)CH{=}CHCH{=}C(Me)CO_2M$$

(250)

[106] D. A. Dickinson, T. A. Hardy, and H. Hart, *Org. Photochem. Synth.*, 1976, **2**, 62.
[107] H. Lind, H. Loeliger, and T. Winkler, *Tetrahedron Letters*, 1978, 1571.
[108] G. Quinkert, K. R. Schmieder, G. Duerner, K. Hacha, A. Stegk, and D. H. R. Barton, *Chem. Ber.*, 1977, **110**, 3582.

Direct irradiation ($\lambda > 300$ nm) of the coumalate esters (251) in ether gave the lactones (252).[109] Similar behaviour is encountered when the compounds are irradiated in methanol but thermal reactions result in further transformation of the lactones into (253), for example. Coumalic acid (251d) was also photoreactive

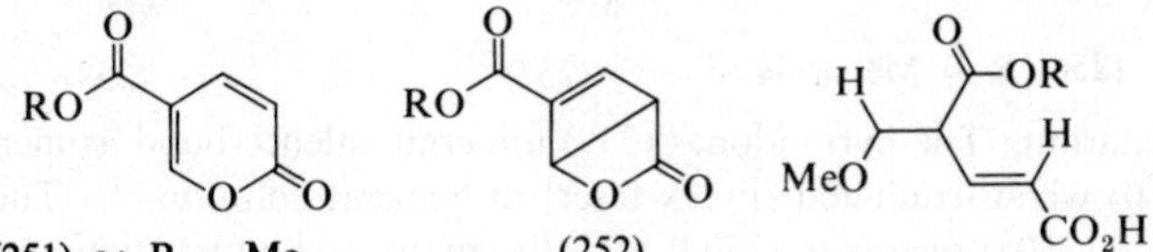

(251) a; R = Me
     b; R = Pr^i
     c; R = CH$_2$Ph
     d; R = H

(252)

(253)

in methanol and gave the products shown in Scheme 19. It is likely that a keten intermediate is involved. In the solid state (KBr discs) all the compounds undergo loss of carbon dioxide when irradiated at 254 nm. Dimerization is encountered when methyl coumalate (251a) is irradiated in KBr at 350 nm. The dimers (254) and (255) obtained are the same as those produced by the benzophenone-sensitized irradiation of methyl coumalate in ether. The authors[109] propose this to be the first conclusive example of an external heavy-atom effect in solid KBr.

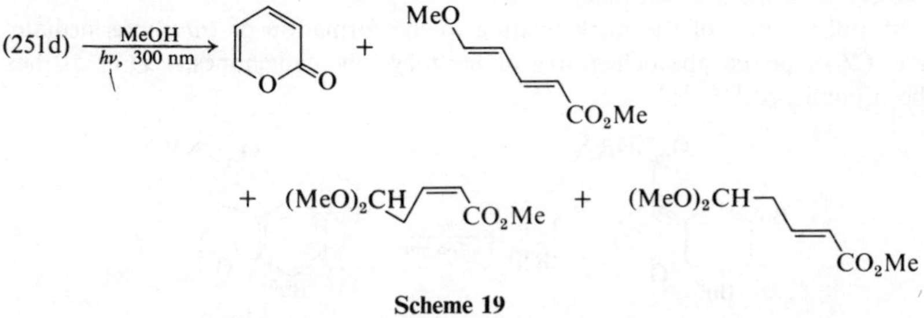

$(251d) \xrightarrow[h\nu,\ 300\ \text{nm}]{\text{MeOH}}$ + MeO ... CO$_2$Me

+ (MeO)$_2$CH CO$_2$Me + (MeO)$_2$CH CO$_2$Me

Scheme 19

Irradiation of (256) in aqueous solutions containing perchloric acid leads to the formation of the pyridinium salts (257).[110] The authors[110] reason that the most likely mechanism for the cyclization involves the singlet state of the pyridones yielding the highly stabilized ylides (258) which would be readily trapped

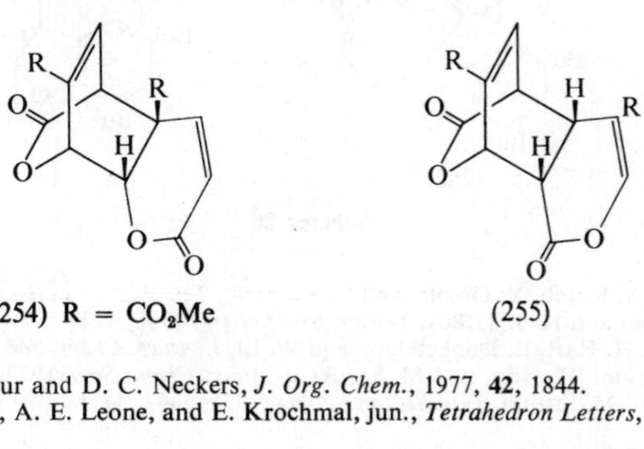

(254) R = CO$_2$Me   (255)

[109] H. Javaheripour and D. C. Neckers, *J. Org. Chem.*, 1977, **42**, 1844.
[110] P. S. Mariano, A. E. Leone, and E. Krochmal, jun., *Tetrahedron Letters*, 1977, 2227.

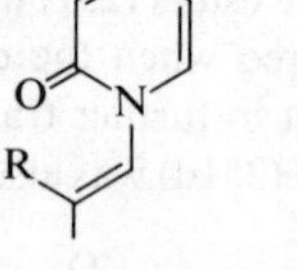

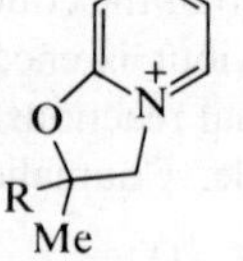

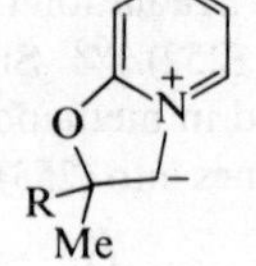

(256) R = Me or H   (257)   (258)

by protonation. The pyrimidones (259) undergo valence-bond isomerization to yield (260) when irradiated (Pyrex filter) in benzene solution.[111] The unstable compound (260a) reverts to (259a) both thermally and on irradiation at 254 nm.

Irradiation ($\lambda > 366$ nm) of benzene solutions of the enones (261) produces the isomers (262).[112] The route envisaged is outlined in Scheme 20 whereby the

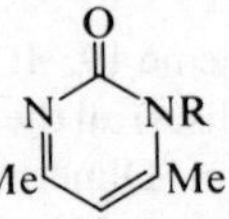

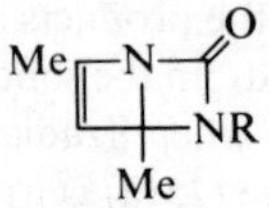

(259) a; R = Ph   (260)
      b; R = Me

photoexcited carbonyl group adds to the benzene ring, yielding the intermediate spiro-biradical (263). Subsequent bonding and ring opening of the cyclopropane moiety yield the product (262).

A full account of the work relating to the formation of *trans*-intermediate, *e.g.* (264), in the photochemistry of benzocycloheptadienones, *e.g.* (265), has been published.[113—115]

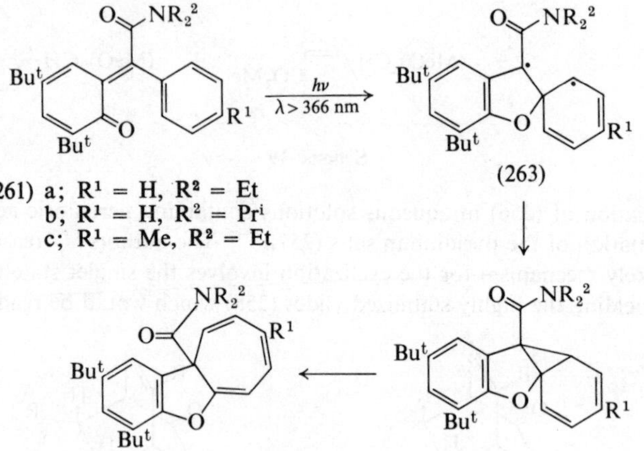

(261) a; $R^1 = H$, $R^2 = Et$
      b; $R^1 = H$, $R^2 = Pr^i$
      c; $R^1 = Me$, $R^2 = Et$

(263)

(262)

Scheme 20

[111] T. Nishio, A. Katoh, Y. Omote, and C. Kashima, *Tetrahedron Letters*, 1978, 1543.
[112] W. Verboom and H. J. T. Bos, *Tetrahedron Letters*, 1978, 1229.
[113] M. Suzuki, H. Hart, E. Dunkelblum, and W. Li, *J. Amer. Chem. Soc.*, 1977, **99**, 5083.
[114] E. Dunkelblum, H. Hart, and M. Suzuki, *J. Amer. Chem. Soc.*, 1977, **99**, 5074.
[115] H. Hart and M. Suzuki, *Tetrahedron Letters*, 1975, 3447, 3451.

(264)   (265)

Boron trihalides complex with eucarvone (266).[116] In line with the arguments put forward for protonated eucarvone,[117] it is reasoned that irradiation of boron trihalide complexes also induces a reversal of the excited-state energies so that the $\pi\pi^*$ state becomes lower in energy. The products (267)—(269), as complexes with the boron halide, are formed on irradiation at $-90\ °C$ and are produced by cationic processes. In the experiments with protonated substrate (267) and (269) were formed as the major products of the reaction.[117] It is thus clear that there is an overall similarity between the boron trihalide and the protonation experiments.

(266)   (267)   (268) a; $R^1 = Pr^i$, $R^2 = H$ (269)
                          b; $R^1 = H$, $R^2 = Pr^i$

Liu et al.[118] report further on their investigations into the photochemical isomerization of isomers of vitamin A and related carotenoids. Thus direct irradiation of the pentanone (270a) at 360 nm in acetonitrile affords products identified as the 9-*cis*-isomer (271a) and the 7-*cis*-isomer (272a). Direct irradiation of the dehydroretinal (270b) at wavelengths greater than 380 nm, also in acetonitrile, gave two products the major of which was identified as (271b). Irradiation at longer wavelength ($\lambda > 480$ nm) enhanced the formation of the minor product, thereby permitting the isolation of this isomer and its identification as (272b). All-*trans*-retinal (273) undergoes photochemical cyclization in acetonitrile solution to yield the hexahydronaphthalene derivative (274a).[119]

(270) a; X = O
      b; X = CHCHO

(271) a; X = O
      b; X = CHCHO

(272) a; X = O
      b; X = CHCHO

[116] R. F. Childs and Y.-C. Hor, *Canad. J. Chem.*, 1977, **55**, 3501.
[117] K. E. Hine and R. F. Childs, *J. Amer. Chem. Soc.*, 1973, **95**, 6116.
[118] R. S. H. Liu, A. E. Asato, and M. Denny, *J. Amer. Chem. Soc.*, 1977, **99**, 8095.
[119] K. Tsukida, M. Ito, and A. Kodama, *J. Nutr. Sci. Vitaminol.*, 1977, **23**, 375 (*Chem. Abs.*, 1978, **88**, 89 881).

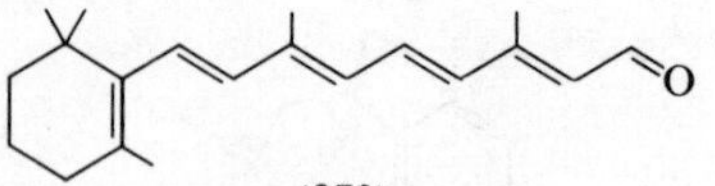

(273)

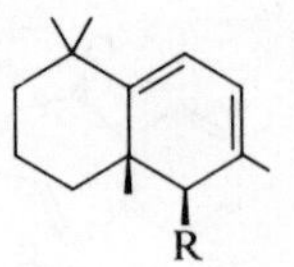

(274) a; R =

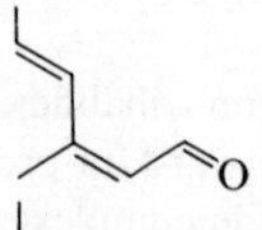

b; R =

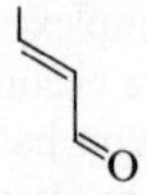

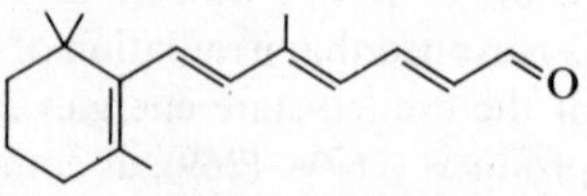

(275)

All-*trans*-β-ionylidenecrotonaldehyde (275) affords the analogous product (274b) under the same conditions.

Cross-conjugated Dienones.—A review dealing with the photochemical transformations of cross-conjugated cyclohexadienones has been published.[120] The dienones (276) undergo photoconversion into the lumiketones (277) when irradiated with 254 nm light in dioxan solution.[121] When light of 300 nm and acetic acid as solvent are used, the dienone (270b) is converted solely into the enone (278). A similar photorearrangement is reported for the dienone (279) which is converted into (280) in 40 % yield.[122]

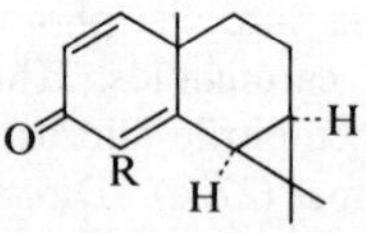

(276) a; R = Me
      b; R = H

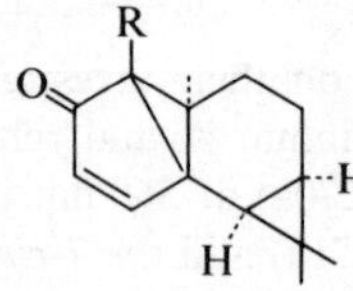

(277)

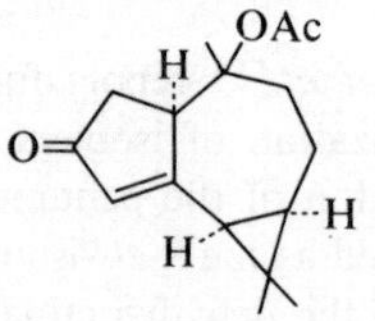

(278)

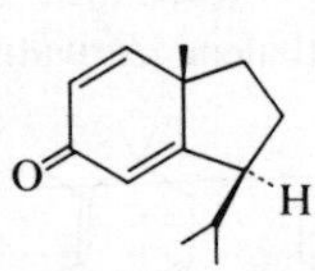

(279) 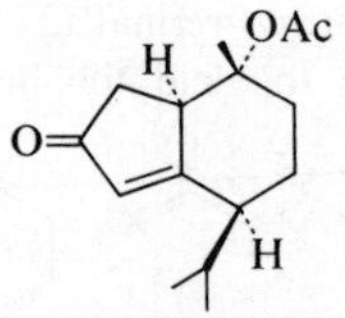   (280)

Irradiation of the pyrone (281) in water with added sodium cyanoborohydride affords the four products (282)—(285), one of which (282) is the mould metabolite terrein.[123] The reactions are shown in Scheme 21 whereby excitation affords the intermediate (286) which collapses to the diketone (287). This is not isolated but undergoes selective reduction *in situ* to afford the mixture of products.

[120] D. I. Schuster, *Accounts Chem. Res.*, 1978, **11**, 65.
[121] D. Caine, H. Deutsch, and J. T. Gupton, *J. Org. Chem.*, 1978, **43**, 343.
[122] D. Caine and A. S. Frobese, *Tetrahedron Letters*, 1977, 3107.
[123] D. H. R. Barton and L. A. Hulshof, *J.C.S. Perkin I*, 1977, 1103.

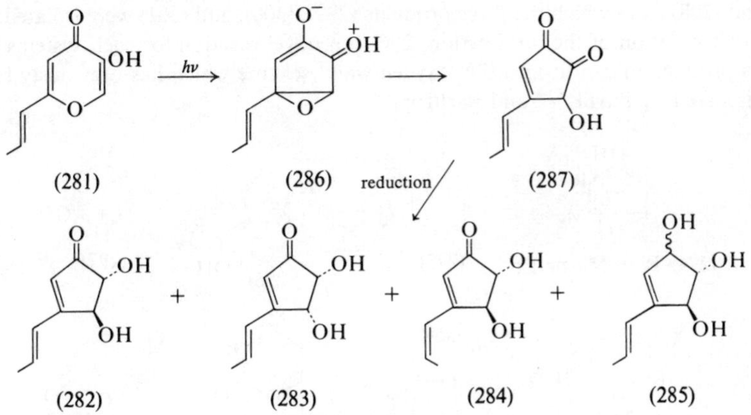

Scheme 21

Barltrop, Day, and Samuel[124] previously reported that irradiation of the γ-pyrone (288) in trifluoroethanol yielded the α-pyrone (289) and the alcohol adduct (290). A further examination of this reaction has identified two other mechanistically significant products (291) and (292).[125] The epoxide (292) is photolabile and is converted into the pyrone (289) and the cyclopentenedione (291). The isolation of the cyclopentadiene epoxide confirms the previous postulates concerning the mechanism of rearrangement of pyrones (and raises the question whether an analogous intermediate may also be involved in the

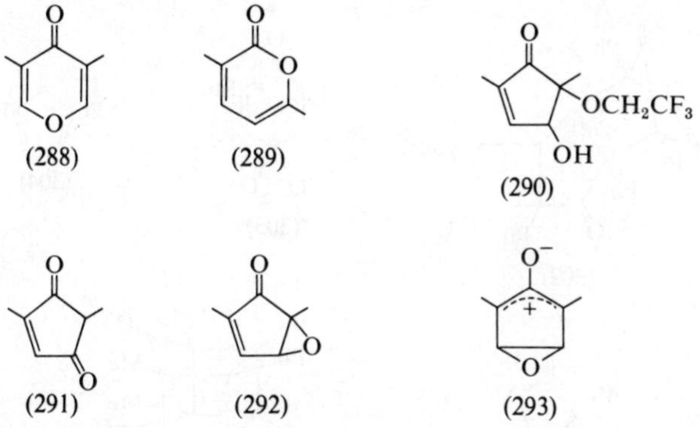

above synthesis of terrein), but the demonstration that the epoxide is not involved in the formation of the alcohol addition product (290) suggests that this latter is formed by interception of the zwitterionic intermediate (293): *cf.* ref. 123. Irradiation of the pyrylium cations (294) [prepared by dissolution of the pyrones (295) in 96% sulphuric acid] affords the three cations (296), (297),

124 J. A. Barltrop, A. C. Day, and C. J. Samuel, *J.C.S. Chem. Comm.*, 1976, 822.
125 J. A. Barltrop, A. C. Day, and C. J. Samuel, *J.C.S. Chem. Comm.*, 1977, 598. 1977, **99**, 2816.

and (298) from which the three products (299), (300), and (301) were isolated.[126] The formation of the furyl cation (296) is a novel reaction for such systems but its production can fit into the 'oxygen-walk' scheme which has previously been discussed by Pavlik [127] and Barltrop.[128]

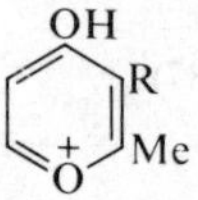

(294) R = Me or Et 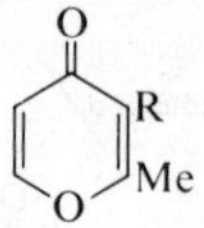   (295) 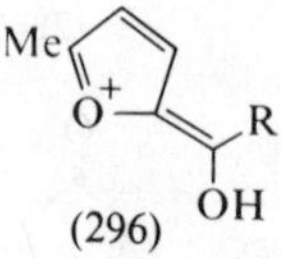   (296) 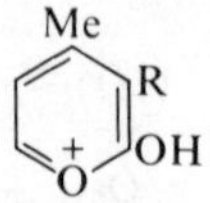   (297)

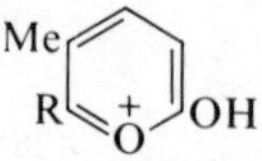

(298) 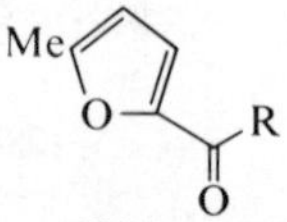   (299) 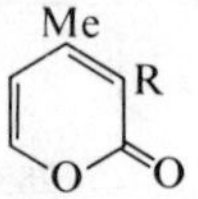   (300) 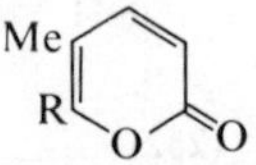   (301)

The cyclopentanedienone trimer (302) is formed when the thiopyran (303) is irradiated (sunlight) either in solution or in the solid.[129] The production of this trimer is thought to provide good evidence for the formation, by elimination of SO_2 from the pyran (303), of the previously unknown cyclopentadienone (304). This compound is obviously prone to Diels–Alder reaction and trimerizes to (302). The previously unknown cyclopentadienone (305) has been isolated and undergoes photochemical dimerization in the crystal to afford the cage compound (306).[130]

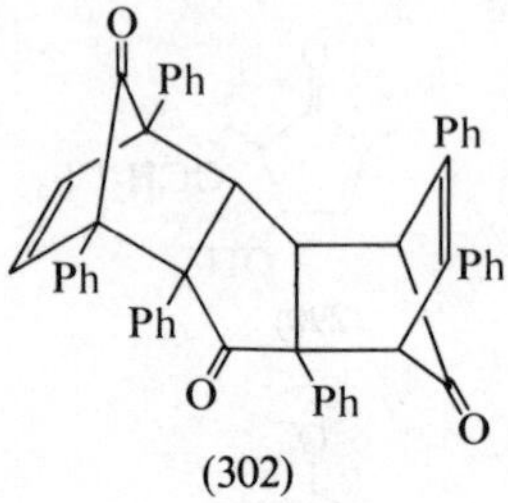

(302) 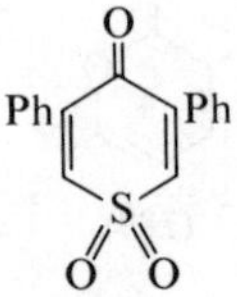   (303) 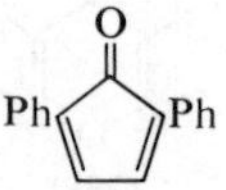   (304)

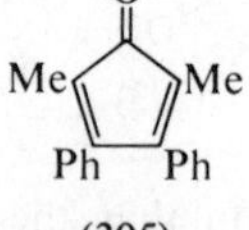

(305) 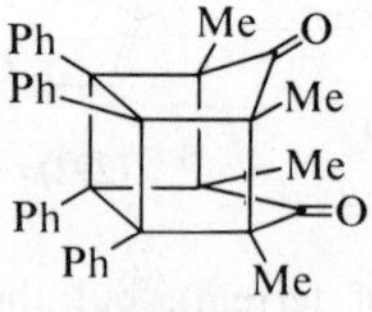   (306)

[126] J. W. Pavlik, D. R. Bolin, K. C. Bradford, and W. G. Anderson, *J. Amer. Chem. Soc.*, 1977, **99**, 2816.
[127] J. W. Pavlik and E. L. Clennan, *J. Amer. Chem. Soc.*, 1973, **95**, 1697; J. W. Pavlik and J. Kwong, *ibid.*, p. 7914.
[128] J. A. Barltrop and A. C. Day, *J.C.S. Chem. Comm.*, 1975, 177; J. A. Barltrop, R. Carder, A. C. Day, J. R. Harding, and C. J. Samuel, *ibid.*, p. 729.
[129] W. Ried and H. Bopp, *Angew. Chem. Internat. Edn.*, 1977, **16**, 653.
[130] B. Fuchs and M. Pasternak, *J.C.S. Chem. Comm.*, 1977, 537.

Pavlik and Seymour [131] have established that the valence-bond isomerization of the azepinones (307a) and (307b) in ether to the bicyclic amides (308) and (309) arises from the triplet state. When methanol was used as solvent, a further product was obtained which is thought to arise from decarbonylation of (307) to the imine (310). The route to this product and the excited state involved in its production have not yet been established. Ushiyama *et al.*[132] have shown that direct irradiation (presumably a singlet-state reaction) of the dienones (311) brings about decarbonylation and formation of highly reactive trienes. Triplet-sensitized irradiation of the dienones (311) results in valence-bond isomerization to yield (312) and structural rearrangement to (313). These products are produced in equal amounts. The triplet-sensitized cyclization to (312) is in line with the work of Pavlik and Seymour.[131]

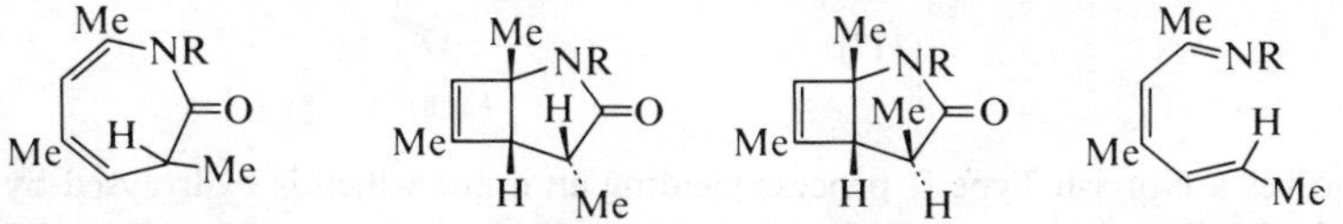

(307) a; R = H
       b; R = Me

(308)

(309)

(310)

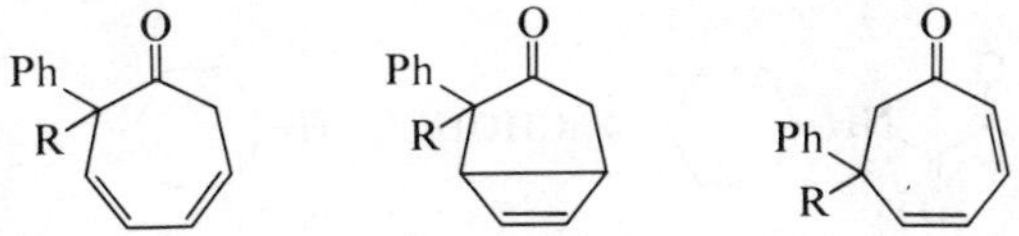

(311) R = H or Me

(312)

(313)

5 1,2-, 1,3-, and 1,4-Diketones

The photochemistry ($\lambda > 300$ nm) of several cyclopropyl-1,2-diketones [*e.g.* (314)] has been reported to involve rupture of the bond between the two carbonyl groups, decarbonylation, and formation of alkyl radicals.[133]

Ryang *et al.*[134] previously reported that the photoaddition of biacetyl to olefins such as 2,3-dimethylbut-2-ene affords unsaturated ethers (315) *via* the formation of a biradical (316) followed by a hydrogen abstraction to afford the isolated product. Shima *et al.*[135] have examined the photoreactions of 1-phenylpropane-1,2-dione with the same olefin and have observed the exclusive formation of the oxetan (317). They[135] conclude that the biradical (318) is a weaker hydrogen abstracter than (316).

Irradiation of the ketoamide (319) in ethanol with added mineral acid results in the formation of cyclohexanone in 77% yield.[136] The reaction presumably

131 J. W. Pavlik and C. A. Seymour, *Tetrahedron Letters*, 1977, 2555.
132 H. Ushiyama, H. Hagiwara, and K. Sato, *Chem. Letters*, 1977, 925 (*Chem. Abs.*, 1977, **87**, 200 879).
133 J. Kelder, H. Cerfontain, and F. W. M. van der Wielen, *J.C.S. Perkin II*, 1977, 710.
134 H.-S. Ryang, K. Shima, and H. Sakurai, *Tetrahedron Letters*, 1970, 1094; *J. Amer. Chem. Soc.*, 1971, **93**, 5270; *J. Org. Chem.*, 1973, **38**, 2860.
135 K. Shima, S. Takeo, K. Yokoyama, and H. Yamaguchi, *Bull. Chem. Soc. Japan*, 1977 **50**, 761.
136 U. Zehavi, *J. Org. Chem.*, 1977, **42**, 2821.

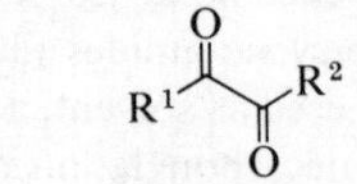

(314) R¹ = R² = cyclopropyl
       R¹ = cyclopropyl, R² = Me
       R¹ = cyclopropyl, R² = Ph

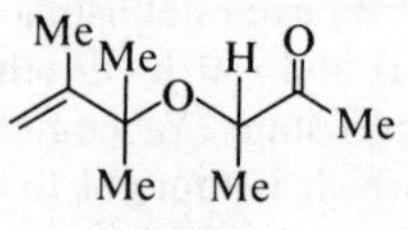

(315)

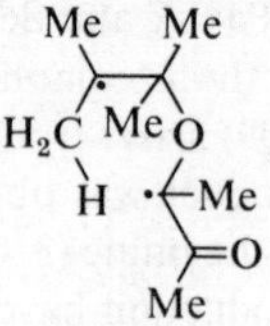

(316)

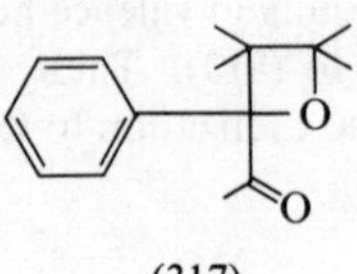

(317)

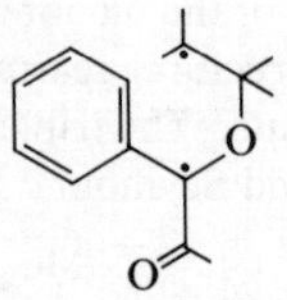

(318)

involves a Norrish Type II process yielding an imine which is hydrolysed by the acid. Omission of the acid gives only a low yield of the ketone, and other non-volatile products (320) and (321) are obtained. The related compound (322) is apparently photostable, possibly owing to a photoenolization process which was not detected. The scope of the reaction has also been studied.[136] Other

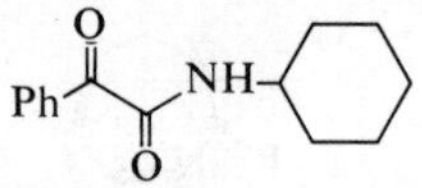

(319)

(320)

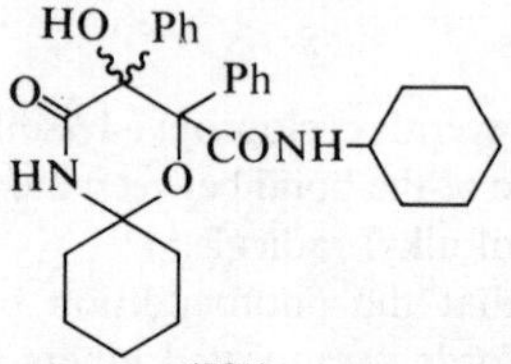

(321)

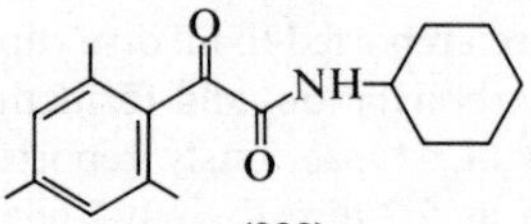

(322)

workers have studied other keto-amides (323). Thus the main products from irradiation of the amides (323a—d) in methanol are the oxazolidinones (324).[137] These are produced by Norrish Type II hydrogen abstraction, affording the biradical (325): 1,4-hydrogen migration within this leads to the biradical (326) from which the enol (327) and subsequently the product (324) are produced. The formation of the oxazolidinones is favoured in alcoholic solvents and suppressed in aprotic solvents such as benzene. It seems likely that the products arise from the triplet state of the amide. In aprotic solvents, the amides (323e and f) yield the lactams (328). Again a triplet state and Norrish Type II reaction are involved.

137 H. Aoyama, T. Hasegawa, M. Watabe, H. Shiraishi, and Y. Omote, *J. Org. Chem.*, 1978, **43**, 419.

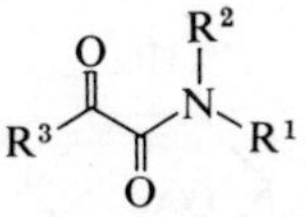

(323) a; $R^1 = R^2 = Et$, $R^3 = Me$
     b; $R^1 = R^2 = Pr^n$, $R^3 = Me$
     c; $R^1 = R^2 = Pr^i$, $R^3 = Me$
     d; $R^1 = R^2 = Et$, $R^3 = Ph$
     e; $R^1 = R^2 = CH_2Ph$, $R^3 = Me$
     f; $R^1 = R^2 = Pr^i$, $R^3 = Ph$

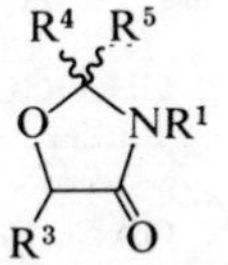

(324) a; $R^4 = Me$, $R^5 = H$
     b; $R^4 = Et$, $R^5 = H$
     c; $R^4 = R^5 = Me$
     d; $R^4 = Me$, $R^5 = H$

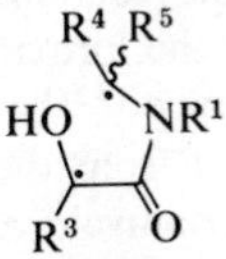

(325)

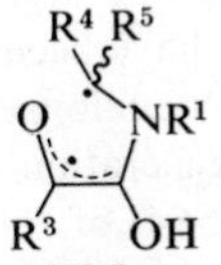

(326)

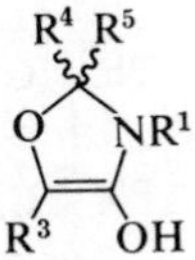

(327)

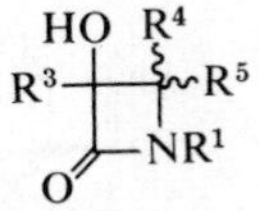

(328) e; $R^5 = Ph$, $R^4 = H$
     f; $R^4 = R^5 = Me$

Tidwell[138] and his group have described the photochemistry of the oxalates (329)—(331) in cyclohexane. The products formed (Scheme 22) are the result of free-radical reactions in which the radicals are produced by the double decarboxylation of the esters. Binkley[139, 140] has reported a mild method for the oxidation of some carbohydrates which utilizes the photolysis of pyruvates in dry benzene. An example of the method is recorded in Scheme 23.

$(PhCH_2OC)_2 \xrightarrow{h\nu} PhCH_3 + PhCH_2CH_2Ph + PhCH_2C_6H_{11}$
(329)   (24%)   (28%)   (24%)

$(Bu^t_2CHOC)_2 \xrightarrow{h\nu} Bu^t_2CH_2$
(330)   (16%)

$(Bu^t_3COC)_2 \xrightarrow{h\nu} Bu^t_3CH$
(331)   (38%)

Scheme 22

[138] S. Icli, V. J. Nowlan, P. M. Rahimi, C. Thankachan, and T. T. Tidwell, *Canad. J. Chem.*, 1977, **55**, 3349.
[139] R. W. Binkley, *Carbohydrate Res.*, 1976, **48**, C1.
[140] R. W. Binkley, *J. Org. Chem.*, 1977, **42**, 1216.

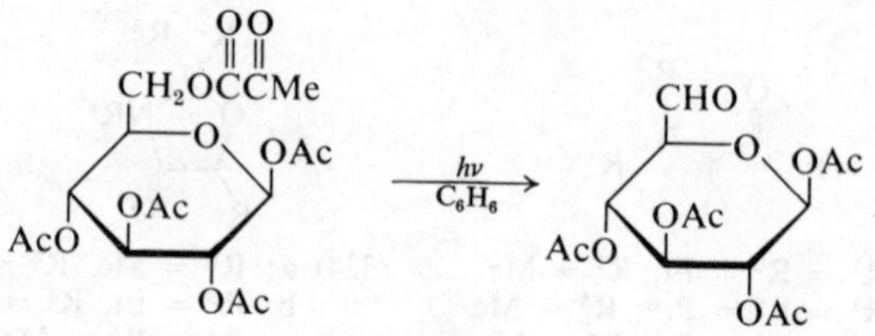

Scheme 23

A full account of the results of the photolysis of the dione (332a) and the derivative (332b) has been reported.[141, 142] The reactions which these compounds undergo have been explained in terms of the formation of the bis-keten (333a) and the *N*-analogue (333b). Jung and Lowe [143] have made use of the photochemical generation of the bis-ketens (334) from the cyclobutenediones (335) in the synthesis of compounds related to adriamycin, a noted chemotherapeutic reagent. A review dealing with the synthesis and reactions of cyclobutenediones has been published.[144]

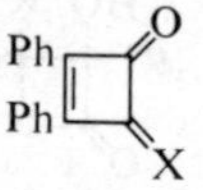

(332) a; X = O

     b; X = N—

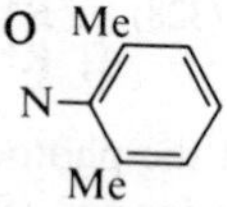

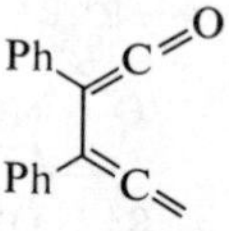

(333)

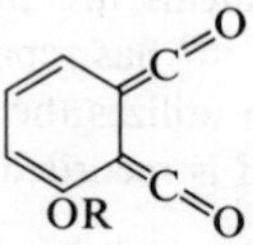

(334) R = Me or CH₂OMe   (335)

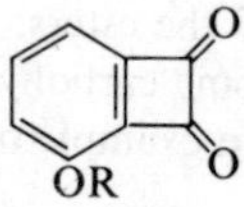

The Diels–Alder products of benzvalene with *o*-benzoquinone or tetrachloro-*o*-benzoquinone are (336). Both of these products undergo photochemical double decarbonylation to afford the dienes (337).[145] Double photochemical decarbonylation is also reported for the diones (338) and (339) [146] and for the diketone (340).[147] In the last example decarbonylation produces the *o*-xylene derivative (341) which in the presence of *N*-methylmaleimide affords the *endo*-adduct (342) exclusively. In the absence of trapping agents decarbonylation of (340) affords two hydrocarbons which have not yet been fully identified.[147]

[141] N. Obata and T. Takizawa, *Bull. Chem. Soc. Japan*, 1977, **50**, 2017.
[142] N. Obata and T. Takizawa, *Chem. Comm.*, 1971, 587.
[143] M. E. Jung and J. A. Lowe, *J. Org. Chem.*, 1977, **42**, 2371.
[144] A. H. Schmidt and W. Ried, *Synthesis*, 1978, 1.
[145] M. Christl, H.-J. Luddeke, A., Nagyrevi-Neppel, and G. Freitag, *Chem. Ber.*, 1977, **110**, 3745.
[146] K. Sakanishi, T. Shigeshima, and H. Hachiya, *Susuka Kogyo Koto Semmon Gakko Kiyo*, 1975, **8**, 269 (*Chem. Abs.*, 1978, **88**, 74 103).
[147] R. N. Warrener, P. A. Harrision, and R. A. Russell, *Tetrahedron Letters*, 1977, 2031.

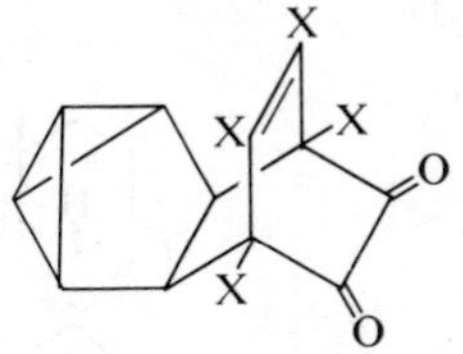

(336) X = H or Cl

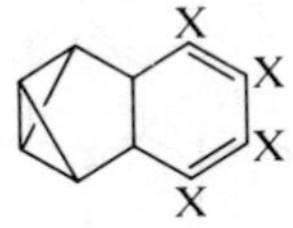

(337)

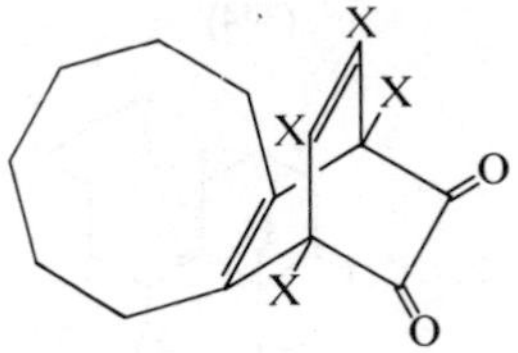

(338) X = Cl or H

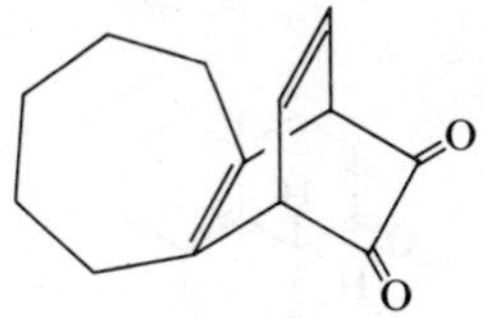

(339)

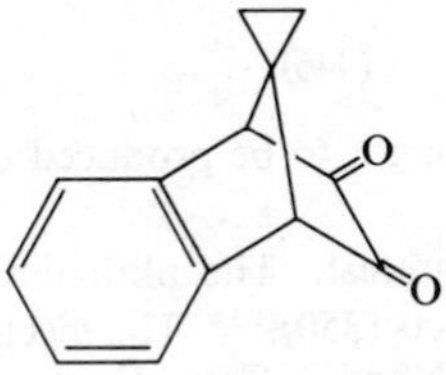

(340)

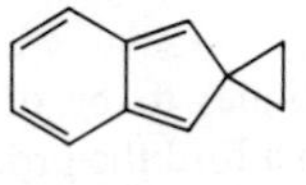

(341)

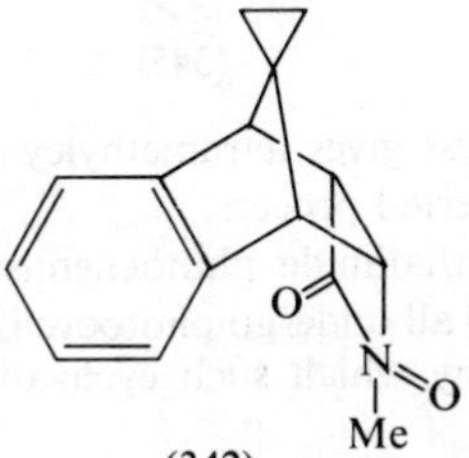

(342)

The exposure of heptane solutions of dibenzoylmethane, benzoylacetone, and ethylbenzoyl acetone to sunlight results in a displacement of the enol–keto equilibrium towards the ketone.[148] A full account of the photochemistry of the enol form of some 1,3-dicarbonyl compounds has been published.[149, 150]

Photoaddition of the enedione (343) to cyclopentene, tetrahydropyran, and 2,3-dimethylbutadiene yields the novel [4 + 2] adducts (344)—(346).[151] A triplet state is thought to be involved in the photochemical addition. The more usual [2 + 2] cycloaddition products were also obtained and this addition mode can also be accommodated in a mechanism involving a triplet excited state.

Diones [*e.g.* (347)] have been shown to rearrange to (348) when irradiated (Corex filter) in ether.[152] The reaction occurs by a Norrish Type I cleavage and rebonding. A quantitative study of the direct and xanthone-sensitized irradiation of tetramethylcyclobutane-1,3-dione has been reported.[153] The triplet-state decomposition of the dione leads exclusively to dimethylketen. The singlet-state

[148] P. Markov and I. Petkov, *Tetrahedron*, 1977, **33**, 1013.
[149] D. Veierov, T. Bercovici, E. Fischer, Y. Mazur, and A. Yogev, *J. Amer. Chem. Soc.*, 1977, **99**, 2723.
[150] D. Veierov, T. Bercovici, E. Fischer, Y. Mazur, and A. Yogev, *J. Amer. Chem. Soc.*, 1973, **95**, 8173.
[151] G. R. Lenz, *J.C.S. Chem. Comm.*, 1977, 700.
[152] H. Hart, C.-T. Peng, and E. Shih, *J. Org. Chem.*, 1977, **42**, 3635.
[153] R. Spafford and M. Vala, *J. Photochem.*, 1978, **8**, 61 (*Chem. Abs.*, 1978, **88**, 67 815).

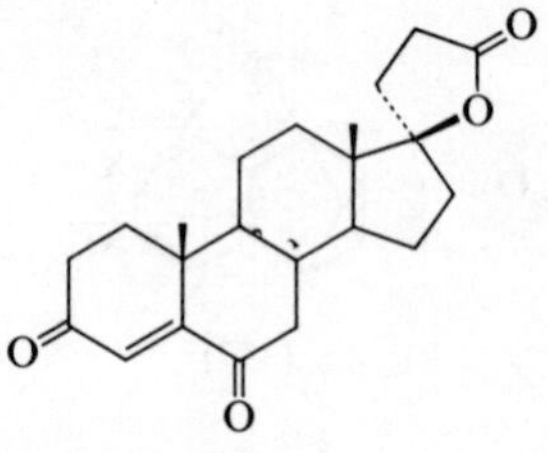

(343)

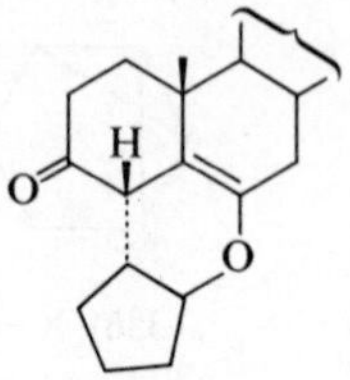

(344)

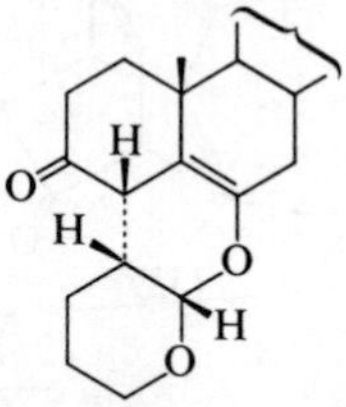

(345)

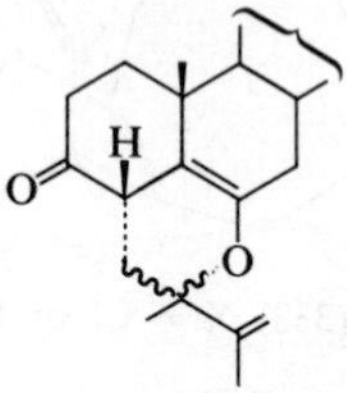

(346)

process gives tetramethylcyclopropanone which is thought to be produced by a concerted process.

Phthalimide photochemistry continues to be of interest. The phthalimides (349) all undergo photocyclization to afford the products (350).[154] The mechanism by which such cyclizations occur is akin to a Norrish Type II process.

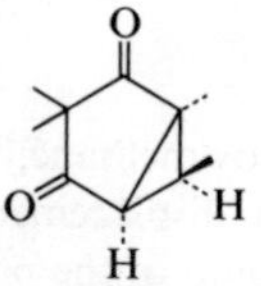

(347)

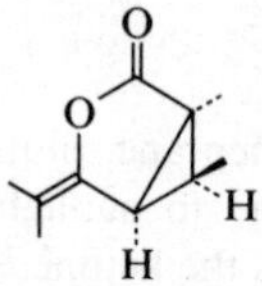

(348)

However, the incorporation of a side-chain nitrogen slightly modifies the process and an electron transfer is probably involved. This question has been studied by Coyle *et al.*[155] for the phthalimides (351): the evidence from emission studies suggests that an electron transfer mechanism is involved. Irradiation of the phthalimide derivative (352) in acetone gave the cyclized product (353) and the dihydro-compound (354).[156]

Attempts to study the effect of quenchers upon these photoreactions of *N*-alkylphthalimides have led to the discovery of a further reaction. The work reported here was simplified by the use of the phthalimide (355) where Norrish Type II reactions were avoided.[157] Irradiation of (355) with butadiene in acetonitrile gave the adducts (356a and b). The addition reaction also occurs with isoprene, yielding (356c) and (357a), and with 2,3-dimethylbutadiene to

[154] M. Machida, H. Takechi, and Y. Kanaoka, *Heterocycles*, 1977, **7**, 273 (*Chem. Abs.*, 1978, **88**, 50 823).
[155] J. D. Coyle, G. L. Newport, and A. Harriman, *J.C.S. Perkin II*, 1978, 133.
[156] M. Terashima, K. Seki, K. Koyama, and Y. Kanaoka, *Chem. and Pharm. Bull.* (*Japan*), 1977, **25**, 1591 (*Chem. Abs.*, 1978, **88**, 22 690).
[157] P. H. Mazzocchi, M. J. Bowen, and N. K. Narain, *J. Amer. Chem. Soc.*, 1977, **99**, 7063.

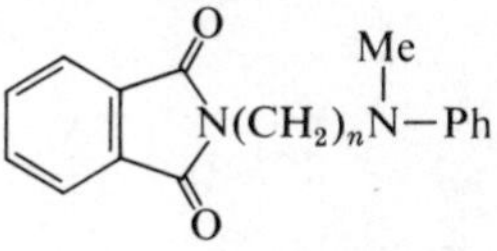

(349) n = 1—6, 10, or 12

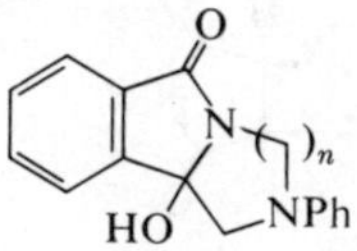

(350)

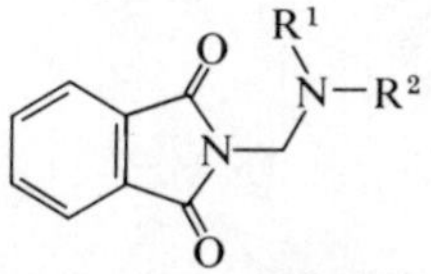

(351) $R^1 = R^2$ = Me
$R^1 = R^2 = Pr^n$
$R^1 = R^2 = CH_2Ph$
R^1 = Me, R^2 = Ph
R^1—$R^2 = (CH_2)_4$
R^1—$R^2 = (CH_2)_2NMe(CH_2)_2$
R^1—$R^2 = (CH_2)_2O(CH_2)_2$

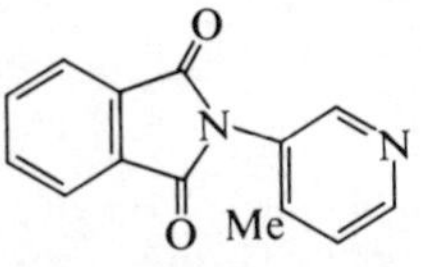

(352)

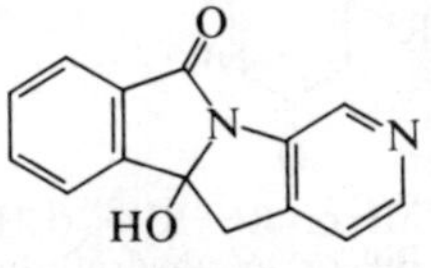

(353)

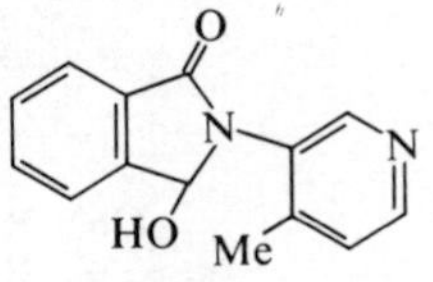

(354)

afford (357b). The addition reaction always occurs such that the terminal methylene group of the diene becomes attached to the nitrogen of the phthalimide. The authors [157] suggest that a mechanism for the formation of the compounds (356) and (357) might involve the polar biradical (358) and further study is on hand to verify this postulate.

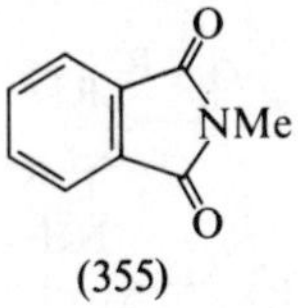

(355)

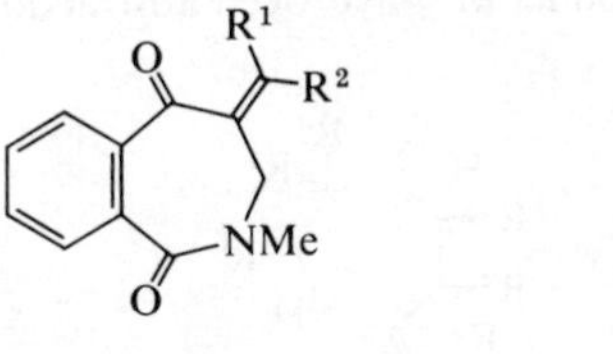

(356) a; R^1 = H, R^2 = Me
b; R^1 = Me, R^2 = H
c; $R^1 = R^2$ = Me

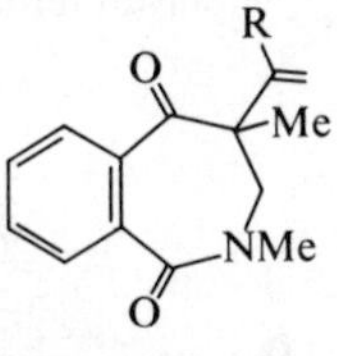

(357) a; R = H
b; R = Me

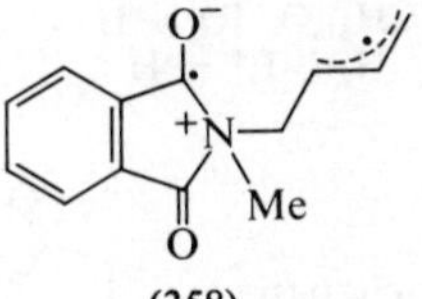

(358)

Norrish Type II reactivity is also encountered in simpler imides such as succinimides (359) and glutarimides (360).[158] Irradiation of these compounds results in hydrogen abstraction from the alkyl side-chain by one of the imide carbonyl groups. The resultant biradical undergoes cyclization to afford a

[158] Y. Kanaoka, H. Okajima, and Y. Hatanaka, *Heterocycles*, 1977, **8**, 339 (*Chem. Abs.*, 1978, **88**, 105 106).

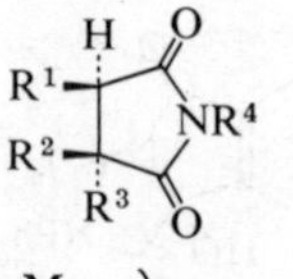

(359) $R^1 = R^3 = H$, $R^2 = Me$
$R^1 = H$, $R^2 = R^3 = Me$
$R^1-R^2 = (CH_2)_2$, $R^3 = H$
$R^1-R^2 = (CH_2)_4$, $R^3 = H$

R^4 = cyclo-octyl; cyclododecyl, Et, Bu^i, tetrahydrofurfuryl, or cyclopentyl

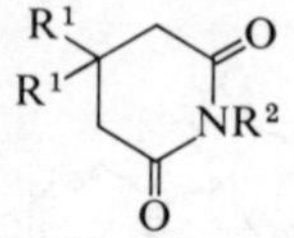

(360) $R^1 = Me$ or $R^1-R^1 = (CH_2)_4$
$R^2 = Bu^i$ or tetrahydrofurfuryl

cyclobutanol intermediate which ring-opens to yield the products (361) from the succinimides and (362) from the glutarimides.[158] Changing the substituents in the side-chain of the succinimides as in (363) changes the pattern of reactivity, and photochemical excitation results in δ- rather than γ-hydrogen abstraction, affording (364).[159] Somewhat similar behaviour is encountered with compounds bearing thio-substituents (365a and b), irradiation of which yields (366a and b), the product of γ-hydrogen abstraction, together with the products (367a and b)

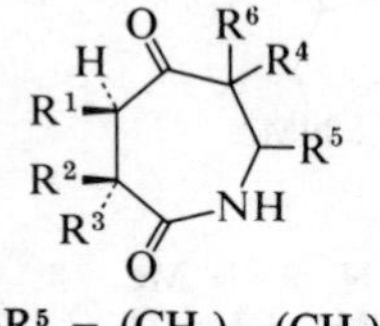

(361) $R^4-R^5 = (CH_2)_6$, $(CH_2)_{10}$, or
$R^4 = R^5 = H$
$R^4 = R^6 = Me$, $R^5 = H$
$R^4-R^5 = (CH_2)_3O$, $R^6 = H$
$R^4-R^5 = (CH_2)_3$, $R^6 = H$

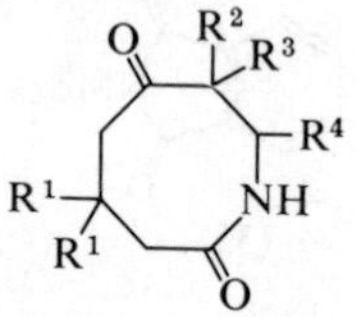

(362) $R^3 = R^2 = Me$
R^4-R^2 = tetrahydrofurfuryl, $R^3 = H$

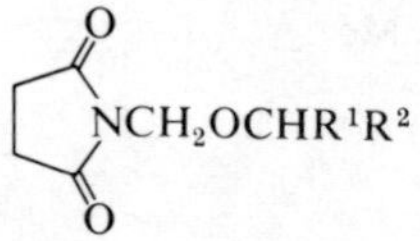

(363)

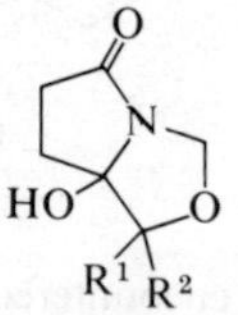

(364)

$R^1 = R^2 = Me$ or $R^1-R^2 = (CH_2)_{4-7}$

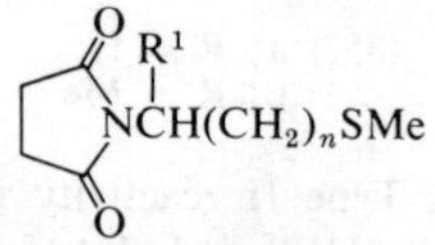

(365) a; $R^1 = CO_2Me$, $n = 2$
b; $R^1 = H$; $n = 2$
c; $R^1 = H$; $n = 1$

[159] H. Nakai, Y., Sato, T. Mizoguchi, M. Yamazaki, and Y. Kanaoka, *Heterocycles*, 1977, **8**, 345 (*Chem. Abs.*, 1978, **88**, 74 330).

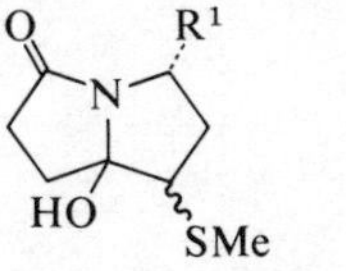

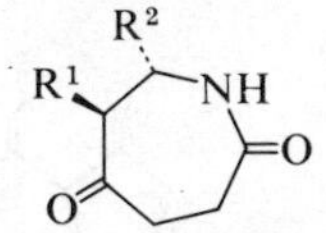

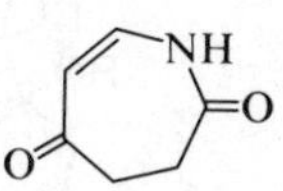

(366) a; $R^1 = CO_2Me$ (367) a; $R^1 = CH_2SMe$, $R^2 = CO_2Me$ (368)
      b; $R^1 = H$ b; $R^1 = CH_2SMe$, $R^2 = H$
                               c; $R^1 = SMe$, $R^2 = H$

of the more conventional δ-H-abstraction. These compounds are produced by ring closure of the biradical to a hydroxycyclobutanol which ring-opens to (367). Similar reactivity is encountered with (365c) which affords both (367c) and the elimination product (368). The oxetan (369) is formed when a solution of the alkenyl imide (370a) in acetonitrile is irradiated using 254 nm light.[160] When methanol with a trace of acid is used as solvent, the product (371) is obtained

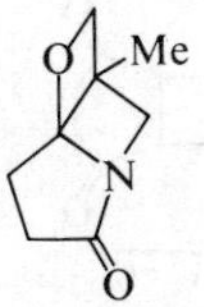

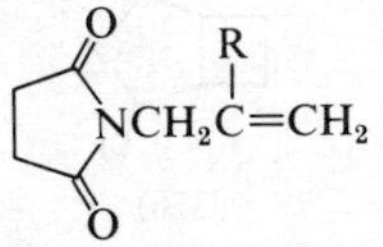

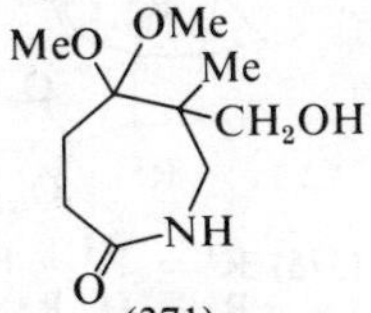

(369)   (370) a; R = Me   (371)
                    b; R = H

as a result of ring opening of the oxetan intermediate and acetalization. Other alkenyl imides (370b) and (372) are also photoreactive in methanol–H$^+$ and yield the products shown in Scheme 24.

The formation of the bicyclic lactams (373) from irradiation of the keto-amides (374) in benzene solution with Pyrex-filtered light has been interpreted in terms of δ-hydrogen transfer by the cycloalkanone C=O function. This is reasoned to be preferred to an electron transfer mechanism.[161]

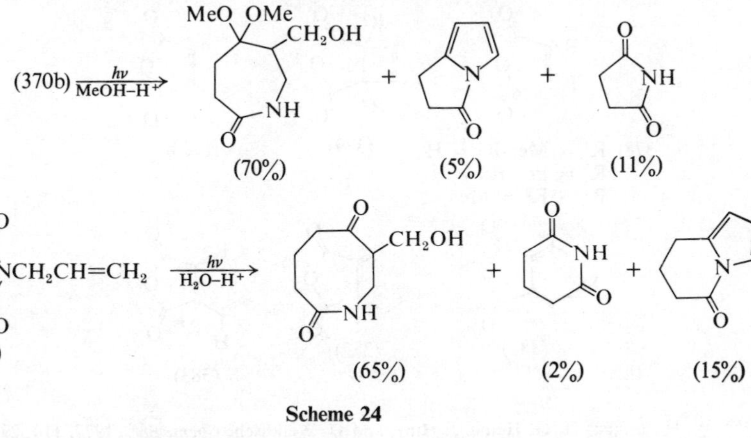

(370b) $\xrightarrow[\text{MeOH–H}^+]{h\nu}$

(70%)   +   (5%)   +   (11%)

(372) $\xrightarrow[\text{H}_2\text{O–H}^+]{h\nu}$

(65%)   +   (2%)   +   (15%)

Scheme 24

[160] K. Maruyama and Y. Kubo, *J. Org. Chem.*, 1977, **42**, 3215.
[161] T. Hasegawa, M. Inoue, H. Aoyama, and Y. Omote, *J. Org. Chem.*, 1978, **43**, 1005.

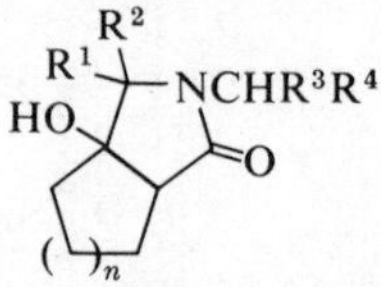

(373)

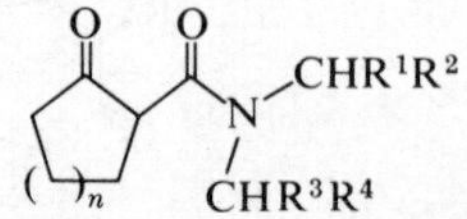

(374) $R^1 = R^3 = H$, $R^2 = R^4 = Ph$, $n = 1$
$R^1 = R^2 = R^3 = R^4 = Me$, $n = 1$
$R^1 = R^2 = R^3 = R^4 = Me$, $n = 2$

The spirocyclohexanes (375) can be prepared in good yield by the photo-addition of methylenecyclopropane to maleic anhydride and its derivatives.[162] The mechanism of the addition is proposed to involve the addition of triplet maleic anhydride at the terminal C-atom of the methylenecyclopropane. The syntheses of the anhydrides (376)[163] and (377)[164] by the irradiation of maleic

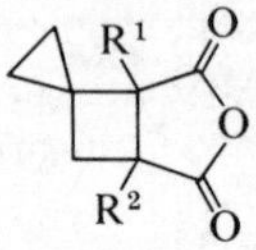

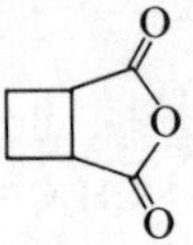

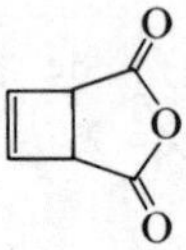

(375) $R^1 = R^2 = H$
$R^1 = H$, $R^2 = Me$
$R^1 = Me$, $R^2 = H$
$R^1 = R^2 = Me$
$R^1 = R^2 = Cl$

(376)

(377)

anhydride in the presence of ethylene and acetylene, respectively, have been reported. The benzophenone-sensitized photoaddition of acetylene to the anhydrides (378) yields the [2 + 2] adducts (379): other adducts (380)—(382) are also produced.[165] The formation of these products is evidence for the involvement of a biradical (383) which can either ring-close to (379) or else undergo intramolecular disproportionation to afford the olefins (380)—(382).

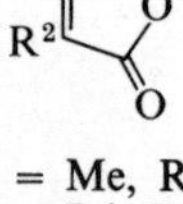

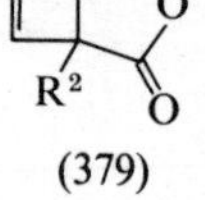

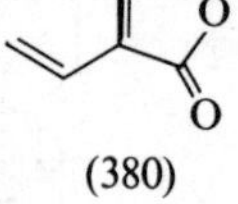

(378) $R^1 = Me$, $R^2 = H$
$R^1 = Pr^i$, $R^2 = H$
$R^1 = R^2 = Me$

(379)

(380)

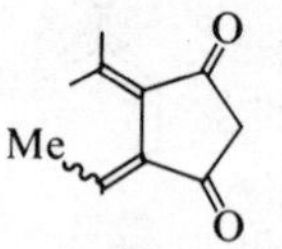

(381)

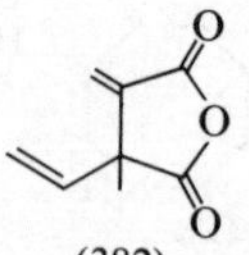

(382)

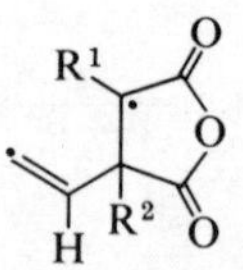

(383)

[162] W. Hartmann, H.-G. Heine, J. Hinz, and D. Wendisch, *Chem. Ber.*, 1977, **110**, 2986.
[163] J. J. Bloomfield and D. C. Owsley, *Org. Photochem. Synth.*, 1976, **2**, 32.
[164] J. J. Bloomfield, D. C. Owsley, and R. Srinivasan, *Org. Photochem. Synth.*, 1976, **2**, 36.
[165] H.-D. Scharf and J. Mattay, *Annalen*, 1977, 7720.

The adduct (384) is produced when the anhydride (385) is irradiated in an ethereal solution containing cyclohexa-1,4-diene.[166] Irradiation of 4-methyl-1,2,3,6-tetrahydrophthalic anhydride in the presence of maleic anhydride gives the [2 + 2] adduct (386).[167]

(384)   (385)   (386)

The photodimerization of methylmaleic anhydrides in carbon tetrachloride has been reported.[168] The influence of solvent on the photochemistry of maleic anhydride has also been studied.[169] The authors[169] suggest that the type of reaction undergone by maleic anhydride depends upon the nature of the complex it can form in the solvent in question.

The photoaddition of dibromomaleic anhydride (387) to benzo[b]thiophen affords the two products (388) and (389).[170] The ratio of these was dependent upon the solvent polarity, the cycloadduct (388) being favoured in more polar solvents. The reaction involves a triplet exciplex.

(387)   (388)   (389)

Low yields of products are obtained when the maleimides (390) were irradiated with thiophen in acetone.[171] The products were the known dimers of the imides and the thienyl-maleimides (391). An extra product (392) was isolated from the reaction of (390b).[171]

Irradiation of the anhydride (393) in an argon matrix at 254 nm affords a complex of tetramethylcyclobutadiene with phthalic anhydride (394) which has an absorption maximum at 525 nm.[172] This is identical with the species produced by the irradiation of the anhydride (395) under similar conditions.

The diones (396) undergo photocleavage in acetonitrile solution ($\lambda > 310$ nm) to afford the isocyanates (397) together with N_2 and CO.[173] When the irradiations are carried out in ether, alkylation of the diones takes place to afford the

[166] B. Deppisch, H.-G. Fritz, I. Schneider, and H. Musso, *Chem. Ber.*, 1978, **111**, 1497.
[167] I. I. Yukel'son, L. V. Kontsova, N. N. Chernousova, and T. M. Evsyukova, *Izvest. Vyssch. Uchebn. Zaved. Khim. khim. Tekhnol.*, 1978, **21**, 40 (*Chem. Abs.*, 1978, **88**, 137 233).
[168] P. Boule and J. Lemaire, *Compt. rend.*, 1977, **285**, C, 3059.
[169] P. Boule, J. Roche, and J. Lemaire, *J. Chim. phys.*, 1977, **74**, 593 (*Chem. Abs.*, 1977, **87**, 117 312).
[170] T. Matsuo and S. Mihara, *Bull. Chem. Soc. Japan*, 1977, **50**, 1797.
[171] H. Wamhoff and H.-J. Hupe, *Tetrahedron Letters*, 1978, 125.
[172] G. Maier, H. P. Reisenauer, and H.-A. Reitag, *Tetrahedron Letters*, 1978, 121.
[173] H. Wamhoff and K. Wald, *Chem. Ber.*, 1977, **110**, 1699.

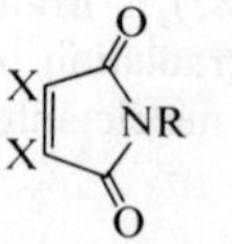

(390) a; R = H, X = Cl
      b; R = Me, X = Br
      c; R = H, X = I

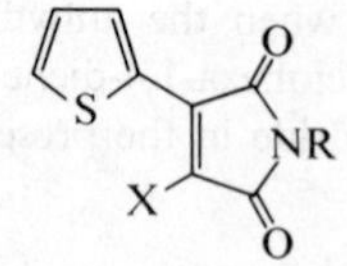

(391) a; 5.8%
      b; 4.5%
      c; 15%

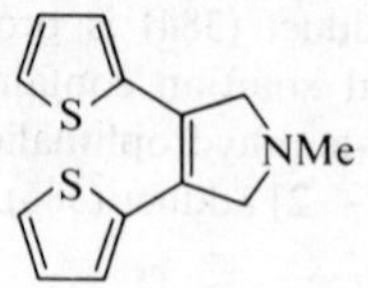

(392)

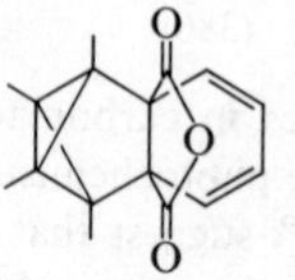

(393) 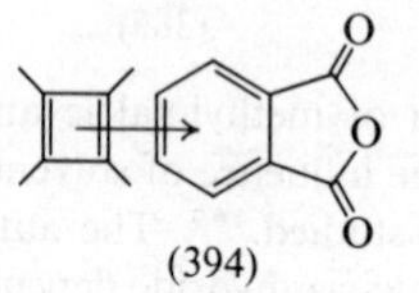   (394) 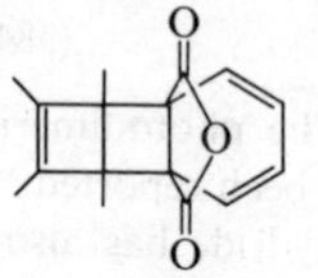   (395)

products (398), presumably by a hydrogen- abstraction pathway. Another free-radical reaction path is also recognized in the photoarylation of anisole to afford (399).[173]

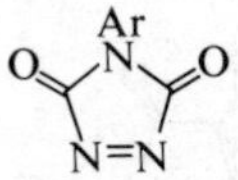

Ar—N=C=O
(397)

(396) a; Ar = Ph
      b; Ar = 4-ClC$_6$H$_4$
      c; Ar = 3,4-Cl$_2$C$_6$H$_3$

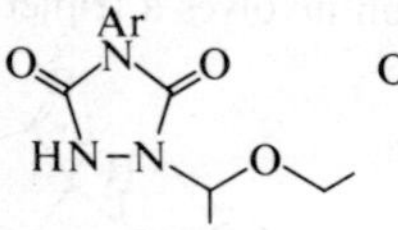

(398)

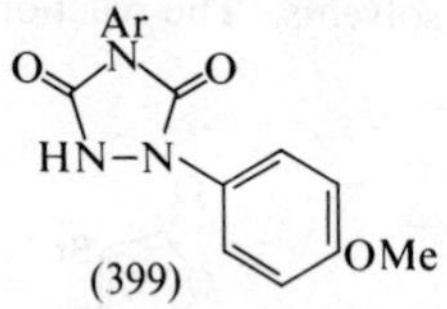

(399)

Caldwell and Creed[174] have studied the influence of quenchers (oxygen or di-t-butyl nitroxide) upon the formation of cyclobutanes in the cycloaddition of dimethyl fumarate or maleate to phenanthrene. They[174] observe that the formation of cyclobutane products is enhanced by 5—20% in the presence of the quencher. The analysis of the reaction system points to the interception by the quencher of a biradical intermediate such as (400).

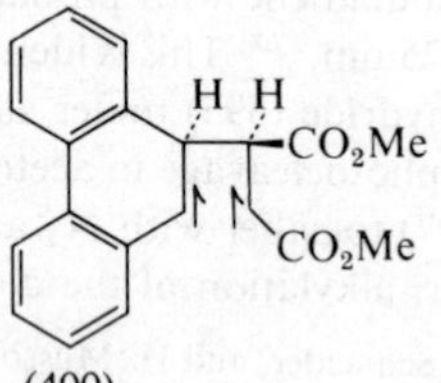

(400)

The dione (401) undergoes photoaddition of diphenylacetylene to give the [2 + 2] adduct (402) and the diketone (403), a result of oxeten formation followed by subsequent ring opening.[175] Intramolecular [2 + 2] addition is

[174] R. A. Caldwell and D. Creed, *J. Amer. Chem. Soc.*, 1977, **99**, 8360.
[175] M. Oda, H. Oikawa, N. Fukazawa, and Y. Kitahara, *Tetrahedron Letters*, 1977, 4409.

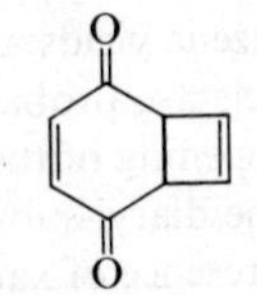

(401) 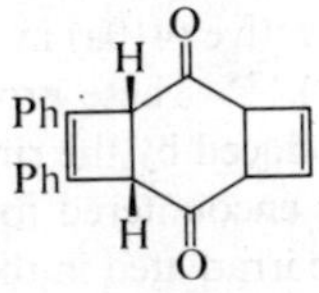   (402) 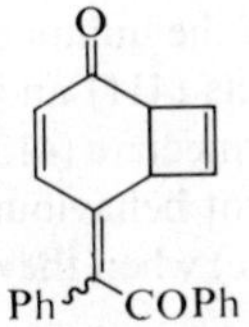   (403)

encountered in the photochemical formation of the highly strained cage compound (404) from (405).[176] The cage compounds (406) are produced when the dienediones (407) are irradiated in the solid phase.[177] The conversion of (406b) into (407b) could also be carried out in solution. The easy formation of the cage compounds suggests that the stable conformation of the precursors is as shown in (407). However, when (407a) is irradiated in benzene the cage compound (406a) is accompanied by the oxetan (408). This is thought to arise from another conformation of (407a), namely (409) where the carbonyl groups are now suitably aligned for the oxetan formation.

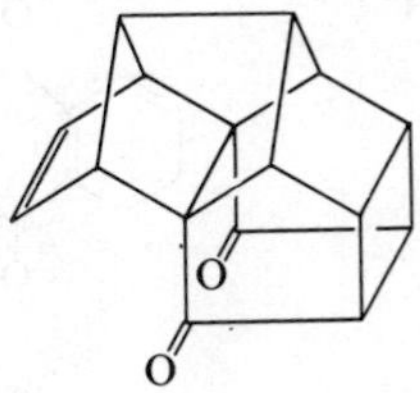

(404) 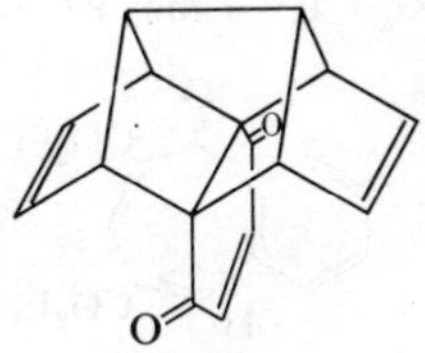   (405)

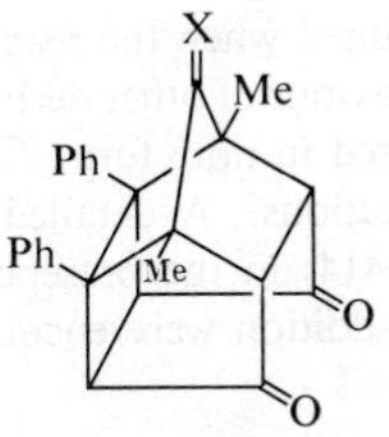

(406) a; X = O
      b; X = H$_2$

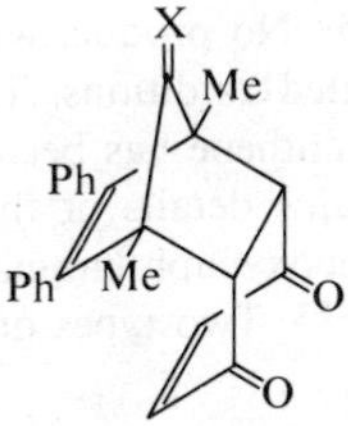

(407)

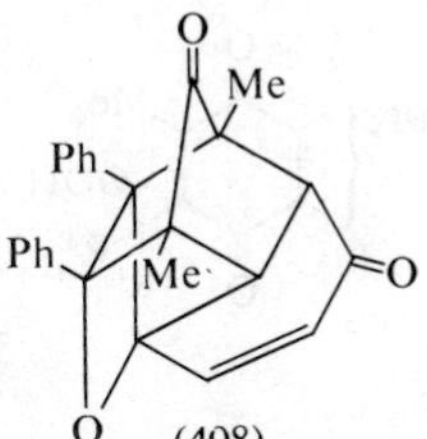

(408) 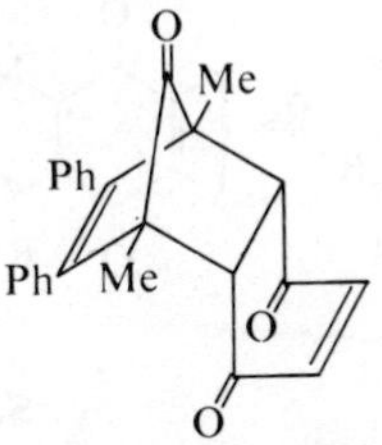   (409)

[176] L. A. Paquette, R. A. Snow, J. L. Muthard, and T. Cynkowski, *J. Amer. Chem. Soc.*, 1978, **100**, 1600.

[177] R. N. Warrener, I. W. McCay, and M. N. Paddon-Row, *Austral. J. Chem.*, 1977, **30**, 2189.

Irradiation of the quinone derivative (410a) in benzene yields a mixture of the isomeric products (411) and (412).[178] These products are probably formed *via* a biradical intermediate (413) produced by the ring opening of the cyclopropane moiety. Different behaviour was encountered for the diaryl-substituted derivatives (410b and c) when they were irradiated in the presence of xanthene and air.

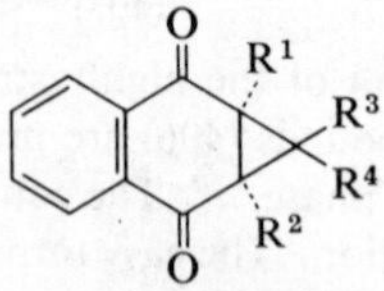

(410) a; $R^1 = R^2 = CO_2Et$, $R^3 = R^4 = H$
b; $R^1 = Me$, $R^2 = H$, $R^3 = R^4 = Ph$
c; $R^1 = Me$, $R^2 = H$, $R^3-R^4 = $ fluorenyl
d; $R^1 = Me$, $R^2 = R^3 = R^4 = H$
e; $R^1 = R^2 = Me$, $R^3 = R^4 = H$
f; $R^1 = R^2 = R^3 = R^4 = H$
g; $R^1 = R^4 = Me$, $R^2 = R^3 = H$

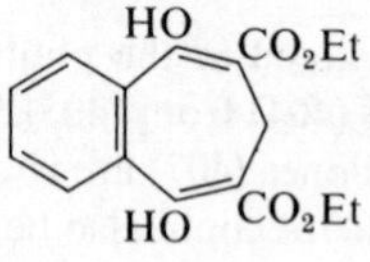

(411)

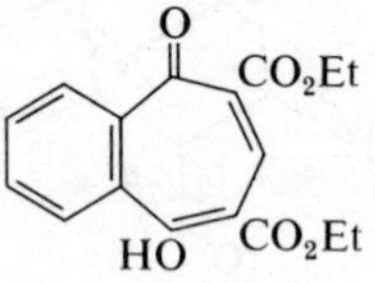

(412) 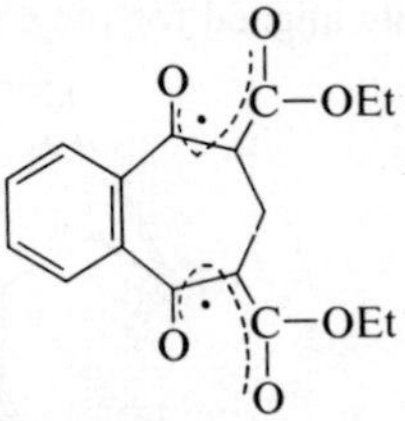　　　　　　　　(413)

This led to the production of the fragmentation products formed by the path shown in Scheme 25. No product was obtained when the reaction was carried out under deoxygenated conditions. The behaviour of other derivatives (410d—g) in the presence of xanthene has been reported in note form,[179] and the above publication[178] contains details of these reactions. A detailed account of the photochemistry of epoxynaphthaquinones (414) in the presence of olefins has been published.[180, 181] Two types of cycloaddition were encountered (Scheme

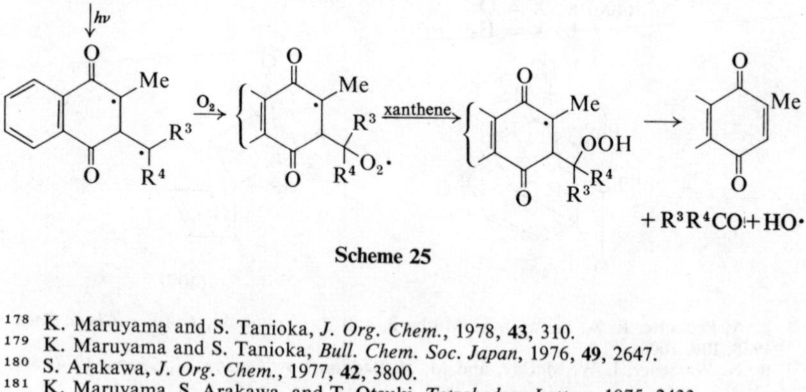

Scheme 25

[178] K. Maruyama and S. Tanioka, *J. Org. Chem.*, 1978, **43**, 310.
[179] K. Maruyama and S. Tanioka, *Bull. Chem. Soc. Japan*, 1976, **49**, 2647.
[180] S. Arakawa, *J. Org. Chem.*, 1977, **42**, 3800.
[181] K. Maruyama, S. Arakawa, and T. Otsuki, *Tetrahedron Letters*, 1975, 2433.

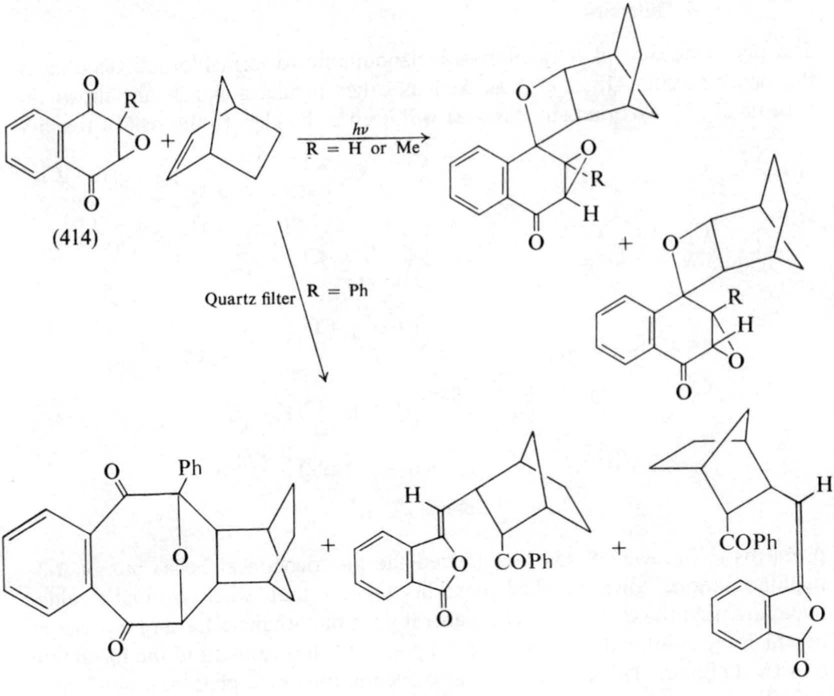

Scheme 26

26), *viz.* formation of spiro-oxetans by [2 + 2] cycloaddition, and additions of 1,3-dipolar type. The behaviour of the quinones with hydrogen donors has also been studied.[182]

Experimental details of two examples of the conversion of the quinone–diene Diels–Alder adducts (415a) and (415b) into tricyclic compounds (416) and (417), respectively, have been reported.[183, 184]

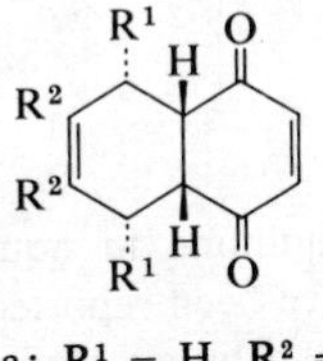

(415) a; R^1 = H, R^2 = Me
       b; R^1 = Me, R^2 = H
(416) 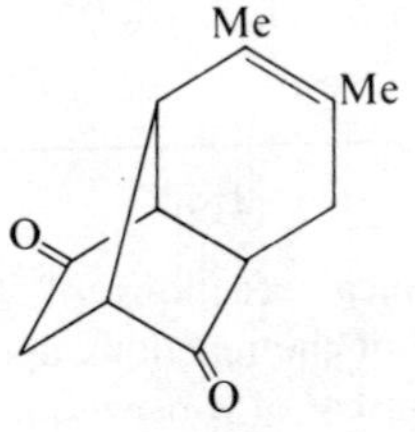
(417) 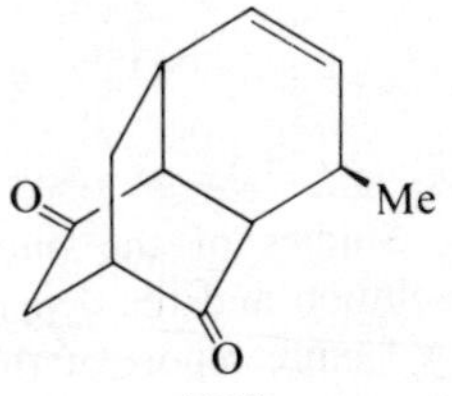

[182] K. Maruyama and S. Arakawa, *J. Org. Chem.*, 1977, **42**, 3793.
[183] J. R. Scheffer and K. S. Bhandari, *Org. Photochem. Synth.*, 1976, **2**, 54 (*Chem. Abs.*, 1977, **87**, 5466).
[184] J. R. Scheffer and R. E. Gayler, *Org. Photochem. Synth.*, 1976, **2**, 80 (*Chem. Abs.*, 1977, **87**, 5467).

6 Quinones

The photoaddition of tetrachloro-*o*-benzoquinone to tetrachloroethylene gives the benzodioxen (418; 17%) as well as other products which are shown in Scheme 27.[185] (Workers in this area will need to be alert to the risk of toxicity

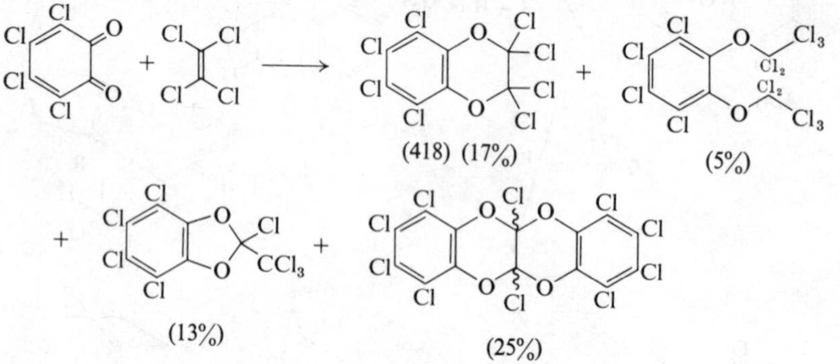

(418) (17%)   (5%)

(13%)   (25%)

Scheme 27

problems.) Takuwa[186] has investigated the photochemical behaviour of 1,2-naphthoquinone towards aldehydes: he reports that when aromatic aldehydes are used the main product is generally the naphthalene-1,2-diol monoester in which the combination of the radical pair (419) has resulted in the formation of a C—O bond. This behaviour is also encountered with propenal, but-2-enal, *trans*-hexen-2-al, 1-cyclopentenecarbaldehydes, and 1-cyclohexenecarbaldehyde. On the other hand, saturated carboxaldehydes or non-conjugated unsaturated aldehydes afford mixtures of naphthalene-1,2-diol monoesters and 3-acyl-naphthalene-1,2-diols. In this latter reaction the combination within the biradical (419) leads to both C—O and C—C bond formation. Takuwa[186] suggests that the difference in products formed in the reactions results from a difference in the nucleophilicity of the acyl radicals.

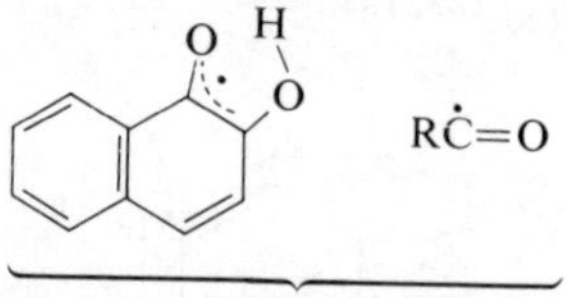

(419)

Studies of the photochemical reactions of 1,4-benzoquinone in aqueous solution and the dependence of the reactions upon pH have been reported.[187] A lecture report of the chemistry of *p*-benzoquinone excited by an argon ion laser has been published.[188] The report summarizes, amongst other reactions,

[185] S. Kumamoto, K. Somekawa, and S. Ide, *Kagoshima Daigaku Kogakubu Kenkyu Hokoku*, 1976, **18**, 101 (*Chem. Abs.*, 1977, **87**, 135 217).
[186] A. Takuwa, *Bull. Chem. Soc. Japan*, 1977, **50**, 2973.
[187] H. Thielemann, *Z. Chem.*, 1977, **17**, 109 (*Chem. Abs.*, 1977, **87**, 38 485).
[188] R. M. Wilson, S. W. Wunderly, J. G. Kalmbacher, and W. Brabender, *Ann. New York Acad. Sci.*, 1976, **267**, 201 (*Chem. Abs.*, 1977, **86**, 170 394).

the work carried out by Wilson *et al.*[189] during the past few years on the photo-addition of *p*-benzoquinone to cyclo-octatetraene.

The aminoquinones (420) undergo intramolecular photochemical hydrogen abstraction reactions leading to the formation of the cyclized products (421).[190]

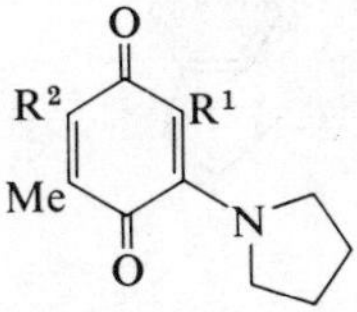

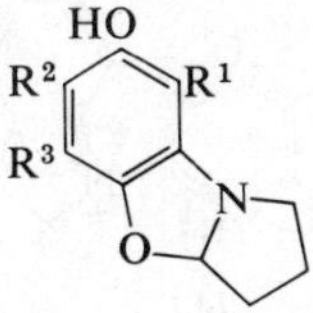

(420) a; R^1 = Me, R^2 = H
     b; R^1 = MeCO; R^2 = H
     c; R^1 = MeCO, R^2 = N⟨⟩

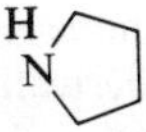

(421) a; R^1 = R^3 = Me, R^2 = H
     b; R^1 = COMe, R^2 = H, R^3 = Me
     c; R^1 = Me, R^2 = 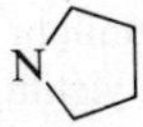 N⟨⟩; R^3 = COMe

The most likely path involves Norrish Type II behaviour to produce a biradical intermediate which conceivably could either undergo electron transfer to give (422) or else produce an intermediate spirocyclopropane formation (423). This latter path has been used by others[191] to explain the reaction of alkyl-substituted 1,4-benzoquinones. A similar mechanism could be involved in the formation of the cyclized products (424) produced by the irradiation of the naphthoquinones

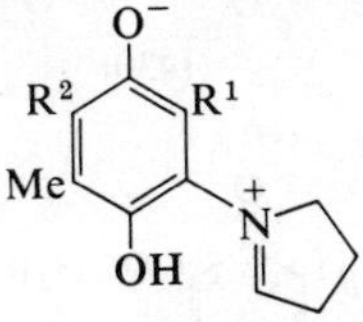

(422)

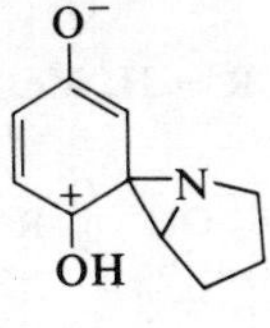

(423)

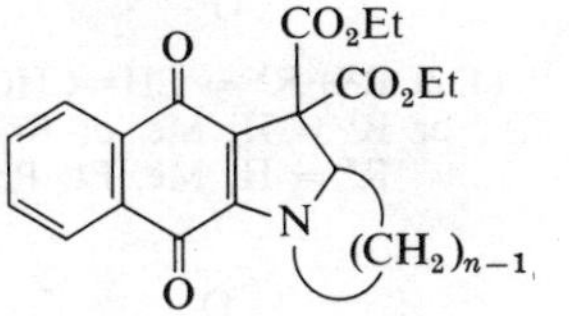

(424)

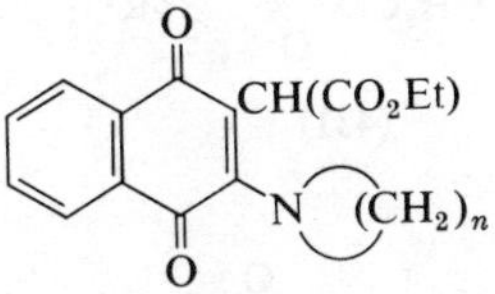

(425)

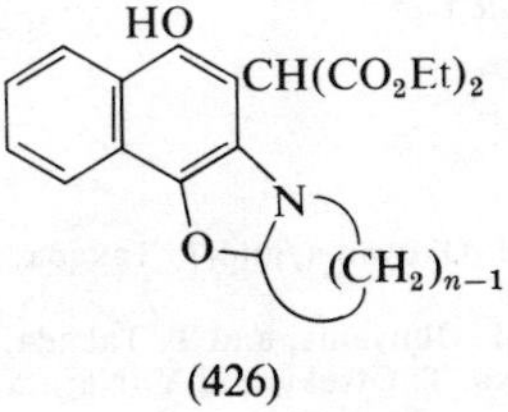

(426)

[189] R. M. Wilson and S. W. Wunderly, *J.C.S. Chem. Comm.*, 1974, 461; *J. Amer. Chem. Soc.*, 1974, **96**, 7350; E. J. Gardner, R. H. Squire, R. C. Elder, and R. M. Wilson, *ibid.*, 1973, **95**, 1693.
[190] K. J. Falci, R. W. Franck, and G. P. Smith, *J. Org. Chem.*, 1977, **42**, 3317.
[191] S. Farid, *Chem. Comm.*, 1971, 73.

(424) ($n = 4$ or 5). The intermediate oxazolines (426) are involved in this reaction.[192, 193] Intramolecular hydrogen abstraction results in the formation of the transient zwitterion (427) when (428) is irradiated.[194]

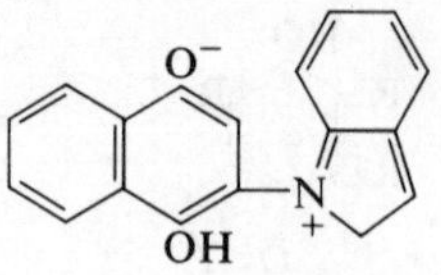

(427) 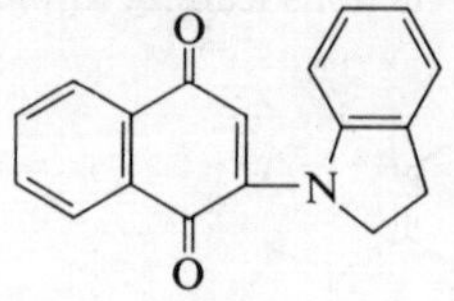 (428)

The cyclobutane-type dimers of the quinone (429) undergo photo-ring-opening in sunlight to afford the bis-quinone derivatives (430).[195] Regiospecific dimerization yielding (431) results when the quinones (432) are irradiated.[196]

Charge-transfer absorption has been observed in solutions of the quinone (433) and furan.[197] Irradiation into this charge-transfer band at 488 nm leads to the

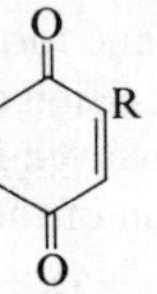

(429) R = H, Me, or Ph 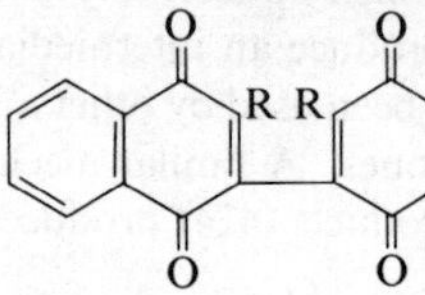 (430)

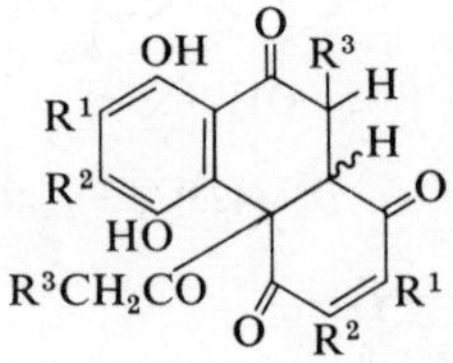

(431)

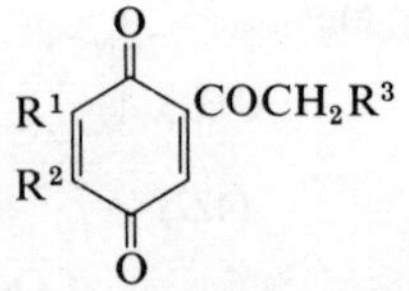

(432) R^1—R^2 = CH=CHCH=CH
or R^1 = H, Me, or Br,
R^3 = H, Me, Et, Pr, or Bu

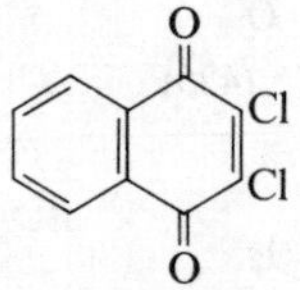

(433) 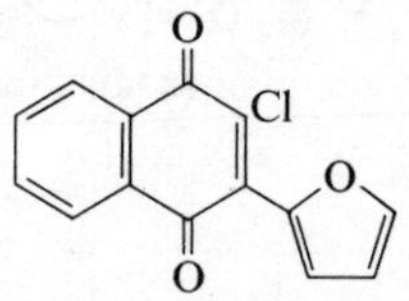 (434)

[192] M. Akiba, Y. Kosugi, M. Okuyama, and T. Takada, *Heterocycles*, 1977, **6**, 1113 (*Chem. Abs.*, 1977, **87**, 151 943).
[193] M. Akiba, Y. Kosugi, M. Okuyama, and T. Takada, *J. Org. Chem.*, 1978, **43**, 181.
[194] K. Maruyama, T. Kozuka, T. Otsuki, and Y. Naruta, *Chem. Letters*, 1977, 1125 (*Chem. Abs.*, 1978, **88**, 37 557).
[195] F. J. C. Martin, J. Dekker, A. M. Viljoen, and D. P. Venter, *S. African J. Chem.*, 1977, **30**, 89 (*Chem. Abs.*, 1978, **88**, 5871).
[196] Y. Miyagi, K. Kitamura, K. Maruyama, and Y. L. Chow, *Chem. Letters*, 1978, 33 (*Chem. Abs.*, 1978, **88**, 104 550).
[197] K. Maruyama and T. Otsuki, *Bull. Chem. Soc. Japan*, 1977, **50**, 3429.

formation of the product (434) *via* formation of a radical pair and the evolution of HCl. Several examples of the reaction are reported.[197]

Royleanone (435a), a naturally occurring quinone, undergoes photochemical transformation by irradiation into the 270 nm band. The products (from side-chain cyclization or isomerization and from the formation of a cyclic ether) are thought to arise from the presence of a loosely associated but highly polarized complex in which photoexcitation brings about double hydrogen transfer with the formation of a hydroquinone molecule (436; isolated as the triacetate) and a biradical (437) capable of yielding the products (439)—(441) *via* the intermediate

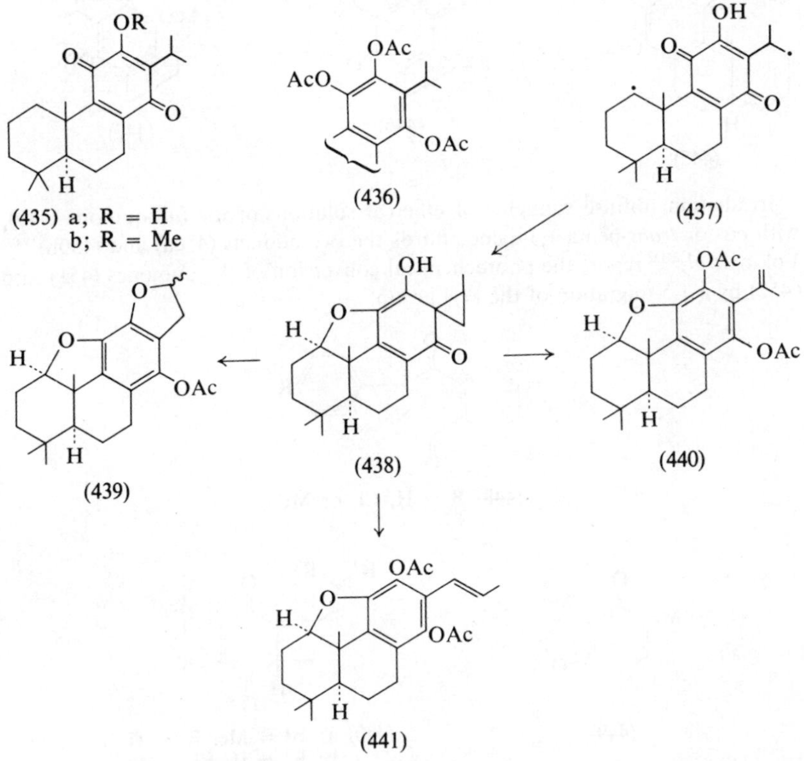

(435) a; R = H
      b; R = Me

(436)

(437)

(439)

(438)

(440)

(441)

(438).[198] The photochemistry of the acetoxy-derivative (442) was also studied in both benzene and dimethoxyethane. The four main products from the reaction were identified as (443) and (444) after the reaction mixture had been acetylated. Hydrogen-abstraction reactions were encountered in the photolysis of the methyl ether (435b) to give the products (445)—(447).

Irradiation of the quinones (448) in the presence of hydrogen donors has been described.[199]

[198] O. E. Edwards and P.-T. Ho, *Canad. J. Chem.*, 1978, **56**, 733.
[199] K. Maruyama, T. Iwai, T. Otsuki, Y. Naruta, and Y. Miyagi, *Chem. Letters*, 1977, 1127 (*Chem. Abs.*, 1978, **88**, 37 587).

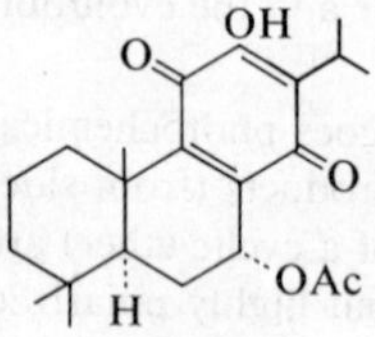

(442) 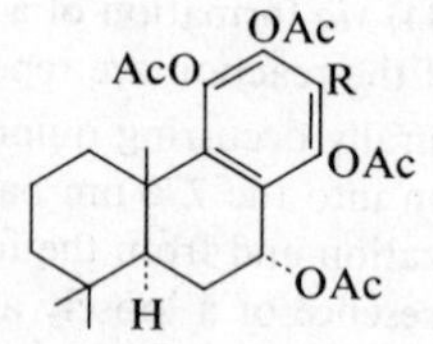   (443) R = C(Me)=CH$_2$ or CH=CHMe 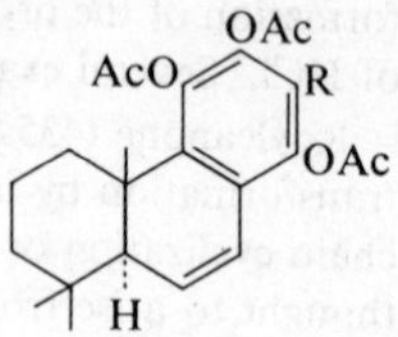   (444)

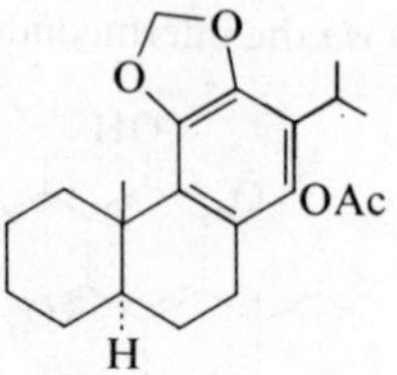

(445) 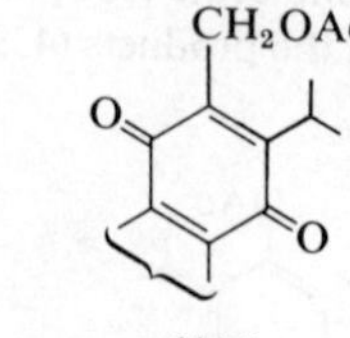   (446) 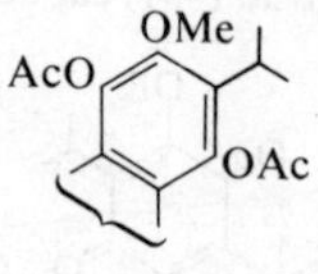   (447)

Irradiation (diffuse sunlight) of ethereal solutions of the furoquinone (449), with *cis*- or *trans*-penta-1,3-diene affords the two adducts (450a) and (450b).[200] Fokin *et al.*[201] report the photochemical conversion of the quinones (451) into (452) by a 1,5-migration of the aryl group.

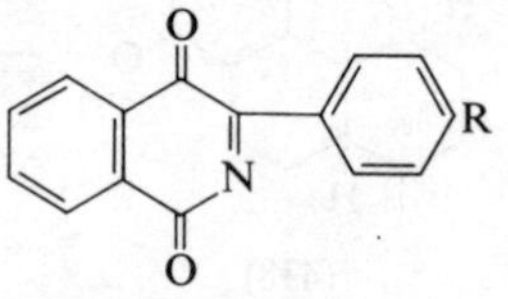

(448) R = H, Cl, or Me

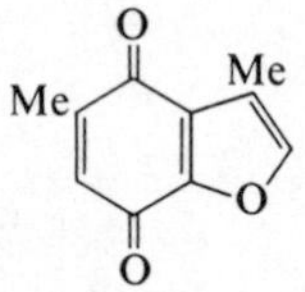

(449)

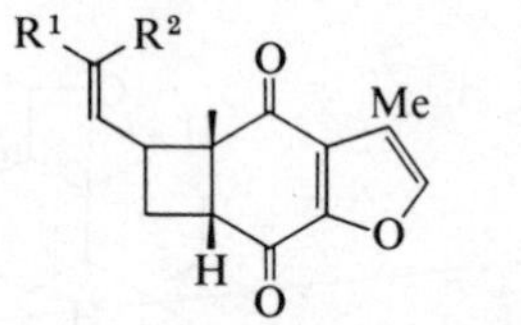

(450) a; R^1 = Me, R^2 = H
      b; R^1 = H, R^2 = Me

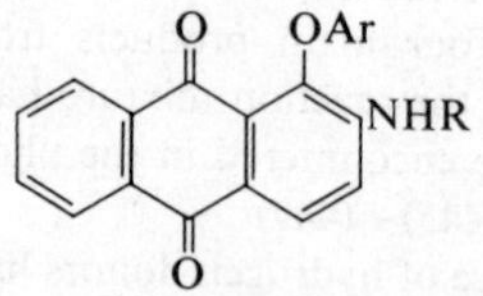

(451) 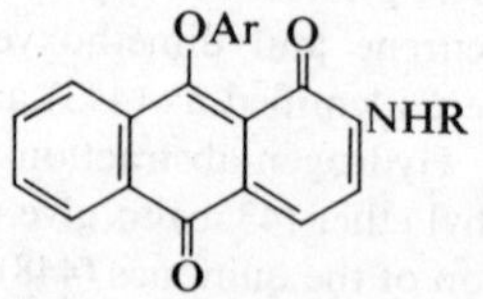   (452)

[200] F. Bohlmann and H.-J. Förster, *Chem. Ber.*, 1977, **110**, 2016.
[201] E. P. Fokin, S. A. Russkikh, and L. S. Klimenko, *Zhur. org. Khim.*, 1977, **13**, 2010 (*Chem. Abs.*, 1978, **88**, 22 446).

3
Photochemistry of Olefins, Acetylenes, and Related Compounds

BY W. M. HORSPOOL

1 Reactions of Alkenes

Addition Reactions.—A detailed study of the photo-behaviour of allyl aryl ethers in hydrocarbon and alcoholic solvents has been reported.[1] Irradiation of isoeugenol (1) has been shown to lead to dehydro-dimers, *cis–trans* isomerization of the double bond, and solvent addition to give products of type (2).[2] Addition of solvent to styryl systems is not common although it has been reported by Miyamoto *et al.*[3] for addition to (3), and also by Baker and Horspool[4] for additions to ferrocenyl olefins (4).

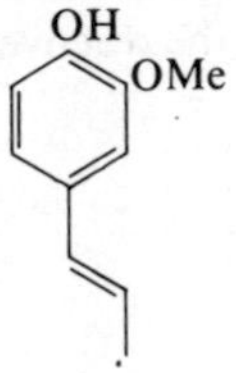

OH
OMe

OH
OMe

MeO Et

R

R
Fc

(1) 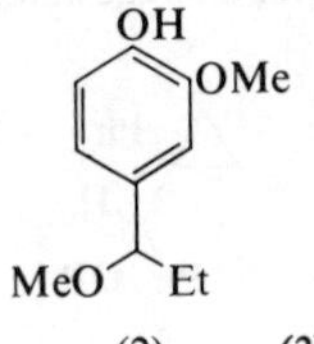   (2) 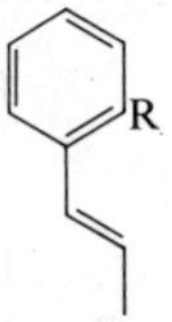   (3) R = OMe, OH, or NH$_2$ 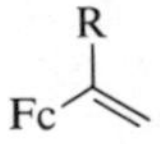   (4) R = H or Me
Fc = C$_5$H$_5$FeC$_5$H$_4$

Arnold and his co-workers[5] have continued their investigations into reactions involving electron transfer processes. Thus irradiation of 1,1-diphenylethylene in the presence of electron-donating sensitizers such as 1-methoxynaphthalene, 1,4-dimethoxynaphthalene, or 1-methylnaphthalene results in formation of the radical-anion (5) which, in the presence of an alcohol, affords an ether, probably by the route shown in Scheme 1. The same technique has been used for the irradiation of other olefins.[6] Thus when olefins (6) and (7) are irradiated along with 1-cyanonaphthalene or methyl *p*-cyanobenzoate as an electron acceptor a radical-anion is produced. This can be trapped byNC$^-$ as the nitriles (8) and (9).

[1] H.-R. Waespe, H. Heimgartner, H. Schmid, H.-J. Hansen, H. Paul, and H. Fischer, *Helv. Chim. Acta*, 1978, **61**, 401.
[2] G. Leary, *Austral. J. Chem.*, 1977, **30**, 1133.
[3] N. Miyamoto and H. Nozaki, *Tetrahedron*, 1973, **29**, 3819; N. Miyamoto, M. Kawanisi, and H. Nozaki, *Tetrahedron Letters*, 1971, 2565; N. Miyamoto, K. Otimoto, and H. Nozaki, *ibid.*, 1972, 2895.
[4] C. Baker and W. M. Horspool, *Chem. Comm.*, 1971, 615.
[5] D. R. Arnold and A. J. Maroulis, *J. Amer. Chem. Soc.*, 1977, **99**, 7355.
[6] A. J. Maroulis, Y. Shigemitsu, and D. R. Arnold, *J. Amer. Chem. Soc.*, 1978, **100**, 535.

$$(Ph_2C-CH_2)^{\cdot-} \xrightarrow{ROH} Ph_2\dot{C}-CH_3$$

$$(5)$$

$$Ph_2\dot{C}Me + D^{\cdot+} \longrightarrow Ph_2\overset{+}{C}Me + D$$

$$Ph_2\overset{+}{C}Me + RO^- \longrightarrow Ph_2CMe$$
$$\underset{OR}{|}$$

Scheme 1

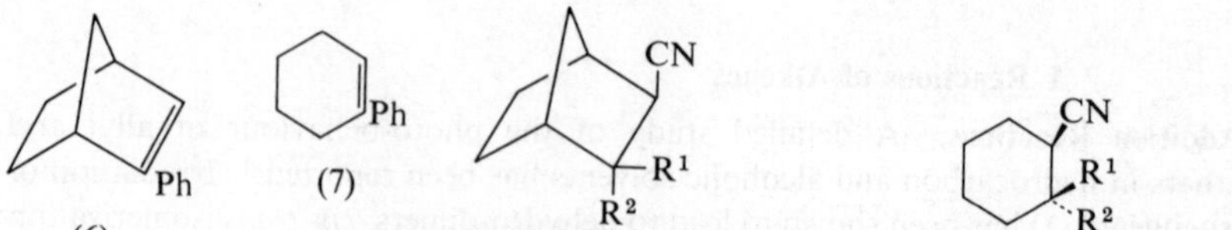

(6)   (7)   (8) R^1 = Ph, R^2 = H (10%)   (9) R^1 = Ph, R^2 = H (20%)
        R^1 = H, R^2 = Ph (29%) R^1 = H, R^2 = Ph (19%)

A full account of the results of the photoaddition of methanol and acetic acid to 1,2-diphenylcyclobutene has been published:[7, 8] the products are shown in Scheme 2. As a result of the quenching of the fluorescence of 1,2-diphenyl-cyclobutene by methanol and by acetic acid, a singlet is suggested as the reactive

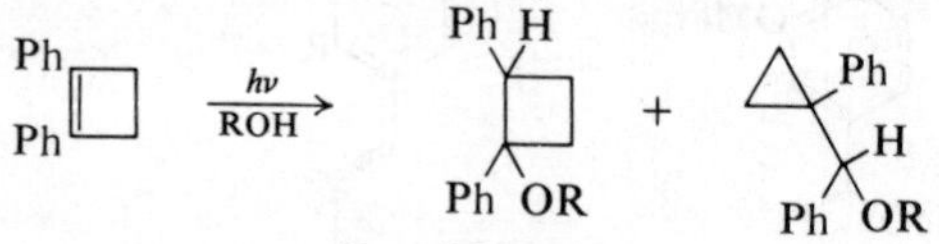

$$Ph \xrightarrow[ROH]{h\nu} \quad + \quad$$

R = Me or Ac

Scheme 2

state. The fluorescence of 1,2-diphenylcyclobutene is also quenched by amines in hexane with a rate close to a diffusion-controlled rate.[9] Irradiation of the olefin with diethylamine at 300 nm gives three products (10)—(12) by a mechanism involving a charge-transfer complex. The authenticity of this mechanism

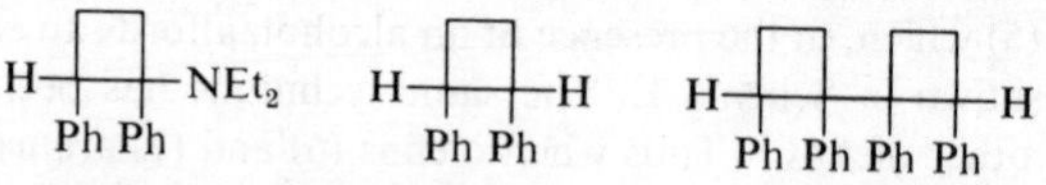

(10)   (11)   (12)

was checked by the use of deuteriated amine. The fluorescence of *trans*-stilbene is also quenched by tertiary and secondary aliphatic amines, but, surprisingly, not by primary amines.[10] Furthermore, irradiation of *trans*-stilbene in the presence of such secondary and tertiary amines gives addition products as illustrated in Scheme 3. The formation of these products implies the intermediacy

[7]   M. Sakuragi, H. Sakuragi, and M. Hasegawa, *Bull. Chem. Soc. Japan*, 1977, **50**, 1562.
[8]   M. Sakuragi and M. Hasegawa, *Chem. Letters*, 1974, 29.
[9]   M. Sakuragi and H. Sakuragi, *Bull. Chem. Soc. Japan*, 1977, **50**, 1802.
[10] F. D. Lewis and T.-I. Ho, *J. Amer. Chem. Soc.*, 1977, **99**, 7991.

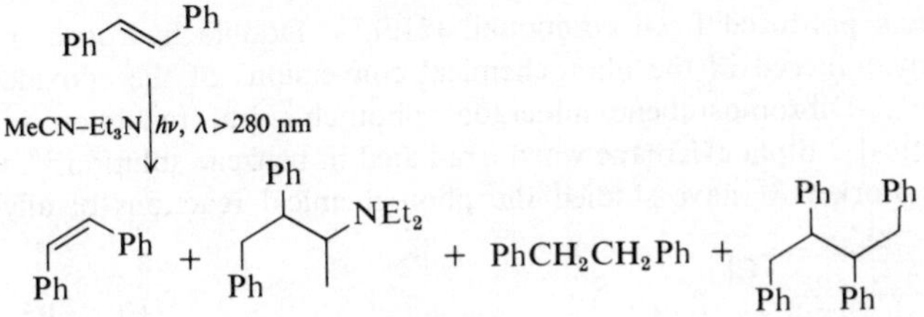

Scheme 3

of free radicals, and the authors[10] suggest that there is some similarity between the reaction described here and the peroxide-induced thermal reaction between amines and alkenes. A study of the fluorescence quenching data indicates that *trans*-stilbene and amines exhibit exciplex emission at higher amine concentrations. From this and other information, the reversible formation of an exciplex is proposed.

Warrener *et al.*[11] have used photochemical fragmentation of the compounds (13) to obtain the azetines (14). The azetine is photochemically reactive and undergoes photochemical hydration to yield the unstable adduct (15), which subsequently ring-opens to yield (16). A full account of the photoaddition of alcohols to phospholenes [*e.g.* (17)] has been reported.[12,13] The phospholene (17) adds a proton on irradiation in methanol to yield the tertiary ion (18). The direction of addition has been interpreted in terms of a polarized double bond. The products formed from the intermediate (18) are the olefin (19) and the ether (20).

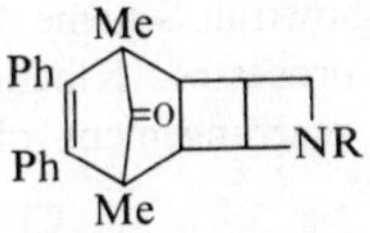

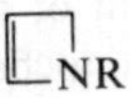

(14)

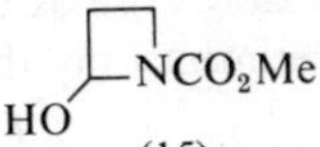

(15)

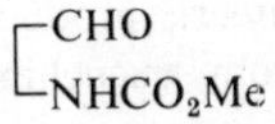

(16)

(13) a; R = p-$MeC_6H_4SO_2$
b; R = CO_2Me

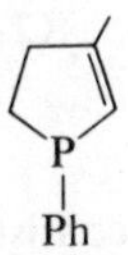

(17)

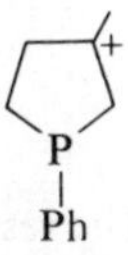

(18)

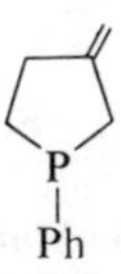

(19)

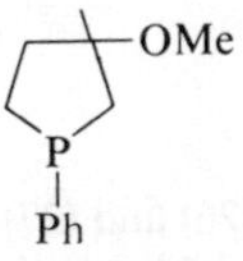

(20)

Halogeno-olefins.—Double hydrogen transfer is used to account for the sensitized conversion of (21a and b) into (22a and b).[14] The products of the reaction are stable to the irradiative conditions, in contrast with the light-sensitive

[11] R. N. Warrener, G. Kretschmer, and M. N. Paddon-Row, *J.C.S. Chem. Comm.*, 1977, 806.
[12] H. Tomioka, K. Sugiura, S. Takata, Y. Hirano, and Y. Izawa, *J. Org. Chem.*, 1977, **42**, 3070.
[13] H. Tomioka and Y. Izawa, *Tetrahedron Letters*, 1973, 5059.
[14] H. Parlar, M. Mansour, and S. Gab, *Tetrahedron Letters*, 1978, 1597.

products produced from compound (21c).[15] Double hydrogen transfers are also encountered in the photochemical conversions of the epoxides (23) into (24). 1,2-Dibromostilbene undergoes photochemical reduction to *meso*-1,2-dibromo-1,2-diphenylethane when irradiated in benzene solution.[16] Cristol and his co-workers[17] have studied the photochemical reactions of allyl chlorides

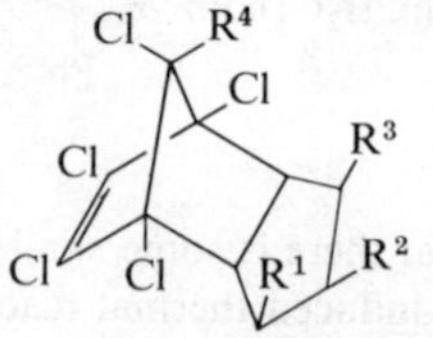

(21) a; $R^1 = R^2 = R^3 = R^4 = H$
      b; $R^1 = R^2 = R^3 = H, R^4 = Cl$
      c; $R^1 = R^3 = CO_2Me, R^2 = R^4 = Cl$

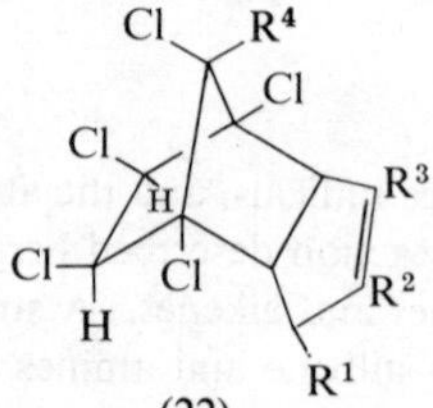

(22)

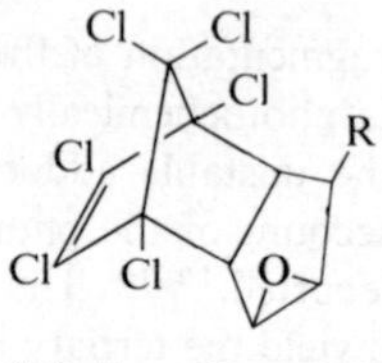

(23) R = H or Cl

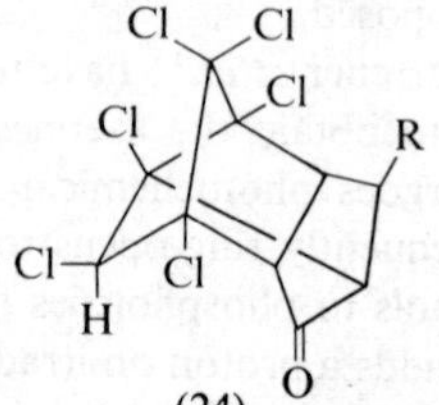

(24)

in greater detail. In particular they have examined the acetone- and benzene-sensitized reactions of 2-methyl-3-chloroprop-1-ene (25). In moderately concentrated solutions, the reaction affords the products shown in Scheme 4. The dimeric products probably arise as a result of radical processes. Kinetic data from quenching experiments on the photochemical rearrangement of allyl

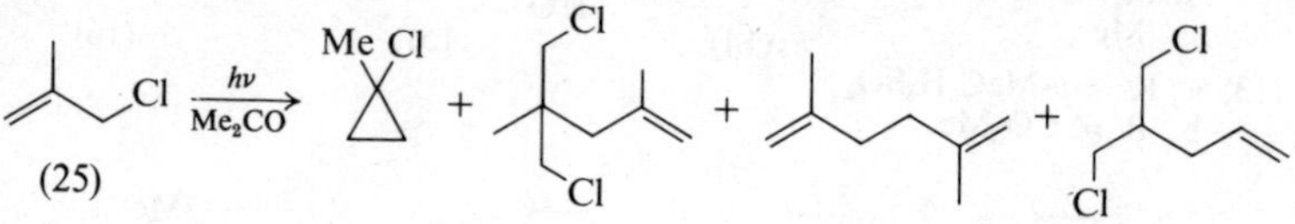

Scheme 4

chlorides (26) and (27) have been obtained.[18] The results are consistent with the formation of two or more intermediates in the excited state. One of these intermediates leads to the cyclopropane derivatives usually encountered in such systems. However, this intermediate is different for each allylic chloride. The second intermediate is thought to interconnect with the three allylic isomers. The photoaddition induced by production of free radicals of alcohols and ethers to tetrachloroethylene affords the adducts (28).[19]

[15] H. Parlar and F. Korte, *Chemosphere*, 1977, **6**, 665; H. Parlar, S. Gab, E. S. Lahaniatis, and F. Korte, *Chem. Ber.*, 1975, **108**, 3692.
[16] W. Schroth and H. Bahn, *Z. Chem.*, 1977, **17**, 56 (*Chem. Abs.*, 1977, **87**, 67 918).
[17] S. J. Cristol, R. J. Daughenbaugh, and R. J. Opitz, *J. Amer. Chem. Soc.*, 1977, **99**, 6347.
[18] S. J. Cristol and R. P. Micheli, *J. Amer. Chem. Soc.*, 1978, **100**, 850.
[19] T. Yumoto, *Nippon Kagaku Kaishi*, 1978, 134 (*Chem. Abs.*, 1978, **88**, 104 366).

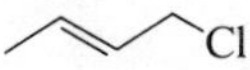

(26) 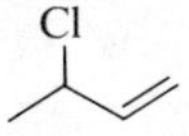   (27) $RO-CCl=CCl_2$

                                        (28)

Group Migration Reactions.—Izawa *et al.*[20] suggest that the products (29) and (30) formed by the acid-catalysed photoaddition of methanol to t-butylbenzene are the result of rearrangement of (31). It needs to be understood that this compound is the initial product of (probably thermal) acid-catalysed addition of methanol to t-butylbenzvalene. It is thought to undergo the sensitized photochemical change outlined in Scheme 5 involving 1,3-shifts of one of the

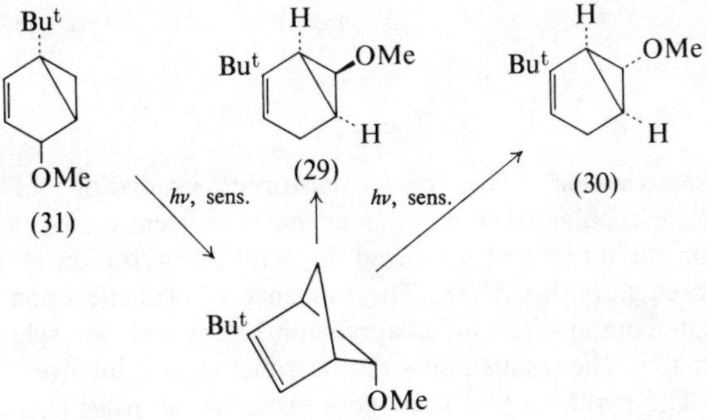

Scheme 5

cyclopropyl bonds. A full account of the direct irradiation of cyclohexene, cycloheptene, and cyclo-octene in the gas and liquid phases has been published.[21, 22] In the gas phase cycloheptene, for example, yields hepta-1,6-diene and vinylcyclopentane when the irradiations are carried out at low gas pressures. At higher gas pressures methylenecyclohexane and bicyclo[4,1,0]heptane are formed *via* a carbene intermediate (32).

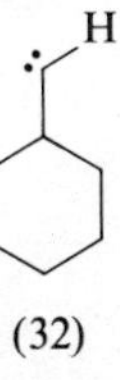

(32)

The cation (33) is formed when 10-methyloctalone is dissolved in concentrated sulphuric acid: it is quite stable, and octalone can be recovered unchanged from the acid solution. When the acid solutions are irradiated (300 nm), however, an efficient conversion to the cation (34) takes place, and the cyclopentenone (35)

20 Y. Izawa, H. Tomioka, T. Kagami, and T. Sato, *J.C.S. Chem. Comm.*, 1977, 780.
21 Y. Inoue, S. Takamuku, and H. Sakurai, *J.C.S. Perkin II*, 1977, 1635.
22 Y. Inoue, S. Takamuku, and H. Sakurai, *J.C.S. Chem. Comm.*, 1975, 577.

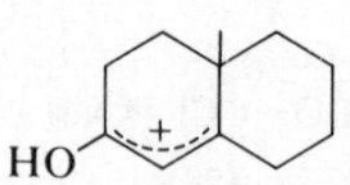

(33) 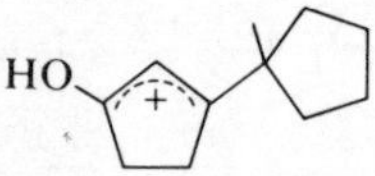   (34) 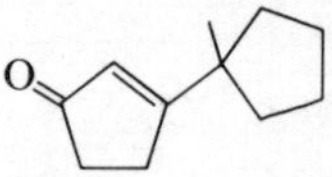   (35)

can be isolated.[23] If the cation (34), produced by irradiation at 300 nm, is
irradiated at 254 nm a second rearrangement takes place (Scheme 6) yielding
the ion (36) from which the ketone (37) can be isolated.

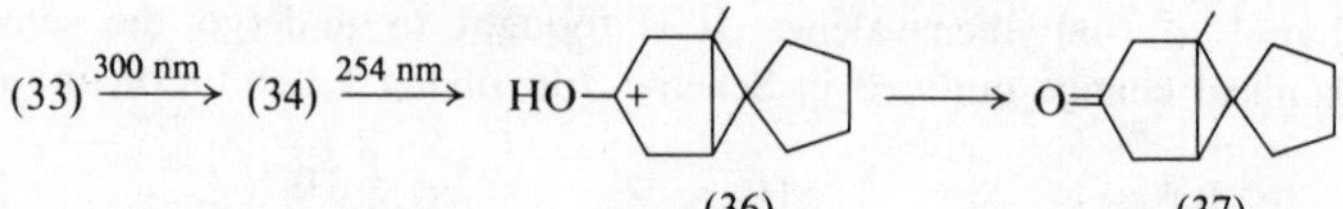

(36)   (37)

Scheme 6

cis–trans Isomerization.—The triplet-sensitized irradiation of 4-nitro-4'-
methoxystilbene in polar and non-polar solvents has been studied.[24] The results
obtained from the direct and sensitized *cis–trans* isomerization of 2-styrylthio-
phen have been published.[25, 26] The influence of azulene upon the photostationary state compositions for β-styrylnaphthalene and 3-styrylquinoline has
been evaluated.[27] The results imply that a triplet state is involved in the direct
irradiation. The pyridone (38) undergoes exclusive *cis–trans* isomerization on

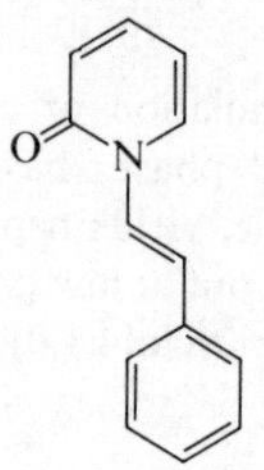

(38) 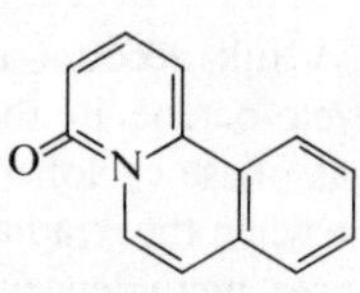   (39)

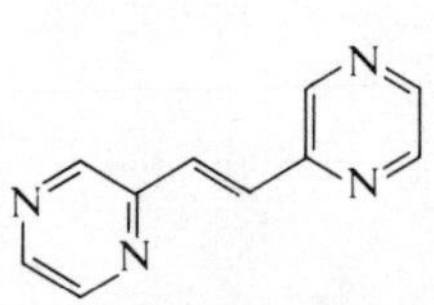

(40) 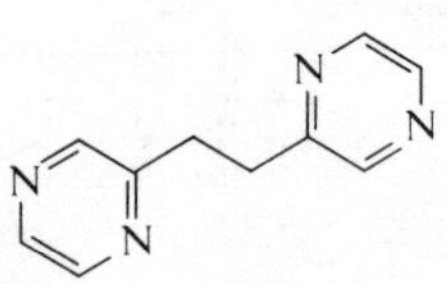   (41)

[23] D. G. Cornell and N. Filipescu, *J. Org. Chem.*, 1977, **42**, 3331.
[24] H. Goerner and D. Schulte-Frohlinde, *J. Photochem.*, 1978, **8**, 91 (*Chem. Abs.*, 1978, **88**, 120 363).
[25] L. L. Costanzo, S. Pistara, G. Scarlata, and G. Condorelli, *Chimica e Industria*, 1977, **59**, 121 (*Chem. Abs.*, 1977, **87**, 167 239).
[26] L. L. Costanzo, S. Pistara, G. Condorelli, and G. Scarlata, *J. Photochem.*, 1977, **7**, 297 (*Chem. Abs.*, 1977, **87**, 167 260).
[27] G. Gennari, G. Cauzzo, and G. Galiazzo, *J. Phys. Chem.*, 1977, **81**, 1551.

direct irradiation in neutral media, but with added HCl photocyclization occurs to yield (39).[28] The dipyrimidylethylene (40) undergoes photoreduction to (41) when irradiated in protic solvents.[29]

2 Reactions involving Cyclopropane Rings

Zimmerman *et al.*[30] have sought to explain why it is that in a series of di-π-methane reactions the rate of rearrangement varies with the system, although consistent substituent-control of regioselectivity is observed. In a study of the photochemical di-π-methane reactions of the dienes (42) Pratt and his co-workers[31] have established that the migratory preference in the biradical (43) is dihalogenovinyl > diphenylvinyl > dicyanovinyl. Thus the rearrangement of the dienes (42) affords the products shown in Scheme 7. Further study has

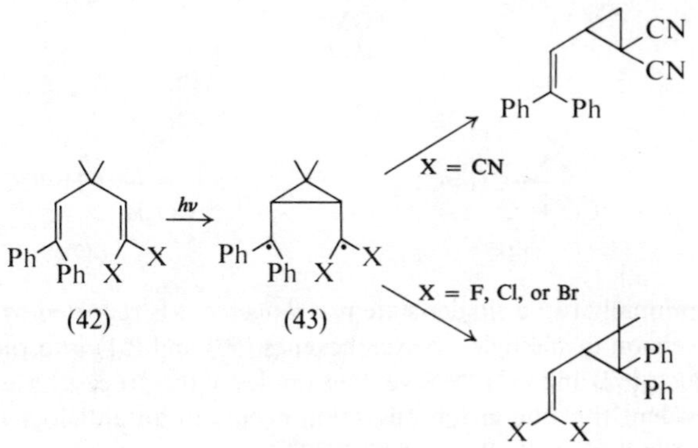

Scheme 7

shown that the dicyanodiene and the difluorodiene undergo di-π-methane rearrangement from the singlet state. On the other hand, the dichlorodiene undergoes the reaction from the triplet state. Secondary reaction processes were also uncovered in this study and it was shown that the difluorodiene yielded the adduct (44) when benzophenone sensitization was employed.

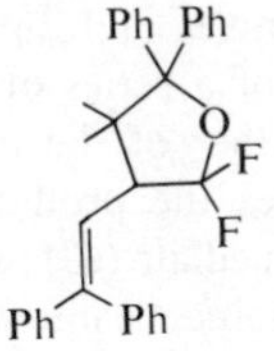

(44)

[28] P. S. Mariano, E. Krochmal, jun., and A. Leone, *J. Org. Chem.*, 1977, **42**, 1122.
[29] S.-C. Shim and J.-S. Chae, *Taehan Hwahak Hoechi*, 1977, **21**, 102 (*Chem. Abs.*, 1977, **87**, 200 487).
[30] H. E. Zimmerman, W. T. Gruenbaum, R. T. Klun, M. G. Steinmetz, and T. R. Welter, *J.C.S. Chem. Comm.*, 1978, 228.
[31] D. W. Alexander, A. C. Pratt, D. H. Rowley, and A. E. Tipping, *J.C.S. Chem. Comm.*, 1978, 101.

Zimmerman *et al.*[32] have demonstrated that the photoreaction of (45), an example of an arylvinylmethane reaction, occurs to give (46) with total stereo-specificity (within experimental error). The authors[32] point out that care has to be taken in the evaluation of such reactions since over-irradiation can result in the racemization of product (46) by formation of (47). The photochemical racemization of (−)-*trans*-(48) into (±)-*cis,trans*-(49) has been reported to occur under triplet-photosensitized conditions.[33]

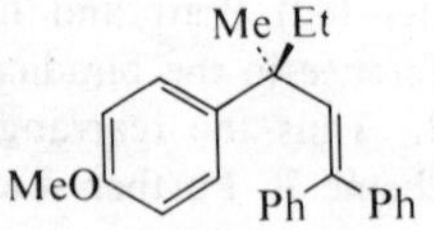

(45)

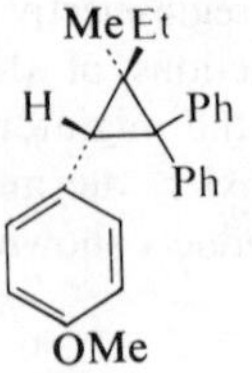

(46)

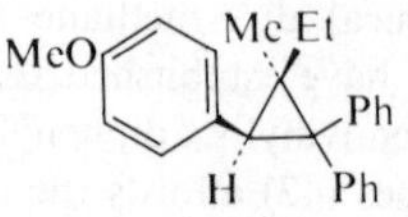

(47)

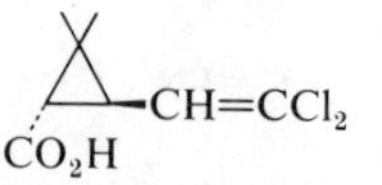

(48)

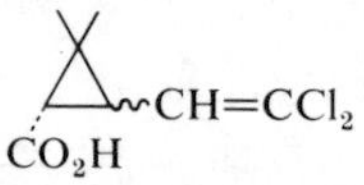

(49)

An exceptionally rapid singlet-state rearrangement is reported to be operative in the conversion of methylenebicyclohexenes (50) and (51) into the spirohepta-diene products (52) and (53).[34] Several examples of the process have been studied and it is evident that the group migration occurs in an anticlockwise direction as illustrated in Scheme 8. Product scrambling, where it occurs, can be accounted for by overshooting and reversal, again as illustrated in Scheme 8 for formation of the product (56) *via* intermediates (54) and (55).

A primary photoproduct (57) is formed when the pyran (58) is irradiated in t-butyl alcohol.[35] This primary product is itself photolabile and yields a product of ring opening. The formation of (57) involves the 1,2-migration of a phenyl group (59), followed by ring closure. Although no definitive experiment has as yet been carried out, the authors[35] consider that the reaction most likely occurs from the triplet state. Other examples of 1,2-phenyl migration reactions have also been provided by the study of a series of 1,1-diphenylindenes (60). The products of the irradiation are either of the two isomeric products (61) and (62).[36] The step which determines the product in each case is the thermal migration of hydrogen in the intermediate (63) produced by a 1,2-photochemical aryl migration. The formation of isoindene intermediates has also been examined by McCullough and his group.[37, 38] They have published a full account of the

[32] H. E. Zimmerman, T. P. Gannett, and G. E. Keck, *J. Amer. Chem. Soc.*, 1978, **100**, 323.
[33] T. Umemura and N. Itaya, Ger. Offen., 2 628 477/1976 (*Chem. Abs.*, 1977, **87**, 102 010).
[34] H. E. Zimmerman and T. P. Cutler, *J.C.S. Chem. Comm.*, 1978, 232.
[35] D. Gravel, C. Leboeuf, and S. Caron, *Canad. J. Chem.*, 1977, **55**, 2373.
[36] W. A. Pettit and J. W. Wilson, *J. Amer. Chem. Soc.*, 1977, **99**, 6372.
[37] K. Kamal de Foneska, C. Manning, J. J. McCullough, and A. J. Yarwood, *J. Amer. Chem. Soc.*, 1977, **99**, 8257.
[38] J. J. McCullough and A. J. Yarwood, *J.C.S. Chem. Comm.*, 1975, 485.

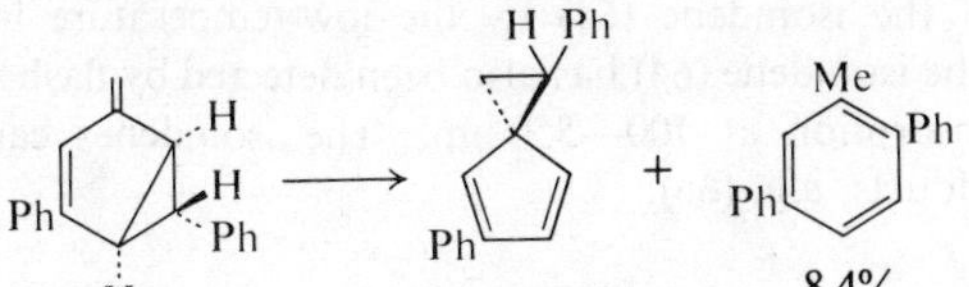

(50) → (52) 91.6% + 8.4%

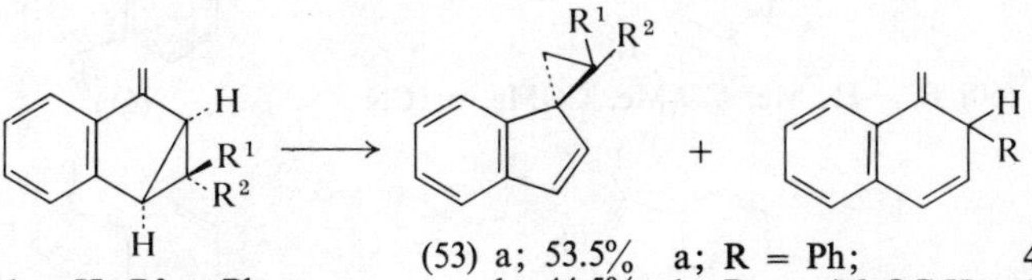

(51) a; R¹ = H, R² = Ph
b; R¹ = H, R² = p-MeOC₆H₄
c; R¹ = Ph, R² = H
d; R¹ = p-MeOC₆H₄, R² = H

(53) a; 53.5%
b; 44.5%
c; 10.6%
d; —

a; R = Ph;   46.5%
b; R = p-MeOC₆H₄;   55.5%
c; R = Ph;   89.4%
d; R = p-MeOC₆H₄;   100%

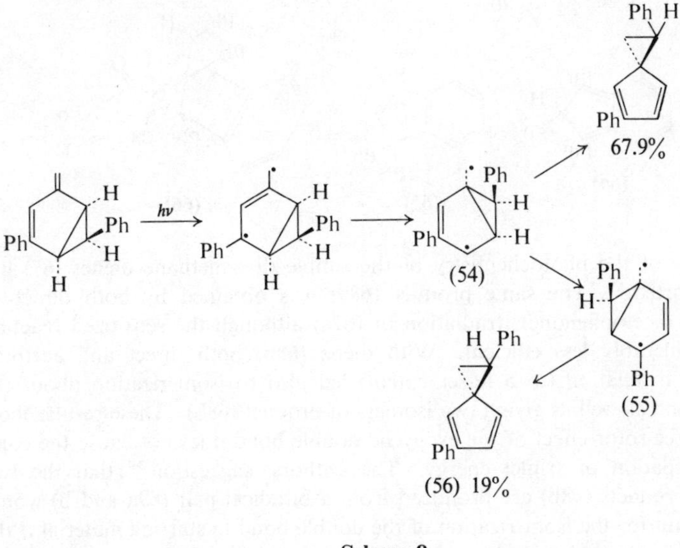

(54) → 67.9%
(55)
(56) 19%

Scheme 8

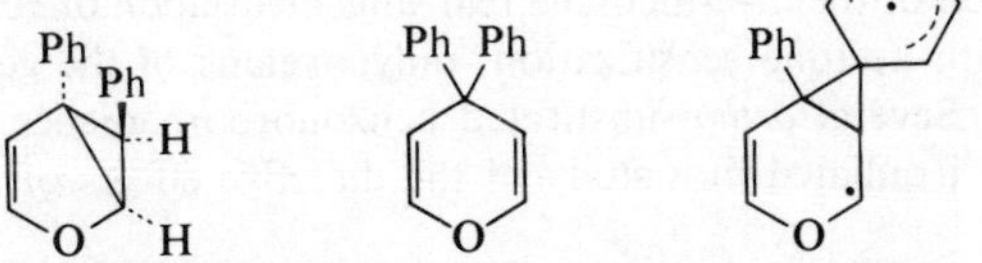

(57)   (58)   (59)

preparation of the isoindene (64) by the low-temperature irradiation of the indene (65). The isoindene (64) has also been detected by flash spectrometry and exhibits an absorption at 400—550 nm. The isoindenes can be trapped as Diels–Alder adducts, *e.g.* (66).

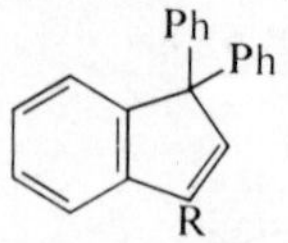

(60) R = H, Me, CO$_2$Me, COMe, or CN

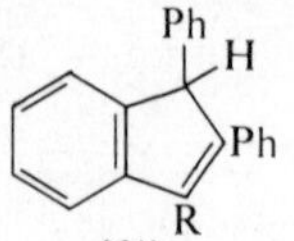

(61)

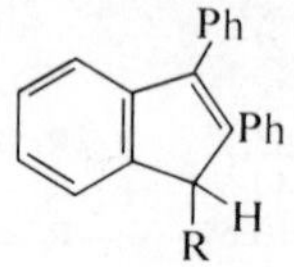

(62)

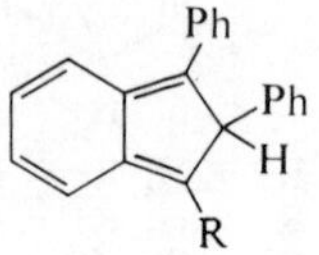

(63)

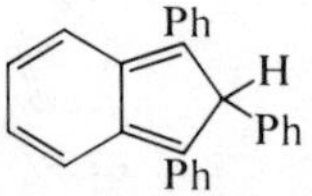

(64)

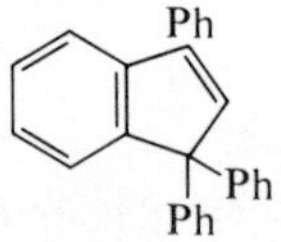

(65)

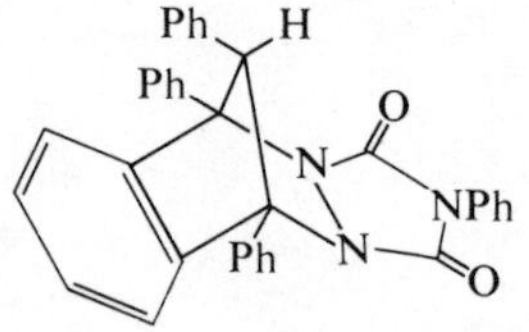

(66)

A study of the photochemistry of the simple di-π-methane dienes (67) has been reported.[39] The same product (68a) was obtained by both direct or sensitized (acetophenone) irradiation of (67a) although the sensitized reaction was considerably less efficient. With diene (68b) both direct and acetone-sensitized irradiation (to a lesser extent) led also to isomerization about the double bond as well as giving two isomers of product (68b). These results show that the free-rotor effect of the exocyclic double bond does not cause the complete dissipation of triplet energy. The authors' suggestion[39] that the two isomeric products (68b) are produced from a biradical pair (69a and b) would also account for the isomerization of the double bond in starting material if the radical pair also decayed back to (68b). The triene (67c) undergoes direct irradiative conversion into (68c) and (70), whereas acetophenone-sensitized irradiation of (67c) gives only one monomeric product (68c). At least part of the triene product (70) arises from secondary photolysis of the di-π-methane product (68c). The incorporation of a heteroatom into the system, as in (67d) led to suppression of the di-π-methane rearrangement upon direct and sensitized irradiation. With acetone sensitization, only oxetans of the general type (71) were produced. Several *ortho*-substituted benzonorbornadienes (72) have been synthesized and irradiated in a study of the directive effects of substituents on

[39] R. G. Weiss and G. S. Hammond, *J. Amer. Chem. Soc.*, 1978, **100**, 1172.

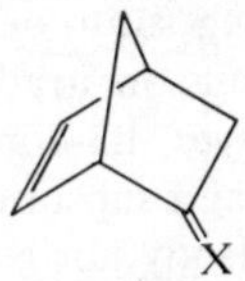

(67) a; X = CH$_2$
b; X = CHMe
c; X = CHCH=CMe$_2$
d; X = NPr^i

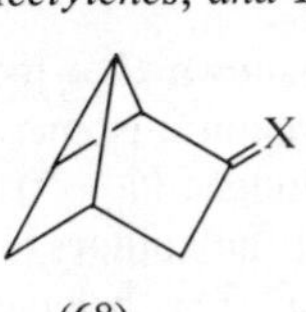

(68)

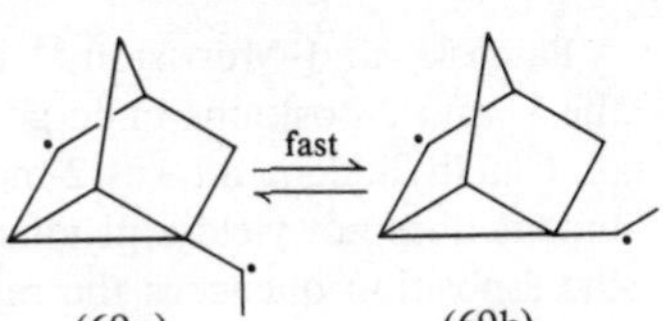

(69a)

fast

(69b)

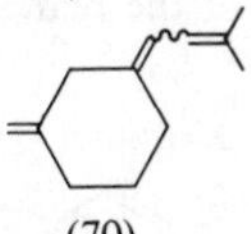

(70)

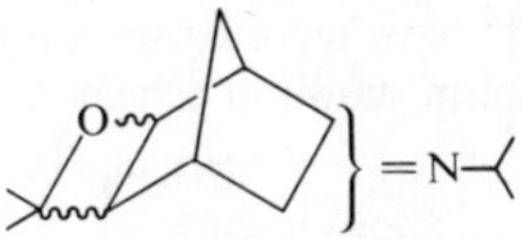

(71)

} = N—⟨

the di-π-methane reaction.[40] The results indicate that the triplet-state di-π-methane rearrangements of these compounds proceed with a high preference for benzo-vinyl bridging *ortho* to the substituent.[40] The one exception to this is the fluoro-compound (72; R = F) which affords products by *ortho*-bridging (50%) and *meta*-bridging (50%). A study was carried out for the tricyclotriene (73). A full account of the control by *meta*-substituents upon the mode of ring closure of benzonorbornadienes has been reported.[41, 42]

The hydrocarbons (74) undergo photochemical di-π-methane rearrangement to afford the products (75).[43]

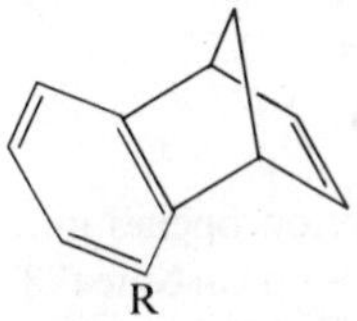

(72) R = CN, NO$_2$, F, OMe, or COMe 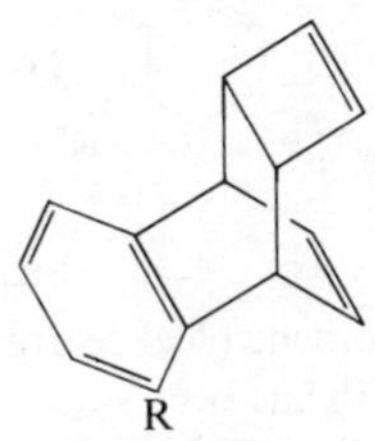   (73) R = CN or OMe

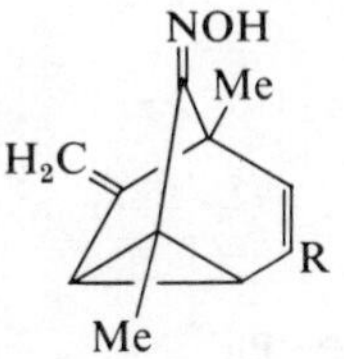

(74) a; R = H
b; R = Me

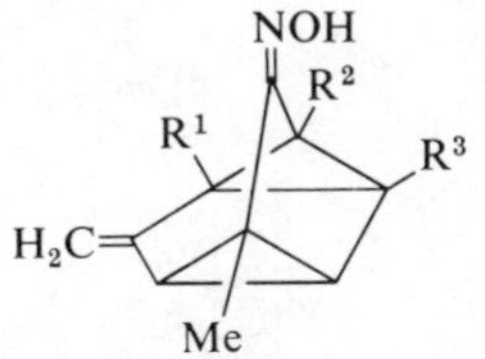

(75) a; R^1 = R^2 = H, R^3 = Me
and R^1 = Me, R^2 = R^3 = H
b; R^1 = R^3 = Me, R^2 = H

[40] R. A. Snow, D. M. Cottrell, and L. A. Paquette, *J. Amer. Chem. Soc.*, 1977, **99**, 3734.
[41] L. A. Paquette, D. M. Cottrell, and R. A. Snow, *J. Amer. Chem. Soc.*, 1977, **99**, 3723.
[42] L. A. Paquette, D. M. Cottrell, R. A. Snow, K. B. Gifkins, and J. Clardy, *J. Amer. Chem. Soc.*, 1975, **97**, 3275.
[43] M. Nitta, O. Inoue, and M. Tada, *Chem. Letters*, 1977, 1065 (*Chem. Abs.*, 1978, **88**, 50 361).

Palensky and Morrision [44] have shown that indenes having substituents at the 1- and 2-positions undergo a structural isomerization. Thus the irradiation of 1-methylindene affords 2-methylindene ($\Phi = 0.03$). However, 1,3-dimethyl-indene does not yield a product and the authors [44] reason that a substituent in the 3-position quenches the reaction. This behaviour is evidently not general, for Padwa *et al.*[45] have observed the rearrangement of 1-phenyl-2-methyl-3-benzylindene (76) into 1-methyl-2-phenyl-3-benzylindene (77). Palensky and Morrison [44] have reported several examples of the rearrangement and suggest the mechanism shown in Scheme 9.

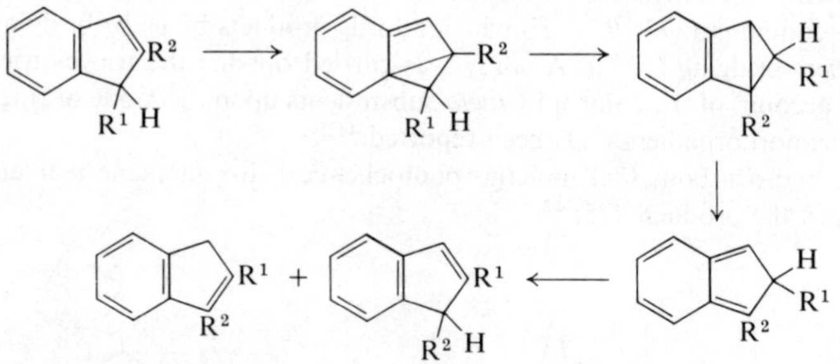

Scheme 9

The photochemical rearrangement of vincyclopropenes into cyclopentadienes (Scheme 10) has been suggested to involve either a carbene (78) [46] or a 'housane-type' biradical (79) [47] intermediate. Padwa *et al.*[48] have examined the process further and have presented information supportive of a carbene process. Thus

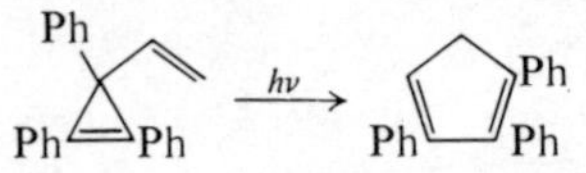

Scheme 10

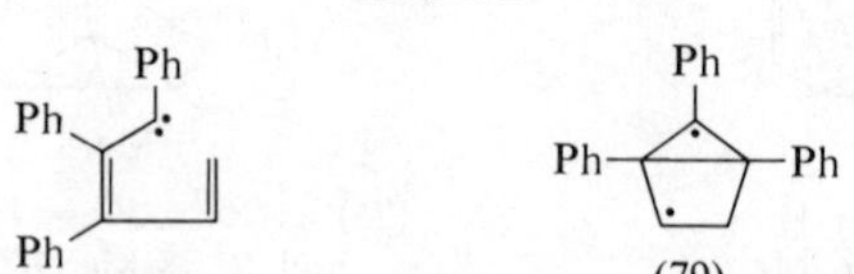

[44] F. J. Palensky and H. A. Morrison, *J. Amer. Chem. Soc.*, 1977, **99**, 3507.
[45] A. Padwa, R. Loza, and D. Getman, *Tetrahedron Letters*, 1977, 2847.
[46] A. Padwa, T. J. Blacklock, D. Getman, and N. Hatanaka, *J. Amer. Chem. Soc.*, 1977, **99**, 2344.
[47] H. E. Zimmerman and S. M. Aasen, *J. Amer. Chem. Soc.*, 1977, **99**, 2342.
[48] A. Padwa, R. Loza, and D. Getman, *Tetrahedron Letters*, 1977, 2847.

irradiation of the cyclopropene (80) in benzene affords the indene (81) as the principal primary product. This compound is accompanied by (82) which can be converted photochemically into (81). When the irradiation is carried out in methanol neither (81) nor (82) is formed but several methoxy ethers are produced instead, consistent with the trapping of a carbene by solvent. A carbene is also involved in the irradiation of cyclopropene (83) which in methanol yields (84) and (85).

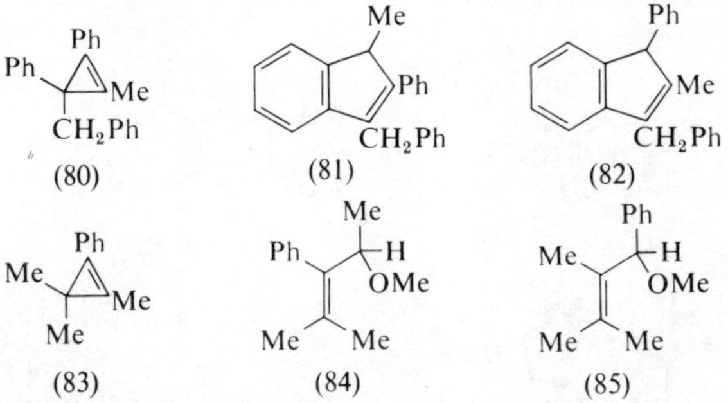

A study of the copper-catalysed photorearrangement and photofragmentation of methylenecyclopropanes has been published: see Scheme 11.[49] The kinetics of the thermal and photochemical reactions of the bicyclohexene derivative (86) have been studied.[50]

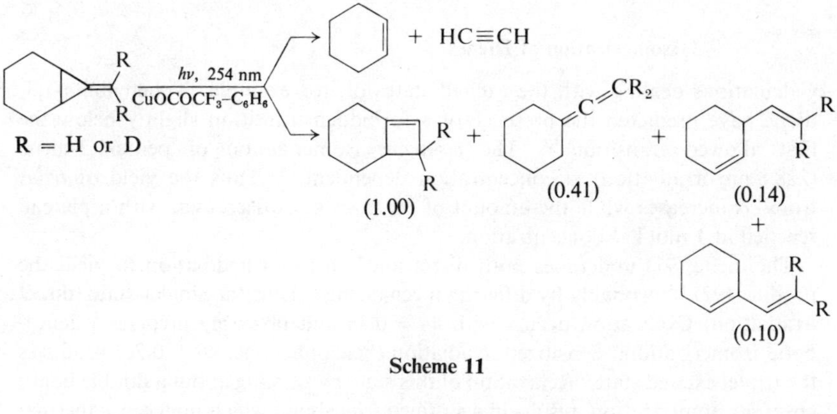

Scheme 11

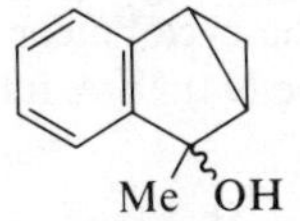

(86)

[49] R. G. Salomon, A. Sinha, and M. F. Salomon, *J. Amer. Chem. Soc.*, 1978, **100**, 520.
[50] V. Wertheimer, A. M. Glatz, and A. C. Razus, *Rev. Roumaine Chim.*, 1977, **22**, 1505 (*Chem. Abs.*, 1978, **88**, 104 354).

Irradiation (Pyrex filter) of the alkaloid crinamine (87) in methanol provides a photoproduct identified as (88). The authors [51] suggest that the reaction proceeds by a $[2\pi + 2\sigma]$ process yielding a cyclopropane intermediate (89). The formation of the cyclopropane (89), if indeed it is an intermediate, could be envisaged as arising by a 1,2-migration of bond *a* in (87) to afford a biradical (90) which could ring-close to (89) or abstract hydrogen from the solvent. If a cyclopropane is involved, the authors [51] suggest that it undergoes an unprecedented type of photoreduction to the observed product (88).

(87)   (88)

(89)   (90)

3 Isomerization of Dienes

Calculations dealing with the excited states of, for example, 1-phenylbuta-1,3-diene have predicted the presence of a forbidden transition slightly below the first allowed transition.[52] The *cis–trans* isomerization of penta-1,3-diene (228.8 nm irradiation) is concentration dependent.[53] Thus the yield of *trans* from *cis* increases while the amount of *cis* from *trans* decreases, with a plateau reached at 1 mol l^{-1} concentration.

The diene (91) undergoes both direct and sensitized irradiation to yield the product (92),[54] probably by differing mechanisms. Thus the singlet-state (direct irradiation) Cyclization occurs with $\Phi = 0.38$ and probably involves valence-bond isomerization. Sensitized irradiation (benzophenone, $\Phi = 0.20$) produces the triplet excited state. Relaxation of this state by twisting about a double bond, *cis–trans* isomerization, results in a strained *trans*-bond which undergoes thermal ring closure to afford (92). The cyclobutene derivative (93) is prepared by the irradiation (quartz) of the diene (94).[55] A full report of the work describing the

51 Y. Tsuda, M. Kaneda, S. Takagi, M. Yamaki, and Y. Iitaka, *Tetrahedron Letters*, 1978, 1199.
52 P. J. Baldry and J. A. Barltrop, *Chem. Phys. Letters*, 1977, **46**, 430.
53 P. Vanderlinden and S. Boue, *Bull. Soc. chim. belges*, 1977, **86**, 785 (*Chem. Abs.*, 1978, **88**, 61 772).
54 J. S. Swenton and G. L. Smyser, *J. Org. Chem.*, 1978, **43**, 165.
55 D. Davalian, P. J. Garratt, and M. M. Mansuri, *J. Amer. Chem. Soc.*, 1978, **100**, 980.

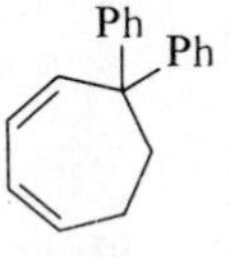

(91) 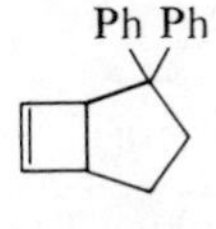   (92) 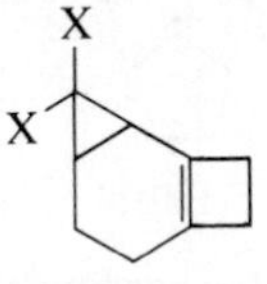   (93)   X = Cl or Br 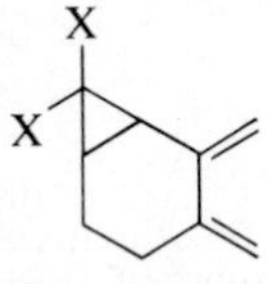   (94)

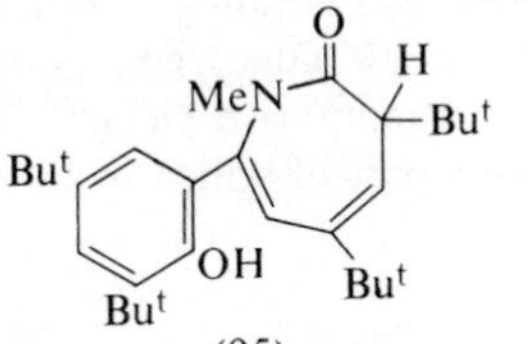

(95) 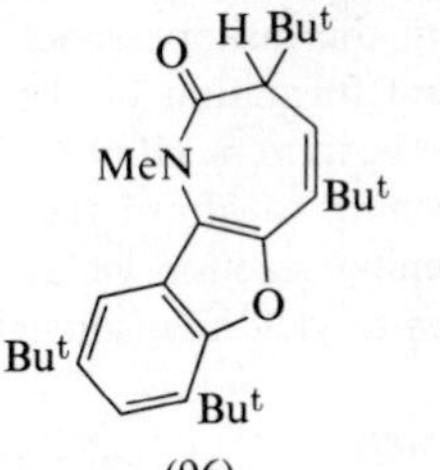   (96)

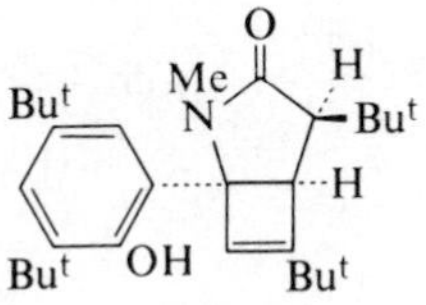

(97) 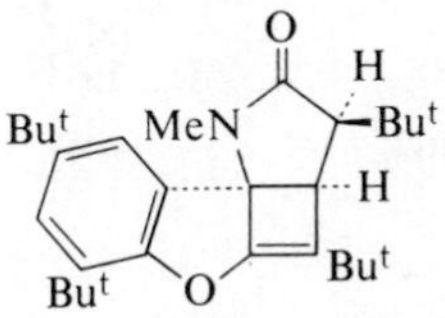   (98)

reactions of (95) and (96), including the photochemical ring closure to (97) and (98), has appeared.[56, 57]

The reactions depicted in Scheme 12 have been further studied.[58, 59] The authors[58] report that the process provides a stereoselective route to homoallylic alcohols.

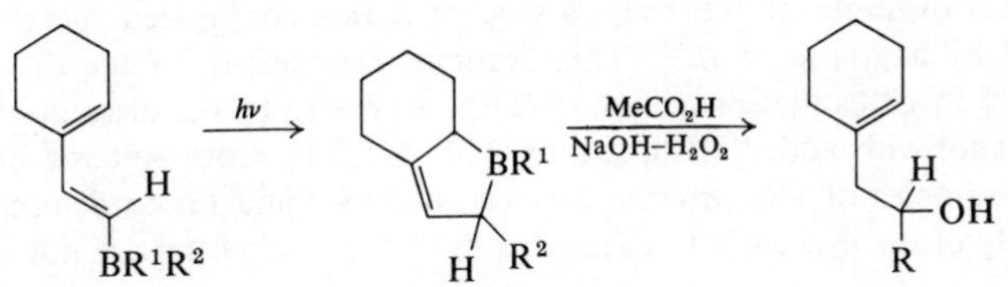

Scheme 12

The ^{13}C-labelled cyclopentadiene (99) undergoes transposition of carbon atoms to afford (100)[60] *via* the formation of bicyclopentene (101; 'housene'), a photoreaction previously reported by van Tamelen.[61] The migration of the C-atom presumably involves a degenerate 1,3-carbon shift, yielding (101). Ring

[56] H.-D. Becker and K. Gustafsson, *J. Org. Chem.*, 1977, **42**, 2966.
[57] H.-D. Becker and K. Gustafsson, *Tetrahedron Letters*, 1976, 1705.
[58] G. Zweifel, S. J. Blacklund, and T. Leung, *J. Amer. Chem. Soc.*, 1977, **99**, 5192.
[59] G. M. Clark, K. G. Hancock, and G. Zweifel, *J. Amer. Chem. Soc.*, 1971, **93**, 1308.
[60] G. D. Andrews and J. E. Baldwin, *J. Amer. Chem. Soc.*, 1977, **99**, 4851.
[61] E. E. van Tamelen, L. E. Ellis, and J. I. Brauman, *J. Amer. Chem. Soc.*, 1967, **89**, 5073; *ibid.*, 1971, **93**, 6145.

(99)   (100)   (101)   (101)   (102)

opening of this yields the cyclopentadiene (100). The reaction mixture also contained the tricyclopentane (102) as a minor product.[60]

Irradiation ($\lambda > 300$ nm) of the dihydropyridine (103) in anhydrous ethanol results in the disappearance of the starting material with a quantum yield of 0.213 and formation of the iminodiene (104).[62] Photoisomerization of (105) gives the hexatriene (106).[63] The cyclohexadiene (107) undergoes photochemical ring opening to afford the dienone (108) in quantitative yield.[64] The initial ring-opening reaction yields a triene (109) one part of which is an enol which ketonizes to yield the isolated product.

(103)   (104)   (105)   (106)

(107)   (108)   (109)

A novel example of the ring closure of a non-conjugated diene has been reported by Mariano *et al.*[65] This example, reminiscent of the di-π-methane process,[66] involves the conversion of (110), generated by the dissolution of (111) in methanol with added $HClO_4$, into (112; 44%) by Corex-filtered irradiation. A few examples of this reaction are recorded (Scheme 13) to demonstrate the generality of the cyclization. Exact details of the mechanism are not yet known

(110)   (111)   (112)

[62] M. Guay and F. Lamy, *Canad. J. Chem.*, 1978, **56**, 302.
[63] E. Baggiolini, J. J. Partridge, and M. R. Uskokovic, U.S.P., 4 021 423/1976 (*Chem. Abs.*, 1977, **87**, 102 541).
[64] Y. Fujimoto, T. Shimizu, and T. Tatsuno, *Japan. Kokai*, 76/125 355 (*Chem. Abs.*, 1977, **87**, 56 463).
[65] P. S. Mariano, J. I. Stavinoha, and R. Swanson, *J. Amer. Chem. Soc.*, 1977, **99**, 6781.
[66] S. S. Hixson, P. S. Mariano, and H. E. Zimmerman, *Chem. Rev.*, 1973, **73**, 531.

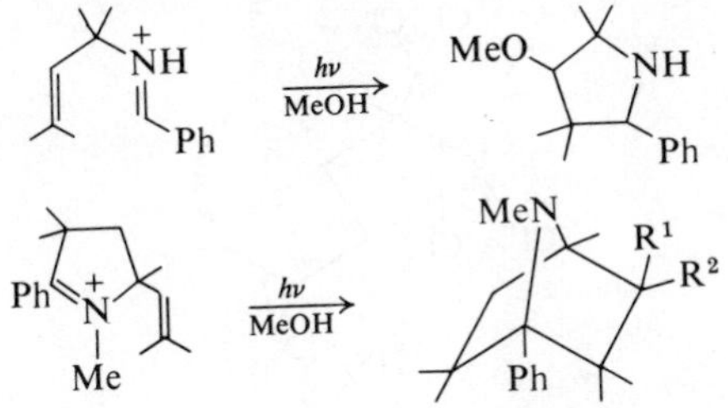

Scheme 13

although several possibilities were considered. Direct irradiation of the anil (113) results in slow decomposition.[67] However, xanthone sensitization leads to rapid *Z–E* isomerization with a photostationary state of 23 : 77. Accompanying this is a slow hydrogen transfer process yielding (114). Hancock *et al.*[68] have previously reported the results of the irradiation of the parent enone.

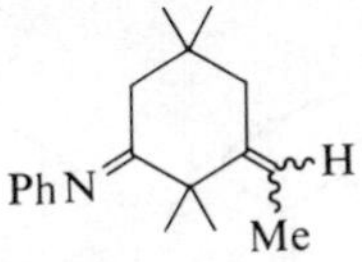

(113) 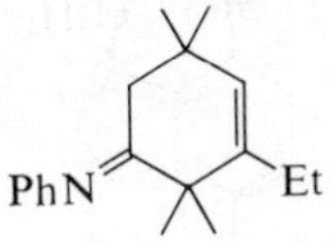   (114)

The picoline isomer (115), produced by the irradiation (254 nm) of the picoline (116) in aqueous NaOH, is also photochemically reactive and rearomatizes to (117) when irradiated in diethyl ether.[69] The rearomatization path is thought to involve N—C bond fission to afford the intermediate shown in Scheme 14.

A study of the photoreactions of Aldrin (118) in the gas phase has been reported.[70]

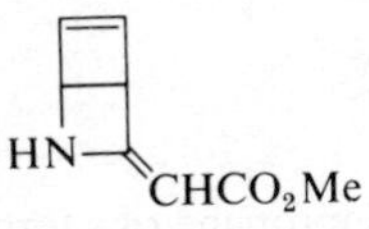

(115) 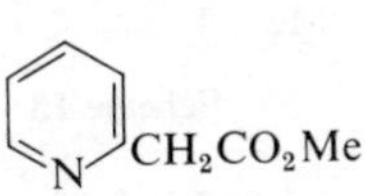   (116) 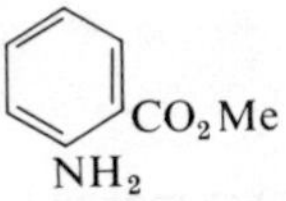   (117)

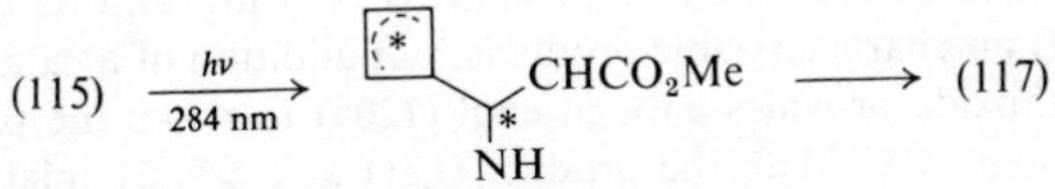

Scheme 14

[67] K. G. Hancock, J. D. Condie, and A. J. Barkovich, *J. Org. Chem.*, 1977, **42**, 2794.
[68] K. G. Hancock and R. O. Grider *Tetrahedron Letters* 1971, 4281; *ibid.*, 1972, 1367.
[69] Y. Ogata and K. Takagi, *J. Org. Chem.*, 1978, **43**, 944.
[70] V. Saravanja-Boanic, S. Gaeb, K. Hustert, and F. Korte, *Chemosphere*, 1977, **6**, 21 (*Chem. Abs.*, 1977, **87**, 38 981).

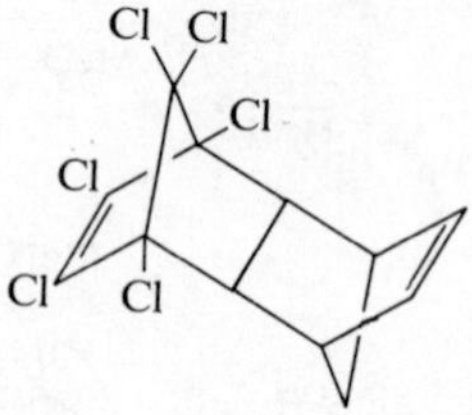

(118)

4 Reactions of Trienes and Higher Polyenes

A review of the photochemical reactions of alkenes and polyenes has been published.[71] Calculations have been carried out concerning the excited states in triene systems with a particular bias towards what is described as 'sudden polarization'.[72] The work is aimed at a possible mechanism for the formation of bicyclo[3,1,0]hex-2-enes. One such example is the cyclization of the triene [73] illustrated in Scheme 15. Bonacic–Koutecky argues that a charged intermediate (Scheme 15) could account for both products. Thus if H-abstraction by the

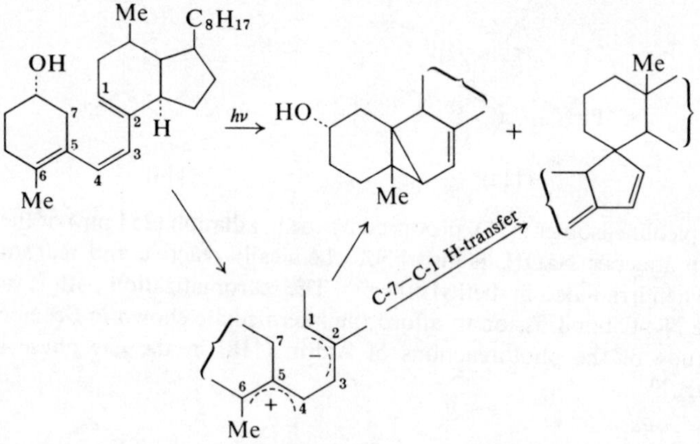

Scheme 15

carbanion moiety C-1, C-2, and C-3 is faster than cyclopropane ring formation, the spiro-product is preferred. Further studies of the photochemistry of diphenyl-polyenes in monolayer assemblies have been reported.[74]

The enol content of the chromanone (119) is very low and its irradiation in methanol yields no characterizable products, but addition of a catalytic amount of sodium methoxide provides enough enol (120a) to make the photoreaction clean and efficient in yielding the product (121) as a 97:21 mixture with the

[71] G. Kaupp, *Angew. Chem. Internat. Edn.*, 1978, **17**, 150.
[72] V. Bonacic-Koutecky, *J. Amer. Chem. Soc.*, 1978, **100**, 396.
[73] E. Havinga, *Chimica*, 1976, **30**, 27.
[74] F. H. Quina, D. Moebius, F. A. Carroll, F. R. Hopf, and D. G. Whitten, *Z. phys. Chem.* (*Frankfurt*), 1976, **101**, 151 (*Chem. Abs.*, 1977, **86**, 163 514).

open-chair tautomer (122).[75] The involvement of the enol in the photochemical reactions is beyond doubt and is further confirmed by the failure of (123) to undergo photoreaction. The enol ether (120b) also undergoes photoreaction to afford the ring-opened compounds (124) and (125). Padwa and Owens[75] implicate the *o*-quinoneallide intermediate (126) in the reactions. This is produced by the ring opening of the enols (120a and b).

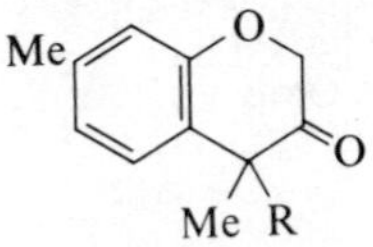

(119) R = H
(123) R = Me

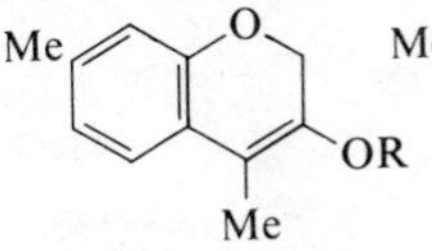

(120) a; R = H
b; R = Me

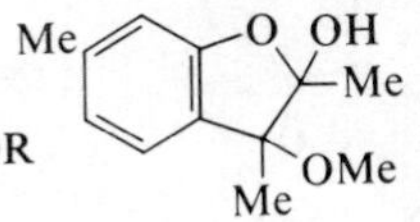

(121)

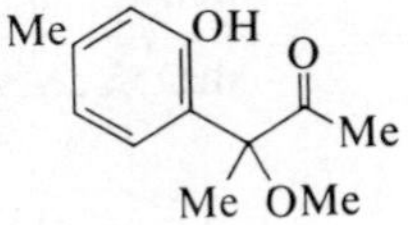

(122)

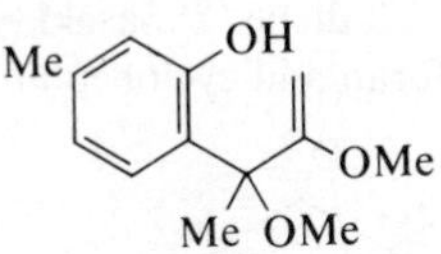

(124)

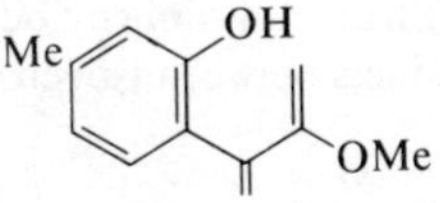

(125)

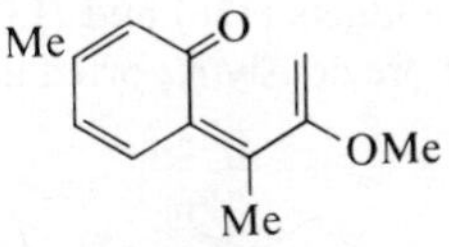

(126)

The scope and mechanism of the photocyclization of the diarylethylenes (127) into (128) in the presence of amines have been investigated.[76] The photochemical cyclization of several formyl enamides (129) into protoberberines (130) has been described.[77]

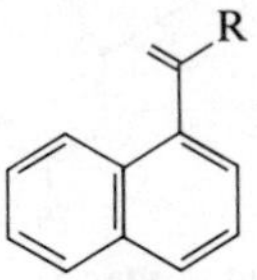

(127)

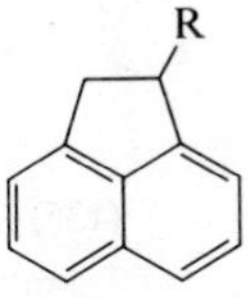

(128)

R = 1-naphthyl, 1-pyrenyl, 9-phenanthryl, 5-chrysenyl, or 3-fluoranthenyl

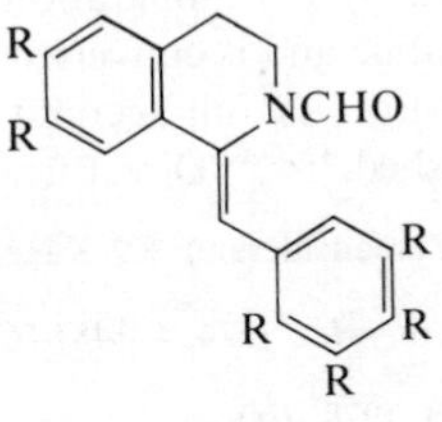

(129)

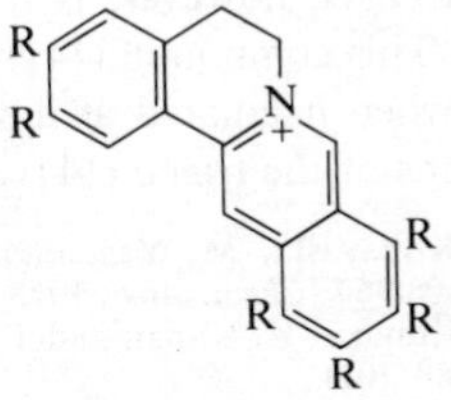

(130)

[75] A. Padwa and W. Owens, *J. Org. Chem.*, 1977, **42**, 3076.
[76] R. Lapouyade, R. Koussini, and H. Bouas-Laurent, *J. Amer. Chem. Soc.*, 1977, **99**, 7374.
[77] G. R. Lenz, *J. Org. Chem.*, 1977, **42**, 1117.

Further patents concerning the transformation of vitamin D derivatives have been published.[78] Irradiation of the furanone derivative (131) results in the formation of β-apolignan and the bicyclohexene derivative (132).[79]

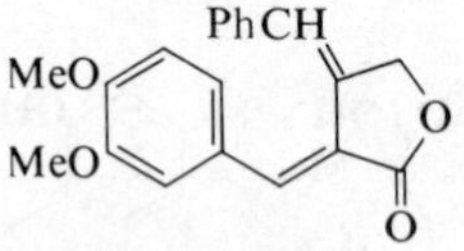

(131) 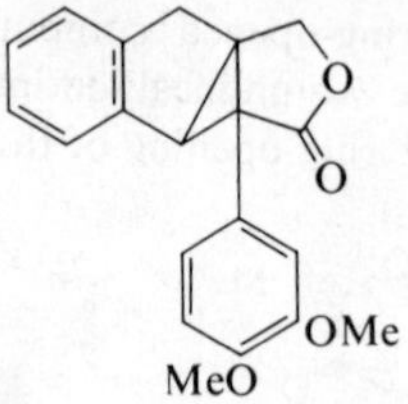   (132)

Adducts are formed when isobenzofuran (133) is irradiated in the presence of dienes. Thus the adducts (134) and (135) are formed from cyclohexa-1,3-diene and adducts (136) and (137) from 1,4-diphenylbuta-1,3-diene.[80] Sasaki *et al.*[81] have previously reported adducts between isobenzofuran and cycloheptatriene.

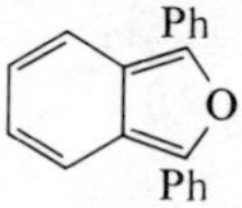

(133) 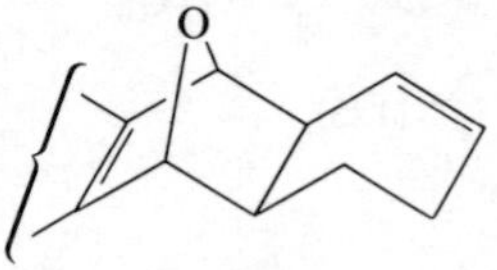   (134) 11% 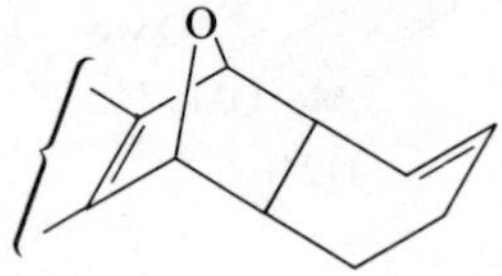   (135) 6%

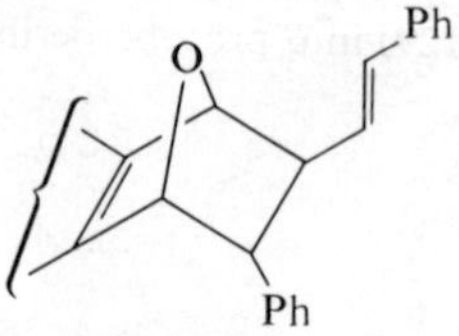

(136) 78%

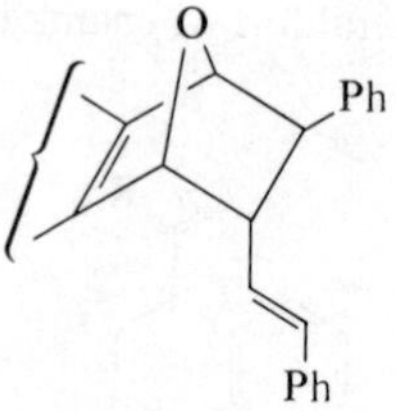

(137) 14%

When (138) is irradiated (low-pressure Hg lamp) in methanol solution three products (139), (140), and (141) are formed.[82] The products (139) and (140) are produced by intramolecular [2 + 2] addition in the two possible modes. Compound (141), however, is formally obtained by a 1,3-migration of bond *a* in (138). This compound (141) is itself photolabile and is converted into (142) and (143) when irradiated at ambient temperatures. A full account of the photo-reactions of the triene (144) has been published.[83, 84] Direct irradiation of this

[78] T. Kobayashi, M. Yasumura, T. Sato, H. Yamauchi, and K. Kagei, *Japan. Kokai*, 77/108 964 (*Chem. Abs.*, 1978, **88**, 62 535).
[79] T. Momose, K. Kanai, and T. Nakamura, *Heterocycles*, 1976, **4**, 1481 (*Chem. Abs.*, 1977, **86**, 88 707).
[80] G. Kaupp and E. Teufel, *J. Chem. Research (S)*, 1978, 100.
[81] T. Sasaki, K. Kanematsu, and K. Hayakawa, *Tetrahedron Letters*, 1974, 343.
[82] M. Kato, T. Chikamoto, and T. Miwa, *Bull. Chem. Soc. Japan*, 1977, **50**, 1082.
[83] L. A. Paquette, M. J. Kukla, S. V. Ley, and S. G. Traynor, *J. Amer. Chem. Soc.*, 1977, **99**, 4756.
[84] L. A. Paquette and M. J. Kukla, *J.C.S. Chem. Comm.*, 1973, 409.

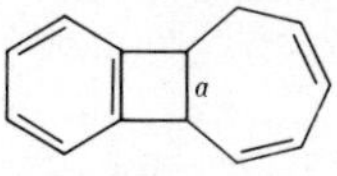

(138)

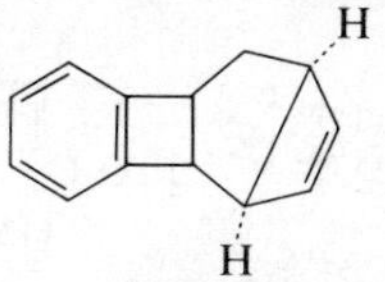

(139)

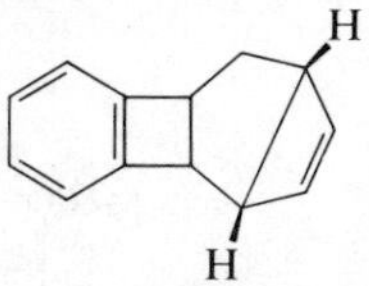

(140)

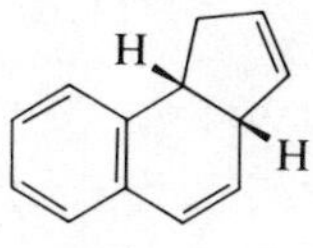

(141)

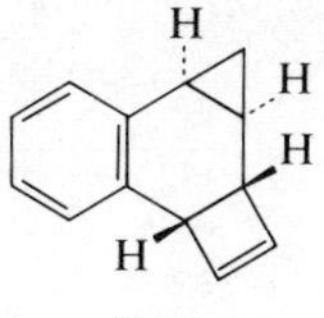

(142)

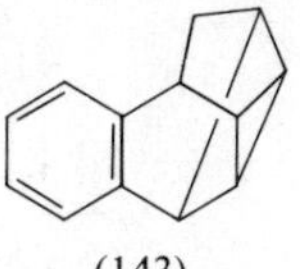

(143)

compound yields a 1 : 1 mixture of the isomers (145) and (146). Acetone-sensitized irradiation of this mixture converts it into two products (147) and (148).

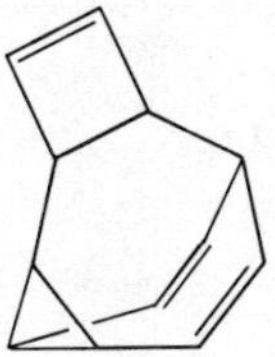

(144)

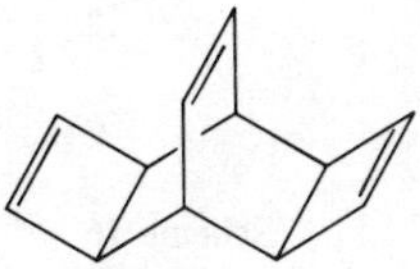

(145)

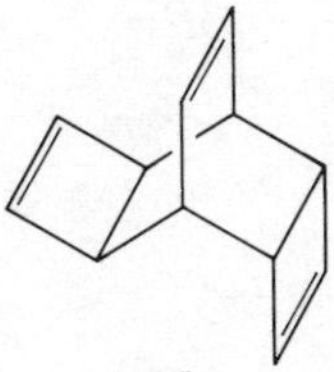

(146)

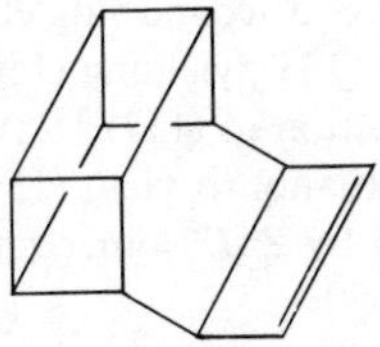

(147)

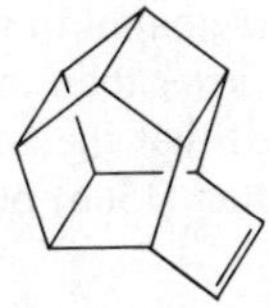

(148)

A study of the photochemical hydrogen migration processes in substituted cycloheptatrienes has been reported.[85] The irradiation of cycloheptatrienes (149) or (150) at 310 nm in cyclohexane results in the formation of the products (151) and (152) shown in Scheme 16.[86] This valence-bond isomerization is in accord with the interpretation based on the polarization of the first excited singlet state by the substituents which Gorman and his co-workers reported some years ago.[87] The interconversion of the cycloheptatrienes (149) and (150) arises by hydrogen migration processes. In contrast to the valence-bond isomerizations brought about by irradiation at 310 nm, irradiation at 254 nm affords four products, the two previous compounds (151) and (152) and two new products (153)

[85] W. Abraham, E. Henke, and D. Kreysig, *Tetrahedron Letters*, 1978, 345.
[86] V. C. Freestone and A. A. Gorman, *Tetrahedron Letters*, 1977, 3389.
[87] A. R. Brember, A. A. Gorman, R. L. Leyland, and J. B. Sheridan, *Tetrahedron Letters*, 1970, 2511.

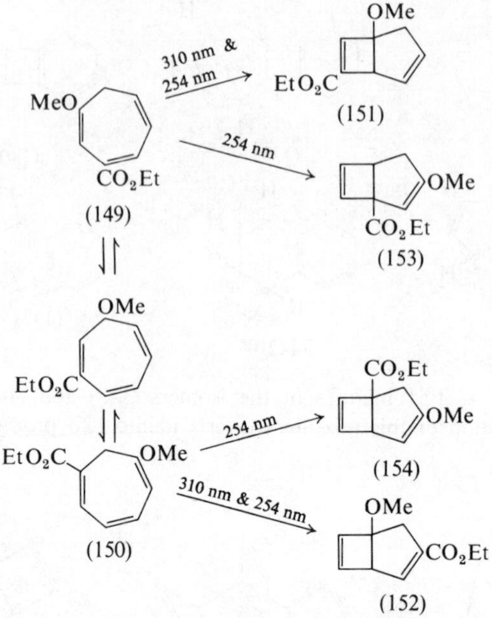

Scheme 16

and (154). The authors[86] suggest that this different mode of cyclization is not a violation of their earlier proposals[87] but involves a second singlet state.

Photoaddition of methanol to the oxazepine (155), yielding (156a) and (156b) in a ratio of 3 : 1, is faster than the ring contraction to (157).[88] A further study of the process showed that the addition of methanol to yield (156a) was highly stereoselective. Product (156b) probably arises by *Z–E* isomerization of (156a).

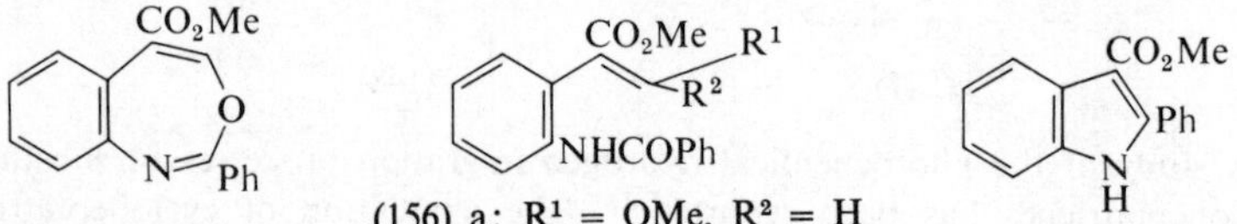

(155) (156) a; R^1 = OMe, R^2 = H
b; R^1 = H, R^2 = OMe (157)

The diazepines (158) undergo photochemical rearrangement to the isomeric species (159) *via* ring closure to the intermediate valence-bond isomers (160) and di-π-methane rearrangement.[89] The tendency of some benzoxepins (161) to undergo valence-bond isomerization to (162) has been studied.[90] The reaction has been analysed in terms of frontier orbital interactions. Irradiation of the benzthiepins (163) results in the formation of the tricyclic compounds (164) as a

88 M. Somei, R. Kitamura, H. Fujii, K. Hashiba, S. Kawai, and C. Kaneko, *J.C.S. Chem. Comm.*, 1977, 899.
89 G. Reissenweber and J. Sauer, *Tetrahedron Letters*, 1977, 4389.
90 H. Hofmann and P. Hofmann, *Annalen*, 1977, 1597.

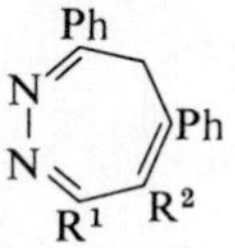

(158) R^1 = Ph, R^2 = Ph
  R^1 = R^2 = *p*-tolyl
  R^1 = R^2 = Me

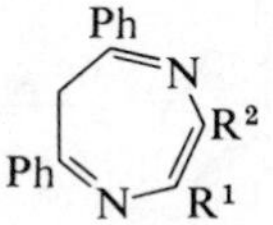

(159)

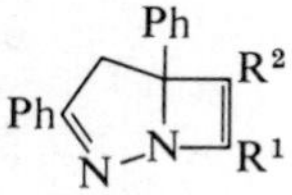

(160)

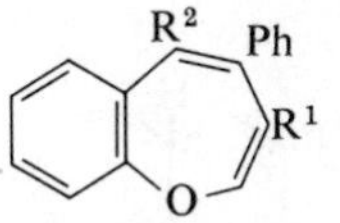

(161)

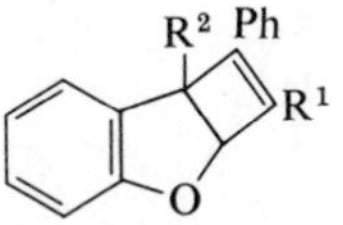

(162)

R^1 = R^2 = H; R^1 = OMe, R^2 = H; R^1 = OAc, R^2 = H;
R^1 = H, R^2 = Cl; R^1 = H, R^2 = MeO; R^1 = Me, R^2 = OAc;
R^1 = Me, R^2 = OMe; R^1 = OAc, R^2 = Cl; R^1 = R^2 = OAc;
R^1 = CN, R^2 = OMe; R^1 = CN, R^2 = OAc

result of valence-bond isomerization.[91] The products (164) are themselves photoreactive, and secondary irradiation converts them into the isomeric compounds (165). The most likely route for the formation of these isomers is *via* a biradical mechanism (a 1,3-bond migration is also possible), whereby bond fission yields the intermediate (166) which subsequently undergoes bond formation to afford the products (Scheme 17).

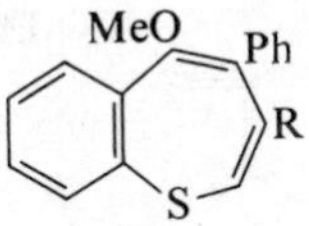

(163)

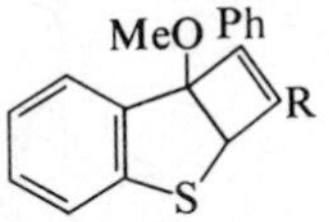

(164)

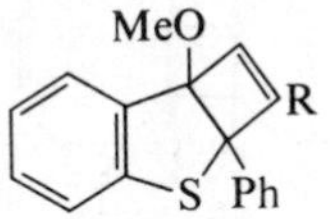

(165)

R = CN, H, OAc, or OMe

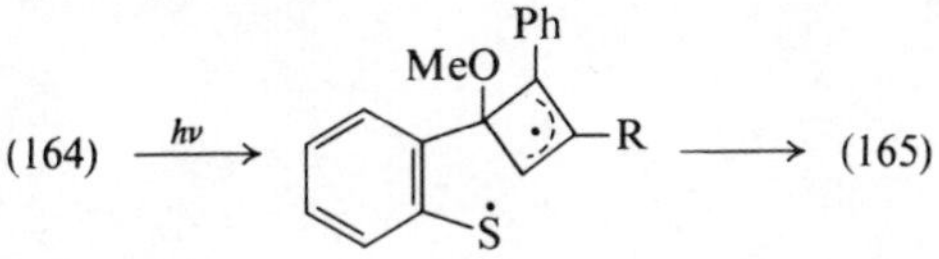

$$(164) \xrightarrow{h\nu} \quad (166) \quad \longrightarrow (165)$$

(166)

Scheme 17

Exploratory photochemistry of trienes related to the hetero-[11]annulenes has been carried out.[92] The three derivatives (167) studied all undergo ready photofragmentation to yield the products shown in Scheme 18. The related compound (168a) does not undergo fragmentation, but direct irradiation at −78 °C causes isomerization to the three products (169), (170), and (168b).

[91] H. Hofmann and H. Gaube, *Annalen*, 1977, 1847.
[92] A. G. Anastassiou, E. Reichmanis, S. J. Girgenti, and M. Schaefer-Ridder, *J. Org. Chem.*, 1978, **43**, 315.

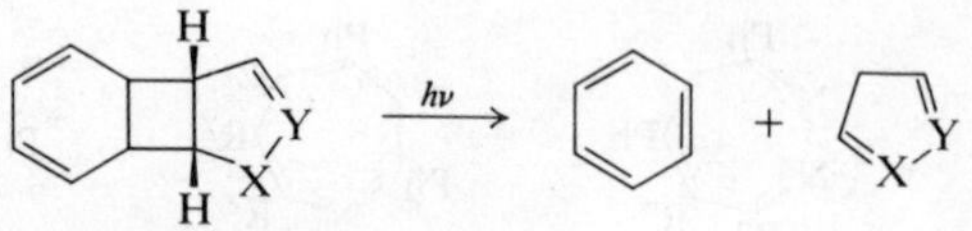

(167) a; X = NCO₂Et, Y = CH
      b; X = O, Y = CH
      c; X = NCO₂Et, Y = N

Scheme 18

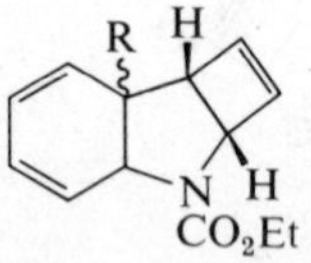

(168) a; R = α-H
      b; R = β-H

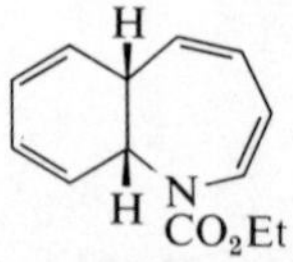

(169)

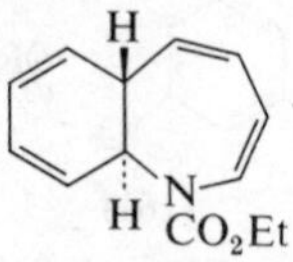

(170)

Sensitized irradiation of (168a) at room temperature yields a dimer. The adduct (171) undergoes photochemical ring opening to the macrocycle (172) when irradiated at −105 to −70 °C in pentane.[93] This process would appear to be less favourable on entropy grounds than photodissociation by cleavage of the C_4 ring. The tribenztriene (173) undergoes photochemical isomerization to (174) when irradiated in pyridine solution.[94]

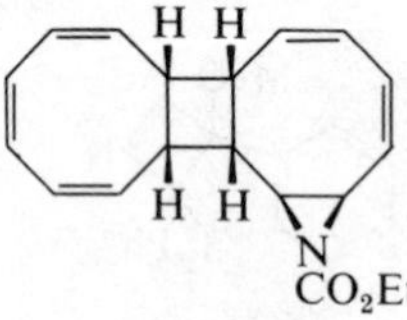

(171)

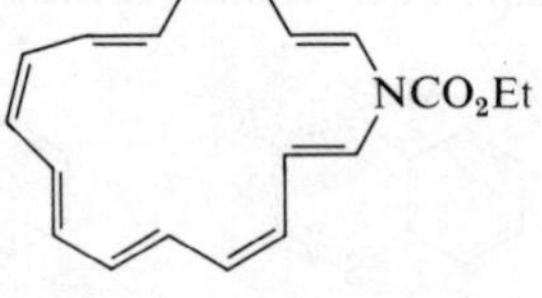

(172)

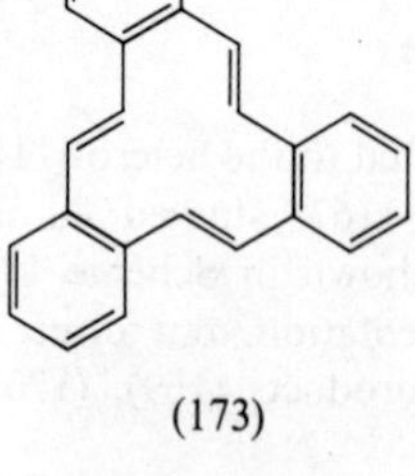

(173)

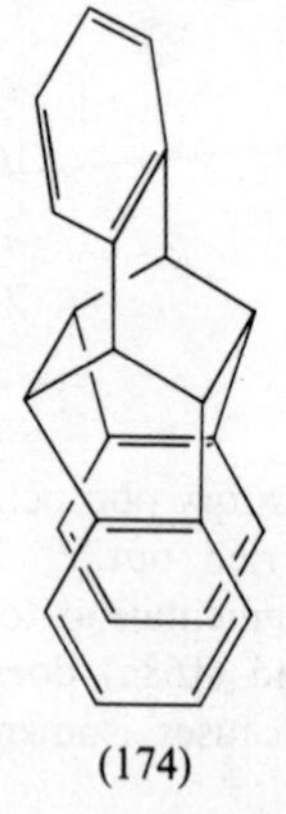

(174)

[93] H. Roettele, G. Heil, and G. Schroeder, *Chem. Ber.*, 1978, **111**, 84.
[94] M. W. Tausch, M. Elian, A. Bucur, and E. Cioranescu, *Chem. Ber.*, 1977, **110**, 1744.

5 [2 + 2] Intramolecular Additions

Further use of Cu[I] trifluoromethanesulphonate ('copper triflate') has been made in photochemical synthesis. Thus irradiation of the alcohol (175) as a complex with the triflate leads to the formation of the bicyclic products (176a) and (176b) in a ratio of 3 : 2.[95] It is apparent, from spectroscopic studies, that the copper

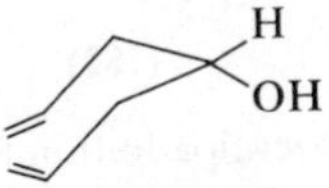

(175)

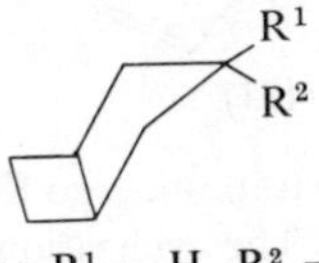

(176) a; R[1] = H, R[2] = OH
     b; R[1] = OH, R[2] = H

co-ordinates symmetrically with the two double bonds and that there is no participation of the OH groups. It is also suggested that the cyclization path involves a two-step process.

Conventional [2 + 2] cycloaddition is reported in the acetone-sensitized irradiation of the diene (177) to yield (178; 55%).[96] Becker[97] suggested that

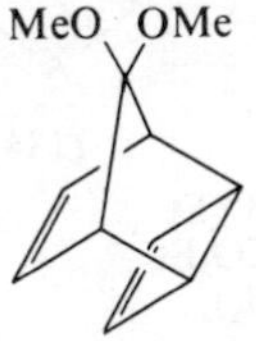

(177)

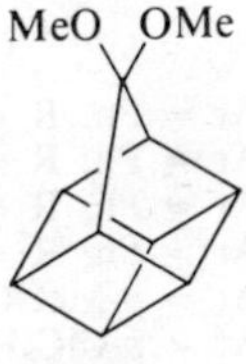

(178)

the diene (179) was inert to cage formation because of inefficient transfer of energy from the sensitizer to the diene. However, Hirao *et al.*[98] have shown that toluene does transfer energy to the diene (179), and they suggest (but without direct evidence) that this leads to the *trans*-isomer which in the presence of acetic acid affords the esters (180). These authors[98] reason that the most likely cause of the inertness in (179) is ring flexibility, for the less flexible diene (181) does undergo [2 + 2] intramolecular cycloaddition to yield (182). Several dienes

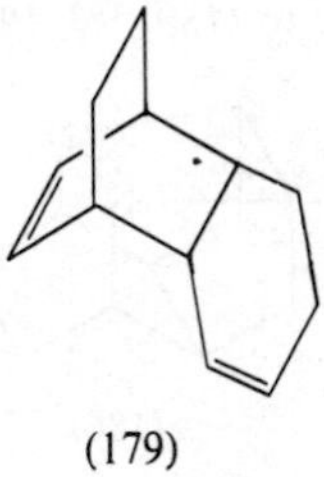

(179)

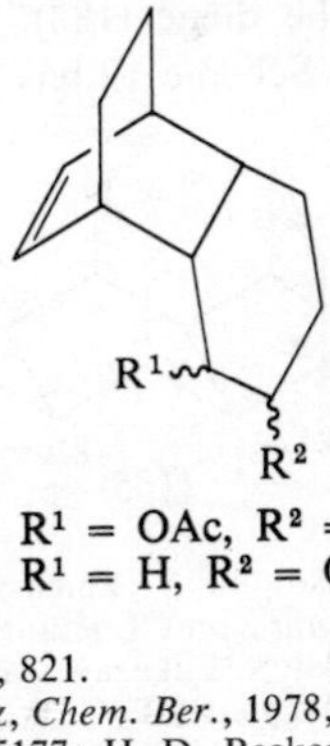

(180) R[1] = OAc, R[2] = H
      R[1] = H, R[2] = OAc

[95] J. Th. M. Evers and A. Mackor, *Tetrahedron Letters*, 1978, 821.
[96] R. W. Hoffmann, H. R. Kurz, J. Becherer, and M. T. Reetz, *Chem. Ber.*, 1978, **111**, 1264.
[97] H.-D. Becker and A. Konar, *Tetrahedron Letters*, 1972, 5177; H.-D. Becker, *Annalen*, 1973, 1675.
[98] K. Hirao, S. Unno, and O. Yonemitsu, *J.C.S. Chem. Comm.*, 1977, 577.

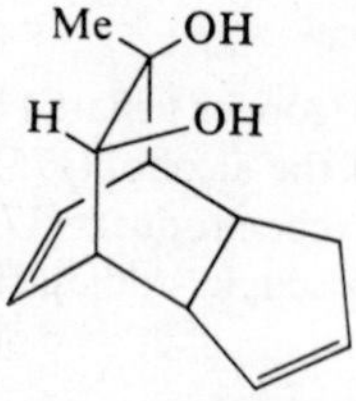

(181)

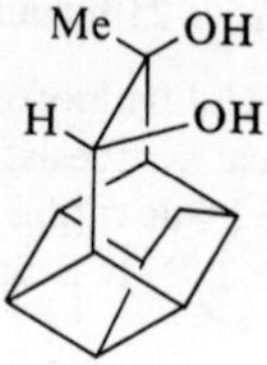

(182)

(183) have been shown to undergo [2 + 2] photocycloaddition to yield the cage compounds (184).[99] The cyclizations were not affected by quenchers or sensitizers but the authors[99] did not conclude whether a singlet or a triplet state was involved in the intramolecular process. The quantum yields for the cyclizations

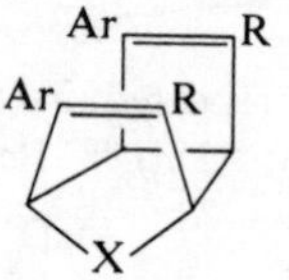

(183) a; Ar = Ph, R = Me, X = $(CH_2)_2$
b; Ar = Ph, R = Me, X = $(CH_2)_3$
c; Ar = Ph, R = Me, X = CH=CHNCO_2Et
d; Ar = Ph, R = Me, X = CH=N—NCO_2Et
e; Ar = p-MeOC_6H_4, R = Me, X = $(CH_2)_2$
f; Ar = p-ClC_6H_4, R = Me, X = $(CH_2)_2$
g; Ar = Ph, R = CO_2Me, X = $(CH_2)_2$

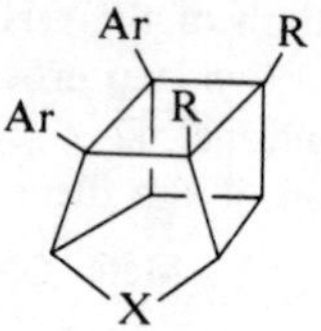

(184)

are all in excess of 0.4. The interest in the formation of these compounds arises from a study of the possibility of using them as a means of storing light energy, and towards this end the thermal reversions have also been studied.

The photoelectron spectrum of the compound (185) clearly shows that there is strong interaction between the N=N and C=C bonds.[100] Irradiation of this compound (185) in acetonitrile solution gives a high yield of the novel cage compound (186). Cage compound (187) is formed by acetone-sensitized irradiation of the diene (188).[101] An application of the photochemical conversion shown in Scheme 19 has been made as a route to (189).[102, 103]

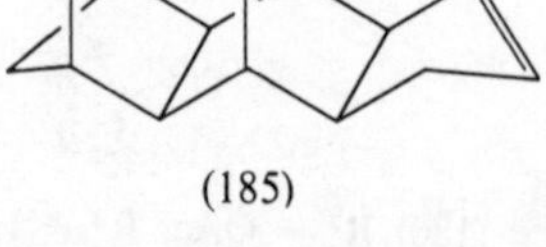

(185)

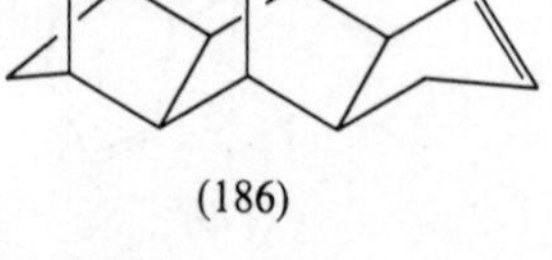

(186)

[99] T. Mukai and Y. Yamashita, *Tetrahedron Letters*, 1978, 357.
[100] W. Berning and S. Hunig, *Angew. Chem. Internat. Edn.*, 1977, **16**, 777.
[101] M. Oda, N. Fukazawa, and Y. Kitahara, *Tetrahedron Letters*, 1977, 3277.
[102] L. A. Paquette, T. G. Wallis, K. Hirotsu, and J. Clardy, *J. Amer. Chem. Soc.*, 1977, **99**, 2815.
[103] L. A. Paquette, T. G. Wallis, T. Kempe, G. G. Christoph, J. P. Springer, and J. Clardy, *J. Amer. Chem. Soc.*, 1977, **99**, 6946.

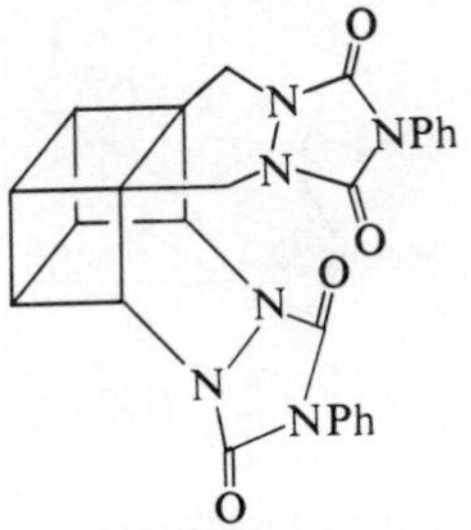

(187)

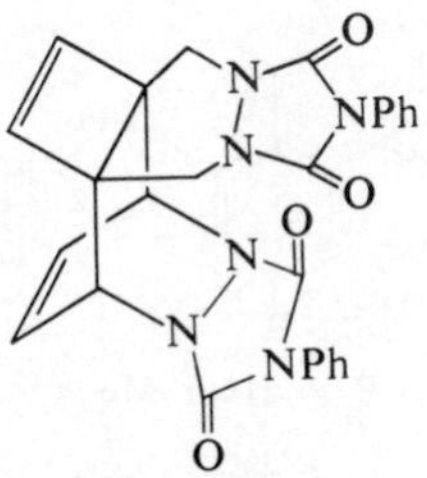

(188)

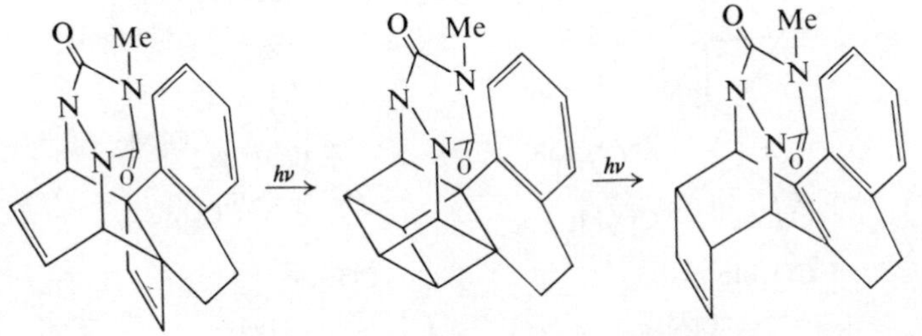

Scheme 19

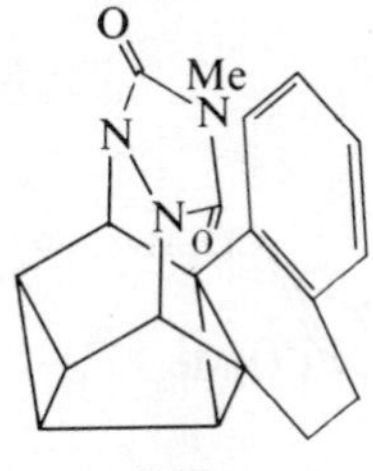

(189)

An investigation of the role of the Cu$^+$ in the CuCl-assisted isomerization of norbornadiene to quadricyclane has shown that a 1:1 CuCl–norbornadiene complex is formed: *cf.* ref. 95.[104] The absence of dimerization of the norbornadiene suggests that the copper complexes only one molecule of diene. The use of copper(I) phosphine complexes for the photosensitization of the norbornadiene-quadricyclane conversion has been studied.[105]

Direct irradiation of the benzoxanorbornene derivatives (190) in ether solution affords the oxepines (191).[106] The transformation presumably involves [2 + 2] cycloaddition to yield the intermediate (192) which thermally rearranges to (191). Full experimental details for the synthesis of (193) by photochemical [2 + 2] addition within (194) has been reported.[107] The direct irradiation ($\lambda > 280$ nm)

104 D. P. Schwendiman and C. Kutal, *J. Amer. Chem. Soc.*, 1977, **99**, 5677.
105 P. A. Grutsch and C. Kutal, *J. Amer. Chem. Soc.*, 1977, **99**, 6460.
106 W. Tochtermann, H. Timm, and J. Diekmann, *Tetrahedron Letters*, 1977, 4311.
107 W. Eberbach and H. Prinzbach, *Org. Photochem. Synth.*, 1976, **2**, 104 (*Chem. Abs.*, 1977, **87**, 22 525).

(190) R = H or Me   (191)   (192)

of methylene chloride solutions of (195) affords a high yield of the rearrangement product (196). Prinzbach and his co-workers [108] argue that this cycloaddition could be an example of a $\pi_2^s + \pi_2^a + \pi_2^a + \sigma_2^s$ reaction.

(193)   (194)

(195)   (196)

6 Dimerization, Intermolecular Cycloaddition, and Cycloaddition of Acetylenes

Benzophenone-sensitized irradiation of 4-vinylpyridine and 2-vinylpyridine in benzene or t-butyl alcohol affords, in each case, a single [2 + 2] photodimer of the *trans*-head-to-head type. 2-Vinylquinoline behaves in a like manner.[109] Lewis and Johnson [110] report that the dimerization of *trans*-stilbene competes efficiently with isomerization at concentrations >0.1 mol l^{-1} The solid-state photochemical [2 + 2] dimerization of (197) affords (198) in 90% yield.[111] An earlier publication dealing with the photochemical dimerization of (199) in

[108] H. Prinzbach, K.-H. Lehr, H. Babsch, and H. Fritz, *Tetrahedron Letters*, 1977, 4199.
[109] T. Nakano, A. Martin, C. Rivas, and C. Perez, *J. Heterocyclic Chem.*, 1977, **14**, 921.
[110] F. D. Lewis and D. E. Johnson, *J. Photochem.*, 1977, **7**, 421 (*Chem. Abs.*, 1977, **87**, 200 479).
[111] D. Donati, M. Fiorenza, E. Moschi, and P. Sarti-Fantoni, *J. Heterocyclic Chem.*, 1977, **14**, 951.

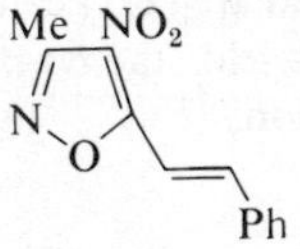

(197)

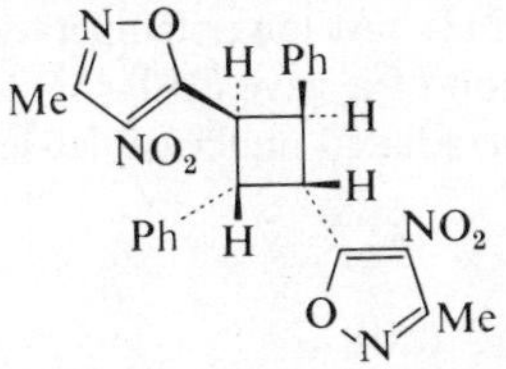

(198)

benzene reported [112] the formation of the cyclobutane-type dimer (200) in addition to several other products. A reinvestigation [113] of the process by the same workers has reported the isolation of four other cyclobutane-type dimers (201)—(204).

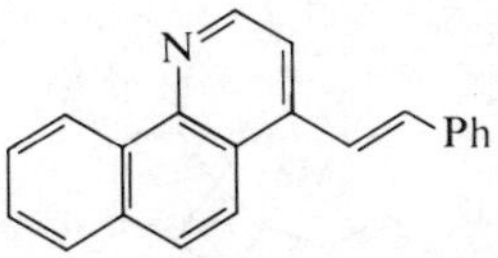

(199)

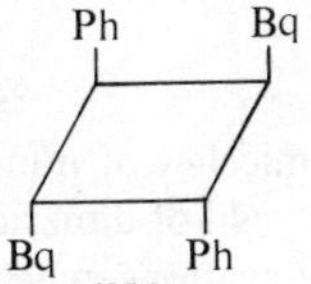

(200) 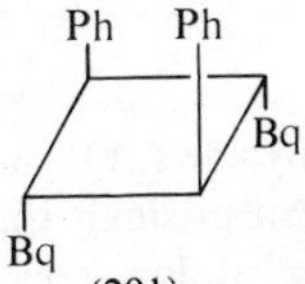   (201) 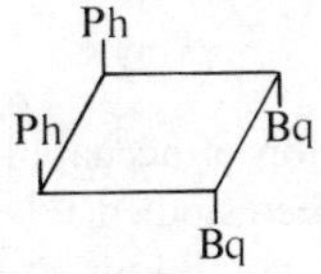   (202) 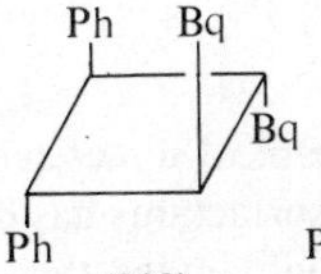   (203) 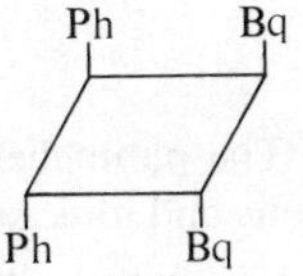   (204)

Bq = 4-Benzoquinolyl

Irradiation of the diazo-ketone (205) induces loss of nitrogen to afford a carbene intermediate, which can either undergo insertion to afford the tricyclic ketone (206) or give the cyclopropene (207). This cyclopropene (207) undergoes photochemical dimerization to yield (208).[114] Full experimental details of the processes are given. The photosensitized dimerization of 1,2-dimethyl-3,4-dimethylenecyclobutene yields (209). Pyrolysis at 380 °C yielded the novel

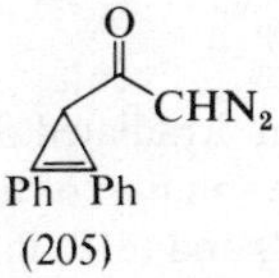

(205) 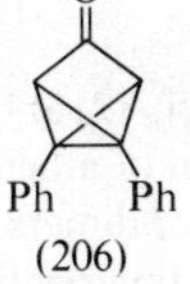   (206) 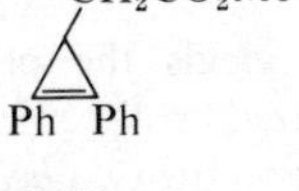   (207) 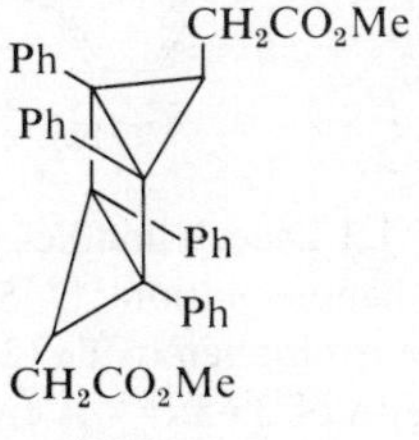   (208)

[112] F. Andreani, R. Andrisano, G. Salvadori, and M. Tramontini, *J. Chem. Soc. (C)*, 1971, 1007.
[113] F. Andreani, R. Andrisano, G. Salvadori, and M. Tramontini, *J.C.S. Perkin I*, 1977, 1737.
[114] N. Nakatsuka, and S. Masamune, *Org. Photochem. Synth.*, 1976, **2**, 57 (*Chem. Abs.*, 1977, **87**, 22 524).

compound (211). At lower temperatures the intermediate (211) can be isolated, and irradiation of it gave (210).[115] 1-Methylindene yields the *cis-anti-cis*-dimer (212) when irradiated under triplet-sensitized conditions.[116]

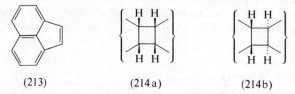

(209)   (210)   (211)

(212)

The photochemical dimerization of acenaphthylene (213) in micelles of non-ionic and ionic surfactants has been studied.[117] Surprisingly the yield of dimeric material was high under these conditions even at low concentrations when similar concentrations of acenaphthylene in benzene solution failed to yield product. The ratios of the *syn-* and *anti*-products (214a and b) was independent of concentration in the range 5—30 mmol l⁻¹. At very low concentrations the *trans*-dimer is favoured. A photochemical study of the acenaphthylene dimerization process in the presence of *para*-substituted bromobenzenes has sought to evaluate the influence of the substituent.[118]

(213)   (214a)   (214b)

1,3-Diacetylindole (215) yields the photodimer (216) when irradiated in ethanol solution.[119] Sensitized or direct irradiation of a benzene solution of the benzthiophen oxide (217) gives the two *head*-to-*head* dimers (218) and (219) in a ratio of 3 : 5.[120] A report on the photochemical dimerization of benzo[*b*]thio-

[115] W. T. Borden, A. Gold, and S. D. Young, *J. Org. Chem.*, 1978, **43**, 486.
[116] J. Dekker, F. Martins, and T. G. Dekker, *S. African J. Chem.*, 1977, **30**, 21 (*Chem. Abs.*, 1977, **87**, 52 439).
[117] Y. Nakamura, Y. Imakura, T. Kato, and Y. Morita, *J.C.S. Chem. Comm.*, 1977, 887.
[118] G. F. Koser and Van-Shau Liu, *J. Org. Chem.*, 1978, **43**, 478.
[119] T. Hino, M. Taniguchi, T. Date, and Y. Iidaka, *Heterocycles*, 1977, **7**, 105 (*Chem. Abs.*, 1978, **88**, 50 688).
[120] M. S. El Faghi El Amoudi, P. Genese, and J. L. Olive, *Tetrahedron Letters*, 1978, 999.

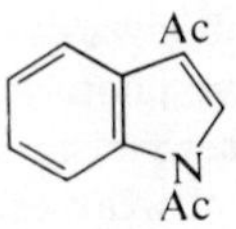

(215)

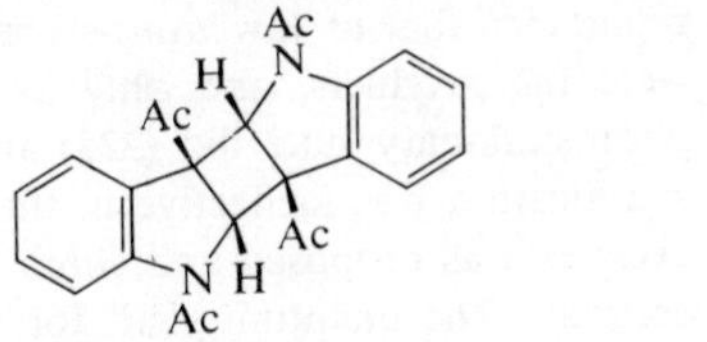

(216)

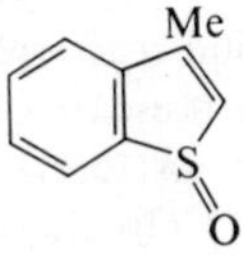

(217)

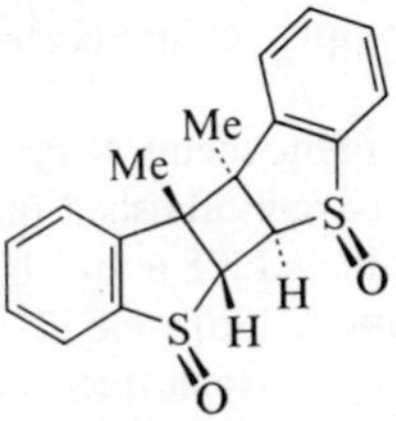

(218)

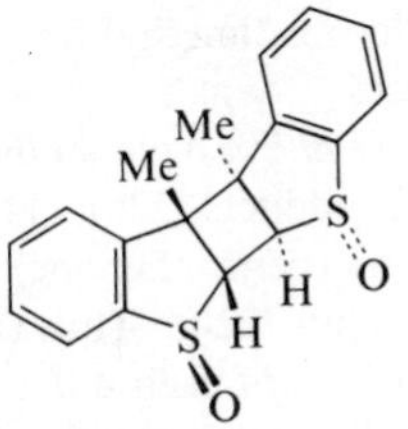

(219)

phen 1,1-dioxides has been published.[121] Irradiation (quartz filter) of the thiophen 1,1-dioxide (220) affords the dimer (221).[122]

Lewis and Hoyle [123] have reported the results of their work dealing with 1,2-diarylethylene–diene exciplexes. Irradiation of *trans*-stilbene (7×10^{-3}

(220)

(221)

mol l^{-1}) in the presence of dimethyl fumarate (7×10^{-2} mol l^{-1}) has previously been reported to yield the two adducts (222) and (223).[124] The photolysis of more concentrated solutions of stilbene (0.2 mol l^{-1}) and dimethyl fumarate (0.22 mol l^{-1}) gives the same products (222) and (223) and also yields the two oxetans (224a and b). (The exact identity of each is unsure.) The present work

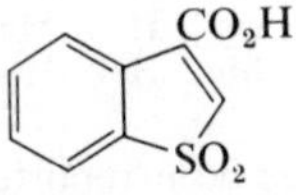

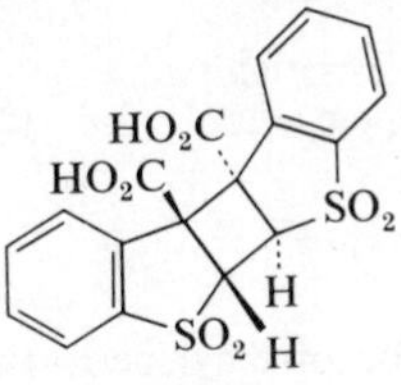

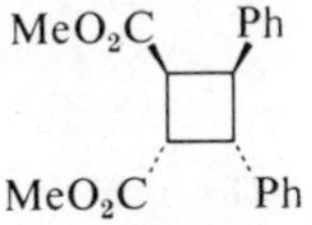

(222)

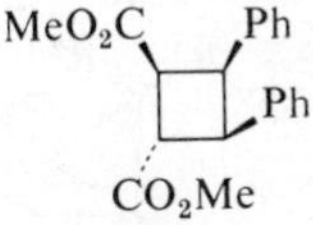

(223)

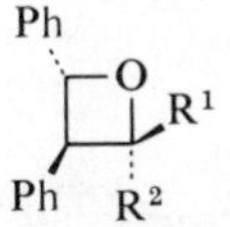

(224) a; R^1 = CH=CHCO$_2$Me, R^2 = OMe
b; R^1 = OMe, R^2 = CH=CHCO$_2$Me

121 G. M. Prokhorov, V. I. Dronov, R. F. Nigmatullina, and F. G. Valuamova, *Khimiya i Fiz.-khimiya Monomerov*, 1975, 204 (*Chem. Abs.*, 1977, **87**, 38 526).
122 W. Davies, B. C. Ennis, C. Mahavera, and Q. N. Porter, *Austral. J. Chem.*, 1977, **30**, 173.
123 F. D. Lewis and C. E. Hoyle, *J. Amer. Chem. Soc.*, 1977, **99**, 3779.
124 B. S. Green and M. Retjo, *J. Org. Chem.*, 1974, **39**, 3284.

reports [125] that at low *trans*-stilbene conversion *cis*-stilbene, (222), and (224a) were the products, and only as photolysis progressed towards the stilbene photostationary state did (223) and (224b) appear in the photolysate. Photosensitization was ineffective in the reaction: a singlet mechanism involving an exciplex was proposed and confirmed by the detection of weak exciplex fluorescence. The quantum yield for the formation of the oxetans increases with increased stilbene concentration but is independent of the concentration of dimethyl fumarate. These observations are in accord with a mechanism whereby the stilbene–dimethyl fumarate exciplex is intercepted by ground-state dimethyl fumarate.

Pac *et al.*[126] have studied the photochemical cycloreversion reactions of the [2 + 2] adducts (225). They [126] have established that in the presence of triplet quencher (benzophenone) the adducts (225 a and b) undergo cycloreversion to yield olefins (226). Thus (225a) and (225b) yield (226a) and (226b) although the proportions of each differ. When the triplet quencher is omitted the reaction is essentially stereospecific, *i.e.* (225a) gives (226a) although a small amount of (226b) is also formed. Similar behaviour is encountered with (225b). Direct irradiation of (225c) and (225d) affords only (226c) and (226d) respectively. Thus in these instances the cycloreversion is completely stereospecific.

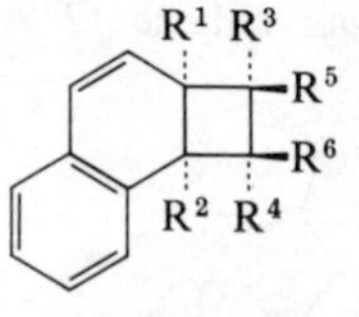

	R^1	R^2	R^3	R^4	R^5	R^6
(226) a;	H	CN	H	H	Me	PhO
b;	H	CN	Me	H	H	PhO
c;	CN	H	H	H	Me	MeO
d;	CN	H	Me	H	H	MeO

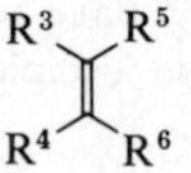

Photoaddition of ethylene to the olefin (227) has been reported to yield the [2 + 2] adduct (228) in reasonable yields.[127] The addition was carried out at −60 °C with acetone sensitization. The scope of the addition process was studied.

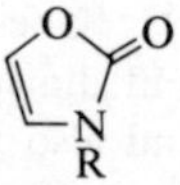

(227) R = H or COMe

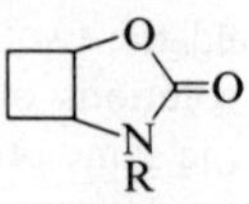

(228)

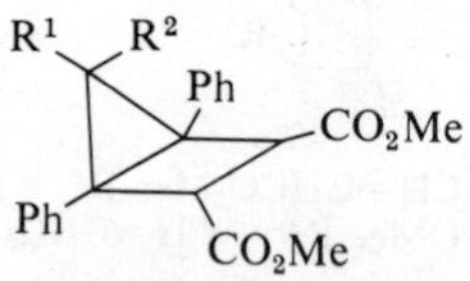

(229) $R^1 = R^2 = Me$
       $R^1 = Ph, R^2 = H$

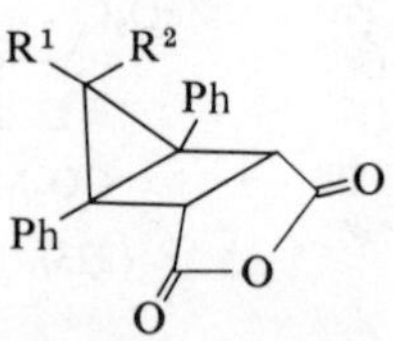

(230) $R^1 = R^2 = Me$
       $R^1 = Ph, R^2 = H$

[125] F. D. Lewis and D. E. Johnson, *J. Amer. Chem. Soc.*, 1978, **100**, 983.
[126] C. Pac, K. Mizuno, and H. Sakurai, *Bull. Chem. Soc. Japan*, 1978, **51**, 329.
[127] K.-H. Scholz, H.-G. Heine, and W. Hartmann, *Tetrahedron Letters*, 1978, 1467.

Direct and sensitized irradiation of 3,3-dimethyl-1,2-diphenylcyclopropene and 1,2,3-triphenylcyclopropene with dimethyl fumarate or maleic anhydride affords the [2 + 2] adducts (229) and (230).[128] Another report on this work has also been published.[129]

Irradiation of indene in the presence of 1-naphthonitrile affords the indene radical-cation which undergoes addition to electron-rich ethylenes such as the vinyl ethers (CH$_2$=CHOR; R = Me, Et, SiMe$_3$, or Ac) to yield the [2 + 2] adducts (231) and (232).[130] Indene also undergoes photoaddition to the perhalogenocycloalkenes (233), yielding the benzotricycloalkenes (234).[131]

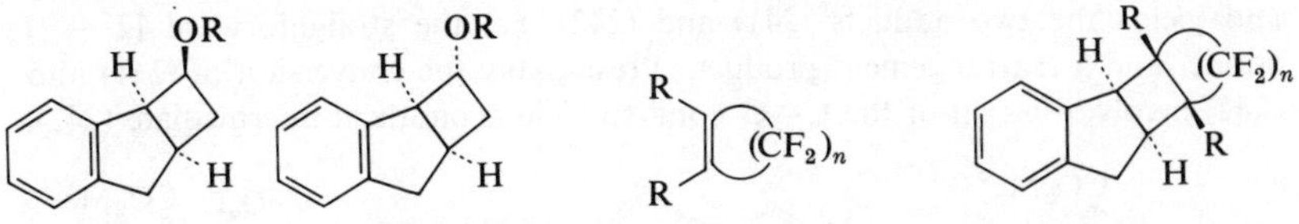

(231)   (232)   (233) R = Cl or F; n = 2 or 3   (234)

Irradiation of the complexes of norbornene and *endo*-dicyclopentadiene with Cu^I triflate in allyl alcohol as solvent yields the products shown in Scheme 20.[132]

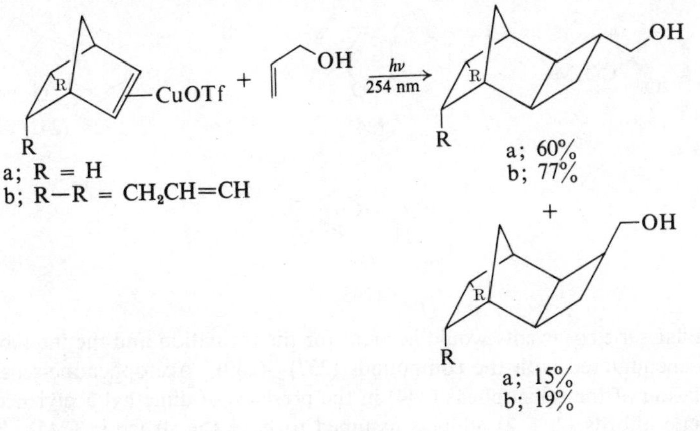

a; R = H
b; R—R = CH$_2$CH=CH

a; 60%
b; 77%

a; 15%
b; 19%

Scheme 20

The photoaddition of diphenylacetylene to *p*-dioxen affords the [2 + 2] adduct (235).[133] This adduct is also photoreactive and undergoes ring opening to yield (236) when irradiation is carried out through a Pyrex filter.

Low yields for four products (237)—(240) were obtained from the acetophenone-sensitized irradiation of benzo[*b*]furan with dimethyl acetylene-

[128] D. R. Arnold and R. M. Morchat, *Afinidad*, 1977, **34**, 276 (*Chem. Abs.*, 1978, **88**, 62 030).
[129] D. R. Arnold and R. M. Morchat, *Canad. J. Chem.*, 1977, **55**, 393.
[130] K. Mizuno, R. Kaji, and Y. Otsuji, *Chem. Letters*, 1977, 1027 (*Chem. Abs.*, 1978, **88**, 36 903).
[131] H. Muramatsu and K. Inukai, *J. Fluorine Chem.*, 1977, **9**, 417 (*Chem. Abs.*, 1977, **87**, 201 142).
[132] R. G. Salomon and A. Sinha, *Tetrahedron Letters*, 1978, 1367.
[133] G. Kaupp and M. Stark, *Angew. Chem. Internat. Edn.*, 1977, **16**, 552.

(235)   (236)

dicarboxylate.[134] Indeed the irradiation of any one of the four products produced the other three. Thus it is obvious that the initial adduct, possessing a
photolabile cyclobutene moiety, undergoes photochemical rearrangement.
Methyl propiolate also undergoes addition, but it behaves more conventionally
and yields the two adducts (241) and (242), *i.e.* the straightforward [2 + 2]
adduct and a rearrangement product. Presumably the conversion of (241) into
(242) involves fission of the C—O bond to yield a biradical intermediate (243).

(237)   (238)   (239)

(240)   (241)   (242)

(243)

A similar series of events would account for the formation and the interconver-
sions encountered with the compounds (237)—(240). Acetophenone-sensitized
irradiation of the selenophens (244) in the presence of dimethyl acetylenedicarboxylate affords [2 + 2] adducts assumed to have the structure (245). These
readily thermally rearrange to the naphthalene derivatives (246).[135] In neither
example of (244) was the primary adduct (245) sufficiently stable for isolation.

(244) R = H or Me   (245)   (246)

[134] A. H. A. Tinnemans and D. C. Neckers, *J. Org. Chem.*, 1977, **42**, 2374.
[135] T. Q. Minh, L. Christiaens, P. Grandclaudon, and A. Lablache-Combier, *Tetrahedron*,
    1977, **33**, 2225.

Cyclobutane derivatives (247) were obtained from the sensitized irradiation of the selenophens (248) in the presence of 1,2-dichloroethylene. The photochemical addition of acetylene to the steroid (249) affords the [2 + 2] adduct (250).[136]

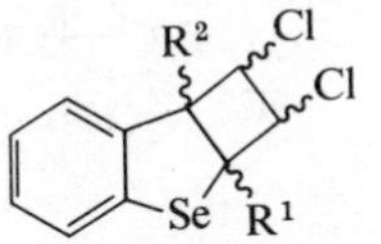

(247) a; $R^1 = R^2 = H$
b; $R^1 = H$, $R^2 = Me$
c; $R^1 = R^2 = Me$
d; $R^1 = H$, $R^2 = AcO$
e; $R^1 = Me$, $R^2 = AcO$

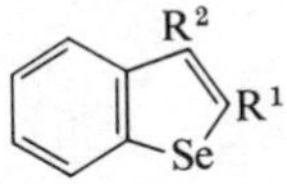

(248)

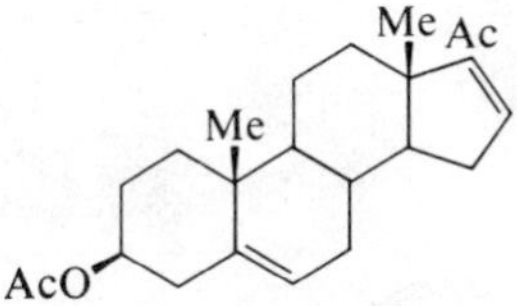

(249) 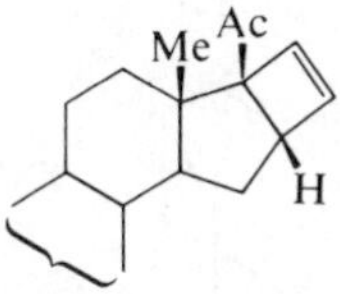　　　　(250)

7 Miscellaneous Reactions

Review articles dealing with the use of photochemical probes of hydrocarbon chain conformations in solution [137] and with the photochemistry of alicyclic compounds [138] have been published. A full paper dealing with the *cis* effect in the photochemical ring fission of substituted cyclobutanes has been published.[139]

The novel tetracyclo-octene derivative (251) has been successfully prepared by the photochemical fragmentation of the adduct (252) using ethereal solutions and Vycor-filtered irradiation at 0 °C.[140] This process yielded 1,4-dimethyl-2,3-diphenylbenzene and the desired compound (252). Irradiation of the ozonide (253) at 77 K in a transparent matrix affords trifluoroacetic anhydride and the cyclobutadiene (254) exclusively.[141] Epling and Yoon [142] have shown that the

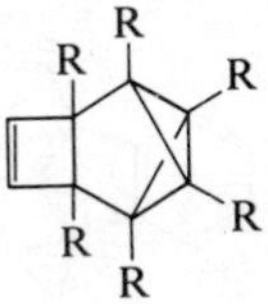

(251) R = CF$_3$ 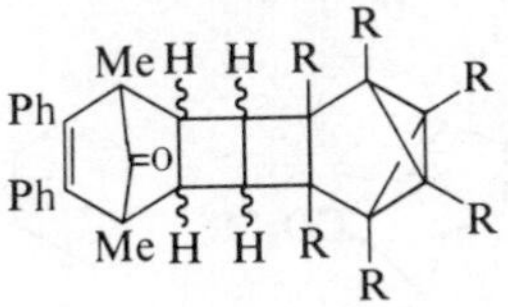　　　　(252)

[136] A. V. Kamernitskii, V. N. Ignatov, S. I. Levina, E. P. Serebryakov, G. V. Nikitina, and V. V. Korkhov, *Khim.-Farm. Zhur.*, 1977, **11**, 96 (*Chem. Abs.*, 1978, **88**, 62 518).
[137] M. A. Winnik, *Accounts Chem. Res.*, 1977, **10**, 173.
[138] M. B. Rubin, in 'Alicyclic Compounds', ed. D. Ginsburg, MTP International Review of Science, Organic Chemistry Series Two, Vol. 5, Butterworths, London, 1976, p. 277 (*Chem. Abs.*, 1977, **86**, 170 106).
[139] G. Kaupp and M. Stark, *Chem. Ber.*, 1977, **110**, 3084.
[140] R. N. Warrener and E. E. Nunn, *Austral. J. Chem.*, 1978, **31**, 221.
[141] S. Masamune, T. Machiguchi, and M. Aratani, *J. Amer. Chem. Soc.*, 1977, **99**, 3524.
[142] G. A. Epling and U. C. Yoon, *Tetrahedron Letters*, 1977, 2471.

(253)   (254)

irradiation of cinchona alkaloids, *e.g.* (255), in methanol affords the fragmentation products (256) and (257) with no sign of dehydroxylation as reported earlier
by Stenberg [143] who had studied the process in 2M-HCl. Further reports of the
photochemical decomposition of tetraphenyl-*p*-dioxin and tetraphenyl-*p*-dithiin
have appeared.[144, 145]

(255)   (256)   (257)

A study of the direct and sensitized *cis–trans* isomerization of the spiro-
cyclopropane (258) has been reported.[146] The photostationary state for the
direct irradiation is 97% *trans*, which is in contrast with the work reported by
Doering and Jones [147] who studied the related spiro-cyclopropane (260) and
observed a corresponding ratio of 1.74 : 1. The quantum yield for the isomerization of *cis*-(259) to *trans*-(258) was 0.0095 and was unaffected by the presence of
air. The isomerization also occurred upon sensitization (acetophenone being the

(258)   (259)   (260)

[143] V. I. Stenberg, E. F. Travecedo, and W. E. Musa, *Tetrahedron Letters*, 1969, 2031.
[144] S. Lahiri, V. Dabral, V. Bhat, E. D. Jemmis, and M. V. George, *Proc. Indian Acad. Sci.,
Sect. A*, 1977, **86**, 1 (*Chem. Abs.*, 1977, **87**, 167 945).
[145] S. Lahiri, V. Dabral, and M. V. George, *Tetrahedron Letters*, 1976, 2259.
[146] G. Ravier, C. Decker, A. M. Braun, and J. Faure, *Tetrahedron Letters*, 1977, 1919.
[147] W. von E. Doering and M. Jones, jun., *Tetrahedron Letters*, 1963, 791.

most efficient; $\Phi = 1.05$). The photostability of the bicyclic compounds (261; $n = 1$) and (261; $n = 2$) is greater than that of (262).[148] Nevertheless both bicyclic compounds afford cyclic ylides when irradiated at 254 nm. The primary products (263; $n = 1$) and (263; $n = 2$) both undergo secondary reactions. A carbonyl-ylide (264) is formed when the epoxy-diene (265) is irradiated (280 nm): subsequent ring closure yields the products (266).[149]

(261) $n = 1$ or 2   (262)   (263) $n = 1$ or 2

(264)   (265)   (266)

The epoxide (267a) undergoes photochemical ring opening in propan-2-ol to afford the alcohol (268a) and the hydroxy-ether (269a). When the epoxide (267b) was irradiated under similar conditions, products analogous to the above were isolated, together with (270) and (271).[150]

(267) a; $R = H$, $n = 1$ or 2   (268) a; $R = H$, 64—74%   (269) a; $R = H$, 13—35%
b; $R = Me$, $n = 1$ or 2   b; $R = Me$, 28—33%   b; $R = Me$, 24—60%

(270) a; —   (271) a; —
b; $R = Me$, 5—20%   b; $R = Me$, 6—21%

Earlier accounts[151, 152] of the photochemistry of 2-iodoadamantane (272a) in methanol interpreted the formation of 2-methoxyadamantane (272b) in terms of the homolytic fission of the I—C bond, electron transfer to yield the adamantyl carbocation (273), and trapping of this ion by solvent. A re-investigation of this reaction in the presence of triethylamine as an acid scavenger

[148] K. Nishiyama, K. Ishikawa, I. Sarkar, D. C. Lankin, and G. W. Griffin, *Heterocycles*, 1977, **6**, 1337 (*Chem. Abs.*, 1978, **88**, 21 680).
[149] W. Eberbach and U. Trostmann, *Tetrahedron Letters*, 1977, 3569.
[150] S. N. Merchant, S. C. Sethi, and H. R. Sonawane, *Indian J. Chem.*, 1977, **15B**, 82 (*Chem. Abs.*, 1977, **87**, 102 001).
[151] P. J. Kropp, G. S. Poindexter, N. J. Pienta, and D. C. Hamilton, *J. Amer. Chem. Soc.*, 1976, **98**, 8135.
[152] R. R. Perkins and R. E. Pincock, *Tetrahedron Letters*, 1975, 943.

shows that the products formed in methanol are (274) and (275) (minor) together
with the ether (272b) and the reduction product (272c).[153] These results show
that the adamantyl carbocation (273), although still readily trapped by solvent,
will also undergo rearrangement and deprotonation to (275) or conversion into
(274). When the reaction is carried out in ether, with triethylamine to scavenge
the acid formed, (274) and (275) are the principal products. These results clearly
illustrate the fate of the carbocation (273) in the absence of acid and show that
the ether (272b) is produced as a secondary product.

(272) a; X = I   (273)   (274)   (275)
      b; X = OMe
      c; X = H

Several products are produced when the di-iodide (276a) is irradiated in
different solvents as shown in the Table.[154] The ionic product (278) is principally
formed although products such as the substitution product (279) were also

Table   *Products from the irradiation of the di-iodide* (276) [154]

		Products, % yield			
Solvent	*Time/h*	(276)	(277)	(278)	*Other*
Ether	12	28	6	51	
Benzene	20	20	trace	62	
Methanol	5	32	3	8	(279) 45
Methylene chloride	1.5	10	trace	39	(280) 28
Methylene bromide	1.5	4	trace	46	(280) 41

(276) a; R = H   (277)   (278)   (279)   (280)   (281)   (282)
      b; R = D

encountered in methanol. The formation of the hydrocarbon product (280) in
methylene chloride and methylene bromide was thought important in that it
suggests a mechanism involving the carbene (281) which subsequently rearranges
to (282) and isomerizes in acid to the product (280). The presence of a carbene
intermediate was substantiated by photolysis of the labelled di-iodide (276b).

[153] P. J. Kropp, J. R. Gibson, J. J. Snyder, and G. S. Poindexter, *Tetrahedron Letters*, 1978,
      207.
[154] N. J. Pienta and P. J. Kropp, *J. Amer. Chem. Soc.*, 1978, **100**, 655.

4
Photochemistry of Aromatic Compounds

1 Introduction

The organization of the subject matter of this Chapter follows that adopted in previous Volumes. In general only publications are covered that describe reactions in which an aromatic ring, rather than substituents alone, is directly involved. Spectroscopic and more physical aspects are reviewed in Part I, Chapters 3 and 4.

A number of reviews pertinent to this area are considered in the appropriate sections. Few really striking reports have been published although in particular areas significant publications which rationalize previous observations have appeared. There does appear to be a fairly general trend to make more use of photochemical procedures in syntheses, particularly intramolecular photo-cyclizations.

2 Isomerization Reactions

The primary photoprocesses in benzene have again received comment and the photoisomerization has been briefly discussed.[1] Mulder has attempted to clear up some of the misunderstanding which he feels exists concerning the application of orbital symmetry arguments to the benzene $\rightleftarrows$ benzvalene interconversion.[2] Previous relevant theoretical reports [3—6] are discussed and the author concludes by advising in general that the use of correlation diagrams which are concerned with potential energy curves or surfaces should be distinguished from the application of the Woodward–Hoffmann rules from which stereochemical deductions may be made, and further that the intended correlation of the ground state of the reactant with the product's excited state or vice versa does not intrinsically predict a high activation energy nor does a low value necessary follow from ground-state correlations. The fact remains that in practice, symmetry correlations have rationalized many puzzling features of benzene photochemistry and have proved to possess predictive value.

The formation of bicyclo[3,1,0]hex-2-ene derivatives from the 254-nm irradiation of benzene in the presence of acidified alcohols was first noted in

1 W. A. Noyes, *Compt. rend.*, 1977, **285**, *C*, 53.
2 J. J. C. Mulder, *J. Amer. Chem. Soc.*, 1977, **99**, 5177.
3 D. Bryce-Smith and A. Gilbert, *Tetrahedron*, 1976, **32**, 1309.
4 I. Haller, *J. Chem. Phys.*, 1967, **47**, 1117.
5 J. J. C. Mulder and L. J. Oosterhoff, *Chem. Comm.*, 1970, 305, 307.
6 E. Farenhorst, *Tetrahedron Letters*, 1966, 6465.

1966,[7] and was later shown to arise by acidolysis of photoproduced benz-valene.[8] Further work on this reaction has been conducted using t-butylbenzene and acidified methanol, the irradiation of which is found to yield a mixture of the *exo*-4-methoxy-1-t-butyl- and *exo*- and *endo*-6-methoxy-2-t-butylbicyclo[3,1,0]-hex-2-enes (1), (2), and (3), respectively.[9] The stereoisomers (2) and (3) are sensitized photoproducts from (1), but the present rearrangement cannot be explained by the earlier proposed cyclohexenyl biradical intermediate (4)[8] which would have led to the 6-methoxy-3-t-butyl bicyclohexene (5) instead of the observed 2-t-butyl isomers (2) and (3). It is proposed that the rearrangement of (1) may occur by a sensitized vinylcyclopropane rearrangement involving the bicyclo[2,1,1]hex-2-ene (6), but a question mark remains in view of the failure to isolate this isomer.

With the exception of perfluoroalkyl benzenes[10] and pyridines,[11] the photo-formation of 'Dewar' (1,4-*para*-bonded) isomers from arenes is a fairly low yield process. It has now, however, been observed that the unfiltered medium-pressure mercury arc lamp irradiation of [6]paracyclophane (7) in [$^2H_{12}$]-cyclohexane leads to quantitative conversion into the Dewar isomer (8).[12] The reverse thermal reaction is described as clean and rapid; indeed only the

[7] E. Farenhorst and A. F. Bickel, *Tetrahedron Letters*, 1966, 5911; K. E. Wilzbach, J. S. Ritscher, and L. Kaplan, *J. Amer. Chem. Soc.*, 1966, **88**, 2881.
[8] L. Kaplan, D. J. Rausch, and K. E. Wilzbach, *J. Amer. Chem. Soc.*, 1972, **94**, 8638.
[9] Y. Izawa, H. Tomioka, T. Kagami, and T. Sato, *J.C.S. Chem. Comm.*, 1977, 780.
[10] M. G Barlow, R N. Haszeldine, and M. J. Kershaw, *J.C.S. Perkin I*, 1975, 2005.
[11] See, for example, R. D. Chambers, and R. Middleton, *J.C.S. Chem. Comm.*, 1977, 154.
[12] S. L. Kammula, L. D. Iroff, M. Jones, J. W. van Straten, W. H. De Wolf, and F. Bickel-haupt, *J. Amer. Chem. Soc.*, 1977, **99**, 5815.

1-chloro- and 1-fluoro-Dewar benzenes exceed (8) in their measured rates of rearomatization.[13] It is interesting to note here that irradiation at wavelengths longer than 290 nm of thoroughly degassed benzene or ether solutions of decamethylanthracene yields the Dewar isomer (9) of which 15% rearomatizes in solution at 40 °C in 30 min.[14]

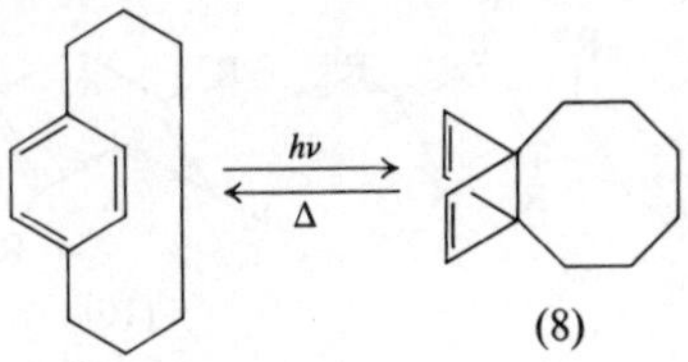

(7)   (8) 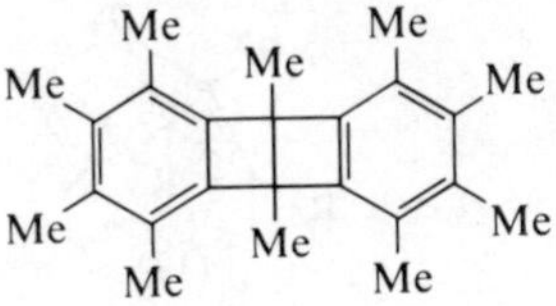   (9)

Chambers and Middleton have reported further on their studies of the photorearrangements of fluorinated alkylpyridines to azaprismanes [*e.g.* (10) and (11)].[15] In each case a 1-azabicyclo[2,2,0]hexadiene derivative [*e.g.* (12)] is formed together with the azaprismanes, the structures of which are deduced from those of the pyridines [*e.g.* (13), (14), and (15)] formed on rearomatization. To account for the structures of the azaprismanes, it is proposed that rearrangement of an initially formed 2-aza-bicyclo[2,2,0]hexadiene derivative to the 1-aza isomer occurs. Thus whereas (12) can yield (11) and thence (15) it would appear that the formation of the azaprismane (10) requires the 1-aza-compound (16) which may be produced by rearrangement of the initially formed 2-aza isomer (17) as shown in Scheme 1.

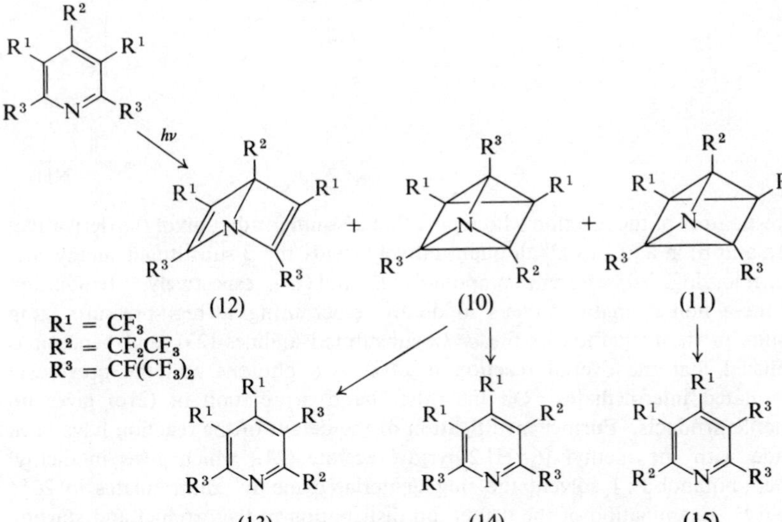

$R^1 = CF_3$
$R^2 = CF_2CF_3$
$R^3 = CF(CF_3)_2$

(12)   (10)   (11)

(13)   (14)   (15)

[13] R. Breslow, J. Napierski, and A. H. Schmidt, *J. Amer. Chem. Soc.*, 1972, **94**, 5906.
[14] H. Hart and B. Ruge, *Tetrahedron Letters*, 1977, 3143.
[15] R. D. Chambers and R. Middleton, *J.C.S. Perkin I*, 1977, 1500.

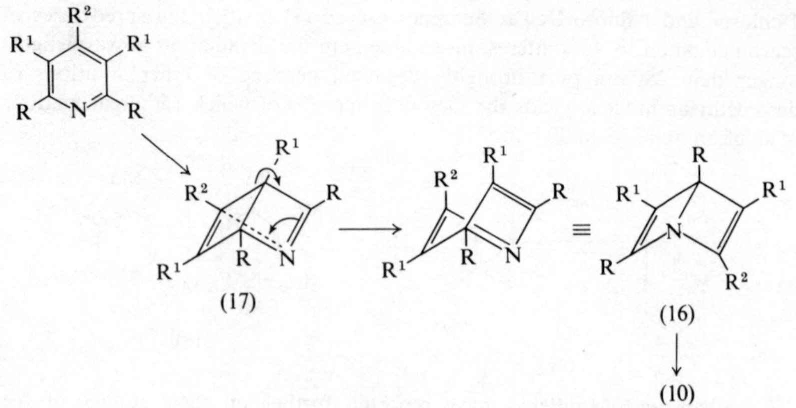

(17)   (16)

                                              ↓

                                             (10)

Scheme 1

The photoisomerization of 2-pyridylacetonitrile to anthranilonitrile was first reported in 1974 and it was tentatively suggested that this intriguing process occurred by way of the bicyclic intermediates (18), (19), and (20).[16] The photolysis of several 2-substituted picolines has now been studied by the original

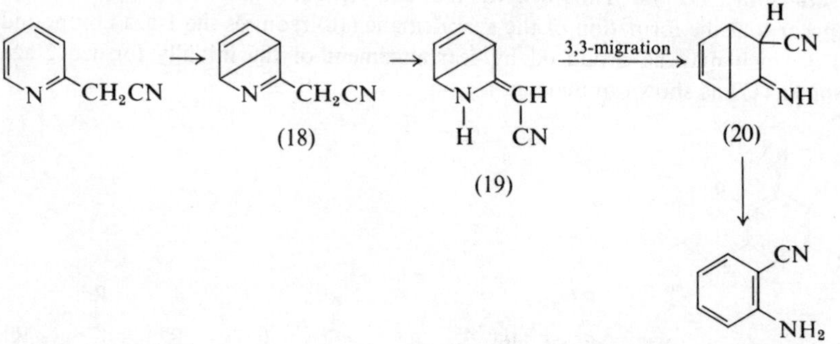

(18)

(19)

(20)

investigators of the reaction who report that 254-nm irradiation of the derivatives (21a and b) in aqueous alkali quantitatively yields the 3-substituted methylene-2-azabicyclo[2,2,0]hex-5-ene compounds (22) and (19), respectively.[17] Irradiation of these non-aromatic isomers in diethyl ether using a high-pressure lamp results in the formation of the *ortho*-substituted anilines (23), and hence it is deduced that the overall reaction involves two photons and the previously postulated intermediates. On the other hand, irradiation of (21c) gives no volatile products. Further investigation of the details of the reaction have been made with the methyl [6-^{2}H]-2-pyridyl acetate (24), which gives in diethyl ether : butanol 1 : 1 solvent the ring deuteriated methyl anthranilates in 20% yield.[18] Examination of the deuterium distribution in the product and starting

[16] Y. Ogata and K. Takagi, *J. Amer. Chem. Soc.*, 1974, **96**, 5933.
[17] Y. Ogata and K. Takagi, *J. Org. Chem.*, 1978, **43**, 944.
[18] K. Takagi and Y. Ogata, *J.C.S. Perkin II*, 1977, 1980.

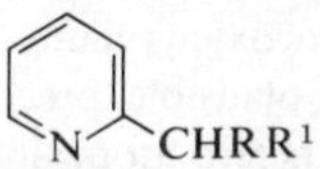

(21) a; R = H, R¹ = CO₂R² (R² = Me or Et)
$$(21)\ a;\ R = H,\ R^1 = CO_2R^2\ (R^2 = Me\ or\ Et)$$

(21) a; R = H, R^1 = CO_2R^2 (R^2 = Me or Et)
    b; R = H, R^1 = CN
    c; R = Me, R^1 = CO_2Me

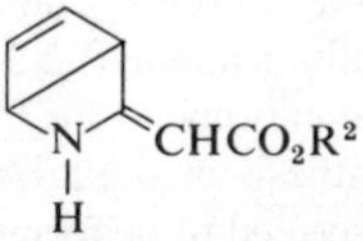

(22) R^2 = Me or Et

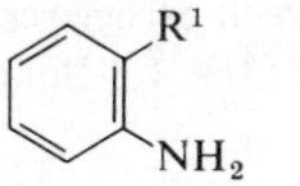

(23) R^1 = CN or CO_2R^2 (R^2 = Me or Et)

material revealed that in the former approximately equal amounts of the [4-²H] and [6-²H] isomers (25) and (26) (*ca.* 40% of each) are produced with 10—20% of the other two deuteriated positional isomers, whereas the latter contained 90% of the original isomer (24). In order to account for these distributions, it is suggested that scisson of the N(1)—C(6) bond of the pyridine occurs to give the biradical intermediate (27), which then undergoes bond formation between C(4 or 6) and (7) as shown in Scheme 2. The reactivity of ethyl methyl-2-pyridyl acetates has been reported to be very dependent upon the position of the methyl substituent.[19] Thus while the 3-, 4-, and 5-methyl isomers all yield the corresponding anthranilates ($\Phi_{disappearance}$0.09—0.25), the 6-isomer has very low photoreactivity and its quantum yield for disappearance is only 0.006. Reasons for this large difference in reactivity are not obvious from the suggested reaction mechanism.

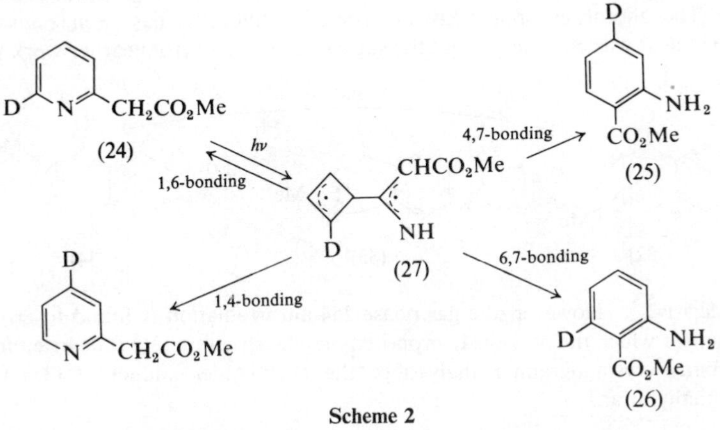

Scheme 2

Photoformation of Dewar-isomers from heteroaromatic compounds is no longer a rare process, and many examples have been reported over the years. 1,3-Bonding leading to benzvalenes in such systems is, on the other hand, rare and is then only generally considered as a process leading to transient intermediates rather than stable products: an example of this type of proposal is

[19] K. Takagi and Y. Ogata, *J.C.S. Perkin II*, 1977, 1148.

found in the case of the rearrangement of hydroxypyrylium compounds where the 'oxoniabenzvalene' (28) was cited as a plausible precursor.[20] The first example of a benzvalene analogue containing hetero atoms in the ring system is now, however, claimed and involves the photoformation of 1,3,4,6-tetrakis-(trifluoromethyl)-2,5-diphosphatricyclo[3,1,0,0^{2,6}]hex-3-ene (29) from the aromatic isomer (30).[21] It is interesting to note here that whereas the thermal synthesis of the aromatic compound (30) had been earlier reported[22] it is now described to be formed also by the irradiation of (31). 1,3-Bonding has also been

observed for 3-oxidopyridazinium betaines.[23] Thus irradiation at wavelengths longer than 290 nm of acetonitrile solutions of, for example (32) yields the fused diaziridine (33) which is sufficiently stable to isolate. Treatment of (33) with water or irradiation of (32) in aqueous solution leads to the ring-opened product (34). The photolysis of tetrakistrifluoromethyl thiophen has been reinvestigated[24] as it was considered that the suggested structure from earlier work was

not decisive.[25] However, the gas phase 254-nm irradiation is found to give an isomer for which the previously proposed bicyclic structure (35) is deduced from an X-ray crystallographic analysis of the Diels–Alder adduct of (35) with tetramethylfuran.

[20] J. W. Pavlik and E. L. Cleannan, *J. Amer. Chem. Soc.*, 1973, **95**, 1697.
[21] Y. Kobayashi, S. Fujino, H. Hamana, I. Kumadaki, and Y. Hanzawa, *J. Amer. Chem. Soc.*, 1977, **99**, 8511.
[22] Y. Kobayashi, I. Kumadaki, A. Ohsawa, and H. Hamana, *Tetrahedron Letters*, 1976, 3715.
[23] Y. Maki, M. Kawamura, H. Okamoto, M. Suzuki, and K. Kaji, *Chem. Letters*, 1977, 1005.
[24] Y. Kobayashi, I. Kumadaki, A. Ohsawa, Y. Sekine, and A. Ando, *Heterocycles*, 1977, **6**, 1587.
[25] H. A. Wiebe, S. Braslavsky, and J. Hiecklen, *Canad. J. Chem.*, 1972, **50**, 2721.

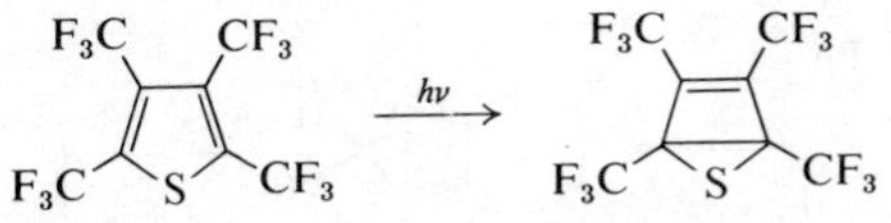

(35)

As in previous years, a number of reports have been published which describe the photorearrangement of benzene derivatives to an aliphatic compound *via* routes other than arene valence isomerization. Again the photoisomer may be isolable but in other cases can be unstable and revert to the starting material, proceed to more stable products, or be trapped by an added reagent. Representative examples of such reports are included here to complete the section on aromatic isomerizations. Direct irradiation of dibenzohomobarrelene (36) causes its rearrangement by a di-π-methane process to yield the isomer (37) and the fluorene (38), which is suggested to be formed *via* the carbene (39).[26] The

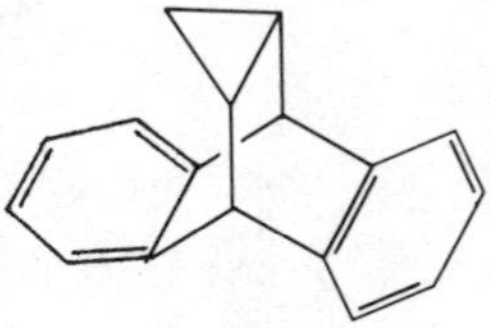

(36)

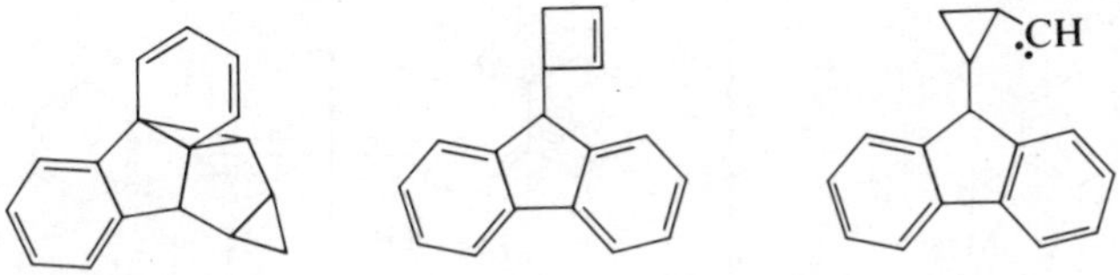

(37)   (38)   (39)

transients formed on the flash photolysis of a series of 1,1-diarylindenes have been assigned as isoindenes.[27] For example, 254-nm irradiation of 1,1,3-triphenylindene at $-70\ °C$ yields an orange solution (λ_{max} 478 nm) which from n.m.r. spectroscopy is considered to contain the isoindene (40). On warming the solution 1,2,3-triphenylindene results and in the presence of dienophiles, adducts, for example (41), are formed. In the conversion of benzisoxazoles (42) into the isomers (43), spectroscopic evidence has been presented which indicates that the intermediate is the azirene (44).[28] It is interesting to note here that the rearrangement for R = methyl or phenyl in (42) occurs in polar solvents, whereas for R = H both polar and non-polar solvents seem to be suitable. Finally, the *ortho* quinomethane (45) is proposed as the reaction intermediate in the photorearrangement-methanol addition reaction of 4,7-dimethyl-3-chromanone (46) in basic methanol to give (47).[29]

[26] H. Hemetsberger, W. Braeuer, and D. Tartler, *Chem. Ber.*, 1977, **110**, 1586.
[27] K. Kamal De Fonseka, C. Manning, J. J. McCullough, and A. J. Yarwood, *J. Amer. Chem. Soc.*, 1977, **99**, 8257.
[28] K. H. Grellmann and E. Tauer, *J. Photochem.*, 1977, **6**, 365.
[29] A. Padwa and W. Owens, *J. Org. Chem.*, 1977, **42**, 3077.

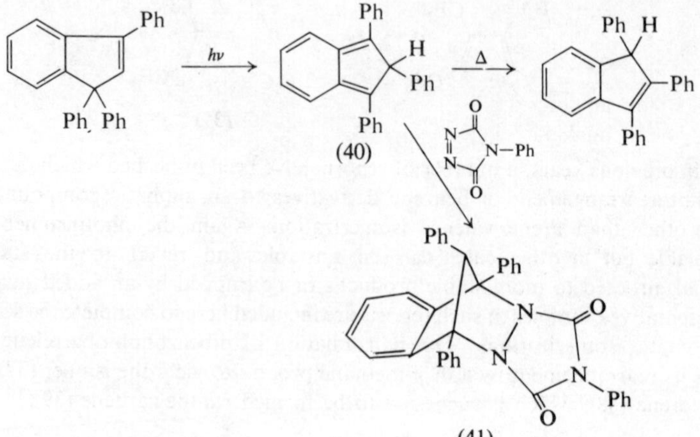

(40)

(41)

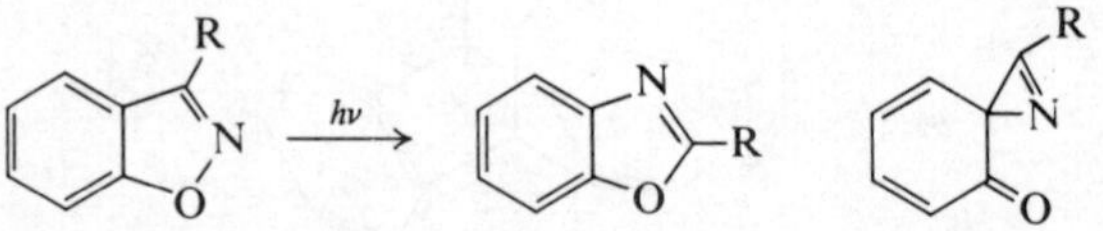

(42) R = H, Me, or Ph   (43)   (44)

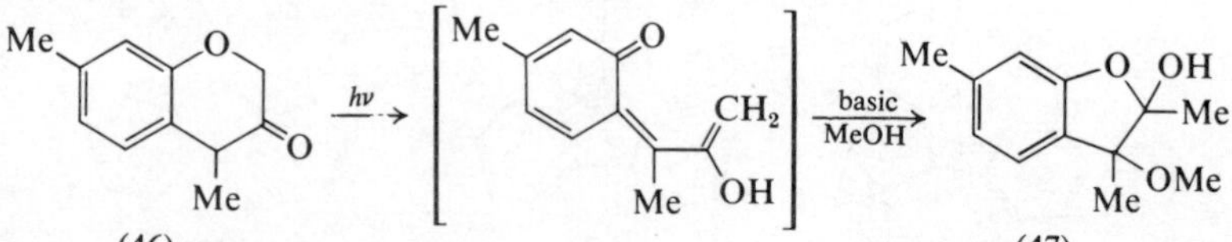

(46)   (45)   (47)

3 Addition Reactions

The second part of the two-part review on the organic photochemistry of benzene has been published and is concerned principally with the photoreactions of benzene with other molecules.[30] Orbital symmetry aspects of the topic are covered and in particular the widely diverse cyclic and acyclic addition reactions are reviewed. An Appendix was added in August 1977, which covered very briefly the literature from the manuscript submission to within a few weeks of its publication. It may seem from this review that there can be little left unknown or unexplored in benzene organic photochemistry, but this year further reports on the details of the various processes have appeared and even the benzene-maleic anhydride photoreaction has received further brief comment.[31]

The 1,2-photocycloaddition of ethylenes to arenes is a general reaction but with benzene in particular, earlier reported examples of the process led exclusively to either an *exo* or an *endo* stereomer of (48). Thus the dienophilic ethy-

[30] D. Bryce-Smith and A. Gilbert, *Tetrahedron*, 1977, **33**, 2459.
[31] P. Boule, J. Roche, and J. Lemaire, *J. Chim. Phys.-Phys. Chim. Biol.*, 1977, **74**, 593.

(48)

lenes maleic anhydride,[32] maleimide,[33] and acrylonitrile[34] gave *exo* isomers, whereas the *endo* configuration was deduced for the 1,2-adducts from *cis*-but-2-ene[35] and *cis*-cyclo-octene[36] and benzene. It now transpires that the reaction is sometimes less specific than originally considered: thus both methyl acrylate and methyl methacrylate undergo 1,2-cycloaddition to benzene to give an approximately 2:1 mixture of stereomers with the ester group *exo* and *endo* respectively to the cyclohexa-1,3-diene moiety in (48).[37] The reasons for this change in specificity are at present unclear but the authors do not discount the possible involvement of triplet rather than, or together with, singlet intermediates – as has so ably been demonstrated for the photoaddition of dimethyl fumarate to phenanthrene.[38]

The 1,3-photocycloaddition of a variety of cyclic and bicyclic alkenes to anisole, and of cyclobutene and cyclopentene to toluene, has been reported to yield essentially only the 1-isomer (49) and in many cases it has been deduced

(49) R = OMe or Me

that this is the *endo* isomer.[39] These observations are rationalized in terms of a preferred *endo* 2,6-orientation of the addends in an intermediate exciplex, but exceptions to this specificity have now been described by other workers.[40] Thus toluene and *cis*-cyclo-octene specifically yield the 5-methyl isomer (50) of which approximately 12% has the *exo* configuration; and whereas the positional isomers obtained from *o*- and *p*-xylene and *cis*-cyclo-octene correspond to those in the corresponding reactions with cyclopentene,[39b] different isomers are formed

[32] D. Bryce-Smith, B. Vickery, and G. I. Fray, *J. Chem. Soc.* (*C*), 1967, 390.
[33] D. Bryce-Smith and M. A. Hems, *Tetrahedron Letters*, 1966, 1895.
[34] R. J. Atkins, G. I. Fray, and A. Gilbert, *Tetrahedron Letters*, 1975, 3087.
[35] K. E. Wilzbach and L. Kaplan, *J. Amer. Chem. Soc.*, 1971, **93**, 2073.
[36] H. M. Tyrrell and A. P. Wolters, *Tetrahedron Letters*, 1974, 4193.
[37] R. J. Atkins, G. I. Fray, A. Gilbert, and M. W. bin Samsudin, *Tetrahedron Letters*, 1977, 3597.
[38] D. Creed and R. A. Caldwell, *J. Amer. Chem. Soc.*, 1974, **96**, 7369; R. A. Caldwell and L. Smith, *J. Amer. Chem. Soc.*, 1974, **96**, 2994; S. Farid, S. E. Hartman, J. C. Doty, and J. L. R. Williams, *J. Amer. Chem. Soc.*, 1975, **97**, 3697.
[39] (*a*) R. Srinivasan, *J. Amer. Chem. Soc.*, 1971, **93**, 3555; (*b*) J. Cornelisse, V. Y. Merritt, and R. Srinivasan, *J. Amer. Chem. Soc.*, 1973, **95**, 6197; (*c*) J. A. Ors and R. Srinivasan, *J. Org. Chem.*, 1977, **42**, 1321.
[40] D. Bryce-Smith, W. M. Dadson, A. Gilbert, B. H. Orger, and H. M. Tyrrell, *Tetrahedron Letters*, 1978, 1093.

from *m*-xylene. Why *cis*-cyclo-octene should be the exception to the general specificity of the reaction is not obvious nor is the underlying feature which dictates exclusive formation of the 5-isomer (50) from toluene and *cis*-cyclo-octene rather than or as well as the 8-isomer, which is the other possible product

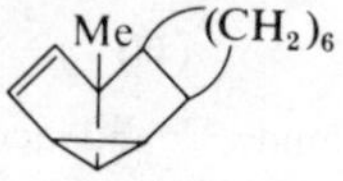

(50)

of 1,3-addition of an alkene to toluene (see Scheme 3). It has previously been pointed out that from an orbital symmetry point of view the ordering of forma-tion of the bonds in 1,3-cycloaddition is unimportant,[41] that is to say whether bond *a or b* are formed prior or subsequent to bonds *c* in (51). Clearly, some-what different intermediates are formed in the two cases and the question which of these is preferred may well decide which isomer is formed from 1,3-cyclo-addition to the substituted benzene. Reference 40 also describes the attempted additions of cyclopentene to a number of other monosubstituted benzenes: of those examined, only 1,3-cycloadducts were observed with fluorobenzene, benzotrifluoride, and chlorobenzene. The reactions appear to be non-selective and with benzotrifluoride the previously unreported vinylic-substituted 1,3-cycloadducts were observed.

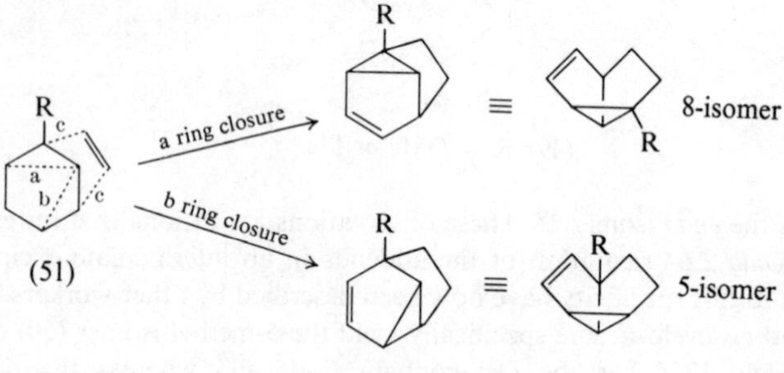

Scheme 3

Two other groups of workers have studied the photoreactions of mono-substituted benzenes with ethylenes: in both cases the addition to the ring yields 1,2-cycloadducts. Cantrell had reported earlier that reaction of benzonitrile with alkenes occurred at both the nitrile function and the nucleus,[42] and although azetines were postulated as intermediates in the former process, it was some time later that Yang and co-workers described these intermediates as stable products from irradiation of such systems.[43] Cantrell has now reported the reaction of benzonitrile with a 'diverse array' of ethylenes in order to elucidate the factor(s) which control the site of reaction (*i.e.* the nitrile function or the aromatic ring).[44]

[41] D. Bryce-Smith, *Pure and Appl. Chem.*, 1973, **34**, 193.
[42] T. S. Cantrell, *J. Amer. Chem. Soc.*, 1972, **94**, 5929.
[43] N. C. Yang, B. Kim, W. Chiang, and T. Hamada, *J.C.S. Chem. Comm.*, 1976, 729.
[44] T. S. Cantrell, *J. Org. Chem.*, 1977, **42**, 4238.

It has been observed that the pathway favoured depends critically on the number and type of electron-donating substituents on the ethylene. Thus ethylenes with four alkyl, two alkoxy, or two alkyl and one alkoxy substituent undergo photoaddition at the nitrile function to yield 2-azabutadienes as the major products together with varying amounts of the azetine precursor. This process is illustrated in Scheme 4 for the photoreaction of 2,3-dimethylbut-2-ene with

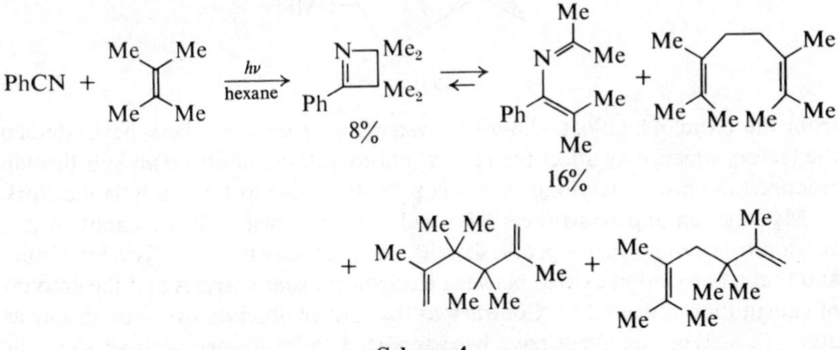

Scheme 4

benzene. With less electron-rich ethylenes 1,2-photocycloaddition occurs to the ring: for example, the irradiation of 2-methylbut-2-ene in benzonitrile yields 62% of a mixture of adducts ($\Phi = 0.17$) of which the 1,2-cycloadduct (52) constitutes 70%. The regioisomer (53), the azetines (54) and (55), and the 2-azabutadienes (56) and (57) are also considered to be formed in this reaction.

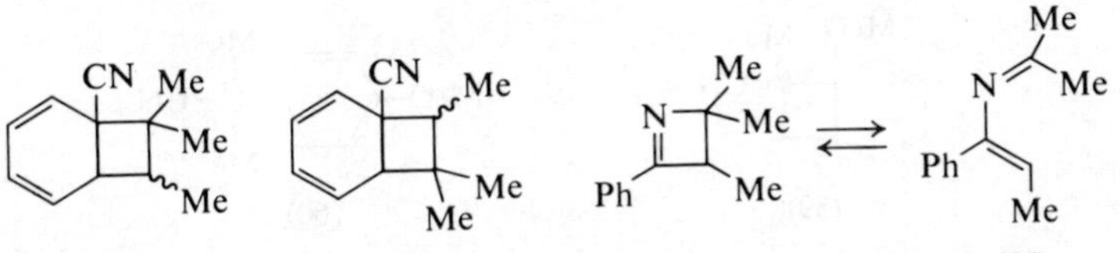

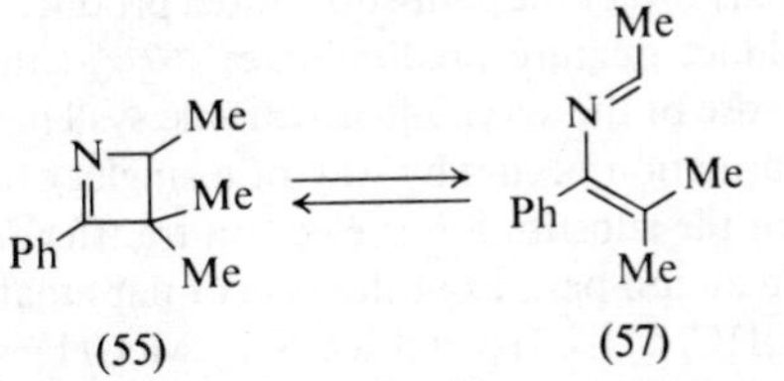

Both types of ethylene are observed to quench strongly the arene fluorescence, from which observation it is deduced that excited nitrile singlets are involved in the processes. The author speculates that the difference in reaction site for the two classes of compound results from the different positions of complexation of the addends in singlet exciplexes. Benzonitrile has not previously been reported to yield 1,3-cycloadducts with ethylenes but the reaction of this arene with 2-methoxypropene is claimed to yield a mixture which may contain such unique compounds.[44] The evidence for their formation arises from the fact that the ratio of vinyl to saturated protons is too low to be accounted for by cyclohexadiene structures. However, it should be noted that this feature may also be reasonably rationalized by the presence of the secondary photoproducts (58)

(58)

from the bicyclo[4,2,0]octa-2,4-diene systems. Japanese workers have studied the factors which may affect the [2 + 2] photocyclo-addition *versus*-substitution reactions in anisole–acrylonitrile systems.[45] Relevant to this study is the work of McCullough and co-workers who had earlier shown with the naphthalene–acrylonitrile system that a protic solvent was necessary for [2 + 2] cycloaddition and that a correlation existed between electron transfer energies and the amount of substitution observed.[46] Contrary to the former observation, with anisole as arene, [2 + 2] cycloadducts have been reported to be formed with acrylonitrile even in acetonitrile.[47] In view of these results the principal objective of the work described in ref. 45 was to examine if the correlation noted with the naphthalene substitution reactions also applied to anisole. Irradiation of 2-cyanopropene with anisole in acetonitrile yields a mixture of [2 + 2] cycloadduct stereomers whereas in methanol solvent the same system leads to 38% of the cycloadduct (59) and 40% of the *para*-substituted product (60) with 5% of the *ortho*-isomer.

(59)   (60)

Likewise crotononitrile with anisole in acetonitrile gives the stereomer mixture, but on the other hand only 6% of the *para*-substituted product results in methanol and again the cycloadduct mixture predominates (67%): thus these reactions are overall similar to those of the acrylonitrile–anisole system.[47] If it is assumed that the cycloadduct formation occurs by way of a singlet exciplex and that the rate-determining step in the substitution is electron transfer from the anisole to the ethylene to produce an ion pair, as in the case of naphthalene[46] the trend of substitution ($CH_2{=}CHCN > CH_2{=}CMeCN > MeCH{=}CHCN$) may be rationalized in terms of electron transfer energies calculated by Weller's equation.[48] Similar results are reported for dimethoxybenzene–acrylonitrile systems: the authors conclude that the higher the reduction potential (or electron affinity) of the ethylene or the lower the oxidation potential of the arene, the more

[45] M. Ohashi, Y. Tanaka, and S. Yamada, *Tetrahedron Letters*, 1977, 3629.
[46] R. M. Bowmann, T. R. Chamberlain, W. C. Huang, and J. J. McCullough, *J. Amer. Chem. Soc.*, 1974, **96**, 692.
[47] M. Ohashi, Y. Tanaka, and S. Yamada, *J.C.S. Chem. Comm.*, 1976, 800.
[48] D. Rehm and A. Weller, *Israel J. Chem.*, 1970, **8**, 259.

substitution occurs. It is further suggested that the rate-determining step for the substitution is transfer of an electron rather than a proton.

The photoaddition of acetylenes to benzene yielding cyclo-octatetraenes *via* initial 1,2-addition is a well-known process;[49] by irradiation of the methyl phenylpropiolate in benzene behind pyrex, it was shown that the addition occurred by excitation of the acetylene.[50] The propiolate–benzene system has been subjected to a detailed examination. The authors confirmed the main result of ref. 50, *viz.* that the reaction proceeds by way of the excited alkyne, but they obtained a more complex product mixture. Thus irradiation of a dilute solution of methyl phenylpropiolate in benzene at wavelengths longer than 290 nm leads to the formation of 85% of the tetracyclic compound (61).[51] In quartz apparatus, irradiation gave a 13 : 7 ratio of (61) and the cyclo-octatetraene (62), respectively, and control experiments showed that the two compounds could be interconverted in quartz apparatus; but in pyrex, (62) gave (61), probably in reflection of the differing absorptivities under the two sets of conditions. Further studies on this system have revealed that the formation of (62) may be photosensitized by xanthone and that (62) is formed in 92% yield from (61) in the presence of this sensitizer using 350 nm radiation.[52] In the former system at incomplete conversions, (61) is detected in the reaction mixture which is consistent with its role as a precursor of the cyclotetraene (62). The authors suggest, as previously noted,[50] that the reaction occurs by way of the 1,2-cycloadduct (63)

$$(61) \qquad (62) \qquad (63)$$

and that the lifetime of this is sufficient for it to absorb a second photon in competition with the residual acetylene in the direct irradiation and hence cyclize to (61). The absorption of a second photon must also compete with the thermal and/or photo valence isomerization of (63) to (62) but surprisingly no (62) is detected at low conversions of the acetylene or at 50 °C, where the isomerization to (62) would be expected to be very efficient. It is concluded that all of the cyclo-octatetraene (62) which is formed from (63) must be efficiently converted into (61) under the reaction conditions. The lack of detection of (61) in the previous study requires comment particularly in view of the fact that additional information is available to the Reporter.[53, 54] Compound (61) is thermally converted into the cyclo-octatetraene (62)[51, 54] [$t_{\frac{1}{2}}$ for (61) = 25.93 min at 187 °C[54]] and since the original isolation of (62) was by preparative g.l.c. at 200 °C it is doubtful if (61) would have survived this procedure: the

49 E. Grovenstein and D. V. Rao, *Tetrahedron Letters*, 1961, 148; D. Bryce-Smith and J. E. Lodge, *J. Chem. Soc.*, 1963, 695.
50 D. Bryce-Smith, A. Gilbert, and J. Grzonka, *Chem. Comm.*, 1970, 498.
51 A. H. A. Tinnemans and D. C. Neckers, *J. Amer. Chem. Soc.*, 1977, **99**, 6459.
52 A. H. A. Tinnemans and D. C. Neckers, *Tetrahedron Letters*, 1978, 1713.
53 D. Bryce-Smith, A. Gilbert, and J. Grzonka, unpublished observations.
54 A. H. A. Tinnemans, personal communication.

authors of ref. 51 and 52 wisely chose column chromatography for their isolation[51] and n.m.r. spectroscopy to assess ratios of products.[54] Another factor
which may be important is that the concentration of the acetylene was much
higher in the former study than in the latter one, so the chance of (63) successfully
competing for the radiation with the acetylene and thereby forming (61) was
very remote: thus its thermal rearrangement to (62) was more likely than in the
more dilute solutions used in the recent study.

The 1,4-1'4'-photocycloaddition of 1,3-dienes to benzene has received the
attention of several groups of workers since the process was first reported in
1967.[55] The reaction of isoprene with benzene which normally yields a complex
mixture of products has now been successfully adapted to a synthesis of 3-
methylene bicyclo[4,2,2]deca-7,9-diene (64).[56] Thus 254-nm irradiation of the
addend in the presence of approximately 0.2% w/v of iodine yields (64) as the
readily isolable main product.

(64)

Reports of the photocycloaddition reactions of hexaflorobenzene have
appeared from two groups of workers. The reactions of C_6F_6 with indene,
1,2-dihydronaphthalene, and cyclopentene have been reported to be stereospecific or stereoselective and involve exclusively the *ortho* mode of attack,[57]
whereas the products from *cis*-cyclo-octene as the addend reflect both *ortho* and
meta processes,[58] and an acyclic addition occurs with 2,3-dimethylbut-2-ene.[59]
The details of the latter two preliminary reports have been published and the
processes observed with C_6F_6 and simple alkynes also described.[60] Both
ortho- [(65), (66), and (67)] and *meta*- [(68) and (69)] type cycloaddition products
result from *cis*-cyclo-octene and similar products are also reported from the
trans-isomer. Adducts (65), (68), and (69) are primary products and (66) and
(67), as may be expected, result from secondary reactions of (65). The effects
of wavelength, temperature, and reactant concentrations on the process have been
studied. An increase in total quantum yield with decrease in wavelength of exciting

(65)   (66)   (67)

[55] K. Kraft and G. Koltzenburg, *Tetrahedron Letters*, 1967, 4357 and 4723.
[56] J. Ipaktschi and M. N. Iqbal, *Synthesis*, 1977, **9**, 633.
[57] B. Sket and M. Zupan, *J.C.S. Chem. Comm.*, 1976, 1053, and *J.C.S. Chem. Comm.*, 1977,
365.
[58] D. Bryce-Smith, A. Gilbert, and B. H. Orger, *Chem. Comm.*, 1969, 800.
[59] D. Bryce-Smith, B. E. Foulger, A. Gilbert, and P. J. Twitchett, *Chem. Comm.*, 1971, 794.
[60] D. Bryce-Smith, A. Gilbert, B. H. Orger, and P. J. Twitchett, *J.C.S. Perkin I*, 1978, 232

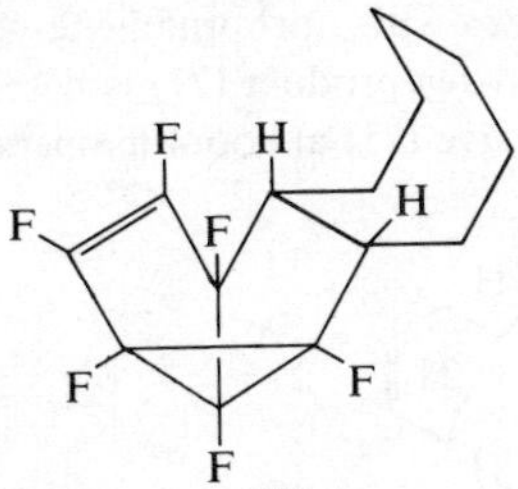

(68)

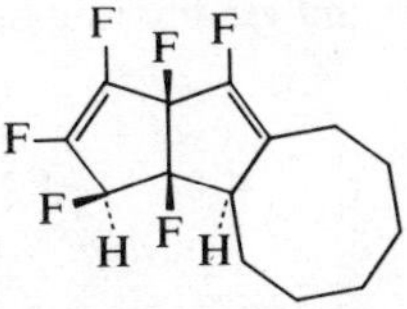

(69)

radiation is noted ($\Phi_{254} = 0.51$; $\Phi_{231} = 0.75$) although the ratio of *ortho* and *meta* adducts remains essentially constant, but an increase in temperature from 60 to 75 °C almost doubles the quantum yields for the former and its derived products while that for the latter is essentially unaffected. The effect of relative concentration of the reactants on the quantum yield of adduct formation in this system is remarkable for although the quantum yield for *meta* addition remains approximately constant over the range 10—90 mol % C_6F_6, the yield of products from *ortho* addition increases by 50% from 10 to 50 mol % C_6F_6 and more than doubles in passing from 50 to 90 mol % C_6F_6. These effects are discussed in the context of the involvement of free and complexed S_1 C_6F_6. Maleic anhydride and maleimide fail to photoadd to C_6F_6 but 2,3-dimethylbut-2-ene gave mainly the 1 : 1 acyclic adduct (70) and its dehydrofluorination product, and propyne and but-2-yne with C_6F_6 gave cyclo-octatetraenes. For the latter alkyne, the precursor *ortho*-cycloadduct (71) is the sole primary product: this readily converts either thermally or photochemically into the cyclo-octatetraene and photochemically into the intramolecular cyclized product (72).[60] The isolation

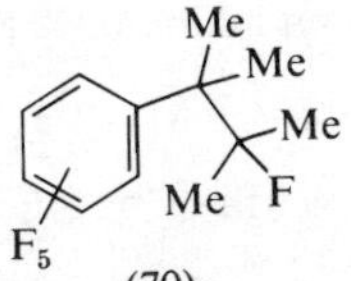

(70)

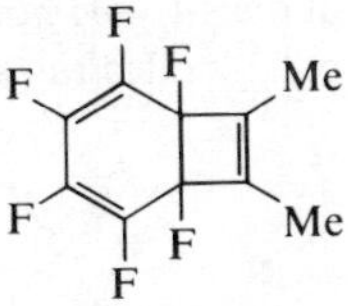

(71)

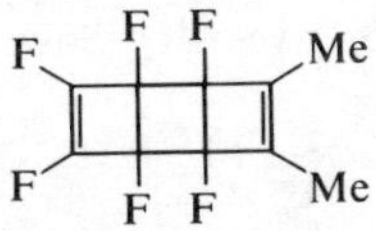

(72)

of a bicyclic intermediate analogous to (71) has been previously described from the photoaddition of phenylacetylenes to C_6F_6.[61] Sket and Zupan have previously contributed a number of significant accounts of photoadditions to C_6F_6 and have now observed that C_6F_6 does give photoadducts with norbornadiene,[62] contrary to a previous report.[30] Thus irradiation of C_6F_6 and norbornadiene, in cyclohexane yields the three 1 : 1 adducts (73), (74), and (75) in respective yields of 47, 33, and 20%. Heating (73) to 150 °C causes partial decomposition to starting materials and also gives (76) which is presumably the primary photoadduct and which on irradiation yields (75). Similar heat treatment of (74) and (75) gives both (73) and (76). From the interrelationship of the adducts, it is

61 B. Sket and M. Zupan, *J. Amer. Chem. Soc.*, 1977, **99**, 3504.
62 B. Sket and M. Zupan, *Tetrahedron Letters*, 1977, 2811.

suggested that C_6F_6 reacts exclusively *exo* with norbornadiene with *syn* attack being 80% and *anti* attack 20%: the *exo-syn*-product (77) is described as being very unstable and readily ring-opens to give (73) at room temperature.

(73)

(74)   (75)   (76)   (77)

The photoaddition of aliphatic amines to benzene has been the subject of several preliminary communications,[63] and the full details of the reactions with primary, secondary, and tertiary aliphatic amines have now been published.[64] In all cases, the reactions arise from the S_1 state of benzene. Primary amines give the products (78), (79), and (80) of 1,2-, 1,3-, and 1,4-addition, respectively to the N—H bond, the analogous product (79) from secondary amines is not observed: this is because (79) arises from (78) by way of hexa-1,3,5-triene intermediates and the 1,2-adduct from secondary amines is photostable – surprisingly so in view of the lability of (78). The addition reactions of tertiary amines to benzene involve attack at the α-C—H bond in the amine: triethylamine, for example, gives (81). Solvents of high dielctric constant inhibit these additions,

(78)   (79)   (80)   (81)

but the reaction with tertiary amines proceeds only slowly in the absence of a proton source such as methanol: primary and secondary amines appear to act as their own proton donor in these photoadditions. The mechanism of the reaction to account for the formation of the adducts and neutral products and the effects of solvents are discussed in terms of the involvement of charge-transfer

[63] M. Bellas, D. Bryce-Smith, and A. Gilbert, *Chem. Comm.*, 1967, 862; D. Bryce-Smith, A. Gilbert, and C. Manning, *Angew. Chem.*, 1974, **86**, 350; D. Bryce-Smith, M. T. Clarke, A. Gilbert, G. Klunklin, and C. Manning, *Chem. Comm.*, 1971, 916; and D. Bryce-Smith, A. Gilbert, and G. Klunklin, *J.C.S. Chem. Comm.*, 1973, 330.

[64] M. Bellas, D. Bryce-Smith, M. T. Clarke, A. Gilbert, G. Klunklin, S. Krestonosich, C. Manning, and S. Wilson, *J.C.S. Perkin I*, 1977, 2571.

processes and both polar and radical intermediates. Scheme 5 summarizes these mechanistic conclusions for the case of benzene and triethylamine.[64] The photoreactions of anthracene with aromatic amines which produce 9,10-dihydroanthracene derivatives [*e.g.* (82) from aniline] have been the subject of

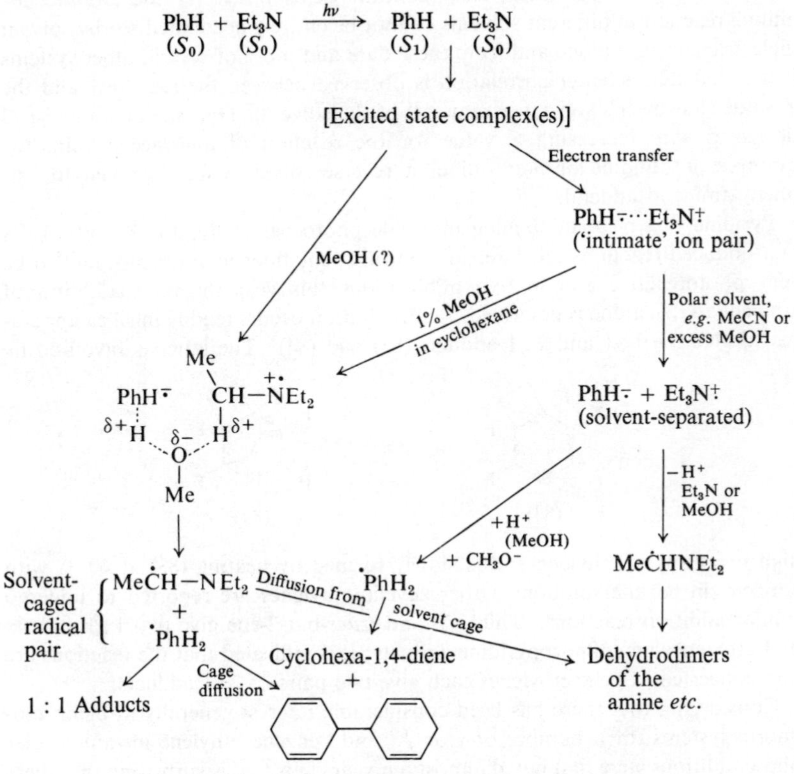

$$PhH + Et_3N \xrightarrow{h\nu} PhH + Et_3N$$
$$(S_0) \quad (S_0) \qquad\qquad (S_1) \quad (S_0)$$

Scheme 5

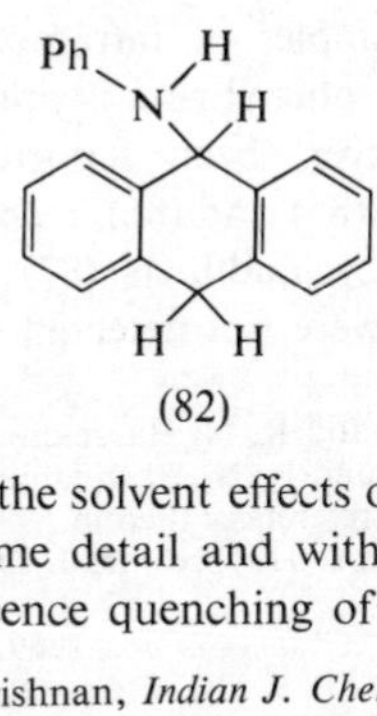

(82)

several reports recently,[65] and the solvent effects on the kinetics of the process have now been described in some detail and with a range of amines.[66] It is found that whereas the fluorescence quenching of anthracene by primary and

65 S. Vaidyanathan and V. Ramakrishnan, *Indian J. Chem.*, 1975, **13**, 257, and references therein.
66 S. Vaidyanathan and V. Ramakrishnan, *Indian J. Chem.*, 1977, **B15**, 269.

secondary aromatic amines is mainly dependent on the viscosity of the medium, the photoreaction which yields the 1 : 1 adduct is controlled by solvent polarity. It is further reported that the solvent effect on the rate of reaction depends on the amine and in some cases is higher in non-polar media, but for other addends is higher in polar solvents. The quantum yields noted for the anthracene–aniline reaction in different solvents cannot be entirely accounted for by solvent dielectric constant alone and from these data and those of several other systems it is noted that a better correlation is observed between the reactivity and the Z value (Kosower's solvent parameter) of the solvent. Thus the quantum yield decreases with increasing Z value for the reaction of anthracene with, for example, *p*-toluidine or *o*-anisidine; a reverse solvent effect is noted for diphenylamine as addend.

Pyridine is a most unobliging molecule photochemically, and reports of its light-induced reactions are rare. In contrast pentafluoropyridine appears to be very photoreactive and in two publications this year the photoaddition of ethylene to this arene is described.[67,68] Addition occurs readily in silica apparatus to give the 1 : 1 and 2 : 1 adducts (83) and (84). The latter is favoured by

(83)   (84)

high pressures of ethylene and is slowly formed by heating (83) at 40 °C with ethylene in hexane solution. Other acyclic ethylenes are reported to undergo similar addition reactions. Thus *cis*- and *trans*-but-2-ene give two 1 : 1 and six 2 : 1 ethylene : pentafluoropyridine adducts; it is indicated that the reactions are non-concerted. Cyclic ethylenes each give two pairs of 2 : 1 adducts.

Considering that there has been considerable interest generally in bichromophoric systems for a number of years,[69] and benzene–ethylene intramolecular photoadditions were first noted almost ten years ago,[71] it is surprising that there are few reports concerned with the intramolecular photoaddition reactions of simple bichromophoric molecules which have a phenyl ring, although the literature abounds with examples of intramolecular cyclization of arenes in which the aromaticity of the phenyl ring is preserved. Irradiation of phenethyl vinyl ether has now, however, been reported to yield the 2,4- and 2,5-intramolecular cycloadducts (85) and (86), respectively, with relative efficiencies of 1 : 15.[72] The *ortho*-type-cycloadducts (87) expected from consideration of intermolecular analogues[73] were not detected. Molecular models suggest that

[67] M. G. Barlow, D. E. Brown, and R. N. Haszeldine, *J.C.S. Chem. Comm.*, 1977, 669.
[68] M. G. Barlow, D. E. Brown, and R. N. Haszeldine, *J.C.S. Perkin I*, 1978, 363.
[69] See, for example, ref. 70 and references therein.
[70] F. C. De Schryver, N. Boens, J. Huybrechts, J. Daemen, and M. De Brackelaire, *Pure Appl. Chem.*, 1977, **49**, 237.
[71] H. Morrison and W. I. Feree, *Chem. Comm.*, 1969, 268.
[72] A. Gilbert and G. N. Taylor, *J.C.S. Chem. Comm.*, 1978, 129.
[73] D. Bryce-Smith, A. Gilbert, B. H. Orger, and H. M. Tyrrell, *J.C.S. Chem. Comm.*, 1974, 334.

additions of *ortho*- and *meta*-type might have been expected to be favoured, and provide no explanation for the normally disfavoured *para*-type cycloaddition to give (86).[72]

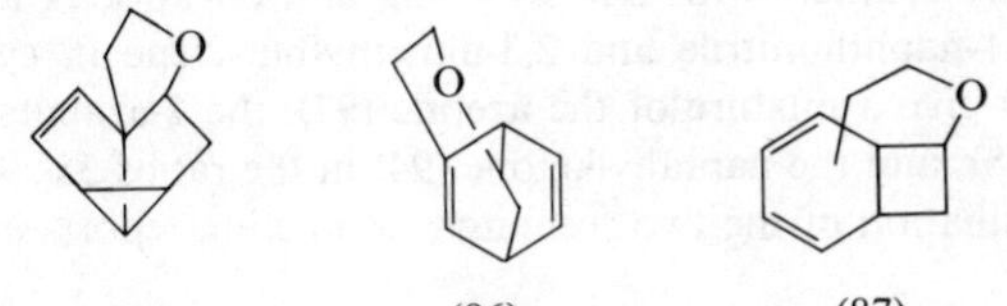

(85)   (86)   (87)

Interest in the photoreactions of naphthonitriles continues at a high level and, as has been noted earlier in this section, the reactions of ethylene with aryl nitriles may lead to attack at either or both of the two chromophores.[44] McCullough and co-workers report, however, that the products from 1- and 2-naphthonitriles and 2,3-dimethylbut-2-ene are solvent dependent and do not apparently involve cycloaddition to the nitrile function.[74] The process has been studied in hexane, benzene, dimethoxyethane, methanol, and acetonitrile, and in the first three solvents with 313-nm radiation, the two nitriles yield the cyclo-butanes (88) and (89), respectively as the exclusive products. In methanol, the

(88)   (89)

1-isomer yields products of reductive dimerization, the exact structures of which are not determined, whereas the 2-naphthonitrile yields the solvent-incorporated adducts (90) and (91), and 1,2-dihydro-2-naphthonitrile. Acetonitrile as solvent

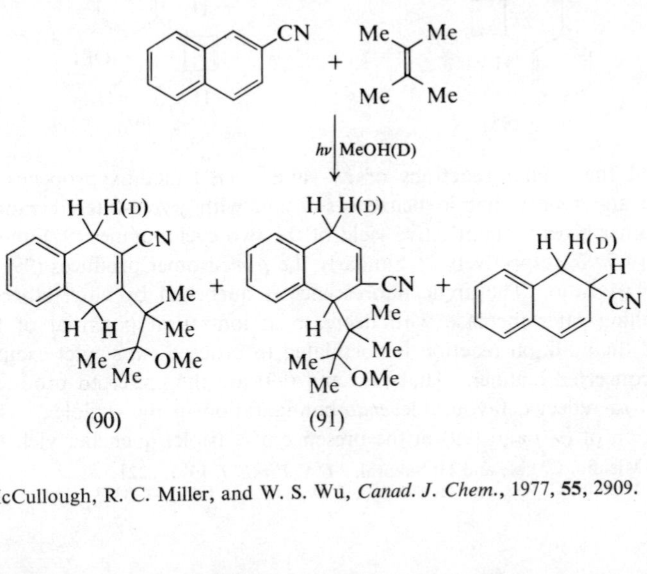

(90)   (91)

[74] J. J. McCullough, R. C. Miller, and W. S. Wu, *Canad. J. Chem.*, 1977, **55**, 2909.

produces complex mixtures from both arenes. It is proposed that the cyclobutane formation in non-polar solvents involves exciplex intermediates, whereas in polar solvents electron transfer occurs and results in photoreduction. Interestingly this report conflicts with that by Yang and co-workers last year[47] who reported that 1-naphthonitrile and 2,3-dimethylbut-2-ene in cyclohexane and under nitrogen give a mixture of the azetine (92), the 2-azabutadiene (93), the cyclobutane (88), and the naphthylketone (94) in the ratios 35 : 9 : 28 : 1 respectively. Reconciliation of the two seemingly conflicting reports may possibly be

found in the different experimental conditions used. Thus Yang and co-workers used Corex-filtered light, noted that the cyclobutanes were labile under these conditions, and proceeded to high reactant conversions, whereas the work reported in ref. 74 is for low reactant conversions and involves pyrex-filtered radiation under which conditions the cyclobutanes are stable. Hence in the study of McCullough and co-workers the azetine (92) and 2-azabutadiene (93) may have been formed but their concentrations relative to that of (88) under the particular conditions were so low that they were not detected.

It has been previously reported that irradiation of alkyl vinyl ethers and of 1-naphthonitrile yield exclusively the *endo* cyclobutane adduct (95) using 313-nm radiation,[75] but that at wavelengths longer than 280 nm and longer irradiation periods, the cyclobutene (96) predominates. The same workers have now

studied the similar reactions of *cis*- and *trans*-1-methoxypropenes with this nitrile and report that in benzene solution with pyrex-filtered radiation the *cis*-isomer gives a quantitative yield of the two cyclobutanes (97) and (98) in a ratio of 7 : 3, respectively.[76] Similarly the *trans*-isomer produces (99) and (100) in a 4 : 1 ratio. The arene fluorescence is quenched by vinyl ethers and the quenching rates increase with decrease in ionization potential of the ether. Again the addition reaction is postulated to proceed *via* singlet exciplexes and in a concerted manner. That (97) and (99) are the preferred products is suggested to reflect a favourable *endo* configuration in the exciplex. The photoreversion of (97) and (99) in the presence of a triplet quencher yields the vinyl

[75] K. Mizuno, C. Pac, and H. Sakurai, *J.C.S. Perkin I*, 1975, 2221.
[76] K. Mizuno, C. Pac, and H. Sakurai, *J. Org. Chem.*, 1977, **42**, 3313.

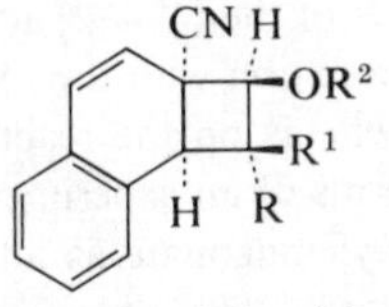

(97) R = H, R^1 = R^2 = Me
(99) R = R^2 = Me, R^1 = H

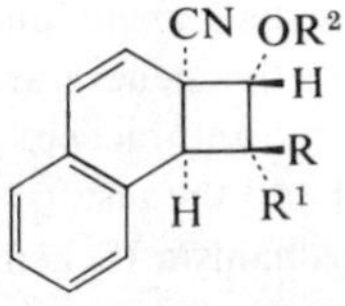

(98) R = H, R^1 = R^2 = Me
(100) R = R^2 = Me, R^1 = H

ethers with > 95% *cis* and *trans* geometries, respectively: this aspect of the study has received more detailed comment in another report by the same authors.[77]

Thanks to the efforts of Caldwell and Farid and their co-workers, it is now well established that the photoaddition of phenanthrene to dimethyl fumarate involves both singlet and triplet exciplexes.[38] The benzophenone-sensitized process is considered to involve the triplet species which collapse to a biradical (101) and thence on to the products (102) and (103). It is now reported that this biradical can be intercepted, for di-t-butylnitroxide and molecular oxygen enhance the cyclobutane formation by interaction with (101).[78] The enhancement of product formation suggests to the authors that the interception of (101)

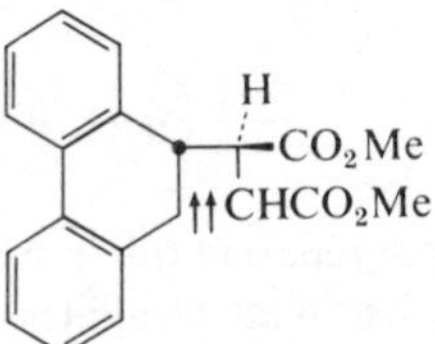

(101)

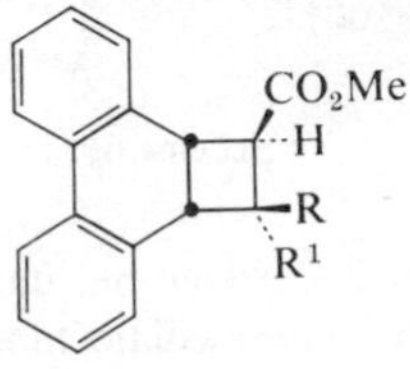

(102a) R = H, R^1 = CO$_2$Me
(102b) R = CO$_2$Me, R^1 = H

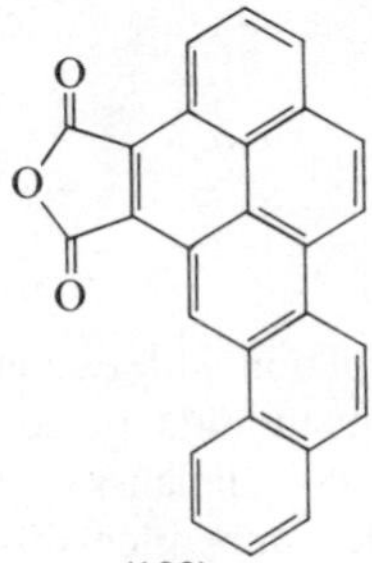

(103)

occurs preferentially in a structure predisposed to the formation of the cyclobutane. These observations are important in a much wider context than this particular case and the potential synthetic utility of interception in altering product yields in reactions with biradical intermediates is noted. The effect of solvent, sensitizer, light, and aryl substituent on the photoaddition of maleic anhydride to phenanthrene,[79] has received further comment.[80] In general the yields of adduct are higher for alkylphenanthrenes than for the parent hydrocarbon (in contrast with corresponding additions to benzenes[80a]) and whereas benzophenone is a good sensitizer, acetone inhibited the reaction.

The photo **Diels-Alder** reactions of polynuclear arenes have received further attention and anthracene is reported to react non-stereospecifically with electron-deficient alkenes, styrenes, and 1,3-dienes.[81] The fluorescence of the arene is

[77] C. Pac, K. Mizuno, and H. Sakurai, *Bull. Chem. Soc. Japan*, 1978, **51**, 329.
[78] R. A. Caldwell and D. Creed, *J. Amer. Chem. Soc.*, 1977, **99**, 8360.
[79] D. Bryce-Smith and B. Vickery, *Chem. and Ind.*, 1961, 429.
[80] (a) L. D. Melikadze, E. G. Lekveishvihi, and E. V. Kartvelishvihi, *Soobshch. Akad. Nauk Gruz S.S.R.*, 1977, **85**, 353; (b) D. Bryce-Smith and A. Gilbert, *J. Chem. Soc.*, 1965, 918; *Chem. Comm.*, 1968, 19.
[81] G. Kaupp, *Justus Liebig's Ann. Chem.*, 1977, **2**, 254.

quenched by the addend, and preparative yields of the [4 + 2] adducts ([4 + 4] in the case of the dienes), are observed in competition with the formation of the photodimer of anthracene. Polar and steric effects on the reaction have been investigated and the results are discussed in terms of an experimentally detected two-step mechanism (Scheme 6). For dimethyl muconate as addend, exciplex

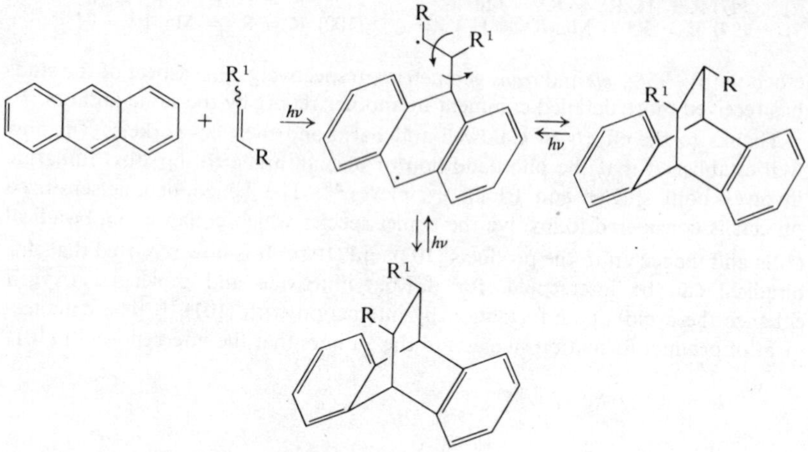

Scheme 6

formation with excited arene is detected but unlike those reported for so many photoaddition processes, this intermediate hinders rather than facilitates the process. Polansky and co-workers have reported on the photoreactions of maleic anhydride with picene[82] and chrysene.[83] The thermal reaction of the latter system in the presence of iodine has been reported to give (103) only,[84] but from the corresponding light-induced process a second isomer (104) is also formed; the ratio of (103) to (104) being approximately 1.5. Irradiation of chloroform solutions of maleic anhydride and chrysene behind a solidex filter yields the dehydro 1 : 1 adduct (105), but irradiation of solutions degassed under

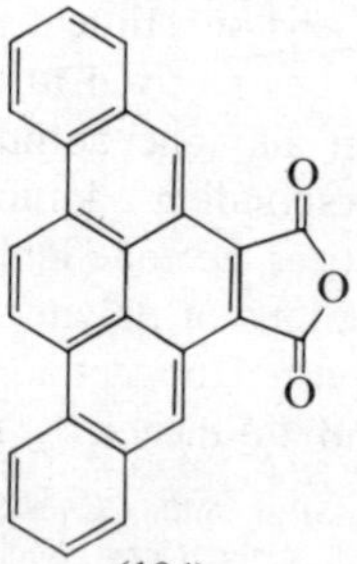

(104)

[82] H. Karpf, O. E. Polansky, and M. Zander, *Tetrahedron Letters*, 1978, 339.
[83] H. Karpf and O. E. Polansky, *Tetrahedron Letters*, 1978, 2069.
[84] E. Clar, *Tetrahedron Letters*, 1973, 3471.

argon gives a product whose mass spectrum suggests the presence of (106).[83] The authors note that the primary product may be (107), the 1 : 1 adduct, which would be a very unstable compound and readily dehydrogenate to (106) and (105).

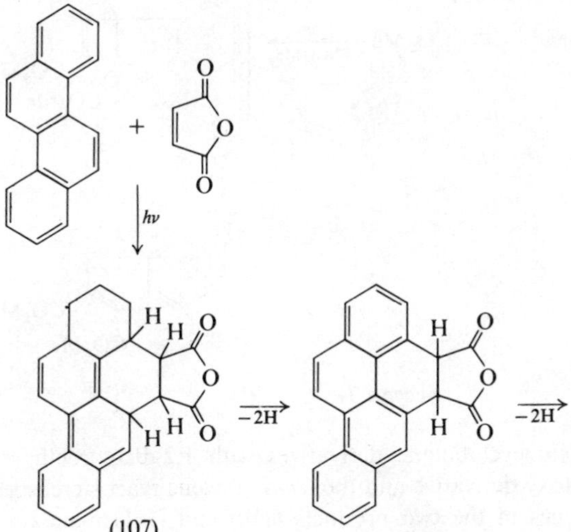

(107)   (106)   (105)

The photoaddition reactions of dimethyl acetylenedicarboxylate and methyl propiolate to benzo[b]furan and of the former alkyne to benzo[b]selenophene have both been described within the year. Tinnemans and Neckers report that acetophenone-sensitized addition of the dicarboxylate to the furan gives the four cyclobutane-type adducts (108a—d) in which the carbomethoxy groups occupy vicinal positions.[85] It is noteworthy here that with benzophenone as sensitizer only oxetans are formed. The authors suggest that (108b) is formed from the initially produced adduct (108a) or from (108c), and the latter is shown to arise both from (108a) and (108b) *via* an intermediate whose structure is postulated to be that of a 1,2-cyclobutenespiro[2,5]octadiene (109). The corresponding sensitized reaction with methyl propiolate yields two cyclobutane 1 : 1 adducts, one with the carbomethoxy group attached to the 2-position of the benzo[b]furan nucleus and the other having an unrearranged structure: it is suggested that the excited state of the arene is highly polarized. Acetophenone

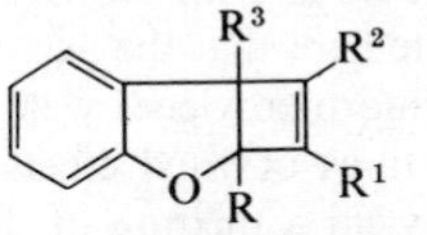

(108) a; R = R^3 = H, R^1 = R^2 = CO$_2$Me   9%
      b; R = R^1 = CO$_2$Me, R^2 = R^3 = H   8%
      c; R = R^1 = H, R^2 = R^3 = CO$_2$Me   7%
      d; R = R^3 = CO$_2$Me, R^1 = R^2 = H   1.5%

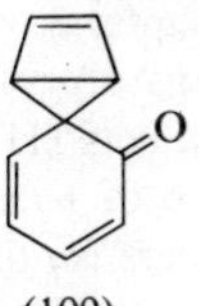

(109)

[85] A. H. A. Tinnemans and D. C. Neckers, *J. Org. Chem.*, 1977, **42**, 2374.

is also used as a sensitizer for photoadditions of acetylenedicarboxylic ester to benzo[b]selenophene and its 3-methyl derivative.[86] In this case the primary product is unstable and not isolated (see Scheme 7), but various derivatives of

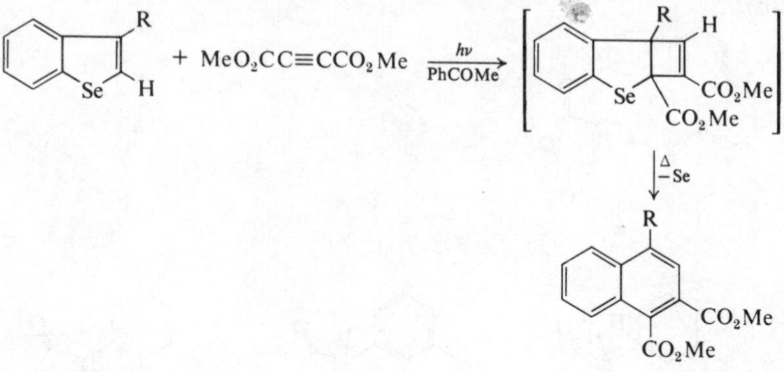

Scheme 7

the selenophene do yield cyclobutane derivatives with 1,2-dichloroethylenes. For example, the 3-acetoxy derivative and the *trans* ethylene react stereospecifically in that the chlorines in the two products (110) and (111) maintain the *trans* relationship. Irradiation of dibromomaleic anhydride and benzo[b]-thiophen is reported to yield both the cyclobutane derivative (112) and the substituted product (113).[87] The ratio of addition to substitution decreases with

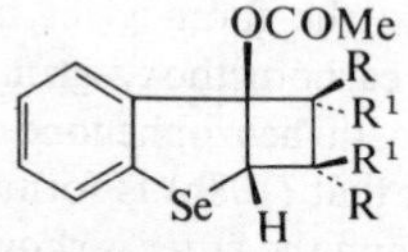

(110) R = H, R^1 = Cl
(111) R = Cl, R^1 = H

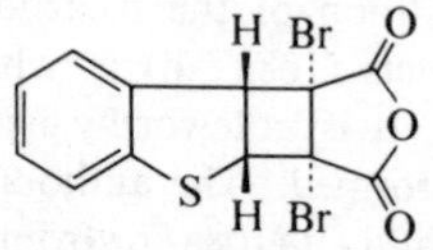

(112)

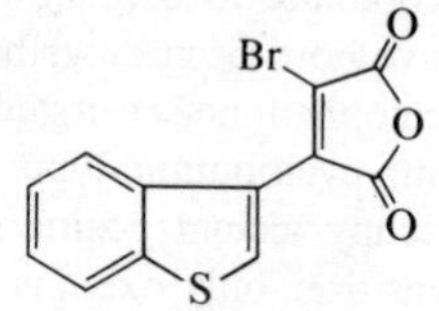

(113)

increase in the solvent polarity and this is accounted for by the proposal that the substitution proceeds *via* a more polar transition state than the addition. The kinetic parameters of the process have been obtained under a variety of conditions and the authors say these indicate a common triplet exciplex. The [2 + 2] photoaddition reactions of ethylenes and indenes are considered in Part III, Chapter 3, but it is pertinent to note here that the acetophenone-sensitized additions of cyclic fluoro- and chlorofluoro-ethylenes yield predominantly the *anti* adducts (114),[88] and that the reactions of vinyl ethers and indene in the presence of 1-naphthanitrile efficiently yield a mixture of the two cyclobutane stereomers *via* the indene cation-radical.[89]

[86] T. Q. Minh, L. Christiaens, P. Grandclaudon, and A. Lablache-Combier, *Tetrahedron*, 1977, **33**, 2225.
[87] T. Matsuo and S. Mihara, *Bull. Chem. Soc. Japan*, 1977, **50**, 1797.
[88] H. Kimoto, H. Muramatsu, and K. Inukai, *J. Fluorine Chem.*, 1977, **9**, 417.
[89] K. Mizuno, R. Kaji and Y. Otsuji, *Chem. Letters*, 1977, 1027.

(114) R = Cl or F
    n = 2 or 3

4 Substitution Reactions

As in previous years, the great diversity of reactions which have been discussed and systems which have been studied under the general heading of photosubstitution of aromatic compounds makes their order of presentation here somewhat arbitrary. It is widely accepted that many nucleophilic photosubstitution processes occur from the triplet $\pi\pi^*$ state of the arene.[90] Russian workers have supported this finding to some extent as they report that comparison of the calculated (INDO) charge densities of fluorobenzene, phenol, aniline, furan, and pyrrole in the ground and triplet states with the experimental reactivity of these molecules indicates that the substitution process may well proceed *via* triplet σ-complexes.[91] The photohydrolysis of *m*-nitroanisole is a well-studied reaction but the process has now been investigated under very high pressure (3000 kg/cm²) using 366-nm radiation.[92] The quantum yield for formation of the *m*-nitrophenolate ion in alkaline media decreases with increase in pressure and this is explained by the difference of the pressure effects on the elimination step of hydroxide anion and that of the methoxide anion from the excited σ-complex. In previous reports, Beugelmans and co-workers have described the effect of 18-crown-6 ether on photonucleophilic substitution reactions,[93] and have now examined the photocyanation of arenes using tetrabutylammonium cyanide in anhydrous acetonitrile or dichloromethane and also under phase-transfer conditions (CH₂Cl₂-KCN in H₂O), and have compared the results with those obtained with the crown ether.[94] The reactions were studied with naphthalene, biphenyl, and phenanthrene and their 1-, 4-, and 9-nitro derivatives, respectively. For naphthalene and 9-nitrophenanthrene, the photocyanation is more efficient with tetrabutylammonium cyanide in acetonitrile than with potassium cyanide and the crown ether, but for 1-nitronaphthalene and 4-nitrobiphenyl the process is much less efficient. In the presence of dichloromethane, the cyanation only occurs with the nitro compounds under both sets of conditions. The yields under the phase-transfer conditions with catalytic amounts of tetrabutylammonium cyanide are appreciably higher than those in anhydrous solvent and, as the authors point out, the former reactions are more convenient for product isolation.

[90] See E. Havinga and J. Cornelisse, *Chem. Rev.*, 1975, **75**, 353; J. Cornelisse, *Pure Appl. Chem.*, 1975, **41**, 433, and references therein.
[91] I. A. Abronin, L. I. Belenkii, and G. M. Zhidomirov, *Izvest. Akad. Nauk S.S.S.R. Ser. Khim.*, 1977, **3**, 588.
[92] H. Aomi, M. Sasaki, and J. Osugi, *Rev. Phys. Chem. Japan*, 1977, **44**, 111.
[93] R. Beugelmans, M. T. Le Goff, J. Pusset, and G. Roussi, *Tetrahedron Letters*, 1976, 2305, and *J.C.S. Chem. Comm.*, 1976, 377.
[94] R. Beugelmans, H. Ginsburg, A. Lecas, M. T. Le Goff, J. Pusset, and G. Roussi, *J.C.S. Chem. Comm.*, 1977, 885.

Photosubstitution reactions of arylcyanides continue to attract interest. Two years ago it was reported that *p*-dicyanobenzene reacts photochemically with triethylamine to give the *para* α-C-substituted product (115).[95] The same workers have now extended their studies to primary and secondary amines and unlike the reactions with benzene[64] or anthracene[96] but as with those of diethylamine with pyridine and chlorobenzene,[97] attack at the α-position rather than N—H of the amine is reported.[98] Thus irradiation of *p*-dicyanobenzene with, for example, cyclohexylamine in acetonitrile yields a mixture of *p*-cyclohexylbenzo-nitrile (11%), (116) (10%), and a small amount of benzonitrile. Neither charge-transfer absorption nor exciplex emission is noted from these systems and the

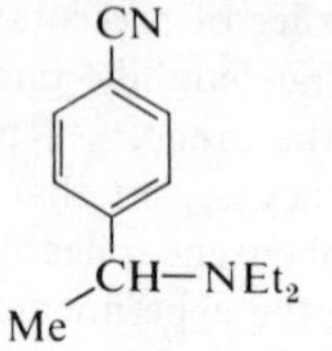

(115)

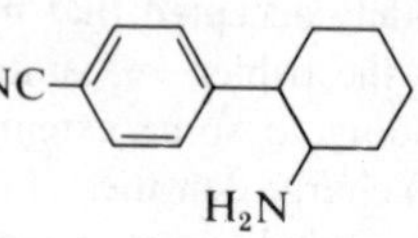

(116)

authors conclude that electron transfer from the amine to the excited arene occurs and that the reaction proceeds *via* the encounter complexes or exciplexes so frequently postulated for other arene–amine systems.[99] *o*- and *p*-Dicyano-benzene also undergo a novel ureiodomethylation process on irradiation of their acetonitrile solutions with tetramethyl urea.[100] Thus (117) and (118) are formed in 27 and 19% respectively from the *o*- and *p*-arene isomers together with varying amounts of the corresponding toluonitrile and the urea-derived product (119). In a non-polar solvent the reaction is completely suppressed and in acetonitrile solution the urea quenches the arene fluorescence. From these results it is suggested that the reaction again involves electron transfer followed by a proton

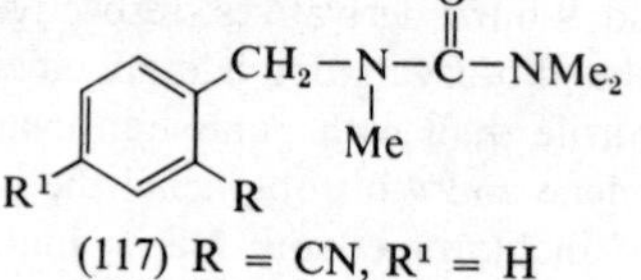

(117) R = CN, R¹ = H
(118) R = H, R¹ = CN

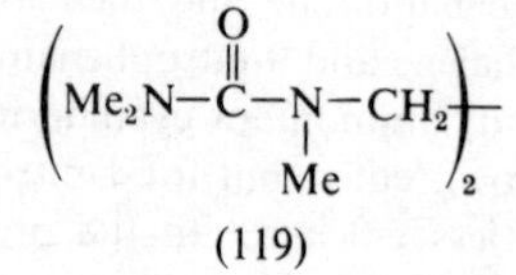

(119)

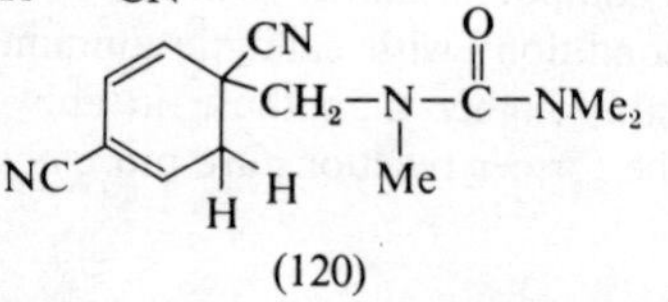

(120)

[95] K. Tsujimoto, K. Miyake, and M. Ohashi, *J.C.S. Chem. Comm.*, 1976, 386.
[96] N. C. Yang and J. Libman, *J. Amer. Chem. Soc.*, 1973, **95**, 5783.
[97] D. Bryce-Smith, A. Gilbert, and S. Krestonosich, *Tetrahedron Letters*, 1977, 385.
[98] M. Ohashi and K. Miyake, *Chem. Letters*, 1977, **6**, 615.
[99] See Volumes 1–9 in this series for previous relevant studies.
[100] Y. Tsujimoto, M. Hayashi, Y. Nishimura, T. Miyamoto, and Y. Odaira, *Chem. Letters*, 1977, **6**, 677.

transfer and coupling of the caged radicals to give the (non-detected) intermediate 1 : 1 adduct (120) which eliminates hydrogen cyanide; precedent is cited for the facility of this latter step. Although the toluene 1,2,4,5-tetracyanobenzene system yields the substituted product (121),[101] benzene is inert as the arene in the system but addition of catalytic amounts of methanol, as with benzene and tertiary aliphatic amines,[64] provides the necessary proton source and reaction proceeds.[102] Irradiation using a pyrex-filtered medium pressure lamp gives the products (122), (123), (124), and (125) in respective yields of 41, 6, 5, and 4%, whereas 254-nm radiation yields (124) and (125) together with low yields of

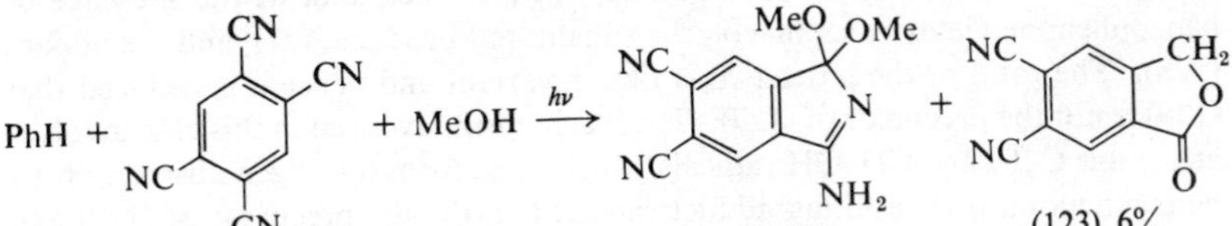

(122) 41%

(123) 6%

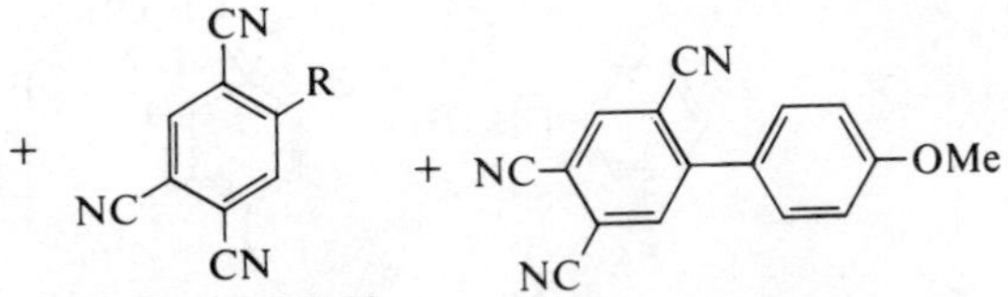

(121) R = —CH₂Ph
(124) R = Ph 5%

(125) 4%

other compounds including (123). The process was extended to other arenes, and phenanthrene, for example, was examined in an attempt to elucidate the mechanism. The 9,10-dihydro derivative (126) was formed in this case and it is postulated that the reaction proceeds by proton transfer from the methanol to the tetracyanobenzene radical anion produced by electron transfer from the

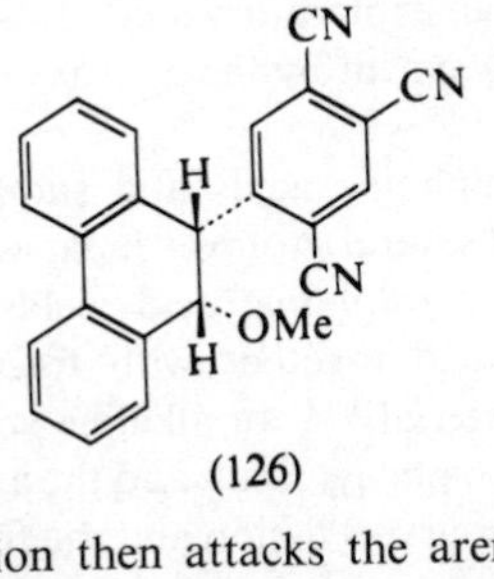

(126)

arene. The methoxide anion then attacks the arene radical-cation, following which coupling of the radicals and elimination of hydrogen cyanide yields the 1 : 1 : 1 addition–elimination products. The alternative to this mechanism involves coupling of the radical-ions to give a zwitterion which then reacts with the methanol. That formation of (126) appears to be stereospecific seems to indicate a preference for this latter mechanism. It is of interest to note here that

[101] T. Yamasaki, T. Yonezawa, and M. Ohashi, *J.C.S. Perkin I*, 1975, 735.
[102] S. Yamada, Y. Kimura and M. Ohashi, *J.C.S. Chem. Comm.*, 1977, 667.

the involvement of the phenanthrene radical-cation has also been postulated in the photosensitized dimerization of α-methylstyrene which yields 1,4-dimethyl-1-phenyl-1,2,3,4-tetrahydronaphthalene and 1,4-dimethyl-1-methoxy-1,4-diphenylbutane in acetonitrile and acetonitrile–methanol solutions respectively in the presence of *m*-dicyanobenzene.[103] In this case, the reaction is suggested to proceed by electron transfer from the styrene to the phenanthrene radical-cation.

Evidence for an addition–elimination mechanism in the normal and cine-substitution of fluorobenzenes by amines has been earlier presented,[104] and this type of process has now been discussed for the photohydroxyalkylation of C_6F_6.[105] Irradiation (300—350 nm) of C_6F_6 in methanol in the presence of benzophenone yields benzpinacol, the substituted product (127), and the adduct (128). The ratio of the latter two is time-invariant and hence it is deduced that (128) is not the precursor of (127). The mechanism favoured in this case involves attack on C_6F_6 by $\cdot CH_2OH$ radicals which arise from hydrogen abstraction by benzophenone; the resulting adduct radical (129) is the precursor of (127) and (128).

(127)   (128)   (129)

The photoreaction of *N,N*-dimethylaniline with chloromethanes gives a complex mixture of products arising, it is suggested, by attack of solvent radicals formed in the solvent exciplex or charge-transfer complex with the aromatic amine.[106] Similar unselectivity is noted in the irradiation of nitromethane with benzene which yields anisole, nitrobenzene, *o*- and *p*-nitrophenols, biphenyl, phenol, and traces of toluene.[107] Nitrophenols themselves are photolabile in aqueous solution and from an exploratory study it is suggested that the reaction largely involves nitrite displacement by the hydroxyl anion from the solvent and the formation of nitric acid.[108]

The irradiation of arylsulphonic acids and sulphonates continues to be a fruitful topic of research for several groups. Japanese workers have shown that in aqueous solution benzene sulphonic acid yields sulphurous acid, sulphur dioxide, and/or sulphuric acid together with traces of biphenyl and much sulphur-containing tarry material.[109] In alkaline solution, on the other hand, benzene (16%) and traces of biphenyl result and the authors postulate reasonably a mechanism involving initial aryl-S fission and the formation of Ph· and ·SO₃H

[103] T. Asanuma, T. Gotoh, A. Tsuchida, M. Yamamoto, and Y. Nisijima, *J.C.S. Chem. Comm.*, 1977, 485.
[104] D. Bryce-Smith, A. Gilbert, and S. Krestonosich, *J.C.S. Chem. Comm.*, 1976, 405.
[105] M. Zupan, B. Sket, and B. Pahor, *Tetrahedron Letters*, 1977, 4541.
[106] T. Latowski, M. Przytarska, and B. Zelent, *Roczniki Chem.*, 1977, **51**, 995.
[107] S. Paszyc, H. Kozubek, and B. Marciniak, *Bull. Acad. polon. Sci. Sér. Sci. chim.*, 1977, **25**, 951.
[108] M. I. O. Ishag and P. G. N. Moseley, *Tetrahedron*, 1977, **33**, 3141.
[109] Y. Ogata, K. Takagi, and S. Yamada, *Bull. Chem. Soc. Japan*, 1977, **50**, 2205.

radicals. Alkylbenzenesulphonic acids in the presence of equimolar amounts of sodium hypochlorite in aqueous solution are reported to undergo photolytic oxidation to give the corresponding alkylbenzenes, alkylphenols, aryl alkyl alcohols, and aryl alkyl aldehydes; but under the same conditions, benzene-sulphonic acid is completely photolysed to carbon dioxide and sulphuric acid.[110] Photodesulphonation has also been examined in the case of anthrahydroquinone sulphonic acids, a process which in aqueous alkaline solution is accompanied by photoreduction to give 9-hydroxyanthracene sulphonic acid, the water acting as the proton donor.[111] In contrast with the competing desulphonation process, photoreduction of anthrahydroquinone sulphonic acid occurs with a quantum yield which remains constant as the wavelength of the exciting light is varied between 254 and 434 nm, is the same for mono- and di-anions, and does not seemingly depend on the position, or indeed the presence of, the SO_3H group. Desulphonation of the sodium salts of arenesulphonic acids $ArSO_3Na$ in dimethyl sulphoxide is reported to occur on 254-nm irradiation.[112] For Ar = phenyl or 2- and 4-tolyl, the hydrocarbon is the sole product, but for Ar = 9-anthryl, traces of anthraquinone are formed, and 1- and 2-naphthols result from the 1- and 2-naphthalenesulphonates, together with naphthalene. The authors suggest that the desulphonation proceeds from the $T_1(\pi\pi^*)$ state but that desulphonylation may occur from the $S_1(n\pi^*)$ state. They note that in contrast with these findings, desulphonylation in aqueous solution involves the $T_1(n\pi^*)$ state, whereas the desulphonation arises from the $T_2(\pi\pi^*)$ state of $ArSO_3Na$.

As usual, numerous accounts have described photoreactions of aryl halides which formally lead to substitution of the halogen; such reactions include the continuing attempts to degrade the polyhalogeno arenes which now pollute the environment. A variety of chlorinated *p*-terphenyls have been photolysed (300 nm) in organic media, and replacement of the chlorine by hydrogen is found to be the major process.[113] Substitution products are reported to be much more prominent with the terphenyls than in the corresponding reactions of poly-chlorinated biphenyls, and quantum yields for the reaction are substantially higher for *ortho* chlorinated compounds than for those in which the *ortho* positions are unsubstituted. A triplet state reaction is deduced from the enhance-ment of the process in the presence of xenon and, less convincingly, from the very inefficient sensitization by benzophenone. Addition of triethylamine was found to promote an electron transfer pathway for the dehalogenation reaction, but only for those compounds which were least photolabile. The same group of workers have also reported further on the photoreactions of various other halogenated arenes including biphenyl and naphthalenes.[114, 115] Photodegrada-tion of 1,1-bis(3,4-dichlorophenyl)ethane and 1-(2,3-dichlorophenyl)-1-(3,4-

110 M. Nakamura and Y. Ogata, *Bull. Chem. Soc. Japan*, 1977, **50**, 2396.
111 O. P. Studzinskii, N. I. Rtishchev, and A. V. Eltsov, *Zhur. org. Khim.*, 1977, **13**, 160
112 N. Suzuki, K. Ito, A. Inoue, and Y. Izawa, *Chem. and Ind.*, 1977, **278**, 399.
113 B. Chittim, S. Safe, N. J. Bunce, L. O. Ruzo, K. Olie, and O. Hutzinger, *Canad. J. Chem.*, 1978, **56**, 1253.
114 G. Sundstrom, K. Olie, O. Hutzinger, S. Safe, and B. Chittim, *Chemosphere*, 1977, **6**, 103.
115 L. O. Ruzo, G. Sundstrom, O. Hutzinger, and S. Safe, *Rec. Trav. Chim.*, 1977, **96**, 249.

dichlorophenyl)ethane in cyclohexane yields mainly reductive dechlorination products, but in polar solvents a photonucleophilic substitution pathway is also followed.[114] The addition of heavy atom lithium salts (LiBr, LiI) is reported to affect the irradiation of methanol solutions of halogenated naphthalenes and biphenyls.[115] Thus the triplet decay rate is increased relative to that of the reaction, so an overall decrease in photodegradation is noted. The effect of iodide is greater than that of bromide, whereas the rates of reaction for chloride, or in the absence of added halide ion, increase in order of decreasing bond strength of the C—X bonds. Irradiation of 2,2'-dichlorobiphenyl in the presence of nitrous oxide has also been reported.[116] In this case, O(^3P) is generated from the photolysis of the nitrous oxide and products reflecting hydroxylation (130) and nitration–hydroxylation (131) are observed. Photolysis of 2,4,5,6-tetrachloro isophthalonitrile (132) in methanol under nitrogen leads to the formation of the dechlorination and methyl substitution products (133) and (134), respectively.[117]

(130)   (131)   (132) R = Cl
                                                                                  (133) R = H
                                                                                  (134) R = Me

Eltsov and co-workers report that the photosubstitution of chloro by cyano or methylsulphano is inhibited in the presence of a nitro substituent.[118] This Russian group, who have contributed much to the area of photosubstitution reactions, also note that the displacement of chloro by cyano in chloroanilines is faster for the *p*- than the *o*- and *m*-isomers.[119] With dihaloanilines, as may be expected, the heavier halogen is substituted before the lighter one and in these compounds halogens *ortho* to the amino group are more reactive than those in the *meta* and *para* positions. The photosubstitution of halogen by selenocyanate ion has also been reported for the aromatic compounds ArX where Ar = 4-methoxy phenyl, 2-amino-1-naphthyl, and 3-cyano-4-aminophenyl, and X = Cl; and for Ar = 2-(dimethylamino)-5-pyrimidyl and X = Br.[120] In all cases, irradiation of the arene in aqueous t-butanol in the presence of sodium or potassium selenocyanate gave the corresponding ArSeCN. Loss of iodine from *o*-di-iodobenzene to give a benzyne intermediate is certainly not new, but anisole is formed from the reaction in methanol using sunlight; phenyl halides under the same conditions give benzene.[121]

[116] V. Saravanja-Bozanic, S. Gaeb, K. Hustert, and F. Korte, *Chemosphere*, 1977, **6**, 21.
[117] R. W. Binkley, G. L. Kirstner, V. C. Opaskar, and P. Olynyk, *Chemosphere*, 1977, **6**, 163.
[118] V. V. Yunnikov, A. N. Frolov, O. V. Kulbitskaya, and A. V. Eltsov, *Zhur. org. Khim.*, 1977, **13**, 361.
[119] A. N. Frolov, V. V. Yunnilov, O. V. Kulbitskaya, and A. V. Eltsov, *Zhur. org. Khim.*, 1977, **13**, 603.
[120] A. N. Frolov, A. V. Eltsov, E. V. Smirnov, and O. V. Kulbitskaya, *Zhur. org. Khim.*, 1977, **13**, 2007.
[121] S. Rajan, *Indian J. Chem.*, 1977, **B15**, 653.

Bunce and Ravanal have commented on the irradiation of chlorobenzene in cyclohexane to give chlorocyclohexene and benzene, and have examined the possible involvement of exciplexes in the photolysis of chlorobenzene by comparing the photochemistry of the systems (135) and (136) with that of chlorobenzene and toluene and chlorobenzene and triethylamine.[122] It is concluded

$$Cl-\langle\!\!\!\langle\ \rangle\!\!\!\rangle-(CH_2)_3-R$$

(135) R = Ph
(136) R = NMe$_2$

that in contrast with the situation with 1-chloronaphthalene,[123] exciplex/ excimer formation in the case of chlorobenzene leads to protection against photodecomposition; in other words, loss of the chlorine from complexed chlorobenzene is less efficient than from the uncomplexed arene.

Photosubstitution in heteroaromatic compounds continues to attract interest. Thus irradiation of 2-bromopyridine in liquid ammonia causes a $S_{RN}1$ reaction with the potassium enolates of acetone, 2,4-dimethyl-3-pentanone, cyclohexanone, and pinacolone to give, for example, in the first case, the ketone (137).[124] In competition experiments, the 2-pyridyl radicals formed in the

(137)

process show a preference for combination with tertiary rather than primary enolates. The pyridine nucleus itself is prone to substitution, and the irradiation of pyridine 2-carboxylate in acidic methanol solution is reported to yield the 6-methoxylation product.[125] Interestingly, the reaction is promoted by the addition to the solution of pyridines bearing either electron-donor or -acceptor groups in the 4-position, whereas addition of the 2- or 3-isomers either has no effect or inhibits the process. The reaction is suggested to proceed in some curious way *via* a specific excimer of the aryl carboxylate and the 4-substituted pyridine but details of the mechanism are not reported. The photoalkylation of aromatic amino-acids with chloroacetamide has been studied as a model for the photomodification of proteins.[126] Irradiation (254 nm) of tryptophan or tyrosine in aqueous solution in the presence of chloroacetamide leads to alkylation [*e.g.* formation of (138)] of both arenes with a quantum yield of 0.1. Nucleophilic photosubstitution reactions of some derivatives of imidazole and pyrazole have been reported.[127] With nitro-derivatives of 1-methyl imidazole and 1-methyl pyrazole, the reaction is strongly dependent on the position of the

[122] N. J. Bunce and L. Ravanal, *J. Amer. Chem. Soc.*, 1977, **99**, 4150.
[123] N. J. Bunce, P. Pilon, L. O. Ruzo, and D. J. Sturch, *J. Org. Chem.*, 1976, **41**, 3023.
[124] A. P. Komin and J. F. Wolfe, *J. Org. Chem.*, 1977, **42**, 2481.
[125] Y. Miyazawa, R. Hasegawa, E. Tobita, T. Furihata, T. Sugiyama, and A. Sugimori, *Chem. Letters*, 1977, 1155.
[126] T. Hamada and O. Yonemitsu, *Chem. Pharm. Bull.*, 1977, **25**, 271.
[127] C. Oldenhof and J. Cornelisse, *Rec. Trav. Chem.*, 1978, **97**, 35.

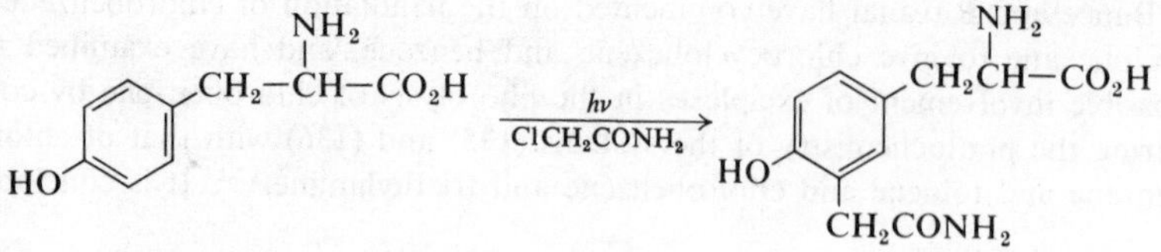

(138)

nitro group, and for reaction with cyano or methoxy anions, the 5-nitro and 3-nitro derivatives are respectively the most reactive for the imidazole and the pyrazole. It is noted that some reactions proceed by normal bimolecular aromatic photosubstitution processes whereas others resemble the photo-reactions of methoxybenzenes and naphthalenes in which a radical-cation is involved. Photonucleophilic substitution of nitrophenazines (139) in the presence of primary amines is reported to give good yields of 6- and 9-alkylamino nitrophenazines (140) with minor amounts of the 4-substituted products.[128] The regioselectivity of the process is discussed and compared with the orientation rules proposed by Havinga and Cornelisse,[129] and an addition elimination mechanism (Scheme 8) is proposed as the reaction pathway. The photoreaction

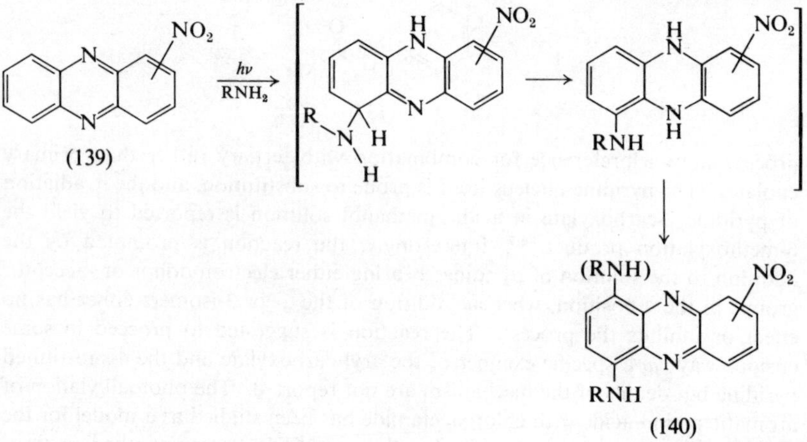

Scheme 8

of amines with 2-nitrophenazine-10-oxide has also been studied.[130] Under the reaction conditions, formation of a σ-anionic Meisenheimer complex (141) was demonstrated by absorbance, conductance, and n.m.r. spectroscopic measurements. The two reactions which occur on irradiation are reduction to 2-nitrophenazine and substitution of the nitro by an amino group.

[128] A. Albini, G. F. Bettinetti, E. Fasani, and G. Minoli, *J.C.S. Perkin I*, 1978, 299.
[129] Reference 90 and E. Havinga and J. Cornelisse, *Pure Appl. Chem.*, 1976, **47**, 1.
[130] G. Minoli, A. Albini, G. F. Bettinetti, and S. Pietra, *J.C.S. Perkin II*, 1977, 1661.

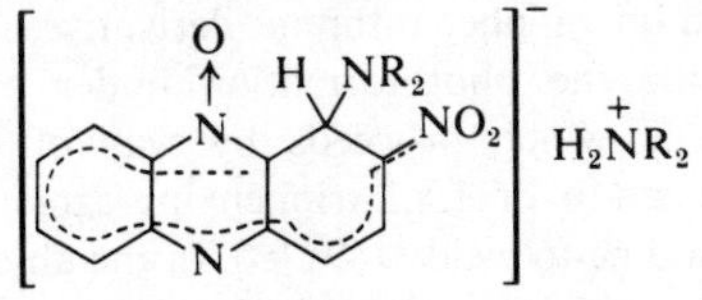

(141)

5 Intramolecular Cyclization Reactions

The numerous reports which have appeared during the year describing reactions in this broad area of research have for ease of presentation been classified where possible into the following types: stilbene-phenanthrene, arenes separated by groups other than C=C, 1,4-diarylbutadienes, ethylene-arene systems, and cyclization by loss of halogen acid. Accounts of other studies are presented at the end of this section.

The stilbene-into-phenanthrene conversion is widely recognized to proceed by way of the *trans*-4a,4b-dihydrophenanthrene intermediate (142), and proof of this was provided from a study of the stilbenediol (143) which yields the diketone (144).[131] A further investigation of compounds of type (143) in which the

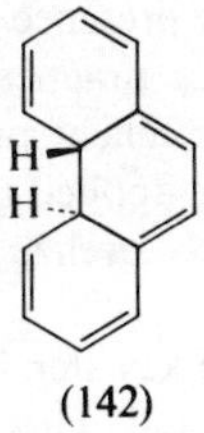

(142) 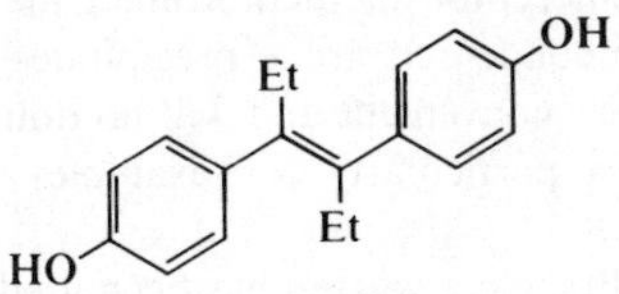 (143) 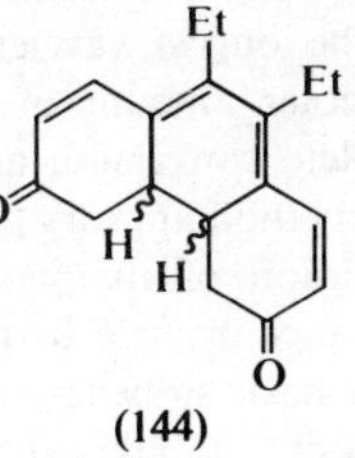 (144)

stilbene moiety is held *cis* [*i.e.* (145)] has been reported and in methanol solution with 254-nm radiation the 4a,4b-dihydrophenanthrene derivative (146) is formed with no side-reactions and with a quantum yield in excess of 0.85.[132] Other

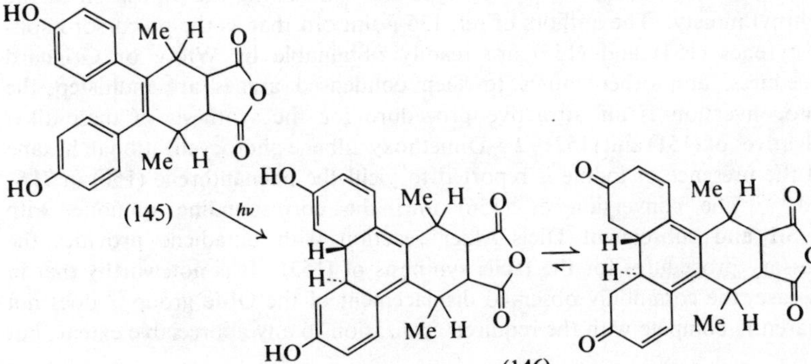

(145) (146)

[131] T. D. Doyle, N. Filipescu, W. R. Benson, and D. Banes, *J. Amer. Chem. Soc.*, 1970, **92**, 6371; T. J. H. M. Cuppen and W. H. Laarhovem, *J. Amer. Chem. Soc.*, 1972, **94**, 5914.
[132] M. Maienthal, W. R. Bensons, E. B. Sheinin, T. D. Doyle, and N. Filipescu, *J. Org. Chem.*, 1978, **43**, 972.

accounts of the formation of phenanthrene derivatives from rigidly held *cis*-stilbene moieties include the photoformation under oxygen of (147) from diphenylcyclobutenedione which proceeds by way of diphenylmaleic anhydride,[133] and the cyclization of 1,4,5-triphenylpyrazole (148) under nitrogen and in the presence of iodine to yield (149).[134] In the absence of iodine, (148) is stable but the benzophenone sensitized reaction induces ring cleavage and the formation of 3-anilino-2,3-diphenylacrylonitrile (150).

(147)   (148)   (149)   (150)

In view of the low yields and by-product formation in many stilbene cyclizations, the reactions of several stilbenes have been investigated in the presence of such π acceptors as tetracyanoethylene, tetracyanoquinodimethane, chloranil, and bromanil: extremely rapid reactions have been observed.[135] Irradiation into the longest wavelength absorption band of the stilbene in the presence of an equimolar amount of the π-acceptor in dichloromethane gives practically a complete conversion and the condensed arene precipitates in excellent purity. This method appears to be very convenient and will no doubt be applied to the cyclization of many stilbenes particularly for examples which cyclize with reluctance or yield by-products.

The basic stilbene–phenanthrene conversion has been used as a key step in the synthesis of alkyl-substituted benzo[c]phenanthrenes (151) and chrysenes (152),[136] and of 7,12-benz[a]anthraquinones (153).[137] Yields for the production of (151) from (154) vary between 66 and 89%, but the product is free from isomeric methyl derivatives although approximately 1% of benz[a]anthracenes is formed from all preparations by cyclization involving the β-position of the naphthyl moiety. The authors of ref. 136 point out that as the precursor naphthylstyrenes (154) and (155) are readily obtainable by Wittig or Grignard procedures, and other routes to such condensed arenes are multistep, the photoconversion is an attractive procedure for the synthesis of these alkyl derivatives of (151) and (152). 2,5-Dimethoxystilbene photocyclization in hexane and the presence of iodine is reported to yield the phenanthrene (156) in 71% yield.[137] The conversion of (156) into the corresponding quinone with Me_3SiI and subsequent Diels–Alder reaction with butadiene provides the necessary procedures for the facile synthesis of (153). It is noteworthy that in this case, the commonly observed displacement of the OMe group[99] does not apparently compete with the required cyclization to any appreciable extent; but

[133] N. Obata and T. Takizawa, *Bull. Chem. Soc. Japan*, 1977, **50**, 2017.
[134] J. Grimshaw and D. Mannus, *J.C.S. Perkin I*, 1977, 2096.
[135] J. Bendig, M. Beyermann, and D. Kreysig, *Tetrahedron Letters*, 1977, 3659.
[136] D. L. Nagel, R. Kupper, K. Antonson, and L. Wallcave, *J. Org. Chem.*, 1977, **42**, 3626.
[137] B. I. Rosen and W. P. Weber, *J. Org. Chem.*, 1977, **42**, 3463.

(154) R = H, Me, or OMe   (151)

(155) R = Me or Et   (152)

(153) R's = H or OMe   (156)

for the stilbene (157), elimination of the methoxy group does occur to yield 2,3-dimethylphenanthrene.[138] The process has, however, been further investigated and low yields of 1-methoxy-3,4-dimethylphenanthrene have been obtained.[139] The problems of separation of the methoxy compound from the reaction mixture limit the value of this procedure in synthesis (maybe π-acceptors which induce a more specific reaction are required here[135]). From the ratio of methoxyphenanthrene to the product following elimination, it is estimated that the attack at the methoxy-substituted position occurs approximately half as frequently as that at the unsubstituted site. The photocyclization of a number of unsymmetrically substituted aryl ethylenes of the type $Ar_2C{=}CAr_2$ has been studied both in cyclohexane solution in air and as a solid film in air in direct sunlight.[140] Although the phenanthrene was formed, this did not apparently react further for no dibenzo[*gp*]chrysene (158) was observed: it is, however, significant that

(157)   (158)

[138] Footnote 6 in F. B. Mallory and C. S. Wood, *J. Org. Chem.*, 1964, **29**, 3373.
[139] E. N. Marvell, J. K. Reed, W. Ganzler, and H. Tong, *J. Org. Chem.*, 1977 **42**, 3783.
[140] R. H. Mitchell, L. Mazuch, B. Shell, and P. R. West, *Canad. J. Chem.*, 1978, **56**, 1246.

under the experimental conditions authentic chrysene underwent oxidative and/ or degradative processes.

In 1971 the first example of photocyclization of a stilbene having one ethylenic electron-withdrawing substituent [*viz.* (159)] to the 9,10-dihydrophenanthrene derivative (160) was reported.[141] Other examples of the process have been described,[142] but the radical mechanism proposed involving the intermediate (161) does not explain why other stilbenes having the one ethylenic substituent appropriate for radical stabilization did not yield products analogous to (160). Further mechanistic details of the (159) into (160) conversion have been obtained from flash photolysis and studies of the reaction in CH_3OD solution.[143] Thus evidence for the involvement of the dihydrophenanthrene intermediate (162) has been obtained; the fate of this intermediate depends on the presence of ethylenic substituents and the reaction conditions. If the substituent is enolizable and the solvent is a proton source, the 9,10-dihydrophenanthrene is formed under anaerobic conditions *via* the intermediate (163) (see Scheme 9) and under oxidative conditions (162) is at least partly converted into the phenanthrene. But other ethylenic substituents may not affect the more general course to phenanthrenes, and 9,10-dihydrophenanthrene is reported to be formed in 12% yield from irradiation of α,β-dichlorostilbene in cyclohexane in the presence of iodine.[144]

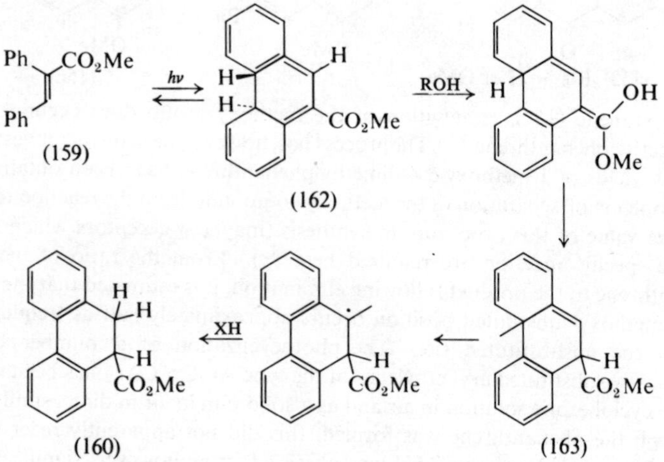

Scheme 9

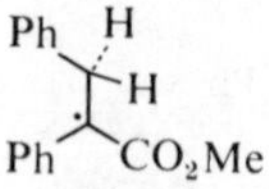

(161)

[141] R. Srinivasan and J. N. C. Hsu, *J. Amer. Chem. Soc.*, 1971, **93**, 2816.
[142] R. Srinivasan and J. N. C. Hsu, *Euchem. Research Conference*, 1975, Ghent, Belgium.
[143] P. H. G. Op Het Veld and W. H. Laarhoven, *J. Amer. Chem. Soc.*, 1977, **99**, 7221.
[144] W. Schroth and H. Bahn, *Z. Chem.*, 1977, **17**, 56.

The possibility of the synthesis of optically active helicene by irradiation of suitable 1,2-diarylethylenes in chiral solvents was outlined last year,[145] and more details of the process have now been published.[146] The irradiation of 2-styrylbenzo[*c*]phenanthrene (164) has been studied in eleven different chiral

(164)

solvents and gives non-racemic hexahelicene with optical yields between 0.2 and 2.0%, the highest being from (*S*)-(+)ethyl mandelate. The role of the chiral solvent is ascribed to its influence on the equilibrium between enantiomeric conformations of *cis*-(164), and this leads to an excess of one conformer. The size and sign of the effect can be related to the size and position of large apolar residues at the chiral centre of the solvent molecule, and it is thus concluded that the solvent behaves as a matrix in which the reaction takes place.

Full details of the earlier preliminary account[147] of the photocyclization reaction of substituted divinylbiphenyl derivatives have been described,[148] and compared with those recently reported by Laarhoven and co-workers.[149] The formation from (165) of the cyclobutane (166) and the tetrahydropyrene (167)[150] as the kinetically and thermodynamically controlled products respectively has been confirmed, but compounds (168), (169), (170), and (171) even at very short irradiation periods only yield the tetrahydropyrene, and [2 + 2] intramolecular cycloadducts are not observed; this latter process is, however, the main reaction for (172), which yields (173). Again it is noted that free-valence indices of

(165) R = R³ = Ph, R¹ = R² = H
(168) R = R¹ = R² = R³ = R⁴ = H
(169) R = R³ = CN, R¹ = R² = H
(170) R = R³ = CO₂Me, R¹ = R² = H
(171) R = R³ = H, R¹ = R² = Ph
(172) R = R² = Ph, R¹ = R³ = H

145   W. H. Laarhoven and T. J. H. M. Cuppen, *J.C.S. Chem. Comm.*, 1977, 47.
146   W. H. Laarhoven and T. J. H. M. Cuppen, *J.C.S. Perkin II*, 1978, 315.
147   A. Padwa and A. Mazzu, *Tetrahedron Letters*, 1974, 4471.
148   A. Padwa, C. Doubleday, and A. Mazzu, *J. Org. Chem.*, 1977, **42**, 3271.
149   P. H. G. Op Het Veld, J. C. Langendam, and W. H. Laarhoven, *Tetrahedron Letters*, 1975, 231, and *J.C.S. Perkin I*, 1977, 268.
150   W. H. Laarhoven and T. J. H. M. Cuppen, *J. Chem. Soc. (C)*, 1972, 2074.

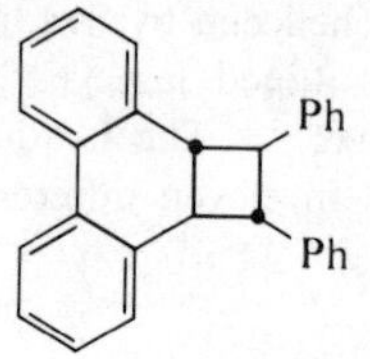

(166) 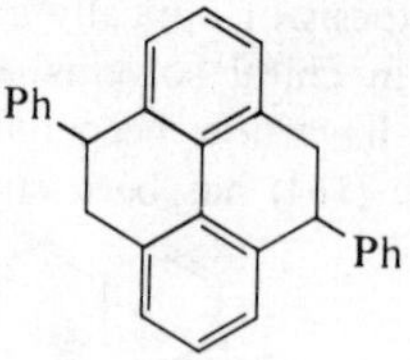   (167) 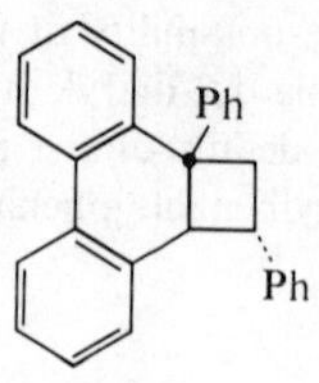   (173)

terminal atoms involved in the cyclizations in the first excited state serve as a guide to predict the direction of cyclization. This report also describes the photoreactions of several 2-vinyl-2'-acylbiphenyls: for example, the 9,10-dihydrophenanthrene (174) is formed from (175), whereas (176) yields the analogue of (174) and (177).

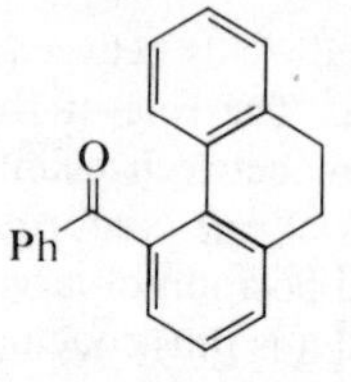

(174) 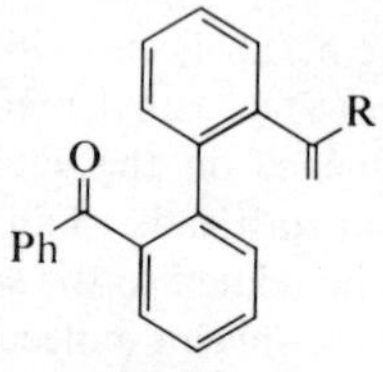   (175) R = H   (177)
                         (176) R = Ph

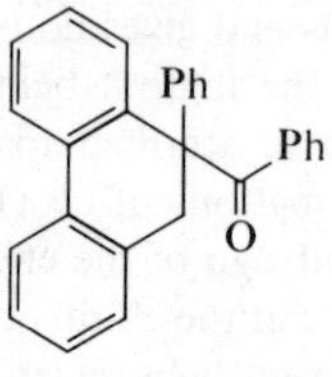

Stilbazoles are known to undergo cyclization to yield benzo[f]quinolines[151] and the reaction has been extended to a number of sulphur-containing derivatives (178).[152] Irradiations were performed in t-butanol–benzene mixtures in the presence of oxygen using a Corex 9700 filter which allowed excitation of the 2-stilbazole but not the cyclized product. Derivatives (178c, d, and g) give the benzo[f]quinolines (179) in respective yields of 21, 19, and 26%, but (178a, b,

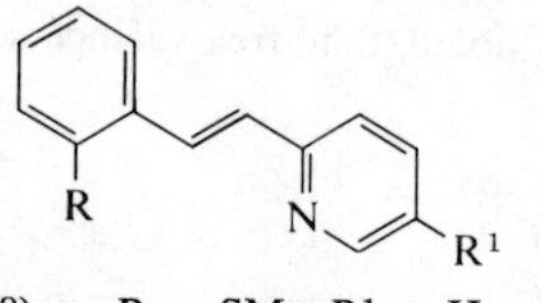

(178) a; R = SMe, R¹ = H
      b; R = SOMe, R¹ = H
      c; R = SO₂Me, R¹ = H
      d; R = H, R¹ = CO₂Me
      e; R = SMe, R¹ = CO₂Me
      f; R = SOMe, R¹ = CO₂Me
      g; R = SO₂Me, R¹ = CO₂Me

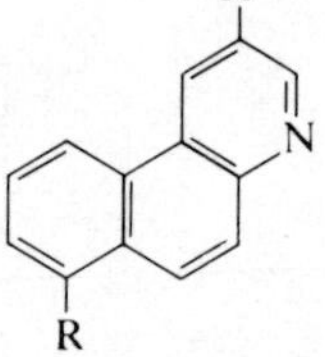

(179)

e, and f) suffered extensive decomposition and no cyclized products could be isolated. The cyclization of azabenzene to benzo[c]cinnoline only occurs in acid media,[153] but it is now reported that the analogous process with 2-phenylazopyridine to yield 4-pyrido[c]cinnoline (180) also occurs in neutral media and a

[151] P. L. Kumler and R. A. Dybas, *J. Org. Chem.*, 1970, **35**, 125 and 3825.
[152] N. R. Beller, D. C. Neckers, and E. P. Papadopoulos, *J. Org. Chem.*, 1977, **42**, 3514.
[153] C. P. Joshua and V. N. R. Pillai, *Indian J. Chem.*, 1975, **B13**, 1018.

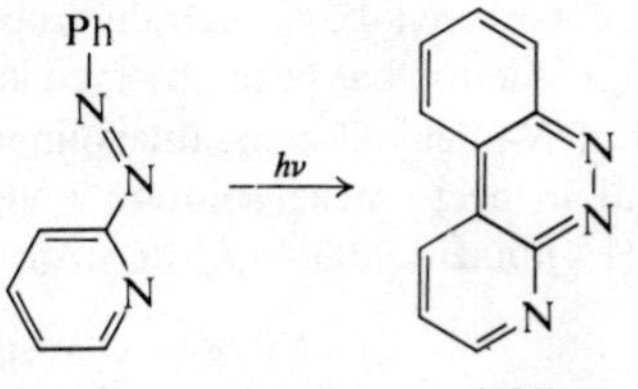

(180)

mechanism similar to that for the azabenzene process is also proposed in the latter reaction.[154]

Two groups report on the photocyclization reactions of diphenylquinoline derivatives. Glinka has observed that the reaction of 3,4-diphenylquinoline which yields dibenzo[*i,k*]phenanthridine (181) in the presence of dilute hydrochloric acid does not occur in neutral media,[155] and Mortelmans and Van Binst

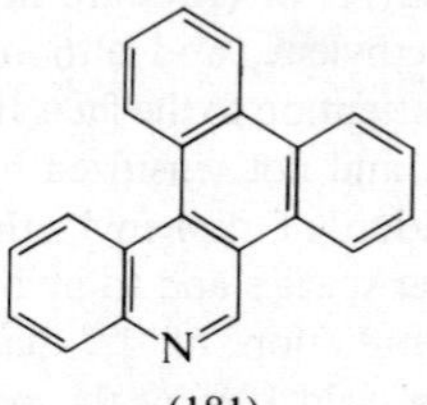

(181)

have studied the related process with 1,2-diphenylquinolinium perchlorates (182) which on irradiation in methanol solution and the presence of iodine yield (183) in 50% yield.[156] By way of contrast, the photoreaction of 1,2-diphenyl-1,2,3,4-tetrahydroquinoline (184) leads to attack of the 1-phenyl group onto the quino-

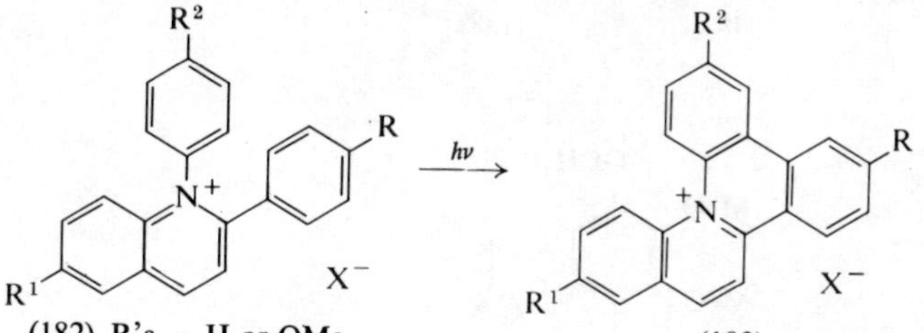

(182) R's = H or OMe   (183)

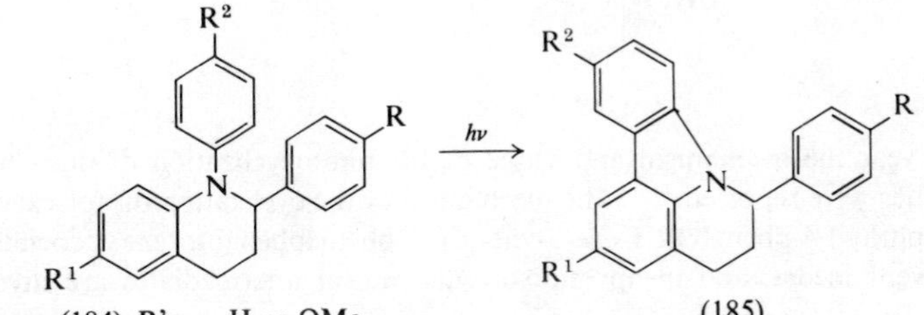

(184) R's = H or OMe   (185)

[154] V. N. R. Pillai and E. Purushothaman, *Current Sci.*, 1977, **46**, 381.
[155] J. Glinka, *Roczniki Chem.*, 1977, **51**, 577.
[156] C. Mortelmans and G. van Binst, *Tetrahedron*, 1978, **34**, 363.

line and the formation of 6-phenyl-4,5,6,7-tetrahydropyrido[3,2,1-*j,k*]carbazole (185). A somewhat similar reaction has been observed in what has been described as a new photoreaction of *N*-phenyl-2-benzothiazolinone (186): in the presence of ethyl vinyl ether and in acetonitrile solution under argon this yields the previously unreported [1,4]thiazino[2,3,4-*j,k*]carbazole system (187) as well as

parent carbazole.[157] Derivatives of (187) are not formed with cyclohexene or acrylonitrile as the added ethylene, and although the mechanism is not yet elucidated the authors draw attention to the facts that the process is regioselective, not quenched by piperylene, and not sensitized by acetone.

The photodehydrocyclization of diphenyl ether to dibenzofuran has been proposed to arise from triplet species and to involve 1*a*,9*a*-dihydrodibenzofuran as intermediate,[158] and irradiation of 1,7-diarylheptanones [*e.g.* (188)] in alkaline media is reported to yield 10% of the cyclized product (189).[159]

Last year the mechanism and scope of the photocyclization of diarylbut-1-ene-3-ynes were reported.[160] The mechanism of the cyclization of, for example, 1-(α-naphthyl)-4-phenylbut-1-ene-3-yne to 1-phenylphenanthrene depends on the solvent media, and in aprotic solvents radical intermediates are involved

[157] L. R. Sousa and J. G. Bucher, *Tetrahedron Letters*, 1978, 2267.
[158] K. P. Zeller and G. Gauglitz, *Z. Naturforsch.*, 1977, **32**, 285.
[159] D. A. Whiting and A. F. Wood, *Tetrahedron Letters*, 1978, 2335.
[160] A. H. A. Tinnemanns and W. H. Laarhoven, *J.C.S. Perkin II*, 1976, 1111 and 1115.

whereas an ionic pathway is operative in alcoholic solvents. The effect of atmosphere has been studied on the cyclization of 1-(9-phenanthryl)-4-phenylbut-1-ene-3-yne (190) which under nitrogen yields the expected product 1-phenyl-triphenylene (191).[161] Under an oxygen atmosphere, however, and in non-polar solvents such as benzene, hexane, or diethyl ether, (192) is the main product, but (191) is still the product in methanol even under oxygen. The authors suggest that the formation of (192) probably involves the radical pathway and from quenching and sensitization studies they conclude that at least part of the reaction involves triplet species (see Scheme 10). Further studies have been

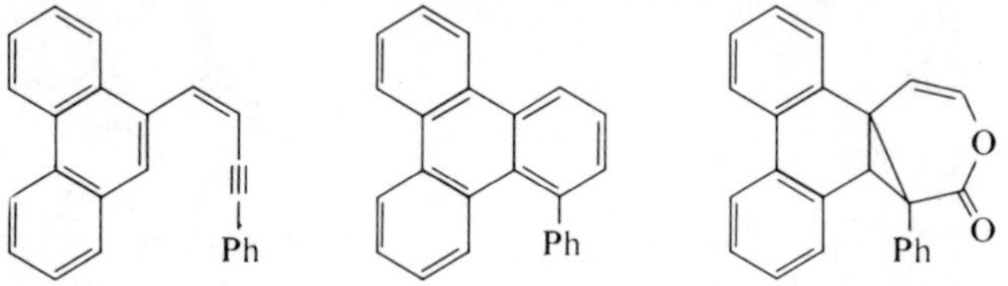

(190)   (191)   (192)

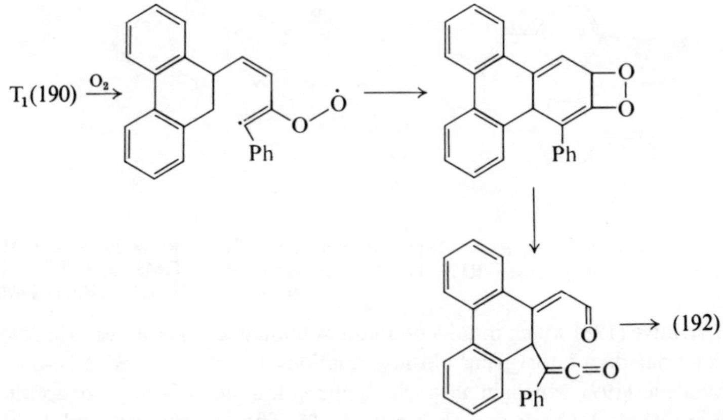

$T_1(190) \xrightarrow{O_2}$

Scheme 10

reported on the cyclization of both 3,3′-diphenyl-2,2′-bi-1*H*-indene-1,1′-dione (193) and 2,3-dibenzylidene butyrolactones (194).[162] The dione (193) yields (195) by way of the now well-established dihydro-compound,[163] and irradiation of (194) in dimethylformamide under nitrogen and in the presence of 1,4-diazabicyclo[2.2.2]octane gives selectively the β-apolignans (196).[164] The photochromic properties of the fulgides (197) have received comment from two groups of workers. Heller and co-workers have reported widely on such over-crowded molecules as (197), and have now studied the details of irradiation of

[161] R. J. F. M. van Arendonk and W. H. Laarhoven, *Tetrahedron Letters*, 1977, 2629.
[162] See F. Toda and Y. Todo, *J.C.S. Chem. Comm.*, 1976, 848; T. Momose, T. Nakamura, and K. Kanai, *Heterocycles*, 1977, 6, 277, for previous pertinent studies of these molecules.
[163] F. Toda and Y. Todo, *Bull. Chem. Soc. Japan*, 1977, 3000.
[164] T. Momose, K. Kanai, T. Nakamura, and Y. Kuni, *Chem. Pharm. Bull.*, 1977, 25, 2755.

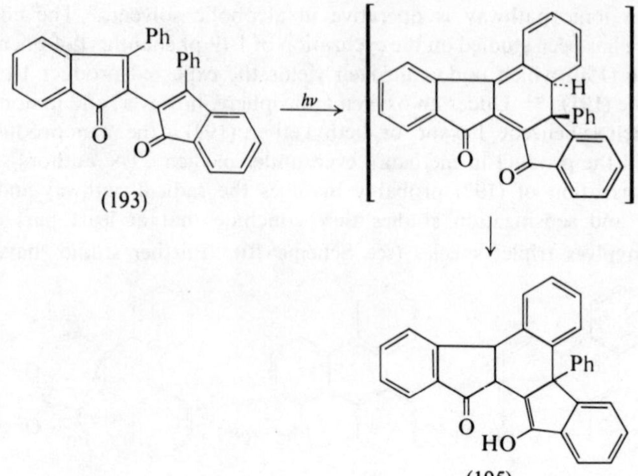

(193)

(195)

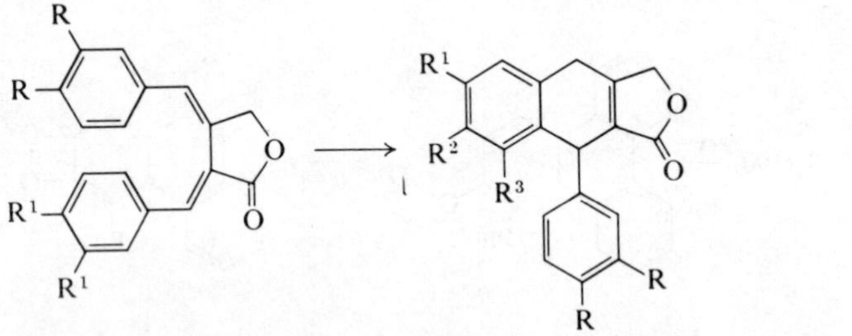

(194) R = H, R¹ = OMe
and R = OMe, R¹ = H

(196) a; R¹ = R² = R³ = H, R = OMe
b; R¹ = R² = OMe, R = R³ = H
c; R = R¹ = H, R² = R³ = OMe

the derivative (198) which in toluene solution undergoes conrotatory ring closure and thermal disrotatory ring closure reactions to give the red 1,8a-dihydro-naphthalene (199) which in turn yields the colourless 1,2-dihydronaphthalene (200) in a facile 1,5-hydrogen shift process.[165] The dimethoxyphenyl-derivative (201) undergoes reversible photochemical ring closure to the deep blue solvato-

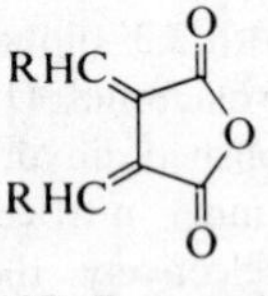

(197) R = aryl

chromatic compound (202) which does not undergo 1,5-hydrogen shifts at ambient temperature but on irradiation yields the 1,4-dihydronaphthalene (203) by a 1,7-hydrogen shift. Ilge and Paetzold have reported the reaction of several derivatives of (197) including R = *p*-anisyl, piperonyl, 4-acetoxy-3-methoxy-

165 P. J. Darcy, R. J. Hart, and H. G. Heller, *J.C.S. Perkin I*, 1978, 571.

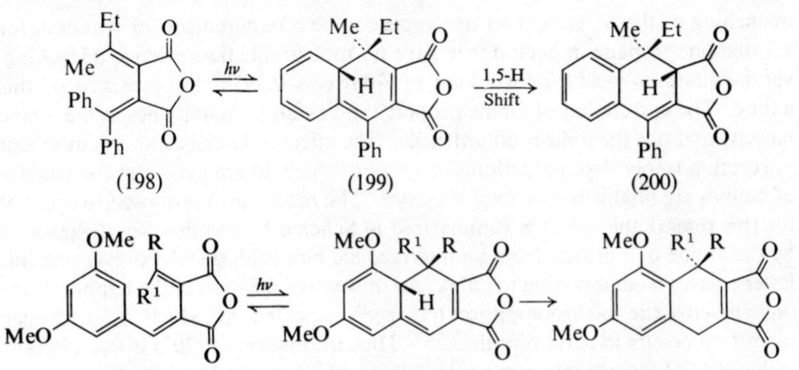

(198)   (199)   (200)

(201)   (202)   (203)

phenyl, 3,4-dimethoxyphenyl, 3,4,5-trimethylphenyl, *p*-nitrophenyl, or 3,5-dimethoxyphenyl.[166] The effects of light intensity, wavelength, and temperature on formation of the coloured 1,8-dihydronaphthalenes were all investigated.

This year many examples have been reported which involve intramolecular cyclization of an ethylene onto an aryl group. Lapouyade and co-workers have previously shown that 1-(α-naphthyl)-1-phenylethylene (204) yields 1-phenylacenaphthylene (205) in the presence of iodine and oxygen, and that in the absence of the oxidant 1-phenylacenaphthene (206) is also formed; the process involves singlet intermediates.[167] The effect of amines on the process has now been studied and the enhancement in the formation of (206) is such that it is almost the exclusive product.[168] With triethylamine as catalyst the quantum yield is increased by approximately 40-fold but the bulk of the amine alkyl group is important as n-tripropylamine is far less efficient in promoting the reaction. It is noteworthy that (204) produces a fluorescent exciplex with the amine; but

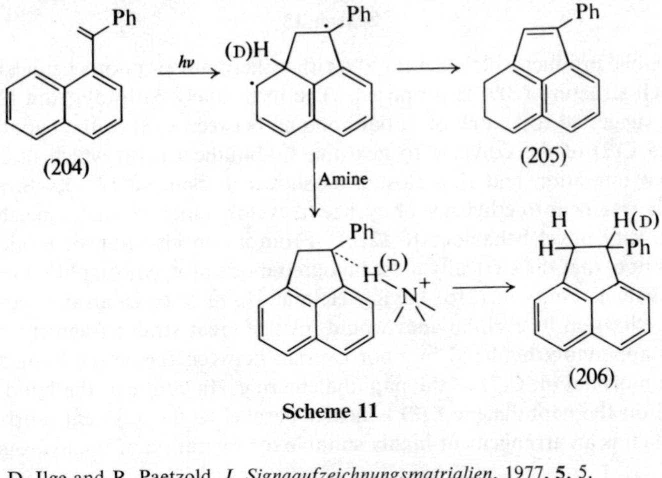

(204)   (205)

(206)

Scheme 11

[166] H. D. Ilge and R. Paetzold, *J. Signaaufzeichnungsmatrialien*, 1977, **5**, 5.
[167] R. Lapouyade, R. Koussini, and J. C. Rayez, *J.C.S. Chem. Comm.*, 1975, 676.
[168] R. Lapouyade, R. Koussini, and H. Bouas-Laurent, *J. Amer. Chem. Soc.*, 1977, **99**, 7374.

quenching of the S_1 state does not appear to be a requirement for reaction, for 1,3-diaminopropane in acetonitrile gave no measurable fluorescence quenching, yet the quantum yield for formation of (206) was 0.39 in the presence of this amine. The correlation of amine property and catalytic activity lies in the amine basicity and not the ionization potential. The effect is described as enhancement of reaction rather than induction, for systems which do not cyclize in the absence of amines are unaffected by their presence. The mechanism proposed to account for this remarkable effect is summarized in Scheme 11 and involves abstraction by the amine of a proton from an intermediate biradical, thereby preventing this latter species from reverting to (204). An unusual cyclization of the naphthalene–diphenylethylene bichromophore has been reported in which 1,2-hydrogen migration occurs in a 1,4-biradical.[169] Thus irradiation of (207) in cyclohexane yields 60 % of the one major product (208) in which, unlike the 4-phenyl analogue of (207)[170] there is no migration of deuterium. Short periods of irradiation

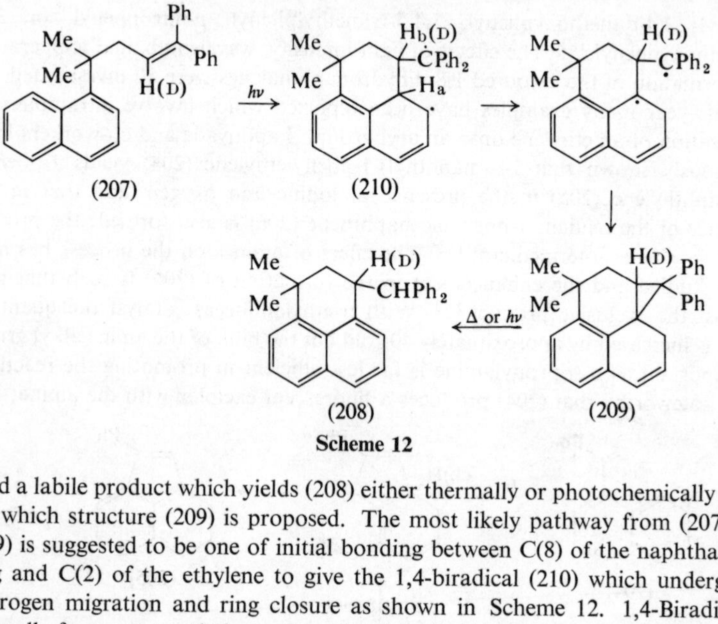

Scheme 12

yield a labile product which yields (208) either thermally or photochemically and for which structure (209) is proposed. The most likely pathway from (207) to (209) is suggested to be one of initial bonding between C(8) of the naphthalene ring and C(2) of the ethylene to give the 1,4-biradical (210) which undergoes hydrogen migration and ring closure as shown in Scheme 12. 1,4-Biradicals normally fragment to ethylenes or cyclize to cyclobutanes, so such a mechanism ascribes very novel behaviour to (210). From a consideration of models, the authors note that the sterically most favoured mode of vinyl—naphthyl bonding is that which results in (210) having H_a and H_b in a *trans* arrangement and hence cyclization to cyclobutanes would involve great strain; fragmentation of (210) is apparently hindered by poor overlap between the newly formed bond and the *p*-orbital on C(7) of the naphthalene ring. In contrast, the bond to Ha in (210) on the naphthalene C(8) is nearly parallel to the adjacent *p*-orbital on C(7) which is an arrangement highly suitable for migration of the hydrogen.

[169] S. S. Hixson and J. C. Tausta, *J. Org. Chem.*, 1977, **42**, 2191.
[170] S. S. Hixson and J. C. Tausta, *J. Amer. Chem. Soc.*, 1975, **97**, 3230.

Intramolecular reaction of the α-(1-cyclohexenyl) cinnamic esters (211) provides examples of arene–ethylene cyclization in which the intervening moiety is itself an ethylene.[171] In methanol solution and in the presence of iodine and air the 5,6,7,8-tetrahydrophenanthrenes (212) are formed as the only products; the yields (62—88%) are very much better than those of diphenylbutadiene cyclizations.[172] Irradiation of (211a—c) under nitrogen gives results which depend on the aromatic substitution. Thus whereas (211c), for example, yields (213) in 86.5% yield, (211a) gives 57% of (214), 24% of the derived acid, and 5% of (212a).

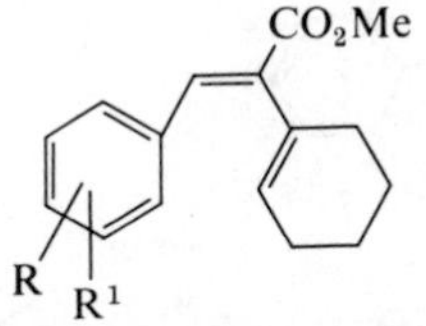

(211) a; R = H, R¹ = CF₃(4-)
       b; R = R¹ = Cl (2,4-)
       c; R = R¹ = OMe (3,5-)

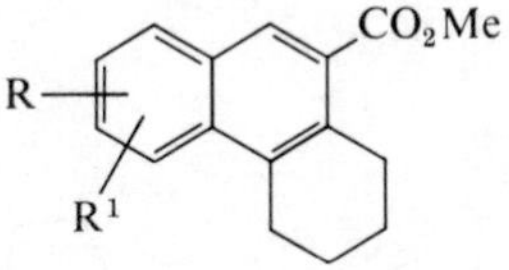

(212)

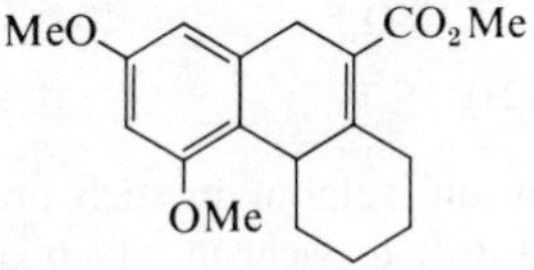

(213)

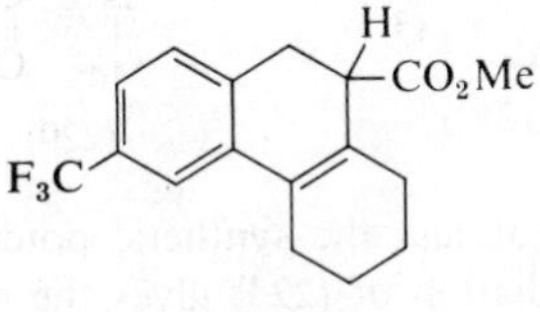

(214)

Analogous intramolecular cyclizations in which the two chromophores are separated by nitrogen, oxygen, and sulphur have all been reported this year, and lead to the corresponding indoles, benzofurans, and benzothiophens respectively. Using circularly polarized light, irradiation of N-methyl-N-aryl enamines [*e.g.* (215)] gives optically active N-methyl dihydroindoles in optical yields of 0.2%[173] which are of the same order as those described for the cyclization leading to optically active hexahelicene.[146] The enamine cyclization has been used as a key step in the synthesis of (±) cryptaustoline; but its application may be limited, for irradiation of ethereal solutions of the precursor (216) only gives low yields of (217) and (218).[174] Schultz and Ching Chiu have investigated the possibility of using this type of reaction to construct the indoline part of certain alkaloids by studying the process with the 2-anilinocyclohex-2-enones (219) and (220).[175] Both compounds apparently cyclize smoothly, and the *cis*-fused indolines (221) and (222) are formed in yields of 90 and 71%, respectively. In the case of (219), 7% of the *trans*-isomer of (221) is also formed, but treatment of this with sodium carbonate in methanol yields the *cis*-isomer. The same group of workers report on heteroatom-directed photoarylation in this type of system

171 R. Srinivasan, V. Y. Merritt, J. N. C. Hsu, P. H. G. Op Het Veld, and W. H. Laarhoven, *J. Org. Chem.*, 1978, **43**, 980.
172 R. J. Hayward and C. C. Leznoff, *Tetrahedron*, 1971, **27**, 2085.
173 J. F. Nicoud and H. B. Kagan, *Israel J. Chem.*, 1977, **15**, 78.
174 I. Ninomiya, J. Yasui, and T. Kiguchi. *Heterocycles*, 1977, **6**, 1855.
175 A. G. Schultz and I. Ching Chiu, *J.C.S. Chem. Comm.*, 1978, 29.

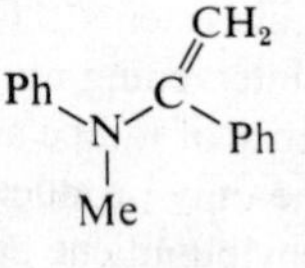

(215) 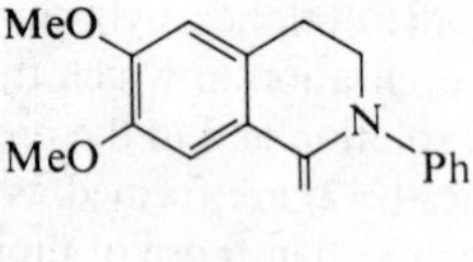   (216)

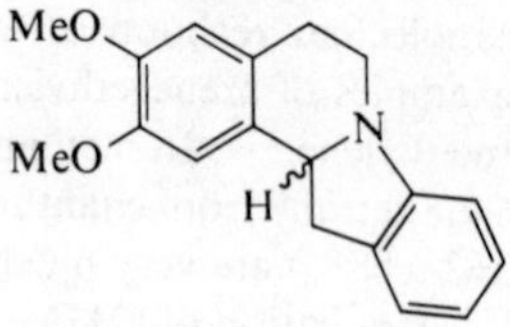

(217)

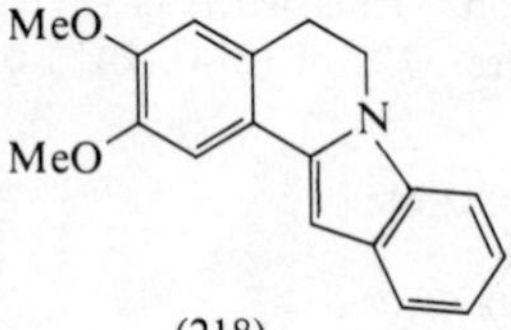

(218)

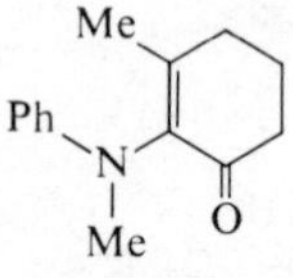

(219) 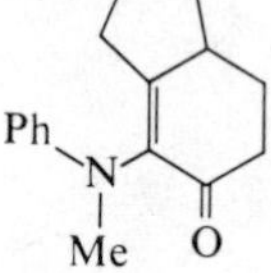   (220) 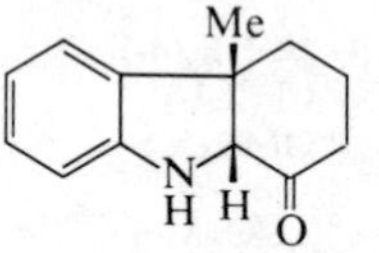   (221) 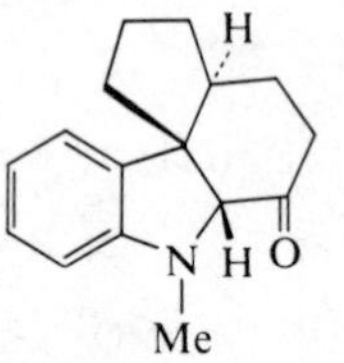   (222)

and discuss the synthetic potential of oxygen and sulphur in such processes. Irradiation of (223) gives the derivative (224) in high yield in which only the *cis*-decalone ring fusion is observed.[176] The stereochemistry of heteroatom-directed photoarylation is demonstrated by the reaction of (225) which in

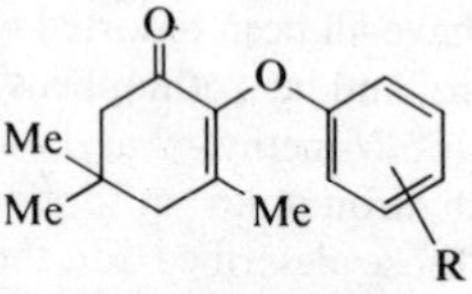

(223) R = H, OMe, Me, CO₂Me,
          CO₂H, COMe, NMe₂

(–OMe, CN), (–OMe, ⟨S S⟩)

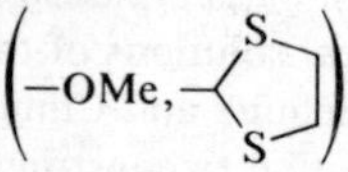

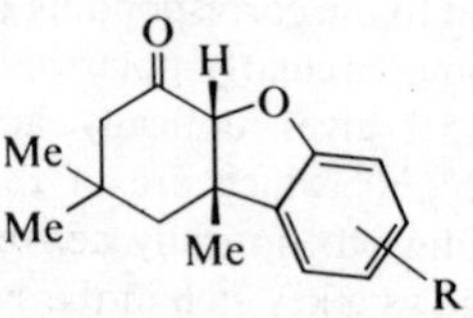

(224)

benzene solution yields only the *trans* product (226), whereas in methanol–benzene solution both (226) and the *cis* isomer (227) are formed. These observations are rationalized by the proposal that the intermediate carbonyl ylide (228) in benzene solvent undergoes suprafacial 1,4-hydrogen migration, whereas in protic solution (methanol) competitive protonation of the ylide occurs and (227) also results. The effect of sulphur on the cyclization has been studied with a range of compounds [*e.g.* formation of (229) from (230)] and the reaction appears to be quite general, proceeding with high chemical and quantum efficiencies, and

[176] A. G. Schultz, R. D. Lucci, W. Y. Fu, M. H. Berger, J. Erhardt, and W. K. Hagmann, *J. Amer. Chem. Soc.*, 1978, **100**, 2150.

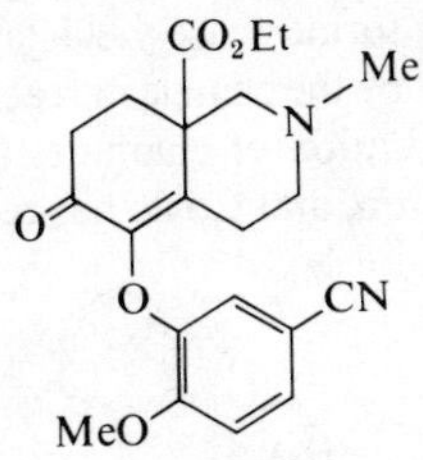

(225)

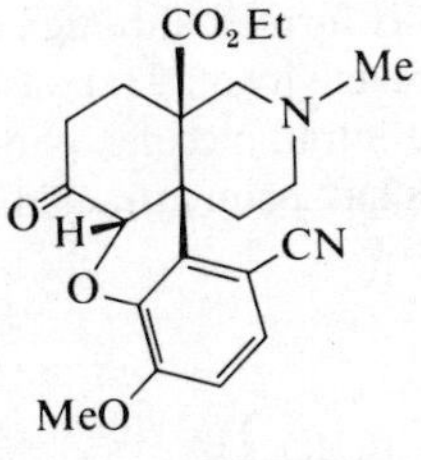

(226)

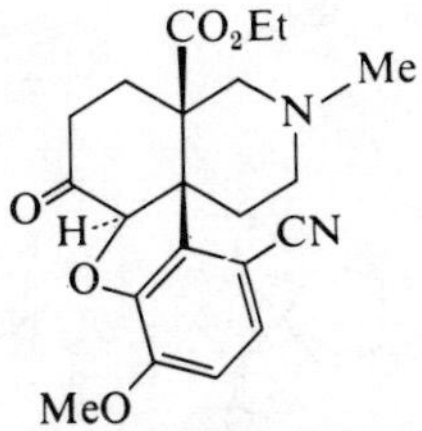

(227)

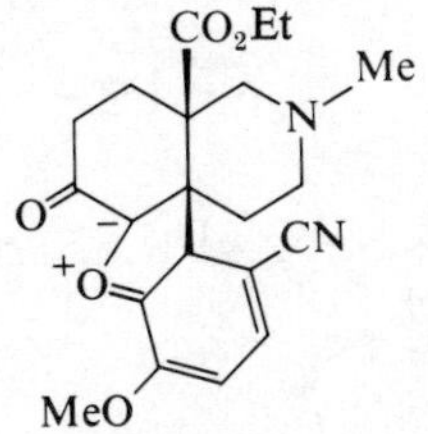

(228)

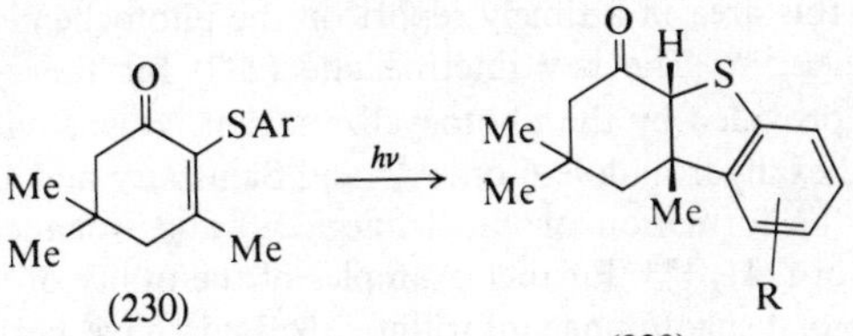

(230)

(229)

Ar = Ph, *o*-, *m*- and *p*-tolyl,
      p-OHC$_6$H$_4$ 2-naphthyl,
      2-Me-1-naphthyl, 2-quinolyl,
      3-indolyl, 1-Me-3-indolyl,
      and 2-benzothiazolyl.

compatible with a wide variety of functional groups within the molecular system.[177]

The photocyclization of 3-benzyl-1-cycloalkenyl ketones (231) to give (232) in the presence of boron trifluoride was reported two years ago,[178] and the reaction has now been studied with the 1-cycloalkenyl phenethyl ketones (233).[179] The cyclohexenyl compound (233; *n* = 2) yields (234) in benzene in the presence of BF$_3$, methanol, or trifluoroacetic acid. In the absence of an acid catalyst, no cyclization is observed but a small amount of fragmentation occurs. But for (233) with *n* = 1, in both the presence and absence of acid catalyst, the spiro

[177] A. G. Schultz, W. Y. Fu, R. D. Lucci, B. G. Kurr, K. M. Lo, and M. Boxer, *J. Amer. Chem. Soc.*, 1978, **100**, 2140.

[178] M. Tada, H. Saiki, K. Miura, and H. Shinozaki, *Bull. Chem. Soc. Japan*, 1976, **49**, 1666, and *J.C.S. Chem. Comm.*, 1975, 55.

[179] M. Tada, H. Saiki, H. Mizutani, and J. Miura, *Bull. Chem. Soc. Japan*, 1978, **51**, 343.

compound (235) is formed although in the former case yields are poor and no difference in the reactivity of (231) was noted by the change in reaction conditions.

The use of the intramolecular photocyclization of enamides (236) in natural product synthesis has again attracted attention, and Coyle has recently reviewed

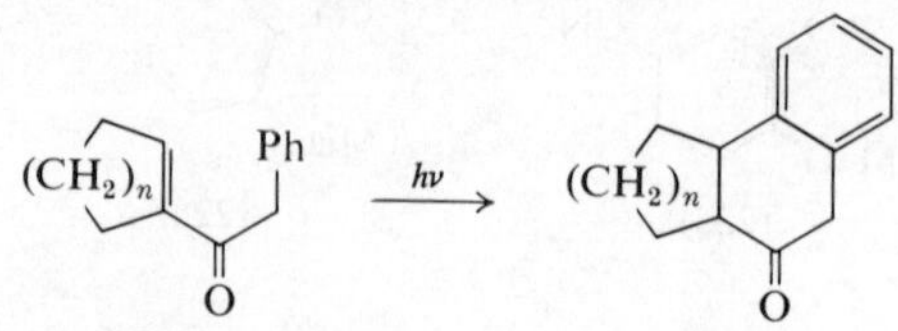

(231)
n = 1 or 2

(232)

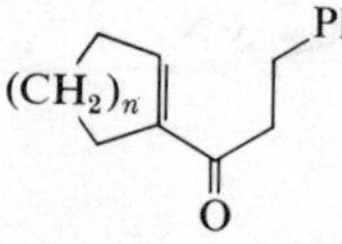

(233) 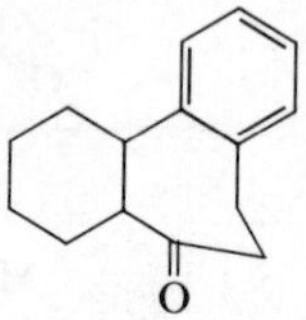   (234) 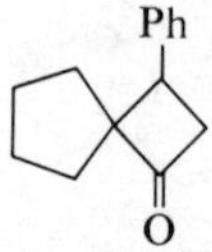   (235) 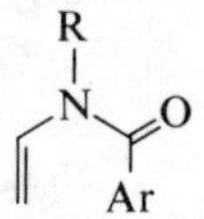   (236)

some of the work in this area in a timely report on the photochemistry of car-boxylic acid derivatives.[180] The key intermediate (237) for the synthesis of γ-lycorane has been provided by the photocyclization of N-benzoyl derivatives (238) of 1,2,3,3a,4,5-hexahydroindole-6-one,[181] and Sainsbury and Uttley comment further[182] on the formation of nauclefine (239) and isonauclefine (240) from the irradiation of (241).[183] Further examples of the utility of this cycliza-tion in the synthesis of benzo[b]phenanthridine alkaloids have been described

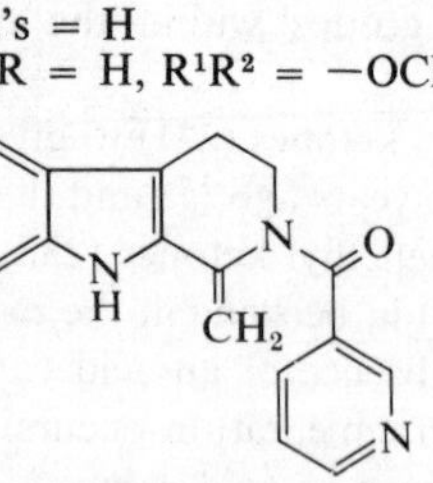

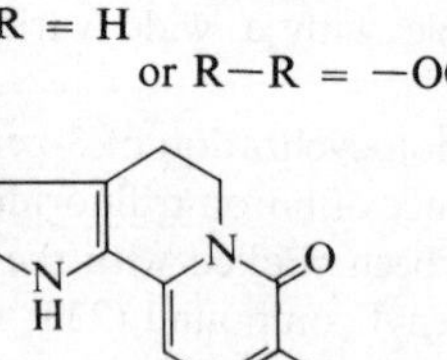

(237) R's = H
R = H, R¹R² = —OCH₂O—

(238) R = H
or R—R = —OCH₂O—

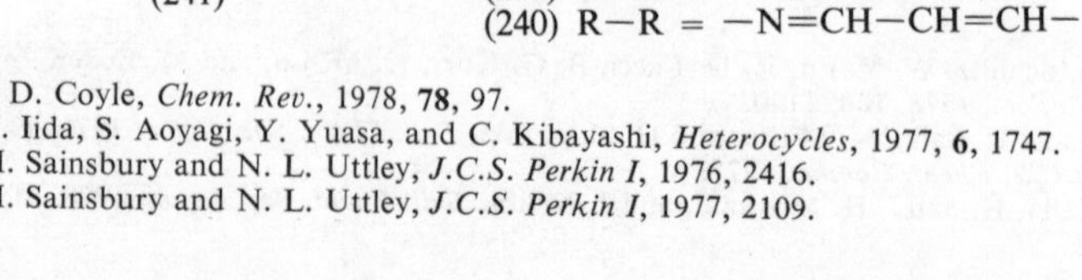

(241)

(239) R—R = —CH=CH—N=CH—
(240) R—R = —N=CH—CH=CH—

[180] J. D. Coyle, *Chem. Rev.*, 1978, **78**, 97.
[181] H. Iida, S. Aoyagi, Y. Yuasa, and C. Kibayashi, *Heterocycles*, 1977, **6**, 1747.
[182] M. Sainsbury and N. L. Uttley, *J.C.S. Perkin I*, 1976, 2416.
[183] M. Sainsbury and N. L. Uttley, *J.C.S. Perkin I*, 1977, 2109.

by Japanese workers for (242) which in methanol yields exclusively the *trans* product (243).[184] In a somewhat different approach to the problem, Soetens and Pandit have used the cyclization of the 4-substituted isoquinolinone (244) as an entry into this ring system.[185]

(242)   (243)   (244)

Photocoupling of two moieties by the loss of halogen acid is a very general intramolecular reaction and has again found a use in natural product synthesis. An example of such a use is the synthesis of ($\pm$)laurolitsine (245) from the irradiation of the hydrochloride (246),[186] and reports of several similar applications have appeared during the year.[187] It is of interest to note that not in all cases does the cyclization involve loss of HX: for example, (247) is also formed

(246)   (245)   (247)

from (246). The conditions of the reaction generally involve pyrex-filtered radiation and the hydrochloride of the amine, although in some cases the reactions are conducted in aqueous solution in the presence of sodium bisulphite. The elimination–cyclization has also been used as a route to phenanthridones (248) from the amides (249).[188] The amounts of products (248) are increased by the presence of a *p*-hydroxy group in (249) and the trend in the yields directly reflects the C—X bond energies. Loss of hydrogen halide resulting in arene–ethylene cyclization has also been observed and has led to a novel synthesis of hexahydroapoerysopine dimethyl ether (250).[189] Thus irradiation of either the 6'-iodo or 7-bromo derivatives (251) and (252) in dioxan or acetonitrile

[184] I. Ninomiya, O. Yamamoto, T. Kiguchi, T. Naito, and H. Ishii, *Heterocycles*, 1977, **6**, 1730.

[185] H. P. Soetens and U. K. Pandit, *Heterocycles*, 1977, **8**, 181.

[186] S. Rajeswari, H. Suguna, and B. R. Pai, *Indian J. Chem.*, 1977, **B15**, 592.

[187] H. Suguna and B. R. Pai, *Indian J. Chem.*, 1977, **B15**, 416 T. R. Govindachari, K. Nagarajan, S. Rajeswari, H. Suguna, and B. R. Pai, *ibid.*, 873; G. Manikumar, B. R. Pai, and H. Suguna, *ibid.*, 740.

[188] S. Lalitha, S. Rajeswari, B. R. Pai, and H. Suguna, *Indian J. Chem.*, 1977, **B15**, 180.

[189] H. Iida, T. Takarai, and C. Kibayashi, *J.C.S. Chem. Comm.*, 1977, 644 *J. Org. Chem.*, 1978, **43**, 975.

(249)   (248)

a, R = R² = R³ = H, R¹ = I
b, R = R² = R³ = H, R¹ = Br
c, R = R² = H, R¹ = Br, R³ = OH
d, R = OH, R¹ = R³ = H, R² = Br

solution in the presence of triethylamine gives (253) in yields of 50 and 38%, respectively; this is readily converted into (250) by lithium aluminium hydride reduction and hydrogenation.

(250)   (251) R = I, R¹ = H   (253)
                         (252) R = H, R¹ = Br

Again this year a number of reports have been published which describe arene photocyclization processes which cannot be easily classified either in this section or elsewhere in the Volume. Their inclusion at the end of this section in no way reflects the Reporter's opinion of their substance. Verboom and Bos have studied the photoisomerization of the α-aryl-γ-oxo-α,β-unsaturated carboxamide (254) which leads to a ring enlargement and formation of the cycloheptatriene derivative (255) in 85% yield.[190] A mechanism is suggested which involves attack of the oxygen atom of the ketone onto C(1) of the arene ring to give (256) which undergoes ring closure to the norcaradiene derivative (257) followed by ring opening to (255). Two groups report extensively on the ring-opening reaction of 3-phenyl- and 3-vinyl-substituted cyclopropenes; pertinent to the present section is the formation of the indenes which result from the process. Thus irradiation of (258) in benzene solution gives 75% of (259),[191] and the vinyl compound (260) yields both the indene (261) and the cyclopentadiene derivative (262).[192]

Photocyclization which involves *ortho* substituents of benzene derivatives occurs for a wide variety of compounds and several examples of the process

[190] W. Verboom and H. J. T. Bos, *Tetrahedron Letters*, 1978, 1229.
[191] A. Padwa, T. J. Blacklock, D. Getman, N. Hatanaka, and R. Loza, *J. Org. Chem.*, 1978, **43**, 1481.
[192] H. E. Zimmermann and S. M. Aasen, *J. Org. Chem.*, 1978, **43**, 1493.

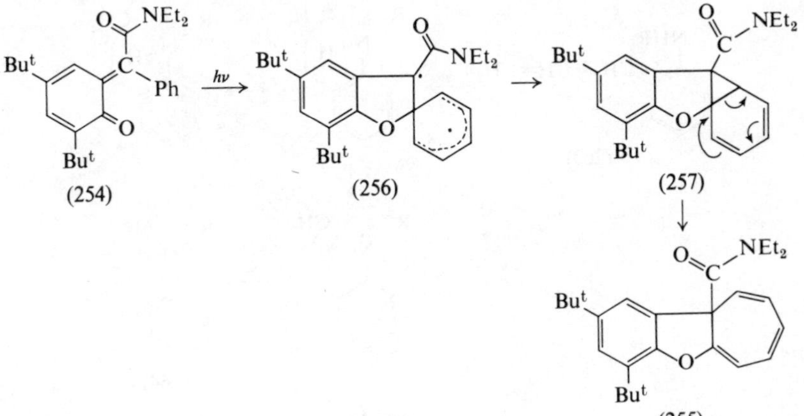

(254) (256) (257)

(255)

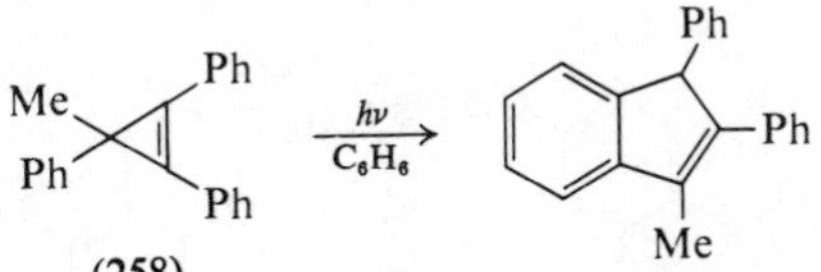

(258) (259)

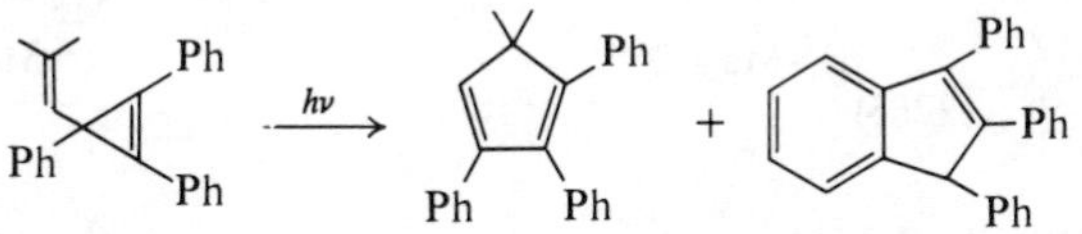

(260) (262) (261)

have been recently reported. The reactions of *o*-, *m*-, and *p*-arylanisoles and of *o*-allylanilines (263) have been studied in detail.[193] Irradiation of the latter compounds (R = H, Me, or Et) in benzene solution gives 40—80% of the 2-methylindolines (264). A similar procedure in methanol or ethanol solvent results in reaction of the alcohol with the ethylenic bond in competition to the formation of (264), and *NN*-dialkylamino derivatives of (263) yield 2-cyclo-propyl-*NN*-diarylanilines. Similarly the allylanisoles give cyclopropylanisoles, but the naphthalene analogues of the anisoles and t-amines are inert. The mechanism of the process is discussed in terms of intramolecular electron transfer (Scheme 13) and the singlet and triplet energies of the aromatic and ethylenic chromophores. The processes are neither quenched by piperylene nor sensitized by acetone, and it is suggested that the transformation starts in the singlet manifold of the arene. Photo-oxidative cyclization of morusin (265) to give high yields of morusin hydroperoxide (266) occurs in chloroform or benzene solution in either bright sunlight or from irradiation using a high-pressure lamp.[194] The

193 U. Koch-Pomeranz, H. Schmid, and H. J. Hansen, *Helv. Chim. Acta*, 1977, **60**, 768.
194 T. Nomura, T. Fakai, S. Yamada, and M. Katayanagi, *Chem. and Pharm. Bull. (Japan)*, 1977, **25**, 1155.

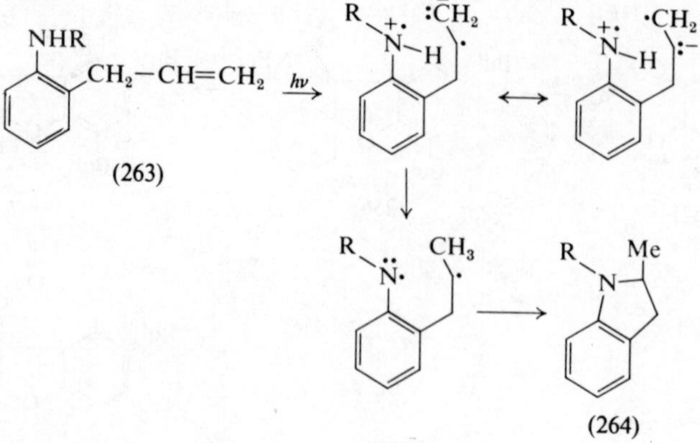

Scheme 13

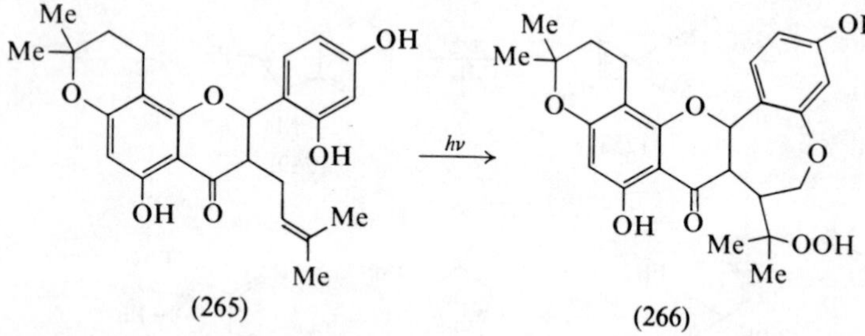

(265)   (266)

results show that both a free hydroxy-group at C(2) and an isolated ethylenic bond are necessary structural features for this photoreaction to occur.

Whereas thermally the 4-phenyl-3-(2-allylphenyl)sydnone (267) yields the 1,3-dipolar intramolecular cycloadduct (268) in 60% yield,[195] irradiation causes deoxygenation to give 30% of the cyclized product (269). The involvement of *o*-nitro groups in arene cyclizations has been previously noted by Döpp,[196] and is now reported for the photoreaction of the 2-nitrobenzylidene hydrazide (270) which yields a 1:1 mixture of (271) and (272).[197] The former product is described as resulting from a $[_{\pi}2_s + {_{\sigma}}2_s]$ cycloaddition between the nitro and N—H groups whereas the latter is considered to result from a $[_{\pi}2_a + {_{\sigma}}2_a]$ process. Involvement of a nitro group in an arene cyclization has also been noted in the case of photolysis of (273) which gives 14—18% yields of (274) and (275).[198] Cyclizations of the 2-(2,6-dichlorobenzylidene hydrazino)pyrimidine (276) and related compounds have been described; whereas under nitrogen the *anti* isomer of (276) simply yields the *syn* compound in benzene solution, in the

[195] H. Meier, H. Heimgartner, and H. Schmid, *Helv. Chim. Acta*, 1977, **60**, 1087.
[196] See, for example, D. Döpp and E. Brugger, *Chem. Ber.*, 1973, **106**, 2166.
[197] N. E. Alexandrou and J. Stephanidou-Stephanatou, *Chem. Chron.*, 1977, **6**, 357.
[198] S. Senda, K. Hirota, M. Suzuki, and M. Takahashi, *Chem. and Pharm. Bull.* (*Japan*), 1977, **25**, 563.

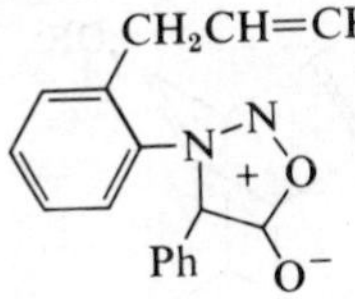

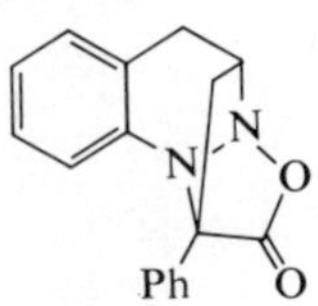

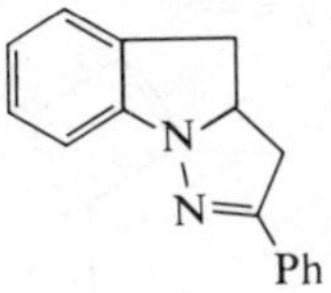

(267)   (268)   (269)

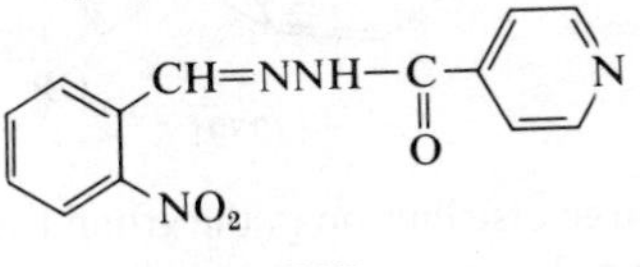

(270) 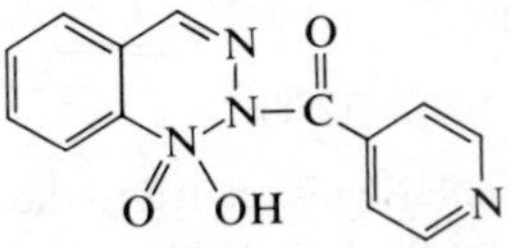   (271)

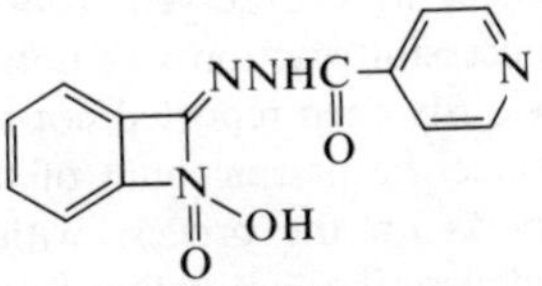

(272)

presence of oxygen, photosensitized auto-oxidation occurs to give the 3-aryl-1,2,4-triazolo[4,3-*a*]pyrimidine (277).[199]

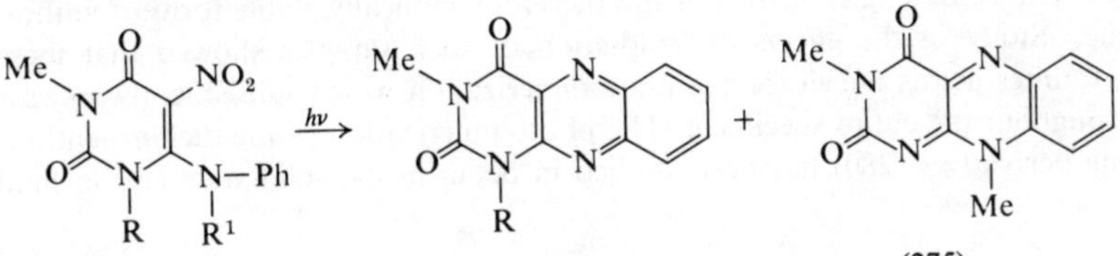

(273) R's = H or Me   (274)   (275)

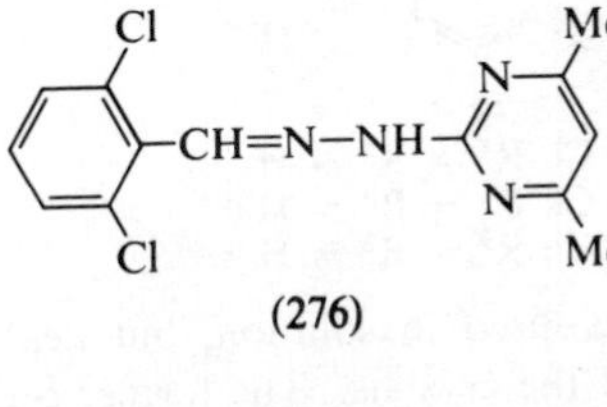

(276) 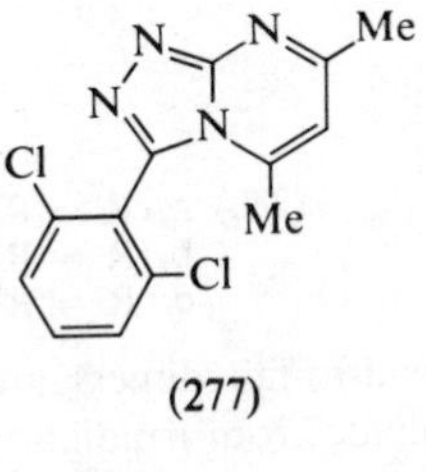   (277)

6 Dimerization Reactions

Sasse and co-workers have continued their studies of the various factors which may affect the photodimerization of naphthalene derivatives and report on photodimers related to that of 2-methoxynaphthalene.[200] Seven naphthalene compounds were examined and the dimer structures are based on the ring systems of (278) and/or (279). With the exception of the products from methyl 6-methoxy-2-naphthoate, head-to-tail dimers were observed exclusively, and this

199 T. Tsujikawa and M. Tatsuta, *Chem. Pharm. Bull.*, 1977, **25**, 3137; *Heterocycles*, 1977, **6**, 423.
200 T. Teitei, D. Wells, T. H. Spurling, and W. H. F. Sasse, *Austral. J. Chem.*, 1978, **31**, 85.

(278)　　　　　　　　(279)

specific orientation is related to the charge distribution in the ground state of the molecules on the reasonable assumption that pairwise association occurs before excitation. Although no dimers were observed from 2,6-dimethoxy and 2,3-dimethoxynaphthalenes, a successful reaction was obtained with 2-isopropoxy-napththalene which had previously been reported not to photodimerize.[201]

Although it is 110 years since the first account of the photodimerization of anthracene appeared,[202] reports on the process with the parent compound continue to be published and describe fascinating features of the reaction. A new phase of crystalline anthracene has been reported and its importance in the solid-state photodimerization discussed.[203] The new phase is readily produced at room temperature by application of a compressive force perpendicular to the (001) planes of single crystals of the thermodynamically stable form of anthracene. Study of the effects of irradiation of such samples showed that these crystallites act as nuclei for the photodimerization which subsequently spreads throughout the entire specimen. The photodimerization of the dichloroanthracene derivatives (280) has been studied in detail in the solid state and in fluid

(280) a, R = R^3 = Cl, R^1 = R^2 = H
　　　b, R = R^2 = Cl, R^1 = R^3 = H
　　　c, R = R^1 = Cl, R^2 = R^3 = H

solution.[204] Head-to-tail dimers are favoured in solution, but head-to-head isomers predominate from irradiation of the crystals. The former observation is rationalized in terms of the solvating molecules allowing the arene monomers to take up freely all possible mutual orientations. The preferred formation of head-to-tail dimers is considered to arise from the repulsion between the chlorine atoms in the head-to-head activated complex. In the solid state, such repulsion is compensated for by crystal forces, and it is demonstrated that in this phase the reaction occurs at disorder sites in which regions it is argued there is a high

[201] J. S. Bradshaw, N. B. Nielsen, and D. P. Rees, *J. Org. Chem.*, 1968, **33**, 259.
[202] J. Fritzsche, *J. Prakt. Chem.*, 1866, **101**, 377.
[203] G. M. Parkinson, M. J. Goringe, S. Ramdas, J. O. Williams, and J. M. Thomas, *J.C.S. Chem. Comm.*, 1978, 134.
[204] J. P. Desvergne, F. Chekpo, and H. Bouas-Laurent, *J.C.S. Perkin II*, 1978, 84.

degree of head-to-head mutual orientation of the monomers reflecting the ideal β-type crystal structure. Photodissociation of anthracene dimers to give sandwich excimers has been previously noted[205] and has been described again by two groups. Ferguson and Puza note that the retroprocess from dianthracene and related products yields appreciable amounts of excimers or exciplex products dependent upon temperature and solvent viscosity,[206] and Haller and Perkampus have irradiated thin films of anthracene at 366-nm light to give the dimer which in turn with 254-nm radiation at 77 K produces a sandwich excimer having characteristic luminescence.[207] Warming the mixture to room temperature without irradiation results in the formation of dianthracene, so the overall process is reversible.

For several years now there has been interest in bichromophoric molecules, and in particular those which involve polynuclear aromatic hydrocarbons as one or both of the two chromophores. De Schryver and co-workers have studied many examples of such systems and have now presented the details of much of their research in this area against a background of their previous studies and those of other workers.[70] It is shown that the intramolecular interactions occur even if the connecting chain of the chromophores contains more than the 'magical' three units. In the case where an intramolecular interaction is not possible, high molecular weight products may result from the corresponding intermolecular process. Such considerations are well demonstrated in the intra- and inter-molecular dimerization of molecules in which the two chromophores are anthracene moieties. Thus, for example, whereas intramolecular dimerization (both head-to-head and head-to-tail) of (281) occurs, irradiation of a 0.2M solution of alkylene bis-9-anthroate (282) in dichloromethane gives rise to

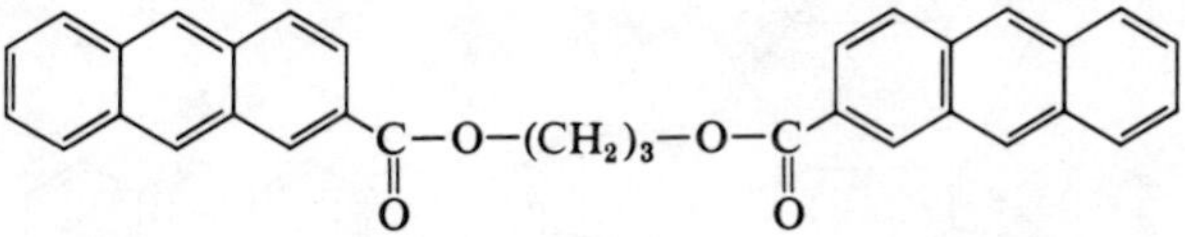

(281)

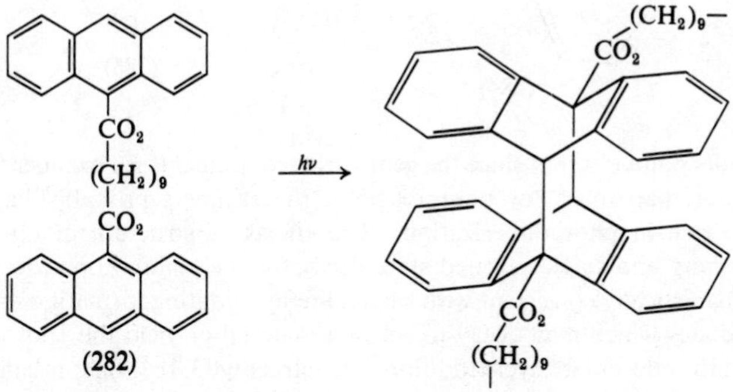

(282)

[205] See, for example, J. Ferguson and S. E. H. Miller, *Chem. Phys. Letters*, 1975, **36**, 635; P. C. Subudhi, N. Kanamaru, and E. C. Lim, *Chem. Phys. Letters*, 1975, **32**, 503.
[206] J. Ferguson and M. Puza, *Chem. Phys. Letters*, 1978, **53**, 215.
[207] W. Haller and H. H. Perkampus, *Ber. Bunsengesellschaft phys. Chem.*, 1978, **82**, 200.

polymerization by intermolecular $[4 + 4]$ π addition of the anthracene moieties. An intramolecular cyclodimerization of the above type has been used in the synthesis of a crown ether by irradiation of the bichromophoric system (283).[208] The isomer, although formed with a high quantum yield (0.32 and 0.27 in benzene and diethyl ether, respectively), readily reverts to (283) ($t_{\frac{1}{2}}$ *ca.* 3 min). When the photoreaction is carried out in the presence of lithium perchlorate, however, the 'cation-locked' isomer (284) is formed and is stable up to 206—210 °C. On the other hand, only the starting material is recovered from irradiation of (285).

The photoisomerization of linked athracenes is described as a demanding test of molecular geometry for intermediates in photodimerization.[209] A number of molecules of the type (286) have been studied, and it is noted that the fluorescence is as expected dependent upon X and is greater for $X = C_2H_4$ than for CH_2. Emission was observed from an excimer of (286) ($X = C_2H_4$, $R = H$), whereas irradiation of (286) with $X = CH_2$ or CHOH and R's = H or OMe resulted in only the normal anthracene fluorescence. The quantum yield for photodimerization of (286) was found to be dependent on the nature of the R-substituents and

$$9\text{-Anthryl}-X-CH_2(CH_2-X-CH_2)_2-CH_2-X-9\text{-Anthryl}$$
$$(283) \ X = O$$
$$(285) \ X = CH_2$$

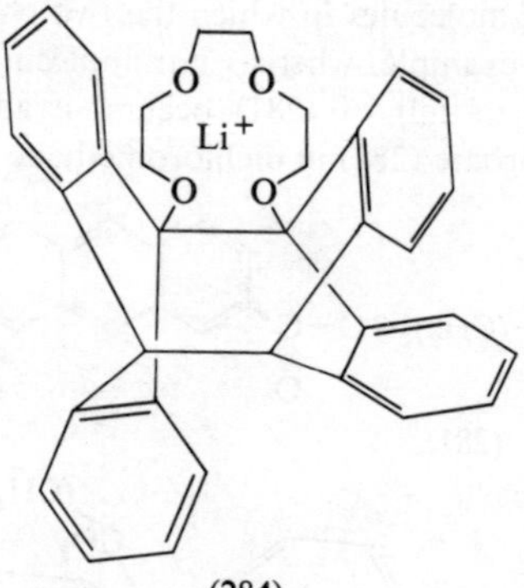

(284)

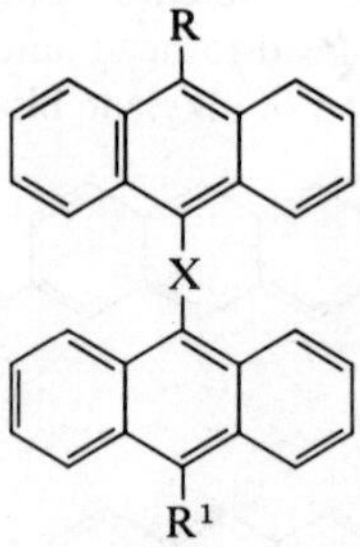

(286)

the authors point out that since the geometric requirements for excimer formation are greater than those for photoreaction, the excimers probably have a subordinate role in photodimerization. The effects of structure of (286) on the photoactivity and singlet excited state decay for the bichromophoric molecules are mechanistically consistent with singlet lifetime limiting formation of biradical intermediates which may decay to starting material or yield the photoisomer.

The intramolecular cycloaddition of tetraethyl[3,3](1,4)naphthaleno(9,10)-anthracenophane-2,2,15,15-tetracarboxylate (287) and tetraethyl[3,3]paracyclo-(9,10)anthracenophane-2,2,11,11-tetracarboxylate (288) has been reported.[210]

[208] J. P. Desvergne and H. Bouas-Laurent, *J.C.S. Chem. Comm.*, 1978, 403.
[209] W. R. Bergmark and G. Jones, *Nouv. J. Chem.*, 1977, **1**, 271.
[210] T. Shinmyozu, T. Inazu, and T. Yoshino, *Chem. Letters*, 1978, 405.

Compound (287) yields (289), and (290) similarly results from irradiation of (288). It is interesting to note here that heating (288) at 100 °C causes a Diels–Alder reaction and the formation of (291): this is specially noteworthy as it represents one of the few thermal addition reactions of the benzene nucleus.

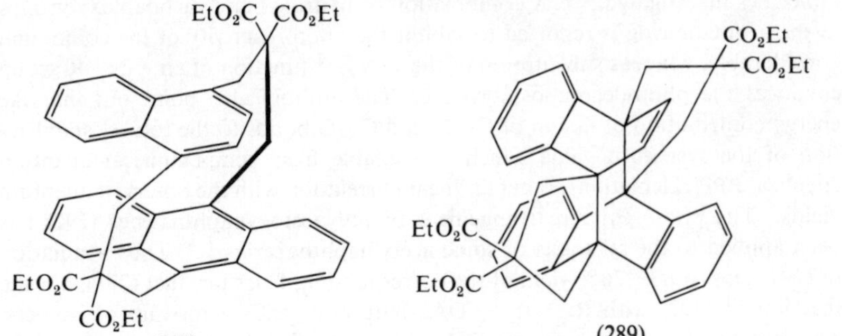

(287)

(289)

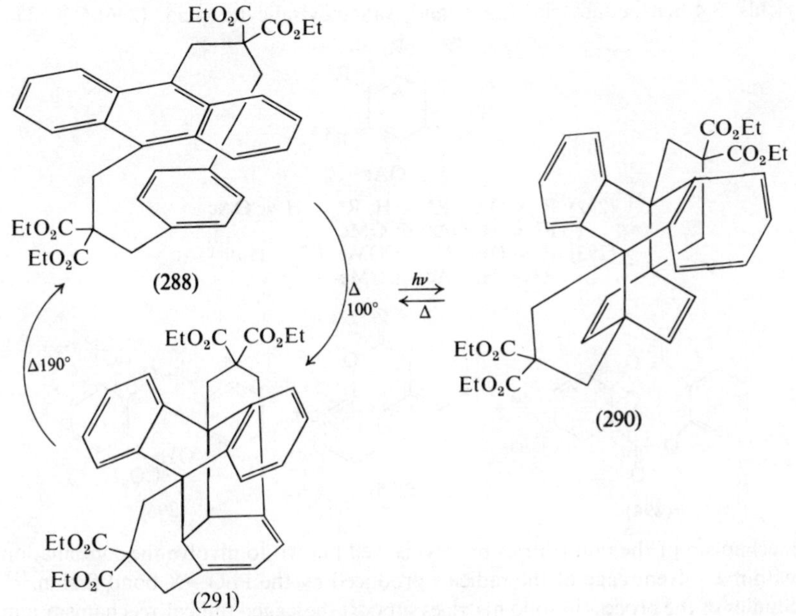

(288)

(290)

(291)

7 Lateral-nuclear Rearrangements

The theoretical treatment published last year[211] of the photo-Fries rearrangement has been extended, and the energy transfer tendency of any π-system to quench or sensitize the reaction of N-phenylurethane has been systematized in terms of the aggregate geometry of the urethane and added compound.[212]

[211] A. Mehlhorn, B. Schwenzer, and K. Schwetlick, *Tetrahedron*, 1977, **33**, 1489.
[212] A. Mehlhorn, B. Schwenzer, H. J. Brückner, and K. Schwetlick, *Tetrahedron*, 1978, **34**, 481.

The same workers have also commented on the structure-limited variation of the quantum yield of the photo-Fries from a quantitative study of the reaction with thirteen *N*-arylamides, *N*-arylurethanes, *N*-arylureas, and *o*-arylesters, and report a rather sensitive dependence of the quantum yield on the topology of the molecules investigated.[213] A combination of hydroxyl and carboalkoxy groups in the aromatic ring is reported to inhibit the photoreactivity of the compound considerably, whereas substitution of the —N—H function of an —N—R group enhances the photodecomposition rate. The authors also point out that the energy contribution of fission of C—N and C—O bonds to the total destabilization of the system, a value which is available from semi-empirical quantum chemical PPP calculations, is in fair linear correlation with the observed quantum yields. The photo-Fries rearrangement of polyacetoxynaphthalenes (292) has been applied to the synthesis of some acetylnaphthazarins.[214] Thus irradiation of (292) gives a 60—76% yield of the corresponding Fries product (293), but the reaction of (292) with R = R' = OAc naturally yields a mixture of isomers. The process has also been studied with difunctional molecules. Thus the disalicylide (294) on irradiation in benzene solution with a high-pressure lamp yields 3,4-benzocoumarin (295) and salicyloylsalicylic acid (296).[215] The

(292) R = OAc, R^1 = H, R^2 = H or OAc
       R^3 = H, OAc, or OMe
(293) R = OH, R^1 = COMe, R^2 = H or OAc
       R^3 = H, OAc, or OMe

(294)   (295)   (296)

mechanism of the photo-Fries process is well known to involve the combination within a solvent cage of the radicals produced by the PhO—X bond fission.[216] Studies of the process in solid matrices support the 'caged radical' mechanism, and as shown in Scheme 14 only a small rotation of the phenoxy radical intermediate is required to give the rearrangement from the phenyl ester.[217] The total quantum

[213] J. Stumpe, A. Mehlhorn, and K. Schwetlick, *J. Photochem.*, 1978, **8**, 1.
[214] C. Escobar, F. Farina, R. Martinez-Utrilla, and M. C. Paredes, *J. Chem. Res. (S)*, 1977, **11**, 266.
[215] V. D. Gaitonde and B. D. Hosangadi, *Indian J. Chem.*, 1977, **B15**, 83.
[216] J. W. Meyer and G. S. Hammond, *J. Amer. Chem. Soc.*, 1972, **94**, 2219; C. E. Kalmus and D. M. Hercules, *ibid.*, 1974, **96**, 449.
[217] S. K. L. Li and J. E. Guillet, *Macromolecules*, 1977, **10**, 840 J. E. Guillet, *Pure Appl. Chem.*, 1977, **49**, 249.

yield for reaction of poly(phenylacrylate) both in solution ($\Phi = 0.25$) and the solid phase ($\Phi = 0.23$ air) at temperatures below the glass transition of the polymer are comparable to those of phenyl acetate ($\Phi \simeq 0.4$) and there appears to be no dependence of quantum yield on the wavelength of excitation. The activation energy for formation of the *para* product from the poly(phenylacrylate) is slightly larger (1.8 kcal mol^{-1}) than that of the *ortho* isomer (1.2 kcal mol^{-1}) which is considered to reflect the slightly larger spatial requirements.

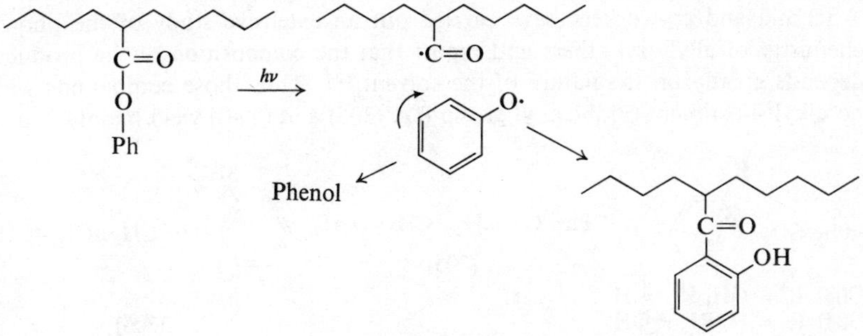

Scheme 14

A type of Fries reaction followed by hydrogen chloride elimination appears to be involved in the photoformation of the new heterocyclic ring system 5H[1]benzoselenino[2,3-*b*]pyridine (297) from the selenonicotinate (298).[218] The reaction proceeds in 25% yield in benzene solution under nitrogen, but the product is accompanied by 57% of di(4-methylphenyl)diselenide. The first telluro-photo-Fries process has been described and involves the irradiation of (299) under conditions similar to those described in ref. 218 to yield various products, including (300) and (301).[219] Irradiation of phenyl benzenesulphonate

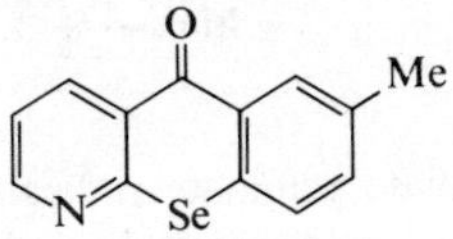

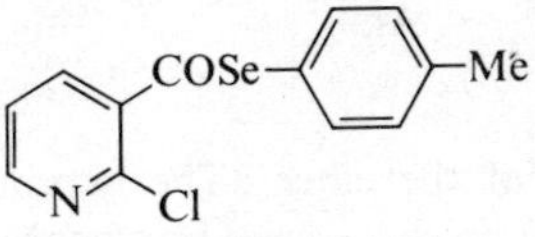

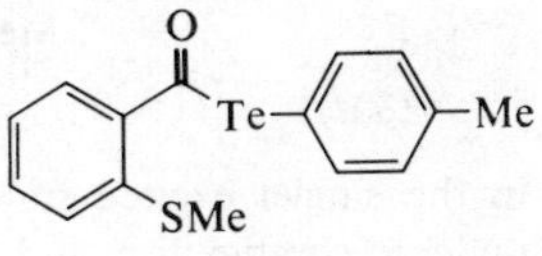

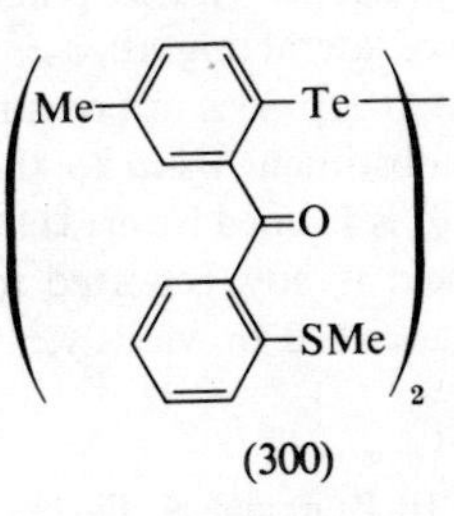

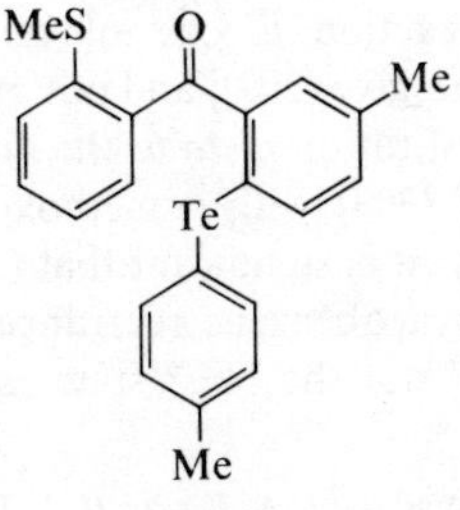

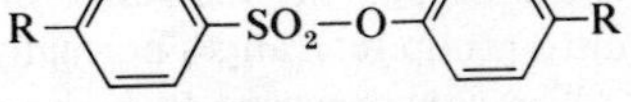

[218] B. Pakzad, K. Praefcke, and H. Simon, *Angew. Chem.*, 1977, **89**, 329.
[219] G. Höhne, W. Lohner, K. Praefcke, U. Schulze, and H. Simon, *Tetrahedron Letters*, 1978, 613.

(302a) in ethanol solution yields *o*- and *p*-hydroxyphenyl phenylsulphones (303) and (304) as well as phenol, small amounts of diphenyl ether, and much polymeric material.[220] Quenching and sensitization studies of the reaction indicate the involvement of an excited singlet state or a short-lived triplet. The methyl group is an effective blocking substituent, for irradiation of (302b) only yields the *ortho*- and *para*-cresol. The rearrangment proceeds predominantly by an intramolecular pathway since mixed irradiations of (302a and b) yield essentially non-mixed products.

Schmid and co-workers have carried out an extensive study of the photochemistry of allyl aryl ethers and report that the composition of the products depends greatly on the nature of the solvent.[221] Thus those compounds with no alkyl substituent on the allyl group [*i.e.* (305) and (306)] yield mainly 2- and

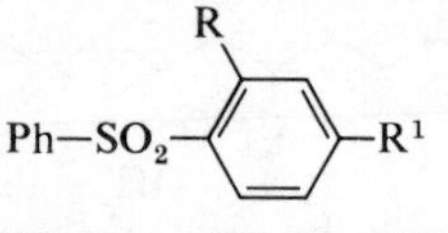

$$Ph-SO_2-\underset{R^1}{\overset{R}{\bigcirc}}$$

(303) R = OH, R^1 = H
(304) R = H, R^1 = OH

$$Ph-O-CH_2-CH = CH_2$$
(305)

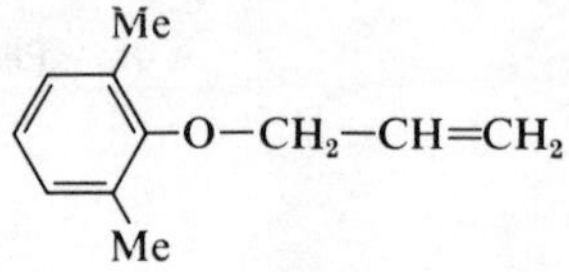

(306)

4-allylated phenols in protic solvents, but in hydrocarbon media, the isomeric 3-allylated compound is formed in significant amounts. The results from the irradiation of these compounds and of those with alkylated allyl groups [*e.g.* (307)] suggest that all products result from homolytic cleavage of the C—O bond

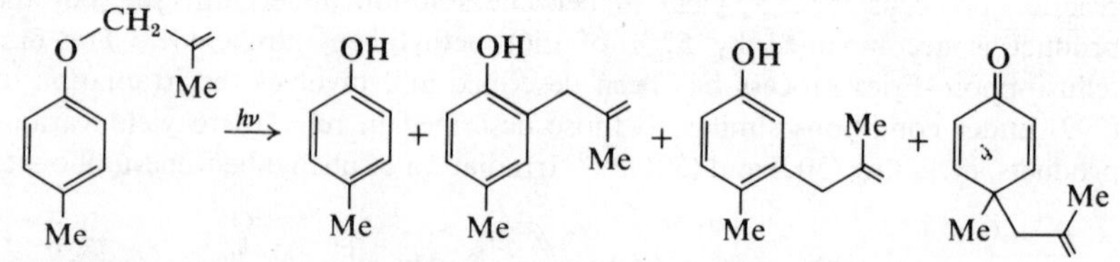

(307)

in the singlet excited state of the ether. The intermediate geminate radicals undergo combination and no products arise by a concerted mechanism. In such compounds as (308), however, Ph—OR bond cleavage appears to result on irradiation, and the overall reaction is one of intramolecular nucleophilic substitution (the photo-Smiles) to give (309), and not one of lateral migration.[222] It is interesting to note that substitution *meta* to the nitro group does not occur, and although the derivative of (308) with a methoxy substituent *meta* to the nitro group rearranges normally, it is significant that (310) is formed from (311).

The azoxybenzene to hydroxyazobenzene rearrangement is now reported to occur *via* the $^1n\pi^*$ state[223] and not the $^1\pi\pi^*$ state as described previously.[224]

[220] Y. Ogata, K. Takagi, and S. Yamada, *J.C.S. Perkin II*, 1977, 1629.
[221] H. R. Waespe, H. Heimgartner, H. Schmid, H. J. Hansen, H. Paul, and H. Fischer, *Helv. Chim. Acta*, 1978, **61**, 406.
[222] K. Mutai, S. Kanno, and K. Kobayashi, *Tetrahedron Letters*, 1978, 1273.
[223] N. J. Bunce, J. P. Schoch, and M. C. Zerner, *J. Amer. Chem. Soc.*, 1977, **99**, 7986.
[224] R. Tanikaga, *Bull. Chem. Soc. Japan*, 1968, **41**, 1664, 2151.

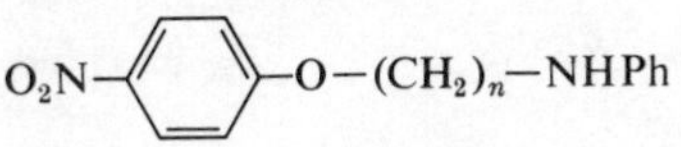

O_2N—⟨benzene ring⟩—O—$(CH_2)_n$—$NHPh$

(308) n = 2 to 5

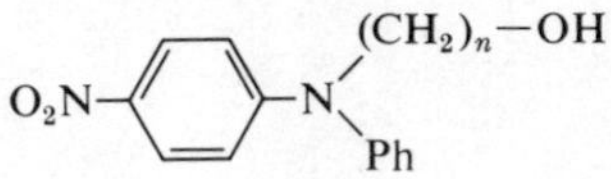

O_2N—⟨benzene ring⟩—N

$(CH_2)_n$—OH / Ph

(309)

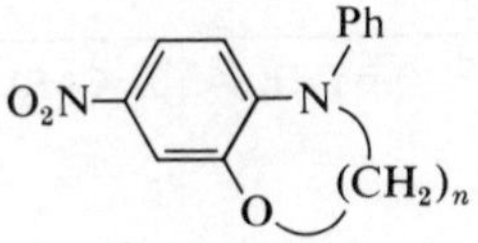

O_2N—⟨benzene ring⟩—N

Ph / $(CH_2)_n$ / O

(310)

O_2N—⟨benzene ring⟩—OMe / O—$(CH_2)_n$—$NHPh$

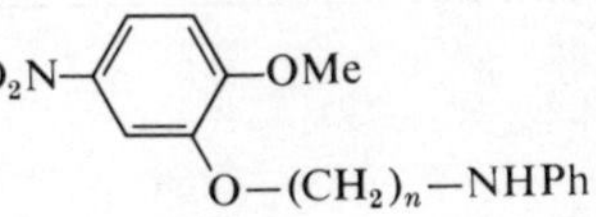

(311)

The theoretical model for the reaction predicts an electrophilic role for oxygen during the rearrangement, and experiments with substituted azoxybenzenes support this proposal. Thus among isomeric compounds with different substituents on the two phenyl rings, the more efficient reaction is that in which oxygen migrates to the ring bearing the more electron-donating substituents.

5
Photo-reduction and -oxidation

BY H. A. J. CARLESS

1 Conversion of C=O into C—OH

There is still much that remains to be understood in the apparently simple photoreduction of ketones. Typical examples are provided by the photoreduction of the heterocyclic analogues of the cyclohexanone (1a) in methanol or propan-2-ol, which give all the products shown in Scheme 1.[1] For the oxygen and nitrogen heterocycles (1b) and (1c), the symmetrical pinacols (2b) and (2c) are the main products, whereas the sulphur analogue (1d) gives mainly alcohol (3d). The cyclohexanone (1a) produces a large amount of mixed pinacol (4a) on

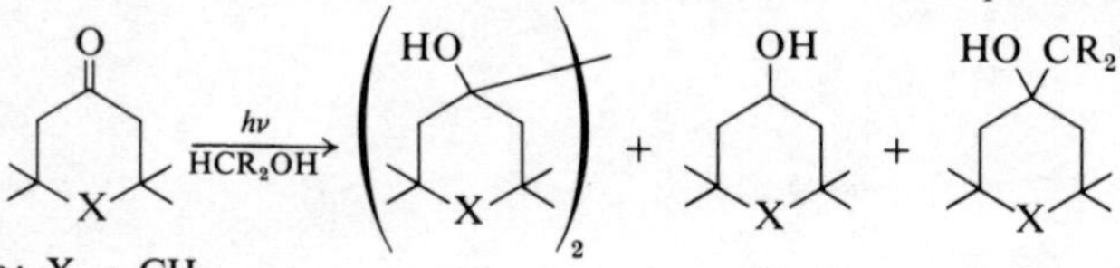

(1) a; X = CH$_2$
    b; X = O
    c; X = NH
    d; X = S

R = H or Me

Scheme 1

irradiation in methanol, but alcohol (3a) is the predominant product on irradiation in propan-2-ol. The products can easily be understood in terms of hydrogen abstraction by excited carbonyl compound from alcohol, followed by various radical reactions. However, it is not possible at present to rationalize the relative importance of the different pathways, which reflect a delicate balance of radical reactivities. Irradiation of the pyranos-3-ulose derivatives (5) in methanol leads to a similar reduction to yield mainly the mixed pinacols (42—55%), thus presenting another route to branched-chain sugars.[2]

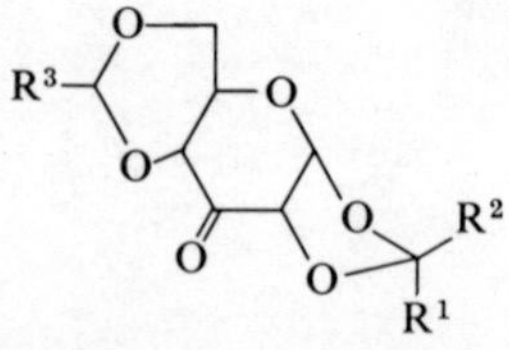

(5) a; R^1 = R^2 = R^3 = Me
    b; R^1 = H, R^2 = Et, R^3 = Ph

[1] N. Nemeroff, M. M. Joullié, and G. Preti, *J. Org. Chem.*, 1978, **43**, 331.
[2] P. M. Collins, V. R. N. Munasinghe, and N. N. Oparaeche, *J.C.S. Perkin I*, 1977, 2423.

The complexities of ketone photoreduction are again highlighted by a detailed kinetic study of the photoreduction of 3,3,5-trimethylcyclohexanone in propan-2-ol, which yields 3,3,5-trimethylcyclohexanol and acetone as the predominant products.[3] In this case, the quantum yield of photoreduction increases with increasing initial ketone concentration, a phenomenon which cannot be explained on a simple mechanism of reduction. The authors can analyse their results in terms of reduction from both propan-2-ol and a ketone–propan-2-ol complex, although the nature of this complex is left for interpretation. A daunting problem still remains, that the rate constant for hydrogen abstraction from the complex needs to be *ca.* 2×10^3 times that from alcohol alone.

An interesting example of a highly selective solid-state photoreaction occurs on irradiation of the 1 : 2 complex of acetone and deoxycholic acid (6), which produces the adducts (7)—(9), as shown in Scheme 2.[4] The *X*-ray crystal structure of the complex gives a detailed picture of the geometric requirements for hydrogen abstraction by the acetone from (6), followed by combination with

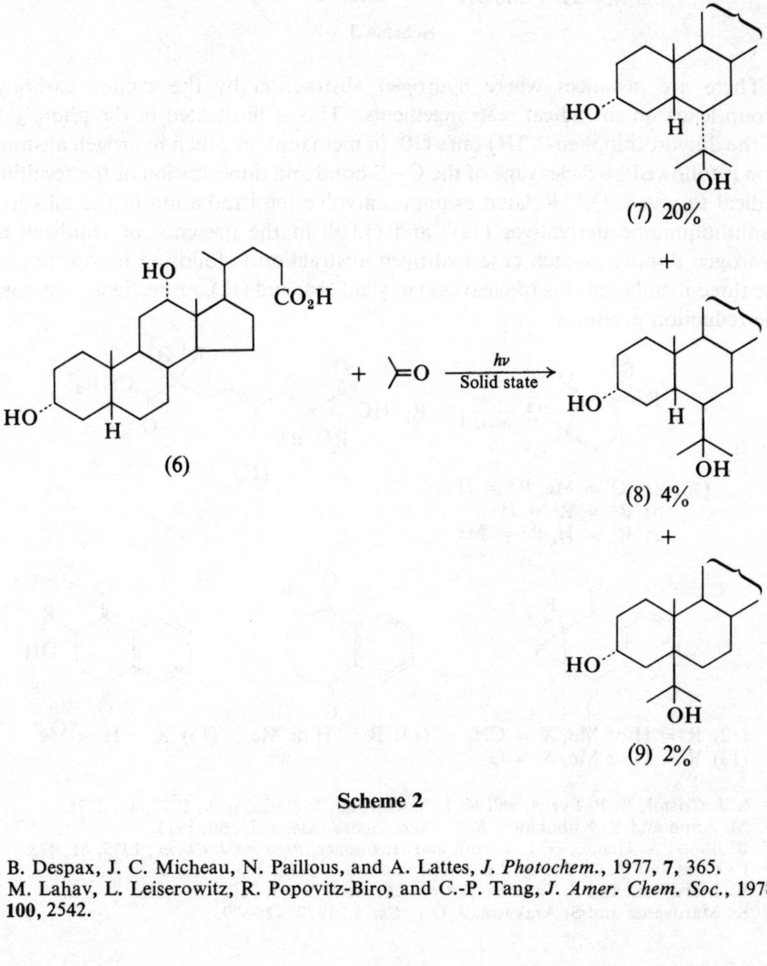

(7) 20%

+

(8) 4%

+

(9) 2%

Scheme 2

[3] B. Despax, J. C. Micheau, N. Paillous, and A. Lattes, *J. Photochem.*, 1977, 7, 365.
[4] M. Lahav, L. Leiserowitz, R. Popovitz-Biro, and C.-P. Tang, *J. Amer. Chem. Soc.*, 1978, 100, 2542.

the ketyl radical (Me$_2$ĊOH), which results only in functionalization of the C-5 and C-6 positions of (6). Irradiation of substituted ureas in acetone solution leads to hydrogen abstraction followed by radical combinations producing ureido-alcohols and bis-ureas.[5] The quantum yield of methane formation on irradiation of acetone in water ($\Phi = 0.06$) is more than an order of magnitude greater than in heptane ($\Phi = 0.003$) or methanol ($\Phi = 0.004$) as solvent.[6] This is because reversible hydrogen abstraction becomes dominant in the latter two solvents (Scheme 3). The slow tautomerization of the resulting enol of acetone[7] back to its original ketonic form results in little overall reaction.

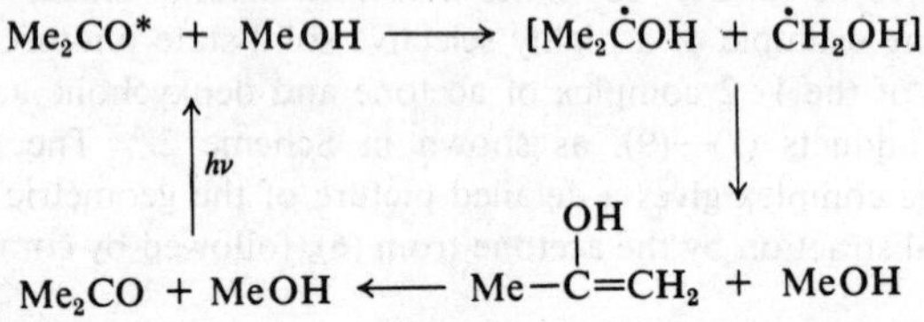

$$\text{Me}_2\text{CO}^* + \text{MeOH} \longrightarrow [\text{Me}_2\dot{\text{C}}\text{OH} + \dot{\text{C}}\text{H}_2\text{OH}]$$

$$\text{Me}_2\text{CO} + \text{MeOH} \longleftarrow \text{Me}-\overset{\overset{\displaystyle\text{OH}}{|}}{\text{C}}=\text{CH}_2 + \text{MeOH}$$

Scheme 3

There are instances where hydrogen abstraction by the excited carbonyl group leads on to radical rearrangements. This is illustrated in the photolysis of the dihydrothiophen-3(2H)-ones (10) in methanol, in which hydrogen abstraction is followed by β-cleavage of the C—S bond and dimerization of the resulting radical to give (11).[8] Related examples involve the irradiation of the dihydronaphthoquinone derivatives (12)[9] and (13)[10] in the presence of xanthene as hydrogen donor; in each case hydrogen abstraction is followed by opening of the three-membered ring (β-cleavage) to yield (14) and (15), respectively, amongst the reduction products.

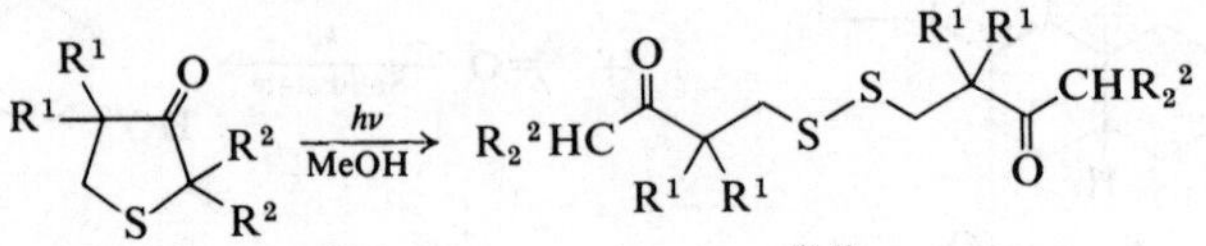

(10) a; R^1 = Me, R^2 = H
     b; R^1 = R^2 = H
     c; R^1 = H, R^2 = Me

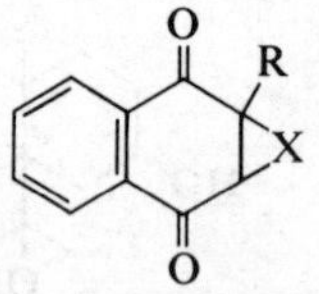

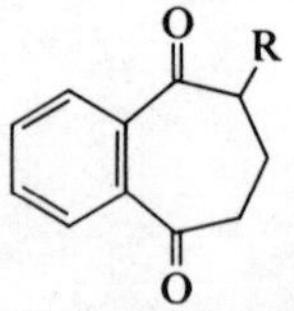

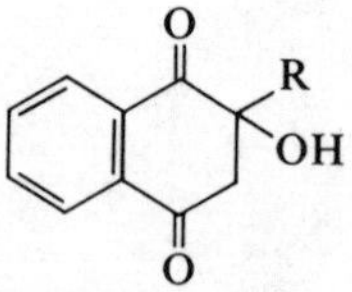

(11)

(12) R = H or Me, X = CH$_2$ (14) R = H or Me   (15) R = H or Me
(13) R = H or Me, X = O

[5] S. J. Cristol, R. P. Evans, and K. L. Lockwood, *J. Org. Chem.*, 1977, **42**, 2378.
[6] M. Anpo and Y. Kubokawa, *Bull. Chem. Soc. Japan*, 1977, **50**, 1913.
[7] B. Blank, A. Henne, G. P. Laroff, and H. Fischer, *Pure Appl. Chem.*, 1975, **41**, 475.
[8] P. Yates and Y. C. Toong, *J.C.S. Chem. Comm.*, 1978, 205.
[9] K. Maruyama and S. Tanioka, *J. Org. Chem.*, 1978, **43**, 310.
[10] K. Maruyama and S. Arakawa, *J. Org. Chem.*, 1977, **42**, 3793.

The photolysis of benzophenone carboxylic esters of long chain alkanols leads to intramolecular hydrogen abstraction and C—C coupling at various positions along the alkanol carbon chain. Breslow and co-workers[11] have now published a full account of their use of this technique to probe the conformation of alkyl chains in solution. Calculated values of rate constants of hydrogen abstraction by triplet $n\pi^*$ benzophenone from various substrates (principally amines) have been compared with the experimental values.[12] The activation energy for hydrogen abstraction by triplet benzophenone from ethanol is 2.2 kcal mol^{-1} in the low temperature range 93—131 K; at slightly higher temperatures, the resulting ketyl radical is formed by escape from the solvent cage with an energy of activation of 12 kcal mol^{-1} in the range 125—155 K.[13] Photoreduction of benzophenone in water occurs by electron transfer from hydroxide ion to $n\pi^*$ excited ketone.[14] The photoreduction of 2-benzoylbenzoic acid in propan-2-ol produces equal quantities (measured by ^{13}C n.m.r.) of the racemic and *meso*-diastereoisomers of the pinacol-like product.[15] There are several reports of the photoreduction of benzophenone in alcohols leading to hydrogen abstraction, followed by useful reactions of the alcohol-derived radical. For example, the benzophenone-induced photo-addition of methanol to carbohydrate enones can be stereo- and regio-specific.[16] Benzophenone-induced addition of species such as diols, aldehydes and ketals is also possible.[17] Hydrogen abstraction and combination occurs entirely from the 2-position of dioxolan derivatives (16),

$$O \underset{2}{\diagup} O$$

$$CH_2R$$

(16) a; R = Et
     b; R = OH

giving a 'masked' acylation of enones. Irradiation of carbonyl compounds in reducing solvents leads to *cis–trans* isomerization of pent-3-en-2-one by a radical chain process,[18] whilst irradiation of hexafluorobenzene in methanol in the presence of benzophenone leads to hydroxymethyl radical ($\dot{C}H_2OH$) addition and substitution of hexafluorobenzene.[19]

Following on the work reported last year of Schuster and co-workers[20] on the photoreduction of Michler's ketone [4,4′-bis(dimethylamino)benzophenone], Brown and Porter[21] have examined the triplet lifetime of Michler's ketone in cyclohexane, propan-2-ol, and ethanol, and have found it to vary only slightly.

[11] R. Breslow, J. Rothbard, F. Herman, and M. L. Rodriguez, *J. Amer. Chem. Soc.*, 1978, **100**, 1213.
[12] G. D. Abbott and D. Phillips, *Mol. Photochem.*, 1977, **8**, 289.
[13] H. Murai, M. Jinguji, and K. Obi, *J. Phys. Chem.*, 1978, **82**, 38.
[14] S. Hashimoto, M. Yasuda, and M. Mouri, *Nippon Kagaku Kaishi*, 1978, 79.
[15] M. Pfau, F. Gobert, J.-C. Gramain, and M.-F. Lhomme, *J.C.S. Perkin I*, 1978, 509.
[16] B. Fraser-Reid, N. L. Holder, D. R. Hicks, and D. L. Walker, *Canad. J. Chem.*, 1977, **55**, 3978.
[17] B. Fraser-Reid, R. C. Anderson, D. R. Hicks, and D. L. Walker, *Canad. J. Chem.*, 1977, **55**, 3986.
[18] A. Rioual, A. Deflandre, and J. Lemaire, *Canad. J. Chem.*, 1977, **55**, 3915.
[19] M. Zupan, B. Šket, and B. Pahor, *Tetrahedron Letters*, 1977, 4541.
[20] D. I. Schuster, M. D. Goldstein, and P. Bane, *J. Amer. Chem. Soc.*, 1977, **99**, 187.
[21] R. G. Brown and G. Porter, *J.C.S. Faraday I*, 1977, **73**, 1569.

They conclude that a lowest charge-transfer triplet state exists in either ethanol or cyclohexane, but that thermal population of a slightly higher ($\sim$2000 cm^{-1}) $n\pi^*$ triplet state in cyclohexane accounts for the much increased rate of hydrogen abstraction found in this solvent compared with ethanol.

Several studies of quinone photoreduction have appeared. E.s.r. spectra obtained in a flow system on photoreduction of *p*-benzoquinone (Q) in ethanol provide evidence for both semiquinone radicals ($\dot{Q}$H) and radical anions (Q$^{\cdot-}$).[22] In contrast with the currently accepted view, the authors propose no effective equilibrium between these two species; instead, they believe that the radical $\dot{Q}$H arises by hydrogen abstraction of excited Q from hydroquinone (QH$_2$) formed as a photolysis product. However, studies by Russian workers[23] of the photoreduction of *p*-benzoquinone in the presence of propan-2-ol do not support this proposal. The photoreduction of *p*-benzoquinone in aqueous solutions has been investigated,[24] and electron transfer from alcohol to excited quinone has been found to be important in the photoreaction of 2,6-diphenyl-*p*-benzoquinone with alcohols.[25] E.s.r. techniques have also been used to examine the photoreduction of tetrachloro-*p*-benzoquinone in the presence of phenols[26] or ethanol.[27] There is evidence that the photoreduction of duroquinone in the presence of tertiary amines, which was thought to occur by electron transfer to yield quinone radical anions in polar solvents, may in fact proceed in the case of triethylamine by hydrogen abstraction followed by deprotonation of the semi-quinone radical.[28] Wan and his group[29] have used the combined techniques of e.s.r. and CIDEP to investigate the photoreduction of t-butyl-*p*-benzoquinones by propan-2-ol and phenols. 2-t-Butyl-*p*-benzoquinone gives the two semi-quinone radicals (17) and (18) in the primary abstraction process but, curiously,

(17)   (18)

CIDEP shows that formation of radical (18) is five to seven times faster than that of (17), the thermodynamically more stable radical. Elliot and Wan[30] have also used CIDEP to determine rate constants of triplet quenching of some *p*-quinones caused by hydrogen abstraction from propan-2-ol, phenol, 2-methylphenol, and pentachlorophenol. For a given quinone, propan-2-ol is generally about two orders of magnitude less effective a quencher than the phenols.

[22] Y. Kambara and H. Yoshida, *Bull. Chem. Soc. Japan*, 1977, **50**, 1367.
[23] V. M. Kuznets, D. N. Shigorin, A. L. Buchachenko, G. A. Val'kova, A. Z. Yankelevich, and N. N. Shapet'ko, *Izvest. Akad. Nauk S.S.S.R., Ser. khim.*, 1978, 62.
[24] H. Thielemann, *Z. Chem.*, 1977, **17**, 109.
[25] V. M. Kuznets, D. N. Shigorin, and A. L. Buchachenko, *Doklady Akad. Nauk S.S.S.R.*, 1977, **234**, 1112.
[26] B. B. Adeleke and J. K. S. Wan, *Spectroscopy Letters*, 1977, **10**, 871.
[27] Y. Kambara, H. Yoshida, and B. Ranby, *Bull. Chem. Soc. Japan*, 1977, **50**, 2554.
[28] E. Amouyal and R. Bensasson, *J.C.S. Faraday I*, 1977, **73**, 1561.
[29] T. Foster, A. J. Elliot, B. B. Adeleke, and J. K. S. Wan, *Canad. J. Chem.*, 1978, **56**, 869.
[30] A. J. Elliot and J. K. S. Wan, *J. Phys. Chem.*, 1978, **82**, 444.

Quantum yields have been determined for the photoreduction of 9,10-anthraquinone and some derivatives on irradiation in alcohols and amines.[31] Japanese workers[32] have reinvestigated a report[33] that the photoreduction of 2-piperidinoanthraquinone (19) occurs from the intramolecular charge-transfer state. Irradiation of (19) in a propan-2-ol–water mixture at long wavelengths

(19)

($\lambda > 420$ nm, which excites only the charge-transfer band) does not lead to efficient production of the quinone radical anion ($\Phi = 0.003$), whereas reaction occurs efficiently ($\Phi = 1.0$) on irradiation at 313 or 365 nm. The triplet quencher cyclohexa-1,3-diene does inhibit the reaction, and consequently the authors[32] believe an upper excited $n\pi^*$ triplet state is responsible for reduction.

$$RCO_2H + CH_3OH \xrightarrow{h\nu} R\dot{C}(OH)_2 + \dot{C}H_2OH \qquad (1)$$

Photoreduction [reaction (1)] is one of the major processes which are detected by e.s.r. spectroscopy on irradiation of aliphatic carboxylic acids (formic, acetic, propionic, and malonic acids) in methanol.[34] A CIDEP study of the direct and the benzophenone-sensitized photoreduction of biacetyl with triethylamine shows the complexity of the sensitized system.[35] The primary radical $Me\dot{C}HNEt_2$, which results from the rapid reaction of triplet benzophenone with triethylamine, transfers its polarization *via* reaction with ground-state biacetyl to the secondary biacetyl radical anions. CIDNP results on the photoreduction of pyruvic acid ($MeCOCO_2H$) by ethanol, propan-2-ol, and acetaldehyde in acetonitrile solution give some detailed information on various radical reactions which follow the initial hydrogen abstraction by excited $n\pi^*$ triplet pyruvic acid from the substrate.[36] Thus, hydrogen exchange is noted between ketyl radicals and ground-state pyruvic acid, and as the concentration of ethanol is lowered, competing abstraction by excited pyruvic acid from product acetaldehyde becomes important. A CIDNP study has been made of the photoreduction of furil (20) by pentachlorophenol.[37]

(20)

[31] R. Mitzner, D. Frosch, and H. Dorst, *Z. phys. Chem. (Leipzig)*, 1977, **258**, 845.
[32] H. Inoue, K. Kawabe, N. Kitamura, and M. Hida, *Chem. Letters*, 1977, 987.
[33] A. K. Davies, J. F. McKellar, and G. O. Phillips, *Proc. Roy. Soc.*, 1971, **A323**, 69.
[34] T. Kaiser, L. Grossi, and H. Fischer, *Helv. Chim. Acta*, 1978, **61**, 223.
[35] K. A. McLauchlan, R. C. Sealy, and J. M. Wittmann, *J.C.S. Faraday II*, 1977, **73**, 926.
[36] G. L. Closs and R. J. Miller, *J. Amer. Chem. Soc.*, 1978, **100**, 3483.
[37] H. M. Vyas and J. K. S. Wan, *Mol. Phys.*, 1977, **34**, 887.

2 Reduction of Nitrogen-containing Compounds

There have been several recent publications on the photoreduction of aromatic nitro-compounds. For example, the photoreduction of 4-substituted nitro-benzenes (21) and (22) in propan-2-ol has been examined by CIDNP techniques.[38] Polarized spectra are observed when the substituent is electron-withdrawing, as in (21), but not when it is electron-releasing, as in (22). These results are

(21) R = CN, NO$_2$, or CHO
(22) R = H, Me, NH$_2$, or OMe

qualitatively interpreted in terms of the increased rate of hydrogen abstraction by triplet nitrobenzene as the electron-withdrawing power of the substituent increases. CIDNP Studies of the hydrogen abstraction by nitroaromatics from phenols have also been reported.[39] Nanosecond flash photolysis has been used to determine rate constants for hydrogen abstraction of the triplet excited states of 1,2- and 1,8-dinitronaphthalenes from tributyltin hydride (3.4×10^8 and 1.6×10^8 l mol^{-1} s^{-1}, respectively),[40] and of triplet 3-nitropyrene from diphenylamine (2.9×10^9 l mol^{-1} s^{-1}).[41] Some synthetic applications of the photoreduction of nitro-compounds are evident in the selective hydrogen abstraction by photoexcited nitrobenzenes from the benzylic positions of methoxy-substituted aromatics,[42] and in the reductive cyclization of the nitro-uracil derivatives (23) to produce the pyrrolo-derivatives (24) in low yield (8—16%).[43] The photoreduction of 1- and 2-nitroanthraquinones in alcohols is reported to give the corresponding amines, *via* the hydroxylamines.[44]

Further examples of the deoxygenation of heterocyclic aromatic *N*-oxides have been noted in the deoxygenation on irradiation of 2,3-diarylquinoxaline

(23)   (24)
R = H, Me, or OMe

[38] N. Levy and M. D. Cohen, *Mol. Photochem.*, 1977, **8**, 155.
[39] K. A. Muszkat, *Chem. Phys. Letters*, 1977, **49**, 583.
[40] C. Capellos and K. Suryanarayanan, *Internat. J. Chem. Kinetics*, 1977, **9**, 399.
[41] R. Scheerer and A. Henglein, *Ber. Bunsengesellschaft phys. Chem.*, 1977, **81**, 1234.
[42] J. Libman, *J.C.S. Chem. Comm.*, 1977, 868.
[43] S. Senda, K. Hirota, M. Suzuki, and M. Takahashi, *Chem. and Pharm. Bull. (Japan)*, 1977, **25**, 563.
[44] A. V. El'tsov, O. P. Studzinskii, Yu. K. Levental, and M. V. Florinskaya, *Zhur. org. Chim.*, 1977, **13**, 1061.

1,4-dioxides in methanol,[45] and of pteridine N-oxides in aqueous solutions.[46] The photochemically-induced deoxygenation of 2-nitrophenazine 10-oxide by triethylamine occurs quite efficiently in acetonitrile or methanol ($\Phi \simeq 0.2$).[47] A new procedure for the deoxygenation of aromatic N-oxides consists of their irradiation in dichloromethane solution in the presence of triphenylphosphine (1.2—5 mole equivalents).[48]

Photoreduction of the C=N chromophore is a process analogous to the photoreduction of carbonyl compounds. Thus, irradiation of the N-acetylimine (25) in propan-2-ol [reaction (2)] has been studied by e.s.r.[49] An intermediate

$$Ph_2C{=}NCOMe + Me_2CHOH \xrightarrow{\ h\nu\ } Ph_2CHNHCOMe + Me_2CO \quad (2)$$
(25)

radical which was detected has been assigned the structure $MeCH(OH)\overset{\cdot}{C}H_2$, and this was taken as evidence for the exceptional β-hydrogen abstraction of the excited imine from propan-2-ol. A β-hydrogen abstraction by the excited C=N bond of oxazole (26) from ethanol was a plausible route to explain the unusual photomethylation of (26) in ethanol to give (27), but ^{14}C-labelling experiments

(26) R = H
(27) R = Me

have now provided evidence for a pathway involving α-hydrogen abstraction by excited imino-group from the alcohol, followed by rearrangement and elimination.[50] Hydrogen abstraction by the imine nitrogen atom also occurs on irradiation of arylimines (28) in propan-2-ol, followed by dimerization of the resultant radical.[51] Photolysis of the azaquinones (29) in acetonitrile in the presence of a hydrogen donor leads to reduction of the C=N bond, besides the formation of other products.[52]

As a model system for solar energy conversion, the photoreduction of the paraquat dication (1,1'-dimethyl-4,4'-bipyridylium) from cysteine as donor in micelles is sensitized by a surfactant derivative of tris(2,2'-bipyridyl)ruthenium-(II).[53] Japanese workers have investigated the photoreduction of acridine in

[45] A. A. Jarrar, *J. Heterocyclic Chem.*, 1978, **15**, 177.
[46] F. L. Lam and T.-C. Lee, *J. Org. Chem.*, 1978, **43**, 167.
[47] S. Pietra, G. F. Bettinetti, A. Albini, E. Fasani, and R. Oberti, *J.C.S. Perkin II*, 1978, 185.
[48] C. Kaneko, M. Yamamori, A. Yamamoto, and R. Hayashi, *Tetrahedron Letters*, 1978, 2799.
[49] N. Toshima, K. Aoki, and H. Hirai, *Bull. Chem. Soc. Japan*, 1977, **50**, 1235.
[50] M. Maeda, M. Kawashige, and M. Kojima, *J.C.S. Chem. Comm.*, 1977, 511.
[51] K. N. Mehrotra and B. P. Giri, *Indian J. Chem.*, 1977, **15B**, 1106.
[52] K. Maruyama, T. Iwai, T. Otsuki, Y. Naruta, and Y. Miyagi, *Chem. Letters*, 1977, 1127.
[53] K. Kalyanasundaram, *J.C.S. Chem. Comm.*, 1978, 628.

$$R^1 \diagdown C=NCHR^3Ph$$
$$R^2 \diagup$$

(28) a; R^1 = Ph, R^2 = R^3 = H
    b; R^1 = Ph, R^2 = Me, R^3 = H
    c; R^1 = p-MeOC$_6$H$_4$, R^2 = H, R^3 = Ph
    d; R^1 = p-ClC$_6$H$_4$, R^2 = H, R^3 = Ph
    e; R^1 = p-O$_2$NC$_6$H$_4$, R^2 = H, R^3 = Ph

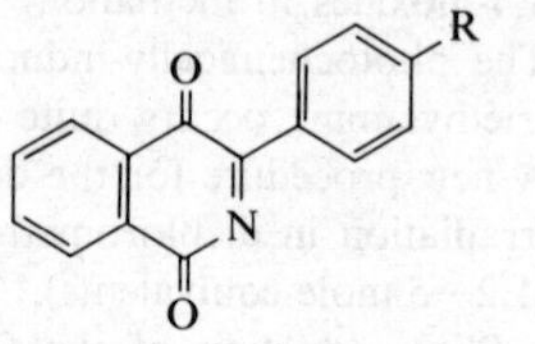

(29) R = H, Me, or Cl

triethylamine[54,55] and diethylamine.[55] Complex mixtures of products are reported from the reduction of azobenzene on irradiation in formamide or dimethylformamide.[56]

3 Miscellaneous Reductions

Further examples of the photoreduction of the carbon—carbon double bond of alkenes and enones have been found. Thus, hydrogen abstraction by the alkene double bond is a significant reaction on irradiation of *trans*-stilbene in acetonitrile in the presence of various secondary and tertiary amines (*e.g.* Scheme 4),[57]

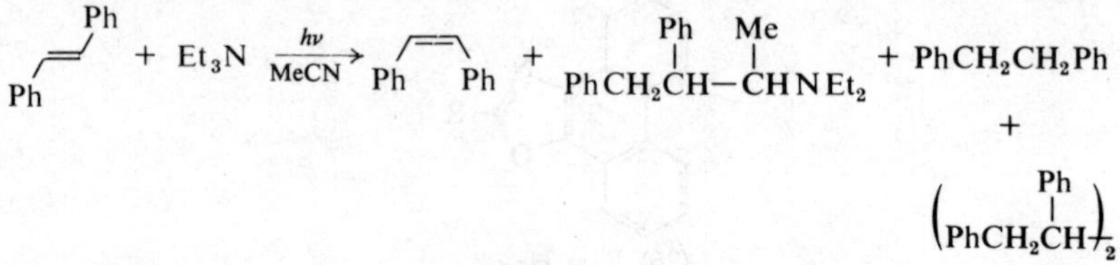

$$\text{Ph} \diagup + \text{Et}_3\text{N} \xrightarrow[\text{MeCN}]{h\nu} \text{Ph} \diagup \text{Ph} + \underset{PhCH_2CH-CHNEt_2}{\overset{Ph \quad Me}{|\quad\quad|}} + PhCH_2CH_2Ph$$

$$+$$

$$\left(\underset{PhCH_2CH}{\overset{Ph}{|}}\right)_2$$

Scheme 4

and also in the photoreduction of *trans*-1,2-bis(pyrazyl)ethylene in protic solvents to the corresponding ethane.[58] Acetone-sensitized photolysis of the dicyclopentadiene epoxide (30) results in efficient reduction of the norbornenyl double bond by a free radical mechanism to give the products shown in Scheme 5.[59] At low concentrations of (30) in acetone the formation of reduction product (31) is favoured, whilst at higher concentrations acetonyl addition products (32) and (33) predominate. Reduction of (E)-$\alpha\beta$-dibromostilbene on irradiation in benzene is reported to give *meso*-PhCHBrCHBrPh (48%).[60] In alcoholic solutions, photolysis of maleimides (34)—(36) results in abstraction of hydrogen by the double bond of triplet maleimide, followed very rapidly by addition of the solvent fragment to the C=C bond of a ground-state maleimide molecule.[61]

54 K. Okutsu and M. Kobayashi, *Josai Shika Daigaku Kiyo*, 1976, **5**, 279 (*Chem. Abs.*, 1977, **87**, 133 478).
55 K. Okutsu and M. Kobayashi, *Josai Shika Daigaku Kiyo*, 1975, **4**, 357 (*Chem. Abs.*, 1977, **87**, 133 479).
56 I. Cepciansky, M. Adamek, and D. Pavlaskova, *Chem. prùmysl*, 1977, **27**, 398.
57 F. D. Lewis and T.-I. Ho, *J. Amer. Chem. Soc.*, 1977, **99**, 7991.
58 S.-C. Shim and J.-S. Chae, *Taehan Hwahak Hoechi*, 1977, **21**, 102 (*Chem. Abs.*, 1977, **87**, 200 487).
59 D. R. Paulson, A. S. Murray, and E. J. Fornoret, *J. Org. Chem.*, 1978, **43**, 2010.
60 W. Schroth and H. Bahn, *Z. Chem.*, 1977, **17**, 56.
61 P. B. Ayscough, T. H. English, G. Lambert, and A. J. Elliot, *J.C.S. Faraday I*, 1977, **73**, 1302.

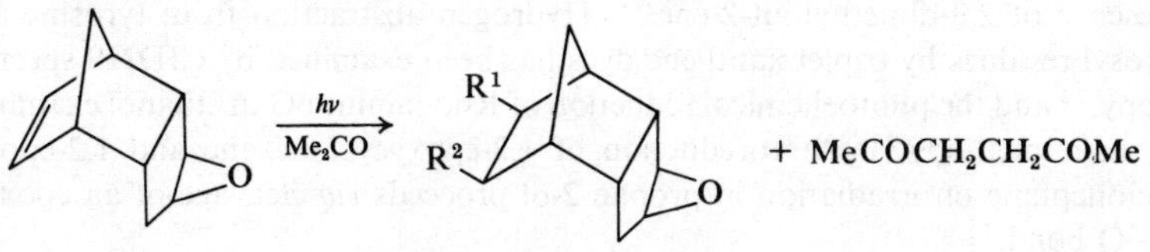

(30) (31) R^1 = R^2 = H
      (32) R^1 = MeCOCH$_2$, R^2 = H
      (33) R^1 = H, R^2 = MeCOCH$_2$

Scheme 5

When base is present, electron transfer results only in the observation of the maleimide radical anion. Irradiation of maleic anhydride (37) in methanol, propan-2-ol[62] or ethers[63] produces a similar hydrogen abstraction from solvent. There is evidence for the abstraction of hydrogen by the β-carbon atom upon irradiation of 2(5*H*)-furanone (38) in hydrocarbon solvents.[64,65] *NN*-Dibenzyl and *NN*-di-isopropyl αβ-unsaturated amides give a related, but

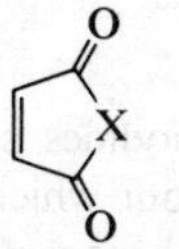

(38)

(34) X = NH
(35) X = NMe
(36) X = NEt
(37) X = O

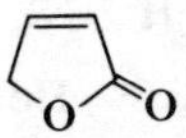

intramolecular, hydrogen abstraction by the β-carbon atom followed by cyclization to azetidinones in the former case.[66] Irradiation of uridine and its derivatives in methanol leads to reduction of the uracil 5,6-double bond.[67]

In carbohydrate chemistry, two groups have successfully exploited the high-yield reduction of sugar acetates[68,69] and pivalates[69] on irradiation in aqueous hexamethylphosphoric triamide (HMPT) [reaction (3)] as a route to deoxy-sugars.

$$R^1CO_2R^2 \xrightarrow[\text{HMPT,H}_2\text{O}]{h\nu} R^1CO_2H + R^2H \quad (3)$$

Photolysis of benzyl acetates in the presence of aliphatic amines gives a similar reductive cleavage.[70] For example, 254 nm irradiation of *p*-cyanobenzyl acetate and triethylamine in acetonitrile yields *p*-toluonitrile (26%) and 4,4′-dicyano-bibenzyl (16%). Reduction of the electron-poor aromatic ring is the major reaction occurring on irradiation of 2-cyanonaphthalene in methanol in the

62 J. Hellebrand and P. Wuensche, *Z. phys. Chem.* (*Leipzig*), 1977, **258**, 481.
63 H. Kasperski, J. Hellebrand, and H. K. Roth, *Z. phys. Chem.* (*Leipzig*), 1978, **259**, 167.
64 T. W. Flechtner, *J. Org. Chem.*, 1977, **42**, 901.
65 B. H. Toder, S. J. Branca, and A. B. Smith, *J. Org. Chem.*, 1977, **42**, 904.
66 T. Hasegawa, M. Watabe, H. Aoyama, and Y. Omote, *Tetrahedron*, 1977, **33**, 485.
67 J.-L. Fourrey and P. Jouin, *Tetrahedron Letters*, 1977, 3397.
68 J. P. Pete and C. Portella, *Synthesis*, 1977, 774.
69 P. M. Collins and V. R. Z. Munasinghe, *J.C.S. Chem. Comm.*, 1977, 927.
70 M. Ohashi, K. Tsujimoto, and Y. Furukawa, *Chem. Letters*, 1977, 543.

presence of 2,3-dimethylbut-2-ene.[71] Hydrogen abstraction from tyrosine and tyrosyl residues by triplet xanthene dyes has been examined by CIDNP spectroscopy,[72] and the photochemical reduction of Rhodamine 6G in ethanol examined by e.s.r. methods.[73] Photoreduction of 1,2-epoxycyclohexane and 1,2-epoxycycloheptane on irradiation in propan-2-ol proceeds *via* cleavage of an epoxide C—O bond.[74]

Photolysis of bromoalkanes and iodoalkanes affords ionic, as well as radical, intermediates in solution, *via* homolytic carbon—halogen bond cleavage followed by electron transfer in the radical pair. Thus, 254 nm irradiation of 2-iodoadamantane (39) in ether produces adamantane (40) by a radical process, with the reactive 2,4-dehydroadamantane (41) and protoadamantene (42) formed

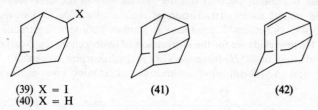

(39) X = I   (41)   (42)
(40) X = H

by an ionic process.[75] However, geminal di-iodides such as (43) can give carbenoid behaviour as well as the ionic behaviour which leads to (44); consequently, labelled (43) produces methylenecyclohexane (45) with retention of deuterium, as shown in Scheme 6.[76] Irradiation of *gem*-di-iodides in the presence

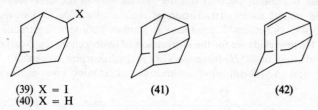

(43)   (44)   (45)

Scheme 6

of alkenes presents a convenient photochemical procedure for cyclopropane formation. Radical reduction is more prominent for bromoalkanes than iodoalkanes, and photoreduction of *gem*-dibromocyclopropanes in ether solutions leads to the corresponding monobromocyclopropanes in 67—88% yield.[77] Reaction presumably does not go further because the product absorbs much less strongly at the irradiation wavelength used (254 nm). A method which is claimed to be superior to conventional chemical methods for the reduction of

[71] J. J. McCullough, R. C. Miller, and W.-S. Wu, *Canad. J. Chem.*, 1977, **55**, 2909.
[72] K. A. Muszkat, *J.C.S. Chem. Comm.*, 1977, 872.
[73] V. V. Ryl'kov and S. V. Litke, *Khim. Vys. Energ.*, 1977, **11**, 373.
[74] S. N. Merchant, S. C. Sethi, and H. R. Sonawane, *Indian J. Chem.*, 1977, **15B**, 82.
[75] P. J. Kropp, J. R. Gibson, J. J. Snyder, and G. S. Poindexter, *Tetrahedron Letters*, 1978, 207.
[76] N. J. Pienta and P. J. Kropp, *J. Amer. Chem. Soc.*, 1978, **100**, 655.
[77] N. Shimizu and S. Nishida, *Chem. Letters*, 1977, 839.

bridgehead bromides involves the tributyltin hydride-reduction catalysed by u.v. light (254 nm).[78]

Environmental concern has encouraged several studies of the photoreactions of polyhalogenated aromatic compounds, such as polychlorinated biphenyls (PCB),[79,80] chlorinated terphenyls,[81,82] and chlorinated diphenyl ethers.[83] The major reaction occurring on irradiation in hydrocarbons is reductive dechlorination, with *ortho*-chlorine atoms being most efficiently replaced.[81] Dechlorination of the pesticide aldrin (46) to (47) occurs on direct irradiation in cyclohexane, and an experiment using [^{2}H$_{12}$]cyclohexane shows that the vinyl radical generated has a distinct geometry for hydrogen abstraction: it undergoes intermolecular abstraction from the cyclohexane to yield (48), rather than abstraction from the adjacent methylene bridge of the aldrin.[84]

(46) R = Cl
(47) R = H
(48) R = D

Irradiation of adamantane-2-thione in cyclohexane using 254 nm light gives rise to hydrogen abstraction by the S_2 $\pi\pi^*$ state of the thione.[85] The excited thione is a very indiscriminate abstractor of hydrogen from alkanes, showing little preference for abstraction from tertiary rather than primary C—H bonds. The reduction of phenacyl sulphonium salts by 1,4-dihydropyridines is enhanced by light, especially in the presence of dyes.[86]

4 Singlet Oxygen

This section mentions only those reports of singlet molecular oxygen ($^1O_2$, $^1\Delta_g$ which are relevant to organic chemistry. It is very noticeable that the importance of $^1O_2$ in synthetic organic chemistry is increasing rapidly. There are several interesting articles in a book on $^1O_2$ reactions with organic compounds and polymers, which is based on papers given at a 1976 EUCHEM conference in Sweden.[87] Rigaudy[88] has reviewed photo-oxygenation reactions, while

[78] T.-Y. Luh and L. M. Stock, *J. Org. Chem.*, 1977, **42**, 2790.
[79] G. Sundstrom, K. Olie, O. Hutzinger, S. Safe, and B. Chittim, *Chemosphere*, 1977, **6**, 103.
[80] D. Kotzias, W. Klein, and F. Korte, *Chemosphere*, 1977, **6**, 99.
[81] B. Chittim, S. Safe, N. J. Bunce, L. O. Ruzo, K. Olie, and O. Hutzinger, *Canad. J. Chem.*, 1978, **56**, 1253.
[82] B. Chittim and S. Safe, *Chemosphere*, 1977, **6**, 269.
[83] A. Norstrom, K. Andersson, and C. Rappe, *Chemosphere*, 1977, **6**, 241.
[84] L. A. Levy, *J.C.S. Chem. Comm.*, 1978, 574.
[85] K. Y. Law, P. de Mayo, and S. K. Wong, *J. Amer. Chem. Soc.*, 1977, **99**, 5813.
[86] D. M. Hedstrand, W. H. Kruizinga, and R. M. Kellogg, *Tetrahedron Letters*, 1978, 1255.
[87] 'Singlet Oxygen. Reactions with Organic Compounds and Polymers', ed. B. Rånby and J. F. Rabek, Wiley–Interscience, 1978.
[88] J. Rigaudy, *Actual. Chim.*, 1974, 7.

Matsuura[89] has published a fascinating report on biomimetic oxygenation in which he gives examples of the photo-oxidation of biologically relevant compounds such as phenols, indoles, and unsaturated fatty acids. Articles of environmental relevance are those which discuss the role of $^1O_2$ in photochemical air pollution[90] and in natural waters.[91]

Several examples have been noted of reactions which involve the generation of $^1O_2$. Methylene blue adsorbed on silica gel has been used as a heterogeneous sensitizer in photo-oxidation, being reportedly more stable than in a homogeneous system.[92] The efficiency of generation of $^1O_2$ from the thermal decomposition of 9,10-diphenylanthracene peroxide and rubrene peroxide has been measured.[93] It has been shown that $^1O_2$ formation can be sensitized by the excimers of aromatic compounds such as naphthalene and pyrene, as well as by their monomers.[94] The action of dehydrating agents (*e.g.* benzoyl isocyanate) on hydrogen peroxide can produce $^1O_2$ in yields of up to 98%.[95] Formation of $^1O_2$ by the oxidation of superoxide radical anion is a process which may be of biological significance. This reaction can be brought about by the use of chemical oxidizing agents such as dibenzoyl peroxide or dilauroyl peroxide, according to reaction (4).[96]

$$2O_2^{\cdot -} + (RCOO)_2 \longrightarrow 2O_2 + 2RCO_2^- \qquad (4)$$

Monroe[97] has measured rate constants for $^1O_2$ quenching by 22 aliphatic amines in chloroform, and has noted that for unbranched alkylamines a correlation exists between quenching rate and amine ionization potential. However, substitution of the carbon α to nitrogen reduces the quenching rate below that predicted from ionization potential data, indicating that $^1O_2$ quenching is subject to steric hindrance. The rate constants for $^1O_2$ addition (including any physical quenching component) to 13 alkenes in chloroform have been determined, as well as the rate constants for $^1O_2$ quenching by anthracene and some substituted anthracenes.[98] Quenching of $^1O_2$ by compounds (*e.g.* heptane, ethylbenzene) which are models for polyolefins has been examined,[99] and the lifetime and reactivity of $^1O_2$ in an aqueous micellar system have been investigated.[100]

5 Oxidation of Aliphatic and Alicyclic Unsaturated Systems

The dispute continues as to whether a perepoxide intermediate [*e.g.* (49) in Scheme 7] is involved in the photo-oxidation of alkenes to dioxetans and allylic hydroperoxides. *Ab initio* studies of the reaction of $^1O_2$ with ethylene suggest

[89]   T. Matsuura, *Tetrahedron*, 1977, **33**, 2869.
[90]   M. Sukigawa, *Seisan Kenkyu*, 1977, **29**, 87 (*Chem. Abs.*, 1978, **88**, 94 135).
[91]   R. G. Zepp, N. L. Wolfe, G. L. Baughman, and R. C. Hollis, *Nature*, 1977, **267**, 421.
[92]   D. Brkic, P. Forzatti, I. Pasquon, and F. Trifiro, *J. Mol. Catal.*, 1977, **3**, 173.
[93]   B. Stevens and R. D. Small, *J. Phys. Chem.*, 1977, **81**, 1605.
[94]   D. M. Shold, *J. Photochem.*, 1978, **8**, 39.
[95]   J. Rebek, S. Wolf, and A. Mossman, *J. Org. Chem.*, 1978, **43**, 180.
[96]   W. C. Danen and R. L. Arudi, *J. Amer. Chem. Soc.*, 1978, **100**, 3944.
[97]   B. M. Monroe, *J. Phys. Chem.*, 1977, **81**, 1861.
[98]   B. M. Monroe, *J. Phys. Chem.*, 1978, **82**, 15.
[99]   P. Bortolus, S. Dellonte, and G. Beggiato, *European Polymer J.*, 1977, **13**, 185.
[100] A. A. Gorman and M. A. J. Rodgers, *Chem. Phys. Letters*, 1978, **55**, 52.

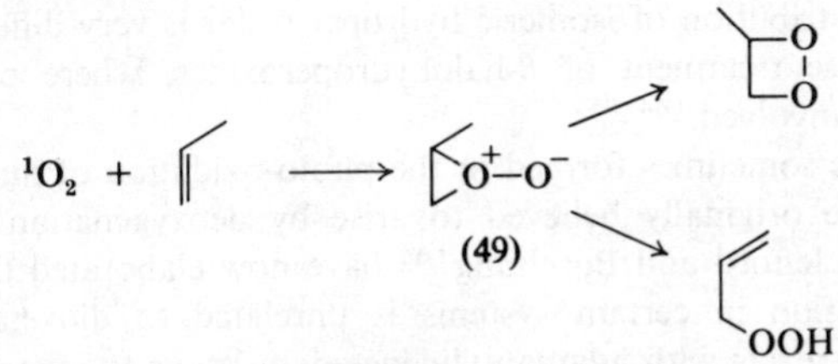

Scheme 7

that a perepoxide analogous to (49) lies at higher energy than 1,2-dioxetan or the 1,4-biradical ($\cdot CH_2CH_2OO\cdot$).[101] The authors therefore argue that perepoxides must be unimportant in $^1O_2$ chemistry and they criticize the predictions of perepoxide stability recently published by Dewar and co-workers.[102] However, the findings of McCapra and Beheshti[103] are in direct contrast with these predictions of Harding and Goddard.[101] Thus, $^1O_2$ reacts with the highly hindered alkene, camphenylidene-adamantane (50) to give not only the expected dioxetan (52), but also the rearranged 1,2-dioxolan (53). The ratio of (52) : (53)

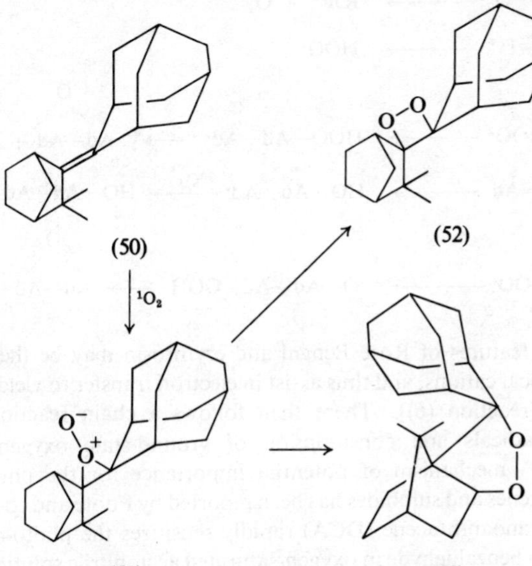

varies from 7 : 3 in dichloromethane to 1 : 9 in methanol. The formation of (53) is most readily understood as arising from the rearrangement of a perepoxide such as (51). To cloud the issue further, it is reported that the formation of allylic hydroperoxides in the reaction of $^1O_2$ with isopropylidenecyclohexane and isopropylidenecyclopentane does not proceed *via* perepoxide intermediates,

101 L. B. Harding and W. A. Goddard, *J. Amer. Chem. Soc.*, 1977, **99**, 4520.
102 M. J. S. Dewar and W. Thiel, *J. Amer. Chem. Soc.*, 1975, **97**, 3978.
103 F. McCapra and I. Beheshti, *J.C.S. Chem. Comm.*, 1977, 517.

because the distribution of isomeric hydroperoxides is very different from that formed by base treatment of β-halohydroperoxides, where perepoxides are believed to be involved.[104]

Epoxides are sometimes formed in the photo-oxidation of hindered alkenes, and these were originally believed to arise by deoxygenation of perepoxide intermediates. Jefford and Boschung[105] have now elaborated their claim that epoxide formation in certain systems is unrelated to dioxetan production. Singlet oxygen reacts with adamantylidene-adamantane to give only the corresponding dioxetan and epoxide. In several solvents, using methylene blue as sensitizer, dioxetan is the predominant product. However, when erythrosin or Rose Bengal is used as sensitizer, epoxide becomes the main product (70—95 %). The authors propose a complex reaction pathway to epoxide, involving both $^1O_2$ and radicals, as outlined in reactions (5)—(10) where RB is Rose Bengal and Ad=Ad is the alkene. There appears to be substantial supporting evidence for this route.

$$RB \xrightarrow{h\nu} {^1RB} \longrightarrow {^3RB} \xrightarrow{^3O_2} RB + {^1O_2} \tag{5}$$

$$RB + {^1O_2} \longrightarrow RB^{\cdot +} + O_2^{\cdot -} \tag{6}$$

$$O_2^{\cdot -} + H^+ \longrightarrow HOO^{\cdot} \tag{7}$$

$$Ad{=}Ad + HOO^{\cdot} \longrightarrow HOO{-}Ad{-}Ad^{\cdot} \longrightarrow \overset{\displaystyle O}{\overset{\displaystyle /\backslash}{Ad{-}Ad}} + HO^{\cdot} \tag{8}$$

$$HO^{\cdot} + Ad{=}Ad \longrightarrow HO{-}Ad{-}Ad^{\cdot} \xrightarrow{^3O_2} HO{-}Ad{-}Ad{-}OO^{\cdot} \tag{9}$$

$$HO{-}Ad{-}Ad{-}OO^{\cdot} \longrightarrow {\cdot}O{-}Ad{-}Ad{-}OOH \longrightarrow \overset{\displaystyle O}{\overset{\displaystyle /\backslash}{Ad{-}Ad}} + HOO^{\cdot} \tag{10}$$

The unusual features of Rose Bengal and erythrosin may be their ability to form stable radical cations, and thus assist in electron transfer to yield superoxide radical anion [reaction (6)]. There then follows a chain reaction involving hydroperoxy radicals and consumption of ground-state oxygen ($^3O_2$). A related non-$^1O_2$ mechanism of potential importance for the photosensitized oxidation of alkenes and sulphides has been reported by Foote and co-workers.[106] Thus, 9,10-dicyanoanthracene (DCA) rapidly sensitizes the photo-oxidation of *trans*-stilbene to benzaldehyde in oxygen-saturated acetonitrile solution, although this reaction does not occur easily with $^1O_2$. The former process apparently involves DCA radical anion, which reduces ground-state oxygen to superoxide radical anion.

The mechanism of the photo-oxidation of 4-methyl-2,3-dihydropyrans such as (54) to yield dioxetan (55) and hydroperoxide (56) has been examined in

[104] K. R. Kopecky, W. A. Scott, P. A. Lockwood, and C. Mumford, *Canad. J. Chem.*, 1978, **56**, 1114.
[105] C. W. Jefford and A. F. Boschung, *Helv. Chim. Acta*, 1977, **60**, 2673.
[106] J. Eriksen, C. S. Foote, and T. L. Parker, *J. Amer. Chem. Soc.*, 1977, **99**, 6455.

(54)   (55)   (56)

detail.[107] Photo-oxidation of suitably tritium-substituted derivatives of (54) allowed the determination of tritium kinetic isotope effects at the α-, β-, and γ-positions of the dihydropyran ring. The interpretation is that for both products, the transition state of the reaction has a weaker C^γ—H bond, a strengthened C^β—H bond, and an almost unaltered C^α—H bond. These results, which are in conflict with a concerted ene-mechanism, would be understandable on the basis of a perepoxide route, although they do not prove such a mechanism. Photo-oxidation of 2-methoxynorbornene (57) in acetonitrile gives dioxetan (58) and its dicarbonyl cleavage product (59), whereas photo-oxidation in methanol largely forms (60),[108] thus providing evidence by trapping for an intermediate such as the perepoxide of (57) or its zwitterionic structure (61).[109] Once again, Harding and Goddard[110] dissent from this view that reaction of

(57)   (58)   (59)

(60)   (61)

$^1O_2$ with methoxy-substituted alkenes occurs *via* a perepoxide intermediate. Instead, these authors claim to rationalize the regioselectivity of Conia's recent examples[111] in terms of a biradical mechanism and a consideration of the anomeric effect. Detection of the chemiluminescence, which accompanies the decomposition of dioxetans, has been used in a molecular-beam study of the activation energy required for the reaction of $^1O_2$ with methyl vinyl ether and 1,1-diethoxyethylene,[112] and in the investigation of the reaction of $^1O_2$ with various alkenes.[113]

[107] A. A. Frimer, P. D. Bartlett, A. F. Boschung, and J. G. Jewett, *J. Amer. Chem. Soc.*, 1977, **99**, 7977.
[108] C. W. Jefford and C. G. Rimbault, *J. Amer. Chem. Soc.*, 1978, **100**, 295.
[109] For a related example, see C. W. Jefford and C. G. Rimbault, *Tetrahedron Letters*, 1977, 2375.
[110] L. B. Harding and W. A. Goddard, *Tetrahedron Letters*, 1978, 747.
[111] G. Rousseau, P. Le Perchec, and J. M. Conia, *Tetrahedron Letters*, 1977, 2517.
[112] K. T. Alben, A. Auerbach, W. M. Ollison, J. Weiner, and R. J. Cross, jun., *J. Amer. Chem. Soc.*, 1978, **100**, 3274.
[113] D. J. Bogan, R. S. Sheinson, and F. W. Williams, *A.C.S. Symposium Series*, 1977, No. 56 (State-to-State Chem., Symp.), p. 127.

The dioxetans (62),[114] (63), and (64)[115] can be prepared by photosensitized oxidation of the corresponding alkenes at $-78\ °C$. The latter two dioxetans produce remarkably high yields of excited singlet N-methylacridone (and hence

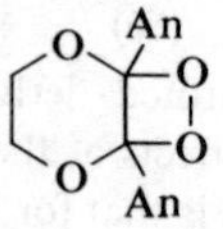

(62) An = 2-anthryl

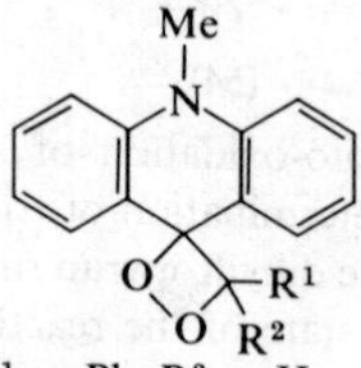

(63) R[1] = Ph, R[2] = H
(64) R[1]–R[2] = adamantylidene

chemiluminescence) on thermal decomposition. McCapra[116] suggests that such dioxetans, which are substituted by electron-donating groups, can give chemi-luminescence *via* an electron-transfer mechanism, and that it is the reverse electron transfer which causes efficient chemiluminescence (Scheme 8, where

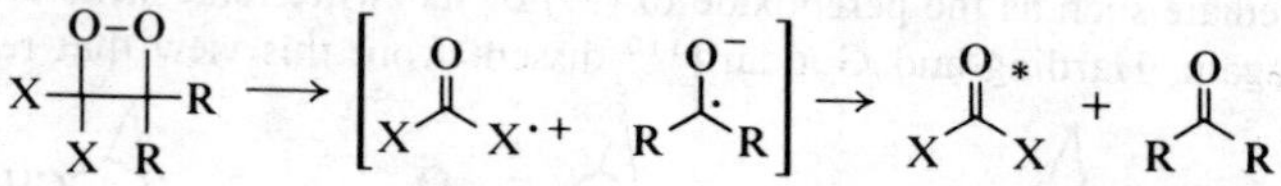

Scheme 8

X = donor). Such electron-donor character is probably also responsible for the very large increases in rate of decomposition and chemiluminescence effi-ciency on adsorption of these dioxetans onto silica gel.[114, 115]

A perepoxide or zwitterion is also reasoned to be an intermediate in the reaction of $^1O_2$ with the double bond of ketenes, which allows the isolation of α-peroxylactones such as (65).[117] It appears that chemical generation of $^1O_2$ gives better yields than dye-sensitized photo-oxidation for this reaction.

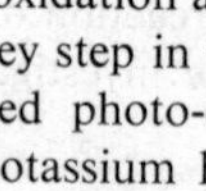

(65) a; R = Me
     b; R = Ph
     c; R = CF₃

Several synthetic uses of photo-oxidation at the carbon—carbon double bond by $^1O_2$ have been described. A key step in the synthesis of progesterone from stigmasterol involves dye-sensitized photo-oxidation of an aldehyde to the degraded ketone in methanolic potassium hydroxide solution (Scheme 9).[118] Reaction presumably occurs *via* the dioxetan formed from the enolate ion. A

[114] K. A. Zaklika, P. A. Burns, and A. P. Schaap, *J. Amer. Chem. Soc.*, 1978, **100**, 318.
[115] F. McCapra, I. Beheshti, A. Burford, R. A. Hann, and K. A. Zaklika, *J.C.S. Chem. Comm.*, 1977, 944.
[116] F. McCapra, *J.C.S. Chem. Comm.*, 1977, 946.
[117] N. J. Turro, Y. Ito, M.-F. Chow, W. Adam, O. Rodriquez, and F. Yany, *J. Amer. Chem. Soc.*, 1977, **99**, 5836.
[118] P. Sundararaman and C. Djerassi, *J. Org. Chem.*, 1977, **42**, 3633.

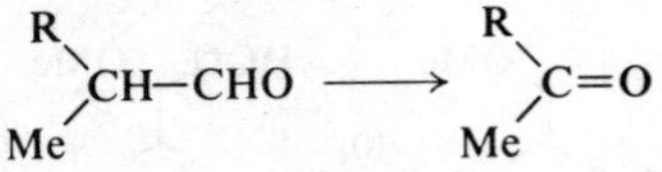

$$\underset{Me}{\overset{R}{>}}CH-CHO \longrightarrow \underset{Me}{\overset{R}{>}}C=O$$

Scheme 9

similar photo-oxidation of aldehydes in the presence of piperidine or morpholine gives the degraded aldehyde and the *N*-formyl derivative of the amine.[119] Photosensitized oxidation of the bicyclic thioenol ethers (66) produces ketothiolactones in reasonable yields,[120] most likely by way of cleavage of an intermediate dioxetan. In the photo-oxidation of the sesquiterpene thujopsene, the role of singlet oxygen has been confirmed by *in situ* dioxetan reduction with dimethyl sulphide to a *cis*-diol.[121] Dioxetans are also believed to be involved in the photo-oxidation of 2'-hydroxychalcones by $^1O_2$.[122]

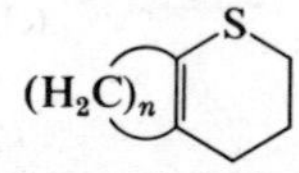

$$(H_2C)_n \overset{S}{\bigcirc}$$

(66) *n* = 4, 5, 6, or 10

When an alkene contains suitably-situated allylic hydrogen atoms, $^1O_2$ may produce an ene-reaction in which allylic hydroperoxide is formed. For example, photo-oxidation of 2-methylnorbornadiene (67), followed by reduction, produces the *exo* and *endo* allylic alcohols (68) and (69) in the ratio 7.3 : 1.0.[123] A comparison can be made with the rates and stereochemistry of $^1O_2$ attack on

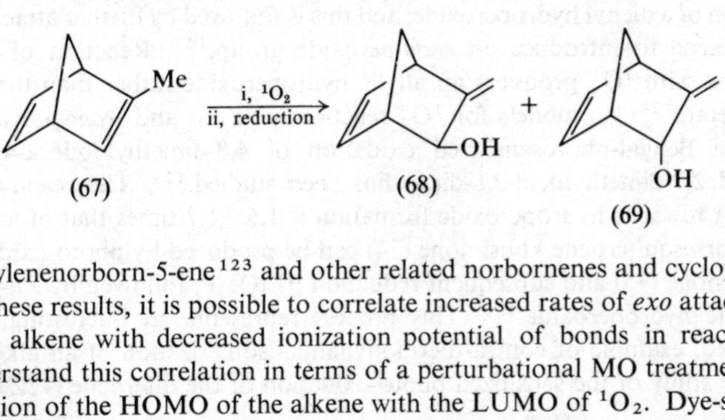

(67) $\xrightarrow[\text{ii, reduction}]{\text{i, }^1O_2}$ (68)   +   (69)

2-methylenenorborn-5-ene[123] and other related norbornenes and cyclopentenes. From these results, it is possible to correlate increased rates of *exo* attack on the bicyclic alkene with decreased ionization potential of bonds in reaction and to understand this correlation in terms of a perturbational MO treatment of the interaction of the HOMO of the alkene with the LUMO of $^1O_2$. Dye-sensitized photo-oxidation of enol ethers presents a general method for the one-carbon homologation of carbonyl compounds to $\alpha\beta$-unsaturated aldehydes and esters, as shown for cyclohexanone in Scheme 10.[124] In the photochemical step, an ene reaction of $^1O_2$ with the enol ether (70) gives a hydroperoxide (71), which is easily reduced to unsaturated aldehyde or dehydrated to ester. The reaction of ketene trimethylsilyl acetals (72) with $^1O_2$ gives, in a trimethylsilyl-analogue of

119 W. Ando and K. Watanabe, *Chem. Letters*, 1977, 947.
120 H. C. Araújo and J. R. Mahajan, *Synthesis*, 1978, 228.
121 H. Takeshita, T. Hatsui, and I. Shimooda, *Tetrahedron Letters*, 1978, 2889.
122 H. M. Chawla, S. S. Chibber, and R. Saigal, *Indian J. Chem.*, 1977, **15B**, 975.
123 C. W. Jefford and C. G. Rimbault, *J. Org. Chem.*, 1978, **43**, 1908.
124 G. Rousseau, P. Le Perchec, and J. M. Conia, *Synthesis*, 1978, 67.

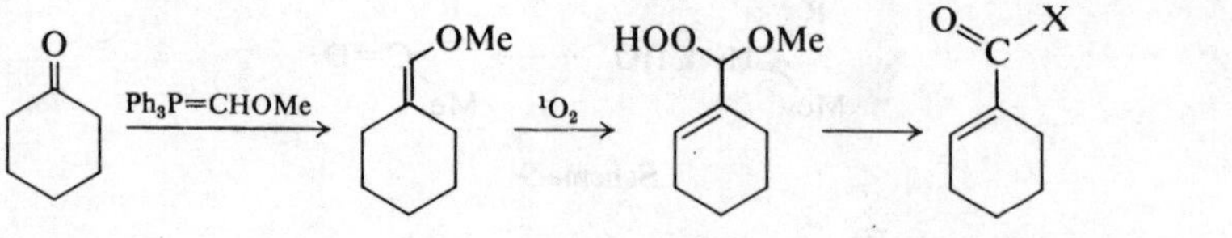

(70)   (71)   X = H or OMe

Scheme 10

the hydrogen shift in the ene-reaction, silylperoxy derivatives (73). Desilylation with methanol provides a route to α-hydroperoxy acids[125] and α-hydroperoxy esters.[126]

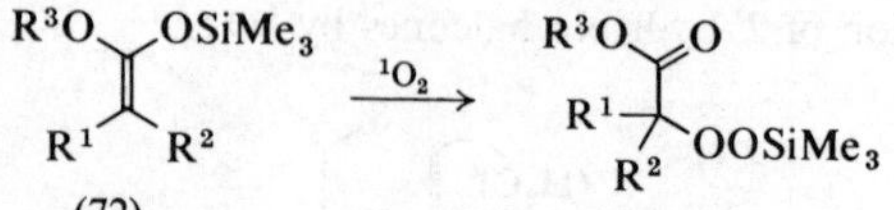

(72)   (73)

a; $R^1 = R^2 = Bu^t$, $R^3 = SiMe_3$
b; $R^1 = H$, $R^2 = Bu^t$, $R^3 = SiMe_3$
c; $R^1 = R^2 = Me$, $R^3 = SiMe_3$
d; $R^1 = H$, $R^2 = Bu^t$, $R^3 = Me$
e; $R^1 = R^2 = R^3 = Me$
f; $R^1 = R^2 = Ph$, $R^3 = Me$

The first step in the photo-oxidation of 1,4-diphenylcyclohexa-1,4-diene is the formation of a dienyl hydroperoxide, and this is followed by further attack of $^1O_2$ on the diene to introduce an *endo*-peroxide group.[127] Reaction of bicyclo-butylidene with $^1O_2$ produces an allylic hydroperoxide rather than the hoped-for dioxetan.[128] As models for $^1O_2$ reaction with *cis*- and *trans*-polyisoprene, the Rose Bengal-photosensitized oxidation of 4,8-dimethyldodeca-4,8-diene, 2,6-, and 2,7-dimethylocta-2,6-diene has been studied.[129] The reactivity of a *trans* unit towards hydroperoxide formation is 1.6—1.7 times that of a *cis* unit.

The norsesquiterpene khusimone (74) can be produced by photo-oxidation of isokhusimone (75) and subsequent reduction [(EtO)₃P, followed by Zn–HCl] of the allylic hydroperoxide.[130] This process represents an interesting, though specialized, example of contra-thermodynamic isomerization of an alkene. An extensive study of the sensitized photo-oxidation of the diterpene (12Z)-abienol

(74)   (75)

[125] W. Adam, A. Alzérreca, J.-C. Liu, and F. Yany, *J. Amer. Chem. Soc.*, 1977, **99**, 5768.
[126] W. Adam and J. del Fierro, *J. Org. Chem.*, 1978, **43**, 1159.
[127] I. Saito, K. Tamoto, A. Katsumura, H. Sugiyama, and T. Matsuura, *Chem. Letters*, 1978, 127.
[128] L. K. Bee, J. W. Everett, and P. J. Garratt, *Tetrahedron*, 1977, **33**, 2143.
[129] C. Tanielian and J. Chaineaux, *J. Photochem.*, 1978, **9**, 19.
[130] G. Büchi, A. Hauser, and J. Limacher, *J. Org. Chem.*, 1977, **42**, 3323.

shows that ene processes predominate, yielding allylic hydroperoxides.[131] A naturally-occurring ene-process with $^1O_2$ may account for the fact that a marine diterpene has been found to be an allylic hydroperoxide.[132] Dye-sensitized photo-oxidation of the sesquiterpene α-cyclocostunolide also yields an allylic hydroperoxide.[133] Erythrosin-sensitized photo-oxidation of methyl oleate produces methyl 9- and 10-hydroperoxyoctadecenoates in the ratio 49.5 : 50.5, whereas autoxidation yields mainly 8- and 11-hydroperoxy isomers.[134] Photo-oxidation of the 3-methylbut-2-enyl group to a 3-hydroperoxy-3-methylbutenyl group plays a part in the synthesis of the coumarin avicennol[135] and (at the 3-position) in the photo-oxidation of the flavonoid morusin.[136—138]

Singlet oxygen reacts with *cisoid* conjugated dienes in a [2 + 4] reaction to yield 1,4-*endo*-peroxides. The finding that the remaining double bond in the product can be reduced by di-imide without affecting the peroxidic link makes accessible a wide variety of saturated *endo*-peroxides.[139—141] For example, the biological precursors of prostaglandins are *endo*-peroxides with the 2,3-dioxa-bicyclo[2,2,1]heptane skeleton, and Scheme 11 shows the formation of the

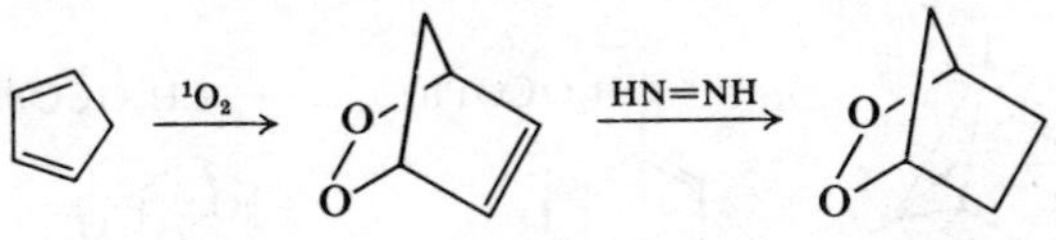

(76)

Scheme 11

parent ring system (76) from cyclopentadiene.[139] In exactly similar manner, photo-oxidation of 6,6-disubstituted fulvenes and di-imide reduction, all at −78 °C, allows the trapping of the unstable intermediate *endo*-peroxides.[142] Photo-oxidation of α-pyrone at 0 °C has allowed the isolation of an unstable *endo*-peroxide (77), which is again stabilized by di-imide reduction.[143] Analogously, photo-oxidation of 1,4-diphenyl-2-benzopyran-3-one at 0 °C gives the *endo*-peroxide (78) which decarboxylates to a novel *o*-xylylene peroxide, itself a source of chemiluminescence.[144]

131   I. Wahlberg, K. Karlsson, M. Curvall, T. Nishida, and C. R. Enzell, *Acta Chem. Scand.*, 1978, **B32**, 203.
132   B. M. Howard, W. Fenical, J. Finer, K. Hirotsu, and J. Clardy, *J. Amer. Chem. Soc.*, 1977, **99**, 6440.
133   S. A. Nadgouda, G. K. Trivedi, and S. C. Bhattacharyya, *Indian J. Chem.*, 1978, **16B**, 16.
134   H. W. S. Chan and G. Levett, *Chem. and Ind.*, 1977, 692.
135   R. D. H. Murray and I. T. Forbes, *Tetrahedron*, 1978, **34**, 1411.
136   T. Nomura and T. Fukai, *Heterocycles*, 1977, **8**, 443.
137   T. Nomura, T. Fukai, S. Yamada, and M. Katayanagi, *Chem. and Pharm. Bull. (Japan)*, 1977, **25**, 1155.
138   T. Nomura and T. Fukai, *Heterocycles*, 1978, **9**, 635.
139   W. Adam and H. J. Eggelte, *J. Org. Chem.*, 1977, **42**, 3987.
140   W. Adam and H. J. Eggelte, *Angew. Chem. Internat. Edn.*, 1977, **16**, 713.
141   W. Adam, A. J. Bloodworth, H. J. Eggelte, and M. E. Loveitt, *Angew. Chem. Internat. Edn.*, 1978, **17**, 209.
142   W. Adam and I. Erden, *Angew. Chem. Internat. Edn.*, 1978, **17**, 210.
143   W. Adam and I. Erden, *Angew. Chem. Internat. Edn.*, 1978, **17**, 211.
144   J. P. Smith and G. B. Schuster, *J. Amer. Chem. Soc.*, 1978, **100**, 2564.

(77)   (78)

Reaction between the bicyclobutane-bridged diene (79) and $^1O_2$ cannot be detected at room temperature or $-70\,°C$.[145] However, the corresponding *endo*-peroxide is believed to be formed in a reversible equilibrium, because it can be diverted by silver ions or acetic acid to yield an aromatic peroxide or a homofulvene peroxide, respectively. An important step in the synthesis of the anti-tumour agent, senepoxide, involves photo-oxidation of the cyclohexadiene (80) to yield the *endo*-peroxide (81).[146] (For a closely related approach to such *endo*-peroxides, see ref. 160.) Another example in which an *endo*-peroxide is

(79)   (80)   (81)

formed by $^1O_2$ attack at the less hindered side of a diene occurs on methylene blue-sensitized photo-oxidation of the terpene α-pyronene (1,5,5,6-tetramethylcyclohexa-1,3-diene).[147] The structure of baccatin, a naturally-occurring nortriterpene *endo*-peroxide, has been established.[148] Another method has been reported for the conversion of dienes to furans, which involves photo-oxidation, treatment of the resultant *endo*-peroxide with base, and dehydration.[149] Photo-oxidation of the furanoid ring of some furanosesquiterpenes has been observed.[150,151] Photo-oxygenation of a diene unit is a reaction common to the synthesis of the alkaloid pontevedrine,[152] and to the reactions of 1,2-diazepins[153] and cyclic α-keto-sulphonamides[154] with $^1O_2$.

Azide radical production is implicated in the dye-sensitized photo-oxidation of spirocyclic vinylcyclopropanes in the presence of sodium azide, which yields

[145] R. F. Heldeweg, H. Hogeveen, and E. P. Schudde, *J. Org. Chem.*, 1978, **43**, 1912.
[146] G. W. Holbert and B. Ganem, *J. Amer. Chem. Soc.*, 1978, **100**, 352.
[147] W. Cocker, K. J. Crowley, and K. Srinivasan, *J.C.S. Perkin I*, 1978, 159.
[148] B. Saha, D. B. Naskar, D. R. Misra, B. P. Pradhan, and H. N. Khastgir, *Tetrahedron Letters*, 1977, 3095.
[149] B. Harirchian and P. D. Magnus, *Synthetic Comm.*, 1977, **7**, 119.
[150] K. Naya, N. Nogi, Y. Makiyama, H. Takashina, and T. Imagawa, *Bull. Chem. Soc. Japan*, 1977, **50**, 3002.
[151] Z. Horii, E. Yoneda, T. Tanaka, and C. Iwata, *Chem. and Pharm. Bull.* (*Japan*), 1977, **25**, 2782.
[152] L. Castedo, R. Estévez, J. M. Saá, and R. Suau, *Tetrahedron Letters*, 1978, 2179.
[153] T. Tsuchiya, H. Arai, H. Hasegawa, and H. Igeta, *Chem. and Pharm. Bull* (*Japan*), 1977, **25**, 2749.
[154] E. Fanghaenel, E. Do Nascimento, R. Radeglia, G. Lutze, and H. Baerhold, *Z. Chem.*, 1977, **17**, 135.

products different from those produced by $^1O_2$ itself.[155] Other photo-oxidation reactions of alkenes which do not appear to involve $^1O_2$ include photo-oxidations in the presence of pyridine and ferric chloride to give α-chlorinated ketones,[156] and photo-oxidations in pyridine in the presence of uranyl acetate to give β-hydroxyhydroperoxides.[157]

6 Oxidation of Aromatic Compounds

Reactions of $^1O_2$ with simple aromatic compounds are rare but, in contrast with an earlier report,[158] the photo-oxidation of hexamethylbenzene can be induced by $^1O_2$.[159] Thus, dye-sensitized photo-oxidation of hexamethylbenzene in various solvents leads to the uptake of two moles of oxygen, giving the product (82). This *endo*-peroxy hydroperoxide can be understood to have arisen from a

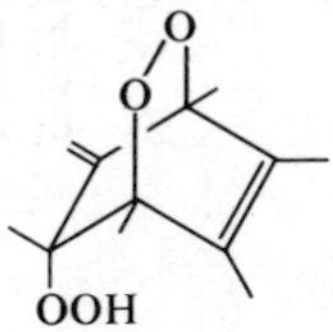

(82)

[2 + 4] addition of $^1O_2$ to a diene unit, and an ene reaction of a methylalkene fragment. However, no evidence could be found for an intermediate species in the reaction, and so it is not certain which step occurs first.

Further examples of the 1,4-cycloaddition of $^1O_2$ to phenyl-conjugated alkenes have been published, in which one of the double bonds of the aromatic ring acts as part of the diene system. Thus, ββ-dimethylstyrenes (83) give favourable yields of adducts (84) and (85) formed by 1,4-attack of $^1O_2$ followed by a further $^1O_2$ attack on the diene system produced.[160] Thermal rearrangement of the

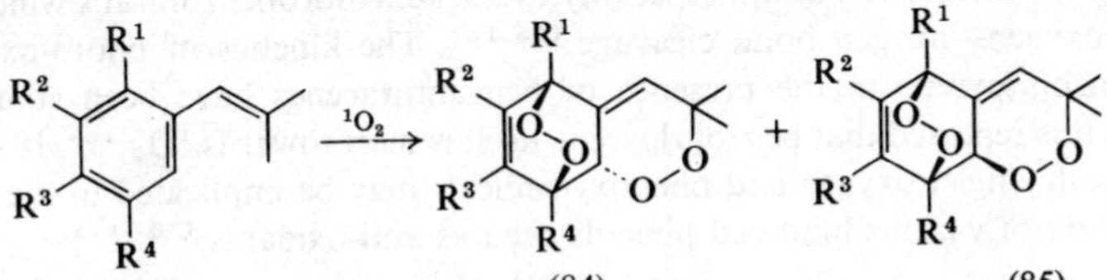

(83) (84) (85)

a; $R^1 = R^2 = R^4 = H, R^3 = Bu^t$
b; $R^1 = R^2 = R^3 = R^4 = H$
c; $R^1 = R^2 = R^4 = H, R^3 = Me$
d; $R^1 = Me, R^2 = R^3 = R^4 = H$
e; $R^1 = R^3 = H, R^2 = R^4 = Me$
f; $R^1 = CH_2OCOPh, R^2 = R^3 = R^4 = H$

155 T. Hatsui and H. Takeshita, *Chem. Letters*, 1977, 603.
156 E. Murayama, A. Kohda, and T. Sato, *Chem. Letters*, 1978, 161.
157 E. Murayama and T. Sato, *Tetrahedron Letters*, 1977, 4079.
158 H. H. Wasserman, P. S. Mariano, and P. M. Keehn, *J. Org. Chem.*, 1971, **36**, 1765.
159 C. J. M. van den Heuvel, H. Steinberg, and T. J. de Boer, *Rec. Trav. chim.*, 1977, **96**, 157.
160 M. Matsumoto, S. Dobashi, and K. Kuroda, *Tetrahedron Letters*, 1977, 3361.

bis(*endo*-peroxide) (84f) to diepoxide can be used in an ingenious synthesis of the anti-tumour agent, crotepoxide (86). Related attack by [2 + 4] cycloaddition of $^1O_2$ occurs at the vinyl-aromatic systems of 9-phenylphenanthrenes[161] and 2-vinylbenzofurans.[162] The production of ring-hydroxylated aromatics on photo-sensitized oxidation of fluorenylidene derivatives (87) in methanol, exemplified by the formation of (88) and (89) from (87a), may likewise be the result of formation and rearrangement of an *endo*-peroxide intermediate.[163]

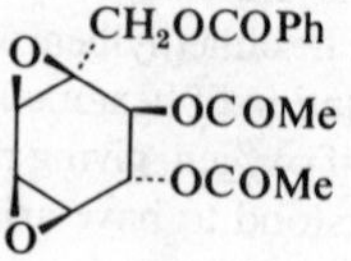

(86)

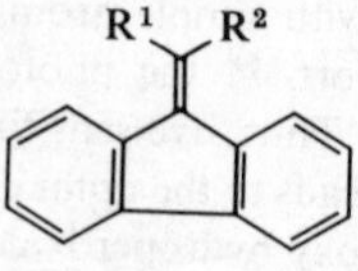

(87) a; $R^1 = R^2 = Me$
b; $R^1-R^2 = (CH_2)_5$
c; $R^1 = R^2 = H$
d; $R^1 = H, R^2 = Me$

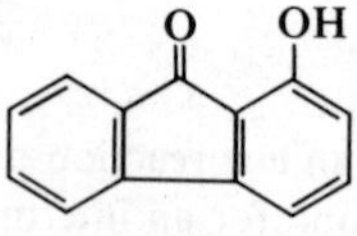

(88) 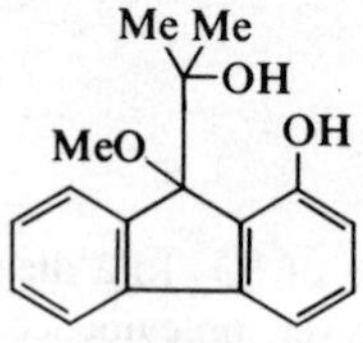   (89)

A superior method of photosensitized oxidation has been described for the preparation of anthracene 9,10-*endo*-peroxide.[164] Photo-oxygenation of 5,12-diphenylnaphthacene gives a mixture of isomeric 5,12-*endo*-peroxide (66%) and 6,11-*endo*-peroxide (29%).[165] Such *endo*-peroxides may undergo further re-arrangements on irradiation, especially to the benzofurobenzofurans which arise from oxygen—oxygen bond cleavage.[166, 167] The kinetics of photo-oxidation of benzo[*a*]pyrene in the presence of benzanthracenes have been studied,[168] whilst it is reported that benzo[*a*]pyrene itself is inert towards $^1O_2$.[169] It appears that both singlet oxygen and phenoxy-radicals may be implicated in the photo-oxidation of various hindered phenols used as anti-oxidants.[170—172]

[161] M. Matsumoto, S. Dobashi, and K. Kondo, *Bull. Chem. Soc. Japan*, 1978, **51**, 185.
[162] M. Matsumoto, S. Dobashi, and K. Kondo, *Bull. Chem. Soc., Japan*, 1977, **50**, 3026.
[163] W. Ando and S. Kohmoto, *J.C.S. Chem. Comm.*, 1978, 120.
[164] J. Rigaudy, J. Baranne-Lafont, A. Defoin, and N. K. Cuong, *Tetrahedron*, 1978, **34**, 73.
[165] J. Rigaudy and D. Sparfel, *Bull. Soc. chim. France*, 1977, 742.
[166] M. K. Logani, W. A. Austin, and R. E. Davis, *Tetrahedron Letters*, 1978, 511.
[167] J. Rigaudy, C. Breliere, and P. Scribe, *Tetrahedron Letters*, 1978, 687.
[168] L. Paalme and M. Gubergrits, *Eesti N.S.V. Tead. Akad. Toim., Keem., Geol.*, 1977, **26**, 239 (*Chem. Abs.*, 1978, **88**, 97 320).
[169] Y. Ioki and C. Nagata, *J.C.S. Perkin II*, 1977, 1172.
[170] L. V. Samsonova, L. Taimr, and J. Pospisil, *Angew. Macromol. Chem.*, 1977, **65**, 197.
[171] L. Taimr and J. Pospisil, *J. Polymer Sci., Polymer Symposia*, 1976, **57** (Degradation Stab. Polyolefins), 213.
[172] T. Kurechi and H. Senda, *Eisei Kagaku*, 1977, **23**, 267 (*Chem. Abs.*, 1978, **88**, 190 281).

7 Oxidation of Nitrogen-containing Compounds

Two mechanisms can be prevalent in the dye-sensitized photo-oxidation of amines: either a Type I (radical hydrogen abstraction) or a Type II (singlet oxygen) pathway. The sensitized photo-oxidation of *NN*-dimethylbenzylamine gives benzaldehyde and *N*-benzyl-*N*-methylformamide as the major products.[173] The product ratio depends strongly on the particular photosensitizer used, since benzaldehyde is the main product from Type I radical-producing sensitizers like benzophenone, whilst significant amounts of *N*-benzyl-*N*-methylformamide are produced by dye sensitizers such as methylene blue. The photosensitized oxidation of tyramine in aqueous solutions is mainly a Type II process, with an estimated rate constant for reaction of $^1O_2$ with tyramine of 2.8×10^8 l mol^{-1} s^{-1} at pH 10.[174] Dye-sensitized photo-oxidation of aziridines has been studied.[175] Thus, either of the aziridines (90) yields a mixture of benzoic acid, *N*-cyclohexylbenzamide, and benzamide. Such reactions may occur *via* $^1O_2$ attack on the azomethine ylide which results from aziridine carbon—carbon bond cleavage.

There is increasing interest in the use of semiconductor photoelectrodes to induce chemical reactions. The cadmium sulphide sensitized photo-oxidation of leuco-Crystal Violet (91) to the Crystal Violet cation in oxygenated acetonitrile

(90) a; $R^1 = Ph, R^2 = H$
     b; $R^1 = H, R^2 = Ph$

(91)

solution occurs by a mechanism of oxidation of leuco-Crystal Violet by a photo-generated hole in the semiconductor.[176] The superoxide radical anion ($O_2^{\cdot-}$) is produced, and can give competitive oxidation. The zinc oxide catalysed photo-oxidation of aniline to azobenzene is believed to involve formation of the superoxide radical anion as the oxidizing species.[177]

NN-Dimethylhydrazones are of increasing importance in synthetic chemistry, and a photochemical method has now been described for the regeneration of carbonyl compounds from such derivatives.[178] Dye-sensitized photo-oxidation of *NN*-dimethylhydrazones, followed by reduction with triphenylphosphine or dimethyl sulphide and hydrolysis, yields the parent carbonyl compound in reasonable yield (48—88%) (*e.g.* Scheme 12). Oxidation *via* an ene-type process is likely to be involved since adamantanone-*NN*-dimethylhydrazone, where an ene-reaction is not possible, remains unchanged under the reaction conditions.

[173] K. Inoue, I. Saito, and T. Matsuura, *Chem. Letters*, 1977, 607.
[174] G. R. Seely, *Photochem. and Photobiol.*, 1977, **26**, 115.
[175] V. Bhat and M. V. George, *Tetrahedron Letters*, 1977, 4133.
[176] F. D. Saeva, G. R. Olin, and J. R. Harbour, *J.C.S. Chem. Comm.*, 1978, 417.
[177] M. A. Hema, V. Ramakrishnan, and J. C. Kuriacose, *Indian J. Chem.*, 1977, **15B**, 947.
[178] E. Friedrich, W. Lutz, H. Eichenauer, and D. Enders, *Synthesis*, 1977, 893.

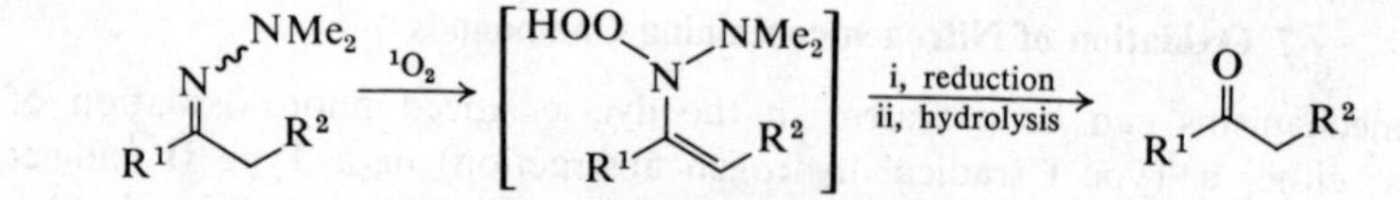

a; $R^1 = H$, $R^2 = C_{10}H_{21}$
b; $R^1-R^2 = (CH_2)_4$
c; $R^1 = Me$, $R^2 = CH_2CO_2Et$

Scheme 12

There have been several reports of the photo-oxidation of alkaloids. Dye-sensitized photo-oxidation of berberinium (92) chloride in methanol affords the berberine phenol betaine (93).[179] The further photo-oxidation of (93) has been separately studied.[180] The analogous betaine (94) can also be photo-oxidized.[181]

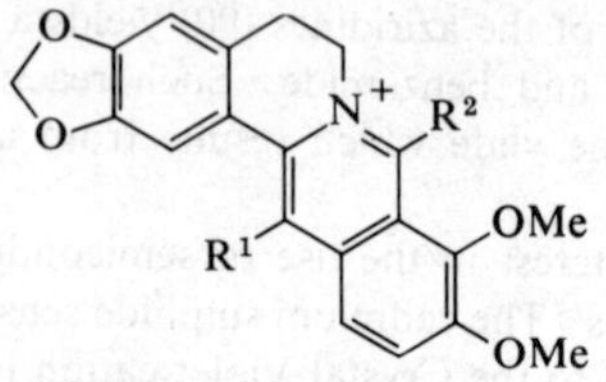

(92) $R^1 = R^2 = H$
(93) $R^1 = O^-$, $R^2 = OMe$
(94) $R^1 = O^-$, $R^2 = H$

A series of alkaloid tertiary amines undergo photo-oxidation sensitized by Rose Bengal in the presence of potassium cyanide to yield α-aminonitriles.[182,183] The reaction, which can often occur in the absence of oxygen, is believed to involve trapping of an iminium ion by a cyanide ion. Direct and dye-sensitized photo-oxidations of the isoquinoline alkaloid glaucine have been noted.[184]

Singlet oxygen reacts with the betaine (95) to yield the trione (96), probably by way of oxygen addition across the 2- and 6-positions of the ring.[185] Photo-oxidation of some phenothiazine drugs has been studied,[186,187] and may be related to the allergic skin reactions sometimes noticed on phenothiazine treatment. Chlorpromazine, for example, reacts with $^1O_2$ mainly by cleavage of the N-alkyl side chain, with a slower photo-oxidation at the phenothiazine sulphur

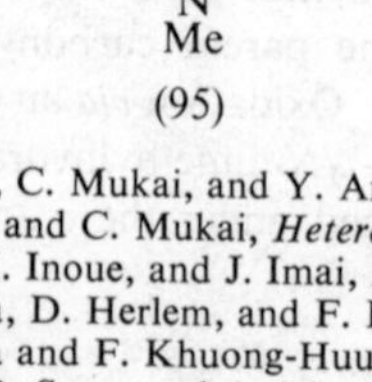

(95) 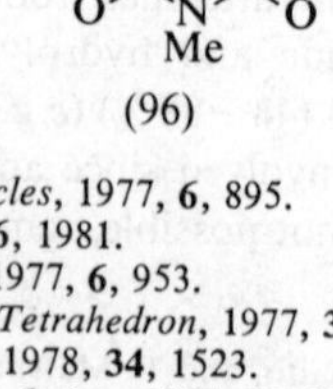   (96)

179 M. Hanaoka, C. Mukai, and Y. Arata, *Heterocycles*, 1977, **6**, 895.
180 M. Hanaoka and C. Mukai, *Heterocycles*, 1977, **6**, 1981.
181 Y. Kondo, H. Inoue, and J. Imai, *Heterocycles*, 1977, **6**, 953.
182 J. Santamaria, D. Herlem, and F. Khuong-Huu, *Tetrahedron*, 1977, **33**, 2389.
183 J. Santamaria and F. Khuong-Huu, *Tetrahedron*, 1978, **34**, 1523.
184 L. Castedo, R. Suau, and A. Mourino, *Anales de Quim.*, 1977, **73**, 290.
185 Y. Tamura, M. Akita, H. Kiyokawa, L.-C. Chen, and H. Ishibashi, *Tetrahedron Letters*, 1978, 1751.
186 I. Rosenthal, T. Bercovici, and A. Frimer, *J. Heterocyclic Chem.*, 1977, **14**, 355.
187 E. Pawelczyk and B. Marciniec, *Polish J. Pharmacol. Pharm.*, 1977, **29**, 137.

atom.[186] Azetidine carboxylic esters can be converted into enol silyl ethers which, as aminoketene acetals, undergo cleavage of the carbon—carbon double bond by $^1O_2$ to form β-lactams.[188] Photo-oxidation of the steroidal enamide 17β-acetoxy-4-aza-androsta-1,5-dien-3-one has been reported.[189]

The photochemical conversion of primary and secondary nitro-groups into carbonyl compounds has become possible with the finding that $^1O_2$ reacts with the nitronate salt made by dissolving the corresponding nitro-compound in base.[190] In the presence of oxygen, tetramethyl-2-tetrazene ($Me_2NN{=}NNMe_2$) undergoes photo-oxidative addition to alkenes to yield 1-amino-2-hydro-peroxides that can be reduced to the corresponding aminoalcohols.[191] *N*-Nitrosodimethylamine can give a related photo-oxidative addition to produce aminonitro-oxyalkanes.[192] Besides acting as a triplet quencher, ground-state oxygen can also act as a biradical trapping agent, forming bicyclic peroxides when the photochemical decomposition of azo compounds is carried out in the presence of oxygen (Scheme 13).[193] Typical oxygen pressures for optimum

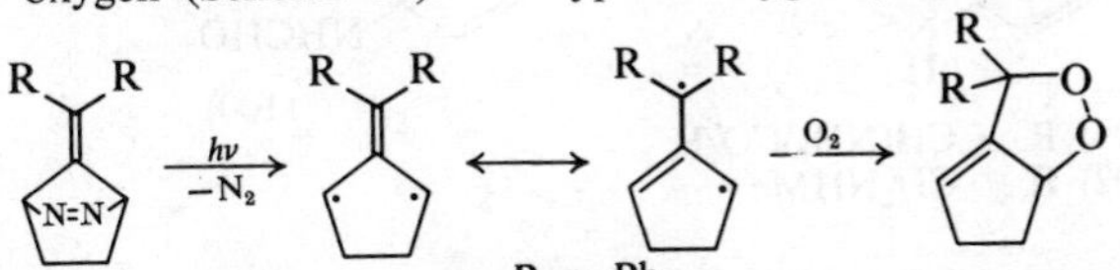

a; R = Ph
b; R = Me

Scheme 13

trapping are *ca.* 103×10^4 N m^{-2}. Benzophenone-sensitized photolysis of 2,3-diazabicyclo[2,2,1]heptane under these conditions produces the dioxabicyclo-heptane (76) in an interesting alternative route to that described earlier.[139] The stability of some 1-phenylazo-2-napththols containing aminomethyl groups towards $^1O_2$ has been examined.[194]

Photo-oxidation of the pyrrole ring has been used in a synthetic approach to the mitomycins. Thus, Rose Bengal-sensitized photo-oxidation of the pyrrolo-indoles (97) has been shown to lead to hydroperoxides (98), which are easily transformed to the corresponding alcohols.[195] Horsey and Whitten[196] have

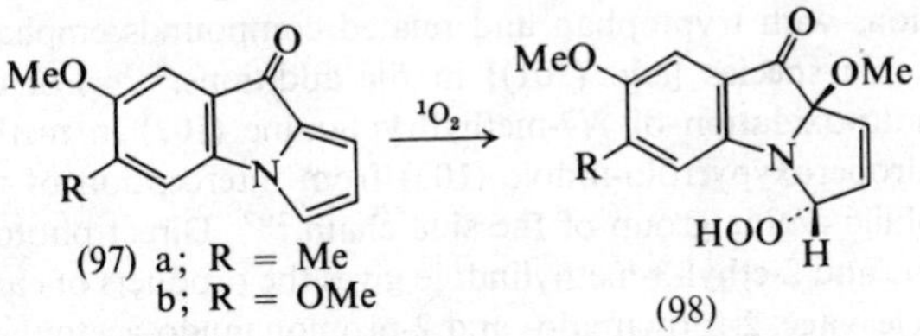

(97) a; R = Me
b; R = OMe

(98)

[188] H. H. Wasserman, B. H. Lipschutz, and J. S. Wu, *Heterocycles*, 1977, **7**, 315.
[189] J. A. Vallet, A. Cánovas, J. Boix, and J.-J. Bonet, *Helv. Chim. Acta*, 1978, **61**, 1165.
[190] J. R. Williams, L. R. Unger, and R. H. Moore, *J. Org. Chem.*, 1978, **43**, 1271.
[191] L. J. Magdzinski and Y. L. Chow, *J. Amer. Chem. Soc.*, 1978, **100**, 2444.
[192] K. S. Pillay and Y. L. Chow, *J.C.S. Perkin II*, 1977, 93.
[193] R. M. Wilson and F. Geiser, *J. Amer. Chem. Soc.*, 1978, **100**, 2225.
[194] J. Griffiths and C. Hawkins, *J. Appl. Chem. Biotechnol.*, 1977, **27**, 558.
[195] T. Kametani, T. Ohsawa, M. Ihara, and K. Fukumoto, *J.C.S. Perkin I*, 1978, 460; *cf.* T. Kametani, T. Ohsawa, K. Takahashi, M. Ihara, and K. Fukumoto, *Heterocycles*, 1976, **4**, 1637.
[196] B. E. Horsey and D. G. Whitten, *J. Amer. Chem. Soc.*, 1978, **100**, 1293.

examined the way in which environment governs the photo-oxidation of proto-porphyrin IX as the dimethyl ester and the bis(dihydrocholesterol) ester: the ratio of oxidation products and the rate of reaction vary according to whether the protoporphyrin is part of a micelle, a monolayer, or in solution. There is apparently a high reactivity in monolayers or assemblies yet, surprisingly, a low reactivity in micelles. The efficiencies of various porphyrins as photosensitizers for the oxidation of methionine have been investigated.[197]

Tryptophan (99) is one of the amino acids most susceptible to photo-oxidative attack, and this reaction is often implicated in the photodynamic action on biological substances which occurs under the influence of sensitizers, light, and oxygen. The oxidative product from (99) is N'-formylkynurenine (100),[198] a

(99) R = CH(NH$_2$)CO$_2$H
(102) R = CH$_2$NHMe

(100)

particularly good photosensitizer. It has now been proven, in agreement with an earlier proposal,[199] that superoxide radical anion $(O_2{}^{\cdot -})$ is formed in the u.v. (315—400 nm) irradiation of tryptophan.[200] The enzyme superoxide dismutase (SOD) is a good probe for the formation of $O_2{}^{\cdot -}$, because of its rapid reaction to yield hydrogen peroxide (reaction 11).

$$2O_2{}^{\cdot -} + 2H^+ \xrightarrow{\text{SOD}} H_2O_2 + O_2 \qquad (11)$$

Fortunately, tryptophan residues are absent from SOD, so that it can be used *in situ* to detect the photogeneration of $O_2{}^{\cdot -}$ by analysis of the hydrogen peroxide formed. Therefore, the increased amounts of hydrogen peroxide formed on irradiation of oxygenated aqueous solutions of tryptophan in the presence of SOD clearly implicate $O_2{}^{\cdot -}$ production. A review which summarizes the results of $^1O_2$ reactions with tryptophan and related compounds emphasizes the role of a zwitterionic species [*e.g.* (101)] in the additions.[201] For example, dye-sensitized photo-oxidation of N^b-methyltryptamine (102) in methanol at 0 °C gives the hydroperoxypyrrolo-indole (103) from interception of the zwitterion by a nucleophilic amine group of the side chain.[202] Direct photo-oxidation of 3-methylindole and 2-ethyl-3-methylindole gives the products of carbon—carbon double bond cleavage, 2-formamido- and 2-propionamido-acetophenone, respectively, in a process analogous to the photo-oxidation of (99) to (100).[203]

[197] G. Cauzzo, G. Gennari, G. Jori, and J. D. Spikes, *Photochem. and Photobiol.*, 1977, **25**, 389.
[198] W. E. Savige, *Proc. Internat. Wolltextil Forschungskonf., 5th*, 1975, **2**, 570 (*Chem. Abs.*, 1977, **87**, 136 344).
[199] P. Walrant, R. Santus, and L. I. Grossweiner, *Photochem. and Photobiol.*, 1975, **22**, 63.
[200] J. P. McCormick and T. Thomason, *J. Amer. Chem. Soc.*, 1978, **100**, 312.
[201] I. Saito, T. Matsuura, M. Nakagawa, and T. Hino, *Accounts Chem. Res.*, 1977, **10**, 346.
[202] M. Nakagawa and T. Hino, *Heterocycles*, 1977, **6**, 1675.
[203] N. A. Evans, *Austral. J. Chem.*, 1977, **30**, 1127.

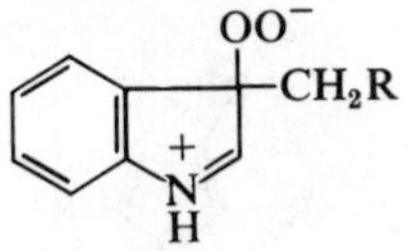

(101) 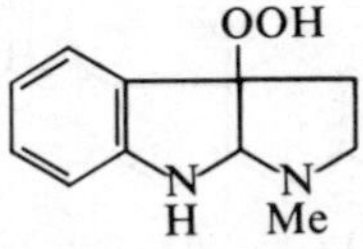 (103)

Some reported photo-oxidations of nitrogen-containing compounds do not involve $^1O_2$ production. Shimizu and Bartlett[204] have previously pointed out such a mechanism of photo-oxidation of alkenes using α-diketones as sensitizers, in which epoxides are the main products. This type of reaction has been applied to the benzil- or biacetyl-sensitized photo-oxidation of pyrimidines (104), which yields products probably derived from the intermediate epoxides (105) (Scheme 14).[205] U.v. irradiation of steroidal isoxazolidines (106) in various solvents in

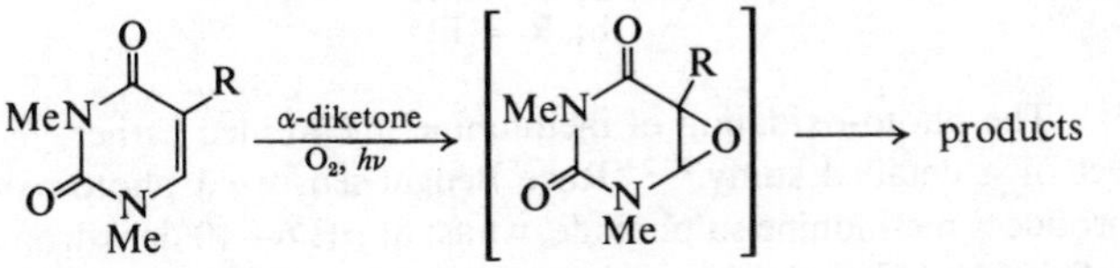

(104) (105) R = Me or H

Scheme 14

the presence of oxygen gives a mixture of nitro- and azoxy-compounds arising from isoxazolidine ring cleavage and rearrangement.[206] Again, the reaction has been shown not to involve $^1O_2$ but rather to occur by an electron transfer from isoxazolidine to oxygen.

Methylene blue-sensitized photo-oxidation of the alkoxyoxazoles (107) to diacylcarbamates by $^1O_2$ has been reported.[207] In some cases, *e.g.* (107a, b), the novel heterocyclic ring system of a 3*H*-1,2,4-dioxazole (108) can be isolated as the main reaction product.[207, 208]

This year there have been several instances of photo-oxidation of amino-acids and related materials. Phenylalanine is not appreciably photo-oxidizable, but substitution into the aromatic ring of electron-donating groups such as hydroxyl (as in tyrosine) leads to rapid photo-oxidation.[209] Both singlet oxygen and hydrogen-abstraction mechanisms are involved in the eosin-sensitized photo-oxidation of hydroxy-substituted phenyl-alanines. Cyclic dipeptides undergo a radical oxidation to mono- and bis-hydroperoxides on irradiation in the presence of benzophenone.[210] The photo-oxidation of ornithine and the related diamine putrescine (1,4-diaminobutane), catalysed by titanium dioxide, have been

204 N. Shimizu and P. D. Bartlett, *J. Amer. Chem. Soc.*, 1976, **98**, 4193.
205 H.-S. Ryang and S. Y. Wang, *J. Amer. Chem. Soc.*, 1978, **100**, 1302.
206 L. Lorenc, I. Juranić, and M. L. Mihailović, *J.C.S. Chem. Comm.*, 1977, 749.
207 M. L. Graziano, M. R. Iesce, A. Carotenuto, and R. Scarpati, *J. Heterocyclic Chem.*, 1977, **14**, 261.
208 M. L. Graziano, M. R. Iesce, A. Carotenuto, and R. Scarpati, *Synthesis*, 1977, 572.
209 F. Rizzuto and J. D. Spikes, *Photochem. and Photobiol.*, 1977, **25**, 465.
210 J. Häusler, R. Jahn, and U. Schmidt, *Chem. Ber.*, 1978, **111**, 361.

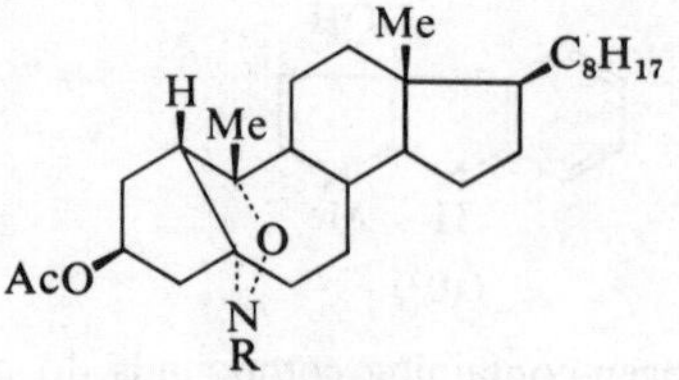

(106) a; R = H
      b; R = Me
      c; R = Ac

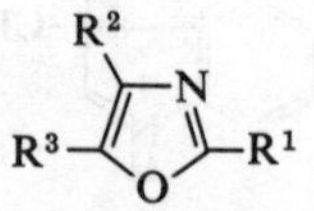

(107) a; R^1 = Ph, R^2 = Me, R^3 = OMe
      b; R^1 = Ph, R^2 = Me, R^3 = OEt
      c; R^1 = OEt, R^2 = OMe, R^3 = Ph

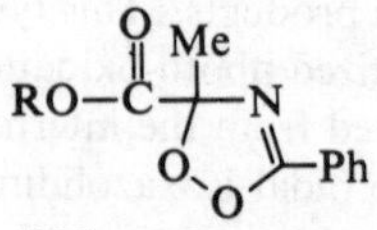

(108) a; R = Me
      b; R = Et

studied.[211] The photo-oxidation of methionine, mentioned earlier,[197] has been the subject of a detailed study.[212] Rose Bengal-sensitized photo-oxidation at pH ⩽ 6 produces methionine sulphoxide, whilst at pH 6—10 dehydromethionine [(110) in Scheme 15] and hydrogen peroxide are produced. Above pH 9, a process leading directly to methionine sulphoxide and hydrogen peroxide becomes important. All the above reactions seem to ensue from $^1O_2$ attack on the sulphur atom of methionine, to yield the persulphoxide (109) as shown in Scheme 15.

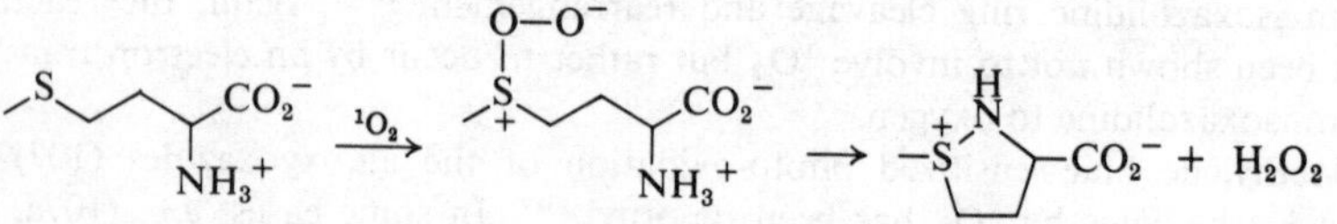

Scheme 15

8 Miscellaneous Oxidations

Binkley[213] has previously reported a mild photochemical method for the oxidation of alcohols to carbonyl compounds which involves the preparation and photolysis of the pyruvate ester of the alcohol, yielding acetaldehyde, carbon monoxide, and the carbonyl compound. Binkley et al.[214] have extended this method to the photo-oxidation of nucleosides such as 5′-O-tritylthymidine, 5′-O-benzoylthymidine, and their respective 3′-epimers. Kinetic parameters for the free-radical photo oxidation of benzyl alcohol have been established.[215]

[211] M. B. Patil and G. D. Kalyankar, *Marathwada Univ. J. Sci., Nat. Sci.*, 1976, **15**, 47 (*Chem. Abs.*, 1978, **89**, 24 751).
[212] P. K. Sysak, C. S. Foote, and T.-Y. Ching, *Photochem. and Photobiol.*, 1977, **26**, 19.
[213] R. W. Binkley, *J. Org. Chem.*, 1976, **41**, 3030.
[214] R. W. Binkley, D. G. Hehemann, and W. W. Binkley, *Carbohydrate Res.*, 1977, **58**, C10.
[215] A. N. Shendrik, N. P. Mytsyk, and I. A. Opeida, *Kinetika i Kataliz*, 1977, **18**, 1077 (*Chem. Abs.*, 1977, **87**, 167 238).

Certain catalysed photo-oxidations of alcohols have been found. For example, irradiation of a homogeneous solution in propan-2-ol of a rhodium–tin chloride catalyst markedly increases the rate of production of hydrogen and acetone, with an energy conversion efficiency of up to 23% being reached.[216] The heterogeneous photo-oxidation of methylbutanols on titanium dioxide is reported.[217] A photoelectrochemical method of oxidation involves irradiation of catalytic amounts of quinones in the presence of a substrate (*e.g.* propan-2-ol or toluene) and a graphite anode.[218] The photochemical reaction yields oxidized substrate (acetone or bibenzyl) and hydroquinone; the anode is held at such a potential that the hydroquinone is selectively oxidized back to quinone.

Bartlett[219] has reviewed the free radical aspects of photo-oxidation. The direct photo-oxidation of biacetyl in decane solution proceeds by attack of the triplet biacetyl on oxygen leading to the formation of biacetyl peroxide, carbon dioxide, acetic anhydride, and acetic acid.[220] The dye-sensitized photo-oxidation of α-ketocarboxylic acids has been the subject of controversy. The initial report[221] was that $^1O_2$ attacked the α-keto-acid to give carbon dioxide and the corresponding peroxy-acid, which instantaneously caused decarboxylation of the starting α-keto-acid. Davidson[222] pointed out that triplet excited dye was responsible for a reaction leading to decarboxylation even in the absence of oxygen. However, the original group have now regained face by showing at least some involvement of $^1O_2$ in the reaction of α-keto-acids (111).[223]

$$\underset{\text{(111)}}{} R-\overset{\overset{\displaystyle O}{\|}}{C}-CO_2H$$

$$\text{a; R} = HO_2C(CH_2)_2$$
$$\text{b; R} = Me$$
$$\text{c; R} = Et$$
$$\text{d; R} = MeCH_2CH_2$$

With the increasing scale of oil spillage onto the seas, a study of the photochemical oxidation of floating hydrocarbon oils is of obvious importance. Samples of isopropylbenzene, toluene, and mineral oil containing various concentrations of 1-hydroxynaphthalene, 1-naphthaldehyde, or 1-nitronaphthalene have been floated on water as thin layers and subjected to photo-oxidation.[224] Identification of some of the photo-oxidation products has allowed the proposal of reaction mechanisms. The xanthone-sensitized photo-oxidation of n-hexadecane on water produces hexadecanol in reasonable quantum yield ($\Phi = 0.16$—0.60).[225] The vapour phase photo-oxidation of alkyltoluenes to alkylbenzaldehydes occurs at a titanium dioxide surface on u.v. irradiation in the presence

216 S. Shinoda, H. Moriyama, Y. Kise, and Y. Saito, *J.C.S. Chem. Comm.*, 1978, 348.
217 A. Walker, M. Formenti, P. Meriaudeau, and S. J. Teichner, *J. Catalysis*, 1977, **50**, 237.
218 J. M. Bobbitt and J. P. Willis, *J. Org. Chem.*, 1977, **42**, 2347.
219 P. D. Bartlett, *A.C.S. Symposia Series*, 1978, No. 69 (Organic Free Radicals), p.15.
220 M. Bouchy and J. C. Andre, *Mol. Photochem.*, 1977, **8**, 345.
221 C. W. Jefford, A. F. Boschung, T. A. B. M. Bolsman, R. M. Moriarty, and B. Melnick, *J. Amer. Chem. Soc.*, 1976, **98**, 1017.
222 R. S. Davidson, *Tetrahedron Letters*, 1976, 4181.
223 C. W. Jefford, A. Exarchou, and P. A. Cadby, *Tetrahedron Letters*, 1978, 2053.
224 W. H. K. Sanniez and N. Pilpel, *J.C.S. Faraday I*, 1978, **74**, 123.
225 H. D. Gesser, T. A. Wildman, and Y. B. Tewari, *Environ. Sci. Technol.*, 1977, **11**, 605.

of oxygen.[226] The photodehydrogenation of thiochromanones (112) occurs on irradiation (350 nm) to yield thiochromones (113) by an unspecified mechanism.[227] The reaction yield increases with increasing solvent polarity, highest yields (40—55%) being reported on irradiation in methanol or t-butanol.

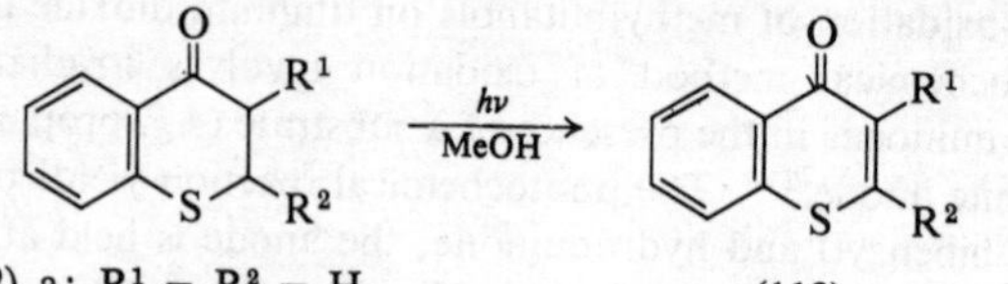

$$(112) \quad a; \ R^1 = R^2 = H$$
$$b; \ R^1 = H, R^2 = Me$$
$$c; \ R^1 = Me, R^2 = H$$

$$(113)$$

Sinnreich *et al.*[228] have recently observed the direct photo-oxidation of aliphatic sulphides to the corresponding sulphoxides; on the basis of the inertness of di-t-butyl sulphide, they proposed that the reaction involved an intermediate arising from hydrogen abstraction at an α-carbon atom. However, Japanese workers[229] now argue against this route, because diphenyl sulphide can be directly photo-oxidized to diphenyl sulphoxide in several solvents. The direct photo-oxidations of methyl phenyl sulphide, pentamethylene sulphide, and di-n-butyl sulphide also yield the corresponding sulphoxides. The rate of sensitized photo-oxidation of 1,3-dithiol-2-thiones (114) increases with increasing electron-density of the heterocyclic ring.[230]

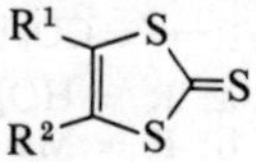

$$(114) \quad a; \ R^1 = H,$$
$$R^2 = p\text{-}XC_6H_4 \text{ where } X = H, MeO, Me, Br, or Cl$$
$$b; \ R^1\text{-}R^2 = (CH_2)_4$$
$$c; \ R^1\text{-}R^2 = (CH_2)_5$$

Organic contaminants in waste water can be removed by photolytic oxidation in the presence of aqueous sodium hypochlorite. In this connection, the products from the photo-oxidation of alkylbenzene sulphonic acids[231] and ethylene glycol dimethyl ether[232] have been investigated, and reaction mechanisms postulated.

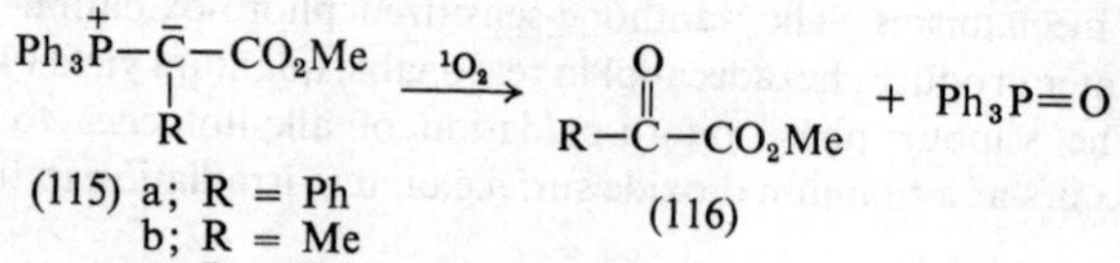

$$(115) \quad a; \ R = Ph$$
$$b; \ R = Me$$
$$c; \ R = Pr^i$$
$$d; \ R = H$$

$$(116)$$

[226] M.-N. Mozzanega, J. M. Herrmann, and P. Pichat, *Tetrahedron Letters*, 1977, 2965.
[227] A. Couture, A. Lablache-Combier, and T. Q. Minh, *Tetrahedron Letters*, 1977, 2873.
[228] D. Sinnreich, H. Lind, and H. Batzer, *Tetrahedron Letters*, 1976, 3541.
[229] T. Tezuka, H. Miyazaki, and H. Suzuki, *Tetrahedron Letters*, 1978, 1959.
[230] E. Fanghaenel and G. Lutze, *J. prakt. Chem.*, 1977, **319**, 875.
[231] M. Nakamura and Y. Ogata, *Bull. Chem. Soc. Japan*, 1977, **50**, 2396.
[232] Y. Ogata, K. Takagi, and T. Suzuki, *J.C.S. Perkin II*, 1978, 562.

Phosphorus ylides (115a—c) undergo a reaction with $^1O_2$, generated by dye-sensitized photo-oxidation, to give the α-ketoesters (116a—c) in high yield (89—100%).[233] Only in the case of ylide (115d) does the α-ketoester product, methyl glyoxylate, react further with the phosphorus ylide to give a mixture of dimethyl fumarate and maleate.

[233] C. W. Jefford and G. Barchietto, *Tetrahedron Letters*, 1977, 4531.

6
Photoreactions of Compounds containing Heteroatoms other than Oxygen

BY S. T. REID

1 Nitrogen-containing Compounds

Rearrangement.—Further examples of *cis–trans* isomerization about the carbon—nitrogen double bond have been reported. *Z*-Isomers were prepared and detected at low temperature by irradiation of *E*-azomethines of 1,5-dimethyl-2-phenyl-4-formyl-Δ^4-pyrazolin-3-one.[1] Similarly, isomerization was observed on irradiation in benzene of *E*-2-(2,6-dichlorobenzylidenehydrazino)pyrimidine and related hydrazones, although in the presence of oxygen a competing photo-sensitized oxidation occurred to give 3-aryl-1,2,4-triazolo[4,3-*a*]pyrimidines.[2,3] Partial quantum yields for the $E \rightleftharpoons Z$ photoisomerization of benzophenonan-thraldazine, adsorbed on silica gel, have been determined.[4] Isomerization was also observed on direct and triplet-sensitized irradiation of *E*-3-ethoxyiminobutan-2-one[5] and on benzophenone-sensitized irradiation of 4-ethoxyimino-2,6-dimethylheptan-3-one.[6]

Photochemically induced cyclization in the 2*H*-azepin-2-ones (1) has been shown to be more complex than originally thought. In addition to the previously reported photoproducts (2), the stereoisomers (3) were also formed.[7] Moreover, irradiation of the tetramethyl derivative (1; R = Me) in methanol gave the imine (4) by loss of carbon monoxide, a reaction which is not surprising in view of the known decarbonylation of cyclohepta-3,5-dienones. An analogous electro-cyclic process has been observed in benzofurano-annelated azepinones.[8] The photorearrangement of pyrid-2-ones, but not the diaza-analogues has been extensively studied. The pyrimidin-2-one (5), however, has now been shown to give on irradiation ($\lambda > 300$ nm) in benzene the diazabicyclo[2,2,0]hex-5-ene (6), an isomer which is unexpectedly stable at room temperature.[9] The process can be reversed quantitatively by irradiation at 254 nm. Electrocyclic ring cleavage occurs readily in 1,2-dipropyl-3,5-diethyl-1,2-dihydropyridine (7) to give the iminodiene (8).[10] The dihydropyridine serves as a model for analogous dihydro

[1] W. Freyer and G. Tomaschewski, *J. prakt. Chem.*, 1977, **319**, 895.
[2] T. Tsujikawa and M. Tatsuta, *Chem. and Pharm. Bull. (Japan)*, 1977, **25**, 3137.
[3] T. Tsujikawa and M. Tatsuta, *Heterocycles*, 1977, **6**, 423.
[4] D. Fassler and W. Guenther, *Z. Chem.*, 1978, **18**, 69.
[5] P. Baas and H. Cerfontain, *J.C.S. Perkin II*, 1977, 1351.
[6] P. Baas and H. Cerfontain, *Tetrahedron Letters*, 1978, 1501.
[7] J. W. Pavlik and C. A. Seymour, *Tetrahedron Letters*, 1977, 2555.
[8] H.-D. Becker and K. Gustafsson, *J. Org. Chem.*, 1977, **42**, 2966.
[9] T. Nishio, A. Katoh, Y. Omote, and C. Kashima, *Tetrahedron Letters*, 1978, 1543.
[10] M. Guay and F. Lamy, *Canad. J. Chem.*, 1978, **56**, 302.

compounds which cross-link polypeptide chains in elastin, and the photo-transformation could in principle be used as a specific method for breaking this link. In contrast to this result, however, the azahexatrienes, previously proposed as intermediates in the photorearrangement of ethyl 2-cyano-1,2-dihydro-quinoline-1-carboxylates, have now been shown not to be involved.[11] Photocyclization of the triazahexatrienes (9) to give the 8-arylaminotheophyllines (10) and, in the presence of oxygen, the azalumazin-7-ones (11) has been reported,[12] and rearrangement of 2-dicyanomethylene-1,3-indanedione monophenylhydra-zone (12) to the 5*H*-indeno[1,2-*c*]pyridazine (13) occurs on treatment with Lewis bases or on irradiation;[13] in the latter case, the corresponding anionic inter-mediate has been detected by flash photolysis.

Interest has been shown in nitrogen-containing systems capable of undergoing processes equivalent to the di-π-methane rearrangement. One example which can be rationalized in this way is the photochemically induced conversion of the

[11] M. Ikeda, S. Matsugashita, and Y. Tamura, *Heterocycles*, 1978, **9**, 281.
[12] F. Yoneda and M. Higuchi, *Chem. and Pharm. Bull. (Japan)*, 1977, **25**, 2794.
[13] A. M. Braun, M. Wittmer-Metz, J. Faure, J. Schultz, and H. Junek, *J. Chim. Phys. Phys.-Chim. Biol.*, 1977, **74**, 1054.

(9)   (10)

(11)

(12)   (13)

diazabicycloheptadienes (14) into the 1,4-diazepines (15).[14] A different explanation is offered to account for the formation in 44% yield of the 7-azabicyclo-[2,2,1]heptane (16) from the pyrrolinium salt (17) on irradiation in methanol;[15] a pathway involving intramolecular $[\pi2 + \pi2]$ cycloaddition and the formation of a short-lived aziridinium salt followed by heterolytic cleavage of the carbon—nitrogen bond and solvent capture has been proposed. An apparently unrelated

(14) $R^1 = R^2 = $ Ph, *p*-tolyl, or Me   (15)
     $R^1 = $ Ph, $R^2 = $ Me
     $R^1 = $ Me, $R^2 = $ Ph

(17)   (16)

[14] G. Reissenweber and J. Sauer, *Tetrahedron Letters*, 1977, 4389.
[15] P. S. Mariano, J. L. Stavinoha, and R. Swanson, *J. Amer. Chem. Soc.*, 1977, **99**, 6781.

photocyclization is observed in 1-vinylpyrid-2-ones (18) leading, in the presence of perchloric acid, to the formation of crystalline oxazolo[3,2-*a*]pyridinium perchlorates (19).[16] Evidence has been advanced supporting the generation of intermediate azomethine ylides (20) by electrocyclic ring closure of the excited singlet state of the pyridones.

(18) R = H or Me   (20)   (19) X = ClO$_4^-$

Low temperature irradiation in pentane of the aziridine (21) gave an *N*-(ethoxycarbonyl)aza[17]annulene with the configuration proposed in (22).[17]

(21)   (22)

Acetone-sensitized irradiation of the azonine 1,2-oxide (23) gave a labile compound which underwent further thermal rearrangement at 20 °C to the *trans*-isomer (24).[18] The structure of the unstable intermediate is tentatively claimed

(23)   (25)   (24)

(26)

$$R = $$

[16] P. S. Mariano, A. A. Leone, and E. Krochmal, *Tetrahedron Letters*, 1977, 2227.
[17] H. Röttele, G. Heil, and G. Schröder, *Chem. Ber.*, 1978, **111**, 84.
[18] A. G. Anastassiou, S. J. Girgenti, R. C. Griffith, and E. Reichmanis, *J. Org. Chem.*, 1977, **42**, 2651.

to be the oxonin (25), the photogeneration and thermal rearrangement both being allowed by orbital symmetry considerations. Rearrangement arising by ring expansion in the aziridine (26) has also been described.[19]

Numerous examples of nitrogen-containing analogues of the stilbene-to-phenanthrene cyclization have been reported, although in general little evidence has been presented for dihydrophenanthrene-type intermediates. Phenanthro-[9,10-*d*]oxazoles have been obtained by direct irradiation of 4,5-diphenyloxazoles in the presence of oxygen or iodine.[20] An analogous cyclodehydrogenation of 1,4,5-triphenylpyrazole (27) to give 1-phenyl-1*H*-phenanthro[9,10-*c*]-pyrazole (28) has been observed, but in the presence of benzophenone rearrangement is preferred leading to the formation of 3-anilino-2,3-diphenylacrylonitrile (29).[21] This observation suggests that 3-aminoacrylonitriles may well be intermediates in the previously reported photorearrangement of pyrazoles to imidazoles. Similar cyclizations have been reported in quinoline derivatives.[22,23]

The analogous cyclization of azobenzene to benzo[*c*]cinnoline is known to occur only in strong acid. 2-Phenylazopyridine, however, has been reported to undergo cyclization to 4-pyrido[*c*]cinnoline both in acid and in neutral solution.[24] The acid-catalysed photocyclization of 2-phenylazobenzo[*c*]cinnoline has been re-examined and affords not the benzo[1,2-*c*:4,5-*c'*]dicinnoline as previously reported but the isomeric benzo[1,2-*c*:4,3-*c'*]dicinnoline.[25] Electrochemical studies of the photoreactions of the triphenyldiazenium cation (30) in acetonitrile or 2,6-lutidine showed that initial cyclization to the dihydrobenzocinnolinium cation (31) occurred, followed by dehydrogenation to give the benzocinnolinium cation (32).[26]

[19] D. W. Jones, *J.C.S. Chem. Comm.*, 1978, 404.
[20] V. N. R. Pillai and M. Ravindran, *Indian J. Chem.*, 1977, **15B**, 1043.
[21] J. Grimshaw and D. Mannus, *J.C.S. Perkin I*, 1977, 2096.
[22] M. Onda, Y. Harigaya, and T. Suzuki, *Chem. and Pharm. Bull (Japan)*, 1977, **25**, 2935.
[23] J. Glinka, *Roczniki Chem.*, 1977, **51**, 577.
[24] V. N. R. Pillai and E. Purushothaman, *Current Sci.*, 1977, **46**, 381.
[25] J. W. Barton and R. B. Walker, *Tetrahedron Letters*, 1978, 1005.
[26] G. Cauquis and G. Reverdy, *Tetrahedron Letters*, 1977, 3267.

A new asymmetric photocyclization of N-arylenamines with circularly polarized light has been reported.[27] The optical yield for the formation of N-methyl-2-phenyldihydroindole (33) from the enamine (34) is estimated to be 0.20%, a value which is of the same order of magnitude as that found in the corresponding asymmetric synthesis of helicenes. The photocyclization of an enamine has been employed in the total synthesis of (±)-cryptoaustoline.[28] The analogous photocyclization of 2-anilinocyclohex-2-enone (35) to the *cis*-fused indoline (36) in high yield provides a new approach to the synthesis of the *Aspidosperma* alkaloids.[29]

New synthetic applications of the photocyclization of enamides have been reported, and in particular the synthesis of benzo[c]phenanthridine alkaloids has been reviewed.[30] A typical example is the formation of the *trans*-lactam (37) on irradiation of the N-benzoylenamine (38).[31] These reactions are viewed as proceeding *via* an intermediate arising by photocyclization, followed by a thermal [1,5]hydrogen shift. Cyclization is frequently accompanied by oxidation, as in the preparation of nauclefine[32] and a thia-analogue,[33] and as in the photocyclization of the enamide (39) to give a precursor (40) of (±)-lycoran.[34]

Further examples of the generation of nitrile ylides from azirines and their reactions have been described. The first example of a [$_\pi 6 + _\pi 4$] addition of a

[27] J. F. Nicoud and H. B. Kagan, *Israel J. Chem.*, 1977, **15**, 78.
[28] I. Ninomiya, J. Yasui, and T. Kiguchi, *Heterocycles*, 1977, **6**, 1855.
[29] A. G. Schultz and I.-C. Chiu, *J.C.S. Chem. Comm.*, 1978, 29.
[30] I. Ninomiya, O. Yamamoto, T. Kiguchi, T. Naito, and H. Ishii, *Heterocycles*, 1977, **6**, 1730.
[31] C.-C. Wei and S. Teitel, *Heterocycles*, 1977, **8**, 97.
[32] M. Sainsbury and N. L. Uttley, *J.C.S. Perkin I*, 1977, 2109.
[33] A. Shafiee and A. Rashidbaigi, *J. Heterocyclic Chem.*, 1977, **14**, 1317.
[34] H. Iida, S. Aoyagi, Y. Yuasa, and C. Kibayashi, *Heterocycles*, 1977, **6**, 1747.

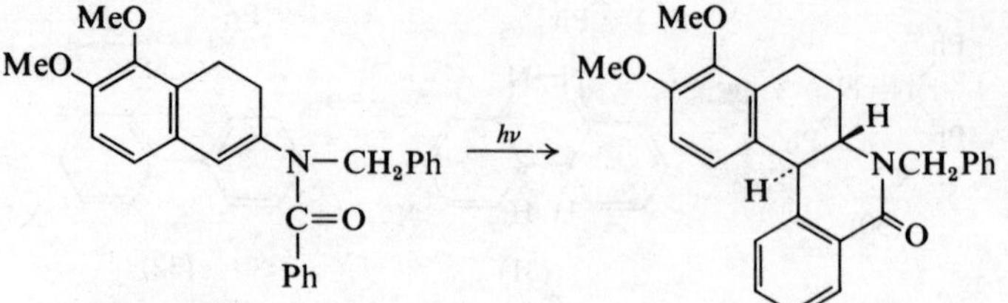

(38)

(37)

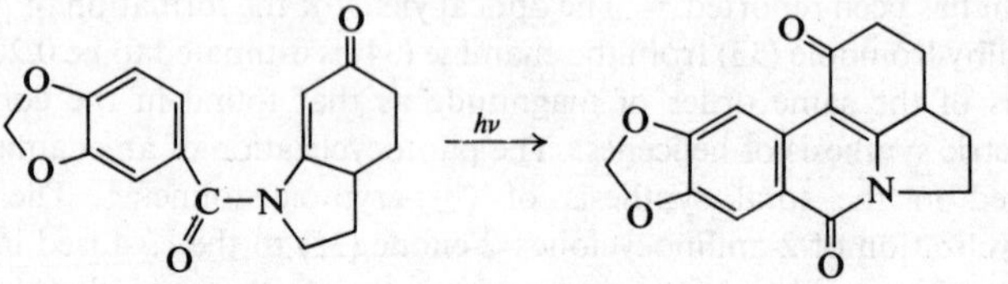

(39)

(40)

nitrile ylide in which the ylide functions as the 4π component has been discovered.[35] Irradiation of a mixture of 3-phenyl-2,2-dimethyl-2*H*-azirine (41) and 6,6-dimethylfulvene (42) in cyclohexane gave the $[\pi6 + \pi4]$ adduct (43) and the $[\pi4 + \pi2]$ adduct (44). The adduct (43) underwent a rapid [1,5]hydrogen shift to give the isomer (45). The intramolecular addition of photochemically generated

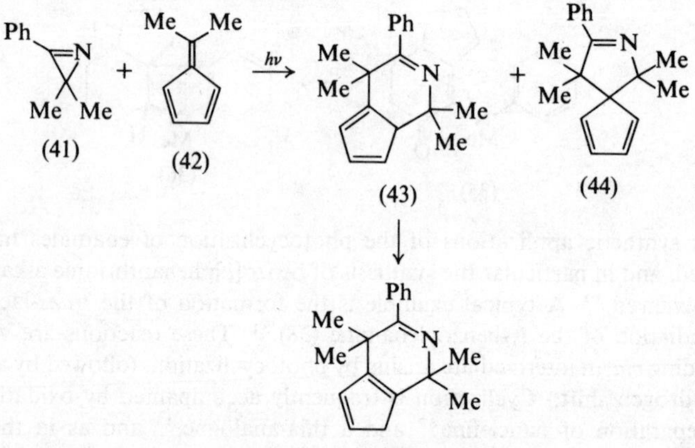

(41)

(42)

(43)

(44)

(45)

nitrile ylides to alkenes is increasingly attracting attention. Irradiation in benzene of 3-(2′-vinyl-2-biphenyl)-2*H*-azirine (46), for example, gave as the only identifiable photoproduct 1*H*-phenanthro[9,10-*b*]pyrrole (47);[36] a pathway involving intramolecular 1,3-dipolar addition of the nitrile ylide to the alkene, followed by oxidation, is proposed (Scheme 1). Irradiation of the bis-2*H*-azirine (48) in the presence of dimethyl acetylenedicarboxylate or dimethyl fumarate gave the

[35] A. Padwa and F. Nobs, *Tetrahedron Letters*, 1978, 93.
[36] A. Padwa, H. Ku, and A. Mazzu, *J. Org. Chem.*, 1978, **43**, 381.

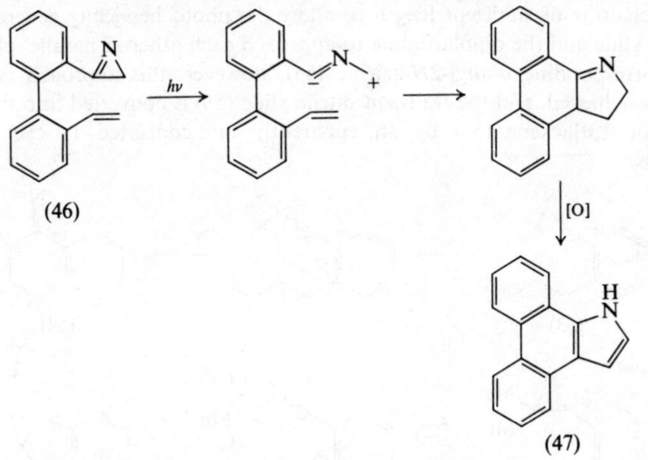

(46)

(47)

Scheme 1

adducts (49) and (50), respectively, derived from the transient azabicyclo[3,1,0]-hexene (51) *via* a second azomethine ylide (52) as outlined in Scheme 2.[36, 37] A more detailed examination of the spatial requirements associated with intra-molecular cycloaddition reactions of nitrile ylides in a series of *o*-alkenylphenyl-substituted 2*H*-azirines has been described.[38] Irradiation in benzene of the *o*-butenylphenyl-2*H*-azirine (53) gave the 2*H*-benz[*g*]indole (54); in this case, the

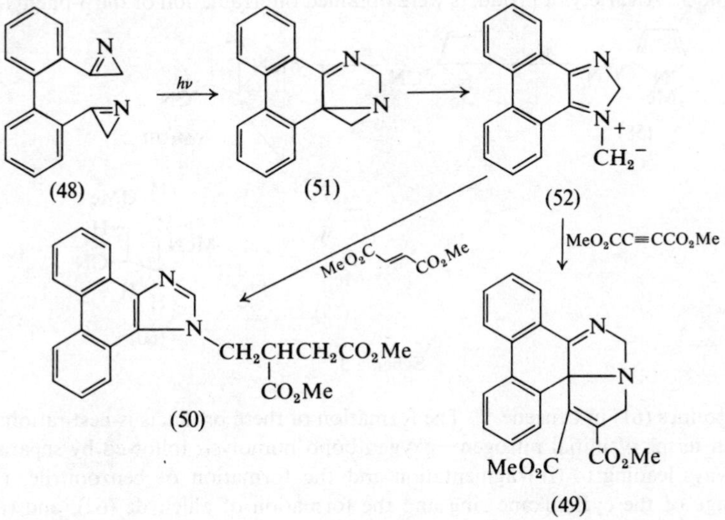

(48)

(51)

(52)

(50)

(49)

Scheme 2

[37] A. Padwa and H. Ku, *J.C.S. Chem. Comm.*, 1977, 551.
[38] A. Padwa and A. Ku, *J. Amer. Chem. Soc.*, 1978, **100**, 2181.

alkyl chain is of sufficient length to allow the photochemically generated bent nitrile ylide and the dipolarophile to approach each other in parallel planes. In the corresponding o-allyl-2H-azirine (55), however, this approach cannot be readily achieved, and the transient nitrile ylide (56) is converted into the benzo-bicyclo[3,1,0]hexene (57) by an apparently non-concerted 1,1-cycloaddition process.

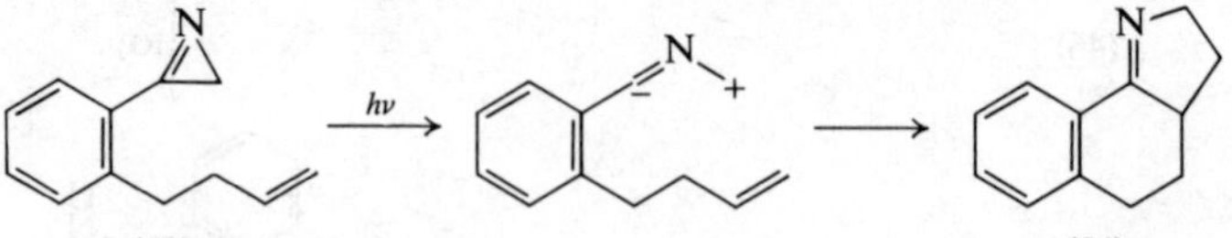

Further efforts have been made to clarify the mechanism of photorearrangement of five-membered heterocycles. Irradiation of 2-cyano-N-methylpyrrole (58) in methanol affords the corresponding 3-cyanopyrrole (59) together with the methanol adduct (60).[39] Compelling evidence has been presented that these products arise by a transposition mechanism involving 2,5-bonding followed by a thermal 'walk' rearrangement of the aziridine nitrogen atom as outlined in Scheme 3. A variety of products were obtained on irradiation of the 3-phenyl-2-

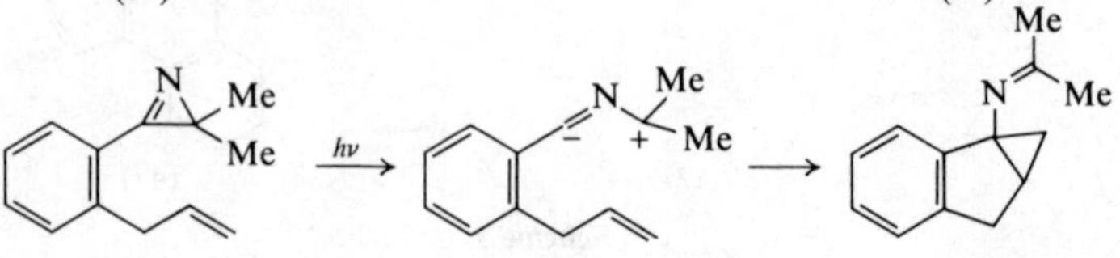

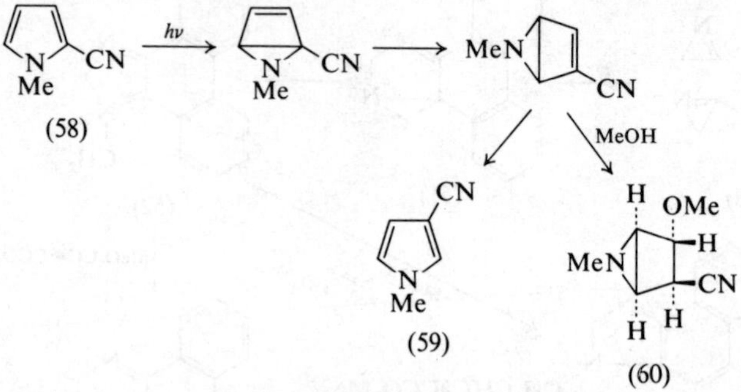

Scheme 3

isoxazolines (61) in benzene.[40] The formation of these products is best rationalized in terms of initial nitrogen—oxygen bond homolysis followed by separate pathways leading to (i) fragmentation and the formation of benzonitrile, (ii) cleavage of the cycloalkane ring and the formation of aldehyde (62), and (iii) azirine formation (isolated in the case of $n = 2$) and subsequent intramolecular addition of nitrile ylide to the aldehyde group to give oxazoles (63) and (64). The

[39] J. A. Barltrop, A. C. Day, and R. W. Ward, *J.C.S. Chem. Comm.*, 1978, 131.
[40] O. Seshimoto, T. Kumagai, K. Shimizu, and T. Mukai, *Chem. Letters*, 1977, 1195.

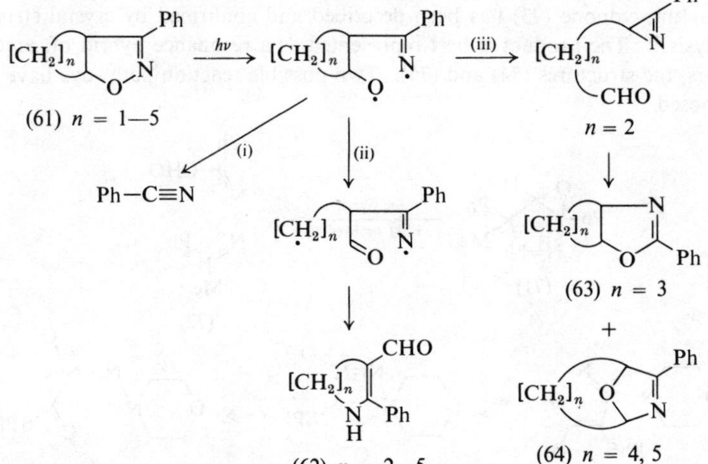

Ph
[CH$_2$]$_n$ ——$h\nu$——> [CH$_2$]$_n$ ——(iii)——> [CH$_2$]$_n$

(61) $n = 1$—5

(i) Ph—C≡N

(ii)

[CH$_2$]$_n$

(62) $n = 2$—5

CHO
n = 2

(63) $n = 3$
+
(64) $n = 4, 5$

product composition is dependent on cycloalkane ring size. A further example of a reversible photochemical Dimroth rearrangement in 5-amino-1,2,3-triazoles has been reported.[41] A diaziridine has been suggested as an intermediate in this transformation but further work is clearly required to substantiate this proposal. The photochemistry of azoles has been reviewed.[42]

Azirines (65) have been shown by spectra to be intermediates in the photo-rearrangement of benzisoxazoles (66) to benzoxazoles (67).[43] The formation of the diaziridines (68) on irradiation of the 3-pyrazolidone azomethinimines (69) in methanol is the result of an initial cyclization to the bicyclic diaziridines (70)

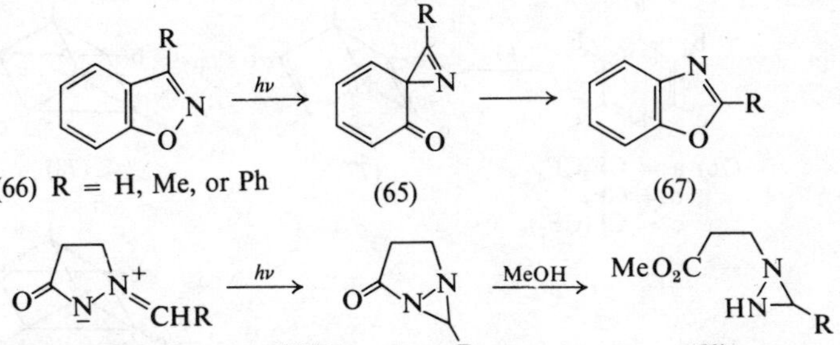

(66) R = H, Me, or Ph   (65)   (67)

(69) R = Ph, p-MeC$_6$H$_4$, or p-ClC$_6$H$_4$ (70)   (68)

followed by reaction with solvent;[44] cyclization only occurs when R = aryl. Extensive conjugation of the oxiran with the azo system in the 1-pyrazolenine (71) is responsible for photorearrangement to the isomeric aldehydes (72) instead of

[41] Y. Ogata, K. Takagi, and E. Hayashi, *Bull. Chem. Soc. Japan*, 1977, **50**, 2505.
[42] M. I. Knyazhanskii, P. V. Gilyanovskii, and A. O. Osipov, *Khim. geterotsikl. Soedinenii*, 1977, 1455.
[43] K. H. Grellmann and E. Tauer, *J. Photochem.*, 1977, **6**, 365.
[44] M. Schulz, G. West, U. Mueller, and D. Henke, *J. prakt. Chem.*, 1976, **318**, 946.

photoelimination of nitrogen.[45] A novel rearrangement of 3-morpholino-4-phenylthiosydnone (73) has been described and confirmed by crystal structure analysis.[46] The product is best represented as a resonance hybrid of, amongst others, the structures (74) and (75). Two possible reaction pathways have been proposed.

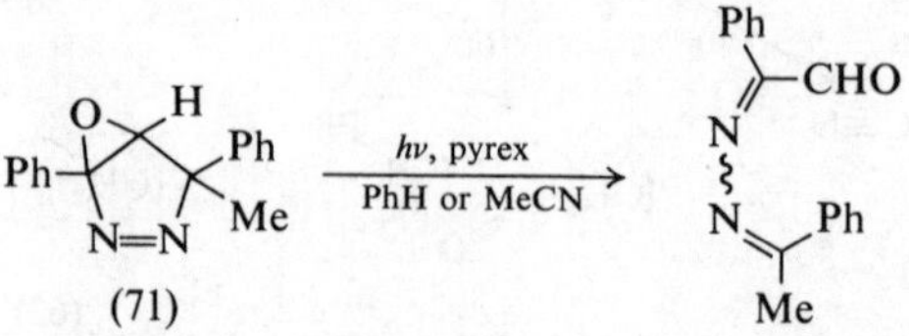

(71) → (72)

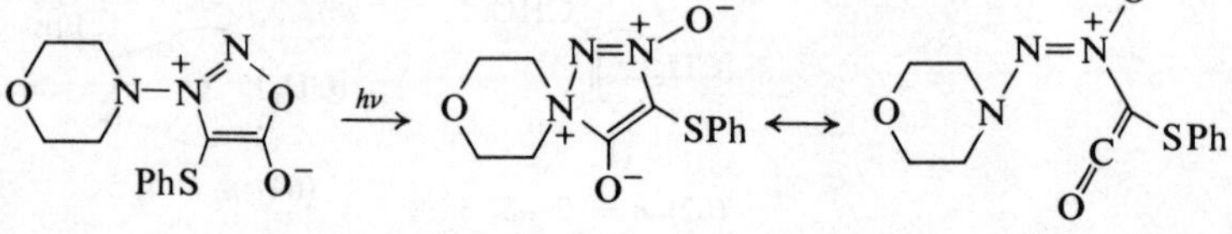

(73)   (74)   (75)

A variety of reports on the photorearrangement of six-membered nitrogen containing heterocycles have been published. Irradiation of highly substituted perfluoroalkylpyridines affords in each case a 1-azabicyclo[2,2,0]hexadiene and two thermally stable azaprismanes.[47] Thus, the pyridine (76), for example, is converted (in $CF_2Cl,CFCl_2$ at 254 nm) into the *p*-bonded isomer (77) and the azaprismanes (78) and (79). Azaprismane (78) appears to be formed directly from isomer (77), whereas azaprismane (79) is claimed to arise by initial formation

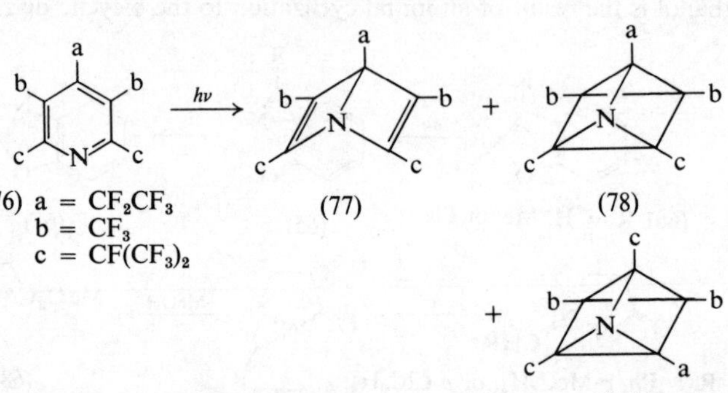

(76) a = CF_2CF_3
     b = CF_3
     c = $CF(CF_3)_2$

(77)   (78)

(79)

of a 2-azabicyclo[2,2,0]hexadiene followed by rapid rearrangement to the corresponding 1-azabicyclo[2,2,0]hexadiene and subsequent prismane formation. These further reactions of bicyclo[2,2,0]hexadienes are considered to be excited state reactions. Irradiation of the perfluoroalkyl aza-cyclohexadiene (80) gave

[45] L. E. Friedrich, N. L. de Vera, and Y. P. Lam, *J. Org. Chem.*, 1978, **43**, 34.
[46] H. Gotthardt, F. Reiter, A. Gieren, and V. Lamm, *Tetrahedron Letters*, 1978, 2331.
[47] R. D. Chambers and R. Middleton, *J.C.S. Perkin I*, 1977, 1500.

as the major photoproduct the azahexatriene (81) by an electrocyclic ring-opening process;[48] fluorine migration and retro-Diels–Alder reactions were also observed in closely related systems.

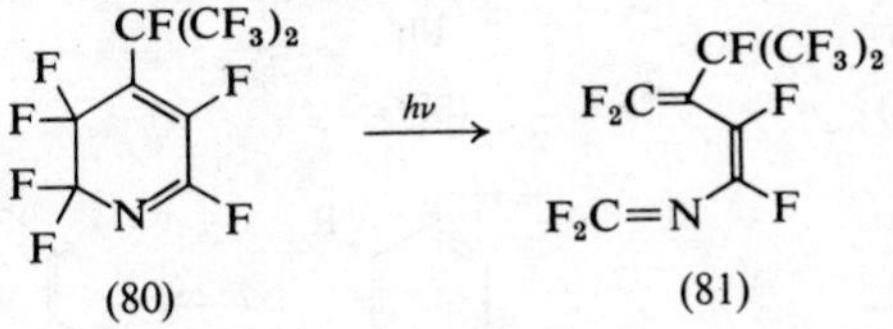

Full details of the photorearrangement of substituted 2-methylpyridines (82) to *o*-substituted anilines (83) have now been published.[49] Deuterium labelling studies have unambiguously established the intermediacy of the biradical (84) in this transformation,[50] and a separate investigation in aqueous alkali has allowed the isolation of the 3-substituted methylene-2-azabicyclo[2,2,0]hex-2-enes (85).[51] The process has been shown to be a two-photon one and follows the pathway outlined in Scheme 4. The conversion of 1-benzyl-2-imino-1,2-dihydropyrimi-dine into 2-benzylaminopyrimidine is another example of a photo-Dimroth type rearrangement.[52]

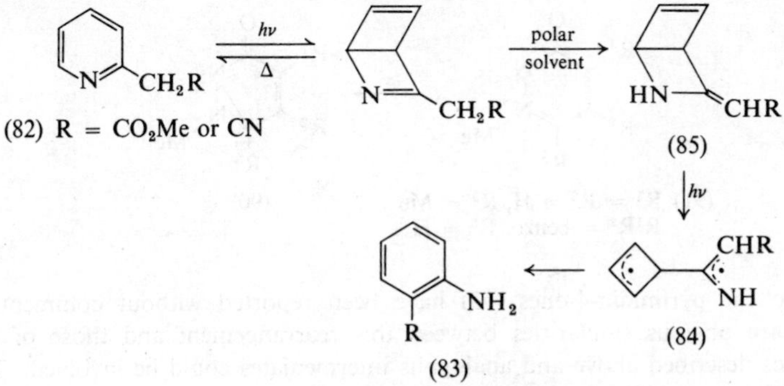

Scheme 4

3-Oxido-1-phenylpyridinium (86; R = H) has previously been reported to give the valence isomer (87; R = H) *via* a photochemically allowed disrotatory ring closure. A different reaction, observed on irradiation of the same betaine and its 6-methyl derivative (86; R = Me) to give the pyrid-2-ones (88), almost certainly involves the valence isomers (87) as intermediates;[53] a pathway *via* a ketene (89) has been proposed (Scheme 5). In a similar fashion, diaziridines (90),

[48] R. D. Chambers, R. D. Hercliffe, and R. Middleton, *J.C.S. Chem. Comm.*, 1978, 305.
[49] K. Takagi and Y. Ogata, *J.C.S. Perkin II*, 1977, 1148.
[50] K. Takagi and Y. Ogata, *J.C.S. Perkin II*, 1977, 1980.
[51] Y. Ogata and K. Takagi, *J. Org. Chem.*, 1978, **43**, 944.
[52] K. Takagi and Y. Ogata, *J.C.S. Perkin II*, 1977, 1410.
[53] T. Laerum and K. Undheim, *Acta Chem. Scand.*, 1978, **B32**, 68.

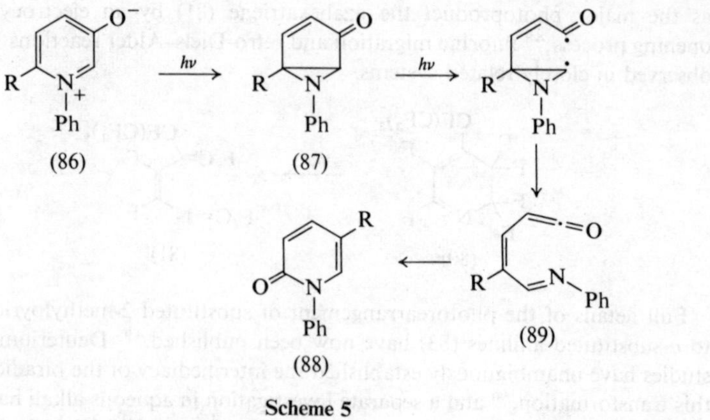

Scheme 5

sufficiently stable to be isolated, were obtained on irradiation of the corresponding 3-oxidopyridazinium betaines (91).[54] A phototransformation of a betaine
is also implicated in the conversion of 13-oxidonorcoralynium into a phthalide-
isoquinoline.[55] The photorearrangement of the fused zwitterionic pyridazines

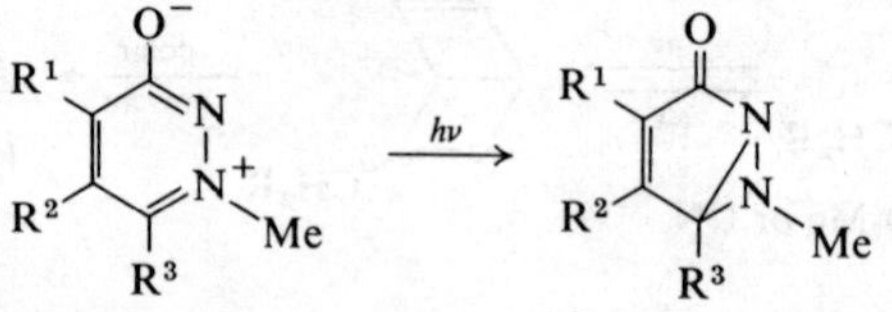

(91) R¹ = R² = H, R³ = Me   (90)
     R¹R² = benzo, R³ = H

(92) to the pyrimidin-4-ones (93) have been reported without comment;[56]
there are obvious similarities between this rearrangement and those of the
betaines described above and analogous intermediates could be involved. The
pyrimidin-4-one (93; R = H) underwent further reaction on irradiation in
methanol to give the 4-oxospiro[4,5]dec-2-ene (94).

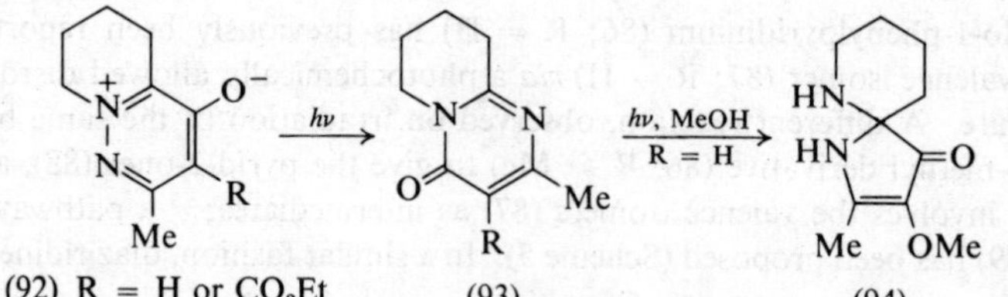

(92) R = H or CO₂Et   (93)   (94)

[54] Y. Maki, M. Kawamura, H. Okamoto, M. Suzuki, and K. Kenji, *Chem. Letters*, 1977, 1005.
[55] J. Imai and Y. Kondo, *Heterocycles*, 1977, **6**, 959.
[56] T. Yamazaki, M. Nagata, S.-I. Hirokami, and S. Miyakoshi, *Heterocycles*, 1977, **8**, 377.

A series of *N*-[ω-(*p*-nitrophenoxy)alkyl]anilines undergoes a photo-Smiles rearrangement leading to the corresponding *N*-(ω-hydroxyalkyl)-*p*-nitrodiphenylamines on irradiation in methanol or acetonitrile.[57]

The study of the photorearrangement of nitrones and heteroaromatic *N*-oxides is one that has again consumed much time and effort. In general, irradiation of cyclic and acyclic nitrones results in the initial formation of oxaziridines. Irradiation of the nitrone (95), for example, afforded a single isomeric oxaziridine (96) which reverted to the original nitrone thermally at room temperature.[58] Several 3-oxo-1-pyrroline 1-oxides (97) undergo photorearrangement to isomeric aryl azetidinones (98) and oxazinones (99).[59] The ratio of products is dependent on the nature of the 2-substituent, but in all cases the azetidinones can be isolated in moderate yield. The ring contraction and expansion processes are believed to take place *via* intermediate oxaziridines (100), and in one case, R = Ph, the

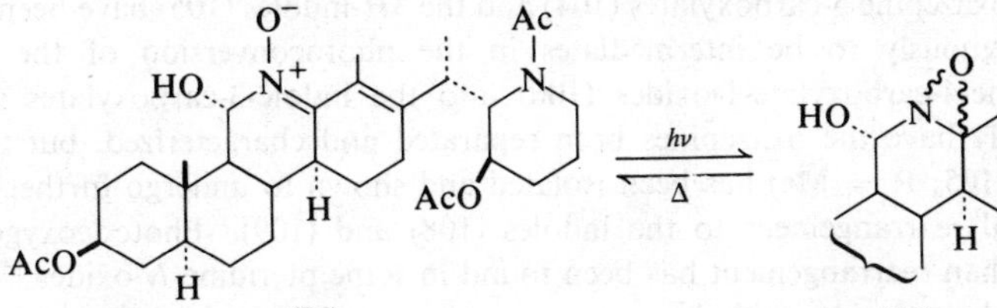

(95)　　　　　　　(96)

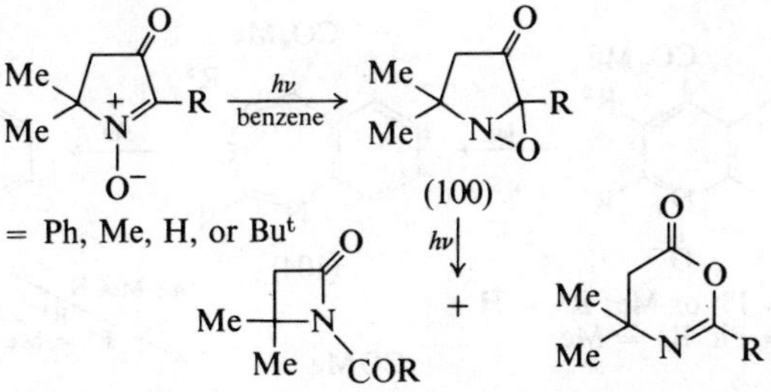

(97) R = Ph, Me, H, or Bu^t

(100)

(98)　　　　　(99)

oxaziridine has been isolated and further converted photochemically into products. It is also reported that 2-cyano-1-pyrroline 1-oxides undergo clean photorearrangement at 254 nm in benzene to give the cyanoformyl azetidines *via* the oxaziridines,[60] whereas in certain 2-acyl-1-pyrroline 1-oxides, fragmentation and not rearrangement is observed.[61]

The intermediacy of oxaziridines in heteroaromatic *N*-oxide photorearrangements is not so clearly established. A reinvestigation of pyridine *N*-oxide (101), in fact, has shown that photochemically induced ring cleavage occurs readily to afford an unstable unsaturated isocyanide with structure (102) or (103);[62] this

[57] K. Mutai, S. Kanno, and K. Kobayashi, *Tetrahedron Letters*, 1978, 1273.
[58] H. Suginome, T. Mizuguchi, and T. Masamune, *Bull. Chem. Soc. Japan*, 1977, **50**, 987.
[59] D. St. C. Black and A. B. Boscacci, *Austral. J. Chem.*, 1977, **30**, 1109.
[60] D. St. C. Black, N. A. Blackman, and A. B. Boscacci, *Tetrahedron Letters*, 1978, 175.
[61] D. St. C. Black and A. B. Boscacci, *Austral. J. Chem.*, 1977, **30**, 1353.
[62] O. Buchardt, J. J. Christensen, C. Lohse, J. J. Turner, and I. R. Dunkin, *J.C.S. Chem. Comm.*, 1977, 837.

species has been trapped in matrices at cryogenic temperatures or with base at room temperature. Oxaziridines are probably intermediates in the photo-rearrangement of imidazole *N*-oxides to imidazolones[63] and in the photo-rearrangement of 4-substituted quinoline-*N*-oxides to the corresponding quinol-2-ones;[64] further photoreaction of these quinolones in propionic acid affords 4-substituted *cis*-3-ethyl-2-oxo-1,2,3,4-tetrahydroquinolines as the major photo-products.

The oxazepine 5-carboxylates (104) and the 3*H*-indoles (105) have been shown unambiguously to be intermediates in the photoconversion of the methyl quinoline-4-carboxylate-1-oxides (106) into the indole-3-carboxylates (107).[65] Not only have the oxazepines been separated and characterized, but the 3*H*-indole (105; R = Me) has been isolated and shown to undergo further photo-chemical rearrangement to the indoles (108) and (109). Photodeoxygenation rather than rearrangement has been found in some pteridine *N*-oxides[66] and in 2-nitrophenazine 10-oxide.[67]

(106) R¹ = Ph or Me; R² = H
R¹ = Ph, R² = Me

The 1,4,6,7-dioxadiozocin structure, originally assigned to the photo-isomer of pyridazine 1,2-dioxide (110), has been shown by crystal structure analysis to be incorrect;[68] the photoproduct is 3*a*,6*a*-dihydroisoxazolo[5,4-*d*]isoxazole (111) and is claimed without any real evidence to arise *via* the *bis*-iminoxy-radical (112) as shown in Scheme 6.

[63] R. Bartnik and G. Mloston, *Roczniki Chem.*, 1977, **51**, 1747.
[64] A. Ide, Y. Mori, K. Matsumori, and H. Watanabe, *Bull. Chem. Soc. Japan*, 1977, **50**, 1959.
[65] R. Kitamura, H. Fujii, K. Hashiba, M. Somei, and C. Kaneko, *Tetrahedron Letters*, 1977, 2911.
[66] F. L. Lam and T. Lee, *J. Org. Chem.*, 1978, **43**, 167.
[67] S. Pietra, G. F. Bettinetti, A. Albini, E. Fasani, and R. Oberti, *J.C.S. Perkin II*, 1978, 185.
[68] H. Arai, A. Ohsawa, K. Saiki, H. Igeta, A. Tsuji, T. Akimoto, and Y. Iitaka, *J.C.S. Chem. Comm.*, 1977, 856.

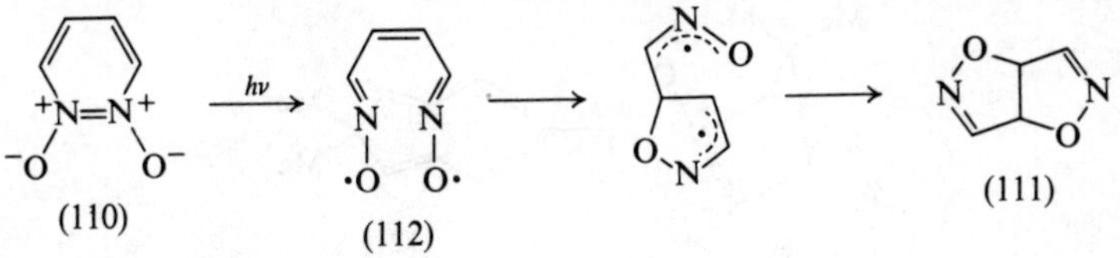

Scheme 6

The photorearrangement of 2,3-substituted quinoxaline-1,4-dioxides (113) to the corresponding benzimidazolones (114) has been carefully examined.[69] Surprisingly, evidence indicates that the dioxaziridines (115) are intermediates in this transformation. The previously proposed stepwise pathway involving the mononitrone (116) cannot be correct as this species is converted on irradiation directly into the quinoxaline dione (117).

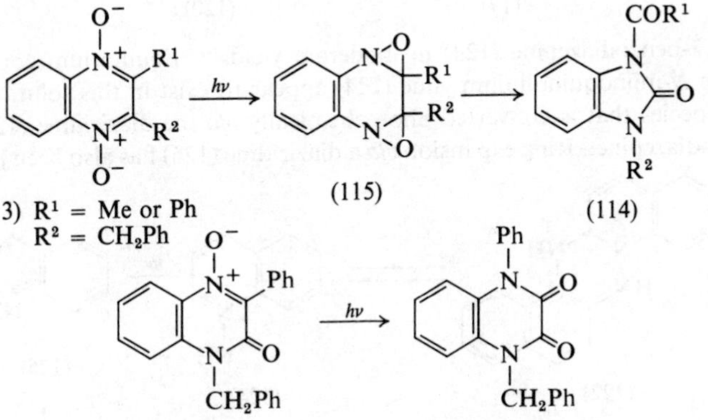

(113) R^1 = Me or Ph
      R^2 = CH$_2$Ph

(115)

(114)

(116)

(117)

Irradiation at $-80\ ^\circ$C of the azoxy compound (118) gave the thermally unstable *endo*- and *exo*-oxadiaziridines (119) and (120).[70] In contrast, however, irradiation of the same azoxy compound in an EPA glass matrix at 77 K led to the formation of the elusive 2,2-dimethyl-2*H*-indene (121). Evidence has been presented for a mechanism involving electrophilic attack by oxygen in the photorearrangement of azoxybenzene to 2-hydroxyazobenzene.[71]

Heteroaromatic *N*-ylides undergo photorearrangements similar to those recorded for *N*-oxides. The rearrangements and related fragmentations of *N*-ylides have been reviewed,[72] and an account of the application of the photorearrangement of pyridinium *N*-ylides to the synthesis of diazepines has been published.[73] Irradiation in dichloromethane–acetic acid solution of the *N*-iminoquinolinium dimer (122) afforded the previously unknown fully unsaturated

[69] A. A. Jarrar and Z. A. Fataftah, *Tetrahedron*, 1977, **33**, 2127.
[70] W. R. Dolbier, K. Matsui, J. Michl, and D. V. Horák, *J. Amer. Chem. Soc.*, 1977, **99**, 3876.
[71] N. J. Bunce, J.-P. Schoch, and M. C. Zerner, *J. Amer. Chem. Soc.*, 1977, **99**, 7986.
[72] J. Streith, *Pure Appl. Chem.*, 1977, **49**, 305.
[73] J. Streith, *Heterocycles*, 1977, **6**, 2021.

1*H*-1,2-benzodiazepine (123) in moderate yield.[74] Equilibrium concentrations of the *N*-iminoquinolinium ylide (124) appear to exist in this solution, and it is this species that is converted photochemically *via* the diaziridine (125) into the benzodiazepine. Ring expansion *via* a diaziridine (126) has also been proposed to

account for the conversion of a series of *N*-(3-quinazolinio)-amidates (127) into the corresponding 4-acyl-5-alkoxy-4,5-dihydro-3*H*-1,3,4-benzotriazepines (128) on irradiation in ethanol or benzyl alcohol.[75] Various other products are also obtained, and stable *N*-6-phenanthridinylbenzimidic acids are the products of irradiation of *N*-(5-phenanthridinio)benzamidates.[76] Ring expansion is not observed, however, on irradiation of *N*-(benzoylimino)pyrimidine (129);[77] the major product in methanol is 4,6-dimethyl-2-(hydroxymethyl)pyrimidine (130), derived presumably by a competing fragmentation involving initial loss of benzoyl nitrene.

[74] T. Tsuchiya, J. Kurita, and V. Snieckus, *J. Org. Chem.*, 1977, **42**, 1856.
[75] J. Fetter, K. Lempert, G. Barta-Szalai, J. Møller, and L. Párkányi, *Acta Chim. Acad. Sci. Hung.*, 1977, **94**, 233.
[76] B. Agai, K. Lempert, and J. Hegedus-Vajda, *Acta Chim. Acad. Sci. Hung.*, 1976, **91**, 91.
[77] F. Roeterdink and H. C. van der Plas, *Rec. Trav. chim.*, 1977, **96**, 156.

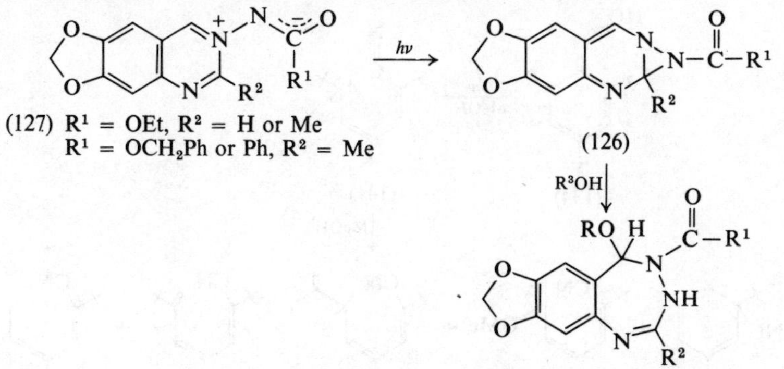

(127) R^1 = OEt, R^2 = H or Me
R^1 = OCH$_2$Ph or Ph, R^2 = Me

(126)

(128) R^3 = Et or CH$_2$Ph

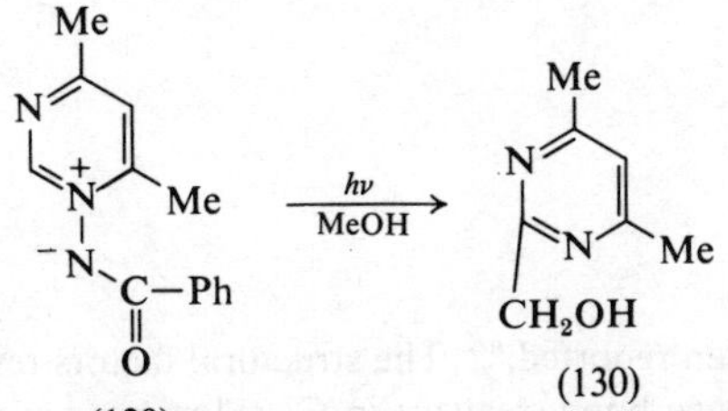

(129)

(130)

The stereochemistry of the photo-Beckmann rearrangement in several cholestanone oximes has been the subject of detailed study.[78] In all cases, the chirality of the migrating group is retained; two isomeric lactams (131) and (132) were obtained, for example, on irradiation of 5α-cholestan-4-one oxime (133) in

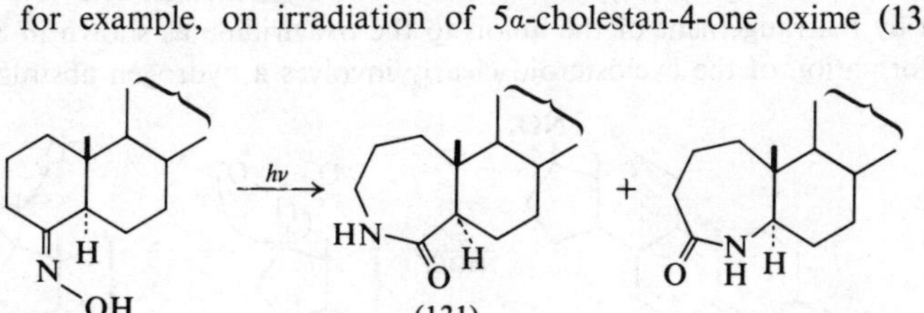

(133)   (131)   (132)

methanol. In this, as in other cases, the lactam arising by migration of the more highly substituted carbon was obtained in greater yield. These results support a pathway in which singlet excited oxime is converted into an oxaziridine followed by reorganization to a lactam, the complete reaction proceeding in a fully concerted manner. A different mode of reaction was observed in 2,4-cyclohexadienone oximes;[79] irradiation of oxime (134) in methanol gave products (135) to (140) arising, it is claimed, by a novel heterolytic α-cleavage to give the cation (141) as intermediate. A di-π-methane type photorearrangement involving an

[78] H. Suginome and F. Yagihashi, *J.C.S. Perkin I*, 1977, 2488.
[79] R. Okazaki, M. Watanabe, and N. Inamoto, *Tetrahedron Letters*, 1977, 4515.

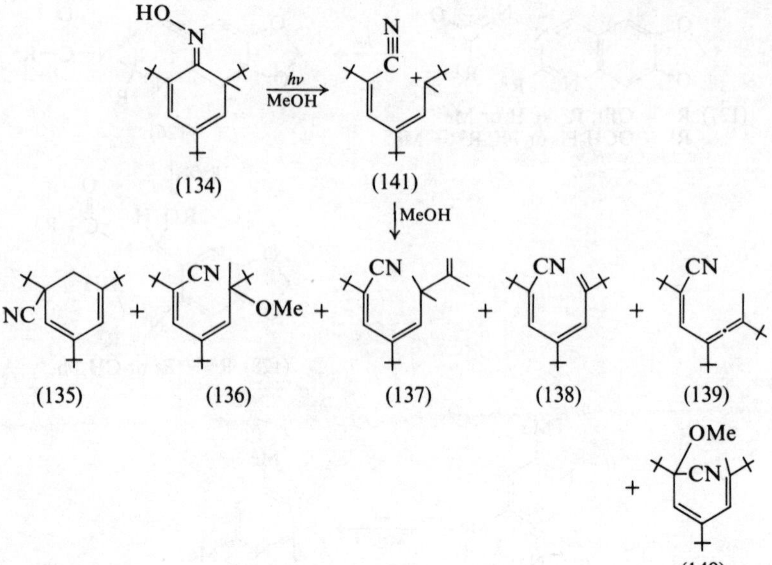

(134) → (141)

(135) + (136) + (137) + (138) + (139)

+ (140)

oxime group has been reported.[80] The structural factors responsible for triplet-state nitrogen—oxygen bond cleavage in *O*-acyloximes have been clarified.[81]

Examples of photorearrangement and photofragmentation of nitro compounds continue to be reported. The major products of irradiation of 3β-acetoxy-17β-nitro-5α-androstane (142) in ethanol–NaOEt are the hydroxamic acid (143) and the corresponding 17,18-cyclosteroid.[82] The hydroxamic acid is believed to arise by rearrangement of the anion to the oxaziridine as shown in Scheme 7; the formation of the cyclosteroid clearly involves a hydrogen abstraction from

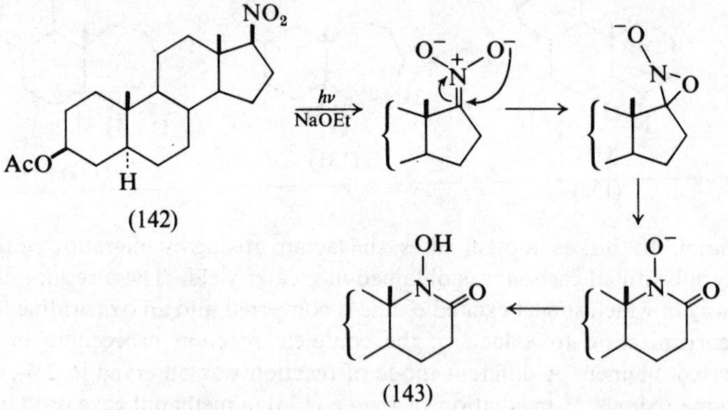

Scheme 7

[80] M. Nitta, O. Inoue, and M. Tada, *Chem. Letters*, 1977, 1065.
[81] H. Sakuragi, M. Yoshida, H. Kinoshita, K. Utena, K. Tokumaru, and M. Hoshino, *Tetrahedron Letters*, 1978, 1529.
[82] S. H. Iman and B. A. Marples, *Tetrahedron Letters*, 1977, 2613.

C-18, but details of this process remain uncertain. Rearrangement of *E*-1-nitro-cyclo-octene to the strained *Z*-isomer can be effected photochemically.[83] Additional details concerning the photoreactions of 3β-acetoxy-6-nitrocholest-5-ene (144) have been reported.[84] All products appear to arise by initial rearrangement involving intramolecular hydrogen abstraction from C-4 to give the *aci*-nitro tautomer (145). Three distinct pathways are recognized. (i) deconjugation to give the isomers (146), (ii) isoxazole (147) formation, and (iii) αβ-unsaturated ketone (148) formation. It is equally possible that the ketone arises by photolysis of the corresponding enol nitrite (149) rather than photofragmentation of the *aci*-nitro compound; enol nitrites are almost certainly involved in the pyrex-filtered

(144)   (145)   (146)

(149)   (148)   (147) R = H or Ac

photorearrangement of 1-(2-pyridyl)- and 1-(3-pyridyl)-2-nitropropene to 1-(2-pyridyl)- and 1-(3-pyridyl)-1-hydroxyiminopropan-2-one in acetone solution.[85]

Examples of intramolecular hydrogen abstraction in aryl nitro compounds are again recorded. Irradiation of *o*-nitrobenzaldehyde in the presence of alkenes gave alkenylaryl nitroxide radicals *via o*-nitrosobenzoic acid.[86] Intramolecular hydrogen abstraction is also responsible for the conversion of 2'-*O*-(2-nitrobenzyl)guanosine (150) into guanosine (151).[87] The *o*-nitrobenzyl substituent has similarly been used as a photolabile protecting group in tyrosine.[88] The photochromic behaviour of *o*-nitrobenzylpyridines may be the result of a reversible hydrogen abstraction of this type.[89,90] The equivalent intermolecular hydrogen abstraction process using nitrobenzene has been used for benzylic hydroxylation.[91,92]

[83] K. Yokoyama, M. Kato, and R. Noyori, *Bull. Chem. Soc. Japan*, 1977, **50**, 2201.
[84] J. T. Pinhey, E. Rizzardo, and G. C. Smith, *Austral. J. Chem.*, 1978, **31**, 97.
[85] R. G. Hunt and S. T. Reid, *J.C.S. Perkin I*, 1977, 2462.
[86] W. G. Filby and K. Guenther, *Z. Naturforsch.*, 1977, **32b**, 693.
[87] E. Ohtsuka, S. Tanaka, and M. Ikehara, *Synthesis*, 1977, 453.
[88] B. Amit, E. Hazum, M. Fridkin, and A. Patchornik, *Internat. J. Pept. Protein Res.*, 1977, **9**, 91.
[89] D. Klemm and E. Klemm, *Z. phys. Chem. (Leipzig)*, 1977, **258**, 1179.
[90] D. Klemm and E. Klemm, *Z. Chem.*, 1977, **17**, 416.
[91] J. Libman, *J.C.S. Chem. Comm.*, 1977, 868.
[92] J. Libman and E. Berman, *Tetrahedron Letters*, 1977, 2191.

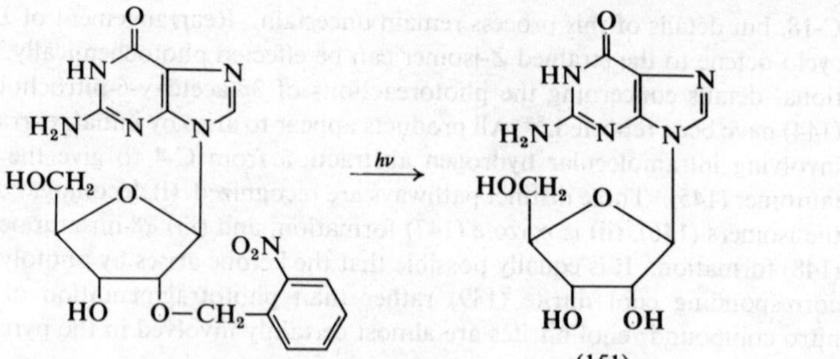

(150)   (151)

An unusual rearrangement is claimed on irradiation of the 2-nitrobenzylidene hydrazide (152) giving rise to an equilibrium mixture of photoproducts (153) and (154).[93] In contrast, a photoreductive cyclization is observed in the 5-nitro-6-styryluracil derivatives (155) leading to the formation of the 1,3-dimethyl-6-phenylpyrrolo[3,2-*d*]pyrimidines (156) in yields of 8—16%.[94]

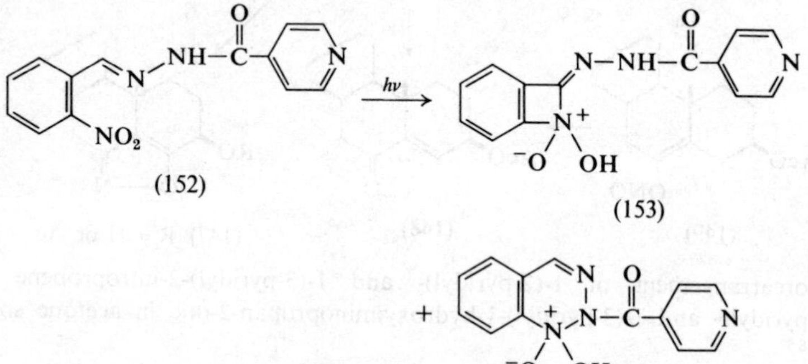

(152)

(153)

+

(154)

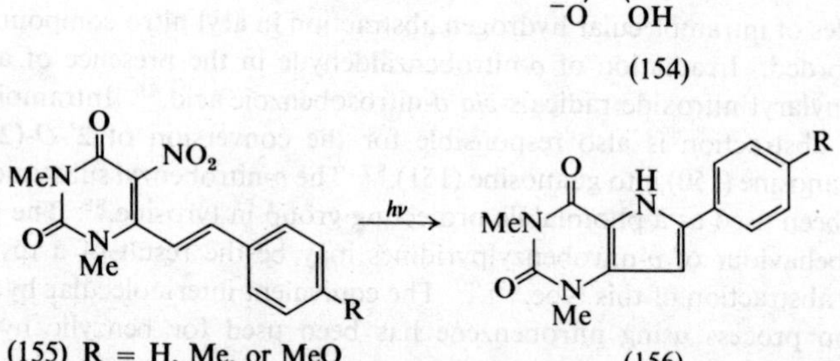

(155) R = H, Me, or MeO   (156)

Rearrangements in nitrogen-containing carbonyl compounds originating from excitation of the carbonyl group merit brief discussion in this section as well as in Part III, Chapter 1. A Norrish Type I cleavage is responsible for the synthetically useful conversion of the keto-imino ethers (157), into the bicyclo-

[93] N. E. Alexandrou and J. Stephanidou-Stephanatou, *Chem. Chron.*, 1977, **6**, 357 (*Chem. Abs.*, 1977, **87** 38 491).
[94] S. Senda, K. Hirota, M. Suzuki, and M. Takahashi, *Chem. and Pharm. Bull.* (*Japan*), 1977, **25**, 563.

[n,1,0]alkanes (158).[95] The stereochemistry of the corresponding rearrangement of 2-ethoxypyrrolin-5-one has been the subject of a separate investigation.[96] Type I processes are also implicated in the ring contraction of 6-methyl-2-phenyltetrahydropyridazin-3-one to 1-anilino-5-methylpyrrolidin-2-one [97] and in the ring expansion of *N*-vinyl-2-pyrrolidone to 4-azacyclohepten-1-one,[98] and the ketene-ketenimine (159) has been shown to be the primary photoproduct of the keto-imine (160).[99]

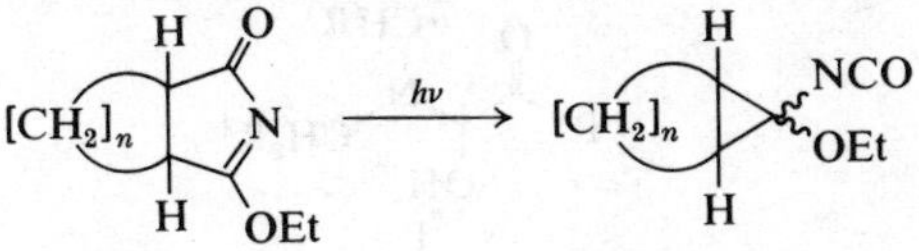

(157)   (158)

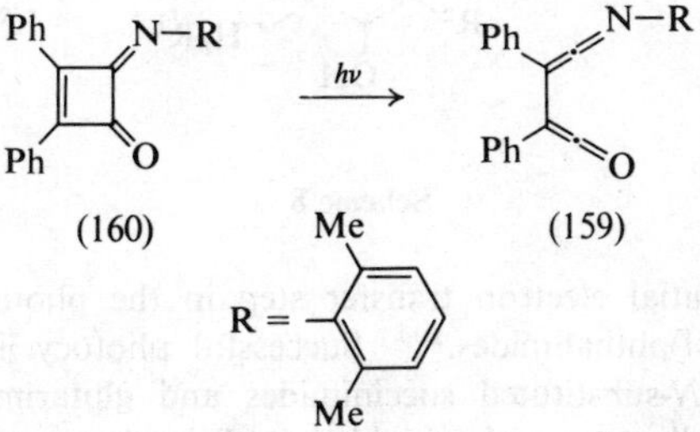

(160)   (159)

R =

Type II processes are typified by the conversion in aprotic solvents of the *N'N*-dibenzyl-α-oxoamides (161; R[1] = Ph) by 1,5-hydrogen transfer (γ-hydrogen abstraction) into the β-lactams (162).[100] Surprisingly, photoreaction of the corresponding *NN*-dialkyl derivatives (161; R[1] = Me or Et) in methanol takes a different course leading to the oxazolidin-4-ones (163); an additional 1,4-hydrogen transfer is postulated to account for this transformation (Scheme 8). The syntheses from *N*-substituted phthalimides of a 6*H*-pyrido[4',3':4,5]-pyrrolo[2,1-*a*]isoindol-6-one [101] and a isoindolo[2,1-*a*]benz[*c,d*]indol-5-one [102] have been accomplished using Type II processes involving δ-hydrogen abstraction. Macrocyclic synthesis by remote photocyclization of *N*-(ω-methylanilino)-alkylphthalimides (164) has been reported,[103] and luminescence data point to the

[95] G. C. Crockett and T. H. Koch, *J. Org. Chem.*, 1977, **42**, 2721.
[96] J. M. Burns, M. E. Ashley, G. C. Crockett, and T. H. Koch, *J. Amer. Chem. Soc.*, 1977, **99**, 6924.
[97] S. N. Ege, M. L. C. Carter, D. F. Ortwine, S.-S. P. Chou, and J. F. Richman, *J.C.S. Perkin I*, 1977, 1252.
[98] H. Shizuka, T. Ogiwara, and T. Morita, *Bull. Chem. Soc. Japan*, 1977, **50**, 2067.
[99] N. Obata and T. Takizawa, *Bull. Chem. Soc. Japan*, 1977, **50**, 2071.
[100] H. Aoyama, T. Hasegawa, M. Watabe, H. Shiraishi, and Y. Omote, *J. Org. Chem.*, 1978, **43**, 419.
[101] M. Terashima, K. Seki, K. Koyama, and Y. Kanaoka, *Chem. and Pharm. Bull.* (*Japan*), 1977, **25**, 1591.
[102] M. Terashima, K. Koyama, and Y. Kanaoka, *Chem. and Pharm. Bull.* (*Japan*), 1978, **26**, 630.
[103] M. Machida, H. Takechi, and Y. Kanaoka, *Heterocycles*, 1977, **7**, 273.

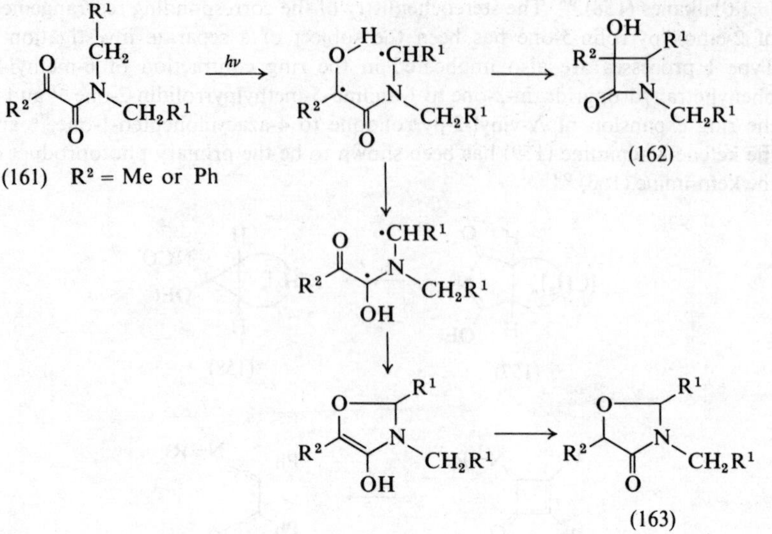

(161) R^2 = Me or Ph

(162)

(163)

Scheme 8

operation of an initial electron transfer step in the photocyclization of *N*-(dialkylaminomethyl)phthalimides.[104] Successful photocyclizations have also been reported in *N*-substituted succinimides and glutarimides.[105,106] *NN*-Dialkyl-2-oxocycloalkanone-carboxamides (165) undergo photocyclization to give the bicyclic lactams (166) by a process involving δ-hydrogen abstraction;[107] there is no precedent for this in simple 2-substituted cycloalkanones.

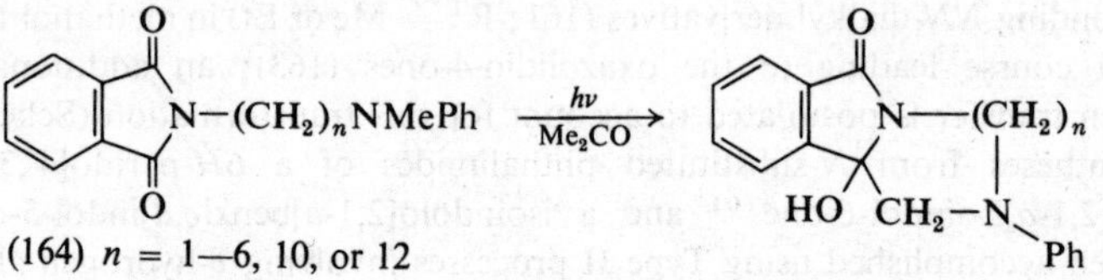

(164) n = 1—6, 10, or 12

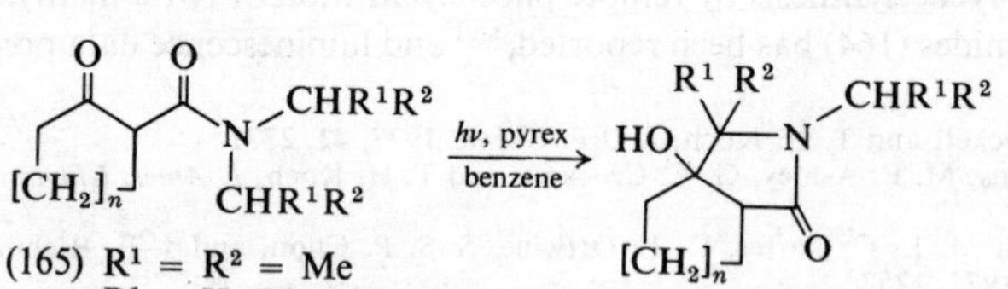

(165) R^1 = R^2 = Me
        R^1 = H, R^2 = Ph
        n = 1 or 2

(166)

[104] J. D. Coyle, G. L. Newport, and A. Harriman, *J.C.S. Perkin II*, 1978, 133.
[105] H. Nakai, Y. Sato, T. Mizoguchi, M. Yamazaki, and Y. Kanaoka, *Heterocycles*, 1977, **8**, 345.
[106] Y. Kanaoka, H. Okajima, and Y. Hatanaka, *Heterocycles*, 1977, **8**, 339.
[107] T. Hasegawa, M. Inoue, H. Aoyama, and Y. Omote, *J. Org. Chem.*, 1978, **43**, 1005.

Other miscellaneous rearrangements include the conversion of the bis-benzaldehyde (167) into (±)-*cis*-alpinigenine (168),[108] and oxetan (169) formation on irradiation of *N*-(2-methylallyl)succinimide (170).[109] Reports of the photocyclization of aminoquinones have been published.[110—114]

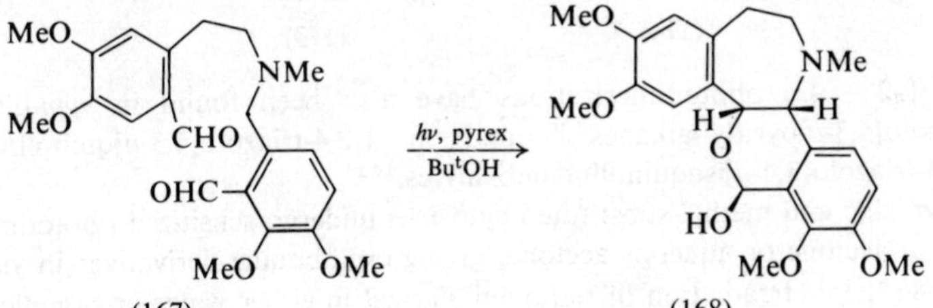

(167)   (168)

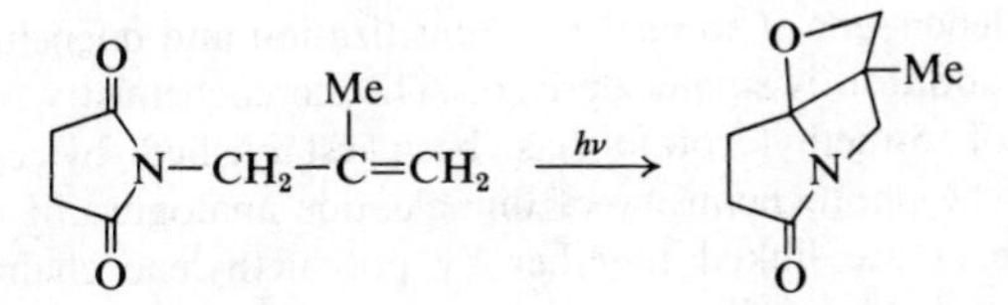

(170)   (169)

Addition.—[$_\pi2 + _\pi2$] Cycloaddition reactions reported include the photo-dimerizations of 3-methyl-4-nitro-5-stryrylisoxazole,[115] *trans*-4-styrylbenzo[*h*]-quinoline,[116] and 2- and 4-vinylpyridine.[117] Photodimerization of 1,3-diacetyl-indole (171) in ethanol gave the cyclobutane (172), the structure of which was established by *X*-ray analysis.[118] The unsymmetrical dimer (173) was the product of irradiation of 8-methyl-*s*-triazolo[4,3-*a*]pyridine (174) in tetrahydrofuran,[119]

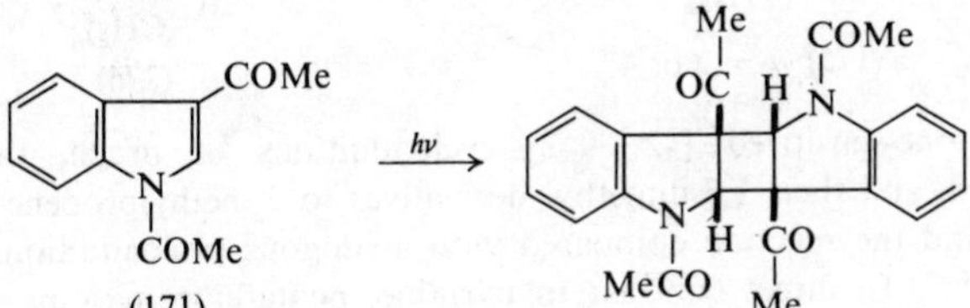

(171)   (172)

[108] S. Prabhakar, A. M. Lobo, and I. M. C. Oliveira, *J.C.S. Chem. Comm.*, 1977, 419.
[109] K. Maruyama and Y. Kubo, *J. Org. Chem.*, 1977, **42**, 3215.
[110] M. Akiba, Y. Kosugi, M. Okuyama, and T. Takada, *Heterocycles*, 1977, **6**, 1113.
[111] M. Akiba, M. Okuyama, Y. Kosugi, and T. Takada, *Heterocycles*, 1977, **6**, 1773.
[112] K. J. Falci, R. W. Franck, and G. P. Smith, *J. Org. Chem.*, 1977, **42**, 3317.
[113] M. Akiba, Y. Kosugi, M. Okuyama, and T. Takada, *J. Org. Chem.*, 1978, **43**, 181.
[114] K. Maruyama, T. Kozuka, and T. Otsuki, *Bull. Chem. Soc. Japan*, 1977, **50**, 2170.
[115] D. Donati, M. Fiorenza, E. Moschi, and P. Sarti-Fantoni, *J. Heterocyclic Chem.*, 1977, **14**, 951.
[116] F. Andreani, R. Andrisano, G. Salvadori, and M. Tramontini, *J.C.S. Perkin I*, 1977, 1737.
[117] T. Nakano, A. Martin, C. Rivas, and C. Perez, *J. Heterocyclic Chem.*, 1977, **14**, 921.
[118] T. Hino, M. Taniguchi, T. Date, and Y. Iidaka, *Heterocycles*, 1977, **7**, 105.
[119] K. T. Potts, E. G. Brugel, and W. C. Dunlap, *Tetrahedron*, 1977, **33**, 1247.

(174)   (173)

and $[\pi 2 + \pi 2]$ photodimerizations have also been found in *sym*-di(3-*sym*-triazolo[4,3-*a*]pyridyl)alkanes[130] and in 1,2,4-triazolo[4,3-*a*]quinoline and 1,2,4-triazolo[3,4-*a*]isoquinoline derivatives.[121]

Cytosine and methyl-substituted cytosines undergo sensitized photodimerization in acetone or aqueous acetone, giving cyclobutane derivatives in yields of 14—86%.[122] Irradiation of tetramethyluracil in either water or organic media affords the *cis,anti*- and the *trans,anti*-dimers with product distribution being essentially independent of solvent.[123] Sensitization and quenching studies indicate that the addition is singlet derived. The stereochemistry of the *trans,anti*-photodimer of 5-methylorotate has been established by crystal structure analysis.[124] The photochemistry of dinucleotide analogues in which purine or pyrimidine bases are linked together by polymethylene chains has been reviewed.[125] Irradiation (290 nm) in aqueous acetone of 1,1'-di- and 1,1'-tetra-methylenebis(5-ethyl)uracils (175) gave the *cis,syn*- adducts (176).[126] A *cis,syn*-configuration has also been established by X-ray analysis for the intramolecular photoadduct of 1,1'-trimethylene-dipropylbisthymine.[127]

(175) $n = 2$ or 4   (176)

The acetone-sensitized $[\pi 2 + \pi 2]$ cycloadditions of uracil, thymine, 6-methyluracil, and their 1,3-dimethyl derivatives to 2-methylpropene have been described and the reaction compared with analogous cycloadditions of cyclohexenones.[128] In direct contrast to pyridine, pentafluoropyridine (177) adds photochemically to ethylene to give the 1:1- and 1:2-adducts (178) and (179) in comparable yields.[129] The mechanism of the photoaddition of *N*-methyl-

[120] K. T. Potts, W. C. Dunlap, and E. G. Brugel, *Tetrahedron*, 1977, **33**, 1253.
[121] K. T. Potts, W. C. Dunlap, and F. S. Apple, *Tetrahedron*, 1977, **33**, 1263.
[122] H. Taguchi, B.-S. Hahn, and S. Y. Wang, *J. Org. Chem.*, 1977, **42**, 4127.
[123] J. C. Otten, C. S. Yeh, S. Byrn, and H. Morrison, *J. Amer. Chem. Soc.*, 1977, **99**, 6353.
[124] C. P. Huber, C. I. Birnbaum, M. L. Post, E. Kulikowska, L. Gajewska, and D. Shugar, *Canad. J. Chem.*, 1978, **56**, 824.
[125] K. Golankiewicz, *Heterocycles*, 1977, **7**, 429.
[126] K. Golankiewicz and H. Koroniak, *Roczniki Chem.*, 1976, **50**, 2041.
[127] E. Skrzypczak-Jankun, H. Maluszynska, Z. Kaluski, and K. Golankiewicz, *Acta Cryst.*, 1977, **B33**, 1624.
[128] A. J. Wexler, J. A. Hyatt, P. W. Raynolds, C. Cottrell, and J. S. Swenton, *J. Amer. Chem. Soc.*, 1978, **100**, 512.
[129] M. G. Barlow, D. E. Brown, and R. N. Haszeldine, *J.C.S. Chem. Comm.*, 1977, 669; *J.C.S. Perkin I*, 1978, 363.

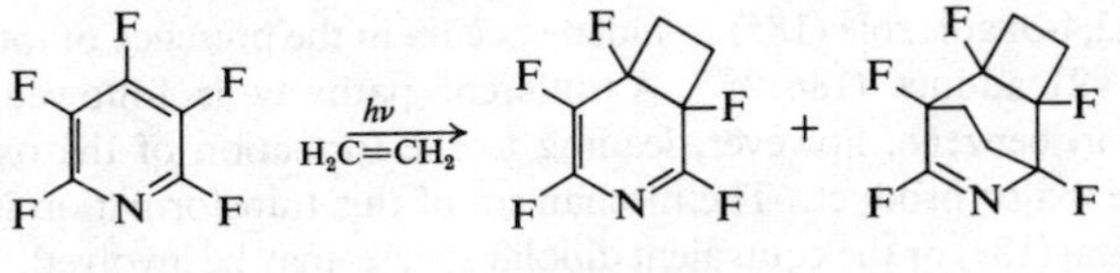

(177)   (178)   (179)

pyrid-2-one to methyl acrylate has been discussed,[130] and details of the photo-cycloaddition of 4-oxazoline-2-one to a variety of alkenes and acetylene have been published.[131] The novel β-lactam system (180) has been obtained by irradiation in dioxan of the dehydrovaline acrylamide (181);[132] the photoreaction takes an alternative path when the acrylamide is not protected as the thiazoline. Nitrogen-containing cage molecules have also been prepared by intramolecular [$_\pi2 + _\pi2$] cycloadditions.[133]

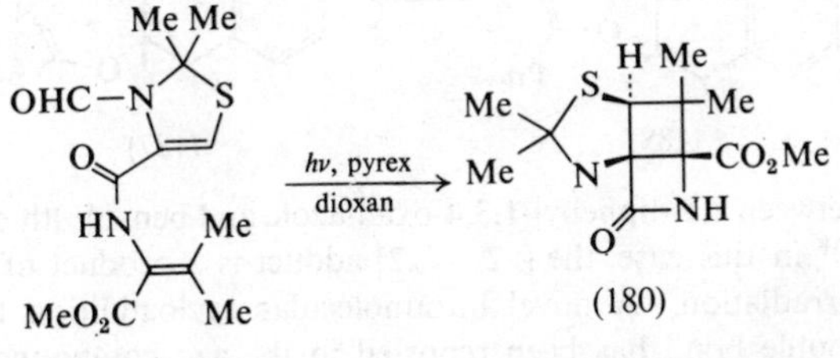

(181)   (180)

Two [$_\pi4 + _\pi4$] dimers (182) and (183) were obtained from *N*-methylisoindole (184) on irradiation in solution;[134] no transient photoisomers were observed, and the photodimers underwent reversion to the monomer photochemically and thermally. A [$_\pi4 + _\pi4$] adduct was also formed from 2-methyl-*sym*-triazolo-[1,5-*a*]pyridine and pyrid-2-one on irradiation;[135] both species are known separately to undergo such dimerizations.

Cycloadditions to the carbon—nitrogen double bond are still relatively rare. The previously documented photoaddition of 3-ethoxyisoindolenone to tetra-methylethylene and *cis*-but-2-ene has now been shown to be quenched by di-t-butyl nitroxide, *cis*-piperylene, and biacetyl.[136] Photoaddition of 2,5-

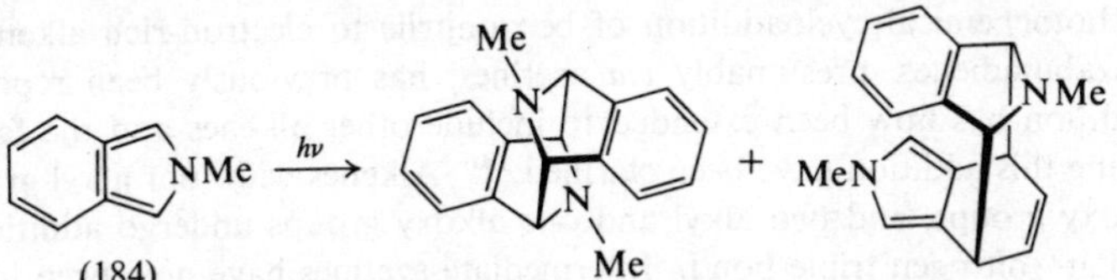

(184)   (182)   (183)

130 K. Somekawa and S. Kumamoto, *Nippon Kagaku Kaishi*, 1977, 1489 (*Chem. Abs.*, 1978, **88**, 73 832).
131 K.-H. Scholz, H.-G. Heine, and W. Hartmann, *Tetrahedron Letters*, 1978, 1467.
132 P. K. Sen, C. J. Veal, and D. J. Young, *J.C.S. Chem. Comm.*, 1977, 678.
133 T. Mukai and Y. Yamashita, *Tetrahedron Letters*, 1978, 357.
134 W. Rettig and J. Wirz, *Helv. Chim. Acta*, 1978, **61**, 444.
135 T. Nagano, M. Hirobe, and T. Okamoto, *Tetrahedron Letters*, 1977, 3891.
136 D. R. Anderson, J. S. Keute, T. H. Koch, and R. H. Moseley, *J. Amer. Chem. Soc.*, 1977, **99**, 6332.

diphenyl-1,3,4-oxadiazole (185) to indene occurs in the presence of iodine to give the [$_\pi2 + _\pi2$] adduct (186).[137] A different pathway is followed on direct irradiation in benzene, however, leading to the formation of the oxadiazepine (187) as the major product. The mechanism of this transformation is not clear, but a biradical (188) or the equivalent dipolar species may be involved. Analogous

reactions between 2,5-diphenyl-1,3,4-oxadiazole and benzo[*b*]thiophen have been reported;[138] in this case, the [$_\pi2 + _\pi2$] adduct is a product of benzophenone-sensitized irradiation. A novel intramolecular cycloaddition to a nitrogen—nitrogen double bond has been reported in the azo compound (189).[139] The efficiency of this phototransformation is attributed to the parallel orientation of the two unsaturated functions, the yield of adduct (190) being virtually quantitative with no competing elimination of nitrogen.

The photochemical cycloaddition of benzonitrile to electron-rich alkenes to give 2-azabutadienes, presumably *via* azetines, has previously been reported. This addition has now been extended to include other alkenes and the factors controlling this addition have been clarified.[140] Alkenes with four alkyl groups, two alkoxy groups, and two alkyl and one alkoxy groups undergo addition to the carbon—nitrogen triple bond. Intermediate azetines have now been identified as, for example, in the addition of benzonitrile to cycloalkene (191) to afford adducts (192) to (194). The alkenes strongly quench benzonitrile fluorescence indicating the intermediacy of excited singlet species. Further examples of the photoaddition of *sym*-triazolo[4,3,*b*]pyridazine (195) to alkenes have been

[137] K. Oe, M. Tashiro and O. Tsuge, *J. Org. Chem.*, 1977, **42**, 1496.
[138] K. Oe, M. Tashiro and O. Tsuge, *Bull. Chem. Soc. Japan*, 1977, **50**, 3281.
[139] W. Berning and S. Hünig, *Angew. Chem. Internat. Edn.*, 1977, **16**, 777.
[140] T. S. Cantrell, *J. Org. Chem.*, 1977, **42**, 4239.

(191)   (192)   (193)   (194)

described.[141,142] Addition to 2,5-dihydrofuran (196), for example, results in the formation of the pyrrolo[1,2-*b*]-*sym*-triazoles (197) and (198). The mechanism of this photoreaction has been discussed previously.

(195)   (196)   (197)   (198)

N-Methylphthalimide (199) undergoes an unusual addition to butadiene on irradiation in acetonitrile to afford the azepine-2,5-diones (200) and (201).[143] The transformation, formally a [$_\pi 2 + _\sigma 2$] process, is thought to involve an intermediate aziridine (202). The addition of 2-phenyl-3-ethoxycarbonyl-Δ^2-pyrroline-4,5-dione to cyclopentadiene has been described[144] and developed as a new route to 3,4-dihydropyrid-2-ones.[145]

(199)   (202)   (200) R^1 = H, R^2 = Me
                     (201) R^1 = Me, R^2 = H

Examples of the photoaddition of solvent molecules to nitrogen-containing systems have again been described, but there is little novelty in many of these reports and the subject will not be reviewed in depth. Factors affecting the photoaddition of methanol to uridine have been studied,[146] and the preparation of *trans*-pyrimidine 1,2-diols by irradiation of thymine and its simple derivatives in the presence of hydrogen peroxide has been reported.[147] 1,3,4-Oxadiazole (203) is converted on irradiation in methanol into the ylide (204) by heterolytic

141 G. E. Mass and J. S. Bradshaw, *J. Heterocyclic Chem.*, 1977, **14**, 81.
142 J. S. Bradshaw, J. E. Tueller, S. L. Baxter, G. E. Mass, and J. T. Carlock, *J. Heterocyclic Chem.*, 1977, **14**, 411.
143 P. H. Mazzocchi, M. J. Bowen, and N. K. Narain, *J. Amer. Chem. Soc.*, 1977, **99**, 7063.
144 T. Sano, Y. Horiguchi, Y. Tsuda, and Y. Itatani, *Heterocycles*, 1978, **9**, 161.
145 Y. Tsuda, M. Kaneda, Y. Itatani, T. Sano, Y. Horiguchi, and Y. Iitaka, *Heterocycles*, 1978, **9**, 153.
146 J.-L. Fourrey and P. Jouin, *Tetrahedron Letters*, 1977, 3397.
147 B. S. Hahn and S. Y. Wang, *Biochem. Biophys. Res. Comm.*, 1977, **77**, 947.

addition of methanol to the carbon—nitrogen double bond followed by cyclo-elimination of the ester (205).[148] The ylide is trapped by oxadiazole to give adduct (206). Photoaddition of methanol is also accompanied by ring cleavage

in prenazone[149] and in 2-phenyl-3,1-benzoxazepine-5-carboxylate.[150] Photo-reactions involving the formation of hydroxyalkyl radicals from alcohols and their subsequent addition to 3-aminopyrido[4,3-*e*]as-triazine,[151] to 1,N^4-dimethylcytosine,[152] and to a benzo[*c*]phenanthridinium salt[153] have been described. An α-hydroxyalkyl radical is also believed to be implicated in the photomethylation of phenanthro[9,10-*d*]oxazole in ethanol.[154] Radicals photo-chemically derived from ethers have been reported to add to carbon—nitrogen double bonds in theophylline[155] and in 2-aza-3-aryl-1,4-naphthoquinones.[156] Photoamidation with formamide of a furanose occurs readily[157] and the adduct (207) is formed *via* a novel photoureidomethylation of 1,1-diphenylethylene with

[148] O. Tsuge, K. Oe, and M. Tashiro, *Chem. Letters*, 1977, 1207.
[149] J. Reisch, K. G. Weidmann, and J. Triebe, *Arch. Pharm.*, 1977, **310**, 827.
[150] M. Somei, R. Kitamura, H. Fujii, K. Hashiba, S. Kawai, and C. Kaneko, *J.C.S. Chem. Comm.*, 1977, 899.
[151] P. Benko and L. Pallos, *Acta Chim. Acad. Sci. Hung.*, 1976, **91**, 327.
[152] K. I. Ekpenyong, R. B. Meyer, and M. D. Shetlar, *Tetrahedron Letters*, 1978, 1619.
[153] H. Ishii, T. Ishikawa, K. Hosoya, and N. Takao, *Chem. and Pharm. Bull.* (*Japan*), 1978, **26**, 166.
[154] M. Maeda, M. Kawashige, and M. Kojima, *J.C.S. Chem. Comm.*, 1977, 511.
[155] A. Erndt, A. Para, and A. Kostuch, *Roczniki Chem.*, 1977, **51**, 2421.
[156] K. Maruyama, T. Iwai, T. Otsuki, Y. Naruta, and Y. Miyagi, *Chem. Letters*, 1977, 1127.
[157] A. Rosenthal and M. Ratcliffe, *Carbohydrate Res.*, 1977, **54**, 61.

tetramethylurea.[158] Irradiation of *N*-acetyldiphenylmethyleneamine (208) in cyclopentene gave the adduct (209).[159]

The addition of several secondary and tertiary amines to singlet stilbene has been reported;[160] in aprotic solvents, for example, triethylamine adds to stilbene to give the products listed in Scheme 9 by a process in which electron transfer precedes proton transfer. Addition of secondary amines to 1,2-diphenylcyclo-butene has also been noted.[161]

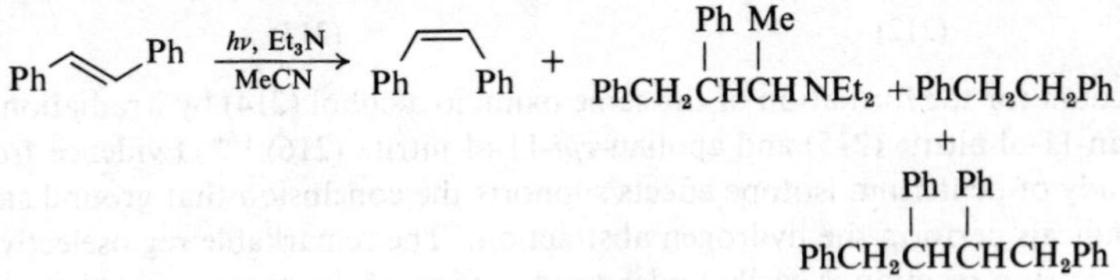

Scheme 9

Miscellaneous Reactions.—The gas-phase photodecomposition of heterocyclic compounds containing nitrogen, oxygen, and sulphur[162] and the photochemistry of cyanates[163] have both been the subject of recent reviews. Renewed interest in the photolysis of nitrites has been shown. On irradiation, 3β-acetoxyoleanan-12α-yl nitrite (210) is converted into the oxime (211) by the normal Barton reaction pathway of nitrogen—oxygen bond homolysis followed by hydrogen abstraction

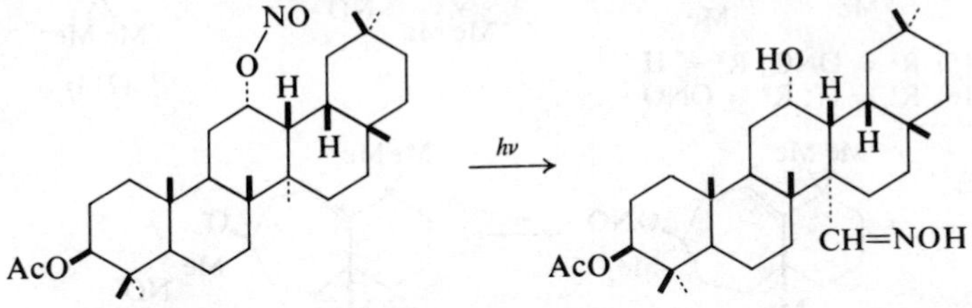

(210)   (211)

from C-27 and radical recombination.[164] In the nitrite ester of 3β-acetoxy-oleanan-13β-ol (212), however, formation of the 13β-oxyl radical is followed not by hydrogen abstraction from C-28 as expected but by collapse *via* fission of the 13,18-bond to give the hydroxynitrone (213). The mechanism of hydroxynitrone formation and the factors promoting β-scission in the oxy-radical have been discussed in detail elsewhere.[165] An analogous reversible β-scission is proposed

[158] T. Miyamoto, Y. Tsujimoto, T. Tsuchinaga, Y. Nishimura, and Y. Odaira, *Tetrahedron Letters*, 1978, 2155.

[159] N. Toshima, S. Asao, and H. Hirai, *Bull. Chem. Soc. Japan*, 1978, **51**, 578.

[160] F. D. Lewis and T.-I. Ho, *J. Amer. Chem. Soc.*, 1977, **99**, 7991.

[161] M. Sakuragi and H. Sakuragi, *Bull. Chem. Soc. Japan*, 1977, **50**, 1802.

[162] S. Braslavsky and J. Heichlen, *Chem. Rev.*, 1977, **77**, 473.

[163] R. Jahn and V. Schmidt, in 'Chemistry of Cyanates and their Thio Derivatives', Vol. 1, eds. S. Patai, Wiley, 1977, p. 343.

[164] R. B. Boar, L. Joukhadar, M. de Luque, J. F. McGhie, D. H. R. Barton, D. Arigoni, H. G. Brunner, and R. Giger, *J.C.S. Perkin I*, 1977, 2104.

[165] H. Suginome, S. Sugiura, N. Yonekura, T. Masamune, and E. Osawa, *J.C.S. Perkin I*, 1978, 612.

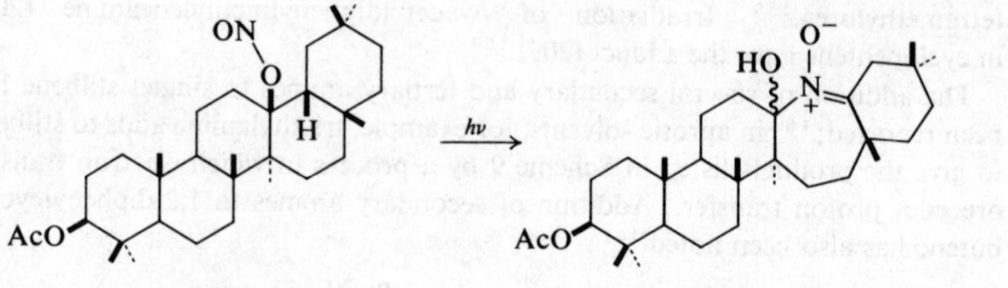

(212)   (213)

to account for the formation of the same oximino alcohol (214) by irradiation of apollan-11-ol nitrite (215) and apollan-*epi*-11-ol nitrite (216).[166] Evidence from the study of deuterium isotope effects supports the conclusion that ground state oxy-radicals perform the hydrogen abstraction. The remarkable regioselectivity of the Barton reaction is easily understood in view of the observation that even slight changes in the molecular geometry can markedly alter the degree of bonding in the transition state. β-Scission in the oxy-radical is also involved in ring cleavage of widdrol nitrite (217) to the four isomeric oximes (218)[167] and in the

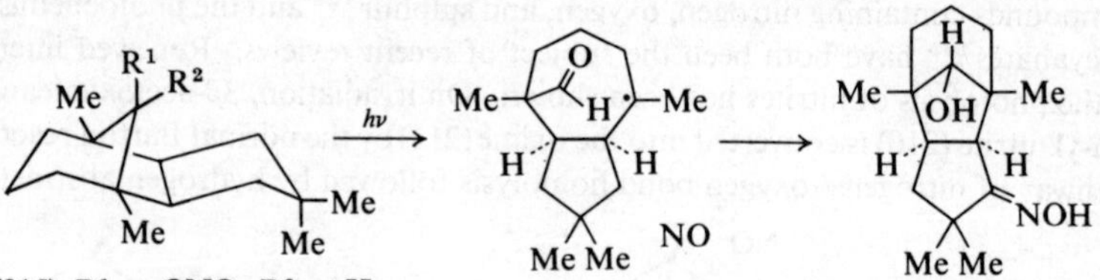

(215) R¹ = ONO; R² = H
(216) R¹ = H; R² = ONO

(214)

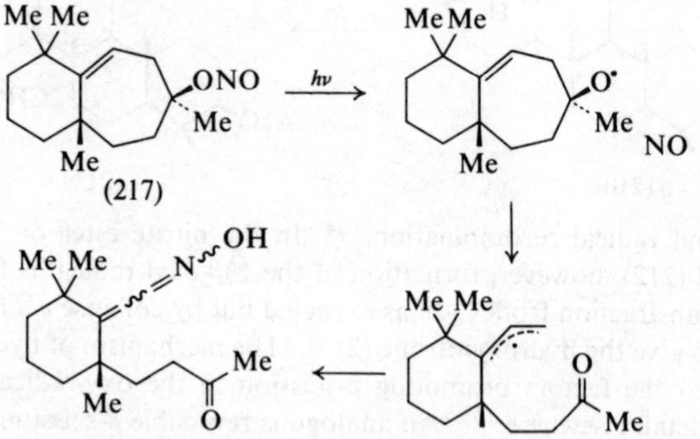

(217)

(218)

conversion of 5α-androstane-3α,17β-diol 3-acetate 17-nitrite into two isomeric hydroxamic acids.[168] A hydroxamic acid (219) was also a product of irradiation of 3β-acetoxy-28-nitrosyloxylupane (220), but the additional formation of

[166] A. Nickon, R. Ferguson, A. Bosch, and T. Iwadare, *J. Amer. Chem. Soc.*, 1977, **99**, 4518.
[167] H. Takahashi, M. Ito, and H. Suginome, *Chem. Letters*, 1977, 241.
[168] H. Suginome, N. Yonekura, T. Mizuguchi, and T. Masamune, *Bull. Chem. Soc. Japan*, 1977, **50**, 3010.

the nitrone (221) as the major photoproduct cannot be easily explained.[169] The photonitrosation of cyclohexane has been reviewed.[170]

Chemical and spectroscopic evidence has been reported for the generation of iminyl and 2-isoxazolinyl radicals by nitrogen—oxygen bond homolysis in the photoreactions of the 2-isoxazoline (222);[171,172] the formation of products (223) to (226) has been rationalized in terms of a radical pair mechanism as outlined in Scheme 11. Initial nitrogen—oxygen bond homolysis is also probably implicated in the photochemical conversion of a 5-phenyl-1,2,4-oxadiazoline into a benzimidazole derivative.[173] Heterolytic nitrogen—oxygen bond cleavage has been proposed both in the conversion in water of N-alkyl-2,1-benzisoxazolium perchlorates into 3-acyl-p-N-alkylaminophenols[174] and in the phototransformations of benzisoxazolo[2,3-a] pyridinium tetrafluoroborates.[175]

Successive nitrogen—nitrogen bond homolysis and ring cleavage are responsible for the photoisomerization of N-cyclopropyl-N-nitrosotosylamide (227) to the C-nitrosoimine (228).[176] β-Hydrogen elimination is preferred in the corresponding cyclopentyl derivative after homolysis, resulting in the formation of an imine. Analogous transformations have been reported in N-cycloalkyl-N-halosulphonamides,[177] and the common transient obtained on flash photolysis of N-chloro-and N-nitroso-piperidine is claimed to be the piperidinium radical.[178]

[169] J. Protiva, M. Budesinsky, and A. Vystrcil, *Coll. Czech. Chem. Comm.*, 1977, **42**, 1220.
[170] M. Fischer, *Angew. Chem. Internat. Edn.*, 1978, **17**, 16.
[171] H. Saiki, T. Miyashi, T. Mukai, and Y. Ikegami, *Tetrahedron Letters*, 1977, 4619.
[172] T. Mukai, H. Saiki, T. Miyashi, and Y. Ikegami, *Heterocycles*, 1977, **6**, 1599.
[173] M. Vaultier and R. Carrie, *J.C.S. Chem. Comm.*, 1978, 356.
[174] N. F. Haley, *J. Org. Chem.*, 1977, **42**, 3929.
[175] R. A. Abramovitch and M. N. Inbasekaran, *J.C.S. Chem. Comm.*, 1978, 149.
[176] E. E. J. Dekker, J. B. F. N. Engberts, and Th. J. de Boer, *Rec. Trav. chim.*, 1977, **96**, 230.
[177] E. E. J. Dekker, J. B. F. N. Engberts, and Th. J. de Boer, *Rec. Trav. chim.*, 1978, **97**, 39.
[178] A. J. Cessna, S. E. Sugamori, R. W. Yip, M. P. Lau, R. S. Snyder, and Y. L. Chow, *J. Amer. Chem. Soc.*, 1977, **99**, 4044.

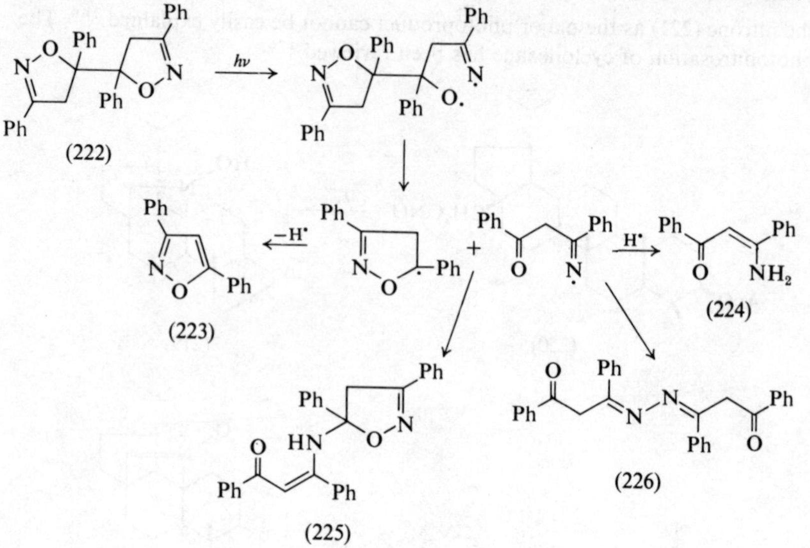

Scheme 11

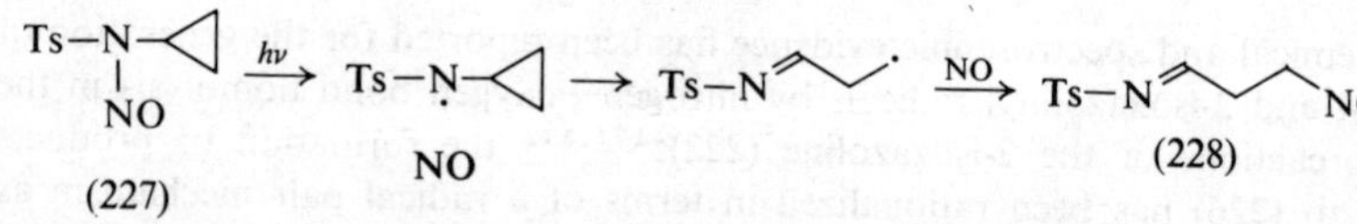

The formation of oxazolines (229) and amides (230) on irradiation of the amidates (231) in 1-methoxy-2-methylprop-1-ene presumably occurs by way of intermediate nitrenes generated by nitrogen—nitrogen bond heterolysis.[179] Aryl nitrenes or aryl nitrenium ions have also been proposed as intermediates in the low-temperature photoreactions of 2-alkylindazoles.[180]

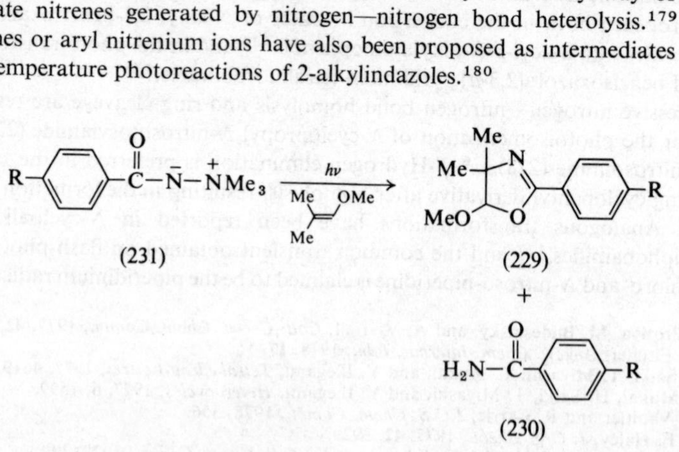

[179] V. P. Semenov, A. K. Khusainova, A. N. Studenikov, and K. A. Ogloblin, *Zhur. obshchei Khim.*, 1977, **13**, 963.
[180] W. Heinzelmann, *Helv. Chim. Acta*, 1978, **61**, 618.

A wide variety of transformations arising by carbon—nitrogen bond homolysis have been described. The photochemical cleavage of the *N*-(benzyloxycarbonyl) group in amino-sugar derivatives has been reported [181] along with other examples of the cleavage of amides and lactams.[182—184] Irradiation of (+)-glaucine methiodide (232) in methanol gave, in a reaction not without precedent, the phenanthrene (233).[185] Carbonium ions have been postulated as intermediates in this transformation, but radicals cannot be completely excluded. The photo-dealkylation in good yield of *N*-8 substituted lumazines has been described,[186]

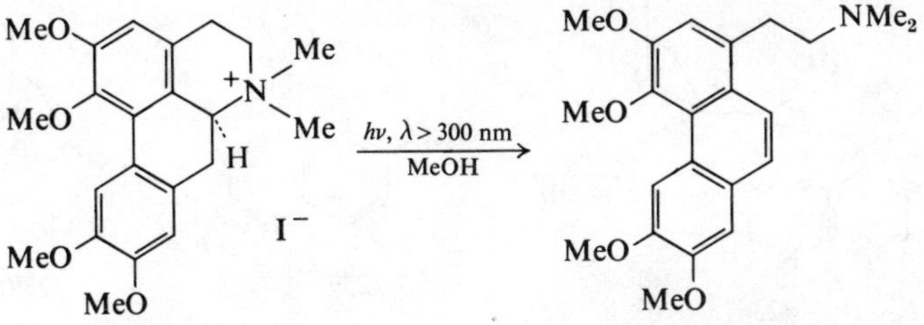

(232) $\xrightarrow[\text{MeOH}]{h\nu,\ \lambda > 300\ \text{nm}}$ (233)

and carbon—nitrogen bond homolysis has been reported to occur in (9-acridinyl-methyl) quaternary ammonium salts [187] and in di-t-butyl nitroxide.[188]

Photodegradation of the pterin (234) to the dihydropterin-6-aldehyde (235) proceeds by a pathway formally equivalent to the Norrish Type II process.[189] The detailed mechanism is, however, not clear, the quantum yield being strongly pH dependent. Photochemically induced hydrogen abstraction is implicated in

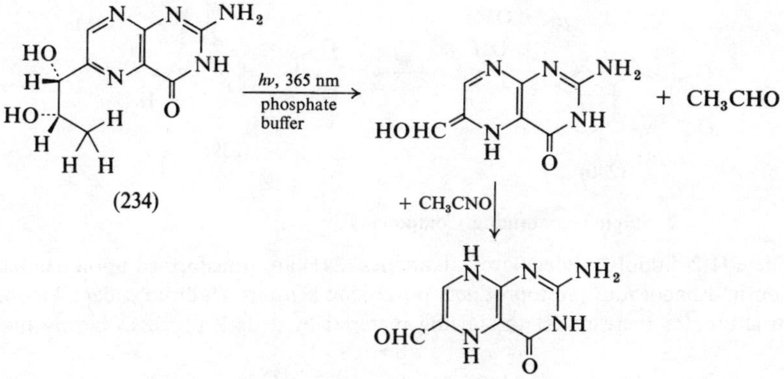

(234) $\xrightarrow[\text{phosphate buffer}]{h\nu,\ 365\ \text{nm}}$ $+ \text{CH}_3\text{CHO}$

$+ \text{CH}_3\text{CNO}$

(235)

[181] S. Hanessian and R. Masse, *Carbohydrate Res.*, 1977, **54**, 142.
[182] U. Zehavi, *J. Org. Chem.*, 1977, **42**, 2821.
[183] K. Shibata, H. Sadaka, M. Matsui, and Y. Takase, *Nippon Kagaku Kaishi*, 1978, 75.
[184] J. Reisch, K. G. Weidmann, and J. Triebe, *Arch. Pharm.*, 1977, **310**, 811.
[185] J. B. Bremner and K. N. Winzenberg, *Austral. J. Chem.*, 1978, **31**, 313.
[186] V. J. Ram, W.-R. Knappe, and W. Pfleiderer, *Tetrahedron Letters*, 1977, 3795.
[187] R. E. Lehr and M. W. Conway, *J. Org. Chem.*, 1977, **42**, 2726.
[188] D. R. Anderson and T. H. Koch, *Tetrahedron Letters*, 1977, 3015.
[189] R. Mengel, W. Pfleiderer, and W.-R. Knappe, *Tetrahedron Letters*, 1977, 2817.

the degradation of riboflavin;[190] the efficiency of this process is solvent dependent owing to conformational effects. Irradiation of quinidine (236) in methanol led to fragmentation and to the formation of 6-methoxyquinoline (237) and 5-vinylquinuclidine-2-carboxaldehyde (238).[191] This is in total contrast with the results of a previous study of 2 M-HCl in which carbon—oxygen bond cleavage was preferred.

(236)   (237)   (238)

The ease of nucleophilic photosubstitution in nitro derivatives of 1-methylimidazole and 1-methylpyrazole depends on the position of the nitro group.[192] In the imidazole series, for example, the 5-nitro derivative is the most reactive isomer, providing a convenient route to 5-cyano- and 5-methoxy-1-methylimidazoles. Dehydrogenation was the major photochemical reaction pathway observed on irradiation of 1,3,5-triphenyl-2-pyrazolines in methanol,[193] and a photoproduct (239) arising by rearrangement and reduction was obtained from the Amaryllidaceae alkaloid crinamine (240).[194]

(240)   (239)

2 Sulphur-containing Compounds

The α-(1,2-dithiol-3-ylidene)cycloalkanones (241) are transformed upon irradiation in ethanol into photoproducts possessing a *trans*-1,2-dithiolylidene ketone structure;[195] these revert to starting material by a dark process obeying first

(241)

[190] M. W. Moore and R. C. Ireton, *Photochem. Photobiol.*, 1977, **25**, 347.
[191] G. A. Epling and U. C. Yoon, *Tetrahedron Letters*, 1977, 2471.
[192] C. Oldenhof and J. Cornelisse, *Rec. Trav. chim.*, 1978, **97**, 35.
[193] J. Lin, D. E. Rivett, and J. F. K. Wilshire, *Austral. J. Chem.*, 1977, **30**, 629.
[194] Y. Tsuda, M. Kaneda, S. Takagi, M. Yamaki, and Y. Iitaka, *Tetrahedron Letters*, 1978, 1199.
[195] C. Lohse and C. Th. Pederson, *Acta Chem. Scand.*, 1977, **31B**, 683.

order kinetics. A series of 4-phenyl-1-benzothiepins (242) have been reported to undergo electrocyclic ring closure to give the expected cyclobutabenzothiophens (243);[196] these are transformed on further irradiation into the isomers (244) by a reaction which is best represented as preceding *via* the biradical (245).

(242) R = H, CN, OAc, or OMe   (243)   (245)

(244)

Reactions of tetrakis(trifluoromethyl) 'Dewar' thiophen, prepared by irradiation of the corresponding thiophen, have been described.[197] Irradiation of bis-6,6'-(1,3-dimethyluracilyl)sulphide (246) is a more complex process, giving rise to the cyclized products (247) and (248) and a third product (249) derived by dimerization of two uracil-6-thiyl radicals followed by oxidation.[198] The mechanism of this cyclization is uncertain although it has been formulated as involving the rearrangement depicted in Scheme 12. The corresponding 1,3-diethyl derivative undergoes cyclization without rearrangement; the reason for this difference in behaviour is obscure.

Sulphur directed photoarylation has been employed in the preparation of a variety of aryl annelated dihydrothiophens from the corresponding aryl vinyl sulphides.[199] Thus, conrotatory cyclization of the 2-thioaryloxyenones (250) gave, *via* the thiocarbonyl ylides (251), the dihydrothiophens (252) in high yield; the efficiency of this process is attributed to the additional stabilization afforded to the ylide by the carbonyl group.

The trimethylsilyl enol ethers (253) of αα-dialkylated α-phenylthioketones undergo a photochemically induced 1,3-phenylthio shift leading to the isomeric enol ethers (254) in high yield.[200] The rearrangement proceeds thermally but much less efficiently. The *S*-benzyl derivative of 4-mercapto-2-methylthiopyrimidine is similarly converted into the 5-benzyl derivative on irradiation,[201] and phenyl benzenesulphonate undergoes a photo-Fries type rearrangement in ethanol to give *o*- and *p*-hydroxyphenyl phenyl sulphones.[202] Competing

[196] H. Hofmann and H. Gaube, *Annalen.*, 1977, 1874.
[197] Y. Kobayashi, I. Kumadaki, A. Ohsawa, Y. Sekine, and A. Ando, *Heterocycles*, 1977, 6, 1587.
[198] T. Itoh, H. Ogura, and K. A. Watanabe, *Tetrahedron Letters*, 1977, 2595.
[199] A. G. Schultz, W. Y. Fu, R. D. Lucci, B. G. Kurr, K. M. Lo, and M. Boxer, *J. Amer. Chem. Soc.*, 1978, 100, 2140.
[200] U. Gerber, U. Widmar, R. Schmid, and H. Schmid, *Helv. Chim. Acta*, 1978, 61, 83.
[201] J.-L. Fourrey, G. Henry, and P. Jouin, *J. Amer. Chem. Soc.*, 1977, 99, 6753.
[202] Y. Ogata, K. Takagi, and S. Yamada, *J.C.S. Perkin II*, 1977, 1629.

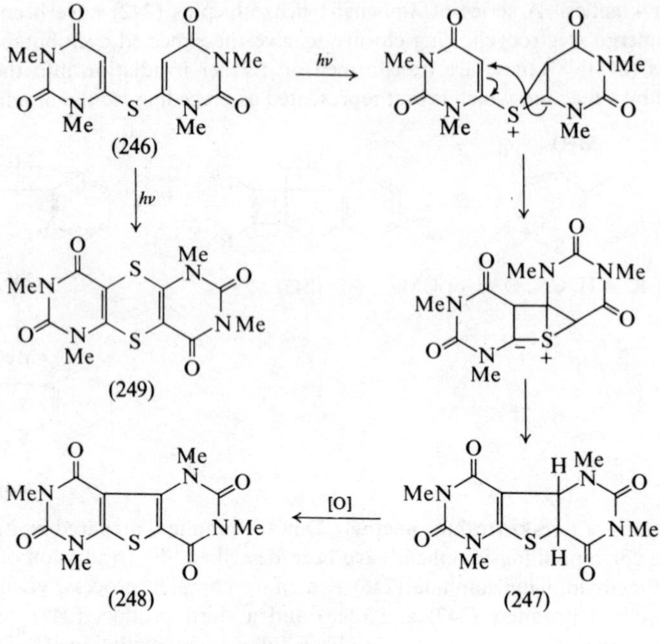

Scheme 12

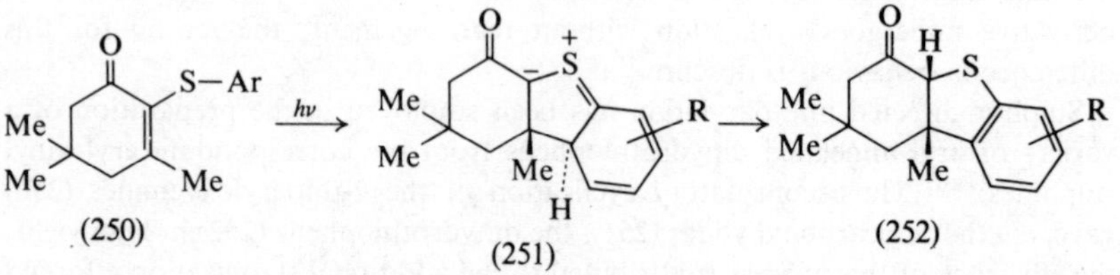

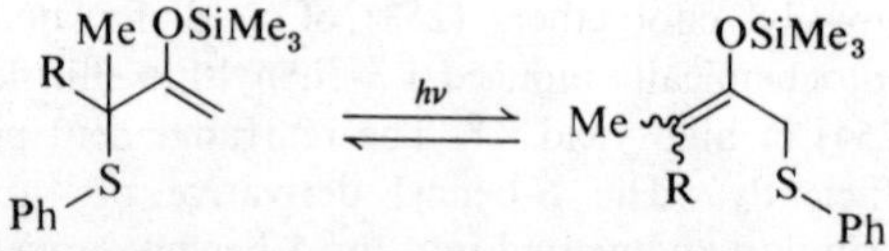

(253) R = Et, $CH_2CH{=}CH_2$, or $CH_2CH{=}CMe_2$

(254)

rearrangements leading to photoproducts (255) and (256) have been reported for the pyrido[2,1-*b*]thiazole (257), the first arising by an apparent ester migration and the second resulting from a 1,3-thio migration.[203] In contrast with this, a 1,2-migration is observed exclusively on irradiation of the sulphonium ylides (258), yielding the isolable isomers (259).[204] Further irradiation induces carbon—

[203] S. Ito, M. Maeda, and M. Kojima, *Heterocycles*, 1977, **8**, 455.
[204] Y. Maki and T. Hiramitsu, *Chem. and Pharm. Bull. (Japan)*, 1977, **25**, 292.

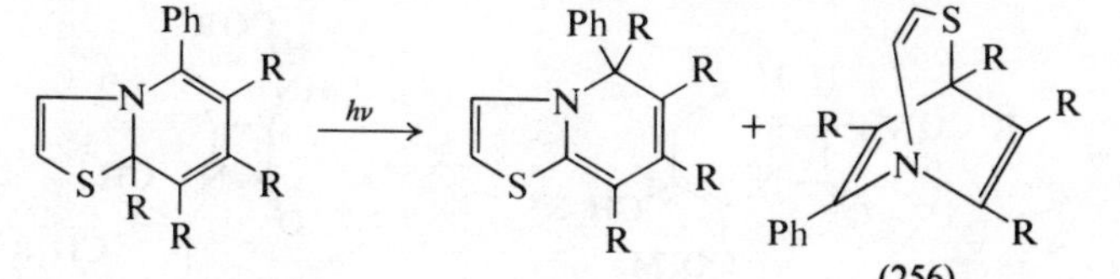

(257) R = CO₂Me (255) (256)

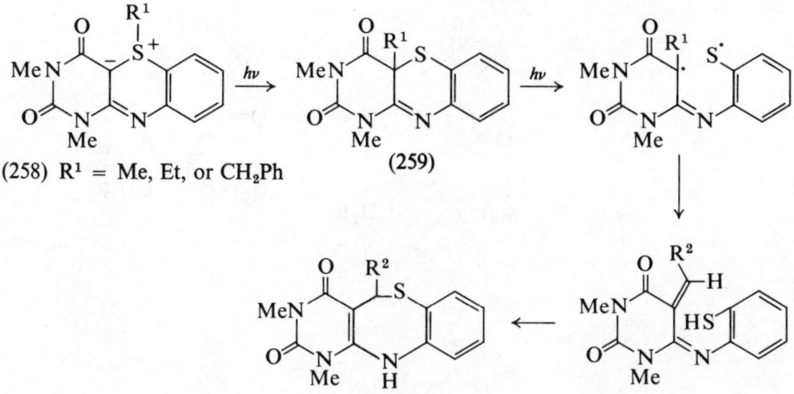

(258) R¹ = Me, Et, or CH₂Ph (259)

(260) R² = H, Me or Ph

Scheme 13

sulphur bond homolysis and gives as the final product the pyrimido-1,4-benzo-[*b*]thiazepines (260); the proposed pathway is outlined in Scheme 13.

The two thiazoles (261) and (262) were obtained on irradiation in methanol of the 3-cephem derivatives (263).[205] A pathway involving initial carbon—sulphur bond homolysis followed by ring cleavage of the β-lactam, elimination of water, and addition of methanol to a common intermediate (264) has been proposed to account for these transformations. The photoreaction was not suppressed by the addition of acetophenone. Photorearrangement of *NN*-bis-*p*-methoxy-phenylbenzenesulphenamide in n-pentane is the result of an initial sulphur—nitrogen bond homolysis.[206] Irradiation of the dihydrothiophen-3(2*H*)-one (265) gave the disulphide (266) as the major product together with a low yield of the alkene (267);[207] preference for β-cleavage in this system unlike the corresponding oxygen analogue must be due to the greater stability of the thiyl radical.

3-Methylbenzo[*b*]thiophen 1-oxide undergoes [π2 + π2] photodimerization in benzene to give the isomeric cyclobutanes (268) and (269).[208] Both photo-cycloaddition and photosubstitution were observed in the photoreaction between benzo[*b*]thiophen and dibromomaleic anhydride;[209] substitution is favoured in

205 Y. Maki and M. Sako, *J. Amer. Chem. Soc.*, 1977, **99**, 5091.
206 T. Anda, M. Nojima, and N. Tokura, *J.C.S. Perkin I*, 1977, 2227.
207 P. Yates and Y. C. Toong, *J.C.S. Chem. Comm.*, 1978, 205.
208 M. S. El Faghi El Amoudi, P. Geneste, and J. L. Olivé, *Tetrahedron Letters*, 1978, 999.
209 T. Matsuo and S. Mihara, *Bull. Chem. Soc. Japan*, 1977, **50**, 1797.

(263) $R^1 = PhCH_2$, $R^2 = H$
       $R^1 = C_4H_3SCH_2$, $R^2 = OAc$

(264)

(262)   +   (261)

(265)   (266)   (267)

solvents of high polarity. Irradiation ($\lambda > 313$ nm) of dibromomaleimide (270) in thiophen gave the 2:1-adduct (271) by further reaction of the $[_\pi 2 + _\pi 2]$ adduct (272);[210] with acetone as solvent, only the dimer of dibromomaleimide and substitution products were obtained. An oxetan was the major product of irradiation of 3-benzoylthiophen in 2,3-dimethylbut-2-ene.[211]

210 H. Wamhoff and H.-J. Hupe, *Tetrahedron Letters*, 1978, 125.
211 T. S. Cantrell, *J. Org. Chem.*, 1977, **42**, 3774.

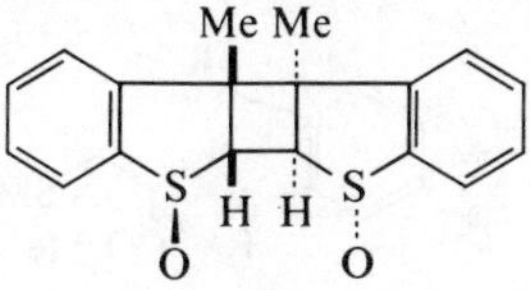

(268)

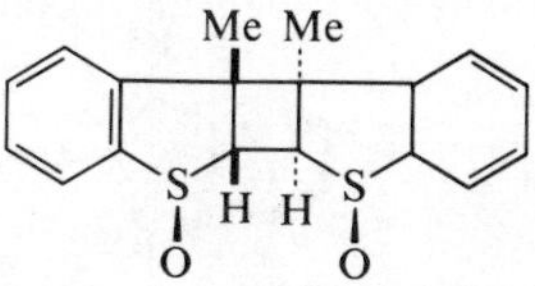

(269)

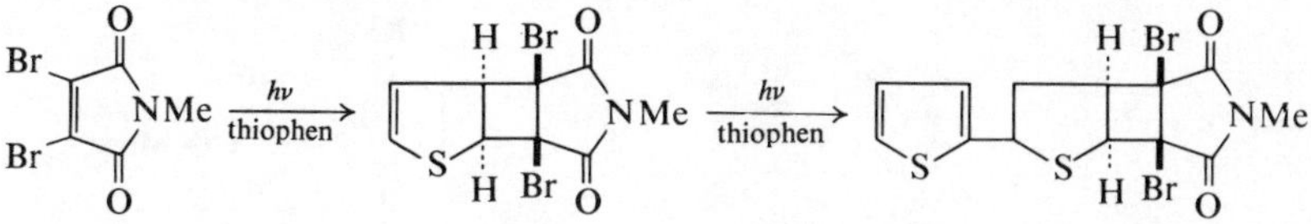

(270)

(272)

(271)

Xanthione (273), on $n \to \pi^*$ excitation ($\lambda = 589$ nm), reacts with acenaphthylene (274) to give the spirothietan (275) in high yield.[212] 2-Iminothietans are similarly formed by [2 + 2] addition of thiones to ketenimines;[213] these are

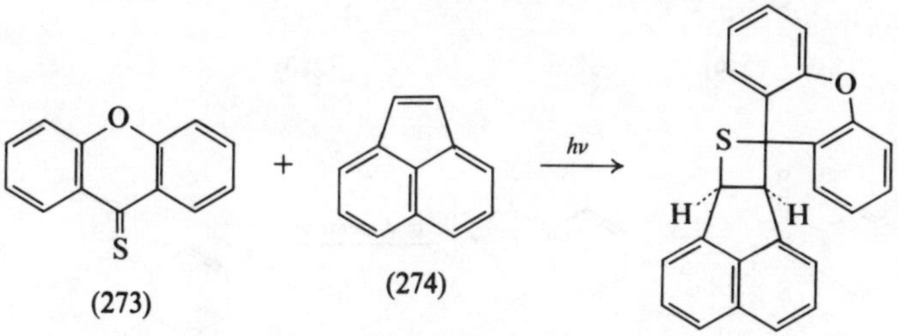

(273)

(274)

(275)

thermally unstable and readily undergo reversion to starting materials. Spirothietans are also presumably intermediates in the photoaddition of indoline-2-thiones (276) to methyl acrylate which, after *S*-methylation, gave the 2-substituted indoles (277) and (278).[214] The major products of irradiation of sulphur dioxide with hexafluorobutadiene are the adducts (279) and (280).[215]

Irradiation of 4,5-benzo-1,2-dithiole-3-thione (281) in the presence of cyclopentene gave the 1:1-adduct (282).[216] The product is envisaged as arising by a sulphur—sulphur bond homolysis as outlined in Scheme 14. In solution the product is in equilibrium with an eight-membered ring dimer (283).

Irradiation of *N*-phenyl-2-benzothiazoline (284) in the presence of the electron-rich alkene, ethyl vinyl ether, gave the previously unreported [1,4]thiazino[2,3,4-*jk*]carbazole ring system (285).[217] The identity of the intermediate species

[212] H. Gotthardt and S. Nieberl, *Chem. Ber.*, 1978, **111**, 1471.
[213] R. G. Visser, J. P. B. Baaij, A. C. Brouwer, and H. J. T. Bos, *Tetrahedron Letters*, 1977, 4343.
[214] C. Marazano, J.-L. Fourrey, and B. C. Das, *J.C.S. Chem. Comm.*, 1977, 742.
[215] N. B. Kazmina, I. L. Knunyants, E. I. Mysov, and G. M. Kuz'yants, *Izvest. Acad. Nauk S.S.S.R., Ser. Khim.*, 1978, 163.
[216] P. de Mayo and H. Y. Ng, *Canad. J. Chem.*, 1977, **55**, 3763.
[217] L. R. Sousa and J. G. Bucher, *Tetrahedron Letters*, 1978, 2267.

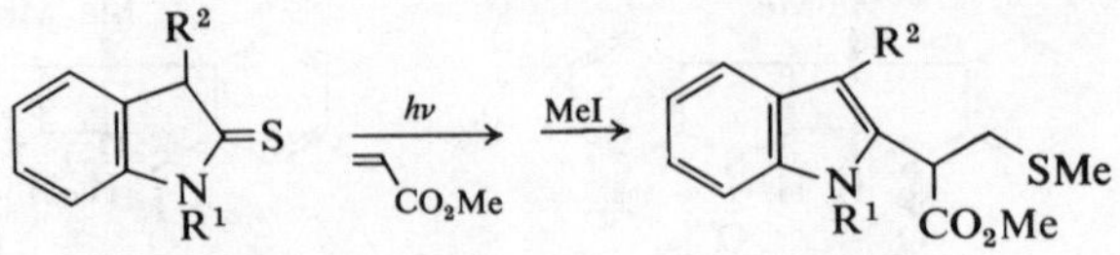

(276) R^1 = Me, R^2 = H
         R^1 = H, R^2 = H or CH$_2$CH$_2$OAc

(277)

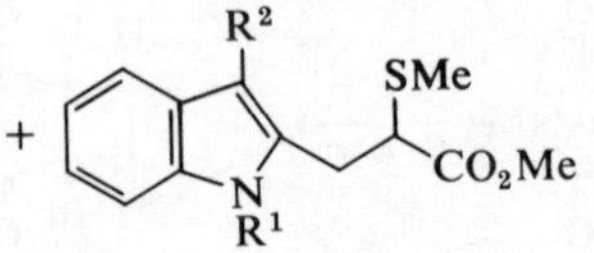

(278)

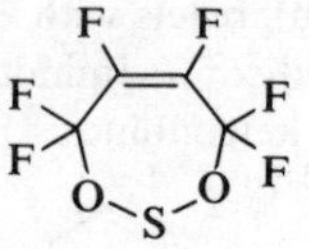

(279) 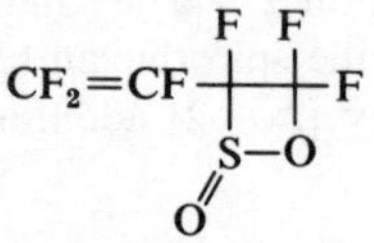   (280)

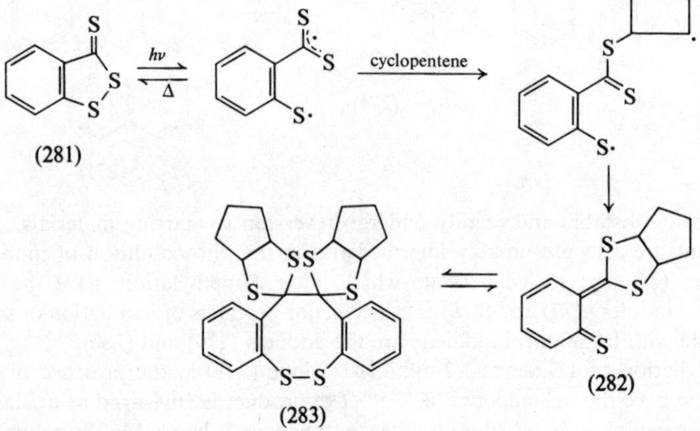

(281)   (282)

(283)

Scheme 14

trapped by the alkene is as yet uncertain; the reaction is not sensitized by acetone or quenched by piperylene although the accompanying formation of carbazole is.

Reactions of photochemically generated thiyl radicals have been described. Hepta-1,6-dienes[218] and hepta-1,6-diynes[219] undergo thiyl radical initiated cyclizations, and the preparation of 4-arylthio-2,6-disubstituted phenols has been effected by a photochemical sulphenylation reaction using 2,6-disubstituted

[218] M. E. Kuehne and R. E. Damon, *J. Org. Chem.*, 1977, **42**, 1825.
[219] M. E. Kuehne and W. H. Parsons, *J. Org. Chem.*, 1977, **42**, 3408.

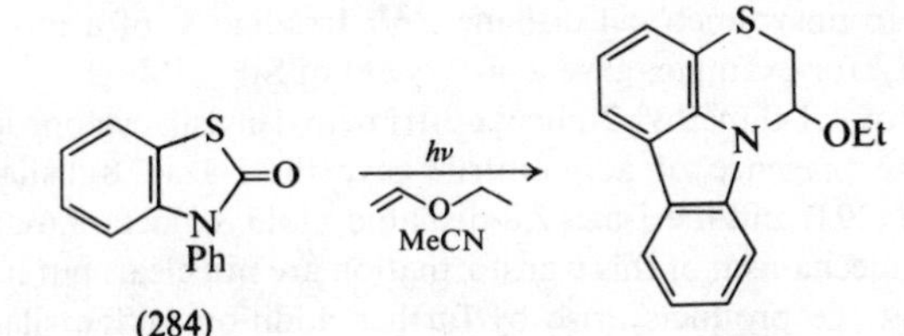

(284) → (285)

phenols and diaryl disulphides.[220] The bisthioethers (286) and (287) are the major products of irradiation of organic disulphides in the presence of tricyclo-[4,1,0,0^{2,7}]heptane (288).[221] Formation of the mono-thioether (289) is preferred when disulphides containing bulky substituents are employed.

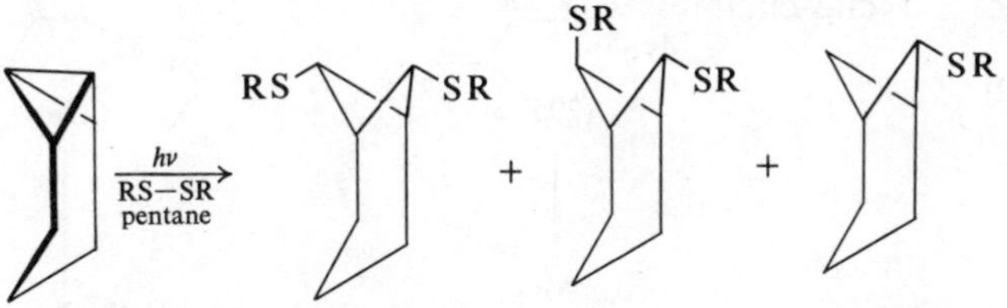

(288) (286) (287) (289)

R = Ph, Me, Et, Bu, Pr^i, or Bu^t

3 Compounds containing other Heteroatoms

Much of the interest in this section has again centred around the photoreactions **of** silicon-containing compounds. Irradiation of triethylsilanethiol in the **presence** of ethynylcarbinols led to the formation of silyl ethers and to addition to the alkyne.[222] Two pathways are operative in the liquid-phase photodegradation of hexamethyldisiloxane (290), the first arising by silicon—carbon bond homolysis and the second involving formation of the unsaturated intermediate (291) which rapidly reacts with free radicals or with polar reagents such as

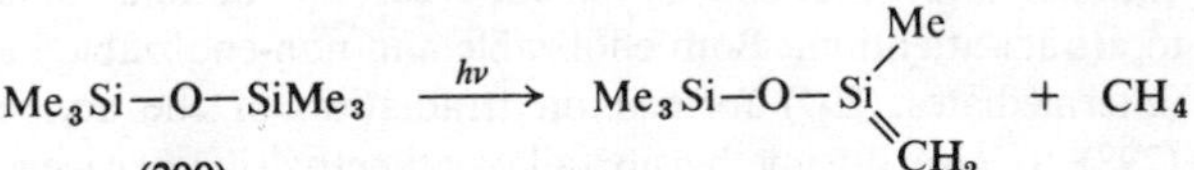

(290) (291)

methanol.[223] Carbon—oxygen bond homolysis is preferred in benzyl-(9-trimethylsilyl-9-fluorenyl) ether.[224] Mercury-sensitized photolysis of silicon—silicon bonds in simple methyl- and fluoro-disilanes has been reported and pro-

[220] T. Fujisawa and T. Kojima, *Bull. Chem. Soc. Japan*, 1977, **50**, 3061.
[221] P. Dietz and G. Szeimies, *Chem. Ber.*, 1978, **111**, 1938.
[222] M. M. Demina, A. S. Medvedeva, N. I. Protsuk, and N. S. Vyazankin, *Zhur. obshchei Khim.*, 1977, **47**, 2292.
[223] H.-P. Schuchmann, A. Ritter, and C. Von Sonntag, *J. Organometallic Chem.*, 1978, **148**, 213.
[224] M. T. Reetz, M. Kliment, and N. Grief, *Chem. Ber.*, 1978, **111**, 1083.

vides a route to unsymmetrical disilanes.[225] Irradiation of a mixture of Si_2F_6 and $(Me_2SiF)_2$, for example, gave a 40% yield of SiF_3SiMe_2F.

Irradiation of 1,1-dimethyl-2-phenyl-3-trimethylsilylsilacyclopropene (292) in benzene in the presence of acrylonitrile gave the 1-aza-2,8-disilabicyclo[3,2,1]-octa-3,6-diene (293) and the 1-aza-2,8-disilabicyclo[3,3,0]octa-3,6-diene (294).[226] Details of the mechanism of this transformation are not clear, but it is reasonable to assume that the products arise by further addition of the silacyclopropene (292) to initially formed adducts (295) and (296). Analogous two-atom insertions of carbonyl groups into silacyclopropanes and silacyclopropenes have been described.[227]

Irradiation of tris(trimethylsilyl)phenylsilane is known to give the highly reactive trimethylsilylphenylsilylene; generation of this silylene in the presence of acetone, diethyl ketone, or cyclohexanone resulted in the formation of the respective 2,2,2-trimethylphenyldisilanyl enol ethers in moderate yield.[228] Photochemically generated trimethylsilylphenylsilylene, unlike methylphenylsilylene, undergoes addition to hex-3-yne to give a silacyclopropene.[229]

The syntheses and reactions of silicon—carbon double-bonded species continue to attract attention. Both enolizable and non-enolizable ketones react with the intermediates (297) formed on irradiation of the arylpentamethyldisilanes (298) to give 2-trimethylsilyl(alkoxydimethylsilyl)benzene derivatives (299) as sole products.[230] Reaction of the same intermediates with alkynes has now been effected,[231] and two new processes have been observed involving nucleophilic attack by dimethyl sulphoxide on the photochemically excited aryldisilane.[232]

[225] K. G. Sharp and P. A. Sutor, *J. Amer. Chem. Soc.*, 1977, **99**, 8046.
[226] H. Sakurai, Y. Kamiyama, and Y. Nakadaira, *J.C.S. Chem. Comm.*, 1978, 80.
[227] D. Seyferth, S. C. Vick, M. L. Shannon, T. F. O. Lim, and D. P. Duncan, *J. Organometallic Chem.*, 1977, **135**, C37.
[228] M. Ishikawa, K. Nakagawa, and M. Kumada, *J. Organometallic Chem.*, 1977, **135**, C45.
[229] M. Ishikawa, K. Nakagawa, and M. Kumada, *J. Organometallic Chem.*, 1977, **131**, C15.
[230] M. Ishikawa, T. Fuchikami, and M. Kumada, *J. Organometallic Chem.*, 1977, **133**, 19.
[231] M. Ishikawa, T. Fuchikami, and M. Kumada, *J. Organometallic Chem.*, 1977, **127**, 261.
[232] H. Okinoshima and W. P. Weber, *J. Organometallic Chem.*, 1978, **149**, 279.

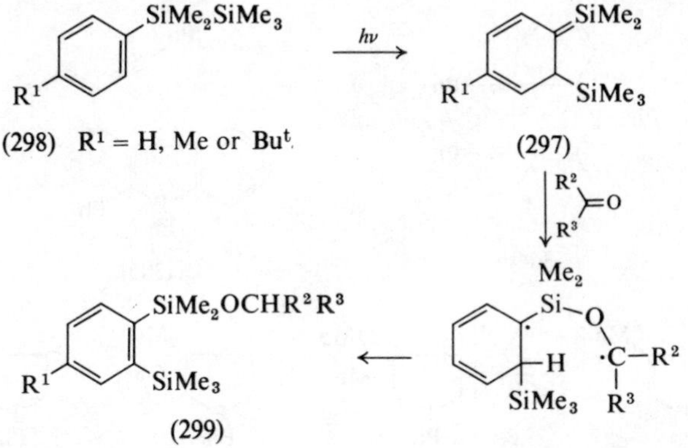

(298)   R¹ = H, Me or Buᵗ.

(297)

(299)

The silacyclopropene (300) was obtained in 75% yield by irradiation of the alkyne (301) in benzene;[233] further reaction with methanol gave the alkenes (302) and (303). Silacyclopropanes have previously been proposed as intermediates in the phototransformations of the corresponding vinyldisilanes. Alkynylsilanes undergo $[\pi 2 + \pi 2]$ cycloaddition reactions with cyclopentenone[234] and with maleic anhydride.[345]

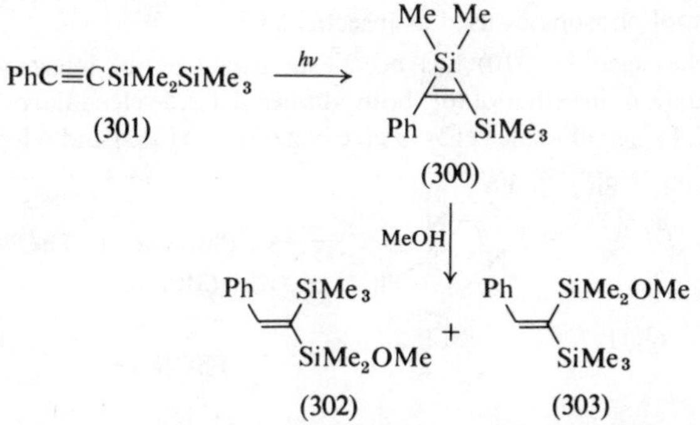

(301)

(300)

(302)   (303)

A trifluoromethylated diphosphabenzvalene has been obtained by pyrex-filtered irradiation of 2,3,5,6-tetrakis(trifluoromethyl)-1,4-diphosphabenzene in perfluoropentane.[236] No nitrogen-containing benzvalenes have so far been isolated from azines or diazines. The triphenylphosphine (304) is photochemically converted in benzene into the phosphindole (305).[237] Photoaddition of methanol to 1-phenyl-3-methyl-2-phospholene (306) affords ethers (307) and (308) and the exocyclic isomer (309);[238] an ionic process involving photoprotonation at C-2 is implicated. The corresponding 2-methyl derivatives were

[233] H. Sakurai, Y. Kamiyama, and Y. Nakadaira, *J. Amer. Chem. Soc.*, 1977, **99**, 3879.
[234] J. Soulie and M. J. Pouet, *Tetrahedron*, 1977, **33**, 2521.
[235] L. Birkofer and D. Eichstaedt, *J. Organometallic Chem.*, 1978, **145**, C29.
[236] Y. Kobayashi, S. Fujino, H. Hamana, I. Kumadaki, and Y. Hanzawa, *J. Amer. Chem. Soc.*, 1977, **99**, 8511.
[237] W. Winter, *Chem. Ber.*, 1977, **110**, 2168.
[238] H. Tomioka, K. Sugiura, S. Takata, Y. Hirano, and Y. Izawa, *J. Org. Chem.*, 1977, **42**, 3070.

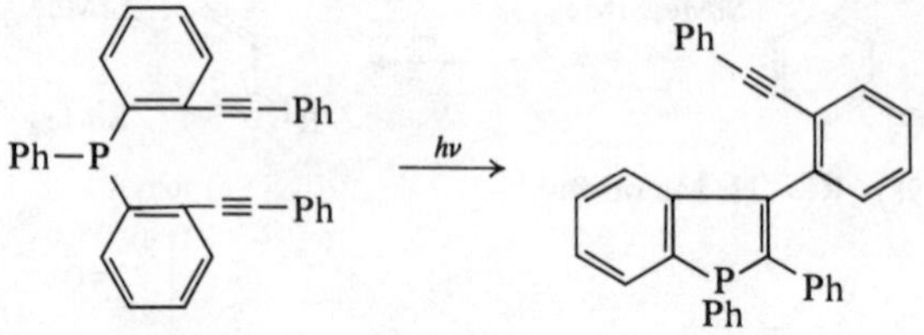

(304)   (305)

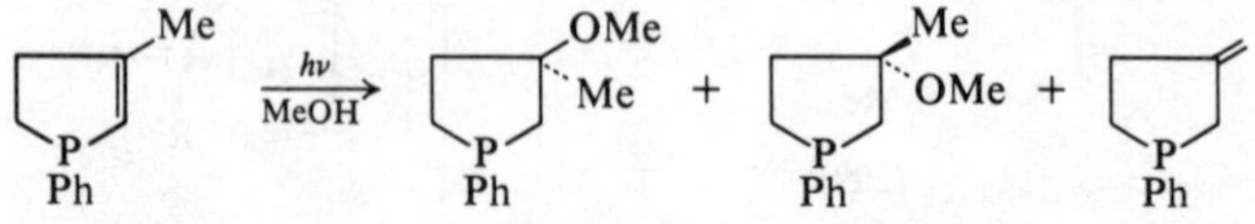

(306)   (307)   (308)   (309)

almost completely inert to photoaddition. *N*-Aryltriphenyliminophosphoranes
are converted on irradiation in inert solvents in nearly quantitative yield into
triphenylphosphine and diaryl azo- compounds.[239] Attempts to trap possible
intermediate nitrenes were unsuccessful. The photochemical behaviour of
dimethyl vinyl phosphate has been studied as a model for the photolytic decomposition of enol phosphates used as insecticides.[240]

Benzonitrile selenide (310) has been established as an intermediate in the
photodegradation in ethanol of both diphenyl-1,2,5-selenadiazole (311) and
diphenyl-1,2,4-selenadiazole (312) to give benzonitrile (313) and selenium.[241, 242]

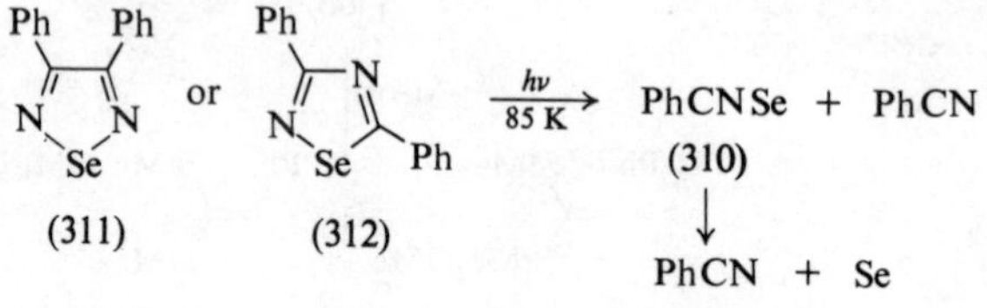

(311)   (312)

$$PhCNSe + PhCN$$
(310)
$$\downarrow$$
$$PhCN + Se$$
(313)

Attempts to trap this intermediate with dimethyl acetylenedicarboxylate were
unsuccessful, but the selenide was characterized spectroscopically at low temperature. It is less stable than the corresponding benzonitrile sulphide and decomposes in EPA glass at 110 K. Photosensitized [$_\pi2 + _\pi2$] cycloaddition of benzo-
[*b*]selenophen and its 3-methyl derivative to dimethyl acetylenedicarboxylate and
to 1,2-dichloroethylene has been reported,[243] and the novel 5*H*-[1]benzo-
selenino[2,3-*b*]pyridin-5-one (314) has been prepared along with the diselenide
(315) by irradiation of the seleno ester (316).[244]

[239] A. S. Yim, M. H. Akhtar, A. M. Unrau, and A. C. Oehlschlager, *Canad. J. Chem.*, 1978,
56, 289.
[240] J. Gignoux, C. Triantaphylides, and G. Peiffer, *Bull. Soc. chim. France*, 1977, 527.
[241] C. L. Pedersen and N. Hacker, *Tetrahedron Letters*, 1977, 3981.
[242] C. L. Pedersen, N. Harrit, M. Poliakoff, and I. Dunkin, *Acta Chem. Scand.*, 1977, **31B**,
848.
[243] T. Q. Minh, L. Christiaens, P. Grandclaudon, and A. Lablache-Combier, *Tetrahedron*,
1977, **33**, 2225.
[244] B. Pakzad, K. Praefcke, and H. Simon, *Angew. Chem. Internat. Edn.*, 1977, **16**, 319.

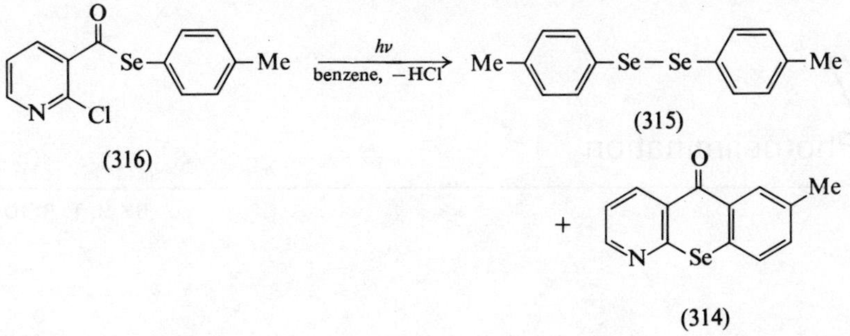

(316)　　　　　　　　　　　　　　(315)

+

(314)

Relevant organoboron phototransformations are briefly considered in this section. Vinylboronic esters undergo photochemically induced $Z \rightleftharpoons E$ isomerization,[245] and further studies of the photocyclization of dialkyl-1,3-dienylboranes to boracyclopent-3-enes have been reported.[246] Photoreaction of 2-amino-pyridine with diphenylboron chloride gave 10-phenyl-8,9-diaza-10-boraphenanthrene.[247] Irradiation of trialkylboranes in propan-2-ol resulted in the formation in good yield of 2-alkyl-1,3,2-dioxaborolanes.[248] Two competing primary photoreactions were observed on irradiation ($\lambda > 350$ nm) of tri-1-naphthylboron;[249] the first, simple boron—carbon cleavage, has already been documented for other arylboranes, but the second leading to the univalent organoboron, naphthylboryne, represents a novel pathway not previously recognized.

[245] A. Hassner and J. A. Soderquist, *J. Organometallic Chem.*, 1977, **131**, C1.
[246] G. Zweifel, S. J. Backlund, and T. Leung, *J. Amer. Chem. Soc.*, 1977, **99**, 5192.
[247] K. D. Mueller and K. Niedenzu, *Inorg. Chim. Acta*, 1977, **25**, L53.
[248] M. Tokuda, V. W. Chung, K. Inagaki, and M. Itoh, *J.C.S. Chem. Comm.*, 1977, 690.
[249] B. G. Ramsey and D. M. Anjo, *J. Amer. Chem. Soc.*, 1977, **99**, 3182.

7
Photoelimination

BY S. T. REID

1 Introduction

This chapter is principally concerned with the photochemically induced fragmentation of organic compounds accompanied by the formation of small molecules such as nitrogen, carbon dioxide, and sulphur dioxide. Photodecompositions resulting in the formation of two or more sizeable fragments are reviewed in the final section. Fragmentations arising by Norrish Type I and Type II reactions of carbonyl-containing compounds are considered in Part III, Chapter 1.

2 Elimination of Nitrogen from Azo-compounds

The mechanism of the photodecomposition of azoalkanes is still a matter of some dispute. Further evidence that the diazenyl radical $MeN_2\cdot$ is an intermediate in the photodecomposition of azomethane comes from an e.s.r. study of the decomposition in a methylcyclohexane matrix.[1] Efficient reversible Z,E-isomerization was detected, however, in azoethane on irradiation ($\lambda = 313$—390 nm) in a hydrocarbon matrix.[2] A new method for the introduction of the trifluoromethyl group has been developed based on the photolysis of trifluoromethyl azo-compounds.[3] Irradiation of azo-compounds (1) in highly viscous solvents such as t-butanol or hexadecane affords the trifluoromethyl derivatives (2) in good yield, presumably by 'cage recombination' of the radicals formed by loss of nitrogen.

$$R-N=N-CF_3 \xrightarrow[-N_2]{h\nu} R\cdot + \cdot CF_3 \longrightarrow R-CF_3$$

(1)   R = n-C_6H_{13}, n-C_8H_{17}, cyclohexyl,   (2)
cycloheptyl, 1-adamantyl *etc.*

The photoelimination of nitrogen from cyclic azo-compounds including heteroaromatics has been reviewed in detail.[4] The photoelimination of nitrogen from 1-pyrazolines continues to provide a useful synthetic route to cyclopropane derivatives. *Exo-* and *endo-*tricyclo[3,2,1,0^{2,4}]oct-6-enes have been prepared

[1] V. I. Pergushov, A. K. Vorob'ev, and V. S. Gurman, *Doklady Akad. Nauk S.S.S.R.*, 1977, **233**, 423.
[2] V. I. Pergushov and O. N. Bormot'ko, *Chem. Phys. Letters*, 1977, **51**, 269.
[3] P. Gölitz and A. de Meijere, *Angew. Chem. Internat. Edn.*, 1977, **16**, 854.
[4] H. Meier and K.-P. Zeller, *Angew. Chem. Internat. Edn.*, 1977, **16**, 835.

stereospecifically in this way by irradiation of the appropriate 1-pyrazolines,[5] and a stepwise elimination of nitrogen from the spiro-pyrazoline (3) to give 1,1,4,4-tetramethylspiro[2,2]pentane *via* 1,1,6,6-tetramethyl-4,5-diazospiro[2,4]-hept-4-ene has been observed.[6] Elimination of nitrogen from the excited singlet pyrazoline (4) affords bicyclo[2,1,0]pentane (5), whereas oxygen-trapping of the azo-derived triplet biradical generated by benzophenone-sensitized irradiation provides a new and versatile route to the peroxide (6).[7]

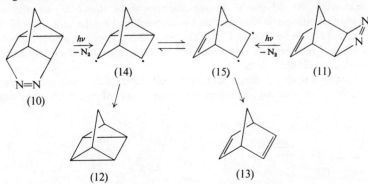

The effect of ring size on the efficiency of the photodecomposition of $\alpha\alpha'$-tetramethyl monocyclic azoalkanes has been examined.[8] Quantum yields for the photoelimination of nitrogen from azoalkanes (7), (8), and (9) are 0.52,

0.98, and 0.008, respectively. Fluorescence is also observed in azoalkanes (7) and (8). The apparent relative stability of the 1,2-diazacyclohexene (9) is tentatively ascribed to competing photoisomerization to the thermally unstable *trans*-isomer. Irradiation of the pyrazoline (10) and the diazetine (11) led to the loss of nitrogen and the formation in each case of quadricyclane (12) and nor-bornadiene (13).[9] These decompositions provide an independent pathway for the generation of quadricyclenyl and norbornadienyl biradicals (14) and (15);

[5] C. Dietrich-Buchecker, D. Martina, and M. Franck-Neumann, *J. Chem. Research (S)*, 1978, 78.
[6] M. Schneider, O. Schuster, and H. Rau, *Chem. Ber.*, 1977, **110**, 2180.
[7] R. M. Wilson and F. Geiser, *J. Amer. Chem. Soc.*, 1978, **100**, 2225.
[8] P. S. Engel, R. A. Hayes, L. Keifer, S. Szilagyi, and J. W. Timberlake, *J. Amer. Chem. Soc.*, 1978, **100**, 1876.
[9] N. J. Turro, W. R. Cherry, M. F. Mirbach, and M. J. Mirbach, *J. Amer. Chem. Soc.*, 1977, **99**, 7388.

norbornadiene is the preferred product from the singlet biradical, and quadri-cyclane that from the triplet biradical.

The products obtained on photodecomposition of 2,3-diazabicyclo[3,1,0]hex-2-enes differ fundamentally from those arising from other bicyclic pyrazolines. Compelling evidence has now been presented that for the 1-pyrazoline (16), photodecomposition takes place *via* the diazo-compound (17) and the carbene derived therefrom by loss of nitrogen; diazo-compound (17) is, in fact, the sole product of irradiation at $-50\,^{\circ}C$ in tetrahydrofuran.[10]

Examples of the photoelimination of nitrogen from cyclic azo-alkanes of varying ring size have also been described. Photolysis of 3-methyl-3-chloro-diazirine in the gas phase is reported to give nitrogen, chloroethylene, acetylene, 1,1-dichloroethane, and HCl.[11] Competing pathways have been observed in the photodecomposition of the cycloalkanespirodiazirines (18), the first leading to the diazocycloalkanes (19) and the second to products derived from the cyclo-alkylidenes (20).[12] Fluorescence ($\Phi_F = 0.76$) is favoured over elimination of

nitrogen on direct or sensitized irradiation of the azo-alkane (21) at room temperature.[13] At 80—90 °C in the vapour phase, however, quantitative elimination of nitrogen is observed and is accompanied by the formation of the novel cyclopentadiene dimers (22), (23), and (24). Low-temperature photo-elimination of nitrogen has been employed in the synthesis of the oddly named pterodactylane (25).[14]

3H-Pyrazoles readily undergo photochemically induced elimination of nitrogen to yield the corresponding cyclopropenes, often by way of detectable vinyl diazo

[10] M. Schneider and B. Csacsko, *Angew. Chem. Internat. Edn.*, 1977, **16**, 867.
[11] H. M. Frey and D. E. Penny, *J.C.S. Faraday I*, 1977, **73**, 2010.
[12] G. F. Bradley, W. B. L. Evans, and I. D. R. Stevens, *J.C.S. Perkin II*, 1977, 1214.
[13] N. J. Turro, K.-C. Liu, W. Cherry, J.-M. Liu, and B. Jacobson, *Tetrahedron Letters*, 1978, 555.
[14] H.-D. Martin and M. Hekman, *Tetrahedron Letters*, 1978, 1183.

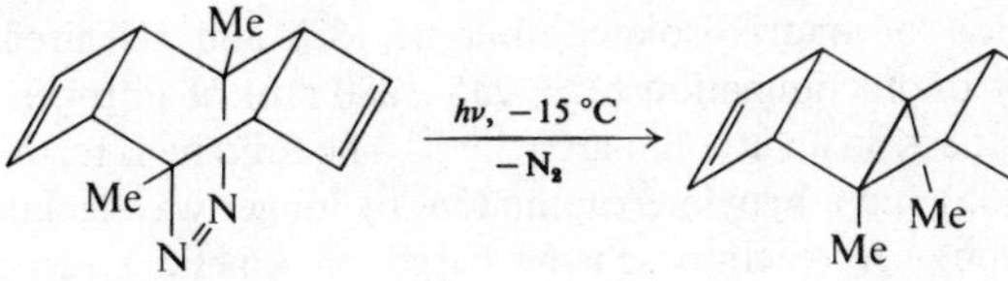

(25)

intermediates. The sulphinyl-3*H*-pyrazole (26), for example, is converted on irradiation into the cyclopropene (27), which can react further as a vinylcarbene with furan or ethyl vinyl ether.[15] Similarly, highly strained spiro-cyclopropenes have been prepared by photoelimination of nitrogen from spiro-3*H*-pyrazoles.[16]

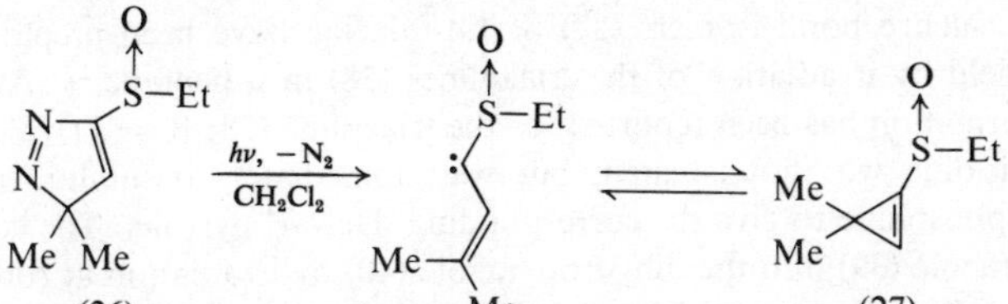

In a detailed study of the mechanism of the phototransformation of the diaza[2,2]spirenes (28) to the benzocyclopropenes (29), no evidence could be obtained for the presumed indazole intermediates.[17] Formation of the cyclobuta[*c*]cycloheptan-1-one (30) rather than the expected cyclopropene on irradiation of the pyrazoles (31) has been explained in terms of successive cyclization and hydrogen migration in the initially formed biradical intermediates (32).[18]

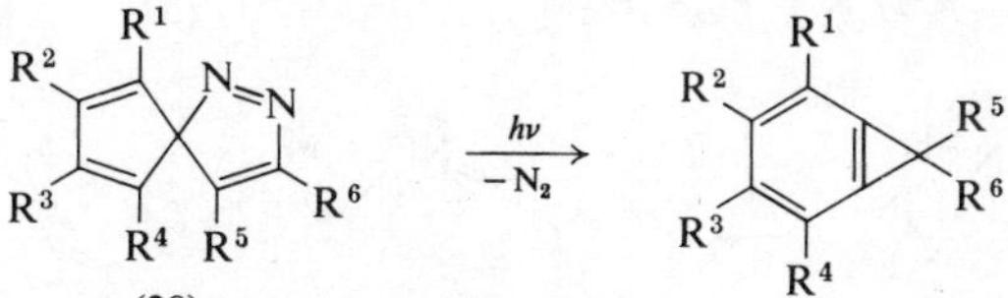

(28)

R^1	R^2	R^3	R^4	R^5, R^6
Ph	H	H	Ph	$-(CH_2)_6-$
Ph	H	H	Ph	CO_2Me
Ph	Ph	Ph	Ph	CO_2Me
Ph	Ph	(benzo)		CO_2Me

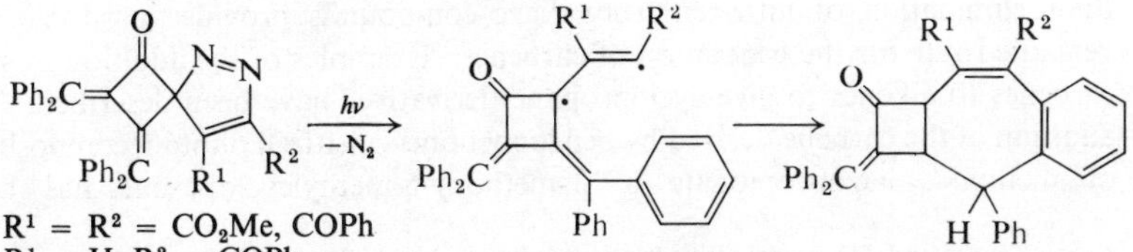

(29)

(31) $R^1 = R^2 = CO_2Me$, COPh
$R^1 = H$, $R^2 = COPh$

(32)   (30)

[15] M. Franck-Neumann and J.-J. Lohmann, *Angew. Chem. Internat. Edn.*, 1977, **16**, 323.
[16] H. Dürr, S. Fröhlich, B. Schley, and H. Weisgerber, *J.C.S. Chem. Comm.*, 1977, 843.
[17] E. Lüddecke, H. Rau, H. Dürr, and H. Schmitz, *Tetrahedron*, 1977, **33**, 2677.
[18] K. Ueda and F. Toda, *Bull. Chem. Soc. Japan*, 1978, **51**, 649.

The preparation of matrix-isolated thiirene (33) and selenirene has been accomplished by photoelimination (λ = 235—280 nm) of nitrogen from 1,2,3-thiadiazole (34) at 8 K in an argon matrix.[19, 20] The thiirene is further converted into thioketene (35) and ethynylmercaptan (36) by longer wavelength irradiation (λ = 330—370 nm). A reaction scheme based on kinetic evidence has been proposed to account for the products arising by photoelimination of nitrogen from 2-oxo-5,5-polymethylene-2,5-dihydro-1,3,4-oxadiazoles.[21]

$$\underset{(34)}{\text{\includegraphics}} \xrightarrow[-N_2]{h\nu} \underset{(33)}{\text{\includegraphics}} \xrightarrow{h\nu} \underset{(35)}{H_2C=C=S} + \underset{(36)}{HC\equiv C-SH}$$

The valence bond isomers (37) of 1,4-thiazine have been prepared in up to 55% yield by irradiation of the triazolines (38) in n-pentane.[22] An analogous transformation has been reported for the triazoline (38; R = H);[23] in this case, the aziridine was not isolated, but was immediately desulphurized with triphenylphosphine to give the corresponding 'Dewar' pyrrole. The conversion of the tetrazole (39) into the dihydropyrazole (40) by irradiation at room temperature is believed to proceed by way of an intramolecular 1,3-dipolar addition in the photochemically generated nitrilimine (41).[24]

(38) R = Ph, C$_6$H$_{11}$ (37)

(39)   (41)   (40)

3 Elimination of Nitrogen from Diazo-compounds

Photoelimination of nitrogen from diazo-compounds provides a simple and versatile route for the generation of carbenes. Examples of the addition of such carbenes to alkenes to give cyclopropane derivatives have been described. The addition of the carbene derived by benzophenone-sensitized photodecomposition of methyl α-diazopropionate to 1-methoxy-5-methylcyclohexene has been

[19] A. Krantz and J. Laureni, *Ber. Bunsengesellschaft Phys. Chem.*, 1978, **82**, 13.
[20] A. Krantz and J. Laureni, *J. Amer. Chem. Soc.*, 1977, **99**, 4842.
[21] D. Daniil, G. Gauglitz, and H. Meier, *Photochem. and Photobiol.*, 1977, **26**, 225.
[22] Y. Kobayashi, I. Kumadaki, A. Ohsawa, and A. Ando, *J. Amer. Chem. Soc.*, 1977, **99**, 7350.
[23] Y. Kobayashi, A. Ando, and I. Kumadaki, *J.C.S. Chem. Comm.*, 1978, 509.
[24] H. Meier and H. Heimgartner, *Helv. Chim. Acta*, 1977, **60**, 3035.

employed in a synthesis of menthofuran.[25] 8-*Endo*-phenyl substituted bicyclo-[5,1,0]octa-2,4-dienes (42) are formed together with minor amounts of the *exo*-isomers on irradiation of the diazo-compounds (43) in cycloheptatriene.[26] *Exo*-isomer formation is, however, favoured in the addition of photochemically generated phenylsulphonylcarbene to cyclohexene.[27] Competing formation of cyclopropane (44) and alkene (45) was observed on photodecomposition of diphenyldiazomethane (46) in 2-methylpropene.[28] This reaction is dramatically

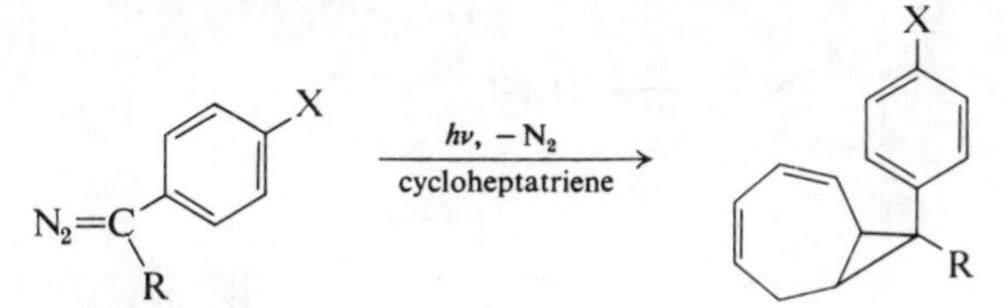

(43) R = PO(OMe)$_2$; X = H, OMe
    R = PO(C$_6$H$_5$)$_2$, CO$_2$Me, Ph; X = H

(42)

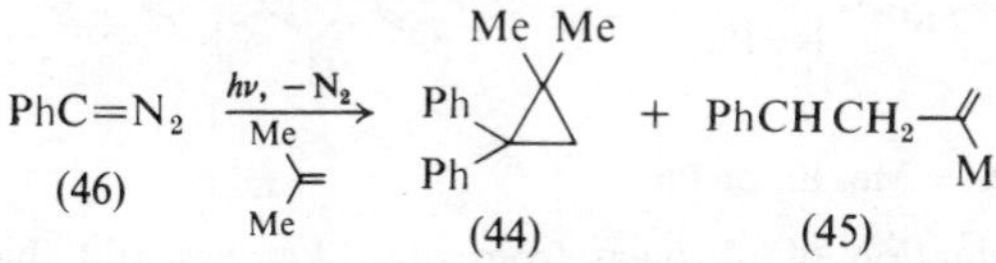

temperature dependent, the alkene being virtually the only product at -190 °C. The alkene has been shown to arise by a triplet-derived hydrogen-abstraction radical-recombination pathway. The major products of the reaction of photochemically generated biscarbomethoxycarbene with vinyl sulphides are the corresponding stable sulphonium ylides and not the cyclopropane derivatives, which are formed in low yield only.[29]

Insertion into the O—H bond rather than a C—H bond is the normal fate of a carbene generated in an alcohol. Thus, the principal product of the decomposition of ethyl 2-diazo-2-toluene-*p*-sulphonylacetate in methanol is the ether (47).[30]

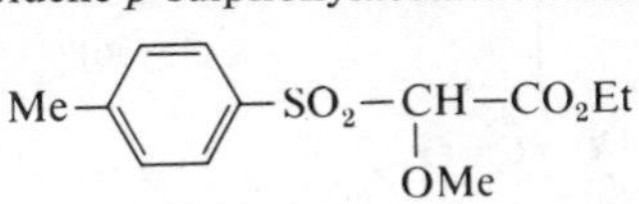

(47)

A detailed study of the photodecomposition of phenyldiazomethane in propan-2-ol, however, has revealed that at -196 °C, C—H bond insertion is preferred;[31] O—H bond insertion leading to 2-propyl benzyl ether occurs in 77% yield at 20 °C. The O—H bond insertion is believed to be characteristic of singlet carbene, whereas C—H bond insertion is the result of an abstraction-recombination reaction of triplet carbene. In contrast, photodecomposition of

[25] E. Wenkert, M. E. Alonso, B. L. Buckwalter, and K. J. Chou, *J. Amer. Chem. Soc.*, 1977, **99**, 4778.
[26] G. Maas and M. Regitz, *Angew. Chem. Internat. Edn.*, 1977, **16**, 711.
[27] R. A. Abramovitch, V. Alexanian, and J. Roy, *J.C.S. Perkin I*, 1977, 1928.
[28] R. A. Moss and J. K. Huselton, *J. Amer. Chem. Soc.*, 1978, **100**, 1314.
[29] W. Ando, H. Higuchi, and T. Migita, *J. Org. Chem.*, 1977, **42**, 3365.
[30] V. Chowdhry and F. H. Westheimer, *J. Amer. Chem. Soc.*, 1978, **100**, 309.
[31] H. Tomioka and Y. Izawa, *J. Amer. Chem. Soc.*, 1977, **99**, 6128.

4-diazoimidazole-5-carboxamide (48) in methanol at 10—20 °C produces the carboxamide (49) arising by C—H bond insertion together with the ether (50).[32] A mechanism involving insertion by singlet carbene has been proposed. Photoelimination of nitrogen from the 6-anilino-5-diazouracils (51) gave the indolo[2,3-*d*]pyrimidines (52).[33] The photodecompositions of methyl phenyl-α-diazopropylphosphinate[34] and of diethyl mercurybis(diazoacetate)[35] have also been studied.

(48)   (49)   (50)

(51) R = Me, Et, or Ph   (52)

The photoelimination of nitrogen from diazo-ketones and the fate of the resulting α-oxocarbene are still areas of major photochemical interest. The ability of α-oxocarbenes to undergo rearrangement to alternative α-oxocarbenes *via* oxirenes has previously been demonstrated by carbon labelling or product isolation studies. It has now been shown that in the photodecomposition of ^{13}C-labelled diazo-ketone (53) to give the 2-phenylpropionic acids (54) and

[32] U. G. Kang and H. Shechter, *J. Amer. Chem. Soc.*, 1978, **100**, 651.
[33] Y. Sakuma and F. Yonedo, *Heterocycles*, 1977, 6, 1911.
[34] R. D. Gareev and A. N. Pudovik, *Zhur. obshchei Khim.*, 1978, **48**, 226.
[35] T. B. Patrick and T.-T. Wu, *J. Org. Chem.*, 1978, **43**, 1506.

(55), 60—68% of the label has migrated *via* oxiren (56).[36] Surprisingly in the diazo-ketone (57), there is no detectable migration. These findings point to the existence of a complicated dependence of the carbene–carbene rearrangement on partial rate constants. A reinvestigation of the photodecomposition of 3-diazoheptan-4-one has revealed that although there are considerable discrepancies between the new observations and the earlier findings in regard to product composition, the existence of an oxocarbene–oxiren equilibrium is undisputed.[37] In fact, 85—88% oxiren involvement is indicated.

The formation of ketone (58) together with ketone (59) on photoelimination of nitrogen from 2-diazo-12-methylcyclododecan-1-one (60) provides the first evidence for the involvement of an oxiren in the decomposition of a cyclic diazo-ketone (Scheme 1).[38]

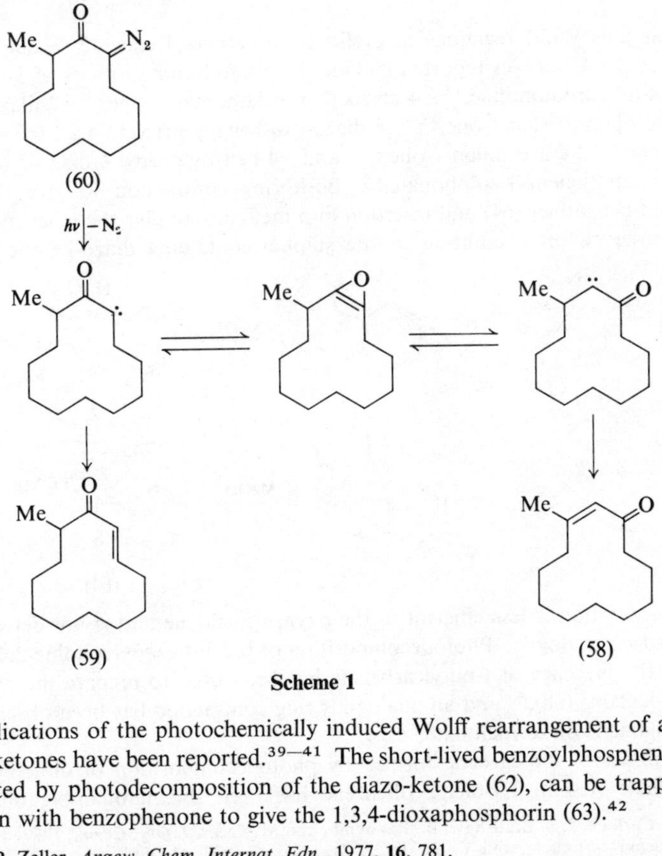

Scheme 1

Applications of the photochemically induced Wolff rearrangement of acyclic diazo-ketones have been reported.[39—41] The short-lived benzoylphosphene (61), generated by photodecomposition of the diazo-ketone (62), can be trapped by reaction with benzophenone to give the 1,3,4-dioxaphosphorin (63).[42]

36 K.-P. Zeller, *Angew. Chem. Internat. Edn.*, 1977, **16**, 781.
37 R. A. Cormier, K. M. Freeman, and D. M. Schnur, *Tetrahedron Letters*, 1977, 2231.
38 U. Timm, K.-P. Zeller, and H. Meier, *Chem. Ber.*, 1978, **111**, 1549.
39 M. Payard, J. Paris, J. Couquelet, and P. Tronche, *Bull. Soc. chim. France*, 1977, 505.
40 S. J. Branca, R. L. Lock, and A. B. Smith, *J. Org. Chem.*, 1977, **42**, 3165.
41 J. A. Katzenellenbogen, H. N. Meyers, H. J. Johnson, R. J. Kempton, and K. E. Carlson, *Biochemistry*, 1977, **16**, 1964.
42 M. Regitz, W. Illger, and G. Mass, *Chem. Ber.*, 1978, **111**, 705.

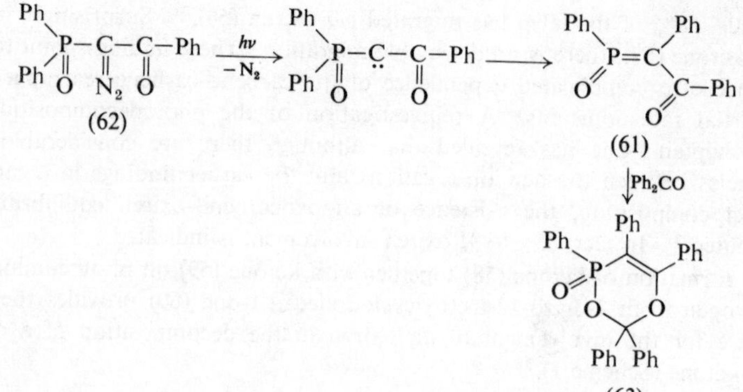

(62)   (61)

(63)

Analogous Wolff reactions in cyclic diazo-ketones frequently result in ring contraction. Examples reported include the photodecompositions of 3-diazo-4-oxo-3,4-dihydroquinoline,[43] 4-diazoethanoadamantan-3-one,[44] 2-diazocyclopenteno[b]biphenylen-1-one,[45] 8-diazo-*exo*-benzo[c]tricyclo[4,2,1,0[2,5]]non-3-en-7-one,[46] 2-diazo-indan-1-one,[47] and 4-benzoylphenyl-6-diazo-5-oxo-5,6-dihydronaphthalene-1-sulphonate.[48] Both ring contraction to give the first reported benzothiet (64) and insertion into methanol to give the ether (65) have been observed on irradiation of the sulphur-containing diazo-ketone (66).[49]

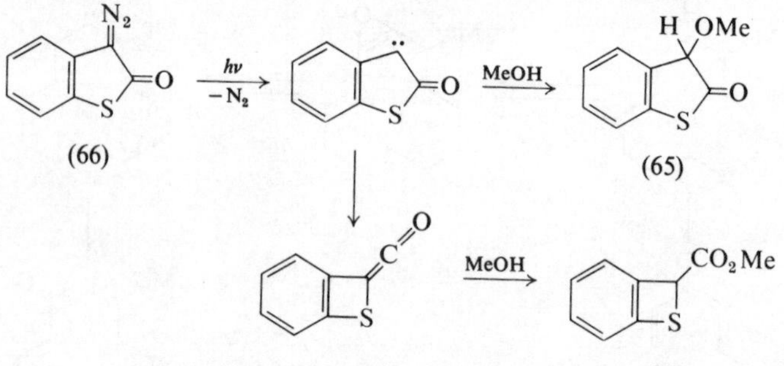

(66)   (65)

(64)

Ring contraction is less efficient in the oxygen analogue and is not detected in the nitrogen analogue. Photodecomposition of the 3-diazopyrrolidine-2,4-dione (67) in the presence of t-butylcarbazate has been used to prepare the *cis-* and *trans-β*-lactams (68),[50] and an analogous ring contraction has been observed in isopropylidene diazo-malonate.[51]

The dimer of the carbene formed by photodecomposition of dimethyl diazomalonate (69) in 1,4-dichlorobenzene has been shown unambiguously by

[43] J. T. Carlock, J. S. Bradshaw, B. Stanovnik, and M. Tišler, *J. Org. Chem.*, 1977, **42**, 1883.
[44] T. Sasaki, S. Eguchi, and Y. Hirako, *J. Org. Chem.*, 1977, **42**, 2981.
[45] P. R. Buckland and J. F. W. McOmie, *Tetrahedron*, 1977, **33**, 1797.
[46] L. Enescu and M. Avram, *Rev. Roumaine Chim.*, 1977, **22**, 1061.
[47] R. J. Spangler, J. H. Kim, and M. P. Cava, *J. Org. Chem.*, 1977, **42**, 1697.
[48] J. Pacansky and D. Johnson, *J. Electrochem. Soc.*, 1977, **124**, 862.
[49] E. Voigt and H. Meier, *Chem. Ber.*, 1977, **110**, 2242.
[50] J. E. T. Corrie, J. R. Hlubucek, and G. Lowe, *J.C.S. Perkin I*, 1977, 1421.
[51] S. L. Kammula, H. L. Tracer, P. B. Shevlin, and M. Jones, *J. Org. Chem.*, 1977, **42**, 2931.

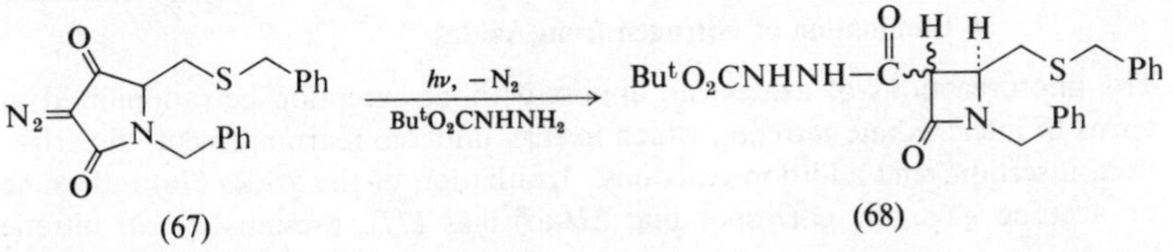

(67)   (68)

X-ray structure analysis to be the dihydrofuran (70);[52] it is believed to arise by cycloaddition of the α-oxocarbene (71) to the methoxyketene (72) formed in turn from the carbene by Wolff rearrangement. Cycloaddition of the carbene generated from 2-diazoacenaphthenone to a variety of nitriles to give acenaphtho[1,2-*d*]oxazoles has also been reported.[53]

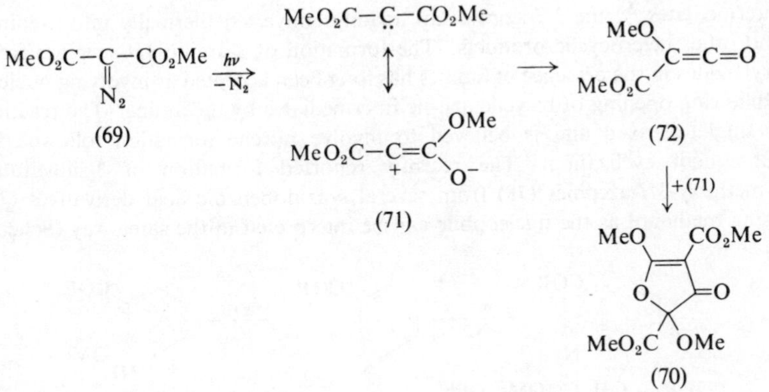

(69)   (71)   (72)

(70)

The photochemically induced valence isomerization of α-diazocarbonyl compounds to give diazirines, a process which had previously been believed to be restricted to α-diazo-amides, has now been observed in the α-diazo-ketone (73).[54] The reaction is photoreversible and both species (73) and (74) undergo slow Wolff rearrangements to the cyclobutane carboxylic acid (75) in aqueous dioxan. A similar isomerization occurs in the corresponding saturated α-diazo-ketone thereby eliminating the possibility of π-bond participation in this process.

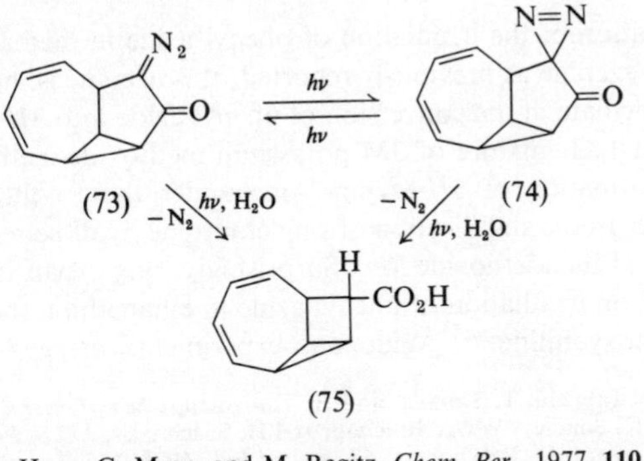

(73)   (74)

(75)

52 K. Eichorn, R. Hoge, G. Maas, and M. Regitz, *Chem. Ber.*, 1977, **110**, 3272.
53 O. Tsuge and M. Koga, *Heterocycles*, 1977, **6**, 411.
54 T. Miyashi, T. Nakajo, and T. Mukai, *J.C.S. Chem. Comm.*, 1978, 442.

4 Elimination of Nitrogen from Azides

The photoreactions of azides can almost without exception be rationalized in terms of intermediate nitrenes, which in turn undergo rearrangement, dimerization, insertion, and addition reactions. Irradiation of the azides (76) in hexane or acetone gave the corresponding 2*H*-azirines (77), presumably *via* nitrene

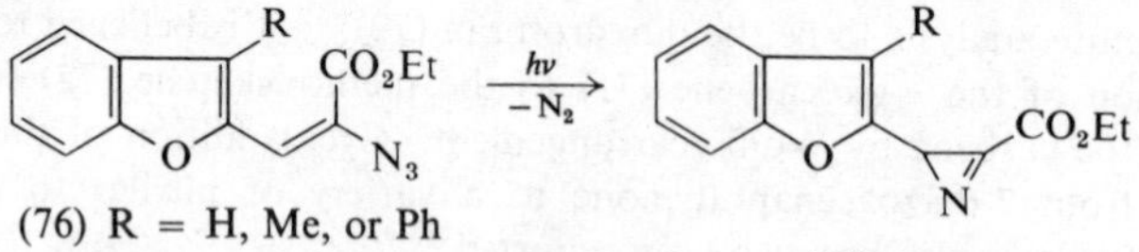

(76) R = H, Me, or Ph

(77)

intermediates;[55] the 2*H*-azirines are readily converted thermally into azepines and other heterocyclic products. The formation of 2-amino-3*H*-azepines from aryl azides in the presence of amines has long been accepted as involving nucleophilic ring opening of bicyclic azirine intermediates by the amine. The reaction is singlet derived and is believed to involve nitrene formation followed by subsequent cyclization. The recently reported formation of 3-substituted 2-methoxy-3*H*-azepines (78) from several *o*-azidobenzoic acid derivatives (79) using methanol as the nucleophile can be interpreted in the same way (Scheme 2).[56]

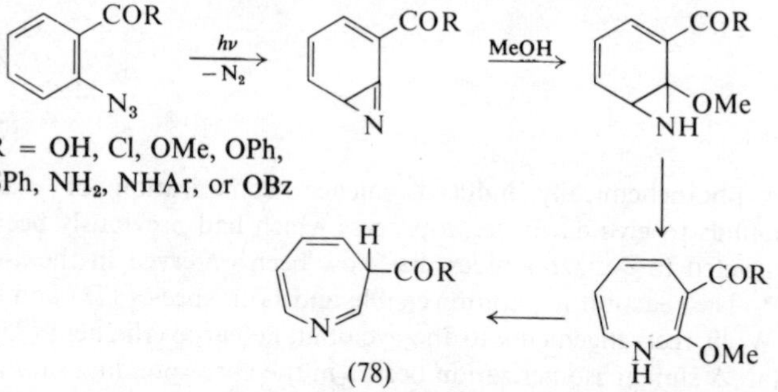

(79) R = OH, Cl, OMe, OPh,
SPh, NH$_2$, NHAr, or OBz

(78)

Scheme 2

A reinvestigation of the irradiation of phenyl azide in methanol did not give 2-methoxy-3*H*-azepine as previously reported; this azepine is, however, believed to be an intermediate in the conversion of phenyl azide into 3*H*-azepin-2-one on irradiation in a 1:1 mixture of 3M potassium methoxide-methanol:dioxan.[57] The ease of formation of 3*H*-azepin-2-one under these conditions has been attributed either to the stabilization of singlet nitrene by dioxan or to the greater nucleophilicity of the methoxide ion. Surprisingly, ring expansion to an azepine is not observed on irradiation of phenyl azide in ethanethiol, the major product being *o*-thioethoxyaniline.[58] Analogous experiments using 8-azidoquinoline,

[55] I. Isomura, H. Taguchi, T. Tanaka, and H. Taniguchi, *Chem. Letters*, 1977, 401.
[56] R. Purvis, R. K. Smalley, W. A. Strachan, and H. Suschitzky, *J.C.S. Perkin I*, 1978, 191.
[57] E. F. V. Scriven and D. R. Thomas, *Chem. and Ind.*, 1978, 385.
[58] S. E. Carroll, B. Nay, E. F. V. Scriven, H. Suschitzky, and D. R. Thomas, *Tetrahedron Letters*, 1977, 3175.

2-azidonaphthalene, and *p*-tolyl azide have clearly established that a 'nitrogen walk' is involved in this process as illustrated for *p*-tolyl azide (80) in Scheme 3. Bicyclic α-azidoarenes such as 5-azido- and 8-azido-quinolines have been successfully converted on photoelimination of nitrogen into azepines, but only in the presence of primary aliphatic amines.[59]

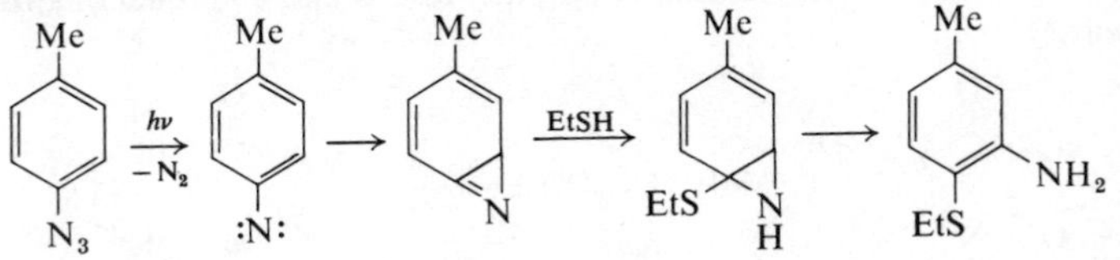

Scheme 3

An important recent study suggests, however, that the singlet photochemistry of phenyl azide (81) may have to be reinterpreted.[60] Irradiation ($\lambda = 216$ nm) in an argon matrix at 8 K gave as primary photoproduct a compound believed on the basis of spectroscopic evidence to be 1-azacyclohepta-1,2,4,6-tetraene (82); the same ketenimine was obtained by photodecomposition under identical conditions of the triazole (83) *via* 2-diazomethylpyridine (84). It is now proposed

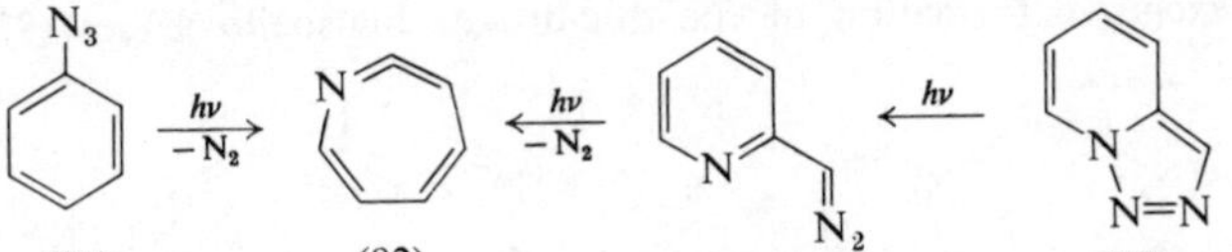

that it is this ketenimine and not the isomeric azirine which is trapped by nucleophiles to give azepines. Firm evidence supporting singlet nitrene as a precursor of the ketenimine is lacking.

The first example of a ring expansion in an azidopyrimidine has been reported;[61] photoelimination of nitrogen from 6-azido-1,3-dimethylthymine (85) in the presence of methylamine affords in 40% yield the 1,3,5-triazepine (86).

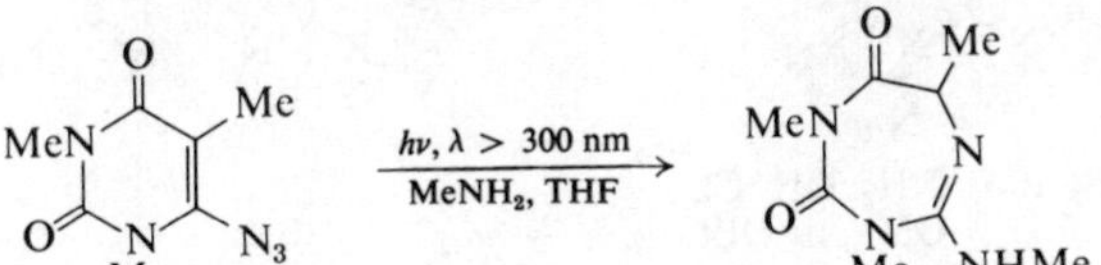

This result contrasts with an earlier study of the photodecomposition of 6-azido-1,3-dimethyluracil in primary or secondary amines which gave 6-alkylamino-5-amino-1,3-dimethyluracils. That photoreaction has now been extended to

[59] B. Nay, E. F. V. Scriven, and H. Suschitzky, *Synthesis*, 1977, 757.
[60] O. L. Chapman and J.-P. Le Roux, *J. Amer. Chem. Soc.*, 1978, **100**, 282.
[61] S. Senda, K. Hirota, T. Asao, K. Maruhashi, and N. Kitamura, *Tetrahedron Letters*, 1978, 1531.

provide a synthetic route to lumazines and fervenulins (7-azalumazines);[62] irradiation of 6-azido-1,3-dimethyluracil (87) and *N*-methylglycine ethyl ester (88) in tetrahydrofuran, for example, gave a 72% yield of 7,8-dihydro-1,3,8-trimethyllumazin-6(5*H*)-one (89). 5-Acylamino-6-chloro-1,3-dimethyluracils are similarly formed on irradiation of 6-azido-1,3-dimethyluracil in the presence of acyl halides, thus providing another approach to C-5 and C-6 functionalization in uracils.[63]

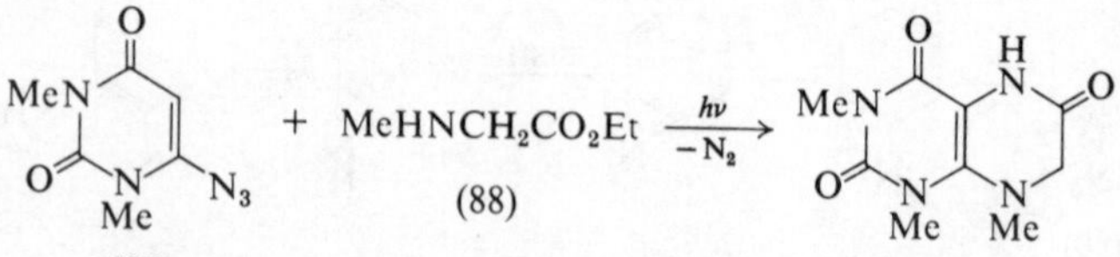

(87)   + MeHNCH$_2$CO$_2$Et $\xrightarrow[-N_2]{h\nu}$

(88)

(89)

The photochemically induced conversion of 2-azidobiphenyl into carbazole *via* 2-nitrenobiphenyl has been the subject of much study. Further evidence that the process involves the concerted cyclization of singlet nitrene has been reported.[64] 2-Methyl-2'-nitrenobiphenyl gives 4-methylcarbazole in high yield under singlet-promoting conditions, but increasing amounts of phenanthridine are produced under conditions favouring the formation of triplet nitrenes. Similarly, irradiation of the 6-(2-azidophenylthio)uracils (90) resulted in the almost exclusive formation of the dihydro-10-thiaisoalloxazines (91).[65] The

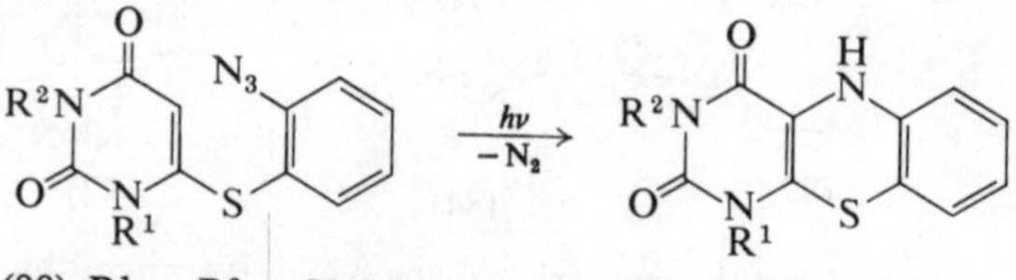

(90) R^1 = R^2 = H, Me
     R^1 = H, R^2 = Me

(91)

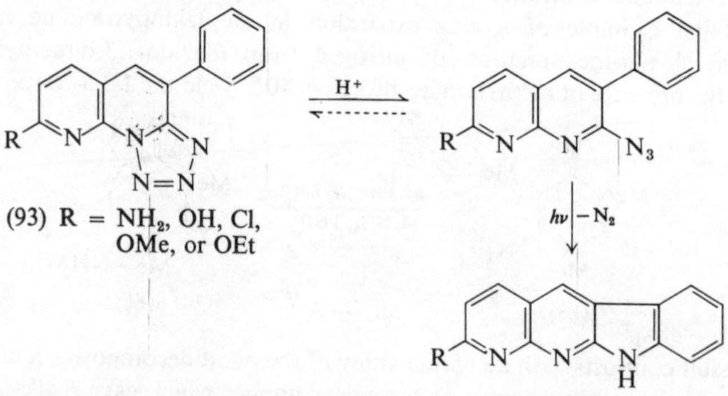

(93) R = NH$_2$, OH, Cl,
     OMe, or OEt

(92)

[62] S. Senda, K. Hirota, T. Asao, and K. Maruhashi, *J. Amer. Chem. Soc.*, 1977, **99**, 7358.
[63] S. Senda, K. Hirota, T. Asao, and K. Maruhashi, *J.C.S. Chem. Comm.*, 1978, 367.
[64] J. M. Lindley, I. M. McRobbie, O. Meth-Cohn, and H. Suschitzky, *J.C.S. Perkin I*, 1977, 2194.
[65] T. Hiramitsu and Y. Maki, *J.C.S. Chem. Comm.*, 1977, 557.

6*H*-indolo[2,3-*b*][1,8]naphthyridines (92) were produced in high yield on irradiation in trifluoroacetic acid of the 4-phenyltetrazolo[1,5-*a*][1,8]naphthyridines (93).[66] The tetrazole–azide equilibrium is known to be displaced towards the azide by trifluoroacetic acid.

The reactions of singlet and triplet 1,3,5-triazinylnitrenes, generated by photoelimination of nitrogen from the 2-azido-4-methoxy-6-(1-naphthyl)-1,3,5-triazines (94), with acetone, acetonitrile, and dimethyl sulphoxide have been described.[67] The singlet nitrene undergoes addition to acetone to afford the oxadiazole (95), whereas triplet nitrene abstracts hydrogen forming the aminotriazine (96). A 1,1-cycloadduct was also obtained with acetonitrile, whereas with dimethyl sulphoxide the products were the ylide (97), the aminotriazine (96), and polymer.

(94) R^1 = H; R^2 = H, OH, OMe, Me, or Cl
R^1 = OH, OMe, or Me; R^2 = H

Acyl nitrenes are also readily prepared by photoelimination of nitrogen from the corresponding azides. Irradiation of benzoyl azide in the presence of anthracene affords phenyl isocyanate, derived by Curtius rearrangement of the acyl nitrene, and the nitrene insertion product 9-benzoylaminoanthracene.[68] Similarly, photodecomposition of aroyl azides in the presence of $Me_2C{=}CHOMe$ gave approximately equal amounts of the corresponding isocyanates and the

66 A. Da Settimo, G. Primofiore, V. Sauterini, G. Biagi, and L. D'Amico, *J. Org. Chem.*, 1977, **42**, 1725.
67 T. Goka, H. Shizuka, and K. Matsui, *J. Org. Chem.*, 1978, **43**, 1361.
68 K. Ichimura, *Bull. Chem. Soc. Japan*, 1977, **50**, 3063.

adducts (98).[69] Examples of the addition of photochemically generated acyl nitrenes to alkenes to give *N*-acylaziridines have also been described.[70, 71]

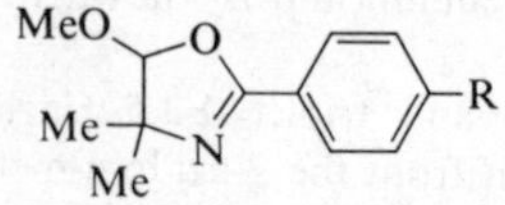

(98) R = H, OMe, or Cl

Ethoxycarbonylnitrene, obtained by irradiation of ethyl azidoformate, has been found to insert preferentially into the α-C—H bonds of acyclic ethers.[72] Insertion and abstraction products were obtained by sensitized photodecomposition of methanesulphonyl azide in hydrocarbons,[73] and irradiation of biphenyl-2-sulphonyl azide (99) in the presence of di-t-butyl sulphide gave the sultam (100) in 3% yield, together with 2-biphenyl-t-butyl sulphone and 2-biphenyl disulphide.[74]

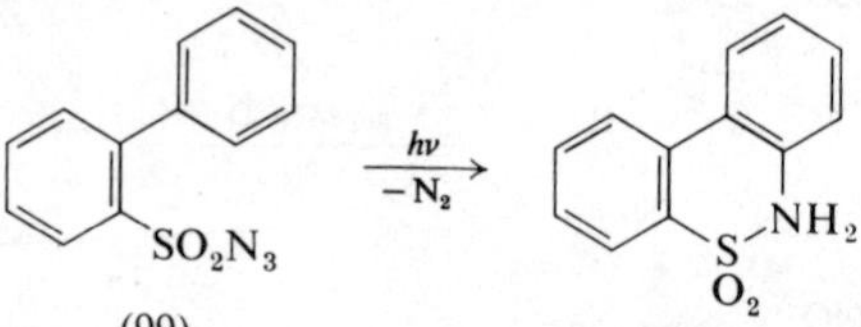

(99)   (100)

5 Photodecomposition of other Compounds having N—N Bonds

On irradiation, the sodium or lithium salts of toluene-*p*-sulphonylhydrazones are converted into the corresponding diazo-compounds which can either eliminate nitrogen and give products derived from the resulting carbene or protonate in protoic solvents such as methanol and give products derived from the diazonium ion. Thus, irradiation of the lithium salt (101) gave 5,5-dimethylnorbornene (102);[75] deuterium labelling studies indicate a preference for *exo* C-3 to C-2

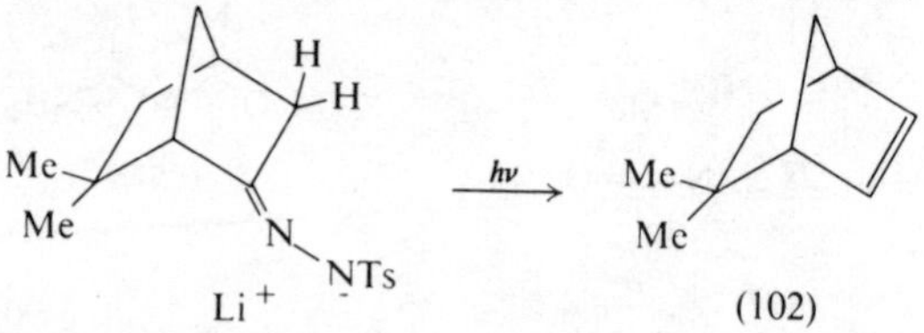

(101)   (102)

[69] V. P. Semenov, A. N. Studenikov, A. D. Bespalov, and K. A. Ogloblin, *Zhur. org. Khim.*, 1077, **13**, 2202.

[70] V. P. Semenov, A. N. Studenikov, A. P. Prosypkina, and K. A. Ogloblin, *Zhur. org. Khim.*, 1977, **13**, 2207.

[71] V. P. Semenov, A. P. Prosypkina, O. F. Gavrilova, and K. A. Ogloblin, *Khim. Geterotsikl Soedinenii*, 1977, 464.

[72] N. Torimoto, T. Shingaki, and T. Nagai, *Bull. Chem. Soc. Japan*, 1977, **50**, 1517.

[73] N. Torimoto, T. Shingaki, and T. Nagai, *J. Org. Chem.*, 1978, **43**, 631.

[74] R. A. Abramovitch, T. Chellathurai, I. T. McMaster, T. Takaya, C. I. Azogu, and D. P. Vanderpool, *J. Org. Chem.*, 1977, **42**, 2914.

[75] P. K. Freeman, T. A. Hardy, J. R. Balyeat, and L. D. Westcott, *J. Org. Chem.*, 1977, **42**, 3356.

hydrogen migration in the intermediate carbene. The carbocyclic aromatic carbene, 4,9-methano[11]annulenylidene (103), has been generated both by photodecomposition of the corresponding ketone tosylhydrazone salt (104) and by decomposition of the sodium salt of the tosylhydrazone of 3-formyl-1,6-methano[10]annulene (105).[76] In the latter case, the reaction is believed to

(104)   (103)

(105)

involve a carbene–carbene rearrangement proceeding with a high degree of regioselectivity. In the absence of sufficiently reactive species, the annulenylidene dimerizes to give a fulvalene, but in the presence of styrene or dimethyl fumarate, spirocyclopropane adducts are obtained. An analogous series of reactions for 3,8-methano[11]annulenylidene has been reported, and the photochemical generation and reactions of *anti*-tricyclo[6,1,0,0^{2,4}]non-5-en-7-ylidene and *anti,anti*-tetracyclo[7,1,0,0^{2,4}0^{5,7}]decan-8-ylidene have been described.[77]

Photodecomposition of the cyclobutanone tosylhydrazone salt (106) in methanol to give products (107) to (111) is the result of protonation of the initially formed diazo-compound as outlined in Scheme 4.[78] Other examples of the generation of diazonium ions by irradiation of tosylhydrazone salts in methanol have been reported.[79, 80] Irradiation of the quinone tosylhydrazone (112) gave the benzoxazoline (113) in good yield; subsequent photodecomposition and elimination of nitrogen led to the formation of the 1*H*-pyrrolo[1,2-*a*]-indole (114) containing the basic skeleton of the mitomycins.[81] Reports of the photochemical decomposition of aryl diazonium salts have been published.[82—84]

The novel photoelimination of nitrogen from the indazolo[2,3-*b*]pyridazine (115) to give the naphthalene (116) in addition to 3-(2-acetamidophenyl)-pyridazine is believed to proceed *via* the carbene (117) formed by loss of nitrogen from intermediate (118) or (119).[85] The influence of substituents on the efficiency of this process is not clear.

[76] W. M. Jones, R. A. LaBar, U. H. Brinker, and P. H. Gebert, *J. Amer. Chem. Soc.*, 1977, **99**, 6379.
[77] M. Oda, Y. Ito, and Y. Kitahara, *Tetrahedron Letters*, 1978, 977.
[78] R. D. Miller, *Tetrahedron Letters*, 1977, 3309.
[79] W. Kirmse and U. Richarz, *Chem. Ber.*, 1978, **111**, 1883.
[80] W. Kirmse and U. Richarz, *Chem. Ber.*, 1978, **111**, 1895.
[81] M. Akiba, Y. Kosugi, and T. Takaday, *Heterocycles*, 1977, **6**, 1125.
[82] H. Baumann, H. G. O. Becker, and S. Thiele, *J. prakt. Chem.*, 1977, **319**, 140.
[83] H. G. O. Becker, G. Hoffmann, and G. Israel, *J. prakt. Chem.*, 1977, **319**, 1021.
[84] R. A. Bartsch, N. F. Haddock, and D. W. McCann, *Tetrahedron Letters*, 1977, 3779.
[85] H. Hasegawa, H. Arai, and H. Igeta, *Chem. and Pharm. Bull. (Japan)*, 1977, **25**, 192.

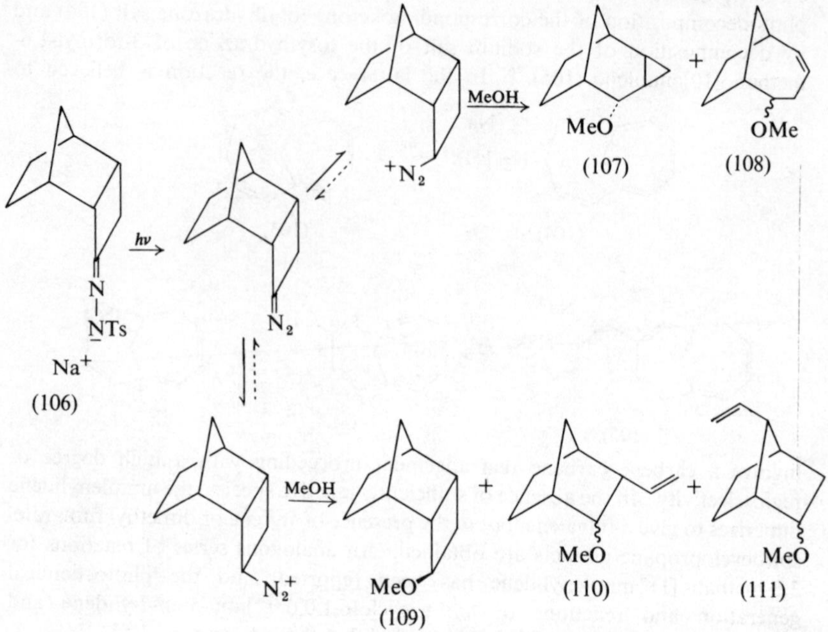

Scheme 4

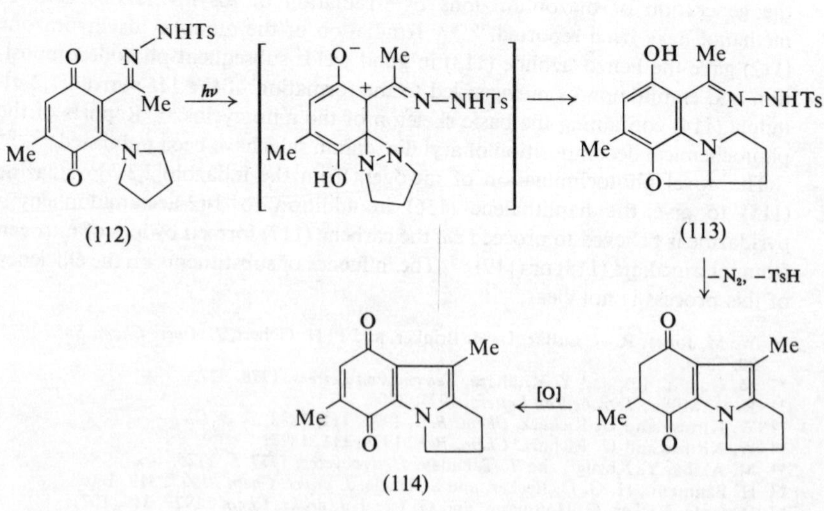

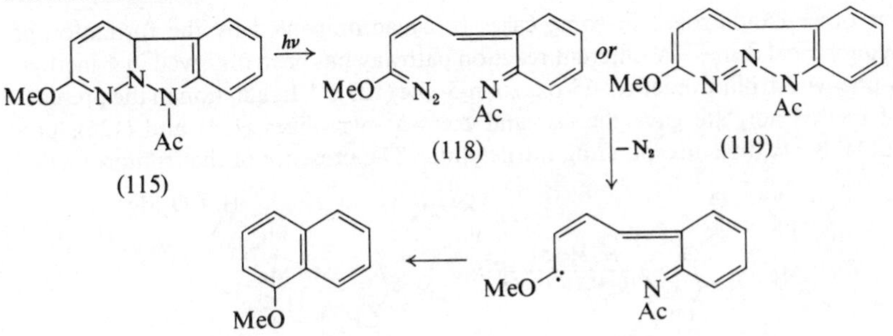

MeO (115) → (118) *or* (119)

−N₂ ↓

(116) ← (117)

6 Photoelimination of Carbon Dioxide

Elimination of carbon dioxide is one of the pathways considered in a recent review of the photochemistry of carboxylic acid derivatives.[86] Carbon dioxide is obtained by irradiation of acetic acid in argon or nitrogen matrices at 10 K,[87] and e.s.r. spectral evidence for alkyl and acyl radicals in the photodecomposition of aliphatic carboxylic acids and esters in aqueous and non-aqueous solvents has been described.[88] Photodecarboxylation of luciferin and certain analogues has been observed in methanol solution,[89] and various radical-derived products including carbon dioxide have been obtained from the photodecomposition of peracetic acid.[90]

The generation of alkyl radicals on irradiation of certain oxalate esters is accompanied by the formation of carbon dioxide,[91] and the photoelimination of carbon dioxide from [2-¹³C]bicyclopyranone and [6-¹³C]bicyclopyranone has been studied in an argon matrix at 8 K with a view to resolving the controversy regarding the infrared spectrum of matrix-isolated cyclobutadiene.[92]

Irradiation of the oxadiazole (120) in the presence of α-methylstyrene (121) results in the evolution of carbon dioxide and the formation of the α-sytryl-methylene hydrazone (122).[93] Analogous addition reactions have been reported

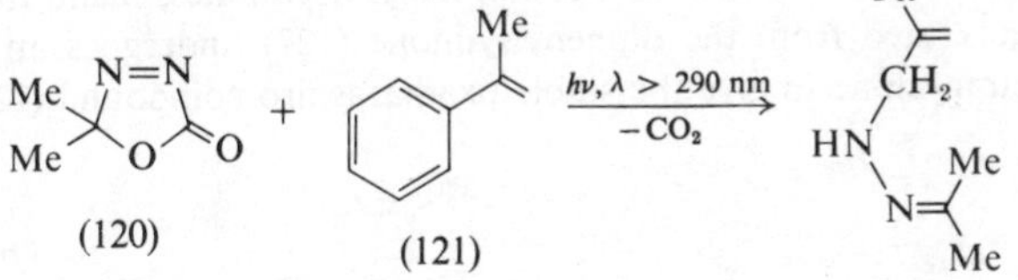

(120)   +   (121) $\xrightarrow[-CO_2]{h\nu, \lambda > 290\ nm}$ (122)

[86] J. D. Coyle, *Chem. Rev.*, 1978, **78**, 97.
[87] J. L. Wilkerson and W. A. Guillory, *J. Photochem.*, 1977, **7**, 251.
[88] T. Kaiser, L. Grossi, and H. Fischer, *Helv. Chim. Acta*, 1978, **61**, 223.
[89] N. Suzuki, M. Sato, H. Yokoyama, H. Morikawa, and T. Goto, *Agric. and Biol. Chem. (Japan)*, 1977, **41**, 217.
[90] Y. Ogata and K. Tomizawa, *J. Org. Chem.*, 1978, **43**, 261.
[91] S. Icli, V. J. Nowlan, P. M. Rahimi, C. Thankachan, and T. T. Tidwell, *Canad. J. Chem.*, 1977, **55**, 3349.
[92] R. G. S. Pong, B.-S. Huang, J. Laureni, and A. Krantz, *J. Amer. Chem. Soc.*, 1977, **99**, 4153.
[93] D. Daniil and H. Meier, *Tetrahedron Letters*, 1977, 3155.

for other oxadiazoles, in some cases being accompanied by the formation of symmetrical azines. A different reaction pathway has been observed in 4-methyl-4-phenyl-2-trifluoromethyl-Δ^2-oxazolin-5-one (123).[94] Irradiation in the presence of methyl acrylate gave the *cis*- and *trans*-Δ^1-pyrrolines (124) and (125), presumably *via* the corresponding nitrile ylide. The presence of the trifluoromethyl

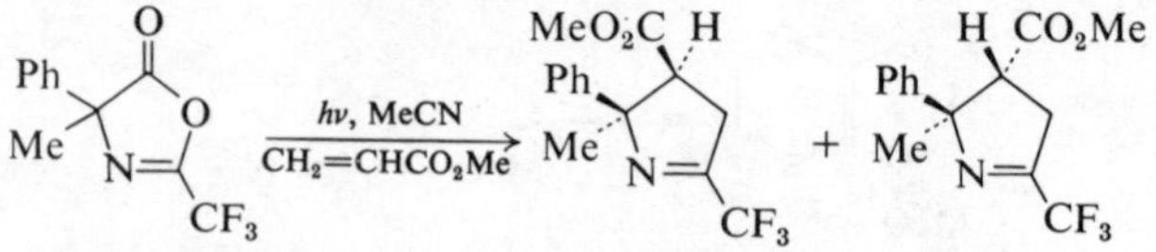

group is apparently essential for ylide formation, for the methyl analogue, 2,4-dimethyl-4-phenyl-Δ^2-oxazolin-5-one, is converted on irradiation both in the presence and absence of dipolarophiles into *N*-(1-methylbenzylidene)acetamide and carbon monoxide.

A detailed study of the photoelimination of carbon monoxide from 3,4-disubstituted sydnones and the subsequent reactions of the photochemically generated nitrile-imines has been reported.[95] Nitrile-imine formation has been shown to be the result of initial bond formation between N-2 and C-4 (Scheme 5).

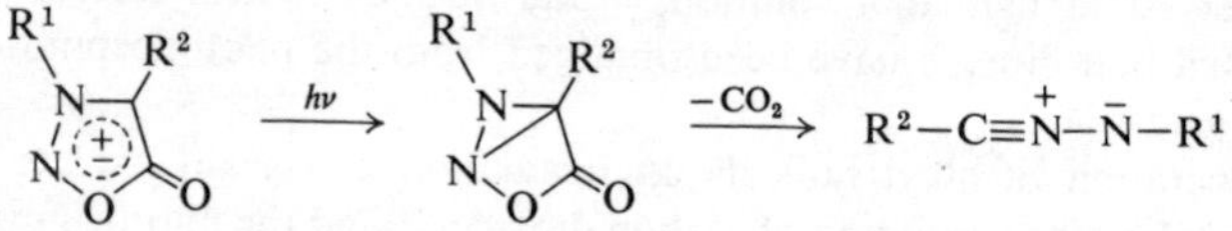

Scheme 5

2,4,5-Trisubstituted 1,2,3-triazoles are formed from 1,2-bisazoethylenes which arise in turn by head-to-head dimerization of nitrile-imines. The nitrile-imines can also be trapped by dipolarophiles to give pyrazoles, and by carboxylic acids, to give *N'*-acylhydrazides. *Z,E*-Photoisomerization competes with carbon dioxide elimination in the photoreactions of 4-styrylsydnones,[96] and the nitrile-imine (126) generated from the diphenylsydnone (127) undergoes an unusual addition to tetracyclone to give the stable oxadiazaspiro-compound (128).[97]

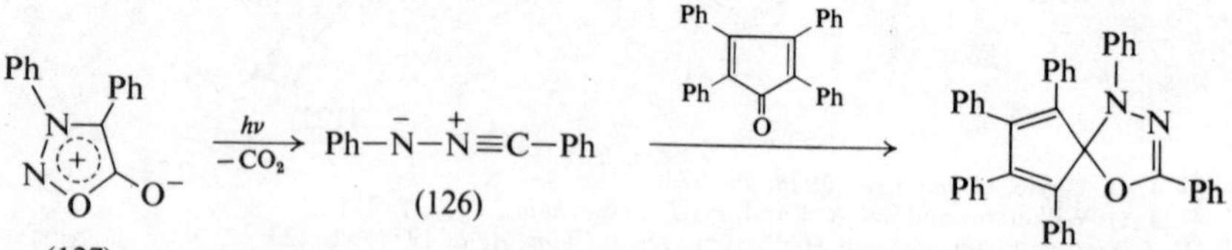

[94] M. R. Johnson and L. R. Sousa, *J. Org. Chem.*, 1977, **42**, 2439.
[95] M. Märky, H. Meier, A. Wunderli, H. Heimgartner, H. Schmid, and H.-J. Hansen, *Helv. Chim. Acta*, 1978, **61**, 1477.
[96] B. M. Neumann, H. G. Henning, D. Gloyna, and J. Sauer, *J. prakt. Chem.*, 1978, **320**, 81.
[97] S. K. Kar, *Indian J. Chem.*, 1977, **15B**, 184.

7 Fragmentation of Organosulphur Compounds

Irradiation of simple alkyl sulphides in a dilute rigid glass of 3-methylpentane at 77 K results in carbon–sulphur bond cleavage and the formation of 'hot' radical pairs, which either react with each other to give products or abstract hydrogen from the solvent to give thiols.[98] Quantum yields for the formation of hydrogen and methane in the photolysis of partially deuteriated methanethiol (CH$_3$SD) have been measured at 185 and 254 nm.[99] Thioketals are converted in high yield into the parent ketone by benzophenone-sensitized irradiation in the presence of oxygen.[100] In contrast, photodecomposition of the fluorenone diaryldithioketal (129) affords the 9-(arylthio)fluoren-9-yl radical, which dimerizes to give the 9,9′-bifluorenyl derivative (130).[101]

(129)   (130)

Carbon—sulphur bond homolysis is involved in the phototransformation of 2-phenyl-Δ^2-thiazolines to the corresponding nitriles and thiirane; a concerted pathway is preferred for this decomposition since the addition of n-octanethiol did not inhibit the formation of nitriles.[102] Preferential carbon—sulphur bond homolysis also occurs on irradiation of 2-(phenylthio)cycloalkanones in alcoholic solution, whereas Type I reactions are found in the corresponding 2-(alkoxycarbonyl)- and 2-cyano-cycloalkanones.[103] The photodecomposition of 3-acetyl-2,4-dioxothiolane has been described.[104]

Photoinduced desulphurization of cyclic sulphides with triethyl phosphite has again been used in the synthesis of a wide variety of cyclophanes;[105–110] [2](1,3)azuleno[2]paracyclophane (131), for example, has been obtained in this way in 70% yield from the azulene (132).[111] Desulphurization of tetraphenyl-*p*-dithin to give tetraphenylthiophen has also been reported.[112]

98   F. C. Adam and A. J. Elliot, *Canad. J. Chem.*, 1977, **55**, 1546.
99   D. Kamra and J. M. White, *J. Photochem.*, 1977, **7**, 171.
100 T. T. Takahashi, C. Y. Nakamura, and J. Y. Satoh, *J.C.S. Chem. Comm.*, 1977, 680.
101 K. Praefcke and H. Simon, *Z. Naturforsch.*, 1977, **32b**, 1172.
102 N. Suzuki, K. Kuroyanagi, and Y. Izawa, *Chem. and Ind.*, 1977, 313.
103 M. Tokuda, Y. Watanabe, and M. Itoh, *Bull. Chem. Soc. Japan*, 1978, **51**, 905.
104 K. Saito and T. Sato, *Chem. Letters*, 1978, 307.
105 M. W. Haenel, *Tetrahedron Letters*, 1977, 4191.
106 M. Matsumoto, T. Otsubo, Y. Sakata, and S. Misumi, *Tetrahedron Letters*, 1977, 4425.
107 H. Machida, H. Tatemitsu, Y. Sakata, and S. Misumi, *Tetrahedron Letters*, 1978, 915.
108 T. Otsubo, T. Kohda, and S. Misumi, *Tetrahedron Letters*, 1978, 2507.
109 P. J. Jessup and J. A. Reiss, *Austral. J. Chem.*, 1977, **30**, 843.
110 T. Otsubo, H. Horita, and S. Misumi, *Synthetic Comm.*, 1976, **6**, 591.
111 Y. Fukazawa, M. Aoyagi, and S. Itô, *Tetrahedron Letters*, 1978, 1067.
112 S. Lahiri, V. Dabral, V. Bhat, E. D. Jemmis, and M. V. George, *Proc. Indian Acad. Sci.*, 1977, **86A**, 1.

(132)   (131)

Photochemically induced homolysis of the disulphide bond, initiated by the benzothiazole nucleus, is responsible for the transformation of disulphide (133) into the two cephams (134) and (135) and 2-mercaptobenzothiazole.[113] The photoreaction of thiocyanogen with mono- and di-substituted alkenes to give

(133)

(134)   (135)

$\alpha\beta$-dithiocyanates and allylic isothiocyanates is believed to proceed by way of a radical chain mechanism initiated by dissociation of thiocyanogen $(SCN)_2$ into thiocyanato radicals $\cdot SCN$.[114] E.s.r. evidence for the production of arene-sulphinyl radicals $ArSO\cdot$ on photolysis of diaryl sulphoxides and S-aryl arene-sulphonates has been presented,[115] but the photodecomposition of 1,3-oxa-thiolan-2-one to oxiran and COS is claimed to exclude participation of radicals and to fragment *via* the excited singlet state.[116]

Numerous examples of the photodecomposition of sulphones accompanied by the elimination of sulphur dioxide have been reported. Particular interest has been shown in benzyl sulphones; a comparison of quantum yields and multiplicity for the photoextrusion of SO_2, and CO_2, and CO from benzyl sulphones, benzyl esters, and dibenzyl ketones in dioxan or benzene solution reveals that sulphur dioxide extrusion is an efficient process and provides a useful means of

[113] E. M. Gordon and C. M. Cimarusti, *Tetrahedron Letters*, 1977, 3425.
[114] R. G. Guy and J. J. Thompson, *Tetrahedron*, 1978, **34**, 541.
[115] B. C. Gilbert, B. Gill, and M. D. Sexton, *J.C.S. Chem. Comm.*, 1978, 79.
[116] H. Chandra and K. S. Sidhu, *Indian J. Chem.*, 1977, **15B**, 823.

forming new carbon bonds.[117] Both singlet and triplet species are involved in this photodecomposition; evidence for a partial solvent cage effect on direct excitation was observed, whereas on triplet excitation the radicals appear to escape almost entirely from the cage. The formation of bibenzyl, 4-chloro-bibenzyl, and 4,4′-dichlorobibenzyl in a ratio of 1.0 : 1.8 : 0.9 on photodecomposition of benzyl 4-chlorobenzyl sulphone in benzene has been separately reported.[118] Photochemically induced sulphur dioxide extrusion has been used as a route to cyclophanes,[119] and sulphinic acids have been obtained as alternative products on irradiation of sulphones in alcohols.[120] Photoelimination by daylight of sulphur dioxide from 3,5-diphenyl-4*H*-thiopyran-4-one 1,1-dioxide (136) gave the previously unknown 2,5-diphenylcyclopentadienone, isolated as the trimer (137) or trapped by suitable dienophiles.[121] The corresponding 2,6-diphenyl and 3,5-dimethyl sulphones are stable to daylight.

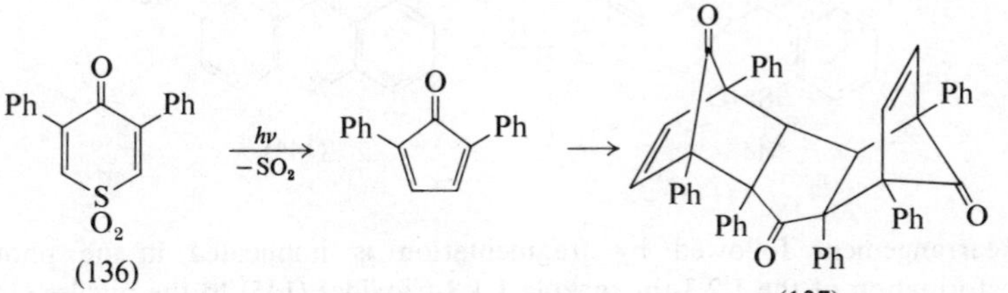

(136)   (137)

A transient detected in the flash photolysis of aromatic sulphones has been attributed to the arenesulphonyl radical.[122] Isomeric *p*-tolylpyridines were obtained by irradiation of di-*p*-tolyl sulphone in pyridine,[123] and sulphur dioxide is a product of irradiation of benzenesulphonic acid in aqueous solution.[124] The γ-sultine (138) has been reported to undergo photochemically induced elimination of sulphur dioxide *via* a biradical intermediate to give phenyl-cyclopropane (139).[125]

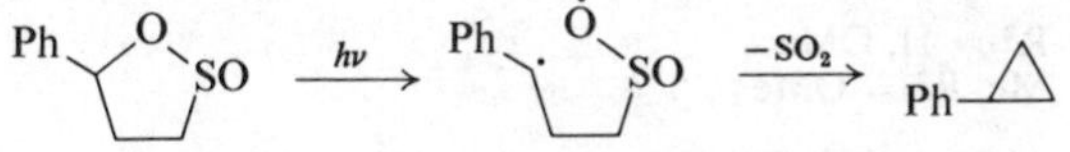

(138)   (139)

On irradiation in benzene, the 2-sulphinyl-substituted aryl benzoates (140) are converted into the corresponding benzoic acid derivatives (141);[126] a cyclic biradical intermediate (142) is proposed. Analogous reactions have been reported for the thiobenzoate (143; X = S) and for the selenobenzoate (143;

[117] R. S. Givens and B. Matuszewski, *Tetrahedron Letters*, 1978, 861.
[118] G. E. Robinson and J. M. Vernon, *J.C.S. Perkin I*, 1977, 1682.
[119] R. S. Givens and P. L. Wylie, *Tetrahedron Letters*, 1978, 865.
[120] R. F. Langler, Z. A. Marini, and J. A. Pincock, *Canad. J. Chem.*, 1978, **56**, 903.
[121] W. Ried and H. Bopp, *Angew. Chem. Internat. Edn.*, 1977, **16**, 653.
[122] H. T. Ho, O. Ito, M. Iino, and M. Matsuda, *J. Phys. Chem.*, 1978, **82**, 314.
[123] T. Nakabayashi, T. Horii, S. Kawamura, and M. Hamada, *Bull. Chem. Soc. Japan*, 1977, **50**, 2491.
[124] Y. Ogata, K. Takagi, and S. Yamada, *Bull. Chem. Soc. Japan*, 1977, **50**, 2205.
[125] T. Durst, J. C. Huang, N. K. Sharma, and D. J. H. Smith, *Canad. J. Chem.*, 1978, **56**, 512.
[126] R. Lüdersdorf, J. Martens, B. Pakzad, and K. Praefcke, *Annalen*, 1977, 1992.

$X = Se$), but in addition these give, *via* a novel aromatic photosubstitution, the thioxanthone (144; $X = S$) and the selenoxanthone (144; $X = Se$). The photodecomposition reactions of mixed thioanhydrides have also been studied.[127]

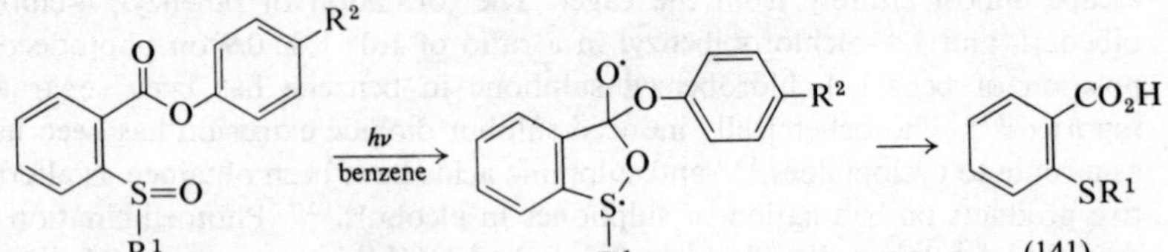

(140) $R^1 = Me$; $R^2 = Me$ or OMe
      $R^1 = p\text{-}MeC_6H_4$; $R^2 = Me$

(141)

(142)

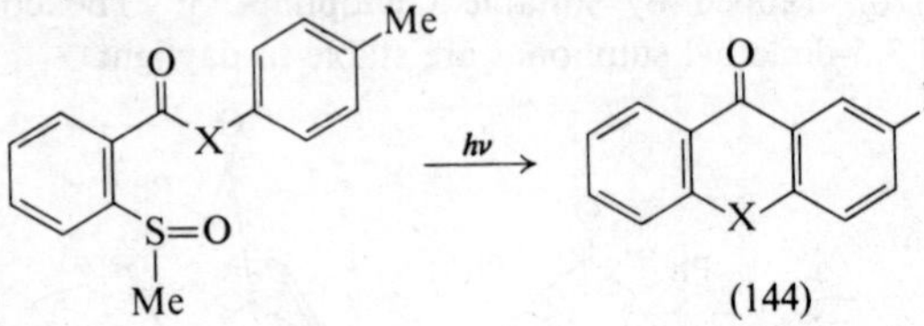

(143)

(144)

Rearrangement followed by fragmentation is implicated in the photo-transformation of the 1,2,3-thiadiazole 1,1,3-trioxides (145) to the nitriles (146) and the 1,3,2,4-dioxathiazole 2-oxides (147).[128] Analogous reactions have been

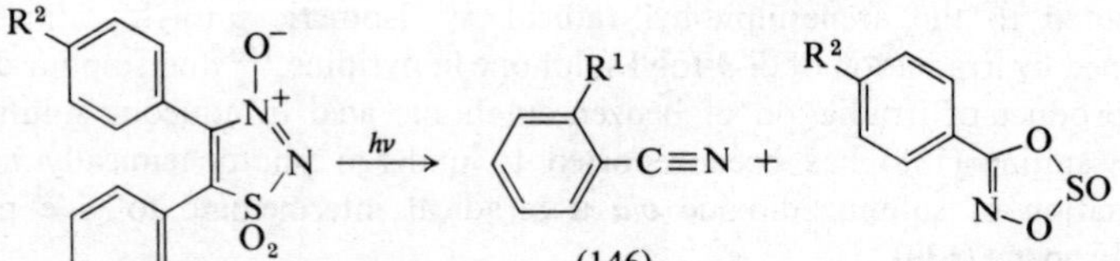

(145) $R^1 = R^2 = H$, OMe
      $R^1 = Me$, $R^2 = OMe$

(146)

(147)

reported for 1,2,3-thiadiazole 1,1,2-trioxides.[129] Valence bond tautomerism followed by nitrogen—nitrogen bond homolysis and hydrogen abstraction has been proposed to account for the photofragmentation of the novel mesoionic 2-vinyl-1,3,4-thiadiazole (148) to *N*-methylthiobenzamide (149).[130] Singlet

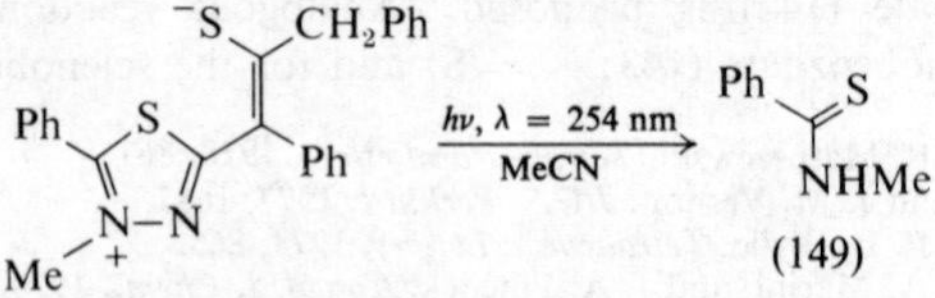

(148)

$\xrightarrow{h\nu, \lambda = 254 \text{ nm}}{MeCN}$

(149)

[127] K. Beelitz and K. Praefcke, *Chem.-Ztg.*, 1978, **102**, 67.
[128] G. Trickes, H. P. Braun, and H. Meier, *Annalen*, 1977, 1347.
[129] G. Trickes, U. Pleucken, and H. Meier, *Z. Naturforsch.*, 1977, **32b**, 956.
[130] R. Mukherjee and M. M. Moriarty, *Indian J. Chem.*, 1977, **15B**, 499.

nitrenes have been obtained by the photodecomposition of *N*-acyl-*S,S*-diphenylsulphimides,[131] and the photoelimination of atomic sulphur from isothiocyanates has been studied.[132] The formation of β-thiolpenicillanates (150) by irradiation of a mixture of βββ-trichloroethyl 6-diazopenicillanate (151) and thiols has been shown to arise *via* intermediate azo-sulphides (152) which undergo photofragmentation with loss of nitrogen.[133] Protected sugar derivatives having one free hydroxy-group may be deoxygenated at this position by irradiation of the corresponding dimethylthiocarbamic ester.[134]

8 Miscellaneous Decomposition and Elimination Reactions

Fragmentation and elimination reactions which cannot be included in any of the above categories are briefly reviewed in this section. It has not proved possible to classify these processes, but analogous reactions are grouped together.

Details of two studies of the liquid phase photodecomposition of 1,4-dioxan have been published;[135,136] the major products at 184.9 nm are ethylene, hydrogen, and carbon monoxide, formed with quantum yields of 0.55, 0.26, and 0.03, respectively. Tetrakis(trifluoromethyl)[4]annulene (153) is the first symmetrically substituted derivative of [4]annulene for which full spectral characterization is available:[137] it is prepared together with trifluoroacetic anhydride by irradiation of the ozonide (154) in a transparent matrix at 77 K. The photodecomposition of cyclohexa-2,4-dienone peroxides has been investigated,[138]

[131] N. Furukawa, T. Nishio, and M. Fukumura, *Chem. Letters*, 1978, 209.
[132] R. Jahn and U. Schmidt, *Monatsh. Chem.*, 1978, **109**, 161.
[133] J. C. Sheehan, T. J. Commons, and Y. S. Lo, *J. Org. Chem.*, 1977, **42**, 2224.
[134] R. H. Bell, D. Horton, D. M. Williams, and E. Winter-Mihaly, *Carbohydrate Res.*, 1977, **58**, 109.
[135] J. Kiwi, *J. Photochem.*, 1977, **7**, 237.
[136] J. J. Houser and B. A. Sibbio, *J. Org. Chem.*, 1977, **42**, 2145.
[137] F. J. Palensky and H. A. Morrison, *J. Amer. Chem. Soc.*, 1977, **99**, 3507.
[138] H. Lind, H. Loeliger, and T. Winkler, *Tetrahedron Letters*, 1978, 1571.

and further examples of the photochemically induced generation of carbonyl ylides from stilbene oxides have been reported.[139,140]

The photoreactions of some *gem*-chloronitrosoalkanes have been explained in terms of a primary process involving carbon—nitrogen bond homolysis with the formation of α-chloroalkyl radicals and nitric oxide.[141] Quantum yields at 580—680 nm were found to be dependent on the structure of the nitrosoalkane, the viscosity and type of solvent and the presence of oxygen, but independent of the wavelength of light. Carbon—nitrogen bond homolysis is also responsible for the photoreactions of 2-chloro-2-nitrosoadamantane in methanol.[142] Products of irradiation of nitrosyl cyanide include nitrogen, carbon monoxide, carbon dioxide, nitrous oxide, and cyanogen isocyanate;[143] these arise by carbon—nitrogen bond homolysis and by the reaction of nitrosoalkanes with free radicals and nitric oxide.

The photochemical cleavage of cyclobutanes has been the subject of further discussion. The cycloreversion of substituted 1,2,2a,8b-tetrahydrocyclobuta[a]-naphthalenes to naphthalene and alkenes proceeds stereospecifically from the singlet state, presumably by a $[_\delta 2_s + {}_\delta 2_s]$ concerted pathway.[144] A mixture of Z- and E-alkenes is obtained on triplet-sensitized decomposition with stereo-retention being favoured. Cleavage of cyclobutanes with phenyl or acrylic ester substituents is reported to take place *via* short-lived 1,4-biradicals,[145] and photofragmentation of the cyclobutane (155) has been employed in the synthesis of 1,2,3,4,5,6-hexakis(trifluoromethyl)tetracyclo[4,2,0,0^{2,4}0^{3,5}]oct-7-ene (156).[146] Irradiation of the heterocycles (157), synthesized as potential precursors of hetero[11]annulenes, gave only benzene and the parent hetero-cycles (158).[147] An analogous route involving photodecarbonylation of the adduct (159), followed by fragmentation of the cyclobutane (160), has been employed in the synthesis of N-methoxycarbonyl-2-azetine (161).[148]

Photochemically induced intramolecular eliminations of HCl, HBr, and HI have again been widely used in the synthesis of heterocycles and alkaloids. Hexahydroapoerysopine (162), for example, is obtained on irradiation of either

[139] K. Nishiyama, K. Ishikawa, I. Sarkar, D. C. Lankin, and G. W. Griffin, *Heterocycles*, 1977, **6**, 1337.
[140] R. Huisgen, V. Markowski, and H. Hermann, *Heterocycles*, 1977, **7**, 61.
[141] D. Forrest, B. G. Gowenlock, and J. Pfab, *J.C.S. Perkin II*, 1978, 12.
[142] A. H. M. Kayen and Th. J. de Boer, *Rec. Trav. chim.*, 1977, **96**, 237.
[143] C. M. Keary, B. G. Gowenlock, and J. Pfab, *J.C.S. Perkin II*, 1978, 242.
[144] C. Pac, K. Mizuno, and H. Sakurai, *Bull. Chem. Soc. Japan*, 1978, **51**, 329.
[145] G. Kaupp and M. Stark, *Chem. Ber.*, 1977, **110**, 3084.
[146] R. N. Warrener and E. E. Nunn, *Austral. J. Chem.*, 1978, **31**, 221.
[147] A. G. Anastassiou, E. Reichmanis, S. J. Girgenti, and M. Schaefer-Ridder, *J. Org. Chem.*, 1978, **43**, 315.
[148] R. N. Warrener, G. Kretschmer, and M. N. Paddon-row, *J.C.S. Chem. Comm.*, 1977, 806.

(155) R=CF₃ → (156) [structures]

(157) X = O or NCO₂Et → (158)

(159) → (160) → (161)

the 6'-iodo- or the 7-bromo-derivative of 1,2,3,3a,4,5-hexahydro-*N*-(3,4-dimethoxyphenethyl)indol-6-one (163) in dioxan.[149,150] Similarly, photo-dehydrochlorination of the chlorobenzo[*b*]thiophens (164) affords the diazanaphthalenes (165).[151] Numerous other examples of this approach to the synthesis of heterocycles[152,153] and alkaloids[154—161] have been reported.

Intermolecular eliminations of HX of potential synthetic value have also been described. Examples include the formation of 8-(8-adenosyl)guanosine (166)

[149] H. Iida, T. Takarai, and C. Kibayashi, *J.C.S. Chem. Comm.*, 1977, 644.
[150] H. Iida, T. Takarai, and C. Kibayashi, *J. Org. Chem.*, 1978, **43**, 975.
[151] M. Terashima, K. Seti, K. Itoh, and Y. Kanaoka, *Heterocycles*, 1977, **8**, 421.
[152] S. Lalitha, S. Rajeswari, and H. Suguna, *Indian J. Chem.*, 1977, **15B**, 180.
[153] S. Kobayashi, M. Kihara, and Y. Nakauchi, *Yakugaku Zasshi*, 1978, **98**, 161 (*Chem. Abs.*, 1978, **88**, 190 574).
[154] L. Castedo, R. Estévez, J. M. Saá, and R. Suau, *Tetrahedron Letters*, 1978, 2179.
[155] W. J. Begley and J. Grimshaw, *J.C.S. Perkin I*, 1977, 2325.
[156] T. R. Govindachari, K. Nagarajan, S. Rajeswari, H. Suguna, and B. R. Pai, *Helv. Chim. Acta*, 1977, **60**, 2138.
[157] B. R. Pai, H. Suguna, S. Natarajan, and G. Manikumar, *Heterocycles*, 1977, **6**, 1993.
[158] H. Suguna and B. R. Pai, *Indian J. Chem.*, 1977, **15B**, 416.
[159] S. Rajeswari, H. Suguna, and B. R. Pai, *Indian J. Chem.*, 1977, **15B**, 592.
[160] G. Manikumar, B. R. Pai, and H. Suguna, *Indian J. Chem.*, 1977, **15B**, 740.
[161] T. R. Govindachari, K. Nagarajan, S. Rajeswari, H. Suguna, and B. R. Pai, *Indian J. Chem.*, 1977, **15B**, 873.

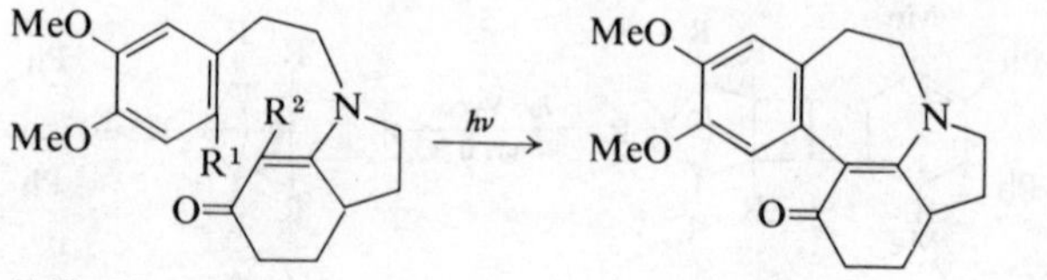

(163) R^1 = I, R^2 = H
       R^1 = H, R^2 = Br

(162)

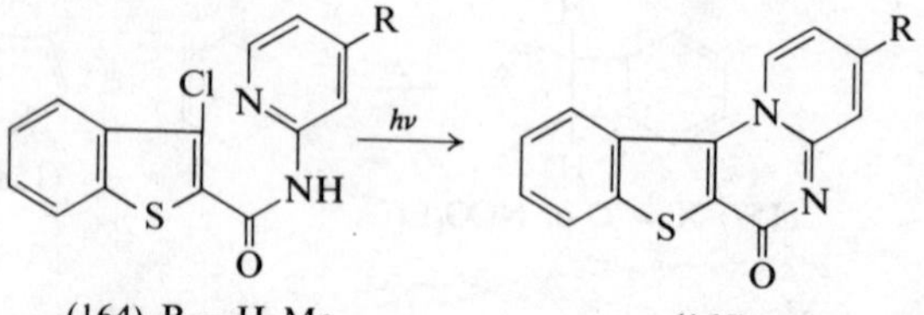

(164) R = H, Me

(165)

by irradiation (254 nm) of an aqueous solution of 8-bromoadenosine and guanosine,[162] the introduction of indolyl groups at the 5-position of uridine,[163] and the photoalkylation of aromatic amino-acids with chloracetamide.[164]

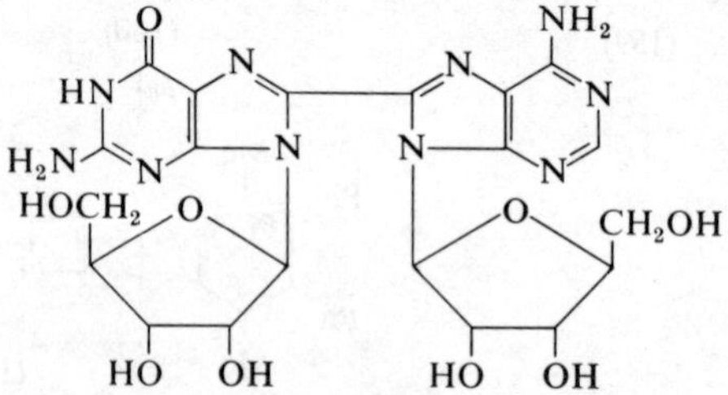

(166)

Photoelimination of HBr is proposed as the first step in the unusual conversion in 55% yield of *N*-aryl-2-nitrobenzhydrazonoyl bromides (167) into 3-aryl-4-oxo-3,4-dihydro-1,2,3-benzotriazines (168).[165] 2-Aryl-4-oxidobenzotriazinium 1-oxides (169) are also believed to be intermediates in this transformation, and the proposed pathway is outlined in Scheme 6.

Many other decomposition reactions arising by carbon—halogen bond homolysis have been described, but these are essentially radical processes having no special photochemical significance, and so are not reviewed in detail in this Report. A few publications are worth of note. Recent studies have shown that bromo- and iodo-alkanes afford ionic as well as radical intermediates on irradiation, by a process involving homolysis followed by electron transfer. 1,1-Di-iodides are now reported to exhibit similar behaviour.[166] In addition, carbenoids have also been proposed as intermediates, cyclopropanes are produced in yields of up to 80% on irradiation in the presence of alkenes. Irradiation of a series of

[162] S. N. Bose and R. J. H. Davies, *Biochem. Soc. Trans.*, 1977, **5**, 282.
[163] I. Saito, S. Ito, and T. Matsuura, *Tetrahedron Letters*, 1978, 2585.
[164] T. Hamada and O. Yonemitsu, *Chem. and Pharm. Bull. (Japan)*, 1977, **25**, 271.
[165] Y. Maki and T. Furuta, *Synthesis*, 1978, **10**, 382.
[166] N. J. Pienta and P. J. Kropp, *J. Amer. Chem. Soc.*, 1978, **100**, 655.

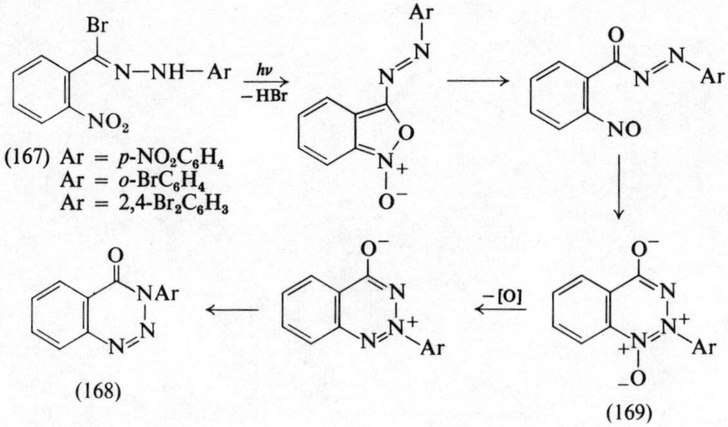

Scheme 6

1,1-dibromocyclopropanes in ether gave the corresponding monobromocyclo-propanes in good to moderate yield.[167] It is now reported that the fate of the 2-adamantyl cation, generated by irradiation of 2-iodoadamantane in ether or methanol, is deprotonation with the formation of a mixture of 2,4-dehydro-adamantane and protoadamantene.[168] The γ- and δ-lactams have been obtained by intramolecular cyclization of amido-radicals generated in turn by photolysis of *N*-iodo-amides.[169]

[167] N. Shimizu and S. Nishida, *Chem. Letters*, 1977, 839.
[168] P. J. Kropp, J. R. Gibson, J. J. Snyder, and G. S. Poindexter, *Tetrahedron Letters*, 1978, 207.
[169] S. A. Glover and A. Goosen, *J.C.S. Perkin I*, 1977, 1348.

Scheme II

Part IV

POLYMER PHOTOCHEMISTRY

By N. S. ALLEN and J. F. McKELLAR (late)*

* During the proofing of this chapter John McKellar died suddenly on 3rd May, 1979 following a long period of illness.

'*It is with regret that we record the death of a fellow reporter John F. McKellar who died suddenly on 3rd May 1979 after a long period of illness. He was an outstanding research worker having published over 150 original research papers and review articles mainly in the field of Applied Photochemistry. Our deepest sympathy goes to his wife and two children.*'

G. O. Phillips
A. K. Davies
N. S. Allen

Professor Bryce-Smith and all the Reporters of 'Photochemistry' would like to be associated with the above expression of regret from the colleagues of Dr. McKellar.

1 Introduction

The format of this Report is essentially the same as that used in the previous volume. Also, because of the large number of papers reviewed, detailed discussion is restricted to a few.

Polymer photochemistry is still an active field of research and continues to expand in new directions, mainly those having commercial relevance. This is particularly so for photopolymerization where in the last year some twenty review articles have appeared in the literature, many of which deal with the commercial applications.

2 Photopolymerization

Photoinitiation of Addition Polymerization.—Several review articles covering various aspects of photoinitiated addition polymerization have appeared.[1—16] General accounts of the role of various initiators in the photopolymerization of organic coatings[1] and monomeric vapours[2] have appeared. The kinetics of photopolymerization of various acrylic and styrene monomers have also been reviewed.[3] Many other review articles, mainly of a technological interest, discuss the u.v. curing of polyester systems for glass fibre reinforcement,[4, 5] coatings,[6, 7] paints,[8] and packaging.[9] Also, three review articles have appeared discussing recent developments in photopolymerizable systems for coatings and inks.[10—13] The differences between u.v. and peroxide curing techniques,[14] the uses of photopolymerizable systems,[15] and the status of the field in the U.S.A.[16] have also been discussed.

1 S. E. Young, *Prog. Org. Coatings*, 1976, **4**, 225.
2 J. F. Kinstle and M. K. Dehnke, *J. Rad. Curing*, 1977, **4**, 2.
3 Yu. L. Spirin and A. K. Chaiko, *Kinet. Mekh. Reakts. Obrag. Polim.*, 1977, 26.
4 K. Koch, *ZIS Mitt.*, 1977, **19**, 888.
5 G. Niederstadt, J. P. Schik, and R. Siegberg, *Kunstharg-Nachr.*, 1975, **34**, 1.
6 H. Motoki, *Mokuzai to Gijutsu*, 1977, **29**, 1.
7 K. Fuhr, *Polymer Paint. Colour. J.*, 1977, 167, 672, 705.
8 T. Ishihara, *Toso To Toryo*, 1976, **277**, 29.
9 R. B. Mesrobian, *Internat. J. Radiation Phys. Chem.*, 1977, **9**, 307.
10 M. Barbier, J. Vlahakis, R. Ouellette, R. Pikul, and R. Rice, *Electrotechnology (Ann Arbor, Mich.)*, 1978, **2**, 28.
11 M. T. Nowak and C. B. Rybny, *Amer. Inkmaker*, 1977, 25.
12 G. Berner, R. Kirchmayr, and G. Rist, *J. Oil Colour Chemists' Assoc.*, 1978, **61**, 105.
13 M. A. Parrish, *J. Oil Colour Chemists' Assoc.*, 1977, **60**, 474.
14 J P. Schik, *Plastica*, 1978, **31**, 4.
15 T. Koshiishi, *Plastics Age*, 1977, **23**, 82.
16 J. Pelgrims, *J. Oil Colour Chemists' Assoc.*, 1978, **61**, 114.

Bamford and his co-workers have used *N*-acetyldibenz[*b,f*]azepine groupings for chain extending poly(methyl methacrylate) and polystyrene.[17] For example, the photopolymerization of styrene using $Mn_2(CO)_{10}$–*N*-bromoacetyldibenzazepine as initiators gives polymer molecules with dibenzazepine units at each end, structure (1). Further irradiation of this polymer in the presence of benzil or benzophenone results in chain extension through photocyclodimerization of the central ethylenic double bond of the dibenzazepine moieties.

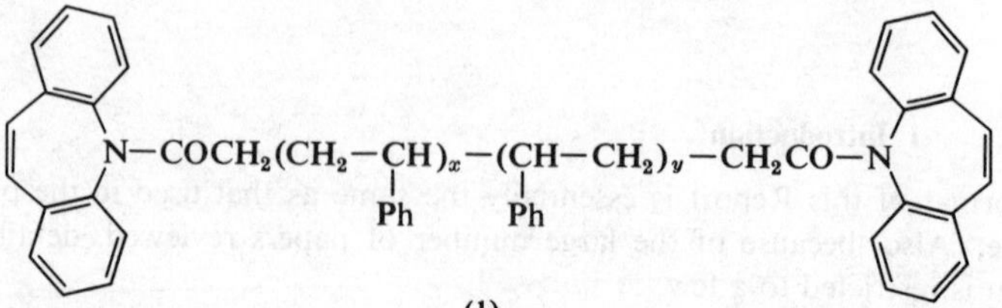

(1)

Ledwith and co-workers[18] have studied the involvement of exciplexes in the photopolymerization of methyl methacrylate by fluorenone–amine systems. Interestingly, the amine radical cation species and not those of the ketone were found to be the more important photoinitiators. This was due to the ability of the radical anion derived from the ketone to dimerize to form the pinacol, as shown in Scheme 1.

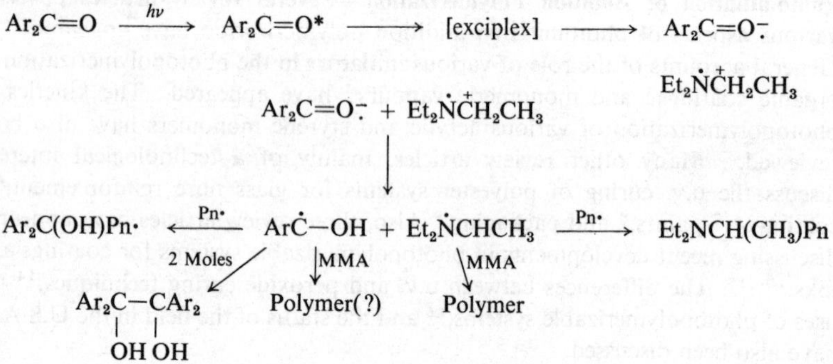

Scheme 1

The photoinitiated polymerization of acrylate monomers with ^{14}C-labelled benzoin methyl ethers has been studied by Carlblom and Pappas.[19] Both the benzoyl and α-methoxybenzyl radicals, produced on photolysis of benzoin methyl ether, were found to be equally effective in both initiation and hydrogen abstraction. The primary photochemical processes in the benzoin-initiated photopolymerization of vinyl monomers have been investigated using laser

[17] A. K. Alimoglu, C. H. Bamford, A. Ledwith, and S. U. Mullik, *Macromolecules*, 1977, **10**, 1081.
[18] A. Ledwith, J. A. Bosley, and M. D. Purbrick, *J. Oil Colour Chemists' Assoc.*, 1978, **61**, 95
[19] L. H. Carlblom and S. P. Pappas, *J. Polymer Sci. Polymer Chem.*, 1977, **15**, 1381.

flash photolysis.[20] In this case the benzoyl and α-hydroxy benzyl radicals derived from the photolysis of benzoin were found to decay by second-order kinetics due to radical–radical recombination to form the corresponding pinacol derivatives (Scheme 2). Second-order kinetics were found for the decay of ketyl radicals produced in the flash photolysis of benzophenone in methyl

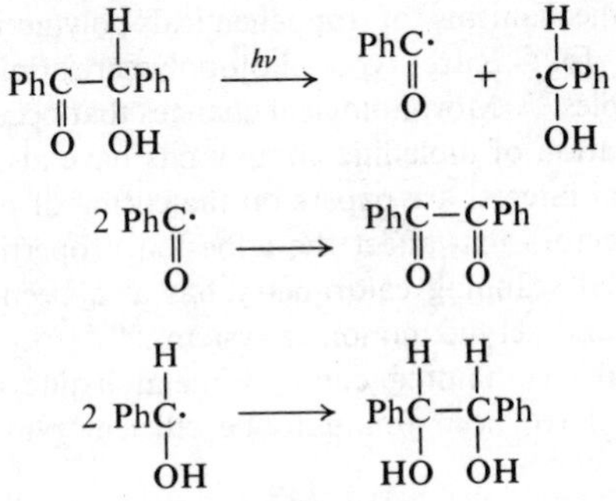

$$
\begin{array}{ccc}
\underset{\substack{| \\ \text{O} \quad \text{OH}}}{\text{PhC}-\text{CPh}} & \xrightarrow{h\nu} & \underset{\text{O}}{\text{PhC}\cdot} + \underset{\substack{| \\ \text{OH}}}{\cdot\text{CPh}}
\end{array}
$$

$$
2\ \underset{\text{O}}{\text{PhC}\cdot} \longrightarrow \underset{\text{O}\quad\text{O}}{\text{PhC}-\text{CPh}}
$$

$$
2\ \underset{\substack{| \\ \text{OH}}}{\text{PhC}\cdot} \longrightarrow \underset{\substack{| \quad | \\ \text{HO}\quad\text{OH}}}{\text{PhC}-\text{CPh}}
$$

Scheme 2

methacrylate.[21] The rate of photopolymerization of methyl methacrylate initiated by benzophenone has not, unexpectedly, been found to decrease on introducing electron donating substituents in the 4 or 4,4′-positions of the benzene rings.[22] Electron-acceptor substituents had the opposite effect.

Armand *et al.*[23] have investigated the photopolymerization of methyl acrylate initiated by alkoxyacetophenone compounds of structures (2) and (3) respectively.

$$
\underset{\substack{| \\ \text{O}\quad\text{OMe}}}{\text{Ph}-\text{C}-\overset{\text{OMe}}{\underset{}{\text{C}}}-\text{Ph}}
\qquad\qquad
\underset{\substack{| \\ \text{O}\quad\text{OEt}}}{\text{Ph}-\text{C}-\overset{\text{OEt}}{\underset{}{\text{C}}}-\text{H}}
$$

(2)   (3)

Whereas compound (2) was photolyzed by a Norrish Type I process, compound (3) was photolysed by both Norrish Type I and II processes.

The photopolymerization of styrene initiated by 1,3-dioxolane and αα′-azobisisobutyronitrile has been studied.[24] Both initiators were found to exhibit similar rates of photoinitiation. The photoinitiated copolymerization of unsaturated monomers using naphthalene as the initiator has also been investigated.[25] Other studies of interest include the photopolymerization of

[20] R. Kuhlman and W. Schnabel, *Polymer*, 1977, **18**, 1163.
[21] V. M. Granchak, V. P. Sherstyuk, and I. I. Dilung, *Doklady Akad. Nauk S.S.R.*, 1977, **235**, 611.
[22] V. M. Granchak, V. P. Sherstyuk, and I. I. Dilung, *Teor. i eksp. Khim.*, 1977, **13**, 257.
[23] B. Armand, R. Kirchmayr, and G. Rist, *Helv. Chim. Acta*, 1978, **61**, 305.
[24] T. Ouchi and Y. Komatsu, *J. Macromol. Sci., Chem. Edn.*, 1977, **A11**, 483.
[25] J. Barton, I. Capek, D. Lath, and E. Lathova, *Chem. Zvesti.*, 1977, **31**, 265.

pentaerythritol tetracinnamate,[26] hexaethylidencyclohexane,[27] *p*-vinyldiphenyl-ether,[28] phenylacetylene,[29] cyclohexene-*N*-ethylmaleimide,[30] styrene,[31] and the photocopolymerization of *p*-vinylbenzophenone and vinylferrocene,[32] *p*-chlorophenyl methacrylate and ethylene glycol dimethacrylate,[33] and maleic anhydride and vinyl monomers.[34] Styrene and acrylic acid have also been photocopolymerized by block copolymerization for use as a reverse osmosis membrane.[35] The mechanisms of topochemical polymerizations have been described using the four-centre type photopolymerization of diolefins and diacetylenes as examples.[36] Morphological changes that occur in the four-centre type photopolymerization of diolefinic compounds have also been examined.[37] Of more technological interest are papers on the curing of polyester resins with visible light[38] and factors that affect the adhesion properties of photocurable systems.[39] Differential scanning calorimetry has also been used to study the photocuring of ester and acrylic monomer systems.[40,41]

Diaryliodonium salts containing complex metal halide anions and having general structure (4) have been found to be efficient photoinitiators for the

$$Ar_2I^+MX_n^-$$

(4)

cationic polymerization of a variety of monomer systems.[42,43] The polymerization rate was found to be dependent on the nature of the acidic component, their reactivity increasing in the order $SbF_6^- > AsF_6^- > PF_6^- > BF_4^-$.

Davies *et al.*[44] have studied the photopolymerization of 2-vinylnapthalene using zinc chloride as the photoinitiator. It is postulated that an exciplex is formed as the initiating species (Scheme 3).

[26] F. Danusso, P. Ferruti, A. Moro, G. Tieghi, and M. Zocchi, *Polymer*, 1977, **18**, 161.
[27] O. A. Yuzhakova, I. V. Isakov, E. E. Rider, G. N. Gerasimov, and A. D.Abkin, *Vysokomol. Soedineniya, Ser. B*, 1977, **19**, 431.
[28] V. V. Sinitsyn, M. A. Butatov, and A. V. Sukin, *Deposited Doc.*, 1974, *VINITI* 240-75, 15.
[29] L. M. Svanidze, E. A. Mushina, A. M. Sladkov, N. I. Sirotkin, A. K. Artemov, I. R. Golding, T. G. Samedova, G. N. Bondarenko, and B. E. Davydov, *Vysokomol. Soedineniya, Ser. B*, 1977, **19**, 51.
[30] H. Heusinger, E. Wipfelder, J. Heinzl, and H. Zott, *Proc. Tihany. Symp. Radiat. Chem.*, 1977, **4**, 613.
[31] H. J. Tiller, U. Wagner, P. Fink, and K. Meyer, *Plaste Kautschuk*, 1977, **24**, 236.
[32] G. Sanchez, G. Weill, and R. Kroesel, *Makromol. Chem.*, 1978, **179**, 131.
[33] I. V. Andreeva, M. M. Koton, and V. N. Artemeva, *Vysokomol. Soedineniya Ser. B*, 1978, **20**, 127.
[34] H. K. Roth, J. Hellebrand, and P. Wuensche, *Plaste. Kautschuk*, 1977, **24**, 736.
[35] W. Kawai, *J. Macromol. Sci., Chem.*, 1977, **A11**, 1027.
[36] G. Wegner, *Pure Appl. Chem.*, 1977, **49**, 443.
[37] H. Nakanishi, M. Hasegawa, H. Kirihura, and T. Yunuzi, *Nippon Kagaku Kaishi*, 1977, **7**, 1046.
[38] B. G. Dixon, D. M. Longenecker, and G. G. Greth, *Soc. Plast. Ind.*, 32nd *Amer. Tech. Corp.*, 1977, **5-D**, 1.
[39] V. D. McGinniss and A. Kah, *Polymer Eng. Sci.*, 1977, **17**, 478.
[40] F. R. Wight and G. W. Hicks, *Soc. Plast. Eng. Technol. Paper*, 1977, **23**, 500.
[41] J. E. Moore, *Chem. Prop. Crosslinked Polym.* (*Proc. Amer. Chem. Soc. Symp.*), 1977, 535.
[42] J. V. Crivello and J. H. W. Lam, *J. Polymer Sci., Part C*, 1976, **56**, 383.
[43] J. V. Crivello, J. H. W. Lam, and C. N. Volante, *J. Radiations Curing*, 1977, **7**, 2.
[44] D. H. Davies, D. C. Phillips, and J. D. B. Smith, *J. Polymer Sci. Polymer Chem.*, 1977, **15**, 2673.

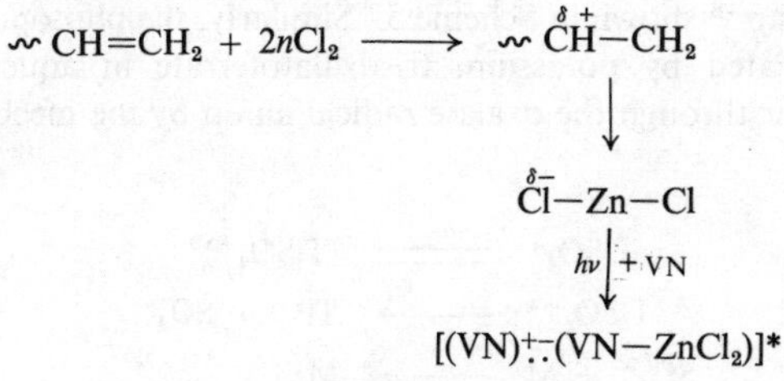

$$\text{\small\textasciitilde\textasciitilde} CH{=}CH_2 + 2nCl_2 \longrightarrow \text{\small\textasciitilde\textasciitilde} \overset{\delta\,+}{C}H{-}CH_2$$

$$\downarrow$$

$$\overset{\delta\,-}{C}l{-}Zn{-}Cl$$

$$h\nu \downarrow + \text{VN}$$

$$[(\text{VN})^{+}_{\cdot}\!\cdot(\text{VN}{-}ZnCl_2)]^*$$

Scheme 3

The cationic polymerization of vinylcarbazole and styrene has been investigated using pulse radiolysis.[45] The cationic species responsible for the initiation and propagation reactions were detected directly by this method. Hayashi *et al.*[46] have examined the photopolymerization of α-methylstyrene and styrene in the presence of electron acceptors such as tetracyanobenzene. The polymerization was found to proceed by three different mechanisms, namely radical, ion-pair, and free ion formation through an excited electron donor–acceptor complex. Takeishi *et al.*[47] have found that *N*-laurylpyridinium azide is an effective initiator for the photopolymerization of styrene in an aqueous medium. A free radical inhibitor, catechol, did not inhibit the polymerization, whereas a nitrene scavenger, pyridine, did. From these results the mechanisms (Scheme 4) involving triplet nitrene (*c*) were proposed to account for photopolymerization and inhibition by pyridine.

The photopolymerization of methyl methacrylate initiated by thallium ions in aqueous sulphuric acid is believed to occur through the sulphate radical anion

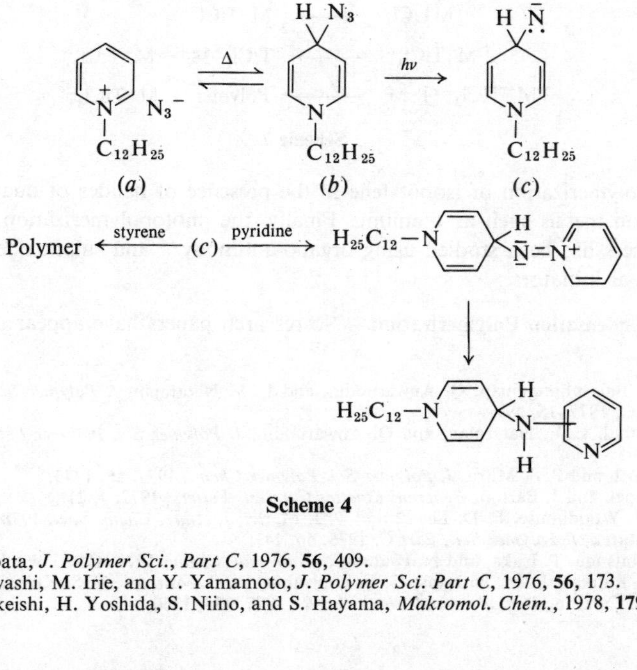

Scheme 4

[45] Y. Tabata, *J. Polymer Sci.*, *Part C*, 1976, **56**, 409.

[46] K. Hayashi, M. Irie, and Y. Yamamoto, *J. Polymer Sci. Part C*, 1976, **56**, 173.

[47] M. Takeishi, H. Yoshida, S. Niino, and S. Hayama, *Makromol. Chem.*, 1978, **179**, 1387.

by the mechanism[48] shown in Scheme 5. Similarly, the photopolymerization of acrylamide initiated by potassium trisoxalatoferrate in aqueous solution is believed to occur through the oxalate radical anion by the mechanism[49] shown in Scheme 6.

$$TlSO_4^+ \; \overset{h\nu}{\rightleftharpoons} \; (TlSO_4^+)^*$$

$$TlSO_4^{+*} \; \longrightarrow \; Tl^{2+} + SO_4^{\cdot-}$$

$$SO_4^{\cdot-} + M \; \longrightarrow \; M\cdot$$

Scheme 5

$$[Fe^{3+}(C_2O_4^{2-})_3]^{3-} \; \overset{h\nu}{\longrightarrow} \; [Fe^{2+}(C_2O_4^{2-})_2(C_2O_4^{\cdot-})]^* \quad \text{(i)}$$

$$[Fe^{2+}(C_2O_4^{2-})_2(C_2O_4^{\cdot-})]^* \; \longrightarrow \; Fe^{2+}(C_2O_4^{2-})_2 + C_2O_4^{\cdot-} \quad \text{(ii)}$$

$$C_2O_4^{\cdot-} + M \; \longrightarrow \; M\cdot \quad \text{(iii)}$$

Scheme 6

The photopolymerization of methyl methacrylate using quinoline–bromine as the initiator has been found to be markedly dependent on the nature of the solvent.[50] Capek and Barton[51] have found that zinc salts catalyse the photopolymerization of acrylonitrile initiated by aromatic hydrocarbons. Of particular interest is a study on the photopolymerization of tetrahydrofuran initiated by copper and silver salts.[52] This is believed to be the first time tetrahydrofuran has been photopolymerized. Marek[53] suggests Scheme 7 to account for the

$$M + TiCl_4 \; \rightleftharpoons \; M\!:\!TiCl_4$$

$$MTiCl_4 \; \longrightarrow \; M^+_{\cdot}TiCl_4^{\cdot-}$$

$$2\, M^+_{\cdot}TiCl_4^{\cdot-} \; \longrightarrow \; TiCl_4^- M^+ - M^+ TiCl_4^{\cdot-}$$

$$2\, M^+ TiCl_4^{\cdot-} + M \; \longrightarrow \; \text{Polymer} + M^+ TiCl_4^{\cdot-}$$

Scheme 7

photopolymerization of isobutylene in the presence of halides of quadrivalent transition metals such as titanium. Finally, the photopolymerization of vinyl monomers has been studied using organo-antimony[54] and sulphoxide[55] compounds as initiators.

Photocondensation Polymerization.—No research papers have appeared on this topic.

[48] V. S. Balasubramanian, Q. Anwaruddin, and L. V. Natarajan, *J. Polymer Sci. Polymer Letters*, 1977, **15**, 599.
[49] K. Sahul, L. V. Natarajan, and Q. Anwaruddin, *J. Polymer Sci. Polymer Letters*, 1977, **15**, 605.
[50] P. Gosh and P. S. Mitra, *J. Polymer Sci. Polymer Chem.*, 1977, **15**, 1743.
[51] I. Capek and J. Barton, *Reaction Kinetics Catalysis Letters*, 1977, **7**, 21.
[52] M. E. Woodhouse, F. D. Lewis, and T. J. Marks, *J. Amer. Chem. Soc.*, 1978, **100**, 996.
[53] M. Marek, *J. Polymer Sci., Part C*, 1976, **56**, 149.
[54] H. Matsuda, T. Isaka, and N. Iwamoto, *Makromol. Chem.*, 1978, **179**, 539.
[55] S. R. Rafikov, E. M. Battalov, G. V. Leplyanin, Yu. I. Murinov, V. S. Kolosintsyn, and Yu. E. Nikitin, *Doklady Akad. Nauk S.S.R.*, 1977, **235**, 1360.

Photografting.—A number of workers[56–58] have successfully photografted various vinyl monomers such as styrene, methyl methacrylate, and the vinyl-pyridines on to cellulose using both inorganic and organic initiator systems. Acrylic monomers have also been successfully photografted on to wool[59] and poly(phenyl vinyl sulphide).[60] A method has been developed to photograft fluoro-olefin vapours on to aromatic polyamide fabrics to improve their flame resistance.[61] Finally, thermally regenerable ion-exchange resins have been developed by a photografting process using polymers of poly(triallylamine hydrochloride) and copolymers of propyldiallylamine hydrochloride and allyl benzoin methyl ether and allyl benzoin methyl ether with acrylic monomers.[62]

Photochemical Cross-linking.—The preparation of a variety of photocross-linkable resins has been reviewed by Panda.[63] Various sensitizers for use with photocross-linkable polymers have also been reviewed.[64]

The photocross-linking of poly-*m*-(vinyloxyethoxy)styrene containing pendant electron donor groups has been successfully carried out with several different acceptors.[65] For various initiator and quencher systems the mechanism of photo-cross-linking was found to be cationic. In the presence of an acceptor (A) the mechanism in Scheme 8 is believed to operate.

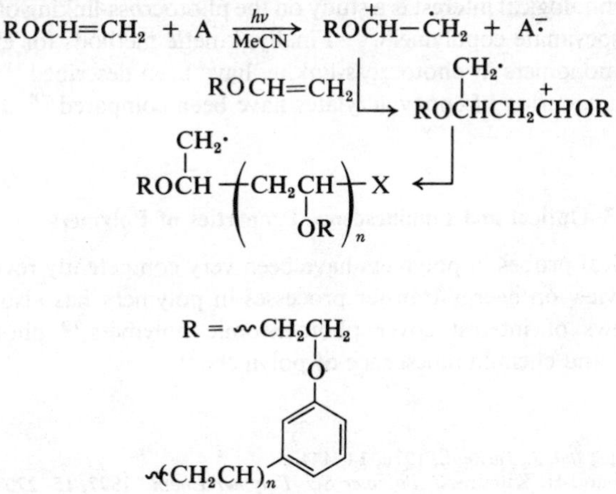

$$ROCH=CH_2 + A \xrightarrow[\text{MeCN}]{hv} RO\overset{+}{C}H-\dot{C}H_2 + A^{\bar{\cdot}}$$

$$ROCH=CH_2 \downarrow \longrightarrow \begin{array}{c} CH_2\cdot \\ | \\ ROCHCH_2\overset{+}{C}HOR \end{array}$$

$$\begin{array}{c} CH_2\cdot \\ | \\ ROCH \end{array} \left(CH_2CH \right)_n X \longleftarrow$$

$$\begin{array}{c} | \\ OR \end{array}$$

$$R = \text{\textasciitilde} CH_2CH_2$$

Scheme 8

[56] J. L. Garnett, in 'Cellulose Chemistry and Technology', *Amer. Chem. Soc., Symp. Ser.* 48, ed. J. C. Arthur, 1977, No. 23, 334.

[57] M. Sirocka and M. Marian, *Pr. Wydz. Nauk Tech. Bydgoskie Tow Nauk.*, Ser. A, 1976, **11**, 61.

[58] A. Takahashi and S. Takahashi, *Seni Gakkaishi*, 1977, **33**, T508.

[59] H. L. Needles ahd K. W. Alger, *Proc. Internat. Wolltextil-Forschungskonf.*, 5th, 1975, **3**, 394.

[60] S. Kondo, T. Nakamatsu, K. Tsuda, and T. Otsu, *J. Macromol. Sci., Chem. Edn.*, 1977, **A11**, 719.

[61] M. Toy, R. S. Stringham, and F. S. Dawn, *J. Appl. Polymer Sci.*, 1977, **21**, 2583.

[62] M. B. Jackson and W. H. F. Sasne, *J. Macromol. Sci., Chem. Edn.*, 1977, **A11**, 1137.

[63] S. P. Panda, *J. Sci. and Ind. Res.*, 1976, **35**, 560.

[64] D. P. Specht and S. Y. Farid, *Res. Discl.*, 1977, **161**, 65.

[65] G. B. Bulter and W. I. Ferric, *J. Polymer Sci., Part C*, 1977, **56**, 397.

Photocross-linkable terpolymers based on 4- and 4'-glycidoxychalcone, styrene, and maleic anhydride have been prepared.[66] On irradiation, whereas both the 4- and 4'-glycidoxychalcone-g-styrene co-maleic anhydride terpolymers cross-linked, only the 4'-terpolymer underwent an isomerization. Steric hindrance in the case of the 4-terpolymer prevented the isomerization. Maleic anhydride has also been used for the photocross-linking of olefin polymers[67] and homopolymers of 2-hydroxy-3-crotonyloxypropyl methacrylate.[68] A study has also been made of the curing of unsaturated polyesters with styrene containing fumarate groups.[69]

Bordin and Williams[70] have prepared some forty different photocross-linkable styrylpyridinium-substituted vinyl polymers by condensing poly[1-(1,6-dimethylpyridinium-3-yl)ethylene methyl sulphate] with various aromatic aldehydes using piperidine as the catalyst. A spectral analysis of the polymers showed that they may be designed for maximum absorption anywhere in the wavelength region 270—540 nm and are therefore particularly useful for photo-imaging systems. In addition, Borden[71] has developed photocross-linkable polymer systems based on 2,5-bis-(4-hydroxy-3-methoxybenzylidene)cyclopentanone. Other photocross-linkable systems that have been studied include phenoxy-resin esters of cinnamylidene–acetic acid[72] and poly(alkoxyacrylates).[73] Of more technological interest is a study on the photocross-linking of acrylamide–polyester–isocyanate copolymers.[74] Finally, kinetic methods for evaluating the activity of monomers in photocross-linking have been described[75] and electron beam and u.v. curing of epoxy acrylates have been compared.[76]

3 Optical and Luminescence Properties of Polymers

Photochemical probes in polymers have been very competently reviewed.[77] An excellent review on energy-transfer processes in polymers has also appeared.[78] Other reviews of interest cover photochromic polymers,[79] photoconductive polymers,[80] and chemiluminescence of polymers,[81]

[66] S. P. Panda, *Ind. J. Technol.*, 1976, **14**, 444.
[67] S. Tazuke and H. Kimura, *J. Polymer Sci. Polymer Chem.*, 1977, **15**, 2707.
[68] N. Ito, *Kobunshi Ronbunshu*, 1977, **34**, 625.
[69] R. Trauzeddel, W. Koenig, J. Sobottka, and E. T. Lasarenko, *Wiss. Z. Techn. Hochsch. Leipzig*, 1977, **1**, 59.
[70] D. G. Borden and L. R. Williams, *Makromol. Chemie*, 1977, **178**, 3035.
[71] D. G. Borden, *J. Appl. Polymer Sci.*, 1978, **22**, 239.
[72] T. Yamaoka, K. Ueno, T. Tsnoda, and K. Torige, *Polymer*, 1977, **18**, 81.
[73] Y. Nakashima and K. Nagoya-Shi, *Kenkyusho Kenkyu Hokoku*, 1976, **55**, 32.
[74] V. G. Matyushova, A. V. Shevchuk, V. F. Matyushev, and V. M. Tremut, *Plast. Massy*, 1977, **6**, 13.
[75] L. D. Taran, E. V. Khoina, N. N. Kalibabchuk, L. P. Paskal, V. G. Syromyatnikov, and M. A. Tsepenyuk, *Sposoby Zapisi Inform. na Besserebyan Nositchyakh*, 1977, **8**, 99.
[76] J. Kumanotani, T. Koshio, T. Yagi, and M. Gotoda, *Radiation Phys. Chem.*, 1977, **9**, 851.
[77] J. L. R. Williams and R. C. Daly, *Prog. Polymer Sci.*, 1977, **5**, 61.
[78] N. J. Turro, *Pure Appl. Chem.*, 1977, **49**, 405.
[79] M. Kryszewski and B. Nadolski, *Pure Appl. Chem.*, 1977, **49**, 511.
[80] J. M. Pearson, *Pure Appl. Chem.*, 1977, **49**, 463.
[81] G. C. Mendenhall, *Angew. Chem. Internat. Edn.*, 1977, **16**, 225.

Several optically active polymers based on 2-phenylvinyl alkyl thioethers have been synthesized [82] while optically active poly(*trans*-2-methylpentadiene) has been synthesized. [83]

The luminescence of polymers continues to attract much interest; an excellent review article has appeared recently on the use of luminescence spectroscopy for studying energy transfer and molecular mobility in polymer systems. [84] The degree of polarization of the fluorescence of poly(vinylchloride), [85] poly-(acryloyl-β-cyclodextrin), [86] agarose, [87] 9,10-bis(bromomethyl) anthracene–isoprene copolymer, [88] polyacrylics, [89, 90] vinylamine–vinylpyrrolidinone copolymer, [91] and polystyrene [92] has been related to the degree of orientation and segmental mobility of the polymer chains. Of these studies the work on polystyrene is particularly interesting. [92] Here an autocorrelation function of the type given in equation (1) has been proposed to describe the motions of a bond

$$M_2(t) = e^{-t/\theta}e^{t/p} \text{ erfc } (t/p)^{\frac{1}{2}} \tag{1}$$

internal to the chain. In equation (1) $M_2(t)$ is the orientation autocorrelation function, erfc is the error function complement, p is the relaxation time, θ is the relaxation time reflecting the effects of possible departures from motions permitted by an ideal tetrahedral lattice, and t is the orientation time. Depolarization of the phosphorescence from poly(methyl acrylate) containing acenaphthylene and 1-vinylnaphthalene as phosphorescent probes has been found to occur at the onset of segmental motion of the polymer chains at 298 K. [93] This value was found to be in good agreement with those obtained by dielectric and relaxation techniques. Experimental studies on the relationship between thermoluminescence and molecular relaxation processes in polymers have also been carried out. [94]

The temperature dependence of the phosphorescence from enones formed in the thermal oxidation of polybutadiene has been studied. [95] Discontinuities in the Arrhenius plots were ascribed to rotational barriers in the enones as well as those of the polymer matrix. Laser flash photolysis has been used to study the steric effects and conformation of macromolecules in solution. [96] Of particular interest is a study describing the possibility of using Förster resonance energy transfer to determine the end-to-end distribution between chromophores

[82] V. Kurešević and D. Fles, *J. Polymer Sci. Polymer Chem.*, 1977, **15**, 847.
[83] M. Miyata and K. Takemoto, *Polymer J.*, 1977, **9**, 111.
[84] J. E. Guillet, *Pure Appl. Chem.*, 1977, **49**, 249.
[85] S. Hibi, M. Maeda, H. Kubota, and T. Muira, *Polymer*, 1977, **18**, 143.
[86] A. Harada, M. Furue, and S. Nozakura, *Macromolecules*, 1977, **10**, 676.
[87] A. Hayashi, K. Kinoshita, and M. Kuwano, *Polymer J.*, 1977, **9**, 219.
[88] D. Kuffer, *Rev. Gen. Caoutch. Plast.*, 1977, **54**, 99.
[89] E. V. Anufrieva, O. A. Belozerova, V. D. Pantov, and I. M. Papisov, *Vysokomol. Soedineniya, Ser. B*, 1977, **19**, 409.
[90] Yu. V. Brestkin, E. S. Edilyan, A. V. Novoselova, S. Ya. Frenkel, and G. Mann, *Faserforsch. Textiltech.*, 1977, **28**, 519.
[91] E. V. Anufrieva, E. F. Panarin, V. D. Pantov, G. V. Semisotnov, and M. V. Solovskii, *Vysokomol. Soedineniya, Ser. A*, 1977, **19**, 1329.
[92] B. Veleur and L. Monnerie, *J. Polymer Sci. Polymer Phys.*, 1976, **14**, 11, 29.
[93] H. Rutherford and I. Soutar, *J. Polymer Sci. Polymer Phys.*, 1977, **15**, 2213.
[94] A. Linkens and J. Vanderschueren, *J. Electrostatics*, 1977, **3**, 149.
[95] S. W. Beavan and D. Phillips, *European Polymer J.*, 1977, **13**, 825.
[96] J. Fauré, J. P. Fouassier, D. J. Lougnot, and R. Salvin, *European Polymer J.*, 1977, **13**, 891.

attached to a polymer chain.[97] Other studies of interest include the use of fluorescence and phosphorescence spectroscopy to study the delamination [98] and curing [99] of glass-reinforced polyester and melamine formaldehyde resins, respectively.

Photocarrier generation and photoconductivity in polymer systems have been investigated. Photocarrier generation in N-vinylcarbazole (VCz)–1-vinylnaphthalene (VNa) copolymers has been postulated to occur *via* an upper excited singlet state of the carbazole moiety as shown [100] in Scheme 9. It is also proposed that photocarrier generation occurs by the same mechanism as that of the

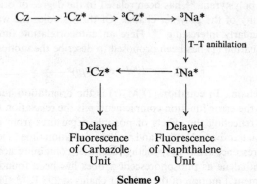

Scheme 9

vinylcarbazole homopolymer. Bromination of poly-(N-vinylcarbazole) significantly reduces photocarrier generation in the polymer,[101] while photocarrier generation in poly(ethylene terephthalate) is believed to occur through $\pi\pi^*$ excitations at wavelengths $\leqslant 320$ nm.[102] The photoconductive properties of polystyrene [103] and polyrhodanines [104] have been examined. In the latter case the activation energy for photoconduction was found to increase with an increase in the number of methylene groups in the chain separating the rhodanine groups.

Excimer and exciplex formation have been observed in a number of polymer systems. Excimer formation in the polystyrenes continues to attract much interest. For example, the activation energies for excimer formation in isotactic polystyrenes have been found to be smaller than those for the atactic polymers.[105] In another paper by the same workers excimer formation was found to be dependent on the molecular weight of the polymer.[106] In this case excimer form-

[97] I. Ohmine, R. Silbey, and J. M. Dentch, *Macromolecules*, 1977, **10**, 862.
[98] T. Fujii and Y. Izumoto, *Zairyo*, 1977, **26**, 1167.
[99] D. J. Morantz, C. S. Bilen, and R. C. Thompson, in 'Proceedings of the Eighth International Symposium on the Reactivity of Solids', Plenum, New York, 1976, 499.
[100] M. Yokoyama, M. Hanabata, T. Tamamura, T. Nakano, and Hiroshi Mikawa, *J. Chem. Phys.*, 1977, **67**, 1742.
[101] C. H. Griffiths, K. Okumura, and A. Van Laeken, *J. Polymer Sci. Polymer Phys.*, 1977, **15**, 1627.
[102] Y. Takai, T. Osawa, T. Mizutani, and M. Ieda, *J. Polymer Sci. Polymer Phys.*, 1977, **15**, 945.
[103] S. Morita and M. Shen, *J. Polymer Sci. Polymer Phys.*, 1977, **15**, 981.
[104] R. Hirohashi, Y. Toda, and Y. Hishihi, *Polymer*, 1977, **18**, 601.
[105] T. Ishii, T. Handa, and S. Matsunaga, *Makromol. Chem.*, 1977, **178**, 2351.
[106] T. Ishii, T. Handa, and S. Matsunaga, *Macromolecules*, 1978, **11**, 40.

ation increases with increasing molecular weight up to 10^4 due to stronger interactions between the phenyl rings. The composition of styrene–acrylonitrile copolymers has been determined by measuring the ratios of excimer to monomer fluorescence for the styrene.[107] The ratio was found to decrease with an increase in the number of acrylonitrile sequences along the chain; with below 40% styrene units no excimer fluorescence was observed. A linear relationship was obtained between the relative intensity of excimer emission and the number of styrene sequences (Figure 1).

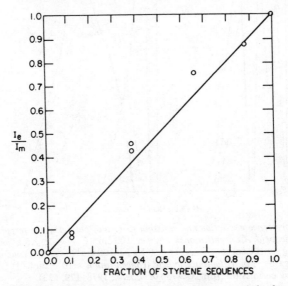

Figure 1 *Correlation between relative excimer fluorescence and the fraction of styrene sequences in SAN copolymers*
(Reproduced by permission from *J. Appl. Polymer Sci.*, 1977, **15**, 2013)

The use of time-resolved fluorescence spectroscopy has shown that excimers in poly(1-vinylnaphthalene) dissociate appreciably to the monomer at room temperature.[108] Interestingly, Ferreira and Porter[109] have attributed excimer formation by perylene in poly(methyl methacrylate) to an excitation transfer process through a Förster inductive resonance mechanism. Excimer formation has been observed in poly-2-phenyl-5-(*p*-vinyl)phenyloxazole.[110] Allen and McKellar[111] have shown that the fluorescence emission from poly(ethylene terephthalate) occurs from an associated ground-state dimer formed between terephthalate units and not an excimer as was previously thought.[112] In tri-fluoroacetic acid solution, poly(ethylene terephthalate) exhibits longer wave-length excitation and emission spectra (Figure 2). On dilution, however, these

[107] L. Alexandru and A. C. Somersall, *J. Polymer Sci. Polymer Chem.*, 1977, **15**, 2013.
[108] K. P. Ghiggino, R. D. Wright, and D. Phillips, *Chem. Phys. Letters*, 1978, **53**, 552.
[109] J. O. Ferreira and G. Porter, *J.C.S. Faraday II*, 1977, **73**, 340.
[110] I. McInnally, I. Soutar, and W. Steedman, *J. Polymer Sci. Polymer Chem.*, 1977, **15**, 2511.
[111] N. S. Allen and J. F. McKellar, *Makromol. Chem.*, 1978, **179**, 523.
[112] D. H. Phillips and J. C. Schug, *J. Chem. Phys.*, 1969, **50**, 3297.

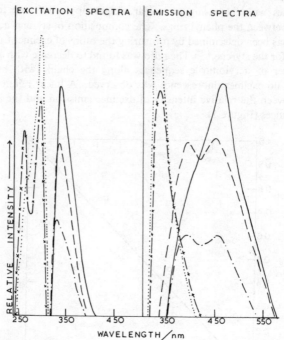

EXCITATION SPECTRA   EMISSION SPECTRA

RELATIVE INTENSITY

WAVELENGTH/nm

Figure 2 *Fluorescence excitation and emission spectra of poly(ethylene terephthalate) in trifluoroacetic acid. Concentrations:* 5 g *per* 100 cm³ (————); 2.5 g *per* 100 cm³ (— — —); 1 g *per* 100 cm³ (—·—·—); < 0.5 g *per* 100 cm³ (—×—×) *and of dimethyl terephthalate* (·····) *in trifluoroacetic acid (concentration* ≈ 10^{-4} mol l^{-1})
(Reproduced by permission from *Makromol. Chem.*, 1978, **179**, 523)

spectra gradually disappear until only monomer spectra are observed. Fluorescence emission has also been observed from associated ground-state dimers in the poly-(1 and 2-vinylnaphthalenes).[113] A fluorescence study of poly-(*N*-vinylcarbazole) containing dimethyl terephthalate has shown that the formation of an 'exterplex' consisting of two carbazolyl groups and one molecule of DMPT[114] can be represented by structure (5). Some Russian workers[115] have reported on an anomalous phosphorescence emission with a mean lifetime of 25 seconds from poly-(*N*-vinylcarbazole). The origin of the emission, however, remains uncertain. Charge transfer complex formation has been observed in polymers containing carbazole groups[116] and polyspiropyran.[117]

Energy transfer in polymer systems has been extensively studied. Energy transfer from a styrene–phenylacetylene copolymer to perylene in poly(methyl

[113] M. Irie, T. Kamijo, M. Alkarva, T. Takemura, K. Hayashi, and H. Baba, *J. Phys. Chem.*, 1977, **81**, 1571.
[114] C. E. Hoyle and J. E. Guillet, *Macromolecules*, 1978, **11**, 221.
[115] I. R. Azizov and A. L. Kartuzhauskii, *Zhur. nauch. priklad. Fotograf. Kinemat.*, 1977, **22**, 289.
[116] S. Iwatsuki and K. Arai, *Makromol. Chem.*, 1977, **178**, 2307.
[117] A. D. Grishina and V. D. Ermakova, *Vysokomol. Soedineniya, Ser. A*, 1977, **19**, 2228.

$$[Cz - Cz^+ - DMPT^-]*$$

$$(5)$$

methacrylate) has been studied using fluorescence depolarization.[118] Application of the Förster theory gave a value of 78 nm for the critical transfer distance R_0. Only the emission from the acceptor depolarized with an increase in the efficiency of energy transfer while the emission from the donor showed no such effect. Intramolecular energy transfer by energy migration along the polymer chain was ruled out in favour of a long range intermolecular energy transfer process. The efficiency of triplet energy migration in a styrene–vinylbenzophenone copolymer has been found to increase with the vinylbenzophenone content as a result of exchange interactions between the benzophenone groups.[119] Vikesland and Wilkinson[120] have studied energy transfer between the lowest triplet state of triphenylene to the transition metal complexes ferrocene and ruthenocene in poly(methyl methacrylate). It was found that a comparison of the change in relative lifetimes and emission yields with acceptor concentrations was not sufficient to determine the mechanism of energy transfer when transition-metal complexes are involved. Other studies involving triplet–triplet energy transfer include an examination of oligmers in polyacrylophenone,[121] 1-methyl-naphthalene in polyacrylophenone,[122] and anthracene and benzil end-groups in polystyrene.[123] Finally, the quenching of triplet anthracene by molecular oxygen has been found to be unaffected by the presence of a polymeric environment.[124]

Singlet molecular oxygen annihilation luminescence has been investigated from polystyrene doped with fluorescent molecules.[125] The emitting transition in $^1O_2$ annihilation luminescence was concluded to be the most efficient radiative transition of either the fluorescent chromophore or an oxygen–organic chromophore complex. Singlet energy migration has been observed in an ethylene–propylene–diene terpolymer containing pendant anthracene groups.[126] Stern–Volmer quenching has been observed in poly-(4-vinylpyridine) containing 9-(methacryloyloxymethyl)anthracene.[127]

The role of energy transfer in the photostabilization of polymers has been extensively studied. Allen *et al.*[128] found a correlation between the photo-stabilizing effect of alkali metal halides in nylon 6,6 and their ability to quench the fluorescence emission from the polymer. The quenching and photostabilizing efficiencies of halide ions were found to increase in the order $F^- > Cl^- > Br^- > I^-$. In the same study a correlation was found between the photostabilizing

118 J. R. MacCallum and L. Rudkin, *Nature*, 1977, **266**, 338.
119 C. David, D. Baeyens-Volant, P. Macedo de Abreu, and G. Geuskens, *European Polymer J.*, 1977, **13**, 841.
120 J. P. Vikesland and F. Wilkinson, *J.C.S. Faraday II*, 1977, **73**, 1082.
121 R. Salvin and J. Meybeck, *J. Photochem.*, 1977, **7**, 411.
122 R. Salvin, J. Meybeck, and J. Faure, *Makromol. Chem.*, 1977, **178**, 2275.
123 K. Horie and I. Mita, *Polymer J.*, 1977, **9**, 201.
124 G. Ramme and J. F. Rabek, *European Polymer J.*, 1977, **13**, 855.
125 R. D. Kenner and A. U. Khan, *J. Chem. Phys.*, 1977, **67**, 1605.
126 J. P. Fonassier, D. J. Lougnot, and J. Faure, *Makromol. Chem.*, 1978, **179**, 437.
127 Yu. Ya. Gotlib, N. R. Pavlov, Yu. E. Kirsh, and V. A. Kabanov, *Vysokomol. Soedineniya, Ser. A*, 1977, **19**, 1150.
128 N. S. Allen, J. F. McKellar, and D. Wilson, *J. Photochem.*, 1977, **7**, 319.

efficiency of aromatic and metal salicylates and their ability to reduce the lifetime and intensity of the phosphorescence from the impurity α,β-unsaturated carbonyl groups in the polymer. In Table 1 it is seen that salicyclic acid, which is a non-stabilizer, exhibits no quenching, phenyl salicylate, a weak stabilizer, is a mild quencher, whereas all the metal salicylates, which are good stabilizers, exhibit a strong quenching effect.

Table 1 *Effect of salicylates on nylon 6,6 phosphorescence* $(Ex\lambda_{max} = 290\ nm)$

Additive	Phosphorescence emission intensity 410 nm (relative)	Phosphorescence emission lifetimes τ/sec
Nylon 6,6 control	1.0	2.20
Salicylic acid	1.0	2.20
Phenyl salicylate	0.75	1.20
Sodium salicylate	<0.10	<0.10
Magnesium salicylate	<0.10	<0.10
Aluminium salicylate	<0.10	<0.10
Manganese salicylate	<0.10	<0.10
Copper salicylate	<0.10	<0.10
Chromium salicylate	<0.10	<0.10

In another study the same workers found that for nickel oxime chelates to be effective triplet quenchers and photostabilizers the nickel and oxime ligand must be combined to form the chelate.[129] Phillips and co-workers[130] have shown that the quenching of excited states in polysulphones is efficient. Interestingly, Pivovarov[131] found that various light stabilizers quench the excimer fluorescence of polystyrene by an inductive-resonance mechanism.

The degradation of polymer systems has been extensively studied using luminescence spectroscopy. Geuskens and David[132] have shown that carbonyl groups can photosensitize the decomposition of hydroperoxide groups in polymers. Evidence for this was shown by the fact that cumene hydroperoxide quenches the excimer fluorescence of polystyrene. Allen *et al.* have established the identity of the luminescent impurities as α,β-unsaturated carbonyl groups in the commercially important polyolefins, polypropylene,[133] polyethylene,[134] and poly-(4-methylpent-1-ene).[135] The fluorescence was attributed to the presence of impurity enone groups and the phosphorescence to dienone groups in all three polymers. The importance of these chromophores in the mechanism of polymer photo-oxidation will be discussed later. Finally, the fluorescence from poly-caproamide has been attributed to the perturbing influence of defects in the

[129] N. S. Allen, J. Homer, and J. F. McKellar, *J. Appl. Polymer Sci.*, 1978, **22**, 611.
[130] F. Abdul-Rasoul, C. L. R. Catherall, J. S. Hargreaves, J. M. Mellor, and D. Phillips, *European Polymer J.*, 1977, **13**, 1019.
[131] A. P. Pivovarov, *Vysokomol. Soedineniya, Ser. B*, 1978, **20**, 54.
[132] G. Genskens and C. David, *Pure Appl. Chem.*, 1977, **49**, 479.
[133] N. S. Allen, J. Homer, and J. F. McKellar, *J. Appl. Polymer Sci.*, 1977, **21**, 2261.
[134] N. S. Allen, J. Homer, J. F. McKellar, and D. G. M. Wood, *J. Appl. Polymer Sci.*, 1977, **21**, 3147.
[135] N. S. Allen and J. F. McKellar, *J. Appl. Polymer Sci.*, 1978, **22**, 325.

chemical structure of the polymer.[136] The yellow products produced on photo-degradation of polystyrenes have been found to quench the normal and excimer fluorescence emissions from the polymers.[137] Other studies of interest cover the use of luminescence for evaluating the mechanical degradation of glass-reinforced plastics[138] and the thermal degradation of cotton.[139]

A number of studies have appeared on the role of energy transfer in the dye-sensitized photodegradation of polymers. Allen *et al.*[140] observed triplet–singlet resonance energy transfer from the impurity α,β-unsaturated carbonyl groups in nylon 6,6 and a photoactive disperse dye, 3-methoxybenzanthrone. Strong long-lived fluorescence was observed from the dye when the polymer was excited at the wavelengths where only the phosphorescent impurity chromophores absorb light. A value of 4.6 nm was obtained for the critical transfer distance R_0. Roberts and co-workers[141,142] have observed weak triplet–singlet resonance energy transfer between the phosphorescent ester carbonyl groups in poly(ethylene terephthalate) and a non-photoactive disperse dye, 1-amino-2-(2-methoxyethoxy)-4-hydroxy-9,10-anthraquinone. They conclude that energy transfer is not important in the dye-sensitized photodegradation. Ershov and Dovbii[143] have also observed energy transfer from the fluorescent species in polycaproamide and some acid dyes.

The thermoluminescence and chemiluminescence of polymers have been actively studied. The thermoluminescence from γ-irradiated polyethylene is dependent on both the formation of crystallites[144] and the degree of pre-oxidation of the polymer.[145,146] The thermoluminescence of poly(methyl methacrylate) irradiated with X-rays has been investigated.[147] Two glow peaks with maxima at 136 and 368 K were observed. The first peak at 136 K was associated with electron traps and the second at 368 K with the macroradical structure (6).

Thermoluminescence[148] and thermally[149] stimulated current analyses have been used to study the behaviour of electron traps in poly(ethyleneterephthalate).

$$\begin{array}{c} CH_3 \\ | \\ \text{\textasciitilde}CH_2C\cdot \\ | \\ CH_3 \end{array}$$

(6)

[136] E. V. Dovbii, Yu. A. Ershov, B. O. Glotov, and G. D. Litovenko, *Zhur. priklad. Spektroskopii*, 1977, **26**, 856.

[137] R. B. Fox and T. R. Rice, *Amer. Chem. Soc., Div. Polymer Chem., Polymer Preprints*, 1977, **18**, 385.

[138] U. Krauja, J. Laizans, Z. Upitis, and M. Tutane, *Mekh. Polim.*, 1977, **2**, 316.

[139] B. D. Semak and I. I. Shiiko, *Izvest. V.U.Z. Teknol. Legk. Prom.-sti.*, 1977, **4**, 41.

[140] N. S. Allen, J. F. McKellar, and D. Wilson, *J. Photochem.* 1977, **7**, 405.

[141] R. G. Merrill and C. W. Roberts, *J. Appl. Polymer Sci.*, 1977. **21**, 2745.

[142] C. W. Roberts and P. R. Cheung, *Ind. Manag. Text. Sci. (Clemson Univ., U.S.A.)*, 1977, 13.

[143] Yu. A. Ershov and Eu. V. Dovbii, *J. Appl. Polymer Sci.*, 1977, **21**, 1511.

[144] S. Nakamura and M. Ieda, *J. Appl. Phys.*, 1977, **48**, 179.

[145] S. Nakamura and G. Saura, *J. Appl. Phys.*, 1977, **48**, 3626.

[146] Y. Suzuoki, K. Yasuda, T. Mizutani, and M. Ieda, *Japan J. Appl. Phys.*, 1977, **16**, 1339.

[147] S. Radhakrishna and M. R. K. Murthy, *J. Polymer Sci. Polymer Phys.*, 1977, **15**, 987.

[148] M. P. Padhye and P. S. Tamhane, *Angew. Makromol. Chem.*, 1978, **67**, 79.

[149] Y. Takai, K. Mori, T. Mizutani, and M. Ieda, *Japan J. Appl. Phys.*, 1977, **16**, 1937.

Further studies on chemiluminescence include an examination of anthracene in polyethylene,[150] phenolic antioxidants in polypropylene,[151] polybutadienes,[152] and a 9,10-bis(methacryloyloxymethyl)anthracene–methyl methacrylate copolymer.[153]

4 Photochemical Processes in Polymeric Materials

Synthetic Polymers.—Several review articles have appeared on the general aspects of photo-oxidation and photodegradation of commercial polymers.[155—158] Tsuji[155] has extensively reviewed e.s.r. studies on polymer photodegradation and Scott[156] has reviewed his own work on the role of hydroperoxides in the photo-oxidation of polymers.

Polyolefins. The photodegradation and photo-oxidation of commercial polyolefins are still an active area of research in polymer photochemistry. Scott[159,160] has discussed the influence of thermal history on the rate of photo-oxidation of polyolefins and maintains that hydroperoxides are the major photoinitiators. In a recent review Allen and McKellar[161] conclude that hydroperoxides and α,β-unsaturated carbonyl groups play a dual role in initiating the photo-oxidation of the polyolefins.

Recently, Frank[162] has shown that the active wavelengths responsible for maximum photo-oxidation of polypropylene occur in the wavelength region 300—320 nm and not at 370 nm as previously determined.[163] Chakraborty and Scott[164] have found that heating prior thermally oxidized polyethylene film in an inert atmosphere for several hours improves the subsequent light stability of the polymer (see Figure 3). Before this treatment the film contained a substantial amount of vinylidene hydroperoxide and photo-oxidized rapidly. After treatment the vinylidene hydroperoxide concentration was reduced to virtually zero. The overall reaction (Scheme 10) was proposed to account for the formation and subsequent photolysis of the vinylidene hydroperoxide groups in polyethylene.

However, the fact that luminescent α,β-unsaturated carbonyl groups have now been found to be present in the three commercially important polyolefins,

[150] N. A. Krinitsyna, G. Ya. Bazarskii, A. A. Kachan, and V. A. Shrubovich, *Vysokomol. Soedineniya, Ser. B*, 1977, **19**, 449.
[151] L. Matisová-Rychla, P. Ambrović, N. Kulickova, and J. Holcik, *J. Polymer Sci., Part C*, 1976, **57**, 181.
[152] C. D. Mendenhall, R. A. Nathan, M. A. Birts, and M. A. Golub, *Amer. Chem. Soc., Div. Polymer Chem., Polymer Preprints*, 1977, **18**, 362.
[153] S. R. Rafikov, V. N. Korobeinikova, G. V. Leplyanin, V. P. Kazakov, and V. I. Ionov, *Zhur. priklad. Spektroskopii*, 1978, **28**, 151.
[154] K. Tsuji, *Polymer-Plast. Technol. Eng.*, 1977, **9**, 1.
[155] G. Scott, in 'Developments in Polymer Degradation', ed. N. Grassie, Applied Science Publishers, London, 1977, Ch. 7, p. 205.
[156] W. J. McGill, *South African J. Sci.*, 1977, **73**, 43.
[157] D. Phillips, *Polymer News*, 1977, **3**, 246.
[158] H. Lind, *Kunststk-Plast.*, 1977, **24**, 27.
[159] G. Scott, *J. Polymer Sci., Part C*, 1976, **57**, 357.
[160] G. Scott, *Amer. Chem. Soc., Div. Polymer Chem., Polymer Preprints*, 1977, **18**, 365.
[161] N. S. Allen and J. F. McKellar, *Brit. Polymer J.*, 1977, **9**, 302.
[162] H. P. Frank, *J. Polymer Sci., Part C*, 1976, **57**, 311.
[163] R. C. Hirt and N. Z. Searle, *Appl. Polymer Symposia*, 1967, **4**, 61.
[164] K. B. Chakraborty and G. Scott, *European Polymer J.*, 1977, **13**, 731.

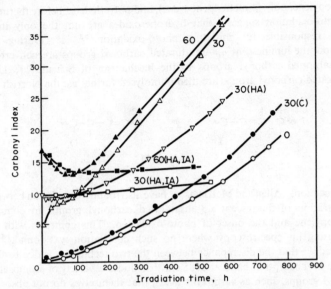

Figure 3 *Change in carbonyl index of LDPE films during u.v. irradiation. (Numbers on curves indicate processing times at 150 °C in air; HA, heated in argon; IA, irradiated in argon; C closed mixer)*
(Reproduced by permission from *European Polymer J*, 1977, **13**, 731)

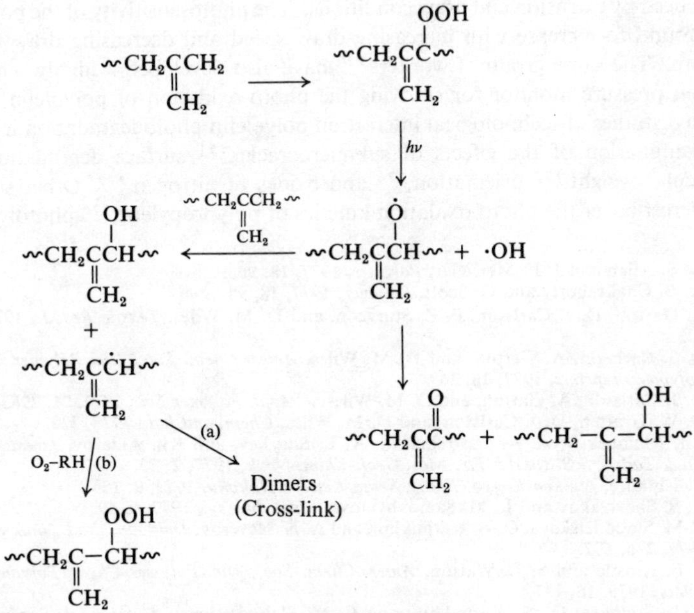

Scheme 10

polypropylene, poly-(4-methylpent-1-ene), and polyethylene, before thermal and/ or photo-oxidation suggests that hydroperoxides are not the only impurity species responsible for initiating photo-oxidation.[133—135] During photo-oxidation, the luminescent α,β-unsaturated carbonyl groups are converted into β,γ-unsaturated carbonyl groups by the mechanism of Scheme 11. The β,γ-unsaturated carbonyl groups are then photolysed further *via* the Norrish Type I

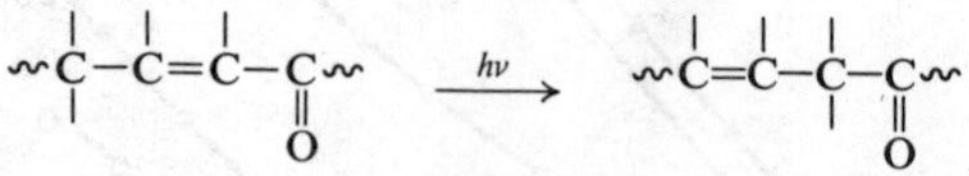

Scheme 11

and II reactions. Allen and McKellar[165] have also found a correlation between the concentration of fluorescent α,β-unsaturated carbonyl groups in commercial polypropylene and the onset of photo-oxidation. This contrasts with earlier work using i.r. spectroscopy where no such correlation was found.[166] It is suggested[165] that the discrepancy between the two techniques is due to the fact that the i.r. absorption band at 1685 cm^{-1} includes a variety of α,β-unsaturated carbonyl groups, such as carboxylic acids, that themselves do not absorb light above 290 nm. Consequently, these groups would not be expected to participate in the sunlight-induced oxidation of the polymer.

Carlsson and co-workers[167—169] have examined the effect of morphology and fibre production conditions on the photo-oxidation of polypropylene. They found that the sensitivity of the polymer to photo-oxidation was markedly influenced by extrusion and draw conditions. The photosensitivity of the polymer was found to increase with increasing draw speed and decreasing draw temperature. The same group of workers[170] have also developed a highly sensitive oxygen pressure monitor for studying the photo-oxidation of polyolefin films. Related studies of technological interest on polyolefin photodegradation include an examination of the effects of submicrocracks,[171] surface degradation,[172] molecular weight,[173] orientation,[174] and oxides of nitrogen.[175] Other studies of interest cover the photo-oxidation kinetics of polypropylene,[176] photolysis of

[165] N. S. Allen and J. F. McKellar, *Polymer*, 1977, **18**, 968.
[166] K. B. Chakraborty and G. Scott, *Polymer*, 1977, **18**, 98.
[167] A. Garton, D. J. Carlsson, P. Z. Sturgeon, and D. M. Wiles, *Textile Res. J.*, 1977, **47**, 423.
[168] D. J. Carlsson, A. Garton, and D. M. Wiles, *Amer. Chem. Soc., Div. Polymer Chem., Polymer Preprints*, 1977, **18**, 369.
[169] D. J. Carlsson, A. Garton, and D. M. Wiles, *J. Appl. Polymer Sci.*, 1977, **21**, 2963.
[170] D. W. Grattan, D. J. Carlsson, and D. M. Wiles, *Chem. and Ind.*, 1978, 228.
[171] Kh. Akimbekov, A. A. Yastrebinskii, A. L. Slutskev, and Kh. Aidarov, *Izvest. Akad. Nauk Tadzh. S.S.R. Otd. Fiz.-Mat. Geol.-Khim. Nauk*, 1977, **2**, 75.
[172] Y. Ishikure, *Ign-ken Kogyo Gijutsa Senta Kenkyu Hokoku*, 1977, **9**, 104.
[173] A. S. Shevlyakov and L. M. Samokhvalova, *Plast. Massy.*, 1977, **9**, 49.
[174] E. M. Slobodetskaya, O. N. Karpukhin, and A. N. Neverov, *Doklady Akad. Nauk S.S.R.*, 1977, **236**, 677.
[175] J. F. Kinstle and S. L. Watson, *Amer. Chem. Soc., Div. Polymer Chem. Polymer Preprints*, 1975, **16**, 137.
[176] V. L. Bodneva, O. N. Karpukhin, and E. M. Slabodetskaya, *Vysokomol. Soedineniya, Ser. B*, 1977, **19**, 276.

polyethylene in the far-u.v.,[177] and photo-oxidation of ethylene-α-olefin copolymers.[178]

Polystyrenes and Related Polymers. Weir[179] has examined the photosensitized degradation of polystyrene in solution by hydroxy-radicals produced from the photodecomposition of hydrogen peroxide. Hydroxy-radical addition was found to occur at the phenyl groups of the polymer to give phenols and dialdehydes. Kubica[180] has found that the addition of low molecular weight acetophenone compounds to polystyrene determines the photo-oxidation kinetics of the polymer in the early stages of irradiation. Other workers have found that the polymerization catalyst, azo-bis-isobutyronitrile (AIBN), photosensitizes the degradation of poly-(α-methylstyrene).[181]

MacCallum and Ramsay[182] have recently proposed a three-stage mechanism for the photo-oxidation of styrene–copolymer systems. The first stage involves the formation of hydroperoxides from charge-transfer complexes between the oxygen and phenyl units, the second involves the breakdown of the hydroperoxides to form ketonic groups, and the third the production of oxygen-consuming radicals by the excited states of the ketones, Schemes 12, 13, and 14, respectively.

Scott and Tahan[183] have examined the photo-oxidation of acrylonitrile and methyl methacrylate butadiene–styrene copolymers. Differences in the rates of photo-oxidation of the copolymers were attributed to differences in the initial concentration of unsaturated groups. Unsaturated carbonyl groups were found to enhance the rate of photocross-linking.

Poly(vinyl halides). Impact modified poly(vinyl chloride) has been found to photo-oxidize in two stages, *viz.* an initial rapid but short stage followed by a longer period of slower degradation.[184] The fast initial stage is believed to be caused by the rapid photo-oxidation of the polybutadiene phase. The polybutadiene was also found to accelerate discoloration of the polymer. Interestingly, colour formation was found to be linearly related to the formation of unsaturated carbonyl groups in the polymers. Prior thermal treatment of both the poly(vinyl chloride) and impact modified polymer was found to accelerate the rate of photo-oxidative degradation.[184] Other workers have obtained similar results.[185] The photo-oxidation of poly(vinyl chloride) has been shown to take place only at the surface of the material (<0.005 cm).[186,187] For

177 I. G. Gorodetskii, E. A. Grigoryan, Yu. I. Dorofuv, V. E. Skurat, and V. L. Talroze, *Tezisy Doklady Vses Konf. Spektroskopii Vak. Ul'trafioleta, Vzaimodeistoga Izluch. Veshchestron. 4th*, (1975) 146.
178 T. N. Guseinov, G. M. Aliev, D. Kh. Ragimova, Z. L. Efendieva, and M. N. Elchiev, *Izvest. Akad. Nauk Az. S.S.S.R. Ser. Fiz.-Tekh. Mat. Nauk*, 1977, **4**, 96.
179 N. A. Weir, *European Polymer J.*, 1978, **14**, 9.
180 J. Kubica, *Makromol. Chemie*, 1977, **178**, 3017.
181 T. Ikeda, S. Okamura, and H. Yamaoka, *J. Polymer Sci., Polymer Chem.*, 1977, **15**, 2971.
182 J. R. MacCullum and D. A. Ramsay, *European Polymer J.*, 1977, **13**, 945.
183 G. Scott and M. Tahan, *European Polymer J.*, 1977, **13**, 981.
184 G. Scott and M. Tahan, *European Polymer J.*, 1977, **13**, 989.
185 J. D. Isner and J. W. Summers, *Soc. Plast. Eng., Amer. Tech. Conf. (Montreal)*, 1977, **23**, 237.
186 L. D. Strelhova, G. T. Fedoseyeva, and K. S. Minsker, *Vysokomol. Soedineniya, Ser. A*, 1976, **18**, 2064.
187 F. Mori, M. Koyama, and Y. Oko, *Angew. Makromol. Chem.*, 1977, **64**, 89.

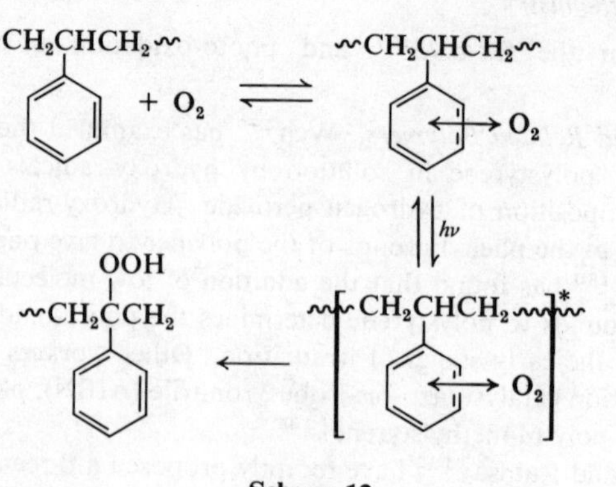

Scheme 12

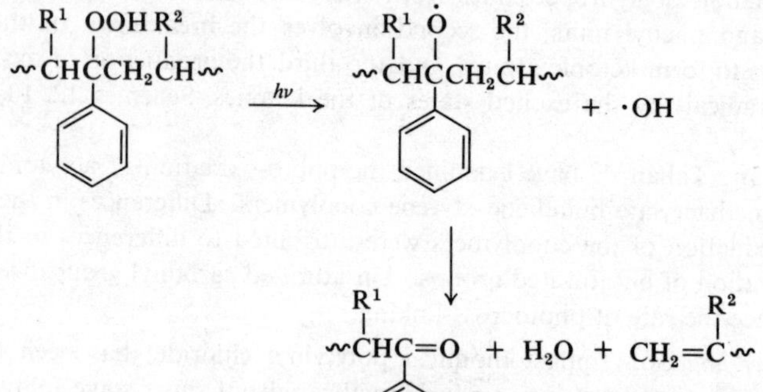

Scheme 13

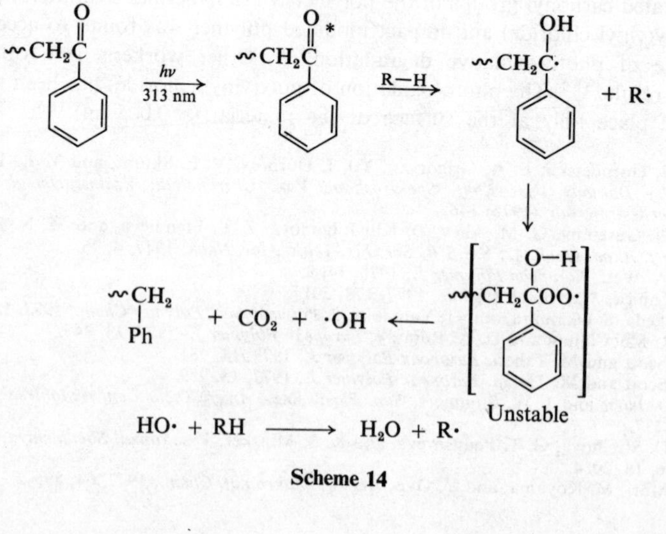

Scheme 14

example, the carbonyl group concentration was found to be higher in the surface layer of the polymer sample than the interior.[187] The photosensitized dehydrochlorination of modified poly(vinyl chloride), treated with halogens, has been shown to occur through a radical mechanism involving polyene–halogen complexes.[188]

Polyacrylates. The photodegradation of poly(methyl methacrylate) has been investigated using a modified flash photolysis set-up.[189] The gaseous products of photodegradation were analysed using a gas chromatograph. The excited ester groups are believed to form radicals which initiate degradation of the polymer. In a vacuum a significant amount of methanol and monomer was detected as products, whereas under nitrogen a significant quantity of carbon particles was formed as an additional product. The rate of photo-oxidation of methacrylate–methacrylic acid copolymer coatings has been found to increase when applied to aluminium foil.[190] This result was attributed to higher internal stress in the coating than in unsupported film. The application of an external stress to the unsupported film was found to increase its rate of photo-oxidation.

Polyamides. Some Russian workers have recently reported that peroxide radicals are unimportant in the photo-oxidation kinetics of commercial polyamides.[191] During irradiation peroxide radicals were found to accumulate at a lower rate than that of oxygen absorption. The same workers[192] also examined in detail the primary and secondary photolysis mechanisms of polyamides. Changes in the mechanical properties of Kapron fibres during irradiation have been examined[193,194] and the light stability of *N*-chloro-nylon 6,6 has been investigated.[195]

Polyesters. Day and Wiles[196] have examined the effect of flame retardants on the photochemical deterioration of polyesters. The presence of bromine was found to accelerate the rate of photo-oxidation of the polymer while the flame resistance remained unaffected. Polyester fibres prepared from ethylene glycol-1,1'-ferrocene–dicarboxylic acid–terephthalic acid copolymer have been found to have greater photostability than that of poly(ethylene terephthalate).[197]

Rubbers. The photo-oxidation of various polydienes has been found to occur by a chain reaction involving the degenerate branching of hydroperoxides.[198]

[188] V. V. Kolesnikova, O. V. Kolninov, V. K. Milinchuk, and S. Ya. Pshezhetskii, *Vysokomol. Soedineniya, Ser. A,* 1976, **18**, 2431.
[189] H. Kambe, H. Watanabe, S. Nagatomo, and Y. Itoh, *Polymer,* 1977, **18**, 1063.
[190] P. I. Zubov and E. A. Kanevskaya, *Doklady Akad. Nauk S.S.S.R.,* 1977, **233**, 148.
[191] E. V. Vichutinskaya, G. G. Makarov, A. L. Margolin, and L. M. Postnikov, *Vysokomol. Soedineniya Ser. B,* 1977, **19**, 441.
[192] A. L. Margolin, L. M. Postnikov, and V. Ya. Shlyapintokh, *Vysokomol. Soedineniya Ser. A,* 1977, **19**, 1954.
[193] L. A. Uvarova, *Khim. Volokna,* 1977, **5**, 73.
[194] A. E. Mirhesnikov, M. P. Nosov, V. F. Putyatin, and V. I. Pavlov, *Izvest. V. U. Z. Teknol. Legk. Prom.-sti.,* 1977, **4**, 446.
[195] A. Banihashemi and R. C. Schulz, *Makromol. Chem.,* 1978, **179**, 855.
[196] M. Day and D. M. Wiles, *Textile Res. J.,* 1978, **48**, 32; 'Proceedings of the Symposium on Textile Flammability', 1977, **5**, 141.
[197] A. I. Volozhin, L. Yu. Verkhovodka, S. S. Gusev, P. N. Zernov, V. A. Levdanskii, and K. A. Bogdanov, *Prepr. Mezhdunar Simp. Khim. Voloknam,* 2nd, 1977, **2**, 21.
[198] V. B. Ivanov, M. N. Kuznetsova, L. G. Angert, and V. Ya. Shlyapintokh, *Vysokomol. Soedineniya, Ser. A,* 1978, **20**, 465.

The rates of photo-oxidation of the rubbers were found to be a linear function of the irradiation dose. N.m.r. spectroscopy has shown that during the photosensitized oxidation of poly(buta-1,4-diene) nearly all the shifted double bonds have a *trans*-configuration.[199] The photodegradation of poly(cyclohexa-1,3-diene) and poly(cyclopentadiene) has been investigated.[200]

Epoxy-resins. The photo-oxidation of epoxy-resin systems has been investigated by a number of workers. Epoxy-novolac resins of general structure (7) have been found to be significantly more unstable to light than epoxy-bisphenol A-resins of general structure (8).[201] The rate of the epoxy-novolac resins was found to be dependent on the conditions of cure. Photoinitiation was shown to occur

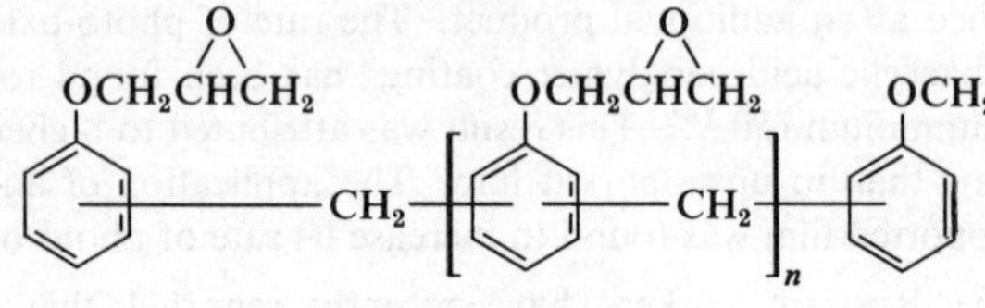

(7)

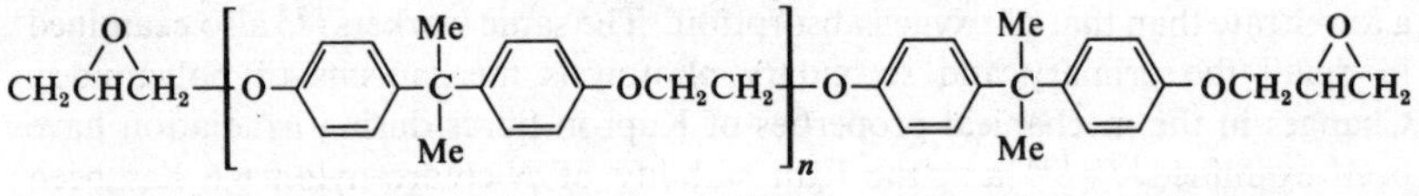

(8)

through aromatic carbonyl groups formed by oxidation of the methylene linkages of the novolac resin. Vacuum curing was found to improve the light stability of the resin. Other workers have obtained similar results.[202] The photodegradation of epoxy-resins has been examined by quantitative spectroscopic techniques.[203] The mechanical changes that occur during the photodegradation of epoxy-resins have been investigated.[204]

Natural Fibres. Bendak[205] has reviewed the mechanisms of the photochemical degradation of wool. Holt and Milligan[206] have shown that carbonyl groups may be indirectly involved in the photoyellowing of wool. The same workers[207] have produced further evidence to show that tryptophan is involved in the process of photoyellowing. As an example tryptophan-depleted wool was found to photoyellow much less than normal wool. The kinetics of decay of free radical

[199] M. A. Golub, R. V. Gemmer, and M. L. Rosenberg, *Amer. Chem. Soc.*, *Div. Polymer Chem.*, *Polymer Preprints*, 1977, **18**, 357.
[200] S. F. Naumkov, D. D. Yurina, A. I. Fleisher, B. G. Gerasimov, and B. V. Erofeev, *Vesti Akad. Nauk S.S.S.R.*, *Ser. Khim. Nauk*, 1977, **4**, 5.
[201] G. A. George, R. E. Sacher, and J. F. Sprouse, *J. Appl. Polymer Sci.*, 1977, **21**, 2241.
[202] D. Roylance and M. Roylance, *Polymer Eng. Sci.*, 1978, **18**, 249.
[203] A. M. Noskov and N. I. Novikov, *Zhur. priklad. Spektroskopii*, 1977, **26**, 1063.
[204] L. D. Kandpal, A. K. Tewari, A. K. Saxena, and K. J. Balakrishna, *Indian J. Technol.*, 1976, **14**, 500.
[205] A. Bendak, *Amer. Dyestuff. Rep.*, 1976, 37.
[206] L. A. Holt and B. Milligan, *Textile Res. J.*, 1977, **47**, 620.
[207] L. A. Holt, B. Milligan, and W. E. Savige, *J. Textile Inst.*, 1977, **68**, 124.

species formed during irradiation of wool keratin has been studied.[208] The nature of the radicals, however, was not considered. Treatment of wool with perchloroethylene has been found to increase the light stability of the fibre owing to the presence of traces of anti-oxidants present in the solvent.[209] Changes in the mechanical properties of wool fibre during photodegradation have also received attention.[210] A copolymer of β-benzyl-L-aspartate and β-p-phenylazobenz1-L-aspartate has been shown to undergo a *cis–trans* isomerization during irradiation.[211]

The photodegradation of cellulose with light of wavelengths <300 nm has been studied.[212] Chain scission was found to obey first-order kinetics and the presence of water was found to retard the photodegradation. The photodegradation of flame-, water-, weather-, and mildew-resistant cotton fabric has been examined.[213]

Miscellaneous Polymers. Poly(dimethacrylimide) has been shown to undergo a photochemical elimination reaction to give isocyanic acid by the mechanism shown in Scheme 15. The rate of photodegradation of atactic and syndiotactic poly(t-butylvinylketone) has been found to be greater than that of the isotactic polymer.[215] The difference was attributed to steric effects on the Norrish Type II photodegradation mechanism. This process was found to be more favourable in the case of the atactic and syndiotactic polymers, for example Scheme 16. Nanosecond and picosecond laser flash photolysis studies have provided valuable information on transient intermediates formed during the photodegradation of poly(methyl-[216] and phenyl-[217] vinyl ketone) polymers. Other studies of interest on polymer photodegradation include polyacrylophenones,[218, 219] poly(vinyl-2-pyridine),[220] N-vinylcarbazole–t-butyl vinyl ketone copolymers,[221] isopropyl vinyl ketone–styrene–maleic anhydride terpolymers,[222] poly-(N-vinylcarbazole),[223] polyurethanes,[224, 225] polyvinyl alcohol,[226, 227] and

208 A. Shatkay and I. L. Weatherall, *J. Polymer Sci. Polymer Chem.*, 1977, **15**, 1735.
209 J. A. Rippon, M. A. Rushforth, and D. J. Westmoreland, *J. Textile Inst.*, 1977, **68**, 158.
210 I. S. Polikarpov, *Izvest V. U. Z. Tekhnol. Legk. Prom.-sti.*, 1977, **2**, 166.
211 A. Ueno, J. Anzai, T. Osa, and Y. Kadoma, *J. Polymer Sci. Polymer Letters*, 1977, **15**, 407.
212 T. Yurugi and H. Iwata, *Nippon Kagaku Kaishi*, 1977, **4**, 544.
213 B. M. Reagan and C. M. Chiasson, *Textile Res. J.*, 1977, **47**, 637.
214 H. Hiraoka, *Macromolecules*, 1977, **10**, 719.
215 H. Tanaka and T. Otsu, *J. Polymer Sci. Polymer Chem.*, 1977, **15**, 2613.
216 D. Lindenau, S. W. Beavan, G. Beck, and W. Schnabel, *European Polymer J.*, 1977, **13**, 819.
217 J. Fauré, *Pure Appl. Chem.*, 1977, **49**, 487.
218 I. Luka, J. Pilka, M. Kulicova, and P. Hrdlovic, *J. Polymer Sci. Polymer Chem.*, 1977, **15**, 1645.
219 R. Salvin and J. Meybeck, *Macromol. Chem.*, 1977, **178**, 2887.
220 J. P. Quaegebeur, A. L. Combier, and C. Chachaty, *Makromol. Chem.*, 1977, **178**, 1507.
221 J. L. Acosta, R. Sastre, and J. Latasa, *Rev. Plast. Mod.*, 1977, **33**, 351.
222 J. L. Acosta, P. M. Garcia, R. M. Garcia, R. Sastre, and Y. J. Fontan, *Anal. de Quim.*, 1977, **73**, 325.
223 A. N. Faidysh, V. V. Slobadyanik, V. N. Yashchuk, and L. N. Fedorova, *Dopovidi Akad. Nauk Ukrain R.S.R.*, Ser. B, *Geol. Khim. Biol. Nauki*, 1977, **8**, 714.
224 W. N. Huang and J. J. Aklonis, *J. Macromol. Sci. Phys. Edn.*, 1977, **B13**, 291.
225 T. Yamagata, T. Ishii, Y. Takanaka, and T. Handa, *Nippon Kagaku Kaishi*, 1977, **10**, 1496.
226 V. S. Che vonenko and S. Ya. Pshezhitskii, *Vysokomol. Soedineniya*, Ser. A, 1977, **19**, 783.
227 H. Kawakami, N. Mori, S. Mori, O. Matsuura, M. Azuchi, M. Ikezawa, T. Sacki, and M. Tsuchiya, *Seni. Gakkaishi*, 1977, **33**, T515.

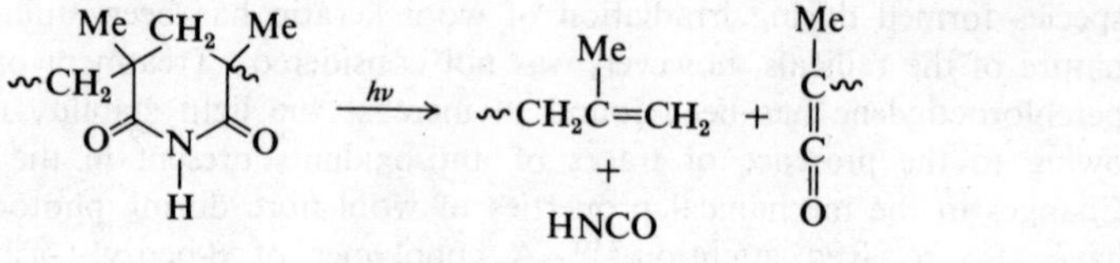

Scheme 15

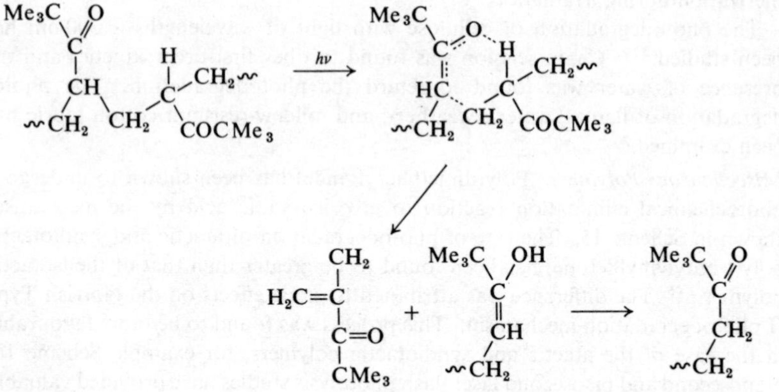

Scheme 16

polyethers.[228] Poly(phenyl acrylate) has been shown to undergo a Photo–Fries rearrangement,[229] and polymers containing cinnamylidene acetate have been shown to undergo a photoreversible cyclization reaction.[230] The photodegradation of polymers has been studied using far-u.v. radiation.[231] A mass spectrometric technique has been developed for studying the gaseous products from the photolysis of polymers.[232] Finally, studies of a more technological interest include a comparison of the natural and artificial weathering of polymers[233] and an examination of the mechanical changes that occur during weathering.[234]

Photodegradable Polymers.—The controlled photochemical degradation of polymers has been reviewed by Nenkov and Kabaivonov[235] and Weir.[236] The use

[228] I. K. Chakhov, V. I. Dindoin, and I. S. Filatov, *Konstruktsion. Polimery pri Nizk. Temperaturakh.*, 1976, 16.

[229] S. K. L. Li and J. E. Guillet, *Macromolecules*, 1977, **10**, 840.

[230] H. Tanaka and K. Honda, *J. Polymer Sci. Polymer Chem.*, 1977, **15**, 2685.

[231] B. I. Khruschch, Yu. A. Ershov, A. N. Lyulichev, and V. F. Udovenko, *Khim. Vysokikh. Energii*, 1977, **11**, 332.

[232] I. G. Gorodetskii, V. E. Skurat, V. L. Talroze, and S. A. Yakovlev *Tezisy Doklady Vses. Konf. Spektrosk. Vak. Ul'trafioleta Vzaimodeistoga. Izluch. Veshchest, 4th,* 1975, 20.

[233] K. Berger and E. Mancioli, *Conv. Ital. Sci. Macromol. (Atti)*, 3rd, 1977, **48**.

[234] M. P. Vershinina, N. G. Kvachadze, and E. E. Tomashevskii, *Fiz. Tverd. Tela (Leningrad)*, 1977, **19**, 2350.

[235] G. Nenkov and V. Kabaivanov, *Plaste Kautschuk*, 1977, **24**, 601.

[236] N. A. Weir, *Review Const.*, 1975, **1**, 21.

of photodegradable polyethylene film for agricultural use has been discussed by Gilead.[237]

Photosensitive Polymers. Guillet and co-workers[238] have described some aspects of photolysis of polyethylene containing in-polymerized ketone groups. The quantum yields of Norrish Type I and II reactions were found to be strongly dependent on the structure of the ketone carbonyl group. The synthesis of some novel photodegradable poly(o-nitrobenzaldehyde acetals) has been described.[239] These polymers have good thermal stability but very poor light stability. They are believed to photodegrade by the mechanism of Scheme 17.

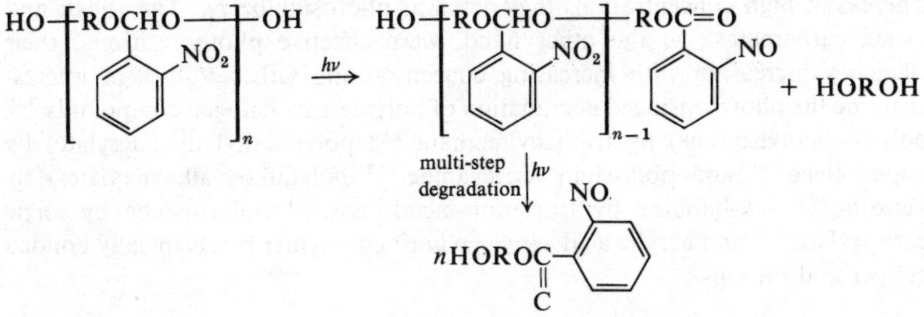

$$
\text{HO}-\left[\text{ROCHO}\underset{\substack{\text{NO}_2}}{}\right]_n-\text{OH} \xrightarrow{h\nu} \text{HO}-\left[\text{ROCHO}\underset{\substack{\text{NO}_2}}{}\right]_{n-1}-\text{ROC}=\text{O}\;\underset{\text{NO}}{} + \text{HOROH}
$$

$$
\xrightarrow[\text{degradation}]{\text{multi-step}\;\Big\downarrow h\nu}\;\; \text{NO}
$$

$$
n\,\text{HOROC}\underset{\substack{\|\\ \text{C}}}{}
$$

Scheme 17

Photodegradable poly(oxime esters) and poly(oxime urethanes) have been prepared[240] and the rate of photodegradation of poly-(2,6-dimethyl-1,4-phenylene oxide) has been found to increase by incorporating 30% of polystyrene into the polymer.[241]

Photoactive Additives. The photosensitized degradation of polymer systems by benzophenone has received much attention. Asquith and co-workers[242] have developed a photodegradable polypropylene film with a long indoor shelf-life by coating the polymer with a film of poly(vinyl acetate) containing benzophenone. The mechanism by which the benzophenone sensitizes the photodegradation of the polypropylene is as yet uncertain. On the other hand, laser flash photolysis indicates that hydrogen atom abstraction is the major primary photochemical process in the benzophenone-sensitized photodegradation of poly-(α-methyl-styrene).[243] Similar conclusions were reached in other studies on the benzo-phenone-sensitized photodegradation of polystyrene[244] and butadiene–isoprene copolymer.[245] The benzophenone-sensitized photodegradation of poly(methyl

[237] D. Gilead, *Soc. Plast. Eng. Technol. Paper*, 1977, **23**, 51.
[238] F. Sitek, J. E. Guillet, and M. Heskins, *J. Polymer Sci., Part C*, 1976, **57**, 343.
[239] C. C. Petropoulos, *J. Polymer Sci. Polymer Chem.*, 1977, **15**, 1637.
[240] S. Hong, *Sonl. Tackakhyo Nonmunijup, Uiyakhe*, 1975, **25**, 119.
[241] Z. Slama and J. Majer, *Plaste Kautschuk*, 1977, **24**, 424.
[242] R. S. Asquith, K. L. Gardner, T. G. Gechan, and G. M. McNally, *J. Polymer Sci. Polymer Letters*, 1977, **15**, 435.
[243] H. Yamaoka, T. Ikeda, and S. Okamura, *Macromolecules*, 1977, **10**, 717.
[244] C. David, W. Piret, M. Sakaguchi, and G. Geuskens, *Makromol. Chem.*, 1978, **179**, 181.
[245] E. S. Turkina, M. A. Rybalchenko, and L. P. Gilyazetdinov, *Lakokras Mater. Ikh. Primen.*, 1977, **3**, 17.

methacrylate) has also been studied.[246] An interesting report has appeared on the photosensitized degradation of polyethylene by ferric chloride.[247] During irradiation, the presence of the ferric chloride was found to suppress alkyl radical formation while enhancing the formation of double bonds in the polymer. These processes were explained by the fact that the ferric chloride increases the rate of photodegradation of the polymer by the Norrish Type II process. This reduces radical formation while at the same time enhancing the formation of double bonds. The effect of transition-metal NN'-diethyldiselenocarbamates on the photo-oxidation of polyethylene has been examined.[248] At low concentrations, zinc and chromium carbamates behaved as photosensitizers, whereas at high concentrations they acted as photostabilizers. The cobalt and nickel carbamates, on the other hand, were effective photosensitizers, their efficiency increasing with increasing concentration. Other studies of interest examine the photosensitized degradation of polymers by halogen compounds,[249] poly-(α-methylstyrene) by triphenylmethane,[250] poly(methyl methacrylate) by naphthalene[251] and potassium ferricyanide,[252] poly(alkoxyalkylacrylates) by benzoin,[253] silk-habutae by transition-metal ions,[254] polyethylene by ferric carboxylate,[255] and acrylic acid–ethylene graft copolymer by chemically bonded copper and tin ions.[256]

5 Photostabilization Processes in Polymeric Materials

Allen and McKellar[161] have recently reviewed the mechanisms of photostabilization of commercial polyolefins. It is concluded that commercial light stabilizers may be multifunctional in their modes of operation. These would depend on the type of photoinitiation mechanism (or mechanisms) that is (or are) operative and this (or these) in turn would depend on the manufacturing and processing history of the polymer. Another review article discusses the importance of stabilizer mobility in the photostabilization of polymers.[257] Other reviews have covered recent developments in the photostabilization of polyolefins[258–260] and poly(vinyl chloride).[261] Suitable u.v. stabilizers have been recommended for the protection of various plastics.[262]

[246] O. N. Karpukhin and A. V. Kutsenova, *Vysokomol. Soedineniya, Ser. A*, 1977, **19**, 1047.
[247] Y. Ogiwara, Y. Kimura, Z. Osawa, and H. Kubota, *J. Polymer Sci. Polymer Chem.* 1977, **15**, 1667.
[248] Z. Osawa and K. Nakano, *J. Polymer Sci., Part C*, 1976, **57**, 267.
[249] J. J. Windle and B. Freedman, *J. Appl. Polymer Sci.*, 1977, **21**, 2225.
[250] T. Ikeda, H. Shimizu, H. Yamaoka, and S. Okamura, *Macromolecules*, 1977, **10**, 1417.
[251] M. Ya. Mel'nikov and N. V. Fok, *Khim. vysok. Energii*, 1977, **11**, 277.
[252] E. Ya. Davydov, G. B. Pariiskii, and D. Ya. Toptygin, *Vysokomol. Soedineniya, Ser. A*, 1977, **19**, 977.
[253] Y. Nakashima, *Nagoya-Shi Kogyo Kenkyusho Kenkyu Hokoku*, 1976, **55**, 32.
[254] F. Shimizu, K. Jyoko, and I. Sakaguchi, *Nippon Sanshigaku Zasshi*, 1976, **45**, 314.
[255] G. Le Brasseur, *Off. Plast. Caoutch*, 1977, **24**, 779.
[256] S. S. Demchenko, A. A. Kachan, and L. L. Chevoyatsova, *Sint. Fiz.-Khim., Polim.*, 1977, **20**, 48.
[257] W. L. Hawkins, *Polymer News*, 1977, **4**, 81.
[258] W. L. Hawkins, *J. Polymer Sci., Part C*, 1976, **57**, 319.
[259] M. Fedlke, *Plaste Kautschuk*, 1977, **24**, 393.
[260] J. Masek, P. Cancik, J. Holcik, M. Karvas, and J. Durmis, *Plaste Kautschuk*, 1977, **24**, 405.
[261] H. O. Wirth and H. Andreas, *Pure Appl. Chem.*, 1977, **49**, 627.
[262] R. W. Kuchkuda, *Plast. Eng.*, 1977, **33**, 32.

The most outstanding u.v. stabilizer, carbon black, has been shown to operate primarily as a screen in polyethylene.[263] Its photostabilizing efficiency was found to be proportional to its specific surface. The photostability of the o-hydroxybenzophenone stabilizers has been attributed to ready deactivation of both their singlet and triplet states.[264] The new hindered amine stabilizers have been investigated by many workers.[265—271] A hindered piperidine derivative (9) (Tinuvin 770) has attracted much interest since it has been shown to be

$$\text{HN} \overset{\text{Me Me}}{\underset{\text{Me Me}}{\bigcirc}} \text{O}-\overset{\text{O}}{\overset{\|}{\text{C}}}-(CH_2)_8-\overset{\text{O}}{\overset{\|}{\text{C}}}-\text{O} \overset{\text{Me Me}}{\underset{\text{Me Me}}{\bigcirc}} \text{NH}$$

(9)

superior to many other conventional types.[265, 266, 268] Allen and McKellar[270] found that it inhibited the photolysis of the α,β-unsaturated carbonyl impurity groups in commercial polypropylene. This may be one mechanism by which the stabilizer operates in the polyolefins. Chakraborty and Scott[271] found that the light stabilizing efficiency of Tinuvin 770 in polyethylene increases with increasing processing time. This effect has been attributed to the ability of the stabilizer to destroy hydroperoxides formed during the processing operation. This process is believed to involve the redox reaction between the hindered amine and hydroperoxides, (Scheme 18) The same workers[272] have also recently observed synergism and antagonism between certain u.v. stabilizers and anti-oxidants in polyethylene. For example, the nickel and zinc dialkyl-dithiocarbamates were found to form synergistic combinations with a 2-hydroxybenzophenones stabilizer. In this case the dithiocarbamates are believed to protect the 2-hydroxybenzophenone by decomposing the hydroperoxides formed during processing and/or photo-oxidation. Prolonged processing was found to reduce the light stabilizing action of the 2-hydroxybenzophenone due to destruction of the stabilizer by hydroperoxides. On the other hand, the metal dithiocarbamates were found to be antagonistic in the presence of a phenolic anti-oxidant. Here, the phenolic anti-oxidant is believed to photosensitize the decomposition of the dithiocarbamates. Interestingly, copolymerization of 2-hydroxybenzophenone stabilizers with acrylonitrile–butadiene–styrene ter-

[263] E. Kovacs and Z. Wolkober, *J. Polymer Sci.*, Part C, 1976, **57**, 171.
[264] W. Klopffer, *J. Polymer Sci.*, Part C, 1976, **57**, 205; R. N. Nurmukhametov, O. I. Betin, and D. N. Shigorin, *Doklady Akad. Nauk S.S.S.R.*, 1977, **234**, 1128.
[265] J. J. Usilton and A. R. Patel, *Amer. Chem. Soc., Div. Polymer Chem., Polymer Preprints*, 1977, **18**, 393.
[266] V. La Rosa, *Pliplasti Plast. Pinf.*, 1977, **25**, 40.
[267] L. G. M. Ogandzhanyan, R. A. Petrosyan, N. M. Beileryan, K. A. Ordukhanyan, and R. V. Bagdasaryan, *Armyan khim. Zhur.*, 1977, **30**, 452.
[268] D. Kh. Khalikov, I. G. Shanyavskii, I. Ya. Kalontarov, and G. S. Sanyukovich, *Vysokomol. Soedineniya, Ser. A*, 1977, **19**, 1132.
[269] V. B. Ivanov, S. G. Burkova, Yu. L. Morozov, and V. Ya. Shlyapintokh, *Vysokomol. Soedineniya, Ser. B*, 1977, **19**, 359.
[270] N. S. Allen and J. F. McKellar, *Chem. and Ind.*, 1977, 537.
[271] K. B. Chakraborty and G. Scott, *Chem. and Ind.*, 1978, 237.
[272] K. B. Chakraborty and G. Scott, *European Polymer J.*, 1977, **13**, 1007.

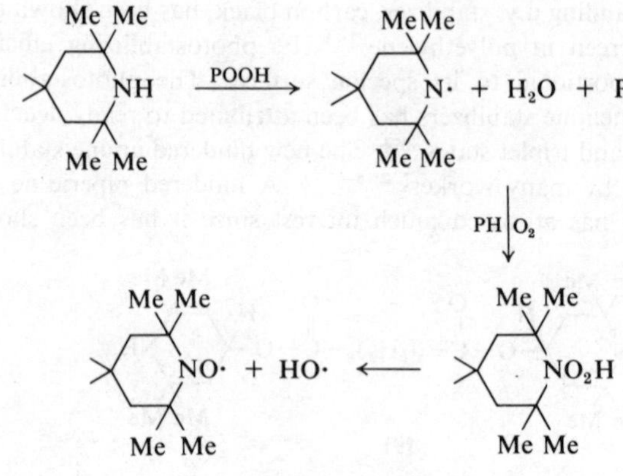

Scheme 18

polymer has been found to improve markedly the photostability of the polymer compared with that of the monomeric stabilizer.[273]

The photostabilization of poly(vinyl chloride) has been extensively studied.[274—276] In particular Rabek et al.[276] have examined the role of tin thermo-stabilizers in the photo-oxidation of PVC. It was found that the presence of the tin stabilizer inhibited colour formation but accelerated the cross-linking process. The reaction in Scheme 19 was proposed to account for the reduction

Scheme 19

in colour formation. Cross-linking may occur through recombination of the macroradical species (I). The photostabilizing action of phenyl allyl mercury in butyl rubber has been attributed to its ability to terminate propagating free radical species formed from the scission of polymeric hydroperoxides.[277] The photo-oxidation of phenolic anti-oxidants continues to be an active field of study.[278—281] Of particular interest are the studies by Pospisil and co-workers,[278, 279] which show that singlet oxygen is a reactive intermediate in the photosensitized oxidation of phenolic anti-oxidants.

[273] K. M. Oo and M. Tahan, *European Polymer J.* 1977, **13**, 915.
[274] Y. Oki, F. Mori, and M. Koyama, *Kinzoku Hyoman Gijutsu*, 1977, **28**, 177.
[275] M. Milke, *Plastverarbirtev*, 1975, **26**, 73; G. Menzel, *ibid.*, 1975, **26**, 259.
[276] J. F. Rabek, G. Canback, and B. Ranby, *J. Appl. Polymer Sci.*, 1977, **21**, 2211.
[277] R. Chandra and H. L. Bhatnagar, *J. Ind. Chem. Soc.*, 1976, **33**, 1108.
[278] L. Taimr and J. Pospisil, *J. Polymer Sci., Part C*, 1976, **57**, 213; J. Lerchova, L. Kotulak, J. Ratschova, J. Pilar, and J. Pospisil, *ibid.*, 229.
[279] L. V. Samsonova, L. Taimr, and J. Pospisil, *Angew. Makromol. Chem.*, 1977, **65**, 197.

Allen and McKellar[129] have shown that for nickel oxime chelates to be both effective photostabilizers in polypropylene and triplet quenchers the nickel and the oxime ligand must react together to form the chelate. On the other hand, other workers have shown that oximes inhibit the photolysis of styrene–phenyl vinyl ketone copolymers by quenching the triplet state of the ketone unit.[282] The photostabilization of polystyrene by the 2-hydroxybenzophenones and 2-(2'-hydroxyphenyl)benzotriazoles has been shown to be due mainly to screening and excited state quenching.[283]

Other studies of more technological interest include the photostabilization of cellulose acetate,[284] cotton fabric,[285] polychloroprene,[286] copolyester ether elastomers,[287] poly(methyl methacrylate),[288] poly(formaldehyde),[289] poly-(vinyl alcohol),[290] high density polyethylene,[291] polypropylene,[292] poly-caproamides,[293] and flame-retardant polystyrene.[294] A quantitative method for the determination of the stabilizing efficiency of u.v. stabilizers has been developed using e.s.r. spectroscopy.[295]

6 Photochemistry of Dyes and Pigments

The lightfastness and photochemical properties of commercial dyes on polymer substrates continue to be an active field of study. A new theoretical model has been proposed for the light fading dyes.[296] Expressions were derived which show that the rate of fading varies with $1/a^2$ (where a = radius of particle) for large dye particles. For smaller dye particles, the rate tends towards $1/a$ and for very small particles is independent of particle size. Singlet oxygen is believed to be important in the photofading of azo-dyes.[297] For example, 4-aryl-azo-1-naphthols have been found to undergo photoreaction to give 1,4-naphthoquinone

[280] J. Kowal and B. Waligora, *Makromol. Chem.*, 1978, **179**, 707, 713.
[281] O. N. Karpukhin and E. M. Slobodetskaya, *Vysokomol. Soedineniya, Ser B*, 1977, **19**, 48.
[282] M.Tsunooka, T. Nishino, and M. Tanaka, *Chem. Letters (Japan)*, 1977, 1107.
[283] M. N. Volkrub, T. A. Rubtsova, and A. F. Lukovnikov, *Vysokomol. Soedineniya, Ser B*, 1977, **19**, 762.
[284] R. Khalmirzaeva, E. I. Berenshtein, and B. I. Aikhodzhaev, *Atsetatn Valokna*, 1977, **7**, 79; Z. A. Rumyantseva, V. S. Polyanskaya, A. I. Galchenko, O. A. In, V. I. Sobolev, K. M. Makhkamov, and I. Yu. Kalontarov, *Doklady Akad. Nauk Tadzh., S.S.S.R.*, 1977, **20**, 31; I. Yu. Kalontarov, K. M. Makhkamov, O. A. In, and R. F. Smirnov, *ibid.*, p. 34.
[285] H. Watamoto and H. Tonami, *Nippon Kagaku Kaishi*, 1977, **6**, 876.
[286] L. G. Melik-Ogandzhanyan, R. A. Petrosyan, N. M. Beileryan, K. A. Ordukhanyan, and R. V. Bagdesuryan, *Armyan khim. Zhur.*, 1977, **30**, 458.
[287] R. P. Kane, *J. Elastomers Plast.*, 1977, **9**, 416.
[288] A. A. Khim, A. S. Bark, and M. A. Askarov, *Vysokomol. Soedineniya, Ser B*, 1977, **19**, 255.
[289] V. I. Nepochatykh, S. K. Bokova, A. V. Yudin, and V. G. Oleinik, *Khim. Volakna*, 1977, **3**, 28.
[290] B. N. Narzullaev, E. Baimuratov, A. V. Zakhavehuk, and D. S. Saidov, *Izvest. V. U. Z. Tekhnol. Tekst. Prom-sti*, 1977, **4**, 66.
[291] A. A. Ozhafarov, R. G. Namelov, E. S. Namelova, M. F. Ganieva, and N. E. Parsegova, *Azerb. khim. Zhur.*, 1976, **4**, 71.
[292] R. I. Mamedov, G. D. Aliev, Sh. M. Novruzov, and S. I. Sadykh-Zade, *Plaste. Massy*, 1977, **6**, 18.
[293] V. V. Bykov and G. N. Smirnov, *Izvest. V. U. Z. Khim. i khim. Tekhnol.*, 1977, **20**, 724.
[294] R. Bradley, J. Farber, and L. Testa, 'Proceedings of the International Symposium on Flammability and Fire Retardants, 1977, 14.
[295] T. Taplik, K. Kuehl, and M. Fedtke, *Plaste Kautschuk*, 1978, **25**, 23.
[296] C. H. Giles, D. J. Walsh, and R. S. Sinclair, *J. Soc. Dyers and Colourists*, 1977, **93**, 349.
[297] J. Griffiths and C. Hawkins, *J.C.S., Perkin II*, 1977, 747.

(Scheme 20). The catalytic fading of yellow azo-dyes in the presence of blue, violet, or red anthraquinonoid dyes is also believed to involve singlet oxygen.[298] The photofading of rhodamine dyes has been found to be dependent on the nature of the substrate.[299] In another study, the fading of an acid azo-dye was found to be accelerated in the presence of tryptophan, tyrosine, and cysteine.[300]

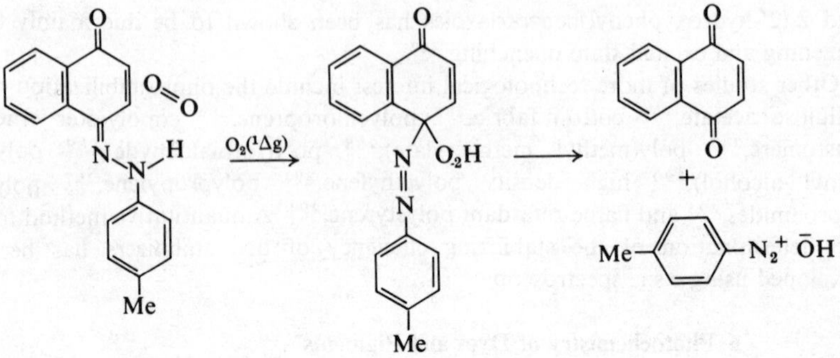

Scheme 20

Energy transfer from the amino-acid to the dye is believed to be responsible for the accelerated fading. Further evidence has appeared to show that the photofading of azo-dyes on cellulose and poly(ethylene terephthalate) occurs by an oxidation-type reaction.[301] The quantum efficiencies of the photofading of some disperse dyes in textile substrates have been determined.[302]

Electron paramagnetic resonance spectroscopy has been used to measure the photofading kinetics of several types of reactive dyes on cellulose.[303] The accumulation of paramagnetic centres was found to be proportional to the degree of photofading. The photofading of malachite green and crystal violet has been studied using thin-layer chromatography.[304] Other studies of interest cover the photofading of methylene blue in PVC[305] and fibre-reactive dyes in polymers.[306] The photofading of fluorescent dyes[307] and brightening agents[307-309] has also received some attention. In one study[309] fluorescent brightening agents of the pyrazoline type were found to photosensitize the degradation of the heterocyclic residues, such as tryptophan, in wool: note that

[298] M. W. Rembold and H. E. A. Kramer, *J. Soc. Dyers and Colourists*, 1978, **94**, 12.
[299] N. A. Evans and P. J. Waters, *Proc.-Internat. Wolltextil-Forschungskonf.*, 5th, 1975, **3**, 567.
[300] J. Barnes, A. Fritz, and C. H. Nicholls, *Proc.-Internat. Wolltextil-Forschungskonf.*, 5th, 1975, **3**, 557.
[301] C. D. Shah and R. Srinivasan, *J. Appl. Chem. and Biotechnol.*, 1977, **27**, 429.
[302] C. D. Shah and R. Srivivasan, *Textile Res. J.*, 1977, **47**, 625.
[303] Ya. O. Grinberg, A. A. Gombkoto, and G. E. Krichevskii, *Kolor Ert.*, 1977, **19**, 14.
[304] R. Bangert, W. Aichele, E. Schollmeyer, B. Weimann, and H. Herlinger, *Melliand Textilber.* 1977, **58**, 399.
[305] A. D. Grishina, *Vysokomol. Soedineniya, Ser B*, 1977, **19**, 411.
[306] J. Gombkoto, G. E. Krichevskii, and A. Vig, *Zhur. priklad. Spektroskopii*, 1977, **26**, 712.
[307] J. H. Anderson, G. Bruce Guise, and G. Caird Ramsay, *J. Soc. Dyers and Colourists*, 1977, **93**, 275.
[308] Y. Yamashita and K. Yoshida, *Yuki Gosei Kagaku Kyokaishi*, 1977, **35**, 285.
[309] N. A. Evans, L. A. Holt, and B. Milligan, *Austral. J. Chem.*, 1977, **30**, 2277.

the tryptophan residues are believed to be responsible for the tendency of wool to become yellow in light.

In recent studies, Allen *et al.*[310,311] have presented evidence to show that electron abstraction may be important in the phototendering of polar substrates such as nylon 6,6. The fluorescence from a well-known phototendering dye, 2-piperidinoanthraquinone, was efficiently quenched by model amide systems.[311] The lightfastness and phototendering properties of 3-methoxybenzanthrone in poly(ethylene terephthate) (PET) and nylon 6,6 were also examined.[311] This dye was found to have low lightfastness in nylon 6,6 compared with that in PET. It also sensitized the photodegradation of nylon 6,6 but had no effect on PET. An examination of the fluorescence properties of the dye in different solvents indicated that in polar environments such as nylon 6,6 there is an increased probability of dye–substrate interaction. The fluorescence of the dye was found to be quenched by amines indicating that the mechanism of phototendering may involve electron abstraction from the substrate. The effect of reactive dyes,[312] metal phthalocyamines,[313] and metal-containing disperse dyes[314] on the photodegradation of polyamides has also been studied.

Lemaire and co-workers[315] have studied the titanium dioxide- and zinc oxide-sensitized photodegradation of polyethylene as a function of wavelength and temperature. An activation energy of about 18.0 kcal mol^{-1} was obtained for the sensitized photodegradation of the polymer by both types of pigment. Maximum degradation occurred with wavelengths >300 nm. The addition of a phenolic anti-oxidant was found to reduce significantly the photosensitizing action of the pigments. Allen *et al.*[316] have found a correlation between the intensity of the i.r. emission at 815 nm from rutile pigments and their photoactivity in polyethylene. Rutile pigments with surface coatings of Al_2O_3–SiO_2 and silicone were found to exhibit less emission at 815 nm than uncoated rutile and were correspondingly less photoactive. The coated rutile pigments were also found to exhibit a new emission band at 1015 nm. Mlinac and co-workers[317] have examined the photodegradation of polyethylene-containing coloured pigments. The protective action of the pigments in the polymer was found to increase in the order cadmium yellow < ultramarine blue < copper phthalocyanine < chrome orange. The cadmium yellow sensitized the photodegradation of the polymer, whereas all the other pigments acted as photostabilizers. The effect of titanium dioxide on the weathering of paint fibres has been studied.[318] A useful study for the technologist has appeared on the photofading of various

310   N. S. Allen, D. Wilson, and J. F. McKellar, *Makromol. Chem.*, 1978, **179**, 269.
311   N. S. Allen, J. F. McKellar, and S. Protopapas, *J. Appl. Chem. and Biotechnol.*, 1977, **27**, 269.
312   G. D. Korodenko, G. F. Pugachevskii, B. D. Semak, and I. I. Shiiko, *Fiz.-khim. Mekh. Materialov*, 1977, **13**, 125.
313   V. V. Kalacheva, A. N. Bykov, and G. N. Smirnova, *Izvest. V. U. Z. Khim. i khim. Tekhnol.* 1977, **20**, 724.
314   L. Jonikaite, A. Kaziliunas, and A. Paulauskas, *Chem. Chem. Technol. Tech. Makslu. Isvystymo. Resp. Ju. Result. Panandojimo Korf. Medziaga*, 1975, 109.
315   P. Laurenson, R. Arnand, J. Lemaire, J. Quemner, and G. Roche, *European Polymer J.*, 1978, **14**, 129.
316   N. S. Allen, D. J. Bullen, and J. F. McKellar, *Chem. and Ind.*, 1977, 797.
317   M. Mlinac, J. Rolich, and M Bravar, *J. Polymer Sci.*, Part C, 1977, **57**, 161.
318   R. Tsuji, *Toyota Chico Kenkyusho R & D, Rebyu*, 1977, **13**, 22.

yellow pigments in latex paints.[319] The photofading of methylene blue pigments in polymeric coatings has been investigated[320] and electron microscopy has been used to study the weathering of pigmented polystyrene.[321] Finally, the u.v. screening properties of various pigments in polystyrene have been determined.[322]

7 Appendix: Review of the Patent Literature

Photopolymerizable Systems.—Patents of interest concerning photopolymerizable and photocurable systems are listed below for the U.K., U.S.A., Germany, and Japan.

United Kingdom

1 461 255	1 474 715	1 480 425

United States

4 022 674	4 048 034	4 073 777	4 045 427	4 054 498	4 037 021	4 024 297
4 020 034	4 049 457	4 017 652	4 040 922	4 065 587	4 043 887	4 008 138
4 024 296	4 040 923	4 028 204	4 047 963	4 026 939	4 045 426	4 052 280
4 042 476	4 054 721	4 035 321	4 054 719	4 029 561	4 042 386	4 069 123
4 040 923	4 065 267					

West Germany

2 545 290	2 731 396	2 702 385	2 643 701	2 625 538	2 602 419	2 712 354
2 639 742	2 705 307	2 646 416	2 704 927	2 706 575	2 705 612	2 559 718
2 643 702	2 706 639	2 621 095	2 702 660	2 643 701	2 616 408	2 733 037
2 613 098	2 558 482	2 600 318	2 706 549	2 724 087	2 728 780	2 065 976
2 649 836	2 733 038	2 704 368	2 632 010	2 656 998	2 727 109	2 352 524
2 647 890	2 706 638	4 039 414				

East Germany

121 329

Japan

77 15 529	77 71 585	77 103 201	77 40 328	77 108 480	77 141 622	77 106 864
77 24 296	77 73 990	77 108 479	77 34 714	77 109 536	77 47 888	77 117 386
76 109 992	77 134 693	77 22 099	77 68 267	77 106 901	77 112 103	77 117 939
76 48 794	77 105 936	77 22 100	77 103 487	77 126 440	77 53 011	77 42 192
77 58 790	77 102 393	77 52 987	77 107 087	77 125 182	77 39 742	77 111 935
77 22 373	77 99 105	77 64 590	77 98 742	77 45 597	77 93 448	77 126 474
77 62 392	77 110 737	77 60 882	77 92 246	77 48 578	77 31 909	77 129 790
77 68 283	77 110 103	77 58 782	77 80 237	77 64 589		

[319] A. M. Keay, *J. Coatings Technol.*, 1977, **49**, 31.
[320] A. D. Grishna, *Sposoby Zapisi Inform. na. Besserebryan Nositelyakh*, 1977, **8**, 56.
[321] V. Masarik, B. Dolezel, L. Adamirova, and J. Pitrev, *Plaste Kautschuk*, 1977, **24**, 431.
[322] P. Marvuglio, R. F. Sharrock, and R. J. Kennedy, *J. Oil. Colour Chemists' Assoc.*, 1978, **61**, 79.

Table A1 *Prodegradants and u.v. sensitizers*

Class and general formula	Specifications	Applications	Ref.
Halogen-containing compositions			
N-bromosuccinimide	—	1% by weight used for polyolefins or polystyrene	323
N-halobenzenesulphonamide	—	1% by weight used for polyolefins or polystyrene	324
Haloalkyl disulphides	—	3—5% by weight used for polyolefins	325
Other halogenated compositions of interest	—	—	326—330
Carbonyl-containing compositions			
Butadiene—methyl vinyl ketone copolymer	—	Mixtures of 0.6735 g with 25 g of polystyrene or polyethylene were milled for 10 min at 125 °C	331
Chloromethyl vinyl ketone	—	Copolymerized with PVC to give a photodegradable resin	332

323 U.S. Dept. Agriculture, 12 April 1977, US 4 017 668.
324 U.S. Dept. Agriculture, 12 April 1977, US 4 017 667.
325 U.S. Dept. Agriculture, 12 April 1977, US 4 017 666.
326 U.S. Dept. Agriculture, 8 March 1977, US 4 011 375.
327 U.S. Dept. Agriculture, 15 March 1977, US 4 012 565.
328 U.S. Dept. Agriculture, 25 April 1977, US 4 020 249.
329 Owens Illinois, Inc., 24 May 1977, US 4 025 580.
330 U.S. Dept. Agriculture, 2 Aug. 1977, US 4 039 731.
331 B.P. Chemicals Ltd., 31 Dec. 1976, GB 1 459 782.
332 Mitsubishi Monsanto Chem. Co., 20 April 1977, JA 77 49 286.

Table A1

Class and general formula	Specification	Application	Ref.
Dinaphthyl phthalate	—	0.05% by weight in polypropylene	333
Other carbonyl-containing compositions of interest	—	—	334, 335
Miscellaneous prodegradents	—	—	336—338

Patents references: BE Belgium, BZ Brazil, CA Canada, CZ Czechoslovakia, EG East Germany, FR France, GB Great Britain, JA Japan, SA South Africa, SP Spain, SU Soviet Union, SW Switzerland, US United States, WG West Germany.

333 Exxon Res. and Eng. Co., 19 July 1977, US 4 037 034.
334 Patronato de Investigación Científica y Tecnica 'Juan de la Cierra', Consejo Superior de Investigaciones Científicas, 1 March 1977, SP 438 711.
335 Biodegradable Plasts. Inc., 26 July 1977, US 4 038 227.
336 Exxon Res. and Eng. Co., 16 Aug. 1977, US 4 042 765.
337 Gulf Res. and Dev. Co., 16 Aug. 1977, US 4 042 763.
338 Asahi Chemical Ind. Co. Ltd., 5 Jan. 1977, JA 77 00 054.

Table A2 *Photostabilizers*

Class and general formula	Specification	Application	Ref.
Organophosphorus compounds			
(11)	Reacted with a C_{26-52} monoepoxyalkane and a mixture of 1,2-epoxydodecane and glycidyl ether of bisphenol A	For PVC and polyolefins	339
(12)	R = stearyl	For polypropylene	340
Other organo-phosphorus compounds of interest	—	—	341—348
poly-(*p*-phenylene sulphide)	—	$\geq 1\%$ blended into polyolefins	349

339 Hoechst A.-G., 1 Dec. 1977, WG 2 621 323.
340 Hoechst A.-G., 2 June 1977, WG 2 552 796.
341 Goodrich B.F. & Co., 24 May 1977, US 3 909 491.
342 Ciba-Geigy A.-G., 14 July 1977, WG 2 656 748.
343 Kurashiki Spinning Co. Ltd., 5 Aug. 1977, JA 77 93 463.
344 Du Pont De Nemours E. I. and Co. Ltd., 23 Aug. 1977, US 4 043 974.
345 American Cyanamid Co. Ltd., 20 Dec. 1977, US 4 064 106.
346 Hoechst A.-G., 12 Jan. 1978, WG 2 630 257.
347 Hoechst A.-G., 26 Jan. 1978, WG 2 633 392.
348 Hoechst A.-G., 26 Jan. 1978, WG 2 633 393.
349 Phillips Petroleum Co., 24 May 1977, US 4 025 582.

Table A2 (cont)

Class and general formula	Specificaton	Application	Ref.
Other organo-sulphur compounds of interest	—	—	350—352
Sterically hindered phenols and benzoic esters			
(13)	Where R = alkyl, cycloalkyl, or aralkyl. Mixed with benzotriazole light stabilizers	For polyurethane rubbers	353
(14)	—	For polyesters	354
Other phenols and benzoic esters of interest	—	—	355—362
Organometallic compounds Dimethyl bis(lauryl mercaptide) (I) and Methylin tris(lauryl mercaptide) (II)	For use with 2-hydroxybenzophenones to give synergistic effect	0.20 % of (I) and 0.05 % of (II) for use in polyolefins	363
Other organometallic compounds of interest	—	—	364—366

Benzotriazole compounds			
Benzotriazole–styrene esters 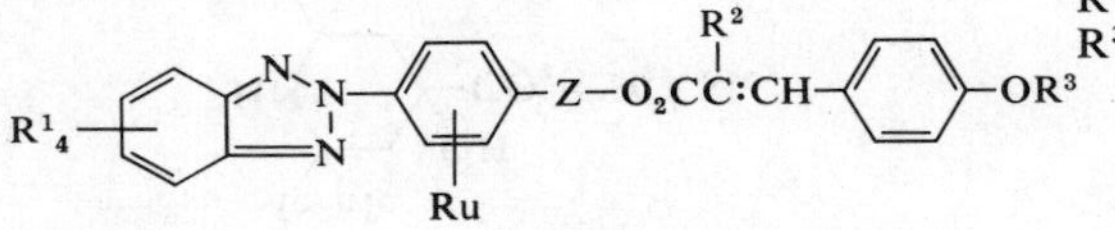 (15)	Where R^1 = H, optional substituent; R^2 = cyano, acyl, or alkanesulphonyl; R^3 = lower alkyl	For use in polymers and organic compositions	367
Benzotriazolylphenol aromatic esters 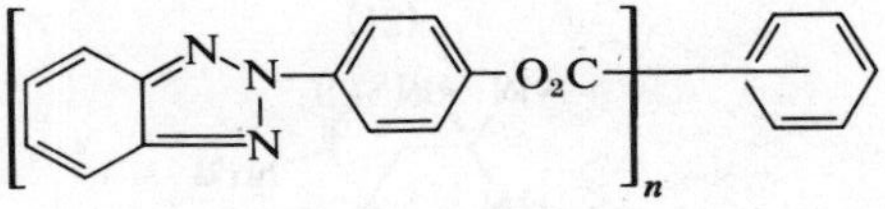 (16)	Where n = 1—3	0.5% by weight for polyesters	368

350 Tenneco Chem. Inc., 24 Nov. 1977, WG 2 719 527.
351 Goodrich B.F. Co. Ltd., 25 Oct. 1977, US 4 055 540.
352 Hitachi Ltd., 10 June 1977, JA 77 69 949.
353 Bayer A.-G., 25 Aug. 1977, WG 2 065 975.
354 Eastman Kodak Co., 26 April 1977, US 4 020 080.
355 Phillips Petroleum Co., 24 May 1977, US 4 025 488
356 Ciba-Geigy Corp., 25 Aug. 1977, US 4 038 250.
357 Ciba-Geigy Corp., 24 May 1977, US 4 025 487.
358 Ciba-Geigy A.-G., 1 Dec. 1977, WG 2 721 398.
359 Eastman Kodak Co. Ltd., 26 April 1977, US 4 020 041.
360 Ciba-Geigy A.-G., 14 July 1977, WG 2 656 769.
361 Sandoz Ltd., 14 June 1977, US 4 029 684.
362 Ciba-Geigy Corp., 26 July 1977, US 4 038 249.
363 Cincinnati Milacron Chem. Inc., 31 March 1977, WG 2 639 086.
364 Uniroyal Inc., 19 April 1977, US 4 018 808.
365 A. Petrik, 15 Oct. 1976, CZ 165 579.
366 Sandoz Ltd., 29 Oct. 1976, SW 581 090.
367 Eastman Kodak Co., 24 Jan. 1978, US 4 070 339.
368 Eastman Kodak Co., 17 May 1977, US 4 024 153.

Table A2 (cont)

Class and general formula	Specification	Application	Ref.
Other benzotriazoles of interest	—	—	369–372
(17)	Where R^1 = H or Me, R^2 = H or Me, R^3 = CN, CO_2Et, or Ac and R^4 = CN, alkoxycarbonyl. Ac, or Ph	For PVC, polytyrene, and polyurethanes	373
Cyclic amines			
(18)	Where R^1 = allyl, R^2 = O or R^1 = $PhOCH_2CH_2$, R^2 = O or R^1 = $BzOCH_2CH_2$, R^2 = O or R^1 = allyl, R^2 = S	For polypropylene 0.25% by weight	374
1,3,8-Triazaspiro-[4,5]-decane derivatives			
(19)	—	For polypropylene	375

Other cyclic amines of interest

369 General Electric Co., 5 May 1977, WG 2 648 367.
370 Eastman Kodak Co., 24 Jan. 1978, US 4 070 337.
371 Eastman Kodak Co., 9 Aug. 1977, US 4 041 011.
372 L. Filacek, I. Holeci, J. Novak, and J. Zaberaysky, 15 Dec. 1976, CZ 166 511.
373 Bayer A.-G., 29 Sept. 1977, WG 2 612 314.
374 Sankyo Co. Ltd., 28 April 1977, WG 2 265 271.
375 Ciba-Geigy A.-G., 1 Dec. 1977, WG 2 719 133.
376 Ciba-Geigy A.-G., 10 Nov. 1977, WG 2 716 100.
377 Ciba-Geigy A.-G., 24 Nov. 1977, WG 2 718 458.
378 American Cyanamid Co., 4 Oct. 1977, US 4 052 361.
379 Mitsubishi Rayon Co. Ltd., 8 Aug. 1977, JA 77 94 351.
380 Ciba-Geigy A.-G., 10 Nov. 1977, WG 2 717 087.
381 Bayer A.-G., 2 April 1977, WG 2 545 647.
382 A. V. Vasilev, T. S. Simonemko, B. I. Golovenko, G. Kh. Richmond, A. G. Liakumovich, V. S. Gumerova, A. F. Rukosov, V. S. Shamaev, and L. P. Steparova, 5 Sept. 1977, SU 450 483.
383 American Cyanamid Co. Ltd., 6 Sept. 1977, US 4 046 736.
384 Ciba-Geigy Corp., 7 June 1977, US 4 028 334.
385 Phillips Petroleum Co., 4 Oct. 1977, US 4 052 351.
386 Chimosa Chimica Organica SpA., 30 June 1977, WG 2 636 130.
387 G. N. Antyina, L. N. Smirnov, G. I. Shilova, M. V. Gorclik, F. M. Egidis, A. S. Rakhmanova, L. A. Skripko, A. B. Shapiro, and E. G. Rosantsev, 25 June 1977, SU 562 561.
388 Bayer A.-G., 18 May 1977, WG 2 551 499.
389 Ciba-Geigy A.-G., 7 July 1977, WG 2 656 766.
390 Ciba-Geigy A.-G., 23 June 1976, WG 2 727 385.
391 Sankyo Co. Ltd. and Ciba-Geigy A.-G., 10 Dec. 1976, JA 76 143 674.
392 Sankyo Co. Ltd. and Ciba-Geigy A.-G., 10 Dec. 1976, JA 76 143 673.
393 Sankyo Co. Ltd., 21 Sept. 1976, CA 997 353.
394 Adeka Argus Chem. Co. Ltd., 19 Feb. 1977, JA 77 22 578.
395 Teijin Ltd., 8 April 1977, JA 77 44 869.
396 Phillips Petroleim Co., 12 July 1977, US 4 035 323.
397 Chimosa Chimica Organica SpA., 26 May 1977, WG 2 636 143.
398 Sankyo Co. Ltd., 12 May 1977, WG 2 651 511.
399 Asahi Chem. Ind. Co. Ltd., 5 Feb. 1977, JA 77 15 549.
400 Ciba-Geigy A.-G., 1 Dec. 1977, WG 2 719 131.
401 American Cyanamid Co., 20 Dec. 1977, US 4 064 102.
402 Goodrich B.F. Co., 29 Dec. 1977, WG 2 726 653.
403 Hoechst A.-G., 18 Aug. 1977, WG 2 606 026.
403 Hoechst A.-G., 18 Aug. 1977, WG 2 606 026.
404 Hoechst A.-G., 1 Sept. 1977, WG 2 606 819.
405 Hoechst A.-G., 28 July 1977, WG 2 602 673.

Table A2 (*cont.*)

Class and general formula	Specification	Application	Ref.
Miscellaneous compounds			
  (20) Bis-[5,5-bis(phenoxymethyl)]-1,3-dioxan-2-yl]	—	For polyethylene	406
  (21) N-(Benzimidazol-2-yl)arylcarboxamides	Where R^1 = aromatic radical with 1—3 nuclei, ring-substituted aromatic radical, or aromatic heterocyclic radical; R^2 = H, Me, aliphatic or aromatic acryl	For plastics, fibres, and sun-tan oil	407
Other miscellaneous photostabilizers	—	—	408—445

406 Ufa-Scientific Res. Inst. of Petroleum, 30 Sept. 1977, SU 574 446.
407 Merck and Co. Inc., 8 March 1977, US 4 011 236.
408 Dynamit Nabel A.-G., 18 Aug. 1977, WG 2 605 325.
409 V. Hart, J. Gallovic, and A. Jusko, 15 Feb. 1977, CZ 167 652.
410 V. Hart, J. Gallovic, and A. Jusko, 15 Feb. 1977, CZ 167 659.
411 American Cyanamid Co., 14 July 1977, BZ 76 06 261.
412 Eastman Kodak Ltd., 23 Aug. 1977, US 4 043 973.
413 Vistron Corp., 17 May 1977, US 4 024 206.
414 Peritech Int. Corp., 9 March 1977, GB 1 464 919.
415 Showa Denko K. K., 30 Nov. 1977, JA 76 138 744.
416 Suva Seikosha Co. Ltd., 28 March 1977, JA 77 39 736.
417 Goodrich B.F. Co., 28 June 1977, US 4 032 505.
418 Deutsche Gold and Silber Scheideanstalt Roesslev, 4 May 1977, US 3 954 702.

419 Eastman Kodak Co., 14 June 1977, US 4 029 670.
420 Furakawa Electric Co. Ltd., 5 Aug. 1977, JA 77 93 462.
421 Ciba-Geigy A.-G., 12 Jan. 1978, WG 2 730 449.
422 Ciba-Geigy A.-G., 12 Jan. 1978, WG 2 730 503.
423 American Cyanamid Co., 3 Jan. 1978, US 4 066 614.
424 Adeka Argus Chem. Co. Ltd., 2 June 1977, JA 77 66 551.
425 American Cyanamid Co., 27 Aug. 1977, SA 75 06 750.
426 Shell International Research Maatschappij B.V., 5 May 1977, WG 2 648 452.
427 Phillips Petroleum Co., 24 May 1977, US 4 025 582.
428 Daiichi Kogyo Seiyaku Co. Ltd., 21 Feb. 1977, JA 77 23 156.
429 Teijin Ltd., 15 March 1977, JA 77 33 945.
430 Adeka Argus Chem. Co. Ltd., 23 Aug. 1977, JA 77 100 543.
431 M. Shultz, H. Wegwart, O. Wolniak, R. Schlimper, and H. Krause, 2nd Feb. 1977, EG 124 054.
432 Sankyo Co. Ltd., 10 Dec. 1976, JA 76 143 691.
433 Teijin Ltd., 10 March 1977, JA 77 32 046.
434 Rhone-Paulene S.A., 24 March 1977, WG 2 641 955.
435 Hoechst A.-G., 14 April 1977, WG 2 545 292.
436 Mitsubishi Chem. Co. Ltd., 27 Oct. 1977, JA
437 V. E. B. Leuna-Werke 'Walter Ulbricht', 14 Sept. 1977, EG 127 223.
438 Eastman Kodak Co., 13 Dec. 1977, US 4 062 800.
439 Daiichi Kogyo Seiyaku Co. Ltd., 21 July 1977, JA 77 87 452.
440 Tryobo Co. Ltd. and Yoshitomi Pharmaceutical Ind. Ltd., 16 Sept. 1977, JA 77 36 535.
441 Akzo G.m.b.H., 13 Oct. 1977, WG 2 612 669.
442 General Electric Co., 24 May 1977, CA 1 011 030.
443 Goodyear Tire and Rubber Co., 13 Sept. 1977, US 4 048 227.
444 Sankyo Organic Chemicals Ltd., 6 July 1977, JA 77 50 349.
445 Asahi Chem. Ind. Co. Ltd., 29 June 1977, JA 77 77 163.

Table A3 Optical brighteners

Class and general formula	Specification	Application	Ref.
Stilbenes (22)	Where $R^1 = H$, $R^2 = p\text{-}CO_2Me$ or $R^1 = 3\text{-}CN$, $R^2 = H$, and sulpho-derivatives of $R^1 = R^2 = H$	For polyesters, polyamides, and cotton	446
Other stilbenes of interest	—	—	447—455
Triazinyl-stilbenes (23)	Where $R = N(CH_2CH_2OH)_2$, $Z = O$ or $R =$ morpholino, $Z = O$	For polyamides	456

Other triazinyl-stilbenes of interest	—	457
Triazol-stilbenes	For lacquers and plastics	458

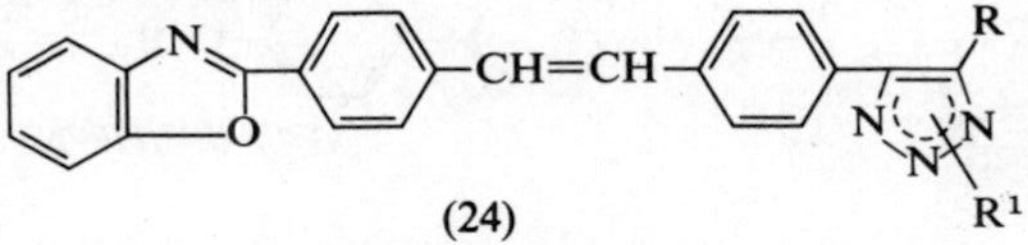

(24)

Where R^1 = CN, 2-benzoxazolyl, 2-benzimidazolyl, CO_2Me, CO_2Et, or $CONH_2$; R^2 = H, Me, Et, Ph_3C, CH_2CONH_2, CH_2CO_2H, CH_2Bz, or CH_2CO_2Me

Other triazol-stilbenes of interest	—	459—461
Coumarins	For polyesters	462

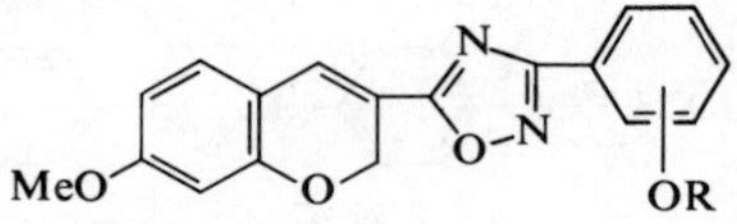

(25)

Where R = CH_2CF_3, CHF_2, or CF_3

446 Sandoz G.m.b.H., 12 May 1977, WG 2 647 179.
447 Showa Chemical Ind. Ltd., 8 Dec. 1977, WG 2 721 730.
448 Ciba-Geigy A.-G., 18 July 1977, BE 850 473.
449 Ciba-Geigy A.-G., 19 Jan. 1977, WG 2 730 246.
450 Ciba-Geigy A.-G., 2 June 1977, WG 2 652 891.
451 J. Pirkl and C. Fisar, 15 March 1977, CZ 168 100.
452 M. Ottova, 15 Oct. 1976, CZ 165 151.
453 J. Pirkl, 15 March 1977, CZ 168 152.
454 Showa Chem. Ind. Ltd., 1 July 1977, JA 77 78 235.
455 Hoechst A.-G., 15 Sept. 1977, WG 2 709 924.
456 Sumitomo Chemical Co. Ltd., 4 Feb. 1977, JA 77 04 576.
457 J. Pirkl and C. Fisar, 15 Jan. 1977, CZ 166 880.
458 Hoechst A.-G., 2 Aug. 1977, US 4 039 531.
459 Sandoz G.m.b.H., 7 April 1977, WG 2 265 268.
460 Bayer A.-G., 21 July 1977, WG 2 601 469.
461 Ciba-Geigy A.-G., 6 Oct. 1977, WG 2 712 409.
462 Produits Chimiques Uginc Khulman, 30 July 1977, FR 2 296 636.

Table A3 (*cont.*)

Class and general formula	Specification	Application	Ref.
Other coumarins of interest	—	—	463
Triazol-coumarins (26)	Where R^1 = Me, Et, Bu, or Ph CH$_2$; R^2 = H, or Cl	For acetate, polyamide, and polyester fibres	464
Other triazol-coumarins of interest	—	—	465
Pyrazolines (27)	Where R^1 = Cl, Ph, MeO, or H; R^2 = H, Cl, or MeO; R^3 = H, Me, or Cl; R^4 = H, CH$_2$CH$_2$NMe$_2$, 2-morpholino-ethyl, 2-piperidimethyl, CHMeCH$_2$NMe$_2$, CH$_2$CH$_2$CH$_2$NMe$_2$, CH$_2$CH$_2$CH$_2$N$^+$-Me$_3$MeSO$_4^-$, or Me	For acetate and acrylic fibres	466
Other pyrazolines of interest	—	—	467—469
Naphthalimides	Where R^1 = Et, R^2 = Et, n = 2 or R^1 = Me, R^2 = MeO, n = 1, or R^1 = Me, R^2 = H, n = 1	For nylon 6 fibres	470

Other naphthalimides of interest — 471

Miscellaneous optical brighteners
Benzoxazole derivatives

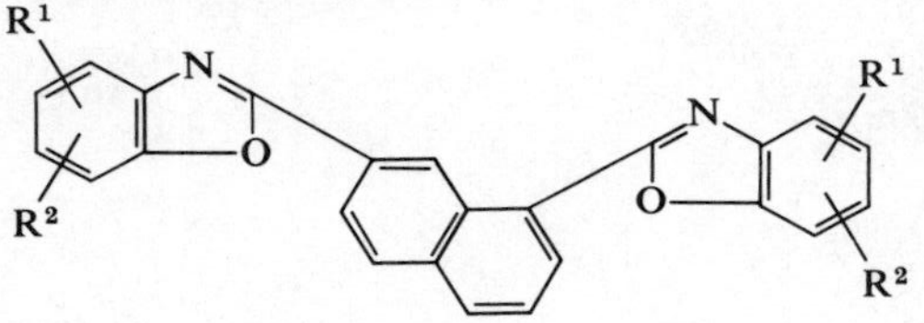

(29)

Where R^1 and R^2 = H, halogen, or a non-chromophoric alkyl, aralkyl carboxy, carboxylate ester or alkylsulphonyl group or (R^1R^2) is a condensed benzene ring — 472

473—485

Other benzoxazole derivatives —

463 Ciba-Geigy A.-G., 3 Nov. 1977, WG 2 717 599.
464 Ciba-Geigy A.-G., 6 Oct. 1977, WG 2 712 408.
465 Ciba-Geigy A.-G., 6 Oct. 1977, WG 2 712 496.
466 Sandoz Ltd., 13 May 1977, SW 587 828.
467 BASF, A.-G., 3 Feb. 1977, WG 2 534 180.
468 BASF, A.-G., 12 May 1977, WG 2 550 548.
469 Gesellschaft für Kernforschung G.m.b.H., 27 Jan. 1977, WG 2 531 495.
470 Mitsubishi Chemical Ind. Co. Ltd., 15 Aug. 1977, JA 77 31 465.
471 Nippon Kayaku Co. Ltd., 3 Feb. 1977, JA 77 04 680.
472 Hoechst A.-G., 1 Dec. 1977, WG 2 621 169.
473 R. Buettner, R. Schrot, and W. Dohrn, 25 May 1977, EG 125 855.
474 Eastman Kodak Co., 20 Sept. 1977, US 4 049 621.
475 Ciba-Geigy A.-G., 31 Aug. 1977, SW 590 960.
476 Sandoz G.m.b.H., 11 Aug. 1977, WG 2 703 864.
477 Mitsui Toatsu Chemicals Inc., 6 June 1977, JA 77 68 229.
478 Mitsubishi Chemical Ind. Co. Ltd., 3 June 1977, JA 77 20 485.
479 Ciba-Geigy A.-G., 15 Sept. 1977, WG 2 709 636.
480 M. Glieman, A. Kurtz, B. Schmidt, M. Dunke, T. Andrich, H. Huscol, and E. Peter, 15 Oct. 1977, EG 122 403.
481 Hoechst A.-G., 5 April 1977, US 4 016 195.
482 Ciba-Geigy A.-G., 8 June 1977, WG 2 653 599.
483 R. Buettner, R. Schrot, and W. Dohrn, 25 May 1977, EG 125 855.
484 Hoechst A.-G., 12 Jan. 1978, WG 2 629 703.
485 Ciba-Geigy A.-G., 12 Jan. 1978, WG 2 729 986.

Part V

PHOTOCHEMICAL ASPECTS OF SOLAR ENERGY
CONVERSION

By M. D. ARCHER

1 Photochemistry

The proceedings of the 1st International Conference on the Photochemical Conversion and Storage of Solar Energy have become available in the course of the year.[1] The eight chapters comprise the invited lectures given at that Conference, held in August 1976. A second, thorough review has been published by the National Swedish Board for Energy Source Development.[2]

Bolton[3,4] has considered the maximum efficiency realisable in a photochemical energy storage system, making allowance for an energy loss of 40% incurred in the formation of the product from the initially produced excited state, and has concluded that 15—16% conversion efficiency is a realistic limit. Archer[5] has considered the maximum conversion efficiency in a dual photochemical system containing two dyes with Gaussian absorption bands, without making allowance for such an energy loss, and has calculated a value of 55%. Using quasi-thermodynamic arguments, the upper limit of efficiency of any photovoltaic solar cell with a single uniform energy gap has been calculated to be 31% for AM1 sunlight.[6]

A novel method of concentrating solar radiation has been proposed,[7—9] based upon a stack of transparent sheets of material doped with fluorescent dyes. The concentrated light emerges from the edges of the slabs. Suitably designed collectors can separate and concentrate various wavebands, which implies a potential improvement in conversion efficiency for single threshold devices.

Valence Isomerization and Energy Storage.—Laird[10] has reviewed the progress that has been made in the past three years in solar energy storage *via* valence isomerization reactions and concludes that the norbornadiene–quadricyclene system remains the most promising one. Kutal and co-workers[11] found that several Cu^I complexes sensitized the photochemical reaction of norbornadiene

[1] 'Solar Power and Fuels', ed. J. R. Bolton, Academic Press, New York, 1977.
[2] 'Solar Energy—Photochemical Conversion and Storage', ed. S. Claessen and L. Engström, National Swedish Board for Energy Source Development, Liber Tryck, Stockholm, 1977.
[3] J. R. Bolton, *J. Solid State Chem.*, 1977, **22**, 3.
[4] J. R. Bolton, *Solar Energy*, 1978, **20**, 181.
[5] M. D. Archer, *Solar Energy*, 1978, **20**, 167.
[6] C. D. Mathers, *J. Appl. Phys.*, 1977, **48**, 3181.
[7] P. B. Mauer and G. D. Turechek, *Res. Discl.*, 1977, **162**, 43.
[8] A. Goetzberger and W. Greibel, *Appl. Phys.*, 1977, **14**, 123.
[9] B. A. Swartz, T. Cole, and A. H. Zewail, *Optics Letters*, 1977, **1**, 73.
[10] T. Laird, *Chem. and Ind.*, 1978, 186.
[11] C. Kutal, D. P. Schwendiman, and P. Grutsch, *Solar Energy*, 1977, **19**, 651.

through formation of a π-bonded complex with a charge-transfer absorption band.[12]

Solar heat storage using chemical reactions has been discussed by Ervin.[13]

Photochemical Decomposition of Water.—Photochemical [14,15] and thermal methods[16] of water decomposition have been reviewed.

Mann *et al.*[17] have reported the production of hydrogen by irradiation with 546 nm light of a dinuclear rhodium(I) complex in 12M-HCl:

$$[Cl^+Rh^I(bridge)_4Rh^I(H^+)] \xrightarrow{h\nu} [Cl^-Rh^{1.5}(bridge)_4Rh^{1.5}(H\cdot)]^{2+*}$$

$$\downarrow \text{fast}$$

$$[Cl^-Rh^{II}(bridge)_4Rh^{II}Cl^-]^{2+} + H_2$$

The bridging ligand is 1,3-di-isocyanopropane. The quantum yield of $[Rh_2(bridge)_4Cl_2]^{2+}$ production was 0.004 ± 0.002 at 546 nm, and the reverse thermal reaction was extremely slow. It is postulated that axial metal–ligand interactions are responsible for the reluctance of the intermediate to return to the Rh^I—Rh^I ground state.

Lehn and Sauvage[18] have reported a more complex system which functions as the reductive half of a photochemical water-splitting cycle. They found that 'appreciable quantities' of hydrogen were produced catalytically by the visible-light irradiation of a neutral aqueous solution containing triethanolamine (TEA), the tris-(2,2'-bipyridyl)ruthenium(II)$^{2+}$ ion, a similar rhodium(III) complex, and colloidal platinum. No data on quantum yields were presented but each molecule of ruthenium produced about 300 molecules of H_2, so that the reaction is certainly catalytic with respect to ruthenium. The tentatively proposed mechanism involves light absorption by the ruthenium complex followed by photo-oxidation of TEA. The ruthenium complex is thought to be converted into its original state by reaction with the rhodium complex, which stores electrons and protons *via* hydride formation. Finally, the colloidal platinum catalyses the decomposition of the hydride with formation of hydrogen. Clearly, much mechanistic work remains to be performed on this system. Interesting and encouraging though the observation of ruthenium(II)-sensitized hydrogen production from water is, it should be noted that the oxidation of the TEA is irreversible and therefore the system could not be used in its present form in a regenerative photochemical water-splitting cycle.

Davis *et al.*[19] have considered the general requirements for hydrogen formation by photoredox reactions of inorganic ions and have pointed out that the

[12] C. Kutal and D. P. Schwendiman, *Inorg. Chem.*, 1977, **16**, 719.
[13] G. Ervin, *J. Solid State Chem.*, 1977, **22**, 51.
[14] T. Ohta and T. N. Veziroglu, *Internat. J. Hydrogen Energy*, 1976, **1**, 255.
[15] E. A. Fletcher and R. L. Moen, *Science*, 1977, **197**, 1050.
[16] E. Bilgen, M. Ducarroir, M. Foex, F. Sibieude, and F. Trombe, *Internat. J. Hydrogen Energy*, 1977, **2**, 251.
[17] K. R. Mann, N. S. Lewis, V. M. Miskowski, D. K. Erwin, G. S. Hammond, and H. B. Gray, *J. Amer. Chem. Soc.*, 1977, **99**, 5525.
[18] J. M. Lehn and J. P. Sauvage, *Nouveau J. Chim.*, 1977, **1**, 449.
[19] D. D. Davis, G. K. King, K. L. Stevenson, E. R. Birnbaum, and J. H. Hageman, *J. Solid State Chem.*, 1977, **22**, 63.

photochemical oxidation of aqueous solutions of chloro- or bromo-complexes of copper(I) produces hydrogen in quite good quantum yield, as shown in Table 1. The Cu^{II} species thus produced cause very little internal masking.

Table 1 *Quantum yields and energy conversion efficiencies for photo-oxidation of Cu^I in 1M-HCl or -HBr.*

	HCl		HBr	
λ/nm	$\Phi_{Cu^{II}}$	$Q_\lambda/\%^a$	$\Phi_{Cu^{II}}$	$Q_\lambda/\%^a$
275	0.65	6.5	0.31	3.1
280	0.65	6.6	0.23	2.3
301	0.44	4.9	0.15	1.6
313	0.34	3.9	0.08	0.9

[a] The monochromatic energy conversion efficiency, Q_λ, is defined as $Q_\lambda = 100\Phi_\lambda \Delta G°_{298}\lambda/hc$.

Leutwyler and Schumacher[20] have investigated the photochemical/thermal cleavage of water at silver ion-exchanged zeolite. This material photo-oxidizes adsorbed water to oxygen on irradiation, with formation of molecularly dispersed silver particles which can reduce the zeolite water to hydrogen at temperatures above 600 °C.

2 Photoelectrochemistry at Semiconductor Electrodes

Nozik[21] has reviewed the performance of electrode materials used in photoregenerative cells and cells for the photoelectrolysis of water. Unique and stringent materials problems must be solved before either of these can be commercially successful. Useful summaries of work published up until the end of 1976 are given in tabular form. A more recent general review[22] covers work done on semiconductor anodes up until the end of 1977.

The flat-band potential of a semiconductor, *i.e.* the potential at which there is no space-charge layer within the semiconductor, is an important quantity since it is one of the parameters that determines open-circuit voltage. Butler and Ginley[23] have extended their method for the calculation of flat-band potentials from the atomic electronegativities of the constituent atoms to all metal oxides. As shown in Figure 1, the calculated and experimental values agree to within *ca.* 0.2 V.

A considerable amount of work has been done on the factors that determine the photoelectrochemical stability of semiconductor electrodes. Gerischer,[24] in an important paper on the cathodic reduction of semiconductors by electrons and their oxidation by holes, has used a thermodynamic approach and the concept of quasi-Fermi levels of holes and electrons in illuminated semiconductors to provide simple criteria for thermodynamic stability. Figure 2 gives, as an example of his approach, the positions of the band edges and the Fermi

20 S. Leutwyler and E. Schumacher, *Chimia (Switz.)*, 1977, **31**, 475.
21 A. J. Nozik, *J. Crystal Growth*, 1977, **39**, 200.
22 W. A. Gerrard and L. M. Rouse, *J. Vacuum Sci. Technol.*, 1978, **15**, 1155.
23 M. A. Butler and D. S. Ginley, *J. Electrochem. Soc.*, 1978, **125**, 228.
24 H. Gerischer, *J. Electroanalyt. Chem. Interfacial Electrochem.*, 1977, **82**, 133.

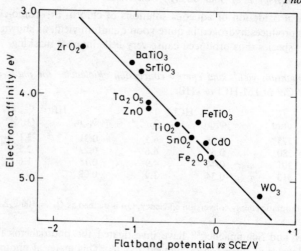

Figure 1 *The electron affinity calculated using atomic electronegativities versus the measured flat-band potentials for the labelled semiconductors corrected to their respective points of zero zeta potential*
(Reproduced with permission from *J. Electrochem. Soc.*, 1978, **125**, 228)

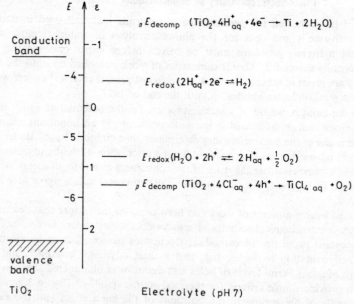

Figure 2 *Energy correlation between band edges and the Fermi energies of electrode reactions in aqueous solution of pH = 7 for TiO₂. E = energies with respect to the vacuum level; ε = energies with respect to the normal hydrogen electrode*
(Reproduced with permission from *J. Electroanalyt. Chem. Interfacial Electrochem.*, 1977, **82**, 133)

energies for various electrode reactions at titanium dioxide. The cathodic decomposition potential, $_nE_{decomp}$, is above (on the diagram) the conduction band edge and the material is therefore not electrochemically reducible. However, $_pE_{decomp}$ is well above the valence band and it is therefore not surprising that the material is subject to slow anodic dissolution under illumination. Fortunately, the oxygen potential, for the overall reaction though possibly not for the first step, is above $_pE_{decomp}$ and oxygen evolution remains the faster reaction.

Bard and Wrighton[25] have used a similar approach to define conditions for thermodynamic stability of an n-type semiconductor in contact with a redox couple, A^+/A. If A is to prevent photoanodic dissolution of the electrode, then E_{redox} must be more negative (on the normal electrochemical scale) than the valence band edge. However, the product A^+ must not be capable of oxidizing the semiconductor either, so it is also necessary that E_{redox} is more negative than the anodic dissolution potential of the semiconductor, $_nE_{decomp}$. Similar criteria may be formulated for p-type electrodes. Bard and Wrighton also point out that surface states unrepresented in these simple thermodynamic considerations may play an important role, and this is certainly borne out by the experimental work of Memming,[26] who has shown by elegant work with ring-disc electrodes that charge transfer to a redox couple can compete very efficiently with semiconductor decomposition even though the simple thermodynamic picture indicates that the necessary electron transfer across the interface between full and empty states has a very low probability. This was certainly true of n-type semiconductors, though p-type electrodes showed simpler behaviour. Fujishima et al.,[27] also working with ring-disc electrodes, have reported data for the relative rates of anodic dissolution and oxidation of the reductant in solution for CdS, ZnO, and TiO_2 in the presence of various redox couples; a smooth correlation is found between the degree to which anodic dissolution is suppressed and the standard redox potential of the reducing agent.

The quasi-thermodynamic approach of Gerischer and others has been criticized by Williams and Nozik[28] on the grounds that observed high quantum efficiencies of charge injection must necessarily mean that tunnelling times are considerably shorter than hole–electron recombination time constants and that therefore the concept of local detailed balance between the rates of creation and annihilation of charge carriers in the semiconductor must be invalid near the interface with the electrolyte.

Photoelectrolysis of Water.—Photoelectrolysis of water at semiconductor electrodes has been thoroughly reviewed.[29] Discussion of the origin of the over-potential terms in the detailed energy balance equation and of dye and impurity sensitization and heterostructure electrodes (e.g. n-TiO_2 films on narrow band-gap semiconductors) is included. Wrighton and co-workers[30] have also reviewed

[25] A. J. Bard and M. S. Wrighton, *J. Electrochem. Soc.*, 1977, **124**, 1706.
[26] R. Memming, *J. Electrochem. Soc.*, 1978, **125**, 117.
[27] A. Fujishima, T. Inoue, T. Watanabe, and K. Honda, *Chem. Letters*, 1978, 357.
[28] F. Williams and A. J. Nozik, *Nature*, 1978, **271**, 137.
[29] H. P. Maruska and A. K. Ghosh, *Solar Energy*, 1978, **29**, 443.
[30] M. S. Wrighton, P. T. Wolczanski, and A. B. Ellis, *J. Solid State Chem.*, 1977, **22**, 17.

the basic principles of semiconductor photoelectrochemistry as applied to photoelectrolysis and have reported the successful use of partially platinized n-type semiconducting $SrTiO_3$ or $KTiO_3$ to photoelectrolyse water. The bare titanate serves as a normal photoanode and the platinum-coated dark portions as a cathode at which the overpotential for hydrogen evolution is small. A slow deterioration in the performance of these short-circuited photoelectrolysis 'cells' is probably connected with slow changes in the platinum coating. Bockris and co-workers[31] have continued their work on photoelectrolysis cells containing two semiconductor electrodes, and found that the n-TiO_2/p-GaP combination was the most efficient of those that they tested, with a solar conversion efficiency of 0.1%.

Titanium Dioxide and Titanate Anodes. Bockris and Uosaki[32] have compared experimental results for single-crystal and thin-film polycrystalline titanium dioxide films with their theoretical model and conclude that surface states must be involved in the charge-transfer process. Nobe *et al.*[33] have also examined single crystals, and find significant differences in the quantum efficiency of charge injection and the wavelength of maximum photoresponse for single TiO_2 crystals with surface orientations parallel and perpendicular to the c-axis.

The pressures of hydrogen and oxygen at the two electrodes exert a considerable, although often ignored, influence on cell performance. Significant pressure produces an appreciable back e.m.f. in the cell, so that to generate hydrogen at 1 atm pressure an auxiliary voltage of at least 0.2 V is necessary.[34] However, contrary to what has often been considered self-evident, cells with a small external bias can have a higher net energy conversion efficiency than unbiased ones. The same point is made by Ghosh and Maruska[35] in their study of aluminium-doped titanium dioxide. They found the diffusion length of holes to be increased from 1.2 μm in undoped TiO_2 to 4.5 μm in the sample doped with 0.05 weight % Al. This leads to an improved net sunlight conversion efficiency of 1.3% in an externally biased cell. Chromium doping is also effective, sensitizing the spectral response of TiO_2 to well past 550 nm. Augustyni *et al.*[36] also found that doping of polycrystalline TiO_2 with Al produced excellent results. They also investigated the effects of doping with Sr, Ga, Eu, and B.

Fleischauer and Allen[37] have examined the sensitization of thin-TiO_2 films to visible light and have demonstrated two types of sensitization. Firstly, non-absorbing sensitizers such as 4,5-dihydroxy-m-benzenedisulphonic acid (tiron) were found to exert a current-multiplying effect upon the anodic photocurrent, which enables hydrogen to be produced at the counter-electrode in the 400—500 nm region. Secondly rhodamine B was found to sensitize the effect of the tiron to wavelengths of about 500 nm. The tiron is, however, presumably irreversibly consumed in both these processes.

[31] K. Ohashi, J. McCann, and J. O'M. Bockris, *Internat. J. Energy Res.*, 1977, **1**, 259.
[32] J. O'M. Bockris and K. Uosaki, *Adv. Chem. Ser.*, 1977, No. 163, p. 33.
[33] K. Nobe, G. L. Bauerle, and M. Braun, *J. Appl. Electrochem.*, 1977, **7**, 379.
[34] H. Ruefer, P. Wuerfel, and W. Ruppel, *J. Appl. Phys.*, 1978, **49**, 2531.
[35] A. K. Ghosh and H. P. Maruska, *J. Electrochem. Soc.*, 1977, **124**, 1516.
[36] J. Augustyni, J. Hinden, and C. Stalder, *J. Electrochem. Soc.*, 1977, **124**, 1063.
[37] P. D. Fleischauer and J. K. Allen, *J. Phys. Chem.*, 1978, **82**, 432.

n-TiO_2 can be spectrally 'sensitized' by polypyridine–ruthenium(II)$^{2+}$ complexes.[38] These are reported, from experiments using chopped light, to decrease the anodic photocurrent produced by greater than band-gap light, simply by absorbing the light, but to increase the photocurrent with less than band-gap light. The photoanodic process is the production of $[Ru(bipy)_3]^{3+}$ from $[*Ru(bipy)_3]^{2+}$ in a quantum yield of approximately unity in the range 420—550 nm.

Frank and Bard[39] have demonstrated the photoassisted oxidations of a number of compounds, including hydroquinone, p-aminophenols, I^-, Br^-, Cl^-, Fe^{2+}, Ce^{3+}, and CN^-, at polycrystalline TiO_2. Many of these compounds compete for photogenerated holes with the solvent water with high current efficiencies. Ethanol, cyanide, and ethyl formate are oxidized by a current doubling mechanism.

Some interesting work[40] on the photoelectron spectroscopy of clean oxygen-deficient TiO_2 crystal surfaces has shown the existence of Ti^{3+} in the dark. While water is adsorbed in molecular form on stoicheiometric TiO_2 surfaces it adsorbs dissociatively on the oxygen-deficient surface, changing the Ti^{3+} to a different oxidation state. Illumination by band-gap light regenerates the Ti^{3+}, however.

Two other brief studies of the photoelectrochemistry of TiO_2 have been published.[41,42]

Other Electrodes for Photoanodic Oxygen Evolution. Tributsch[43] has continued his study of MoS_2 crystals, reported in a preliminary way last year. Both photocathodic and photoanodic effects can be observed in water, with p-type and n-type crystals respectively. The predominant photoanodic process on irradiation in the 400—715 nm waveband is the production of Mo^{VI} and sulphate, possibly *via* a Mo^V—OH intermediate, and the current efficiency for photoanodic oxygen generation was found to be less than 1%. Clearly, the oxidation of crystal-bound sulphur to sulphate must be prevented if the object is the photoelectrochemical dissociation of water. The use of tris-(2,2'-bipyridyl)ruthenium(II)$^{2+}$ ions as redox catalysts in wet acetonitrile is reported to be partially successful. $MoSe_2$, with a band-gap of 1.4 eV, acts very similarly to MoS_2, selenic acid being produced photoanodically in water.[44]

Tungsten trioxide electrodes[45] show photoelectrochemical behaviour typical of n-type semiconductors. Annealing of WO_3 layers in an oxygen atmosphere leads to a strong red shift in the photosensitivity and a change in the electrode capacitance.

Monocrystalline ZnO, serving as a photoanode in KCl solution, is reported to be transformed to the polycrystalline material.[46]

[38] W. D. R. Clark and N. Sutin, *J. Amer. Chem. Soc.*, 1977, **99**, 4676.
[39] S. N. Frank and A. J. Bard, *J. Amer. Chem. Soc.*, 1977, **99**, 4667.
[40] W. J. Lo, Y. W. Chung, and G. A. Somorjai, *Surface Sci.*, 1978, **71**, 199.
[41] Z. A. Rotenberg, T. V. Dzhavrishvili, Yu. V. Pleskov, and A. L. Asatiani, *Elektrokhimiya*, 1977, **13**, 1803.
[42] K. Miyatani and I. Sato, *Japan. J. Appl. Phys.*, 1977, **16**, 1879.
[43] H. Tributsch, *Z. Naturforsch.*, 1977, **32a**, 972.
[44] H. Tributsch, *J. Electrochem. Soc.*, 1978, **125**, 1086.
[45] W. Gissler, *J. Electrochem. Soc.*, 1977, **124**, 1710.
[46] L. L. Larina and E. M. Trukhan, *Zhur. fiz. Khim.*, 1978, **52**, 805.

Tsubomura *et al.*[47,48] have examined the sensitization of the oxidation of iodide ion at sintered zinc oxide by Rose Bengal and rhodamine B, and the resonance Raman spectrum of Rose Bengal adsorbed on ZnO.

Rhodamine B has been chemically attached to SnO_2 and TiO_2 surfaces by an ester linkage.[49] The photocurrent for the oxidation of hydroquinone at these derivatized surfaces was greatly increased compared with earlier chemically modified electrodes involving amide linkages.

_p_-Type Semiconductors for Photocathodic Evolution of Hydrogen from Water.—
Tamura *et al.*[50] have investigated the quantum yield of the cathodic photo-current at *p*-GaP (also of *n*-TiO_2) as a function of hole concentration. Their results are shown in Figure 3. The optimal carrier concentration, *ca.* 2×10^{16} cm^{-3}, is much greater than that predicted from the requirement that the

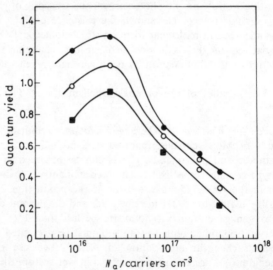

Figure 3 *The quantum yield of the cathodic photocurrent at* p-GaP *electrodes in* 0.5M-H_2SO_4 *as a function of the acceptor concentration of the electrode:* ●, −1.0 V; ○, −0.5 V; ■, −0.2 V *versus S.C.E.*
(Reproduced with permission from *J. Electroanalyt. Cehm. Interfacial Electrochem.*, 1977, **80**, 357)

space-charge layer thickness should equal the optical absorption depth. They conclude that a critical field strength is required in the space-charge layer to cause separation of hole–electron pairs.

Horowitz[51] has presented capacitance measurements which give reproducible values for the flat-band potential, E_{FB}, of etched *p*-GaP. The variation with pH

[47] M. Matsumura, Y. Nomura, and H. Tsubomura, *Bull. Chem. Soc. Japan*, 1977, **50**, 2533.
[48] H. Yamada, T. Amamiya, and H. Tsubomura, *Chem. Phys. Letters*, 1978, **56**, 591.
[49] M. Fujihira, N. Ohishi, and T. Osa, *Nature*, 1977, **268**, 226.
[50] H. Tamura, H. Yoneyama, C. Iwakura, H. Sakamoto, and S. Murakami, *J. Electroanalyt. Chem. Interfacial Electrochem.*, 1977, **80**, 357.
[51] G. Horowitz, *J. Appl. Phys.*, 1978, **49**, 3571.

is given by the relationship $E_{FB} = 0.72 - 0.059$ pH ± 0.03 V (S.C.E.). Although such electrodes are photoelectrochemically stable, they are not well suited to solar energy conversion since the wavelength of maximum photoresponse is 410 nm.[52]

Bockris and co-workers[53] have examined silver-doped p-ZnTe and undoped p-CdTe as possible photocathodes for the production of hydrogen from water. The flat-band potentials in alkaline solution were found to be -0.79 ± 0.02 V and $+0.21 \pm 0.02$ (N.H.E.) for ZnTe and CdTe respectively. Thus ZnTe is unsuitable since the onset of hydrogen evolution occurs at a potential more negative than that at which oxygen evolution occurs at any presently known n-type photoanode. CdTe, however, appeared to be stable and is therefore, with its band-gap of 1.5 eV, a suitable photocathodic material.

Regenerative Photoelectrochemical Cells.—Heller et al.[54] have discussed the use of a technique they call photocurrent spectroscopy, in which one pumping beam and a less intense probing beam irradiate the electrode simultaneously, to examine the nature of the carrier recombination centres in cells with n-type CdS, CdSe, CdTe, and GaAs photoanodes and chalcogenide anion solutions. Poor short-wavelength response is shown to be due to surface or near-surface recombination centres, while lowered long-wavelength response is associated with an absorption length that exceeds the carrier diffusion length.

Cells containing CdX photoanodes and X^{2-} in solution, where X is a chalcogen, are sometimes classified as 'stable' on the basis of constancy of weight of the electrode and virtual constancy with time of the output current. However, these two criteria may be insensitive to changes that nevertheless in the long run would affect cell performance. CdSe and CdS electrodes under illumination in sulphide/polysulphide solutions, to take an example of such a 'stable' system, show significant changes in photocurrent spectra and Mott–Schottky plots indicative of a restructuring of the surface to polycrystalline or amorphous CdS.[55] How greatly such a restructuring would affect cell performance probably depends on how far the transformation reactions penetrate into the photoanode.

CdS. Minoura and co-workers have investigated the behaviour of CdS films formed by chemical spray deposition on conductive glass and find behaviour similar to that obtained with a single-crystal electrode.[56, 57] They have also discovered that the (000$\bar{1}$) (sulphur) face of single-crystal n-CdS shows a cathodic photocurrent in sulphide solutions at negative potentials, in addition to the usual anodic photocurrent at more positive potentials. They ascribe this cathodic photocurrent to reduction of polysulphide ion *via* surface states within the forbidden gap of CdS.[58]

[52] A. Bourrasse and G. Horowitz, *J. Phys. (Paris), Letters*, 1977, **38**, 291.
[53] K. Ohashi, K. Uosaki, and J. O'M. Bockris, *Internat. J. Energy Res.*, 1977, **1**, 25.
[54] A. Heller, K. C. Chang, and B. Miller, *J. Amer. Chem. Soc.*, 1978, **100**, 684.
[55] H. Gerischer and J. Gobrecht, *Ber. Bunsengesellschaft phys. Chem.*, 1978, **82**, 520.
[56] H. Minoura, T. Nakamura, Y. Ueno, and M. Tsuiki, *Chem. Letters*, 1977, **8**, 913.
[57] H. Minoura, H. Okada, and M. Tsuiki, *Nippon Kagaku Kaishi*, 1977, **10**, 1443.
[58] H. Minoura and M. Tsuiki, *Chem. Letters*, 1978, 205.

Comparatively little work on the prevention of semiconductor corrosion by the use of organic solvents has been reported this year, but a CdS electrode in combination with iodide ions in acetonitrile has been found to be stable when the concentration of NaI in solution exceeds 0.2 mol l⁻¹.[59]

Wagner and Shay[60] have grown n-Cds films on (111)A GaAs substrates, producing a heterojunction which enhances the performance of the CdS photo-anode in a regenerative cell. The solar conversion efficiency is increased by a factor of about 1.5 relative to a simple n-CdS photoelectrochemical solar cell, as shown in Figure 4.

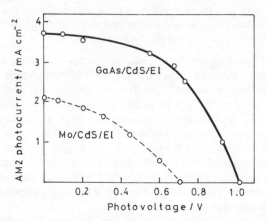

Figure 4 *AM2 photocurrent–photovoltage characteristics of cells with* Mo/n-CdS *and* n-GaAs/n-CdS *anodes*, Pt *counter-electrodes, and electrolyte containing* 0.1M-Na₂S + 0.2M-S + 0.2M-NaOH
(Reproduced with permissionm from *Appl. Phys. Letters*, 1977, **31**, 446)

CdSe. Work on *n*-CdSe in sulphide/polysulphide solutions has continued. Although such cells show good performance (solar conversion efficiencies in excess of 9%),[61, 62] Gerischer's findings, quoted above,[55] should be borne in mind.

Miller and co-workers have considered the use of polycrystalline rather than single-crystal semiconductors, with particular reference to CdSe.[63] Pressure-sintered CdSe electrodes show conversion efficiencies approaching three-quarters of those of single-crystal specimens. They point out that little deterioration in performance as a result of using polycrystalline anodes is to be expected provided that the crystallite size is at least comparable to the light absorption depth. The latter is of the order of 1 μm for direct band-gap semi-conductors, whereas a grain size of 10—20 μm can be produced by sintering.

[59] K. Nakatani, S. Matsudaira, and H. Tsubomura, *J. Electrochem. Soc.*, 1978, **125**, 406.
[60] S. Wagner and J. L. Shay, *Appl. Phys. Letters*, 1977, **31**, 446.
[61] M. S. Wrighton, A. B. Ellis, and S. W. Kaiser, *Adv. Chem. Ser.*, 1977, No. 163, p. 71.
[62] W. A. Gerrard and J. R. Owen, *Materials Res. Bull.*, 1977, **12**, 677.
[63] B. Miller, A. Heller, M. Robbins, S. Menezes, K. C. Chang, and J. Thomson, *J. Electro-chem. Soc.*, 1977, **124**, 1019.

Bard and co-workers[64] have investigated the performance of solid solutions of CdS and CdSe; studies of the photoassisted oxidation of sulphide show that the flat-band potentials of the mixtures were shifted to negative potentials with respect to CdS and CdSe, which is advantageous. When $x_{CdSe} = 0.9$, a power conversion efficiency of 9% at a cell voltage of 202 mV (irradiated with 577 nm light of intensity 1 mW cm^{-2}) was obtained.

CdTe. CdTe has a band-gap (1.52 eV) that is very suitable for direct solar energy conversion, and it is unique among II–VI materials in that it can readily be made either *n*- or *p*-type, which may indicate that its band structure is different from those of the other cadmium chalcogenides.

Menezes *et al.*[65] have investigated the cell

polycrystalline *n*-CdTe | SnCl$_2$ (saturated solution) | Pt

and found that the initial AM1 solar conversion efficiency of 4.8% declined on a time-scale of hours, despite the photoanode material appearing to be stable. However, traces of indium (the doping material) were detected in solution.

Using a thin film of *n*-CdTe deposited on titanium, the cell

polycrystalline *n*-CdTe | 1M-NaOH + 1M-Na$_2$S + 1M-S | Pt

was found to have the following characteristics under xenon light (50 mW cm^{-2}): $V_{oc} = 0.34$ V, $i_{sc} = 2.6$ mA cm^{-2}, $\eta = 0.4\%$.[66] An electrode of *n*-CdTe immersed in 0.1M-NaOH has been found, perhaps unsurprisingly, to be unstable under irradiation.[67]

Other Materials. Tributsch has investigated the suitability of MoS$_2$[68] and MoSe$_2$[69] as photoelectrodes in regenerative solar cells. The cathodic photocurrent drawn from a *p*-MoS$_2$ electrode in contact with Fe$^{3+/2+}$ remained stable for at least a day, but the photocathodic behaviour of *p*-type MoS$_2$ was always found to be inferior to the photoanodic behaviour of *n*-type MoS$_2$ because of the relatively poor *p*-character of the available material.[68] *n*-MoSe$_2$, which tends to show superficial oxidation to Mo^{VI} sulphate on irradiation in water, is stabilized by I$^-$/I$_2$, Fe^{2+}/Fe^{3+}, and quinhydrone, and the cell

n-MoSe$_2$ | I$^-$ (saturated solution) + I$_2$ | Metal

exhibited remarkable stability, operating over a period of 150 days with an output current that remained stable to within 15%. The solar conversion efficiency was approximately 1%.[69]

Wrighton and co-workers[70] have produced a novel stable regenerative cell based on an *n*-type Si photoanode in contact with the ferrocene/ferricenium

[64] R. N. Noufi, P. A. Kohl, and A. J. Bard, *J. Electrochem. Soc.*, 1978, **125**, 375.
[65] C. Menezes, F. Sanchez-Sinencio, J. S. Helman, R. Williams, and J. Dresner, *Appl. Phys. Letters*, 1977, **31**, 16.
[66] W. J. Danaher and L. E. Lyons, *Nature*, 1978, **271**, 137.
[67] Y. G. Chai, W. W. Anderson, and L. B. Anderson, *I.E.E.E. Trans. Electron Devices*, 1977, **ED24**, 492.
[68] H. Tributsch, *Ber. Bunsengesellschaft phys. Chem.*, 1977, **81**, 361.
[69] H. Tributsch, *Ber. Bunsengesellschaft phys. Chem.*, 1978, **82**, 169.
[70] K. D. Legg, A. B. Ellis, J. M. Bolts, and M. S. Wrighton, *Proc. Nat. Acad. Sci. U.S.A.*, 1977, **74**, 4116.

redox couple in ethanol. The solar conversion efficiency was estimated to be *ca.* 1%.

3 Photoelectrochemistry at Metal Electrodes

Photogalvanic Cells.—Albery and Archer[71] have continued their theoretical analysis of photogalvanic cell efficiency by considering the performance of thin-layer concentration and differential cells. In the former, the two electrodes are identical and the maximum conversion efficiency is very low ($<0.2\%$), but in the latter at least one electrode is selective in its redox activity and this arrangement is capable of much better performance (7—18% efficiency, depending on the assumptions made about the threshold energy and $\Delta G°$ value of the photoredox reaction). This series of papers is summarized in a short, separate paper.[72]

Several studies of the iron–thionine photogalvanic system at semiconductor electrodes have been published this year. Cadmium stannate, Cd_2SnO_4, sputtered on glass, forms thin films of excellent transparency and conductivity. However, it is electrochemically unstable, as well as being irreversible both to $Fe^{3+/2+}$ and to thionine/leucothionine. Its use as an underlay for an SnO_2 anode increases short-circuit currents somewhat, presumably by lowering the internal resistance of the cell.[73] SnO_2 itself is fairly selective for thionine/leucothionine, but gives rise to an undesirable ohmic loss when used as the anode material.[74] Tsubomura and co-workers[75] have made a comparative study of the behaviour of three semiconductor anodes and platinum, with the results summarized in Table 2. The unusual observation of an anodic photocurrent at a *p*-type electrode (GaP) is ascribed to hole tunnelling from the semiconductor to reduced thionine in solution, despite the unfavourable electric field in the space-charge layer.

Table 2 *Anodic short-circuit photocurrent density* (i_{sc}) *and open-circuit voltage* (V_{oc}) *between an illuminated electrode and a dark platinum electrode*

Illuminated electrode[a]	$i_{sc}/\mu A\ cm^{-2}$	V_{oc}/mV
SnO_2	2.8	−170
TiO_2	1.5	−155
p-GaP	12[b]	−100[b]
Pt	1.9	−135

[a] Illuminated by 500 W xenon lamp. Wavelengths shorter than 550 nm were cut off with a filter.   [b] Illumination with wavelengths longer than 600 nm.

[71] W. J. Albery and M. D. Archer, *J. Electroanalyt. Chem. Interfacial Electrochem.*, 1978, **86**, 1, 19.
[72] W. J. Albery and M. D. Archer, *Nature*, 1977, **270**, 399.
[73] D. Hall, *J. Electrochem. Soc.*, 1977, **124**, 804.
[74] D. E. Hall, W. D. K. Clark, J. A. Eckert, N. N. Lichtin, and P. D. Wildes, *Bull. Amer. Ceram. Soc.*, 1977, **56**, 408.
[75] Y. Suda, Y. Shimoura, T. Sakata, and H. Tsubomura, *J. Amer. Chem. Soc.*, 1978, **82**, 268.

Shigehara *et al.*[76] have compared conventional iron–thionine thin layer cells, containing the components in aqueous solution, with cells containing various gels, membranes, and polymer-bound thionine. Moorthy and co-workers[77] have investigated the effect of a temperature differential between the illuminated and dark half-cells. Increasing the temperature of the illuminated half-cell has a deleterious effect, presumably because it increases the rate of the thermal back reaction between Fe^{III} and semi- and leuco-thionine. Increasing the temperature of the dark half-cell is, however, beneficial, presumably because it increases the rate of mass transport to the electrodes. Lichtin and co-workers[78] have examined the kinetics of the Fe^{III}–leucothionine reaction in water–acetonitrile, and have interpreted their results in terms of complex formation between the two reactants.

Photogalvanic effects of aqueous solutions of tris-(2,2'-bipyridyl)ruthenium-(II)$^{2+}$ at gold have been ascribed to formation of $[Ru(bipy)_3]^+$ by reductive quenching.[79]

Pigment-covered Metal Electrodes.—Platinum electrodes coated with a 50 nm thick layer of vanadyl phthalocyanine are reported to display substantial pH-dependent photoelectrochemical effects.[80]

Small photogalvanic effects are observed on irradiating aqueous suspensions of chromatophores from *Rhodospirillum rubrum*.[81]

4 Photosynthesis

The proceedings of a conference on biological solar energy conversion, held in 1976, have been published.[82] Sauer[83] has reviewed current knowledge of the structure and function of photosynthetic membranes, particularly those of photosynthetic bacteria, and Clayton[84] has provided a general review of solar energy conversion by photosynthesis.

5 Photovoltaic Cells

Several useful reviews have been published this year. The greater part of one issue of the *Journal of Crystal Growth* has been devoted to thirteen review papers on materials for photovoltaic cells,[85] and a shorter French review of the same subject has appeared.[86] Wagner[87] has also reviewed the current status of photovoltaic cells, including approaches to the development of inexpensive cells.

[76] K. Shigehara, M. Nishimura, and E. Tsuchida, *Bull. Chem. Soc. Japan*, 1977, **50**, 3397.
[77] P. V. Kamat, M. D. Karkhanavala, and P. N. Moorthy, *Solar Energy*, 1978, **20**, 173.
[78] T. L. Osif, N. N. Lichtin, and M. Z. Hoffman, *J. Phys. Chem.*, 1978, **82**, 1778.
[79] A. Lottaz and A. von Zelewsky, *Chimia (Switz.)*, 1977, **31**, 180.
[80] V. A. Ilatovskii, I. B. Dmitriev, and G. G. Komissarov, *Zhur. fiz. Khim.*, 1978, **52**, 121.
[81] T. Erabi, H. Hiura, M. Hayashi, M. Yamada, T. Endo, J. Yamashita, M. Tanaka, and T. Horio, *Chem. Letters*, 1978, 341.
[82] 'Biological Solar Enery Conversion', ed. A. Mitsui, S. Miyachi, A. San Pietro, and S. Tamura, Academic Press, New York, 1977.
[83] K. Sauer, *Accounts Chem. Res.*, 1978, **11**, 257.
[84] R. K. Clayton, *Adv. Chem. Ser.*, 1977, No. 163, p. 93.
[85] *J. Crystal Growth*, 1977, **39** (1): issue entitled 'Materials and Energy'.
[86] M. Rodot, M. Barbe, and J. Dixmier, *Rev. Phys. Appl.*, 1977, **12**, 1223.
[87] S. Wagner, *Adv. Chem. Ser.*, 1977, No. 163, p. 109.

Fabre[88] has analysed the different junction types in solar cells and discussed the effects of the physical characteristics of the base material, especially its band structure, on the cell performance. Hovel[89] has reviewed solar cells for terrestrial applications, including some discussion of cell arrays. Finally, many valuable papers have appeared in a special issue of *I.E.E.E. Trans. Electron Devices*.[90] These papers, and those of ref. 85, are not abstracted individually in this Report, although they span the same subject matter as do the following sections.

Inorganic Homojunction Cells.—The characteristics of silicon used in solar cells[91] and the prospects for single-crystal cells[92] have recently been reviewed (in French). The most efficient single-crystal solar cells are the n^+/p cells originally developed for space applications (AM1 efficiency *ca.* 18%). However, the relatively high sheet resistance of the n^+ layer in this cell is disadvantageous. Bae and D'Aiello[93] have fabricated p^+/n cells by boron diffusion, producing lower values of sheet resistance and an AM1 conversion efficiency of 14—17% without using the expensive fine grid top electrode required by the n^+/p cells.

The electronic properties of amorphous silicon continue to attract interest[94, 95] and a Schottky barrier solar cell of *ca.* 1 μm *a*-Si on platinum of solar conversion efficiency 5.5% ($V_{oc} = 0.86$ V, $i_{sc} = 14$ mA cm^{-2}) has been fabricated.[96]

Thin-film homojunction CuInSe$_2$ of conversion efficiency 3.3% (tungsten–halogen illumination) has been reported.[97] The minority carrier lifetime in the *p*-doped material appears to limit the performance. The same parameter limits efficiency of photovoltaic CdTe homojunctions made by implantation of As$^+$ ions into *n*-type undoped CdTe crystals, which also have a solar efficiency of 3%.[98] Conversion efficiencies of 17% were obtained from single-crystal $n^+/p/p^+$ homojunction GaAs cells, prepared by chemical vapour deposition without the use of the usual Ga$_{1-x}$Al$_x$As layers.[99]

Inorganic Heterojunction Cells.—Further work has been carried out on the preparation of graded band-gap Ga$_{1-x}$Al$_x$As/GaAs solar cells, and on the analytical techniques used to profile these graded layers.[100] Highly efficient large-area (>10 cm^2) cells of this type have been made[101, 102] by vapour-phase

88   E. Fabre, *Acta Electron.*, 1977, **20**, 117.
89   H. T. Hovel, *Solar Energy*, 1977, **19**, 605.
90   *I.E.E.E. Trans. Electron Devices*, 1977, **ED24** (4): special issue on photovoltaic devices.
91   J. J. Brissot, *Acta Electron.*, 1977, **20**, 101.
92   J. Michel, *Acta Electron.*, 1977, **20**, 133.
93   M. S. Bae and R. V. D'Aiello, *Appl. Phys. Letters*, 1977, **31**, 285.
94   A. R. Moore, *Appl. Phys. Letters*, 1977, **31**, 762.
95   D. L. Staetler and C. R. Wronski, *Appl. Phys. Letters*, 1977, **31**, 292.
96   D. E. Carlson, C. R., Wronski, J. Pankove, P. J. Zanzucchi, and D. L. Staetler, *R.C.A. Rev.*, 1977, **38**, 211; D. E. Carlson, J. I. Pankove, C. R. Wronski, and P. J. Zanzucchi, *Thin Solid Films*, 1977, **45**, 43.
97   L. L. Kazmerski and G. A. Sanborn, *J. Appl. Phys.*, 1977, **48**, 3178.
98   M. Chu, A. L. Fahrenbruch, and R. H. Bube, *J. Appl. Phys.*, 1978, **49**, 322.
99   C. O. Bozler and J. C. C. Fan, *Appl. Phys. Letters*, 1977, **31**, 629.
100   R. Sahai, J. S. Harris, D. D. Edwall, and F. H. Eisen, *J. Electron Mater.*, 1977, **6**, 645.
101   N. J. Nelson, K. K. Johnson, R. L. Moon, H. A. Vander Plas, and L. W. James, *Appl. Phys. Letters*, 1978, **83**, 26.
102   R. D. Dupuis, P. D. Dapkus, R. D. Yingling, and L. A. Moudy, *Appl. Phys. Letters*, 1977, **31**, 201.

epitaxy, a technique which is capable of producing much larger areas than is the liquid-phase epitaxial process generally employed to grow these structures. These large-area cells also function well in concentrated sunlight (conversion efficiency 19% at ×933 suns[101]).

CdS/Cu$_x$S solar cells continue to pose some stability problems,[103] but the non-stoicheiometry of thin CdS and Cu$_x$S films is now better understood from work with electron diffraction[104] and Auger spectroscopy.[105] The performance of some other heterojunction cells containing n-CdS is summarized in Table 3.

Metal–Semiconductor and Metal–Insulator–Semiconductor Cells.—Recent data are summarized in Table 3. Noteworthy developments this year are the fabrication of many of these diodes into relatively large area (several cm²) cells without loss in performance. Use of a large-area device avoids the loss of a significant fraction of the photogenerated charge carriers around the periphery of the cell, which occurs in very small-area (<0.02 cm²) cells.[106] The production of MIS cells based on amorphous silicon, a material readily produced in thin but large-area films, seems particularly promising, particularly if low-cost metals such as nickel and aluminium are usuable as the metallic component.

Theoretical work on MS[107] and MIS[108, 109] cells continues. The influence of interfacial states in MIS is attracting attention:[110–113] these have an electro-static role in that they store charge: for a p-type semiconductor, donor-like hole trapping states are desirable and acceptor-like negatively charged states very unfavourable. Interfacial states also have a kinetic role in that they provide a path between the various carrier reservoirs.

Organic Semiconductors.—There has been a significant breakthrough this year, with three reports of cells based on merocyanine and phthalocyanine dyes reaching overall solar conversion efficiencies of the order of 1%. (The previous best was 0.02% in the cell based on hydroxysquarylium reported last year.) A team at Exxon[114] have fabricated thin film cells of the type

Electrode of high work function (metallic or semiconducting)	Merocyanine dye (vapour deposited or solution cast)	Semitransparent Al electrode

where the merocyanine dye is of structure (1) or (2) in a film of thickness 10—500 nm, absorbing 15% and 50% respectively of incident sunlight. A cell containing dye (1) had a sunlight engineering efficiency of 0.7% (V_{oc}, = 1.2 V

103 H. Matsumoto, N. Nakayama, K. Yamaguchi, and S. Ikegami, *Japan. J. Appl. Phys.*, 1977, **16**, 1283.
104 B. G. Caswell, G. J. Russell, and J. Woods, *J. Phys. (D)*, 1977, **10**, 1345.
105 R. Hill and I. A. S. Edwards, *Vacuum*, 1977, **27**, 277.
106 R. B. Godfrey and M. A. Green, *Appl. Phys. Letters*, 1977, **31**, 705.
107 R. J. Soukup and L. A. Akers, *J. Appl. Phys.*, 1978, **49**, 4031.
108 A. K. Ghosh, C. Fishman, and T. Feng, *J. Appl. Phys.*, 1978, **49**, 3490.
109 H. C. Card, *Solid-State Electronics*, 1977, **20**, 971.
110 P. T. Landsberg and C. Klimpke, *Proc. Roy. Soc.*, 1977, **A354**, 101.
111 L. C. Olsen, *Solid-State Electronics*, 1977, **20**, 741.
112 P. Viktorovitch and G. Kamarinos, *J. Appl. Phys.*, 1977, **48**, 3060.
113 P. Viktorovitch, G. Kamarinos, P. Even, and E. Fabre, *Phys. Stat. Solidi (A)*, 1978, **48**, 137.
114 D. L. Morel, A. K. Ghosh, T. Feng, E. L. Stogryn, P. E. Purwin, R. F. Shaw, and C. Fishman, *Appl. Phys. Letters*, 1978, **32**, 495.

Table 3 *Performance of some inorganic heterojunction, MS, and MIS cells*

Cell	V_{oc}/V	i_{sc}/mA cm^{-2}	η/%	Illumination
Heterojunction cells				
Polycrystalline n-CdS/monocrystalline p-CdTe	0.63	16.1	7.9	85 mW cm^{-2} solar simulation[a]
Epitaxial n-CdS/monocrystalline p-CdTe	0.67	20.1	10.5	AM1.3 sunlight[b]
	0.69	23.7	6.0	AMO sunlight[b]
Epitaxial n-CdS/monocrystalline p-Si	0.36	8.0	1.9	100 mW cm^{-2} solar simulation[c]
MS cells				
polycrystalline n-ITO/monocrystalline n-Si	0.48	32	10.2	AM1 sunlight[d]
Anodically formed SnO$_2$/monocrystalline n-Si			ca. 1	Tungsten lamp[e]
MIS cells				
Al/oxide/polycrystalline p-Si	0.52	22	9	AM1 sunlight[f]
Al/SiO$_2$/mono- or poly-crystalline p-Si with TiO$_x$ dielectric	0.49	21	8	AM1 sunlight[g]
with SiO$_2$ dielectric	0.52	33.3	13	AM1 sunlight[g]
In$_2$O$_3$ antireflection coating/Pt/SiO$_2$/monocrystalline n-Si	0.41	29.2	8.8	Tungsten lamp[h]
Ni/TiO$_x$/amorphous Si	0.67	7.9	4.8	60 mW cm^{-2} solar simulation[i]
Ni or Au/oxide/amorphous Si	0.40			AM1 sunlight[j]
Oxidized polycrystalline n-GaAs/5 μm W/graphite	0.57	17.7	5.5	AM1 sunlight[k]

V_{oc} = open-circuit voltage; i_{sc} = short-circuit current density; η = overall solar power conversion efficiency.

[a] K. W. Mitchell, A. L. Fahrenbruch, and R. H. Bube, *J. Appl. Phys.*, 1977, **48**, 4365. [b] K. Yamaguchi, N. Nakayama, H. Matsumoto, and S. Ikegami, *Japan. J. Appl. Phys.*, 1977, **16**, 1203. [c] F. M. Livingstone, W. M. Tsang, A. J. Barlow, R. M. de la Rue, and W. Duncan, *J. Phys.* (*D*), 1977, **10**, 1959. [d] J. C. Manifacier and L. Szepessy, *Appl. Phys. Letters*, 1977, **31**, 459. [e] A. Usami and S. Ishihara, *Opt. Quantum Electronics*, 1977, **9**, 357. [f] D. R. Lillington and W. G. Townesnd, *Appl. Phys. Letters*, 1977, **31**, 471. [g] P. Van Halen, R. P. Mertens, R. J. Van Overstraeten, R. E. Thomas, and J. Van Meerbergen, *I.E.E.E. Trans. Electron Devices*, 1978, **ED25**, 507. [h] H. Matsunami, S. Matsumoto and T. Tanaka, *Japan. J. Appl. Phys.*, 1977, **16**, 1491. [i] J. I. B. Wilson, J. McGill, and S. Kinmond, *Nature*, 1978, **272**, 152. [j] J. I. B. Wilson and P. Robinson, *Solid-State Electronics*, 1978, **21**, 489. [k] S. S. Chu, T. L. Chu, and H. T. Yang, *Appl. Phys. Letters*, 1978, **32**, 557.

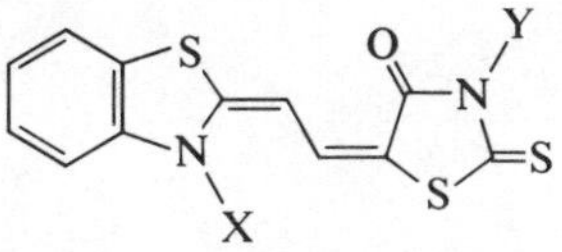

(1)

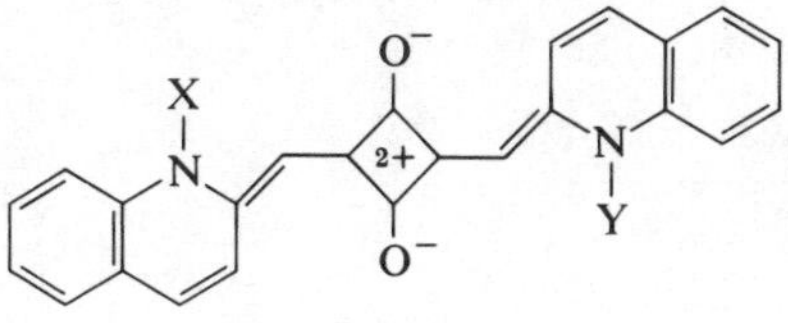

(2)

$i_{so} = 1.8$ mA cm^{-2} for a cell of 1 cm^2 area). The open-circuit voltage, V_{oc}, is determined by the difference in the work functions of the two electrodes but the quantum efficiency of current generation is less straightforward, approaching 100% at only high photon energies and averaging only about 35% in the main absorption region of the dye. The short-circuit current is reported to diminish by 10% in 8 h of exposure to white light.

Cells of conversion efficiency *ca.* 0.5% have been made by barrier layer cells containing thin copper phthalocyanine[115] and indium phthalocyanine.[116]

These much improved performances are probably due in part to the successful use of thin organic films in which the probability of hole–electron recombination or exciton annihilation in the bulk is minimized: improved purity may have a role to play here, though water and oxygen can both increase cell outputs, possibly at the expense of long-term stability. Techikawa and Faulkner,[117] in a most interesting paper comparing the electrochemical and solid-state behaviour of thin films of phthalocyanines, found dramatic changes in resistance in the cell Au | ZnPC (α-phase) | Au resulting from exposure to moist air.

Corker and Lundstrom[118] have examined both the d.c. and the a.c. electrical properties of metal | chlorophyll *a* | Hg sandwich cells and found evidence for a Schottky barrier of width *ca.* 45 nm at the Cr | Chl junction which is responsible for the photovoltaism of cells based on Cr.

Rosseinsky and co-workers[119] found small photovoltaic effects (105 and 30 mV) in heterodiscs of methylene blue: metal phosphate, where the metal was Mn^{II} and Fe^{II} respectively. The Fe^{II} discs were unstable, and discs containing Co^{II} phosphate were photovoltaically inactive. Photovoltaic activity has also been observed in a cell consisting of multilayered films of purple membrane (bacteriorhodopsin) and lipid sandwiched between two palladium electrodes.[120]

[115] C. W. Tang, *Res. Discl.*, 1977, **162**, 71.
[116] V. A. Benderskii, M. I. Alyanov, M. I. Fedorov, and L. M. Fedorov, *Doklady Akad. Nauk S.S.S.R.*, 1978, **239**, 856.
[117] H. Tachikawa and L. R. Faulkner, *J. Amer. Chem. Soc.*, 1978, **100**, 4379.
[118] G. A. Corker and I. Lundstrom, *Photochem. and Photobiol.*, 1977, **26**, 139; *J. Appl. Phys.*, 1978, **49**, 686.
[119] D. R. Rosseinsky, R. E. Malpas, and T. E. Booty, *Transition Metal. Chem.*, 1978, **3**, 254.
[120] S.-B. Hwang, J. I. Korenbrot, and W. Stoeckenius, *Biochim. Biophys. Acta*, 1978, **509**, 300.

Part VI

CHEMICAL ASPECTS OF PHOTOBIOLOGY

By G. BEDDARD

1 Introduction

The emphasis of this Report is on the light harvesting and charge separation in photosynthesis and on the first few intermediates found in the proton pump of the purple membrane and in the visual process. Only papers from mid-1976 to the Autumn of 1978 have been reviewed.

2 Photosynthesis

Introduction.—In photosynthesizing organisms several hundred chlorophyll (Chl) molecules and various accessory pigments act collectively to transfer the absorbed light energy into a specialized Chl-containing complex called the reaction centre (r.c.) at which charge separation occurs. In the r.c. the primary electron donor is postulated to be a specialized Chl dimer. This transfers an electron to the primary acceptor, a plastoquinone in many systems. In green plants the electron is replaced in the primary donor (P) by the oxidation of water and the electron from the acceptor eventually results in the reduction of carbon dioxide. Furthermore, in green plants two photosystems called PS II and PS I co-operate to facilitate the generation of oxidants and reductants. In photosynthetic bacteria, only one photosystem is present since compounds with a smaller redox potential than water are used.

Figure 1 shows in schematic form the intermediates involved in the Z scheme electron transport pathway of green plants. A similar scheme for bacteria is shown in Figure 6.

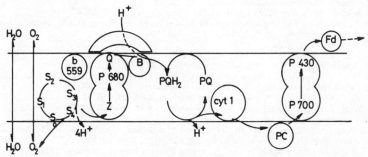

Figure 1 *The electron transport chain of green plant photosynthesis arranged as two proton pumps (the protons shown are not meant to indicate any particular stoicheiometry)* (Reproduced by permission from 'Current Topics in Bioenergetics–Photosynthesis Part A', ed. D. Sanadi and L. Vernon, Academic Press, London, 1977, Vol. 7, p. 216)

Many reviews dealing with the primary processes of photosynthesis have appeared. These include PS I,[1—5] PS II,[2—4, 6—8] bacterial reaction centres,[2, 9—11] antennae systems and picosecond fluorescence,[4, 9—19] Chl triplet states and e.s.r.,[20, 21] delayed fluorescence,[22] *in vitro* photosynthesis and properties of Chl-like compounds,[16, 23—29] and books and conference reports on photosynthesis.[14, 30—33]

Models for Photosynthetic Reaction Centres.—Various lines of evidence relating to the molecular organization of chlorophylls in photosynthetic reaction centres[1, 2] of both bacteria and green plants[29, 30] suggest that special chlorophyll dimers ('special pairs') act as the primary electron donors.

In Photosystem I (PS I) the reaction centre has a characteristic absorbance at $\sim$700 nm due to an unknown species P700. A model of P700 has been synthesized by Boxer and Closs[34] where two pyrochlorophyll macrocycles are joined by a ten-carbon link to form a bis(pyrochlorophyllide-a) ethylene glycol diester. Only in polar and hydrogen-bonding solvents does the molecule fold

1 B. Ke in 'Current Topics in Bioenergetics – Photosynthesis Part A', ed. D. Sanadi and L. Vernon, Academic Press, London, 1977, Vol. 7, p. 76.
2 R. Blankenship and W. Parson, *Ann. Rev. Biochem.*, 1978, **47**, 635.
3 P. Loach, *Photochem. and Photobiol.*, 1977, **26**, 87.
4 J. Barber, *Developments Bioenerg. and Biomembr.*, 1977, **1**, 459.
5 E. Land, *U.V. Spectrum Group Bulletin*, 1977, **5**, 56.
6 D. Knaff, *Photochem. and Photobiol.*, 1977, **26**, 327.
7 D. Knaff and R. Malkin in ref. 1, p. 139.
8 H. Boehme, *Z. Naturforsch.*, 1978, **33c**, 161.
9 D. Holten and M. Windsor. *Ann. Rev. Biophys. Bioeng.*, 1978, **7**, 189.
10 M. Siebert in ref. 1, p. 39.
11 *Photochem. and Photobiol.*, Dec. 1978.
12 K. Sauer, *Accounts Chem. Res.*, 1978, **11**, 257.
13 G. Busch and P. Rentzepis, *Trends Biochem. Sci.*, 1977, **2**, N253.
14 'Topics in Photosynthesis – Primary Processes in Photosynthesis', ed. J. Barber, Elsevier, Amsterdam, 1977, Vol. 2.
15 N. Boardman, in 'The Molecular Biology of Plant Cells', ed. H. Smith Oxford University Press, Oxford 1977, vol. 14, p. 85.
16 A. Glaser, *Mol. Cell Biochem.*, 1977, **18**, 125.
17 W. Butler, *Ann. Rev. Plant Physiol.*, 1978, **29**, 345.
18 G. Harnischfeger, *Adv. Bot. Res.*, 1977, **5**, 1.
19 W. Parson, *Chem. and Biochem. Appl. Lasers*, 1974, **1**, 339.
20 A. Levanon and J. Norris, *Chem. Rev.*, 1978, **3**, 185.
21 T. Moore, *Photochem. and Photobiol.*, 1977, **26**, 75.
22 J. Amesz and A. Van Gorkom, *Ann. Rev. Plant Physiol.*, 1978, **29**, 47.
23 G. Porter, *Proc. Roy. Soc. (London)*, 1978, **A362**, 281; G. Porter and M. Archer, *Inter-Interdisiplin. Sci. Rev.*, 1976, **1**, 119.
24 G. Seely in ref. 1, p. 3.
25 M. Calvin, *Accounts Chem. Res.*, 1978, **11**, 369.
26 R. Fugate and P-s. Song, *Photochem. and Photobiol.*, 1976, **24**, 629.
27 H. Witt, 'Living Syst. Energy Converters – Proc. European Conference', ed. R. Buvet, M. Allen, and J. Massue, North Holland, Amsterdam, 1976, p. 185.
28 T. Lee and R. Fugate, *Photochem. and Photobiol.*, 1978, **27**, 803.
29 J. Katz, J. Norris, L. Shipman, M. Thurnauer, and M. Wasielewski, *Ann. Rev. Biophys. Bioeng.*, 1978, **7**, 393.
30 'Chlorophyll Proteins, Reaction Centres and Photosynth. Membranes', ed. J. Olson and G. Hind, Brookhaven Symposia in Biology, 28, 1976, BNL 50530.
31 'The Science of Photobiology', ed. K. Smith, Plenum Press, New York, 1977.
32 'Research in Photobiology; Proceedings of the Seventh International Congress on Photobiology', ed. A. Castellani, Plenum Press, New York, 1977.
33 'Photochemistry and Photobiology Review, 1976', ed. K. Smith, Plenum Press, New York, 1976, Vol. 1.
34 S. Boxer and G. Closs, *J. Amer. Chem. Soc.*, 1976, **98**, 5406.

to form a bridged sandwich complex. The proposed structure is shown in Figure 2. Neither the $(H_2\text{-pyrochl-CO}_2\text{CH}_2-)_2$ nor the $Mg\text{-pyrochl-CO}_2\text{CH}_3$ have absorbances at 695 nm where the folded dimer absorbs. The flexible chain linkage reduces the $T\Delta S$ term rather than forcing the molecule into a particular geometry.

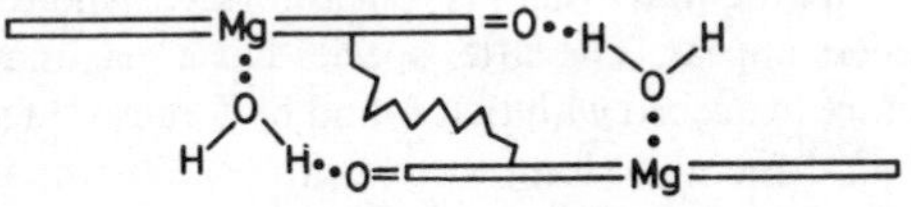

Figure 2 *The proposed bridged sandwich complex structure of P700 in polar and hydrogen bonding solvents*

A similar compound has been synthesized by Wasielewski *et al.*[35] who demonstrated that the radical cation of the dimer is similar to that observed *in vivo*. Flash photolysis studies on the pyrochlorophyllide dimers demonstrates[36] that, upon excitation, the folded species forms an excited singlet and then a triplet state. This folded triplet then converts into an unfolded triplet and decays to an unfolded ground state, from which the folded ground state is regenerated. There is no evidence for a $(Chl^+\text{--}Chl^-)$ species, possibly due to the high symmetry of the system. A dimeric bacteriochlorophyll compound has also been synthesized[37] but, unlike the chlorophyllide analogue, the absorption peaks (780 and 812 nm) do not correspond to the *in vivo* r.c.'s absorption (865 nm). Since the e.s.r. properties of the dimers are similar to those *in vivo*, Wasielewski *et al.*[37] suggest that the optical and e.s.r. properties monitor different Chl molecules. The synthesis of several bridged porphyrins has been reported.[38, 39] All the covalently bonded dimers synthesized show remarkably different properties in the open and closed configurations.

A bis-Chl cyclophane in which the two Chl rings lie on top of one another has been synthesized.[40] The absorption and emission spectra indicate that no interaction occurs between the rings but upon photo-oxidation an e.s.r. signal similar to that of $P700^+$ is observed. Kagan *et al.*,[41] however, did note some interaction between the rings in similar compounds.

Stimulated emission at 735 nm from several Chl dimers has been observed but only from the folded forms.[42] The fluorescence lifetime of the folded dimers is slightly less than that of the open ones.

There has been considerable controversy over the nature of the dimeric chlorophyll water species. Essentially water-free chlorophyll can be obtained

[35] M. Wasielewski, M. Studier, and J. Katz, *Proc. Nat. Acad. Sci. U.S.A.*, 1976, **73**, 4282.
[36] N. Periasamy, H. Lindschitz, G. Closs, and S. Boxer, *Proc. Nat. Acad. Sci. U.S.A.*, 1978, **75**, 2563.
[37] M. Wasielewski, V. Smith, B. Cope, and J. Katz, *J. Amer. Chem. Soc.*, 1977, **99**, 4172.
[38] J. Collman, C. Elliot, T. Halbert, and B. Torrog, *Proc. Nat. Acad. Sci. U.S.A.*, 1977, **74**, 18.
[39] A. Ogoshi, A. Sugimoto, and Z. Yoshida, *Tetrahedron Letters*, 1977, **2**, 169.
[40] M. Wasielewski, W. Svec, and B. Cope, *J. Amer. Chem. Soc.*, 1978, **100**, 1961; L. Shipman and J. Katz, *J. Phys. Chem.*, 1977, **81**, 577.
[41] N Kajan, D Mauzerall, and R. Merrifield, *J. Amer. Chem. Soc.*, 1978, **99**, 5484.
[42] J. Hindman, R. Kugel, M. Wasielewski, and J. Katz, *Proc. Nat. Acad. Sci., U.S.A.* 1978, **75**, 2076.

by vigorous drying of Chl and solvent,[43] although under most circumstances water is bonded onto the Mg of the Chl. Some workers,[44] however, claim that the ratio of water to the Chl cannot be reduced to a value of less than 1 : 1, except in extreme circumstances. In a series of papers Fong and co-workers[44—50] have detailed the conditions necessary for different types of complexes to be formed. In organic solvents with a little water $(Chl–H_2O)_2$ is formed, whereas in excess water $(Chl–2H_2O)_n$ polymers appear. The latter species has a maximum absorption at 743 nm and a structure in accord with that found by Strause[51] for chlorophyllide species. Cotton *et al.*[43] found no long wavelength ($\sim$700 nm) absorbing species in dried solution but such an absorption appears on adding water or other species with OH, NH_2, NH, or SH groups; some Chl self aggregation also occurs. Calculations of the electronic spectra of hydrated Chl stacks[40,49,54] place most of the oscillator strength in the lowest energy transition.

Three forms of $(Chl–H_2O)_2$ species are proposed where linkages are between the carbomethoxy-groups at C-7 and C-10 of each Chl or between the C-9 keto groups.[45] A comparison of absorption[45] and e.s.r.[53] spectra has led to the assignment by Fong of an *endo* C-10 keto carbonyl bonded $(Chl–H_2O)_2$ as the analogue of P700, the PS I reaction centre. This has been criticized by Katz and co-workers[43,54] who favoured an *exo* C-9 ester carbonyl complex. Simultaneous measurements of zero field splitting and fluorescence parameters on Chl–water adducts, at 4.2 K, provide evidence in favour of an ester carbonyl complex (the Shipman–Katz model) rather than the keto carbonyl complex of Fong. Figure 3 gives diagrammatic representations of the two types of Chl–water dimers.

These data on the Chl–water species and those of the bridged Chl species support the proposal that the 'special pair' dimer geometry can account for the redox, spin localization, and absorption properties observed *in vivo*. Additionally, the absorption properties of any *in vivo* Chl dimer species would be greatly influenced by the presence of additional chromophores and by Chl protein interactions.[43] A number of the comparisons between *in vivo* and *in vitro* species can only be made at low temperatures and raise questions of the validity of Chl dimeric species as models of the reaction centre. Possibly, interactions with the protein can sufficiently alter the Chl dimer properties to match the *in vivo* r.c. properties at room temperature.

Properties of Chlorophyll *in vitro*.—The spectroscopy of Chl *in vivo*, and in

[43] T. Cotton, P. Loach, J. Katz, and K. Ballschmitter, *Photochem. and Photobiol.*, 1978, **27**, 735.
[44] J. Brace, F. Fong, D. Karweik, V. Koester, A. Sheperd, and N. Winograd, *J. Amer. Chem. Soc.*, 1978, **100**, 5203.
[45] F. Fong, V. Koester, and L. Galloway, *J. Amer. Chem. Soc.*, 1977, **99**, 2372.
[46] F. Fong, V. Koester, and J. Polles, *J. Amer. Chem. Soc.*, 1976, **98**, 6406.
[47] F. Fong and W. Wassam, *J. Amer. Chem. Soc.*, 1977, **99**, 2375.
[48] L. Galloway, J. Roettger, D. Fruge, and F. Fong, *J. Amer. Chem. Soc.*, 1978, **100**, 4635.
[49] F. Fong and W. Wassam, *J. Amer. Chem. Soc.*, 1977, **99**, 2375.
[50] F. Fong, J. Polles, L. Galloway, and D. Fruge, *J. Amer. Chem. Soc.*, 1977, **99**, 5802.
[51] C. Strause, *Proc. Inorg. Chem.*, 1976, **21**, 159.
[52] R. Kooyan, T. Schaafsma, and J. Kleibeuker, *Photochem. and Photobiol.*, 1977, **26**, 235.
[53] F. Fong, A. Hoff, and F. Brinkman, *J. Amer. Chem. Soc.*, 1978, **100**, 619.
[54] J. Katz, in ref. 30.

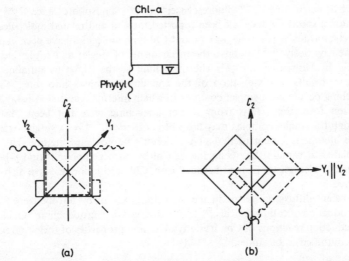

Figure 3 *Simplified representation of proposed structures of the two types of Chl–water dimers: (a) structure related to the one proposed by Fong with y-axes highly inclined; (b) structure related to the one proposed by Shipman et al. with the y-axes little inclined* (Reproduced by permission from *Biochim. Biophys. Acta*, 1977, **462**, 73)

synthetic systems, has been reviewed by Brown[55] and the properties of Chl in bilayer membranes has also been discussed in detail.[56] Photosensitive[57] and asymmetrical lipid[58] bilayers have been produced. Photoelectron spectroscopy has been found to be a sensitive detector of assymetry in BChl bilayers.[59]

Non-linear absorption of Chl *in vivo*,[41, 60—62] and *in vitro*,[63] and stimulated emission from folded Chl dimers,[42] and Chl *in vitro*,[64—66] but not *in vivo*, have been reported. The triplet states,[67—71] energy levels,[72] dipole strengths,[73]

[55] J. Brown, *Photochem. and Photobiol.*, 1977, **26**, 319.
[56] *Photochem. and Photobiol.*, 1976, Vol. 24.
[57] C. Chen and D. Berns, *Photochem. and Photobiol.*, 1976, **24**, 255.
[58] S. Brody and S. Brody, *Photochem. and Photobiol.*, 1977, **26**, 57.
[59] J. Amend, W. Sistrom, and O. Griffith, *Biophys. J.*, 1978, **21**, 195.
[60] D. Leupold, S. Mory, and O. Hoffmann, *Acta Phys. Chem.*, 1977, **23**, 33.
[61] D. Leupold, B. Voight, S. Mory, P. Hoffmann, and B. Hieke, *Biophys. J.*, 1978, **21**, 177.
[62] D. Leupold, B. Voight, S. Mory, R. Koenig, and P. Hoffmann, in 'Lasers in Chemistry', Proc. Conf., ed. M. A. West, Elsevier, Amsterdam, 1977, p. 299.
[63] B. Hieke, P. Hoffmann, D. Leupold, S. Mory, and J. Schoffe, *Photosynthetica*, 1979, **13**, 37.
[64] J. Hindman, R. Kugel, A. Svirmikas, and J. Katz, *Proc. Nat. Acad. Sci. U.S.A.*, 1977, **74**, 5.
[65] D. Leupold, P. Mory, B. Hieke, and R. Konig, *Chem. Phys. Letters*, 1977, **45**, 567.
[66] P. Mathis, J. Breton, A. Vermeglio, and M. Yates, *F.E.B.S. Letters*, 1976, **63**, 171.
[67] N. Andreeva, G. Zakharova, V. Shubin, and A. Chibisov, *Chem. Phys. Letters*, 1978, **53**, 317.
[68] W. Haegele, D. Schmid, and H. Wolff, *Z. Naturforsch.*, 1978, **33a**, 94.
[69] K. Solov'ev, A. Gradyushko, and M. Tsvirko, *Tezisy Doklady Vses. Soveshch. Lyumin 22nd*, 1975, 102.
[70] R. Burgner and A. Ponte-Goncalves, *Chem. Phys. Letters*, 1977, **46**, 275.
[71] R. Avaarma, *Chem. Phys. Letters*, 1977, **46**, 279.
[72] A. Mau and M. Puza, *Photochem. and Photobiol.*, 1977, **25**, 601.
[73] L. Shipman, *Photochem. and Photobiol.*, 1977, **26**, 287.

π-electron structure,[74—76] fluorescence,[40,77—80] phosphorescence,[81,82] and vibrational spectra [83] have all been reported for Chl and related molecules.

Electron transfer reactions of Chl and Chl-type molecules have been expertly reviewed by Seely [84,85] as have the mechanisms of singlet and triplet electron transfer by Holten *et al.*[86—87] The quenching of the triplet by quinones, for example, results in dissociation of the encounter complex into ions. Singlet quenching occurs *via* a singlet complex in which internal conversion is favoured over ion formation. Ions from singlet quenching have not been observed although the singlet complex may have been observed.[86—88] A singlet exciplex, whose lifetime is similar to that of the donor's fluorescence lifetime, has been postulated to explain the quenching of Chl by nitro compounds and *p*-benzoquinone.[89] Complexes formed between iron, Chl, and quinones have also been observed.[90]

Transient diffusional terms in the quenching of Chl by nitrobenzene [91] and the production of cation radicals,[92,93] are observed during quenching. Oxidation and reduction reactions of Chl, its derivatives and properties of their aggregates, have continued to be studied.[78,—89,94—117]

[74] H. Nagae and S. Yomosa, *J. Phys. Soc. Japan*, 1977, **42**, 988.
[75] D. Spangler, G. Maggiora, L. Shipman, and R. Christoffersen, *J. Amer. Chem. Soc.*, 1977, **99**, 7470.
[76] D. Spangler, G. Maggiora, L. Shipman, and R. Christoffersen, *J. Amer. Chem. Soc.*, 1977, **99**, 7478.
[77] V. Raskin, G. Gurevich, and V. Yunevich, *Biosint. Sostoyanie Khlorofillov Rast*, 1975, ed. A. Shlyk, p. 19 (*Chem. Abs.*, **88**, 34663).
[78] V. Gheorghe, L. Tugulea, and E. Vasile, *An. Univ. Bucuresti Fiz.*, 1977, **26**, 59.
[79] M. Kaplanova, *Sb. Ved. Pr. Vys. Sk. Chemickotechnol. Pardubice*, 1975, **34**, 101 (Chem. Abs., **86**, 148058).
[80] F. Litvin, I. Stadnichuk, and V. Kruglov, *Biofizika*, 1978, **23**, 450.
[81] L. Boguslavskii, L. Zhuravlev, M. Kandelaki, and K. Shengeliya, *Doklady Akad. Nauk S.S.S.R.*, 1978, **240**, 1453.
[82] A. Krasnovskii, *Acta Phys. Chem.*, 1977, **23**, 147.
[83] M. Lutz, J. Kleo, R. Gilet, M. Henry, R. Plus, and J. Leicknam, 'Proceedings of the International Conference on Stable Isotopes, 2nd', ed. R. Klein and P. Klein, Springfield 1975, p. 462.
[84] G. Seely, *Photochem. and Photobiol.*, 1978, **27**, 639.
[85] G. Seely, in 'Current Topics in Bioenergetics', ed. L. Vernon and G. Seely, Academic Press, New York, 1978, Vol. 7.
[86] D. Holten, M. Gouterman, M. Windsor, and M. Rockley, *Photochem. and Photobiol.*, 1976, **23**, 415.
[87] M. Gouterman and D. Holten, *Photochem. and Photobiol.*, 1977, **25**, 85; D. Holton, M. Gouterman, W. Parson, and M. Windsor, *Photochem. and Photobiol.*, 1978, **28**, 951.
[88] D. Huppert, P. Rentzepis, and G. Tollin, *Biochim. Biophys. Acta*, 1976, **440**, 356.
[89] S. Bondarev, G. Gurinovich, B. Dzhagarov, and K. Salokhiddinov, *Tezisy Doklady Vses. Soveshch. Lyumin*, 22nd, 1975, 103.
[90] G. Tollin and F. Rizzuto, *Photochem. and Photobiol.*, 1978, **27**, 487.
[91] G. Beddard, S. Carlin, L. Harris, G. Porter, and C. Tredwell, *Photochem. and Photobiol.*, 1978, **27**, 433.
[92] A. Kostikov, N. Sadovnikova, and V. Evstigneev, *Biofizika*, 1976, **21**, 803.
[93] E. Elfimov, V. Voznyak, T. Kantseva, and V. Evstigneev, *Biofizika*, 1977, **22**, 1004.
[94] G. Olovyanishnikova and V. Voznyak, *Deposited Doc.*, 1973, *Viniti*, 7176, 68.
[95] I. Sheshukova, *Deposited Doc.*, 1973, *Viniti*, 7176, 88.
[96] A. Rasquain, G. Houssier, and C. Sironval, *Biochim. Biophys. Acta*, 1977, **462**, 622.
[97] E. Sagun and B. Dzhagarov, *Tezisy Doklady Vses. Soveshch. Lyumin.* 22nd, 1975, 18.
[98] V. Kreslavskii, O. Shadrikov, V. Romanov, and Yu. Stolovitskii, *Deposited Doc.*, 1973, *Viniti*, 7176, 42.

The interaction of Chl with lipids has been studied in detail.[118-122] Coupling between Chl and monogalactosyl diglyceride has been observed,[120,122,123] confirming earlier suggestions that such interactions could prevent self-quenching[124] during energy migration. Energy transfer between Chl[125] and from Chl to its dimers in micelles[115] and between several pheophytins in solution[116] have been measured.

Alternative Schemes of Photosynthetic Electron Transport.—A revised 'up-conversion' theory of photosynthesis has been proposed by Fong[126] in which the transfer of electrons through the reaction centre occurs *via* a long-lived triplet state. This 'up-conversion' theory has little experimental support and has also been strongly criticized elsewhere.[127,128]

Another alternative scheme to the Z scheme has been proposed by Arnold. Here two separate one-photon events are hypothesized and a new scheme for the flow of electrons is presented.[129,130]

99   B. Kiselev, V. Evstigneev and I. Tsygankova, *Doklady Acad. Nauk S.S.S.R.*, 1976, **230**, 726.
100 G. Cauzzo, G. Gennari, G. Jori, and J. Spikes, *Photochem. and Photobiol.*, 1977, **25**, 389.
101 J. Nakata, T. Imura, and K. Kawabe, *J. Phys. Soc. Japan*, 1977, **42**, 146.
102 K. Vacek, J. Naus, M. Svabova, E. Vavrinec, M. Kaplanova, and J. Hala, *Studia Biophys.*, 1977, **62**, 201.
103 K. Iriyama and M. Yoshiura, *Colloid Polymer Sci.*, 1977, **255**, 133.
104 G. Beddard, R. Davidson, and A. Trethowey, *Nature*, 1977, **267**, 373.
105 L. Boguslavski, A. Volkov, and M. Kanelaki, *Biofizika*, 1976, **21**, 808.
106 N. Gudkov and Yu. Stolovifskii, *Deposited Doc.*, 1973, *Viniti*, 7176, 11.
107 M. Sarzhevskaya and A. Losev, *Biokhimiya*, 1977, **42**, 2105.
108 Yu. Stolovitskii and V. Evstigneev, *Tezisy Doklady Vses. Konf. 9th*, 1976, 20.
109 B. Kiselev, Yu. Kozlov, and V. Evstigneev, in ref. 108, p. 16.
110 V. Blok, *Zhur. fiz. Khim.*, 1978, **52**, 245.
111 N. Samoshina, N. Novikova, and L. Nekrasov, *Vestnik Moskov. Univ. Ser. 2: Khim.*, 1978, **19**, 85.
112 R. Tamkivi, R. Avarmaa, *Izvest. Akad. Nauk S.S.S.R., Ser. fiz.*, 1978, **42**, 568.
113 E. Zen'kevich, G. Kochubeev, A. Losev, and G. Gurinovich, *Zhur. priklad. Spektroskopii*, 1977, **27**, 834.
114 A. Peshkin, T. Slavnova, and A. Chibisov, *Izvest. Akad. Nauk S.S.S.R., Ser. fiz.*, 1978, **42**, 388.
115 E. Lehoczki and L. Szalay, *Acta Phys. Chem.*, 1977, **23**, 161.
116 E. Zen'kevich, A. Losev, G. Kochubeev, and G. Gurinovich, *Izvest. Akad. Nauk S.S.S.R., Ser. fiz.*, 1978, **42**, 573.
117 W. Stillwell and H. Tien, *Biochim. Biophys. Acta*, 1977, **461**, 239.
118 C. Jones and R. Mackay, *J. Phys. Chem.*, 1978, **82**, 63.
119 N. Murata and N. Sato, *Plant Cell. Physiol.*, 1978, **19**, 401.
120 I. Ivnitskaya, G. Yakovenko, S. Manuil'skaya, and I. Dilung, *Dopovidi Akad. Nauk Ukrain R.S.R. Ser. B*, 1977, **7**, 637.
121 S. Otavin, *Dopovidi Akad. Nauk Ukrain R.S.R. Ser. B.*, 1978, **2**, 165.
122 A. Vecher, R. Koval'chuk, A. Mas'ko, and V. Reshetnikov, *Ret. Dokl. Soobsch – Mendeleevsk. S'ezd Obshchei Priklad. Khim.* 11th, 1975, **6**, 115.
123 L. De Kok, P. Van Hasselt, and P. Kuiper, *Physiol. Plant*, 1978, **43**, 7.
124 G. Beddard and G. Porter, *Nature*, 1976, **260**, 366.
125 K. Vacek, J. Naus, M. Svabova, E. Vavrinec, M. Kaplanova and, J. Hala, *Mol. Spectroscopy Dense Phases, Proc. European Congr. Mol. Spectroscopy 12th*, 1975, 463.
126 F. Fong, *J. Amer. Chem. Soc.*, 1976, **98**, 7840.
127 J. Warden, *Proc. Nat. Acad. Sci. U.S.A.*, 1976, **73**, 2773.
128 Govindjee and J. Warden, *J. Amer. Chem. Soc.*, 1977, **99**, 8088.
129 D. Arnon and R. Chain, *Plant Cell Physiol.*, 1977, **129**, 1977.
130 W. Arnold, *Proc. Nat. Acad. Sci. U.S.A.*, 1976, **73**, 4502.

3 Chlorophyll Protein Complexes

The photosynthetic membranes of green algae and higher plants can be separated into several different Chl protein complexes. Two of these complexes, the P700–Chl-a protein complex and the light-harvesting Chl-a/b protein have been well characterized. The former represents the pigment molecules forming Photosystem I (10% of total Chl); the latter is the major antennae component (55%) of higher plants[131–133] and is present in all Chl-b containing organisms. Also surrounding the reaction centres is a third Chl-a protein which acts as an antennae for PS I. Figure 4 gives a diagrammatic representation of these complexes.

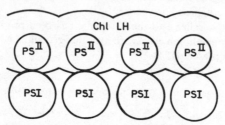

Figure 4 *Tripartite model represented as individual PS II units embedded in a matrix of light-harvesting Chl-a/b complex. Each PS II unit is connected to a PS I unit* (Reproduced by permission from *Proc. Nat. Acad. Sci. U.S.A.*, 1977, **74**, 3382)

The P700 protein complex has two parts, a trimer of 48 kD polypeptides each containing seven Chls, and a trimer of two 48 kD polypeptides with seven Chls each and one 46 kD polypeptide containing the reaction centre and a number of Chl molecules.[132] In the Chl-a/b protein, pigment/protein ratios indicate that three Chl-a and three Chl-b, one carotene, and less than one lipid molecule are present in each 30 kD polypeptide particle. Many of these units are aggregated to form the light-harvesting (LH) complex[132] (see ref. 134 for a review). Recently three new Chl protein complexes have been prepared,[135] two are similar to the Chl-a/b and are probably oligomers of this complex, the third is a Chl-a complex and this may be associated with PS II. The binding of Chl-a/b protein to ethylene diamine sepharose 4B indicates that a tetramer and not a trimer is the intrinsic structural unit in the chloroplast membrane.[136]

The topology of the proteins involved and their organization in the membrane is still an open question, however, but the schemes such as that in Figure 4 are now widely accepted.

The interaction of Chl in photosynthetic membranes at 35 K has been studied by Lutz,[137] using resonance Raman spectroscopy. Most Chl-a molecules are

[131] J. Thornber, R. Alberte, F. Hunter, J. Shiozawa, and S. Kan, in ref. 30, p. 132.
[132] J. Anderson and J. Barrett, 'Chlorophyll Organisation in the Reaction Centre and Energy Transfer Processes', Ciba Foundation 61, 1979, p. 81.
[133] R. Alberte and J. Thornber, *F.E.B.S. Letters*, 1978, **91**, 126.
[134] N. Boardman, J. Anderson, and D. Goodchild, 'Current Topics in Bioenergetics', ed. D. Sanadi and L. Vernon, Academic Press, New York, 1978, Vol. 8, p. 36.
[135] F. Henriques and R. Park, *Biophys. Res. Commn.*, 1978, **81**, 1113.
[136] E. Gross, *Biochemistry*, 1978, **17**, 806.
[137] M. Lutz, *Biochim. Biophys. Acta*, 1977, **460**, 408.

bound, possibly to the proteins, *via* hydrogen bonds at the 9-C=O group, and the Mg atom in both Chl-a and -b is bonded to a fifth ligand, possibly water. The spectra indicate that at least five environmental sub-species of Chl-a exist, which are related to the species observed in absorption. In Chl-b the picture is not so clear but the 3-C=O as well as the 9-C=O groups are bonded. No evidence for Chl oligomers, as proposed by Katz and co-workers[138] for the antennae systems, was found. In addition to carbonyl and Mg bonding, the phytyl chain and ester carbonyl can interact with the lipid and protein environment to ensure bonding of the Chl to the polypeptide chains and thus determine distance and relative orientation of the Chl matrix.

4 Fluorescence from Photosynthetic Systems

The fluorescence from photosynthetic systems can be characterized into 'prompt' and 'variable' fluorescence in the presence of continuous illumination. Virtually all the fluorescence is emitted by the Chl in the light-gathering systems and none from the reaction centre, excepting luminescence caused by charge recombination in these centres. The variable fluorescence monitors how the system responds when photochemical traps are becoming closed and energy is redistributed from PS II antennae to PS I antennae. This redistribution ensures that both PS I and PS II receive a similar number of excitations as is required for efficient electron transport, according to the Z scheme in Figure 1. This redistribution process is sometimes called spill-over.[139–141] To explain the kinetic properties of the fluorescence which are associated with the energy redistribution, Butler and co-workers[142, 143] have proposed a tripartite model which allows an analysis of the initial photon distribution and its subsequent evolution; this is shown in Figure 4.

Energy can be absorbed by any of the pigments in the light-harvesting or antennae systems and eventually arrives at the reaction centres. From the proposed topology of the photosynthetic unit, energy in the LH complex may pass into PS II or PS I antennae and that absorbed in PS II antennae into PS I antennae, but has only a small probability of going into the LH complex.

Fluorescence Decay Times.—Considerable effort has been put into measuring fluorescence decay times *in vivo*. It has been found that excited state annihilation occurs between excited Chl molecules in the light-gathering pigments when high intensity laser sources are used to excite the fluorescence, and leads to anomalously short lifetimes. Annihilation occurs in *Chlorella*,[144–146] chloro-

[138] L. Shipman, T. Cotton, J. Norris, and J. Katz, *J. Amer. Chem. Soc.*, 1976, **98**, 8222.
[139] J. Barber, 'Topics in Photosynthesis – The Intact Chloroplast', ed. J. Barber, Elsevier, Amsterdam, 1976, Vol. 1, p. 89.
[140] A. Ley and W. Butler, *Proc. Nat. Acad. Sci. U.S.A.*, 1976, **73**, 3957.
[141] A. Ley and W. Butler, *Biochim. Biophys. Acta*, 1977, **462**, 290.
[142] W. Butler and R. Strasser, *Proc. Nat. Acad. Sci. U.S.A.*, 1977, **74**, 3382.
[143] W. Butler, in ref. 30, p. 338.
[144] G. Porter, J. Synowiec, and C. Tredwell, *Biochim. Biophys. Acta*, 1977, **459**, 329.
[145] C. Tredwell, G. Porter, J. Synowiec, J. Barber, G. Searle, and L. Harris, in ref. 62, p. 304.
[146] J. Barber, C. Tredwell, and G. Porter, *U.V. Spectrom. Group Bulletin*, 1977, **5**, 65.

plasts,[147—154] photosynthetic bacteria,[155,156] and red algae.[157,158] Such effects do not occur with low intensity sources and much longer and more accurate lifetimes are measured under these conditions.[159—162]

The migratory nature of the excitation before trapping makes possible the annihilation and the probable quenching processes are $S + S \rightarrow T + T$; $S + T \rightarrow T + T$ (or $S + T$); $S + S \rightarrow$ ions and $S + T \rightarrow$ ions.[144,152,163] Triplets can be formed by a single laser pulse or from pulse trains.[144,152]

Lower bounds to the extent of excitation migration in the membrane of 200 Å, an energy transfer rate of $>10^{11}\,s^{-1}$, and a diffusion coefficient of $>10^{-3}\,cm^2\,s^{-1}$ have been estimated.[147,148,163] It was assumed that the excitation diffuses freely among the pigments before trapping, that the pigments are in a regular two-dimensional array, and that diffusion is sufficiently fast to randomize the initial excitation distribution before annihilation occurs. If the last condition is not fulfilled, then time-dependent diffusion terms need to be included in the analysis.[149]

Experiments at low temperatures indicate that most annihilation occurs in the LH complex.[50] The intensity of a 735 nm emission in PS I, which occurs only at low temperatures,[164] is quenched in unison with the 685 and 695 nm emission from the Chl-a/b (LH) complex, but the lifetime of the 735 nm emission is unchanged and is 1.5 ns.[153,165,166] This red emission therefore monitors the energy transferred to PS I but is probably not PS I antennae emission.[167] The species emitting at 735 nm also traps energy as efficiently as P700 at $-196\,°C$, but at room temperature does not compete with P700.[167] Using intense microsecond pulses or mode-locked pulse trains of the same envelope, the 735 nm emission is quenched more than the 685 nm, indicating a preferential build up

147 C. Swenberg, N. Geacintov, and M. Pope, *Biophys. J.*, 1976, **16**, 1447.
148 N. Geacintov, J. Breton, C. Swenberg, and G. Paillotin, *Photochem. and Photobiol.*, 1977, **26**, 629.
149 G. Beddard and G. Porter, *Biochim. Biophys. Acta*, 1977, **459**, 329.
150 N. Geacintov, J. Breton, C. Swenberg, A. Campillo, R. Hyer, and S. Shapiro, *Biochim. Biophys. Acta*, 1977, **461**, 306.
151 J. Breton and N. Geacintov, *F.E.B.S. Letters*, 1976, **69**, 89.
152 N. Geacintov and J. Breton, *Biophys. J.*, 1977, **17**, 1.
153 A. Campillo and S. Shapiro, in 'Picosecond Phenomena; Advances in Chemical Physics', ed. C. Shank, E. Ippen, and S. Shapiro, Springer. Verlag, Berlin, 1978, vol. 4, p. 140.
154 W. Yu, F. Pellegrino, and R. Alfano, *Biochim. Biophys. Acta*, 1977, **460**, 171.
155 A. Campillo, R. Hyer, T. Monger, W. Parson, and S. Shapiro, *Proc. Nat. Acad. Sci. U.S.A.*, 1977, **74**, 1997.
156 T. Monger, R. Cogdell, and W. Parson, *Biochim. Biophys. Acta*, 1976, **449**, 136.
157 G. Porter, C. Tredwell, G. Searle, and J. Barber, *Biochim. Biophys. Acta*, 1978, **501**, 232.
158 G. Searle, J. Barber, G. Porter, and C. Tredwell, *Biochim. Biophys. Acta*, 1978, **501**, 246.
159 G. Beddard, G. Fleming, G. Porter, and C. Tredwell, in ref. 153 p. 149.
160 F. Moya, Govindjee, C. Vernotte, and J. Briantais, *F.E.B.S. Letters*, 1977, **75**, 13.
161 K. Sauer and G. Brewington, *Energy Res. Abs.*, No. 15 918, 1978, Vol. 3, Report 1977, LBL-6179 (*Chem. Abs.*, **89**, 39581).
162 K. Sauer and G. Brewington, 'Proceedings of the International Congress on Photosynthesis', ed. D. Hall, J. Coombs, and T. Goodwin, Biochemical Society, London, 1977, p. 409.
163 G. Paillotin, in ref. 162, p. 33.
164 Z. Abilov and R. Gasanov, *Studia Biophys.*, 1978, **68**, 155.
165 A. Campillo, S. Shapiro, N. Geacintov, and C. Swenberg, *F.E.B.S. Letters*, 1977, **83**, 316.
166 G. Searle, J. Barber, L. Harris, G. Porter, and C. Tredwell, *Biochim. Biophys. Acta*, 1977, **459**, 390.
167 K. Satoh and W. Butler, *Biochim. Biophys. Acta*, 1978, **502**, 103.

of long-lived quenchers in PS I.[150] Furthermore, the 735 nm emission was shown to have a 150 ps rise-time,[153,165] which is much slower than the instrumental response. Campillo *et al.*[153,165] suggest that this may be due to a few Chls in the LH complex being initially associated with PS I while the rest are only loosely associated and transfer to PS I takes some time. In similar experiments in the author's laboratory, however, no rise time in the 735 nm emission was noticed.[168]

The observation of lasing from Chl compounds[42,64,169] at high excitation intensities *in vitro* has been proposed as an alternative explanation for the observed *in vivo* fluorescence quenching, but no lasing from any *in vivo* system has been observed[61,65,152] and is unlikely in view of the efficient transport of energy to the photochemical traps.

Most research groups are now in general agreement on the mean decay time for *Chlorella*, chloroplasts, and PS I particles at room temperature.[150,153,159,161,162,165,166,170–173] Yu *et al.*[154] observe slightly shorter decay times for chloroplasts, possibly due to slight annihilation effects. The lifetimes at room temperature and when dark-adapted are 490 ps for *Chlorella*, 453 ps for chloroplasts,[160] and 113 ps for PS I particles.[159] When photochemistry is blocked with DCMU and continuous illumination, the PS II reaction centres become closed and quench the fluorescence less efficiently, and a decay of 1.5 ns in *Chlorella* and 1.3 ns in chloroplasts is measured.[159,166 172] Only Harris *et al.*[172] and Searle *et al.*[166] analysed the decays of *Chlorella* and chloroplasts according to the equation:

$$I(t) = \exp(-at - bt^{\frac{1}{2}})$$

which describes the effects of non-random distributions of excitations before quenching. They found that the $(-at)$ term is small compared with the non-linear $(-bt^{\frac{1}{2}})$ term. Campillo *et al.*,[150,156,165] however, found only exponential decays. With the extra precision available from photon counting over that of the streak camera, the decays can be shown to be more accurately described by two exponential terms.[159,161,162,170,171] Mean lifetimes measured by the phase method are in agreement with those calculated by other methods.[160] Furthermore, a nearly linear relationship observed between mean lifetime and fluorescence yield upon continuous illumination was proposed to support the hypothesis that the energy can migrate freely from one photosynthetic unit to another, *i.e.* the Chls are analogous to a lake in which excitation energy can move.[160,174] Data from fluorescence induction experiments[175–177] and some lifetime data[161,162] support a model in which limited migration between a few

168 C. Tredwell, personal communication.
169 J. Hindman, R. Kugel, A. Smirmikas, and J. Katz, *Chem. Phys. Letters*, 1978, **53**, 197.
170 G. Beddard, G. Fleming, G. Porter, G. Searle, and J. Synowiec, *Biochim. Biophys. Acta*, 1979, **545**, 165.
171 G. Beddard, G. Fleming, G. Synowiec, and G. Porter, *Biochem. J.*, 1978, **6**, 1385.
172 L. Harris, G. Porter, J. Synowiec, C. Tredwell, and J. Barber, *Biochim. Biophys. Acta*, 1976, **449**, 329.
173 J. Barber, G. Searle, and C. Tredwell, *Biochim. Biophys. Acta*, 1978, **501**, 174.
174 W. Williams, in ref. 14, p. 100.
175 M. Hipkins, *Biochim. Biophys. Acta*, 1978, **502**, 161.
176 S. Malkin and J. Barber, *Biochim. Biophys. Acta*, 1978, **502**, 524.
177 M. Hipkins, *Biochim. Biophys. Acta*, 1978, **502**, 514.

PSUs is possible. In chloroplasts the two decays of 453 ps and 1300 ps remain constant as the system moves from a state with traps open to one with some traps closed. The proportion of the 1.3 ns decay, due to units with closed reaction centres, increases upon illumination.[159,170,171] These data mean that there is little or no energy transfer from one PSU to another, in contradiction with data from less direct experiments.[175—177] Clearly more work is needed to solve this fundamental problem to the limits of energy migration.

The rates of migration and of trapping of the excitation have still to be measured and this prevents a comparison with various theoretical models.[163,178—182] Calculations from a random walk model yielded decays similar to those observed *in vivo*, and a time-dependent trapping rate was observed at short times, indicating that the initial excitation distribution was not randomized before trapping.[183] Quantum yield calculations and the ratio of Chl-a : Chl-b expected *in vivo* have been calculated from a random walk on a square lattice.[184]

Role of Bivalent Cations.—During the past decade much evidence has accumulated which indicates that bivalent cations such as Mg^{2+} can regulate the distribution of energy between PS II and PS I (see ref. 139 for a review of this topic). In the absence of Mg^{2+}, at room temperature and at 77 K, energy transfer from PS II to PS I has been demonstrated.[142,185—188] This spill-over of energy was enhanced by closing the reaction centres.

With added bivalent cations and continuous illumination, both the background and variable PS II fluorescence are enhanced[189] and PS I fluorescence is decreased.[190] The absorption spectrum of the chloroplasts was slightly altered.[189,191] Arntzen and co-workers[190,192—194] have shown that the Chl-a/b protein is essential for the cation-regulated process to occur, but the mechanisms of the induced fluorescence changes are still unclear. A change in the rate of energy transfer from PS II to PS I,[142,173] caused by a cation-induced conformational change in the membrane altering distances between certain molecules, and so changing the rate of Förster transfer, seems to be a possible mechanism for this effect.[191,192,195,196] Both an increase in light scattering

[178] G. Paillotin, *J. Theor. Biol.*, 1976, **58**, 237.
[179] T. Markvart, *J. Theor. Biol.*, 1978, **72**, 117.
[180] S. Kuprin, A. Kokushkin, and A. Tikhonov, *Biofizika*, 1977, **22**, 161.
[181] K. Colbow, 'Physics of Biological Membranes' (Lect. Summer Inst.), 1974, p. 430.
[182] S. Haan and R. Zwanzig, *J. Chem. Phys.*, 1978, **68**, 1879.
[183] J. Altmann, G. Beddard, and G. Porter, *Chem. Phys. Letters*, 1978, **58**, 54.
[184] C. Swenberg, R. Pominijani, and N. Geacintov, *Photochem. and Photobiol.*, 1976, **24**, 601.
[185] K. Satoh, R. Strasser, and W. Butler, *Biochim. Biophys. Acta*, 1976, **440**, 337.
[186] R. Strasser and W. Butler, *Biochim. Biophys. Acta*, 1977, **462**, 295.
[187] W. Butler and R. Strasser, *Biochim. Biophys. Acta*, 1977, **462**, 283.
[188] R. Strasser and W. Butler, *Biochim. Biophys. Acta*, 1977, **460**, 230.
[189] B. Henkin and K. Sauer, *Photochem. and Photobiol.*, 1977, **26**, 277.
[190] J. Burke, C. Ditto, and C. Arntzen, *Arch. Biochem. Biophys.*, 1978, **187**, 252.
[191] E. Gross and J. Grenier, *Arch. Biochem. Biophys.*, 1978, **187**, 387.
[192] R. Lieberman, S. Bose, and C. Arntzen, *Biochim. Biophys. Acta*, 1978, **502**, 417.
[193] S. Bose, J. Burke, and C. Arntzen, *Dev. Bioeng. Biomembrane*, 1977, **1**, 245.
[194] C. Arntzen and C. Ditto, *Biochim. Biophys. Acta*, 1976, **449**, 259.
[195] J. Biggins and J. Svejkovsky, *F.E.B.S. Letters*, 1978, **89**, 201.
[196] T. Krendeleva, G. Kukarskikh, N. Nizovskaya, V. Pashchenko, K. Timofeev, G. Tulbu, and Yu. Khitrov, *Biokhimiya*, 1977, **42**, 1965.

regulated by cations[191] and the orientation by cations of long wavelength-absorbing species[195] support this hypothesis. A change in the effective absorption cross-section of the pigment array affecting the initial distribution of excitation has also been proposed.[189]. This would be brought about by associating more Chl into the PS II system. Cations have also been observed to have a direct effect on the PS II reaction centre.[190, 193] Clearly the action of cations is complex and they seem to act at several places in the PSU. It is not clear that the observations made in the presence of cations are not the result of several processes.

5 Orientation of Chl in the Membrane

A number of interesting papers on the orientation of Chl molecules in isolated Chl-a/b proteins,[197] PS I particles,[198] and chloroplast lamellae[199—202] have appeared. The spectral properties of many of the Chl proteins have been reported.[203—206] From fluorescence polarization measurement on the Chl-a/b protein,[197] it is concluded that depolarization occurs with a single jump and that energy transfer is fast compared with the fluorescence decay. Direct absorption into Chl-b leads to complete depolarization but this is not the case for Chl-a absorption, indicating that two Chl-a species are present.[197] The CD spectra showed a doublet for Chl-b which indicates that there is exciton coupling between the molecules. A trimeric pigment complex of C_3 symmetry is proposed for the Chl-a/b protein.[197] In PS I particles containing between 40 and 100 pigment molecules, the Chls absorbing at 680 nm have their Q_y transitions parallel to the long axis of β carotene, and at about 18° to the plane of the membrane,[198, 207, 208] (it was suggested that β carotene protects the Chl from oxidation). Unfortunately, the possible presence of antennae Chl absorbing at 700 nm makes the determination of the relative orientation of the two Chl molecules, postulated to be in the PS I particle (P700) reaction centre, ambiguous.[209] The P700 pigments are either parallel or perpendicular to the plane of the membrane.[207, 208] The Chl-a in chloroplasts has been found to have its transition moment from 55° to 90° to the normal of the membrane and that of Chl-b at ~50° to the normal.[199, 200] From these measurements it is inferred that the Q_y transition becomes nearer to being parallel to the plane of the membrane the nearer the Chl is to the reaction centre.[199, 200, 207, 208, 210]

[197] R. Van Metter, *Biochim. Biophys. Acta*, 1977, **462**, 642.
[198] W. Junge, H. Schaffernicht, and N. Nelson, *Biochim. Biophys. Acta*, 1977, **462**, 73.
[199] G. Paillotin and J. Breton, *Biophys. J.*, 1977, **18**, 63.
[200] G. Garab and J. Breton, *Acta Phys. Chem.*, 1977, **23**, 135.
[201] A. Gegliano, N. Geacintov, and J. Breton, *Biochim. Biophys. Acta*, 1977, **461**, 460.
[202] F. Tjerneld, B. Norden, H. Akerland, R. Anderson, and P. Albertsson, *Lin. Dichoism. Spectrosc. Proc. Nobel. Workshop. Chem. Appl. Polaris. Spectosc.*, 1976, 224.
[203] R. Strasser and W. Butler, *Biochim. Biophys, Acta.*, 1977, **461**, 307.
[204] K. Satoh and W. Butler, *Plant Physiol.*, 1978, **61**, 373.
[205] J. Brown, *Photochem. and Photobiol.*, 1977, **26**, 519.
[206] J. Brown, *Dev. Bioenerg. Biomembr.*, 1977, **1**, 297.
[207] J. Breton, *Biochim. Biophys. Acta*, 1976, **459**, 66.
[208] W. Junge and H. Schaffernicht, in ref. 162, p. 21.
[209] J. Katz, J. Norris, L. Shipman, M. Thurnauer, and M. Wasielewski, *Ann. Rev. Biophys. Bioenerg.*, 1978, **7**, 393.
[210] J. Breton and N. Geacintov, *Biochim. Biophys. Acta*, 1977, **459**, 58.

6 Fluorescence from Photosynthetic Bacteria

Time-resolved fluorescence studies of the light-harvesting pigments in photosynthetic bacteria have not been as numerous as those of green plants even though the bacteria have a relatively simple structure, which makes them especially interesting. They contain only a single photosystem and the size of the photosynthetic apparatus is often small, containing only 100 BChl molecules per reaction centre. The minimal functional unit of *Rhodopseudomonas sphaeroides*, the B800—850 light-harvesting pigment–protein complex, has one carotenoid and three BChl molecules. One BChl is responsible for the 800 nm band and the other two for the 850 nm band.[211] Sauer and Austen[212] found intermediate exciton couplings between the BChl molecules and weak interactions between the LH complexes either in aggregates or in the membrane. These couplings are used to explain the fluorescence depolarization and yield. A number of other studies have been made on the light-harvesting complexes.[213—217]

Fluorescence Decay Times.—Campillo *et al.*[155] have measured the emission from chromatophores of four strains of *R. sphaeroides* using frequency-doubled picosecond pulses from a Nd laser at 530 nm. They examined two wild strains which have 100 ps decay times, a mutant without a reaction centre, and a mutant lacking any carotenes, which have 1.1 ns and 300 ps decay times, respectively. The 1.1 ns in the reaction centreless mutant represents the decay time of the antennae BChl quenched by adventitious quenchers, such as O_2, in and near the antennae.

As with higher plants, excitation annihilation processes occur but at higher excitation intensities than plants and half quenching occurs at different intensities depending upon the type of antennae. At a given excitation intensity, quenching is greater the longer the lifetime (and hence migration) in the unquenched state. At moderate redox potential the annihilation quenching curves, which decrease monotonically in chloroplasts and *Chlorella*, pass through a maxima.[218] This is due to the light initially closing the traps and hence reducing the fluorescence quenching, the annihilation soon overcomes this increase and a quenching is observed. Why this effect is not observed in green plants is unknown. An analysis of the annihilation processes is consistent with a lake model of photosynthetic units.

Pashenko *et al.*[219] have measured a decay of 200 ps in *R. sphaeroides* 1760—1 cells and chromatophores, and Godik and Borisov,[220] using the phase method, measured 1.1—1.5 ns in antennae complexes of *R. rubrum*. The antennae decay

[211] R. Cogdell and A. Crofts, *Biochim. Biophy.. Acta*, 1978, **502**, 409.
[212] K. Sauer and L. Austen, *Biochemistry*, 1978, **17**, 2011.
[213] P. Cuendet and H. Zuber, *Ber. Deutsch. Bot. Gesellschaft*, 1977, **30**, 493.
[214] A. Moskalenko and Yu. Erokhin, *F.E.B.S. Letters*, 1978, **87**, 254.
[215] S. Morito and T. Miyazaki, *J. Biochem. (Tokyo)*, 1978, **83**, 1715.
[216] Yu. Erokhin, V. Chuguno, A. Moskalenko, Z. Makhneva, and I. Agrikova, *Itog. Issled. Mekh. Fotosint. Tezisy Doklady Vses. Konf.*, 1976, 12.
[217] R. Feick and G. Drews, *Biochim. Biophys. Acta*, 1978, **501**, 499.
[218] T. Monger and W. Parson, *Biochim. Biophys. Acta*, 1977, **460**, 393.
[219] V. Paschenko, A. Kononenko, S. Protasov, A. Rubin, L. Rubin, and N. Uspenskaya, *Biochim. Biophys. Acta*, 1977, **461**, 403.
[220] A. Godik and A. Borisov, *F.E.B.S. Letters*, 1977, **82**, 355.

times were found to vary upon changing the oxidation–reduction state of the reaction centres. In the active state, a decay of 30—60 ps and a trapping rate of 2×10^{10} s^{-1} was measured. When the reaction centre was closed, efficient fluorescence quenching also occurred. In the state PA$^-$ a 3.0 ns decay was observed, which is longer than from the antennae alone and may be due to free Chl.[220] Emission from reaction centres after 694 nm excitation has a fast (15 ps) and a slow 200 ps component which lengthens in proportion to oxidation of the reaction centre,[219] the 200 ps decay is thought to come from B pheophytin, in good agreement with the lifetime of state P^f (see p. 000). A rise time of the emission of up to 350 ps was also observed (dependent upon the excitation and observation wavelengths) in oxidized reaction centres. This was attributed to slow energy migration (10^9 s^{-1}) into the reaction centre. This result must be considered as tentative, however, as the rise time exceeded the decay time.[219]

Measurements made by phase fluorometry on sub-chromatophore complexes of *Chlorobium limicola* show that two decay times are present: one of about 2 ns represents BChl not functionally connected to the traps and the other has a lifetime variable with the oxidation of the reaction centre of between 40 ± 20 ps and 400 ± 200 ps.[221] Further details of the structure of the interesting BChl protein from this bacteria and exciton interactions between BChl have been reported.[222] It seems that the excitation spreads over several BChl in the complex.

Excited state annihilation has been observed after ns pulse excitation at 694 nm, and 834 nm in five different bacteria;[156,218] BChl and carotenoid triplet states were produced. The rate of ^3T carotene formation was 3×10^7 s^{-1} and it was formed only from ^3T BChl. This formation is 10^3 times faster than from other processes, such as O$_2$ quenching. The carotenoid serves to dissipate the excess energy from the antennae system but the triplet mechanism only operates when the reaction centres are saturated and so protects the antennae from oxidation of the BChl.[156,223] Carotene triplets can also quench the BChl fluorescence directly. A linear (Stern–Volmer) relationship was found between the reciprocal fluorescence yield and ^3T carotene in *R. sphaeroides* and indicates that the excitation energy may migrate over several photosynthetic units, *i.e.* a lake model.[218] The data are explained by a model in which B870 BChl proteins are adjacent to one another and connect the traps. The B800—850 BChl protein complexes surround these other pigments.

An unusual fluorescence at 450—600 nm has been observed in *R. sphaeroides* with a yield of 10^{-9} upon excitation with 694 nm or 868 nm laser pulses. Possible two-photon effects are discussed.[224]

7 Photosynthetic Accessory Pigments

Biliproteins are globular proteins with covalently linked tetra pyrrol chromophores, and occur as photosynthetic accessory pigments in red, blue, green, and

221 A. Borisov, Z. Fetisova, and V. Godik, *Biochim. Biophys. Acta*, 1977, **461**, 500.
222 R. Fenna and B. Matthews, in ref. 30, p. 170.
223 G. Renger and Ch. Wolff, *Biochim. Biophys. Acta*, 1977, **460**, 47.
224 M. Kung and D. DeVault, *Biochim. Biophys. Acta*, 1978, **501**, 217.

cryptomonad algae. The protein structure has been studied in detail[225—227] and the phycobiliproteins have been reviewed.[228,229] These algae lack the Chl-a/b protein of plants but use instead phycobilisomes (PBS), which are prolate hemiellipsoids measuring 300 Å high and 450 Å wide. *Porphyridium cruentum*, which supposedly evolved in regions of the sea where there was relatively little light that could be absorbed by Chl, developed the pigments *B* and *b* phycoethryin (PE), R-phycocyanin (RPC), and allophycocyanin (APC) to absorb light in the 500—600 nm region.

A structural model consists of an APC core surrounded by RPC with PE on the periphery.[140,230,231] The APC channels energy into the Chl which is in the thylakoid membrane. The energy levels of all these pigments are such that Förster energy migration between like pigments and energy migration along the pathway: PE → RPC → APC → Chl are very favourable. Most energy from the accessory pigments is transferred into PS II. Ley and Butler[140,141] have demonstrated that spill-over of energy into PS I then occurs. The *in vitro*[232,233] and *in vivo*[157,158,233—237] fluorescence polarization, yields, lifetimes, and Förster R_0 values are reported for several of the pigments.

As the pigments PE, RPC, APC, and Chl each fluoresce in well defined spectral regions, Porter and co-workers,[157,158] using picosecond spectroscopy, were able to time-resolve the various energy transfer processes in the intact algae and in isolated intact phycobilisomes (Figure 5). The decay times for the species PE, RPC, APC, and Chl are 70, 90, 118, and 175 ps, respectively. When the PS II trap is closed, the Chl emission becomes complex with components of 110 and 810 ps. Rise times for the fluorescence are observed as 0, 12, 24, and 50 ps for the same pigments, see Figure 5. These finite rise times are a measure of the rate at which energy is transferred to the next pigment in the pathway. In the isolated PBS the APC is the last pigment in the chain and has a lifetime of ~4 ns but takes 150 ps to reach its full intensity. As the decay is so long, this rise time mirrors the energy transfer from the previous pigment RPC. In the intact algae where the Chl quenches the APC fluorescence and also in the PBS the PE and RPC fluorescences are described by complex multiexponential decays.[157]

Using fluorescence emission and excitation spectra, Grabowski and Gantt[233,234] calculated transfer times of 280 ps from PE to APC and they also calculate similar times from random walk models of energy transfer in regular

[225] V. Williams and A. Glazer, *J. Biol. Chem.*, 1978, **253**, 202.
[226] P. Freidenreich, G. Apell, and A. Glazer, *J. Biol. Chem.*, 1978, **253**, 212.
[227] D. Bryant, C. Hixson, and A. Glazer, *J. Biol. Chem.*, 1978, **253**, 220.
[228] E. Gantt, *Photochem. and Photobiol.*, 1977, **26**, 685.
[229] H. Koest, W. Ruediger, DHEW Publ. (NIH) (U.S.) Report NIH 77, 1977, 1100 (*Chem. Abs.*, **89**, 38136).
[230] E. Gantt, C. Lipschultz, and B. Zilinskas, *Biochim. Biophys. Acta*, 1976, **430**, 375.
[231] E. Gantt, C. Lipschultz, and B. Zilinskas, in ref. 30, p. 347.
[232] D. Barber and J. Richards, *Photochem. and Photobiol.*, 1977, **25**, 365.
[233] J. Grabowski and E. Gantt, *Photochem. and Photobiol.*, 1978, **28**, 39.
[234] J. Grabowski and E. Gantt, *Photochem. and Photobiol.*, 1978, **28**, 47.
[235] R. MacColl and D. Berns, *Photochem. and Photobiol.*, 1978, **27**, 343.
[236] M. Mimura and Y. Fujita, *Biochim. Biophys. Acta*, 1977, **459**, 376.
[237] M. Mimura and Y. Fujita, *Plant Cell Physiol.*, 1977, **3**, 23; 'Photosynthetic Organelles', ed. S. Miyachi, S. Katoh, Y. Fujita, and K. Shibatu,

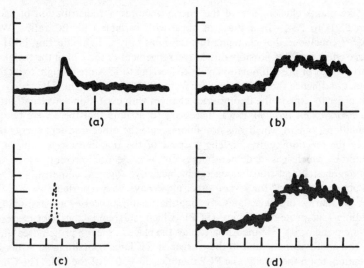

Figure 5 *Energy transfer from phycoerythrin to allophycocyanin in phycobilisomes:*
(a) phycoerythrin fluorescence; (b) allophycocyanin fluorescence; (c) laser pulse profile;
(d) rise of allophycocyanin fluorescence on a faster time scale. (a) and (b) time scale
120 ps cm^{-1}
(Adapted from *Proc. Roy. Soc.*, 1978, **A362**, 281)

lattices.[178,238] These transfer times, although of the correct order of magnitude, are still two to three times longer than the directly measured values and illustrate the difficulties of inferring kinetic information from steady-state measurements.

In cryptomonad biliproteins (PC and RPE) each protein contains two chromophores, each on a different polypeptide chain. Only chromophores from one α chain are responsible for fluorescence, those on the β chain transfer energy efficiently to the α chain and then to Chl-a.[235] Chlorophyll c_2 is also present but may not be an intermediate in the energy transfer chain.

Energy distribution in red algae of the Florideae and Bangides is similar to that in *P. cruentum*, *i.e.* PBS → PS II → PS I. The PS II to PS I transfer depends upon the state of the reaction centres.[239,240] Some energy distribution may result from conformational changes in the protein(s).

No picosecond studies have been made on the blue green alga, *Anacystis nidulans*, although the energy transfer processes have been studied.[241,242] In *Anacystis* the light-harvesting function is performed by PC and Chl. APC may be by-passed in the transfer chain.[24] Sixty per cent of PC transfers energy to only a few (16%) of the Chls associated with PS II, the remaining Chl and PC is associated with PS I. This is almost the reverse of the situation in green plants

238 Z. Bay and R. Pearlstein, *Proc. Nat. Acad. Sci., U.S.A.*, 1963, **50**, 1071.
239 A. Reid, B. Hessenberg, H. Metzler, and R. Ziegler, *Biochim. Biophys. Acta*, 1977, **459**, 175.
240 A. Reid and B. Reinhardt, *Biochim. Biophys. Acta*, 1977, **460**, 25.
241 K. Csatordy, J. Hammans, and J. Goedheer, *Biochem. Biophys. Res. Comm.*, 1978, **81**, 571.
242 R. Wang, C. Stevens, and J. Myers, *Photochem. and Photobiol.*, 1977, **25**, 103.

but like plants efficient use of the energy requires a redistribution of energy from PS II to PS I. In a study of algae with variable Chl : PC ratios, Wang *et al.*[242] concluded that changing this ratio has only a slight effect on PS II but a large effect on PS I, possibly due to rearrangement of the Chl in the thylakoid. Short periods of pre-illumination prior to freezing to 77 K are thought to increase contact and hence the energy transfer from the PBS to the Chl.[243,244]

In order to explain the fluorescence changes with excitation wavelength in the alga *Phormidium luridum*, it was necessary to assume that there were two Chl populations, one of which was non-fluorescent; the other transfers energy from PC to the reaction centre.[245] The purpose of the non-fluorescent Chl, or the method by which it is made non-fluorescent, is unclear. If present, it is presumably aggregated and transfers energy inefficiently. Pigment composition[246,247] and action spectra[248] for several other algae have been reported.

In the marine dinoflagellates, the light-harvesting processes are assisted by a peridinin–Chl–protein complex (PCP), which surrounds the light-harvesting Chl-a/c protein;[249–251] the Chl-a/c takes the place of the a/b protein (see Figure 4). The photosynthetic unit in these organisms is large containing 600 Chls and, depending upon the family, the PCP contains 30—50% of the Chl. The Chl-a/c has 20—40%, the Chl-a light-harvesting 20%, and both reaction centres combined 10% of the total Chl.[249,252]

The fluorescence decay times at 675 nm of the PCP complexes are between 4 and 5 ns, similar to the lifetimes for free Chl.[253] In the *Glenodinium* sp., *G. polyhedra*, and *Amphidinium rhynocephadeum*, the peridinin : Chl ratio is 4 : 1 and in *A. carterae* it is 9 : 2.[253] The chromophores are accommodated in a hydrophobic cleft in the protein as the fluorescence is unquenched by 2M-KI. The surface of the protein is highly polar. Energy absorbed by the carotene is transferred with nearly unit efficiency to the Chl, and CD and fluorescence polarization spectra lead to a structure of two pairs of peridinin molecules on opposite sides of the Chl macrocycle with ~12 Å between the dimer pairs. Exciton resonance interactions are thought to lengthen the peridinin singlet state lifetime, estimated to be ≪1 ps, to make energy transfer more probable. The Förster critical distance is estimated as 8 Å and it seems unlikely that dipole–dipole interactions alone can be responsible for energy transfer; exciton couplings may also contribute.

8 Photosynthetic Bacteria

In the bacterial reaction centres, when the BChl complex P (a BChl dimer) is excited to the singlet state (P*) an electron is released to an intermediate acceptor

243 G. Harnischfeger and G. Codd, *Biochim. Biophys. Acta*, 1978, **502**, 507.
244 G. Harnischfeger and G. Codd, *Brit. Phycol. J.*, 1974, **12**, 225.
245 E. Tel-or and S. Malkin, *Biochim. Biophys. Acta*, 1977, **459**, 157.
246 B. Zilinskas, B. Zimmerman, and E. Gantt, *Photochem. and Photobiol.*, 1978, **27**, 587.
247 B. Gray, J. Cosner, and E. Gantt, *Photochem. and Photobiol.*, 1976, **24**, 299.
248 A. Larkum and S. Wayrauch, *Photochem. and Photobiol.*, 1977, **25**, 65.
249 B. Prézelin and R. Alberte, *Proc. Nat. Acad. Sci. U.S.A.*, 1978, **75**, 1801.
250 H. Siegelman, J. Kycia, and F. Haxo, Report, 1976, BNL 21729, p. 17, *ERDA Energy Res. Abs.*, 1977, Vol. 2, No. 3845.
251 H. Siegelman, J. Kycia, and F. Haxo, in ref. 30, p. 72.
252 P-S. Song, P. Koka, B. Prézelin, and F. Haxo, *Biochemistry*, 1976, **15**, 4422.
253 P. Koka and P-S. Song, *Biochim. Biophys. Acta*, 1977, **495**, 220.

(I), possibly a BPh, and is then transferred to a 'primary' acceptor X, a quinone iron complex.

Structure of the Reaction Centre.—Reaction centres in *R. sphaeroides* and *R. viridis* have been found to contain only four BChl, two BPh, a quinone (ubiquinone or menaquinone), and a closely associated non-heme iron and a carotene. *Viridis* r.c.'s also retain the secondary donor cytochrome *c*. In most reaction centres these pigments are contained in two of three polypeptides which have a molecular weight of 20—24 K dalton.[2, 254—256] From linear dichroism and fluorescence polarization studies of the reaction centres[257, 258] it was found that the 870 nm band corresponds to a single transition, but at 800 nm absorption changes were attributed to a band shift and to a bleaching of a BChl dimer band at 805 nm. The angle between the Q_y transitions in this dimer is less than 25°, whereas the angle between the Q_y BPh transitions is 55° or 125° but both are almost perpendicular to the 870 nm absorption band. The carotene is nearly perpendicular to this band and the other BChl molecules.[258]

The inhomogeneous nature of the back electron transfer, which was not properly accounted for by earlier workers,[259, 260] led to the low polarization value within the 870 nm band, which was inconsistent with the constant dichroic ratio[261] and fluorescence polarization.[262] At low temperature the Q_x transition of PBh near 540 nm is split into two bands with similar polarizations,[259] which are probably not caused by exciton interactions but by orientational differences of the two BPh molecules.

Evidence from energy transfer experiments[263] has led to the suggestion that all the BChl is monomeric but far more evidence has accumulated for two dimeric and two monomeric BChl.[2, 9, 257, 258, 262]

Resonance Raman spectra of the carotenes from several bacterial species have rather similar spectra. The carotenes are found to be in a di-*cis* conformation involving one methylated and one unmethylated double bond in the reaction centre, and are all *trans* in the chromatophores.[264] The *cis* form of the carotenes may facilitate their close approach to the reaction centre and so assist in the protective role against photo-oxidation. In carotene-depleted reaction centres, to which several different carotenes are added, the carotenes appear to be in a mono-*cis* form.[265]

Evidence for two types of reaction centres in *R. rubrum* has been obtained from flash studies of P^+ [the $(BChl)_2^+$ dimer] and from cytochrome oxidation in

254 B. Clayton and R. Clayton, *Biochim. Biophys. Acta*, 1978, **501**, 470.
255 R. Clayton and B. Clayton, *Biochim. Biophys. Acta*, 1978, **501**, 478.
256 T. Trosper, D. Benson, and J. Thornber, *Biochim. Biophys. Acta*, 1977, **460**, 318.
257 C. Rafferty and R. Clayton, *Biochim. Biophys. Acta*, 1978, **502**, 51.
258 A. Vermeglio, J. Breton, G. Paillotin, and R. Cogdell, *Biochim. Biophys. Acta*, 1978, **501**, 514.
259 T. Mar and G. Gingras, *Biochim. Biophys. Acta*, 1976, **440**, 609.
260 V. Shuvalov, A. Asadov, and I. Krakhmaleva, *F.E.B.S. Letters*, 1977, **76**, 240.
261 A. Vermeglio and R. Clayton, *Biochim. Biophys. Acta*, 1976, **449**, 500.
262 T. Ebrey and R. Clayton, *Photochem. and Photobiol.*, 1969, **10**, 109.
263 T. Mar and G. Gingras, *Biochim. Biophys. Acta*, 1977, **460**, 239.
264 M. Lutz, T. Agalidis, G. Hervo, R. Cogdell, and F. Reiss-Husson, *Biochim. Biophys. Acta*, 1978, **503**, 287.
265 F. Boucher, M. Van Der Rest, and G. Gingras, *Biochim. Biophys. Acta*, 1977, **461**, 339.

whole cells.[266] In one type, 300 BChl and one cytochrome c_{428} are attached to the reaction centres. The other type has only 30 BChl molecules and shares one cytochrome c_{420} between two centres. A second cytochrome shared by six of the centres is involved in cyclic electron transport from the primary acceptor to c_{420}. In *Chromatium vinosum*,[267] cytochrome c_{551} acts as a mobile electron carrier between two reaction centres. This cytochrome donates electrons to two c_{555} cytochromes that act as parallel electron donors to *P*870. Cytochrome c_{552} also donates to *P*870. Of these cytochromes only c_{551} is extrinsic to the membrane,[267] and *P*870, the reaction centre, is located just inside the membrane.[268—271]

Picosecond Studies on *Rhodopseudomonas sphaeroides*.—Two of the four BChl molecules in the reaction centre appear to interact sufficiently well with one another to facilitate the release of an electron after excitation (see refs. 2, 9, and 272 for details of work in this area).

Detailed work on *R. sphaeroides* has shown that after excitation the BChl reaction centre complex transfers an electron to an intermediate (I) in less than 10 ps. A radical pair state $P^{+}I^{-}$, called P^{f}, is formed and subsequently transfers an electron in $\sim$200 ps to the ubiquinone acceptor (X). Lowering the redox potential reduces the acceptor X and the state P^{f} now lives for 15 ns as electron transfer is frustrated. As P^{f} decays a state P^{R}, possibly the triplet, is formed which has a lifetime of 6 μs at room temperature.[9] Assuming that 530 nm light excites only BPh, an energy transfer rate of 10^{11} s^{-1} between BPh and the BChl dimer is calculated and P^{f} is formed within 5 ps.[273] Both spectral and kinetic evidence favours the assignment of the intermediate I as one of the two BPh molecules. At 1250 nm, where BPh^{-} has no absorption, the species (BChl)$_2^{+}$, also called P^{+}, has been observed to form in <5 ps and the absorption is unchanged until P is produced again.[9, 273] Ubiquinone (X) has been found to have no effect on the formation of the 1250 nm band, indicating the presence of an intermediate electron acceptor.[274] Figure 6 shows schemes for bacterial reaction centre reactions at various redox potentials. P^{f} has been proposed to be (BChl$^{\pm}$BChl BPh$^{-\cdot}$)X and P^{R} (BChl—BChl)^{T} BPh X$^{-\cdot}$

Picosecond Studies on *Rhodopseudomonas viridis*.—Picosecond studies of the transients in isolated *R. viridis* reaction centres[275—277] reveal that as in *R. sphaeroides* the electron is transferred from the initially excited state to an intermediate electron carrier I in <10 ps. The radical pair state (P^{+}I^{-}), or P^{f},

[266] R. Van Grondelle, L. Duysens, and H. Van der Waal, *Biochim. Biophys. Acta*, 1976, **449**, 169.
[267] R. Van Grondelle, L. Duysens, J. Van der Waal, and H. Van der Waal, *Biochim. Biophys. Acta*, 1977, **461**, 188.
[268] B. De Grooth and J. Amesz, *Biochim. Biophys. Acta*, 1977, **462**, 237.
[269] B. De Grooth and J. Amesz, *Biochim. Biophys. Acta*, 1977, **462**, 247.
[270] K. Matsuura and M. Nashimura, *Biochim. Biophys. Acta*, 1977, **462**, 700.
[271] M. Symons and C. Swysen, *Biochim. Biophys. Acta*, 1977, **462**, 706.
[272] D. Hollen, M. Windsor and W. Parson, in ref. 153, p. 126.
[273] F. Moscowitz and H. Malley, *Photochem. and Photobiol.*, 1978, **27**, 55.
[274] T. Netzel, P. Dutton, K. Petty, E. Degenkolb, and P. Rentzepis, *Adv. Mol. Relax. Interact. Process*, 1977, **11**, 217.

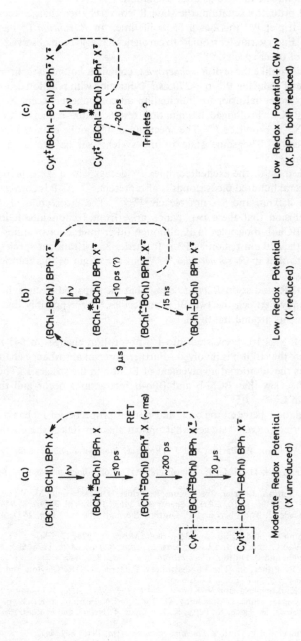

Figure 6 *Scheme of early events in bacterial photosynthesis at room temperature.* (a) *Corresponds to physiological conditions. RET denotes reverse electron transfer. Lifetimes given are exponential lifetimes* (Reproduced by permission from Ann. Rev. Biophys. Bioeng., 1978, **7**, 189)

subsequently formed, transfers an electron to the acceptor X in 230 ps.[275] Monitoring at 1250 nm in *Ch. vinosum* and at 1301 nm in *R. viridis*, the oxidation was found to be complete in <10 ps after excitation.[276]

If X is reduced prior to excitation, the state P^f converts after charge recombination into the triplet P^R; this has a 55 μs lifetime. In *R. rubrum* P^R has a 2.5 μs lifetime.[279] Energy transfer from P^R to carotenes has also been observed[279] as has the triplet of other species.[280]

R.c.'s of *R. viridis*, unlike those of *R. sphaeroides*, contain a bound cytochrome which has the role of reducing P^+ in ~270 ns.[275] Starting with reaction centres at moderate redox potential prior to illumination all components except X are reduced, *i.e.* Cyt PIX⁻. Prolonged illumination results in the reduction of I; the state Cyt PI⁻X⁻ is produced.[278] The spectrum of I⁻ can be obtained from the difference spectra. The same state of the system can be produced by chemical reduction.[281,282]

With I reduced, most of the excited complex P* decays with a 20 ps lifetime but a decay of several hundred picoseconds is also present.[275] As P is converted into P* bands at 850 nm and 960 nm bleach.[275,278] The picosecond results support the conclusion that these two bands arise from exciton interactions between the two BChl-b molecules in P, although other interactions cannot be ruled out.[283] In the 810 nm region as P* is formed a 35 ps transient species is observed similar to that in *R. sphaeroides*,[9,275] but the origin of this transient is unknown.

E.s.r. and ENDOR studies on *R. viridis* show that the unpaired electron in P^f is not shared equally between the two BChl molecules,[284,285] possibly due to different environments around the BChl.

The Intermediate I.—As I is converted into I⁻, absorption changes at 545 nm and 790 nm suggest that BPh-b is involved. Furthermore, an absorbance change at 830 nm suggests the additional involvement of BChl-b in the species I. These results support the view that BChl-b and BPh-b interactions occur and that both are present in I.[278,286,287]

No direct correlation between the e.s.r. and optical properties of I⁻ has been made, although Prince *et al.*[278] suggest that optical changes reflect the possible

275 D. Holten, M. Windsor, W. Parson, and J. Thornber, *Biochim. Biophys. Acta*, 1978, **501**, 112.
276 T. Netzel, P. Rentzepis, D. Tiede, R. Prince, and P. Dutton, *Biochim. Biophys. Acta*, 1977, **460**, 467.
277 J. Thornber, P. Dutton, J. Fajer, W. Parson, R. Prince, D. Tiede, and M. Windsor, Report, 1977, BNL 23306, p. 16, *ERDA, Energy Res. Abs.*, 1978, Vol. 3, No. 75 914.
278 R. Prince, D. Tiede, J. Thornber, and P. Dutton, *Biochim. Biophys. Acta*, 1977, **462**, 467.
279 V. Godik, V. Samuilov, and A. Borisov, *Mol. Biol.* (*Moscow*), 1978, **12**, 290.
280 N. Holmes, R. Van Grondelle, and L. Duysens, *Biochim. Biophys. Acta*, 1978, **503**, 26.
281 R. Prince, J. Leigh, and P. Dutton, *Biochim. Biophys. Acta*, 1976, **440**, 662.
282 P. Dutton and R. Prince, in 'The Photosynthetic Bacteria', ed. R. Clayton and W. Sistrom, Plenum Press, New York, 1977, p. 525.
283 V. Shuvalov, I. Krakhmaleva, and V. Klimov, *Biochim. Biophys. Acta*, 1976, **449**, 597.
284 J. Fajer, M. Davis, D. Brune, L. Spaulding, D. Borg, and A. Forman, in ref. 30, p. 74.
285 P. Dutton, R. Prince, D. Tiede, K. Petty, K. Kaufman, T. Netzel, and P. Rentzepis, in ref. 30, p. 213.
286 D. Tiede, R. Prince, and P. Dutton, *Biochim. Biophys. Acta*, 1976, **449**, 447.
287 V. Shuvalov and V. Klimov, *Biochim. Biophys. Acta*, 1976, **440**, 587.

BPh interactions with BChl, while the two e.s.r. signals reflect interactions of BPh with the Q^- Fe complex. One signal comes from close association of these two species, the other from a weak association. Fajer *et al.*,[288] however, concluded from e.s.r. and ENDOR studies on I^- trapped in isolated reaction centres that in BChl-b containing bacteria such as *R. viridis*, I is BPh-b. They also suggest that in BChl-a containing species BPh-a fulfils a similar role.

When using 880 nm excitation and monitoring at 750 nm in *R. rubrum*, and during a 15 ps instrumental response, Shuvalov *et al.*[289] noticed an absorbance increase which they attributed to a new species, $P^+ BChl_{800}^-$. A bleaching at 750 nm reflects the formation, within 35 ps, of $P^+ BPh^-$. These transient species have not been observed by other workers but nevertheless cast doubt on previous suggestions of the nature of the species I. According to Shuvalov the electron transfer sequence is: $(BChl)_2 \rightarrow BChl \rightarrow BPh \rightarrow QFe$. The difference in this work to that of other picosecond absorption studies is that the $(BChl)_2$ in the reaction centre is excited directly at 880 nm. At 530 nm, light from a doubled Nd laser causes excitation of BPh; this may make the early time history of the reactions confusing as excitation has first to reach the $(BChl)_2$. Picosecond studies with higher time resolution than 15 ps, and where the BChl absorbs, are clearly needed.

Because of the involvement of BPh-a or BPh-b in the intermediate, I, several *in vitro* studies of these species have been made. Mg^{290} and ^1H hyperfine interactions have been calculated[291] and obtained from ENDOR experiments,[292,293] and e.s.r. properties[293-296] have also been obtained. A new Raman technique was also used to identify these cations.[297]

Effects of Magnetic Fields on the Reaction Centres.—When the state P^f is formed the two electrons of the radical pair are relatively far apart and are weakly coupled together. The relative orientation of their spins may change rapidly before P^f decays. P^f probably forms in its singlet state but the return of the electron in a triplet state places the radical pair in the BChl triplet state. P^R or $^TPIX^-$ is formed when the acceptor is reduced. Spin polarization is consistent with a radical pair mechanism forming the triplet state. Other intramolecular models have also been reported.[298] A radical pair method is also supported by a decrease in fluorescence intensity with magnetic fields of up to 300 G.[299]

[288] J. Fajer, M. Davis, D. Brune, A. Foreman, and J. Thornber, *J. Amer. Chem. Soc.*, 1978, **100**, 1918.
[289] V. Shuvalov, A. Klevanik, A. Sharkov, J. Matveetz, and P. Krukov, *F.E.B.S. Letters*, 1978, **91**, 135.
[290] M. Thurnauer, M. Bowman, B. Cope, and J. Norris, *J. Amer. Chem. Soc.*, 1978, **100**, 1965.
[291] J. Chang and T. Das, *Biochim. Biophys. Acta*, 1978, **502**, 61.
[292] A. Hoff and K. Moebius, *Proc. Nat. Acad. Sci. U.S.A.*, 1978, **75**, 2296.
[293] D. Borg, A. Forman, and J. Fajer, *J. Amer. Chem. Soc.*, 1976, **98**, 6889.
[294] J. Fajer, A. Forman, M. Davis, L. Spaulding, D. Brune, and R. Felton, *J. Amer. Chem. Soc.*, 1977, **99**, 4134.
[295] V. Voznyak, I. Proskuryakov, E. Elfimov, V. Kim, V. Evstigneev, *Itogin Issled. Mekh. Fotosint. Tezisy Doklady Vses. Konf.*, *9th*, 1976, 25.
[296] E. Nissani, A. Scherz, and H. Levanon, *Photochem. and Photobiol.*, 1977, **25**, 93.
[297] T. Cotton and R. Van Duyne, *Biochim. Biophys. Res. Comm.*, 1978, **82**, 424.
[298] A. Ponte-Goncalves and W. Spendel, *Chem. Phys. Letters*, 1978, **54**, 611.
[299] H. De Vries and A. Hoff, *Chem. Phys. Letters*, 1978, **55**, 395.

Spin echoes have been observed by monitoring *in vivo* fluorescence changes, and so allow spin memory times to be determined.[300]

By adding a weak magnetic field the probability of forming a triplet state can be changed, since the radical pair oscillates between the singlet and only one of the triplet state (T_0) sublevels. Consequently, the yield of P^R is reduced.[301] The iron in the Fe–quinone complex also exerts an effect on the triplet formation in a magnetic field.[302]. The magnitude of the magnetic fields needed to observe an effect differs between reaction centres and chromatophores, and between different reaction centre preparations,[303] and may reflect orientational or distance changes between the members of P^f in these systems.[302]

Fast laser flash e.p.r. spectroscopy has been used to resolve discrepancies between the optical and e.s.r. measurements of the triplets in *R. sphaeroides* at 5 K.[304] Decay rates of the individual levels of the triplet state have been determined *in vivo*[305] and *in vitro*.[306] It was also found that a simple exciton model was inadequate to derive the geometry of the reaction centre BChl dimer from the zero field splittings and decay rates, as was previously suggested.[307,308] The normals to the planes of the two macrocycles have, nevertheless, been proposed to be at 48° to one another but rotated by 78°.[307,308]

At 77 K an efficient charge recombination $(P^+I^-X^- \rightarrow P^*IX^-)$ occurs, when the primary acceptor is reduced, which generates delayed fluorescence in the antennae BChl.[309] The fluorescence and triplet yields were measured, and compared favourably with predictions of a model in which the excitation energy can migrate freely in the antennae, and has a fractional probability of trapping at the reaction centre.

Properties of the Primary Acceptor X.—The 'primary' acceptor X is now thought to be a ubiquinone or a menaquinone closely coupled to a non-heme iron. Removal of the quinone results in a loss of photochemical activity, which can be restored by adding ubi- or related quinones.[2,310] In *C. vinosum*, menaquinone takes the place of the ubiquinone, which could be removed without affecting its activity.[311] In *R. rubrum* activity was retained after less than one ubiquinone was present per reaction centre, and as menaquinone was also absent, the nature of the acceptor is unknown.[312]

It is now known that two ubiquinones are involved in the electron transfer in the plastoquinone pool. The non-heme iron also seems to be involved in this process.[278,313] In *R. sphaeroides*, on each odd numbered flash, the ubi-

[300] N. Nishi, J. Schmidt, A. Hoff, and J. Van Der Waals, *Chem. Phys. Letters*, 1978, **56**, 205.
[301] R. Blankenship, T. Schaafsma, and W. Parson, *Biochim. Biophys. Acta*, 1977, **461**, 297.
[302] A. Hoff, H. Rademaker, and R. Grondelle, *Biochim. Biophys. Acta*, 1977, **460**, 457.
[303] H. Werner, K. Shulten, and A. Weller, *Biochim. Biophys. Acta*, 1978, **502**, 255.
[304] P. Gast and A. Hoff, *F.E.B.S. Letters*, 1978, **85**, 183.
[305] A. Hoff and H. De Vries, *Biochim. Biophys. Acta*, 1978, **503**, 94.
[306] J. Kleibeuker, R. Platenkamp, and T. Schaafsma, *Chem. Phys.*, 1978, **27**, 51.
[307] R. Clarke, R. Connors, H. Frank, and J. Hoch, *Chem. Phys. Letters*, 1977, **45**, 523.
[308] R. Clarke and R. Connors, *Chem. Phys. Letters*, 1976, **42**, 69.
[309] R. Van Grondelle, N. Holmes, H. Rademaker, and L. Duysens, *Biochim. Biophys. Acta*, 1978, **503**, 10.
[310] C. Bering and P. Loach, *Photochem. and Photobiol.*, 1977, **26**, 607.
[311] G. Feher and M. Okamura, in ref. 30, p. 183.
[312] L. Morrison, J. Rundquist, and P. Loach, *Photochem. and Photobiol.*, 1977, **26**, 607.
[313] C. Wraight, *Biochim. Biophys. Acta*, 1977, **459**, 525.

semiquinone was formed and on the next flash the fully reduced quinone was generated.[313—315] Both a duplex[310,313,316] and a series model[314,317] have been proposed to explain this effect. Studies on the kinetics of the electron transfer, by Vermeglio and Clayton,[317] have shown that the $t_{\frac{1}{2}}$ for transfer from the first (Q_1) to the second (Q_2) quinone was 200 µs and did not depend upon the state of Q_2. The electrons are then transferred two at a time into the quinone pool. A possible scheme, with Q_p as the quinone pool, is shown:[317]

$$Q_1 \quad Q_2 \quad Q_p \xrightarrow{hv} Q_1^- \quad Q_2 \quad Q_p \longrightarrow Q_1 \quad Q_2^- \quad Q_p \xrightarrow{hv}$$

$$Q_1^- \quad Q_2^- \quad Q_p \longrightarrow Q_1 \quad Q_2^{2-} \quad Q_p^2 \longrightarrow Q_1 \quad Q_2 \quad Q_p^{2-}$$

The reduction of Q_1 causes the BPh Q_x transitions to red shift and the reduction of Q_2 causes a blue shift. At 750 nm there is an isobestic point in the region of the Q_y transition. These shifts imply a close relationship between the quinone and a BPh.[317] By chemically removing the loosely bound quinones (Q_2, Q_p), charge recombination between Q_1^- and its donor is possible. This recombination is biphasic and, to account for this, one quinone is thought to serve sequentially two donors (a duplex model).[312,316]

Other Photosynthetic Bacteria.—The acceptor in *Thiocapsa pfennigii* is similar to other purple bacteria.[318] The e.p.r. signals show a BChl dimer similar to that of *R. viridis*; however, two pairs of cytochromes donate electrons to the reaction centre dimer. In *R. gelatinosa*,[319] e.p.r. signals were observed to be similar to those of other bacteria studied, but at low temperatures I$^-$ was observed to undergo electron tunnelling to the oxidized cytochrome *c* which had not been previously observed.

Field-induced Absorption Changes.—In photosynthetic bacteria absorbance changes, which cannot be ascribed directly to oxidation or reduction reactions in the reaction centre, have been observed. It is generally accepted that these changes, in the carotene and BChl absorptions, reflect the generation of a potential across the thylakoid or chromatophore membrane.[268,269,320—322] The kinetics of the absorbance changes indicate that there is a 0.25 nm red shift per electron transfer per reaction centre in 10—30% of the most red-shifted forms of one of the two identical pools of the carotene sphaeroidene. The BChl shift

314 A. Vermeglio, *Biochim. Biophys. Acta*, 1977, **459**, 516.
315 Y. Barouch and R. Clayton, *Biochim. Biophys. Acta*, 1977, **462**, 785.
316 L. Morrison and P. Loach, *Photochem. and Photobiol.*, 1978, **27**, 751.
317 A. Vermeglio and R. Clayton, *Biochim. Biophys. Acta*, 1977, **461**, 159.
318 R. Prince, *Biochim. Biophys. Acta*, 1978, **501**, 195.
319 R. Prince, L. Dutton, B. Clayton, and R. Clayton, *Biochim. Biophys. Acta*, 1978, **502**, 354.
320 K. Seiwe and R. Reich, *F.E.B.S. Letters*, 1977, **80**, 30.
321 K. Seiwe and R. Reich, *Z. Naturforsch.*, 1977, **32c**, 161.
322 G. Borisovich, P. Knoks, A. Kononenko, A. Rubin, and E. Vosary, *Biochim. Biophys. Acta*, 1978, **13**, 67.

is *ca.* one third that of the carotene.[268—271] Alternative explanations of these shifts, besides generation of a potential, have however also been presented.[323, 324] Chls and carotenes have also been observed to produce electrochromic shifts *in vitro.*[321]

In addition to the delocalized field effects, local field effects were also observed by Heathcote *et al.*[325] In purified reaction centres they found that the absorption band of a specific carotenoid associated with the reaction centre was shifted by 1 nm only when the BChl of the BChl dimer (P870) was oxidized. The state of the electron acceptor had no effect on the shift.

9 Photosystem I

In Photosystem I (PS I) an absorbance decrease at 700, 680, and 430 nm and an increase at 820 nm, accompany oxidation of the primary electron donor, P700, in the reaction centre. E.s.r., ENDOR, and CD spectra of P700$^+$ are characteristic of the cation of a chlorophyll dimer. The 820 nm absorption band is also characteristic of this species. From a study of the dichroism of P700 (measured at 660 nm), it was concluded that the P700 dimer was tilted out of the plane of the membrane; it was also suggested that there is a heterogeneous population of P700 traps.[207]

The fluorescence from P700-enriched particles (8 Chl/P700) was found to exhibit a weak variable fluorescence on top of the steady fluorescence; this variable emission depends upon the redox state of the reaction centre.[326] The fluorescence was not from P700 but from the other Chls, *i.e.* Chl-a at 672 nm and 685 nm, which fluoresce at 679 and 694 nm, respectively. The 685 nm form is most closely associated with P700. The oxidized donor and acceptor were observed to quench the variable fluorescence, which is probably emitted by Chl 685.[326]

Chlorophyll anion radicals, identified with a 50 μs absorbance change,[327] and Chl triplet states[296, 328—331] have both been observed in PS I particles. The evidence presented for a radical pair mechanism in the initial charge separation seems overwhelming;[332] thus the triplets observed may be from Chl in the antennae associated with the reaction centre.

The primary acceptor is no longer thought to be bound ferredoxin,[333] but rather an unidentified species X. This species reacts reversibly with P700 upon oxidation, and without changing the e.s.r. spectrum of bound ferredoxin. This

[323] N. Holmes and A. Crofts, *Biochim. Biophys. Acta*, 1977, **461**, 141.
[324] N. Holmes and A. Crofts, *Biochim. Biophys. Acta*, 1977, **459**, 492.
[325] P. Heathcote, A. Vermeglio, and R. Clayton, *Biochim. Biophys. Acta*, 1977, **461**, 358.
[326] I. Ikegami, *Biochim. Biophys. Acta*, 1976, **449**, 245.
[327] N. Zakharova and A. Chibisov, *Doklady Akad. Nauk S.S.S.R.*, 1977, **237**, 473.
[328] V. Shuvalov, *Biochim. Biophys. Acta*, 1976, **430**, 113.
[329] A. Chibisov, N. Fakharova, A. Peshkin, and T. Slavnova, *Studia Biophys.*, 1978, **69**, 29.
[330] S. Van der Bent and T. Schaafsma, *Biochem. Biophys. Res. Comm.*, 1976, **71**, 1147.
[331] A. Hoff, J. Govindjee, and J. Romijn, *F.E.B.S. Letters*, 1977, **73**, 191.
[332] C. Dismukes, A. McGuire, R. Blankenship, and K. Sauer, *Biophys. J.*, 1978, **21**, 239.
[333] R. Malkin, and A. Bearden, *Proc. Nat. Acad. Sci. U.S.A.*, 1971, **68**, 16.

e.p.r. spectrum is observed when the ferredoxin is reduced.[334—337] X^- is observed to form in parallel with $P700^+$ and relegates the bound ferredoxin to the role of a secondary acceptor.

The CIDEP emission associated with the initial charge separation is now thought to arise from radical pairs,[332] rather than triplets, in P700;[338] the charge separation arises from the P700 singlet state. The triplet mechanism cannot account for both emission and enhanced absorption among the hyperfine states of $P700^+$. The counter radical of $P700^+$ is thought to be oriented in the membrane because of the anisotropy in the g-tensor. This anisotropy is characteristic of both X^- and bound ferredoxin. The amplitude of the polarized signal calculated for oriented samples and assuming that X^- is the counter radical are too small by a factor of 10. This difficulty can be resolved if an intermediate acceptor A is postulated between P700 and P430.[332] Spin polarization is then developed by both A_1^-X and A_1X^- [339] which are formed successively. Further evidence for an intermediate electron carrier was provided by flash experiments on SDS and Triton solubilized P700–Chl protein complexes.[339—342] At room temperature, a 10 μs decay component of the 820 nm absorption is due to the back reaction of $P700^+$ with its reduced primary acceptor, postulated as A^-.[340, 341] In previous experiments with Triton solubilized particles, where a laser pulse was superimposed upon red illumination, a 3 μs and a 250 μs back reaction were observed[339, 341, 343] due to $P700^+$ reaction with A_1^-, and a new second intermediate acceptor, A_2^-. In SDS and Triton solubilized particles these species were observed only because the FeS proteins were denatured or removed, or a reductant added to eliminate reactions with the secondary acceptors. Below 110 K the electron tunnels from the intermediate acceptors to P700. These two acceptors (A_1, A_2) lie between P700 and the P430 iron sulphur protein.[340, 341, 343] Kinetic measurements on a picosecond time scale are probably necessary to establish the time course of the electron transfers, as was the case for bacterial reaction centres.

E.p.r. spectra of PS I particles, at physiological[344] and cryogenic temperatures,[345] indicated the presence of two electron transport components (iron sulphur centres A and B), which act after X in the electron transport scheme. Redox titrations confirm that centre A has an $E_m = -550$ mV and Centre B, $E_m = -585$ mV. P700 oxidation is reversible only when B is reduced, indicating its participation as an electron acceptor. The data indicate that B lies before A and that P700, X, and B and A are present in quantitative amounts in the particles. Species X is still considered by some workers, however, to be the

334 A. McIntosh and J. Bolton, *Biochim. Biophys. Acta*, 1976, **430**, 555; see also ref. 382.
335 M. Evans, C. Sihra, and R. Cammak, *Biochem. J.*, 1976, **158**, 71.
336 M. Evans, C. Sihra, R. Bolton, and R. Cammak, *Nature*, 1975, **256**, 668.
337 B. Ke, E. Dolan, K. Sugahara, F. Hawkbridge, S. Demeter, and E. Shaw, *Plant Cell Physiol.*, 1977, **3**, 187.
338 R. Blankenship and A. McGuire, *Proc. Nat. Acad. Sci. U.S.A.*, 1975, **72**, 4943.
339 K. Sauer, S. Acker, P. Mathis, and J. Van Best, in 'Bioenergetics of Biomembranes', ed. L. Packer, G. Papageorgiou and A. Trebst, Elsevier, North Holland, 1977, p. 351.
340 P. Mathis, K. Sauer, and R. Remy, *F.E.B.S. Letters*, 1978, **88**, 275.
341 K. Sauer, P. Mathis, S. Acker, and J. Van Best, *Biochim. Biophys. Acta*, 1978, **503**, 120.
342 S. Demeter and B. Ke, *Biochim. Biophys. Acta*, 1977, **462**, 770.
343 K. Sauer, S. Acker, P. Mathis, and J. Van Best, *Dev. Bioenerg. Biomembr.*, 1977, **1**, 351.
344 D. Arnold, H. Tsujimoto, and T. Hijama, *Proc. Nat. Acad. Sci. U.S.A.*, 1977, **74**, 3826.

primary acceptor.[345—348] The relationship between centres A and B and soluble ferredoxin and $NADP^+$ has been investigated by e.p.r.[344]

An alternative electron transport scheme has been proposed with water reducing ferrodoxin and $NADP^+$ rather than the commonly accepted idea of P700 sequentially reducing both species.[344, 349]

A linearity between signals from $P700^+$ and FeS centres, as the FeS protein is denatured, implied that all 10—12 four FeS pairs are in the same protein.[350] Attempts to generate the FeS complex in the protein failed, since only artificial clusters formed, and the photochemical activity was not returned.[350]

A flavoprotein, ferredoxin NADP reductase (FNR), which accepts two electrons and acts between ferredoxin and NADP has been identified by flash photolysis.[351] A band at 470 nm indicates that *in vivo* FNR forms a complex with ferredoxin.[351]

10 Photosystem II

The light reaction in Photosystem II (PS II) leads to oxygen evolution from water and the reduction of plastoquinone. The primary donor is called P680 (sometimes $Chl-a)_{II}$ and on excitation a bleaching occurs at $\sim$680 and 435 nm and an absorbance increases at 820 nm. As in PS I, the absorbance changes are possibly due to the radical cations of a Chl-a dimer.[352—354] The reduced acceptor has an absorbance at 320 nm, is sometimes called X320, and is a specialized plastoquinone (PQ) molecule. The reduction of the acceptor occurs within 1 µs and decays with $t_{\frac{1}{2}} = 600$ ms. An additional role for PQ has been proposed whereby it also acts on the oxidized side of P680, but in an unknown manner.[355] Removal of PQ to less than one PQ per 750 Chl results in a loss of P680 activity; this is restored by reconstituting the reaction centre with PQ but not with β-carotene.[356] Absorbance changes in the 520—550 nm range are due to the electrochromic response of the pigment molecules to the electric field, caused by charge separation, and is a convenient measure of the redox state.[1, 357—362]

345 P. Heathcote, D. Williams-Smith, C. Sihra, and M. Evans, *Biochim. Biophys. Acta*, 1978, **503**, 333.
346 M. Evans, P. Heathcote, and D. Williams-Smith, *Dev. Bioenerg. Biomembr.*, 1977, **1**, 217.
347 P. Heathcote, D. Williams-Smith, and M. Evans, *Biochem. J.*, 1978, **170**, 378.
348 D. Williams-Smith, P. Heathcote, C. Sihra, and M. Evans, *Biochem. J.*, 1978, **170**, 365.
349 D. Arnon, in 'Encyclopedia of Plant Physiology, New Series', ed. A. Pirson and H. Zimmerman, Springer-Verlag, Berlin, 1975, Vol. 5, p. 7.
350 J. Golbeck and B. Kok, *Arch. Biochem. Biophys.*, 1978, **188**, 233.
351 B. Bouges-Bocquet, *F.E.B.S. Letters*, 1978, **85**, 340.
352 J. Van Best and P. Mathis, *Biochim. Biophys. Acta*, 1975, **503**, 178.
353 P. Mathis, J. Haveman, and M. Yates, in ref. 30, p. 267.
354 G. Renger, H. Eckerk, and H. Buchwald, *F.E.B.S. Letters*, 1978, **90**, 10.
355 D. Sadewasser and R. Dilley, *Biochim. Biophys. Acta*, 1978, **501**, 208.
356 D. Knaff, R. Malkin, J. Myron, and M. Stoller, *Biochim. Biophys. Acta*, 1977, **459**, 402.
357 P. Loach, *Photochem. and Photobiol.*, 1977, **26**, 87.
358 Y. Yamamoto and M. Nishimura, *Plant Cell Physiol.*, 1977, **18**, 55.
359 W. Vredenberg and A. Schapendonk, *F.E.B.S. Letters*, 1978, **91**, 90.
360 A. Schapendonk and W. Vredenberg, *Biochim. Biophys. Acta*, 1977, **462**, 613.
361 R. Malkin, *F.E.B.S. Letters*, 1978, **87**, 329.
362 J. Duniec and S. Thorne, *J. Bioenerg. Biomembr.*, 1977, **9**, 223.

The electron transfer scheme for the reaction centres can be shown diagrammatically as:

$$H_2O \atop O_2 \Big) \ Z \longrightarrow Y \longrightarrow \overset{\overset{\displaystyle D}{\downarrow}}{P680} \longrightarrow Q_1 \longrightarrow Q_2 \longrightarrow PQ$$

where Z is the secondary electron donor and Q the acceptor, a plastoquinone. The recovery of the bleaching due to $P680^+$ occurs by the back reaction with Q^-, or by the reduction of $P680^+$ with secondary electron donors. The back reaction is complex with at least 200 and 20 ms ($t_{\frac{1}{2}}$) components at room temperature. Reduction of $P680^+$ by a donor, at a low pH in the intra-thylakoid space, proceeds mainly by kinetics[363] with $t_{\frac{1}{2}} = 35$ µs. At high pH this reaction is more rapid, $t_{\frac{1}{2}} < 1$ µs.[354,363,364] The primary acceptor behaves similarly.[365] Additionally, the water-splitting enzyme (Y) has to be intact to observe the fast kinetics, which depend upon the charge accumulation on Y.[354,364,366] Under normal illumination conditions, $P680^+$ is reduced with three different half lives of <3 µs, 35 µs, and 200 µs.[354]

In untreated chloroplasts Van Best and Mathis[352] observed a 25—45 ns component at 820 nm, due to $P680^+$ reduction by the normal electron donor. When hydroxylamine is added, the reduction is biphasic and much slower, with half lives of 20 µs and 200 µs, and is due to $P680^+$ reduction by accessory donors. Renger *et al.*[354] suggest that the nanosecond component may only be due to a priming reaction as Van Best[352] used dark-adapted chloroplasts. Renger *et al.* find no evidence for a component of <1 µs in chloroplasts pre-illuminated with a single flash. The nanosecond component,[352] however, correlates well with the fast fluorescence induction observed after a single flash in *Chlorella*.[367]

The reduction of $P680^+$ has also been correlated with delayed light emission from charge recombination between P680 and Q^-.[357,368—371] $P680^+$ and Q^- are both effective fluorescence quenchers.[357,369,371]

When the reaction centres are closed before the flash, a strong 0.7 µs luminescence is observed, possibly due to $P680^+$ reduction by a secondary and adventitious acceptor (W^-).[368] Under normal conditions this acceptor (W) must function poorly compared with Q (the usual acceptor), so that photosynthesis can function efficiently. Acceptor (W) is proposed to be a molecule in the secondary donor complex.[368] Hydroxylamine removed this fast luminescence by inhibiting electron transfer to P680. In tris-treated chloroplasts a 120 µs component in the luminescence decay was due to the back reaction between

363 G. Renger, M. Glazer, and H. Buchwald, *Biochim. Biophys. Acta*, 1976, **461**, 392.
364 M. Glazer, Ch. Wolff, and G. Renger, *Z. Naturforsch.*, 1976, **31c**, 712.
365 B. Diner and P. Joliot, *Biochim. Biophys. Acta*, 1976, **423**, 479.
366 J. Haveman and P. Mathis, *Biochim. Biophys. Acta*, 1976, **440**, 346.
367 D. Mauzerall, *Proc. Nat. Acad. Sci. U.S.A.*, 1972, **69**, 1358.
368 J. Van Best and L. Duysens, *Biochim. Biophys. Acta*, 1977, **459**, 187.
369 P. Jursinic and Govindjee, *Biochim. Biophys. Acta*, 1977, **462**, 253.
370 P. Jursinic and Govindjee, *Photochem. and Photobiol.*, 1978, **27**, 61.
371 P. Jursinic and Govindjee, *Photochem. and Photobiol.*, 1977, **26**, 617.

P680[+] and Q[−]. Haveman and Mathis[366] found that, without background illumination and even after two flashes, the absorption at 820 nm, due to P680[+], was not observed. This indicates that at least two positive charges are stored on the donor side of PS II in tris-washed chloroplasts and, from this conclusion, it is suggested that e.p.r. signal II_f, observed by Babcock and Sauer,[372] is not the immediate donor to P680[+] but is at least one step removed from this role.[366]

Unlike the microsecond, the millisecond luminescence was sensitive to the effects of light-induced membrane potention, in fact, 128 mV are generated in chloroplasts by a single flash.[370]

On the acceptor side of P680, the existence of a two electron carrier has been inferred from the periodicity of two in the number of electrons reaching PS I (after short light flashes), from the delayed luminescence of PS II, and from the absorbance changes at 310 nm. Witt et al.[373] proposed that two plastoquinones act in parallel and disproportionate to form the fully reduced quinone. Two electrons are then transferred to the plastoquinone pool. Alternatively, the two plastoquinones may act in series. Recent flash[374] and steady-state studies[375] reveal complex absorption changes, between 300 and 360 nm, and provide support for the hypothesis that the two PQ acceptors (A_1, A_2) act in series and without H[+] uptake, in agreement with Haenhil.[376] The transfer from acceptor A_1 to A_2 is pH sensitive, indicating either a complex mechanism or that the reactions may involve a protein. The rate of electron transfer from $A_1{}^-$ to A_2 is the same as from $A_1{}^-$ to $A_2{}^-$. A_2 may be two plastoquinones operating between their oxidized and semireduced forms.[374]

The time evolution (μs to ms) of the increase in fluorescence yield, after a 3 μs saturating flash, has yielded information about a 'double hit', processes in which each reaction centre has two acceptors, Q_1, Q_2. The P680 is initially reduced by the secondary donor complex Y, and after a second 'hit' on the reaction centre by an auxiliary donor D[377] (see p. 000). These acceptors, Q_1, Q_2, are not those observed by Duysens et al.[368] as described above. Donor D is responsible for the slow rise in the fluorescence yield in hydroxylamine-washed chloroplasts, as the function of Z is impaired by Mn removal. The species D[+]P680 Q[−] has a 100 ms lifetime.[378] In untreated chloroplasts, D[+] is far more stable indicating that D, as well as Z, may be on the main electron transport route *via* a new carrier Z_2 which acts in parallel to Z. Both Z and Z_2 would then accumulate two charges.[377] The donor D may only be an auxiliary donor, however, as its role is not yet well defined.

A transient e.p.r. signal, observed at 10—100 K, similar to that of P700[+], has recently been assigned to P680[+]. This signal decays in 30 ms, at 100 K, due to its reduction by the secondary donor cytochrome b_{559}. The e.p.r. signal indicates that P680[+] is a chlorophyll dimer radical species.[379] The reaction centres are

[372] G. Babcock and K. Sauer, *Biochim. Biophys. Acta*, 1975, **376**, 329.
[373] H. Stiehl and T. Witt, *Z. Naturforsch.*, 1969, **246**, 1588.
[374] P. Mathis and J. Haveman, *Biochim. Biophys. Acta*, 1977, **461**, 167.
[375] M. Pulles, H. Van Gorkom, and J. Willemsen, *Biochim. Biophys. Acta*, 1976, **449**, 536.
[376] W Haenhil, *Biochim. Biophys. Acta*, 1976, **440**, 506.
[377] P. Joliot and A. Joliot, *Biochim. Biophys. Acta*, 1977, **462**, 559.
[378] A. Joliot, *Biochim. Biophys. Acta*, 1977, **460**, 142.
[379] J. Visser, C. Rijgersberg, and P. Gast, *Biochim. Biophys. Acta*, 1977, **460**, 36.

found to be similar at 5 and 80 K, but to be heterogeneous with respect to the secondary electron donors.[380,381] McIntosh and Bolton[382] have observed CIDEP emission in deuteriated algae and attribute this to the triplet rather than the radical pair mechanism.

11 Bacteriorhodopsin

At low oxygen tension, or in the absence of suitable metabolic carbon, the bacterium *Halobacterium halobium* develops purple patches of bacteriorhodopsin (bR) whose light-adapted form absorbs at 568 nm.[383,384] This pigment acts as a light-driven proton pump and converts light energy, *via* a series of intermediates (called *K, L, M, N, O*), into chemical energy and subsequently into a proton gradient across a cell membrane. This proton gradient is used to synthesize ATP or serves as a driving force for amino acid transport.

The chromophore, retinal, is attached to the protein, apo bacteriorhodopsin, *via* an ε amino group of the lysine residue.[385] The resultant bacteriorhodopsin is a compact globular protein consisting of seven helical segments 35—40 Å long inserted into the bilayer membrane. The protein consists of 75% of the weight of the membrane. The 568 nm transition dipole, which lies along the all-*trans* polyene chain, is between 63°[386] and 72°[387] to the normal of the plane of the membrane. The bacteriorhodopsin is oriented in the membrane such that its carboxy terminus is on the cytoplasmic side of the membrane.[388] It is arranged in a two-dimensional crystalline array with P_3 symmetry so that the proteins are grouped into trimers with their centres 15 Å apart.[389]

The bR exists in two forms, light- and dark-adapted. The dark-adapted form absorbs maximally at 560 nm and consists of equal amounts of all-*trans* and 11-*cis* bR isomers.[390,391] Pettei *et al.*,[392] however, consider that the dark-adapted form is all-*trans* and contains no *cis* molecules. The light-adapted form (bR$_{570}$ or sometimes PM) has an all-*trans* configuration and absorbs at 570 nm.

Light absorption by these pigments induces a photochemical cycle represented by the general scheme,[383,384] shown in Figure 7. The subscripts refer to the absorption maxima in nm. The cycling time of the light-adapted form is about 10 ms at 36 °C. Several additional intermediates to the ones shown have been

[380] C. Rijgersberg and J. Amesz, *Biochim. Biophys. Acta*, 1978, **502**, 152.
[381] F. Bonnet, C. Vernotte, J. Briantaiss and A. Etienne, *Biochim. Biophys. Acta*, 1977, **461**, 151.
[382] A. McIntosh and J. Bolton, *Nature*, 1976, **263**, 443.
[383] R. Lozier, R. Bogomolni, and W. W. Stoeckenius, *Biophys. J.*, 1975, **15**, 955.
[384] D. Oesterhelt and W. Stoeckenius, *Proc. Nat. Acad. Sci. U.S.A.*, 1973, **70**, 2853.
[385] T. Schreckenbach, B. Walckhoff, and D. Oesterhelt, *European J. Biochem.*, 1977, **76**, 499.
[386] R. Korenstein and B. Hess, *F.E.B.S. Letters*, 1978, **89**, 15.
[387] M. Heyn, R. Cherry, and M. Mueller, *J. Mol. Biol.*, 1977, **117**, 607.
[388] G. Gerber, C. Gray, D. Widenauer, and H. Khorana, *Proc. Nat. Acad. Sci. U.S.A.*, 1977, **74**, 5426.
[389] R. Henderson and P. Unwin, *Nature*, 1975, **257**, 28.
[390] A. Maeda and T. Yoshizawa, *J. Biochem. (Tokyo)*, 1977, **82**, 1599.
[391] B. Becher, F. Tokunaga, and T. Ebrey, *Biochemistry*, 1978, **17**, 2293.
[392] M. Pettei, A. Yudd, K. Nakanishi, R. Henselman, and W. Stoekenius, *Biochemistry*, 1977, **16**, 1955.

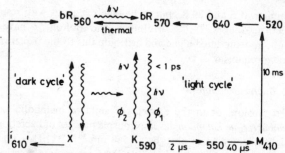

Figure 7 *General scheme of a photochemical cycle induced by light absorption of pigments*

proposed,[393-395] some of which are stable for several seconds. The quantum yield of the primary process, bR_{570} to K_{590}, is temperature independent and equals 0.25 ± 0.05 with $\Phi_1/\Phi_2 = 0.4$.[396-399] Species M and N have recently been demonstrated to be kinetically distinct species as the action spectra of the flash-induced photovoltages match their absorption spectra.[400,401] Resonance Raman and absorption spectra also identify the L and M intermediates.[402-404] The dark-adapted *cis* form is not involved in the normal photocycle[405,406] but undergoes a cycle involving a transient X, and a transient absorbing at 610 nm, which forms in 20 μs and has $t_{\frac{1}{2}} = 40$ ms. Either the initially formed transient X or the *cis* bR_{560} can convert into either bR_{570} or K at high light intensities (see Figure 7). The quantum yield for the light conversion from bR_{560} to bR_{570} equals 0.035.[406]

Using picosecond flash photolysis, Ippen *et al.*[407] and Kaufmann *et al.*,[408] have observed transients due to the formation of species K. This species formed in ~1 ps[407] and lasted for 4 μs at room temperature[383] (see Figure 8). Kaufmann *et al.* also observed a transient forming in <6 ps, but at 580 nm the species recovers in 20 ps, possibly due to ground-state repopulation.[408] The fluorescence from bR_{570} has been measured to decay with a lifetime of 40 ps at low tempera-

[393] P. Kyrukov, Yu. Lazarev, V. Letokhov, Yu. Matveets, E. Terpugov, L. Chekulaeva, and A. Sharkov, *Biofizika*, 1978, **23**, 171.
[394] F. Litvin and S. Balashov, *Biofizika*, 1977, **22**, 1111.
[395] A. Shkrob and A. Rodionov, *Bioorg. Khim.*, 1978, **4**, 500.
[396] C. Goldschmidt, O. Kalinsky, T. Rosenfeld, and M. Ottolenghi, *Biophys. J.*, 1977, **17**, 179.
[397] C. Goldschmidt, M. Ottolenghi, and R. Korenstein, *Biophys. J.*, 1976, **16**, 839.
[398] B. Becher and T. Ebrey, *Biophys. J.*, 1977, **17**, 185.
[399] R. Lozier and W. Niederberger, *Fed. Proc.*, 1977, **36**, 1805.
[400] S.-B. Hwang, J. Korenbrot, and W. Stoeckenius, *Biochim. Biophys. Acta*, 1978, **509**, 300.
[401] S.-B. Hwang, J. Korenbrot, and W. Stoeckenius, *Dev. Bioenerg. Biomembr.*, 1977, **1**, 137.
[402] J. Terner, A. Campion, and M. El-Sayed, *Proc. Nat. Acad. Sci. U.S.A.*, 1977, **74**, 5212.
[403] J. Spoonhower, *Diss. Abs. Internat. B*, 1978, **38**, 3012.
[404] A. Campion, M. El-Sayed, and J. Terner, *Proc. Soc. Photo-opt. Instrum. Eng.*, 1977, **113**, 128.
[405] K. Ohno, Y. Takeuchi, and M. Yoshida, *Biochim. Biophys. Acta*, 1977, **462**, 575.
[406] O. Kalinsky, C. Goldschmidt, and M. Ottolenghi, *Biophys. J.*, 1977, **19**, 185.
[407] E. Ippen, C. Shank, A. Lewis, and M. Marcus, *Science*, 1978, **200**, 1279.
[408] K. Kaufmann, V. Sundstrom, T. Tamane, and P. Rentzepis, *Biophys. J.*, 1978, **22**, 121.

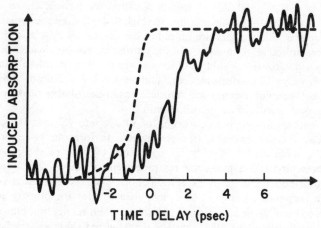

Figure 8 *Induced absorption at* 615 nm *plotted against the delay between the probe and excitation pulses*; (–––) *experimentally determined instantaneous response of the system*; (· · ·) *computed response for a* 1 ps *exponential rise*
(Reproduced by permission from *Science*, 1978, **200**, 1279)

tures,[409] and of 15 ± 3 ps at physiological temperatures, by Hirsch *et al.*[410], using an up-conversion light-gate technique. This fluorescence appeared instantaneously within the picosecond time resolution of the technique. The fluorescence quantum yield was 2.4×10^{-5} and the intrinsic lifetime was calculated as 6 ns leading to an expected 0.14 ps decay time, approximately 100 times shorter than the measured decay time. The emission from bR does not originate from the ^1B_u state but possibly from the forbidden state ^1A_g, proposed to exist in many polyenes.[411] (Fluorescence at *ca.* 330 nm due to protein fluorescence in bR, and which varies with the light cycle has also been observed.[412])

These data suggest that the transient absorption and the fluorescence arise from different states of the bR molecule. The temperature dependence of the fluorescence quantum yield, and the temperature independence of the bR $\rightarrow K$ photochemical conversion and of the transient species, which still forms in <20 ps at 68 K,[413,414] support the idea that the fluorescence state is different from that undergoing photochemical transformation.[415] No fluorescence has been observed from intermediate K and no energy transfer from K to a fluorescing state of bR_{540} occurs.[415] The strong interaction between bR molecules suggests, however, that some energy transfer occurs between them, with an

[409] R. Alfano, R. Yu, B. Govindjee, B. Becher, and T. Ebrey, *Biophys. J.*, 1976, **16**, 1399.
[410] M. Hirsch, M. Marcus, A. Lewis, H. Mahr, and N. Frigo, *Biophys. J.*, 1976, **16**, 1399.
[411] B. Hudson and B. Kohler, *Ann. Rev. Phys. Chem.*, 1974, **25**, 437.
[412] R. Bogomolni, L. Stubbs, and J. Lanyi, *Biochemistry*, 1978, **17**, 1037.
[413] A. Lewis, M. Marcus, E. Ippen, C. Shank, M. Hirsch, and H. Mahr, *Biophys. J.*, 1977, **17**, 759.
[414] K. Kaufmann, P. Rentzepis, W. Stoeckenius, and A. Lewis, *Biochem. Biophys. Res. Comm.*, 1976, **68**, 1109.
[415] R. Govindjee, B. Becher, and T. Ebrey, *Biophys. J.*, 1978, **22**, 67.

efficiency of 0.47,[416] and also to species K within the trimeric cluster of pigments.[418] Contrary to this observation, the high (0.45) fluorescence polarization observed over a wide temperature range has been explained by a lack of energy transfer or molecular rotation.[417] If the chromophores are indeed oriented, then little depolarization would be observed even if energy transfer occurred.

There is as yet no satisfactory explanation for the significant differences observed between fluorescence and transient absorption studies of the primary photochemical process in bacteriorhodopsin.

Three mechanisms for the conversion of bR_{540} into K_{590} in the primary process have been considered, (i) isomerization to form an 11-*cis* or 14-*cis* retinylidene molecule,[392, 399, 419—422] (ii) change of state of the protonated Schiff's base, and (iii) a molecular rearrangement of the protein induced by electron distribution change in the chromophore.[423, 424] Some recent evidence has made proposals (i) and (ii) seem less attractive for the primary process.

Firstly, combining the 13-*cis* isomer with the protein results in a blue shift of the absorption relative to bR_{570}, rather than the red shirts expected for the formation of intermediate K.[425] Secondly, time-resolved resonance Raman spectroscopy[426] indicates that the state of protonation changes in 10—30 μs, rather than the few ps, observed for the transient absorption changes. The third proposal, that of an electron distribution change as a result of light absorption, can explain the observed picosecond transients. Lewis[423, 427] has suggested that a protein conformational transition occurs, followed by the movement of protons from one amino acid to another. A similar explanation has been proposed by Konishi and Packer[424, 428, 429] and is shown in Figure 9. The various states in Figure 9 are (i) dark adapted, (ii) the light-adapted state where the chromophore moves closer to the tryptophan, and (iii) upon light absorption an electron transfers to the amino acid giving rise to the observed transients,[430] and a proton is also released from the Schiff base nitrogen to a neighbouring low pK_a group. There is then a proton conformational change, step (iv). The electron is returned from the tryptophan and increases the electron density on the Schiff base. This species is the 412 nm intermediate (M), which accepts a proton from a nearby tyrosine group, and a proton is released from the negatively charged group R in Figure 9. Tyrosine has previously been shown to be involved

[416] T. Ebrey, B. Becher, B. Mao, P. Kilbride, and B. Honig, *J. Mol. Biol.*, 1977, **112**, 377.
[417] V. Sineschekov and F. Litvin, *Biochim. Biophys. Acta*, 1977, **462**, 450.
[418] J. Hurley and T. Ebrey, *Biophys. J.*, 1978, **22**, 49.
[419] T. Rosenfeld, B. Honig, M. Ottolenghi, J. Hurley, and T. Ebrey, *Pure Appl. Chem.*, 1977, **49**, 391.
[420] B. Aton, A. Doukas, R. Callender, B. Becher, and T. Ebrey, *Biochemistry*, 1977, **16**, 2995.
[421] J. Hurley, T. Ebrey, B. Honig, and M. Ottolenghi, *Nature*, 1977, **270**, 540.
[422] C. Shulten and P. Tavan, *Nature*, 1978, **272**, 85.
[423] A. Lewis, *Proc. Nat. Acad. Sci. U.S.A.*, 1978, **75**, 2.
[424] T. Konishi and L. Packer, *F.E.B.S. Letters*, 1978, **92**, 1.
[425] N. Sperling, P. Caul, C. Rafferty, and N. Dencher, *Biophys. Struct. Mech.*, 1977, **3**, 79.
[426] M. Marcus and A. Lewis, *Science*, 1977, **195**, 1328.
[427] S. Rackovsky and A. Lewis, in ref. 153, p. 330.
[428] T. Konishi and L. Packer, *Biochem. Biophys. Res. Comm.*, 1976, **72**, 1437.
[429] T. Konishi and L. Packer, *F.E.B.S. Letters*, 1977, **80**, 455.
[430] T. Konishi and L. Packer, *F.E.B.S. Letters*, 1977, **79**, 369.

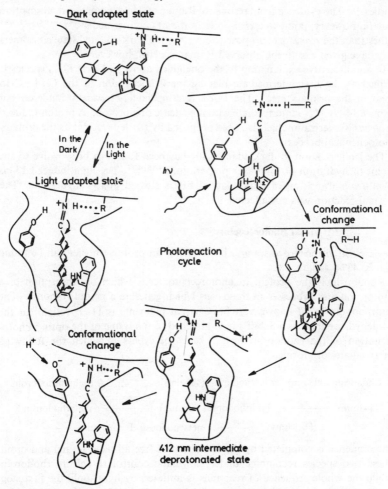

Figure 9 *Mechanism of proton translocation by bacteriorhodopsin*
(Reproduced by permission from *F.E.B.S. Letters*, 1978, **92**, 1)

with the reprotonation of the Schiff base[431] in step (v). In the last steps the
phenolate which is formed accepts a proton from the medium and the light-
adapted state (ii) is reformed.

In general agreement with this scheme, time-resolved resonance raman spectra
of *bacteriorhodopsin* suggest that deprotonation occurs on a microsecond
time scale and between the 570 nm and the M_{412} intermediate,[426] possibly in the
K_{550} to M_{412} step.[404,432,433] Similar conclusions have been drawn from u.v.

[431] R. Bogomolni, R. Renthal, and J. Lanyi, *Biophys. J.*, 1978, **21**, 183a.
[432] A. Campion, M. El-Sayed, and J. Terner, *Biophys. J.*, 1977, **20**, 369.
[433] A. Campion, J. Terner, and M. El-Sayed, *Nature*, 1977, **265**, 659.

studies.[391] The evidence from resonance Raman and low temperature absorption spectra, however, point to a *cis* isomer in the M_{412} intermediate[402,404,422,434] rather than the *trans* form proposed above. In the dark-adapted state *cis* isomers may have given rise to the observed Raman spectra, however.[406]

In an alternative explanation to the one given above, a 14-*cis* K_{590} species is formed as the first intermediate placing the isomerization at the C-14—C-15 bond in the primary step. The 14-*cis* intermediate is sterically hindered but forms a 14-*cis* M_{412} intermediate with the release of a proton. A proton-induced *cis*–*trans* isomerization has also been proposed by Hurley *et al.*[421] as the primary photochemical process.

The isomerization in dark-adapted bR has been found to be sensitive to the extent of hydration of thin-layer preparations.[435,436] The light-induced 13-*cis* to all-*trans* transition of dark-adapted bR is slowed at low hydration. The thermal reaction was not affected.

12 Visual Photoreceptors

Reviews on vision processes and bacteriorhopsin proton transfer can be found in refs. 437—443.

Rhodopsin is the protein in photoreceptor cells which on absorption of a photon initiates processes in these cells which generate a neural response. The chromophore in all known vertebrate visual pigments is 11-*cis* retinal (or the 3,4-dehydroretinal) as a Schiff's base attached to the lysine of the opsin. Photo-excitation of the rhodopsin starts a series of events in which the following intermediates are formed:

$$\text{Rhodopsin (498 nm) [or isorhodopsin (485 nm)]} \underset{\longleftarrow}{\overset{ps}{\longrightarrow}} \text{prelumirhodopsin}$$

$$(543 \text{ nm}) \xrightarrow{\text{ns}} \text{lumirhodopsin (497 nm)} \xrightarrow{\mu s} \text{metarhodopsin I}$$

$$(481 \text{ nm}) \xrightarrow{\text{ms}} \text{metarhodopsin II (380 nm)}$$

The sequence is completed by the formation of free all-*trans* retinal and opsin. These two species recombine under enzymatic control to reform rhodopsin. While the whole sequence of reactions is initiated by light, only the first step, the formation of prelumirhodopsin (sometimes called bathorhodopsin), requires light. All other reactions are thermally activated.

The photochemistry of *cis*–*trans* isomerization, the photophysics of retinals[444,445] and their Schiff base,[441] have been studied as models for the

[434] J. Jurley, B. Becher, and T. Ebrey, *Nature*, 1978, **272**, 87.
[435] R. Korenstein and B. Hess, *Nature*, 1977, **270**, 184.
[436] R. Korenstein and B. Hess, *F.E.B.S. Letters*, 1977, **82**, 7.
[437] S. Ostroy, *Biochim. Biophys. Acta*, 1977, **463**, 91.
[438] W. Stone and F. Dratz, *Photochem. and Photobiol.*, 1977, **26**, 79.
[439] H. Suzuki and H. Kobayashi, *Photochem. and Photobiol.*, 1978, **27**, 815.
[440] J. Wolken, *Jerusalem Symp. Quant. Chem. Biol.*, 1977, **10**, 175.
[441] W. Kuehne, *Vision Res.*, 1977, **17**, 1273.
[442] M. Ostrovskii, *Membrane Transport Processes*, 1978, **2**, 217.
[443] B. Wiggert, D. Bergsma, M. Lewis, and G. Chader, *J. Neurochem*, 1977, **29**, 947.
[444] R. Hochstrasser and D. Narva, *Photochem. and Photobiol.*, 1977, **26**, 595.
[445] R. Hochstrasser, D. Narva, and A. Nelson, *Chem. Phys. Letters*, 1976, **43**, 15.

visual processes. The protonated Schiff bases show wavelength-dependent quantum yields for isomerization,[446] whereas rhodopsin has a higher and wavelength-independent isomerization yield. From this and other arguments it is clear that the protein plays a vital role in controlling the photochemistry *in vivo*.

The 'classical' *cis–trans* isomerization has been questioned recently as the primary step in the visual transduction processes, by the observation of pre-lumirhodopsin formation in <6 ps[447,448] or <9 ps.[449] A proton shift mechanism, between a retinal proton and the Schiff base linkage, mediated by a nearby amino-acid residue in the opsin, has been proposed as the primary processes after light absorption.[447,448,450,451] The isomerization then occurs in one of the dark reactions.

Picosecond absorption measurements on rod outer segments from bovine retinas, in clear glasses at low temperatures, clearly show that only below 20 K does the formation of prelumirhodopsin, from the excited singlet of rhodopsin, become resolvable.[447] This formation takes 9 ps at 20 K and is still surprisingly fast (36 ps at 4 K) when monitored at 570 nm. Upon deuteriation of the protonated Schiff base an isotope effect (K_H/K_D) ~7 was observed; the prelumi formation now took 257 ps at 4 K. Any *cis–trans* isomerization might be expected to be inhibited at this temperature. Experiments on the protonated 11-*cis* retinylidene Schiff base in methanol solution show that the isomerization at room temperature takes longer than 6 ps and that prelumirhodopsin does not contain all-*trans* retinal.[452] No evidence for a hypsorhodopsin, a proposed precursor of prelumirhodopsin, was obtained in the 530 nm picosecond experiments,[447,448] perhaps not surprisingly as the hypso intermediate is supposed to form by excitation at >540 nm.[437]

A marked non-Arrheniuus behaviour was observed with the formation rate of prelumirhodopsin. As 0 K is approached, the rate became temperature independent instead of vanishing as it would if the reaction followed a classical Arrheniuus behaviour.

In experiments at room temperature with nanosecond time resolution, Bensasson *et al.*[453] also found prelumirhodopsin to be the first transient formed. The light-driven equilibrium rhodopsin ⇌ prelumirhodopsin ⇌ isorhodopsin was also observed. Isorhodopsin is the 9-*cis* isomer and is formed during the 530 nm laser pulse.

These kinetic studies can be explained by models not involving isomerization as the primary step. Firstly, van der Meer[450,454] suggested that a two proton

[446] T. Rosenfeld, B. Honig, M. Ottolenghi, J. Hurley, and T. Ebrey, *Pure Appl. Chem.*, 1977, **49**, 341.

[447] K. Peters, M. Applebury, and P. Rentzepis, *Proc. Nat. Acad. Sci. U.S.A.*, 1977, **74**, 3119.

[448] G. Busch, M. Applebury, A. Lamola, and P. Rentzepis, *Proc. Nat. Acad. Sci. U.S.A.*, 1972, **69**, 2802.

[499] B. Green, R. Monger, R. Alfano, B. Aton, and R. Callender, *Nature*, 1977, **269**, 179.

[450] K. Van Der Meer, J. Mulder, and J. Lugtenberg, *Photochem. and Photobiol.*, 1976, **24**, 363.

[451] R. Mathies and L. Stryer, *Proc. Nat. Acad. Sci. U.S.A.*, 1976, **73**, 2169.

[452] D. Huppert, P. Rentzepis, and D. Kliger, *Photochem. and Photobiol.*, 1977, **25**, 193.

[453] R. Bensasson, E. Land, and T. Truscott, *Photochem. and Photobiol.*, 1977, **26**, 601.

[454] M. Fransen, W. Luyten, J. Van Thuijl, and J. Lugtenberg, *Nature*, 1976, **260**, 726.

shift occurs to form the prelumi species. An imidazole or similar group in the opsin was proposed to fit into the bend of the *cis* retinal and allow a proton to be transferred from C-18 to it. A second proton is then transferred to the Schiff base nitrogen. Kropf,[455] however, has shown that a prelumirhodopsin forms with 5-desmethyl retinal (which does not contain C-18) as the chromophore. It is possible then that C-4 of the ionone ring is used as a proton source.[447] Secondly, very large changes in dipole moments (10—15 D) have been observed, by Mathies and Stryer,[451] in *cis* retinal and all-*trans* retinal and its Schiff base with n-butylamine. Excitation of either molecule is accompanied by an electron shift to the Schiff nitrogen and a positive charge being associated on or near the ionone ring. The initially formed excited state is far from its equilibrium excited state configuration; similarly a nearby amino-acid residue is far from its equilibrium, and the interaction produced could initiate conformational changes in the protein.[423,451,456,457] These changes then produce the prelumi intermediate. It has also been suggested that nearby polar groups in the opsin could cause electron transfer to the retinal.[458]

Peters *et al.*[447] suggest that the finite rate of prelumi formation at 4 K and also the deuterium effect provide evidence for tunnelling of a proton between an amino-acid side-group and the Schiff base nitrogen. Lewis,[423] however, argues that, as the Raman spectra of rhodopsin and prelumirhodopsin have similar CN bond frequencies, the proton must be associated with the opsin matrix and not the Schiff base, and that this must control the onset of absorption. Some authors have, however, reaffirmed the *cis–trans* isomerization mechanism from a consideration of similar data.[446]

[455] A. Kropf, *Nature*, 1976, **264**, 92.
[456] L. Salem and P. Bruchmann, *Nature*, 1975, **258**, 526.
[457] L. Salem, *Science*, 1976, **191**, 822.
[458] J. Simons, *Proc. Nat. Acad. Sci. U.S.A.*, 1977, **74**, 3375.

Author Index

671